Periodic Table of The Elements

Legend:
- Metals
- Nonmetals
- Metalloids

1A (1)	2A (2)	3B (3)	4B (4)	5B (5)	6B (6)	7B (7)	8B (8)	8B (9)	8B (10)	1B (11)	2B (12)	3A (13)	4A (14)	5A (15)	6A (16)	7A (17)	8A (18)
1 H 1.008																1 H 1.008	2 He 4.0026
3 Li 6.941	4 Be 9.0122											5 B 10.81	6 C 12.011	7 N 14.007	8 O 15.9994	9 F 18.9984	10 Ne 20.1797
11 Na 22.9898	12 Mg 24.3050											13 Al 26.9815	14 Si 28.085	15 P 30.9738	16 S 32.06	17 Cl 35.45	18 Ar 39.948
19 K 39.0983	20 Ca 40.078	21 Sc 44.9559	22 Ti 47.867	23 V 50.9415	24 Cr 51.9961	25 Mn 54.9380	26 Fe 55.845	27 Co 58.9332	28 Ni 58.6934	29 Cu 63.546	30 Zn 65.38	31 Ga 69.723	32 Ge 72.63	33 As 74.9216	34 Se 78.96	35 Br 79.904	36 Kr 83.798
37 Rb 85.4678	38 Sr 87.62	39 Y 88.9059	40 Zr 91.224	41 Nb 92.9064	42 Mo 95.94	43 Tc (98)	44 Ru 101.07	45 Rh 102.9055	46 Pd 106.42	47 Ag 107.8682	48 Cd 112.411	49 In 114.818	50 Sn 118.710	51 Sb 121.760	52 Te 127.60	53 I 126.9045	54 Xe 131.293
55 Cs 132.9055	56 Ba 137.327	57 La 138.9055 *	72 Hf 178.49	73 Ta 180.9479	74 W 183.84	75 Re 186.207	76 Os 190.23	77 Ir 192.217	78 Pt 195.084	79 Au 196.9666	80 Hg 200.59	81 Tl 204.38	82 Pb 207.2	83 Bi 208.9804	84 Po (209)	85 At (210)	86 Rn (222)
87 Fr (223)	88 Ra (226)	89 Ac (227) **	104 Rf (265)	105 Db (268)	106 Sg (271)	107 Bh (270)	108 Hs (277)	109 Mt (268)	110 Ds (281)	111 Rg (280)	112 Cn (285)	113 Uut (284)	114 Fl (289)	115 Uup (288)	116 Lv (293)	117 Uus (294)	118 Uuo (294)

*Lanthanide Series

58 Ce 140.116	59 Pr 140.9076	60 Nd 144.242	61 Pm (145)	62 Sm 150.36	63 Eu 151.964	64 Gd 157.25	65 Tb 158.9253	66 Dy 162.500	67 Ho 164.9303	68 Er 167.259	69 Tm 168.9342	70 Yb 173.054	71 Lu 174.9668

** Actinide Series

90 Th 232.0381	91 Pa 231.0359	92 U 238.0289	93 Np (237)	94 Pu (244)	95 Am (243)	96 Cm (247)	97 Bk (247)	98 Cf (251)	99 Es (252)	100 Fm (257)	101 Md (258)	102 No (259)	103 Lr (262)

Note: Atomic masses are IUPAC values (2009, up to four decimal places). More accurate values for some elements are given in the International Table of Atomic Weights on the facing page.

Chemistry

for CHM 1046

Selections from the 10th edition of Whitten, Chemistry

Kenneth W. Whitten | Raymond E. Davis | M. Larry Peck |
George G. Stanley

CENGAGE
Learning·

Australia • Brazil • Japan • Korea • Mexico • Singapore • Spain • United Kingdom • United States

Chemistry: for CHM 1046, Selections from the 10th edition of Whitten, Chemistry

Chemistry, Tenth Edition
Kenneth W. Whitten | Raymond E. Davis | M. Larry Peck | George G. Stanley

© 2014, 2010 Cengage Learning. All rights reserved.

Student Solutions Manual: Chemistry, Tenth Edition
Kenneth W. Whitten | Raymond E. Davis | M. Larry Peck | George G. Stanley

© 2014 Cengage Learning. All rights reserved.

Senior Project Development Manager:
Linda deStefano

Market Development Manager:
Heather Kramer

Senior Production/Manufacturing Manager:
Donna M. Brown

Production Editorial Manager:
Kim Fry

Sr. Rights Acquisition Account Manager:
Todd Osborne

For product information and technology assistance, contact us at
Cengage Learning Customer & Sales Support, 1-800-354-9706

For permission to use material from this text or product,
submit all requests online at **cengage.com/permissions**
Further permissions questions can be emailed to
permissionrequest@cengage.com

This book contains select works from existing Cengage Learning resources and was produced by Cengage Learning Custom Solutions for collegiate use. As such, those adopting and/or contributing to this work are responsible for editorial content accuracy, continuity and completeness.

Compilation © 2013 Cengage Learning
ISBN-13: 978-1-285-90826-7

ISBN-10: 1-285-90826-0

Cengage Learning
5191 Natorp Boulevard
Mason, Ohio 45040
USA
Cengage Learning is a leading provider of customized learning solutions with office locations around the globe, including Singapore, the United Kingdom, Australia, Mexico, Brazil, and Japan. Locate your local office at:
international.cengage.com/region.

Cengage Learning products are represented in Canada by Nelson Education, Ltd.
For your lifelong learning solutions, visit **www.cengage.com/custom.**
Visit our corporate website at **www.cengage.com.**

Printed at CLDPC, USA, 05-18

Brief Contents

Excerpted from:
Chemistry, Tenth Edition
Kenneth W. Whitten | Raymond E. Davis | M. Larry Peck | George G. Stanley

Excerpted from:
Student Solutions Manual: Chemistry, Tenth Edition
Kenneth W. Whitten | Raymond E. Davis | M. Larry Peck | George G. Stanley

Liquids and Solids

13

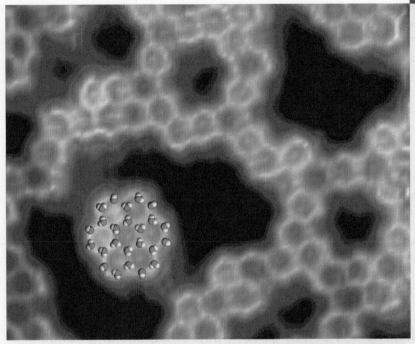

Interfacial force microscopy (a form of scanning tunneling microscopy) gives an image of water wetting an extended surface of a metal such as palladium or ruthenium. Theoretical calculations based on these images give new clues for how water wets such surfaces. A proposed water rosette on the metal surface is represented as a space-filling model.

Courtesy of Professor Miquel Salmeron, Lawrence Berkeley National Laboratory

OBJECTIVES

After you have studied this chapter, you should be able to

▶ Describe the properties of liquids and solids and how they differ from gases

▶ Understand the kinetic–molecular description of liquids and solids, and show how this description differs from that for gases

▶ Use the terminology of phase changes

▶ Understand various kinds of intermolecular attractions and how they are related to physical properties such as vapor pressure, viscosity, melting point, and boiling point

▶ Describe evaporation, condensation, and boiling in molecular terms

▶ Calculate the heat transfer involved in warming or cooling without change of phase

▶ Calculate the heat transfer involved in phase changes

▶ Describe melting, solidification, sublimation, and deposition in molecular terms

▶ Interpret *P* versus *T* phase diagrams

▶ Describe the regular structure of crystalline solids

▶ Describe various types of solids

▶ Relate the properties of different types of solids to the bonding or interactions among particles in these solids

▶ Visualize some common simple arrangements of atoms in solids

▶ Carry out calculations relating atomic arrangement, density, unit cell size, and ionic or atomic radii in some simple crystalline arrangements

▶ Describe the bonding in metals

▶ Explain why some substances are conductors, some are insulators, and others are semiconductors

The molecules of most gases are so widely separated at ordinary temperatures and pressures that they do not interact with one another significantly. The physical properties of gases are reasonably well described by the simple relationships in Chapter 12. In liquids and solids, the so-called **condensed phases**, the particles are close together so they interact much more strongly. Although the properties of liquids and solids can be described, they cannot be adequately explained by simple mathematical relationships. Table 13-1 and Figure 13-1 summarize some of the characteristics of gases, liquids, and solids.

13-1 Kinetic–Molecular Description of Liquids and Solids

The properties listed in Table 13-1 can be qualitatively explained in terms of the kinetic–molecular theory of Chapter 12. We saw in Section 12-13 that the average kinetic energy of a collection of gas molecules decreases as the temperature is lowered. As a sample of gas is cooled and compressed, the rapid, random motion of gaseous molecules decreases. The molecules approach one another, and the intermolecular attractions increase. Eventually, these increasing intermolecular attractions overcome the reduced kinetic energies. At this point the gas changes to a liquid; this process is called condensation (liquefaction). The temperatures and pressures required for condensation vary from gas to gas, because different kinds of molecules have different attractive forces.

In the liquid state the forces of attraction among particles are great enough that disordered clustering occurs. The particles are so close together that very little of the volume occupied by a liquid is empty space. As a result, it is very hard to compress a liquid. Particles in liquids have sufficient energy of motion to overcome partially the attractive forces among them. They are able to slide past one another so that liquids assume the shapes of their containers up to the volume of the liquid.

▶ *Inter*molecular attractions are those between different molecules or ions. *Intra*molecular attractions are those between atoms within a single molecule or ion.

Liquids diffuse into other liquids with which they are *miscible*. For example, when a drop of red food coloring is added to a glass of water, the water becomes red throughout after diffusion is complete. The natural diffusion rate is slow relative to gases at normal temperatures. Because the average separations among particles in liquids are far less than those in gases, the densities of liquids are much higher than the densities of gases (see Table 12-1).

▶ The *miscibility* of two liquids refers to their ability to mix and produce a homogeneous solution.

Cooling a liquid lowers its molecular kinetic energy, so its molecules slow down even more. If the temperature is lowered sufficiently, at ordinary pressures, stronger but shorter-range attractive interactions overcome the reduced kinetic energies of the

Table 13-1 Some Characteristics of Solids, Liquids, and Gases

Solids	Liquids	Gases
1. Have definite shape (resist deformation)	1. Have no definite shape (assume shapes of containers)	1. Have no definite shape (fill containers completely)
2. Are nearly incompressible	2. Have definite volume (are only very slightly compressible)	2. Are compressible
3. Usually have higher density than liquids	3. Have high density	3. Have low density
4. Are not fluid	4. Are fluid	4. Are fluid
5. Diffuse only very slowly through solids	5. Diffuse through other liquids	5. Diffuse rapidly
6. Have an ordered arrangement of particles that are very close together; particles usually have only vibrational motion	6. Consist of disordered clusters of particles that are quite close together; particles have random motion in three dimensions	6. Consist of extremely disordered particles with much empty space between them; particles have rapid, random motion in three dimensions

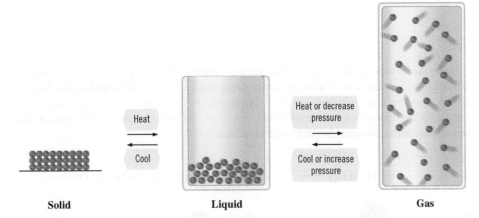

Figure 13-1 Representations of the kinetic–molecular description of the three phases of matter.

molecules to cause *solidification*. The temperature required for *crystallization* at a given pressure depends on the nature of short-range interactions among the particles and is characteristic of each substance.

Most solids have ordered arrangements of particles with a very restricted range of motion. Particles in the solid state cannot move freely past one another so they only vibrate about fixed positions. Consequently, solids have definite shapes and volumes. Because the particles are so close together, solids are nearly incompressible and are very dense relative to gases. Solid particles do not diffuse readily into other solids. However, analysis of two blocks of different solids, such as copper and lead, that have been pressed together for a period of years shows that each block contains some atoms of the other element. This demonstrates that solids do diffuse, but very slowly (Figure 13-2).

▶ *Solidification* and *crystallization* refer to the process in which a liquid changes to a solid. Crystallization is the more specific term, referring to the formation of a very ordered solid material.

13-2 Intermolecular Attractions and Phase Changes

We have seen (see Section 12-15) how the presence of strong attractive forces between gas molecules can cause gas behavior to become nonideal when the molecules get close together at higher pressures. In liquids and solids the molecules are much closer together than in gases. As a result, properties of liquids, such as boiling point, vapor pressure, viscosity, and heat of vaporization, depend markedly on the strengths of the intermolecular attractive forces. These forces are also directly related to the properties of solids, such as melting point and heat of fusion. As the basis for our study of these condensed phases, let us discuss the types of attractive forces that can exist between molecules and ions.

Figure 13-2 A representation of diffusion in solids.

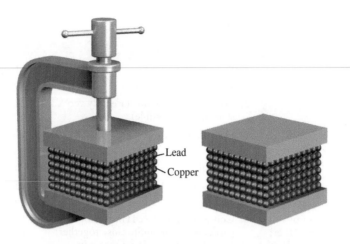

When blocks of two different metals are clamped together for a long time, a few atoms of each metal diffuse into the other metal.

*Inter*molecular forces refer to the forces *between* individual particles (atoms, molecules, ions) of a substance. These forces are quite weak relative to *intra*molecular forces, that is, covalent and ionic bonds *within* compounds. For example, 927 kJ of energy is required to decompose one mole of water vapor into H and O atoms. This reflects the strength of intramolecular forces (covalent bonds).

$$H{-}\overset{..}{\underset{..}{O}}{-}H(g) \longrightarrow 2H\cdot(g) + \cdot\overset{..}{\underset{..}{O}}\cdot(g) \quad \text{(absorbs 927 kJ/mol H}_2\text{O)}$$

Only 40.7 kJ is required to convert one mole of liquid water into steam at 100°C.

$$H_2O(\ell) \longrightarrow H_2O(g) \quad \text{(absorbs 40.7 kJ/mol H}_2\text{O)}$$

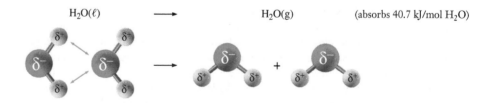

This reflects the lower strength of the intermolecular forces of attraction between the water molecules, compared to the covalent bonds within the water molecules. The attractive forces between water molecules are mainly due to *hydrogen bonding*.

If it were not for the existence of intermolecular attractions, condensed phases (liquids and solids) could not exist. These are the forces that hold the particles close to one another in liquids and solids. As we shall see, the effects of these attractions on melting points of solids parallel those on boiling points of liquids. High boiling and melting points are associated with substances that have strong intermolecular attractions. Let us consider the effects of the general types of forces that exist among ionic, covalent, and monatomic species.

STOP & THINK
It is important to be able to tell whether a substance is ionic, nonpolar covalent, or polar covalent. You should review the discussion of bonding in Chapters 7 and 8.

Ion–Ion Interactions

According to Coulomb's Law, the *force of attraction* between two oppositely charged ions is directly proportional to the charges on the ions, q^+ and q^-, and inversely proportional to the square of the distance between them, d.

$$F \propto \frac{q^+ q^-}{d^2}$$

▶ When oppositely charged ions are close together, d (the denominator) is small, so F, the attractive force between them, is large.

Energy has the units of force × distance, $F \times d$, so the *energy of attraction* between two oppositely charged ions is directly proportional to the charges on the ions and inversely proportional to the distance of separation.

$$E \propto \frac{q^+ q^-}{d}$$

Ionic compounds such as NaCl, CaBr$_2$, and K$_2$SO$_4$ exist as extended arrays of discrete ions in the solid state. As we shall see in Section 13-16, the oppositely charged ions in these arrays are quite close together. As a result of these small distances, d, the energies of attraction in these solids are substantial. Most ionic bonding is strong, and as a result most ionic compounds have relatively high melting points (Table 13-2). At high enough temperatures, ionic solids melt as the added heat energy overcomes the potential energy associated with the attraction of oppositely charged ions. The ions in the resulting liquid samples are free to move about, which accounts for the excellent electrical conductivity of molten ionic compounds.

▶ The covalent bonding *within* a polyatomic ion such as NH$_4^+$ or SO$_4^{2-}$ is very strong, but the forces that hold the *entire substance* together are ionic. Thus a compound that contains a polyatomic ion is an ionic compound (see Section 7-12).

For most substances, the liquid is less dense than the solid. Melting a solid nearly always produces greater average separations among the particles. This means that the forces (and energies) of attractions among the ions in an ionic liquid are less than in the solid state because the average d is greater in the melt. However, these energies of attraction are still much greater in magnitude than the energies of attraction among neutral species (molecules or atoms).

The product $q^+ q^-$ increases as the charges on ions increase. Ionic substances containing multiply charged ions, such as Al^{3+}, Mg^{2+}, O^{2-}, and S^{2-} ions, *usually* have higher melting and boiling points than ionic compounds containing only singly charged ions, such as Na$^+$, K$^+$, F$^-$, and Cl$^-$. For a series of ions of similar charges, the closer approach of smaller ions results in stronger interionic attractive forces and higher melting points (compare NaF, NaCl, and NaBr in Table 13-2).

STOP & THINK
The simple concepts of ion size and magnitudes of the ionic charges combined with Coulomb's Law can give you very good guidance on understanding a number of solid-state properties ranging from melting point trends to solubility.

Dipole–Dipole Interactions

Permanent dipole–dipole interactions occur between polar covalent molecules because of the attraction of the $\delta+$ atoms of one molecule to the $\delta-$ atoms of another molecule (see Section 7-11).

Electrostatic forces between two ions decrease by the factor $1/d^2$ as their separation distance, d, increases. But dipole–dipole forces vary as $1/d^4$. Because of the higher power of d in the denominator, $1/d^4$ diminishes with increasing d much more rapidly than does

Table 13-2 Melting Points of Some Ionic Compounds

Compound	mp (°C)	Compound	mp (°C)	Compound	mp (°C)
NaF	993	CaF$_2$	1423	MgO	2800
NaCl	801	Na$_2$S	1180	CaO	2580
NaBr	747	K$_2$S	840	BaO	1923
KCl	770				

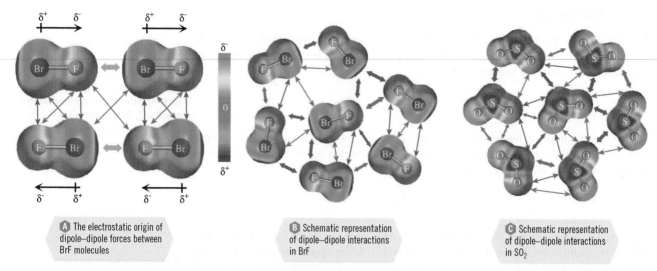

B Schematic representation of dipole–dipole interactions in BrF

C Schematic representation of dipole–dipole interactions in SO₂

Figure 13-3 Dipole–dipole interactions among polar molecules. Each polar molecule is shaded with regions of highest negative charge ($\delta-$) in red and regions of highest positive charge ($\delta+$) colored blue. Attractive forces are shown as blue arrows, and repulsive forces are shown as red arrows. Stronger attractions and repulsions are indicated by thicker arrows. Molecules tend to arrange themselves to maximize attractions by bringing regions of opposite charge together while minimizing repulsions by separating regions of like charge.

$1/d^2$. As a result, dipole forces are effective only over very short distances. Furthermore, for dipole–dipole forces, q^+ and q^- represent only "partial charges," so these forces are weaker than ion–ion forces. Average dipole–dipole interaction energies are approximately 4 kJ per mole of bonds. They are much weaker than ionic and covalent bonds, which have typical energies of about 400 kJ per mole of bonds. Substances in which permanent dipole–dipole interactions affect physical properties include bromine fluoride, BrF, and sulfur dioxide, SO_2. Dipole–dipole interactions are illustrated in Figure 13-3. All dipole–dipole interactions, including hydrogen bonding (discussed in the following section), are somewhat directional. An increase in temperature causes an increase in translational, rotational, and vibrational motion of molecules. This produces more random orientations of molecules relative to one another. Consequently, dipole–dipole interactions become less important as temperature increases. All these factors make compounds having only dipole–dipole interactions more volatile than ionic compounds.

Hydrogen Bonding

Hydrogen bonds are a special case of strong dipole–dipole interaction. They are not really chemical bonds in the formal sense.

> Strong hydrogen bonding occurs among polar covalent molecules containing H bonded to one of the three small, highly electronegative elements—F, O, or N.

Like ordinary dipole–dipole interactions, hydrogen bonds result from the electrostatic attractions between $\delta+$ atoms of one molecule, in this case H atoms, and the $\delta-$ atoms of another molecule. The small sizes of the F, O, and N atoms, combined with their high electronegativities, concentrate the electrons of these molecules around these $\delta-$ atoms. This causes an H atom bonded to one of these highly electronegative elements to become quite positive. The $\delta+$ H atom is attracted to a lone pair of electrons on an F, O, or N atom other than the atom to which it is covalently bonded (Figure 13-4). The molecule that contains the hydrogen-bonding $\delta+$ H atom is often referred to as the *hydrogen-bond donor*; the $\delta-$ atom to which it is attracted is called the *hydrogen-bond acceptor*.

Recently, careful studies of light absorption and magnetic properties in solution and of the arrangements of molecules in solids have led to the conclusion that the same kind of attraction occurs (although more weakly) when H is bonded to carbon. In some cases, very weak C—H···O "hydrogen bonds" exist. Similar observations suggest the existence of weak hydrogen bonds to chlorine atoms, such as O—H···Cl. However, most chemists usually restrict usage of the term "hydrogen bonding" to compounds in which H is covalently bonded to F, O, or N, and we will do likewise throughout this text.

Typical hydrogen-bond energies are in the range 15 to 20 kJ/mol, which is four to five times greater than the energies of other dipole–dipole interactions. As a result, hydrogen bonds exert a considerable influence on the properties of substances. Hydrogen bonding is responsible for the unusually high melting and boiling points of compounds such as water, methyl alcohol, and ammonia (see Figure 13-4) compared with other compounds of similar molecular weight and molecular geometry (Figure 13-5). Hydrogen bonding between amino acid subunits, for example, is very important in establishing the three-dimensional structures of proteins.

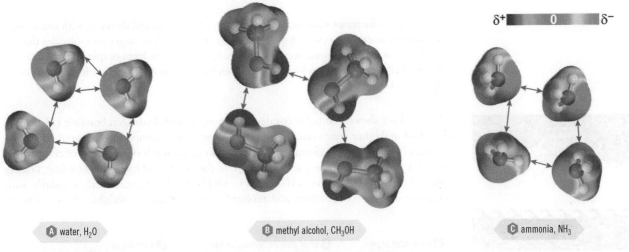

$\delta+$ 0 $\delta-$

Ⓐ water, H₂O   Ⓑ methyl alcohol, CH₃OH   Ⓒ ammonia, NH₃

Figure 13-4 Hydrogen bonding (indicated by blue arrows) in (a) water, H_2O; (b) methyl alcohol, CH_3OH; and (c) ammonia, NH_3. Hydrogen bonding is a special case of very strong dipole interaction. The hydrogen bonding is due to the electrostatic attraction of the $\delta+$ charged hydrogen of one molecule and the $\delta-$ charged oxygen or nitrogen atom of another.

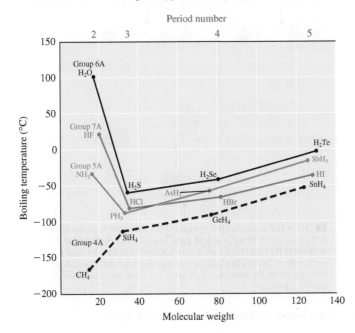

Figure 13-5 Boiling points of some hydrides as a function of molecular weight. The unusually high boiling points of NH_3, H_2O, and HF compared with those of other hydrides of the same groups are due to hydrogen bonding. The electronegativity difference between H and C is small, and there are no unshared pairs on C; thus, CH_4 is not hydrogen bonded. Increasing molecular weight corresponds to increasing number of electrons; this makes the electron clouds easier to deform and causes increased dispersion forces, accounting for the increase in boiling points for the nonhydrogen-bonded members of each series.

Dispersion Forces

▶ Dispersion forces are often called **London forces**, after the German-born physicist Fritz London (1900–1954). He initially postulated their existence in 1930, on the basis of quantum theory. Yet another name is **van der Waals** attractive forces. You should be familiar with all three names.

Dispersion forces are weak attractive forces that are important only over *extremely* short distances because they vary as $1/d^7$. These forces are present between all types of molecules in condensed phases but are weak for small molecules. Dispersion forces are the only kind of intermolecular forces present among symmetrical nonpolar substances such as SO_3, CO_2, O_2, N_2, Br_2, H_2, and monatomic species such as the noble gases. Without dispersion forces, such substances could not condense to form liquids or solidify to form solids. Condensation of some substances occurs only at very low temperatures and/or high pressures.

Dispersion forces result from the attraction of the positively charged nucleus of one atom for the electron cloud of an atom in nearby molecules. This induces *temporary* dipoles in neighboring atoms or molecules. As electron clouds become larger and more diffuse, they are attracted less strongly by their own (positively charged) nuclei. Thus, they are more easily distorted, or *polarized*, by adjacent atoms or molecules. Dispersion forces are depicted in Figure 13-6. They exist in all substances.

> Polarizability increases with increasing numbers of electrons and therefore with increasing sizes of molecules. Therefore, dispersion forces are generally stronger for molecules that are larger or that have more electrons. For molecules that are large or quite polarizable the total effect of dispersion forces can be even higher than dipole–dipole interactions or hydrogen bonding.

Figure 13-5 shows that polar covalent compounds with hydrogen bonding (H_2O, HF, NH_3) boil at higher temperatures than analogous polar compounds without hydrogen bonding (H_2S, HCl, PH_3). Symmetrical, nonpolar compounds (CH_4, SiH_4) of comparable molecular weight boil at lower temperatures. In the absence of hydrogen bonding, boiling points of analogous substances (CH_4, SiH_4, GeH_4, SnH_4) increase fairly regularly with increasing number of electrons and molecular size (molecular weight). This is due to

For a physical analogy to dispersion forces think about Velcro. Each little hook and loop in Velcro represents a very weak interaction. But adding them all together can make the total attraction very strong.

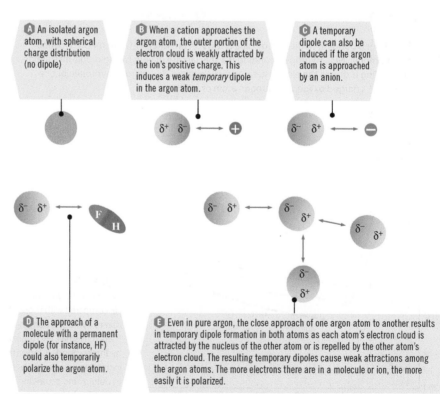

A An isolated argon atom, with spherical charge distribution (no dipole)

B When a cation approaches the argon atom, the outer portion of the electron cloud is weakly attracted by the ion's positive charge. This induces a weak *temporary* dipole in the argon atom.

C A temporary dipole can also be induced if the argon atom is approached by an anion.

D The approach of a molecule with a permanent dipole (for instance, HF) could also temporarily polarize the argon atom.

E Even in pure argon, the close approach of one argon atom to another results in temporary dipole formation in both atoms as each atom's electron cloud is attracted by the nucleus of the other atom or is repelled by the other atom's electron cloud. The resulting temporary dipoles cause weak attractions among the argon atoms. The more electrons there are in a molecule or ion, the more easily it is polarized.

Figure 13-6 An illustration of how a temporary dipole can be induced in an atom.

increasing effectiveness of dispersion forces of attraction in the larger molecules and occurs even in the case of some polar covalent molecules. The increasing effectiveness of dispersion forces, for example, accounts for the increase in boiling points in the sequences $HCl < HBr < HI$ and $H_2S < H_2Se < H_2Te$, which involve nonhydrogen-bonded polar covalent molecules. The differences in electronegativities between hydrogen and other nonmetals *decrease* in these sequences, and the increasing dispersion forces override the decreasing permanent dipole–dipole forces. The *permanent* dipole–dipole interactions therefore have very little effect on the boiling point trend of these compounds.

Let's compare the magnitudes of the various contributions to the total energy of interactions in some simple molecules. Table 13-3 shows the permanent dipole moments and the energy contributions for five simple molecules. The contribution from dispersion forces is substantial in all cases. The permanent dipole–dipole energy is greatest for substances in which hydrogen bonding occurs. The variations of these total energies of interaction are closely related to molar heats of vaporization. As we shall see in Section 13-9, the heat of vaporization measures the amount of energy required to overcome the attractive forces that hold the molecules together in a liquid.

The properties of a liquid or a solid are often the result of many forces. The properties of an ionic compound are determined mainly by the very strong ion–ion interactions, even though other forces may also be present. In a polar covalent compound that contains N — H, O — H, or F — H bonds, strong hydrogen bonding is often the strongest force present. If hydrogen bonding is absent in a polar covalent compound, dipole–dipole forces are likely to be the most important. In a slightly polar or nonpolar covalent compound or a monatomic nonmetal, the dispersion forces, though weak, are the strongest ones present, so they determine the overall attraction between molecules. For large molecules, even the very weak dispersion forces can total up to a considerable interactive force.

Table 13-3   Approximate Contributions to the Total Energy of Interaction Between Molecules, in kJ/mol

Molecule	Permanent Dipole Moment (D)	Permanent Dipole–Dipole Energy	Dispersion Energy	Total Energy	Molar Heat of Vaporization (kJ/mol)
Ar	0	0	8.5	8.5	6.7
CO	0.1	≈0	8.7	8.7	8.0
HCl	1.03	3.3*	17.8	21	16.2
NH₃	1.47	13*	16.3	29	27.4
H₂O	1.85	36*	10.9	47	40.7

*Hydrogen-bonded.

The ability of the Tokay gecko to climb on walls and ceilings is due to the dispersion forces between the tiny hairs on the gecko's foot and the surface.

EXAMPLE 13-1 Intermolecular Forces

Identify the types of intermolecular forces that are present in a condensed phase (liquid or solid) sample of each of the following. For each, make a sketch, including a few molecules, that represents the major type of force: (a) water, H_2O; (b) iodine, I_2; (c) nitrogen dioxide, NO_2.

Plan

Any substance exhibits dispersion forces, which are generally weaker than other forces that might be present. To identify other possible forces, we must determine whether the molecule is polar (which can lead to dipole–dipole forces) or whether it can form hydrogen bonds.

Solution

(a) Water, H_2O, is polar, with the H sufficiently positive and the O sufficiently negative to form strong hydrogen bonds in addition to the dispersion forces that are present for any condensed phase. Hydrogen bonds are the strongest intermolecular forces present in this liquid.

(b) Iodine, I_2, is nonpolar. The close approach of two I_2 molecules easily distorts the large electron cloud of each molecule, causing dispersion forces resulting in a temporary weak attraction. Because the molecule is nonpolar, other stronger forces are absent.

(c) The Lewis formula of nitrogen dioxide, NO_2, is shown by the following two resonance formulas:

This molecule contains $N — O$ polar bonds, and the VSEPR approach (see Chapter 8) lets us conclude that the molecule is angular. It is therefore a polar molecule, which we can represent as

Many properties of liquids and solids depend on intermolecular attractions. You must learn to use the electronic structure of a molecule or ion to predict which types of intermolecular forces it can exhibit.

The positive part of one molecule is attracted to the negative parts of its neighbors, so there are dipole–dipole interactions in addition to the usual dispersion forces. Dipole–dipole attractions are the strongest ones between these molecules and can be represented as

You should now work Exercise 124.

The Liquid State

We shall briefly describe several properties of the liquid state. These properties vary markedly among various liquids, depending on the nature and strength of the attractive forces among the particles (atoms, molecules, ions) making up the liquid.

13-3 Viscosity

Viscosity is the resistance to flow of a liquid. Honey, which flows very slowly at room temperature, has a high viscosity while gasoline has a low viscosity. The viscosity of a liquid can be measured with a viscometer such as the one in Figure 13-7.

For a liquid to flow, the molecules must be able to slide past one another. In general, the stronger the intermolecular forces of attraction, the more viscous the liquid is. Substances that have a great ability to form hydrogen bonds, especially involving several hydrogen-bonding sites per molecule, such as glycerine (see structure below), usually have high viscosities. Increasing the size and surface area of molecules generally results in increased viscosity, due to the increased dispersion forces. For example, the shorter-chain hydrocarbon pentane (a free-flowing liquid at room temperature) is less viscous than dodecane (an oily liquid at room temperature). The longer the molecules are, the more they attract one another via dispersion forces, and the harder it is for them to flow.

▶ The *poise* is the unit used to express viscosity. The viscosity of water at 25°C is 0.89 centipoise.

▶ The "weight" of motor oil actually refers to its viscosity. SAE 40 oil ("40-weight") is more viscous than SAE 10 oil.

glycerine, $C_5H_5(OH)_3$
viscosity = 945
   centipoise at 25°C

pentane, C_5H_{12}
viscosity = 0.215
   centipoise at 25°C

dodecane, $C_{12}H_{26}$
viscosity = 1.38
   centipoise at 25°C

$$\begin{pmatrix} \text{stronger} \\ \text{attractive} \\ \text{forces} \end{pmatrix} \longleftrightarrow \begin{pmatrix} \text{higher} \\ \text{viscosity} \end{pmatrix}$$

$$\begin{pmatrix} \text{increasing} \\ \text{temperature} \end{pmatrix} \longleftrightarrow \begin{pmatrix} \text{lower} \\ \text{viscosity} \end{pmatrix}$$

As temperature increases and the molecules move more rapidly, their kinetic energies are better able to overcome intermolecular attractions. Thus, viscosity decreases with increasing temperature, as long as no changes in composition occur.

Honey is a very viscous liquid.

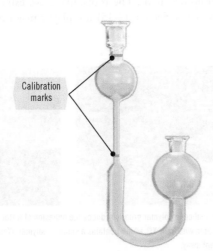

Calibration marks

Figure 13-7 The Ostwald viscometer, a device used to measure viscosity of liquids. The time it takes for a known volume of a liquid to flow through a narrow neck of known size is measured. Liquids with low viscosities flow rapidly.

Charles D. Winters

The surface tension of water supports this water strider. The nonpolar surfaces of its feet also help to repel the water.

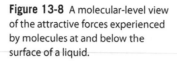

Figure 13-8 A molecular-level view of the attractive forces experienced by molecules at and below the surface of a liquid.

The shape of a soap bubble is due to the inward force (surface tension) that acts to minimize the surface area.

Droplets of mercury lying on a glass surface. The small droplets are almost spherical, whereas the larger droplets are flattened due to the effects of gravity. This shows that surface tension has more influence on the shape of the small (lighter) droplets.

13-4 Surface Tension

Molecules below the surface of a liquid are influenced by intermolecular attractions from all directions. Those on the surface, however, are attracted only toward the interior (Figure 13-8); these attractions pull the surface layer toward the center. The most stable situation is one in which the surface area is minimal. For a given volume, a sphere has the least possible surface area, so drops of liquid tend to assume spherical shapes. **Surface tension** is a measure of the inward forces that must be overcome to expand the surface area of a liquid.

13-5 Capillary Action

All forces holding a liquid together are called **cohesive forces**. The forces of attraction between a liquid and another surface are **adhesive forces**. The partial positive charges on the H atoms of water hydrogen bond strongly to the partial negative charges on the oxygen atoms at the surface of the glass. As a result, water *adheres* to glass, or is said to *wet* glass. As water creeps up the side of the glass tube, its favorable area of contact with the glass increases. The surface of the water, its **meniscus**, has a concave shape (Figure 13-9). On the other hand, mercury does not wet glass because its cohesive forces are much stronger than its attraction to glass. Thus, its meniscus is convex. **Capillary action** occurs when one end of a capillary tube, a glass tube with a small bore (inside diameter), is immersed in a liquid. If adhesive forces exceed cohesive forces, the liquid creeps up the sides of the tube until a balance is reached between adhesive forces and the weight of liquid. The smaller the bore, the higher the liquid climbs. Capillary action helps plant roots take up water and dissolved nutrients from the soil and transmit them up the stems. The roots, like glass, exhibit strong adhesive forces for water. Osmotic pressure (see Section 14-15) also plays a major role in this process.

Coating glass with a silicone polymer greatly reduces the adhesion of water to the glass. The left side of each glass has been treated with Rain-X®, which contains a silicone polymer. Water on the treated side forms droplets that are easily swept away.

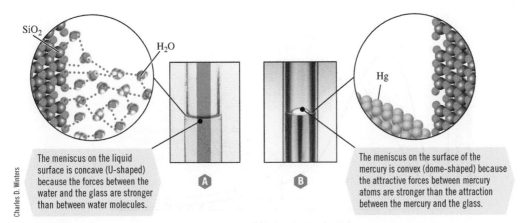

The meniscus on the liquid surface is concave (U-shaped) because the forces between the water and the glass are stronger than between water molecules.

The meniscus on the surface of the mercury is convex (dome-shaped) because the attractive forces between mercury atoms are stronger than the attraction between the mercury and the glass.

Figure 13-9 The meniscus, as observed in glass tubes (a) with water (concave) and (b) with mercury (convex).

13-6 Evaporation

Evaporation, or **vaporization**, is the process by which molecules on the surface of a liquid break away and go into the gas phase (Figure 13-10). Kinetic energies of molecules in liquids depend on temperature in the same way as they do in gases. The distribution of kinetic energies among liquid molecules at two different temperatures is shown in Figure 13-11. To break away from the attractions of its neighbors, a molecule must possess at least some minimum kinetic energy. Figure 13-11 shows that at a higher temperature, a greater fraction of molecules possess at least that minimum energy. The rate of evaporation increases as temperature increases.

Only the higher-energy molecules can escape from the liquid phase. The average molecular kinetic energy of the molecules remaining in the liquid state is thereby lowered, resulting in a lower temperature in the liquid. The liquid would then be cooler than its surroundings, so it absorbs heat from its surroundings. The cooling of your body by evaporation of perspiration is a familiar example of the cooling of the surroundings by evaporation of a liquid. This is called "cooling by evaporation."

A molecule in the vapor may strike the liquid surface and be captured there. This process, the reverse of evaporation, is called **condensation**. As evaporation occurs in a closed container, the volume of liquid decreases and the number of gas molecules above the surface increases. Because more gas phase molecules can then collide with the surface, the rate of condensation increases. The system composed of the liquid and gas molecules of

The dew on this spider web was formed by condensation of water vapor from the air.

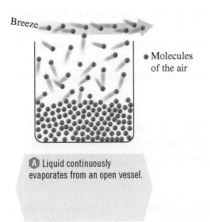

A Liquid continuously evaporates from an open vessel.

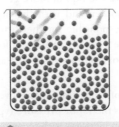

B In a closed container, equilibrium between liquid and vapor is established in which molecules return to the liquid at the same rate as they leave it.

C A bottle in which liquid–vapor equilibrium has been established. Note that droplets have condensed.

Figure 13-10

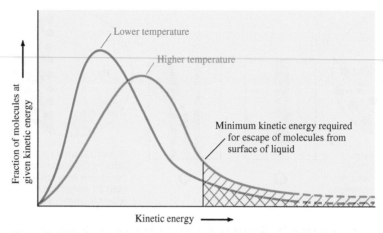

Figure 13-11 Distribution of kinetic energies of molecules in a liquid at different temperatures. At the lower temperature, a smaller fraction of the molecules have the energy required to escape from the liquid, so evaporation is slower and the equilibrium vapor pressure (Section 13-7) is lower.

▶ As an analogy, suppose that 30 students per minute leave a classroom, moving into the closed hallway outside, and 30 students per minute enter it. The total number of students in the room would remain constant, as would the total number of students in the hallway.

the same substance eventually achieves a **dynamic equilibrium** in which the rate of evaporation equals the rate of condensation in the closed container.

$$\text{liquid} \underset{\text{condensation}}{\overset{\text{evaporation}}{\rightleftharpoons}} \text{vapor (gas)}$$

The two opposing rates are not zero, but are equal to one another—hence we call this "dynamic," rather than "static," equilibrium. Even though evaporation and condensation are both continuously occurring, *no net change occurs* because the rates are equal.

If the vessel were left open to the air, however, this equilibrium could not be reached. Molecules would diffuse away, and slight air currents would also sweep some gas molecules away from the liquid surface. This would allow more evaporation to occur to replace the lost vapor molecules. Consequently, a liquid can eventually evaporate entirely if it is left uncovered. This situation illustrates **LeChatelier's Principle**:

▶ This is one of the guiding principles that allows us to understand chemical equilibrium. It is discussed further in Chapter 17.

> A system at equilibrium, or changing toward equilibrium, responds in the way that tends to relieve or "undo" any stress placed on it.

In this example the stress is the removal of molecules in the vapor phase. The response is the continued evaporation of the liquid to replace molecules in the vapor.

13-7 Vapor Pressure

Vapor molecules cannot escape when vaporization of a liquid occurs in a closed container. As more molecules leave the liquid, more gaseous molecules collide with the walls of the container, with one another, and with the liquid surface, so more condensation occurs. This condensation is responsible for the formation of liquid droplets that adhere to the sides of the vessel above a liquid surface and for the eventual establishment of equilibrium between liquid and vapor (see Figure 13-10b and c).

▶ Note the terminology: vapor pressure is measured as a *gas pressure* in equilibrium with a liquid, but we call it the vapor pressure *of the liquid*.

> The partial pressure of vapor molecules above the surface of a liquid at equilibrium at a given temperature is the **vapor pressure (vp)** of the liquid at that temperature. Because the rate of evaporation increases with increasing temperature, vapor pressures of liquids *always* increase as temperature increases.

Vapor pressures can be measured with manometers (Figure 13-12). Easily vaporized liquids are called **volatile** liquids, and they have relatively high vapor pressures. The most volatile liquid in Table 13-4 is diethyl ether, whereas water is the least volatile.

Stronger cohesive forces tend to hold molecules in the liquid state. Methanol molecules are strongly linked by hydrogen bonding, whereas diethyl ether molecules are not, so methanol has a lower vapor pressure than diethyl ether. The very strong hydrogen bonding in water accounts for its unusually low vapor pressure (see Table 13-4) and high boiling point. Dispersion forces generally increase with increasing molecular size, so substances composed of larger molecules have lower vapor pressures.

We can understand the order of vapor pressures of some of the liquids cited in Table 13-4 and Figure 13-13 by considering the strengths of their intermolecular attractions. Water has the lowest vapor pressure (strongest cohesive forces) because each molecule has two hydrogen atoms to act as hydrogen-bond donors and each molecule can accept hydrogen bonds from two other molecules. Low-molecular-weight alcohols, such as methanol or ethanol, each have only one potential hydrogen-bond donor, so their average cohesive forces are weaker than those in water and their vapor pressures are higher. In benzene and diethyl

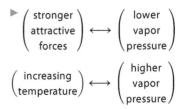

$$\begin{pmatrix} \text{stronger} \\ \text{attractive} \\ \text{forces} \end{pmatrix} \longleftrightarrow \begin{pmatrix} \text{lower} \\ \text{vapor} \\ \text{pressure} \end{pmatrix}$$

$$\begin{pmatrix} \text{increasing} \\ \text{temperature} \end{pmatrix} \longleftrightarrow \begin{pmatrix} \text{higher} \\ \text{vapor} \\ \text{pressure} \end{pmatrix}$$

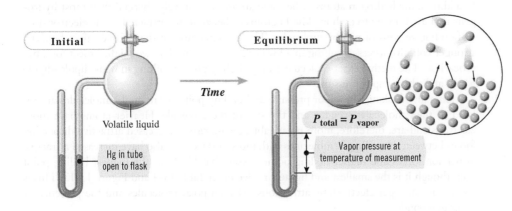

Figure 13-12 A representation of the measurement of vapor pressure of a liquid at a given temperature. The container is evacuated before the liquid is added. At the instant the liquid is added to the container, there are no molecules in the gas phase so the pressure is zero. Some of the liquid then vaporizes until equilibrium is established. The difference in heights of the mercury columns is a measure of the vapor pressure of the liquid at that temperature.

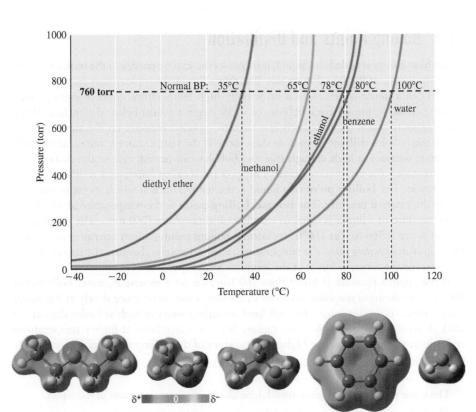

Figure 13-13 Plots of the vapor pressures of the liquids in Table 13-4. Notice that the increase in vapor pressure is *not* linear with temperature. Each substance exists as a liquid for temperatures and pressures to the left of its curve, and as a gas at conditions to the right of the curve. The *normal* boiling point of a liquid is the temperature at which its vapor pressure is equal to one atmosphere. The electrostatic charge potential surface for each molecule is also shown. These demonstrate the important role of polarity in reducing vapor pressure and increasing boiling points. Note the presence of partial positive charge on hydrogen atoms that are attached to the electronegative oxygen atoms. These can engage in hydrogen bonding to oxygen atoms in other molecules that have a build-up of partial negative charge.

▶ As long as some liquid remains in contact with the vapor, the pressure does not depend on the volume or surface area of the liquid.

Table 13-4   Vapor Pressure (in torr) of Some Liquids

	MW	0°C	25°C	50°C	75°C	100°C	125°C
diethyl ether	74 g/mol	185	470	1325	2680	4859	
methanol	32 g/mol	29.7	122	404	1126		
ethanol	46 g/mol	13	63	258	680		
benzene	78 g/mol	27.1	94.4	271	644	1360	
water	18 g/mol	4.6	23.8	92.5	300	760	1741

ether, the hydrogen atoms are all bonded to carbon, so strong hydrogen bonds are not possible. Electrons can move easily throughout the delocalized π-bonding orbitals of benzene, however, so benzene is quite polarizable and exhibits significant dispersion forces. In addition, the hydrogen atoms of benzene are more positively charged than most hydrogens that are bonded to carbon. The H atoms of benzene are attracted to the electron-rich π-bonding regions of nearby molecules. The accumulation of these forces gives benzene stronger cohesive forces, resulting in a lower vapor pressure than we might expect for a hydrocarbon. The diethyl ether molecule is only slightly polar, resulting in weak dipole–dipole forces and a high vapor pressure.

Generally speaking, the vapor pressure and boiling points of nonpolar molecules are related to their molecular weights and the size of the molecule. Heavier atoms have more electrons and are, therefore, more polarizable, giving rise to increased attractive dispersion forces between them. Bigger molecules with more surface area also have increased attractive dispersion forces and higher boiling points. Note that H_2O has the highest boiling point even though it is the smallest and lightest molecule in Table 13-4 and Figure 13-13. This is due to the stronger electrostatic attractions between polar molecules and the presence of hydrogen bonds.

13-8 Boiling Points and Distillation

When heat energy is added to a liquid, it increases the kinetic energy of the molecules, and the temperature of the liquid increases. Heating a liquid always increases its vapor pressure. When a liquid is heated to a sufficiently high temperature under a given applied pressure (usually atmospheric), bubbles of vapor begin to form below the surface. If the vapor pressure inside the bubbles is less than the applied pressure on the surface of the liquid, the bubbles collapse as soon as they form. If the temperature is raised sufficiently, the vapor pressure is high enough that the bubbles can persist, rise to the surface, and burst, releasing the vapor into the air. This process is called *boiling* and is different from evaporation. The **boiling point** of a liquid is the temperature at which its vapor pressure equals the external pressure. The **normal boiling point** is the temperature at which the vapor pressure of a liquid is equal to exactly one atmosphere (760 torr). The vapor pressure of water is 760 torr at 100°C, its normal boiling point. As heat energy is added to a pure liquid *at its boiling point*, the temperature remains constant, because the energy is used to overcome the cohesive forces in the liquid to form vapor.

If the applied pressure is lower than 760 torr, say on a mountain, water boils below 100°C. The chemical reactions involved in cooking food occur more slowly at the lower temperature, so it takes longer to cook food in boiling water at high altitudes than at sea level. A pressure cooker cooks food rapidly because water boils at higher temperatures under increased pressures. The higher temperature of the boiling water increases the rate of cooking.

▶ As water is being heated, but before it boils, small bubbles may appear in the container. This is not boiling, but rather the formation of bubbles of dissolved gases such as CO_2 and O_2 whose solubilities in water decrease with increasing temperature.

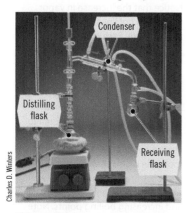

Charles D. Winters

Figure 13-14 A laboratory setup for distillation. During distillation of an impure liquid, nonvolatile substances remain in the distilling flask. The liquid is vaporized and condensed before being collected in the receiving flask.

The vapor pressure of any given liquid depends *only* on the temperature of the liquid.

The boiling point of any given liquid depends *only* on the external pressure.

Different liquids have different cohesive forces, so they have different vapor pressures and boil at different temperatures. A mixture of liquids with sufficiently different boiling points can often be separated into its components by **distillation**. In this process the mixture is heated slowly until the temperature reaches the point at which the most volatile liquid boils off. If this component is a liquid under ordinary conditions, it is subsequently recondensed in a water-cooled condensing column (Figure 13-14) and collected as a distillate. After enough heat has been added to vaporize all of the most volatile liquid, the temperature again rises slowly until the boiling point of the next substance is reached, and the process continues. Any nonvolatile substances dissolved in the liquid do not boil, but remain in the distilling flask. Impure water can be purified and separated from its dissolved salts by distillation. Compounds with similar boiling points, especially those that interact very strongly with one another, are not effectively separated by simple distillation, but require a modification called fractional distillation (see Section 14-10).

13-9 Heat Transfer Involving Liquids

Heat must be added to a liquid to raise its temperature (see Section 1-14). The **specific heat** (J/g·°C) or **molar heat capacity** (J/mol·°C) of a liquid is the amount of heat that must be added to the stated mass of liquid to raise its temperature by one degree Celsius. If heat is added to a pure liquid under constant pressure, the temperature rises until its boiling point is reached. Then the temperature remains constant until enough heat has been added to boil away all the liquid. The **molar heat** (or **enthalpy**) **of vaporization** (ΔH_{vap}) of a liquid is the amount of heat that must be added to one mole of the liquid at its boiling point to convert it to vapor with no change in temperature. Heats of vaporization can also be expressed as energy per gram. For example, the heat of vaporization for water at its boiling point is 40.7 kJ/mol, or 2.26×10^3 J/g:

$$\frac{?\,J}{g} = \frac{40.7\,kJ}{mol} \times \frac{1000\,J}{kJ} \times \frac{1\,mol}{18.0\,g} = 2.26 \times 10^3\,J/g$$

▶ The specific heat and heat capacity of a substance change somewhat with its temperature. For most substances, this variation is small enough to ignore.

Like many other properties of liquids, heats of vaporization reflect the strengths of intermolecular forces. Heats of vaporization generally increase as boiling points and intermolecular forces increase and as vapor pressures decrease. Table 13-5 illustrates this. The high heats of vaporization of water, ethylene glycol, and ethyl alcohol are due mainly to the strong hydrogen-bonding interactions in these liquids (see Section 13-2). The very high value for water makes it quite effective as a coolant and, in the form of steam, as a source of heat.

▶ Molar heats of vaporization (also called molar *enthalpies* of vaporization) are often expressed in kilojoules rather than joules. The units of heat of vaporization do *not* include temperature. This is because boiling occurs with *no change in temperature.*

Liquids can evaporate even below their boiling points. The water in perspiration is an effective coolant for our bodies. Each gram of water that evaporates absorbs 2.41 kJ of heat from the body. We feel even cooler in a breeze because perspiration evaporates faster, so heat is removed more rapidly.

▶ The heat of vaporization of water is higher at 37°C (normal body temperature) than at 100°C (2.41 kJ/g compared to 2.26 kJ/g).

Table 13-5 Heats of Vaporization, Boiling Points, and Vapor Pressures of Some Common Liquids

Liquid	Molecular Weight (g/mol)	Vapor Pressure (torr at 20°C)	Boiling Point at 1 atm (°C)	Heat of Vaporization at Boiling Point	
				J/g	kJ/mol
diethyl ether, $CH_3CH_2OCH_2CH_3$	74	442.	34.6	351	26.0
carbon tetrachloride, CCl_4	153	85.6	76.8	213	32.8
ethanol, CH_3CH_2OH	46	43.9	78.3	855	39.3
benzene, C_6H_6	78	74.6	80.1	395	30.8
water, H_2O	18	17.5	100.	2260	40.7
ethylene glycol, $HOCH_2CH_2OH$	62	0.1	197.3	984	58.9

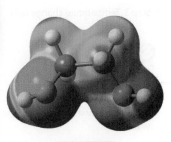

ethylene glycol,
$HOCH_2CH_2OH$

Condensation is the reverse of evaporation. The amount of heat that must be removed from a vapor to condense it (without change in temperature) is called the **heat of condensation**.

$$\text{liquid} + \text{heat} \underset{\text{condensation}}{\overset{\text{evaporation}}{\rightleftharpoons}} \text{vapor}$$

The heat of condensation of a liquid is equal in magnitude to the heat of vaporization. It is released by the vapor during condensation.

Because 2.26 kJ must be absorbed to vaporize one gram of water at 100°C, that same amount of heat must be released to the environment when one gram of steam at 100°C condenses to form liquid water at 100°C. In steam-heated radiators, steam condenses and releases 2.26 kJ of heat per gram as its molecules collide with the cooler radiator walls and condense there. The metallic walls conduct heat well. They transfer the heat to the air in contact with the outside walls of the radiator. The heats of condensation and vaporization of non–hydrogen-bonded liquids, such as benzene, have smaller magnitudes than those of hydrogen-bonded liquids (see Table 13-5). They are, therefore, much less effective as heating and cooling agents.

▶ Because of the large amount of heat released by steam as it condenses, burns caused by steam at 100°C are much more severe than burns caused by liquid water at 100°C.

EXAMPLE 13-2 Heat of Vaporization

Calculate the amount of heat, in joules, required to convert 180. grams of water at 10.0°C to steam at 105.0°C.

Plan

The total amount of heat absorbed is the sum of the amounts required to (1) warm the liquid water from 10.0°C to 100.0°C, (2) convert the liquid water to steam at 100.0°C, and (3) warm the steam from 100.0°C to 105.0°C.

Step 1:
180. g $H_2O(\ell)$ at 10.0°C $\xrightarrow[\text{(temp. change)}]{\text{warm the liquid}}$ 180. g $H_2O(\ell)$ at 100.0°C $\xrightarrow[\text{(phase change)}]{\text{boil the liquid}}$

Step 3:
180. g $H_2O(g)$ at 100.0°C $\xrightarrow[\text{(temp. change)}]{\text{warm the steam}}$ 180. g $H_2O(g)$ at 105.0°C

Steps 1 and 3 involve the specific heats of water and steam, 4.18 J/g·°C and 2.03 J/g·°C, respectively (see Appendix E), whereas step 2 involves the heat of vaporization of water (2.26×10^3 J/g).

Solution

▶ Steps 1 and 3 of this example involve warming with *no* phase change. Such calculations were introduced in Section 1-13.

▶ Step 1: Temperature change only

1. $\underline{?}$ J = 180. g $\times \dfrac{4.18\ \text{J}}{\text{g} \cdot °\text{C}} \times (100.0°\text{C} - 10.0°\text{C}) = 6.77 \times 10^4$ J $= 0.677 \times 10^5$ J

▶ Step 2: Phase change only

2. $\underline{?}$ J = 180. g $\times \dfrac{2.26 \times 10^3\ \text{J}}{\text{g}}$ $= 4.07 \times 10^5$ J

▶ Step 3: Temperature change only

3. $\underline{?}$ J = 180. g $\times \dfrac{2.03\ \text{J}}{\text{g} \cdot °\text{C}} \times (105.0°\text{C} - 100.0°\text{C}) = 1.8 \times 10^3$ J $= 0.018 \times 10^5$ J

Total amount of heat absorbed = 4.76×10^5 J

🛑 **TOP & THINK**
Far more heat is associated with the phase change than with the warming of liquid or steam.

You should now work Exercises 50 and 51.

Distillation is not an economical way to purify large quantities of water for public water supplies. The high heat of vaporization of water requires a large amount of energy, making it too expensive to vaporize large volumes of water.

Problem-Solving Tip Temperature Change or Phase Change?

A problem such as Example 13-2 can be broken down into steps so that each involves *either* a temperature change *or* a phase change, but not both. A temperature change calculation uses the specific heat of the substance (steps 1 and 3 of Example 13-2); remember that each different phase has its own specific heat. Note also that because we are using ΔT (change in temperature) we do not need to convert °C to kelvin. A phase change always takes place with *no change* in temperature, so that calculation does not involve temperature (step 2 of Example 13-2).

EXAMPLE 13-3 Heat of Vaporization

Compare the amount of "cooling" experienced by an individual who drinks 400. mL of ice water (0.0°C) with the amount of "cooling" experienced by an individual who "sweats out" 400. mL of water. Assume that the sweat is essentially pure water and that all of it evaporates. The density of water is very nearly 1.00 g/mL at both 0.0°C and 37.0°C, average body temperature. The heat of vaporization of water is 2.41 kJ/g at 37.0°C.

Plan

In the case of drinking ice water, the body is cooled by the amount of heat required to raise the temperature of 400. mL (400. g) of water from 0.0°C to 37.0°C. The amount of heat lost by perspiration is equal to the amount of heat required to vaporize 400. g of water at 37.0°C.

Solution

Raising the temperature of 400. g of water from 0.0°C to 37.0°C requires

$$\underline{?} = (400.\text{ g})(4.18\text{ J/g}\cdot°\text{C})(37.0°\text{C}) = 6.19 \times 10^4 \text{ J, or } 61.9 \text{ kJ}$$

Evaporating (i.e., "sweating out") 400. mL of water at 37°C requires

$$\underline{?} = (400.\text{ g})(2.41 \times 10^3 \text{ J/g}) = 9.64 \times 10^5 \text{ J, or } 964 \text{ kJ}$$

Thus, we see that "sweating out" 400. mL of water removes 964 kJ of heat from one's body, whereas drinking 400. mL of ice water removes only 61.9 kJ. Stated differently, sweating removes (964/61.9) or 15.6 times more heat than drinking ice water!

You should now work Exercise 55.

As Example 13-3 illustrates, perspiration is more effective at cooling the body than drinking cold fluids. For health reasons, however, it is important to replace the water lost by perspiration.

ENRICHMENT

The Clausius–Clapeyron Equation

We have seen (see Figure 13-13) that vapor pressure increases with increasing temperature. Let us now discuss the quantitative expression of this relationship.

When the temperature of a liquid is changed from T_1 to T_2, the vapor pressure of the liquid changes from P_1 to P_2. These changes are related to the molar heat of vaporization, ΔH_{vap}, for the liquid by the **Clausius–Clapeyron equation**.

$$\ln\left(\frac{P_2}{P_1}\right) = \frac{\Delta H_{vap}}{R}\left(\frac{1}{T_1} - \frac{1}{T_2}\right)$$

Although ΔH_{vap} changes somewhat with temperature, it is usually adequate to use the value tabulated at the normal boiling point of the liquid (see Appendix E) unless more precise values are available. The units of R must be consistent with those of ΔH_{vap}.

▶ The Clausius–Clapeyron equation is used for three types of calculations: (1) to predict the vapor pressure of a liquid at a specified temperature, as in Example 13-4; (2) to determine the temperature at which a liquid has a specified vapor pressure; and (3) to calculate ΔH_{vap} from measurement of vapor pressures at different temperatures.

EXAMPLE 13-4 Vapor Pressure Versus Temperature

The normal boiling point of ethanol, C_2H_5OH, is 78.3°C, and its molar heat of vaporization is 39.3 kJ/mol (see Appendix E). What would be the vapor pressure, in torr, of ethanol at 50.0°C?

Plan

The normal boiling point of a liquid is the temperature at which its vapor pressure is 760 torr, so we designate this as one of the conditions (subscript 1). We wish to find the vapor pressure at another temperature (subscript 2), and we know the molar heat of vaporization. We use the Clausius–Clapeyron equation to solve for P_2.

Solution

$$P_1 = 760 \text{ torr} \qquad \text{at} \qquad T_1 = 78.3°C + 273.2 = 351.5 \text{ K}$$

$$P_2 = \underline{?} \qquad \text{at} \qquad T_2 = 50.0°C + 273.2 = 323.2 \text{ K}$$

$$\Delta H_{vap} = 39.3 \text{ kJ/mol} \qquad \text{or} \qquad 3.93 \times 10^4 \text{ J/mol}$$

We solve for P_2.

$$\ln\left(\frac{P_2}{760 \text{ torr}}\right) = \frac{3.93 \times 10^4 \text{ J/mol}}{\left(8.314 \dfrac{\text{J}}{\text{mol} \cdot \text{K}}\right)} \left(\frac{1}{351.5 \text{ K}} - \frac{1}{323.2 \text{ K}}\right)$$

$$\ln\left(\frac{P_2}{760 \text{ torr}}\right) = -1.18 \qquad \text{so} \qquad \left(\frac{P_2}{760 \text{ torr}}\right) = e^{-1.18} = 0.307$$

$$P_2 = 0.307(760 \text{ torr}) = 233 \text{ torr} \qquad \text{(Note: lower vapor pressure at lower temperature)}$$

You should now work Exercise 58.

Table 13-6 General Effects of Intermolecular Attractions on Physical Properties of Liquids

Property	Volatile Liquids (weak intermolecular attractions)	Nonvolatile Liquids (strong intermolecular attractions)
cohesive forces	low	high
viscosity	low	high
surface tension	low	high
specific heat	low	high
vapor pressure	high	low
rate of evaporation	high	low
boiling point	low	high
heat of vaporization	low	high

We have described many properties of liquids and discussed how they depend on intermolecular forces of attraction. The general effects of these attractions on the physical properties of liquids are summarized in Table 13-6. "High" and "low" are relative terms. Table 13-6 is intended to show only very general trends. Example 13-5 illustrates the use of intermolecular attractions to predict boiling points.

EXAMPLE 13-5 Boiling Points Versus Intermolecular Forces

Predict the order of increasing boiling points for the following: H_2S; H_2O; CH_4; H_2; KBr.

Plan

We analyze the polarity and size of each substance to determine the kinds of intermolecular forces that are present. In general, the stronger the intermolecular forces, the higher is the boiling point of the substance.

Solution

KBr is ionic, so it boils at the highest temperature. Water exhibits strong hydrogen bonding and boils at the next highest temperature. Hydrogen sulfide is the only other polar covalent substance in the list, so it boils below H_2O but above the other two substances. Both CH_4 and H_2 are nonpolar. The larger CH_4 molecule is more easily polarized than the very small H_2 molecule, so dispersion forces are stronger in CH_4. Thus, CH_4 boils at a higher temperature than H_2.

$$H_2 < CH_4 < H_2S < H_2O < KBr$$

increasing boiling points

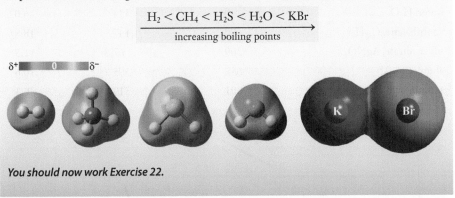

You should now work Exercise 22.

The Solid State

13-10 Melting Point

The **melting point (freezing point)** of a substance is the temperature at which its solid and liquid phases coexist in equilibrium.

$$\text{solid} \underset{\text{freezing}}{\overset{\text{melting}}{\rightleftharpoons}} \text{liquid}$$

The *melting point* of a solid is the same as the *freezing point* of its liquid. It is the temperature at which the rate of melting of a solid is the same as the rate of freezing of its liquid under a given applied pressure.

The **normal melting point** of a substance is its melting point at one atmosphere pressure. Changes in pressure have very small effects on melting points, but they have large effects on boiling points.

13-11 Heat Transfer Involving Solids

When heat is added to a solid below its melting point, its temperature rises. After enough heat has been added to bring the solid to its melting point, additional heat is required to convert the solid to liquid. During this melting process, the temperature remains constant at the melting point until all of the substance has melted. After melting is complete, the continued addition of heat results in an increase in the temperature of the liquid, until the boiling point is reached. This is illustrated graphically in the first three segments of the heating curve in Figure 13-15.

The **molar heat** (or **enthalpy**) **of fusion** (ΔH_{fus}; kJ/mol) is the amount of heat required to melt one mole of a solid at its melting point. Heats of fusion can also be expressed on a per gram basis. The heat of fusion depends on the *inter*molecular forces of attraction in the solid state. These forces "hold the molecules together" as a solid. Heats of fusion are *usually* higher for substances with higher melting points. Values for some common compounds are shown in Table 13-7. Appendix E has more values.

▶ Melting is always endothermic. The term "fusion" means "melting."

Table 13-7 Some Melting Points and Heats of Fusion

Substance	Melting Point (°C)	Heat of Fusion	
		J/g	*kJ/mol*
methane, CH_4	−182	58.6	0.92
ethanol, CH_3CH_2OH	−117	109	5.02
water, H_2O	0	334	6.02
naphthalene, $C_{10}H_8$	80.2	147	18.8
silver nitrate, $AgNO_3$	209	67.8	11.5
aluminum, Al	658	395	10.6
sodium chloride, NaCl	801	519	30.3

Charles D. Winters

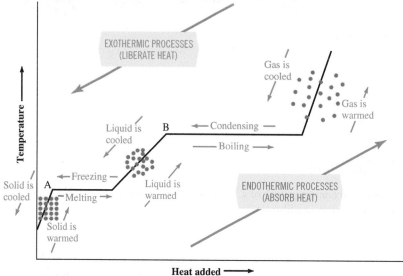

During any phase change,

liquid ⇌ gas

and

solid ⇌ liquid

the temperature remains constant.

Figure 13-15 A typical heating curve at constant pressure. When heat energy is added to a solid below its melting point, the temperature of the solid rises until its melting point is reached (point A). In this region of the plot, the slope is rather steep because of the low specific heats of solids [e.g., 2.09 J/g · °C for $H_2O(s)$]. If additional heat energy is added to the solid at its melting point (A), its temperature remains constant until all of the solid has melted, because the melting process requires energy. The length of this horizontal line is proportional to the heat of fusion of the substance—the higher the heat of fusion, the longer the line. When all of the solid has melted, further heating of the liquid raises its temperature until its boiling point is reached (point B). The slope of this line is less steep than that for warming the solid, because the specific heat of the liquid phase [e.g., 4.18 J/g · °C for $H_2O(\ell)$] is usually greater than that of the corresponding solid. If heat is added to the liquid at its boiling point (B), the added heat energy is absorbed as the liquid boils and the temperature remains constant until all of the liquid has boiled. This horizontal line is longer than the previous one, because the heat of vaporization of a substance is always higher than its heat of fusion. When all of the liquid has been converted to a gas (vapor), the addition of more heat raises the temperature of the gas. This segment of the plot has a steep slope because of the relatively low specific heat of the gas phase [e.g., 2.03 J/g · °C for $H_2O(g)$]. Each step in the process can be reversed by removing the same amount of heat.

The **heat** (or **enthalpy**) **of solidification** of a liquid is equal in magnitude to the heat of fusion. It represents removal of a sufficient amount of heat from a given amount (1 mol or 1 g) of liquid to solidify the liquid at its freezing point. For water,

$$\text{ice} \underset{\substack{6.02 \text{ kJ/mol} \\ \text{or } 334 \text{ J/g released}}}{\overset{\substack{6.02 \text{ kJ/mol} \\ \text{or } 334 \text{ J/g absorbed}}}{\rightleftharpoons}} \text{water} \qquad (\text{at } 0°C)$$

EXAMPLE 13-6 Heat of Fusion

The molar heat of fusion, ΔH_{fus}, of Na is 2.6 kJ/mol at its melting point, 97.5°C. How much heat must be absorbed by 5.0 g of solid Na at 97.5°C to melt it?

Plan

Melting takes place at a constant temperature. The molar heat of fusion tells us that every mole of Na, 23 grams, absorbs 2.6 kJ of heat at 97.5°C during the melting process. We want to know the amount of heat that 5.0 grams would absorb. We use the appropriate unit factors, constructed from the atomic weight and ΔH_{fus}, to find the amount of heat absorbed.

Solution

$$\underline{?} \text{ kJ} = 5.0 \text{ g Na} \times \frac{1 \text{ mol Na}}{23 \text{ g Na}} \times \frac{2.6 \text{ kJ}}{1 \text{ mol Na}} = \boxed{0.57 \text{ kJ}}$$

> **S TOP & THINK**
> Melting takes place at constant temperature, so temperature does not appear in this calculation, as the units of ΔH_{fus} show. The mass of Na given (5.0 g) is smaller than the atomic weight of Na (23 g/mol), so the answer (0.57 kJ) is smaller than the molar heat of fusion.

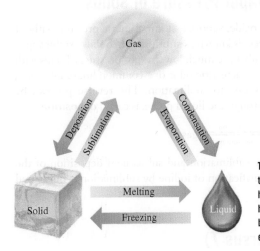

Transitions among the three states of matter. The transitions shown in blue are endothermic (absorb heat); those shown in red are exothermic (release heat). Water vapor is invisible; we can see clouds because they contain tiny droplets of liquid water in equilibrium with water vapor.

EXAMPLE 13-7 Heat of Fusion

Calculate the amount of heat that must be absorbed by 50.0 grams of ice at −12.0°C to convert it to water at 20.0°C. Refer to Appendix E.

Plan

We must determine the amount of heat absorbed during three steps: (1) warming 50.0 g of ice from −12.0°C to its melting point, 0.0°C (we use the specific heat of ice, 2.09 J/g·°C); (2) melting the ice with no change in temperature (we use the heat of fusion of ice at 0.0°C, 334 J/g; and (3) warming the resulting liquid from 0.0°C to 20.0°C (we use the specific heat of water, 4.18 J/g·°C).

▶ Ice is very efficient for cooling because considerable heat is required to melt a given mass of it. However, ΔH_{vap} is generally much greater than ΔH_{fusion}, so evaporative cooling is preferable when possible.

It is important to know the melting and boiling point temperatures for the material in question. We must always treat a phase change as a separate part of the heat calculation.

▶ Step 1: Temperature change only

▶ Step 2: Phase change only

▶ Step 3: Temperature change only

Step 1:
$$50.0 \text{ g H}_2\text{O(s)} \atop \text{at } -12.0°C \xrightarrow[\text{(temp. change)}]{\text{warm the ice}} 50.0 \text{ g H}_2\text{O(s)} \atop \text{at } 0.0°C \xrightarrow[\text{(phase change)}]{\text{melt the ice}}$$

Step 2:

Step 3:
$$50.0 \text{ g H}_2\text{O}(\ell) \atop \text{at } 0.0°C \xrightarrow[\text{(temp. change)}]{\text{warm the liquid}} 50.0 \text{ g H}_2\text{O}(\ell) \atop \text{at } 20°C$$

Solution

1. $50.0 \text{ g} \times \dfrac{2.09 \text{ J}}{\text{g} \cdot °\text{C}} \times [0.0 - (-12.0)]°\text{C} = 1.25 \times 10^3 \text{ J} = 0.125 \times 10^4 \text{ J}$

2. $50.0 \text{ g} \times \dfrac{334 \text{ J}}{\text{g}}$ $\qquad\qquad\qquad\qquad = 1.67 \times 10^4 \text{ J}$

3. $50.0 \text{ g} \times \dfrac{4.18 \text{ J}}{\text{g} \cdot °\text{C}} \times (20.0 - 0.0)°\text{C} = 4.18 \times 10^3 \text{ J} = 0.418 \times 10^4 \text{ J}$

Total amount of heat absorbed $= 2.21 \times 10^4 \text{ J} = 22.1 \text{ kJ}$

Note that most of the heat was absorbed in step 2, melting the ice.

You should now work Exercise 56.

13-12 Sublimation and the Vapor Pressure of Solids

Some solids, such as iodine and carbon dioxide, vaporize at atmospheric pressure without passing through the liquid state. This process is known as **sublimation**. Solids exhibit vapor pressures just as liquids do, but they generally have much lower vapor pressures. Solids with high vapor pressures sublime easily. The characteristic odor of a common household solid, *para*-dichlorobenzene (moth repellent), is due to sublimation. The reverse process, by which a vapor solidifies without passing through the liquid phase, is called **deposition**.

$$\text{solid} \underset{\text{deposition}}{\overset{\text{sublimation}}{\rightleftharpoons}} \text{gas}$$

Some impure solids can be purified by sublimation and subsequent deposition of the vapor (as a solid) onto a cooler surface. Purification of iodine by sublimation is illustrated in Figure 13-16.

13-13 Phase Diagrams (*P* versus *T*)

We have discussed the general properties of the three phases of matter. Now we can describe **phase diagrams** that show the equilibrium pressure–temperature relationships among the different phases of a given pure substance in a closed system. Our discussion of phase diagrams applies only to *closed systems* (e.g., a sample in a sealed container), in which matter does not escape into the surroundings. This limitation is especially important when the vapor phase is involved. Figure 13-17 shows a portion of the phase diagrams for water and carbon dioxide. The curves are not drawn to scale. The distortion allows us to describe the changes of state over wide ranges of pressure or temperature using one diagram.

The curved line from *A* to *C* in Figure 13-17a is a vapor pressure curve obtained experimentally by measuring the vapor pressures of water at various temperatures (Table 13-8). Points along this curve represent the temperature–pressure combinations for which liquid and gas (vapor) coexist in equilibrium. At points above *AC*, the stable form of water is liquid; below the curve, it is vapor.

Figure 13-16 Sublimation can be used to purify volatile solids. Iodine, I_2, sublimes readily to form purple I_2 vapor. The high vapor pressure of the solid I_2 in the lower mixture causes it to sublime when heated. Crystals of purified iodine are formed when the vapor is deposited to form solid on the cooler (upper) portion of the apparatus.

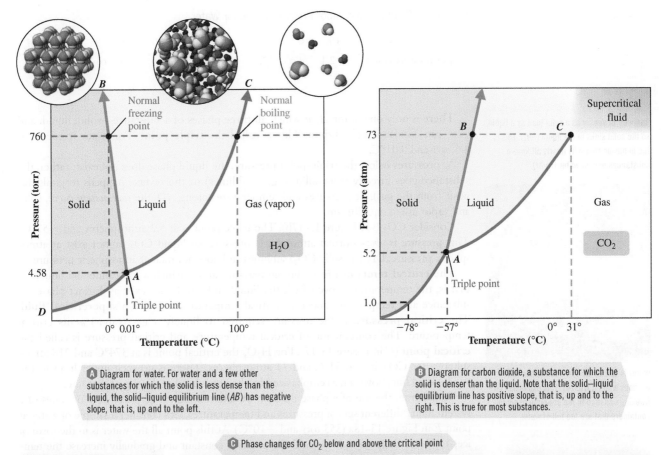

A Diagram for water. For water and a few other substances for which the solid is less dense than the liquid, the solid–liquid equilibrium line (*AB*) has negative slope, that is, up and to the left.

B Diagram for carbon dioxide, a substance for which the solid is denser than the liquid. Note that the solid–liquid equilibrium line has positive slope, that is, up and to the right. This is true for most substances.

C Phase changes for CO_2 below and above the critical point

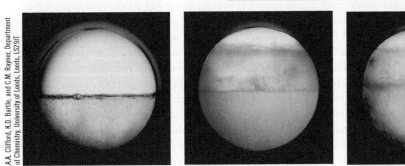

The separate phases of CO_2 are seen through the window in a high-pressure vessel.

As the sample warms and the pressure increases, the meniscus becomes less distinct.

As the temperature continues to increase, it is more difficult to distinguish the liquid and vapor phases.

Above the critical *T* and *P*, distinct liquid and vapor phases are no longer in evidence. This homogeneous phase is "supercritical CO_2."

Figure 13-17 Phase diagrams (not to scale).

Line *AB* represents the liquid–solid equilibrium conditions. We see that this line has a negative slope (up and to the left). Water is one of the very few substances for which this is the case. The negative slope indicates that increasing the pressure sufficiently on the surface of ice causes it to melt. This is because ice is *less dense* than liquid water in the vicinity of the liquid–solid equilibrium. The network of hydrogen bonding in ice is more extensive than that in liquid water and requires a greater separation of H_2O molecules. This causes ice to float in liquid water. Almost all other solids are denser than their corresponding liquids; they would have positive slopes associated with line *AB*. The stable form of water at points to the left of *AB* is solid (ice). Thus *AB* is called a *melting curve*.

▶ The CO_2 in common fire extinguishers is liquid. As you can see from Figure 13-17b, the liquid must be at some pressure greater than 10 atm for temperatures above 0°C. It is ordinarily at about 65 atm (more than 900 lb/in.²), so these cylinders must be *handled with care*.

Table 13-8 Points on the Vapor Pressure Curve for Water

temperature (°C)	−10.	0	20	30	50	70	90	95	100	101
vapor pressure (torr)	2.1	4.6	17.5	31.8	92.5	234	526	634	760	788

Benzene is *denser* as a solid than as a liquid, so the solid sinks in the liquid (*left*). This is the behavior shown by nearly all known substances except water (*right*).

Water is one of the few compounds that expands when it freezes. This expansion on freezing can lead to the breaking of sealed containers that are too full of water.

▶ Phase diagrams are obtained by combining the results of heating curves measured experimentally at different pressures.

▶ A fluid *below* its critical temperature may properly be identified as a liquid or as a gas. *Above* the critical temperature, we should use the term "fluid."

There is only one point, A, at which all three phases of a substance—solid, liquid, and gas—can coexist at equilibrium. This is called the **triple point**. For water it occurs at 4.6 torr and 0.01°C.

At pressures below the triple-point pressure, the liquid phase does not exist; rather, the substance goes directly from solid to gas (sublimes) or the reverse happens (crystals deposit from the gas). At pressures and temperatures along AD, the *sublimation curve*, solid and vapor are in equilibrium.

Consider CO_2 (see Figure 13-17b). The triple point is at 5.2 atmospheres and −57°C. This pressure is *above* normal atmospheric pressure, so liquid CO_2 cannot exist at atmospheric pressure. Dry ice (solid CO_2) sublimes and does not melt at atmospheric pressure.

The **critical temperature** is the temperature above which a gas cannot be liquefied, that is, the temperature above which the liquid and gas do not exist as distinct phases. A substance at a temperature above its critical temperature is called a **supercritical fluid**. The **critical pressure** is the pressure required to liquefy a gas (vapor) *at* its critical temperature. The combination of critical temperature and critical pressure is called the **critical point** (C in Figure 13-17). For H_2O, the critical point is at 374°C and 218 atmospheres; for CO_2, it is at 31°C and 73 atmospheres. There is no such upper limit to the solid–liquid line, however, as emphasized by the arrowhead at the top of that line.

To illustrate the use of a phase diagram in determining the physical state or states of a system under different sets of pressures and temperatures, let's consider a sample of water at point E in Figure 13-18a (355 torr and −10°C). At this point all the water is in the form of ice, $H_2O(s)$. Suppose that we hold the pressure constant and gradually increase the temperature—in other words, trace a path from left to right along EH. At the temperature F at which EH intersects AB (the melting curve) some of the ice melts. If we stopped here, equilibrium between solid and liquid water would eventually be established, and both phases would be present. If we added more heat, all the solid would melt with no temperature change. Remember that all phase changes of pure substances occur at constant temperature.

Once the solid is completely melted, additional heat causes the temperature to rise. Eventually, at point G (355 torr and 80°C), some of the liquid begins to boil; liquid, $H_2O(\ell)$, and vapor, $H_2O(g)$, are in equilibrium. Adding more heat at constant pressure vaporizes the rest of the water with no temperature change. Adding still more heat warms the vapor (gas) from G to H. Complete vaporization would also occur if, at point G and before all the liquid had vaporized, the temperature were held constant and the pressure were decreased to, say, 234 torr at point I. If we wished to hold the pressure at 234 torr and condense some of the vapor, it would be necessary to cool the vapor to 70°C, point J, which lies on the vapor pressure curve, AC. To state this in another way, the vapor pressure of water at 70°C is 234 torr.

Suppose we move back to solid at point E (355 torr and −10°C). If we now hold the temperature at −10°C and reduce the pressure, we move vertically down along EK. At a pressure of 2.1 torr we reach the sublimation curve, at which point the solid passes directly into the gas phase (sublimes) until all the ice has sublimed. An important application of this phenomenon is in the freeze-drying of foods. In this process a water-containing food is cooled below the freezing point of water to form ice, which is then removed as a vapor by decreasing the pressure.

Let's clarify the nature of the fluid phases (liquid and gas) and of the critical point by describing two different ways that a gas can be liquefied. A sample at point W in the phase diagram of Figure 13-18b is in the vapor (gas) phase, below its critical temperature. Suppose we compress the sample at constant T from point W to point Z. We can identify a definite pressure (the intersection of line WZ with the vapor pressure curve AC) where the transition from gas to liquid takes place. If we go *around* the critical point by the path $WXYZ$, however, no such clear-cut transition takes place. By this second path, the density

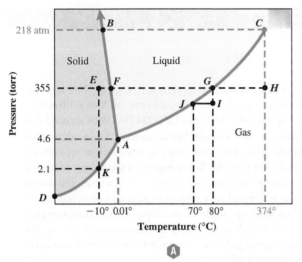

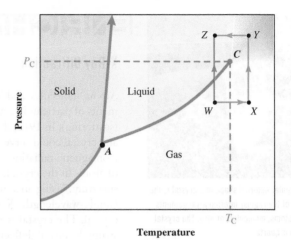

A   **B**

Figure 13-18 Some interpretations of phase diagrams. (a) The phase diagram of water. Phase relationships at various points in this diagram are described in the text. (b) Two paths by which a gas can be liquefied. (1) Below the critical temperature. Compressing the sample at *constant* temperature is represented by the vertical line *WZ*. Where this line crosses the vapor pressure curve *AC*, the gas liquefies; at that set of conditions, *two distinct phases*, gas and liquid, are present in equilibrium with each other. These two phases have different properties, for example, different densities. Raising the pressure further results in a completely liquid sample at point *Z*. (2) Above the critical temperature. Suppose that we instead first warm the gas at constant pressure from *W* to *X*, a temperature above its critical temperature. Then, holding the temperature constant, we increase the pressure to point *Y*. Along this path, the sample increases *smoothly* in density, with no sharp transition between phases. From *Y*, we then decrease the temperature to reach final point *Z*, where the sample is clearly a liquid.

and other properties of the sample vary in a continuous manner; there is no definite point at which we can say that the sample changes from gas to liquid (see Figure 13-17c).

13-14 Amorphous Solids and Crystalline Solids

We have already seen that solids have definite shapes and volumes, are not very compressible, are dense, and diffuse only very slowly into other solids. They are generally characterized by compact, ordered arrangements of particles that vibrate about fixed positions in their structures.

Some noncrystalline solids, called **amorphous solids**, have no well-defined, ordered structure. Examples include rubber, some kinds of plastics, glass, and amorphous sulfur. Some amorphous solids are called "glasses" because, like liquids, they flow, although *very* slowly. The irregular structures of glasses are intermediate between those of freely flowing liquids and those of crystalline solids; there is only short-range order. Crystalline solids such as ice and sodium chloride have well-defined, sharp melting temperatures. Particles in amorphous solids are irregularly arranged, so intermolecular forces among their particles vary in strength within a sample. Melting occurs at different temperatures for various portions of the same sample as the intermolecular forces are overcome. Unlike crystalline solids, glasses and other amorphous solids do not exhibit sharp melting points, but soften over a temperature range.

The shattering of a crystalline solid can produce fragments having the same (or related) interfacial angles and structural characteristics as the original sample. The shattering of a cube of rock salt often produces several smaller cubes of rock salt. Such cleaving occurs preferentially along the crystal lattice planes between which the interionic or intermolecular forces of attraction are weakest. Amorphous solids with irregular structures, such as glasses, shatter irregularly to yield pieces with curved edges and irregular angles.

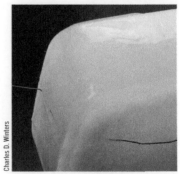

A weighted wire cuts through a block of ice. The ice melts under the high pressure of the wire, and then refreezes behind the wire.

▶ One test for the purity of a crystalline solid is the sharpness of its melting point. Impurities alter the intermolecular forces and cause melting to occur over a considerable temperature range.

ENRICHMENT

The regular external shape of a crystal is the result of the regular internal arrangements of its atoms, molecules, or ions. The crystal shown is quartz.

▶ William and Lawrence Bragg are the only father and son to receive the Nobel Prize, which they shared in physics in 1915.

▶ The lattice planes are planes within the crystal containing regularly ordered arrangements of particles.

X-Ray Diffraction

Atoms, molecules, and ions are much too small to be seen with the eye. The arrangements of particles in crystalline solids are determined indirectly by X-ray diffraction (scattering). In 1912, the German physicist Max von Laue (1879–1960) showed that any crystal could serve as a three-dimensional diffraction grating for incident electromagnetic radiation with wavelengths approximating the internuclear separations of atoms in the crystal. Such radiation is in the X-ray region of the electromagnetic spectrum. Using an apparatus such as that shown in Figure 13-19, a monochromatic (single-wavelength) X-ray beam is defined by a system of slits and directed onto a crystal. The crystal is rotated to vary the angle of incidence θ. At various angles, strong beams of deflected X-rays hit a photographic plate. Upon development, the plate shows a set of symmetrically arranged spots due to deflected X-rays. Different crystals produce different arrangements of spots.

In 1913, the English scientists William (1862–1942) and Lawrence (1890–1971) Bragg found that diffraction photographs are more easily interpreted by considering the crystal as a reflection grating rather than a diffraction grating. The analysis of the spots is somewhat complicated, but an experienced crystallographer can determine the separations between atoms within identical layers and the distances between layers of atoms. The more electrons an atom has, the more strongly it scatters X-rays, so it is also possible to determine the identities of individual atoms using this technique.

Figure 13-20 illustrates the determination of spacings between layers of atoms. The X-ray beam strikes parallel layers of atoms in the crystal at an angle θ. Those rays colliding with atoms in the first layer are reflected at the same angle θ. Those passing through the first layer may be reflected from the second layer, third layer, and so forth. A reflected beam results only if all rays are in phase.

For the waves to be in phase (interact constructively), the difference in path length must be equal to the wavelength, λ, times an integer, n. This leads to the condition known as the **Bragg equation**.

$$n\lambda = 2d \sin \theta \qquad \text{or} \qquad \sin \theta = \frac{n\lambda}{2d}$$

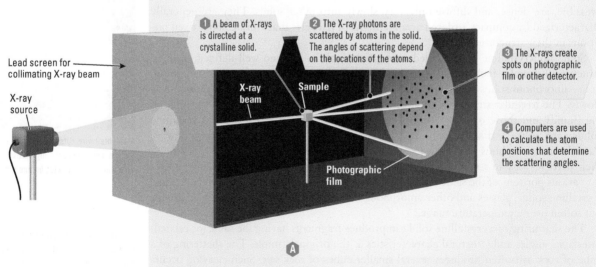

❶ A beam of X-rays is directed at a crystalline solid.

❷ The X-ray photons are scattered by atoms in the solid. The angles of scattering depend on the locations of the atoms.

❸ The X-rays create spots on photographic film or other detector.

❹ Computers are used to calculate the atom positions that determine the scattering angles.

Lead screen for collimating X-ray beam

X-ray source

X-ray beam

Sample

Photographic film

Ⓐ

Figure 13-19 (a) X-ray diffraction by crystals (schematic).

It tells us that for X-rays of a given wavelength λ, atoms in planes separated by distances d give rise to reflections at angles of incidence θ. The reflection angle increases with increasing order, $n = 1, 2, 3, \ldots$.

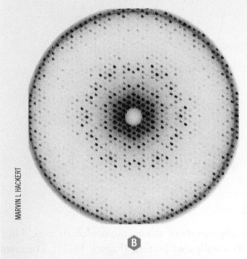

MARVIN L. HACKERT

B

Figure 13-19 (b) A photograph of the X-ray diffraction pattern from a crystal of the enzyme histidine decarboxylase (MW ≈ 37,000 amu). The crystal was rotated so that many different lattice planes with different spacings were moved in succession into diffracting position (see Figure 13-20).

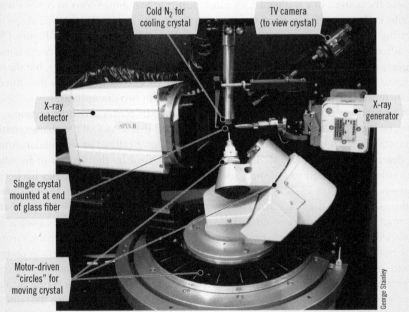

Cold N₂ for cooling crystal

TV camera (to view crystal)

X-ray detector

X-ray generator

APEX II

Single crystal mounted at end of glass fiber

Motor-driven "circles" for moving crystal

George Stanley

A modern single-crystal X-ray diffractometer system.

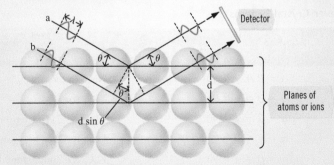

a

λ

b

θ θ

Detector

d

Planes of atoms or ions

θ

d sin θ

Figure 13-20 Reflection of a monochromatic beam of X-rays by two lattice planes (layers of atoms) of a crystal.

Figure 13-21 Patterns that repeat in two dimensions. Such patterns might be used to make wallpaper. We must imagine that the pattern extends indefinitely (to the end of the wall). In each pattern two of the many possible choices of unit cells are outlined. Once we identify a unit cell and its contents, repetition by translating this unit generates the entire pattern. Any crystal is an analogous pattern in which the contents of the three-dimensional unit cell consist of atoms, molecules, or ions. The pattern extends in *three* dimensions to the boundaries of the crystal, usually including many thousands of unit cells.

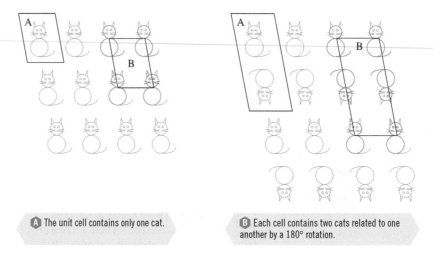

Ⓐ The unit cell contains only one cat.

Ⓑ Each cell contains two cats related to one another by a 180° rotation.

13-15 Structures of Crystals

All crystals contain regularly repeating arrangements of atoms, molecules, or ions. They are analogous (but in three dimensions) to a wallpaper pattern (Figure 13-21). Once we discover the pattern of a wallpaper, we can repeat it in two dimensions to cover a wall. To describe such a repeating pattern we must specify two things: (1) the size and shape of the repeating unit and (2) the contents of this unit. In the wallpaper pattern of Figure 13-21a, two different choices of the repeating unit are outlined. Repeating unit A contains one complete cat; unit B, with the same area, contains parts of four different cats, but these still add up to one complete cat. From whichever unit we choose, we can obtain the entire pattern by repeatedly translating the contents of that unit in two dimensions.

In a crystal the repeating unit is three-dimensional; its contents consist of atoms, molecules, or ions. The smallest unit of volume of a crystal that shows all the characteristics of the crystal's pattern is its **unit cell**. We note that the unit cell is just the fundamental *box* that describes the arrangement. The unit cell is described by the lengths of its edges— *a*, *b*, *c* (which are related to the spacings between layers, *d*)—and the angles between the edges—α, β, γ (Figure 13-22). Unit cells are stacked in three dimensions to build a lattice, the three-dimensional *arrangement* corresponding to the crystal. It can be proven that unit cells must fit into one of the seven crystal systems (Table 13-9). Each crystal system is distinguished by the relations between the unit cell lengths and angles *and* by the symmetry of the resulting three-dimensional patterns. Crystals have the same symmetry as their constituent unit cells because all crystals are repetitive multiples of such cells.

For a mathematical description, we can replace each repeat unit in the crystal by a point (called a *lattice point*) placed at the same place in the unit. All such points have the same environment and are indistinguishable from one another. The resulting three-dimensional array

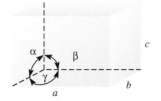

Figure 13-22 A representation of a unit cell.

Table 13-9 The Unit Cell Relationships for the Seven Crystal Systems*

System	Unit Cell		Example (common name)
	Lengths	*Angles*	
cubic	$a = b = c$	$\alpha = \beta = \gamma = 90°$	NaCl (rock salt)
tetragonal	$a = b \neq c$	$\alpha = \beta = \gamma = 90°$	TiO_2 (rutile)
orthorhombic	$a \neq b \neq c$	$\alpha = \beta = \gamma = 90°$	$MgSO_4 \cdot 7H_2O$ (epsomite)
monoclinic	$a \neq b \neq c$	$\alpha = \gamma = 90°$; $\beta \neq 90°$	$CaSO_4 \cdot 2H_2O$ (gypsum)
triclinic	$a \neq b \neq c$	$\alpha \neq \beta \neq \gamma \neq 90°$	$K_2Cr_2O_7$ (potassium dichromate)
hexagonal	$a = b \neq c$	$\alpha = \beta = 90°$; $\gamma = 120°$	SiO_2 (silica)
rhombohedral	$a = b = c$	$\alpha = \beta = \gamma \neq 90°$	$CaCO_3$ (calcite)

*In these definitions, the sign $\neq$ means "is not necessarily equal to."

Figure 13-23 Shapes of unit cells for the seven crystal systems and a representative mineral of each system.

of points is called a *lattice*. It is a simple but complete description of the basic geometric pattern of the crystal structure.

The unit cells shown in Figure 13-23 are the simple, or primitive, unit cells corresponding to the seven crystal systems listed in Table 13-9. Each of these unit cells corresponds to *one* lattice point. As a two-dimensional representation of the reasoning behind this statement, look at the unit cell marked "B" in Figure 13-21a. Each corner of the unit cell is a lattice point, and can be imagined to represent one cat. The cat at each corner is shared among four unit cells (remember—we are working in two dimensions here). The unit cell has four corners, and in the corners of the unit cell are enough pieces to make one complete cat. Thus, unit cell B contains one cat, the same as the alternative unit cell choice marked "A." Now imagine that each lattice point in a three-dimensional crystal represents

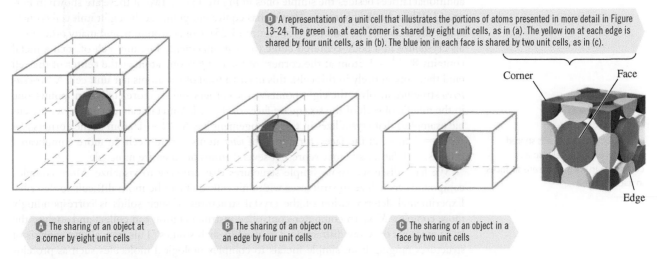

Figure 13-24 Representation of the sharing of an object (an atom, ion, or molecule) among unit cells. The fraction of each sphere that "belongs" to a single unit cell is shown in red.

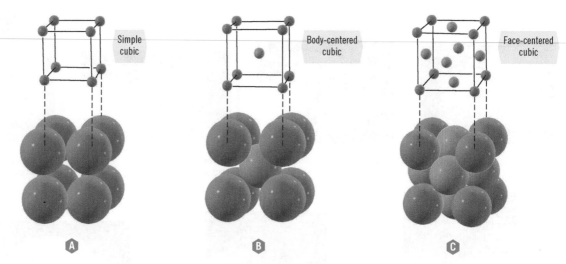

Simple cubic

Body-centered cubic

Face-centered cubic

A

B

C

Figure 13-25 Unit cells for (a) simple cubic, (b) body-centered cubic, and (c) face-centered cubic. The spheres in each figure represent *identical* atoms or ions; different colors are shown *only* to help you visualize the spheres in the center of the cube in body-centered cubic (b) and in face-centered cubic (c) forms.

an object (a molecule, an atom, and so on). Such an object at a corner (Figure 13-24a) is shared by the eight unit cells that meet at that corner. Each unit cell has eight corners, so it contains eight "pieces" of the object, so it contains $8 \times \frac{1}{8} = 1$ object. Similarly, an object on an edge, but not at a corner, is shared by four unit cells (see Figure 13-24b), and an object on a face is shared by two unit cells (see Figure 13-24c).

Each unit cell contains atoms, molecules, or ions in a definite arrangement. Often the unit cell contents are related by some additional symmetry. (For instance, the unit cell in Figure 13-21b contains *two* cats, related to one another by a rotation of 180°.) Different substances that crystallize in the same type of lattice with the same atomic arrangement are said to be **isomorphous**. A single substance that can crystallize in more than one arrangement is said to be **polymorphous**.

In a *simple*, or *primitive*, lattice, only the eight corners of the unit cell are equivalent. In other types of crystals, objects equivalent to those forming the outline of the unit cell may occupy extra positions within the unit cell. (In this context, "equivalent" means that the same atoms, molecules, or ions appear in *identical environments and orientations* at the eight corners of the cell and, when applicable, at other locations in the unit cell.) This results in additional lattices besides the simple ones in Figure 13-23. Two of these are shown in Figure 13-25b and c. A *body-centered* lattice has equivalent points at the eight unit cell corners *and* at the center of the unit cell (see Figure 13-25). Iron, chromium, and many other metals crystallize in a body-centered cubic (bcc) arrangement. The unit cell of such a metal contains $8 \times \frac{1}{8} = 1$ atom at the corners of the cell *plus* one atom at the center of the cell (and therefore entirely in this cell); this makes a total of *two* atoms per unit cell. A *face-centered* structure involves the eight points at the corners and six more equivalent points, one in the middle of each of the six square faces of the cell. A metal (calcium and silver are cubic examples) that crystallizes in this arrangement has $8 \times \frac{1}{8} = 1$ atom at the corners *plus* $6 \times \frac{1}{2} = 3$ more in the faces, for a total of *four* atoms per unit cell. In more complicated crystals, each lattice site may represent several atoms or an entire molecule.

We have discussed some simple structures that are easy to visualize. More complex compounds crystallize in structures with unit cells that can be more difficult to describe. Experimental determination of the crystal structures of such solids is correspondingly more complex. Modern computer-controlled instrumentation can collect and analyze the large amounts of X-ray diffraction data used in such studies. This now allows analysis of structures ranging from simple metals to complex biological molecules such as proteins and nucleic acids. Most of our knowledge about the three-dimensional arrangements of atoms depends on crystal structure studies.

▶ A crystal of one form of polonium metal has Po atoms at the corners of a simple cubic unit cell that is 3.35 Å on edge (Example 13-8).

▶ Each object in a face is shared between two unit cells, so it is counted $\frac{1}{2}$ in each; there are six faces in each unit cell.

13-16 Bonding in Solids

We classify crystalline solids into categories according to the types of particles in the crystal and the bonding or interactions among them. The four categories are (1) metallic solids, (2) ionic solids, (3) molecular solids, and (4) covalent solids. Table 13-10 summarizes these categories of solids and their typical properties.

Metallic Solids

Metals crystallize as solids in which metal ions may be thought to occupy the lattice sites and are embedded in a cloud of delocalized valence electrons. Nearly all metals crystallize in one of three types of lattices: (1) body-centered cubic (bcc), (2) face-centered cubic (fcc; also called cubic close-packed), and (3) hexagonal close-packed. The latter two types are called close-packed structures because the particles (in this case metal atoms) are packed together as closely as possible. The differences between the two close-packed structures are illustrated in Figures 13-26 and 13-27. Let spheres of equal size represent identical metal atoms, or any other particles, that form close-packed structures. Consider a layer of spheres packed in a plane, *A*, as closely as possible (Figure 13-27a). An identical plane of spheres, *B*, is placed in the depressions of plane *A*. If the third plane is placed with its spheres directly above those in plane *A*, the *ABA* arrangement results. This is the hexagonal close-packed structure (see Figure 13-27a). The extended pattern of arrangement of planes is *ABABAB*.... If the third layer is placed in the alternate set of depressions in the

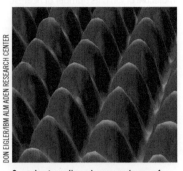

Scanning tunneling microscope image of nickel atoms on the surface of nickel metal.

Table 13-10 **Characteristics of Types of Solids**

	Metallic	**Ionic**	**Molecular**	**Covalent**
Particles of unit cell	Metal ions in "electron cloud"	Anions, cations	Molecules (or atoms)	Atoms
Strongest interparticle forces	Metallic bonds (due to attraction between cations and e^-'s)	Electrostatic	Dispersion, dipole–dipole, and/or hydrogen bonds	Covalent bonds
Properties	Soft to very hard; good thermal and electrical conductors; wide range of melting points (-39 to $3400°C$)	Hard; brittle; poor thermal and electrical conductors; high melting points (400 to $3000°C$)	Soft; poor thermal and electrical conductors; low melting points (-272 to $400°C$)	Very hard; poor thermal and electrical conductors;* high melting points (1200 to $4000°C$)
Examples	Li, K, Ca, Cu, Cr, Ni (metals)	$NaCl$, $CaBr_2$, K_2SO_4 (typical salts)	CH_4 (methane), P_4, O_2, Ar, CO_2, H_2O, S_8	C (diamond), SiO_2 (quartz)

*Exceptions: Diamond is a good conductor of heat; graphite is soft and conducts electricity well.

A Spheres in the same plane, packed as closely as possible. Each sphere touches six others.

B Spheres in two planes, packed as closely as possible. All spheres represent *identical* atoms or ions; different colors are shown *only* to help you visualize the layers. Real crystals have many more than two planes. Each sphere touches six others in its own layer, three in the layer below it, and three in the layer above it; that is, it contacts a total of 12 other spheres (has a coordination number of 12).

Figure 13-26

▶ The **coordination number** of a molecule or ion is its number of nearest neighbors in a crystal. This term is used differently in discussions of coordination chemistry (see Section 25-3).

second layer so that spheres in the first and third layers are not directly above and below each other, the cubic close-packed structure, *ABCABCABC* . . . , results (see Figure 13-27b). In close-packed structures each sphere has a *coordination number* of 12, that is, 12 nearest neighbors. In ideal close-packed structures 74% of a given volume is due to spheres and 26% is empty space. The body-centered cubic structure is less efficient in packing; each sphere has only eight nearest neighbors, and there is more empty space.

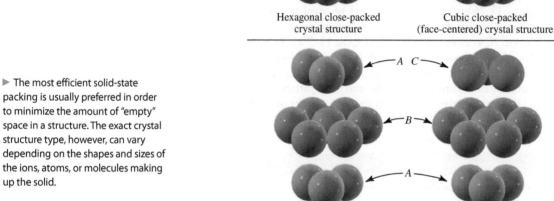

Hexagonal close-packed crystal structure

Cubic close-packed (face-centered) crystal structure

Expanded view

▶ The most efficient solid-state packing is usually preferred in order to minimize the amount of "empty" space in a structure. The exact crystal structure type, however, can vary depending on the shapes and sizes of the ions, atoms, or molecules making up the solid.

Ⓐ In the hexagonal close-packed structure, the first and third layers are oriented in the same direction, so that each atom in the third layer (*A*) lies directly above an atom in the first layer (*A*). All spheres represent identical atoms or ions; different colors are shown only to help you visualize the layers.

Ⓑ In the cubic close-packed structure, the first and third layers are oriented in opposite directions, so that no atom in the third layer (*C*) is directly above an atom in either of the first two layers (*A* and *B*). In both cases, every atom is surrounded by 12 other atoms if the structure is extended indefinitely, so each atom has a coordination number of 12. Although it is not obvious from this figure, the cubic close-packed structure is face-centered cubic. To see this, we would have to include additional atoms and tilt the resulting cluster of atoms.

Figure 13-27 There are two crystal structures in which atoms are packed together as compactly as possible. The diagrams show the structures expanded to clarify the difference between them.

Problem-Solving Tip The Locations of the Nearest Neighbors in Cubic Crystals

The distance from an atom to one of its nearest neighbors in any crystal structure depends on the arrangement of atoms and on the size of the unit cell. For structures (such as metals) that contain only one kind of atom, the nearest neighbors of each atom can be visualized as follows. (Recall that for a cubic structure, the unit cell edge is *a*.) In a simple cubic structure, the nearest neighbors are along the cell edge (a). In a face-centered cubic structure, the nearest neighbors are along the face diagonal (b). In a body-centered cubic structure, they are along the body diagonal (c). The relationships just described hold only for structures composed of a single kind of atom. For other structures, the relationships are more complex.

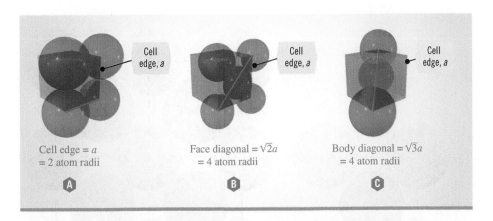

Cell edge = a
= 2 atom radii

A

Face diagonal = $\sqrt{2}a$
= 4 atom radii

B

Body diagonal = $\sqrt{3}a$
= 4 atom radii

C

EXAMPLE 13-8 Nearest Neighbors

In the simple cubic form of polonium there are Po atoms at the corners of a simple cubic unit cell that is 3.35 Å on edge. (a) What is the shortest distance between centers of neighboring Po atoms? (b) How many nearest neighbors does each atom have?

Plan

We visualize the simple cubic cell.

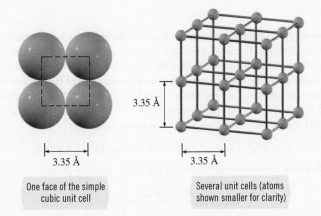

3.35 Å

3.35 Å

3.35 Å

One face of the simple
cubic unit cell

Several unit cells (atoms
shown smaller for clarity)

Solution

(a) One face of the cubic unit cell is shown in the left-hand drawing, with the atoms touching. The centers of the nearest neighbor atoms are separated by one unit cell edge, at the distance 3.35 Å.

(b) A three-dimensional representation of eight unit cells is also shown. In that drawing the atoms are shown smaller for clarity. Consider the atom at the intersection of the eight unit cells. As we can see, this atom has six nearest neighbors, one in each of the unit cell directions. The same would be true of any atom in the structure.

EXAMPLE 13-9 Nearest Neighbors

Silver crystals are face-centered cubic, with a cell edge of 4.086 Å. (a) What is the distance between centers of the two closest Ag atoms? (b) What is the atomic radius of silver in this crystal? (c) How many nearest neighbors does each atom have?

Plan

We reason as in Example 13-8, except that now the two atoms closest to each other are those along the face diagonal.

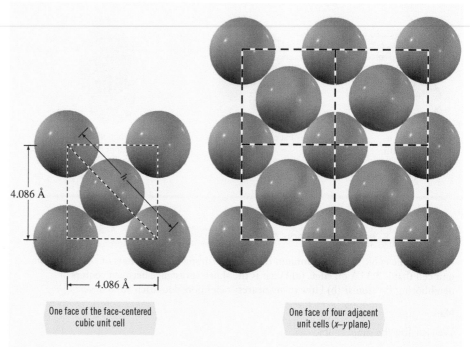

4.086 Å

4.086 Å

One face of the face-centered
cubic unit cell

One face of four adjacent
unit cells (x–y plane)

► In this figure, all atoms are identical; different colors are shown only to help clarify the explanation.

Solution

(a) One face of the face-centered cubic unit cell is shown in the left-hand drawing, with the atoms touching. The nearest neighbor atoms are the ones along the diagonal of the face of the cube. We may visualize the face as consisting of two right isosceles triangles sharing a common hypotenuse, h, and having sides of length a = 4.086 Å. The hypotenuse is equal to *twice* the center-to-center distance. The hypotenuse can be calculated from the Pythagorean theorem, $h^2 = a^2 + a^2$. The length of the hypotenuse equals the square root of the sum of the squares of the sides.

$$h = \sqrt{a^2 + a^2} = \sqrt{2a^2} = \sqrt{2(4.086\text{Å})^2} = 5.778\,\text{Å}$$

The distance between centers of adjacent silver atoms is one half of h, so

$$\text{Distance} = \frac{5.778\,\text{Å}}{2} = 2.889\,\text{Å}$$

(b) The hypotenuse of the unit cell face is four times the radius of the silver atom.

$$\text{Atom radius} = \frac{5.778\,\text{Å}}{4} = 1.444\,\text{Å}$$

(c) To see the number of nearest neighbors, we expand the left-hand drawing to include several unit cells, as shown in the right-hand drawing. Suppose that this is the x–y plane. The atom shown in orange has four nearest neighbors in this plane. There are four more such neighbors in the x–z plane (perpendicular to the x–y plane), and four additional neighbors in the y–z plane (also perpendicular to the x–y plane). This gives a total of 12 nearest neighbors (coordination number 12).

EXAMPLE 13-10 Density and Cell Volume

From data in Example 13-9, calculate the density of metallic silver.

Plan

We first determine the mass of a unit cell, that is, the mass of four atoms of silver. The density of the unit cell, and therefore of silver, is its mass divided by its volume.

Solution

$$\underline{?} \text{ g Ag per unit cell} = \frac{4 \text{ Ag atoms}}{\text{unit cell}} \times \frac{1 \text{ mol Ag}}{6.022 \times 10^{23} \text{ Ag atoms}} \times \frac{107.87 \text{ g Ag}}{1 \text{ mol Ag}}$$

$$= 7.165 \times 10^{-22} \text{ g Ag/unit cell}$$

$$V_{\text{unit cell}} = (4.086 \text{ Å})^3 = 68.22 \text{ Å}^3 \times \left(\frac{10^{-8} \text{ cm}}{\text{Å}}\right)^3 = 6.822 \times 10^{-23} \text{ cm}^3/\text{unit cell}$$

$$\text{Density} = \frac{7.165 \times 10^{-22} \text{ g Ag/unit cell}}{6.822 \times 10^{-23} \text{ cm}^3/\text{unit cell}} = \boxed{10.50 \text{ g/cm}^3}$$

You should now work Exercises 90 and 92.

▶ A handbook gives the density of silver as 10.5 g/cm³ at 20°C.

Data obtained from crystal structures and observed densities give us information from which we can calculate the value of Avogadro's number. The next example illustrates these calculations.

EXAMPLE 13-11 Density, Cell Volume, and Avogadro's Number

Titanium crystallizes in a body-centered cubic unit cell with an edge length of 3.306 Å. The density of titanium is 4.401 g/cm³. Use these data to calculate Avogadro's number.

Plan

We relate the density and the volume of the unit cell to find the total mass contained in one unit cell. Knowing the number of atoms per unit cell, we can then find the mass of one atom. Comparing this to the known atomic weight, which is the mass of one mole (Avogadro's number) of atoms, we can evaluate Avogadro's number.

Solution

We first determine the volume of the unit cell.

$$V_{\text{cell}} = (3.306 \text{ Å})^3 = 36.13 \text{ Å}^3$$

We now convert Å³ to cm³.

$$\underline{?} \text{ cm}^3 = 36.13 \text{ Å}^3 \times \left(\frac{10^{-8} \text{ cm}}{\text{Å}}\right)^3 = 3.613 \times 10^{-23} \text{ cm}^3$$

The mass of the unit cell is its volume times the observed density.

$$\text{Mass of unit cell} = 3.613 \times 10^{-23} \text{ cm}^3 \times \frac{4.401 \text{ g}}{\text{cm}^3} = 1.590 \times 10^{-22} \text{ g}$$

The bcc unit cell contains $8 \times \frac{1}{8} + 1 = 2$ Ti atoms, so this represents the mass of two Ti atoms. The mass of a single Ti atom is

$$\text{Mass of atom} = \frac{1.590 \times 10^{-22} \text{ g}}{2 \text{ atoms}} = 7.950 \times 10^{-23} \text{ g/atom}$$

From the known atomic weight of Ti (47.88), we know that the mass of one mole of Ti is 47.88 g/mol. Avogadro's number represents the number of atoms per mole, and can be calculated as

$$N_{Av} = \frac{47.88 \text{ g}}{\text{mol}} \times \frac{1 \text{ atom}}{7.950 \times 10^{-23} \text{ g}} = 6.023 \times 10^{23} \text{ atoms/mol}$$

You should now work Exercise 96.

Ionic Solids

Most salts crystallize as ionic solids with ions occupying the unit cell. Sodium chloride (Figure 13-28) is an example. Many other salts crystallize in the sodium chloride (face-centered cubic) arrangement. Examples are the halides of Li^+, K^+, and Rb^+, and $M^{2+}X^{2-}$ oxides and sulfides such as MgO, CaO, CaS, and MnO. Two other common ionic structures are those of cesium chloride, CsCl (simple cubic lattice), and zincblende, ZnS (face-centered cubic lattice), shown in Figure 13-29. Salts that are isomorphous with the CsCl structure include CsBr, CsI, NH_4Cl, TlCl, TlBr, and TlI. The sulfides of Be^{2+}, Cd^{2+}, and Hg^{2+}, together with CuBr, CuI, AgI, and ZnO, are isomorphous with the zincblende structure (see Figure 13-29c).

Ionic solids are usually poor electrical and thermal conductors. Liquid (molten) ionic compounds are excellent electrical conductors, however, because their ions are freely mobile.

In certain types of solids, including ionic crystals, particles *different* from those at the corners of the unit cell occupy extra positions within the unit cell. For example, the face-centered cubic unit cell of sodium chloride can be visualized as having chloride ions at the corners and middles of the faces; sodium ions are on the edges between the chloride ions and in the center (see Figures 13-28 and 13-29b). Thus, a unit cell of NaCl contains the following.

$$Cl^- \begin{pmatrix} \text{eight at} \\ \text{corners} \\ (8 \times \frac{1}{8}) \end{pmatrix} + \begin{pmatrix} \text{six in middles} \\ \text{of faces} \\ (6 \times \frac{1}{2}) \end{pmatrix} = 1 + 3 = 4 \; Cl^- \text{ ions/unit cell}$$

$$Na^+ \begin{pmatrix} \text{twelve on} \\ \text{edges} \\ (12 \times \frac{1}{4}) \end{pmatrix} + \begin{pmatrix} \text{one in} \\ \text{center} \\ (1) \end{pmatrix} = 3 + 1 = 4 \; Na^+ \text{ ions/unit cell}$$

The unit cell contains equal numbers of Na^+ and Cl^- ions, as required by its chemical formula. Alternatively, we could translate the unit cell by half its length in any axial direc-

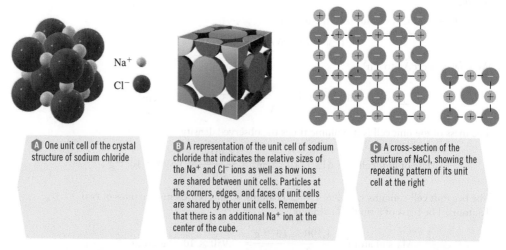

Na^+ ⬤
Cl^- ⬤

Ⓐ One unit cell of the crystal structure of sodium chloride

Ⓑ A representation of the unit cell of sodium chloride that indicates the relative sizes of the Na^+ and Cl^- ions as well as how ions are shared between unit cells. Particles at the corners, edges, and faces of unit cells are shared by other unit cells. Remember that there is an additional Na^+ ion at the center of the cube.

Ⓒ A cross-section of the structure of NaCl, showing the repeating pattern of its unit cell at the right

Figure 13-28 Some representations of the crystal structure of sodium chloride, NaCl. Sodium ions are shown in gray and chloride ions are shown in green.

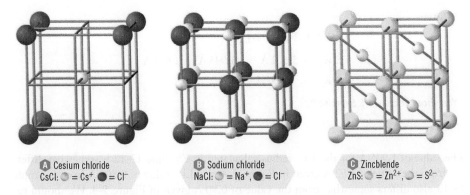

Figure 13-29 Crystal structures of some ionic compounds of the MX type. The gray circles represent cations. One unit cell of each structure is shown. (a) The structure of cesium chloride, CsCl, is simple cubic. It is *not* body-centered, because the point at the center of the cell (Cs$^+$, gray) is not the same as the point at a corner of the cell (Cl$^-$, green). (b) Sodium chloride, NaCl, is face-centered cubic. (c) Zincblende, ZnS, is face-centered cubic, with four Zn^{2+} (gray) and four S^{2-} (yellow) ions per unit cell. The Zn^{2+} ions are related by the same translations as the S^{2-} ions.

tion within the lattice and visualize the unit cell in which sodium and chloride ions have exchanged positions. Such an exchange is not always possible. You should confirm that this alternative description also gives four chloride ions and four sodium ions per unit cell.

Ionic radii such as those in Figure 5-4 and Table 14-1 are obtained from X-ray crystallographic determinations of unit cell dimensions, assuming that adjacent ions are in contact with each other.

EXAMPLE 13-12 Ionic Radii from Crystal Data

Lithium bromide, LiBr, crystallizes in the NaCl face-centered cubic structure with a unit cell edge length of $a = b = c = 5.501$ Å. Assume that the Br$^-$ ions at the corners of the unit cell are in contact with those at the centers of the faces. Determine the ionic radius of the Br$^-$ ion. One face of the unit cell is depicted in Figure 13-30.

Plan

We may visualize the face as consisting of two right isosceles triangles sharing a common hypotenuse, h, and having sides of length $a = 5.501$ Å. The hypotenuse is equal to four times the radius of the bromide ion, $h = 4r_{Br^-}$.

Solution

The hypotenuse can be calculated from the Pythagorean theorem, $h^2 = a^2 + a^2$. The length of the hypotenuse equals the square root of the sum of the squares of the sides.

$$h = \sqrt{a^2 + a^2} = \sqrt{2a^2} = \sqrt{2(5.501 \text{ Å})^2} = 7.780 \text{ Å}$$

The radius of the bromide ion is one fourth of h, so

$$r_{Br^-} = \frac{7.780}{4} = \boxed{1.945 \text{ Å}}$$

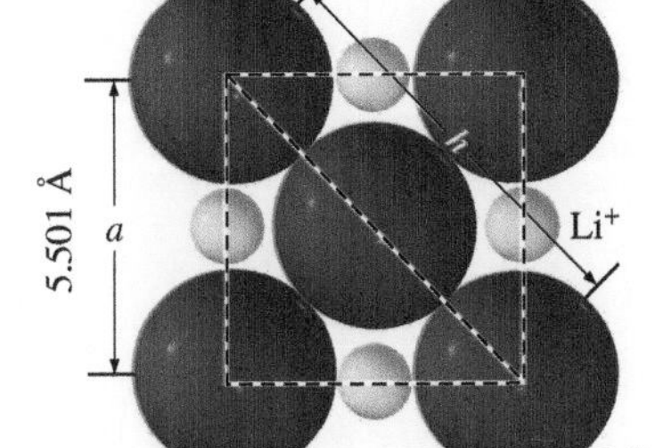

Figure 13-30 One face of the face-centered cubic unit cell of lithium bromide (Example 13-12).

EXAMPLE 13-13 Ionic Radii from Crystal Data

Refer to Example 13-12. Calculate the ionic radius of Li$^+$ in LiBr, assuming anion–cation contact along an edge of the unit cell.

Plan

The edge length, $a = 5.501$ Å, is twice the radius of the Br$^-$ ion plus twice the radius of the Li$^+$ ion. We know from Example 13-12 that the radius for the Br$^-$ ion is 1.945 Å.

Solution

$$5.501 \text{ Å} = 2r_{Br^-} + 2r_{Li^+}$$

$$2r_{Li^+} = 5.501 \text{ Å} - 2(1.945 \text{ Å}) = 1.611 \text{ Å}$$

$$r_{Li^+} = \boxed{0.806 \text{ Å}}$$

You should now work Exercise 88.

The value of 1.945 Å for the Br$^-$ radius calculated in Example 13-12 is somewhat different from the value of 1.82 Å given in Figure 5-4; the Li$^+$ value of 0.806 Å from Example 13-13 also differs somewhat from the value of 0.90 Å given in Figure 5-4. We should remember that the tabulated value in Figure 5-4 is the *average* value obtained from a number of crystal structures of compounds containing the specified ion. Calculations of ionic radii usually assume that anion–anion contact exists, but this assumption is not always true. Calculated radii therefore vary from structure to structure, and we should not place too much emphasis on a value of an ionic radius obtained from any *single* structure determination. We now see that there is some difficulty in determining precise values of ionic radii. Similar difficulties can arise in the determination of atomic radii from molecular and covalent solids or of metallic radii from solid metals.

Molecular Solids

The lattice positions that describe unit cells of molecular solids represent molecules or monatomic elements (sometimes referred to as monatomic molecules). Figure 13-31 shows the unit cells of two simple molecular crystals. Although the bonds *within* molecules are covalent and strong, the forces of attraction *between* molecules are much weaker. They range from hydrogen bonds and weaker dipole–dipole interactions in polar molecules such as H_2O and SO_2 to very weak dispersion forces in symmetrical, nonpolar molecules such as CH_4, CO_2, and O_2 and monatomic elements, such as the noble gases. Because of the relatively weak intermolecular forces of attraction, molecules can be easily displaced. Thus, molecular solids are usually soft substances with low melting points. Because electrons do not move from one molecule to another under ordinary conditions, molecular solids are poor electrical conductors and good insulators.

▶ Dispersion forces are also present among polar molecules.

Covalent Solids

Covalent solids (or "network solids") can be considered giant molecules that consist of covalently bonded atoms in an extended, rigid crystalline network. Diamond (one crystalline

Figure 13-31 The packing arrangement in a molecular crystal depends on the shape of the molecule as well as on the electrostatic interactions of any regions of excess positive and negative charge in the molecules. The arrangements in some molecular crystals are shown here: (a) carbon dioxide, CO_2; (b) benzene, C_6H_6.

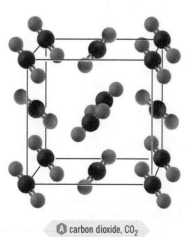

Ⓐ carbon dioxide, CO_2

Ⓑ benzene, C_6H_6

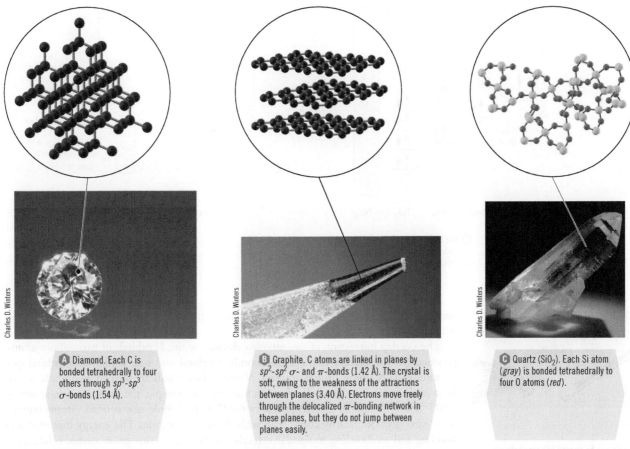

A **Diamond.** Each C is bonded tetrahedrally to four others through sp^3-sp^3 σ-bonds (1.54 Å).

B **Graphite.** C atoms are linked in planes by sp^2-sp^2 σ- and π-bonds (1.42 Å). The crystal is soft, owing to the weakness of the attractions between planes (3.40 Å). Electrons move freely through the delocalized π-bonding network in these planes, but they do not jump between planes easily.

C **Quartz (SiO_2).** Each Si atom (*gray*) is bonded tetrahedrally to four O atoms (*red*).

Figure 13-32 Portions of the atomic arrangements in three covalent solids.

form of carbon) and quartz are examples of covalent solids (Figure 13-32). Because of their rigid, strongly bonded structures, *most* covalent solids are very hard and melt at high temperatures. Because electrons are localized in covalent bonds, they are not freely mobile. As a result, covalent solids are *usually* poor thermal and electrical conductors at ordinary temperatures. (Diamond, however, is a good conductor of heat; jewelers use this property to distinguish gem diamonds from imitations.)

An important exception to these generalizations about properties is *graphite*, an allotropic form of carbon. It has the layer structure shown in Figure 13-32b. The overlap of an extended π-electron network in each plane makes graphite an excellent conductor. The very weak attraction between layers allows these layers to slide over one another easily. Graphite is used as a lubricant, as an additive for motor oil, and in pencil "lead" (combined with clay and other fillers to control hardness).

▶ It is interesting to note that these two allotropes of carbon include one very hard substance and one very soft substance. They differ only in the arrangement and bonding of the C atoms.

13-17 Band Theory of Metals

As described in the previous section, most metals crystallize in close-packed structures. The ability of metals to conduct electricity and heat must result from strong electronic interactions of an atom with its 8 to 12 nearest neighbors. This might be surprising at first if we recall that each Group 1A and Group 2A metal atom has only one or two valence electrons available for bonding. This is too few to participate in bonds localized between it and each of its nearest neighbors.

Bonding in metals is called **metallic bonding**. It results from the electrical attractions among positively charged metal ions and mobile, delocalized electrons belonging to the crystal as a whole. The properties associated with metals—metallic luster, high thermal

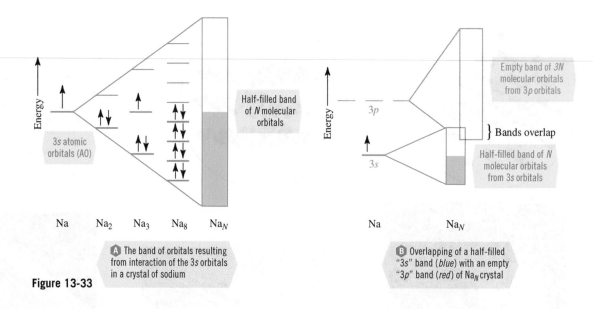

A The band of orbitals resulting from interaction of the 3s orbitals in a crystal of sodium

B Overlapping of a half-filled "3s" band (*blue*) with an empty "3p" band (*red*) of Na$_N$ crystal

Figure 13-33

Metals can be formed into many shapes because of their malleability and ductility.

▶ The alkali metals are those of Group 1A; the alkaline earth metals are those of Group 2A.

and electrical conductivity, and so on—can be explained by the **band theory** of metals, which we now describe.

The overlap interaction of two atomic orbitals, say the 3s orbitals of two sodium atoms, produces two molecular orbitals, one bonding orbital and one antibonding orbital (see Chapter 9). If N atomic orbitals interact, N molecular orbitals are formed. In a single metallic crystal containing one mole of sodium atoms, for example, the interaction (overlap) of 6.022×10^{23} 3s atomic orbitals produces 6.022×10^{23} molecular orbitals. Atoms interact more strongly with nearby atoms than with those farther away. The energy that separates bonding and antibonding molecular orbitals resulting from two given atomic orbitals decreases as the overlap between the atomic orbitals decreases (see Chapter 9). The interactions among the mole of Na atoms result in a series of very closely spaced molecular orbitals (formally σ_{3s} and σ_{3s}^*). These constitute a nearly continuous **band** of orbitals that belongs to the crystal as a whole. One mole of Na atoms contributes 6.022×10^{23} valence electrons (Figure 13-33a), so the 6.022×10^{23} orbitals in the band are half-filled.

The ability of metallic Na to conduct electricity is due to the ability of any of the highest energy electrons in the "3s" band to jump to a slightly higher-energy vacant orbital in the same band when an electric field is applied. The resulting net flow of electrons through the crystal is in the direction of the applied field.

The empty 3p atomic orbitals of the Na atoms also interact to form a wide band of $3 \times 6.022 \times 10^{23}$ orbitals. The 3s and 3p atomic orbitals are quite close in energy, so the fanned-out bands of molecular orbitals overlap, as shown in Figure 13-33b. The two overlapping bands contain $4 \times 6.022 \times 10^{23}$ orbitals and only 6.022×10^{23} electrons. Because each orbital can hold two electrons, the resulting combination of 3s and 3p bands is only one-eighth full.

Overlap of "3s" and "3p" bands is not necessary to explain the ability of Na or of any other Group 1A metal to conduct electricity, as the half-filled "3s" band is sufficient for this. In the Group 2A metals, however, such overlap is important. Consider a crystal of magnesium as an example. The 3s atomic orbital of an isolated Mg atom is filled with two electrons. Thus, without this overlap, the "3s" band in a crystal of Mg is also filled. Mg is a good conductor at room temperature because the highest energy electrons are able to move readily into vacant orbitals in the "3p" band (Figure 13-34).

According to band theory, the highest-energy electrons of metallic crystals occupy either a partially filled band or a filled band that overlaps an empty band. A band within which (or into which) electrons move to allow electrical conduction is called a **conduction band**. The electrical conductivity of a metal decreases as temperature increases. The increase in temperature causes thermal agitation of the metal ions. This impedes the flow of electrons when an electric field is applied.

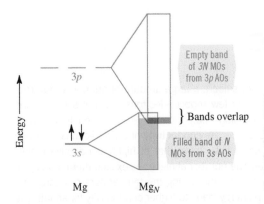

Figure 13-34 Overlapping of a filled "3s" band (*blue*) with an empty "3p" band of Mg$_N$ crystal. The higher-energy electrons are able to move into the "3p" band (*red*) as a result of this overlap. There are now empty orbitals immediately above the filled orbitals, leading to conductivity.

Crystalline nonmetals, such as diamond and phosphorus, are **insulators**—they do not conduct electricity. The reason for this is that their highest energy electrons occupy filled bands of molecular orbitals that are separated from the lowest empty band (conduction band) by an energy difference called the **band gap (E_g)**. In an insulator, this band gap is an energy difference that is too large for electrons to jump to get to the conduction band (Figure 13-35). Many ionic solids are also insulators, but are good conductors in their molten (liquid) state.

Elements that are **semiconductors** have filled bands that are only slightly below, but do not overlap with, empty bands. They do not conduct electricity at low temperatures, but a small increase in temperature is sufficient to excite some of the highest-energy electrons into the empty conduction band.

Let us now summarize some of the physical properties of metals in terms of the band theory of metallic bonding.

1. We have just accounted for the *ability of metals to conduct electricity*.

2. Metals are also *conductors of heat*. They can absorb heat as electrons become thermally excited to low-lying vacant orbitals in a conduction band. The reverse process accompanies the release of heat.

3. Metals have a *lustrous appearance* because the mobile electrons can absorb a wide range of wavelengths of radiant energy as they jump to higher energy levels. Then they emit photons of visible light and fall back to lower levels within the conduction band.

4. Metals are *malleable or ductile* (or both). A crystal of a metal is easily deformed when a mechanical stress is applied to it. All of the metal ions are identical, and they are imbedded in a "sea of electrons." As bonds are broken, new ones are readily formed with adjacent metal ions. The features of the arrangement remain unchanged, and the environment of each metal ion is the same as before the deformation occurred (Figure 13-36). The breakage of bonds involves the promotion of electrons to higher-energy levels. The formation of bonds is accompanied by the return of the electrons to the original energy levels.

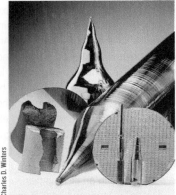

Various samples of elemental silicon. The circle at the lower right is a disk of ultrapure silicon on which many electronic circuits have been etched.

▶ A *malleable* substance can be rolled or pounded into sheets. A *ductile* substance can be drawn into wires. Gold is the most malleable metal known.

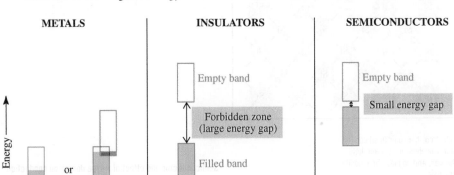

Figure 13-35 Distinction among metals, insulators, and semiconductors. In each case an unshaded area represents a conduction band.

CHEMISTRY IN USE

Semiconductors

A **semiconductor** is an element or a compound with filled bands that are only slightly below, but do not overlap with, empty bands. The difference between an insulator and a semiconductor is only the size of the energy gap, and there is no sharp distinction between them. An *intrinsic* semiconductor (i.e., a semiconductor in its pure form) is a much poorer conductor of electricity than a metal because, for conduction to occur in a semiconductor, electrons must be excited from bonding orbitals in the filled *valence band* into the empty *conduction band*. Figure (a) shows how this happens. An electron that is given an excitation energy greater than or equal to the **band gap** (E_g) enters the conduction band and leaves behind a positively charged *hole* (h^+, the absence of a bonding electron) in the valence band. Both the electron and the hole reside in *delocalized* orbitals, and both can move in an electric field, much as electrons move in a metal. (Holes migrate when an electron in a nearby orbital moves to fill in the hole, thereby creating a new hole in the nearby orbital.) Electrons and holes move in opposite directions in an electric field.

Silicon, a semiconductor of great importance in electronics, has a band gap of 1.94×10^{-22} kJ, or 1.21 *electron volts* (eV). This is the energy needed to create one electron and one hole or, put another way, the energy needed to break one Si—Si bond. This energy can be supplied either thermally or by using light with a photon energy greater than the band gap. To excite one *mole* of electrons from the valence band to the conduction band, an energy of

$$\frac{6.022 \times 10^{23} \text{ electrons}}{\text{mol}} \times \frac{1.94 \times 10^{-22} \text{ kJ}}{\text{electron}} = 117 \text{ kJ/mol}$$

is required. For silicon, a large amount of energy is required, so there are very few mobile electrons and holes (about one electron in a trillion—i.e., 1 in 10^{12}—is excited thermally at room temperature); the conductivity of pure silicon is therefore about 10^{11} times lower than that of highly conductive metals such as silver. The number of electrons excited thermally is proportional to $e^{-E_g/2RT}$. Increasing the temperature or decreasing the band gap energy leads to higher conductivity for an intrinsic semiconductor. Insulators such as diamond and silicon dioxide (quartz), which have very large values of E_g, have conductivities 10^{15} to 10^{20} times lower than most metals.

The electrical conductivity of a semiconductor can be greatly increased by *doping* with impurities. For example, silicon, a Group 4A element, can be doped by adding small amounts of a Group 5A element, such as phosphorus, or a Group 3A element, such as boron. Figure (b) shows the effect of substituting phosphorus for silicon in the crystal structure (silicon has the same structure as diamond, Figure 13-31a). There are exactly enough valence band orbitals to accommodate four of the valence electrons from the phosphorus atom. However, a phosphorus atom has one more electron (and one more proton in its nucleus) than does silicon. The fifth electron enters a higher-energy orbital that is localized in the lattice near the phosphorus atom; the energy of this orbital, called a *donor level*, is just below the conduction band, within the energy gap. An electron in this orbital can easily become *delocalized* when a small amount of thermal energy promotes it into the conduction band. Because the phosphorus-doped silicon contains mobile, *negatively* charged carriers (electrons), it is said to be doped *n-type*. Doping the silicon crystal with boron produces a related, but opposite, effect. Each boron atom contributes only three valence electrons to bonding orbit-

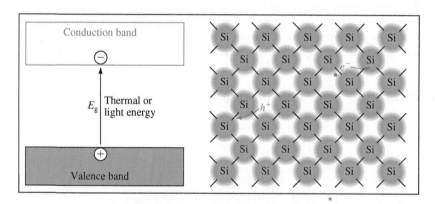

Ⓐ Generation of an electron–hole pair in silicon, an intrinsic semiconductor. The electron (e^-) and hole (h^+) have opposite charges, and so move in opposite directions in an electric field.

Semiconductor and effect of *n*-type doping on conduction properties.

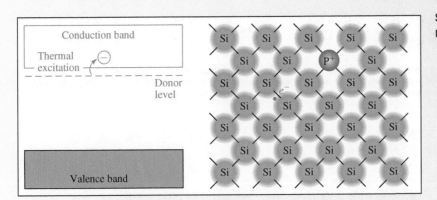

Semiconductor and effect of *n*-type doping on conduction properties.

B *n*-type doping of silicon by phosphorus. The extra valence electron from a phosphorus atom is thermally excited into the conduction band, leaving a fixed positive charge on the phosphorus atom.

als in the valence band, and therefore a *hole* is localized near each boron atom. Thermal energy is enough to separate the negatively charged boron atom from the hole, delocalizing the latter. In this case the charge carriers are the holes, which are *positive,* and the crystal is doped *p-type.* In both *p-* and *n*-type doping, an extremely small concentration of dopants (as little as one part per billion) is enough to cause a significant increase in conductivity. For this reason, great pains are taken to purify the semiconductors used in electronic devices.

The colors of semiconductors are determined by the band gap energy E_g. Only photons with energy greater than E_g can be absorbed. From the Planck radiation formula ($E = h\nu$) and $\lambda\nu = c$, we calculate that the wavelength, λ, of an absorbed photon must be less than hc/E_g. Gallium arsenide (GaAs; $E_g = 1.4$ eV) absorbs photons of wavelengths shorter than 890 nm, which is in the near infrared region. Because it absorbs all wavelengths of visible light, gallium arsenide appears black to the eye. Iron oxide (Fe_2O_3; $E_g = 2.2$ eV) absorbs light of wavelengths shorter than 570 nm; it absorbs both yellow and blue light, and therefore appears red. Cadmium sulfide (CdS; $E_g = 2.6$ eV), which absorbs blue light ($\lambda \leq 470$ nm), appears yellow. Strontium titanate ($SrTiO_3$; $E_g = 3.2$ eV) absorbs only in the ultraviolet ($\lambda \leq 390$ nm). It appears white to the eye because visible light of all colors is reflected by the fine particles.

Even in a doped semiconductor, mobile electrons and holes are both present, although one carrier type is predominant. For example, in a sample of silicon doped with arsenic (*n*-type doping), the concentrations of mobile electrons are slightly less than the concentration of arsenic atoms (usually expressed in terms of atoms/cm³), and the concentrations of mobile holes are extremely low. Interestingly, the concentrations of electrons and holes always follow an equilibrium expression that is entirely analogous to that for the autodissociation of water into H^+ and OH^- ions (see Chapter 18); that is,

$$[e^-][h^+] = K_{eq}$$

where the equilibrium constant K_{eq} depends only on the identity of the semiconductor and the absolute temperature. For silicon at room temperature, $K_{eq} = 4.9 \times 10^{19}$ carriers²/cm⁶.

Doped semiconductors are extremely important in electronic applications. A *p–n junction* is formed by joining *p-* and *n*-type semiconductors. At the junction, free electrons and holes combine, annihilating each other and leaving positively and negatively charged dopant atoms on opposite sides. The unequal charge distribution on the two sides of the junction causes an electric field to develop and gives rise to current rectification (electrons can flow, with a small applied voltage, only from the *n* side to the *p* side of the junction; holes flow only in the reverse direction). Devices such as *diodes* and *transistors,* which form the bases of most analog and digital electronic circuits, are composed of *p–n* junctions.

PROFESSOR THOMAS A. MALLOUK

PENN STATE UNIVERSITY

A crystal can be cleaved into smaller crystals that have the same appearance as the larger crystal.

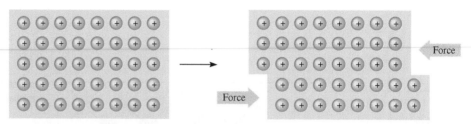

A In a metal, the positively charged metal ions are immersed in a delocalized "cloud of electrons." When the metal is distorted (e.g., rolled into sheets or drawn into wires), the environment around the metal atoms is essentially unchanged, and no new repulsive forces occur. This explains why metal sheets and wires remain intact.

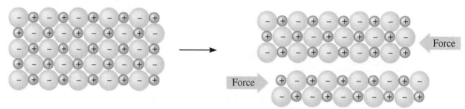

B By contrast, when an ionic crystal is subjected to a force that causes it to slip along a plane, the increased repulsive forces between like-charged ions cause the crystal to break or cleave along a crystal plane.

Figure 13-36

KEY TERMS

Adhesive force   Force of attraction between a liquid and another surface.

Allotropes   Different structural forms of the same element.

Amorphous solid   A noncrystalline solid with no well-defined, ordered structure.

Band   A series of very closely spaced, nearly continuous molecular orbitals that belong to the material as a whole.

Band gap (E_g)   An energy separation between an insulator's highest filled electron energy band and the next higher-energy vacant band.

Band theory of metals   A theory that accounts for the bonding and properties of metallic solids.

Boiling point   The temperature at which the vapor pressure of a liquid is equal to the external pressure; also the condensation point.

Capillary action   The drawing of a liquid up the inside of a small-bore tube when adhesive forces exceed cohesive forces, or the depression of the surface of the liquid when cohesive forces exceed adhesive forces.

Clausius-Clapeyron equation   An equation that relates the change in vapor pressure of a liquid to the change in its temperature and its molar heat of vaporization.

Cohesive forces   All the forces of attraction among particles of a liquid.

Condensation   The process by which a gas or vapor becomes a liquid; liquefaction.

Condensed phases   The liquid and solid phases; phases in which particles interact strongly.

Conduction band   A partially filled band or a band of vacant energy levels just higher in energy than a filled band; a band

within which, or into which, electrons must be promoted to allow electrical conduction to occur in a solid.

Coordination number   In describing crystals, the number of nearest neighbors of an atom or ion.

Critical point   The combination of critical temperature and critical pressure of a substance.

Critical pressure   The pressure required to liquefy a gas (vapor) at its critical temperature.

Critical temperature   The temperature above which a gas cannot be liquefied; the temperature above which a substance cannot exhibit distinct gas and liquid phases.

Crystal lattice   The pattern of arrangement of particles in a crystal.

Crystalline solid   A solid characterized by a regular, ordered arrangement of particles.

Deposition   The direct solidification of a vapor by cooling; the reverse of sublimation.

Dipole–dipole interactions   Interactions between polar molecules, that is, between molecules with permanent dipoles.

Dipole-induced dipole interaction   See *Dispersion forces*.

Dispersion forces   Very weak and very short-range attractive forces between short-lived temporary (induced) dipoles; also known as *London forces* or *van der Waals forces*.

Distillation   The separation of a liquid mixture into its components on the basis of differences in boiling points.

Dynamic equilibrium   A situation in which two (or more) processes occur at the same rate so that no net change occurs.

Enthalpy of fusion   See *Heat of fusion*.

Enthalpy of solidification   See *Heat of solidification*.

Evaporation   The process by which molecules on the surface of a liquid break away and go into the gas phase.

Freezing point See *Melting point*.

Heat of condensation The amount of heat that must be removed from a specific amount of a vapor at its condensation point to condense the vapor with no change in temperature; usually expressed in J/g or kJ/mol; in the latter case it is called the *molar heat of condensation*.

Heat of fusion The amount of heat required to melt a specific amount of a solid at its melting point with no change in temperature; usually expressed in J/g or kJ/mol; in the latter case it is called the *molar heat of fusion*.

Heat of solidification The amount of heat that must be removed from a specific amount of a liquid at its freezing point to freeze it with no change in temperature; usually expressed in J/g or kJ/mol; in the latter case it is called the *molar heat of solidification*.

Heat of vaporization The amount of heat required to vaporize a specific amount of a liquid at its boiling point with no change in temperature; usually expressed in J/g or kJ/mol; in the latter case it is called the *molar heat of vaporization*.

Hydrogen bond A fairly strong dipole–dipole interaction (but still considerably weaker than covalent or ionic bonds) between molecules containing hydrogen directly bonded to a small, highly electronegative atom, such as N, O, or F.

Insulator A poor conductor of electricity and heat.

Intermolecular forces Forces *between* individual particles (atoms, molecules, ions) of a substance.

Intramolecular forces Forces between atoms (or ions) *within* molecules (or formula units).

Isomorphous Refers to crystals having the same atomic arrangement.

LeChatelier's Principle A system at equilibrium, or striving to attain equilibrium, responds in such a way as to counteract any stress placed upon it.

London forces See *Dispersion forces*.

Melting point The temperature at which liquid and solid coexist in equilibrium; also the freezing point.

Meniscus The upper surface of a liquid in a cylindrical container.

Metallic bonding Bonding within metals due to the electrical attraction of positively charged metal ions for mobile electrons that belong to the crystal as a whole.

Molar enthalpy of vaporization See *Molar heat of vaporization*.

Molar heat capacity The amount of heat necessary to raise the temperature of one mole of a substance one degree Celsius with no change in state; usually expressed in kJ/mol · °C. See *Specific heat*.

Molar heat of condensation The amount of heat that must be removed from one mole of a vapor at its condensation point to condense the vapor with no change in temperature; usually expressed in kJ/mol. See *Heat of condensation*.

Molar heat of fusion The amount of heat required to melt one mole of a solid at its melting point with no change in temperature; usually expressed in kJ/mol. See *Heat of fusion*.

Molar heat of vaporization The amount of heat required to vaporize one mole of a liquid at its boiling point with no change in temperature; usually expressed in kJ/mol. See *Heat of vaporization*.

Normal boiling point The temperature at which the vapor pressure of a liquid is equal to one atmosphere pressure.

Normal melting point The melting (freezing) point at one atmosphere pressure.

Phase diagram A diagram that shows equilibrium temperature–pressure relationships for different phases of a substance.

Polymorphous Refers to substances that crystallize in more than one crystalline arrangement.

Semiconductor A substance that does not conduct electricity well at low temperatures but that does at higher temperatures.

Specific heat The amount of heat necessary to raise the temperature of a specific amount of a substance one degree Celsius with no change in state; usually expressed in J/g · °C. See *Molar heat capacity*.

Sublimation The direct vaporization of a solid by heating without passing through the liquid state.

Supercritical fluid A substance at a temperature above its critical temperature. A supercritical fluid cannot be described as either a liquid or gas, but has the properties of both.

Surface tension The result of inward intermolecular forces of attraction among liquid particles that must be overcome to expand the surface area.

Triple point The point on a phase diagram that corresponds to the only pressure and temperature at which three phases (usually solid, liquid, and gas) of a substance can coexist at equilibrium.

Unit cell The smallest repeating unit showing all the structural characteristics of a crystal.

van der Waals forces See *Dispersion forces*.

Vaporization See *Evaporation*.

Vapor pressure The partial pressure of a vapor in equilibrium with its parent liquid or solid.

Viscosity The tendency of a liquid to resist flow; the inverse of its fluidity.

Volatility The ease with which a liquid vaporizes.

EXERCISES

⚫ **Molecular Reasoning** exercises

▲ **More Challenging** exercises

Blue-Numbered exercises are solved in the Student Solutions Manual

General Concepts

1. ⚫ What causes dispersion forces? What factors determine the strengths of dispersion forces between molecules?

2. ⚫ What is hydrogen bonding? Under what conditions can hydrogen bonds be formed?

3. ⚫ Which of the following substances have permanent dipole–dipole forces? (a) GeH_4; (b) molecular $MgCl_2$; (c) PI_3; (d) F_2O.

4. ⚫ Which of the following substances have permanent dipole–dipole forces? (a) molecular $AlBr_3$; (b) PCl_5; (c) NO; (d) SeF_4.

5. ⚫ For which of the substances in Exercise 3 are dispersion forces the only important forces in determining boiling points?

6. ⚫ For which of the substances in Exercise 4 are dispersion forces the only important forces in determining boiling points?

7. ⚫ For each of the following pairs of compounds, predict which compound would exhibit stronger hydrogen bonding. Justify your prediction. It may help to write a Lewis formula for each. (a) water, H_2O, or hydrogen sulfide, H_2S; (b) dichloromethane, CH_2Cl_2 or fluoroamine, NH_2F; (c) acetone, C_3H_6O (contains a $C = O$ double bond) or ethanol, C_2H_6O (contains one $C — O$ single bond).

8. ⚫ Which of the following substances exhibits strong hydrogen bonding in the liquid and solid states? (a) CH_3OH (methanol); (b) PH_3; (c) CH_4; (d) $(CH_3)_2NH$; (e) CH_3NH_2.

9. ⚫ Describe the intermolecular forces that are present in each of the following compounds. Which kind of force would have the greatest influence on the properties of each compound? (a) bromine pentafluoride, BrF_5; (b) acetone, C_3H_6O (contains a central $C = O$ double bond); (c) formaldehyde, H_2CO.

10. ⚫ Describe the intermolecular forces that are present in each of the following compounds. Which kind of force would have the greatest influence on the properties of each compound? (a) ethanol, CH_3CH_2OH (contains one $C — O$ single bond); (b) phosphine, PH_3; (c) sulfur hexafluoride, SF_6.

11. ⚫ Account for the fact that ethylene glycol ($HOCH_2CH_2OH$) is less viscous than glycerine ($HOCH_2CHOHCH_2OH$), but more viscous than ethyl alcohol (CH_3CH_2OH).

12. ⚫ Hydrogen bonding is a very strong dipole–dipole interaction. Why is hydrogen bonding so strong in comparison with other dipole–dipole interactions?

13. ⚫ Which of the following substances exhibits strong hydrogen bonding in the liquid and solid states? (a) H_2S; (b) NH_3; (c) SiH_4; (d) HF; (e) HCl.

14. ⚫ The molecular weights of SiH_4 and PH_3 are nearly the same. Account for the fact that the melting and boiling points of PH_3 ($-133°C$ and $-88°C$) are higher than those of SiH_4 ($-185°C$ and $-112°C$).

15. Give an example of each of the six types of phase changes.

16. ⚫ For each of the following pairs of compounds, predict which would exhibit hydrogen bonding. Justify your prediction. It may help to write a Lewis formula for each. (a) ammonia, NH_3, or phosphine, PH_3; (b) ethylene, C_2H_4, or hydrazine, N_2H_4; (c) hydrogen fluoride, HF, or hydrogen chloride, HCl.

17. ⚫ ▲ Imagine replacing one H atom of a methane molecule, CH_4, with another atom or group of atoms. Account for the order in the normal boiling points of the resulting compounds: CH_4 ($-161°C$); CH_3Br ($3.59°C$); CH_3F ($-78°C$); CH_3OH ($65°C$).

18. ⚫ Suppose you examine samples of copper (a pre-1982 penny is nearly pure copper), rubbing alcohol, and nitrogen (air is mostly nitrogen). Which one is a solid, which one is a liquid, and which one is a gas? How can you tell? What did you observe? How is the solid different from the liquid? How is the liquid different from the gas? Imagine a group of atoms or molecules of each of these substances. What do they look like? Are they in an organized or random arrangement? How far apart are they? How fast are they moving?

19. ⚫ Why does HF have a lower boiling point and lower heat of vaporization than H_2O, even though their molecular weights are nearly the same and the hydrogen bonds between molecules of HF are stronger?

20. ⚫ Consider the following substances: sodium fluoride (NaF), chlorine monofluoride (ClF), hydrogen fluoride (HF), and fluorine (F_2). (a) What type of particles does each substance have? (b) What type of intermolecular attractive forces hold the particles together in each substance? The table below lists the melting point (mp), boiling point (bp), density (D), and solubility in water of each of these substances.

	mp, °C	bp, °C	D, g/L	Solubility
NaF	993	1700	2558 (s)	4.13 g/100 g
ClF	−156	−100	2.43 (g)	violent reaction
HF	−84	20	0.922 (g)	miscible
F_2	−220	−188	1.69 (g)	mild reaction

⚫ **Molecular Reasoning** exercises   ▲ **More Challenging** exercises   Blue-Numbered exercises solved in Student Solutions Manual

Unless otherwise noted, all content on this page is © Cengage Learning.

(c) Based on this information, determine the relative strengths of the forces holding the particles together.
(d) List the intermolecular attractive forces in order of increasing strength.

The Liquid State

21. Use the kinetic–molecular theory to describe the behavior of liquids with changing temperature. Why are liquids denser than gases?

22. ☁ Within each group, assign each of the boiling points to the appropriate substance on the basis of intermolecular forces. (a) Ne, Ar, Kr: $-246°C$, $-186°C$, $-152°C$; (b) NH_3, H_2O, HF: $-33°C$, $-20°C$, $100°C$.

23. Imagine a 250-mL beaker half full of acetone, CH_3COCH_3, sitting in a laboratory in which the air pressure is 760 torr. The acetone is heated to $56°C$, which causes its vapor pressure to increase to 760 torr. At this temperature, bubbles of the vapor form within the body of the liquid, rise to the surface, collapse, and release vaporized molecules. (a) What is this process called? (b) What effect, if any, would each of the following changes have on the temperature at which this process occurs:
(1) adding acetone to the beaker,
(2) removing some acetone from the beaker,
(3) increasing the atmospheric pressure,
(4) decreasing the atmospheric pressure,
(5) increasing the amount of heat being added to the acetone?

24. (a) What is the definition of the normal boiling point? (b) Why is it necessary to specify the atmospheric pressure over a liquid when measuring a boiling point?

25. ☁ Within each group, assign each of the boiling points to the respective substances on the basis of intermolecular forces. (a) N_2, HCN, C_2H_6: $-196°C$, $-89°C$, $26°C$; (b) H_2, HCl, Cl_2: $-35°C$, $-259°C$, $-85°C$.

26. What type of intermolecular forces must be overcome in converting each of the following from a liquid to a gas? (a) CO_2; (b) NH_3; (c) $CHCl_3$; (d) CCl_4.

27. ☁ What factors determine how viscous a liquid is? How does viscosity change with increasing temperature?

28. ☁ What is the surface tension of a liquid? What causes this property? How does surface tension change with increasing temperature?

29. ☁ Dispersion forces are extremely weak in comparison to the other intermolecular attractions. Explain why this is so.

30. ☁ What are some of the similarities of the molecular-level descriptions of the viscosity, surface tension, vapor pressure, and rate of evaporation of a liquid?

31. ☁ What happens inside a capillary tube when a liquid "wets" the tube? What happens when a liquid does not "wet" the tube?

32. ☁ Choose from each pair the substance that, in the liquid state, would have the greater vapor pressure at a given temperature. Base your choice on predicted strengths of intermolecular forces. (a) $BiBr_3$ or $BiCl_3$; (b) CO or CO_2; (c) N_2 or NO; (d) CH_3COOH or $HCOOCH_3$.

33. ☁ Repeat Exercise 32 for (a) C_2H_6 or C_2Cl_6; (b) $F_2C=O$ or CH_3OH; (c) He or H_2.

34. ☁ The temperatures at which the vapor pressures of the following liquids are all 100 torr are given. Predict the order of increasing boiling points of the liquids: butane, C_4H_{10}, $-44.2°C$; 1-butanol, $C_4H_{10}O$, $70.1°C$; diethyl ether, $C_4H_{10}O$, $-11.5°C$.

35. Plot a vapor pressure curve for $GaCl_3$ from the following vapor pressures. Determine the boiling point of $GaCl_3$ under a pressure of 250 torr from the plot:

$t(°C)$	91	108	118	132	153	176	200
vp (torr)	20.	40.	60.	100.	200.	400.	760.

36. Plot a vapor pressure curve for Cl_2O_7 from the following vapor pressures. Determine the boiling point of Cl_2O_7 under a pressure of 125 torr from the plot:

$t(°C)$	-24	-13	-2	10	29	45	62	79
vp (torr)	5.0	10.	20.	40.	100.	200.	400.	760.

37. The vapor pressure of liquid bromine at room temperature is 168 torr. Suppose that bromine is introduced drop by drop into a closed system containing air at 775 torr and room temperature. (The volume of liquid bromine is negligible compared to the gas volume.) If the bromine is added until no more vaporizes and a few drops of liquid are present in the flask, what would be the total pressure? What would be the total pressure if the volume of this closed system were decreased to one half its original value at the same temperature?

38. Which of the following compounds would be expected to form intermolecular hydrogen bonds in the liquid state? (a) CH_3OCH_3 (dimethyl ether); (b) CH_4; (c) HF; (d) CH_3CO_2H (acetic acid); (e) Br_2; (f) CH_3OH (methanol)

39. ▲ ΔH_{vap} is usually greater than ΔH_{fus} for a substance, yet the *nature* of interactions that must be overcome in the vaporization and fusion processes are similar. Why is ΔH_{vap} greater?

40. ▲ The heat of vaporization of water at $100°C$ is 2.26 kJ/g; at $37°C$ (body temperature), it is 2.41 kJ/g. (a) Convert the latter value to standard molar heat of vaporization, $\Delta H°_{vap}$, at $37°C$. (b) Why is the heat of vaporization greater at $37°C$ than at $100°C$?

41. Plot a vapor pressure curve for $C_2Cl_2F_4$ from the following vapor pressures. Determine the boiling point of $C_2Cl_2F_4$ under a pressure of 300. torr from the plot:

$t(°C)$	-95.4	-72.3	-53.7	-39.1	-12.0	3.5
vp (torr)	1.0	10.	40.	100.	400.	760.

42. Plot a vapor pressure curve for $C_2H_4F_2$ from the following vapor pressures. From the plot, determine the boiling point of $C_2H_4F_2$ under a pressure of 200. torr.

$t(°C)$	-77.2	-51.2	-31.1	-15.0	14.8	31.7
vp (torr)	1.0	10.	40.	100.	400.	760.

☁ **Molecular Reasoning** exercises   ▲ **More Challenging** exercises   Blue-Numbered exercises solved in Student Solutions Manual

Phase Changes and Associated Heat Transfer

The following values for water will be useful in some exercises in this section. Values for some other substances appear in Appendix E.

Specific heat of ice	2.09 J/g · °C
Heat of fusion of ice at 0°C	334 J/g
Specific heat of liquid H_2O	4.184 J/g · °C
Heat of vaporization of liquid H_2O at 100°C	2.26×10^3 J/g
Specific heat of steam	2.03 J/g · °C

43. What amount of heat energy, in joules, must be removed to condense 25.4 g of water vapor at 122.5°C to liquid at 23.1°C?

44. What is the total quantity of heat energy transfer required to change 0.50 mol ice at −5°C to 0.50 mol steam at 100°C?

45. Which of the following changes of state are exothermic? Explain. (a) fusion; (b) liquefaction; (c) sublimation; (d) deposition.

46. Is the equilibrium that is established between two physical states of matter an example of static or dynamic equilibrium? How could one demonstrate the type of equilibrium established? Explain your answer.

47. The molar enthalpy of vaporization of methanol is 38.0 kJ/mol at 25°C. How much heat energy transfer is required to convert 250. mL of the alcohol from liquid to vapor? The density of CH_3OH is 0.787 g/mL at 25°C.

48. Suppose 50.0 g of solid bromine at its melting point of −7.2°C is heated, eventually producing gaseous bromine at 100.0°C. Which one of the following steps requires the most heat energy: melting the solid bromine, heating the liquid bromine from its melting point to its boiling point, boiling the bromine, or heating the gaseous bromine from its boiling point to 100.0°C?

> heat of fusion for bromine = 66.15 J/g
> specific heat of liquid bromine = 0.473 J/g · °C
> boiling point of bromine = 58.7°C
> heat of vaporization for bromine = 193.21 J/g
> specific heat of gaseous bromine 0.225 J/g · °C

49. Calculate the amount of heat required to convert 65.0 g of ice at 0°C to liquid water at 100.°C.

50. Calculate the amount of heat required to convert 80.0 g of ice at −15.0°C to steam at 125.0°C

51. Use data in Appendix E to calculate the amount of heat required to warm 165. g of mercury from 25°C to its boiling point and then to vaporize it.

52. If 250. g of liquid water at 100°C and 525 g of water at 30.0°C are mixed in an insulated container, what is the final temperature?

53. If 10.0 g of ice at −10.0°C and 20.0 g of liquid water at 100.°C are mixed in an insulated container, what will the final temperature be?

54. If 180. g of liquid water at 0°C and 18.0 g of steam at 110.°C are mixed in an insulated container, what will the final temperature be?

55. Water can be cooled in hot climates by the evaporation of water from the surfaces of canvas bags. What mass of water can be cooled from 35.0°C to 25.0°C by the evaporation of one gram of water? Assume that $\triangle H_{vap}$ does not change with temperature.

56. (a) How much heat must be removed to prepare 14.0 g of ice at 0.0°C from 14.0 g of water at 25.0°C? (b) Calculate the mass of water at 100.0°C that could be cooled to 23.5°C by the same amount of heat as that calculated in part (a).

Clausius–Clapeyron Equation

57. Toluene, $C_6H_5CH_3$, is a liquid used in the manufacture of TNT. Its normal boiling point is 111.0°C, and its molar heat of vaporization is 35.9 kJ/mol. What would be the vapor pressure (torr) of toluene at 85.00°C?

58. At their normal boiling points, the heat of vaporization of water (100°C) is 40,656 J/mol and that of heavy water (101.41°C) is 41,606 J/mol. Use these data to calculate the vapor pressure of each liquid at 80.00°C.

59. (a) Use the Clausius–Clapeyron equation to calculate the temperature (°C) at which pure water would boil at a pressure of 400.0 torr. (b) Compare this result with the temperature read from Figure 13-13. (c) Compare the results of (a) and (b) with a value obtained from Appendix E.

60. ▲ Show that the Clausius–Clapeyron equation can be written as

$$\ln P = \frac{-\triangle H_{vap}}{RT} + B$$

where B is a constant that has different values for different substances. This is an equation for a straight line. (a) What is the expression for the slope of this line? (b) Using the following vapor pressure data, plot $\ln P$ vs. $1/T$ for ethyl acetate, $CH_3COOC_2H_5$, a common organic solvent used in nail polish removers.

t(°C)	−43.4	−23.5	−13.5	−3.0	+9.1
vp (torr)	1	5	10.	20.	40.

t(°C)	+16.6	+27.0	+42.0	+59.3
vp (torr)	60.	100.	200.	400.

(c) From the plot, estimate $\triangle H_{vap}$ for ethyl acetate.
(d) From the plot, estimate the normal boiling point of ethyl acetate.

Products containing ethyl acetate

61. ▲ Repeat Exercise 60(b) and 60(c) for mercury, using the following data for liquid mercury. Then compare this value with the one in Appendix E.

t(°C)	126.2	184.0	228.8	261.7	323.0
vp (torr)	1	10.	40.	100.	400.

🧪 **Molecular Reasoning** exercises ▲ **More Challenging** exercises Blue-Numbered exercises solved in Student Solutions Manual

Unless otherwise noted, all content on this page is © Cengage Learning.

62. Isopropyl alcohol, C_3H_8O, is marketed as "rubbing alcohol." Its vapor pressure is 100. torr at 39.5°C and 400. torr at 67.8°C. Estimate the molar heat of vaporization of isopropyl alcohol.

63. Using data from Exercise 62, predict the normal boiling point of isopropyl alcohol.

64. Boiling mercury is often used in diffusion pumps to attain a very high vacuum; pressures down to 10^{-10} atm can be readily attained with such a system. Mercury vapor is very toxic to inhale, however. The normal boiling point of liquid mercury is 357°C. What would be the vapor pressure of mercury at 25°C?

Phase Diagrams

65. How many phases exist at a triple point? Describe what would happen if a small amount of heat were added under constant-volume conditions to a sample of water at the triple point. Assume a negligible volume change during fusion.

66. What is the critical point? Will a substance always be a liquid below the critical temperature? Why or why not?

Refer to the phase diagram of CO_2 in Figure 13-17b to answer Exercises 67–70.

67. What phase of CO_2 exists at 1.25 atm pressure and a temperature of (a) −90°C; (b) −60°C; (c) 0°C?

68. What phases of CO_2 are present (a) at a temperature of −78°C and a pressure of 1.0 atm? (b) at −57°C and a pressure of 5.2 atm?

69. List the phases that would be observed if a sample of CO_2 at 10 atm pressure were heated from −80°C to −40°C.

70. How does the melting point of CO_2 change with pressure? What does this indicate about the relative density of solid CO_2 versus liquid CO_2?

71. ▲ You are given the following data for ethanol, C_2H_5OH.

Normal melting point	−117°C
Normal boiling point	78.0°C
Critical temperature	243°C
Critical pressure	63.0 atm

Assume that the triple point is slightly lower in temperature than the melting point and that the vapor pressure at the triple point is about 10^{-5} torr. (a) Sketch a phase diagram for ethanol. (b) Ethanol at 1 atm and 140.°C is compressed to 70. atm. Are two phases present at any time during this process? (c) Ethanol at 1 atm and 270.°C is compressed to 70. atm. Are two phases present at any time during this process?

72. ▲ You are given the following data for butane, C_4H_{10}.

Normal melting point	−138°C
Normal boiling point	0°C
Critical temperature	152°C
Critical pressure	38 atm

Assume that the triple point is slightly lower in temperature than the melting point and that the vapor pressure at the triple point is 3×10^{-5} torr. (a) Sketch a phase diagram for butane. (b) Butane at 1 atm and 140°C is compressed to 40 atm. Are two phases present at any time during this process? (c) Butane at 1 atm and 200°C is compressed to 40 atm. Are two phases present at any time during this process?

Exercises 73 and 74 refer to the phase diagram for sulfur. (The vertical axis is on a logarithmic scale.) Sulfur has two *solid* forms, monoclinic and rhombic.

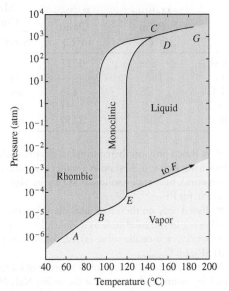

Phase diagram for sulfur

73. ▲ (a) How many triple points are there for sulfur? (b) Indicate the approximate pressure and temperature at each triple point. (c) Which phases are in equilibrium at each triple point?

74. Which physical states should be present at equilibrium under the following conditions? (a) 10^{-1} atm and 110°C; (b) 10^{-5} atm and 80°C; (c) 5×10^3 atm and 160°C; (d) 10^{-1} atm and 80°C; (e) 10^{-5} atm and 140°C; (f) 1 atm and 140°C.

The Solid State

75. ▲ Comment on the following statement: "The only perfectly ordered state of matter is the crystalline state."

76. 🔊 ▲ Ice floats in water. Why? Would you expect solid mercury to float in liquid mercury at its freezing point? Explain.

77. 🔊 ▲ Distinguish among and compare the characteristics of molecular, covalent, ionic, and metallic solids. Give two examples of each kind of solid.

78. 🔊 ▲ Classify each of the following substances, in the solid state, as molecular, ionic, covalent (network), or metallic solids:

	Melting Point (°C)	Boiling Point (°C)	Electrical Conductor	
			Solid	**Liquid**
MoF₆	17.5 (at 406 torr)	35	no	no
BN	3000 (sublimes)	—	no	no
Se₈	217	684	poor	poor
Pt	1769	3827	yes	yes
RbI	642	1300	no	yes

🔊 **Molecular Reasoning** exercises   ▲ **More Challenging** exercises   Blue-Numbered exercises solved in Student Solutions Manual

Unless otherwise noted, all content on this page is © Cengage Learning.

79. 🐾 ▲ Classify each of the following substances, in the solid state, as molecular, ionic, covalent (network), or metallic solids:

	Melting Point (°C)	Boiling Point (°C)	Electrical Conductor	
			Solid	Liquid
CeCl₃	848	1727	no	yes
Ti	1675	—	yes	yes
TiCl₄	−25	136	no	no
NO₃F	−175	−45.9	no	no
B	2300	2550	no	no

80. 🐾 ▲ Based only on their formulas, classify each of the following in the solid state as a molecular, ionic, covalent (network), or metallic solid: (a) SO_2F; (b) MgF_2; (c) W; (d) Pb; (e) PF_5.

81. 🐾 Based only on their formulas, classify each of the following in the solid state as a molecular, ionic, covalent (network), or metallic solid: (a) Au; (b) NO_2; (c) CaF_2; (d) SF_4; (e) $C_{diamond}$.

82. 🐾 Arrange the following solids in order of increasing melting points and account for the order: NaF, MgF_2, AlF_3.

83. 🐾 Arrange the following solids in order of increasing melting points and account for the order: MgO, CaO, SrO, BaO.

Unit Cell Data: Atomic and Ionic Sizes

84. Distinguish among and sketch simple cubic, body-centered cubic (bcc), and face-centered cubic (fcc) lattices. Use CsCl, sodium, and nickel as examples of solids existing in simple cubic, bcc, and fcc lattices, respectively.

85. Describe a unit cell as precisely as you can.

86. Determine the number of ions of each type present in each unit cell shown in Figure 13-29.

87. Refer to Figure 13-29a. (a) If the unit cell edge is represented as a, what is the distance (center to center) from Cs^+ to its nearest neighbor? (b) How many equidistant nearest neighbors does each Cs^+ ion have? What are the identities of these nearest neighbors? (c) What is the distance (center to center), in terms of a, from a Cs^+ ion to the nearest Cs^+ ion? (d) How many equidistant nearest neighbors does each Cl^- ion have? What are their identities?

88. Refer to Figure 13-28. (a) If the unit cell edge is represented as a, what is the distance (center to center) from Na^+ to its nearest neighbor? (b) How many equidistant nearest neighbors does each Na^+ ion have? What are the identities of these nearest neighbors? (c) What is the distance (center to center), in terms of a, from a Na^+ ion to the nearest Na^+ ion? (d) How many equidistant nearest neighbors does each Cl^- ion have? What are their identities?

89. Polonium crystallizes in a simple cubic unit cell with an edge length of 3.36 Å. (a) What is the mass of the unit cell? (b) What is the volume of the unit cell? (c) What is the theoretical density of Po?

90. Calculate the density of Na metal. The length of the body-centered cubic unit cell is 4.24 Å.

91. Gold crystallizes into a face-centered structure, and its density is 19.3 g/cm³. What is the length of an edge of the unit cell in nm?

92. The atomic radius of iridium is 1.36 Å. The unit cell of iridium is a face-centered cube. Calculate the density of iridium.

93. A certain metal has a specific gravity of 10.200 at 25°C. It crystallizes in a body-centered cubic arrangement with a unit cell edge length of 3.147 Å. Determine the atomic weight, and identify the metal.

94. The structure of diamond follows, with each sphere representing a carbon atom. (a) How many carbon atoms are there per unit cell in the diamond structure? (b) Verify, by extending the drawing if necessary, that each carbon atom has four nearest neighbors. What is the arrangement of these nearest neighbors? (c) What is the distance (center to center) from any carbon atom to its nearest neighbor, expressed in terms of a, the unit cell edge? (d) The observed unit cell edge length in diamond is 3.567 Å. What is the C—C single bond length in diamond? (e) Calculate the density of diamond.

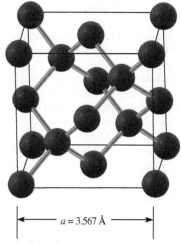

a = 3.567 Å

Structure of diamond

95. The crystal structure of CO_2 is cubic, with a cell edge length of 5.540 Å. A diagram of the cell is shown in Figure 13-31a. (a) What is the number of molecules of CO_2 per unit cell? (b) Is this structure face-centered cubic? How can you tell? (c) What is the density of solid CO_2 at this temperature?

96. A Group 4A element with a density of 11.35 g/cm³ crystallizes in a face-centered cubic lattice whose unit cell edge length is 4.95 Å. Calculate its atomic weight. What is the element?

97. Crystalline silicon has the same structure as diamond, with a unit cell edge length of 5.430 Å. (a) What is the Si—Si distance in this crystal? (b) Calculate the density of crystalline silicon.

98. ▲ (a) What types of electromagnetic radiation are suitable for diffraction studies of crystals? (b) Describe the X-ray diffraction experiment. (c) What must be the relationship between the wavelength of incident radiation and the spacing of the particles in a crystal for diffraction to occur?

🐾 **Molecular Reasoning** exercises   ▲ **More Challenging** exercises   Blue-Numbered exercises solved in Student Solutions Manual

Unless otherwise noted, all content on this page is © Cengage Learning.

99. ▲ (a) Write the Bragg equation. Identify each symbol. (b) X-rays from a palladium source ($\lambda = 0.576$ Å) were reflected by a sample of copper at an angle of 9.408. This reflection corresponds to the unit cell length ($d = a$) with $n = 2$ in the Bragg equation. Calculate the length of the copper unit cell.

100. The spacing between successive planes of platinum atoms parallel to the cubic unit cell face is 2.256 Å. When X-radiation emitted by copper strikes a crystal of platinum metal, the minimum diffraction angle of X-rays is 19.98°. What is the wavelength of the Cu radiation?

101. Gold crystallizes in a fcc structure. When X-radiation of 0.70926 Å wavelength from molybdenum is used to determine the structure of metallic gold, the minimum diffraction angle of X-rays by the gold is 8.6838. Calculate the spacing between parallel layers of gold atoms.

Metallic Bonding and Semiconductors

102. ♣ In general, metallic solids are ductile and malleable, whereas ionic salts are brittle and shatter readily (although they are hard). Explain this observation.

103. ♣ What single factor accounts for the ability of metals to conduct both heat and electricity in the solid state? Why are ionic solids poor conductors of heat and electricity although they are composed of charged particles?

104. Compare the temperature dependence of electrical conductivity of a metal with that of a typical metalloid. Explain the difference.

Mixed Exercises

105. Benzene, C_6H_6, boils at 80.1°C. How much energy, in joules, would be required to change 450.0 g of liquid benzene at 21.5°C to a vapor at its boiling point? (The specific heat of liquid benzene is 1.74 J/g · °C and its heat of vaporization is 395 J/g.)

106. The three major components of air are N_2 (bp = −196°C), O_2 (bp = −183°C), and Ar (bp = −186°C). Suppose we have a sample of liquid air at −200°C. In what order will these gases evaporate as the temperature is raised?

107. ▲ A 20.0-g sample of liquid ethanol, C_2H_5OH, absorbs 6.84×10^3 J of heat at its normal boiling point, 78.0°C. The molar enthalpy of vaporization of ethanol, $\triangle H_{vap}$, is 39.3 kJ/mol. (a) What volume of C_2H_5OH vapor is produced? The volume is measured at 78.0°C and 1.00 atm pressure. (b) What mass of C_2H_5OH remains in the liquid state?

108. Liquid ethylene glycol, $HOCH_2CH_2OH$, is one of the main ingredients in commercial antifreeze. Do you predict its viscosity to be greater or less than that of ethanol, CH_3CH_2OH?

109. ▲ The boiling points of HCl, HBr, and HI increase with increasing molecular weight. Yet the melting and boiling points of the sodium halides, NaCl, NaBr, and NaI, decrease with increasing formula weight. Explain why the trends are opposite. Describe the intermolecular forces present in each of these compounds and predict which has the lowest boiling point.

110. The structures for three molecules having the formula $C_2H_2Cl_2$ are

Describe the intermolecular forces present in each of these compounds and predict which has the lowest boiling point.

111. Are the following statements true or false? Indicate why if a statement is false. (a) The vapor pressure of a liquid will decrease if the volume of liquid decreases. (b) The normal boiling point of a liquid is the temperature at which the external pressure equals the vapor pressure of the liquid. (c) The vapor pressures of liquids in a similar series tend to increase with increasing molecular weight.

112. Are the following statements true or false? Indicate why if a statement is false. (a) The equilibrium vapor pressure of a liquid is independent of the volume occupied by the vapor above the liquid. (b) The normal boiling point of a liquid changes with changing atmospheric pressure. (c) The vapor pressure of a liquid will increase if the mass of liquid is increased.

113. The following are vapor pressures at 20°C. Predict the order of increasing normal boiling points of the liquids, acetone, 185 torr; ethanol, 44 torr; carbon disulfide, CS_2, 309 torr.

114. Refer to Exercise 113. What is the expected order of increasing molar heats of vaporization, $\triangle H_{vap}$, of these liquids at their boiling points? Account for the order.

115. Refer to the sulfur phase diagram in Exercises 73 and 74. (a) Can rhombic sulfur be sublimed? If so, under what conditions? (b) Can monoclinic sulfur be sublimed? If so, under what conditions? (c) Describe what happens if rhombic sulfur is slowly heated from 80°C to 140°C at constant 1-atm pressure. (d) What happens if rhombic sulfur is heated from 80°C to 140°C under constant pressure of 5×10^{-6} atm?

116. The normal boiling point of ammonia, NH_3, is −33°C, and its freezing point is −78°C. Fill in the blanks. (a) At STP (0°C, 1 atm pressure), NH_3 is a _____. (b) If the temperature drops to −40°C, the ammonia will _____ and become a _____. (c) If the temperature drops further to −80°C and the molecules arrange themselves in an orderly pattern, the ammonia will _____ and become a _____. (d) If crystals of ammonia are left on the planet Mars at a temperature of −100°C, they will gradually disappear by the process of _____ and form a _____.

117. Give the correct names for these changes in state: (a) Crystals of *para*-dichlorobenzene, used as a moth repellent, gradually become vapor without passing through the liquid phase. (b) As you enter a warm room from the outdoors on a cold winter day, your eyeglasses become fogged with a film of moisture. (c) On the same (windy) winter day, a pan of water is left outdoors. Some of it turns to vapor, the rest to ice.

118. The normal boiling point of trichlorofluoromethane, CCl_3F, is 24°C, and its freezing point is −111°C. Complete these sentences by supplying the proper terms that describe a state of matter or a change in state. (a) At standard

♣ **Molecular Reasoning** exercises   ▲ **More Challenging** exercises   Blue-Numbered exercises solved in Student Solutions Manual

Unless otherwise noted, all content on this page is © Cengage Learning.

temperature and pressure, CCl₃F is a _____. (b) In an arctic winter at −40°C and 1 atm pressure, CCl₃F is a _____. If it is cooled to −120°C, the molecules arrange themselves in an orderly lattice, the CCl₃F _____ and becomes a _____. (c) If crystalline CCl₃F is held at a temperature of −120°C while a stream of helium gas is blown over it, the crystals will gradually disappear by the process of _____. If liquid CCl₃F is boiled at atmospheric pressure, it is converted to a _____ at a temperature of _____.

Conceptual Exercises

119. The van der Waals constants (see Section 12–15) are $a = 19.01$ L²·atm/mol², $b = 0.1460$ L/mol for pentane, and $a = 18.05$ L²·atm/mol², $b = 0.1417$ L/mol for isopentane.

$$CH_3-CH_2-CH_2-CH_2-CH_3 \qquad CH_3-\underset{\underset{H}{|}}{\overset{\overset{CH_3}{|}}{C}}-CH_2-CH_3$$

pentane   isopentane

(a) Basing your reasoning on intermolecular forces, why would you expect a for pentane to be greater? (b) Basing your reasoning on molecular size, why would you expect b for pentane to be greater?

120. 🫧 What is contained in the bubbles formed during the boiling process?

121. 🫧 How can one convert a sample of a liquid to a vapor without changing its temperature?

122. ▲ Iodine sublimes at room temperature and pressure; water does not. Explain the differences you would expect to observe at room temperature if 7.5 grams of iodine crystals were sealed in a 10.-mL container and 7.5 mL of water were sealed in a similar container.

Iodine sublimes

123. 🫧 Using ball-and-stick representations, draw a sketch similar to Figure 13-3 of four molecules of HBr that indicates dipole–dipole forces between molecules.

124. 🫧 Using space-filling representations, draw a sketch similar to Figure 13-3 of four molecules of HBr that indicates dipole–dipole forces between molecules.

125. Referring to the phase diagram for carbon dioxide shown in Figure 13-17b for approximate values, draw a heating curve similar to that in Figure 13-15 for carbon dioxide at 1 atmosphere pressure. Draw a second heating curve for carbon dioxide at 5 atm pressure. Estimate the transition temperatures.

126. A popular misconception is that "hot water freezes more quickly than cold water." In an experiment, two 100.-mL samples of water, in identical containers, were placed far apart in a freezer at −25°C. One sample had an initial temperature of 78°C, while the other was at 24°C. The second sample took 151 minutes to freeze, and the warmer sample took 166 minutes. The warmer sample took more time but not much. How can you explain their taking about the same length of time to freeze?

127. Consider the portions of the heating curves shown here. Which compound has the highest specific heat capacity?

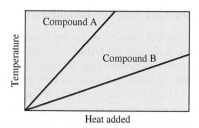

128. An unopened ice-cold can of Coca-Cola® is placed on the kitchen table in the summertime. After a while, drops of water are observed on the outside of the can. Where did they come from?

129. The element potassium (a metal) has a melting point of 63.65°C and a boiling point of 774°C. Suppose a sample of potassium at 100°C is cooled to room temperature (25°C). In what state of matter is the potassium at 100°C? In what state of matter is the potassium at 25°C? What term is used to describe the phase change, if any, that occurs as potassium is cooled from 100°C to 25°C?

Building Your Knowledge

130. More than 150 years ago Pierre Dulong and A. T. Petit discovered a *rule of thumb* that the heat capacity of one mole of a pure solid element is about 6.0 calories per °C (i.e., about 25 J/°C). A 100.2-g sample of an unknown metal at 99.9°C is placed in 50.6 g of water at 24.8°C. The temperature is 36.6°C when the system comes to equilibrium. Assume all heat lost by the metal is absorbed by the water. What is the likely identity of this metal?

131. In a flask containing dry air and some *liquid* silicon tetrachloride, SiCl₄, the total pressure was 988 torr at 225°C. Halving the volume of the flask increased the pressure to 1742 torr at constant temperature. What is the vapor pressure of SiCl₄ in this flask at 225°C?

132. A friend comes to you with this problem: "I looked up the vapor pressure of water in a table; it is 26.7 torr at 300

🫧 **Molecular Reasoning** exercises   ▲ **More Challenging** exercises   Blue-Numbered exercises solved in Student Solutions Manual

Unless otherwise noted, all content on this page is © Cengage Learning.

K and 92,826 torr at 600 K. That means that the vapor pressure increases by a factor of 3477 when the absolute temperature doubles over this temperature range. But I thought the pressure was proportional to the absolute temperature, $P = nRT/V$. The pressure doesn't just double. Why?" How would you help the friend?

133. 🐾 Using as many as six drawings (frames) for each, depict the changes that occur at the molecular level during each of the following physical changes: (a) melting an ice cube; (b) sublimation of an ice cube below 4.6°C and 4.6 torr; and (c) evaporation of a droplet of water at room temperature and pressure.

134. 🐾 Write the Lewis formula of each member of each of the following pairs. Then use VSEPR theory to predict the geometry about each central atom, and describe any features that lead you to decide which member of each pair would have the lower boiling point. (a) CH_3COOH and $HCOCH_3$; (b) NHF_2 and BH_2Cl; (c) CH_3CH_2OH and CH_3OCH_3

135. At its normal melting point of 271.3°C, solid bismuth has a density of 9.73 g/cm³, and liquid bismuth has a density of 10.05 g/cm³. A mixture of liquid and solid bismuth is in equilibrium at 271.3°C. If the pressure were increased from 1 atm to 10 atm, would more solid bismuth melt or would more liquid bismuth freeze? What unusual property does bismuth share with water?

136. 🐾 Although the specific heat of liquid water is usually given as 4.18 J/g°C, it actually varies with temperature, as shown in the graph below. At what temperatures are the deviations from the accepted value the greatest? What is the approximate percent error at these temperatures that have the greatest deviation?

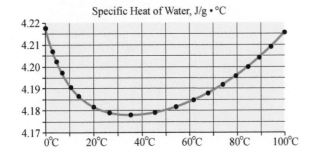

Specific Heat of Water, J/g • °C

Beyond the Textbook

NOTE: *Whenever the answer to an exercise depends on information obtained from a source other than this textbook, the source must be included as an essential part of the answer.*

137. Using *The Merck Index* or *Handbook of Chemistry and Physics* or a suitable site found on the web, look up the temperature at which caffeine sublimes at standard atmospheric pressure.

138. (a) On the web locate the temperature and the pressure at the triple point of carbon dioxide and compare the data that you find with that in Figure 13-17 in the textbook. It is suggested that you follow these steps: Go to **webbook.nist.gov**, and click on "NIST Chemistry WebBook" and then "Formula." Enter CO_2, and click on "Phase Change." If this route does not work, you will need to rely upon one of the search programs to help you find these data. (b) Repeat part (a) for water instead of carbon dioxide.

139. Go to **www.ibiblio.org/e-notes/Cryst/Cryst.htm** using a compatible browser. Click on Face Centered Cubic (FCC) and view the interactive unit cell model. (a) Sketch the atom arrangement in the top layer, middle layer, and bottom layer. What do you observe about the arrangement of the atoms in each layer? Which are the same? (b) Rotate the model to view another face and redraw the atom arrangement on the three layers. Do you observe any difference?

140. Open **www.lyo-san.ca/english/lyophilisation.html.** Using information at this site, at another site, or from reference books, compare the terms: lyophilization, freeze-drying, and sublimation.

🐾 **Molecular Reasoning** exercises   ▲ **More Challenging** exercises   Blue-Numbered exercises solved in Student Solutions Manual

Unless otherwise noted, all content on this page is © Cengage Learning.

Solutions

14

Spreading a soluble salt such as calcium chloride, $CaCl_2$, on icy roadways or sidewalks forms a solution with a lower freezing point than pure water. This causes the ice or snow to melt at a lower temperature and prevents the resulting salt solution from freezing until it reaches the same lower temperature. This lower temperature of melting or freezing is determined by the salt ion concentration (see Section 14-12).

iStockphoto.com/ollo

OBJECTIVES

After you have studied this chapter, you should be able to

▶ Describe the factors that favor the dissolution process

▶ Describe the dissolution of solids in liquids, liquids in liquids, and gases in liquids

▶ Describe how temperature and pressure affect solubility

▶ Express concentrations of solutions in terms of molality and mole fractions

▶ Describe the four colligative properties of solutions and some of their applications

▶ Carry out calculations involving the four colligative properties of solutions: lowering of vapor pressure (Raoult's Law),

boiling point elevation, freezing point depression, and osmotic pressure

▶ Use colligative properties to determine molecular weights of compounds

▶ Describe how dissociation and ionization of compounds affect colligative properties

▶ Recognize and describe colloids: the Tyndall effect, the adsorption phenomenon, and hydrophilic and hydrophobic colloids

A solution is defined as a *homogeneous mixture*, at the molecular level, of two or more substances in which phase separation does not occur. A solution consists of a solvent and one or more solutes, whose proportions can vary from one solution to another. By contrast, a pure substance has a fixed composition. The *solvent* is the medium in which the *solutes* are dissolved. Solutes usually dissolve to give ions or molecules in solution.

Solutions include different combinations in which a solid, liquid, or gas acts as either solvent or solute. Usually the solvent is a liquid. For instance, seawater is an aqueous solution of many salts and some gases such as carbon dioxide and oxygen. Carbonated water is a saturated solution of carbon dioxide in water. Solutions are common in nature and are extremely important in all life processes, in all scientific inquiry, and in many industrial processes. Many naturally occurring fluids contain particulate matter suspended in a solution. For example, blood contains a solution (plasma) with suspended blood cells. Seawater contains dissolved substances as well as suspended solids. The body fluids of all life forms are solutions. Variations in concentrations of our bodily fluids, especially those of blood and urine, give physicians valuable clues about a person's health. Solutions in which the solvent is not a liquid are also common. Air is a solution of gases with variable composition. Dental fillings are solid amalgams, or solutions of liquid mercury dissolved in solid metals. Alloys are solid solutions of solids dissolved in a metal.

It is usually obvious which of the components of a solution is the solvent and which is (are) the solute(s): The solvent is usually the most abundant species present. In a cup of instant coffee, the coffee and any added sugar are considered solutes, and the hot water is the solvent. When both components are liquids, the decision is less clear. Though it is preferable to refer simply to the different components, the solute/solvent terminology is often used. If we mix 10 grams of alcohol with 90 grams of water, we would call alcohol the solute. If we mix 10 grams of water with 90 grams of alcohol, water would be considered the solute. But which is the solute and which is the solvent in a solution of 50 grams of water and 50 grams of alcohol? In such cases, the terminology is arbitrary with no set rules.

The Dissolution Process

14-1 Spontaneity of the Dissolution Process

In Section 6-1, Part 5, we listed the solubility guidelines for aqueous solutions. Now we investigate the major factors that influence solubility.

A substance may dissolve with or without reaction with the solvent. For example, when metallic sodium reacts with water, bubbles of hydrogen are produced and a great deal of

Sodium reacts with water to form hydrogen gas and a solution of sodium hydroxide. A small amount of phenolphthalein indicator added to this solution turns pink due to the presence of sodium hydroxide.

heat is given off. A chemical change occurs in which H_2 and soluble ionic sodium hydroxide, NaOH, are produced.

$$2Na(s) + 2H_2O(\ell) \longrightarrow 2Na^+(aq) + 2OH^-(aq) + H_2(g)$$

If the resulting solution is evaporated to dryness, solid sodium hydroxide, NaOH, is obtained rather than metallic sodium. This, along with the production of hydrogen bubbles, is evidence of a reaction with the solvent. Reactions that involve oxidation state changes are usually considered to be chemical reactions and not dissolutions.

Solid sodium chloride, NaCl, on the other hand, dissolves in water with no evidence of chemical reaction.

$$NaCl(s) \xrightarrow{\text{H}_2\text{O}} Na^+(aq) + Cl^-(aq)$$

Evaporation of the water from the sodium chloride solution yields the original NaCl. In this chapter we will focus on dissolution processes of this type, in which no irreversible reaction occurs between components.

The ease of dissolution of a solute depends on two factors that accompany the process: (1) the change in energy and (2) the change in disorder (called entropy change). In the next chapter we will study in detail how both of these factors influence many kinds of physical and chemical changes. For now, we point out that a process is *favored* by (1) a *decrease in the energy* of the system, which corresponds to an *exothermic process*, and (2) an *increase in the disorder*, or randomness, of the system.

Let us look at the first of these factors. The energy change that accompanies a dissolution process is called the **heat of solution, $\Delta H_{\text{solution}}$**. This change depends mainly on how strongly the solute and solvent particles interact. If a solution gets hotter as a substance dissolves, energy is being released in the form of heat. A negative value of $\Delta H_{\text{solution}}$ designates the release of heat, and the process is called *exothermic*. More negative (less positive) values of $\Delta H_{\text{solution}}$ favor the dissolution process.

In a pure liquid, all the intermolecular forces are between like molecules. When the liquid and a solute are mixed, each molecule then interacts with unlike molecules (or ions) as well as with like molecules. The relative strengths of these interactions help to determine the extent of solubility of a solute in a solvent. The main interactions that affect the dissolution of a solute in a solvent follow.

   a. Weak solute–solute attractions favor solubility.

   b. Weak solvent–solvent attractions favor solubility.

   c. Strong solvent–solute attractions favor solubility.

Figure 14-1 illustrates the interplay of these factors. The intermolecular or interionic attractions among solute particles in the pure solute must be overcome (Step a) to dissolve the solute. This part of the process requires an *absorption* of energy (*endothermic*). Separat-

▶ Ionic solutes that do not react with the solvent undergo solvation. This is a kind of process in which molecules of solvent are attracted, usually in oriented clusters, to the solute particles.

The ocean is a solution of many different dissolved compounds. In this chapter we look at solution formation and solution properties.

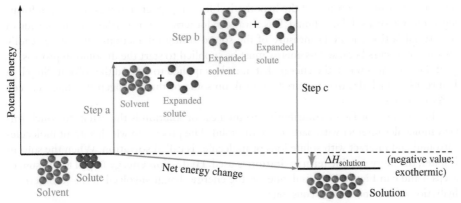

Figure 14-1 A diagram representing the changes in energy content associated with the hypothetical three-step sequence in a dissolution process—in this case, for a solid solute dissolving in a liquid solvent. (Similar considerations would apply to other combinations.) The process depicted here is exothermic. The amount of heat absorbed in Steps a and b is less than the amount released in Step c. In an *endothermic* process (not shown), the heat content of the solution would be *higher* than that of the original solvent plus solute. Thus, the amount of energy absorbed in Steps a and b would be *greater* than the amount of heat released in Step c.

► We can consider these energy changes separately, even though the actual process cannot be carried out in these separate steps.

ing the solvent molecules from one another (Step b) to "make room" for the solute particles also requires the *absorption* of energy (endothermic). Energy is *released*, however, as the solute particles and solvent molecules interact in the solution (Step c, exothermic). The overall dissolution process is exothermic (and favored) if the amount of heat absorbed in hypothetical Steps a and b is less than the amount of energy released in Step c. The process is endothermic (and disfavored) if the amount of energy absorbed in Steps a and b is greater than the amount of heat released in Step c.

Many solids do dissolve in liquids by *endothermic* processes, however. The reason such processes can occur is that the endothermicity can be outweighed by a large increase in disorder of the solute during the dissolution process. The solute particles are highly ordered in a solid crystal, but are free to move about randomly in liquid solutions. Likewise, the degree of disorder in the solvent increases as the solution is formed, because solvent molecules are then in a more random environment. They are surrounded by a mixture of solvent and solute particles.

► The dissolution of NaF in water is one of the few cases in which the dissolving process is not accompanied by an overall increase in disorder. The water molecules become more ordered around the small F⁻ ions. This is due to the strong hydrogen bonding between H_2O molecules and F⁻ ions.

$$O-H\cdots F^-\cdots H-O$$
$$\mid \qquad\qquad\qquad \mid$$
$$H \qquad\qquad\qquad H$$

The amount of heat released on mixing, however, outweighs the disadvantage of this ordering.

Most dissolving processes are accompanied by an overall increase in disorder. Thus, the disorder factor is usually *favorable* to solubility. The determining factor, then, is whether the heat of solution (energy) also favors dissolution or, if it does not, whether it is small enough to be outweighed by the favorable effects of the increasing disorder. In gases, for instance, the molecules are so far apart that intermolecular forces are quite weak. Thus, when gases are mixed, changes in the intermolecular forces are very slight. So the very favorable increase in disorder that accompanies mixing is always more important than possible changes in intermolecular attractions (energy). Hence, gases that do not react with one another can always be mixed in any proportion.

The most common types of solutions are those in which the solvent is a liquid. In the next several sections we consider liquid solutions in more detail.

14-2 Dissolution of Solids in Liquids

The ability of a solid to go into solution depends most strongly on its crystal lattice energy, or the strength of attractions among the particles making up the solid. The **crystal lattice energy** is defined as the energy change accompanying the formation of one mole of formula units in the crystalline state from constituent particles in the gaseous state. This process is always exothermic; that is, crystal lattice energies are always *negative*. For an ionic solid, the process is written as

STOP & THINK
A very negative crystal lattice energy indicates very strong attractions within the solid.

$$M^+(g) + X^-(g) \longrightarrow MX(s) + energy$$

The amount of energy involved in this process depends on the electrostatic attraction between ions in the solid. When these attractions are strong due to higher charges and/or small ion sizes, a large amount of energy is released as the solid forms, increasing its stability.

The reverse of the crystal formation reaction is breaking the crystal into separated gas phase ions.

$$MX(s) + energy \longrightarrow M^+(g) + X^-(g)$$

This process can be considered the hypothetical first step (Step a in Figure 14-1) in forming a solution of a solid in a liquid. It is always endothermic. The smaller the magnitude of the crystal lattice energy (a measure of the solute–solute interactions), the more readily dissolution occurs because less energy must be supplied to start the dissolution process.

If the solvent is water, the energy that must be supplied to expand the solvent (Step b in Figure 14-1) includes that required to break up some of the hydrogen bonding between water molecules.

The third major factor contributing to the heat of solution is the extent to which solvent molecules interact with particles of the solid. The process in which solvent molecules surround and interact with solute ions or molecules is called **solvation**. When the solvent is water, the more specific term **hydration** is used. **Hydration energy** (equal to the sum of Steps b and c in Figure 14-1) is defined as the energy change involved in the (exothermic) hydration of one mole of gaseous ions.

► Hydration energy is also referred to as the **heat of hydration**.

$$M^{n+}(g) + xH_2O(\ell) \longrightarrow M(OH_2)_x^{n+} + \text{energy} \quad \text{(for cation)}$$

$$X^{y-}(g) + rH_2O(\ell) \longrightarrow X(H_2O)_r^{y-} + \text{energy} \quad \text{(for anion)}$$

Hydration is usually highly exothermic for ionic or polar covalent compounds because the polar water molecules interact very strongly with ions and polar molecules. In fact, the only solutes that are appreciably soluble in water either undergo dissociation or ionization or are able to form hydrogen bonds with water.

The overall heat of solution for a solid dissolving in a liquid is equal to the heat of solvation minus the crystal lattice energy.

$$\Delta H_{\text{solution}} = (\text{heat of solvation}) - (\text{crystal lattice energy})$$

Remember that both terms on the right are always negative.

Nonpolar solids such as naphthalene, $C_{10}H_8$, do not dissolve appreciably in polar solvents such as water because the two substances do not attract each other significantly. This is true despite the fact that crystal lattice energies of solids consisting of nonpolar molecules are much less negative (smaller in magnitude) than those of ionic solids. Naphthalene dissolves readily in nonpolar solvents such as benzene because there are no strong attractive forces between solute molecules or between solvent molecules. In such cases, the increase in disorder controls the process. These facts help explain the observation that "*like dissolves like.*"

> The statement "like dissolves like" means that polar solvents dissolve ionic and polar molecular solutes, and nonpolar solvents dissolve nonpolar molecular solutes.

Consider what happens when a piece of sodium chloride, a typical ionic solid, is placed in water. The $\delta+$ parts of water molecules (H atoms) attract the negative chloride ions on the surface of the solid NaCl, as shown in Figure 14-2. Likewise, the $\delta-$ parts of H_2O molecules (O atoms) orient themselves toward the Na^+ ions and solvate them. These attractions help to overcome the forces holding the ions in the crystal, and NaCl dissolves in the H_2O.

$$NaCl(s) \xrightarrow{H_2O} Na^+(aq) + Cl^-(aq)$$

A cube of sugar is slowly dissolving, and the more dense sugar-rich solution can be seen descending through the less dense solvent.

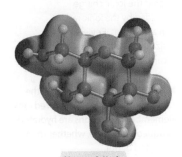

Glucose, $C_6H_{12}O_6$

Water, H_2O

Naphthalene, $C_{10}H_8$

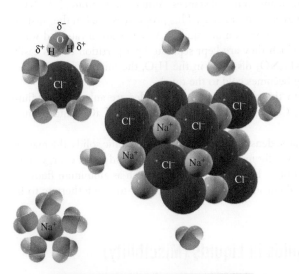

A The role of electrostatic attractions in the dissolution of NaCl in water. The δ^+ H of the polar H_2O molecule helps to attract Cl^- away from the crystal. Likewise, Na^+ is attracted by the δ^- O. Once they are separated from the crystal, both kinds of ions are surrounded by water molecules to complete the hydration process.

B Electrostatic plots illustrating the attraction of the positive portions (*blue*) of the water molecules to the negative ions, and the attraction of the negative portions (*red*) of the water molecules to the positive ions.

Glucose, like all sugars, is quite polar and dissolves in water, which is also very polar. Naphthalene is nonpolar and insoluble in water.

Figure 14-2 Interaction of water molecules with Na^+ cations and Cl^- anions

Table 14-1 Ionic Radii, Charge/Radius Ratios, and Hydration Energies for Some Cations

Ion	Ionic Radius (Å)	Charge/Radius Ratio	Hydration Energy (kJ/mol)
K^+	1.52	0.66	−351
Na^+	1.16	0.86	−435
Li^+	0.90	1.11	−544
Ca^{2+}	1.14	1.75	−1650
Fe^{2+}	0.76	2.63	−1980
Zn^{2+}	0.74	2.70	−2100
Cu^{2+}	0.72	2.78	−2160
Fe^{3+}	0.64	4.69	−4340
Cr^{3+}	0.62	4.84	−4370
Al^{3+}	0.68	4.41	−4750

Solid ammonium nitrate, NH_4NO_3, dissolves in water in a very endothermic process, absorbing heat from its surroundings. It is used in instant cold packs for early treatment of injuries, such as sprains and bruises, to minimize swelling.

S TOP & THINK

The **charge density** (charge/radius ratio) of an ion is the ionic charge divided by the ionic radius in angstroms. Review the sizes of ions in Figure 5-4 carefully.

▶ For simplicity, we often omit the (aq) designations from dissolved ions. Remember that all ions are hydrated in aqueous solution, whether this is indicated or not.

▶ A negative value for heat of hydration indicates that heat is *released* during hydration.

When we write Na^+(aq) and Cl^-(aq), we refer to hydrated ions. The number of H_2O molecules attached to an ion differs depending on the ion. Sodium ions are thought to be hexahydrated; that is, Na^+(aq) probably represents $[Na(OH_2)_6]^+$. Most cations in aqueous solution are surrounded by four to nine H_2O molecules, with six being the most common. Larger cations can generally accommodate more H_2O molecules than smaller cations.

Many solids that are appreciably soluble in water are ionic compounds. Magnitudes of crystal lattice energies generally increase with increasing charge and decreasing size of ions. That is, the size of the lattice energy increases as the ionic charge densities increase and, therefore, as the strength of electrostatic attractions within the crystal increases. Hydration energies vary in the same order (Table 14-1). As we indicated earlier, crystal lattice energies and hydration energies are generally much smaller in magnitude for molecular solids than for ionic solids.

Hydration and the effects of attractions in a crystal oppose each other in the dissolution process. Hydration energies and lattice energies are usually of about the same magnitude for low-charge species, so they often nearly cancel each other. As a result, the dissolution process is slightly endothermic for many ionic substances. Ammonium nitrate, NH_4NO_3, is an example of a salt that dissolves endothermically. This property is used in the "instant cold packs" used to treat sprains and other minor injuries. Ammonium nitrate and water are packaged in a plastic bag in which they are kept separate by a partition that is easily broken when squeezed. As the NH_4NO_3 dissolves in the H_2O, the mixture absorbs heat from its surroundings and the bag becomes cold to the touch.

Many ionic solids dissolve with the release of heat. Examples are anhydrous sodium sulfate, Na_2SO_4; calcium acetate, $Ca(CH_3COO)_2$; calcium chloride, $CaCl_2$; and lithium sulfate hydrate, $Li_2SO_4 \cdot H_2O$.

As the charge-to-size ratio (charge density) increases for ions in ionic solids, the magnitude of the crystal lattice energy usually increases more than the hydration energy. This makes dissolution of solids that contain highly charged ions—such as aluminum fluoride, AlF_3; magnesium oxide, MgO; and chromium(III) oxide, Cr_2O_3—too endothermic to be soluble in water.

14-3 Dissolution of Liquids in Liquids (Miscibility)

The term **miscibility** is used to describe the ability of one liquid to dissolve in another. The three kinds of attractive interactions (solute–solute, solvent–solvent, and solvent–solute) must be considered for liquid–liquid solutions just as they were for solid–liquid solutions. Because solute–solute attractions are usually much weaker for liquid solutes than for solids, this factor is less important and so the mixing process is often exothermic for miscible liquids. Polar liquids tend to interact strongly with and dissolve readily in other polar liquids. Methanol, CH_3OH; ethanol, CH_3CH_2OH; acetonitrile, CH_3CN; and

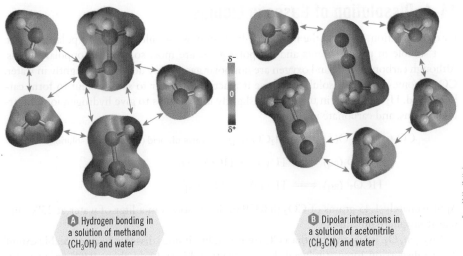

A Hydrogen bonding in a solution of methanol (CH₃OH) and water

B Dipolar interactions in a solution of acetonitrile (CH₃CN) and water

The nonpolar molecules in oil do not attract polar water molecules, so oil and water are immiscible. The polar water molecules attract one another strongly—they "squeeze out" the nonpolar molecules in the oil. Oil is less dense than water, so it floats on water.

Figure 14-3 Molecules tend to arrange themselves to maximize attractions by bringing regions of opposite charge together while minimizing repulsions by separating regions of like charge. In this representation, colors for each molecule range from blue (most positive regions) to red (most negative regions).

sulfuric acid, H_2SO_4, are all polar liquids that are soluble in most polar solvents (such as water). The hydrogen bonding between methanol and water molecules and the dipolar interactions between acetonitrile and water molecules are depicted in Figure 14-3.

Because hydrogen bonding is so strong between sulfuric acid, H_2SO_4, and water, a large amount of heat is released when concentrated H_2SO_4 is diluted with water (Figure 14-4). This can cause the solution to boil and spatter. If the major component of the mixture is water, this heat can be absorbed with less increase in temperature because of the unusually high specific heat of H_2O. For this reason, *sulfuric acid (as well as other concentrated mineral acids) must always be diluted by adding the acid slowly and carefully to water. Water should never be added to the acid.* Then if spattering did occur when the acid is added to water, it would be mainly water that spattered, not the corrosive concentrated acid that initially tends to sink to the bottom of the container.

Nonpolar liquids that do not react with the solvent generally are not very soluble in polar liquids because of the mismatch of forces of interaction. They are said to be *immiscible*. Nonpolar liquids are, however, usually quite soluble (*miscible*) in other nonpolar liquids. Between nonpolar molecules (whether alike or different) there are only dispersion forces, which are weak and easily overcome. As a result, when two nonpolar liquids are mixed, their molecules just "slide between" one another.

▶ Hydrogen bonding and dipolar interactions were discussed in Section 13-2.

▶ When water is added to concentrated acid, the danger is due more to the spattering of the acid itself than to the steam from boiling water.

Figure 14-4 The heat released by pouring 50 mL of sulfuric acid, H_2SO_4, into 50 mL of water increases the temperature by 100°C (from 21°C to 121°C)!

14-4 Dissolution of Gases in Liquids

Based on Section 13-2 and the foregoing discussion, we should expect that polar gases are most soluble in polar solvents and nonpolar gases are most soluble in nonpolar liquids. Although carbon dioxide and oxygen are nonpolar gases, they do dissolve slightly in water. CO_2 is somewhat more soluble because it reacts with water to some extent to form carbonic acid, H_2CO_3. This in turn ionizes slightly in two steps to give hydrogen ions, bicarbonate ions, and carbonate ions.

▶ Carbon dioxide is called an acid anhydride, that is, an "acid without water." As noted in Section 5-9, many other oxides of nonmetals, such as N_2O_5, SO_3, and P_4O_{10}, are also acid anhydrides.

$$CO_2(g) + H_2O(\ell) \rightleftharpoons H_2CO_3(aq) \qquad \text{carbonic acid (exists only in solution)}$$

$$H_2CO_3(aq) \rightleftharpoons H^+(aq) + HCO_3^-(aq)$$

$$HCO_3^-(aq) \rightleftharpoons H^+(aq) + CO_3^{2-}(aq)$$

Approximately 1.45 grams of CO_2 (0.0329 mole) dissolves in a liter of water at 25°C and one atmosphere pressure.

Oxygen, O_2, is less soluble than CO_2 in water, but it does dissolve to a noticeable extent due to dispersion forces (induced dipoles, Section 13-2). Only about 0.041 gram of O_2 (1.3×10^{-3} mole) dissolves in a liter of water at 25°C and 1 atm pressure. This is sufficient to support aquatic life.

Immersed in water, this green plant oxidizes the water to form gaseous oxygen. A small amount of the oxygen dissolves in water, and the excess oxygen forms bubbles of gas. The gaseous oxygen in the bubbles and the dissolved oxygen are in equilibrium with one another.

The hydrogen halides, HF, HCl, HBr, and HI, are all polar covalent gases. In the gas phase the interactions among the widely separated molecules are not very strong, so solute–solute attractions are minimal. Due to the polarity of the hydrogen halides and the hydration of their ions, the dissolution processes in water are very exothermic. The resulting solutions, called hydrohalic acids, contain predominantly ionized HX (X = Cl, Br, I). The ionization involves *protonation* of a water molecule by HX to form a hydrated hydrogen ion, H_3O^+, and halide ion X^- (also hydrated).

▶ Aqueous HCl, HBr, and HI are strong acids (Section 6-1, Part 2, and Section 10-7). Aqueous HF is a weak acid.

A saturated solution of copper(II) sulfate, $CuSO_4$, in water. As H_2O evaporates, blue $CuSO_4 \cdot 5H_2O$ crystals form. They are in dynamic equilibrium with the saturated solution.

HF is only slightly ionized in aqueous solution because of its strong covalent bond. In addition, the polar covalent bond between H and the small F atoms in HF causes very strong hydrogen bonding between H_2O and the largely intact HF molecules.

The only gases that dissolve appreciably in water are those that are capable of hydrogen bonding (such as HF), those that ionize extensively (such as HCl, HBr, and HI), and those that react with water (such as CO_2 or SO_3).

▶ Concentrated hydrochloric acid, HCl(aq), is an approximately 40% solution of HCl gas in water.

14-5 Rates of Dissolution and Saturation

At a given temperature, the rate of dissolution of a solid increases if large crystals are ground to a powder. Grinding increases the surface area, which in turn increases the number of solute ions or molecules in contact with the solvent. When a solid is placed in water, some of its particles solvate and dissolve. The rate of this process slows as time passes because the surface area of the crystals gets smaller and smaller. At the same time, the number of solute particles in solution increases, so they collide with the solid more frequently. Some of these collisions result in recrystallization. The rates of the two opposing processes become equal after some time. The solid and dissolved ions are then in equilibrium with each other.

$$\text{solid} \underset{\text{crystallization}}{\overset{\text{dissolution}}{\rightleftharpoons}} \text{dissolved particles}$$

Such a solution is said to be **saturated**. Saturation occurs at very low concentrations of dissolved species for slightly soluble substances and at high concentrations for very soluble substances. When imperfect crystals are placed in saturated solutions of their ions, surface defects on the crystals are slowly "patched" with no net increase in mass of the solid. Often, after some time has passed, we see fewer but larger crystals. These observations provide evidence of the dynamic nature of the solubility equilibrium. When equilibrium is established, no more solid dissolves without the simultaneous crystallization of an equal mass of dissolved ions.

The solubilities of many solids increase at higher temperatures. **Supersaturated solutions** contain higher-than-saturated concentrations of solute. They can sometimes be prepared by saturating a solution at a high temperature. The saturated solution is then cooled slowly, without agitation, to a temperature at which the solute is less soluble. At this point, the resulting supersaturated solution is *metastable* (temporarily stable). This may be thought of as a state of pseudoequilibrium in which the system is at a higher energy than in its most stable state. In such a case, the solute has not yet become sufficiently organized for crystallization to begin. A supersaturated solution produces crystals rapidly if it is slightly disturbed or if it is "seeded" with a dust particle or a tiny crystal. Under such conditions enough solid crystallizes to leave a saturated solution (Figure 14-5).

▶ All saturated solutions involve dynamic equilibria; for instance, there is a continuous exchange of oxygen molecules across the surface of water in an open container. This allows fish to "breathe" dissolved oxygen.

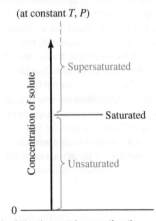

A solution that contains more than the amount of solute normally contained at saturation is said to be supersaturated. A solution that contains less than the amount of solute necessary for saturation is said to be unsaturated.

  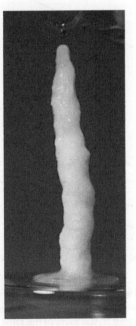

Figure 14-5 A supersaturated solution can be seeded by pouring it very slowly onto a seed crystal. A supersaturated sodium acetate solution was used in these photographs.

Sodium acetate crystals ($NaCH_3COO$) form quickly in a supersaturated solution when a small amount of solute is added.

14-6 Effect of Temperature on Solubility

In Section 13-6 we introduced LeChatelier's Principle, which states that *when a stress is applied to a system at equilibrium, the system responds in a way that best relieves the stress.* Recall that exothermic processes release heat and endothermic processes absorb heat.

$$\text{Exothermic:} \qquad \text{reactants} \longrightarrow \text{products} + heat$$

$$\text{Endothermic:} \qquad \text{reactants} + heat \longrightarrow \text{products}$$

Many ionic solids dissolve by endothermic processes. Their solubilities in water usually *increase* as heat is added and the temperature increases. For example, KCl dissolves endothermically.

$$KCl(s) + 17.2 \text{ kJ} \xrightarrow{\text{H}_2\text{O}} K^+(aq) + Cl^-(aq)$$

Figure 14-6 shows that the solubility of KCl increases as the temperature increases because more heat is available to increase the dissolving process. Raising the temperature (by adding *heat*) causes a stress on the solubility equilibrium. This stress favors the process that *absorbs* heat. In this case, more KCl dissolves.

Some solids, such as anhydrous Na_2SO_4, and many liquids and gases dissolve by exothermic processes. Their solubilities usually decrease as temperature increases. The solubility of O_2 in water decreases (by 34%) from 0.041 gram per liter of water at 25°C to 0.027 gram per liter at 50°C. Raising the temperature of rivers and lakes by dumping heated waste water from industrial plants and nuclear power plants is called **thermal pollution**. A slight increase in the temperature of the water causes a small but significant decrease in the concentration of dissolved oxygen. As a result, the water can no longer support the marine life it ordinarily could.

Solid iodine, I_2, dissolves to a limited extent in water to give an orange solution. This aqueous solution does not mix with nonpolar carbon tetrachloride, CCl_4 (*left*). After the funnel is shaken and the liquids are allowed to separate (*right*), the upper aqueous phase is lighter orange and the lower CCl_4 layer is much more highly colored. This is because iodine is much more soluble in the nonpolar carbon tetrachloride than in water; much of the iodine dissolves preferentially in the lower (CCl_4) phase. The design of the separatory funnel allows the lower (denser) layer to be drained off. Fresh CCl_4 could be added and the process repeated. This method of separation is called *extraction*. It takes advantage of the different solubilities of a solute in two immiscible liquids.

Calcium acetate, $Ca(CH_3COO)_2$, is more soluble in cold water than in hot water. When a cold, concentrated solution of calcium acetate is heated, solid calcium acetate precipitates.

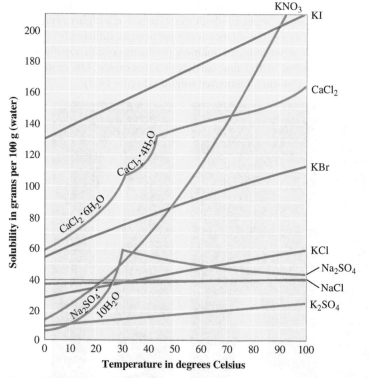

Figure 14-6 A graph that illustrates the effect of temperature on the solubilities of some salts. Some compounds exist either as nonhydrated crystalline substances or as hydrated crystals. Hydrated and nonhydrated crystal forms of the same compounds often have different solubilities because of the different total forces of attraction in the solids. The discontinuities in the solubility curves for $CaCl_2$ and Na_2SO_4 are due to transitions between hydrated and nonhydrated crystal forms.

The dissolution of anhydrous calcium chloride, $CaCl_2$, in water is quite exothermic. Here the temperature increases from 21°C to 88°C. This dissolution process is utilized in commercial instant hot packs for quick treatment of injuries requiring heat.

14-7 Effect of Pressure on Solubility

Changing the pressure has no appreciable effect on the solubilities of either solids or liquids in liquids. The solubilities of gases in all solvents increase, however, as the partial pressures of the gases increase (Figure 14-7). Carbonated water is a saturated solution of carbon dioxide in water under pressure. When a can or bottle of a carbonated beverage is opened, the pressure on the surface of the beverage is reduced to atmospheric pressure, and much of the CO_2 bubbles out of solution. If the container is left open, the beverage becomes "flat" because the released CO_2 escapes.

Henry's Law applies to gases that do not react with the solvent in which they dissolve (or, in some cases, gases that react incompletely). It is usually stated as follows.

> The concentration or solubility of a gas in a liquid at any given temperature is directly proportional to the partial pressure of the gas over the solution.
>
> $$C_{gas} = kP_{gas}$$

P_{gas} is the pressure of the gas above the solution, and k is a constant for a particular gas and solvent at a particular temperature. C_{gas} represents the concentration of dissolved gas; it is usually expressed either as molarity (see Section 3-6) or as mole fraction (see Section 14-8). The relationship is valid at low concentrations and low pressures.

Carbonated beverages can be used to illustrate Henry's Law. When the bottle is opened, the equilibrium is disturbed and bubbles of CO_2 form within the liquid and rise to the surface. After some time, an equilibrium between dissolved CO_2 and atmospheric CO_2 is reached.

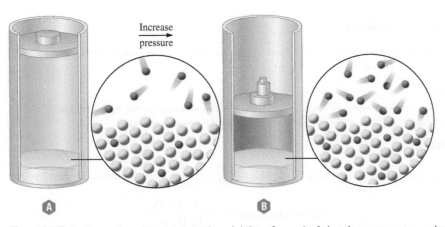

Increase pressure →

Ⓐ Ⓑ

Figure 14-7 An illustration of Henry's Law. The solubility of a gas (*red*) that does not react completely with the solvent (*yellow*) increases with increasing pressure of that gas above the solution.

14-8 Molality and Mole Fraction

We saw in Section 3-6 that concentrations of solutions are often expressed as percent by mass of solute or as molarity. Discussion of many physical properties of solutions is often made easier by expressing concentrations either in molality units or as mole fractions (Sections 14-9 to 14-14).

▶ Molality is based on the amount of *solvent*, not the amount of *solution*.

Molality

> The **molality**, m, of a solute in solution is the number of moles of solute *per kilogram of solvent*.
>
> $$\text{molality} = \frac{\text{number of moles solute}}{\text{number of kilograms solvent}}$$

Each beaker holds the amount of a crystalline ionic compound that will dissolve in 100. grams of water at 100°C. The compounds are (*left-most beaker, then clockwise*): 102 grams of potassium dichromate ($K_2Cr_2O_7$, *red-orange*), 191 grams of cobalt(II) chloride hexahydrate ($CoCl_2 \cdot 6H_2O$, *dark red*), 341 grams of nickel sulfate hexahydrate ($NiSO_4 \cdot 6H_2O$, *green*), 203 grams of copper sulfate pentahydrate ($CuSO_4 \cdot 5H_2O$, *blue*), 39 grams of sodium chloride (NaCl, *white*), and 79 grams of potassium chromate (K_2CrO_4, *yellow*).

EXAMPLE 14-1 Molality

What is the molality of a solution that contains 54.6 g of CH_3OH in 108 g of water?

Plan

We convert the amount of solute (CH_3OH) to moles, express the amount of solvent (water) in kilograms, and apply the definition of molality.

$$\text{g } CH_3OH \longrightarrow \text{mol } CH_3OH$$
$$\text{g } H_2O \longrightarrow \text{kg } H_2O$$
$$\longrightarrow \text{molality}$$

Solution

$$\frac{?\text{ mol}}{\text{kg } H_2O} = \frac{54.6 \text{ g } CH_3OH}{0.108 \text{ kg } H_2O} \times \frac{1 \text{ mol } CH_3OH}{32.0 \text{ g } CH_3OH} = \frac{15.8 \text{ mol } CH_3OH}{\text{kg } H_2O}$$

$$= 15.8 \; m \; CH_3OH$$

You should now work Exercise 28.

STOP & THINK

It is very easy to confuse molality (*m*) with molarity (*M*). Pay close attention to what is being asked.

EXAMPLE 14-2 Molality

How many grams of H_2O must be used to dissolve 50.0 grams of sucrose to prepare a 1.25 m solution of sucrose, $C_{12}H_{22}O_{11}$?

Plan

We convert the amount of solute ($C_{12}H_{22}O_{11}$) to moles, rearrange the molality expression for kilograms of solvent (water), and then express the result in grams.

Solution

$$?\text{ mol } C_{12}H_{22}O_{11} = 50.0 \text{ g } C_{12}H_{22}O_{11} \times \frac{1 \text{ mol } C_{12}H_{22}O_{11}}{342 \text{ g } C_{12}H_{22}O_{11}} = 0.146 \text{ mol } C_{12}H_{22}O_{11}$$

$$\text{molality of solution} = \frac{\text{mol } C_{12}H_{22}O_{11}}{\text{kg } H_2O}$$

Rearranging gives

$$kg\ H_2O = \frac{mol\ C_{12}H_{22}O_{11}}{molality\ of\ solution} = \frac{0.146\ mol\ C_{12}H_{22}O_{11}}{1.25\ mol\ C_{12}H_{22}O_{11}/kg\ H_2O}$$

$$= 0.117\ kg\ H_2O = 117\ g\ H_2O$$

We will calculate several properties of this solution in other examples later in this chapter.

> **S TOP & THINK**
> Use of complete unit factors will help you to solve problems involving solution concentrations in various required units.

Mole Fraction

Recall that in Chapter 12 the **mole fractions**, X_A and X_B, of each component in a mixture containing components A and B were defined as

$$X_A = \frac{no.\ mol\ A}{no.\ mol\ A + no.\ mol.\ B} \quad and \quad X_B = \frac{no.\ mol\ B}{no.\ mol\ A + no.\ mol.\ B}$$

Mole fraction is a dimensionless quantity; that is, it has no units.

EXAMPLE 14-3 Mole Fraction

What are the mole fractions of CH_3OH and H_2O in the solution described in Example 14-1? It contains 54.6 grams of CH_3OH and 108 grams of H_2O.

Plan

We express the amount of both components in moles, and then apply the definition of mole fraction.

Solution

$$\underline{?}\ mol\ CH_3OH = 54.6\ g\ CH_3OH \times \frac{1\ mol\ CH_3OH}{32.0\ g\ CH_3OH} = 1.71\ mol\ CH_3OH$$

$$\underline{?}\ mol\ H_2O = 108\ g\ H_2O \times \frac{1\ mol\ H_2O}{18.0\ g\ H_2O} = 6.00\ mol\ H_2O$$

Now we calculate the mole fraction of each component.

$$X_{CH_3OH} = \frac{no.\ mol\ CH_3OH}{no.\ mol\ CH_3OH + no.\ mol\ H_2O} = \frac{1.71\ mol}{(1.71 + 6.00)\ mol} = 0.222$$

$$X_{H_2O} = \frac{no.\ mol\ H_2O}{no.\ mol\ CH_3OH + no.\ mol\ H_2O} = \frac{6.00\ mol}{(1.71 + 6.00)\ mol} = 0.778$$

You should now work Exercise 30.

> **S TOP & THINK**
> Check: In any mixture the sum of the mole fractions must be 1:
>
> $0.222 + 0.778 = 1$

Colligative Properties of Solutions

Physical properties of solutions that depend on the *number*, but not the *kind*, of solute particles in a given amount of solvent are called **colligative properties**. There are four important colligative properties of a solution that are directly proportional to the number of solute particles present. They are (1) vapor pressure lowering, (2) boiling point elevation, (3) freezing point depression, and (4) osmotic pressure. These properties of a solution depend on the *total concentration of all solute particles*, regardless of their ionic or molecular nature, charge, or size. For most of this chapter, we will consider *nonelectrolyte* solutes (see Section 6-1, Part 1); these substances dissolve to give one mole of dissolved particles for each mole of solute. In Section 14-14 we will see how we must adjust our predictions of colligative properties to account for ion formation in electrolyte solutions.

▶ *Colligative* means "tied together."

14-9 Lowering of Vapor Pressure and Raoult's Law

Many experiments have shown that a solution containing a *nonvolatile* liquid or a solid as a solute always has a lower vapor pressure than the pure solvent (Figure 14-8). The vapor pressure of a liquid depends on the ease with which the molecules are able to escape from the surface of the liquid. When a solute is dissolved in a liquid, some of the total volume of the solution is occupied by solute molecules, so there are fewer solvent molecules *per unit area* at the surface. As a result, solvent molecules vaporize at a slower rate than if no solute were present. The increase in disorder that accompanies evaporation is also a significant factor. Because a solution is already more disordered ("mixed up") than a pure solvent, the evaporation of the pure solvent involves a larger increase in disorder and is thus more favorable. Hence, the pure solvent exhibits a higher vapor pressure than does the solution. The lowering of the vapor pressure of the solution is a colligative property. It is a function of the number, and not the kind, of solute particles in solution. We emphasize that solutions of gases or low-boiling (volatile) liquid solutes can have *higher* total vapor pressures than the pure solvents, so this discussion does not apply to them.

▶ A *vapor* is a gas formed by the boiling or evaporation of a liquid or sublimation of a solid. The *vapor pressure* of a liquid is the pressure (partial pressure) exerted by a vapor in equilibrium with its liquid (see Section 13-7).

The lowering of the vapor pressure of a solvent due to the presence of *nonvolatile, nonionizing* solutes is summarized by **Raoult's Law**.

> The vapor pressure of a solvent in an ideal solution is directly proportional to the mole fraction of the solvent in the solution.

The relationship can be expressed mathematically as

$$P_{solvent} = X_{solvent}P^0_{solvent}$$

where $X_{solvent}$ represents the mole fraction of the solvent in a solution, $P^0_{solvent}$ is the vapor pressure of the *pure* solvent, and $P_{solvent}$ is the vapor pressure of the solvent *in the solution* (Figure 14-9). If the solute is nonvolatile, the vapor pressure of the solution is entirely due to the vapor pressure of the solvent, $P_{solution} = P_{solvent}$.

The *lowering* of the vapor pressure, $\Delta P_{solvent}$, is defined as

$$\Delta P_{solvent} = P^0_{solvent} - P_{solvent}$$

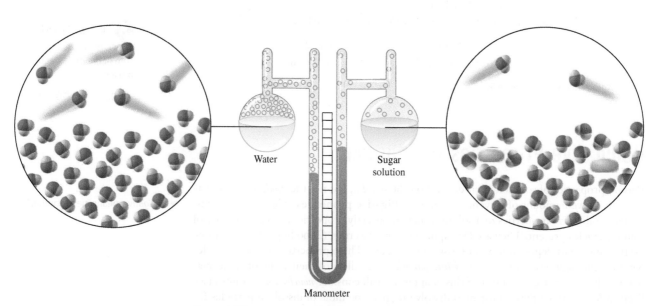

Water   Sugar solution

Manometer

Figure 14-8 Lowering of vapor pressure. If no air is present in the apparatus, the pressure above each liquid is due to water vapor. This pressure is less over the solution of sugar and water.

Thus,

$$\Delta P_{\text{solvent}} = P^0_{\text{solvent}} - (X_{\text{solvent}}P^0_{\text{solvent}}) = (1 - X_{\text{solvent}})P^0_{\text{solvent}}$$

Now $X_{\text{solvent}} + X_{\text{solute}} = 1$, so $1 - X_{\text{solvent}} = X_{\text{solute}}$. We can express the *lowering* of the vapor pressure in terms of the mole fraction of the solute.

$$\Delta P_{\text{solvent}} = X_{\text{solute}}P^0_{\text{solvent}}$$

Solutions that obey this relationship exactly are called **ideal solutions**. The vapor pressures of many solutions, however, do not behave ideally.

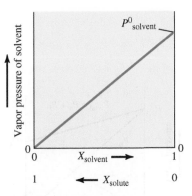

Figure 14-9 Raoult's Law for an ideal solution of a solute in a volatile liquid. The vapor pressure exerted by the liquid is proportional to its mole fraction in the solution.

EXAMPLE 14-4 Vapor Pressure of a Solution of Nonvolatile Solute

Sucrose is a nonvolatile, nonionizing solute in water. Determine the vapor pressure lowering, at 25°C, of the 1.25 *m* sucrose solution in Example 14-2. Assume that the solution behaves ideally. The vapor pressure of pure water at 25°C is 23.8 torr (see Appendix E).

Plan

The solution in Example 14-2 was made by dissolving 50.0 grams of sucrose (0.146 mol) in 117 grams of water (6.50 mol). We calculate the mole fraction of the solute in the solution. Then we apply Raoult's Law to find the vapor pressure lowering, $\Delta P_{\text{solvent}}$.

Solution

$$X_{\text{sucrose}} = \frac{0.146 \text{ mol}}{0.146 \text{ mol} + 6.50 \text{ mol}} = 0.0220$$

Applying Raoult's Law in terms of the vapor pressure lowering,

$$\Delta P_{\text{solvent}} = (X_{\text{solute}})(P^0_{\text{solvent}}) = (0.0220)(23.8 \text{ torr}) = \boxed{0.524 \text{ torr}}$$

You should now work Exercise 38.

> **STOP & THINK**
> The vapor pressure of water in the solution is
> $(23.8 - 0.524)$ torr
> $= 23.3$ torr.
> We could calculate this vapor pressure directly from the mole fraction of the solvent (water) in the solution, using the relationship
>
> $P_{\text{solvent}} = X_{\text{solvent}}P^0_{\text{solvent}}.$

When a solution consists of two components that are very similar, each component behaves essentially as it would if it were pure. For example, the two liquids heptane, C_7H_{16}, and octane, C_8H_{18}, are so similar that each heptane molecule experiences nearly the same intermolecular forces whether it is near another heptane molecule or near an octane molecule, and similarly for each octane molecule. The properties of such a solution can be predicted from a knowledge of its composition and the properties of each component. Such a solution is very nearly ideal.

Consider an ideal solution of two volatile components, A and B. The vapor pressure of each component above the solution is proportional to its mole fraction in the solution.

$$P_A = X_A P^0_A \qquad \text{and} \qquad P_B = X_B P^0_B$$

According to Dalton's Law of Partial Pressures (see Section 12-11), the total vapor pressure of the solution is equal to the sum of the vapor pressures of the two components.

$$P_{\text{total}} = P_A + P_B \qquad \text{or} \qquad P_{\text{total}} = X_A P^0_A + X_B P^0_B$$

This is shown graphically in Figure 14-10. We can use these relationships to predict the vapor pressures of an ideal solution, as Example 14-5 illustrates.

▶ If component B were nonvolatile, then P^0_B would be zero, and this description would be the same as that given earlier for a solution of a nonvolatile, nonionizing solute in a volatile solvent, $P_{\text{total}} = P_{\text{solvent}}$.

EXAMPLE 14-5 Vapor Pressure of a Solution of Volatile Components

At 40°C, the vapor pressure of pure heptane is 92.0 torr and the vapor pressure of pure octane is 31.0 torr. Consider a solution that contains 1.00 mole of heptane and 4.00 moles of octane. Calculate the vapor pressure of each component and the total vapor pressure above the solution.

Plan

We first calculate the mole fraction of each component in the liquid solution. Then we apply Raoult's Law to each of the two volatile components. The total vapor pressure is the sum of the vapor pressures of the components.

Solution

We first calculate the mole fraction of each component in the liquid solution.

$$X_{heptane} = \frac{1.00 \text{ mol heptane}}{(1.00 \text{ mol heptane}) + (4.00 \text{ mol octane})} = 0.200$$

$$X_{octane} = 1 - X_{heptane} = 0.800$$

Then, applying Raoult's Law for volatile components,

$$P_{heptane} = X_{heptane}P^0_{heptane} = (0.200)(92.0 \text{ torr}) = 18.4 \text{ torr}$$

$$P_{octane} = X_{octane}P^0_{octane} = (0.800)(31.0 \text{ torr}) = 24.8 \text{ torr}$$

$$P_{total} = P_{heptane} + P_{octane} = 18.4 \text{ torr} + 24.8 \text{ torr} = 43.2 \text{ torr}$$

You should now work Exercise 40.

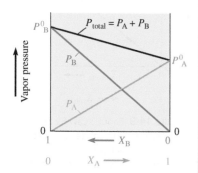

Figure 14-10 Raoult's Law for an ideal solution of two volatile components. The left-hand side of the plot corresponds to pure B ($X_A = 0$, $X_B = 1$), and the right-hand side corresponds to pure A ($X_A = 1$, $X_B = 0$). Of these hypothetical liquids, B is more volatile than A ($P^0_B > P^0_A$).

The vapor in equilibrium with a liquid solution of two or more volatile components has a higher mole fraction of the more volatile component than does the liquid solution.

EXAMPLE 14-6 Composition of Vapor

Calculate the mole fractions of heptane and octane in the vapor that is in equilibrium with the solution in Example 14-5.

Plan

We learned in Section 12-11 that the mole fraction of a component in a gaseous mixture equals the ratio of its partial pressure to the total pressure. In Example 14-5 we calculated the partial pressure of each component in the vapor and the total vapor pressure.

Solution

In the *vapor*

$$X_{heptane} = \frac{P_{heptane}}{P_{total}} = \frac{18.4 \text{ torr}}{43.2 \text{ torr}} = 0.426$$

$$X_{octane} = \frac{P_{octane}}{P_{total}} = \frac{24.8 \text{ torr}}{43.2 \text{ torr}} = 0.574$$

or

$$X_{octane} = 1.000 - 0.426 = 0.574$$

You should now work Exercise 42.

STOP & THINK

Heptane (pure vapor pressure = 92.0 torr at 40°C) is a more volatile liquid than octane (pure vapor pressure = 31.0 torr at 40°C). Its mole fraction in the vapor, 0.426, is higher than its mole fraction in the liquid, 0.200.

Many dilute solutions behave ideally. Some solutions do not behave ideally over the entire concentration range. For some solutions, the observed vapor pressure is greater than that predicted by Raoult's Law (Figure 14-11a). This kind of deviation, known as a *positive deviation*, is due to differences in polarity of the two components. On the molecular level, the two substances do not mix entirely randomly, so there is self-association of each component with local regions enriched in one type of molecule or the other. In a region enriched in A molecules, substance A acts as though its mole fraction were greater than it is in the solution as a whole, and the vapor pressure due to A is greater than if the solution were ideal. A similar description applies to component B. The total vapor pressure is then greater than it would be if the solution were behaving ideally. A solution of acetone (polar) and carbon disulfide (nonpolar) is an example of a solution that shows a positive deviation from Raoult's Law.

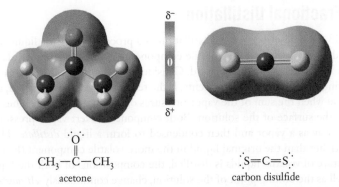

$$CH_3-\overset{\overset{\displaystyle :\!O\!:}{\|}}{C}-CH_3$$
acetone

$$:\!S\!=\!C\!=\!S\!:$$
carbon disulfide

Another, more common type of deviation occurs when the total vapor pressure is less than that predicted (Figure 14-11b). This is called a *negative deviation*. Such an effect is due to unusually strong attractions (such as hydrogen bonding) between *different polar* molecules. As a result, different polar molecules are strongly attracted to one another, so fewer molecules escape to the vapor phase. The observed vapor pressure of each component is thus less than ideally predicted. An acetone–chloroform solution and an ethanol–water solution are two polar combinations that show negative deviations from Raoult's Law.

Figure 14-11
Deviations (*solid lines*) from Raoult's Law for two volatile components. Ideal behavior (Raoult's Law) is shown by the dashed lines.

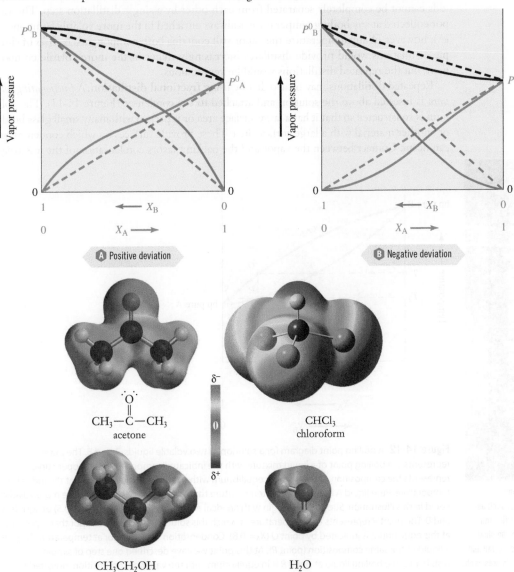

A Positive deviation

B Negative deviation

acetone CH$_3$—C—CH$_3$

CHCl$_3$ chloroform

CH$_3$CH$_2$OH ethanol

H$_2$O water

14-10 Fractional Distillation

In Section 13-8 we described *simple* distillation as a process in which a liquid solution can be separated into volatile and nonvolatile components. But separation of volatile components is not very efficient by this method. Consider the simple distillation of a liquid solution consisting of two volatile components. If the temperature is slowly raised, the solution begins to boil when the sum of the vapor pressures of the components reaches the applied pressure on the surface of the solution. Both components exert vapor pressures, so both are carried away as a vapor and then condensed to form a liquid *distillate*. The resulting distillate is richer than the original liquid in the more volatile component (Example 14-6).

As a mixture of volatile liquids is distilled, the compositions of both the liquid and the vapor, as well as the boiling point of the solution, change continuously. *At constant pressure*, we can represent these quantities in a **boiling point diagram** (Figure 14-12). In such a diagram, the lower curve represents the boiling point of a *liquid mixture* with the indicated composition. The upper curve represents the composition of the *vapor* in equilibrium with the boiling liquid mixture at the indicated temperature. The intercepts at the two vertical axes show the boiling points of the two pure liquids. The distillation of the two liquids is described in the legend for Figure 14-12.

From the boiling point diagram in Figure 14-12, we see that two or more volatile liquids cannot be completely separated from each other by a single distillation step. The vapor collected at any boiling temperature is always enriched in the more volatile component (A); however, at any temperature the vapor still contains both components. A *series* of simple distillations would provide distillates increasingly richer in the more volatile component, but the repeated distillations would be very tedious.

Repeated distillations may be avoided by using **fractional distillation**. A *fractionating column* is inserted above the solution and attached to the condenser (Figure 14-13). The column is constructed so that it has a large surface area or is packed with many small glass beads or another material with a large surface area. These provide surfaces on which condensation can occur. Contact between the vapor and the packing favors condensation of the less vola-

▶ The applied pressure is usually atmospheric pressure. Distillation under vacuum lowers the applied pressure. This allows boiling at lower temperatures than under atmospheric pressure. This technique allows distillation of some substances that would decompose at higher temperatures.

© Ocean/Corbis

Crude oil is separated by fractional distillation into many components such as gasoline, kerosene, fuel oil, paraffin, and asphalt. The distillation towers in chemical plants are typically four to ten stories tall and can process many tons of liquid mixtures each day.

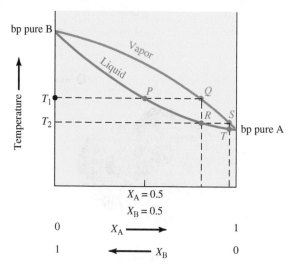

Figure 14-12 A boiling point diagram for a solution of two volatile liquids, A and B. The lower curve represents the boiling point of a liquid mixture with the indicated composition. The upper curve represents the composition of the *vapor* in equilibrium with the boiling liquid mixture at the indicated temperature. Pure liquid A boils at a lower temperature than pure liquid B; hence, A is the more volatile liquid in this illustration. Suppose we begin with an ideal equimolar mixture ($X_A = X_B = 0.5$) of liquids A and B. The point P represents the temperature at which this solution boils, T_1. The vapor that is present at this equilibrium is indicated by point Q ($X_A \approx 0.8$). Condensation of that vapor at temperature T_2 gives a liquid of the same composition (point R). At this point we have described one step of simple distillation. The boiling liquid at point R is in equilibrium with the vapor of composition indicated by point S ($X_A > 0.95$), and so on.

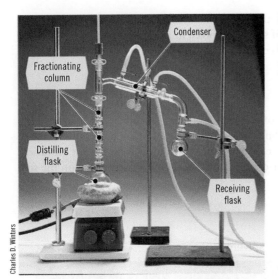

Figure 14-13 A fractional distillation apparatus. The vapor rising in the column is in equilibrium with the liquid that has condensed and is flowing slowly back down the column.

Charles D. Winters

tile component. The column is cooler at the top than at the bottom. By the time the vapor reaches the top of the column, practically all of the less volatile component has condensed and fallen back down the column. The more volatile component goes into the condenser, where it is liquefied and delivered as a highly enriched distillate into the collection flask. The longer the column or the greater the packing, the more efficient is the separation.

14-11 Boiling Point Elevation

Recall that the boiling point of a liquid is the temperature at which its vapor pressure equals the applied pressure on its surface (see Section 13-8). For liquids in open containers, this is atmospheric pressure, and the temperature is the *normal boiling point*. We have seen that the vapor pressure of a solvent at a given temperature is lowered by the presence of a *nonvolatile* solute. Such a solution must then be heated to a higher temperature, $T_{b(\text{solution})}$, than that of the pure solvent T_b, to cause the vapor pressure of the solvent to equal atmospheric pressure (Figure 14-14). In accord with Raoult's Law, the elevation of

▶ When the solute is nonvolatile, only the *solvent* distills from the solution.

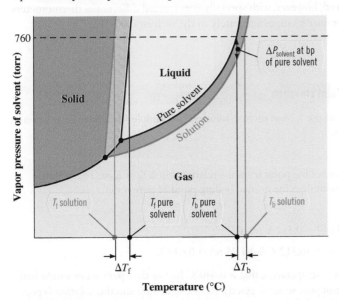

Figure 14-14 Because a *nonvolatile* solute lowers the vapor pressure of a solvent, the boiling point of a solution is higher and the freezing point lower than the corresponding points for the pure solvent. The boiling point elevation, ΔT_b, is positive, and its magnitude is less than the magnitude of the freezing point depression, ΔT_f.

the boiling point of a solvent caused by the presence of a nonvolatile, nonionized solute is proportional to the number of moles of solute dissolved in a given mass of solvent. Mathematically, the boiling point elevation is expressed as

$$\Delta T_b = K_b m$$

The term ΔT_b represents the elevation of the boiling point of the solvent, that is, the boiling point of the solution minus the boiling point of the pure solvent. The m is the molality of the *solute*, and K_b is a proportionality constant called the molal **boiling point elevation constant**. This constant is different for different solvents and *does not* depend on the solute (Table 14-2).

Table 14-2 Some Properties of Common Solvents

Solvent	bp (pure)	K_b (°C/m)	fp (pure)	K_f (°C/m)
water	100*	0.512	0*	1.86
benzene	80.1	2.53	5.5	5.12
acetic acid	118.1	3.07	16.6	3.90
nitrobenzene	210.9	5.24	5.7	7.00
phenol	182	3.56	43	7.40
camphor	207.4	5.61	178.4	40.0

*Exact values by definition.

K_b corresponds to the change in boiling point produced by a one-molal *ideal* solution of a nonvolatile nonelectrolyte. The units of K_b are °C/m.

Elevations of boiling points are usually quite small for solutions of typical concentrations. They can be measured, however, with specially constructed differential thermometers that measure small temperature changes accurately to the nearest 0.001°C.

EXAMPLE 14-7 Boiling Point Elevation

Predict the boiling point of the 1.25 m sucrose solution in Example 14-2. Sucrose is a nonvolatile, nonionized solute.

Plan

We first find the *increase* in boiling point from the relationship $\Delta T_b = K_b m$. The boiling point is *higher* by this amount than the normal boiling point of pure water.

Solution

From Table 14-2, K_b for H_2O = 0.512°C/m, so

$$\Delta T_b = (0.512°C/m)(1.25\ m) = 0.640°C$$

The solution would boil at a temperature that is 0.640°C higher than pure water would boil. The normal boiling point of pure water is exactly 100°C, so at 1.00 atm this solution is predicted to boil at 100°C + 0.640°C = 100.640°C.

You should now work Exercise 46.

14-12 Freezing Point Depression

Molecules of liquids move more slowly and approach one another more closely as the temperature is lowered. The freezing point of a liquid is the temperature at which the forces of attraction among molecules are just great enough to overcome their kinetic energies and thus cause a phase change from the liquid to the solid state. Strictly speaking, the freezing (melting) point of a substance is the temperature at which the liquid and solid phases are in equilibrium. When a dilute solution freezes, it is the *solvent* that begins to solidify first, leaving the solute in a more concentrated solution. Solvent molecules are somewhat more separated in a solution from one another (because of solute particles) than they are in the pure solvent. Consequently, the temperature of a solution must be lowered below the freezing point of the pure solvent to freeze it.

The freezing point depression, $\Delta T_f = T_{f(\text{pure solvent})} - T_{f(\text{solution})}$, of any solution of nonelectrolytes has been found to be equal to the molality of the solute times a proportionality constant called the molal **freezing point depression constant, K_f.**

$$\Delta T_f = K_f m$$

The values of K_f for a few solvents are given in Table 14-2. Each is numerically equal to the freezing point depression of a one-molal *ideal* solution of a nonelectrolyte in that solvent.

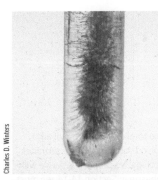

When a solution freezes, the solvent solidifies as the pure substance. For this photo a dye has been added. As the solute freezes along the wall of the test tube, the dye concentration increases near the center.

 STOP & THINK

ΔT_f is the *depression* of the freezing point. It is defined as

$$\Delta T_f = T_{f(\text{pure solvent})} - T_{f(\text{solution})}$$

so it is always *positive*.

 STOP & THINK

Remember to use molality (*m*) and *not* molarity (*M*) when calculating boiling point elevation or freezing point depression.

EXAMPLE 14-8 Freezing Point Depression

When 15.0 grams of ethyl alcohol, C_2H_5OH, is dissolved in 750. grams of formic acid, the freezing point of the solution is 7.20°C. The freezing point of pure formic acid is 8.40°C. Evaluate K_f for formic acid.

Plan

The molality and the depression of the freezing point are calculated first. Then we solve the equation $\Delta T_f = K_f m$ for K_f and substitute values for m and ΔT_f.

$$K_f = \frac{\Delta T_f}{m}$$

Solution

$$\text{molality} = \frac{\text{mol } C_2H_5OH}{\text{kg formic acid}} = \frac{15.0 \text{ g } C_2H_5OH}{0.750 \text{ kg formic acid}} \times \frac{1 \text{ mol } C_2H_5OH}{46.0 \text{ g } C_2H_5OH} = 0.435 \; m$$

$$\Delta T_f = (T_{f[\text{formic acid}]}) - (T_{f[\text{solution}]}) = 8.40°C - 7.20°C = 1.20°C \quad \text{(depression)}$$

Then $K_f = \dfrac{\Delta T_f}{m} = \dfrac{1.20°C}{0.435 \; m} = 2.76°C/m$ for formic acid.

You should now work Exercise 48.

EXAMPLE 14-9 Freezing Point Depression

Calculate the freezing point of the 1.25 *m* sucrose solution in Example 14-2.

Plan

We first find the *decrease* in freezing point from the relationship $\Delta T_f = K_f m$. The temperature at which the solution freezes is *lower* than the freezing point of pure water by this amount.

Solution

From Table 14-2, K_f for $H_2O = 1.86°C/m$, so

$$\Delta T_f = (1.86°C/m)(1.25 \; m) = 2.32°C$$

Lime, CaO, is added to molten iron ore during the manufacture of pig iron. It lowers the melting point of the mixture. The metallurgy of iron is discussed in more detail in Chapter 26.

The solution freezes at a temperature 2.32°C *below* the freezing point of pure water, or

$$T_{f(solution)} = 0.00°C - 2.32°C = -2.32°C$$

You should now work Exercise 56.

▶ The total concentration of all dissolved solute species determines the colligative properties. As we will emphasize in Section 14-14, we must take into account the extent of ion formation in solutions of ionic solutes.

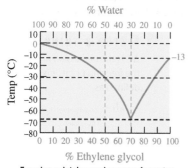

Freezing point depression curve for water and ethylene glycol calculated from the $\Delta T_f = K_f m$ formula. Despite some differences due to nonideal behavior, it matches the experimental data fairly well.

Ethylene glycol, $HOCH_2CH_2OH$, is the major component of "permanent" antifreeze. It depresses the freezing point of water in an automobile radiator and also raises its boiling point. The solution remains in the liquid phase over a wider temperature range than does pure water. This protects against both freezing and boil-over.

You may be familiar with several examples of the effects we have studied. Seawater does not freeze on some days when fresh water does, because seawater contains higher concentrations of solutes, mostly ionic solutes. Spreading soluble salts such as sodium chloride, NaCl, or calcium chloride, $CaCl_2$, on an icy road lowers the freezing point of the ice, causing the ice to melt.

A familiar application is the addition of "permanent" antifreeze, mostly ethylene glycol, $HOCH_2CH_2OH$, to the water in an automobile radiator. Because the boiling point of the solution is elevated, addition of a solute as a winter antifreeze also helps protect against loss of the coolant by summer "boil-over." The amounts by which the freezing and boiling points change depend on the concentration of the ethylene glycol solution. The addition of too much ethylene glycol is counterproductive, however. The freezing point of pure ethylene glycol is −13°C. A solution that is mostly ethylene glycol would have a somewhat lower freezing point due to the presence of water as a solute. Suppose you graph the freezing point depression of water below 0°C as ethylene glycol is added, and also graph the freezing point depression of ethylene glycol below −13°C as water is added. These two curves would intersect at some temperature, indicating the limit of lowering that can occur. Most antifreeze labels recommend a 50:50 mixture by volume (fp = −34°F, bp = 265°F with a 15-psi pressure cap on the radiator), and cite the limit of possible protection with a 70:30 mixture by volume of antifreeze:water (fp = −84°F, bp = 276°F with a 15-psi pressure cap).

14-13 Determination of Molecular Weight by Freezing Point Depression or Boiling Point Elevation

The colligative properties of freezing point depression and, to a lesser extent, boiling point elevation are useful in the determination of molecular weights of solutes. The solutes *must* be nonvolatile in the temperature range of the investigation if boiling point elevations are to be determined. We will restrict our discussion of determination of molecular weight to nonelectrolytes.

EXAMPLE 14-10 Molecular Weight from a Colligative Property

A 1.20-gram sample of an unknown covalent compound is dissolved in 50.0 grams of benzene. The solution freezes at 4.92°C. Calculate the molecular weight of the compound.

Plan

To calculate the molecular weight of the unknown compound, we find the number of moles that is represented by the 1.20 grams of unknown compound. We first use the freezing point data to find the molality of the solution. The molality relates the number of moles of solute to the mass of solvent (known), so this allows us to calculate the number of moles of the unknown compound.

Solution

From Table 14-2, the freezing point of pure benzene is 5.48°C and K_f is 5.12°C/*m*.

$$\Delta T_f = T_{f(pure\ benzene)} - T_{f(solution)} = 5.48°C - 4.92°C = 0.56°C$$

$$m = \frac{\Delta T_f}{K_f} = \frac{0.56°C}{5.12°C/m} = 0.11\ m$$

The molality is the number of moles of solute per kilogram of benzene, so the number of moles of solute in 50.0 g (0.0500 kg) of benzene can be calculated.

$$0.11 \ m = \frac{?\ \text{mol solute}}{0.0500 \ \text{kg benzene}}$$

$$\underline{?} \ \text{mol solute} = (0.11 \ m)(0.0500 \ \text{kg}) = 0.0055 \ \text{mol solute}$$

$$\text{mass of 1.0 mol} = \frac{\text{no. of g solute}}{\text{no. of mol solute}} = \frac{1.20 \ \text{g solute}}{0.0055 \ \text{mol solute}} = 2.2 \times 10^2 \ \text{g/mol}$$

$$\text{molecular weight} = 2.2 \times 10^2 \ \text{amu}$$

You should now work Exercise 61.

EXAMPLE 14-11 Molecular Weight from a Colligative Property

Either camphor ($C_{10}H_{16}O$, molecular weight = 152 g/mol) or naphthalene ($C_{10}H_8$, molecular weight = 128 g/mol) can be used to make mothballs. A 5.2-gram sample of mothballs was dissolved in 100. grams of ethanol, and the resulting solution had a boiling point of 78.90°C. Were the mothballs made of camphor or naphthalene? Pure ethanol has a boiling point of 78.41°C; its K_b = 1.22°C/m.

Plan

We can distinguish between the two possibilities by determining the molecular weight of the unknown solute. We do this by the method shown in Example 14-10, except that now we use the observed boiling point data.

Solution

The observed boiling point elevation is

$$\Delta T_b = T_{b(\text{solution})} - T_{b(\text{solvent})} = (78.90 - 78.41)°\text{C} = 0.49°\text{C}$$

Using ΔT_b = 0.49°C and K_b = 1.22°C/m, we can find the molality of the solution.

$$\text{molality} = \frac{\Delta T_b}{K_b} = \frac{0.49°\text{C}}{1.22°\text{C}/m} = 0.40 \ m$$

The number of moles of solute in the 100. g (0.1000 kg) of solvent used is

$$\left(0.40 \ \frac{\text{mol solute}}{\text{kg solvent}}\right)(0.100 \ \text{kg solvent}) = 0.040 \ \text{mol solute}$$

The molecular weight of the solute is its mass divided by the number of moles.

$$\frac{?\ \text{g}}{\text{mol}} = \frac{5.2 \ \text{g}}{0.040 \ \text{mol}} = 130 \ \text{g/mol}$$

The value 130 g/mol for the molecular weight indicates that naphthalene was used to make these mothballs.

You should now work Exercise 59.

> **S**TOP & THINK
> Remember that grams of *solvent* must be converted into kg when using molalities (*m*).

14-14 Colligative Properties and Dissociation of Electrolytes

As we have emphasized, colligative properties depend on the *number* of solute particles in a given mass of solvent. A 0.100 molal *aqueous* solution of a covalent compound that does not ionize gives a freezing point depression of 0.186°C. If dissociation were complete, 0.100 m NaCl would have a *total* molality of 0.200 m (i.e., 0.100 m Na$^+$ + 0.100 m Cl$^-$). So we might predict that a 0.100 molal solution of this 1:1 strong electrolyte would have a

> $\Delta T_f = K_f m$
> $= (1.86°\text{C}/m)(0.100 \ m)$
> $= 0.186°\text{C}$

freezing point depression of $2 \times 0.186°C$, or $0.372°C$. In fact, the *observed* depression is only $0.348°C$. This value for ΔT_f is about 6% less than we would expect for an effective molality of $0.200\ m$.

In an ionic solution the solute particles are not randomly distributed. Rather, each positive ion has more negative than positive ions near it. Some of the ions undergo **association** in solution (Figure 14-15). At any given instant, some Na^+ and Cl^- ions collide and "stick together." During the brief time that they are in contact, they behave as a single particle. This tends to reduce the effective molality, causing the solution to behave nonideally. The freezing point depression (ΔT_f) is therefore reduced (as well as the boiling point elevation (ΔT_b) and the lowering of vapor pressure).

▶ Ionic solutions are elegantly described by the Debye–Hückel theory, which is beyond the scope of this text.

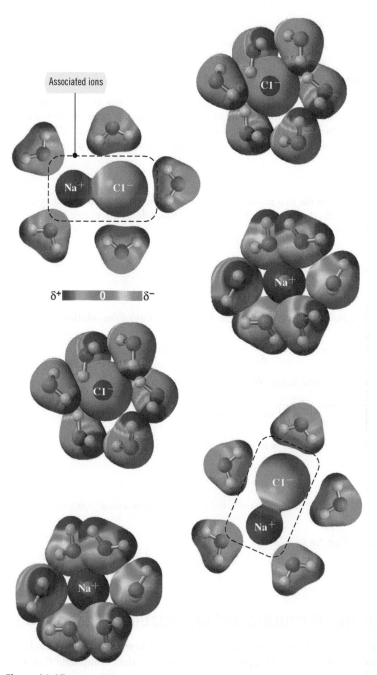

Figure 14-15 Diagrammatic representation of the various species thought to be present in a solution of NaCl in water. This would explain unexpected values for its colligative properties, such as freezing point depression.

A (more concentrated) 1.00 m solution of NaCl might be expected to have a freezing point depression of $2 \times 1.86°C = 3.72°C$, but the observed depression is only 3.40°C. This value for ΔT_f is about 9% less than we would expect for complete dissociation. We see a greater deviation from the depression predicted for an ideal solution in the more concentrated solution. This is because the Na^+ cations and Cl^- anions are closer together and collide more often in the more concentrated solution. Consequently, the ionic association is greater.

One measure of the extent of dissociation (or ionization) of an electrolyte in water is the **van't Hoff factor**, i, for the solution. This is the ratio of the *actual* colligative property to the value that *would* be observed *if no ionic dissociation occurred*.

$$i = \frac{\Delta T_{f(actual)}}{\Delta T_{f(if\ nonelectrolyte)}} = \frac{K_f m_{effective}}{K_f m_{stated}} = \frac{m_{effective}}{m_{stated}}$$

The ideal, or limiting, value of i for a solution of NaCl would be 2, and the value for a 2:1 electrolyte such as Na_2SO_4 would be 3; these values would apply to infinitely dilute solutions in which no appreciable ion association occurs. For 0.10 m and 1.0 m solutions of NaCl, i is somewhat *less than* 2.

$$\text{For } 0.10\ m: \quad i = \frac{0.348°C}{0.186°C} = 1.87 \qquad \text{For } 1.0\ m: \quad i = \frac{3.40°C}{1.86°C} = 1.83$$

Table 14-3 lists actual and ideal values of i for solutions of some strong electrolytes, based on measurements of freezing point depressions.

Many weak electrolytes are quite soluble in water, but they ionize only slightly. The i value and percent ionization for a weak electrolyte in solution can also be determined from freezing point depression data (Example 14-12).

▶ Weak acids and weak bases (see Section 6-1) are weak electrolytes.

Table 14-3 Actual and Ideal van't Hoff Factors, i, for Aqueous Solutions of Nonelectrolytes and Strong Electrolytes

Compound	i for 0.100 m Solution	i for 1.00 m Solution
nonelectrolytes	1.00 (ideal)	1.00 (ideal)
sucrose, $C_{12}H_{22}O_{11}$	1.00	1.00
If 2 ions in solution/formula unit	2.00 (ideal)	2.00 (ideal)
KBr	1.88	1.77
NaCl	1.87	1.83
If 3 ions in solution/formula unit	3.00 (ideal)	3.00 (ideal)
K_2CO_3	2.45	2.39
K_2CrO_4	2.39	1.95
If 4 ions in solution/formula unit	4.00 (ideal)	4.00 (ideal)
$K_3[Fe(CN)_6]$	2.85	—

EXAMPLE 14-12 Colligative Property and Weak Electrolytes

Lactic acid, $C_2H_4(OH)COOH$, is found in sour milk. It is also formed in muscles during intense physical activity and is responsible for the pain felt during strenuous exercise. It is a weak monoprotic acid and therefore a weak electrolyte. The freezing point of a 0.0100 m aqueous solution of lactic acid is $-0.0206°C$. Calculate (a) the van't Hoff factor i and (b) the percent ionization in the solution.

Plan for (a)

To evaluate the van't Hoff factor, i, we first calculate $m_{effective}$ from the observed freezing point depression and K_f for water; we then compare $m_{effective}$ and m_{stated} to find i.

Lactic acid that is formed in muscles is responsible for the discomfort felt during strenuous physical activity.

Solution for (a)

$$m_{effective} = \frac{\Delta T_f}{K_f} = \frac{0.0206°C}{1.86°C/m} = 0.0111 \ m$$

$$i = \frac{m_{effective}}{m_{stated}} = \frac{0.0111 \ m}{0.0100 \ m} = 1.11$$

Plan for (b)

The percent ionization is given by

$$\% \text{ ionization} = \frac{m_{ionized}}{m_{original}} \times 100\% \qquad (\text{where } m_{original} = m_{stated} = 0.0100 \ m)$$

The freezing point depression is caused by the $m_{effective}$, the total *concentration of all dissolved species*—in this case, the sum of the concentrations of HA, H^+, and A^-. We know the value of $m_{effective}$ from part (a). Thus, we need to construct an expression for the effective molality in terms of the amount of lactic acid that ionizes. We represent the molality of lactic acid that ionizes as an unknown, x, and write the concentrations of all species in terms of this unknown.

Solution for (b)

In many calculations, it is helpful to write down (1) the values, or symbols for the values, of initial concentrations; (2) the changes in concentrations due to reaction; and (3) the final concentrations, as shown here. The coefficients of the equation are all ones, so the reaction ratio must be 1:1:1.

Let x = molality of lactic acid that ionizes; then

x = molality of H^+ and lactate ions that have been formed

▶ To simplify the notation, we denote the weak acid as HA and its anion as A^-. The reaction summary used here to analyze the extent of reaction was introduced in Section 11-1.

	HA	$\longrightarrow$	H^+	+	A^-
Start	0.0100 m		0		0
Change	$-x$ m		$+x$ m		$+x$ m
Final	$(0.0100 - x)$ m		x m		x m

The $m_{effective}$ is equal to the sum of the molalities of all the solute particles.

$$m_{effective} = m_{HA} + m_{H^+} + m_{A^-}$$
$$= (0.0100 - x) \ m + x \ m + x \ m$$
$$= (0.0100 + x) \ m$$

This must equal the value for $m_{effective}$ calculated earlier, 0.0111 m.

$$0.0111 \ m = (0.0100 + x) \ m$$

$$x = 0.0011 \ m = \text{molality of the acid that ionizes}$$

We can now calculate the percent ionization.

$$\% \text{ ionization} = \frac{m_{ionized}}{m_{original}} \times 100\% = \frac{0.0011 \ m}{0.0100 \ m} \times 100\% = 11\%$$

This experiment shows that in 0.0100 m solutions, only 11% of the lactic acid has been converted into H^+ and $C_2H_4(OH)COO^-$ ions. The remainder, 89%, exists as nonionized molecules.

You should now work Exercises 78 and 80.

Problem-Solving Tip Selection of a van't Hoff Factor

Use an ideal value for the van't Hoff factor unless the question clearly indicates to do otherwise, as in the previous example and in some of the end-of-chapter exercises. For a strong electrolyte dissolved in water, the ideal value for its van't Hoff factor is listed in Table 14-3. For nonelectrolytes dissolved in water or any solute dissolved in common nonaqueous solvents, the van't Hoff factor is considered to be 1. For weak electrolytes dissolved in water, the van't Hoff factor is a little greater than 1.

14-15 Osmotic Pressure

Osmosis is the spontaneous process by which the solvent molecules pass through a semipermeable membrane from a solution of lower concentration of solute into a solution of higher concentration of solute. Figure 14-16a shows one type of experimental apparatus that demonstrates osmotic pressure. A **semipermeable membrane** such as cellophane separates the sugar solution from pure water. Solvent molecules (water in this case) may pass through the membrane in either direction, but the rate at which they pass into the more concentrated solution is initially greater than the rate in the opposite direction. The initial difference between the two rates is directly proportional to the difference in concentrations across the membrane. Solvent molecules continue to pass through the membrane until the hydrostatic pressure due to the increasing weight of the solution in the column forces the solvent molecules back through the membrane at the same rate at which they enter from the dilute side. The pressure exerted under this condition is called the **osmotic pressure** of the solution.

Osmotic pressure depends on the number, and not the kind, of solute particles in solution; it is therefore a colligative property.

The osmotic pressure of a given aqueous solution can be measured with an apparatus such as that depicted in Figure 14-16a. The solution of interest is placed inside an inverted glass (thistle) tube that has a semipermeable membrane firmly fastened across the bottom. This part of the thistle tube and its membrane are then immersed in a container of pure water. As time passes, the height of the solution in the neck rises until the back pressure it exerts counterbalances the osmotic pressure.

Alternatively, we can view osmotic pressure as the external pressure exactly sufficient to prevent osmosis. The pressure required (Figure 14-17) is equal to the osmotic pressure of the solution.

Like molecules of an ideal gas, solute particles are widely separated in very dilute solutions and do not interact significantly with one another. For very dilute solutions, osmotic pressure, π, is found to follow the equation

$$\pi = \frac{nRT}{V}$$

In this equation n is the number of moles of solute in volume, V (in liters), of the solution. The other quantities have the same meaning as in the ideal gas law. The term n/V is a concentration term. In terms of molarity, M,

$$\pi = MRT$$

Osmotic pressure increases with increasing temperature because T affects the number of solvent–membrane collisions per unit time. It also increases with increasing molarity because M affects the difference in the numbers of solvent molecules hitting the membrane from the two sides, and because a higher M leads to a stronger drive to equalize the concentration difference by dilution and to increase disorder in the solution.

▶ Osmosis is one of the main ways in which water molecules move into and out of living cells. The membranes and cell walls in living organisms allow solvent to pass through. Some of these also selectively permit passage of ions and other small solute particles.

▶ The greater the concentration of solute particles, the greater the height to which the column rises, and the greater the osmotic pressure.

▶ For a solution of an electrolyte,

$$\pi = M_{effective}RT$$

where $M_{effective} = iM_{stated}$

For *dilute aqueous solutions*, the molarity is approximately equal to the molality (because the density of the solution is nearly 1 g/mL or 1 kg/L), so

$$\pi = m_{effective}RT = im_{stated}RT$$

(dilute aqueous solutions)

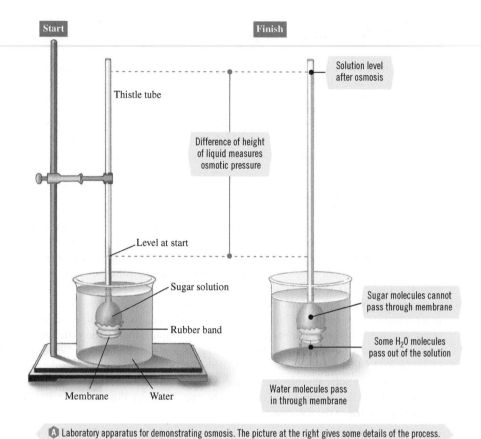

Start

Finish

Thistle tube

Solution level after osmosis

Difference of height of liquid measures osmotic pressure

Level at start

Sugar solution

Sugar molecules cannot pass through membrane

Rubber band

Some H$_2$O molecules pass out of the solution

Membrane   Water

Water molecules pass in through membrane

A Laboratory apparatus for demonstrating osmosis. The picture at the right gives some details of the process.

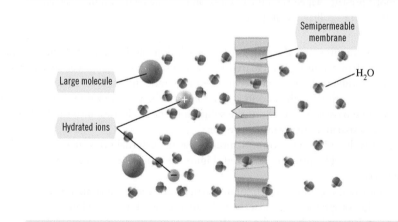

Semipermeable membrane

Large molecule

H$_2$O

Hydrated ions

B A simplified representation of the function of a semipermeable membrane

Figure 14-16

Pressure = π

Equal levels

Solution   Solvent

Membrane

Figure 14-17 The pressure that is just sufficient to prevent solvent flow from the pure solvent side through the semipermeable membrane to the solution side is another measure of the osmotic pressure of the solution.

EXAMPLE 14-13 Osmotic Pressure Calculation

What osmotic pressure would the 1.25 *m* sucrose solution in Example 14-2 exhibit at 25°C? The density of this solution is 1.34 g/mL.

Plan

We note that the approximation $M \approx m$ is not very good for this solution, because the density of this solution is quite different from 1 g/mL or kg/L. Thus, we must first find the molarity of sucrose, and then use the relationship $\pi = MRT$.

Solution

Recall from Example 14-2 that there is 50.0 g of sucrose (0.146 mol) in 117 g of H_2O, which gives 167 g of solution. Using the given density, we find that the volume of this solution is

$$\text{vol solution} = 167 \text{ g} \times \frac{1 \text{ mL}}{1.34 \text{ g}} = 125 \text{ mL or } 0.125 \text{ L}$$

Thus, the molarity of sucrose in the solution is

$$M_{\text{sucrose}} = \frac{0.146 \text{ mol}}{0.125 \text{ L}} = 1.17 \text{ mol/L}$$

Now we can calculate the osmotic pressure.

$$\pi = MRT = (1.17 \text{ mol/L})\left(0.0821 \frac{\text{L} \cdot \text{atm}}{\text{mol} \cdot \text{K}}\right)(298 \text{ K}) = \boxed{28.6 \text{ atm}}$$

You should now work Exercise 84.

STOP & THINK

The osmotic pressure can be quite large and is usually expressed in atm units. The *R* value used here is the one that includes atm units.

Let's compare the calculated values of the four colligative properties for the 1.25 *m* sucrose solution.

vapor pressure lowering	= 0.524 torr	(Example 14-4)
boiling point elevation	= 0.640°C	(Example 14-7)
freezing point depression	= 2.32°C	(Example 14-9)
osmotic pressure	= 28.6 atm	(Example 14-13)

The first of these is so small that it would be hard to measure precisely. Even this small lowering of the vapor pressure is sufficient to raise the boiling point by an amount that could be measured, although with difficulty. The freezing point depression is greater, but still could not be measured very precisely without a special apparatus. The osmotic pressure, on the other hand, is so large that it could be measured much more precisely. Thus, osmotic pressure is often the most easily measured of the four colligative properties, especially when very dilute solutions are used.

Osmotic pressures represent very significant forces. For example, a 1.00 molar solution of a nonelectrolyte in water at 0°C produces an equilibrium osmotic pressure of approximately 22.4 atmospheres (≈ 330 psi).

The use of measurements of osmotic pressure for the determination of molecular weights has several advantages. Even very dilute solutions give easily measurable osmotic pressures. This method is therefore useful in determination of the molecular weights of (1) very expensive substances, (2) substances that can be prepared only in very small amounts, and (3) substances of very high molecular weight that are not very soluble. Because high-molecular-weight materials are often difficult, and in some cases impossible, to obtain in a high state of purity, determinations of their molecular weights are not as accurate as we might like. Nonetheless, osmotic pressures provide a very useful method of estimating molecular weights.

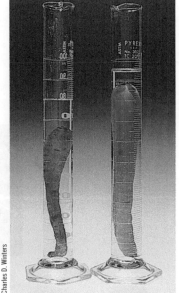

Charles D. Winters

An illustration of osmosis. When a carrot is soaked in a concentrated salt solution, water flows out of the plant cells by osmosis. A carrot soaked overnight in salt solution (*left*) has lost much water and become limp. A carrot soaked overnight in pure water (*right*) is little affected.

EXAMPLE 14-14 Molecular Weight from Osmotic Pressure

Pepsin is an enzyme present in the human digestive tract. A solution of a 0.500-gram sample of purified pepsin in 30.0 mL of aqueous solution exhibits an osmotic pressure of 8.92 torr at 27.0°C. Estimate the molecular weight of pepsin.

▶ An enzyme is a protein that acts as a biological catalyst. Pepsin catalyzes the metabolic cleavage of amino acid chains (called peptide chains) in other proteins.

Plan

As we did in earlier molecular weight determinations (see Section 14-13), we must first find n, the number of moles that 0.500 grams of pepsin represents. We start with the equation $\pi = MRT$. The molarity of pepsin is equal to the number of moles of pepsin per liter of solution, n/V. We substitute this for M and solve for n.

Solution

$$\pi = MRT = \left(\frac{n}{V}\right) RT \qquad \text{or} \qquad n = \frac{\pi V}{RT}$$

We convert 8.92 torr to atmospheres to be consistent with the units of R.

$$n = \frac{\pi V}{RT} = \frac{\left(8.92 \text{ torr} \times \dfrac{1 \text{ atm}}{760 \text{ torr}}\right)(0.0300 \text{ L})}{\left(0.0821 \dfrac{\text{L} \cdot \text{atm}}{\text{mol} \cdot \text{K}}\right)(300 \text{ K})} = 1.43 \times 10^{-5} \text{ mol pepsin}$$

Thus, 0.500 g of pepsin is 1.43×10^{-5} mol. We now estimate its molecular weight.

$$? \text{ g/mol} = \frac{0.500 \text{ g}}{1.43 \times 10^{-5} \text{ mol}} = 3.50 \times 10^{4} \text{ g/mol}$$

The molecular weight of pepsin is approximately 35,000 amu. This is typical for medium-sized proteins.

You should now work Exercise 90.

STOP & THINK

The freezing point depression of this very dilute solution would be only about 0.0009°C, which would be difficult to measure accurately. The osmotic pressure of 8.92 torr, on the other hand, is easily measured.

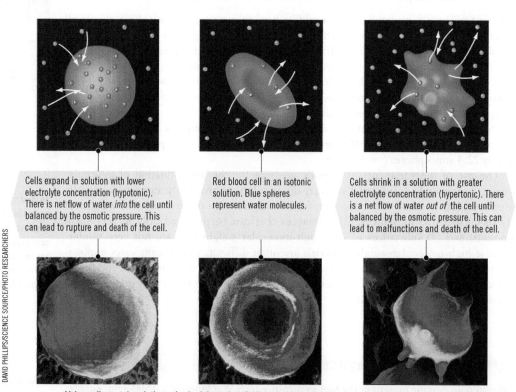

DAVID PHILLIPS/SCIENCE SOURCE/PHOTO RESEARCHERS

Cells expand in solution with lower electrolyte concentration (hypotonic). There is net flow of water *into* the cell until balanced by the osmotic pressure. This can lead to rupture and death of the cell.

Red blood cell in an isotonic solution. Blue spheres represent water molecules.

Cells shrink in a solution with greater electrolyte concentration (hypertonic). There is a net flow of water *out of* the cell until balanced by the osmotic pressure. This can lead to malfunctions and death of the cell.

Living cells contain solutions of salts (electrolytes). When a cell is put in contact with solutions having different electrolyte concentrations, the resulting osmotic pressure will cause a net flow of water to flow into (*left*) or out of (*right*) the cell causing it to expand or contract. The top row shows illustrations, while the bottom row shows actual scanning electron microscope images of red blood cells exhibiting these effects.

Problem-Solving Tip Units in Osmotic Pressure Calculations

The strict definition for osmotic pressure is in terms of molarity, $\pi = MRT$. Osmotic pressure, π, has the units of pressure (atmospheres); M (mol/L); and T (kelvins).

Therefore, the appropriate value of R is $0.0821 \dfrac{L \cdot atm}{mol \cdot K}$. We can balance the units in this equation as follows.

$$atm = \left(\frac{mol}{L}\right)\left(\frac{L \cdot atm}{mol \cdot K}\right)(K)$$

When we use the approximation that $M \approx m$, we might think that the units do not balance. For very dilute aqueous solutions, we can think of M as being *numerically* equal to m, but still having the units mol/L, as shown above.

Sports drinks were developed to counteract dehydration while maintaining electrolyte balance. Such solutions also help to prevent cell damage due to the osmotic pressure that would be caused by drinking large amounts of water.

Colloids

A solution is a homogeneous mixture in which no phase separation occurs and in which solutes are present as individual molecules or ions. This represents one extreme of mixtures. The other extreme is a suspension, a clearly heterogeneous mixture in which solute particles settle out after mixing with a solvent phase. Such a situation results when a handful of sand is stirred into water. **Colloids (colloidal dispersions)** represent an intermediate kind of mixture in which the solute particles, or **dispersed phase**, are suspended in the solvent phase, or **dispersing medium**. The particles of the dispersed phase are so small that settling is negligible. They are large enough, however, to scatter light as it passes through the colloid, making the mixture appear cloudy or even opaque.

Table 14-4 indicates that all combinations of solids, liquids, and gases can form colloids except mixtures of nonreacting gases (all of which are homogeneous and, therefore, true solutions). Whether a given mixture forms a solution, a colloidal dispersion, or a suspension depends on the size of the solute-like particles (Table 14-5), as well as solubility and miscibility.

14-16 The Tyndall Effect

Particles that are too small cannot scatter light. Solute particles in true solutions are below the size limit required for light scattering. Colloidal particles, however, are larger—up to about 10,000 Å (1000 nm or 1 micron)—so they do scatter light. This scattering of light by colloidal particles is called the **Tyndall effect** (Figure 14-18).

The scattering of light from automobile headlights by fogs and mists is an example of the Tyndall effect, as is the scattering of a light beam in a laser show by dust particles in the air in a darkened room. If the air didn't have any suspended solid particles you would not be able to see the laser beam.

14-17 The Adsorption Phenomenon

Much of the chemistry of everyday life is the chemistry of colloids, as one can tell from the examples in Table 14-4. Because colloidal particles are so finely divided, they have a tremendously high total surface area in relation to their volume. It is not surprising, therefore, that an understanding of colloidal behavior requires an understanding of surface phenomena.

Atoms on the surface of a colloidal particle are bonded only to other atoms of the particle on and below the surface. These atoms interact with whatever comes in contact

Figure 14-18 The dispersion of a beam of light by colloidal particles is called the Tyndall effect. The right-most container has distilled water that just barely scatters the green laser beam. The middle container is tap water that has small amounts of particles that do scatter the light to a moderate extent. The left container has 500 mL of distilled water with one drop of milk added. Milk is a colloid, so strong scattering of the laser beam is seen.

CHEMISTRY IN USE

Water Purification and Hemodialysis

Semipermeable membranes play important roles in the normal functioning of many living systems. In addition, they are used in a wide variety of industrial and medical applications. Membranes with different permeability characteristics have been developed for many different purposes. One of these is the purification of water by reverse osmosis.

Suppose we place a semipermeable membrane between a saline (salt) solution and pure water. If the saline solution is pressurized under a greater pressure than its osmotic pressure, the direction of flow can be reversed. That is, the net flow of water molecules will be from the saline solution through the membrane into the pure water. This process is called **reverse osmosis**. The membrane usually consists of cellulose acetate or hollow fibers of a material structurally similar to nylon. This method has been used for the purification of brackish (mildly saline) water. It has the economic advantages of low cost, ease of apparatus construction, and simplicity of operation. Because this method of water purification requires no heat, it has a great advantage over distillation.

The city of Sarasota, Florida, has a large reverse osmosis plant to purify drinking water. It processes more than 4 million gallons of water per day from local wells. Total dissolved solids (mostly salts) are reduced in concentration from 1744 parts per million (ppm) (0.1744% by mass) to 90 ppm. This water is mixed with additional well water purified by an ion exchange system. The final product is more than 10 million gallons of water per day containing less than 500 ppm of total dissolved solids, the standard for drinking water set by the World Health

A reverse osmosis unit used to provide all the fresh water (82,500 gallons per day) for the steamship *Norway*.

Organization. A large reverse osmosis plant in Ashkelon, Israel, processes seawater to produce more than 330,000 cubic meters per day (330 million L/day) of fresh water. This is enough to provide 13% of the country's domestic consumer demand. The Kuwaiti and Saudi water purification plants that were of strategic concern in the Persian Gulf War use reverse osmosis in one of their primary stages.

Human kidneys carry out many important functions. One of the most crucial is the removal of metabolic waste prod-

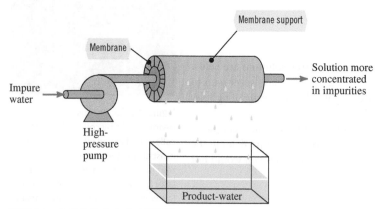

The reverse osmosis method of water purification.

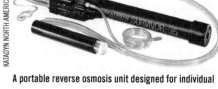

A portable reverse osmosis unit designed for individual use.

with the surface. Colloidal particles often adsorb ions or other charged particles, as well as gases and liquids. The process of **adsorption** involves adhesion of any such species onto the surfaces of particles. For example, a bright red **sol** (solid dispersed in liquid) is formed by mixing hot water with a concentrated aqueous solution of iron(III) chloride (Figure 14-19).

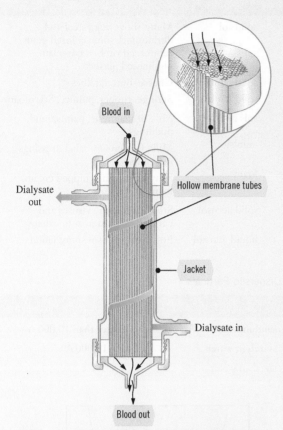

Blood in

Dialysate out

Hollow membrane tubes

Jacket

Dialysate in

Blood out

A schematic diagram of the hollow fiber (or capillary) dialyzer, the most commonly used artificial kidney. The blood flows through many small tubes constructed of a semipermeable membrane material; these tubes are bathed in the dialyzing solution.

The membrane separates the blood from a dialyzing solution, or *dialysate,* that is similar to blood plasma in its concentration of needed substances (e.g., electrolytes and amino acids) but contains none of the waste products. Because the concentrations of undesirable substances are higher in the blood than in the dialysate, they flow preferentially out of the blood and are washed away. The concentrations of *needed* substances are the same on both sides of the membrane, so these substances are maintained at the proper concentrations in the blood. The small pore size of the membrane prevents passage of blood cells. However, Na$^+$ and Cl$^-$ ions and some small molecules do pass through the membrane. A patient with total kidney failure may require up to four hemodialysis sessions per week, at 3 to 4 hours per session. To help hold down the cost of such treatment, the dialysate solution is later purified by a combination of filtration, distillation, and reverse osmosis and is then reused.

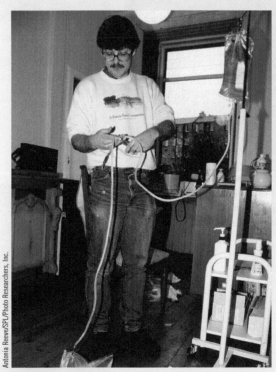

Antonia Reeve/SPL/Photo Researchers, Inc.

A portable dialysis unit.

ucts (e.g., creatinine, urea, and uric acid) from the blood without removal of substances needed by the body (e.g., glucose, electrolytes, and amino acids). The process by which this is accomplished in the kidney involves *dialysis,* a phenomenon in which the membrane allows transfer of both solvent molecules *and* certain solute molecules and ions, usually small ones. Many patients whose kidneys have failed can have dialysis performed by an artificial kidney machine. In this mechanical procedure, called *hemodialysis,* the blood is withdrawn from the body and passed in contact with a semipermeable membrane.

$$2x[Fe^{3+}(aq) + 3Cl^-(aq)] + x(3 + y)H_2O \longrightarrow [Fe_2O_3 \cdot yH_2O]_x(s) + 6x[H^+ + Cl^-]$$

yellow solution   bright red sol

Each colloidal particle of this sol is a cluster of many formula units of hydrated Fe$_2$O$_3$. Each particle attracts positively charged Fe^{3+} ions to its surface. Because each particle is then surrounded by a shell of positively charged ions, the particles repel one another and cannot combine to the extent necessary to cause precipitation.

Freshly made wines are often cloudy because of colloidal particles. Removing these colloidal particles clarifies the wine.

The Tyndall effect is observed as the shafts of light created by the tree's foliage are scattered by the dust particles in the air.

Table 14-4 Types of Colloids

Dispersed (solute-like) Phase		Dispersing (solvent-like) Medium	Common Name	Examples
solid	in	solid	solid sol	Many alloys (e.g., steel and duralumin), some colored gems, reinforced rubber, porcelain, pigmented plastics
liquid	in	solid	solid emulsion	Cheese, butter, jellies
gas	in	solid	solid foam	Sponge, rubber, pumice, Styrofoam
solid	in	liquid	sols and gels	Milk of magnesia, paints, mud, puddings
liquid	in	liquid	emulsion	Milk, face cream, salad dressings, mayonnaise
gas	in	liquid	foam	Shaving cream, whipped cream, foam on beer
solid	in	gas	solid aerosol	Smoke, airborne viruses and particulate matter, auto exhaust
liquid	in	gas	liquid aerosol	Fog, mist, aerosol spray, clouds

Table 14-5 Approximate Sizes of Dispersed Particles

Mixture	Example	Approximate Particle Size
suspension	sand in water	larger than 10,000 Å
colloidal dispersion	starch in water	10–10,000 Å
solution	sugar in water	1–10 Å

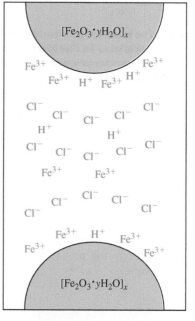

Figure 14-19 Stabilization of a colloid (Fe_2O_3 sol) by electrostatic forces. Each colloidal particle of this orange-red sol is a cluster of many formula units of hydrated Fe_2O_3. Each cluster attracts positively charged Fe^{3+} ions to its surface. (Fe^{3+} ions fit readily into the crystal structure, so they, rather than the Cl^- ions, are preferentially adsorbed.) Each particle is then surrounded by a shell of positively charged ions, so the particles repel one another and cannot combine to the extent necessary to cause actual precipitation. The suspended particles scatter light, making the path of the light beam through the suspension visible.

14-18 Hydrophilic and Hydrophobic Colloids

Colloids are classified as **hydrophilic** ("water loving") or **hydrophobic** ("water hating") based on the surface characteristics of the dispersed particles.

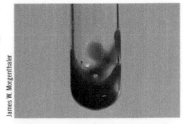

Fe(OH)₃ is a gelatinous precipitate (a gel).

Hydrophilic Colloids

Proteins such as the oxygen-carrier hemoglobin form hydrophilic sols when they are suspended in saline aqueous body fluids such as blood plasma. Such proteins are macromolecules (giant molecules) that fold and twist in an aqueous environment so that polar groups are exposed to the fluid, whereas nonpolar groups are encased (Figure 14-20). Protoplasm and human cells are examples of **gels**, which are special types of sols in which the solid particles (in this case mainly proteins and carbohydrates) join together in a semirigid network structure that encloses the dispersing medium. Other examples of gels are gelatin, jellies, and gelatinous precipitates such as $Al(OH)_3$ and $Fe(OH)_3$.

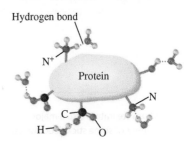

Figure 14-20 Examples of hydrophilic groups at the surface of a giant molecule (macromolecule) that help keep the macromolecule suspended in water.

Hydrophobic Colloids

Hydrophobic colloids cannot exist in polar solvents without the presence of **emulsifying agents**, or **emulsifiers**. These agents coat the particles of the dispersed phase to prevent their coagulation into a separate phase. Milk and mayonnaise are examples of hydrophobic colloids (milk fat in milk, vegetable oil in mayonnaise) that stay suspended with the aid of emulsifying agents (casein in milk and egg yolk in mayonnaise).

Consider the mixture resulting from vigorous shaking of salad oil (nonpolar) and vinegar (polar). Droplets of hydrophobic oil are temporarily suspended in the water. In a short time, however, the very polar water molecules, which attract one another strongly, squeeze out the nonpolar oil molecules. The less dense oil then coalesces and floats to the top. If we add an emulsifying agent, such as egg yolk, and shake or beat the mixture, a stable emulsion (mayonnaise) results.

Oil and grease are mostly long-chain hydrocarbons that are very nearly nonpolar. Our most common solvent is water, a polar substance that does not dissolve nonpolar substances. To use water to wash soiled fabrics, greasy dishes, or our bodies, we must enable the water to suspend and remove nonpolar substances. Soaps and detergents are emulsifying agents that accomplish this. Their function is controlled by the intermolecular interactions that result from their structures.

Solid soaps are usually sodium salts of long-chain organic acids called fatty acids. They have a polar "head" and a nonpolar "hydrocarbon tail." Sodium stearate, a typical soap, is shown here:

Some edible colloids.

▶ "Dry" cleaning does not involve water. The organic solvents that are used in dry cleaning dissolve grease to form true solutions.

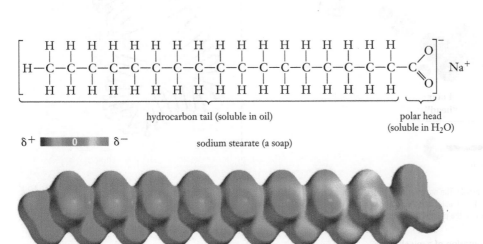

hydrocarbon tail (soluble in oil)

polar head (soluble in H₂O)

$\delta+$ ■■■■■ 0 ■■■■■ $\delta-$

sodium stearate (a soap)

(*Left*) Powdered sulfur floats on pure water because of the high surface tension of water.
(*Right*) When a drop of detergent solution is added to the water, its surface tension is lowered and sulfur sinks. This lowering of surface tension enhances the cleaning action of detergent solutions.

▶ Sodium stearate is also a major component of some stick deodorants.

The stearate ion is typical of the anions in soaps. It has a polar carboxylate head, $\overset{\overset{\text{O}}{\|}}{-\text{C}}-\text{O}^-$, and a long nonpolar tail, $CH_3(CH_2)_{16}-$. The head of the stearate ion is compatible with ("soluble in") water, whereas the hydrocarbon tail is compatible with ("soluble in") oil and grease. Groups of such ions can be dispersed in water because they form **micelles** (Figure 14-21a). Their "water-insoluble" nonpolar tails orient into the interior of a micelle, while their polar heads point to the outside where they can interact

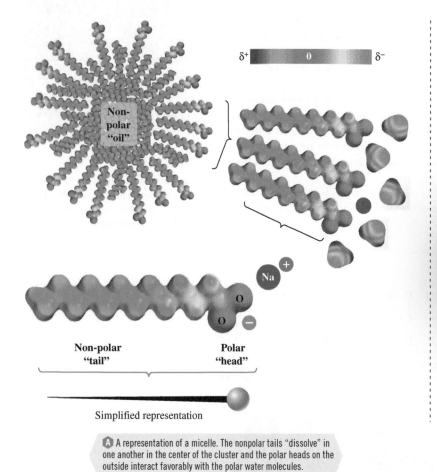

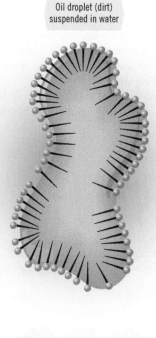

A A representation of a micelle. The nonpolar tails "dissolve" in one another in the center of the cluster and the polar heads on the outside interact favorably with the polar water molecules.

B Attachment of soap or detergent molecules to a droplet of oily dirt to suspend it in water

Figure 14-21 Two-dimensional representation of a micelle.

with the polar water molecules. When sodium stearate is stirred into water, the result is not a true solution. Instead it contains negatively charged micelles of stearate ions, surrounded by the positively charged Na^+ ions. The result is a suspension of micelles in water. These micelles are large enough to scatter light, so a soap–water mixture appears cloudy. Oil and grease "dissolve" in soapy water because the nonpolar oil and grease are preferentially absorbed into the nonpolar interior of micelles (see Figure 14-21b). Micelles form a true emulsion in water, so the oil and grease can be washed away. Sodium stearate is called a **surfactant** (meaning "surface-active agent") or wetting agent because it has the ability to suspend and wash away oil and grease. Other soaps and detergents behave similarly.

"**Hard**" **water** contains Fe^{3+}, Ca^{2+}, and/or Mg^{2+} ions, all of which displace Na^+ from soaps to form precipitates. This removes the soap from the water and puts an undesirable coating ("soap scum") on the bathtub or on the fabric being laundered. Synthetic **detergents** are soap-like emulsifiers that contain sulfonate, $—SO_3^-$, or sulfate, $—OSO_3^-$, instead of carboxylate groups, $—COO^-$. They do not precipitate the ions of hard water due to their larger size and more delocalized negative charge (Coulomb's Law, see Section 13-2) so they can be used in hard water as soap substitutes without forming undesirable scum.

Phosphates were added to commercial detergents for various purposes. They complexed the metal ions that contribute to water hardness and kept them dissolved, controlled acidity, and influenced micelle formation. The use of detergents containing phosphates is now discouraged because they cause **eutrophication** in rivers and streams that receive sewage. This is a condition (not related to colloids) in which an overgrowth of vegetation is caused by the high concentration of phosphorus, a plant nutrient. This overgrowth and the subsequent decay of the dead plants lead to decreased dissolved O_2 in the water, which causes the gradual elimination of marine life. There is also a foaming problem associated with branched alkylbenzenesulfonate (ABS) detergents in streams and in pipes, tanks, and pumps of sewage treatment plants. Such detergents are not **biodegradable**; that is, they cannot be broken down by bacteria.

$$\left[\begin{array}{c} \overset{\displaystyle CH_3}{|} \quad \overset{\displaystyle CH_3}{|} \quad \overset{\displaystyle CH_3}{|} \quad \overset{\displaystyle CH_3}{|} \\ CH_3CHCH_2CHCH_2CHCH_2CH \end{array} \text{—} \bigcirc \text{—} SO_3 \right]^{-} Na^+$$

a sodium branched alkylbenzenesulfonate (ABS)—a nonbiodegradable detergent

Currently used linear-chain alkylbenzenesulfonate (LAS) detergents are biodegradable and do not cause such foaming.

$$\left[CH_3CH_2CH_2CH_2CH_2CH_2CH_2CH_2CH_2CH_2CH_2CH_2 \text{—} \bigcirc \text{—} SO_3 \right]^{-} Na^+$$

sodium lauryl benzenesulfonate
a linear alkylbenzenesulfonate (LAS)—a biodegradable detergent

Humic acids are a complex family of brown-colored natural components of plants (especially leaves and wood) that have multiple carboxylic acid groups and benzene rings. They are readily leached from decaying plant matter and often give streams a distinctive brownish color (Figure 14-22).

© Chinch Gryniewicz, Ecoscene/Corbis

Phosphates and nonbiodegradable detergents are responsible for the devastation of plant life in and along this river.

▶ The two detergents shown here each have a $C_{12}H_{25}$ tail, but the branched one is not biodegradable.

George Stanley

Figure 14-22 The Gooseberry River in northern Minnesota has a distinct brown color due to humic acids leached from decaying plant matter. These humic acids can act as natural detergents or surfactants, producing foams in the vicinity of rapids and waterfalls. The foaming in this case is not caused by synthetic detergents and is not a sign of pollution.

CHEMISTRY IN USE

Why Does Red Wine Go with Red Meat?

Choosing the appropriate wine to go with dinner is a problem for some diners. Experts, however, have offered a simple rule for generations, "serve red wine with red meat and white wine with fish and chicken." Are these cuisine choices just traditions, or are there fundamental reasons for them?

Red wine is usually served with red meat because of a desirable matching of the chemicals found in each. The most influential ingredient in red meat is fat; it gives red meats their desirable flavor. As you chew a piece of red meat, the fat from the meat coats your tongue and palate, which desensitizes your taste buds. As a result, your second bite of red meat is less tasty than the first. Your steak would taste better if you washed your mouth between mouthfuls; there is an easy way to wash away the fat deposits.

Red wine contains a surfactant that cleanses your mouth, removing fat deposits, re-exposing your taste buds, and allowing you to savor the next bite of red meat almost as well as the first. The tannic acid (also called tannin) in red wine provides a soap-like action. Like soap, tannic acid consists of both a nonpolar complex hydrocarbon part as well as a polar one. The polar part of tannic acid dissolves in polar saliva, while the nonpolar part dissolves in the fat film that coats your palate. When you sip red wine, a suspension of micelles forms in the saliva. This micelle emulsion has the fat molecules in its interior; the fat is washed away by swallowing the red wine.

White wines go poorly with red meats because they lack the tannic acid needed to cleanse the palate. In fact, the presence

BRAND X PICTURES/GETTY IMAGES

or absence of tannic acid (not the color of the grapes) is what distinguishes red wines from white wines. Grapes fermented with their skins produce red wines; grapes fermented without their skins produce white wines.

Because fish and chicken have less fat than red meats, they can be enjoyed without surfactants to cleanse the palate. Also, the flavor of tannic acid can overpower the delicate flavor of many fish. The absence of tannic acid in white wines gives them a lighter flavor than red wines, and many people prefer this lighter flavor with their fish or chicken dinners.

RONALD DELORENZO

KEY TERMS

The following terms were defined at the end of Chapter 3: **concentration, dilution, molarity, percent by mass, solute, solution,** and **solvent.** The following terms were defined at the end of Chapter 13: **condensation, condensed phases, evaporation, phase diagram,** and **vapor pressure.**

Adsorption   Adhesion of species onto surfaces of particles.

Associated ions   Short-lived species formed by the collision of dissolved ions of opposite charge.

Biodegradability   The ability of a substance to be broken down into simpler substances by bacteria.

Boiling point elevation   The increase in the boiling point of a solvent caused by dissolution of a nonvolatile solute.

Boiling point elevation constant, K_b   A constant that corresponds to the change (increase) in boiling point produced by a one-molal *ideal* solution of a nonvolatile nonelectrolyte.

Charge density of an ion   The ionic charge divided by the ionic radius in Angstroms.

Colligative properties   Physical properties of solutions that depend on the number, but not the kind, of solute particles present.

Colloid   A heterogeneous mixture in which solute-like particles do not settle out; also known as a *colloidal dispersion.*

Crystal lattice energy   The energy change when one mole of formula units of a crystalline solid is formed from its ions, atoms, or molecules in the gas phase; always negative.

Detergent   A soap-like emulsifier that contains a sulfonate, $-SO_3^-$, or sulfate, $-OSO_3^-$, group instead of a carboxylate, $-COO^-$, group.

Dispersed phase   The solute-like species in a colloid.

Dispersing medium   The solvent-like phase in a colloid.

Dispersion   See *Colloid.*

Distillation   The process in which components of a mixture are separated by boiling away the more volatile liquid.

Effective molality   The sum of the molalities of all solute particles in solution.

Emulsifier   See *Emulsifying agent.*

Emulsifying agent   A substance that coats the particles of a dispersed phase and prevents coagulation of colloidal particles; an emulsifier; also known as an *emulsifier.*

Emulsion   A colloidal dispersion of a liquid in a liquid.

Eutrophication The undesirable overgrowth of vegetation caused by high concentrations of plant nutrients in bodies of water.

Foam A colloidal dispersion of a gas in a liquid.

Fractional distillation The process in which a fractionating column is used in a distillation apparatus to separate components of a liquid mixture that have different boiling points.

Freezing point depression The decrease in the freezing point of a solvent caused by the presence of a solute.

Freezing point depression constant, K_f A constant that corresponds to the change in freezing point produced by a one-molal *ideal* solution of a nonvolatile nonelectrolyte.

Gel A colloidal dispersion of a solid in a liquid; a semirigid sol.

Hard water Water containing Fe^{3+}, Ca^{2+}, or Mg^{2+} ions, which form precipitates with soaps.

Heat of hydration See *Hydration energy (molar) of an ion.*

Heat of solution (molar) The amount of heat absorbed in the formation of a solution that contains one mole of solute; the value is positive if heat is absorbed (endothermic) and negative if heat is released (exothermic).

Henry's Law The concentration or solubility of a gas in a liquid at any given temperature is directly proportional to the partial pressure of the gas over the solution.

Hydration The interaction (surrounding) of solute particles with water molecules.

Hydration energy (molar) of an ion The energy change accompanying the hydration of a mole of gaseous ions.

Hydrophilic colloids Colloidal particles that attract water molecules.

Hydrophobic colloids Colloidal particles that repel water molecules.

Ideal solution A solution that obeys Raoult's Law exactly.

Liquid aerosol A colloidal dispersion of a liquid in a gas.

Micelle A cluster of a large number of soap or detergent molecules or ions, assembled with their hydrophobic tails directed toward the center and their hydrophilic heads directed outward.

Miscibility The ability of one liquid to mix with (dissolve in) another liquid.

Molality (*m*) Concentration expressed as number of moles of solute per kilogram of solvent.

Mole fraction of a component in solution The number of moles of the component divided by the total number of moles of all components.

Osmosis The process by which solvent molecules pass through a semipermeable membrane from a dilute solution into a more concentrated solution.

Osmotic pressure The hydrostatic pressure produced on the surface of a semipermeable membrane by osmosis.

Percent ionization of weak electrolytes The percent of the weak electrolyte that ionizes in a solution of a given concentration.

Raoult's Law The vapor pressure of a solvent in an ideal solution is directly proportional to the mole fraction of the solvent in the solution.

Reverse osmosis The forced flow of solvent molecules through a semipermeable membrane from a concentrated solution into a dilute solution. This is accomplished by application of hydrostatic pressure on the concentrated side greater than the osmotic pressure that is opposing it.

Saturated solution A solution in which no more solute will dissolve at a given temperature.

Semipermeable membrane A thin partition between two solutions through which certain molecules can pass but others cannot.

Soap An emulsifier that can disperse nonpolar substances in water; the sodium salt of a long-chain organic acid; consists of a long hydrocarbon chain attached to a carboxylate group, $-CO_2^-Na^+$.

Sol A colloidal dispersion of a solid in a liquid.

Solid aerosol A colloidal dispersion of a solid in a gas.

Solid emulsion A colloidal dispersion of a liquid in a solid.

Solid foam A colloidal dispersion of a gas in a solid.

Solid sol A colloidal dispersion of a solid in a solid.

Solvation The process by which solvent molecules surround and interact with solute ions or molecules.

Supersaturated solution A (metastable) solution that contains a higher-than-saturation concentration of solute; slight disturbance or seeding causes crystallization of excess solute.

Surfactant A "surface-active agent"; a substance that has the ability to emulsify and wash away oil and grease in an aqueous suspension.

Thermal pollution Introduction of heated waste water into natural waters.

Tyndall effect The scattering of light by colloidal particles.

van't Hoff factor (*i*) A number that indicates the extent of dissociation or ionization of a solute; equal to the actual colligative property divided by the colligative property calculated assuming no ionization or dissociation.

EXERCISES

🦠 **Molecular Reasoning** exercises

▲ **More Challenging** exercises

Blue-Numbered exercises are solved in the Student Solutions Manual

General Concepts: The Dissolving Process

1. Support or criticize the statement "Solutions and mixtures are the same thing."

2. 🦠 Give an example of a solution that contains each of the following: (a) a solid dissolved in a liquid; (b) a gas dissolved in a gas; (c) a gas dissolved in a liquid; (d) a liquid

🦠 **Molecular Reasoning** exercises   ▲ **More Challenging** exercises   **Blue-Numbered** exercises solved in **Student Solutions Manual**

Unless otherwise noted, all content on this page is © Cengage Learning.

dissolved in a liquid; (e) a solid dissolved in a solid. Identify the solvent and the solute in each case.

3. 🐾 There are no *true* solutions in which the solvent is gaseous and the solute is either liquid or solid. Why?

4. 🐾 Explain why (a) solute–solute, (b) solvent–solvent, and (c) solute–solvent interactions are important in determining the extent to which a solute dissolves in a solvent.

5. 🐾 Define and distinguish between dissolution, solvation, and hydration.

6. 🐾 The amount of heat released or absorbed in the dissolution process is important in determining whether the dissolution process is spontaneous, meaning, whether it can occur. What is the other important factor? How does it influence solubility?

7. 🐾 An old saying is that "oil and water don't mix." Explain the molecular basis for this saying.

8. 🐾 Two liquids, A and B, do not react chemically and are completely miscible. What would be observed as one is poured into the other? What would be observed in the case of two completely immiscible liquids and in the case of two partially miscible liquids?

9. 🐾 Consider the following solutions. In each case, predict whether the solubility of the solute should be high or low. Justify your answers. (a) LiCl in *n*-octane, C_8H_{18}; (b) $CaCl_2$ in H_2O; (c) C_8H_{18} in H_2O; (d) $CHCl_3$ in C_6H_{14}; (e) C_8H_{18} in CCl_4

10. 🐾 Consider the following solutions. In each case predict whether the solubility of the solute should be high or low. Justify your answers. (a) HCl in H_2O; (b) HF in H_2O; (c) Al_2O_3 in H_2O; (d) SiO_2 in H_2O; (e) $Na_2(SO_4)$ in hexane, C_6H_{14}

11. 🐾 For those solutions in Exercise 9 that can be prepared in "reasonable" concentrations, classify the solutes as nonelectrolytes, weak electrolytes, or strong electrolytes.

12. 🐾 For those solutions in Exercise 10 that can be prepared in "reasonable" concentrations, classify the solutes as nonelectrolytes, weak electrolytes, or strong electrolytes.

13. 🐾 Both methanol, CH_3OH, and ethanol, CH_3CH_2OH, are completely miscible with water at room temperature because of strong solvent–solute intermolecular attractions. Predict the trend in solubility in water for 1-propanol, $CH_3CH_2CH_2OH$; 1-butanol, $CH_3CH_2CH_2CH_2OH$; and 1-pentanol, $CH_3CH_2CH_2CH_2CH_2OH$.

14. 🐾 (a) Does the solubility of a solid in a liquid exhibit an appreciable dependence on pressure? (b) Is the same true for the solubility of a liquid in a liquid? Why?

15. 🐾 Describe a technique for determining whether or not a solution contains an electrolyte.

16. 🐾 A reagent bottle in the storeroom is labeled as containing a sodium chloride solution. How can one determine whether or not the solution is saturated?

17. ▲ A handbook lists the value of the Henry's Law constant as 3.31×10^{-5} atm^{-1} for ethane, C_2H_6, dissolved in water at 25°C. The absence of concentration units on *k* means that the constant is meant to be used with concentration expressed as a mole fraction. Calculate the mole fraction of ethane in water at an ethane pressure of 0.15 atm.

18. ▲ The mole fraction of methane, CH_4, dissolved in water can be calculated from the Henry's Law constants of 2.42×10^{-5} atm^{-1} at 25°C and 1.73×10^{-5} atm^{-1} at 50.°C. Calculate the solubility of methane at these temperatures for a methane pressure of 10. atm above the solution. Does the solubility increase or decrease with increasing temperature? (See Exercise 17 for interpretation of units.)

19. 🐾 Choose the ionic compound from each pair for which the crystal lattice energy should be the most negative. Justify your choice. (a) LiF or LiBr; (b) KF or CaF_2; (c) FeF_2 or FeF_3; (d) NaF or KF

20. 🐾 Choose the ion from each pair that should be more strongly hydrated in aqueous solution. Justify your choice. (a) Na^+ or Rb^+; (b) Cl^- or Br^-; (c) Fe^{3+} or Fe^{2+}; (d) Na^+ or Mg^{2+}

21. ▲ The crystal lattice energy for LiBr(s) is -818.6 kJ/mol at 25°C. The hydration energy of the ions of LiBr(s) is -867.4 kJ/mol at 25°C (for infinite dilution). (a) What is the heat of solution of LiBr(s) at 25°C (for infinite dilution)? (b) The hydration energy of $Li^+(g)$ is -544 kJ/mol at 25°C. What is the hydration energy for $Br^-(g)$ at 25°C?

22. Why is the dissolving of many ionic solids in water an endothermic process, whereas the mixing of most miscible liquids is an exothermic process?

Concentrations of Solutions

23. Under what conditions are the molarity and molality of a solution nearly the same? Which concentration unit is more useful when measuring volume with burets, pipets, and volumetric flasks in the laboratory? Why?

24. Many handbooks list solubilities in units of (g solute/ 100. g H_2O). How would you convert from this unit to mass percent?

25. The density of a 15.00% by mass aqueous solution of acetic acid, CH_3COOH, is 1.0187 g/mL. What is (a) the molarity? (b) the molality? (c) the mole fraction of each component?

26. Assume you dissolve 45.0 g of camphor, $C_{10}H_{16}O$, in 425 mL of ethanol, C_2H_5OH. Calculate the molality, mole fraction, and weight percent of camphor in this solution. (The density of ethanol is 0.785 g/mL.)

27. ▲ Urea, $(NH_2)_2CO$, is a product of metabolism of proteins. An aqueous solution is 32.0% urea by mass and has a density of 1.087 g/mL. Calculate the molality of urea in the solution.

28. Calculate the molality of a solution that contains 56.5 g of benzoic acid, C_6H_5COOH, in 350 mL of ethanol, C_2H_5OH. The density of ethanol is 0.789 g/mL.

29. Sodium fluoride has a solubility of 3.84 g in 100.0 g of water at 18°C. Express this solute concentration in terms of (a) mass percent, (b) mole fraction, and (c) molality.

30. What are the mole fractions of ethanol, C_2H_5OH, and water in a solution prepared by mixing 55.0 g of ethanol with 45.0 g of water?

31. What are the mole fractions of ethanol, C_2H_5OH, and water in a solution prepared by mixing 55.0 mL of ethanol with 45.0 mL of water at 25°C? The density of ethanol is 0.789 g/mL, and that of water is 1.00 g/mL.

🐾 **Molecular Reasoning** exercises   ▲ **More Challenging** exercises   Blue-Numbered exercises solved in Student Solutions Manual

Unless otherwise noted, all content on this page is © Cengage Learning.

32. The density of an aqueous solution containing 12.50 g K_2SO_4 in 100.00 g solution is 1.083 g/mL. Calculate the concentration of this solution in terms of molarity, molality, percent of K_2SO_4, and mole fraction of solvent.

33. ▲ A piece of jewelry is marked "14 carat gold," meaning that on a mass basis the jewelry is 14/24 pure gold. What is the molality of gold in this alloy—considering the other metal as the solvent?

34. ▲ A solution that is 21.0% fructose, $C_6H_{12}O_6$, in water has a density of 1.10 g/mL at 20°C. (a) What is the molality of fructose in this solution? (b) At a higher temperature, the density is lower. Would the molality be less than, greater than, or the same as the molality at 20°C? Explain.

35. The density of a sulfuric acid solution taken from a car battery is 1.225 g/cm³. This corresponds to a 3.75 M solution. Express the concentration of this solution in terms of molality, mole fraction of H_2SO_4, and percentage of water by mass.

Car battery

Raoult's Law and Vapor Pressure

36. 🌐 In your own words, explain briefly *why* the vapor pressure of a solvent is lowered by dissolving a nonvolatile solute in it.

37. (a) Calculate the vapor pressure lowering associated with dissolving 20.2 g of table sugar, $C_{12}H_{22}O_{11}$, in 400. g of water at 25.0°C. (b) What is the vapor pressure of the solution? Assume that the solution is ideal. The vapor pressure of pure water at 25.0°C is 23.76 torr. (c) What is the vapor pressure of the solution at 100.°C?

38. Calculate the vapor pressure of water over each of the following ethylene glycol ($C_2H_6O_2$) solutions at 22°C (vp pure water = 19.83 torr). Ethylene glycol can be assumed to be nonvolatile.
    (a) $X_{ethylene\ glycol}$ = 0.288
    (b) % ethylene glycol by mass = 39.0%
    (c) 2.42 m ethylene glycol

39. At −100.°C ethane, CH_3CH_3, and propane, $CH_3CH_2CH_3$, are liquids. At this temperature, the vapor pressure of pure ethane is 394 torr and that of pure propane is 22 torr. What is the vapor pressure at −100.°C over a solution containing equal molar amounts of these substances?

40. Using Raoult's Law, predict the partial pressures in the vapor above a solution containing 0.550 mol acetone (P^0 = 345 torr) and 0.550 mol chloroform (P^0 = 295 torr).

41. What is the composition of the vapor above the solution described in Exercise 39?

42. What is the composition of the vapor above the solution described in Exercise 40?

43. Use the following vapor pressure diagram to estimate (a) the partial pressure of chloroform, (b) the partial pressure of acetone, and (c) the total vapor pressure of a solution in which the mole fraction of $CHCl_3$ is 0.30, assuming *ideal* behavior.

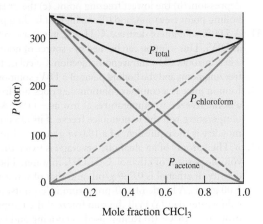

44. Answer Exercise 43 for the *real* solution of acetone and chloroform.

45. A solution is prepared by mixing 50.0 g of dichloromethane, CH_2Cl_2, and 30.0 g of dibromomethane, CH_2Br_2, at 0°C. The vapor pressure at 0°C of pure CH_2Cl_2 is 0.175 atm, and that of CH_2Br_2 is 0.0150 atm. (a) Assuming ideal behavior, calculate the total vapor pressure of the solution. (b) Calculate the mole fractions of CH_2Cl_2 and CH_2Br_2 in the *vapor* above the liquid. Assume that both the vapor and the solution behave ideally.

Boiling Point Elevation and Freezing Point Depression: Solutions of Nonelectrolytes

46. Calculate the boiling point of a 2.97 m aqueous solution of ethylene glycol, a nonvolatile nonelectrolyte.

Ethylene glycol solutions

47. A solution is prepared by dissolving 8.89 g of ordinary sugar (sucrose, $C_{12}H_{22}O_{11}$, 342 g/mol) in 34.0 g of water. Calculate the boiling point of the solution. Sucrose is a nonvolatile nonelectrolyte.

48. What is the freezing point of the solution described in Exercise 46?

🌐 **Molecular Reasoning** exercises   ▲ **More Challenging** exercises   Blue-Numbered exercises solved in Student Solutions Manual

Unless otherwise noted, all content on this page is © Cengage Learning.

49. What is the freezing point of the solution described in Exercise 47?

50. Refer to Table 14-2. Suppose you had a 0.175 *m* solution of a nonvolatile nonelectrolyte in each of the solvents listed there. Which one should have (a) the greatest freezing point depression, (b) the lowest freezing point, (c) the greatest boiling point elevation, and (d) the highest boiling point?

51. A 5.00% solution of dextrose, $C_6H_{12}O_6$, in water is referred to as DW_5. This solution can be used as a source of nourishment when introduced by intravenous injection. Calculate the freezing point and the boiling point of a DW_5 solution.

52. Lemon juice is a complex solution that does not freeze in a home freezer at temperatures as low as −11°C. At what temperature will the lemon juice freeze if its effective molality is the equivalent of a 10.0 *m* glucose solution?

53. ▲ The "proof" of an alcoholic beverage is twice the volume percent of ethanol, C_2H_5OH, in water. The density of ethanol is 0.789 g/mL and that of water is 1.00 g/mL. A bottle of 100-proof rum is left outside on a cold winter day. (a) Will the rum freeze if the temperature drops to −18°C? (b) Rum is used in cooking and baking. At what temperature does 100-proof rum boil?

54. A 3.0-g sample of a nonelectrolyte was isolated from beef fat. The molecular weight of the compound was determined to be 137 amu. The sample was dissolved in 250.0 mL of ethanol. At what temperature should the solution boil? For ethanol: boiling point = 78.41°C, $K_b = 1.22°C/m$, density = 0.789 g/mL.

55. The normal boiling point of benzene is 80.1°C. A 0.85-gram sample of a nonvolatile compound with the molar mass of 185 g/mol is dissolved in 2.75 g of benzene. What is the expected boiling point of this solution?

56. Pure copper melts at 1083°C. Its molal freezing point depression constant is 23°C/*m*. What will be the melting point of a brass made of 12% Zn and 88% Cu by mass?

57. How many grams of the nonelectrolyte sucrose, $C_{12}H_{22}O_{11}$, should be dissolved in 725 g of water to produce a solution that freezes at −1.85°C?

58. An aqueous solution contains 0.180 g of an unknown, nonionic solute in 50.0 g of water. The solution freezes at −0.040°C. What is the molar mass of the solute?

59. A solution was made by dissolving 3.75 g of a nonvolatile solute in 108.7 g of acetone. The solution boiled at 56.58°C. The boiling point of pure acetone is 55.95°C, and $K_b = 1.71°C/m$. Calculate the molecular weight of the solute.

60. The molecular weight of an organic compound was determined by measuring the freezing point depression of a benzene solution. A 0.500-g sample was dissolved in 75.0 g of benzene, and the resulting depression was 0.53°C. What is the approximate molecular weight? The compound gave the following elemental analysis: 37.5% C, 12.5% H, and 50.0% O by mass. Determine the formula and exact molecular weight of the substance.

61. When 0.154 g of sulfur is finely ground and melted with 4.38 g of camphor, the freezing point of the camphor is lowered by 5.47°C. What is the molecular weight of sulfur? What is its molecular formula?

62. ▲ (a) Suppose we dissolve a 6.00-g sample of a mixture of naphthalene, $C_{10}H_8$, and anthracene, $C_{14}H_{10}$, in 360. g

of benzene. The solution is observed to freeze at 4.85°C. Find the percent composition (by mass) of the sample. (b) At what temperature should the solution boil? Assume that naphthalene and anthracene are nonvolatile nonelectrolytes.

Boiling Point Elevation and Freezing Point Depression: Solutions of Electrolytes

63. 🐾 Explain ion association in solution. Can you suggest why the term "ion pairing" is sometimes used to describe this phenomenon?

64. 🐾 You have separate 0.10 *M* aqueous solutions of the following salts: $LiNO_3$, $Ca(NO_3)_2$, and $Al(NO_3)_3$. In which one would you expect to find the highest particle concentration? Which solution would you expect to conduct electricity most strongly? Explain your reasoning.

65. 🐾 What is the significance of the van't Hoff factor, *i*?

66. 🐾 What is the value of the van't Hoff factor, *i*, for the following strong electrolytes at infinite dilution? (a) Na_2SO_4; (b) NaOH; (c) $Al_2(SO_4)_3$; (d) $CaSO_4$.

67. 🐾 Compare the number of solute particles that are present in equal volumes of solutions of equal concentrations of strong electrolytes, weak electrolytes, and nonelectrolytes.

68. 🐾 Four beakers contain 0.010 *m* aqueous solutions of CH_3OH, $KClO_3$, $CaCl_2$, and CH_3COOH, respectively. Without calculating the actual freezing points of each of these solutions, arrange them from lowest to highest freezing point.

69. ▲ A 0.050 *m* aqueous solution of $K_3[Fe(CN)_6]$ has a freezing point of −0.2800°C. Calculate the total concentration of solute particles in this solution, and interpret your results.

70. ▲ A solution is made by dissolving 1.25 grams each of NaCl, NaBr, and NaI in 150. g of water. What is the vapor pressure above this solution at 100°C? Assume complete dissociation of the three salts.

71. The ice fish lives under the polar ice cap where the water temperature is −4°C. This fish does not freeze at that temperature due to the solutes in its blood. The solute concentration of the fish's blood can be related to a sodium chloride solution that would have the same freezing point. What is the minimum concentration of a sodium chloride solution that would not freeze at −4°C? Assume ideal behavior and complete dissociation of NaCl.

72. 🐾 Which solution would freeze at a lower temperature, 0.100 *m* sodium sulfate or 0.100 *m* calcium sulfate? Explain without calculation.

73. 🐾 Two solutions are produced by dissolving 10.0 grams of sodium nitrate in 0.500 kg of water and by dissolving 10.0 grams of calcium nitrate in 0.500 kg of water. Which solution would freeze at a lower temperature? Explain without calculation.

74. 🐾 A series of 1.1 *m* aqueous solutions is produced. Predict without calculation which solution in each pair would boil at the higher temperature. (a) NaCl or LiCl; (b) NaCl or Na_2SO_4; (c) Na_2SO_4 or HCl; (d) HCl or $C_6H_{12}O_6$; (e) $C_6H_{12}O_6$ or CH_3OH; (f) CH_3OH or CH_3COOH; (g) CH_3COOH or KCl.

75. 🐾 Identify without calculation which of the following solutions, each prepared by dissolving 65.0 g of the solute

🐾 **Molecular Reasoning** exercises   ▲ **More Challenging** exercises   Blue-Numbered exercises solved in Student Solutions Manual

Unless otherwise noted, all content on this page is © Cengage Learning.

in 150. mL of water, would display the highest boiling point, and explain your choice. (a) NaCl or LiCl; (b) LiCl or Li_2SO_4; (c) Na_2SO_4 or HCl; (d) HCl or $C_6H_{12}O_6$; (e) $C_6H_{12}O_6$ or CH_3OH; (f) CH_3OH or CH_3COOH; (g) CH_3COOH or KCl.

76. What is the boiling point of a solution composed of 15.0 g urea, $(NH_2)_2CO$, in 0.500 kg water?

77. *Estimate* the freezing point of an aqueous solution that contains 2.0 moles of $CaCl_2$ in 1000. grams of water.

78. A 0.100 *m* acetic acid solution in water freezes at $-0.188°C$. Calculate the percent ionization of CH_3COOH in this solution.

79. ▲ In a home ice cream freezer, we lower the freezing point of the water bath surrounding the ice cream container by dissolving NaCl in water to make a brine solution. A 15.0% brine solution is observed to freeze at $-10.89°C$. What is the van't Hoff factor, *i*, for this solution?

Charles D. Winters

Ice cream

80. CsCl dissolves in water according to

$$CsCl(s) \xrightarrow{H_2O} Cs^+(aq) + Cl^-(aq)$$

A 0.121 *m* solution of CsCl freezes at $-0.403°C$. Calculate *i* and the apparent percent dissociation of CsCl in this solution.

Osmotic Pressure

81. 🐾 What are osmosis and osmotic pressure?

82. ▲ Show numerically that the molality and molarity of 1.00×10^{-4} *M* aqueous sodium chloride are nearly equal. Why is this true? Would this be true if another solvent, say acetonitrile, CH_3CN, replaced water? Why or why not? The density of CH_3CN is 0.786 g/mL at 20.0°C.

83. Show how the expression $\pi = MRT$, where π is osmotic pressure, is similar to the ideal gas law. Rationalize qualitatively why this should be so.

84. What is the osmotic pressure associated with a 0.0200 *M* aqueous solution of a nonvolatile nonelectrolyte solute at 75°C?

85. The osmotic pressure of an aqueous solution of a nonvolatile nonelectrolyte solute is 1.21 atm at 0.0°C. What is the molarity of the solution?

86. Calculate the freezing point depression and boiling point elevation associated with the solution in Exercise 85.

87. Estimate the osmotic pressure associated with 12.5 g of an enzyme of molecular weight 4.21×10^6 dissolved in 1500. mL of ethyl acetate solution at 38.0°C.

88. Calculate the osmotic pressure at 25°C of 1.20 *m* K_2CrO_4 in water, taking ion association into account. Refer to Table 14-3. The density of a 1.20 *m* K_2CrO_4 solution is 1.25 g/mL.

89. Estimate the osmotic pressure at 25°C of 1.20 *m* K_2CrO_4 in water, assuming no ion association. The density of a 1.20 *m* K_2CrO_4 solution is 1.25 g/mL.

90. ▲ Many biological compounds are isolated and purified in very small amounts. We dissolve 11.0 mg of a biological macromolecule with a molecular weight of 2.00×10^4 in 10.0 g of water. (a) Calculate the freezing point of the solution. (b) Calculate the osmotic pressure of the solution at 25°C. (c) Suppose we are trying to use freezing point measurements to *determine* the molecular weight of this substance and that we make an error of only 0.001°C in the temperature measurement. What percent error would this cause in the calculated molecular weight? (d) Suppose we could measure the osmotic pressure with an error of only 0.1 torr (not a very difficult experiment). What percent error would this cause in the calculated molecular weight?

Colloids

91. 🐾 How does a colloidal dispersion differ from a true solution?

92. Distinguish among (a) sol, (b) gel, (c) emulsion, (d) foam, (e) solid sol, (f) solid emulsion, (g) solid foam, (h) solid aerosol, and (i) liquid aerosol. Try to give an example of each that is not listed in Table 14-4.

93. 🐾 What is the Tyndall effect, and how is it caused?

94. 🐾 Distinguish between hydrophilic and hydrophobic colloids.

95. 🐾 What is an emulsifier?

96. 🐾 Distinguish between soaps and detergents. How do they interact with hard water? Write an equation to show the interaction between a soap and hard water that contains Ca^{2+} ions.

97. What is the disadvantage of branched alkylbenzenesulfonate (ABS) detergents compared with linear alkylbenzenesulfonate (LAS) detergents?

Mixed Exercises

98. ▲ An aqueous solution prepared by dissolving 1.56 g of anhydrous $AlCl_3$ in 50.0 g of water has a freezing point of $-1.61°C$. Calculate the boiling point and osmotic pressure at 25°C of this solution. The density of the solution at 25°C is 1.002 g/mL. K_f and K_b for water are 1.86°C/*m* and 0.512°C/*m*, respectively.

99. Dry air contains 20.94% O_2 by volume. The solubility of O_2 in water at 25°C is 0.041 gram O_2 per liter of water. How many liters of water would dissolve the O_2 in one liter of dry air at 25°C and 1.00 atm?

100. (a) The freezing point of a 1.00% aqueous solution of acetic acid, CH_3COOH, is $-0.310°C$. What is the approximate formula weight of acetic acid in water? (b) A 1.00% solution of acetic acid in benzene has a

🐾 **Molecular Reasoning** exercises   ▲ **More Challenging** exercises   Blue-Numbered exercises solved in Student Solutions Manual

Unless otherwise noted, all content on this page is © Cengage Learning.

freezing point depression of 0.441°C. What is the formula weight of acetic acid in this solvent? Explain the difference.

101. An aqueous ammonium chloride solution contains 5.75 mass % NH_4Cl. The density of the solution is 1.0195 g/mL. Express the concentration of this solution in molarity, molality, and mole fraction of solute.

102. ☁ Starch contains $C-C$, $C-H$, $C-O$, and $O-H$ bonds. Hydrocarbons contain only $C-CC$ and $C-H$ bonds. Both starch and hydrocarbon oils can form colloidal dispersions in water. (a) Which dispersion is classified as hydrophobic? (b) Which is hydrophilic? (c) Which dispersion would be easier to make and maintain?

103. ▲ Suppose we put some one-celled microorganisms in various aqueous NaCl solutions. We observe that the cells remain unperturbed in 0.7% NaCl, whereas they shrink in more concentrated solutions and expand in more dilute solutions. Assume that 0.7% NaCl behaves as an *ideal* 1:1 electrolyte. Calculate the osmotic pressure of the aqueous fluid within the cells at 25°C.

104. ▲ A sample of a drug ($C_{21}H_{23}O_5N$, molecular weight = 369 g/mol) mixed with lactose (a sugar, $C_{12}H_{22}O_{11}$, molecular weight = 342 g/mol) was analyzed by osmotic pressure to determine the amount of sugar present. If 100. mL of solution containing 1.00 g of the drug–sugar mixture has an osmotic pressure of 519 torr at 25°C, what is the percent sugar present?

105. A solution containing 4.52 g of a nonelectrolyte polymer per liter of benzene solution has an osmotic pressure of 0.786 torr at 20.0°C. (a) Calculate the molecular weight of the polymer. (b) Assume that the density of the dilute solution is the same as that of benzene, 0.879 g/mL. What would be the freezing point depression for this solution? (c) Why are boiling point elevations and freezing point depressions difficult to use to measure molecular weights of polymers?

106. ☁ On what basis would you choose the components to prepare an ideal solution of a molecular solute? Which of the following combinations would you expect to act most nearly ideally? (a) $CH_4(\ell)$ and $CH_3OH(\ell)$; (b) $CH_3OH(\ell)$ and NaCl(s); (c) $CH_4(\ell)$ and $CH_3CH_3(\ell)$.

107. Physiological saline (normal saline) is a 0.90% NaCl solution. This solution is isotonic with human blood. Calculate the freezing point and boiling point of physiological saline. Assume complete dissociation.

108. A mixture used in parasitology "for clearing large nematodes" consists of 500 mL of glycerol, 300 mL of 95% ethanol (by volume), and 200 mL of water. Calculate the volume of pure ethanol, the total volume of water, the mass of each component, the number of moles of each component, and the mole fraction of each component. (densities: glycerol = 1.26 g/mL, ethanol = 0.789 g/mL, and water = 0.998 g/mL) If you were to pick one component as the solvent, would it be by volume, mass, or number of moles?

Conceptual Exercises

109. Hydration energies for ions generally increase (become more negative) as their charge/radius ratio increases. What effect would you expect each of the following changes to have on the hydration energies if all other factors are

unchanged: (a) decreasing the ionic charge (such as from Cu^{2+} to Cu^{+}); (b) increasing the ionic radius (such as from 1.66 Å for Rb^{+} to 1.81 Å for Cs^{+})?

110. Many gases are only minimally soluble in water because they are not capable of hydrogen bonding with water, they do not ionize extensively, and they do not react with water. Explain the relatively high solubility of each of the following gases in water: H_2S (3.5 g/L), SO_2 (106.4 g/L), and NH_3 (520 g/L).

111. Unlike molarity, the molality of a solution is independent of the temperature of the solution. Examine the defining formulas for molarity and molality. Why does the molarity of a solution change with a change in temperature while the molality does not?

112. What effects does adding a solute to a solvent, thereby producing a solution, have on the following properties of the solution, compared with those of the pure solvent? It (a) _____ the vapor pressure, (b) _____ the boiling point, (c) _____ the freezing point, and (d)_____ the osmotic pressure.

113. Ideally, how many particles (molecules or ions) are produced per formula unit when each of the following substances is dissolved in water: ethylene glycol [$C_2H_4(OH)_2$, a nonelectrolyte], barium hydroxide [$Ba(OH)_2$], potassium phosphate (K_3PO_4), and ammonium nitrate (NH_4NO_3)?

114. In the first five sections of this chapter, the term "dissolution" was used to describe the process by which a solute is dispersed by a solvent to form a solution. A popular dictionary defines dissolution as "decomposition into fragments or parts." Using either of the definitions, compare the following two uses of the term. "It was ruled that there must be dissolution of the estate." "The ease of dissolution of a solute depends on two factors."

115. ▲ Consider two nonelectrolytes A and B; A has a higher molecular weight than B, and both are soluble in solvent C. A solution is made by dissolving x grams of A in 100 grams of C; another solution is made by dissolving the same number of grams, x, of B in 100 grams of C. Assume that the two solutions have the same density. (a) Which solution has the higher molality? (b) Which solution has the higher mole fraction? (c) Which solution has the higher percent by mass? (d) Which solution has the higher molarity?

116. ☁ The actual value for the van't Hoff factor, i, is often observed to be less than the ideal value. Less commonly, it is greater than the ideal value. Give molecular explanations for cases where i is observed to be greater than the ideal value.

117. ▲ A 1.0 m aqueous solution of a monoprotic acid freezes at $-3.35°C$, 2.0 m solution of the acid freezes at $-6.10°C$, and 4.0 m solution of a monoprotic acid freezes at $-8.70°C$. Calculate the van't Hoff factor, i, for the acid at each concentration. Plot i versus concentration, and from the graph determine the value of i for the 3.0 m solution. Describe the general trend of the value of i versus concentration.

118. Using the data in Exercise 117, calculate the percent ionization for the acid in 1.0 m, 2.0 m, and 4.0 m solutions. Describe the general trend of the value of percent ionization versus concentration.

119. ▲ When the van't Hoff factor, i, is included, the boiling point elevation equation becomes $\Delta T_b = iK_bm$. The van't Hoff factor can be inserted in a similar fashion in the

---

☁ **Molecular Reasoning** exercises   ▲ **More Challenging** exercises   Blue-Numbered exercises solved in Student Solutions Manual

Unless otherwise noted, all content on this page is © Cengage Learning.

freezing point depression equation and the osmotic pressure equation. The van't Hoff factor, however, cannot be inserted in the same way into the vapor pressure-lowering equation. Show algebraically that $\Delta P_{solvent} = iX_{solute}P^0_{solvent}$ is not equivalent to $\Delta P_{solvent} = \{(ix\ moles_{solute})/\ [(ix\ moles_{solute}) + (moles_{solvent})]\}P^0_{solvent}$. Which equation for $\Delta P_{solvent}$ is correct?

120. The two solutions shown were both prepared by dissolving 194 g of K_2CrO_4 (1.00 mol) in a 1.00-L volumetric flask. The solution on the right was diluted to the 1-L mark on the neck of the flask, and the solution on the left was diluted by adding 1.00 L of water. Which solution, the one on the left or the one on the right, is (a) more concentrated; (b) a 1.00 m solution; (c) a 1.00 M solution?

Charles D. Winters

Chromate solutions

121. 🜋 Would you expect lowering the freezing point or elevating the boiling point to be the better method to obtain the approximate molecular weight of an unknown? Explain your choice.

122. Rock candy consists of crystals of sugar on a string or stick. Propose a method of making rock candy, and explain each step.

123. Concentrations expressed in units of parts per million and parts per billion often have no meaning for people until they relate these small and large numbers to their own experiences. How many seconds are equal to 1 ppm of a year?

Building Your Knowledge

124. 🜋 ▲ DDT is a toxin still found in the fatty tissues of some animals. DDT was transported into our lakes and streams as runoff from agricultural operations where it was originally used several years ago as an insecticide. In the lakes and streams it did not dissolve to any great extent; it collected in the lake and stream bottoms. It entered the bodies of animals via fatty tissues in their diet; microorganisms collected the DDT, the fish ate the microorganisms, and so on. Fortunately, much of the once large quantities of DDT in lakes and streams has biodegraded. Based on this information, what can you conclude regarding the intermolecular forces present in DDT?

125. 🜋 Draw Figure 14-1, but instead of using colored circles to represent the solvent and solute molecules, draw

ball-and-stick models to represent water as the solvent and acetone, CH_3COCH_3, as the solute. Use dashed lines to show hydrogen bonds. Twelve water molecules and two acetone molecules should be sufficient to illustrate the interaction between these two kinds of molecules.

126. A sugar maple tree grows to a height of about 40 feet, and its roots are in contact with water in the soil. What must be the concentration of the sugar in its sap if osmotic pressure is solely responsible for forcing the sap to the top of the tree at 15°C? The density of mercury is 13.6 g/cm^3, and the density of the sap can be considered to be 1.10 g/cm^3.

127. 🜋 Many metal ions become hydrated in solution by forming coordinate covalent bonds with the unshared pair of electrons from the water molecules to form "AB_6" ions. Because of their sizes, these hydrated ions are unable to pass through the semipermeable membrane described in Section 14-15, whereas water as a trimer, $(H_2O)_3$, or a tetramer, $(H_2O)_4$, can pass through. Anions tend to be less hydrated. Using the VSEPR theory, prepare three-dimensional drawings of $Cu(H_2O)_6{}^{2+}$ and a possible $(H_2O)_3$ that show their relative shapes and sizes.

Beyond the Textbook

NOTE: *Whenever the answer to an exercise depends on information obtained from a source other than this textbook, the source must be included as an essential part of the answer.*

128. Use an internet search engine (such as **http://www.google.com**) to find the circumference of Earth at its equator. What is 1.00 ppm of this length?

129. Use an internet search engine (such as **http://www.google.com**) to locate information on the first winner of the Nobel Prize in Chemistry. With what colligative property do we associate his name?

130. Use an internet search engine (such as **http://www.google.com**) to locate information on the scientist John Tyndall. (a) What colloidal property has John Tyndall's name associated with it? (b) John Tyndall died due to a drug overdose that had been secretly administered by his wife in an effort to treat an ailment that had plagued him for several years. What ailment was his wife attempting to treat?

131. Use an internet search engine (such as **http://www.google.com**) to locate information on edible colloids. (a) Give the addresses of three sites that discuss edible colloids. (b) List eight edible colloids that are not also listed in Table 14-4. (c) Is ice cream an emulsion, a foam, or a sol? (d) Is any part of Spam a colloid?

132. 🜋 Use an internet search engine (such as **http://www.google.com**) to locate information on thixotropy. Describe the relationship between thixotropy and the molecular properties of some colloids.

133. Use an internet search engine (such as **http://www.google.com**) to locate information on fractional distillation of petroleum. List the names and boiling point ranges of six or more distillation fractions usually obtained from most crude petroleum.

Chemical Thermodynamics

15

Thousands of windmills such as these in west Texas are becoming an increasingly important source of electricity. The kinetic energy of wind turns large windmill-powered turbines, which in turn convert their kinetic energy to electrical (potential) energy.

Raymond E. Davis

OBJECTIVES

After you have studied this chapter, you should be able to

► Understand the terminology of thermodynamics, and the meaning of the signs of changes

► Use the concept of state functions

► Carry out calculations of calorimetry to determine changes in energy and enthalpy

► Use Hess's Law to find the enthalpy change, ΔH, for a reaction from other thermochemical equations with known ΔH values

► Use Hess's Law to find the standard enthalpy change, ΔH^0, for a reaction by using tabulated values of standard molar enthalpies of formation

► Use Hess's Law to find the molar enthalpy of formation, ΔH_f^0, given ΔH^0 for a reaction and the known molar enthalpies of formation of the other substances in the reaction

► Use the First Law of Thermodynamics to relate heat, work, and energy changes

► Relate the work done on or by a system to changes in its volume

► Use bond energies to estimate heats of reaction for gas phase reactions; use ΔH values for gas phase reactions to find bond energies

► Understand what is meant by a product-favored (spontaneous) process and by a reactant-favored (nonspontaneous) process

► Understand the relationship of entropy to the dispersal of energy and dispersal of matter (disorder) in a system

► Use tabulated values of absolute entropies to calculate the entropy change, ΔS^0

► Understand how the spontaneity of a process is related to entropy changes—the Second Law of Thermodynamics

► Calculate changes in Gibbs free energy, ΔG, by two methods: (a) from values of ΔH and ΔS and (b) from tabulated values of standard molar free energies of formation; know when to use each type of calculation

► Use ΔG^0 to predict whether a process is product-favored (spontaneous) at constant T and P

► Understand how changes in temperature can affect the spontaneity of a process

► Predict the temperature range of spontaneity of a chemical or physical process

► Some forms of energy are potential, kinetic, electrical, nuclear, heat, and light.

Energy is very important in every aspect of our daily lives. The food we eat supplies the energy to sustain life with all of its activities and concerns. The availability of relatively inexpensive energy is an important factor in our technological society. This is seen in the costs of fuel, heating and cooling of our homes and workplaces, and electricity to power our lights, appliances, and computers. It is also seen in the costs of goods and services we purchase because a substantial part of the cost of production and delivery is for energy in one form or another. We must understand the scientific basis of the storage and use of energy to learn how to decrease our dependence on consumable oil and natural gas as our main energy sources. Such understanding will have profound ramifications, ranging from our daily lifestyles to international relations.

The concept of energy is at the very heart of science. All physical and chemical processes, including those in all living systems, are accompanied by the transfer of energy. Because energy cannot be created or destroyed, we must understand how to do the "accounting" of energy transfers from one body or one substance to another, or from one form of energy to another.

In **thermodynamics** we study the energy changes that accompany physical and chemical processes. These energy changes usually involve *heat*—hence the "thermo-" part of the term. In this chapter we will study the two main aspects of thermodynamics. The first is **thermochemistry**. This is how we *observe*, *measure*, and *predict* energy changes for both physical changes and chemical reactions. The second part of the chapter will address a more fundamental aspect of thermodynamics. There we will learn to use energy changes to tell whether or not a given process can occur under specified conditions to give predominantly products (or reactants) and how to make a process more (or less) favorable.

Heat Changes and Thermochemistry

15-1 The First Law of Thermodynamics

We can define energy as follows.

> Energy is the capacity to do work or to transfer heat.

We classify energy into two general types: kinetic and potential. **Kinetic energy** is the energy of motion. The kinetic energy of an object is equal to one half its mass, m, times the square of its velocity, v.

$$E_{\text{kinetic}} = \tfrac{1}{2}\,mv^2$$

The heavier a hammer is and the more rapidly it moves, the greater its kinetic energy and the more work it can accomplish.

Potential energy is the energy that a system possesses by virtue of its position or composition. The work that we do to lift an object in a gravitational field is stored in the object as potential energy. If we drop a hammer, its potential energy is converted into kinetic energy as it falls, and it could do work on something it hits—for example, drive a nail or break a piece of glass. Similarly, an electron in an atom has potential energy because of the electrostatic force on it that is due to the positively charged nucleus and the other electrons in that atom and surrounding atoms. Energy can take many other forms: electrical energy, radiant energy (light), nuclear energy, and chemical energy. At the atomic or molecular level, we can think of these as either kinetic or potential energy.

The chemical energy in a fuel or food comes from potential energy stored in atoms due to their arrangements in the molecules. This stored chemical energy can be released when compounds undergo chemical changes, such as those that occur in combustion and metabolism.

Reactions that release energy in the form of heat are called **exothermic** reactions. Combustion reactions of fossil fuels are familiar examples of exothermic reactions. Hydrocarbons—including methane (CH_4), the main component of natural gas, and octane (C_8H_{18}), a minor component of gasoline—undergo combustion with an excess of O_2 to yield CO_2 and H_2O. These reactions release heat energy. The amounts of heat energy released at constant pressure are shown for the reactions of one mole of methane and of two moles of octane.

$$CH_4(g) + 2O_2(g) \longrightarrow CO_2(g) + 2H_2O(\ell) + 890 \text{ kJ}$$

$$2C_8H_{18}(\ell) + 25O_2(g) \longrightarrow 16CO_2(g) + 18H_2O(\ell) + 1.090 \times 10^4 \text{ kJ}$$

In such reactions, the total energy of the products is lower than that of the reactants by the amount of energy released, most of which is heat. Some initial activation (e.g., by heat) is needed to get these reactions started. This is shown for CH_4 in Figure 15-1. This activation energy *plus* 890 kJ is released as one mole of $CO_2(g)$ and two moles of $H_2O(\ell)$ are formed.

A process that absorbs energy from its surroundings is called **endothermic**. One such process is shown in Figure 15-2.

Energy changes accompany physical changes, too (see Chapter 13). For example, the melting of one mole of ice at 0°C at constant pressure must be accompanied by the absorption of 6.02 kJ of energy.

$$H_2O(s) + 6.02 \text{ kJ} \longrightarrow H_2O(\ell)$$

This tells us that the total energy of the water is raised by 6.02 kJ in the form of heat during the phase change, even though the temperature remains constant.

As matter falls from a higher to a lower level, its gravitational potential energy is converted into kinetic energy. A hydroelectric power plant converts the kinetic energy of falling water into electrical (potential) energy.

▶ A hydrocarbon is a binary compound of only hydrogen and carbon. Hydrocarbons may be gaseous, liquid, or solid. All burn.

The amount of heat shown in such an equation always refers to the reaction for the number of moles of reactants and products specified by the coefficients. We call this *one mole of reaction* (even though more than one mole of reactants or products may be shown).

 **S** **TOP & THINK**
It is important to specify the physical states of all substances because different physical states have different energy contents.

Figure 15-1 The difference between the potential energy of the reactants—one mole of $CH_4(g)$ and two moles of $O_2(g)$—and that of the products—one mole of $CO_2(g)$ and two moles of $H_2O(\ell)$—is the amount of heat evolved in this *exothermic* reaction at constant pressure. For this reaction, it is 890 kJ/mol of reaction. In this chapter, we see how to measure the heat absorbed or released and how to calculate it from other known heat changes. Some initial activation, by heat, for example, is needed to get this reaction started (see Section 16-6). In the absence of such activation energy, a mixture of CH_4 and O_2 can be kept at room temperature for a long time without reacting. For an *endothermic* reaction, the final level is higher than the initial level.

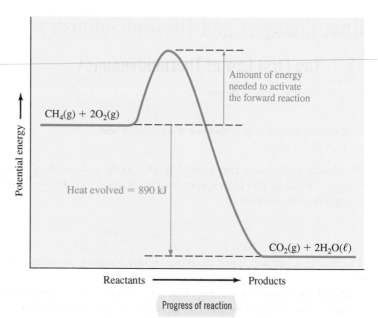

A When solid hydrated barium hydroxide, $Ba(OH)_2 \cdot 8H_2O$, and *excess* solid ammonium nitrate, NH_4NO_3, are mixed, an endothermic reaction occurs.

$$Ba(OH)_2 \cdot 8H_2O(s) + 2NH_4NO_3(s) \longrightarrow Ba(NO_3)_2(s) + 2NH_3(g) + 10H_2O(\ell)$$

The excess ammonium nitrate dissolves in the water produced in the reaction.

B The dissolution process is also very endothermic. If the flask is placed on a wet wooden block, the water freezes and attaches the block to the flask.

Figure 15-2 An endothermic process.

Some important ideas about energy are summarized in the **First Law of Thermodynamics**.

> The combined amount of matter and energy in the universe is constant.

The **Law of Conservation of Energy** is just another statement of the First Law of Thermodynamics.

> Energy can never be created nor destroyed in any process. Energy can be transformed into other forms of energy or converted into matter under special circumstances such as nuclear reactions.

15-2 Some Thermodynamic Terms

The substances involved in the chemical and physical changes that we are studying are called the **system**. Everything in the system's environment constitutes its **surroundings**. The **universe** is the system plus its surroundings. The system may be thought of as the part of the universe under investigation. The First Law of Thermodynamics tells us that energy is neither created nor destroyed; it is only transferred between the system and its surroundings.

The **thermodynamic state of a system** is defined by a set of conditions that completely specifies all the properties of the system. This set commonly includes the temperature, pressure, composition (identity and number of moles of each component), and physical state (gas, liquid, or solid) of each part of the system. Once the state has been specified, all other properties—both physical and chemical—are fixed.

The properties of a system—such as P, V, T—are called **state functions**. The *value* of a state function depends only on the state of the system and not on the way in which the system came to be in that state. A *change* in a state function describes a *difference* between the two states. It is independent of the process or pathway by which the change occurs.

▶ State functions are represented by capital letters. Here P refers to pressure, V to volume, and T to absolute temperature.

For instance, consider a sample of one mole of pure liquid water at 30°C and 1 atm pressure. If at some later time the temperature of the sample is 22°C at the same pressure, then it is in a different thermodynamic state. We can tell that the *net* temperature change is −8°C. It does not matter whether (1) the cooling took place directly (either slowly or rapidly) from 30°C to 22°C, or (2) the sample was first heated to 36°C, then cooled to 10°C, and finally warmed to 22°C, or (3) any other conceivable path was followed from the initial state to the final state. The change in other properties (e.g., the pressure) of the sample is likewise independent of path.

The most important use of state functions in thermodynamics is to describe *changes*. We describe the difference in any quantity, X, as

$$\Delta X = X_{\text{final}} - X_{\text{initial}}$$

When X increases, the final value is greater than the initial value, so ΔX is *positive*; a decrease in X makes ΔX a *negative* value.

You can consider a state function as analogous to a bank account. With a bank account, at any time you can measure the amount of money in your account (your balance) in convenient terms—dollars and cents. Changes in this balance can occur for several reasons, such as deposit of your paycheck, writing of checks, or service charges assessed by the bank. In our analogy these transactions are *not* state functions, but they do cause *changes in* the state function (the balance in the account). You can think of the bank balance on a vertical scale; a deposit of $150 changes the balance by +$150, no matter what it was at the start, just as a withdrawal of $150 would change the balance by −$150. Similarly, we shall see that the energy of a system is a state function that can be changed—for instance, by an energy "deposit" of heat absorbed or work done on the system, or by an energy "withdrawal" of heat given off or work done by the system.

We can describe *differences* between levels of a state function, regardless of where the zero level is located. In the case of a bank balance, the "natural" zero level is obviously the point at which we open the account, before any deposits or withdrawals. In contrast, the zero levels on most temperature scales are set arbitrarily. When we say that the temperature of an ice–water mixture is "zero degrees Celsius," we are not saying that the mixture contains no temperature! We have simply chosen to describe this point on the temperature scale by the number *zero*; conditions of higher temperature are described by positive temperature values, and those of lower temperature have negative values, "below zero." The phrase "15 degrees cooler" has the same meaning anywhere on the scale. Many of the scales that we use in thermodynamics are arbitrarily defined in this way. Arbitrary scales are useful when we are interested only in *changes* in the quantity being described.

Any property of a system that depends only on the values of its state functions is also a state function. For instance, the volume of a given sample of water depends only on tem-

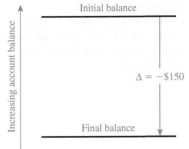

Here is a graphical representation of a $150 decrease in your bank balance. We express the change in your bank balance as $\Delta\$ = \$_{\text{final}} - \$_{\text{initial}}$. Your final balance is *less* than your initial balance, so the result is *negative*, indicating a *decrease*. There are many ways to get this same net change—one large withdrawal or some combination of deposits, withdrawals, interest earned, and service charges. All of the Δ values we will see in this chapter can be thought of in this way.

perature, pressure, and physical state; therefore, volume is a state function. We shall encounter other thermodynamic state functions.

15-3 Enthalpy Changes

Most chemical reactions and physical changes occur at constant (usually atmospheric) pressure.

▶ The symbol q represents the amount of heat absorbed by the system. The subscript p indicates a constant-pressure process.

> The quantity of heat transferred into or out of a system as it undergoes a chemical or physical change at constant pressure, q_p, is defined as the **enthalpy change, ΔH,** of the process.

An enthalpy change is sometimes loosely referred to as a *heat change* or a *heat of reaction*. The enthalpy change is equal to the enthalpy or "heat content," H, of the substances produced minus the enthalpy of the substances consumed.

$$\Delta H = H_{final} - H_{initial} \qquad or \qquad \Delta H = H_{substances\ produced} - H_{substances\ consumed}$$

It is impossible to know the absolute enthalpy (heat content) of a system. *Enthalpy is a state function*, however, and it is the *change in enthalpy* in which we are interested; this can be measured for many processes. In the next several sections, we focus on chemical reactions and the enthalpy changes that occur in these processes. We first discuss the experimental determination of enthalpy changes.

15-4 Calorimetry: Measurement of Heat Transfer

We can determine the energy change associated with a chemical or physical process by using an experimental technique called *calorimetry*. This technique is based on observing the temperature change when a system absorbs or releases energy in the form of heat. The experiment is carried out in a device called a **calorimeter,** in which the temperature change of a known amount of substance (often water) of known specific heat is measured. The temperature change is caused by the absorption or release of heat by the chemical or physical process under study.

▶ A review of calculations involved with heat transfer (see Sections 1-14, 13-9, and 13-11) will help you understand this section.

▶ The polystyrene insulation of the simple coffee-cup calorimeter ensures that little heat escapes from or enters the container.

A "coffee-cup" calorimeter (Figure 15-3) is often used in laboratory classes to measure "heats of reaction" at constant pressure, q_p, in aqueous solutions. Reactions are chosen so that there are no gaseous reactants or products. Thus, all reactants and products remain in the vessel throughout the experiment. Such a calorimeter could be used to measure the amount of heat absorbed or released when a reaction takes place in aqueous solution. We can consider the reactants and products as the system and the calorimeter plus the solution (mostly water) as the surroundings. For an exothermic reaction, the amount of heat evolved by the reaction can be calculated from the amount by which it causes the temperature of the calorimeter and the solution to rise. The heat can be visualized as divided into two parts.

$$\begin{pmatrix} \text{amount of heat} \\ \text{released by reaction} \end{pmatrix} = \begin{pmatrix} \text{amount of heat absorbed} \\ \text{by calorimeter} \end{pmatrix} + \begin{pmatrix} \text{amount of heat} \\ \text{absorbed by solution} \end{pmatrix}$$

The amount of heat absorbed by a calorimeter is sometimes expressed as the *heat capacity* of the calorimeter, in joules per degree. The heat capacity of a calorimeter is determined by adding a known amount of heat and measuring the rise in temperature of the calorimeter and of the solution it contains. This heat capacity of a calorimeter is sometimes called its *calorimeter constant*.

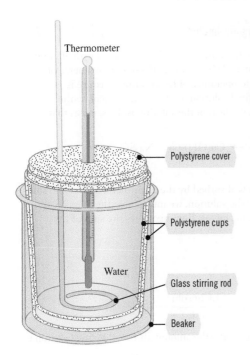

Thermometer

Polystyrene cover

Polystyrene cups

Water

Glass stirring rod

Beaker

Figure 15-3 A coffee-cup calorimeter. The stirring rod is moved up and down to ensure thorough mixing and uniform heating of the solution during reaction. The polystyrene walls and top provide insulation so that very little heat escapes. This kind of calorimeter measures q_p, the heat transfer due to a reaction occurring at constant *pressure*.

EXAMPLE 15-1 Heat Capacity of a Calorimeter

We add 3.358 kJ of heat to a calorimeter that contains 50.00 g of water. The temperature of the water and the calorimeter, originally at 22.34°C, increases to 36.74°C. Calculate the heat capacity of the calorimeter in J/°C. The specific heat of water is 4.184 J/g · °C.

▶ One way to add a measurable amount of heat is to use an electric heater.

Plan

We first calculate the amount of heat gained by the water in the calorimeter. The rest of the heat must have been gained by the calorimeter, so we can determine the heat capacity of the calorimeter.

Solution

$$50.00 \text{ g H}_2\text{O}(\ell) \text{ at } 22.34°C \longrightarrow 50.00 \text{ g H}_2\text{O}(\ell) \text{ at } 36.74°C$$

The temperature change is (36.74 − 22.34)°C = 14.40°C.

$$\underline{?} \text{ J} = 50.00 \text{ g} \times \frac{4.184 \text{ J}}{\text{g} \cdot °C} \times 14.40°C = 3.012 \times 10^3 \text{ J}$$

The total amount of heat added was 3.358 kJ or 3.358×10^3 J. The difference between these heat values is the amount of heat absorbed by the calorimeter.

$$\underline{?} \text{ J} = 3.358 \times 10^3 \text{ J} - 3.012 \times 10^3 \text{ J} = 0.346 \times 10^3 \text{ J, or 346 J absorbed by the calorimeter}$$

To obtain the heat capacity of the calorimeter, we divide the amount of heat absorbed by the calorimeter, 346 J, by its temperature change.

$$\underline{?} \frac{\text{J}}{°C} = \frac{346 \text{ J}}{14.40°C} = 24.0 \text{ J/°C}$$

The calorimeter absorbs 24.0 J of heat for each degree Celsius increase in its temperature.

You should now work Exercise 60.

S TOP & THINK

Note that, because we are using the *change* in temperature (ΔT) in this example, temperatures can be expressed as °C or K. That is because the magnitude of a change of 1°C is equal to 1 K. In this problem ΔT = 14.40°C = 14.40 K. But, when working with mathematical equations that use absolute temperatures T (not ΔT!), it is important to express temperatures in kelvins. In this chapter, we will see mathematical formulas that use T and ΔT, so it is important to keep track of which is being used.

EXAMPLE 15-2 Heat Measurements Using a Calorimeter

A 50.0-mL sample of 0.400 M copper(II) sulfate solution at 23.35°C is mixed with 50.0 mL of 0.600 M sodium hydroxide solution, also at 23.35°C, in the coffee-cup calorimeter of Example 15-1. After the reaction occurs, the temperature of the resulting mixture is measured to be 25.23°C. The density of the final solution is 1.02 g/mL. Calculate the amount of heat evolved. Assume that the specific heat of the solution is the same as that of pure water, 4.184 J/g · °C.

$$CuSO_4(aq) + 2NaOH(aq) \longrightarrow Cu(OH)_2(s) + Na_2SO_4(aq)$$

Plan

The amount of heat released by the reaction is absorbed by the calorimeter *and* by the solution. To find the amount of heat absorbed by the solution, we must know the mass of solution; to find that, we assume that the volume of the reaction mixture is the sum of the volumes of the original solutions.

▶ When *dilute aqueous solutions* are mixed, their volumes are very nearly additive.

Solution

The mass of solution is

$$\underline{?}\text{ g soln} = (50.0 + 50.0)\text{ mL} \times \frac{1.02\text{ g soln}}{\text{mL}} = 102\text{ g soln}$$

The amount of heat absorbed by the calorimeter *plus* the amount absorbed by the solution is

$$\underline{?}\text{ J} = \overbrace{\frac{24.0\text{ J}}{°C} \times (25.23 - 23.35)°C}^{\substack{\text{amount of heat} \\ \text{absorbed by calorimeter}}} + \overbrace{102\text{ g} \times \frac{4.18\text{ J}}{\text{g}·°C} \times (25.23 - 23.35)°C}^{\substack{\text{amount of heat} \\ \text{absorbed by solution}}}$$

$$= 45\text{ J} + 801\text{ J} = 846\text{ J absorbed by solution plus calorimeter}$$

Thus, the reaction must have liberated 846 J, or 0.846 kJ, of heat.

You should now work Exercise 64(a).

The heat released by the reaction of HCl(aq) with NaOH(aq) causes the temperature of the solution to rise.

$C_2H_5OH(\ell) + 3O_2(g)$

Increasing enthalpy

$\Delta H = -1367$ kJ

$2CO_2(g) + 3H_2O(\ell)$

15-5 Thermochemical Equations

A balanced chemical equation, written together with a description of the corresponding heat change, is called a **thermochemical equation**. For example,

$$\underset{\substack{\text{1 mol}}}{C_2H_5OH(\ell)} + \underset{\substack{\text{3 mol}}}{3O_2(g)} \longrightarrow \underset{\substack{\text{2 mol}}}{2CO_2(g)} + \underset{\substack{\text{3 mol}}}{3H_2O(\ell)} + 1367\text{ kJ}$$

is a thermochemical equation that describes the combustion (burning) of one mole of liquid ethanol at a particular temperature and pressure. The coefficients in a balanced thermochemical equation *must* be interpreted as *numbers of moles* of each reactant and product. Thus, 1367 kJ of heat is released when *one* mole of $C_2H_5OH(\ell)$ reacts with *three* moles of $O_2(g)$ to give *two* moles of $CO_2(g)$ and *three* moles of $H_2O(\ell)$. We refer to this amount of reaction as one **mole of reaction**, which we abbreviate "mol rxn." This interpretation allows us to write various unit factors as desired.

$$\frac{1\text{ mol }C_2H_5OH(\ell)}{1\text{ mol rxn}}, \quad \frac{2\text{ mol }CO_2(g)}{1\text{ mol rxn}}, \quad \frac{1367\text{ kJ of heat released}}{1\text{ mol rxn}}, \text{ and so on}$$

The thermochemical equation is more commonly written as

$$C_2H_5OH(\ell) + 3O_2(g) \longrightarrow 2CO_2(g) + 3H_2O(\ell) \qquad \Delta H = -1367\text{ kJ/mol rxn}$$

The negative sign indicates that this is an *exothermic* reaction (i.e., it gives *off* heat).

Charles D. Winters

We always interpret ΔH as the enthalpy change for the reaction as written; that is, as (enthalpy change)/(mole of reaction), where "mole of reaction" means "for the number of moles of each substance shown by the coefficients in the balanced equation."

We can then use several unit factors to interpret this thermochemical equation.

$$\frac{1367 \text{ kJ given off}}{\text{mol of reaction}} = \frac{1367 \text{ kJ given off}}{\text{mol C}_2\text{H}_5\text{OH}(\ell) \text{ consumed}} = \frac{1367 \text{ kJ given off}}{3 \text{ mol O}_2(\text{g}) \text{ consumed}}$$

$$= \frac{1367 \text{ kJ given off}}{2 \text{ mol CO}_2(\text{g}) \text{ formed}} = \frac{1367 \text{ kJ given off}}{3 \text{ mol H}_2\text{O}(\ell) \text{ formed}}$$

▶ To interpret what a value of ΔH means, we must know the balanced chemical equation to which it refers.

The reverse reaction would require the absorption of 1367 kJ under the same conditions.

$$1367 \text{ kJ} + 2\text{CO}_2(\text{g}) + 3\text{H}_2\text{O}(\ell) \longrightarrow \text{C}_2\text{H}_5\text{OH}(\ell) + 3\text{O}_2(\text{g})$$

That is, it is *endothermic*, with $\Delta H = +1367$ kJ.

$$2\text{CO}_2(\text{g}) + 3\text{H}_2\text{O}(\ell) \longrightarrow \text{C}_2\text{H}_5\text{OH}(\ell) + 3\text{O}_2(\text{g}) \qquad \Delta H = +1367 \text{ kJ/mol rxn}$$

It is important to remember the following conventions regarding thermochemical equations:

1. The coefficients in a balanced thermochemical equation refer to the numbers of *moles* of reactants and products involved. In the thermodynamic interpretation of equations, we *never* interpret the coefficients as *numbers of molecules*. Thus, it is acceptable to write coefficients as fractions rather than as integers, when convenient. But most chemists prefer to use the smallest integer values for the coefficients.

2. The numerical value of ΔH (or any other thermodynamic change) is specific to the *number of moles* of substances specified by the balanced equation. This amount of change is called *one mole of reaction*, so we can express ΔH in units of energy/mol rxn. For brevity, the units of ΔH are sometimes written kJ/mol or even just kJ. No matter what units are used, be sure that you interpret the thermodynamic change *per mole of reaction for the balanced chemical equation to which it refers*. If a different amount of material is involved in the reaction, then the ΔH (or other change) must be scaled accordingly.

3. The physical states of all species are important and must be specified. Heat is given off or absorbed when phase changes occur, so different amounts of heat could be involved in a reaction depending on the phases of reactants and products.

4. The value of ΔH usually does not change significantly with moderate changes in temperature.

The launch of the space shuttle requires about 2×10^{10} kilojoules of energy. About one-sixth of this comes from the reaction of hydrogen, H_2, and oxygen, O_2. The rest comes from the explosive decomposition of ammonium perchlorate, NH_4ClO_4, in solid-fuel rockets. The final launch was on July 8, 2011, by the shuttle Atlantis.

EXAMPLE 15-3 Thermochemical Equations

When 2.61 grams of dimethyl ether, CH_3OCH_3, is burned at constant pressure, 82.5 kJ of heat is given off. Find ΔH for the reaction

$$\text{CH}_3\text{OCH}_3(\ell) + 3\text{O}_2(\text{g}) \longrightarrow 2\text{CO}_2(\text{g}) + 3\text{H}_2\text{O}(\ell)$$

Plan

We scale the amount of heat given off in the experiment to correspond to the amount of CH_3OCH_3 shown in the balanced equation.

Solution

$$\frac{?\text{ kJ given off}}{\text{mol rxn}} = \frac{82.5\text{ kJ given off}}{2.61\text{ g CH}_3\text{OCH}_3} \times \frac{46.0\text{ g CH}_3\text{OCH}_3}{\text{mol CH}_3\text{OCH}_3} \times \frac{1\text{ mol CH}_3\text{OCH}_3}{\text{mol rxn}}$$

$$= 1450\text{ kJ/mol rxn}$$

Because heat is given off, we know that the reaction is exothermic so the value of ΔH should be negative.

$$\Delta H = -1450\text{ kJ/mol rxn}$$

You should now work Exercise 21.

EXAMPLE 15-4 Thermochemical Equations

Write the thermochemical equation for the reaction in Example 15-2.

Plan

We must determine *how much* reaction occurred—that is, how many moles of reactants were consumed. We first multiply the volume, in liters, of each solution by its concentration in mol/L (molarity) to determine the number of moles of each reactant mixed. Then we identify the limiting reactant. We scale the amount of heat released in the experiment to correspond to the number of moles of that reactant shown in the balanced equation.

Solution

Using the data from Example 15-2,

$$?\text{ mol CuSO}_4 = 0.0500\text{ L} \times \frac{0.400\text{ mol CuSO}_4}{1.00\text{ L}} = 0.0200\text{ mol CuSO}_4$$

$$?\text{ mol NaOH} = 0.0500\text{ L} \times \frac{0.600\text{ mol NaOH}}{1.00\text{ L}} = 0.0300\text{ mol NaOH}$$

We determine which is the limiting reactant (review Section 3-3).

Required Ratio	**Available Ratio**
$\dfrac{1\text{ mol CuSO}_4}{2\text{ mol NaOH}} = \dfrac{0.500\text{ mol CuSO}_4}{1.00\text{ mol NaOH}}$	$\dfrac{0.0200\text{ mol CuSO}_4}{0.0300\text{ mol NaOH}} = \dfrac{0.667\text{ mol CuSO}_4}{1.00\text{ mol NaOH}}$

▶ NaOH is the limiting reactant.

More CuSO$_4$ is available than is required to react with the NaOH. Thus, 0.846 kJ of heat was given off during the consumption of 0.0300 mol of NaOH. The amount of heat given off per "mole of reaction" is

$$\frac{?\text{ kJ released}}{\text{mol rxn}} = \frac{0.846\text{ kJ given off}}{0.0300\text{ mol NaOH}} \times \frac{2\text{ mol NaOH}}{\text{mol rxn}} = \frac{56.4\text{ kJ given off}}{\text{mol rxn}}$$

Thus, when the reaction occurs *to the extent indicated by the balanced chemical equation*, 56.4 kJ is released. Remembering that exothermic reactions have negative values of ΔH_{rxn}, we write

$$CuSO_4(aq) + 2NaOH(aq) \longrightarrow Cu(OH)_2(s) + Na_2SO_4(aq) \quad \Delta H_{rxn} = -56.4\text{ kJ/mol rxn}$$

You should now work Exercise 64(b).

EXAMPLE 15-5 Amount of Heat Produced

When aluminum metal is exposed to atmospheric oxygen (as in aluminum doors and windows), it is oxidized to form aluminum oxide. How much heat is released by the complete oxidation of 24.2 grams of aluminum at 25°C and 1 atm? The thermochemical equation is

$$4Al(s) + 3O_2(g) \longrightarrow 2Al_2O_3(s) \quad \Delta H = -3352\text{ kJ/mol rxn}$$

Plan

The thermochemical equation tells us that 3352 kJ of heat is released for every mole of reaction, that is, for every 4 moles of Al that reacts. We convert 24.2 g of Al to moles and then calculate the number of kilojoules corresponding to that number of moles of Al, using the unit factors

$$\frac{-3352 \text{ kJ}}{\text{mol rxn}} \quad \text{and} \quad \frac{1 \text{ mol rxn}}{4 \text{ mol Al}}$$

Solution

For 24.2 g Al,

$$\underline{?}\text{ kJ} = 24.2 \text{ g Al} \times \frac{1 \text{ mol Al}}{27.0 \text{ g Al}} \times \frac{1 \text{ mol rxn}}{4 \text{ mol Al}} \times \frac{-3352 \text{ kJ}}{\text{mol rxn}} = -751 \text{ kJ}$$

This tells us that 751 kJ of heat is released to the surroundings during the oxidation of 24.2 grams of aluminum.

You should now work Exercises 16 and 17.

> **STOP & THINK**
> The *sign* tells us that heat was released, but it would be grammatical nonsense to say in words that "−751 kJ of heat was released." As an analogy, suppose you give your friend $5. Your Δ$ is −$5, but in describing the transaction you would not say "I gave her minus five dollars." Instead, you would say, "I gave her five dollars."

> ⚙ **Problem-Solving Tip** Mole of Reaction

Remember that a thermochemical equation can have *different* coefficients (numbers of moles) of *different* reactants or products, which need not equal one. In Example 15-5 one mole of reaction corresponds to 4 moles of Al(s), 3 moles of $O_2(g)$, and 2 moles of $Al_2O_3(s)$.

15-6 Standard States and Standard Enthalpy Changes

The **thermodynamic standard state** of a substance is its most stable pure form under standard pressure (one atmosphere)* and at some specific temperature (25°C or 298 K unless otherwise specified). Examples of elements in their standard states at 25°C are hydrogen, gaseous diatomic molecules, $H_2(g)$; mercury, a silver-colored liquid metal, $Hg(\ell)$; sodium, a silvery-white solid metal, $Na(s)$; and carbon, a grayish-black solid called graphite, C(graphite). We use C(graphite) instead of C(s) to distinguish it from other solid forms of carbon, such as C(diamond). The reaction C(diamond) $\longrightarrow$ C(graphite) would be *exothermic* by 1.897 kJ/mol rxn; C(graphite) is thus more stable than C(diamond). Examples of standard states of compounds include ethanol (ethyl alcohol or grain alcohol), a liquid, $C_2H_5OH(\ell)$; water, a liquid, $H_2O(\ell)$; calcium carbonate, a solid, $CaCO_3(s)$; and carbon dioxide, a gas, $CO_2(g)$. Keep in mind the following conventions for thermochemical standard states.

1. For a *pure* substance in the liquid or solid phase, the standard state is the pure liquid or solid.
2. For a gas, the standard state is the gas at a pressure of *one atmosphere*; in a mixture of gases, its partial pressure must be one atmosphere.
3. For a substance in solution, the standard state refers to *one-molar* concentration.

For ease of comparison and tabulation, we often refer to thermochemical or thermodynamic changes "at standard states" or, more simply, to a *standard change*. To indicate a change at standard pressure, we add a superscript zero, which is read as "naught." If some temperature other than the standard temperature of 25°C (298 K) is specified, we indicate it with a subscript; if no subscript appears, a temperature of 25°C (298 K) is implied.

> ▶ A temperature of 25°C is 77°F. This is slightly above typical room temperature. Notice that these thermodynamic "standard conditions" are not the same as the "standard temperature and pressure (STP)" that we used in gas calculations involving standard molar volume (see Chapter 12).

> ▶ If the substance exists in several different forms, the form that is most stable at 25°C and 1 atm is the standard state.

> ▶ For gas laws (see Chapter 12), standard temperature is taken as 0°C. For thermodynamics, it is taken as 25°C.

*IUPAC has changed the standard pressure from 1 atm to 1 bar. Because 1 bar is equal to 0.987 atm, the differences in thermodynamic calculations are negligible except in work of very high precision. Many tables of thermodynamic data are still based on a standard pressure of 1 atm, so we use that pressure in this text.

▶ This is sometimes referred to as the *standard heat of reaction*. We verbally refer to this as "delta H naught reaction."

The **standard enthalpy change, ΔH^0_{rxn}**, for a reaction

$$\text{reactants} \longrightarrow \text{products}$$

refers to the ΔH when the specified number of moles of reactants, all at standard states, are converted *completely* to the specified number of moles of products, all at standard states.

We allow a reaction to take place, with changes in temperature or pressure if necessary; when the reaction is complete, we return the products to the same conditions of temperature and pressure that we started with, *keeping track of energy or enthalpy changes* as we do so. When we describe a process as taking place "at constant T and P," we mean that the initial and final conditions are the same. Because we are dealing with changes in state functions, the net change is the same as the change we would have obtained hypothetically with T and P actually held constant.

15-7 Standard Molar Enthalpies of Formation, ΔH^0_f

It is not possible to determine the total enthalpy content of a substance on an absolute scale. Because we need to describe only *changes* in this state function, we can define an *arbitrary scale* as follows.

▶ We can think of ΔH^0_f as the enthalpy content of each substance, in its standard state, relative to the enthalpy content of the elements, in their standard states. This is why ΔH^0_f for an element in its standard state is zero.

The **standard molar enthalpy of formation, ΔH^0_f**, of a substance is the enthalpy change for the reaction in which *one mole* of the substance in a specified state is formed from its elements in their standard states. By convention, the ΔH^0_f value for any *element in its standard state* is defined as zero.

Standard molar enthalpy of formation is often called **standard molar heat of formation** or, more simply, **heat of formation**. The superscript zero in ΔH^0_f signifies standard pressure, 1 atmosphere. Negative values for ΔH^0_f describe exothermic formation reactions, whereas positive values for ΔH^0_f describe endothermic formation reactions.

The enthalpy change for a balanced equation that gives a compound from its elements does not necessarily give a molar enthalpy of formation for the compound. Consider the following exothermic reaction at standard conditions.

$$H_2(g) + Br_2(\ell) \longrightarrow 2HBr(g) \qquad \Delta H^0_{rxn} = -72.8 \text{ kJ/mol rxn}$$

We see that *two* moles of HBr(g) are formed in the reaction as written. Half as much energy, 36.4 kJ, is liberated when *one mole* of HBr(g) is produced from its constituent elements in their standard states. For HBr(g), $\Delta H^0_f = -36.4$ kJ/mol. This can be shown by dividing all coefficients in the balanced equation by 2.

▶ The coefficients $\frac{1}{2}$ preceding $H_2(g)$ and $Br_2(\ell)$ do *not* imply half a molecule of each. In thermochemical equations, the coefficients always refer to the number of *moles* under consideration.

$$\tfrac{1}{2}H_2(g) + \tfrac{1}{2}Br_2(\ell) \longrightarrow HBr(g) \qquad \Delta H^0_{rxn} = -36.4 \text{ kJ/mol rxn}$$
$$\Delta H^0_{f\,HBr(g)} = -36.4 \text{ kJ/mol HBr(g)}$$

Standard heats of formation of some common substances are tabulated in Table 15-1. Appendix K contains a more extensive listing.

When referring to a thermodynamic quantity for a *substance*, we often omit the description of the substance from the units. Units for tabulated ΔH^0_f values are given as "kJ/mol"; we must interpret this as "per mole of the substance in the specified state." For instance, for HBr(g) the tabulated ΔH^0_f value of -36.4 kJ/mol should be interpreted as $\dfrac{-36.4 \text{ kJ}}{\text{mol HBr}(g)}$.

Table 15-1 Selected Standard Molar Enthalpies of Formation at 298 K

Substance	ΔH_f^0 (kJ/mol)	Substance	ΔH_f^0 (kJ/mol)
$Br_2(\ell)$	0	HgS(s) red	−58.2
$Br_2(g)$	30.91	$H_2(g)$	0
C(diamond)	1.897	HBr(g)	−36.4
C(graphite)	0	$H_2O(\ell)$	−285.8
$CH_4(g)$	−74.81	$H_2O(g)$	−241.8
$C_2H_4(g)$	52.26	NO(g)	90.25
$C_6H_6(\ell)$	49.03	Na(s)	0
$C_2H_5OH(\ell)$	−277.7	NaCl(s)	−411.0
CO(g)	−110.5	$O_2(g)$	0
$CO_2(g)$	−393.5	$SO_2(g)$	−296.8
CaO(s)	−635.5	$SiH_4(g)$	34.0
$CaCO_3(s)$	−1207.0	$SiCl_4(g)$	−657.0
$Cl_2(g)$	0	$SiO_2(s)$	−910.9

S TOP & THINK
The ΔH_f^0 values of $Br_2(g)$ and C(diamond) are *not equal to 0* at 298 K. This is because the standard states of these elements are $Br_2(\ell)$ and C(graphite), respectively.

EXAMPLE 15-6 Interpretation of ΔH_f^0

The standard molar enthalpy of formation of ethanol, $C_2H_5OH(\ell)$, is −277.7 kJ/mol. Write the thermochemical equation for the reaction for which $\Delta H_{rxn}^0 = -277.7$ kJ/mol rxn.

Plan

The definition of ΔH_f^0 of a substance refers to a reaction in which *one mole* of the substance in a specified state is formed from the elements *in their standard states*. We put one mole of $C_2H_5OH(\ell)$ on the right side of the chemical equation and put the appropriate elements in their standard states on the left. We balance the equation *without changing the coefficient of the product*, even if we must use fractional coefficients on the left.

Solution

$$2C(\text{graphite}) + 3H_2(g) + \tfrac{1}{2}O_2(g) \longrightarrow C_2H_5OH(\ell) \qquad \Delta H = -277.7 \text{ kJ/mol rxn}$$

You should now work Exercise 28.

Problem-Solving Tip   How Do We Interpret Fractional Coefficients?

Remember that we *always* interpret the coefficients in thermochemical equations as numbers of *moles* of reactants or products. The $\tfrac{1}{2}O_2(g)$ in the answer to Example 15-6 refers to $\tfrac{1}{2}$ *mole* of O_2 molecules, or

$$\tfrac{1}{2} \text{ mol } O_2 \times \frac{32.0 \text{ g } O_2}{\text{mol } O_2} = 16.0 \text{ g } O_2$$

It is important to realize that this is *not* the same as one mole of O atoms (though that would also weigh 16.0 g).

Similarly, the fractional coefficients in

$$\tfrac{1}{2} H_2(g) + \tfrac{1}{2} Br_2(\ell) \longrightarrow HBr(g)$$

refer to

$$\tfrac{1}{2} \text{ mol } H_2 \times \frac{2.0 \text{ g } H_2}{\text{mol } H_2} = 1.0 \text{ g } H_2$$

and

$$\tfrac{1}{2} \text{ mol } Br_2 \times \frac{159.8 \text{ g } Br_2}{\text{mol } Br_2} = 79.9 \text{ g } Br_2$$

respectively.

15-8 Hess's Law

In 1840, G. H. Hess (1802–1850) published his **law of heat summation,** which he derived on the basis of numerous thermochemical observations.

> The enthalpy change for a reaction is the same whether it occurs by one step or by any series of steps.

▶ As an analogy, consider traveling from Kansas City (elevation 884 ft above sea level) to Denver (elevation 5280 ft). The change in elevation is (5280 − 884) ft = 4396 ft, regardless of the route taken.

Enthalpy is a state function. Its *change* is therefore independent of the pathway by which a reaction occurs. We do not need to know whether the reaction *does*, or even *can*, occur by the series of steps used in the calculation. The steps must (if only "on paper") result in the overall reaction. Hess's Law lets us calculate enthalpy changes for reactions for which the changes could be measured only with difficulty, if at all. In general terms, Hess's Law of heat summation may be represented as

$$\Delta H^0_{rxn} = \Delta H^0_a + \Delta H^0_b + \Delta H^0_c + \cdots$$

Here a, b, c, . . . refer to balanced thermochemical equations that can be summed to give the equation for the desired reaction.

Consider the following reaction.

$$C(graphite) + \tfrac{1}{2}O_2(g) \longrightarrow CO(g) \qquad \Delta H^0_{rxn} = \underline{?}$$

The enthalpy change for this reaction cannot be measured directly. Even though $CO(g)$ is the predominant product of the reaction of graphite with a *limited* amount of $O_2(g)$, some $CO_2(g)$ is always produced as well. The following reactions do go to completion with excess $O_2(g)$; therefore, ΔH^0 values have been measured experimentally for them. [Pure $CO(g)$ is readily available.]

$$C(graphite) + O_2(g) \longrightarrow CO_2(g) \qquad \Delta H^0_{rxn} = -393.5 \text{ kJ/mol rxn} \qquad (1)$$
$$CO(g) + \tfrac{1}{2}O_2(g) \longrightarrow CO_2(g) \qquad \Delta H^0_{rxn} = -283.0 \text{ kJ/mol rxn} \qquad (2)$$

▶ You are familiar with the addition and subtraction of algebraic equations. This method of combining thermochemical equations is analogous.

We can "work backward" to find out how to combine these two known equations to obtain the desired equation. We want one mole of CO on the right, so we reverse equation (2) [designated below as (−2)]; heat is then absorbed instead of released, so we must change the sign of its ΔH^0 value. Then we add it to equation (1), canceling equal numbers of moles of the same species on each side. This gives the equation for the reaction we want. Adding the corresponding enthalpy changes gives the enthalpy change we seek.

C(graphite) + O₂(g)

−110.5 kJ

CO(g) + ½O₂(g)

−393.5 kJ

−283.0 kJ

CO₂(g)

Above is a schematic representation of the enthalpy changes for the reaction

C(graphite) + ½ O₂(g) ⟶ CO(g).

The Δ H value for each step is based on the number of moles of each substance indicated.

	ΔH^0	
$C(graphite) + O_2(g) \longrightarrow \cancel{CO_2(g)}$	−393.5 kJ/mol rxn)	(1)
$\cancel{CO_2(g)} \longrightarrow CO(g) + \tfrac{1}{2}O_2(g)$	−(−283.0 kJ/mol rxn)	(−2)
$C(graphite) + \tfrac{1}{2}O_2(g) \longrightarrow CO(g)$	$\Delta H^0_{rxn} = -110.5$ kJ/mol rxn	

This equation shows the formation of one mole of $CO(g)$ in its standard state from the elements in their standard states. In this way, we determine that ΔH^0_f for $CO(g)$ is −110.5 kJ/mol.

EXAMPLE 15-7 Combining Thermochemical Equations: Hess's Law

Use the thermochemical equations shown here to determine ΔH^0_{rxn} at 25°C for the following reaction.

$$C(graphite) + 2H_2(g) \longrightarrow CH_4(g)$$

	ΔH^0	
$C(graphite) + O_2(g) \longrightarrow CO_2(g)$	-393.5 kJ/mol rxn	(1)
$H_2(g) + \frac{1}{2}O_2(g) \longrightarrow H_2O(\ell)$	-285.8 kJ/mol rxn	(2)
$CH_4(g) + 2O_2(g) \longrightarrow CO_2(g) + 2H_2O(\ell)$	-890.3 kJ/mol rxn	(3)

▶ These are combustion reactions, for which ΔH^0_{rxn} values can be readily determined from calorimetry experiments.

Plan

(i) We want one mole of C(graphite) as overall reactant, so we write down thermochemical equation (1).

(ii) We want two moles of $H_2(g)$ as overall reactants, so we multiply thermochemical equation (2) by 2 [designated below as 2 × (2)].

(iii) We want one mole of $CH_4(g)$ as overall product, so we reverse thermochemical equation (3) to give (−3).

(iv) Then we add these thermochemical equations term by term. The result is the desired thermochemical equation, with all unwanted substances canceling. The sum of the ΔH^0 values is the ΔH^0 for the desired reaction.

S TOP & THINK
Remember that multiplying a thermochemical equation also multiplies its ΔH^0 by the same factor. Reversing a thermochemical equation changes the sign of its ΔH^0.

Solution

	ΔH^0	
$C(graphite) + \cancel{O_2(g)} \longrightarrow \cancel{CO_2(g)}$	-393.5 kJ/mol rxn)	(1)
$2H_2(g) + \cancel{O_2(g)} \longrightarrow \cancel{2H_2O(\ell)}$	$2(-285.8$ kJ/mol rxn)	2 × (2)
$\cancel{CO_2(g)} + \cancel{2H_2O(\ell)} \longrightarrow CH_4(g) + \cancel{2O_2(g)}$	$+890.3$ kJ/mol rxn)	(−3)
$C(graphite) + 2H_2(g) \longrightarrow CH_4(g)$	$\Delta H^0_{rxn} = -74.8$ kJ/mol rxn)	

S TOP & THINK
We have used a series of reactions for which ΔH^0 values can be easily measured to calculate ΔH^0 for a reaction that cannot be carried out.

$CH_4(g)$ cannot be formed directly from C(graphite) and $H_2(g)$, so its ΔH^0_f value cannot be measured directly. The result of this example tells us that this value is -74.8 kJ/mol.

EXAMPLE 15-8 Combining Thermochemical Equations: Hess's Law

Given the following thermochemical equations, calculate the heat of reaction at 298 K for the reaction of ethylene with water to form ethanol.

$$C_2H_4(g) + H_2O(\ell) \longrightarrow C_2H_5OH(\ell)$$

	ΔH^0	
$C_2H_5OH(\ell) + 3O_2(g) \longrightarrow 2CO_2(g) + 3H_2O(\ell)$	-1367 kJ/mol rxn	(1)
$C_2H_4(g) + 3O_2(g) \longrightarrow 2CO_2(g) + 2H_2O(\ell)$	-1411 kJ/mol rxn	(2)

Plan

We reverse equation (1) to give (−1); when the equation is reversed, the sign of ΔH^0 is changed because the reverse of an exothermic reaction is endothermic. Then we add the thermochemical equations.

Solution

	ΔH^0	
$2\cancel{CO_2(g)} + 3H_2O(\ell) \longrightarrow C_2H_5OH(\ell) + \cancel{3O_2(g)}$	$+1367$ kJ/mol rxn	(−1)
$C_2H_4(g) + \cancel{3O_2(g)} \longrightarrow 2\cancel{CO_2(g)} + 2H_2O(\ell)$	-1411 kJ/mol rxn	(2)
$C_2H_4(g) + H_2O(\ell) \longrightarrow C_2H_5OH(\ell)$	$\Delta H^0_{rxn} = -44$ kJ/mol rxn	

S TOP & THINK
If you reverse a chemical equation, remember to switch the sign of its ΔH^0_{rxn}.

You should now work Exercises 32 and 34.

Problem Solving Tip ΔH_f^0 Refers to a Specific Reaction

The ΔH^0 for the reaction in Example 15-8 is -44 kJ for each mole of $C_2H_5OH(\ell)$ formed. This reaction, however, does *not* involve formation of $C_2H_5OH(\ell)$ *from its constituent elements* because C_2H_4 and H_2O are not elements; therefore, ΔH_{rxn}^0 is *not* ΔH_f^0 for $C_2H_5OH(\ell)$. We have seen the reaction for ΔH_f^0 of $C_2H_5OH(\ell)$ in Example 15-6.

Similarly, the ΔH_{rxn}^0 for

$$CO(g) + \tfrac{1}{2} O_2(g) \longrightarrow CO_2(g)$$

is *not* ΔH_f^0 for $CO_2(g)$.

Another interpretation of Hess's Law lets us use tables of ΔH_f^0 values to calculate the enthalpy change for a reaction. Let us consider again the reaction of Example 15-8.

$$C_2H_4(g) + H_2O(\ell) \longrightarrow C_2H_5OH(\ell)$$

A table of ΔH_f^0 values (see Appendix K) gives $\Delta H_{f\,C_2H_5OH(\ell)}^0 = -277.7$ kJ/mol, $\Delta H_{f\,C_2H_4(g)}^0 = 52.3$ kJ/mol, and $\Delta H_{f\,H_2O(\ell)}^0 = -285.8$ kJ/mol. We may express this information in the form of the following thermochemical equations.

		ΔH^0	
$2C(graphite) + 3H_2(g) + \tfrac{1}{2}O_2(g) \longrightarrow C_2H_5OH(\ell)$		-277.7 kJ/mol rxn	(1)
$2C(graphite) + 2H_2(g) \longrightarrow C_2H_4(g)$		52.3 kJ/mol rxn	(2)
$H_2(g) + \tfrac{1}{2}O_2(g) \longrightarrow H_2O(\ell)$		-285.8 kJ/mol rxn	(3)

We may generate the equation for the desired net reaction by adding equation (1) to the reverse of equations (2) and (3). The value of ΔH^0 for the desired reaction is then the sum of the corresponding ΔH^0 values.

		ΔH^0	
$2C(\text{graphite}) + 3H_2(g) + \tfrac{1}{2}O_2(g) \longrightarrow C_2H_5OH(\ell)$		-277.7 kJ/mol rxn	(1)
$C_2H_4(g) \longrightarrow 2C(\text{graphite}) + 2H_2(g)$		-52.3 kJ/mol rxn	(-2)
$H_2O(\ell) \longrightarrow H_2(g) + \tfrac{1}{2}O_2(g)$		$+285.8$ kJ/mol rxn	(-3)
net rxn: $C_2H_4(g) + H_2O(\ell) \longrightarrow C_2H_5OH(\ell)$		$\Delta H_{rxn}^0 = -44.2$ kJ/mol rxn	

We see that ΔH^0 for this reaction is given by

$$\Delta H_{rxn}^0 = \Delta H_{(1)}^0 + \Delta H_{(-2)}^0 + \Delta H_{(-3)}^0$$

or by

$$\Delta H_{rxn}^0 = \Delta H_{f\,C_2H_5OH(\ell)}^0 - [\Delta H_{f\,C_2H_4(g)}^0 + \Delta H_{f\,H_2O(\ell)}^0]$$

In general terms this is a very useful form of Hess's Law.

$$\Delta H_{rxn}^0 = \Sigma\, n\, \Delta H_{f\,products}^0 - \Sigma\, n\, \Delta H_{f\,reactants}^0$$

The standard enthalpy change of a reaction is equal to the sum of the standard molar enthalpies of formation of the products, each multiplied by its coefficient, n, in the *balanced equation*, minus the corresponding sum of the standard molar enthalpies of formation of the reactants.

▶ The capital Greek letter sigma (Σ) is read "the sum of." The $\Sigma\, n$ means that the ΔH_f^0 value of each product and reactant must be multiplied by its coefficient, n, in the balanced equation. The resulting values are then added.

In effect this form of Hess's Law supposes that the reaction occurs by converting reactants to the elements in their standard states, then converting these to products (Figure 15-4). Few, if any, reactions actually occur by such a pathway. Nevertheless, because H is a state function the ΔH^0 for this *hypothetical* pathway for *reactants* $\longrightarrow$ *products* would be the same as that for any other pathway—including the one by which the reaction actually occurs.

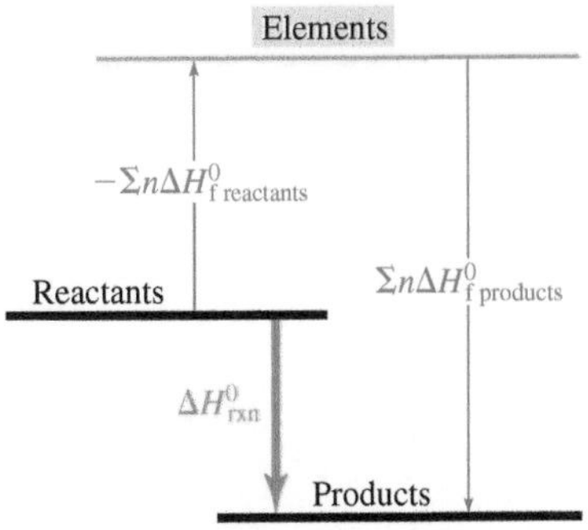

Figure 15-4 A schematic representation of Hess's Law. The red arrow represents the *direct* path from reactants to products. The series of blue arrows is a path (hypothetical) in which reactants are converted to elements, and they in turn are converted to products—all in their standard states.

EXAMPLE 15-9   Using ΔH_f^0 Values: Hess's Law

Calculate ΔH_{rxn}^0 for the following reaction at 298 K.

$$SiH_4(g) + 2O_2(g) \longrightarrow SiO_2(s) + 2H_2O(\ell)$$

Plan

We apply Hess's Law in the form $\Delta H_{rxn}^0 = \Sigma\, n\, \Delta H_f^0{}_{\text{products}} - \Sigma\, n\, \Delta H_f^0{}_{\text{reactants}}$, so we use the ΔH_f^0 values tabulated in Appendix K.

Solution

We can first list the ΔH_f^0 values we obtain from Appendix K:

	$SiH_4(g)$	$O_2(g)$	$SiO_2(s)$	$H_2O(\ell)$
ΔH_f^0, kJ/mol:	34.3	0	−910.9	−285.8

$\Delta H_{rxn}^0 = \Sigma\, n\, \Delta H_f^0{}_{\text{products}} - \Sigma\, n\, \Delta H_f^0{}_{\text{reactants}}$

$\Delta H_{rxn}^0 = [\Delta H_f^0{}_{SiO_2(s)} + 2\, \Delta H_f^0{}_{H_2O(\ell)}] - [\Delta H_f^0{}_{SiH_4(g)} + 2\, \Delta H_f^0{}_{O_2(g)}]$

$$\Delta H_{rxn}^0 = \left[\frac{1\ \text{mol}\ SiO_2(s)}{\text{mol rxn}} \times \frac{-910.9\ \text{kJ}}{\text{mol}\ SiO_2(s)} + \frac{2\ \text{mol}\ H_2O(\ell)}{\text{mol rxn}} \times \frac{-285.8\ \text{kJ}}{\text{mol}\ H_2O(\ell)} \right]$$

$$- \left[\frac{1\ \text{mol}\ SiH_4(g)}{\text{mol rxn}} \times \frac{+34.3\ \text{kJ}}{\text{mol}\ SiH_4(g)} + \frac{2\ \text{mol}\ O_2(g)}{\text{mol rxn}} \times \frac{0\ \text{kJ}}{\text{mol}\ O_2(g)} \right]$$

$\Delta H_{rxn}^0 = -1516.8$ kJ/mol rxn

You should now work Exercise 38.

▶ $O_2(g)$ is an element in its standard state, so its ΔH_f^0 is zero.

Each term in the sums on the right-hand side of the solution in Example 15-9 has the units

$$\frac{\text{mol substance}}{\text{mol rxn}} \times \frac{\text{kJ}}{\text{mol substance}} \quad \text{or} \quad \frac{\text{kJ}}{\text{mol rxn}}$$

For brevity, we shall omit units in the intermediate steps of calculations of this type, and just assign the proper units to the answer. Be sure that you understand how these units arise.

Suppose we measure ΔH_{rxn}^0 at 298 K and know all but one of the ΔH_f^0 values for reactants and products. We can then calculate the unknown ΔH_f^0 value.

EXAMPLE 15-10   Using ΔH_f^0 Values: Hess's Law

Use the following information to determine ΔH_f^0 for PbO(s, yellow).

$$PbO(s, \text{yellow}) + CO(g) \longrightarrow Pb(s) + CO_2(g) \qquad \Delta H_{rxn}^0 = -65.69\ \text{kJ}$$

$\Delta H_f^0 = $ for $CO_2(g) = -393.5$ kJ/mol   and ΔH_f^0 for $CO(g) = -110.5$ kJ/mol

▶ We will consult Appendix K to check the answer only after working the problem.

Plan

We again use Hess's Law in the form $\Delta H^0_{rxn} = \Sigma\, n\, \Delta H^0_{f\ products} - \Sigma\, n\, \Delta H^0_{f\ reactants}$. The standard state of lead is Pb(s), so $\Delta H^0_{f\ Pb(s)} = 0$ kJ/mol. Now we are given ΔH^0_{rxn} and the ΔH^0_f values for all substances *except* PbO(s, yellow). We can solve for this unknown.

Solution

We list the known ΔH^0_f values:

	PbO(s, yellow)	CO(g)	Pb(s)	CO$_2$(g)
ΔH^0_f, kJ/mol:	$\Delta H^0_{f\,PbO_2(s,\,yellow)}$	-110.5	0	-393.5

$$\Delta H^0_{rxn} = \Sigma\, n\, \Delta H^0_{f\ products} \qquad -\Sigma\, n\, \Delta H^0_{f\ reactants}$$

$$\Delta H^0_{rxn} = \Delta H^0_{f\ Pb(s)} + \Delta H^0_{f\ CO_2(g)} - [\Delta H^0_{f\ PbO(s,\,yellow)} + \Delta H^0_{f\ CO(g)}]$$

Substituting values stated in the problem gives

$$-65.69 = 0 \qquad + (-393.5) \quad -[\Delta H^0_{f\ PbO(s,\,yellow)} + (-110.5)]$$

Rearranging to solve for, $\Delta H^0_{f\ PbO(s,\,yellow)}$, we have

$$\Delta H^0_{f\ PbO(s,\,yellow)} = 65.69 - 393.5 + 110.5 = \boxed{-217.3 \text{ kJ/mol of PbO}}$$

You should now work Exercise 44.

Problem-Solving Tip Remember the Values of ΔH^0_f for Elements

In Example 15-10, we were not given the value of ΔH^0_f for Pb(s). We should know without reference to tables that ΔH^0_f for an *element* in its most stable form is exactly 0 kJ/mol, so ΔH^0_f for Pb(s) = 0 kJ/mol. But the element *must* be in its most stable form. Thus, ΔH^0_f for O$_2$(g) is zero, because ordinary oxygen is gaseous and diatomic. We would *not* assume that ΔH^0_f would be zero for oxygen atoms, O(g), or for ozone, O$_3$(g). Similarly, ΔH^0_f is zero for Cl$_2$(g) and for Br$_2$(ℓ), but not for Br$_2$(g). Recall that bromine is one of the few elements that is liquid at room temperature and 1 atm pressure.

15-9 Bond Energies

Chemical reactions involve the breaking and making of chemical bonds. Energy is always required to break a chemical bond (see Section 7-4). Often this energy is supplied in the form of heat.

▶ For all practical purposes, the bond energy is the same as bond enthalpy. Tabulated values of average bond energies are actually average bond enthalpies. We use the term "bond *energy*" rather than "bond *enthalpy*" because it is common practice to do so.

The **bond energy (B.E.)** is the amount of energy necessary to break *one mole* of bonds in a gaseous covalent substance to form products in the gaseous state at constant temperature and pressure.

The greater the bond energy, the more stable (stronger) the bond is, and the harder it is to break. Thus bond energy is a measure of bond strengths.

Consider the following reaction.

$$H_2(g) \longrightarrow 2H(g) \qquad \Delta H^0_{rxn} = \Delta H_{H-H} = +436 \text{ kJ/mol H---H bonds}$$

▶ We have discussed these changes in terms of absorption or release of heat. Another way of breaking bonds is by absorption of light energy (see Chapter 4). Bond energies can be determined from the energies of the photons that cause bond dissociation.

The bond energy of the hydrogen–hydrogen bond is 436 kJ/mol of bonds. In other words, 436 kJ of energy must be absorbed for every mole of H---H bonds that are broken. This endothermic reaction (ΔH^0_{rxn} is positive) can be written

$$H_2(g) + 436 \text{ kJ} \longrightarrow 2H(g)$$

Table 15-2 Some Average Single Bond Energies (kJ/mol of bonds)

H	C	N	O	F	Si	P	S	Cl	Br	I	
436	413	391	463	565	318	322	347	432	366	299	**H**
	346	305	358	485			272	339	285	213	**C**
		163	201	283				192			**N**
			146	190	452	335		218	201	201	**O**
				155	565	490	284	253	249	278	**F**
					222		293	381	310	234	**Si**
						201		326		184	**P**
							226	255			**S**
								242	216	208	**Cl**
									193	175	**Br**
										151	**I**

Table 15-3 Comparison of Some Average Single and Multiple Bond Energies
(kJ/mol of bonds)

Single Bonds		Double Bonds		Triple Bonds	
C—C	346	C=C	602	C≡C	835
N—N	163	N=N	418	N≡N	945
O—O	146	O=O	498		
C—N	305	C=N	615	C≡N	887
C—O	358	C=O	732*	C≡O	1072

*Except in CO_2, where it is 799 kJ/mol.

Some average bond energies are listed in Tables 15-2 and 15-3. We see from Table 15-3 that for any combination of elements, a triple bond is stronger than a double bond, which in turn is stronger than a single bond. Bond energies for double and triple bonds are *not* simply two or three times those for the corresponding single bonds. A single bond is a σ bond, whereas double and triple bonds involve a combination of σ and π bonding. The bond energy measures the effectiveness of orbital overlap, and we should not expect the strength of a π bond to be the same as that of a σ bond between the same two atoms because π bonds have poorer orbital overlap than σ bonds.

We should keep in mind that each of the values listed is the average bond energy from a variety of compounds. The *average C—H bond energy* is 413 kJ/mol of bonds. Average C—H bond energies differ slightly from compound to compound, as in CH_4, CH_3Cl, CH_3NO_2, and so on. Nevertheless, they are sufficiently constant to be useful in estimating thermodynamic data that are not readily available by another approach. Values of ΔH^0_{rxn} estimated in this way are not as reliable as those obtained from ΔH^0_f values for the substances involved in the reaction.

A special case of Hess's Law involves the use of bond energies to *estimate* heats of reaction. Consider the enthalpy diagrams in Figure 15-5. In general terms, ΔH^0_{rxn} is related to the bond energies of the reactants and products in *gas phase reactions* by the following version of Hess's Law.

$$\Delta H^0_{rxn} = \Sigma \text{ B.E.}_{reactants} - \Sigma \text{ B.E.}_{products} \qquad \text{for gas phase reactions only}$$

The net enthalpy change of a reaction is the amount of energy required to break all the bonds in reactant molecules *minus* the amount of energy required to break all the bonds in product molecules. Stated in another way, the amount of energy released when a bond is

STOP & THINK
Remember that this equation involves bond energies of *reactants* minus bond energies of *products*. This is opposite Hess's Law, where it is the sum of the products minus the sum of the reactants.

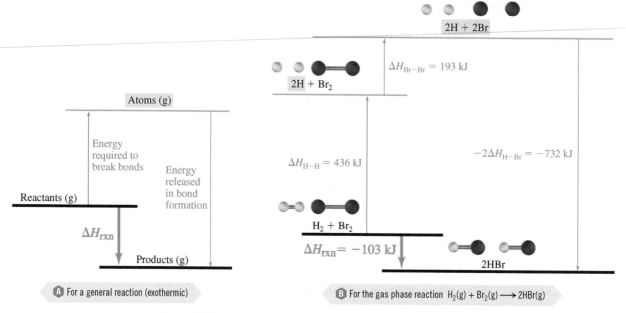

Figure 15-5 A schematic representation of the relationship between bond energies and ΔH_{rxn} for gas phase reactions. As usual for such diagrams, the value shown for each change refers to the number of moles of substances or bonds indicated in the diagram.

▶ Very few real reactions occur by breaking and forming *all* bonds. But because bond energy is a state function, we can assume *any* pathway between reactants and products and get the same result.

formed is equal to the amount absorbed when the same bond is broken. The heat of reaction for a gas phase reaction can be described as the amount of energy released in forming all the bonds in the products minus the amount of energy released in forming all the bonds in the reactants (see Figure 15-5). This heat of reaction can be estimated using the average bond energies in Tables 15-2 and 15-3.

The definition of bond energies is limited to the bond-breaking process *only* and does not include any provision for changes of state. Thus, it is valid only for substances in the gaseous state. The calculations of this section therefore apply *only* when all substances in the reaction are gases. If liquids or solids were involved, then additional information such as heats of vaporization and fusion would be needed to account for phase changes.

EXAMPLE 15-11 Bond Energies

MOLECULAR REASONING

Use the bond energies listed in Table 15-2 to estimate the heat of reaction at 298 K for the following reaction.

$$N_2(g) + 3H_2(g) \longrightarrow 2NH_3(g)$$

Plan

Each NH_3 molecule contains three $N—H$ bonds, so two moles of NH_3 contain six moles of $N—H$ bonds. Three moles of H_2 contain a total of three moles of $H—H$ bonds, and one mole of N_2 contains one mole of $N\equiv N$ bonds. From this we can estimate the heat of reaction.

H—N̈—H
 |
 H

▶ For each term in the sum, the units are $\dfrac{\text{mol bonds}}{\text{mol rxn}} \times \dfrac{\text{kJ}}{\text{mol bonds}}$

Solution

Using the bond energy form of Hess's Law,

$$\Delta H^0_{rxn} = [\Delta H_{N\equiv N} + 3\,\Delta H_{H—H}] - [6\,\Delta H_{N—H}]$$

$$= 945 + 3(436) - 6(391) = \boxed{-93 \text{ kJ/mol rxn}}$$

You should now work Exercise 48.

EXAMPLE 15-12 Bond Energies

Use the bond energies listed in Table 15-2 to estimate the heat of reaction at 298 K for the following reaction.

$$C_3H_8(g) + Cl_2(g) \longrightarrow C_3H_7Cl(g) + HCl(g)$$

$$\underset{\substack{| \quad | \quad | \\ H \; H \; H}}{H-\overset{\substack{H \; H \; H \\ | \quad | \quad |}}{C-C-C}-H} + Cl-Cl \longrightarrow \underset{\substack{| \quad | \quad | \\ H \; H \; H}}{H-\overset{\substack{H \; H \; H \\ | \quad | \quad |}}{C-C-C}-Cl} + H-Cl$$

Plan

Two moles of C—C bonds and seven moles of C—H bonds are the same before and after reaction, so we do not need to include them in the bond energy calculation. The only reactant bonds that are broken are one mole of C—H bonds and one mole of Cl—Cl bonds. On the product side, the only new bonds formed are one mole of C—Cl bonds and one mole of H—Cl bonds. We need to take into account only the bonds that are different on the two sides of the equation. As before, we add and subtract the appropriate bond energies, using values from Table 15-2.

Solution

$$\Delta H^0_{rxn} = [\Delta H_{C-H} + \Delta H_{Cl-Cl}] - [\Delta H_{C-Cl} + \Delta H_{H-Cl}]$$

$$= [413 + 242] - [339 + 432] = -116 \text{ kJ/mol rxn}$$

You should now work Exercises 50 and 52.

> **STOP & THINK**
> We would get the same value for ΔH^0_{rxn} if we used the full bond energy form of Hess's Law and assumed that *all* bonds in reactants were broken and then *all* bonds in products were formed. In such a calculation the bond energies for the unchanged bonds would cancel. Why? Try it!

15-10 Changes in Internal Energy, ΔE

The **internal energy, E,** of a specific amount of a substance represents all the energy contained within the substance. It includes such forms as kinetic energies of the molecules; energies of attraction and repulsion among subatomic particles, atoms, ions, or molecules; and other forms of energy. The internal energy of a collection of molecules is a state function. The difference between the internal energy of the products and the internal energy of the reactants of a chemical reaction or physical change, ΔE, is given by the equation

> ▶ Internal energy is a state function, so it is represented by a capital letter.

$$\Delta E = E_{final} - E_{initial} = E_{products} - E_{reactants} = q + w$$

The terms q and w represent heat and work, respectively. These are two ways in which energy can flow into or out of a system. **Work** involves a change of energy in which a body is moved through a distance, d, against some force, f; that is, $w = fd$.

$$\Delta E = (\text{amount of heat absorbed by system}) + (\text{amount of work done on system})$$

SURROUNDINGS

Heat *absorbed*
by system—
endothermic

Work done
on system

$q > 0$

$q < 0$

SYSTEM

$w > 0$

$w < 0$

Heat *released*
by system—
exothermic

Work done
by system

Sign conventions for *q* and *w*.

Ⓐ Some powdered dry ice (solid CO_2) is placed into a flexible bag, which is then sealed.

Ⓑ As the dry ice absorbs heat from the surroundings, some solid CO_2 sublimes to form gaseous CO_2. The larger volume of the gas causes the bag to expand. The expanding gas does the work of raising a book that has been placed on the bag. Work would be done by the expansion, even if the book were not present, as the bag pushes against the surrounding atmosphere. The heat absorbed by such a process at constant pressure, q_p, is equal to ΔH for the process.

Figure 15-6 A system that absorbs heat and does work.

$$\frac{F}{d^2} \times d^3 = Fd = w$$

$$P \qquad V$$

▶ At 25° the change in internal energy for the combustion of methane is −887 kJ/ mol CH_4. The change in heat content is −890 kJ/ mol CH_4 (see Section 15-1). The small difference is due to work done on the system as it is compressed by the atmosphere.

The following conventions apply to the signs of q and w.

q is positive:	Heat is *absorbed* by the system from the surroundings (endothermic).
q is negative:	Heat is *released* by the system to the surroundings (exothermic).
w is positive:	Work is done *on* the system by the surroundings.
w is negative:	Work is done *by* the system on the surroundings.

Whenever a given amount of energy is added to or removed from a system, either as heat or as work, the energy of the system changes by that same amount. The equation $\Delta E = q + w$ is another way of expressing the First Law of Thermodynamics (see Section 15-1).

The only type of work involved in most chemical and physical changes is pressure–volume work. From dimensional analysis we can see that the product of pressure and volume is work. Pressure is the force exerted per unit area, where area is distance squared, d^2; volume is distance cubed, d^3. Thus, the product of pressure and volume is force times distance, which is work. An example of a physical change (a phase change) in which the system expands and thus does work as it absorbs heat is shown in Figure 15-6. Even if the book had not been present, the expanding system pushing against the atmosphere would have done work for the expansion.

When energy is released by a reacting system, ΔE is negative; energy can be written as a product in the equation for the reaction. When the system absorbs energy from the surroundings, ΔE is positive; energy can be written as a reactant in the equation.

For example, the complete combustion of CH_4 at constant volume at 25°C *releases* energy.

$$CH_4(g) + 2O_2(g) \longrightarrow CO_2(g) + 2H_2O(\ell) + 887 \text{ kJ}$$

indicates release of energy

We can write the *change in energy* that accompanies this reaction as

$$CH_4(g) + 2O_2(g) \longrightarrow CO_2(g) + 2H_2O(\ell) \qquad \Delta E = -887 \text{ kJ/mol rxn}$$

As discussed in Section 15-2, the negative sign indicates a *decrease* in energy of the system, or a *release* of energy by the system.

The reverse of this reaction *absorbs* energy. It can be written as

$$CO_2(g) + 2H_2O(\ell) + 887 \text{ kJ} \longrightarrow CH_4(g) + 2O_2(g)$$

or

indicates absorption of energy

$$CO_2(g) + 2H_2O(\ell) \longrightarrow CH_4(g) + 2O_2(g) \qquad \Delta E = +887 \text{ kJ/mol rxn}$$

If the latter reaction could be forced to occur, the system would have to absorb 887 kJ of energy per mole of reaction from its surroundings.

When a gas is produced against constant external pressure, such as in an open vessel at atmospheric pressure, the gas does work as it expands against the pressure of the atmosphere. If no heat is absorbed during the expansion, the result is a decrease in the internal energy of the system. On the other hand, when a gas is consumed in a process, the atmosphere does work on the reacting system.

Let us illustrate the latter case. Consider the complete reaction of a 2:1 mole ratio of H_2 and O_2 to produce steam at some constant temperature above 100°C and at one atmosphere pressure (Figure 15-7).

$$2H_2(g) + O_2(g) \longrightarrow 2H_2O(g) + \text{heat}$$

Assume that the constant-temperature bath surrounding the reaction vessel completely absorbs all the evolved heat so that the temperature of the gases does not change. The volume of the system decreases by one third (3 mol gaseous reactants → 2 mol gaseous products). The surroundings exert a constant pressure of one atmosphere and do work on the system by compressing it. The internal energy of the system increases by an amount equal to the amount of work done on it.

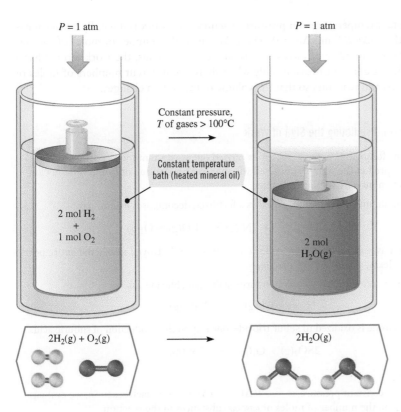

Figure 15-7 An illustration of the one-third decrease in volume that accompanies the reaction of H_2 with O_2 at constant temperature. The temperature is above 100°C.

The work done on or by a system depends on the *external* pressure and the volume. When the external pressure is constant during a change, the amount of work done is equal to this pressure times the change in volume. The work done *on* a system equals $-P\,\Delta V$ or $-P(V_2 - V_1)$.

▶ V_2 is the final volume, and V_1 is the initial volume.

Compression (volume decreases)	Expansion (volume increases)
Work is done *by* the surroundings *on* the system, so the sign of w is positive	Work is done *by* the system *on* the surroundings, so the sign of w is negative
V_2 is less than V_1, so $\Delta V = (V_2 - V_1)$ is negative	V_2 is greater than V_1, so $\Delta V = (V_2 - V_1)$ is positive
$w = -P\,\Delta V$ is positive $(-) \times (+) \times (-) = +$	$w = -P\,\Delta V$ is negative $(-) \times (+) \times (+) = -$
Can be due to a *decrease* in number of moles of gas (Δn negative)	Can be due to an *increase* in number of moles of gas (Δn positive)

We substitute $-P\,\Delta V$ for w in the equation $\Delta E = q + w$ to obtain

$$\Delta E = q - P\,\Delta V$$

In constant-volume reactions, no $P\,\Delta V$ work is done. Volume does not change, so nothing "moves through a distance," and $d = 0$ and $fd = 0$. The change in internal energy of the system is just the amount of heat absorbed or released at constant volume, q_v.

▶ Do not make the error of setting work equal to $V\,\Delta P$.

▶ A subscript v indicates a constant-volume process; a subscript p indicates a constant-pressure process.

$$\Delta E = q_v$$

Figure 15-8 shows the same phase change process as in Figure 15-6, but at constant volume condition, so no work is done.

Solids and liquids do not expand or contract significantly when the pressure changes ($\Delta V \approx 0$). In reactions in which equal numbers of moles of gases are produced and con-

Figure 15-8 A system that absorbs heat at constant volume. Some dry ice [CO_2(s)] is placed into a rigid flask, which is then sealed. As the dry ice absorbs heat from the surroundings, some CO_2(s) sublimes to form CO_2(g). In contrast to the case in Figure 15-6, this system cannot expand ($\Delta V = 0$), so no work is done, and the pressure in the flask increases. Thus, the heat absorbed at constant volume, q_v, is equal to ΔE for the process.

▶ Δn refers to the balanced equation.

sumed at constant temperature and pressure, essentially no work is done. By the ideal gas equation, $P\,\Delta V = (\Delta n)RT$ and $\Delta n = 0$, where Δn equals the number of moles of gaseous products minus the number of moles of gaseous reactants. Thus, the work term w has a significant value at constant pressure only when there are different numbers of moles of gaseous products and reactants so that the volume of the system changes.

EXAMPLE 15-13 Predicting the Sign of Work

For each of the following chemical reactions carried out at constant temperature and constant pressure, predict the sign of w and tell whether work is done *on* or *by* the system. Consider the reaction mixture to be the system.

(a) Ammonium nitrate, commonly used as a fertilizer, decomposes explosively.

$$2NH_4NO_3(s) \longrightarrow 2N_2(g) + 4H_2O(g) + O_2(g)$$

This reaction was responsible for an explosion in 1947 that destroyed nearly the entire port of Texas City, Texas, and killed 576 people.

(b) Hydrogen and chlorine combine to form hydrogen chloride gas.

$$H_2(g) + Cl_2(g) \longrightarrow 2HCl(g)$$

(c) Sulfur dioxide is oxidized to sulfur trioxide, one step in the production of sulfuric acid.

$$2SO_2(g) + O_2(g) \longrightarrow 2SO_3(g)$$

Plan

For a process at constant pressure, $w = -P\,\Delta V = -(\Delta n)RT$. For each reaction, we evaluate Δn, the change in the number of moles of *gaseous* substances in the reaction.

$$\Delta n = (\text{no. of moles of gaseous products}) - (\text{no. of moles of gaseous reactants})$$

Because both R and T (on the Kelvin scale) are positive quantities, the sign of w is opposite from that of Δn; it tells us whether the work is done *on* ($w = +$) or *by* ($w = -$) the system.

The decomposition of NH_4NO_3 produces large amounts of gas, which expands rapidly as the very fast reaction occurs. This explosive reaction was the main cause of the destruction of the Federal Building in Oklahoma City in 1995.

Solution

▶ Here there are no gaseous reactants.

(a) $\Delta n = [2 \text{ mol } N_2(g) + 4 \text{ mol } H_2O(g) + 1 \text{ mol } O_2(g)] - 0 \text{ mol}$

   $= 7 \text{ mol} - 0 \text{ mol} = +7 \text{ mol}$

Δn is positive, so w is negative. This tells us that work is done *by* the system. The large amount of gas formed by the reaction pushes against the surroundings (as happened with devastating effect in the Texas City disaster).

(b) $\Delta n = [2 \text{ mol } HCl(g)] - [1 \text{ mol } H_2(g) + 1 \text{ mol } Cl_2(g)]$

   $= 2 \text{ mol} - 2 \text{ mol} = 0 \text{ mol}$

Thus, $w = 0$, and no work is done as the reaction proceeds. We can see from the balanced equation that for every two moles (total) of gas that react, two moles of gas are formed, so the volume neither expands nor contracts as the reaction occurs.

(c) $\Delta n = [2 \text{ mol } SO_3(g)] - [2 \text{ mol } SO_2(g) + 1 \text{ mol } O_2(g)]$

$= 2 \text{ mol} - 3 \text{ mol} = -1 \text{ mol}$

Δn is negative, so w is positive. This tells us that work is done *on* the system as the reaction proceeds. The surroundings push against the diminishing volume of gas.

You should now work Exercises 77 and 78.

> **STOP & THINK**
>
> When the number of moles of gas *increases*, work is done *by* the system, so *w* is *negative*. When the number of moles of gas *decreases*, work is done *on* the system, so *w* is *positive*.

A **bomb calorimeter** is a device that measures the amount of heat evolved or absorbed by a reaction occurring at constant volume (Figure 15-9). A strong steel vessel (the bomb) is immersed in a large volume of water. As heat is produced or absorbed by a reaction inside the steel vessel, the heat is transferred to or from the large volume of water. Thus, only rather small temperature changes occur. For all practical purposes, the energy changes associated with the reactions are measured at constant volume and constant temperature. No work is done when a reaction is carried out in a bomb calorimeter, even if gases are involved, because $\Delta V = 0$. Therefore,

$$\Delta E = q_v \qquad \text{(constant volume)}$$

> ▶ The "calorie content" of a food can be determined by burning it in excess oxygen inside a bomb calorimeter and determining the heat released. 1 "nutritional Calorie" = 1 kcal = 4.184 kJ.

EXAMPLE 15-14   Bomb Calorimeter

A 1.000-gram sample of ethanol, C_2H_5OH, was burned in a bomb calorimeter whose heat capacity had been determined to be 2.71 kJ/°C. The temperature of 3000 grams of water rose from 24.284°C to 26.225°C. Determine ΔE for the reaction in joules per gram of ethanol, and then in kilojoules per mole of ethanol. The specific heat of water is 4.184 J/g · °C. The combustion reaction is

$$C_2H_5OH(\ell) + 3O_2(g) \longrightarrow 2CO_2(g) + 3H_2O(\ell)$$

> ▶ Benzoic acid, C_6H_5COOH, is often used to determine the heat capacity of a calorimeter. It is a solid that can be compressed into pellets. Its heat of combustion is accurately known: 3227 kJ/mol benzoic acid, or 26.46 kJ/g benzoic acid. Another way to measure the heat capacity of a calorimeter is to add a known amount of heat electrically.

Plan

The amount of heat given off by the system (in the sealed compartment) raises the temperature of the calorimeter and its water. The amount of heat absorbed by the water can be calculated using the specific heat of water; similarly, we use the heat capacity of the calorimeter to find the amount of heat absorbed by the calorimeter. The sum of these two amounts of heat is the total amount of heat released by the combustion of 1.000 gram of ethanol. We must then scale that result to correspond to one mole of ethanol.

Solution

The increase in temperature is

$$\underline{?}\ \degree C = 26.225°C - 24.284°C = 1.941°C \text{ rise}$$

The amount of heat responsible for this increase in temperature of 3000 grams of water is

$$\text{heat to warm water} = 1.941°C \times \frac{4.184 \text{ J}}{g \cdot °C} \times 3000 \text{ g} = 2.436 \times 10^4 \text{ J} = 24.36 \text{ kJ}$$

The amount of heat responsible for the warming of the calorimeter is

$$\text{heat to warm calorimeter} = 1.941°C \times \frac{2.71 \text{ kJ}}{°C} = 5.26 \text{ kJ}$$

The total amount of heat absorbed by the calorimeter *and* by the water is

$$\text{total amount of heat} = 24.36 \text{ kJ} + 5.26 \text{ kJ} = 29.62 \text{ kJ}$$

Combustion of one gram of C_2H_5OH liberates 29.62 kJ of energy in the form of heat, that is

$$\Delta E = q_v = -29.62 \text{ kJ/g ethanol}$$

A disassembled bomb calorimeter

The negative sign indicates that energy is released by the system to the surroundings. Now we may evaluate ΔE in kJ/mol of ethanol by converting grams of C_2H_5OH to moles.

$$\frac{?\ kJ}{mol\ ethanol} = \frac{-29.62\ kJ}{g} \times \frac{46.07g\ C_2H_5OH}{1\ mol\ C_2H_2OH} = -1365\ kJ/mol\ ethanol$$

$$\Delta E = -1365\ kJ/mol\ ethanol$$

This calculation shows that for the combustion of ethanol at constant temperature and constant volume, the change in internal energy is -1365 kJ/mol ethanol.

You should now work Exercises 66 and 67.

The balanced chemical equation involves one mole of ethanol, so we can write the unit factor $\frac{1\ mol\ ethanol}{1\ mol\ rxn}$. Then we express the result of Example 15-14 as

$$\Delta E = \frac{-1365\ kJ}{mol\ ethanol} \times \frac{1\ mol\ ethanol}{1\ mol\ rxn} = -1365\ kJ/mol\ rxn$$

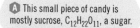
A This small piece of candy is mostly sucrose, $C_{12}H_{22}O_{11}$, a sugar.

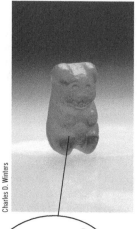

B When the piece of candy is heated together with potassium chlorate, $KClO_3$ (a good oxidizing agent), a highly product-favored reaction occurs.

$$C_{12}H_{22}O_{11}(s) + 12O_2(g) \longrightarrow 12CO_2(g) + 11H_2O(g)$$

If that amount of sucrose is completely metabolized to carbon dioxide and water vapor in your body, the same amount of energy is released, though more slowly.

Oxidation of sugar.

Charles D. Winters

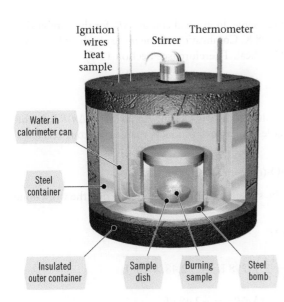

Figure 15-9 A bomb calorimeter measures q_v, the amount of heat given off or absorbed by a reaction occurring at constant *volume*. The amount of energy introduced via the ignition wires is measured and taken into account.

15-11 Relationship Between ΔH and ΔE

The fundamental definition of enthalpy, H, is

$$H = E + PV$$

For a process at constant temperature and pressure,

$$\Delta H = \Delta E + P\,\Delta V \qquad \text{(constant } T \text{ and } P\text{)}$$

From Section 15-10, we know that $\Delta E = q + w$, so

$$\Delta H = q + w + P\,\Delta V \qquad \text{(constant } T \text{ and } P\text{)}$$

At constant pressure, $w = -P\,\Delta V$, so

$$\Delta H = q + (-P\,\Delta V) + P\,\Delta V$$

$$\Delta H = q_p \qquad \text{(constant } T \text{ and } P\text{)}$$

The difference between ΔE and ΔH is the amount of expansion work ($P\,\Delta V$ work) that the system can do. Unless there is a change in the number of moles of gas present, this difference is extremely small and can usually be neglected. For an ideal gas, $PV = nRT$. At constant temperature and constant pressure, $P\,\Delta V = (\Delta n)RT$, a work term. Substituting gives

$$\Delta H = \Delta E + (\Delta n)RT \qquad \text{or} \qquad \Delta E = \Delta H - (\Delta n)RT \qquad \text{(constant } T \text{ and } P\text{)}$$

 Problem-Solving Tip Two Equations Relate ΔH and ΔE—Which One Should Be Used?

The relationship $\Delta H = \Delta E + P\,\Delta V$ is valid for *any* process that takes place at constant temperature and pressure. It is very useful for physical changes that involve volume changes, such as expansion or compression of a gas.

When a chemical reaction occurring at constant T and P results in a change in the number of moles of gas, it is more convenient to use the relationship in the form $\Delta H = \Delta E + (\Delta n)RT$. You should always remember that Δn refers to the change in the number of moles of *gas in the balanced chemical equation*.

STOP & THINK

As usual, Δn refers to the number of moles of *gaseous products* minus the number of moles of *gaseous reactants* in the *balanced chemical equation*.

In Example 15-14 we found that the change in internal energy, ΔE, for the combustion of ethanol is -1365 kJ/mol ethanol at 298 K. Combustion of one mole of ethanol at 298 K and constant pressure releases 1367 kJ of heat. Therefore (see Section 15-5)

$$\Delta H = -1367 \; \frac{\text{kJ}}{\text{mol ethanol}}$$

The difference between ΔH and ΔE is due to the work term, $-P\,\Delta V$ or $-(\Delta n)RT$. In this balanced equation there are fewer moles of gaseous products than of gaseous reactants: $\Delta n = 2 - 3 = -1$.

$$C_2H_5OH(\ell) + 3O_2(g) \longrightarrow 2CO_2(g) + 3H_2O(\ell)$$

Thus, the atmosphere does work on the system (compresses it). Let us find the work done on the system per mole of reaction.

$$w = -P\,\Delta V = -(\Delta n)RT$$

$$= -(-1 \text{ mol})\left(\frac{8.314 \text{ J}}{\text{mol} \cdot \text{K}}\right)(298 \text{ K}) = +2.48 \times 10^3 \text{ J}$$

$$w = +2.48 \text{ kJ} \quad \text{or} \quad (\Delta n)RT = -2.48 \text{ kJ}$$

S TOP & THINK

The positive sign for *w* is consistent with the fact that work is done on the system. The balanced equation involves one mole of ethanol, so this is the amount of work done when one mole of ethanol undergoes combustion.

We can now calculate ΔE for the reaction from ΔH and $(\Delta n)RT$ values.

$$\Delta E = \Delta H - (\Delta n)RT = [-1367 - (-2.48)] = -1365 \text{ kJ/mol rxn}$$

This value agrees with the result that we obtained in Example 15-14. The size of the work term ($+2.48$ kJ) is very small compared with ΔH (-1367 kJ/mol rxn). This is true for many reactions. Of course, if $\Delta n = 0$, then $\Delta H = \Delta E$, and the same amount of heat would be absorbed or given off by the reaction whether it is carried out at constant pressure or at constant volume.

Spontaneity of Physical and Chemical Changes

Another major concern of thermodynamics is predicting *whether* a particular process can occur under specified conditions to give predominantly products. We may summarize this concern in the question "Which would be more stable at the given conditions—the reactants or the products?" A change for which the collection of products is thermodynamically *more stable* than the collection of reactants under the given conditions is said to be **product-favored**, or **spontaneous**, under those conditions. A change for which the products are thermodynamically *less stable* than the reactants under the given conditions is described as **reactant-favored**, or **nonspontaneous**, under those conditions. Some changes are spontaneous under all conditions; others are nonspontaneous under all conditions. The great majority of changes, however, are spontaneous under some conditions but not under others. We use thermodynamics to predict conditions for which the latter type of reactions can occur to give predominantly products.

The concept of spontaneity has a specific interpretation in thermodynamics. A spontaneous chemical reaction or physical change is one that can happen without any continuing outside influence. Examples are the loss of heat by a hot metal to its cooler surroundings, the rusting of a piece of iron, the expansion of a gas into a larger volume, or the melting of ice at room temperature. Such changes have a tendency to occur without being driven by an external influence. We can think of a spontaneous process as one for which products are favored over reactants *at the specified conditions*. The reverse of each of the spontaneous changes just listed is nonspontaneous at the same conditions, that is, it does not occur naturally. We can, however, cause some *nonspontaneous* changes to occur. For example, forcing an electric current through a block of metal can heat it to a temperature higher than that of its surroundings. We can compress a gas into a smaller volume by pressing on it with a piston. But to cause a process to occur in its *nonspontaneous* direction, we must influence it from outside the system; that is, energy must be added to the system.

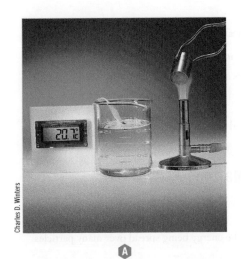

A hot piece of metal (a) is placed into cooler water. Heat is transferred *spontaneously* from the hotter metal to the cooler water (b), until the two are at the same temperature (the condition of *thermal equilibrium*).

Figure 15-10 The diffusion of two gases into one another is spontaneous. (a) A sample of gas in which all molecules of one gas are in one bulb and all molecules of the other gas are in the other bulb. (b) A sample of gas that contains the same number of each kind of molecule as in (a), but with the two kinds randomly mixed in the two bulbs. Sample (b) has greater dispersal of both matter and energy, and is thus more probable.

Although a spontaneous reaction *might* occur rapidly, thermodynamic spontaneity is not related to speed. The fact that a process is spontaneous does not mean that it will occur at an observable rate. It may occur rapidly, at a moderate rate, or very slowly. The rate at which a spontaneous reaction occurs is addressed by kinetics (see Chapter 16).

We now study the factors that influence spontaneity of a physical or chemical change.

15-12 The Two Aspects of Spontaneity

Many product-favored reactions are exothermic. For instance, the combustion (burning) reactions of hydrocarbons such as methane and octane are all exothermic and highly product-favored (spontaneous). The total enthalpy content of the products is lower than that of the reactants. Not all exothermic changes are spontaneous, however, nor are all spontaneous changes exothermic. As an example, consider the freezing of water, which is an exothermic process (heat is released). This process is spontaneous at temperatures below 0°C, but we know it is not spontaneous at temperatures above 0°C. Likewise, we know there are conditions at which the melting of ice, an endothermic process, is spontaneous. Spontaneity is *favored* but not required when heat is released during a chemical reaction or a physical change.

There is another factor that also plays a fundamental role in determining spontaneity. Let's think about two spontaneous processes. Figure 15-10 shows what happens when two gas samples at the same pressure are allowed to mix. The molecules move randomly throughout the two containers to mix the gases (a spontaneous process). We don't expect the more homogeneous sample in Figure 15-10b to spontaneously "unmix" to give the arrangement in Figure 15-10a (a nonspontaneous process).

As a hot metal cools, some of the energy of its vibrating atoms is transferred to the surroundings (a spontaneous process). This warms the surroundings until the two temperatures are equal. We do not expect to observe the reverse process in which energy is transferred from the surroundings to a block of metal, originally at the same temperature, to raise the temperature of the metal (a nonspontaneous process).

▶ Recall that the temperature of a sample is a measure of the average kinetic energy of its particles.

In these examples, we see that *energy and matter tend to become dispersed (spread out)*. We shall see that this dispersal is a fundamental driving force that affects the spontaneity of any process.

Two factors affect the spontaneity of any physical or chemical change:

1. Spontaneity is *favored* when *heat is released* during the change (exothermic).

2. Spontaneity is *favored* when the change causes an *increase in the dispersal of energy and matter.*

The balance of these two effects is considered in Section 15-16.

15-13 Dispersal of Energy and Matter

Dispersal of Energy

The **dispersal of energy** in a system results in the energy being spread over many particles rather than being concentrated in just a few.

To understand this concept, think about a system consisting of just two molecules, A and B, with a total of two units of energy. Denoting one unit of energy with a*, we can list the three ways to distribute these two energy units over the two molecules as

A** (Molecule A has two units of energy, B has none.)

A*B* (Each molecule has one unit of energy.)

B** (Molecule B has two units of energy, A has none.)

Suppose these two molecules are mixed with two other molecules, C and D, that initially have no energy. When collisions occur, energy can be transferred from one molecule to another. Now the energy can be dispersed among the four molecules in ten different ways:

A** B** C** D** A*B* A*C* A*D* B*C* B*D* C*D*

Now there are obviously more ways (ten) the energy can be dispersed than before. In only three of these ways would all of the energy be distributed as before—A**, A*B*, and B**. Put another way, there is only a 3/10 probability that the energy will be restricted to the original molecules, A and B. There are seven ways out of ten, or a probability of 7/10, that at least some of the energy has been transferred to C or D.

What would happen if large numbers of molecules were present, as in any real sample? The probability that the energy is dispersed would be huge, and there would be only an infinitesimally small chance that all of the energy would be concentrated in one or a few molecules. This reasoning leads us to an important conclusion.

If energy can be dispersed over a larger number of particles, it will be.

To see what happens if there is more energy to distribute (as at a higher temperature), let's consider another system with four molecules, but with two of the molecules (A and B) initially having three units of energy and the other two (C and D) initially with no energy, as shown in Figure 15-11a. When these molecules are brought together and allowed to exchange energy by colliding, the energy can be distributed in a total of 84 different ways, as shown in Figure 15-11b. For example, one molecule can have all six units of energy and the other three have no energy, as shown in the leftmost drawing of Figure 15-11b; the six-unit molecule could be any one of the four, so there are four ways to achieve this arrangement. Thus, the probability is only 4/84 (or 1/21) that all of the energy of the system is concentrated in one molecule. There are six ways to distribute the energy to arrive at the rightmost drawing (see Figure 15-11c). Some of the other energy distributions can be achieved in many more ways, leading to a much higher probability for their occurrence. This leads to a broader statement of our earlier conclusion.

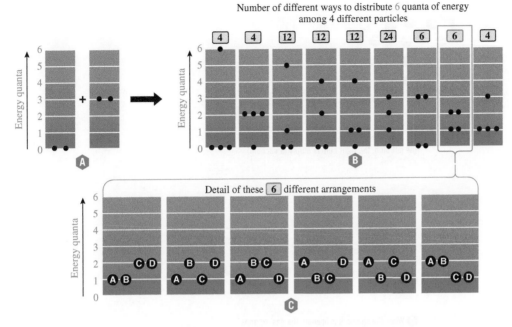

Number of different ways to distribute 6 quanta of energy among 4 different particles

Figure 15-11 Possible ways of distributing six quanta of energy among four molecules. (a) Initially four molecules are separated from each other. Two molecules each have three quanta of energy, and the other two have none. A total of six quanta of energy will be distributed once the molecules interact. (b) Once the molecules begin to interact, there are nine ways to distribute the six available quanta. Each of these arrangements will have multiple ways of distributing the energy among the four molecules. (c) The six different ways to arrange four molecules (A, B, C, and D) such that two molecules have two quanta of energy and the other two have one quanta of energy.

The greater the number of molecules and the higher the total energy of the system, the less likely the energy will be concentrated in a few molecules. Therefore, the energy will be more dispersed.

This generalization provides the molecular explanation behind the Maxwell-Boltzmann distribution of molecular speeds (kinetic energies) that was discussed in Section 12-13 (see Figure 12-9) and in relation to evaporation and vapor pressures of liquids in Sections 13-6 and 13-7.

Dispersal of Matter

Let us apply the idea of dispersal to the arrangements of matter and its molecules. Our experience tells us that a gas originally confined in a bulb (Figure 15-12a) will expand freely into a vacuum (see Figure 15-12b). We don't expect the gas then to concentrate spontaneously into only one bulb. We can consider this expansion from the molecular point of view.

Suppose there are four molecules of gas in the two-part container of Figure 15-13. The probability that a particular molecule is on the left at any given time is $\frac{1}{2}$. A second specific molecule has its own probability of $\frac{1}{2}$ of being on the left, so the probability that *both* of these are on the left at the same time is $\frac{1}{2} \times \frac{1}{2} = \frac{1}{4}$. There are 16 ways that the molecules can be arranged in this container. But in only one of these arrangements are all four molecules in the left-hand portion of the container. The probability of that "concentrated" arrangement is

$$\frac{1}{2} \times \frac{1}{2} \times \frac{1}{2} \times \frac{1}{2} = \left(\frac{1}{2}\right)^4 = \frac{1}{16}$$

We see that there is only a small likelihood that this gas would spontaneously concentrate into the left side of the container.

Similar reasoning shows that the probability of one mole, or 6.0×10^{23} molecules, spontaneously concentrating into the left-hand bulb of Figure 15-12 would be

$$\frac{1}{2} \times \frac{1}{2} \times \ldots \times \frac{1}{2} = \left(\frac{1}{2}\right)^{6.0 \times 10^{23}} = \frac{1}{2^{6.0 \times 10^{23}}}$$

Without writing the number in the denominator explicitly, we can mention that it is larger than the number of molecules in the entire universe! Thus there is virtually no chance that

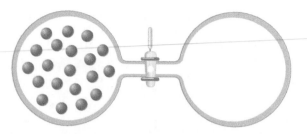

A The stopcock is closed, keeping all of the gas in the left bulb; the right bulb is empty (a vacuum).

B When the stopcock is opened, the gas expands to fill the entire available volume, with half of the gas in each bulb.

Figure 15-12 The expansion of a gas into a vacuum.

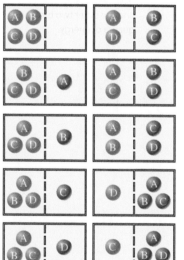

Figure 15-13 The 16 possible ways of arranging four molecules in a two-part container. Only one of these arrangements has all four molecules on the left side, so the probability of that arrangement is 1/16.

all 6.0×10^{23} molecules in a mole of gas would spontaneously concentrate into half the available volume, leaving no molecules in the other half.

The dispersal of matter, as in the expansion of a gas, also often contributes to energy dispersal. Consider the expansion of a gas into a vacuum, as shown in Figure 15-14. When the gas is allowed to occupy the larger volume in Figure 15-14b, its energy levels become closer together than in the smaller volume of Figure 15-14a. This means that there are even more ways for the expanded gas to disperse its energy to arrive at the same total energy. The concept of dispersal of energy thus also predicts that the gas is far more likely to exist in the expanded state of Figure 15-14b, and we would not expect it to concentrate spontaneously to occupy only one chamber as in Figure 15-14a. Similar reasoning helps us to describe the spontaneous mixing of two gases in Figure 15-10. Each gas can have its energy more dispersed in its own more closely spaced energy levels when expanded into both containers. But in addition, the molecules of one gas can transfer energy to the molecules of the other gas by collision. This results in an even larger number of ways for the total energy of the mixture to be dispersed, leading to a much higher probability that the gases will be mixed (see Figure 15-10b) than unmixed (see Figure 15-10a). Thus the spatial dispersal of matter also results in a greater dispersal of energy.

When a soluble substance dissolves in a liquid, the solute particles become dispersed in the solvent (see Sections 14-1 through 14-4). This allows the particles to transfer energy to one another, giving a larger number of ways of distributing the same total energy than if the substances remained in separate phases. In this case, too, dispersal of the two kinds of matter in one another allows more dispersal of energy. In more general terms, we often describe the dispersal of matter as an increase in **disorder**.

To summarize the conclusions of this section:

The final state of a system can be more probable than its initial state (spontaneous, product-favored) in either or both of two ways:

1. Energy can be dispersed over a greater number and variety of molecules.

2. The particles of the system can be more dispersed (more disordered).

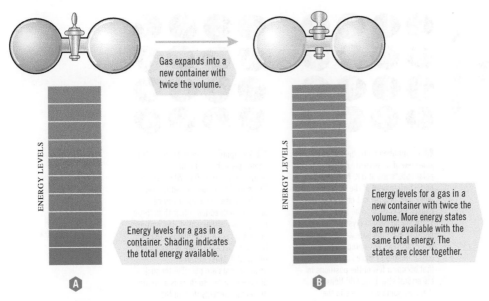

Figure 15-14 Larger gas volumes give more energy levels.

15-14 Entropy, *S*, and Entropy Change, Δ*S*

The dispersal of energy and matter is described by the thermodynamic state function **entropy, *S***. In the most fundamental description, the greater the energy dispersal in a system, the higher is its entropy. We have said that greater disorder (dispersal of matter, both in space and in variety) can lead to greater dispersal of energy, and hence to higher entropy. This connection allows us to discuss many entropy changes in terms of increases or decreases in the disorder and the energy dispersal of the system. Let us first see how entropy is tabulated and how entropy changes are calculated.

The **Third Law of Thermodynamics** establishes the zero of the entropy scale.

> The entropy of a pure, perfect crystalline substance (perfectly ordered) is zero at absolute zero (0 K).

As the temperature of a substance increases, the particles vibrate more vigorously, so the entropy increases (Figure 15-15). Further heat input causes either increased temperature (still higher entropy) or phase transitions (melting, sublimation, or boiling) that also result in higher entropy. The entropy of a substance at any condition is its **absolute entropy**, also called **standard molar entropy**. Consider the absolute entropies at 298 K listed in Table 15-4. At 298 K, *any* substance is more disordered than if it were in a perfect crystalline state at absolute zero, so tabulated S^0_{298} values for compounds and elements are *always positive*. Notice especially that S^0_{298} of an element, unlike its ΔH^0_f, is *not* equal to zero. The reference state for absolute entropy is specified by the Third Law of Thermodynamics. It is different from the reference state for ΔH^0_f (see Section 15-7). The absolute entropies, S^0_{298}, of various substances under standard conditions are tabulated in Appendix K.

▶ Enthalpies are measured only as *differences* with respect to an arbitrary standard state. Entropies, in contrast, are defined relative to an absolute zero level. In either case, the *per mole* designation means *per mole of substance in the specified state*.

Just as for other thermodynamic quantities, the entropy change for a system, ΔS_{system}, is the difference between final and initial states:

$$\Delta S_{system} = S_{system,\ final} - S_{system,\ initial}$$

The **standard entropy change, ΔS^0**, of a reaction can be determined from the absolute entropies of reactants and products. The relationship is analogous to Hess's Law.

▶ The Σ*n* means that each S^0 value must be multiplied by the appropriate coefficient, *n*, from the balanced equation. These values are then added.

$$\Delta S^0_{rxn} = \Sigma\, n\, S^0_{products} - \Sigma\, n\, S^0_{reactants}$$

Table 15-4 Absolute Entropies at 298 K for a Few Common Substances

Substance	S^0 (J/mol · K)
C(diamond)	2.38
C(g)	158.0
$H_2O(\ell)$	69.91
$H_2O(g)$	188.7
$I_2(s)$	116.1
$I_2(g)$	260.6

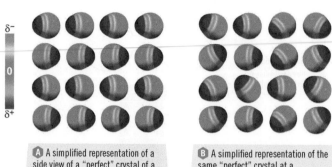

δ^-

0

δ^+

A A simplified representation of a side view of a "perfect" crystal of a polar substance at 0 K. Note the perfect alignment of the dipoles in all molecules in a perfect crystal. This causes its entropy to be zero at 0 K. There are no perfect crystals, however, because even the purest substances that scientists have prepared are contaminated by traces of impurities that occupy a few of the positions in the crystal structure. Additionally, there are some vacancies in the crystal structures of even very highly purified substances such as those used in semiconductors (Section 13-17).

B A simplified representation of the same "perfect" crystal at a temperature above 0 K. Vibrations of the individual molecules within the crystal cause some dipoles to be oriented in directions other than those in a perfect arrangement. The entropy of such a crystalline solid is greater than zero because there is disorder in the crystal. Many different arrangements are possible for such a disordered array, so there is a greater dispersal of energy than in the perfectly ordered crystal.

Figure 15-15

S^0 values are tabulated in units of *J/mol · K* rather than the larger units involving kilojoules that are used for enthalpy changes. The "mol" term in the units for a *substance* refers to a mole of the substance, whereas for a *reaction* it refers to a mole of reaction. Each term in the sums on the right-hand side of the equation has the units

$$\frac{\text{mol substance}}{\text{mol rxn}} \times \frac{\text{J}}{(\text{mol substance}) \cdot \text{K}} = \frac{\text{J}}{(\text{mol rxn}) \cdot \text{K}}$$

The result is usually abbreviated as J/mol · K, or sometimes even as J/K. As before, we will usually omit units in intermediate steps and then apply appropriate units to the result.

We now illustrate this calculation for a phase change and for a chemical reaction.

EXAMPLE 15-15 **Calculation of ΔS^0 for a Phase Change**

Use the values of standard molar entropies in Appendix K to calculate the entropy change for the vaporization of one mole of bromine at 25°C.

$$Br_2(\ell) \longrightarrow Br_2(g)$$

Plan

We use the equation for standard entropy change to calculate ΔS^0 from the tabulated values of standard molar entropies, S^0, for the final and initial states shown in the process.

Solution

We list the ΔS^0_{298} values from Appendix K:

	$Br_2(\ell)$	$Br_2(g)$
S^0, J/mol · K:	152.2	245.4

$$\Delta S^0 = \Sigma\, n\, S^0_{\text{products}} - \Sigma\, n\, S^0_{\text{reactants}}$$

$$= 1(245.4) - 1(152.2) = 93.2 \text{ J/mol} \cdot \text{K}$$

The "mol" designation for J/mol · K refers to a mole of reaction. The reaction as written shows one mole of Br_2, so this is the entropy change for the vaporization of one mole of Br_2. It can be called the *molar entropy of vaporization* of Br_2 at 25°C.

STOP & THINK

The matter in a gaseous sample is far more greatly dispersed than in a liquid. Thus vaporization is always accompanied by an increase in entropy ($\Delta S > 0$).

EXAMPLE 15-16 Calculation of ΔS^0_{rxn} for a Chemical Reaction

Use the values of standard molar entropies in Appendix K to calculate the entropy change at 25°C and one atmosphere pressure for the reaction of hydrazine with hydrogen peroxide. This explosive reaction has been used for rocket propulsion. Do you think the reaction is spontaneous? The balanced equation for the reaction is

$$N_2H_4(\ell) + 2H_2O_2(\ell) \longrightarrow N_2(g) + 4H_2O(g) \quad \Delta H^0_{rxn} = -642.2 \text{ kJ/mol reaction}$$

Plan

We use the equation for standard entropy change to calculate ΔS^0_{rxn} from the tabulated values of standard molar entropies, S^0_{298}, for the substances in the reaction.

Solution

We can list the S^0_{298} values that we obtain from Appendix K for each substance:

	$N_2H_4(\ell)$	$H_2O_2(\ell)$	$N_2(g)$	$H_2O(g)$
S^0, J/mol · K:	121.2	109.6	191.5	188.7

$$\Delta S^0_{rxn} = \Sigma \, n \, S^0_{products} - \Sigma \, n \, S^0_{reactants}$$

$$= [S^0_{N_2(g)} + 4S^0_{H_2O(g)}] - [S^0_{N_2H_4(\ell)} + 2S^0_{H_2O_2(\ell)}]$$

$$= [1(191.5) + 4(188.7)] - [1(121.2) + 2(109.6)]$$

$$\Delta S^0_{rxn} = +605.9 \text{ J/mol} \cdot \text{K}$$

The "mol" designation for ΔS^0_{rxn} refers to a mole of reaction, that is, one mole of $N_2H_4(\ell)$, two moles of $H_2O_2(\ell)$, and so on. Although it may not appear to be, +605.9 J/mol · K is a relatively large value of ΔS^0_{sys}. The positive entropy change favors spontaneity. This reaction is also exothermic (ΔH^0 is negative). As we shall see, this reaction *must* be spontaneous, because both factors are favorable: the reaction is exothermic (ΔH^0_{rxn} is negative) and the disorder of the system increases (ΔS^0_{rxn} is positive).

You should now work Exercise 104.

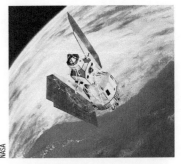

Small booster rockets adjust the course of a satellite in orbit. Some of these small rockets are powered by the N_2H_4–H_2O_2 reaction.

STOP & THINK

In this reaction, liquid reactants are converted into a larger number of moles of gaseous products. Thus it is reasonable that entropy increases ($\Delta S^0 > 0$).

Because changes in the thermodynamic quantity *entropy* may be understood in terms of changes in *energy dispersal* and *molecular disorder*, we can often predict the sign of ΔS_{sys}. The following illustrations emphasize several common types of processes that result in predictable entropy changes for the system.

Phase changes. In a solid, the molecules are in an ordered arrangement, where they can vibrate only around their relatively fixed positions. In a liquid, the molecules are more disordered, so they exchange energy much more freely, and the entropy is higher than in the solid. Similarly, gas molecules are in a much larger volume, and they are far less restrained than in the liquid. In the gas, they move even more randomly, both in direction and in speed; this gives any substance a much higher entropy as a gas than as a liquid or a solid (Figure 15-16). Thus the processes of melting, vaporization, and sublimation are always accompanied by an increase in entropy (Figure 15-17). The reverse processes of freezing, condensation, and deposition always correspond to a decrease in entropy.

For any substance, entropy increases in the order solid < liquid < gas.

Melting, vaporization, and sublimation always have $\Delta S_{system} > 0$.

Freezing, condensation, and deposition always have $\Delta S_{system} < 0$.

Temperature changes. As the temperature of any sample increases, its molecules have increased total kinetic energy; this higher energy can be dispersed among these molecules in more ways, increasing the entropy of the sample. Furthermore, the greater motion of the molecules (translational for gases and liquids, vibrational for solids) corresponds to a state

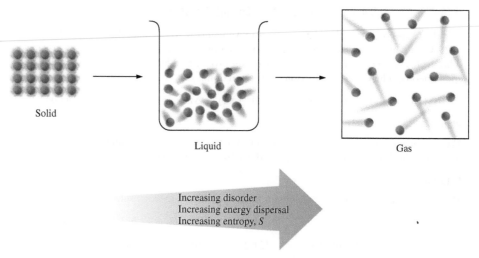

Figure 15-16 As a sample changes from solid to liquid to gas, its particles become increasingly more disordered, allowing greater dispersal of energy, so its entropy increases.

Figure 15-17 The vaporization of bromine, $Br_2(\ell) \rightarrow Br_2(g)$ (*left*) and the sublimation of iodine, $I_2(s) \rightarrow I_2(g)$ (*right*) both lead to an increase in disorder, so $\Delta S_{sys} > 0$ for each process. Which do you think results in the more positive ΔS? Carry out the calculation using values from Appendix K to check whether your prediction was correct.

of greater matter dispersal. The increased dispersal of matter and energy leads to the following result.

> The entropy of any sample increases as its temperature increases.

Volume changes. When the volume of a sample of gas increases, the molecules can occupy more positions, and hence are more randomly arranged (greater dispersal of matter). As pointed out in Section 15-13, the energy levels available to the molecules are closer together in the larger volume, leading to more ways to distribute the same total energy (greater dispersal of energy). Thus a gas has higher entropy at higher volume.

> The entropy of a gas increases as its volume increases.
> For an increase in gas volume, $\Delta S_{system} > 0$.

Mixing of substances, even without chemical reaction. Situations in which particles of more than one kind are "mixed up" are more disordered (greater dispersal of matter) and can exchange energy among both like and unlike particles (greater dispersal of energy). The entropy of the mixture is thus greater than that of the individual substances. This increase in entropy favors processes such as the mixing of gases (see Figure 15-10) and the dissolving of solid and liquid solutes in liquid solvents (Figures 15-18 and 15-19; see also Section 14-2). For example, when one mole of solid NaCl dissolves in water, $NaCl(s) \longrightarrow NaCl(aq)$, the entropy (Appendix K) increases from 72.4 J/mol · K to 115.5 J/mol · K, or $\Delta S^0 = +43.1$ J/mol · K. The term "mixing" can be interpreted rather liberally. For example, the reaction $H_2(g) + Cl_2(g) \longrightarrow 2HCl(g)$ has $\Delta S^0 > 0$; in the reactants, each atom is bonded to an identical atom, a less "mixed-up" situation than in the products, where unlike atoms are bonded together.

> Mixing of substances or dissolving a solid in a liquid causes an increase in entropy, $\Delta S_{system} > 0$.

Changes in the number of particles, as in the dissociation of a diatomic gas such as $F_2(g) \longrightarrow 2F(g)$. Any process in which the number of particles increases results in an increase in entropy, $\Delta S_{sys} > 0$. Values of ΔS^0 calculated for several reactions of this type are given in Table 15-5. As you can see, the ΔS^0 values for the dissociation process $X_2 \longrightarrow 2X$ are all similar for X = H, F, Cl, and N. Why is the value given in Table 15-5 so much larger for X = Br? This process starts with *liquid* Br$_2$. The total process $Br_2(\ell) \longrightarrow 2Br(g)$, for which $\Delta S^0 = 197.6$ J/mol · K, can be treated as the result of *two* processes. The first of these is *vaporization,* $Br_2(\ell) \longrightarrow Br_2(g)$, for which $\Delta S^0 = 93.2$ J/mol · K. The second step is the dissociation of gaseous bromine, $Br_2(g) \longrightarrow 2Br(g)$, for

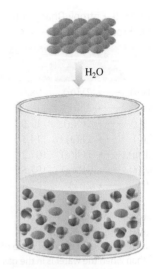

Figure 15-18 As particles leave a crystal to go into solution, they become more disordered and energy dispersal increases. This favors dissolution of the solid.

Figure 15-19 When water, H$_2$O, and propyl alcohol, CH$_3$CH$_2$CH$_2$OH (*left*) are mixed to form a solution (*right*), disorder increases. $\Delta S > 0$ for the mixing of any two molecular substances.

Table 15-5 Entropy Changes for Some Processes $X_2 \longrightarrow 2X$

Reaction	ΔS^0 (J/mol · K)
$H_2(g) \longrightarrow 2H(g)$	98.0
$N_2(g) \longrightarrow 2N(g)$	114.9
$O_2(g) \longrightarrow 2O(g)$	117.0
$F_2(g) \longrightarrow 2F(g)$	114.5
$Cl_2(g) \longrightarrow 2Cl(g)$	107.2
$Br_2(\ell) \longrightarrow 2Br(g)$	197.6
$I_2(s) \longrightarrow 2I(g)$	245.3

▶ Do you think that the reaction $2H_2(g) + O_2(g) \longrightarrow 2H_2O(\ell)$ would have a higher or lower value of ΔS^0 than when the water is in the gas phase? Confirm by calculation.

which $\Delta S^0 = 104.4$ J/mol · K; this entropy increase is about the same as for the other processes that involve *only* dissociation of a gaseous diatomic species. Can you rationalize the even higher value given in the table for the process $I_2(s) \longrightarrow 2I(g)$?

Increasing the number of particles causes an increase in entropy, $\Delta S_{system} > 0$.

Changes in the number of moles of gaseous substances. Processes that result in an increase in the number of moles of gaseous substances have $\Delta S_{sys} > 0$. Example 15-16 illustrates this. There are no gaseous reactants, but the products include five moles of gas. Conversely, we would predict that the process $2H_2(g) + O_2(g) \longrightarrow 2H_2O(g)$ has a negative ΔS^0 value; here, three moles of gas is consumed while only two moles is produced, for a net decrease in the number of moles in the gas phase. You should be able to calculate the value of ΔS^0 for this reaction from the values in Appendix K.

An increase in the number of moles of gaseous substances causes an increase in entropy, $\Delta S_{system} > 0$.

The following guidelines for absolute entropy (standard molar entropy) for individual substances can also be helpful in predicting entropy changes.

Molecular size and complexity. For molecules with similar formulas, absolute entropies usually increase with increasing molecular weight. For instance, the entropies of the gaseous hydrogen halide increase in the order HF (173.7 J/mol · K) < HCl (186.8 J/mol · K) < HBr (198.59 J/mol · K) < HI (206.5 J/mol · K). In more complicated molecular structures, there are more ways for the atoms to move about in three-dimensional space, so there is greater entropy. This trend is especially strong for a series of related compounds. For instance, the absolute entropy of liquid hydrogen peroxide, H_2O_2 (109.6 J/mol · K), is greater than that of liquid water, H_2O (69.91 J/mol · K). Gaseous PCl_5 has a greater absolute entropy (353 J/mol · K) than gaseous PCl_3 (311.7 J/mol · K). Note that molecular complexity and molecular weight play an increasingly important role in the entropy when comparing molecules. For example, gases usually have the highest entropy values when comparing molecules of similar molecular weight. But a solid with a high molecular weight and complexity can have a higher entropy than a lighter gas. Solid sucrose ($C_{12}H_{22}O_{11}$; MW = 342 g/mol) has S^0 = 392 J/mol, whereas $PCl_5(g)$ (MW = 208 g/mol) has a lower entropy of 353 J/mol.

Ionic compounds with similar formulas but different charges. Stronger ionic attractions (due to higher charges or closer approach) hold the ions more tightly in their crystal positions, so they vibrate less, leading to lower entropy. For instance, CaS and KCl have similar formula weights; but CaS (2+ and 2− ionic charges) has stronger ionic attractive forces than KCl (1+, 1− charges), consistent with their relative entropies CaS (56.5 J/mol · K) < KCl (82.6 J/mol · K).

Problem-Solving Tip

Of the factors just listed, *changes in number of particles* and *changes in number of moles of gas* are usually the most important factors in predicting the sign of an entropy change for a reaction.

EXAMPLE 15-17   Predicting the Sign of ΔS^0 for a Chemical Reaction

Without doing a calculation, predict whether the entropy change will be positive or negative when each reaction occurs in the direction it is written.

(a) $C_2H_6(g) + \frac{7}{2}O_2(g) \longrightarrow 3H_2O(g) + 2CO_2(g)$

(b) $3C_2H_2(g) \longrightarrow C_6H_6(\ell)$

(c) $C_6H_{12}O_6(s) + 6O_2(g) \longrightarrow 6CO_2(g) + 6H_2O(\ell)$

Plan

We use the qualitative guidelines described above to predict whether entropy increases or decreases.

Solution

(a) In this reaction, 9/2 moles (4.5 moles) of gaseous reactants produce 5 moles of gaseous products. This increase in number of moles of gas leads to an increase in entropy, $\Delta S^0 > 0$.

(b) Here 3 moles of gaseous reactants produce only liquid product. This decrease in number of moles of gas leads to a decrease in entropy, $\Delta S^0 < 0$.

(c) In this reaction, there are 6 moles of gaseous reactants and the same number of moles of gaseous products, so the prediction cannot be made on this basis. One of the reactants, $C_6H_{12}O_6$, is a solid. The total entropy of one mole of solid and six moles of gas (reactants) is less than the total entropy of six moles of gas and six moles of liquid (products), so we predict an increase in entropy, $\Delta S^0 > 0$.

You should now work Exercise 100.

> **S TOP & THINK**
> You can apply this kind of reasoning to help you judge whether a calculated entropy change has the correct sign.

EXAMPLE 15-18 **Predicting the Order of Absolute Entropies** MOLECULAR REASONING

Rank each group of substances in order of increasing absolute entropy at 25°C. Give reasons for your ranking.

(a) $Hg(\ell)$, $Hg(s)$, $Hg(g)$

(b) $C_2H_6(g)$, $CH_4(g)$, $C_3H_8(g)$

(c) $CaS(s)$, $CaO(s)$

Plan

We use the qualitative comparisons of this section to rank the absolute entropies within each group.

Solution

(a) For any substance, entropy increases in the order solid < liquid < gas, so we predict the absolute entropies to increase as $Hg(s) < Hg(\ell) < Hg(g)$.

(b) Entropy increases with molecular complexity, so we predict the absolute entropies to increase in the order $CH_4(g) < C_2H_6(g) < C_3H_8(g)$.

(c) Both of these ionic compounds contain 2+ and 2− ions, so charge differences alone will not allow us to rank them. But O^{2-} ions are smaller than S^{2-} ions, so they are held closer to the Ca^{2+} ions, giving a stronger attraction in CaO and hence lower entropy. We expect the absolute entropies to increase in the order $CaO(s) < CaS(s)$.

▶ You can check the rankings in Example 15-18 by comparing the S^0 values listed in Appendix K.

You should now work Exercise 98.

15-15 The Second Law of Thermodynamics

We now know that two factors determine whether a reaction is spontaneous under a given set of conditions. The effect of the first factor, the enthalpy change, is that spontaneity is favored (but not required) by exothermicity, and nonspontaneity is favored (but not required) by endothermicity. The effect of the other factor is summarized by the **Second Law of Thermodynamics**.

In any spontaneous change, entropy of the universe increases, $\Delta S_{universe} > 0$.

The Second Law of Thermodynamics is based on our experiences, as seen by the following examples. Gases mix spontaneously. When a drop of food coloring is added to a glass of water, it diffuses until a homogeneously colored solution results. When a truck is driven down the street, it consumes fuel and oxygen, producing carbon dioxide, water vapor, and other emitted substances.

Time

Charles D. Winters

▶ We abbreviate these subscripts as follows: system = sys, surroundings = surr, and universe = univ.

SURROUNDINGS

$S_{surr} \uparrow$ so $\Delta S_{surr} > 0$

Heat   SYSTEM $S_{sys} \downarrow$ so $\Delta S_{sys} < 0$ Heat

Ⓐ Freezing below mp

SURROUNDINGS

$S_{surr} \downarrow$ so $\Delta S_{surr} < 0$

Heat   SYSTEM $S_{sys} \uparrow$ so $\Delta S_{sys} > 0$ Heat

Ⓑ Melting above mp

Figure 15-20 A schematic representation of heat flow and entropy changes for (a) freezing and (b) melting of a pure substance.

The reverse of any spontaneous change is nonspontaneous, because if it did occur, the universe would tend toward a state of lower entropy (greater order, more concentration of energy). This is contrary to our experience. A gas mixture does not spontaneously separate into its components. A colored solution does not spontaneously concentrate all of its color in a small volume. A truck cannot be driven along the street, even in reverse gear, so that it sucks up CO_2, water vapor, and other substances to produce fuel and oxygen.

If the entropy of a system increases during a process, the spontaneity of the process is favored but not required. The Second Law of Thermodynamics says that the entropy of the *universe* (but not necessarily the system) increases during a spontaneous process, that is,

$$\Delta S_{universe} = \Delta S_{system} + \Delta S_{surroundings} > 0 \qquad \text{(spontaneous process)}$$

Of the two ideal gas samples in Figure 15-10, the more ordered arrangement (Figure 15-10a) has lower entropy than the randomly mixed arrangement with the same volume (Figure 15-10b). Because these ideal gas samples mix without absorbing or releasing heat and without a change in total volume, they do not interact with the surroundings, so the entropy of the surroundings does not change. In this case

$$\Delta S_{universe} = \Delta S_{system}$$

If we open the stopcock between the two bulbs in Figure 15-10a, we expect the gases to mix spontaneously, with an increase in the disorder of the system, that is, ΔS_{system} is positive.

$$\text{unmixed gases} \longrightarrow \text{mixed gases} \qquad \Delta S_{universe} = \Delta S_{system} > 0$$

We do not expect the more homogeneous sample in Figure 15-10b to spontaneously "unmix" to give the arrangement in Figure 15-10a (which would correspond to a decrease in ΔS_{system}).

$$\text{mixed gases} \longrightarrow \text{unmixed gases} \qquad \Delta S_{universe} = \Delta S_{system} < 0$$

The entropy of a system can decrease during a spontaneous process or increase during a nonspontaneous process, depending on the accompanying ΔS_{surr}. If ΔS_{sys} is negative (decrease in disorder), then ΔS_{univ} may still be positive (overall increase in disorder) *if* ΔS_{surr} is more positive than ΔS_{sys} is negative. A refrigerator provides an illustration. It removes heat from inside the box (the system) and ejects that heat, *plus* the heat generated by the compressor, into the room (the surroundings). The entropy of the system decreases because the air molecules inside the box move more slowly. But the increase in the entropy of the surroundings more than makes up for that, so the entropy of the universe (refrigerator + room) increases.

Similarly, if ΔS_{sys} is positive but ΔS_{surr} is even more negative, then ΔS_{univ} is still negative. Such a process will be nonspontaneous.

Let's consider the entropy changes that occur when a liquid solidifies at a temperature *below* its freezing (melting) point (Figure 15-20a). ΔS_{sys} is negative because a solid forms from its liquid, yet we know that this is a spontaneous process. A liquid releases heat to its surroundings (atmosphere) as it crystallizes. The released heat increases the motion (more disorder, greater dispersal of matter) of the molecules of the surroundings, so ΔS_{surr} is positive. As the temperature decreases, the ΔS_{surr} contribution becomes more important. When the temperature is low enough (below the freezing point), the positive ΔS_{surr} outweighs the negative ΔS_{sys}. Then ΔS_{univ} becomes positive, and the freezing process becomes spontaneous.

The situation is reversed when a liquid is boiled or a solid is melted (see Figure 15-20b). For example, at temperatures above its melting point, a solid spontaneously melts, the system becomes more disordered (greater dispersal of matter), and ΔS_{sys} is positive. The heat absorbed when the solid (system) melts comes from its surroundings. This decreases the motion of the molecules of the surroundings. Thus, ΔS_{surr} is negative (the surroundings become less disordered). The positive ΔS_{sys} is greater in magnitude than the negative ΔS_{surr}, however, so ΔS_{univ} is positive and the process is spontaneous.

Above the melting point, ΔS_{univ} is positive for melting. Below the melting point, ΔS_{univ} is positive for freezing. At the melting point, ΔS_{surr} is equal in magnitude and opposite in

Table 15-6 Entropy Effects Associated with Melting and Freezing

Change	Temperature	Sign of		(Magnitude of ΔS_{sys}) Compared with (Magnitude of ΔS_{surr})	$\Delta S_{univ} = \Delta S_{sys} + \Delta S_{surr}$	Spontaneity
		ΔS_{sys}	ΔS_{surr}			
1. Melting (solid → liquid)	> mp	+	−	>	> 0	Spontaneous
	= mp	+	−	=	= 0	Equilibrium
	< mp	+	−	<	< 0	Nonspontaneous
2. Freezing (liquid → solid)	> mp	−	+	>	< 0	Nonspontaneous
	= mp	−	+	=	= 0	Equilibrium
	< mp	−	+	<	> 0	Spontaneous

sign to ΔS_{sys}. Then ΔS_{univ} is zero for both melting and freezing; the system is at *equilibrium*. Table 15-6 lists the entropy effects for these changes of physical state.

We have said that ΔS_{univ} is positive for all spontaneous (product-favored) processes. Unfortunately, it is not possible to make direct measurements of ΔS_{univ}. Consequently, entropy changes accompanying physical and chemical changes are reported in terms of ΔS_{sys}. The subscript "sys" for system can be replaced with rxn if we wish to indicate a chemical reaction. The symbol ΔS_{rxn} refers to the change in entropy of a reacting system, just as ΔH_{rxn} refers to the change in enthalpy of the reacting system.

▶ Can you develop a comparable table for boiling (liquid → gas) and condensation (gas → liquid)? (Study Table 15-6 carefully.)

15-16 Free Energy Change, ΔG, and Spontaneity

Energy is the capacity to do work. If heat is released in a chemical reaction (ΔH is negative), *some* of the heat energy may be converted into useful work. If ΔS is negative, some of the energy must be expended to increase the order of the system. If the system becomes more disordered ($\Delta S > 0$), however, more useful energy becomes available than indicated by ΔH alone. J. Willard Gibbs (1839–1903), a prominent 19th-century American professor

Ⓐ The entropy of an organism decreases (unfavorable) when new cells are formed. The energy to sustain animal life is provided by the metabolism of food. This energy is released when the chemical bonds in the food are broken. Exhalation of gases and excretion of waste materials increase the entropy of the surroundings enough so that the entropy of the universe increases and the overall process can occur.

Ⓑ Stored chemical energy can later be transformed by the organism to the mechanical energy for muscle contraction, to the electrical energy for brain function, or to another needed form.

▶ Entropy plays an important role in the energy transfer in cell growth and maintenance in all living organisms.

Dennis Drenner

of mathematics and physics, formulated the relationship between enthalpy and entropy in terms of another function that we now call the **Gibbs free energy, G.** It is defined as

$$G = H - TS$$

The **Gibbs free energy change, ΔG,** at constant temperature and pressure, is

▶ This is often called simply the *Gibbs energy change* or the *free energy change*.

$$\Delta G = \Delta H - T \Delta S \qquad \text{(constant } T \text{ and } P)$$

The amount by which the Gibbs free energy decreases is the *maximum useful energy* obtainable in the form of work from a given process at constant temperature and pressure. It is also the *indicator of spontaneity of a reaction or physical change* at constant T and P. If there is a net decrease of free energy, ΔG is negative and the process is spontaneous (product-favored). We see from the equation that ΔG becomes more negative as (1) ΔH becomes more negative (the process gives off more heat) and (2) ΔS becomes more positive (the process results in greater disorder); *both* factors are favorable so the process *must be* spontaneous (product-favored). If there is a net increase in free energy of the system during a process, ΔG is positive and the process is nonspontaneous (reactant-favored). This means that the reverse process is spontaneous under the given conditions. When $\Delta G = 0$, there is no net transfer of free energy; both the forward and reverse processes are equally favorable. Thus, $\Delta G = 0$ describes a system at *equilibrium*.

The relationship between ΔG and spontaneity may be summarized as follows.

ΔG	Spontaneity of Reaction (constant T and P)
ΔG is positive	Reaction is nonspontaneous (reactant-favored)
ΔG is zero	System is at equilibrium
ΔG is negative	Reaction is spontaneous (product-favored)

—	0	+
$\Delta G < 0$		$\Delta G > 0$
Reaction is spontaneous		Reaction is not spontaneous
Product-favored reaction		Reactant-favored reaction
Forward reaction is favored		Reverse reaction is favored

The free energy content of a system depends on temperature and pressure (and, for mixtures, on concentrations). The value of ΔG for a process depends on the states and the concentrations of the various substances involved. It also depends strongly on temperature, because the equation $\Delta G = \Delta H - T \Delta S$ includes temperature. Just as for other thermodynamic variables, we choose some set of conditions as a standard state reference. The standard state for ΔG^0 is the same as for ΔH^0; 1 atm and the specified temperature, usually 25°C (298 K). Values of standard molar free energy of formation, ΔG^0_f, for many substances are tabulated in Appendix K. For *elements* in their standard states, $\Delta G^0_f = 0$. The values of ΔG^0_f may be used to calculate the standard free energy change of a reaction *at 298 K* by using the following relationship.

$$\Delta G^0_{rxn} = \Sigma n \, \Delta G^0_{f \, products} - \Sigma n \, \Delta G^0_{f \, reactants} \qquad \text{(1 atm and 298 K } only)$$

▶ We want to know which are more stable *at standard conditions*—the *reactants* or the *products*.

The value of ΔG^0_{rxn} allows us to predict the spontaneity of a hypothetical reaction in which the numbers of moles of reactants shown in the balanced equation, all at standard conditions, are *completely* converted to the numbers of moles of products shown in the balanced equation, all at standard conditions.

We must remember that it is ΔG, and not ΔG^0, that is the general criterion for spontaneity. ΔG depends on concentrations of reactants and products in the mixture. For most reactions, there is an *equilibrium mixture* of reactant and product concentrations that is more stable than either all reactants or all products. In Chapter 17 we will study the concept of equilibrium and see how to find ΔG for mixtures.

EXAMPLE 15-19 Spontaneity of Reaction

Diatomic nitrogen and oxygen molecules make up about 99% of all the molecules in reasonably "unpolluted" dry air. Evaluate ΔG^0 for the following reaction at 298 K, using ΔG_f^0 values from Appendix K. Is the reaction spontaneous?

$$N_2(g) + O_2(g) \longrightarrow 2NO(g) \qquad \text{(nitrogen oxide)}$$

Plan

The reaction conditions are 1 atm and 298 K, so we can use the tabulated values of ΔG_f^0 for each substance in Appendix K to evaluate ΔG_{rxn}^0 in the preceding equation. The treatment of units for calculation of ΔG^0 is the same as that for ΔH^0 in Example 15-9.

Solution

We obtain the following values of ΔG_f^0 from Appendix K:

	$N_2(g)$	$O_2(g)$	$NO(g)$
ΔG_f^0, kJ/mol:	0	0	86.57

$$\Delta G_{rxn}^0 = \Sigma\, n\, \Delta G_f^0 \text{ products} - \Sigma\, n\, \Delta G_f^0 \text{ reactants}$$

$$= 2\ \Delta G_{f\ NO(g)}^0 \quad - [\,\Delta G_{f\ N_2(g)}^0 + \Delta G_{f\ O_2(g)}^0\,]$$

$$= 2(86.57) \quad - [0 + 0]$$

$$\Delta G_{rxn}^0 = +173.1 \text{ kJ/mol rxn} \qquad \text{for the reaction as written}$$

Because ΔG^0 is positive, the reaction is nonspontaneous at 298 K under standard state conditions.

You should now work Exercise 111.

▶ For the reverse reaction at 298 K, $\Delta G_{rxn}^0 = -173.1$ kJ/mol rxn. So the reverse reaction is product-favored, although it is very slow at room temperature. The NO formed in automobile engines is oxidized to even more harmful NO_2 much more rapidly than it decomposes to N_2 and O_2. Thermodynamic spontaneity does not guarantee that a process occurs at an observable rate. The oxides of nitrogen in the atmosphere represent a major environmental problem.

The value of ΔG^0 can also be calculated by the equation

$$\Delta G^0 = \Delta H^0 - T\,\Delta S^0 \qquad \text{(constant } T \text{ and } P\text{)}$$

Strictly, this last equation applies to standard conditions; however, ΔH^0 and ΔS^0 often do not vary much with temperature, so the equation can often be used to *estimate* free energy changes at other temperatures.

Problem-Solving Tip Some Common Pitfalls in Calculating ΔG_{rxn}^0

Be careful of these points when you carry out calculations that involve ΔG^0:

1. The calculation of ΔG_{rxn}^0 from tabulated values of ΔG_f^0 is valid *only* for the reaction at 25°C (298 K) and one atmosphere.

2. Calculations with the equation $\Delta G^0 = \Delta H^0 - T\,\Delta S^0$ must be carried out with the temperature in kelvins.

3. The energy term in ΔS^0 is usually in joules, whereas that in ΔH^0 is usually in kilojoules; remember to convert one of these so that units are consistent before you combine them.

EXAMPLE 15-20 Spontaneity of Reaction

Make the same determination as in Example 15-19, using heats of formation and absolute entropies rather than free energies of formation.

Plan

First we calculate ΔH^0_{rxn} and ΔS^0_{rxn}. We use the relationship $\Delta G^0 = \Delta H^0 - T\,\Delta S^0$ to evaluate the free energy change under standard state conditions at 298 K.

Solution

The values we obtain from Appendix K are

	$N_2(g)$	$O_2(g)$	$NO(g)$
ΔH^0_f, kJ/mol:	0	0	90.25
S^0, J/mol · K:	191.5	205.0	210.7

$$\Delta H^0_{rxn} = \Sigma\, n\, \Delta H^0_{f\,products} \quad - \Sigma\, n\, \Delta H^0_{f\,reactants}$$

$$= 2\,\Delta H^0_{f\,NO(g)} \quad - [\Delta H^0_{f\,N_2(g)} + \Delta H^0_{f\,O_2(g)}]$$

$$= [2(90.25) \quad - (0 + 0)] = 180.5 \text{ kJ/mol}$$

$$\Delta S^0_{rxn} = \Sigma\, n\, S^0_{products} \quad - \Sigma\, n\, S^0_{reactants}$$

$$= 2S^0_{NO(g)} \quad - [S^0_{N_2(g)} + S^0_{O_2(g)}]$$

$$= [2(210.7) \quad - (191.5 + 205.0)] = 24.9 \text{ J/mol · K} = 0.0249 \text{ kJ/mol · K}$$

Now we use the relationship $\Delta G^0 = \Delta H^0 - T\,\Delta S^0$, with $T = 298$ K, to evaluate the free energy change under standard state conditions at 298 K.

$$\Delta G^0_{rxn} = \Delta H^0_{rxn} \quad - T\,\Delta S^0_{rxn}$$

$$= 180.5 \text{ kJ/mol} - (298 \text{ K})(0.0249 \text{ kJ/mol · K})$$

$$= 180.5 \text{ kJ/mol} - 7.42 \text{ kJ/mol}$$

$$\Delta G^0_{rxn} = +173.1 \text{ kJ/mol rxn}, \text{ the same value obtained in Example 15-19.}$$

You should now work Exercise 113.

15-17 The Temperature Dependence of Spontaneity

The methods developed in Section 15-16 can also be used to estimate the temperature at which a process is in equilibrium. When a system is at equilibrium, $\Delta G = 0$. Thus,

$$\Delta G_{rxn} = \Delta H_{rxn} - T\,\Delta S_{rxn} \qquad \text{or} \qquad 0 = \Delta H_{rxn} - T\,\Delta S_{rxn}$$

so

$$\Delta H_{rxn} = T\,\Delta S_{rxn} \qquad \text{or} \qquad T = \frac{\Delta H_{rxn}}{\Delta S_{rxn}} \qquad \text{(at equilibrium)}$$

EXAMPLE 15-21 Estimation of Boiling Point

Use the thermodynamic data in Appendix K to estimate the normal boiling point of bromine, Br_2. Assume that ΔH and ΔS do not change with temperature.

Plan

The process we must consider is

$$Br_2(\ell) \longrightarrow Br_2(g)$$

▶ Actually, both ΔH^0_{rxn} and ΔS^0_{rxn} vary with temperature, but usually not enough to introduce significant errors for modest temperature changes. The value of ΔG^0_{rxn}, on the other hand, is strongly dependent on the temperature.

By definition, the normal boiling point of a liquid is the temperature at which pure liquid and pure gas coexist in equilibrium at 1 atm. Therefore, $\Delta G = 0$ for the process as written. We assume that $\Delta H_{rxn} = \Delta H^0_{rxn}$ and $\Delta S_{rxn} = \Delta S^0_{rxn}$. We can evaluate these two quantities, substitute them in the relationship $\Delta G = \Delta H - T \Delta S$, and then solve for the value of T that makes $\Delta G = 0$.

Solution

The required values (see Appendix K) are as follows:

	$Br_2(\ell)$	$Br_2(g)$
ΔH^0_f, kJ/mol:	0	30.91
S^0, J/mol · K:	152.2	245.4

$$\Delta H_{rxn} = \Delta H^0_{f\,Br_2(g)} \quad - \Delta H^0_{f\,Br_2(\ell)}$$

$$= 30.91 \quad - 0 = 30.91 \text{ kJ/mol}$$

$$\Delta S_{rxn} = S^0_{Br_2(g)} \quad - S^0_{Br_2(\ell)}$$

$$= (245.4 \quad - 152.2) = 93.2 \text{ J/mol} \cdot \text{K} = 0.0932 \text{ kJ/mol} \cdot \text{K}$$

We can now solve for the temperature at which the system is in equilibrium, that is, the boiling point of Br_2.

$$\Delta G_{rxn} = \Delta H_{rxn} - T \Delta S_{rxn} = 0 \qquad \text{so} \qquad \Delta H_{rxn} = T \Delta S_{rxn}$$

$$T = \frac{\Delta H_{rxn}}{\Delta S_{rxn}} = \frac{30.91 \text{ kJ/mol}}{0.0932 \text{ kJ/mol} \cdot \text{K}} = 332 \text{ K } (59°C)$$

This is the temperature at which the reactant (liquid) is in equilibrium with the product (gas), that is, the boiling point of Br_2. The experimentally measured value is 58.78°C.

You should now work Exercise 122.

> **S TOP & THINK**
> Remember to use the same energy units in ΔS^0 and ΔH^0.

The free energy change and spontaneity of a reaction depend on both enthalpy and entropy changes. Both ΔH and ΔS may be either positive or negative, so we can group reactions in four classes with respect to spontaneity (Figure 15-21).

	$\Delta G = \Delta H - T \Delta S$		(constant temperature and pressure)
1.	$\Delta H = -$ (favorable)	$\Delta S = +$ (favorable)	Reactions are spontaneous (product-favored) at all temperatures
2.	$\Delta H = -$ (favorable)	$\Delta S = -$ (unfavorable)	Reactions become spontaneous (product-favored) below a definite temperature
3.	$\Delta H = +$ (unfavorable)	$\Delta S = +$ (favorable)	Reactions become spontaneous (product-favored) above a definite temperature
4.	$\Delta H = +$ (unfavorable)	$\Delta S = -$ (unfavorable)	Reactions are nonspontaneous (reactant-favored) at all temperatures

When ΔH and ΔS have opposite signs (classes 1 and 4), they act in the same direction, so the direction of spontaneous change does not depend on temperature. When ΔH and ΔS have the same signs (classes 2 and 3), their effects oppose one another, so changes in temperature can cause one factor or the other to dominate, and spontaneity depends on temperature. For class 2, decreasing the temperature decreases the importance of the *unfavorable* $T \Delta S$ term, so the reaction becomes spontaneous at lower temperatures. For class 3, increasing the temperature increases the importance of the *favorable* $T \Delta S$ term, so the reaction becomes spontaneous at higher temperatures.

The temperature at which $\Delta G^0_{rxn} = 0$ is the temperature limit of spontaneity. The sign of ΔS^0_{rxn} tells us whether the reaction is spontaneous *below* or *above* this limit (Table 15-7). We can estimate the temperature range over which a chemical reaction in class 2 or 3 is spontaneous by evaluating ΔH^0_{rxn} and ΔS^0_{rxn} from tabulated data.

Figure 15-21 A graphical representation of the dependence of ΔG and spontaneity on temperature for each of the four classes of reactions listed in the text and in Table 15-7.

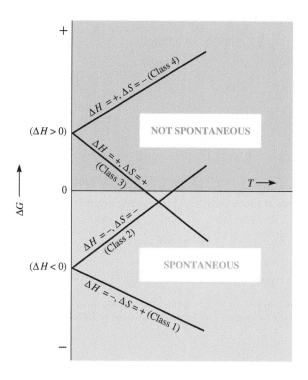

Table 15-7 Thermodynamic Classes of Reactions

Class	Examples	ΔH (kJ/mol)	ΔS (J/mol · K)	Temperature Range of Spontaneity
1	$2H_2O_2(\ell) \longrightarrow 2H_2O(\ell) + O_2(g)$	-196	$+126$	All temperatures
	$H_2(g) + Br_2(\ell) \longrightarrow 2HBr(g)$	-72.8	$+114$	All temperatures
2	$NH_3(g) + HCl(g) \longrightarrow NH_4Cl(s)$	-176	-285	Lower temperatures (< 619 K)
	$2H_2S(g) + SO_2(g) \longrightarrow 3S(s) + 2H_2O(\ell)$	-233	-424	Lower temperatures (< 550 K)
3	$NH_4Cl(s) \longrightarrow NH_3(g) + HCl(g)$	$+176$	$+285$	Higher temperatures (> 619 K)
	$CCl_4(\ell) \longrightarrow C(graphite) + 2Cl_2(g)$	$+135$	$+235$	Higher temperatures (> 574 K)
4	$2H_2O(\ell) + O_2(g) \longrightarrow 2H_2O_2(\ell)$	$+196$	-126	Nonspontaneous, all temperatures
	$3O_2(g) \longrightarrow 2O_3(g)$	$+285$	-137	Nonspontaneous, all temperatures

EXAMPLE 15-22 Temperature Range of Spontaneity

Mercury(II) sulfide is a dark red mineral called cinnabar. Metallic mercury is obtained by roasting the sulfide in a limited amount of air. Estimate the temperature range in which the reaction is product-favored.

$$HgS(s) + O_2(g) \longrightarrow Hg(\ell) + SO_2(g)$$

Plan

We evaluate ΔH^0_{rxn} and ΔS^0_{rxn} and assume that these values are independent of temperature. Each of these values allows us to assess its contribution to spontaneity.

Solution

From Appendix K:

	HgS(s)	O_2	Hg(ℓ)	SO_2(g)
ΔH_f^0, kJ/mol:	−58.2	0	0	−296.8
S^0, J/mol · K:	82.4	205.0	76.0	248.1

$$\Delta H_{rxn}^0 = \Delta H_{f\ Hg(\ell)}^0 + \Delta H_{f\ SO_2(g)}^0 - [\Delta H_{f\ HgS(s)}^0 + \Delta H_{f\ O_2(g)}^0]$$

$$= 0 - 296.8 \qquad - [-58.2 + 0] = -238.6 \text{ kJ/mol}$$

$$\Delta S_{rxn}^0 = S_{Hg(\ell)}^0 + S_{SO_2(g)}^0 - [S_{HgS(s)}^0 + S_{O_2(g)}^0]$$

$$= 76.02 + 248.1 - [82.4 + 205.0] = +36.7 \text{ J/mol} \cdot \text{K}$$

Heating red HgS in air produces liquid Hg. Gaseous SO_2 escapes. Cinnabar, an important ore of mercury, contains HgS.

When ΔH_{rxn}^0 is negative (favorable to spontaneity) and ΔS_{rxn}^0 is positive (favorable to spontaneity), the reaction must be spontaneous (product-favored) at all temperatures; no further calculation is necessary. The reverse reaction is, therefore, nonspontaneous at all temperatures.

The fact that a reaction is spontaneous at all temperatures does not mean that the reaction occurs rapidly enough to be useful at all temperatures. As a matter of fact, Hg(ℓ) can be obtained from HgS(s) by this reaction at a reasonable rate only at high temperatures.

S TOP & THINK

When ΔH^0 and ΔS^0 are both *favorable*, we know without further calculation that ΔG^0 must be *negative* and the reaction must be *spontaneous (product-favored)* at all temperatures. When ΔH^0 and ΔS^0 are both *unfavorable*, we know without further calculation that ΔG^0 must be *positive* and the reaction must be *nonspontaneous (reactant-favored)* at all temperatures.

EXAMPLE 15-23 Temperature Range of Spontaneity

Estimate the temperature range for which the following reaction is product-favored.

$$SiO_2(s) + 2C(\text{graphite}) + 2Cl_2(g) \longrightarrow SiCl_4(g) + 2CO(g)$$

Plan

When we proceed as in Example 15-22, we find that ΔS_{rxn}^0 is favorable to spontaneity, whereas ΔH_{rxn}^0 is unfavorable. Thus, we know that the reaction becomes product-favored *above* some temperature. We can set ΔG^0 equal to zero in the equation $\Delta G^0 = \Delta H^0 - T\Delta S^0$ and solve for the temperature at which the system is *at equilibrium*. This will represent the temperature above which the reaction would be product-favored.

Solution

From Appendix K:

	SiO_2(s)	C(graphite)	Cl_2(g)	$SiCl_4$(g)	CO(g)
ΔH_f^0, kJ/mol:	−910.9	0	0	−657.0	−110.5
S^0, J/mol · K:	41.84	5.740	223.0	330.6	197.6

$$\Delta H_{rxn}^0 = [\Delta H_{f\ SiCl_4(g)}^0 + 2\ \Delta H_{f\ CO(g)}^0] - [\Delta H_{f\ SiO_2(s)}^0 + 2\Delta H_{f\ C(\text{graphite})}^0 + 2\Delta H_{f\ Cl_2(g)}^0]$$

$$= [(-657.0) + 2(-110.5)] - [(-910.9) + 2(0) \qquad + 2(0)]$$

$$= +32.9 \text{ kJ/mol (unfavorable to spontaneity)}$$

$$\Delta S_{rxn}^0 = \Delta S_{SiCl_4(g)}^0 + 2S_{CO(g)}^0 - [\Delta S_{SiO_2(s)}^0 + 2S_{C(\text{graphite})}^0 + 2S_{Cl_2(g)}^0]$$

$$= [330.6 + 2(197.6)] - [41.84 + 2(5.740) + 2(223.0)]$$

$$= 226.5 \text{ J/mol} \cdot \text{K} = 0.2265 \text{ kJ/mol} \cdot \text{K (favorable to spontaneity)}$$

At the temperature at which $\Delta G^0 = 0$, neither the forward nor the reverse reaction is favored and the system is at equilibrium. Let's find this temperature.

$$0 = \Delta G^0 = \Delta H^0 - T\Delta S^0$$

$$\Delta H^0 = T\,\Delta S^0$$

$$T = \frac{\Delta H^0}{\Delta S^0} = \frac{+32.9 \text{ kJ/mol}}{+0.2265 \text{ kJ/mol} \cdot \text{K}} = 145 \text{ K}$$

At temperatures above 145 K, the $T\,\Delta S^0$ term would be greater ($-T\,\Delta S^0$ would be more negative) than the ΔH^0 term, which would make ΔG^0 negative; so the reaction would be spontaneous (produced–favored) above 145 K. At temperatures below 145 K, the $T\,\Delta S^0$ term would be smaller than the ΔH^0 term, which would make ΔG^0 positive; so the reaction would be nonspontaneous below 145 K.

However, 145 K ($-128°C$) is a very low temperature. For all practical purposes, the reaction is product-favored at all but very low temperatures. In practice, it is carried out at 800°C to 1000°C because of the greater reaction rate at these higher temperatures. This gives a useful and economical rate of production of $SiCl_4$, an important industrial chemical.

You should now work Exercises 116 and 120.

KEY TERMS

Absolute entropy (of a substance)   The entropy of a substance relative to its entropy in a perfectly ordered crystalline form at 0 K (where its entropy is zero).

Bomb calorimeter   A device used to measure the heat transfer between system and surroundings at constant volume.

Bond energy   The amount of energy necessary to break one mole of bonds in a gaseous substance to form gaseous products at the same temperature and pressure.

Calorimeter   A device used to measure the heat transfer that accompanies a physical or chemical change.

Dispersal of energy   The degree to which the total energy of a system can be distributed among its particles.

Dispersal of matter   The degree to which the particles of a sample can be distributed in space; also known as *disorder*.

Endothermic process   A process that absorbs heat.

Enthalpy change, ΔH   The quantity of heat transferred into or out of a system as it undergoes a chemical or physical change at constant temperature and pressure.

Entropy (S)   A thermodynamic state function that measures the dispersal of energy and the dispersal of matter (disorder) of a system.

Equilibrium   A state of dynamic balance in which the rates of forward and reverse processes (reactions) are equal; the state of a system when neither the forward nor the reverse process is thermodynamically favored.

Exothermic process   A process that gives off (releases) heat.

First Law of Thermodynamics   The total amount of energy in the universe is constant (also known as the Law of Conservation of Energy); energy is neither created nor destroyed in ordinary chemical reactions and physical changes.

Gibbs free energy (G)   The thermodynamic state function of a system that indicates the amount of energy available for the system to do useful work at constant T and P. It is defined as $G = H - TS$; also known as *free energy*.

Heat of formation   See *Standard molar enthalpy of formation*.

Hess's Law of Heat Summation   The enthalpy change for a reaction is the same whether it occurs in one step or a series of steps.

Internal energy (E)   All forms of energy associated with a specific amount of a substance.

Kinetic energy   The energy of motion. The kinetic energy of an object is equal to one half its mass times the square of its velocity.

Law of Conservation of Energy   Energy cannot be created or destroyed in a chemical reaction or in a physical change; it may be changed from one form to another; see *First Law of Thermodynamics*.

Mole of reaction (mol rxn)   The amount of reaction that corresponds to the number of moles of each substance shown in the balanced equation.

Nonspontaneous change   A change for which the collection of reactants is more stable than the collection of products under the given conditions; also known as *reactant-favored change*.

Potential energy   The energy that a system or object possesses by virtue of its position or composition.

Pressure–volume work   Work done by a gas when it expands against an external pressure, or work done on a system as gases are compressed or consumed in the presence of an external pressure.

Product-favored change   See *Spontaneous change*.

Reactant-favored change   See *Nonspontaneous change*.

Second Law of Thermodynamics   The universe tends toward a state of greater disorder in spontaneous processes.

Spontaneous change   A change for which the collection of products is more stable than the collection of reactants under the given conditions; also known as *product-favored change*.

Standard enthalpy change (ΔH^0)   The enthalpy change in which the number of moles of reactants specified in the balanced chemical equation, all at standard states, is converted completely to the specified number of moles of products, all at standard states.

Standard entropy change (ΔS^0)   The entropy change in which the number of moles of reactants specified in the balanced chemical equation, all at standard states, is converted completely to the specified number of moles of products, all at standard states.

Standard Gibbs free energy change (ΔG^0)   The indicator of spontaneity of a process at constant T and P. If ΔG^0 is negative, the process is product-favored (spontaneous); if ΔG^0 is positive, the reverse process is reactant-favored (nonspontaneous).

Standard molar enthalpy of formation (ΔH_f^0) (of a substance)   The enthalpy change for the formation of one mole of a substance in a specified state from its elements in their standard states; also known as *standard molar heat of formation* or just *heat of formation*.

Standard molar heat of formation   See *Standard molar enthalpy of formation*.

Standard molar entropy (S^0) (of a substance)   The absolute entropy of a substance in its standard state at 298 K.

Standard state (of a substance)   See *Thermodynamic standard state of a substance*.

State function   A variable that defines the state of a system; a function that is independent of the pathway by which a process occurs.

Surroundings   Everything in the environment of the system.

System   The substances of interest in a process; the part of the universe under investigation.

Thermochemical equation   A balanced chemical equation together with a designation of the corresponding value of ΔH_{rxn}. Sometimes used with changes in other thermodynamic quantities.

Thermochemistry   The observation, measurement, and prediction of energy changes for both physical changes and chemical reactions.

Thermodynamics   The study of the energy transfers accompanying physical and chemical processes.

Thermodynamic state of a system   A set of conditions that completely specifies all of the properties of the system.

Thermodynamic standard state of a substance   The most stable state of the substance at one atmosphere pressure and at some specific temperature (25°C unless otherwise specified); also known as *standard state (of a substance)*.

Third Law of Thermodynamics   The entropy of a hypothetical pure, perfect, crystalline substance at absolute zero temperature is zero.

Universe   The system plus the surroundings.

Work   The application of a force through a distance; for physical changes or chemical reactions at constant external pressure, the work done on the system is $-P\Delta V$; for chemical reactions that involve gases, the work done on the system can be expressed as $-(\Delta n)RT$.

EXERCISES

. . . .

🐾 **Molecular Reasoning** exercises

▲ **More Challenging** exercises

Blue-Numbered exercises are solved in the Student Solutions Manual

General Concepts

1. State precisely the meaning of each of the following terms. You may need to review Chapter 1 to refresh your memory concerning terms introduced there. (a) energy; (b) kinetic energy; (c) potential energy; (d) joule.

2. State precisely the meaning of each of the following terms. You may need to review Chapter 1 to refresh your memory about terms introduced there. (a) heat; (b) temperature; (c) system; (d) surroundings; (e) thermodynamic state of system; (f) work.

3. (a) Give an example of the conversion of heat into work. (b) Give an example of the conversion of work into heat.

4. (a) Give an example of heat being given off by the system. (b) Give an example of work being done on the system.

5. (a) Give an example of heat being added to the system. (b) Give an example of work being done by the system.

6. Distinguish between endothermic and exothermic processes. If we know that a reaction is endothermic in one direction, what can be said about the reaction in the reverse direction?

7. According to the First Law of Thermodynamics, the total amount of energy in the universe is constant. Why, then, do we say that we are experiencing a declining supply of energy?

8. 🐾 Use the First Law of Thermodynamics to describe what occurs when an incandescent light is turned on.

9. 🐾 Define enthalpy and give an example of a reaction that has a negative enthalpy change.

10. 🐾 Which of the following are examples of state functions? (a) your bank balance; (b) the mass of a candy bar; (c) your weight; (d) the heat lost by perspiration during a climb up a mountain along a fixed path.

11. 🐾 What is a state function? Would Hess's Law be a law if enthalpy were not a state function?

Enthalpy and Changes in Enthalpy

12. (a) Distinguish between ΔH and ΔH^0 for a reaction. (b) Distinguish between ΔH_{rxn}^0 and ΔH_f^0.

13. A reaction is characterized by $\Delta H_{rxn} = +450$ kJ/mol. Does the reaction mixture absorb heat from the surroundings or release heat to them?

14. For each of the following reactions, (a) does the enthalpy increase or decrease; (b) is $H_{reactants} > H_{products}$ or is $H_{products} > H_{reactants}$; (c) is ΔH positive or negative?
    (i) $Al_2O_3(s) \longrightarrow 2Al(s) + \frac{3}{2} O_2(g)$ (endothermic)
    (ii) $Sn(s) + Cl_2(g) \longrightarrow SnCl_2(s)$ (exothermic)

. . . .

🐾 **Molecular Reasoning** exercises   ▲ **More Challenging** exercises   Blue-Numbered exercises solved in Student Solutions Manual

15. (a) The combustion of 0.0222 g of isooctane vapor, $C_8H_{18}(g)$, at constant pressure raises the temperature of a calorimeter 0.400°C. The heat capacity of the calorimeter and water combined is 2.48 kJ/°C. Find the molar heat of combustion of gaseous isooctane.

$$C_8H_{18}(g) + 12\tfrac{1}{2}O_2(g) \longrightarrow 8CO_2(g) + 9H_2O(\ell)$$

(b) How many grams of $C_8H_{18}(g)$ must be burned to obtain 495 kJ of heat energy?

16. Methanol, CH_3OH, is an efficient fuel with a high octane rating.

$$CH_3OH(g) + \tfrac{3}{2}O_2(g) \longrightarrow CO_2(g) + 2H_2O(\ell)$$
$$\Delta H = -764 \text{ kJ/mol rxn}$$

(a) Find the heat evolved when 115.0 g $CH_3OH(g)$ burns in excess oxygen. (b) What mass of O_2 is consumed when 925 kJ of heat is given off?

17. How much heat is liberated when 0.113 mole of sodium reacts with excess water according to the following equation?

$$2Na(s) + 2H_2O(\ell) \longrightarrow H_2(g) + 2NaOH(aq)$$
$$\Delta H = -368 \text{ kJ/mol rxn}$$

18. What is ΔH for the reaction

$$PbO(s) + C(s) \longrightarrow Pb(s) + CO(g)$$

if 5.95 kJ must be supplied to convert 13.43 g lead(II) oxide to lead?

19. From the data in Appendix K, determine the form that represents the standard state for each of the following elements: iodine, oxygen, sulfur.

20. Why is the standard molar enthalpy of formation, ΔH_f^0 for liquid water different than ΔH_f^0 for water vapor, both at 25°C? Which formation reaction is more exothermic? Does your answer indicate that $H_2O(\ell)$ is at a higher or lower enthalpy than $H_2O(g)$?

21. Methylhydrazine is burned with dinitrogen tetroxide in the attitude-control engines of the space shuttles.

$$CH_6N_2(\ell) + \tfrac{5}{4}N_2O_4(\ell) \longrightarrow$$
$$CO_2(g) + 3H_2O(\ell) + \tfrac{9}{4}N_2(g)$$

The two substances ignite instantly on contact, producing a flame temperature of 3000. K. The energy liberated per 0.100 g of CH_6N_2 at constant atmospheric pressure after the products are cooled back to 25°C is 750. J. (a) Find ΔH for the reaction as written. (b) How many kilojoules are liberated when 87.5 g of N_2 is produced?

NASA

A space shuttle

22. ☁Which is more exothermic, the combustion of one mole of methane to form $CO_2(g)$ and liquid water or the combustion of one mole of methane to form $CO_2(g)$ and steam? Why? (No calculations are necessary.)

23. ☁Which is more exothermic, the combustion of one mole of gaseous benzene, C_6H_6, or the combustion of one mole of liquid benzene? Why? (No calculations are necessary.)

Thermochemical Equations, ΔH_f^0, and Hess's Law

24. ☁ Explain the meaning of the phrase "thermodynamic standard state of a substance."

25. ☁ Explain the meaning of the phrase "standard molar enthalpy of formation." Give an example.

26. ☁ From the data in Appendix K, determine the form that represents the standard state for each of the following elements: (a) chlorine; (b) chromium; (c) bromine; (d) iodine; (e) sulfur; (f) nitrogen.

27. ☁ From the data in Appendix K, determine the form that represents the standard state for each of the following elements: (a) oxygen; (b) carbon; (c) phosphorus; (d) rubidium; (e) mercury; (f) tin.

28. ☁ Write the balanced chemical equation whose ΔH_{rxn}^0 value is equal to ΔH_f^0 for each of the following substances: (a) calcium hydroxide, $Ca(OH)_2(s)$; (b) benzene, $C_6H_6(\ell)$; (c) sodium bicarbonate, $NaHCO_3(s)$; (d) calcium fluoride, $CaF_2(s)$; (e) phosphine, $PH_3(g)$; (f) propane, $C_3H_8(g)$; (g) atomic sulfur, $S(g)$; (h) water, $H_2O(\ell)$.

29. ☁ Write the balanced chemical equation for the formation of one mole of each of the following substances from its elements in the standard state (the equation whose ΔH_{rxn}^0 value is equal to ΔH_f^0 for the substance): hydrogen peroxide [$H_2O_2(\ell)$], calcium fluoride [$CaF_2(s)$], ruthenium(III) hydroxide [$Ru(OH)_3(s)$].

30. ▲ We burn 3.47 g of lithium in excess oxygen at constant atmospheric pressure to form Li_2O. Then we bring the reaction mixture back to 25°C. In this process 146 kJ of heat is given off. What is the standard molar enthalpy of formation of Li_2O?

31. ▲ We burn 7.20 g of magnesium in excess nitrogen at constant atmospheric pressure to form Mg_3N_2. Then we bring the reaction mixture back to 25°C. In this process 68.35 kJ of heat is given off. What is the standard molar enthalpy of formation of Mg_3N_2?

32. From the following enthalpies of reaction,

$$4HCl(g) + O_2(g) \longrightarrow 2H_2O(\ell) + 2Cl_2(g)$$
$$\Delta H = -202.4 \text{ kJ/mol rxn}$$

$$\tfrac{1}{2}H_2(g) + \tfrac{1}{2}F_2(g) \longrightarrow HF(\ell)$$
$$\Delta H = -600.0 \text{ kJ/mol rxn}$$

$$H_2(g) + \tfrac{1}{2}O_2(g) \longrightarrow H_2O(\ell)$$
$$\Delta H = -285.8 \text{ kJ/mol rxn}$$

find ΔH_{rxn} for $2HCl(g) + F_2(g) \longrightarrow 2HF(\ell) + Cl_2(g)$.

☁ **Molecular Reasoning** exercises   ▲ **More Challenging** exercises   Blue-Numbered exercises solved in Student Solutions Manual

Unless otherwise noted, all content on this page is © Cengage Learning.

33. From the following enthalpies of reaction,

$$CaCO_3(s) \longrightarrow CaO(s) + CO_2(g)$$
$$\Delta H = -178.1 \text{ kJ/mol rxn}$$

$$CaO(s) + H_2O(\ell) \longrightarrow Ca(OH)_2(s)$$
$$\Delta H = -65.3 \text{ kJ/mol rxn}$$

$$Ca(OH)_2(s) \longrightarrow Ca^{2+}(aq) + 2OH^-(aq)$$
$$\Delta H = -16.2 \text{ kJ/mol rxn}$$

calculate ΔH_{rxn} for

$$Ca^{2+}(aq) + 2OH^-(aq) + CO_2(g) \longrightarrow$$
$$CaCO_3(s) + H_2O(\ell)$$

34. Given that

$$S(s) + O_2(g) \longrightarrow SO_2(g) \quad \Delta H = -296.8 \text{ kJ/mol}$$
$$S(s) + \tfrac{3}{2}O_2(g) \longrightarrow SO_3(g) \quad \Delta H = -395.6 \text{ kJ/mol}$$

determine the enthalpy change for the decomposition reaction

$$2SO_3(g) \longrightarrow 2SO_2(g) + O_2(g)$$

35. Evaluate ΔH_{rxn}^0 for the reaction below at 25°C, using the heats of formation.

$$Fe_3O_4(s) + CO(g) \longrightarrow 3FeO(s) + CO_2(g)$$

ΔH_f^0, kJ/mol: -1118 -110.5 -272 -393.5

36. Given that

$$2H_2(g) + O_2(g) \longrightarrow 2H_2O(\ell)$$
$$\Delta H = -571.6 \text{ kJ/mol}$$

$$C_3H_4(g) + 4O_2(g) \longrightarrow 3CO_2(g) + 2H_2O(\ell)$$
$$\Delta H = -1937 \text{ kJ/mol}$$

$$C_3H_8(g) + 5O_2(g) \longrightarrow 3CO_2(g) + 4H_2O(\ell)$$
$$\Delta H = -2220. \text{ kJ/mol}$$

determine the heat of the hydrogenation reaction

$$C_3H_4(g) + 2H_2(g) \longrightarrow C_3H_8(g)$$

37. Determine the heat of formation of liquid hydrogen peroxide at 25°C from the following thermochemical equations.

$$H_2(g) + \tfrac{1}{2}O_2(g) \longrightarrow H_2O(g)$$
$$\Delta H^0 = -241.82 \text{ kJ/mol}$$

$$2H(g) + O(g) \longrightarrow H_2O(g)$$
$$\Delta H^0 = -926.92 \text{ kJ/mol}$$

$$2H(g) + 2O(g) \longrightarrow H_2O_2(g)$$
$$\Delta H^0 = -1070.60 \text{ kJ/mol}$$

$$2O(g) \longrightarrow O_2(g)$$
$$\Delta H^0 = -498.34 \text{ kJ/mol}$$

$$H_2O_2(\ell) \longrightarrow H_2O_2(g)$$
$$\Delta H^0 = 51.46 \text{ kJ/mol}$$

38. Use data in Appendix K to find the enthalpy of reaction for
(a) $NH_4NO_3(s) \longrightarrow N_2O(g) + 2H_2O(\ell)$
(b) $2FeS_2(s) + \tfrac{11}{2}O_2(g) \longrightarrow Fe_2O_3(s) + 4SO_2(g)$
(c) $SiO_2(s) + 3C(graphite) \longrightarrow SiC(s) + 2CO(g)$

39. Repeat Exercise 38 for
(a) $CaCO_3(s) \longrightarrow CaO(s) + CO_2(g)$
(b) $2HI(g) + F_2(g) \longrightarrow 2HF(g) + I_2(s)$
(c) $SF_6(g) + 3H_2O(\ell) \longrightarrow 6HF(g) + SO_3(g)$

40. The internal combustion engine uses heat produced during the burning of a fuel. Propane, $C_3H_8(g)$, is sometimes used as the fuel. Gasoline is the most commonly used fuel. Assume that the gasoline is pure octane, $C_8H_{18}(\ell)$, and the fuel and oxygen are completely converted into $CO_2(g)$ and $H_2O(g)$. For each of these fuels, determine the heat released per gram of fuel burned.

41. Write the balanced equation for the complete combustion (in excess O_2) of kerosene. Assume that kerosene is $C_{10}H_{22}(\ell)$ and that the products are $CO_2(g)$ and $H_2O(\ell)$. Calculate ΔH_{rxn}^0 at 25°C for this reaction.
$\Delta H_f^0 C_{10}H_{22}(\ell) = -249.6 \text{ kJ/mol}$
$\Delta H_f^0 CO_2(g) = -393.5 \text{ kJ/mol}$
$\Delta H_f^0 H_2O(\ell) = -285.8 \text{ kJ/mol}$

42. The thermite reaction, used for welding iron, is the reaction of Fe_3O_4 with Al.

$$8Al(s) + 3Fe_3O_4(s) \longrightarrow 4Al_2O_3(s) + 9Fe(s)$$
$$\Delta H^0 = -3350. \text{ kJ/mol rxn}$$

Because this large amount of heat cannot be rapidly dissipated to the surroundings, the reacting mass may reach temperatures near 3000.°C. How much heat is released by the reaction of 27.6 g of Al with 69.12 g of Fe_3O_4?

The thermite reaction

43. When a welder uses an acetylene torch, the combustion of acetylene liberates the intense heat needed for welding metals together. The equation for this combustion reaction is

$$2C_2H_2(g) + 5O_2(g) \longrightarrow 4CO_2(g) + 2H_2O(g)$$

The heat of combustion of acetylene is -1255.5 kJ/mol of C_2H_2. How much heat is liberated when 1.731 kg of C_2H_2 is burned?

44. ▲ Write balanced equations for the oxidation of sucrose (a carbohydrate) and tristearin (a fat). Assume that each reacts with $O_2(g)$, producing $CO_2(g)$ and $H_2O(g)$.

Sucrose, $C_{12}H_{22}O_{11}$ Tristearin, $C_{57}H_{110}O_6$

Use tabulated bond energies to estimate ΔH^0_{rxn} for each reaction in kJ/mol (ignoring phase changes). Convert to kJ/g and kcal/g. Which has the greater energy density?

45. Natural gas is mainly methane, $CH_4(g)$. Assume that gasoline is octane, $C_8H_{18}(\ell)$, and that kerosene is $C_{10}H_{22}(\ell)$. (a) Write the balanced equations for the combustion of each of these three hydrocarbons in excess O_2. The products are $CO_2(g)$ and $H_2O(\ell)$. (b) Calculate at 25°C for each combustion reaction. ΔH^0_f for $C_{10}H_{22}$ is −300.9 kJ/mol. (c) When burned at standard conditions, which of these three fuels would produce the most heat per mole? (d) When burned at standard conditions, which of the three would produce the most heat per gram?

Bond Energies

46. (a) How is the heat released or absorbed in a *gas phase reaction* related to bond energies of products and reactants? (b) Hess's Law states that

$$\Delta H^0_{rxn} = \Sigma\, n\, \Delta H^0_{f\,products} - \Sigma\, n\, \Delta H^0_{f\,reactants}$$

The relationship between ΔH^0_{rxn} and bond energies for a *gas phase reaction* is

$$\Delta H^0_{rxn} = \Sigma\, \text{bond energies}_{reactants} - \Sigma\, \text{bond energies}_{products}$$

It is *not* true, in general, that ΔH^0_f for a substance is equal to the negative of the sum of the bond energies of the substance. Why?

47. (a) Suggest a reason for the fact that different amounts of energy are required for the successive removal of the three hydrogen atoms of an ammonia molecule, even though all N—H bonds in ammonia are equivalent. (b) Suggest why the N—H bonds in different compounds such as ammonia, NH_3; methylamine, CH_3NH_2; and ethylamine, $C_2H_5NH_2$, have slightly different bond energies.

48. Use tabulated bond energies to estimate the enthalpy of reaction for each of the following gas phase reactions.
    (a) $H_2C{=}CH_2 + Br_2 \longrightarrow BrH_2C{-}CH_2Br$
    (b) $H_2O_2 \longrightarrow H_2O + \frac{1}{2}O_2$

49. Use tabulated bond energies to estimate the enthalpy of reaction for each of the following gas phase reactions.
    (a) $N_2 + 3H_2 \longrightarrow 2NH_3$
    (b) $CH_4 + Cl_2 \longrightarrow CH_3Cl + HCl$
    (c) $CO + H_2O \longrightarrow CO_2 + H_2$

50. Use the bond energies listed in Table 15-2 to estimate the heat of reaction for

51. Estimate ΔH for the burning of one mole of butane, using the bond energies listed in Tables 15-2 and 15-3.

$$4\,O{=}C{=}O(g) + 5\,H{-}O{-}H(g)$$

52. (a) Use the bond energies listed in Table 15-2 to estimate the heats of formation of $HCl(g)$ and $HF(g)$. (b) Compare your answers to the standard heats of formation in Appendix K.

53. (a) Use the bond energies listed in Table 15-2 to estimate the heats of formation of $H_2O(g)$ and $O_3(g)$. (b) Compare your answers to the standard heats of formation in Appendix K.

54. Using data in Appendix K, calculate the average P—Cl bond energy in $PCl_3(g)$.

55. Using data in Appendix K, calculate the average P—H bond energy in $PH_3(g)$.

56. Using data in Appendix K, calculate the average P—Cl bond energy in $PCl_5(g)$. Compare your answer with the value calculated in Exercise 54.

57. ✿ ▲ Methane undergoes several different exothermic reactions with gaseous chlorine. One of these forms chloroform, $CHCl_3(g)$.

$$CH_4(g) + 3Cl_2(g) \longrightarrow CHCl_3(g) + 3HCl(g)$$
$$\Delta H^0_{rxn} = -305.2 \text{ kJ/mol rxn}$$

Average bond energies per mole of bonds are: C—H = 413 kJ; Cl—Cl = 242 kJ; H—Cl = 432 kJ. Use these to calculate the average C—Cl bond energy in chloroform. Compare this with the value in Table 15-2.

58. ▲ Ethylamine undergoes an endothermic gas phase dissociation to produce ethylene (or ethene) and ammonia.

$$\Delta H^0_{rxn} = +53.6 \text{ kJ/mol rxn}$$

The following average bond energies per mole of bonds are given: C—H = 413 kJ; C—C = 346 kJ; C=C = 602 kJ; N—H = 391 kJ. Calculate the C—N bond energy in ethylamine. Compare this with the value in Table 15-2.

Calorimetry

59. ✿ What is a coffee-cup calorimeter? How do coffee-cup calorimeters give us useful information?

✿ **Molecular Reasoning** exercises   ▲ **More Challenging** exercises   Blue-Numbered exercises solved in Student Solutions Manual

Unless otherwise noted, all content on this page is © Cengage Learning.

60. A calorimeter contained 75.0 g of water at 16.95°C. A 93.3-g sample of iron at 65.58°C was placed in it, giving a final temperature of 19.68°C for the system. Calculate the heat capacity of the calorimeter. Specific heats are 4.184 J/g · °C for H_2O and 0.444 J/g · °C for Fe.

61. ▲ A student wishes to determine the heat capacity of a coffee-cup calorimeter. After she mixes 100.0 g of water at 58.5°C with 100.0 g of water, already in the calorimeter, at 22.8°C, the final temperature of the water is 39.7°C. (a) Calculate the heat capacity of the calorimeter in J/°C. Use 4.184 J/g · °C as the specific heat of water. (b) Why is it more useful to express the value in J/°C rather than units of J/(g calorimeter · °C)?

62. A coffee-cup calorimeter is used to determine the specific heat of a metallic sample. The calorimeter is filled with 50.0 mL of water at 25.0°C (density = 0.997 g/mL). A 36.5-gram sample of the metallic material is taken from water boiling at 100.0°C and placed in the calorimeter. The equilibrium temperature of the water and sample is 32.5°C. The calorimeter constant is known to be 1.87 J/°C. Calculate the specific heat of the metallic material.

63. A 5.1-gram piece of gold jewelry is removed from water at 100.0°C and placed in a coffee-cup calorimeter containing 16.9 g of water at 22.5°C. The equilibrium temperature of the water and jewelry is 23.2°C. The calorimeter constant is known from calibration experiments to be 1.54 J/°C. What is the specific heat of this piece of jewelry? The specific heat of pure gold is 0.129 J/g · °C. Is the jewelry pure gold?

64. A coffee-cup calorimeter having a heat capacity of 472 J/°C is used to measure the heat evolved when the following aqueous solutions, both initially at 22.6°C, are mixed: 100. g of solution containing 6.62 g of lead(II) nitrate, $Pb(NO_3)_2$, and 100. g of solution containing 6.00 g of sodium iodide, NaI. The final temperature is 24.2°C. Assume that the specific heat of the mixture is the same as that for water, 4.184 J/g · °C. The reaction is

$$Pb(NO_3)_2(aq) + 2NaI(aq) \longrightarrow PbI_2(s) + 2NaNO_3(aq)$$

(a) Calculate the heat evolved in the reaction. (b) Calculate the ΔH for the reaction under the conditions of the experiment.

65. A coffee-cup calorimeter is used to determine the heat of reaction for the acid–base neutralization

$$CH_3COOH(aq) + NaOH(aq) \longrightarrow$$
$$NaCH_3COO(aq) + H_2O(\ell)$$

When we add 20.00 mL of 0.625 M NaOH at 21.400°C to 30.00 mL of 0.500 M CH_3COOH already in the calorimeter at the same temperature, the resulting temperature is observed to be 24.347°C. The heat capacity of the calorimeter has previously been determined to be 27.8 J/°C. Assume that the specific heat of the mixture is the same as that of water, 4.184 J/g · °C, and that the density of the mixture is 1.02 g/mL. (a) Calculate the amount of heat given off in the reaction. (b) Determine ΔH for the reaction under the conditions of the experiment.

66. In a bomb calorimeter compartment surrounded by 945 g of water, the combustion of 1.048 g of benzene, $C_6H_6(\ell)$, raised the temperature of the water from 23.640°C to 32.692°C. The heat capacity of the calorimeter is 891 J/°C. (a) Write the balanced equation for the combustion reaction, assuming that $CO_2(g)$ and $H_2O(\ell)$ are the only products. (b) Use the calorimetric data to calculate ΔE for the combustion of benzene in kJ/g and in kJ/mol.

67. A 2.00-g sample of hydrazine, N_2H_4, is burned in a bomb calorimeter that contains 6.40×10^3 g of H_2O, and the temperature increases from 25.00°C to 26.17°C. The heat capacity of the calorimeter is 3.76 kJ/°C. Calculate ΔE for the combustion of N_2H_4 in kJ/g and in kJ/mol.

68. ▲ A strip of magnesium metal having a mass of 1.22 g dissolves in 100. mL of 6.02 M HCl, which has a specific gravity of 1.10. The hydrochloric acid is initially at 23.0°C, and the resulting solution reaches a final temperature of 45.5°C. The heat capacity of the calorimeter in which the reaction occurs is 562 J/°C. Calculate ΔH for the reaction under the conditions of the experiment, assuming the specific heat of the final solution is the same as that for water, 4.184 J/g · °C.

$$Mg(s) + 2HCl(aq) \longrightarrow MgCl_2(aq) + H_2(g)$$

69. When 3.16 g of salicylic acid, $C_7H_6O_3$, is burned in a bomb calorimeter containing 5.00 kg of water originally at 23.00°C, 69.3 kJ of heat is evolved. The calorimeter constant is 3255 J/°C. Calculate the final temperature.

70. A 6.620-g sample of decane, $C_{10}H_{22}(\ell)$, was burned in a bomb calorimeter whose heat capacity had been determined to be 2.45 kJ/°C. The temperature of 1250.0 g of water rose from 24.6°C to 26.4°C. Calculate ΔE for the reaction in joules per gram of decane and in kilojoules per mole of decane. The specific heat of water is 4.184 J/g · °C.

71. A nutritionist determines the caloric value of a 10.00-g sample of beef fat by burning it in a bomb calorimeter. The calorimeter held 2.500 kg of water, the heat capacity of the bomb is 1.360 kJ/°C, and the temperature of the calorimeter increased from 25.0°C to 56.9°C. (a) Calculate the number of joules released per gram of beef fat. (b) One nutritional Calorie is 1 kcal or 4184 joules. What is the dietary, caloric value of beef fat, in nutritional Calories per gram?

Internal Energy and Changes in Internal Energy

72. (a) What are the sign conventions for q, the amount of heat added to or removed from a system? (b) What are the sign conventions for w, the amount of work done on or by a system?

73. What happens to ΔE for a system during a process in which (a) $q < 0$ and $w < 0$, (b) $q = 0$ and $w > 0$, and (c) $q > 0$ and $w < 0$.

74. Ammonium nitrate, commonly used as a fertilizer, decomposes explosively:

$$2NH_4NO_3(s) \longrightarrow 2N_2(g) + 4H_2O(g) + O_2(g)$$

(This reaction has been responsible for several major explosions.) For this reaction:
(a) Is work (w) positive, negative, or 0?
(b) If $w < 0$, is work done on the system or by the system?

🐚 **Molecular Reasoning** exercises   ▲ **More Challenging** exercises   Blue-Numbered exercises solved in Student Solutions Manual

Unless otherwise noted, all content on this page is © Cengage Learning.

75. A system performs 720. L · atm of pressure–volume work (1 L · atm = 101.325 J) on its surroundings and absorbs 5750. J of heat from its surroundings. What is the change in internal energy of the system?

76. A system receives 96 J of electrical work, performs 257 J of pressure–volume work, and releases 175 J of heat. What is the change in internal energy of the system?

77. For each of the following chemical and physical changes carried out at constant pressure, state whether work is done by the system on the surroundings or by the surroundings on the system, or whether the amount of work is negligible.
(a) $C_6H_6(\ell) \longrightarrow C_6H_6(g)$
(b) $\frac{1}{2}N_2(g) + \frac{3}{2}H_2(g) \longrightarrow NH_3(g)$
(c) $SiO_2(s) + 3C(s) \longrightarrow SiC(s) + 2CO(g)$

78. Repeat Exercise 77 for
(a) $2SO_2(g) + O_2(g) \longrightarrow 2SO_3(g)$
(b) $CaCO_3(s) \longrightarrow CaO(s) + CO_2(g)$
(c) $CO_2(g) + H_2O(\ell) + CaCO_3(s) \longrightarrow$
$ Ca^{2+}(aq) + 2HCO_3{}^-(aq)$

79. Assuming that the gases are ideal, calculate the amount of work done (in joules) in each of the following reactions. In each case, is the work done *on* or *by* the system? (a) A reaction in the Mond process for purifying nickel that involves formation of the gas nickel(0) tetracarbonyl at $50-100°C$. Assume one mole of nickel is used and a constant temperature of 75°C is maintained.

$$Ni(s) + 4CO(g) \longrightarrow Ni(CO)_4(g)$$

(b) The conversion of one mole of brown nitrogen dioxide into colorless dinitrogen tetroxide at 8.0°C.

$$2NO_2(g) \longrightarrow N_2O_4(g)$$

80. Assuming that the gases are ideal, calculate the amount of work done (in joules) in each of the following reactions. In each case, is the work done *on* or *by* the system? (a) The oxidation of one mole of HCl(g) at 200°C.

$$4HCl(g) + O_2(g) \longrightarrow 2Cl_2(g) + 2H_2O(g)$$

(b) The decomposition of one mole of nitric oxide (an air pollutant) at 300.°C.

$$2NO(g) \longrightarrow N_2(g) + O_2(g)$$

81. ▲ When an ideal gas expands at *constant temperature*, there is no change in molecular kinetic energy (kinetic energy is proportional to temperature), and there is no change in potential energy due to intermolecular attractions (these are zero for an ideal gas). Thus for the isothermal (constant temperature) expansion of an ideal gas, $\Delta E = 0$. Suppose we allow an ideal gas to expand isothermally from 2.50 L to 5.50 L in two steps: (a) against a constant external pressure of 3.50 atm until equilibrium is reached, then (b) against a constant external pressure of 2.50 atm until equilibrium is reached. Calculate q and w for this two-step expansion.

Entropy and Entropy Changes

82. ☁ A car uses gasoline as a fuel. Describe the burning of the fuel in terms of chemical and physical changes. Relate your answer to the Second Law of Thermodynamics.

83. State the Second Law of Thermodynamics. Why can't we use ΔS_{univ} directly as a measure of the spontaneity of a reaction?

84. State the Third Law of Thermodynamics. What does it mean?

85. Explain why ΔS may be referred to as a contributor to spontaneity.

86. Suppose you flip a coin. (a) What is the probability that it will come up heads? (b) What is the probability that it will come up heads two times in a row? (c) What is the probability that it will come up heads ten times in a row?

87. ☁ Consider two equal-sized flasks connected as shown in the figure.

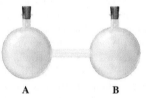

A   B

(a) Suppose you put one molecule inside. What is the probability that the molecule will be in flask A? What is the probability that it will be in flask B?
(b) If you put 100 molecules into the two-flask system, what is the most likely distribution of molecules? Which distribution corresponds to the highest entropy?
(c) Write a mathematical expression for the probability that all 100 molecules in part (b) will be in flask A. (You do not need to evaluate this expression.)

88. ☁ Suppose you have two identical red molecules labeled A and B, and two identical blue molecules labeled C and D. Draw a simple two-flask diagram as in the figure for Exercise 87, and then draw all possible arrangements of the four molecules in the two flasks.
(a) How many different arrangements are possible?
(b) How many of the arrangements have a mixture of unlike molecules in at least one of the flasks?
(c) What is the probability that at least one of the flasks contains a mixture of unlike molecules?
(d) What is the probability that the gases are not mixed (each flask contains only like molecules)?

89. ☁ For each process, tell whether the entropy change of the system is positive or negative. (a) Water vapor (the system) condenses as droplets on a cold windowpane. (b) Water boils. (c) A can of carbonated beverage loses its fizz. (Consider the beverage, but not the can, as the system. What happens to the entropy of the dissolved gas?)

90. ☁ For each process, tell whether the entropy change of the system is positive or negative. (a) A glassblower heats glass (the system) to its softening temperature.

(b) A teaspoon of sugar dissolves in a cup of coffee. (The system consists of both sugar and coffee.) (c) Calcium carbonate precipitates out of water in a cave to form stalactites and stalagmites. (Consider only the calcium carbonate to be the system.)

91. ☁ For each of the following processes, tell whether the entropy of the *universe* increases, decreases, or remains constant: (a) melting one mole of ice to water at 0.°C; (b) freezing one mole of water to ice at 0.°C; (c) freezing one mole of water to ice at −15°C; (d) freezing one mole of water to ice at 0.°C and then cooling it to −15°C.

92. ▲ In which of the following changes is there an increase in entropy?
(a) the freezing of water
(b) the condensation of steam
(c) the sublimation of dry ice, solid CO_2
(d) the separation of salts and pure water from seawater

93. ☁ When solid sodium chloride is cooled from 25°C to 0.°C, the entropy change is −4.4 J/mol · K. Is this an increase or decrease in randomness? Explain this entropy change in terms of what happens in the solid at the molecular level.

94. ☁ When a one-mole sample of argon gas at 0.°C is compressed to one half its original volume, the entropy change is −5.76 J/mol · K. Is this an increase or a decrease in dispersal of energy? Explain this entropy change in terms of what happens in the gas at the molecular level.

95. ☁ Which of the following processes are accompanied by an increase in entropy of the system? (No calculation is necessary.) (a) Dry ice, $CO_2(s)$, sublimes at −78°C and then the resulting $CO_2(g)$ is warmed to 0°C. (b) Water vapor forms snowflakes, $H_2O(s)$. (c) Iodine sublimes, $I_2(s) \longrightarrow I_2(g)$. (d) White silver sulfate, Ag_2SO_4, precipitates from a solution containing silver ions and sulfate ions. (e) A partition is removed to allow two gases to mix.

96. ☁ ▲ Which of the following processes are accompanied by an increase in entropy of the system? (No calculation is necessary.) (a) Solid NaCl is dissolved in water at room temperature. (b) A saturated solution of NaCl is cooled, causing some solid NaCl to precipitate. (c) Water freezes. (d) Carbon tetrachloride, CCl_4, evaporates. (e) The reaction $PCl_5(g) \longrightarrow PCl_3(g) + Cl_2(g)$ occurs. (f) The reaction $PCl_3(g) + Cl_2(g) \longrightarrow PCl_5(g)$ occurs.

97. ☁ For each pair, tell which would have the greater absolute entropy per mole (standard molar entropy) at the same temperature. Give the reasons for your choice.
(a) NaCl(s) or CaO(s)
(b) $Cl_2(g)$ or $P_4(g)$
(c) $AsH_3(g)$ or Kr(g)
(d) $NH_4NO_3(s)$ or $NH_4NO_3(aq)$
(e) Ga(s) or Ga(ℓ)

98. ☁ For each pair, tell which would have the greater absolute entropy per mole (standard molar entropy) at the same temperature. Give the reasons for your choice.
(a) NaF(s) or MgO(s)
(b) Au(s) or Hg(ℓ)
(c) $H_2O(g)$ or $H_2S(g)$
(d) $CH_3OH(\ell)$ or $C_2H_5OH(\ell)$
(e) NaOH(s) or NaOH(aq)

99. ☁ (a) For which change would the entropy change by the greatest amount: (i) condensation of one mole of water vapor to make one mole of liquid water, or (ii) deposition of one mole of water vapor to make one mole of ice? (b) Would the entropy changes for the changes in (a) be positive or negative? Give reasons for your answer.

100. ☁ Without doing a calculation predict whether the entropy change will be positive or negative when each reaction occurs in the direction it is written.
(a) $C_3H_6(g) + H_2(g) \longrightarrow C_3H_8(g)$
(b) $N_2(g) + 3H_2(g) \longrightarrow 2NH_3(g)$
(c) $CaCO_3(s) \longrightarrow CaO(s) + CO_2(g)$
(d) Mg(s) + $\frac{1}{2}O_2(g) \longrightarrow$ MgO(s)
(e) $Ag^+(aq) + Cl^-(aq) \longrightarrow$ AgCl(s)

101. ☁ Without doing a calculation predict whether the entropy change will be positive or negative when each reaction occurs in the direction it is written.
(a) $CH_3OH(\ell) + \frac{3}{2}O_2(g) \longrightarrow CO_2(g) + 2H_2O(g)$
(b) $Br_2(\ell) + H_2(g) \longrightarrow$ 2HBr(g)
(c) Na(s) + $\frac{1}{2}F_2(g) \longrightarrow$ NaF(s)
(d) $CO_2(g) + 2H_2(g) \longrightarrow CH_3OH(\ell)$
(e) $NH_3(g) \longrightarrow N_2(g) + 3H_2(g)$

102. ☁ ▲ Consider the boiling of a pure liquid at constant pressure. Is each of the following greater than, less than, or equal to zero? (a) ΔS_{sys}; (b) ΔH_{sys}; (c) ΔT_{sys}.

103. Use S^0 data from Appendix K to calculate the value of ΔS^0_{298} for each of the following reactions. Compare the signs and magnitudes for these ΔS^0_{298} values and explain your observations.
(a) 2NO(g) + $H_2(g) \longrightarrow N_2O(g) + H_2O(g)$
(b) $2N_2O_5(g) \longrightarrow 4NO_2(g) + O_2(g)$
(c) $NH_4NO_3(s) \longrightarrow N_2O(g) + 2H_2O(g)$

104. Use S^0 data from Appendix K to calculate the value of ΔS^0_{298} for each of the following reactions. Compare the signs and magnitudes for these ΔS^0_{298} values and explain your observations.
(a) 4HCl(g) + $O_2(g) \longrightarrow 2Cl_2(g) + 2H_2O(g)$
(b) $PCl_3(g) + Cl_2(g) \longrightarrow PCl_5(g)$
(c) $2N_2O(g) \longrightarrow 2N_2(g) + O_2(g)$

Gibbs Free Energy Changes and Spontaneity

105. ☁ (a) What are the two factors that favor spontaneity of a process? (b) What is Gibbs free energy? What is change in Gibbs free energy? (c) Most spontaneous reactions are exothermic, but some are not. Explain. (d) Explain how the signs and magnitudes of ΔH and ΔS are related to the spontaneity of a process.

106. Which of the following conditions would predict a process that is (a) always spontaneous, (b) always nonspontaneous, or (c) spontaneous or nonspontaneous depending on the temperature and magnitudes of ΔH and ΔS? (i) $\Delta H > 0$, $\Delta S > 0$; (ii) $\Delta H > 0$, $\Delta S < 0$; (iii) $\Delta H < 0$, $\Delta S > 0$; (iv) $\Delta H < 0$, $\Delta S < 0$

107. Calculate ΔG^0 at 45°C for reactions for which
(a) $\Delta H^0 = 293$ kJ; $\Delta S^0 = -695$ J/K.
(b) $\Delta H^0 = -1137$ kJ; $\Delta S^0 = 0.496$ kJ/K.
(c) $\Delta H^0 = -86.6$ kJ; $\Delta S^0 = -382$ J/K.

☁ **Molecular Reasoning** exercises   ▲ **More Challenging** exercises   Blue-Numbered exercises solved in Student Solutions Manual

Unless otherwise noted, all content on this page is © Cengage Learning.

108. Evaluate ΔS^0 at 25°C and 1 atm for the reaction:

$$SiH_4(g) + 2O_2(g) \longrightarrow SiO_2(s) + 2H_2O(\ell)$$

S^0, J/mol · K:　　204.5　　205.0　　　41.84　　69.91

109. The standard Gibbs free energy of formation is −286.06 kJ/mol for NaI(s), −261.90 kJ/mol for Na⁺(aq), and −51.57 kJ/mol for I⁻(aq) at 25°C. Calculate ΔG^0 for the reaction

$$NaI(s) \xrightarrow{H_2O} Na^+(aq) + I^-(aq)$$

110. ▲ Use the following equations to find ΔG_f^0 for HBr(g) at 25°C.

$Br_2(\ell) \longrightarrow Br_2(g)$	$\Delta G^0 = 3.14$ kJ/mol
$HBr(g) \longrightarrow H(g) + Br(g)$	$\Delta G^0 = 339.09$ kJ/mol
$Br_2(g) \longrightarrow 2Br(g)$	$\Delta G^0 = 161.7$ kJ/mol
$H_2(g) \longrightarrow 2H(g)$	$\Delta G^0 = 406.494$ kJ/mol

111. Use values of standard free energy of formation, ΔG_f^0, from Appendix K to calculate the standard free energy change for each of the following reactions at 25°C and 1 atm.
    (a) $3NO_2(g) + H_2O(\ell) \longrightarrow 2HNO_3(\ell) + NO(g)$
    (b) $SnO_2(s) + 2CO(g) \longrightarrow 2CO_2(g) + Sn(s)$
    (c) $2Na(s) + 2H_2O(\ell) \longrightarrow 2NaOH(aq) + H_2(g)$

112. Make the same calculations as in Exercise 111, using values of standard enthalpy of formation and absolute entropy instead of values of ΔG_f^0.

113. Calculate ΔG^0 at 298 K for the reaction:

$$P_4O_{10}(s) + 6H_2O(\ell) \longrightarrow 4H_3PO_4(s)$$

ΔH_f^0, kJ/mol:　　−2984　　−285.8　　−1281

S^0, J/mol · K:　　　228.9　　69.91　　110.5

Temperature Range of Spontaneity

114. Are the following statements true or false? Justify your answers. (a) An exothermic reaction is always spontaneous. (b) If ΔH and ΔS are both positive, then ΔG will decrease when the temperature increases. (c) A reaction for which ΔS_{sys} is positive is spontaneous.

115. For the reaction

$$C(s) + O_2(g) \longrightarrow CO_2(g)$$

$\Delta H^0 = -393.51$ kJ/mol and $\Delta S^0 = 2.86$ J/mol · K at 25°C. (a) Does this reaction become more or less favorable as the temperature increases? (b) For the reaction

$$C(s) + \tfrac{1}{2}O_2(g) \longrightarrow CO(g)$$

$\Delta H^0 = -110.52$ kJ/mol and $\Delta S^0 = 89.36$ J/mol · K at 25°C. Does this reaction become more or less favorable as the temperature increases? (c) Compare the temperature dependencies of these reactions.

116. (a) Calculate ΔH^0, ΔG^0, and ΔS^0 for the reaction

$$2H_2O_2(\ell) \longrightarrow 2H_2O(\ell) + O_2(g) \text{ at 25°C.}$$

(b) Is there any temperature at which $H_2O_2(\ell)$ is stable at 1 atm?

117. When is it true that $\Delta S = \dfrac{\Delta H}{T}$?

118. 🔬 Dissociation reactions are those in which molecules break apart. Why do high temperatures favor the spontaneity of most dissociation reactions?

119. Estimate the temperature range over which each of the following standard reactions is spontaneous.
    (a) $2Al(s) + 3Cl_2(g) \longrightarrow 2AlCl_3(s)$
    (b) $2NOCl(g) \longrightarrow 2NO(g) + Cl_2(g)$
    (c) $4NO(g) + 6H_2O(g) \longrightarrow 4NH_3(g) + 5O_2(g)$
    (d) $2PH_3(g) \longrightarrow 3H_2(g) + 2P(g)$

120. Estimate the temperature range over which each of the following standard reactions is spontaneous. (a) The reaction by which sulfuric acid droplets from polluted air convert water-insoluble limestone or marble (calcium carbonate) to slightly soluble calcium sulfate, which is slowly washed away by rain:

$$CaCO_3(s) + H_2SO_4(\ell) \longrightarrow CaSO_4(s) + H_2O(\ell) + CO_2(g)$$

(b) The reaction by which Antoine Lavoisier achieved the first laboratory preparation of oxygen in the late eighteenth century: the thermal decomposition of the red-orange powder, mercury(II) oxide, to oxygen and the silvery liquid metal, mercury:

$$2HgO(s) \longrightarrow 2Hg(\ell) + O_2(g)$$

(c) The reaction of coke (carbon) with carbon dioxide to form the reducing agent, carbon monoxide, which is used to reduce some metal ores to metals:

$$CO_2(g) + C(s) \longrightarrow 2CO(g)$$

(d) The reverse of the reaction by which iron rusts:

$$2Fe_2O_3(s) \longrightarrow 4Fe(s) + 3O_2(g)$$

121. Estimate the normal boiling point of pentacarbonyliron(0), Fe(CO)₅, at 1 atm pressure, using Appendix K.

122. (a) Estimate the normal boiling point of water, at 1 atm pressure, using Appendix K. (b) Compare the temperature obtained with the known boiling point of water. Can you explain the discrepancy?

123. Sublimation and subsequent deposition onto a cold surface are a common method of purification of I₂ and other solids that sublime readily. Estimate the sublimation temperature (solid to vapor) of the dark violet solid iodine, I₂, at 1 atm pressure, using the data of Appendix K.

Sublimation and deposition of I₂

124. Some metal oxides can be decomposed to the metal and oxygen under reasonable conditions. Is the decomposition of nickel(II) oxide product-favored at 25°C?

$$2NiO(s) \longrightarrow 2Ni(s) + O_2(g)$$

If not, can it become so if the temperature is raised? At what temperature does the reaction become product-favored?

🔬 **Molecular Reasoning** exercises　　▲ **More Challenging** exercises　　Blue-Numbered exercises solved in Student Solutions Manual

Unless otherwise noted, all content on this page is © Cengage Learning.

125. Calculate ΔH^0 and ΔS^0 for the reaction of tin(IV) oxide with carbon.

$$SnO_2(s) + C(s) \longrightarrow Sn(s) + CO_2(g)$$

(a) Is the reaction spontaneous under standard conditions at 298 K?

(b) Is the reaction predicted to be spontaneous at higher temperatures?

126. Calculate ΔS^0 system at 25°C for the reaction

$$C_2H_4(g) + H_2O(g) \longrightarrow C_2H_5OH(\ell)$$

Can you tell from the result of this calculation whether this reaction is product-favored? If you cannot tell, what additional information do you need? Obtain that information and determine whether the reaction is product-favored.

Mixed Exercises

127. ▲ An ice calorimeter, shown below, can be used to measure the amount of heat released or absorbed by a reaction that is carried out at a constant temperature of 0.°C. If heat is transferred from the system to the bath, some of the ice melts. A given mass of liquid water has a smaller volume than the same mass of ice, so the total volume of the ice and water mixture decreases. Measuring the volume decrease using the scale at the left indicates the amount of heat released by the reacting system. As long as some ice remains in the bath, the temperature remains at 0.°C. In Example 15-2 we saw that the reaction

$$CuSO_4(aq) + 2NaOH(aq) \longrightarrow Cu(OH)_2(s) + Na_2SO_4(aq)$$

releases 846 J of heat at constant temperature and pressure when 50.0 mL of 0.400 M $CuSO_4$ solution and 50.0 mL of 0.600 M NaOH solution are allowed to react. (Because no gases are involved in the reaction, the volume change of the reaction mixture is negligible.) Calculate the change in volume of the ice and water mixture that would be observed if we carried out the same experiment in an ice calorimeter. The density of $H_2O(\ell)$ at 0.°C is 0.99987 g/mL and that of ice is 0.917 g/mL. The heat of fusion of ice at 0°C is 334 J/g.

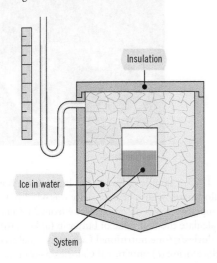

Insulation

Ice in water

System

128. It is difficult to prepare many compounds directly from their elements, so ΔH_f^0 values for these compounds cannot be measured directly. For many organic compounds, it is easier to measure the standard enthalpy of combustion by reaction of the compound with excess $O_2(g)$ to form $CO_2(g)$ and $H_2O(\ell)$. From the following standard enthalpies of combustion at 25°C, determine ΔH_f^0 for the compound. (a) cyclohexane, $C_6H_{12}(\ell)$, a useful organic solvent: $\Delta H_{combustion}^0 = -3920.$ kJ/mol; (b) phenol, $C_6H_5OH(s)$, used as a disinfectant and in the production of thermo-setting plastics: $\Delta H_{combustion}^0 = -3053$ kJ/mol.

129. ▲ Standard entropy changes cannot be measured directly in the laboratory. They are calculated from experimentally obtained values of ΔG^0 and ΔH^0. From the data given here, calculate ΔS^0 at 298 K for each of the following reactions.

(a) $OF_2(g) + H_2O(g) \longrightarrow O_2(g) + 2HF(g)$
    $\Delta H^0 = -323.2$ kJ/mol $\Delta G^0 = -358.4$ kJ/mol

(b) $CaC_2(s) + 2H_2O(\ell) \longrightarrow Ca(OH)_2(s) + C_2H_2(g)$
    $\Delta H^0 = -125.4$ kJ/mol $\Delta G^0 = -145.4$ kJ/mol

(c) $CaO(s) + H_2O(\ell) \longrightarrow Ca(OH)_2(aq)$
    $\Delta H^0 = 81.5$ kJ/mol $\Delta G^0 = -26.20$ kJ/mol

130. ▲ Calculate q, w, and ΔE for the vaporization of 12.5 g of liquid ethanol (C_2H_5OH) at 1.00 atm at 78.0°C, to form gaseous ethanol at 1.00 atm at 78.0°C. Make the following simplifying assumptions: (a) the density of liquid ethanol at 78.0°C is 0.789 g/mL, and (b) gaseous ethanol is adequately described by the ideal gas equation. The heat of vaporization of ethanol is 855 J/g.

131. We add 0.100 g of CaO(s) to 125 g H_2O at 23.6°C in a coffee-cup calorimeter. The following reaction occurs.
$CaO(s) + H_2O(\ell) \longrightarrow Ca(OH)_2(aq)$
                                    $\Delta H^0 = 81.5$ kJ/mol rxn

What will be the final temperature of the solution?

132. (a) The accurately known molar heat of combustion of naphthalene, $C_{10}H_8(s)$, $\Delta H = -5156.8$ kJ/mol $C_{10}H_8$, is used to calibrate calorimeters. The complete combustion of 0.01520 g of $C_{10}H_8$ at constant pressure raises the temperature of a calorimeter by 0.212°C. Find the heat capacity of the calorimeter. (b) The initial temperature of the calorimeter (part a) is 22.102°C; 0.1040 g of $C_8H_{18}(\ell)$, octane (molar heat of combustion $\Delta H = -5451.4$ kJ/mol C_8H_{18},) is completely burned in the calorimeter. Find the final temperature of the calorimeter.

Conceptual Exercises

133. When a gas expands suddenly, it may not have time to absorb a significant amount of heat: $q = 0$. Assume that 1.00 mol N_2 expands suddenly, doing 3000. J of work. (a) What is ΔE for the process? (b) The heat capacity of N_2 is 20.9 J/mol ? °C. How much does its temperature fall during this expansion? (This is the principle of most snow-making machines, which use compressed air mixed with water vapor.)

🐾 **Molecular Reasoning** exercises   ▲ **More Challenging** exercises   Blue-Numbered exercises solved in Student Solutions Manual

Unless otherwise noted, all content on this page is © Cengage Learning.

134. As a rubber band is stretched, it gets warmer; when released, it gets cooler. To obtain the more nearly linear arrangement of the rubber band's polymeric material from the more random relaxed rubber band requires that there be rotation about carbon–carbon single bonds. Based on these data, give the sign of ΔG, ΔH, and ΔS for the stretching of a rubber band and for the relaxing of a stretched rubber band. What drives the spontaneous process?

135. 🐾 (a) The decomposition of mercury(II) oxide has been used as a method for producing oxygen, but this is not a recommended method. Why? (b) Write the balanced equation for the decomposition of mercury(II) oxide. (c) Calculate the ΔH^0, ΔS^0, and ΔG^0 for the reaction. (d) Is the reaction spontaneous at room temperature?

136. ▲ (a) A student heated a sample of a metal weighing 32.6 g to 99.83°C and put it into 100.0 g of water at 23.62°C in a calorimeter. The final temperature was 24.41°C. The student calculated the specific heat of the metal, but neglected to use the heat capacity of the calorimeter. The specific heat of water is 4.184 J/g · °C. What was his answer? The metal was known to be chromium, molybdenum, or tungsten. By comparing the value of the specific heat to those of the metals (Cr, 0.460; Mo, 0.250; W, 0.135 J/g · °C), the student identified the metal. What was the metal? (b) A student at the next laboratory bench did the same experiment, obtained the same data, and used the heat capacity of the calorimeter in his calculations. The heat capacity of the calorimeter was 410. J/°C. Was his identification of the metal different?

137. 🐾 According to the Second Law of Thermodynamics what would be the ultimate state or condition of the universe?

138. For each of the following changes, estimate the signs (positive, negative, or 0) of ΔS and ΔG.
(a) The growth of a crystal from a supersaturated solution
(b) Sugar cube + cup of hot tea $\longrightarrow$
cup of hot, sweetened tea
(c) $H_2O(s) \longrightarrow H_2O(\ell)$

139. Estimate the boiling point of tin(IV) chloride, $SnCl_4$, at one atmosphere pressure:

$$SnCl_4(\ell) \rightleftharpoons SnCl_4(g)$$

$SnCl_4(\ell)$: $\Delta H^0_f = -511.3$ kJ/mol, $S^0 = 258.6$ J/mol · K

$SnCl_4(g)$: $\Delta H^0_f = -471.5$ kJ/mol, $S^0 = 366$ J/mol · K

Building Your Knowledge

140. ▲ Energy to power muscular work is produced from stored carbohydrates (glycogen) or fat (triglycerides). Metabolic consumption and production of energy are described with the nutritional "Calorie," which is equal to 1 kilocalorie. Average energy output per minute for various activities follows: sitting, 1.7 kcal; walking, level, 3.5 mph, 5.5 kcal; cycling, level, 13 mph, 10. kcal; swimming, 8.4 kcal; running, 10. mph, 19 kcal. Approximate energy values of some common foods are also given: large apple, 100. kcal;

8-oz cola drink, 105 kcal; malted milkshake, 500. kcal; $\frac{3}{4}$ cup pasta with tomato sauce and cheese, 195 kcal; hamburger on bun with sauce, 350. kcal; 10.-oz sirloin steak, including fat, 1000. kcal. To maintain body weight, fuel intake should balance energy output. Prepare a table showing (a) each given food, (b) its fuel value, and (c) the minutes of each activity that would balance the kcal of each food.

141. From its heat of fusion, calculate the entropy change associated with the melting of one mole of ice at its melting point. From its heat of vaporization, calculate the entropy change associated with the boiling of one mole of water at its boiling point. Are your calculated values consistent with the simple model that we use to describe order in solids, liquids, and gases?

142. The energy content of dietary fat is 39 kJ/g, and for protein and carbohydrates it is 17 and 16 kJ/g, respectively. A 70.0-kg (155-lb) person utilizes 335 kJ/h while resting and 1250. kJ/h while walking 6 km/h. How many hours would the person need to walk per day instead of resting if he or she consumed 100. g (about $\frac{1}{4}$ lb) of fat instead of 100. g of protein in order to not gain weight?

143. The enthalpy change for melting one mole of water at 273 K is $\Delta H^0_{273} = 6010.$ J/mol, whereas that for vaporizing a mole of water at 373 K is $\Delta H^0_{273} = 40,660$ J/mol. Why is the second value so much larger?

144. A 43.6-g chunk of lead was removed from a beaker of boiling water, quickly dried, and dropped into a polystyrene cup containing 50.0 g of water at 25.0°C. As the system reached equilibrium, the water temperature rose to 26.8°C. The heat capacity of the polystyrene cup had previously been determined to be 18.6 J/°C. Calculate the molar heat capacity and the specific heat of lead.

145. Methane, $CH_4(g)$, is the main constituent of natural gas. In excess oxygen, methane burns to form $CO_2(g)$ and $H_2O(\ell)$, whereas in limited oxygen, the products are $CO(g)$ and $H_2O(\ell)$. Which would result in a higher temperature: a gas–air flame or a gas–oxygen flame? How can you tell?

A methane flame

146. A 0.483-g sample of butter was burned in a bomb calorimeter whose heat capacity was 4572 J/°C, and the temperature was observed to rise from 24.76 to 27.93°C. Calculate the fuel value of butter in (a) kJ/g; (b) nutritional Calories/g (one nutritional Calorie is equal to one kilocalorie); (c) nutritional Calories/5-gram pat.

🐾 **Molecular Reasoning** exercises ▲ **More Challenging** exercises Blue-Numbered exercises solved in Student Solutions Manual

Beyond the Textbook

NOTE: *Whenever the answer to an exercise depends on information obtained from a source other than this textbook, the source must be included, as an essential part of the answer.*

Go to a suitable site and find the answers to the next two questions about Germain Henri Hess.

147. Hess's Law is also known as the Law of _____ _____ _____.

148. Before he became a professor of chemistry at 28 years of age, what career did Germain Hess pursue?

149. Go to a website that describes the events in the life of Josiah Willard Gibbs.
(a) In what way was the doctorate degree granted to Gibbs, from whom Gibbs free energy gets its name, a first? (b) At what university did he spend nearly all his academic career?

150. Use the *Handbook of Chemistry and Physics* or another suitable reference to find the heats of formation of the following hydrocarbons: methane, ethane, propane, and *n*-butane. (a) What are the units of the values you have found? (b) Is there a trend when you compare the formula weight of each hydrocarbon with its heat of formation?

151. Why does $Al_2O_3(s)$ have a lower entropy than $Fe_2O_3(s)$? There are two primary qualitative reasons for this. You may have to use the chemistry library or some other source to get more information on the properties of these two common compounds to answer the question.

152. Steel is made by the high-temperature reaction of iron oxide (Fe_2O_3) with coke (a form of carbon) to produce metallic iron and CO_2. This same reaction can NOT be done with alumina (Al_2O_3) and carbon to make metallic Al and CO_2. Why not? Explain fully and give thermodynamic reasons for your answer. You may need to obtain further information from outside sources on the properties of the substances involved.

● **Molecular Reasoning** exercises   ▲ **More Challenging** exercises   Blue-Numbered exercises solved in Student Solutions Manual

Unless otherwise noted, all content on this page is © Cengage Learning.

Chemical Kinetics

16

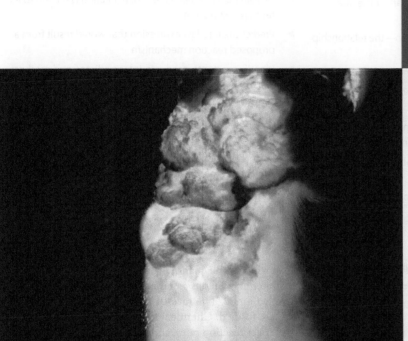

A burning building is an example of a rapid, highly exothermic reaction. Firefighters use basic principles of chemical kinetics to battle the fire. When water is sprayed onto a fire, its evaporation absorbs a large amount of energy; this lowers the temperature and slows the reaction. Other common methods for extinguishing fires include covering them with CO_2 (used in some fire extinguishers), which decreases the supply of oxygen, and backburning (for grass and forest fires), which removes combustible material. In both cases, the removal of a reactant slows (or stops) the reaction.

Lincoln Potter/Tony Stone Images

OBJECTIVES

After you have studied this chapter, you should be able to

▶ Express the rate of a chemical reaction in terms of changes in concentrations of reactants and products with time

▶ Describe the experimental factors that affect the rates of chemical reactions

▶ Use the rate-law expression for a reaction—the relationship between concentration and rate

▶ Use the concept of order of a reaction

▶ Apply the method of initial rates to find the rate-law expression for a reaction

▶ Use the integrated rate-law expression for a reaction—the relationship between concentration and time

▶ Analyze concentration-versus-time data to determine the order of a reaction

▶ Describe the collision theory of reaction rates

▶ Describe the main aspects of transition state theory and the role of activation energy in determining the rate of a reaction

▶ Explain how the mechanism of a reaction is related to its rate-law expression

▶ Predict the rate-law expression that would result from a proposed reaction mechanism

▶ Identify reactants, products, intermediates, and catalysts in a multistep reaction mechanism

▶ Explain how temperature affects rates of reactions

▶ Use the Arrhenius equation to relate the activation energy for a reaction to changes in its rate constant with changing temperature

▶ Explain how a catalyst changes the rate of a reaction

▶ Describe homogeneous catalysis and heterogeneous catalysis

We are all familiar with processes in which some quantity changes with time—a car travels at 40 miles/hour, a faucet delivers water at 3 gallons/minute, or a factory produces 32,000 tires/day. Each of these ratios is called a rate. The **rate of a reaction** describes how fast reactants are used up and products are formed. **Chemical kinetics** is the study of rates of chemical reactions, the factors that affect reaction rates, and the mechanisms (the series of steps) by which reactions occur.

Our experience tells us that different chemical reactions occur at very different rates. For instance, combustion reactions—such as the burning of methane, CH_4, in natural gas and the combustion of isooctane, C_8H_{18}, in gasoline—proceed very rapidly, sometimes even explosively.

$$CH_4(g) + 2O_2(g) \longrightarrow CO_2(g) + 2H_2O(g)$$

$$2C_8H_{18}(g) + 25O_2(g) \longrightarrow 16CO_2(g) + 18H_2O(g)$$

On the other hand, the rusting of iron occurs only very slowly.

Understanding and controlling the rates of reactions is critically important in nearly all areas. In any living system, a huge number of reactions must all interconnect smoothly. Plants and animals use many different ways, all based on the ideas of this chapter, to provide materials from one reaction at rates and in amounts needed for subsequent reactions. Illnesses often disrupt the normal control of reaction rates. The preservation of food by refrigeration uses lowered temperatures to slow the rate of unwanted spoilage reactions. Similarly, chemical engineers use the concepts of chemical kinetics to develop conditions to produce desired materials at a useful and economic rate, while slowing unwanted side reactions and avoiding dangerously high reaction rates that might lead to explosions.

In our study of thermodynamics, we learned to assess whether a particular reaction was spontaneous or not and how much energy was released or absorbed. The question of how rapidly a reaction proceeds is addressed by kinetics. Even though a reaction is thermodynamically spontaneous, it might not occur at a measurable rate.

The reactions of strong acids with strong bases are thermodynamically favored and occur at very rapid rates. Consider, for example, the reaction of hydrochloric acid solution with solid magnesium hydroxide. It is thermodynamically spontaneous at standard state conditions, as indicated by the negative ΔG^0_{rxn} value. It also occurs rapidly.

$$2HCl(aq) + Mg(OH)_2(s) \longrightarrow MgCl_2(aq) + 2H_2O(\ell) \qquad \Delta G^0_{rxn} = -97 \text{ kJ/mol rxn}$$

STOP & THINK

Remember that *spontaneous* is a thermodynamic term for a reaction that releases free energy, which could potentially be used to do work. It does not mean that a reaction is *instantaneous* (extremely fast). Only by studying the kinetics of a spontaneous reaction can we determine how rapidly it will proceed.

▶ This is one of the reactions that occurs in a human digestive system when an antacid containing relatively insoluble magnesium hydroxide neutralizes excess stomach acid.

The reaction of diamond with oxygen is also thermodynamically spontaneous.

$$C(\text{diamond}) + O_2(g) \longrightarrow CO_2(g) \qquad \Delta G^0_{rxn} = -397 \text{ kJ/mol rxn}$$

However, we know from experience that diamonds exposed to the oxygen in air, even over long periods, do not react to form carbon dioxide. The reaction does not occur at an observable rate near room temperature, but diamonds will burn when heated to a high enough temperature in the presence of oxygen.

The reaction of graphite with oxygen is also spontaneous, with a similar value of $\Delta G^0_{rxn} = -394$ kJ/mol rxn. Once it is started, this reaction occurs rapidly. These observations of reaction speeds are explained by kinetics, not thermodynamics.

▶ Recall that the units kJ/mol rxn refer to the numbers of moles of reactants and products in the balanced equation. The units kJ/mol rxn are often shortened to kJ/mol.

Propane burning is a very rapid reaction.

16-1 The Rate of a Reaction

Rates of reactions are usually expressed in units of moles per liter per unit time. If we know the chemical equation for a reaction, its rate can be determined by following the change in concentration of any product or reactant that can be detected quantitatively.

To describe the rate of a reaction, we must determine the concentration of a reactant or product at various times as the reaction proceeds. Devising effective methods for measuring these concentrations is a continuing challenge for chemists who study chemical kinetics. If a reaction is slow enough, we can take samples from the reaction mixture after successive time intervals and then analyze them. For instance, if one reaction product is an acid, its concentration can be determined by titration (see Section 11-2) after each time interval. The reaction of ethyl acetate with water in the presence of a small amount of strong acid produces acetic acid. The extent of the reaction at any time can be determined by titration of the acetic acid.

$$\underset{\text{ethyl acetate}}{CH_3\overset{\overset{\displaystyle O}{\|}}{C}-OCH_2CH_3(aq)} + H_2O(\ell) \xrightarrow{H^+} \underset{\text{acetic acid}}{CH_3\overset{\overset{\displaystyle O}{\|}}{C}-OH(aq)} + \underset{\text{ethanol}}{CH_3CH_2OH(aq)}$$

This approach is suitable only if the reaction is sufficiently slow that the time elapsed during withdrawal and analysis of the sample is negligible. Sometimes the sample is withdrawn and then quickly cooled ("quenched"). This slows the reaction (see Section 16-8) so much that the desired concentration does not change significantly while the analysis is performed.

It is more convenient, especially when a reaction is rapid, to use a technique that continually monitors the change in some physical property of the system. If one of the reactants or products is colored, the increase (or decrease) in intensity of its color might be used to measure a decrease or increase in its concentration. Such an experiment is a special case of *spectroscopic* methods. These methods involve passing light (visible, infrared, or ultraviolet) through the sample. The light should have a wavelength that is absorbed by some substance whose concentration is changing (Figure 16-1). An appropriate light-sensing apparatus provides a signal that depends on the concentration of the absorbing substance. Modern techniques that use computer-controlled pulsing and sensing of lasers have enabled scientists to sample concentrations at very frequent intervals on the order of picoseconds (1 picosecond = 10^{-12} second) or even femtoseconds (1 femtosecond = 10^{-15} second). Such studies have yielded information about very fast reactions, such as energy transfer resulting from absorption of light in photosynthesis.

Ⓐ Blue dye is reacting with bleach, which converts it into a colorless product. The color decreases and eventually disappears. The rate of the reaction could be determined by repeatedly measuring both the color intensity and the elapsed time. The concentration of dye could be calculated from the intensity of the blue color. (Only unreacted dye molecules are shown for clarity.)

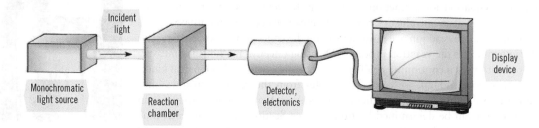

Ⓑ A schematic of a method for determining reaction rates is shown. Light of a wavelength that is absorbed by some substance whose concentration is changing is passed through a reaction chamber. Recording the change in light intensity gives a measure of the changing concentration of a reactant or product as the reaction progresses.

Figure 16-1 Monitoring a reaction via a change in color intensity.

▶ In reactions involving gases, rates of reactions may be related to rates of change of partial pressures. Pressures of gases and concentrations of gases are directly proportional.

$$PV = nRT \quad \text{or} \quad P = \frac{n}{V} RT = MRT$$

where M is molarity.

If the progress of a reaction causes a change in the total number of moles of gas present, the change in pressure of the reaction mixture (held at constant temperature and constant volume) lets us measure how far the reaction has gone. For instance, the decomposition of dinitrogen pentoxide, $N_2O_5(g)$, has been studied by this method.

$$2N_2O_5(g) \longrightarrow 4NO_2(g) + O_2(g)$$

For every two moles of N_2O_5 gas that react, a total of five moles of gas is formed (four moles of NO_2 and one mole of O_2). The resulting increase in pressure can be related by the ideal gas equation to the total number of moles of gas present. This indicates the extent to which the reaction has occurred.

Once we have measured the changes in concentrations of reactants or products with time, how do we describe the rate of a reaction? Consider a hypothetical reaction.

$$aA + bB \longrightarrow cC + dD$$

In this generalized representation, a represents the coefficient of substance A in the balanced chemical equation, b is the coefficient of substance B, and so on. For example, in the equation for the decomposition of N_2O_5, A would represent the reactant N_2O_5, with $a = 2$; C would be the first product NO_2, with $c = 4$; and D the remaining product O_2, with $d = 1$.

▶ The Greek "delta," Δ, stands for "change in," just as it did in Chapter 15.

The amount of each substance present can be given by its concentration, usually expressed as molarity (mol/L) and designated by brackets. The rate at which the reaction proceeds can be described in terms of the rate at which one of the reactants disappears, $-\Delta[A]/\Delta t$ or $-\Delta[B]/\Delta t$, or the rate at which one of the products appears, $\Delta[C]/\Delta t$ or $\Delta[D]/\Delta t$. The reaction rate must be positive because it describes the forward (left-to-right) reaction, which consumes A and B. The concentrations of reactants A and B decrease in the time interval Δt. Thus, $\Delta[A]/\Delta t$ and $\Delta[B]/\Delta t$ would be *negative* quantities. The purpose of the negative sign in the definition when using a reactant is to make the *reaction rate* positive.

Changes in concentration are related to one another by the stoichiometry of the balanced equation. For every a mol/L that [A] decreases, [B] must decrease by b mol/L, [C] must increase by c mol/L, and so on. In the case of the N_2O_5 reaction:

$$2N_2O_5(g) \longrightarrow 4NO_2(g) + O_2(g)$$

NO_2 is produced at a rate $\frac{4}{2}$ or twice as fast as N_2O_5 disappears. O_2, on the other hand, is produced $\frac{1}{2}$ or half as fast as N_2O_5 is consumed.

We wish to describe the rate of reaction on a basis that is the same regardless of which reactant or product we choose to measure. To do this, we can describe the number of *moles of reaction* that occur per liter in a given time. For instance, this is accomplished for reactant A as

$$\left(\frac{1 \text{ mol rxn}}{a \text{ mol A}}\right)\left(\begin{array}{c}\text{rate of decrease} \\ \text{in } [A]\end{array}\right) = -\left(\frac{1 \text{ mol rxn}}{a \text{ mol A}}\right)\left(\frac{\Delta[A]}{\Delta t}\right)$$

The units for rate of reaction are $\dfrac{\text{mol rxn}}{L \cdot \text{time}}$, which we usually shorten to $\dfrac{\text{mol}}{L \cdot \text{time}}$ or $\text{mol} \cdot L^{-1} \cdot \text{time}^{-1}$. The units $\dfrac{\text{mol}}{L}$ represent molarity, M, so the units for rate of reaction can also be written as $\dfrac{M}{\text{time}}$ or $M \cdot \text{time}^{-1}$. Similarly, we can divide each concentration change by its coefficient in the balanced equation. Bringing signs to the beginning of each term, we write the rate of reaction based on the rate of change of concentration of each species.

STOP & THINK
The *relative* reaction rates for the disappearance of reactants and formation of products are based on the reaction stoichiometries.

▶ Time units are usually expressed in seconds and abbreviated as s. Occasionally, minutes (min), hours (h), or years (y) are used for slower reactions.

	in terms of reactants		in terms of products	
rate of reaction $=$	$\dfrac{1}{a}\left(\begin{array}{c}\text{rate of} \\ \text{decrease} \\ \text{in } [A]\end{array}\right)$ $=$	$\dfrac{1}{b}\left(\begin{array}{c}\text{rate of} \\ \text{decrease} \\ \text{in } [B]\end{array}\right)$ $=$	$\dfrac{1}{c}\left(\begin{array}{c}\text{rate of} \\ \text{increase} \\ \text{in } [C]\end{array}\right)$ $=$	$\dfrac{1}{d}\left(\begin{array}{c}\text{rate of} \\ \text{increase} \\ \text{in } [D]\end{array}\right)$
rate of reaction $=$	$-\dfrac{1}{a}\left(\dfrac{\Delta[A]}{\Delta t}\right)$ $=$	$-\dfrac{1}{b}\left(\dfrac{\Delta[B]}{\Delta t}\right)$ $=$	$\dfrac{1}{c}\left(\dfrac{\Delta[C]}{\Delta t}\right)$ $=$	$\dfrac{1}{d}\left(\dfrac{\Delta[D]}{\Delta t}\right)$

This representation gives several equalities, any one of which can be used to relate changes in observed concentrations to the rate of reaction.

The expressions just given describe the *average* rate over some finite time period Δt. The rigorous expressions for the rate at any instant involve the derivatives of concentrations with respect to time.

$$-\frac{1}{a}\left(\frac{d[A]}{dt}\right), \frac{1}{c}\left(\frac{d[C]}{dt}\right), \text{ and so on.}$$

The shorter the time period, the closer $\dfrac{\Delta(\text{concentration})}{\Delta t}$ is to the corresponding derivative.

Problem-Solving Tip Signs and Divisors in Expressions for Rate

As an analogy to these chemical reaction rate expressions, suppose we make sardine sandwiches by the following procedure:

$$2 \text{ bread slices} + 3 \text{ sardines} + 1 \text{ pickle} \longrightarrow 1 \text{ sandwich}$$

As time goes by, the number of sandwiches increases, so Δ(sandwiches) is positive; the rate of the process is given by Δ(sandwiches)/Δ(time). Alternatively, we would count the decreasing number of pickles at various times. Because Δ(pickles) is negative, we must multiply by (-1) to make the rate positive; rate $= -\Delta(\text{pickles})/\Delta(\text{time})$. If we measure the rate by counting slices of bread, we must also take into account that bread slices are consumed *twice as fast* as sandwiches are produced, so rate $= -\frac{1}{2}(\Delta(\text{bread})/\Delta(\text{time}))$. Four different ways of describing the rate all have the same numerical value.

$$\text{rate} = \left(\frac{\Delta(\text{sandwiches})}{\Delta t}\right) = -\frac{1}{2}\left(\frac{\Delta(\text{bread})}{\Delta t}\right) = -\frac{1}{3}\left(\frac{\Delta(\text{sardines})}{\Delta t}\right) = -\left(\frac{\Delta(\text{pickles})}{\Delta t}\right)$$

Consider as a specific chemical example the gas-phase reaction that occurs when we mix 1.000 mole of hydrogen and 2.000 moles of iodine chloride at 230°C in a closed 1.000-liter container.

$$H_2(g) + 2ICl(g) \longrightarrow I_2(g) + 2HCl(g)$$

The coefficients tell us that one mole of H_2 disappears for every two moles of ICl that disappear and for every one mole of I_2 and two moles of HCl that are formed. In other terms, the rate of disappearance of moles of H_2 is one-half the rate of disappearance of moles of ICl, and so on. So we write the rate of reaction as

	in terms of reactants		in terms of products	
rate of reaction $=$	$\begin{pmatrix}\text{rate of}\\\text{decrease}\\\text{in } [H_2]\end{pmatrix}$	$= \dfrac{1}{2}\begin{pmatrix}\text{rate of}\\\text{decrease}\\\text{in } [ICl]\end{pmatrix}$	$= \begin{pmatrix}\text{rate of}\\\text{increase}\\\text{in } [I_2]\end{pmatrix}$	$= \dfrac{1}{2}\begin{pmatrix}\text{rate of}\\\text{increase}\\\text{in } [HCl]\end{pmatrix}$

$$\text{rate of reaction} = -\left(\frac{\Delta[H_2]}{\Delta t}\right) = -\frac{1}{2}\left(\frac{\Delta[ICl]}{\Delta t}\right) = \left(\frac{\Delta[I_2]}{\Delta t}\right) = \frac{1}{2}\left(\frac{\Delta[HCl]}{\Delta t}\right)$$

Table 16-1 lists the concentrations of reactants remaining at 1-second intervals, beginning with the time of mixing ($t = 0$ seconds). The *average* rate of reaction over each 1-second interval is indicated in terms of the rate of decrease in concentration of hydrogen. Verify for yourself that the rate of loss of ICl is twice that of H_2. Therefore, the rate of reaction could also be expressed as rate $= -\frac{1}{2}(\Delta[ICl]/\Delta t)$. Increases in concentration of products could be used instead. Figure 16-2 shows graphically the rates of change of concentrations of all reactants and products.

STOP & THINK

Suppose a driver travels 40 miles in an hour; we describe his average speed (rate) as 40 mi/h. This does not necessarily mean that he drove at a steady speed. He might have stopped at a few traffic signals, made a fuel stop, driven sometimes faster, sometimes slower—his *instantaneous rate* (the rate at which he was traveling at any instant) was quite changeable.

▶ For example, the *average* rate over the interval from 1 to 2 seconds can be calculated as

$$-\frac{\Delta[H_2]}{\Delta t} =$$

$$-\frac{(0.526 - 0.674)\,\text{mol} \cdot L^{-1}}{(2 - 1)\,s}$$

$$= 0.148\,\text{mol} \cdot L^{-1}s^{-1}$$

$$= 0.148\,M \cdot s^{-1}$$

This does *not* mean that the reaction proceeds at this rate during the entire interval.

Table 16-1   Concentration and Rate Data for Reaction of 2.000 *M* ICl and 1.000 *M* H_2 at 230°C

$H_2(g) + 2ICl(g) \longrightarrow I_2(g) + 2HCl(g)$		Average Rate During Time Interval $= -\dfrac{\Delta[H_2]}{\Delta t}$	
[ICl] (mol/L)	[H_2] (mol/L)	$(M \cdot s^{-1})$	Time (t) (seconds)
2.000	1.000		0
		0.326	
1.348	0.674		1
		0.148	
1.052	0.526		2
		0.090	
0.872	0.436		3
		0.062	
0.748	0.374		4
		0.046	
0.656	0.328		5
		0.035	
0.586	0.293		6
		0.028	
0.530	0.265		7
		0.023	
0.484	0.242		8

Figure 16-3 is a plot of the hydrogen concentration versus time, using data from Table 16-1. The initial rate, or the rate at the instant of mixing the reactants, is the negative of the slope at $t = 0$. The *instantaneous* rate of reaction at time t (2.0 seconds, for example) is the negative of the slope of the tangent to the curve at time t. We see that the rate decreases with time; lower concentrations of H_2 and ICl result in slower reaction. Had we plotted concentration of a product versus time, the rate would have been related to the *positive* slope of the tangent at time t.

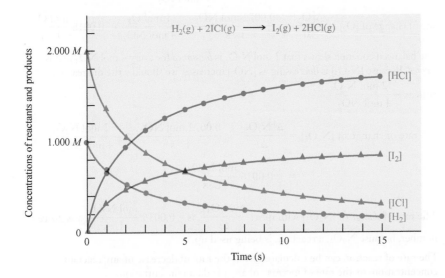

Figure 16-2 Plot of concentrations of all reactants and products versus time in the reaction of 1.000 M H_2 with 2.000 M ICl at 230°C, from data in Table 16-1 (and a few more points).

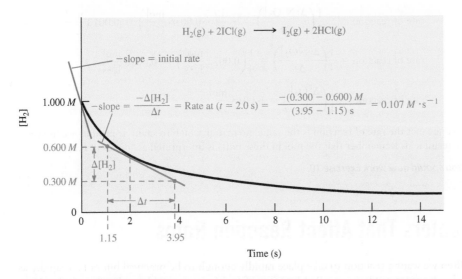

Figure 16-3 Plot of H_2 concentration versus time for the reaction of 1.000 M H_2 with 2.000 M ICl. The instantaneous rate of reaction at any time, t, equals the negative of the slope of the tangent to this curve at time t. The initial rate of the reaction is equal to the negative of the initial slope ($t = 0$). The determination of the instantaneous rate at $t = 2$ seconds is illustrated. (If you do not recall how to find the slope of a straight line, refer to Figure 16-5.)

EXAMPLE 16-1 Rate of Reaction

At some time, we observe that the reaction $2N_2O_5(g) \longrightarrow 4NO_2(g) + O_2(g)$ is forming NO_2 at the rate of 0.0072 $\dfrac{mol}{L \cdot s}$.

(a) What is the rate of change of $[O_2]$, $\dfrac{\Delta[O_2]}{\Delta t}$, in $\dfrac{mol}{L \cdot s}$, at this time?

(b) What is the rate of change of $[N_2O_5]$, $\dfrac{\Delta[N_2O_5]}{\Delta t}$, in $\dfrac{mol}{L \cdot s}$, at this time?

(c) What is the rate of reaction at this time?

Plan

We can use the mole ratios (stoichiometry) from the balanced equation to determine the rates of change of other products and reactants. The rate of reaction can then be derived from any one of these individual rates.

Solution

(a) The balanced equation gives the reaction ratio $\dfrac{1 \text{ mol } O_2}{4 \text{ mol } NO_2}$.

$$\text{rate of change of } [O_2] = \frac{\Delta [O_2]}{\Delta t} = \frac{0.0072 \text{ mol } NO_2}{L \cdot s} \times \frac{1 \text{ mol } O_2}{4 \text{ mol } NO_2} = 0.0018 \frac{\text{mol } O_2}{L \cdot s}$$

(b) The balanced equation shows that 2 mol N_2O_5 is *consumed* for every 4 mol NO_2 that is *formed*. Because $[N_2O_5]$ is decreasing as $[NO_2]$ increases, we should write the reaction ratio as $\dfrac{-2 \text{ mol } N_2O_5}{4 \text{ mol } NO_2}$.

$$\text{rate of change of } [N_2O_5] = \frac{\Delta [N_2O_5]}{\Delta t} = \frac{0.0072 \text{ mol } NO_2}{L \cdot s} \times \frac{-2 \text{ mol } N_2O_5}{4 \text{ mol } NO_2}$$

$$= -0.0036 \frac{\text{mol } N_2O_5}{L \cdot s}$$

▶ The concentration of N_2O_5 is decreasing, so its rate of change is negative.

The rate of *change* of $[N_2O_5]$ with time, $\dfrac{\Delta [N_2O_5]}{\Delta t}$, is $-0.0036 \dfrac{\text{mol } N_2O_5}{L \cdot s}$, a *negative* number, because N_2O_5, a reactant, is being used up.

(c) The rate of reaction can be calculated from the rate of decrease of any reactant concentration or the rate of increase of any product concentration.

$$\text{rate of reaction} = -\frac{1}{2}\left(\frac{\Delta [N_2O_5]}{\Delta t}\right) = -\frac{1}{2}\left(-0.0036 \frac{\text{mol}}{L \cdot s}\right) = 0.0018 \frac{\text{mol}}{L \cdot s}$$

$$\text{rate of reaction} = \frac{1}{4}\left(\frac{\Delta [NO_2]}{\Delta t}\right) = \frac{1}{4}\left(0.0072 \frac{\text{mol}}{L \cdot s}\right) = 0.0018 \frac{\text{mol}}{L \cdot s}$$

$$\text{rate of reaction} = \frac{1}{1}\left(\frac{\Delta [O_2]}{\Delta t}\right) = 0.0018 \frac{\text{mol}}{L \cdot s}$$

> **S TOP & THINK**
>
> Always remember to divide by the balanced reactant or product coefficients in order to convert the individual reactant and product rates into an overall reaction rate. The overall reaction rate is always positive and expressed on a per mole basis.

We see that the rate of reaction is the same, no matter which reactant or product we use to determine it. Remember that the mol in these units is interpreted as "moles of reaction."

You should now work Exercise 10.

Factors That Affect Reaction Rates

Often we want a reaction to take place rapidly enough to be practical but not so rapidly as to be dangerous. The controlled burning of fuel in an internal combustion engine is an example of such a process. On the other hand, we want some undesirable reactions, such as the spoiling of food, to take place more slowly.

Four factors have marked effects on the rates of chemical reactions.

1. Nature of the reactants
2. Concentrations of the reactants
3. Temperature
4. The presence of a catalyst

Understanding the effects of these factors can help us control the rates of reactions in desirable ways. The study of these factors gives important insight into the details of the processes by which a reaction occurs. This kind of study is the basis for developing theories of chemical kinetics. We will now study these factors and the related theories—collision theory and transition state theory.

16-2 Nature of the Reactants

The physical states of reacting substances are important in determining their reactivities. A puddle of liquid gasoline can burn smoothly, but gasoline vapors can burn explosively. Two immiscible liquids may react slowly at their interface, but if they are mixed to provide better contact, the reaction speeds up. White phosphorus and red phosphorus are different solid forms (allotropes) of elemental phosphorus. White phosphorus ignites when exposed to oxygen in the air. By contrast, red phosphorus can be kept in open containers for long periods of time without noticeable reaction.

Two allotropes of phosphorus. White phosphorus (*above*) ignites and burns rapidly when exposed to oxygen in the air, so it is stored under water. Red phosphorus (*below*) reacts with air much more slowly, so it can be stored in contact with air.

Samples of dry solid potassium sulfate, K_2SO_4, and dry solid barium nitrate, $Ba(NO_3)_2$, can be mixed with no appreciable reaction occurring for several years. But if aqueous solutions of the two are mixed, a reaction occurs rapidly, forming a white precipitate of barium sulfate.

$$Ba^{2+}(aq) + SO_4^{2-}(aq) \longrightarrow BaSO_4(s) \qquad \text{(net ionic equation)}$$

Chemical identities of elements and compounds affect reaction rates. Metallic sodium, with its low ionization energy and strong reducing capability, reacts rapidly with water at room temperature; metallic calcium has a higher ionization energy and reacts only slowly with water at room temperature. Solutions of a strong acid and a strong base react rapidly when they are mixed because the interactions involve mainly electrostatic attractions between ions in solution. Reactions that involve the breaking of covalent bonds are usually slower.

The extent of subdivision of solids or liquids can be crucial in determining reaction rates. Large chunks of most metals do not burn. But many powdered metals, with larger surface areas and hence more atoms exposed to the oxygen of the air, burn easily. One pound of fine iron wire rusts much more rapidly than a solid one-pound chunk of iron. Violent explosions sometimes occur in grain elevators, coal mines, and chemical plants in which large amounts of powdered substances are produced. These explosions are examples of the effect of large surface areas on rates of reaction. The rate of reaction depends on the surface area or degree of subdivision. The ultimate degree of subdivision would make all reactant molecules (or ions or atoms) accessible to react at any given time. This situation can be achieved when the reactants are in the gaseous state or in solution.

Powdered iron burns very rapidly when heated in a flame. Iron oxide is formed.

Powdered chalk (mostly calcium carbonate, $CaCO_3$) reacts rapidly with dilute hydrochloric acid because it has a large total surface area. A stick of chalk has a much smaller surface area, so it reacts much more slowly.

16-3 Concentrations of Reactants: The Rate-Law Expression

As the concentrations of reactants change at constant temperature, the rate of reaction changes. We write the **rate-law expression** (often called simply the **rate law**) for a reaction to describe how its rate depends on concentrations; this rate law is experimentally deduced for each reaction from a study of how its rate varies with concentration.

The rate-law expression for a reaction in which A, B, . . . are reactants has the general form

$$\text{rate} = k[A]^x[B]^y \dots$$

The constant k is called the **specific rate constant** (or just the **rate constant**) for the reaction at a particular temperature. The values of the exponents, x and y, and of the rate constant, k, bear no necessary relationship to the coefficients in the *balanced chemical equation* for the overall reaction and must be determined *experimentally*.

The powers to which the concentrations are raised, x and y, are usually small integers or zero but are occasionally fractional or even negative. A power of *one* means that the rate is directly proportional to the concentration of that reactant. A power of *two* means that the rate is directly proportional to the *square* of that concentration. A power of *zero* means that the rate does not depend on the concentration of that reactant, *so long as some of the reactant is present*. The value of x is said to be the order of the reaction with respect to A, and y is the order of the reaction with respect to B. The overall **order of the reaction** is the sum of the reactant orders, $x + y$. Examples of observed rate laws for some reactions follow.

▶ The order of the reaction with respect to a reactant is usually called simply the order of that reactant. The word *order* is used in kinetics in its mathematical meaning.

1. $3NO(g) \longrightarrow N_2O(g) + NO_2(g)$ $\text{rate} = k[NO]^2$

second order in NO; second order overall

2. $2NO_2(g) + F_2(g) \longrightarrow 2NO_2F(g)$ $\text{rate} = k[NO_2][F_2]$

first order in NO_2 and first order in F_2; second order overall

3. $2NO_2(g) \longrightarrow 2NO(g) + O_2(g)$ $\text{rate} = k[NO_2]^2$

second order in NO_2; second order overall

▶ Any number raised to the zero power is one. Here $[H^+]^0 = 1$.

4. $H_2O_2(aq) + 3I^-(aq) + 2H^+(aq) \longrightarrow 2H_2O(\ell) + I_3^-(aq)$ $\text{rate} = k[H_2O_2][I^-]$

first order in H_2O_2 and first order in I^-; zero order in H^+; second order overall

We see that the orders (exponents) in the rate law expression *may* or *may not* match the coefficients in the balanced equation. There is *no* way to predict reaction orders from the balanced overall chemical equation. The orders must be determined experimentally.

▶ More details about values and units of k will be discussed in later sections.

It is important to remember the following points about the specific rate constant, k.

1. Experimental data must first be used to determine the reaction orders and then the value of k for the reaction at appropriate conditions.
2. The value we determine is for a *specific reaction*, represented by a balanced equation.
3. The units of k depend on the *overall order* of the reaction.
4. The value we determine does not change with concentrations of either reactants or products.
5. The value we determine does not change with time (Section 16-4).
6. The value we determine refers to the reaction *at a particular temperature* and changes if we change the temperature (Section 16-8).
7. The value we determine depends on whether a *catalyst* is present and on the nature of the catalyst (Section 16-9).

EXAMPLE 16-2   Interpretation of the Rate Law

For a hypothetical reaction

$$A + B + C \longrightarrow \text{products}$$

the rate law is determined to be

$$\text{rate} = k[A][B]^2$$

What happens to the reaction rate when we make each of the following concentration changes?

(a) We triple the concentration of A without changing the concentration of B or C. (b) We triple the concentration of B without changing the concentration of A or C. (c) We triple the concentration of C without changing the concentration of A or B. (d) We triple all three concentrations simultaneously.

Plan

We interpret the rate law to predict the changes in reaction rate. We remember that changing concentrations of reactants does not change the value of k.

Solution

(a) We see that rate is directly proportional to the *first power* of [A]. We do not change [B] or [C]. Tripling [A] (i.e., increasing [A] by a factor of 3) causes the reaction rate to increase by a factor of $3^1 = 3$ so the reaction rate triples.

(b) We see that rate is directly proportional to the *second power* of [B], that is $[B]^2$. We do not change [A] or [C]. Tripling [B] (i.e., increasing [B] by a factor of 3) causes the reaction rate to increase by a factor of $3^2 = 9$.

(c) The reaction rate is independent of [C] (zero order), so changing [C] causes no change in reaction rate ($3^0 = 1$).

(d) Tripling all concentrations would cause the changes described in (a), (b), and (c) simultaneously. The rate would increase by a factor of 3 due to the change in [A], by a factor of 9 due to the change in [B], and be unaffected by the change in [C]. The result is that the reaction rate increases by a factor of $3^1 \times 3^2 = 3 \times 9 = 27$.

You should now work Exercises 14 and 15.

S TOP & THINK

The concentration change is raised to the power represented by the order of that reactant. A common mistake is to multiply the concentration change by the order and not raise it to that power.

We can use the **method of initial rates** to deduce the rate law from experimentally measured rate data. Usually we know the concentrations of all reactants at the start of the reaction. We can then measure the initial rate of the reaction, corresponding to these initial concentrations. The following tabulated data refer to the hypothetical reaction

$$A + 2B \longrightarrow C$$

at a specific temperature. The brackets indicate the concentrations of the reacting species *at the beginning* of each experimental run listed in the first column—that is, the initial concentrations for each experiment.

▶ In such an experiment, we often keep some initial concentrations the same and vary others by simple factors, such as 2 or 3. This makes it easier to assess the effect of each change on the rate.

Experiment	Initial [A]		Initial [B]		Initial Rate of Formation of C	
1	$1.0 \times 10^{-2} M$	— no change	$1.0 \times 10^{-2} M$	— × 2	$1.5 \times 10^{-6} M \cdot s^{-1}$	— × 2
2	$1.0 \times 10^{-2} M$	— × 2	$2.0 \times 10^{-2} M$	— no change	$3.0 \times 10^{-6} M \cdot s^{-1}$	— × 4
3	$2.0 \times 10^{-2} M$		$2.0 \times 10^{-2} M$		$1.2 \times 10^{-5} M \cdot s^{-1}$	

Because we are describing the same reaction in each experiment, each is governed by the same rate-law expression. This expression has the form

$$\text{rate} = k[A]^x[B]^y$$

Let's compare the initial rates of formation of product (reaction rates) for different experimental runs to see how changes in concentrations of reactants affect the rate of reaction. This lets us calculate the values of x and y, and then k.

We see that the initial concentration of A is the same in experiments 1 and 2; for these trials, any change in reaction rate would be due to different initial concentrations of B. Comparing these two experiments, we see that [B] has been changed by a factor of

$$\frac{2.0 \times 10^{-2}}{1.0 \times 10^{-2}} = 2.0 = [B] \text{ ratio}$$

The rate changes by a factor of

$$\frac{3.0 \times 10^{-6}}{1.5 \times 10^{-6}} = 2.0 = \text{rate ratio}$$

The exponent y can be deduced from

$$\text{rate ratio} = ([B] \text{ ratio})^y$$

$$2.0 = (2.0)^y \qquad \text{so} \qquad y = 1$$

The reaction is first order in [B]. Thus far we know that the rate-law expression is

$$\text{rate} = k[A]^x[B]^1$$

To evaluate x, we observe that the concentrations of [A] are different in experiments 2 and 3. For these two trials, the initial concentration of B is the same, so any change in reaction rate could only be due to different initial concentrations of A. Comparing these two experiments, we see that [A] has been increased by a factor of

$$\frac{2.0 \times 10^{-2}}{1.0 \times 10^{-2}} = 2.0 = [A] \text{ ratio}$$

The rate increases by a factor of

$$\frac{1.2 \times 10^{-5}}{3.0 \times 10^{-6}} = 4.0 = \text{rate ratio}$$

The exponent x can be deduced from

$$\text{rate ratio} = ([A] \text{ ratio})^x$$

$$4.0 = (2.0)^x \qquad \text{so} \qquad x = 2$$

The reaction is second order in [A]. We can now write its rate-law expression as

$$\text{rate} = k[A]^2[B]$$

Now that we know the orders, the specific rate constant, k, can be evaluated by substituting any of the three sets of data into the rate-law expression. Using the data from experiment 1 gives

$$\text{rate}_1 = k[A]_1^2[B]_1 \qquad \text{or} \qquad k = \frac{\text{rate}_1}{[A]_1^2[B]_1}$$

$$k = \frac{1.5 \times 10^{-6} M \cdot s^{-1}}{(1.0 \times 10^{-2} M)^2(1.0 \times 10^{-2} M)} = 1.5\ M^{-2} \cdot s^{-1}$$

At the temperature at which the measurements were made, the rate-law expression for this reaction is

$$\text{rate} = k[A]^2[B] \qquad \text{or} \qquad \text{rate} = (1.5\ M^{-2} \cdot s^{-1})\ [A]^2[B]$$

We can check our result by evaluating k from one of the other sets of data.

S TOP & THINK

Whenever possible pick two experiments where only one reactant has different initial concentrations. We can calculate the order for this reactant from the effect of the concentration change on the initial rate of reaction.

S TOP & THINK

Remember that the specific rate constant k does *not* change with concentration. Only a temperature change or the introduction of a catalyst can change the value of k.

▶ The units of k depend on the overall order of the reaction, consistent with converting the product of concentrations on the right to concentration/time on the left. For any reaction that is third order overall, the units of k are $M^{-2} \cdot \text{time}^{-1}$.

Problem-Solving Tip   Be Sure to Use the Rate of Reaction

The rate-law expression should always give the dependence of the *rate of reaction* on concentrations. The data for the preceding calculation describe the rate of formation of the product C; the coefficient of C in the balanced equation is one, so the rate of reaction is equal to the rate of formation of C. If the coefficient of the measured substance had been 2, then before we began the analysis we should have divided each value of the "initial rate of formation" by 2 to obtain the initial rate of reaction. For instance, suppose we measure the rate of formation of AB in the reaction

$$A_2 + B_2 \longrightarrow 2AB$$

Then

$$\text{rate of reaction} = \frac{1}{2}\left(\frac{\Delta[AB]}{\Delta t}\right) = \frac{1}{2} \text{ (rate of formation of AB)}$$

Then we would analyze how this reaction rate changes as we change the concentrations of reactants.

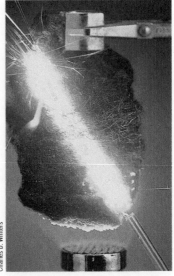

When heated in air, steel wool glows but does not burn rapidly, due to the low O_2 concentration in air (about 21%). When pure oxygen is passed through the center of the steel wool via a porous tube, the steel wool burns vigorously because of the much greater accessibility of O_2 reactant molecules.

An Alternative Method

We can also use a simple algebraic approach to find the exponents in a rate-law expression. Consider a set of rate data similar to that given earlier for the hypothetical reaction

$$A + 2B \longrightarrow C$$

Experiment	Initial [A] (M)	Initial [B] (M)	Initial Rate of Formation of C $(M \cdot s^{-1})$
1	1.0×10^{-2}	1.0×10^{-2}	1.5×10^{-6}
2	1.0×10^{-2}	2.0×10^{-2}	3.0×10^{-6}
3	2.0×10^{-2}	1.0×10^{-2}	6.0×10^{-6}

Because we are describing the same reaction in each experiment, all the experiments are governed by the same rate-law expression,

$$\text{rate} = k[A]^x[B]^y$$

The initial concentration of A is the same in experiments 1 and 2, so any change in the initial rates for these experiments would be due to different initial concentrations of B. To evaluate y, we solve the ratio of the rate-law expressions of these two experiments for y. We can divide the first rate-law expression by the corresponding terms in the second rate-law expression.

$$\frac{\text{rate}_1}{\text{rate}_2} = \frac{k[A]_1^x[B]_1^y}{k[A]_2^x[B]_2^y}$$

▶ It does not matter which way we take the ratio. We would get the same value for y if we divided the second rate-law expression by the first—try it!

The value of k always cancels from such a ratio because it is constant at a particular temperature. The initial concentrations of A are equal, so they too cancel. Thus, the expression simplifies to

$$\frac{\text{rate}_1}{\text{rate}_2} = \left(\frac{[B]_1}{[B]_2}\right)^y$$

The only unknown in this equation is y. We substitute data from experiments 1 and 2 into the equation, which gives us

$$\frac{1.5 \times 10^{-6} M \cdot s^{-1}}{3.0 \times 10^{-6} M \cdot s^{-1}} = \left(\frac{1.0 \times 10^{-2} M}{2.0 \times 10^{-2} M}\right)^y$$

$$0.5 = (0.5)^y \qquad \text{so} \qquad y = 1$$

Thus far, we know that the rate-law expression is

$$\text{rate} = k[A]^x[B]^1$$

Next we evaluate x. In experiments 1 and 3, the initial concentration of B is the same, so any change in the initial rates for these experiments would be due to the different initial concentrations of A. We solve the ratio of the rate-law expressions of these two experiments for x. We divide the third rate-law expression by the corresponding terms in the first rate-law expression.

$$\frac{\text{rate}_3}{\text{rate}_1} = \frac{k[A]_3^x[B]_3^1}{k[A]_1^x[B]_1^1}$$

The value k cancels, and so do the concentrations of B because they are equal. Thus, the expression simplifies to

$$\frac{\text{rate}_3}{\text{rate}_1} = \frac{[A]_3^x}{[A]_1^x} = \left(\frac{[A]_3}{[A]_1}\right)^x$$

$$\frac{6.0 \times 10^{-6}\, M \cdot s^{-1}}{1.5 \times 10^{-6}\, M \cdot s^{-1}} = \left(\frac{2.0 \times 10^{-2}\, M}{1.0 \times 10^{-2}\, M}\right)^x$$

$$4.0 = (2.0)^x \qquad \text{so} \qquad x = 2$$

The power to which [A] is raised in the rate-law expression is 2, so the rate-law expression for this reaction is the same as that obtained earlier.

$$\text{rate} = k[A]^2[B]^1 \qquad \text{or} \qquad \text{rate} = k[A]^2[B]$$

Problem-Solving Tip

Although we will usually work with simple integer values for reactant orders, fractional values can occur. For fractional orders the value cannot be determined quite so simply. In the previous example, if doubling the concentration of [A] changed the rate by a factor of 2.83 we would have:

$$2.83 = (2.0)^x$$

In this case the value of x is not obvious. But we can easily solve for x by taking the log of each side and remembering that the log of A^x is $(x)(\log A)$.

$$\log(2.83) = (x)\log(2.0)$$

Dividing through by $\log(2.0)$ gives

$$x = \log(2.83)/\log(2.0)$$

$$x = 0.452/0.301 = 1.50$$

So in this case the order of [A] is 1.5 or $[A]^{1.5}$.

EXAMPLE 16-3 Method of Initial Rates

Given the following data, determine the rate-law expression and the value of the rate constant for the reaction

$$2A + B + C \longrightarrow D + E$$

▶ The coefficient of E in the balanced equation is 1, so the rate of reaction is equal to the rate of formation of E.

Experiment	Initial [A]	Initial [B]	Initial [C]	Initial Rate of Formation of E
1	0.20 M	0.20 M	0.20 M	$2.4 \times 10^{-6}\, M \cdot min^{-1}$
2	0.40 M	0.30 M	0.20 M	$9.6 \times 10^{-6}\, M \cdot min^{-1}$
3	0.20 M	0.30 M	0.20 M	$2.4 \times 10^{-6}\, M \cdot min^{-1}$
4	0.20 M	0.40 M	0.60 M	$7.2 \times 10^{-6}\, M \cdot min^{-1}$

Plan

The rate law is of the form: rate $= k[A]^x[B]^y[C]^z$. We must evaluate x, y, z, and k. We use the reasoning outlined earlier; in this presentation the first method is used.

▶ The alternative algebraic method outlined previously can also be used.

Solution

Dependence on [B]: In experiments 1 and 3, the initial concentrations of A and C are unchanged. Thus, any change in the rate would be due to the change in concentration of B. But we see that the rate is the same in experiments 1 and 3, even though the concentration of B is different. Thus, the reaction rate is independent of [B], so $y = 0$ (zero order). We can neglect changes in [B] in subsequent reasoning. The rate law can now be simplified to

$$\text{rate} = k[A]^x[C]^z$$

▶ $[B]^0 = 1$

Dependence on [C]: Experiments 1 and 4 involve the same initial concentration of A; thus the observed change in rate must be due entirely to the changed [C]. So we compare experiments 1 and 4 to find z.

▶ It is helpful to set up the ratios with the larger concentration on top to give a ratio > 1.

$$[C] \text{ has increased by a factor of } \frac{0.60}{0.20} = 3.0 = [C] \text{ ratio}$$

The rate changes by a factor of

$$\frac{7.2 \times 10^{-6}}{2.4 \times 10^{-6}} = 3.0 = \text{rate ratio}$$

The exponent z can be deduced from

$$\text{rate ratio} = ([C] \text{ ratio})^z$$
$$3.0 = (3.0)^z \qquad \text{so} \qquad z = 1 \qquad \text{The reaction is first order in [C].}$$

Now we know that the rate law is of the form

$$\text{rate} = k[A]^x[C]$$

Dependence on [A]: We use experiments 1 and 2 to evaluate x, because [A] is changed, [B] does not matter, and [C] is unaltered. The observed rate change is due *only* to the changed [A].

$$[A] \text{ has increased by a factor of } \frac{0.40}{0.20} = 2.0 = [A] \text{ ratio}$$

The rate increases by a factor of

$$\frac{9.6 \times 10^{-6}}{2.4 \times 10^{-6}} = 4.0 = \text{rate ratio}$$

The exponent x can be deduced from

$$\text{rate ratio} = ([A] \text{ ratio})^x$$
$$4.0 = (2.0)^x \qquad \text{so} \qquad x = 2 \qquad \text{The reaction is second order in [A].}$$

From these results we can write the complete rate-law expression.

$$\text{rate} = k[A]^2[B]^0[C]^1 \qquad \text{or} \qquad \text{rate} = k[A]^2[C]$$

We can evaluate the specific rate constant, k, by substituting any of the four sets of data into the rate-law expression we have just derived. Data from experiment 2 give

$$\text{rate}_2 = k[A]_2^2[C]_2$$
$$k = \frac{\text{rate}_2}{[A]_2^2[C]_2} = \frac{9.6 \times 10^{-6} M \cdot \text{min}^{-1}}{(0.40\ M)^2(0.20\ M)} = 3.0 \times 10^{-4} M^{-2} \cdot \text{min}^{-1}$$

The rate-law expression can also be written with the value of k incorporated.

$$\text{rate} = (3.0 \times 10^{-4}\ M^{-2} \cdot \text{min}^{-1})\,[A]^2[C]$$

This expression allows us to calculate the rate at which this reaction occurs with any known concentrations of A and C (provided some B is present). As we shall see presently, changes in temperature alter reaction rates. This value of k is valid *only* at the temperature at which the data were collected.

You should now work Exercises 17 and 18.

S **TOP & THINK**
Because we are comparing only the rate change due to concentration changes in A, the coefficient of 2 for reactant A cancels out in the ratio, so it does not play a role here.

> **Problem-Solving Tip** Check the Rate Law You Have Derived
>
> If the rate law that you deduce from initial rate data is correct, it will not matter which set of data you use to calculate k. As a check, you can calculate k several times, once from each set of experimental concentration and rate data. If the reaction orders in your derived rate law are correct, then all sets of experimental data will give the same value of k (within rounding error); but if the orders are wrong, then the k values will vary considerably.

EXAMPLE 16-4 Method of Initial Rates

Use the following initial rate data to determine the form of the rate-law expression for the reaction

$$3A + 2B \longrightarrow 2C + D$$

Experiment	Initial [A]	Initial [B]	Initial Rate of Formation of D
1	$1.00 \times 10^{-2}\,M$	$1.00 \times 10^{-2}\,M$	$6.00 \times 10^{-3}\,M \cdot min^{-1}$
2	$2.00 \times 10^{-2}\,M$	$3.00 \times 10^{-2}\,M$	$1.44 \times 10^{-1}\,M \cdot min^{-1}$
3	$1.00 \times 10^{-2}\,M$	$2.00 \times 10^{-2}\,M$	$1.20 \times 10^{-2}\,M \cdot min^{-1}$

Plan

The rate law is of the form rate = $k[A]^x[B]^y$. No two experiments have the same initial [B], so let's use the alternative method presented earlier to evaluate x and y.

Solution

The initial concentration of A is the same in experiments 1 and 3. We divide the third rate-law expression by the corresponding terms in the first one

$$\frac{rate_3}{rate_1} = \frac{k[A]_3{}^x[B]_3{}^y}{k[A]_1{}^x[B]_1{}^y}$$

The initial concentrations of A are equal, so they cancel, as does k. Simplifying and then substituting known values of rates and [B],

$$\frac{rate_3}{rate_1} = \frac{[B]_3{}^y}{[B]_1{}^y} \quad \text{or} \quad \frac{1.20 \times 10^{-2}\,M \cdot min}{6.00 \times 10^{-3}\,M \cdot min} = \left(\frac{2.00 \times 10^{-2}\,M}{1.00 \times 10^{-2}\,M}\right)^y$$

$$2.0 = (2.0)^y \quad \text{or} \quad \boxed{y = 1} \quad \text{The reaction is first order in [B].}$$

No two of the experimental runs have the same concentrations of B, so we must proceed somewhat differently. Let's compare experiments 1 and 2. The observed change in rate must be due to the *combination* of the changes in [A] and [B]. We can divide the second rate-law expression by the corresponding terms in the first one, cancel the equal k values, and collect terms.

▶ This is solvable because we have already determined the order of B.

$$\frac{rate_2}{rate_1} = \frac{k[A]_2{}^x[B]_2{}^y}{k[A]_1{}^x[B]_1{}^y} = \left(\frac{[A]_2}{[A]_1}\right)^x \left(\frac{[B]_2}{[B]_1}\right)^y$$

Now let's insert the known values for rates and concentrations and the known [B] exponent (order) of 1.

$$\frac{1.44 \times 10^{-1}\,M \cdot min^{-1}}{6.00 \times 10^{-3}\,M \cdot min^{-1}} = \left(\frac{2.00 \times 10^{-2}\,M}{1.00 \times 10^{-2}\,M}\right)^x \left(\frac{3.00 \times 10^{-2}\,M}{1.00 \times 10^{-2}\,M}\right)^1$$

$$24.0 = (2.00)^x(3.00)$$

$$8.00 = (2.00)^x \quad \text{or} \quad x = 3 \quad \text{The reaction is third order in [A].}$$

The rate-law expression has the form rate = $k[A]^3[B]$.

You should now work Exercise 22.

16-4 Concentration Versus Time: The Integrated Rate Equation

Often we want to know the concentration of a reactant that would remain after some specified time, or how long it would take for some amount of the reactants to be used up.

> The equation that relates *concentration* and *time* is the **integrated rate equation.** We can also use it to calculate the **half-life, $t_{1/2}$,** of a reactant—the time it takes for half of that reactant to be converted into product. The integrated rate equation and the half-life are different for reactions of different order.

We will look at relationships for some simple cases. If you know calculus, you may be interested in the derivation of the integrated rate equations. This development is presented in the Enrichments at the end of this section.

First-Order Reactions

For reactions involving $a\text{A} \longrightarrow$ products that are *first order in* A and *first order overall*, the integrated rate equation is

$$\ln\left(\frac{[\text{A}]_0}{[\text{A}]}\right) = akt \quad \text{(first order)}$$

▶ The variable a represents the coefficient of reactant A in the balanced overall equation.

$[\text{A}]_0$ is the initial concentration of reactant A, and $[\text{A}]$ is its concentration at some time, t, after the reaction begins. Solving this relationship for t gives

$$t = \frac{1}{ak}\ln\left(\frac{[\text{A}]_0}{[\text{A}]}\right)$$

By definition, $[\text{A}] = \frac{1}{2}[\text{A}]_0$ at $t = t_{1/2}$. Thus

$$t_{1/2} = \frac{1}{ak}\ln\frac{[\text{A}]_0}{\frac{1}{2}[\text{A}]_0} = \frac{1}{ak}\ln 2$$

▶ When time $t_{1/2}$ has elapsed, half of the original $[\text{A}]_0$ has reacted, so half of it remains.

$$t_{1/2} = \frac{\ln 2}{ak} = \frac{0.693}{ak} \quad \text{(first order)}$$

This relates the half-life of a reactant in a *first-order reaction* and its rate constant, k. In such reactions, the half-life *does not depend* on the initial concentration of A. This is not true for reactions having overall orders other than first order.

▶ Nuclear decay (Chapter 22) is a very important first-order process. Exercises at the end of that chapter involve calculations of nuclear decay rates.

EXAMPLE 16-5 Half-Life: First-Order Reaction

Compound A decomposes to form B and C in a reaction that is first order with respect to A and first order overall. At 25°C, the specific rate constant for the reaction is 0.0450 s⁻¹. What is the half-life of A at 25°C?

$$\text{A} \longrightarrow \text{B} + \text{C}$$

Plan

We use the equation given earlier for $t_{1/2}$ for a first-order reaction. The value of k is given in the problem; the coefficient of reactant A is $a = 1$.

Solution

$$t_{1/2} = \frac{\ln 2}{ak} = \frac{0.693}{1(0.0450\ \text{s}^{-1})} = 15.4\ \text{s}$$

After 15.4 seconds of reaction, half of the original reactant remains, so $[\text{A}] = \frac{1}{2}[\text{A}]_0$.

You should now work Exercise 37.

Ⓢ TOP & THINK
Many examples, especially those involving rates of radioactive decay, have coefficients of 1. Some textbooks do not show a in the rate expression for first order. It is presumed to be one unless stated otherwise.

Problem-Solving Tip

If the problem you are solving asks for or gives *time*, use the integrated rate equation. If it asks for or gives *rate*, use the rate-law expression.

EXAMPLE 16-6 Concentration Versus Time: First-Order Reaction

The reaction $2N_2O_5(g) \longrightarrow 2N_2O_4(g) + O_2(g)$ obeys the rate law: rate $= k[N_2O_5]$, in which the specific rate constant is 0.00840 s^{-1} at a certain temperature. (a) If 2.50 moles of N_2O_5 were placed in a 5.00-liter container at that temperature, how many moles of N_2O_5 would remain after 1.00 minute? (b) How long would it take for 90% of the original N_2O_5 to react?

Plan

We apply the integrated first-order rate equation.

$$\ln \left(\frac{[N_2O_5]_0}{[N_2O_5]} \right) = akt$$

(a) First, we must determine $[N_2O_5]_0$, the original molar concentration of N_2O_5. Then we solve for $[N_2O_5]$, the molar concentration after 1.00 minute. We must remember to express k and t using the same time units. Finally, we convert molar concentration of N_2O_5 to moles remaining. (b) We solve the integrated first-order equation for the required time.

Solution

(a) The original concentration of N_2O_5 is

> **S**TOP & THINK
> Make sure the units of time are the same—chemists often use seconds (s) in specific rate constants.

$$[N_2O_5]_0 = \frac{2.50 \text{ mol}}{5.00 \text{ L}} = 0.500 \ M$$

The other quantities are

$$a = 2 \qquad k = 0.00840 \text{ s}^{-1} \qquad t = 1.00 \text{ min} = 60.0 \text{ s} \qquad [N_2O_5] = \underline{?}$$

The only unknown in the integrated rate equation is $[N_2O_5]$ after 1.00 minute. Let us solve for the unknown. Because $\ln x/y = \ln x - \ln y$,

$$\ln \frac{[N_2O_5]_0}{[N_2O_5]} = \ln [N_2O_5]_0 - \ln [N_2O_5] = akt$$

$$\ln [N_2O_5] = \ln [N_2O_5]_0 - akt$$

$$= \ln (0.500) - (2)(0.00840 \text{ s}^{-1})(60.0 \text{ s}) = -0.693 - 1.008$$

$$\ln [N_2O_5] = -1.701$$

▶ inv $\ln x = e^{\ln x}$

Be sure that you know how to use your calculator to obtain x by taking the inverse of $\ln x$.

Taking the inverse natural logarithm of both sides gives

$$[N_2O_5] = 1.82 \times 10^{-1} \ M$$

Thus, after 1.00 minute of reaction, the concentration of N_2O_5 is 0.182 M. The number of moles of N_2O_5 left in the 5.00-L container is

$$\underline{?} \text{ mol } N_2O_5 = 5.00 \text{ L} \times \frac{0.182 \text{ mol}}{\text{L}} = 0.910 \text{ mol } N_2O_5$$

(b) Because the integrated *first-order* rate equation involves a *ratio* of concentrations, we do not need to obtain the numerical value of the required concentration. When 90.0% of the original N_2O_5 has reacted, 10.0% remains, or

$$[N_2O_5] = (0.100)[N_2O_5]_0$$

We make this substitution into the integrated rate equation and solve for the elapsed time, t.

$$\ln \frac{[N_2O_5]_0}{[N_2O_5]} = akt$$

$$\ln \frac{[N_2O_5]_0}{(0.100)[N_2O_5]_0} = (2)(0.00840 \text{ s}^{-1})t$$

$$\ln(10.0) = (0.0168 \text{ s}^{-1})t$$

$$2.302 = (0.0168 \text{ s}^{-1})t \quad \text{or} \quad t = \frac{2.302}{0.0168 \text{ s}^{-1}} = \boxed{137 \text{ seconds}}$$

You should now work Exercises 34 and 38.

Problem-Solving Tip Does Your Answer Make Sense?

We know that the amount of N_2O_5 in Example 16-6 must be decreasing. The calculated result, 0.910 mol N_2O_5 after 1.00 minute, is less than the initial amount, 2.50 mol N_2O_5, which is a reasonable result. If our solution had given a result that was *larger* than the original, we should recognize that we must have made some error. For example, if we had *incorrectly* written the equation as

$$\ln \frac{[N_2O_5]}{[N_2O_5]_0} = akt$$

we would have obtained $[N_2O_5] = 1.37 \, M$, corresponding to 6.85 mol N_2O_5. This would be more N_2O_5 than was originally present, which we should immediately recognize as an impossible answer.

Second-Order Reactions

For reactions involving $aA \longrightarrow$ products that are *second order with respect to A and second order overall*, the integrated rate equation is

$$\frac{1}{[A]} - \frac{1}{[A]_0} = akt \qquad \left(\begin{array}{l} \text{second order in A,} \\ \text{second order overall} \end{array} \right)$$

For $t = t_{1/2}$, we have $[A] = \frac{1}{2}[A]_0$, so

$$\frac{1}{\frac{1}{2}[A]_0} - \frac{1}{[A]_0} = akt_{1/2}$$

Simplifying and solving for $t_{1/2}$, we obtain the relationship between the rate constant and $t_{1/2}$.

$$t_{1/2} = \frac{1}{ak[A]_0} \qquad \left(\begin{array}{l} \text{second order in A,} \\ \text{second order overall} \end{array} \right)$$

In this case $t_{1/2}$ *depends on the initial concentration of* A. Figure 16-4 illustrates the different behavior of half-life for first- and second-order reactions.

▶ You should carry out the algebraic steps to solve for $t_{1/2}$.

EXAMPLE 16-7 Half-Life: Second-Order Reaction

Compounds A and B react to form C and D in a reaction that was found to be second order in A and second order overall. The rate constant at 30°C is 0.622 liter per mole per minute. What is the half-life of A when $4.10 \times 10^{-2} \, M$ A is mixed with excess B?

$$A + B \longrightarrow C + D, \quad \text{rate} = k[A]^2$$

Plan

As long as some B is present, only the concentration of A affects the rate. The reaction is second order in [A] and second order overall, so we use the appropriate equation for the half-life.

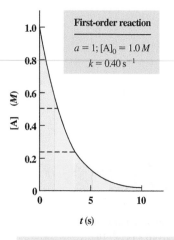

First-order reaction

$a = 1$; $[A]_0 = 1.0\,M$
$k = 0.40\,s^{-1}$

Second-order reaction

$a = 1$; $[A]_0 = 1.0\,M$
$k = 0.40\,M^{-1} \cdot s^{-1}$

A Plot of concentration versus time for a first-order reaction. During the first half-life, 1.73 seconds, the concentration of A falls from 1.00 M to 0.50 M. An additional 1.73 seconds is required for the concentration to fall by half again, from 0.50 M to 0.25 M, and so on. For a first-order reaction, $t_{1/2} = \dfrac{\ln 2}{ak} = \dfrac{0.693}{ak}$; $t_{1/2}$ does not depend on the concentration at the beginning of that time period.

B Plot of concentration versus time for a second-order reaction. The same values are used for a, $[A]_0$, and k as in (a). During the first half-life, 2.50 seconds, the concentration of A falls from 1.00 M to 0.50 M. The concentration falls by half again from 2.50 to 7.50 seconds, so the second half-life is 5.00 seconds. The half-life beginning at 0.25 M is 10.00 seconds. For a second-order reaction, $t_{1/2} = \dfrac{1}{ak\,[A]_0}$; $t_{1/2}$ is inversely proportional to the concentration at the beginning of that time period.

Figure 16-4 Comparison of first and second order decreases in reactant concentration.

Solution

$$t_{1/2} = \frac{1}{ak[A]_0} = \frac{1}{(1)(0.622\,M^{-1} \cdot min^{-1})(4.10 \times 10^{-2}\,M)} = \boxed{39.2\,min}$$

EXAMPLE 16-8   Concentration Versus Time: Second-Order Reaction

The gas-phase decomposition of NOBr is second order in [NOBr], with $k = 0.810\,M^{-1} \cdot s^{-1}$ at 10°C. We start with $4.00 \times 10^{-3}\,M$ NOBr in a flask at 10°C. How many seconds does it take to use up $1.50 \times 10^{-3}\,M$ of this NOBr?

$$2NOBr(g) \longrightarrow 2NO(g) + Br_2(g) \qquad rate = k[NOBr]^2$$

Plan

We first determine the concentration of NOBr that remains after $1.50 \times 10^{-3}\,M$ is used up. Then we use the second-order integrated rate equation to determine the time required to reach that concentration.

Solution

$$\underline{?}\,M\,NOBr\ remaining = (0.00400 - 0.00150)\,M = 0.00250\,M = [NOBr]$$

We solve the integrated rate equation $\dfrac{1}{[NOBr]} - \dfrac{1}{[NOBr]_0} = akt$ for t.

▶ The coefficient of NOBr is $a = 2$.

$$t = \frac{1}{ak}\left(\frac{1}{[NOBr]} - \frac{1}{[NOBr]_0}\right) = \frac{1}{(2)(0.810\,M^{-1} \cdot s^{-1})}\left(\frac{1}{0.00250\,M} - \frac{1}{0.00400\,M}\right)$$

$$= \frac{1}{1.62\,M^{-1} \cdot s^{-1}}(400\,M^{-1} - 250\,M^{-1})$$

$$= 92.6\,s$$

You should now work Exercise 32.

EXAMPLE 16-9   Concentration Versus Time: Second-Order Reaction

Consider the reaction of Example 16-8 at 10°C. If we start with $2.40 \times 10^{-3}\ M$ NOBr, what concentration of NOBr will remain after 5.00 minutes of reaction?

Plan

We use the integrated second-order rate equation to solve for the concentration of NOBr remaining at $t = 5.00$ minutes.

Solution

Again, we start with the expression $\dfrac{1}{[\text{NOBr}]} - \dfrac{1}{[\text{NOBr}]_0} = akt$. Then we put in the known values and solve for [NOBr].

$$\frac{1}{[\text{NOBr}]} - \frac{1}{2.40 \times 10^{-3}\,M} = (2)\,(0.810\ M^{-1} \cdot \text{s}^{-1})\,(5.00\ \text{min})\left(\frac{60\ \text{s}}{1\ \text{min}}\right)$$

$$\frac{1}{[\text{NOBr}]} - 4.17 \times 10^2\ M^{-1} = 486\ M^{-1}$$

$$\frac{1}{[\text{NOBr}]} = 486\ M^{-1} + 417\ M^{-1} = 903\ M^{-1}$$

$$[\text{NOBr}] = \frac{1}{903\,M^{-1}} = 1.11 \times 10^{-3}\ M \qquad \text{(46.2\% remains unreacted)}$$

Thus, 53.8% of the original concentration of NOBr reacts within the first 5 minutes. This is reasonable because, as you can easily verify, the reaction has an initial half-life of 257 seconds, or 4.29 minutes.

You should now work Exercises 33 and 36.

Zero-Order Reaction

For a reaction $a\text{A} \longrightarrow$ products that is zero order, the reaction rate is independent of concentrations. We can write the rate-law expression as

$$\text{rate} = -\frac{1}{a}\left(\frac{\Delta[\text{A}]}{\Delta t}\right) = k$$

The corresponding integrated rate equation is

$$[\text{A}] = [\text{A}]_0 - akt \qquad \text{(zero order)}$$

and the half-life is

$$t_{1/2} = \frac{[\text{A}]_0}{2ak} \qquad \text{(zero order)}$$

Table 16-2 summarizes the relationships that we have presented in Sections 16-3 and 16-4.

Table 16-2 Summary of Relationships for Various Orders of the Reaction $aA \rightarrow$ Products

	Order		
	Zero	*First*	*Second*
Rate-law expression	rate = k	rate = $k[A]$	rate = $k[A]^2$
Integrated rate equation	$[A] = [A]_0 - akt$	$\ln \dfrac{[A]_0}{[A]} = akt$	$\dfrac{1}{[A]} - \dfrac{1}{[A]_0} = akt$
Half-life, $t_{1/2}$	$\dfrac{[A]_0}{2ak}$	$\dfrac{\ln 2}{ak} = \dfrac{0.693}{ak}$	$\dfrac{1}{ak[A]_0}$

Problem-Solving Tip Which Equation Should Be Used?

How can you tell which equation to use to solve a particular problem?

1. You must decide whether to use the rate-law expression or the integrated rate equation. Remember that

the *rate-law expression* relates *rate and concentration*

whereas

the *integrated rate equation* relates *time and concentration.*

When you need to find the *rate* that corresponds to particular concentrations, or the concentrations needed to give a desired rate, you should use the rate-law expression. When *time* is involved in the problem, you should use the integrated rate equation.

2. You must choose the form of the rate-law expression or the integrated rate equation—zero, first, or second order—that is appropriate to the order of the reaction. These are summarized in Table 16-2. One of the following usually helps you decide.

   a. The statement of the problem may state explicitly what the order of the reaction is.
   b. The rate-law expression may be given so that you can tell the order of the reaction from the exponents in that expression.
   c. The units of the specific rate constant, k, may be given; you can interpret these stated units to tell you the order of the reaction.

Order	Units of k
0	$M \cdot \text{time}^{-1}$
1	time^{-1}
2	$M^{-1} \cdot \text{time}^{-1}$

S TOP & THINK
Many students do not realize that it is easier to derive the units of the rate constants for zero-, first-, and second-order reactions than it is to memorize them.

▶ You should test this method using the concentration-versus-time data of Example 16-10, plotted in Figure 16-8b.

One method of assessing reaction order is based on comparing successive half-lives. As we have seen, $t_{1/2}$ for a first-order reaction does not depend on initial concentration. We can measure the time required for different concentrations of a reactant to fall to half of their original values. If this time remains constant, it is an indication that the reaction is first order for that reactant and first order overall (see Figure 16-4a). By contrast, for other orders of reaction, $t_{1/2}$ would change depending on initial concentration. For a second-order reaction, successively measured $t_{1/2}$ values would increase by a factor of 2 as $[A]_0$ decreases by a factor of 2 (see Figure 16-4b). $[A]_0$ is measured at the *beginning of each particular measurement period.*

ENRICHMENT

Calculus Derivation of Integrated Rate Equations

The derivation of the integrated rate equation is an example of the use of calculus in chemistry. The following derivation is for a reaction that is assumed to be first order in a reactant A and first order overall. If you do not know calculus, you can still use the results of this derivation, as we have already shown in this section. For the reaction

$$aA \longrightarrow \text{products}$$

the rate is expressed as

$$\text{rate} = -\frac{1}{a}\left(\frac{\Delta[A]}{\Delta t}\right)$$

For a first-order reaction, the rate is proportional to the first power of [A].

$$-\frac{1}{a}\left(\frac{\Delta[A]}{\Delta t}\right) = k\,[A]$$

In calculus terms, we express the change during an infinitesimally short time dt as the derivative of [A] with respect to time.

$$-\frac{1}{a}\frac{d[A]}{dt} = k\,[A]$$

Separating variables, we obtain

$$-\frac{d[A]}{[A]} = (ak)dt$$

We integrate this equation with limits: As the reaction progresses from time = 0 (the start of the reaction) to time = t elapsed, the concentration of A goes from $[A]_0$, its starting value, to [A], the concentration remaining after time t:

$$-\int_{[A]_0}^{[A]} \frac{d[A]}{[A]} = ak \int_0^t dt$$

The result of the integration is

$$-(\ln[A] - \ln[A]_0) = ak(t - 0) \qquad \text{or} \qquad \ln[A]_0 - \ln[A] = akt$$

Remembering that $\ln(x) - \ln(y) = \ln(x/y)$, we obtain

$$\ln\frac{[A]_0}{[A]} = akt \qquad \text{(first order)}$$

This is the integrated rate equation for a reaction that is first order in reactant A and first order overall.

Integrated rate equations can be derived similarly from other simple rate laws. For a reaction $aA \rightarrow$ products that is second order in reactant A and second order overall, we can write the rate equation as

$$-\frac{d[A]}{adt} = k[A]^2$$

▶ $-\frac{1}{a}\left(\frac{\Delta[A]}{\Delta t}\right)$ represents the average rate over a finite time interval Δt.

$-\frac{1}{a}\left(\frac{d[A]}{dt}\right)$ involves a change over an infinitesimally short time interval dt, so it represents the *instantaneous* rate.

Again, using the methods of calculus, we can separate variables, integrate, and rearrange to obtain the corresponding integrated second-order rate equation.

$$\frac{1}{[A]} - \frac{1}{[A]_0} = akt \qquad \text{(second order)}$$

For a reaction $aA \longrightarrow$ products that is zero order overall, we can write the rate equation as

$$-\frac{d[A]}{adt} = k$$

In this case, the calculus derivation already described leads to the integrated zero-order rate equation

$$[A] = [A]_0 - akt \qquad \text{(zero order)}$$

ENRICHMENT

Using Integrated Rate Equations to Determine Reaction Order

Analysis of the slope and intercept of a straight-line plot can often aid in analyzing data. Recall that the standard equation for a straight line may be written as

$$y = mx + b$$

where y is the variable plotted along the ordinate (vertical axis), x is the variable plotted along the abscissa (horizontal axis), m is the slope of the line, and b is the intercept of the line with the y axis (Figure 16-5).

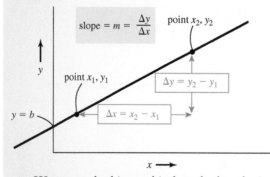

Figure 16-5 Plot of the equation $y = mx + b$, where m and b are constant. The slope of the line (positive in this case) is equal to m; the intercept on the y axis is equal to b.

We can apply this graphical method to the integrated rate equation to help us determine reaction order from concentration-versus-time data. Let us rearrange the integrated first-order rate equation

$$\ln \frac{[A]_0}{[A]} = akt$$

as follows. The logarithm of a quotient, $\ln (x/y)$, is equal to the difference of the logarithms, $\ln x - \ln y$, so we can write

$$\ln [A]_0 - \ln [A] = akt \quad \text{or} \quad \ln [A] = -akt + \ln [A]_0$$

Comparing this rearranged equation to the standard straight-line equation, we see that ln [A] can be interpreted as y, and t as x.

$$\underbrace{\ln [A]}_{\downarrow \atop y} = \underbrace{-\ akt}_{\downarrow\ \downarrow \atop m\ x\ +} + \underbrace{\ln [A]_0}_{\downarrow \atop b}$$

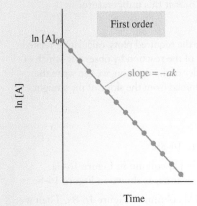

First order

slope = $-ak$

ln [A]₀ → ln [A]

Time

Figure 16-6 Plot of ln [A] versus time for a reaction aA $\longrightarrow$ products that follows first-order kinetics. The observation that such a plot gives a straight line would confirm that the reaction is first order in [A] and first order overall; that is, rate = k[A]. The slope is equal to $-ak$. Because a and k are positive numbers, the slope of the line is always negative. Logarithms are dimensionless, so the slope has the units (time)$^{-1}$. The logarithm of a quantity less than 1 is negative, so data points for concentrations less than 1 molar would have negative values and appear below the time axis.

The quantity $-ak$ is a constant as the reaction proceeds, so it can be interpreted as m. The initial concentration of A is fixed, so ln [A]$_0$ is a constant for each experiment, and ln [A]$_0$ can be interpreted as b. Thus, a plot of ln [A] versus time for a first-order reaction would be expected to give a straight line (Figure 16-6) with the slope of the line equal to $-ak$ and the intercept equal to ln [A]$_0$.

We can proceed in a similar fashion with the integrated rate equation for a reaction that is second order in A and second order overall. We rearrange

$$\frac{1}{[A]} - \frac{1}{[A]_0} = akt \qquad \text{to read} \qquad \frac{1}{[A]} = akt + \frac{1}{[A]_0}$$

Again comparing this with the equation for a straight line, we see that a plot of 1/[A] versus time would be expected to give a straight line (Figure 16-7). The line would have a slope equal to ak and an intercept equal to 1/[A]$_0$.

For a zero-order reaction, we can rearrange the integrated rate equation

$$[A]_0 - [A] = akt \qquad \text{to} \qquad [A] = -\ akt + [A]_0$$

Comparing this with the equation for a straight line, we see that a straight-line plot would be obtained by plotting concentration versus time, [A] versus t. The slope of this line is $-ak$, and the intercept is [A]$_0$.

This discussion suggests another way to deduce an unknown rate-law expression from experimental concentration data. The following approach is particularly useful for any decomposition reaction, one that involves only one reactant.

$$a\text{A} \longrightarrow \text{products}$$

We plot the data in various ways as suggested above. *If* the reaction followed zero-order kinetics, *then* a plot of [A] versus t would give a straight line. But *if* the reaction followed first-order kinetics, *then* a plot of ln [A] versus t would give a straight line whose slope could be interpreted to derive a value of k. *If* the reaction were second order in A and second order overall, *then* neither of these plots would give a straight line, but a plot of 1/[A] versus t would. If none of these plots gave a straight line (within expected scatter due to experimental error), we would know that none of these possibilities is the correct order (rate law) for the reaction. Plots to test for other orders can be devised, as can graphical tests for rate-law expressions involving more than one reactant, but those are subjects for more advanced texts. The graphical approach that we have described is illustrated in the following example.

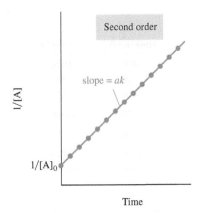

Second order

slope = ak

1/[A]₀ → 1/[A]

Time

Figure 16-7 Plot of 1/[A] versus time for a reaction aA $\longrightarrow$ products that follows second-order kinetics. The observation that such a plot gives a straight line would confirm that the reaction is second order in [A] and second order overall; that is, rate = k[A]^2. The slope is equal to ak. Because a and k are positive numbers, the slope of the line is always positive. Because concentrations cannot be negative, 1/[A] is always positive, and the line is always above the time axis.

▶ It is not possible for *all* of the plots suggested here to yield straight lines for a given reaction. The nonlinearity of the plots may not become obvious, however, if the reaction times used are too short. In practice, all three lines might seem to be straight; we should then suspect that we need to observe the reaction for a longer time.

Time (min)	[A] (mol/L)
0.00	2.000
2.00	1.107
4.00	0.612
6.00	0.338
8.00	0.187
10.00	0.103

EXAMPLE 16-10 Graphical Determination of Reaction Order

We carry out the reaction A → B + C at a particular temperature. As the reaction proceeds, we measure the molarity of the reactant, [A], at various times. The observed data are tabulated in the margin. (a) Plot [A] versus time. (b) Plot ln [A] versus time. (c) Plot 1/[A] versus time. (d) What is the order of the reaction? (e) Write the rate-law expression for the reaction. (f) What is the value of k at this temperature?

Plan

For Parts (a)–(c), we use the observed data to make the required plots, calculating related values as necessary. (d) We can determine the order of the reaction by observing which of these plots gives a straight line. (e) Knowing the order of the reaction, we can write the rate-law expression. (f) The value of k can be determined from the slope of the straight-line plot.

Solution

(a) The plot of [A] versus time is given in Figure 16-8b.

(b) We first use the given data to calculate the ln [A] column in Figure 16-8a. These data are then used to plot ln [A] versus time, as shown in Figure 16-8c.

(c) The given data are used to calculate the 1/[A] column in Figure 16-8a. Then we plot 1/[A] versus time, as shown in Figure 16-8d.

(d) It is clear from the answer to part (b) that the plot of ln [A] versus time gives a straight line. This tells us that the reaction is first order in [A].

(e) In the form of a rate-law expression, the answer to part (d) gives rate = k[A].

Time (min)	[A]	ln [A]	1/[A]
0.00	2.000	0.693	0.5000
2.00	1.107	0.102	0.9033
4.00	0.612	−0.491	1.63
6.00	0.338	−1.085	2.95
8.00	0.187	−1.677	5.35
10.00	0.103	−2.273	9.71

Ⓐ Data for Example 16-10.

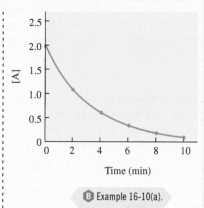

Ⓑ Example 16-10(a).

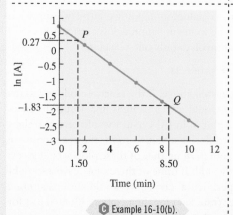

Ⓒ Example 16-10(b).

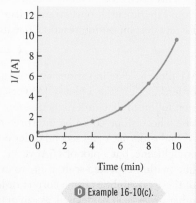

Ⓓ Example 16-10(c).

Figure 16-8 Data conversion and plots for Example 16-10. (a) The data are used to calculate the two columns ln [A] and 1/[A]. (b) Test for zero-order kinetics: a plot of [A] versus time. The nonlinearity of this plot shows that the reaction does not follow zero-order kinetics. (c) Test for first-order kinetics: a plot of ln [A] versus time. The observation that this plot gives a straight line indicates that the reaction follows first-order kinetics. (d) Test for second-order kinetics: a plot of 1/[A] versus time. If the reaction had followed second-order kinetics, this plot would have resulted in a straight line and the plot in Part (c) would not.

(f) We use the straight-line plot in Figure 16-8c to find the value of the rate constant for this first-order reaction from the relationship

$$\text{Slope} = -ak \qquad \text{or} \qquad k = -\frac{\text{slope}}{a}$$

To determine the slope of the straight line, we pick any two points, such as P and Q, on the line. From their coordinates, we calculate

$$\text{Slope} = \frac{\text{change in ordinate}}{\text{change in abscissa}} = \frac{(-1.83) - (0.27)}{(8.50 - 1.50) \text{ min}} = -0.300 \text{ min}^{-1}$$

$$k = -\frac{\text{slope}}{a} = -\frac{-0.300 \text{ min}^{-1}}{1} = 0.300 \text{ min}^{-1}$$

You should now work Exercises 42 and 43.

The graphical interpretations of concentration-versus-time data for some common reaction orders are summarized in Table 16-3.

Problem-Solving Tip Some Warnings About the Graphical Method for Determining Reaction Order

1. When we deal with real experimental data, there is always some error in each of the data points on the plot. For this reason, we should *not* use experimental data points to determine the slope. (Random experimental errors of only 10% can introduce errors of more than 100% in slopes based on only two points.) Rather we should draw the best straight line and then use points on that line to find its slope. Errors are further minimized by choosing points that are widely separated.
2. Remember that the ordinate is the vertical axis and the abscissa is the horizontal one. If you are not careful to keep the points in the same order in the numerator and denominator, you will get the wrong sign for the slope.

Table 16-3 Graphical Interpretations for Various Orders of the Reaction $a\text{A} \longrightarrow$ Products

	Order		
	Zero	*First*	*Second*
Plot that gives straight line	[A] vs. t	ln [A] vs. t	$\dfrac{1}{[\text{A}]}$ vs. t
Direction of straight-line slope	down with time	down with time	up with time
Integrated rate equation	$-ak$	$-ak$	ak
Interpretation of intercept	$[\text{A}]_0$	ln $[\text{A}]_0$	$\dfrac{1}{[\text{A}]_0}$

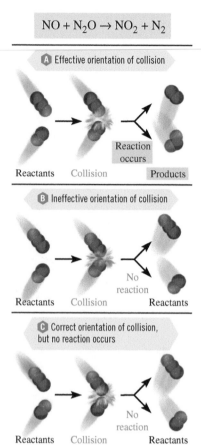

$NO + N_2O \rightarrow NO_2 + N_2$

A Effective orientation of collision

Reactants   Collision   Products

Reaction occurs

B Ineffective orientation of collision

Reactants   Collision   Reactants

No reaction

C Correct orientation of collision, but no reaction occurs

Reactants   Collision   Reactants

No reaction

Figure 16-9 Some possible collisions between N_2O and NO molecules in the gas phase. (a) A collision that could be effective in producing the reaction. (b, c) Collisions that would be ineffective. The molecules must have the proper orientations relative to one another *and* have sufficient energy to react.

16-5 Collision Theory of Reaction Rates

The fundamental notion of the **collision theory of reaction rates** is that for a reaction to occur, molecules, atoms, or ions must first collide. Increased concentrations of reacting species result in greater numbers of collisions per unit time. However, not all collisions result in reaction; that is, not all collisions are **effective collisions**. For a collision to be effective, the reacting species must (1) possess at least a certain minimum energy necessary to rearrange outer electrons in breaking bonds and forming new ones and (2) have the proper orientations toward one another at the time of collision.

> Collisions must occur in order for a chemical reaction to proceed, but they do not guarantee that a reaction will occur.

A collision between atoms, molecules, or ions is not like one between two hard billiard balls. Whether or not chemical species "collide" depends on the distance at which they can interact with one another. For instance, the gas-phase ion–molecule reaction $CH_4^+ + CH_4 \longrightarrow CH_5^+ + CH_3$ can occur with a fairly long-range contact. This is because the interactions between ions and induced dipoles are effective over a relatively long distance. By contrast, the reacting species in the gas reaction $CH_3 + CH_3 \longrightarrow C_2H_6$ are both neutral. They interact appreciably only through very short-range forces between induced dipoles, so they must approach one another very closely before we could say that they "collide."

Recall (Chapter 12) that the average kinetic energy of a collection of molecules is proportional to the absolute temperature. At higher temperatures, more of the molecules possess sufficient energy to react (see Section 16-8).

The colliding molecules must have the proper orientation relative to one another *and* have sufficient energy to react. If colliding molecules have improper orientations, they do not react even though they may possess sufficient energy. Figure 16-9 depicts some possible collisions between molecules of NO and N_2O, which can react to form NO_2 and N_2.

$$NO + N_2O \longrightarrow NO_2 + N_2$$

$$Zn(s) + 2H^+(aq) \longrightarrow Zn^{2+}(aq) + H_2(g)$$

Dilute sulfuric acid reacts slowly with zinc metal (*left*), whereas more concentrated acid reacts rapidly (*right*). The H$^+$(aq) concentration is higher in the more concentrated acid, so more H$^+$(aq) ions collide with Zn per unit time.

Only the collision in Figure 16-9a is in the correct orientation with sufficient kinetic energy to transfer an oxygen atom from the linear N_2O molecule to form the angular NO_2 molecule. For some reactions, the presence of a heterogeneous catalyst (see Section 16-9) can increase the fraction of colliding molecules that have the proper orientations.

16-6 Transition State Theory

Chemical reactions involve the making and breaking of chemical bonds. The energy associated with a chemical bond is a form of potential energy. Reactions are accompanied by changes in potential energy. Consider the following hypothetical, one-step reaction at a certain temperature.

$$A + B_2 \longrightarrow AB + B$$

Figure 16-10 shows plots of potential energy versus the progress of the reaction. In Figure 16-10a the ground state energy of the reactants, A and B_2, is higher than the ground state energy of the products, AB and B. The energy released in the reaction is the difference between these two energies, ΔE. It is related to the change in enthalpy, ΔH^0_{rxn} (see Section 15-11).

Quite often, for a reaction to occur, some covalent bonds must be broken so that others can be formed. This can occur only if the molecules collide *with enough kinetic energy* to overcome the potential energy stabilization of the bonds. According to the **transition state theory**, the reactants pass through a short-lived, high-energy intermediate state, called a **transition state**, before the products are formed. In the transition state, at least one bond is in the process of being broken while a new bond is being formed. This is represented by dashed lines (or dots) between the atoms.

$$A + B—B \longrightarrow A\text{---}B\text{---}B \longrightarrow A—B + B$$

reactants   transition state   products

$A + B_2$ AB_2 $AB + B$

The **activation energy** (or **activation barrier**), E_a, is the kinetic energy that reactant molecules must have to allow them to reach the transition state. If A and B_2 molecules do not possess the necessary amount of energy, E_a, when they collide, a reaction cannot occur. If they do possess sufficient energy to "climb the energy barrier" to reach the transition state, the reaction can proceed. When the atoms go from the transition state arrangement to the product molecules, energy is *released*. If the reaction results in a net *release* of energy (see Figure 16-10a), *more* energy than the activation energy is returned to the surroundings and the reaction is exothermic. If the reaction results in a *net absorption* of energy (see Figure 16-10b), an amount less than E_a is given off when the transition state is converted to products and the reaction is endothermic. The *net* release of energy is ΔE_{rxn}.

For the reverse reaction to occur, some molecules on the right (AB) must have kinetic energy equal to the reverse activation energy, $E_{a\ reverse}$, to allow them to reach the transition state. As you can see from the potential energy diagrams in Figure 16-10,

$$E_{a\ forward} - E_{a\ reverse} = \Delta E_{reaction}$$

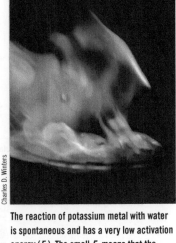

The reaction of potassium metal with water is spontaneous and has a very low activation energy (E_a). The small E_a means that the reaction will be very fast.

▶ Remember that ΔE_{rxn} relates product energy to reactant energy, regardless of the pathway. ΔE_{rxn} is negative when energy is given off; ΔE_{rxn} is positive when energy is absorbed from the surroundings.

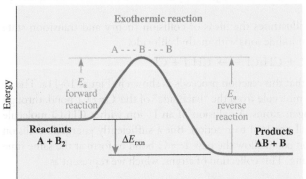

A A reaction that releases energy (exothermic). An example of an exothermic gas-phase reaction is
$$H + I_2 \longrightarrow HI + I$$

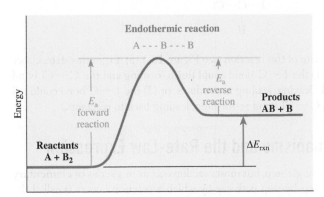

B A reaction that absorbs energy (endothermic). An example of an endothermic gas-phase reaction is
$$I + H_2 \longrightarrow HI + H$$

Figure 16-10 A potential energy diagram. The **reaction coordinate** represents how far the reaction has proceeded *along the pathway* leading from reactants to products. This represents the progress of the reaction.

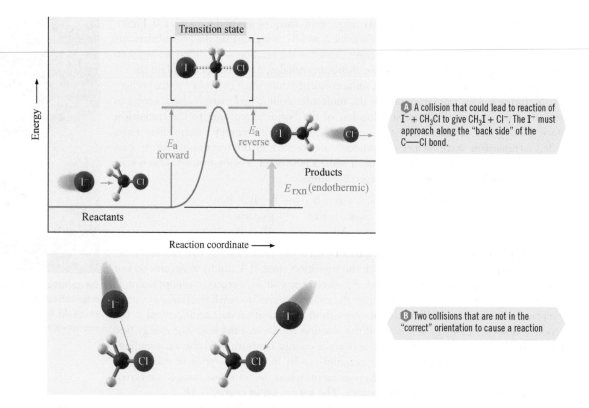

A A collision that could lead to reaction of $I^- + CH_3Cl$ to give $CH_3I + Cl^-$. The I^- must approach along the "back side" of the C—Cl bond.

B Two collisions that are not in the "correct" orientation to cause a reaction

Figure 16-11

▶ The CH_3Cl and CH_3I molecules each have tetrahedral molecular geometry.

As we shall see, increasing the temperature changes the rate by altering the fraction of molecules that can get over a given energy barrier (see Section 16-8). Introducing a catalyst increases the rate by providing a different pathway that has a lower activation energy (see Section 16-9).

As a specific example that illustrates the ideas of collision theory and transition state theory, consider the reaction of iodide ions with methyl chloride.

$$I^- + CH_3Cl \longrightarrow CH_3I + Cl^-$$

Many studies have established that this reaction proceeds as shown in Figure 16-11a. The I^- ion must approach the CH_3Cl molecule from the "back side" of the C—Cl bond, through the middle of the three hydrogen atoms. A collision of an I^- ion with a CH_3Cl molecule from any other direction would not lead to reaction. But a sufficiently energetic collision with the appropriate orientation could allow the new I—C bond to form at the same time that the C—Cl bond is breaking. This collection of atoms, which we represent as

▶ We can view this transition state as though carbon is only partially bonded to I and only partially bonded to Cl. Because we are showing the molecular geometry using wedges, dashes, and regular lines, the bonds that are forming and breaking are indicated by dotted lines and are in the plane of the paper.

$$\begin{array}{ccc} & H & H \\ & | & \diagup \\ I\cdots & C & \cdots Cl \\ & | & \\ & H & \end{array}$$

is what we call the transition state of this reaction (see Figure 16-11a). From this state, either of two things could happen: (1) the I—C bond could finish forming and the C—Cl bond could finish breaking with Cl^- leaving, leading to products, or (2) the I—C bond could fall apart with I^- leaving and the C—Cl bond re-forming, leading back to reactants.

16-7 Reaction Mechanisms and the Rate-Law Expression

Some reactions take place in a single step, but most reactions occur in a series of **elementary** or **fundamental steps**. The step-by-step pathway by which a reaction occurs is called the **reaction mechanism**.

The reaction orders *for any single elementary step* are equal to the coefficients for that step.

In many mechanisms, however, one step is much slower than the others.

An overall reaction can never occur faster than its slowest elementary reaction step.

This slowest step is called the **rate-determining step**. The speed at which the slowest step occurs limits the rate at which the overall reaction occurs.

As an analogy, suppose you often drive a distance of 120 miles at the speed limit of 60 mi/h, requiring 2 hours. But one day there is an accident along the route, causing a slow-down. After passing the accident scene, you resume the posted speed of 60 mi/h. If the total time for this trip was 4 hours, then the *average* speed would be only 120 miles/4 hours, or 30 mi/h. Even though you drove for many miles at the same high speed, 60 mi/h, the overall rate was limited by the slow step, passing the accident scene.

The balanced equation for the overall reaction is equal to the sum of *all* the individual fundamental steps, including any that might follow the rate-determining step. We emphasize again that the rate-law exponents *do not necessarily match* the coefficients of the *overall* balanced equation.

For the general overall reaction

$$a\text{A} + b\text{B} \longrightarrow c\text{C} + d\text{D}$$

the experimentally determined rate-law expression has the form

$$\text{rate} = k[\text{A}]^x[\text{B}]^y$$

The values of x and y are related to the coefficients of the reactants in the slowest (rate-determining) step, influenced in some cases by earlier reaction steps.

Using a combination of experimental data and chemical intuition, we can *postulate* a mechanism by which a reaction could occur. We can never prove that a proposed mechanism is correct. All we can do is postulate a mechanism that is *consistent* with experimental data. We might later detect reaction-intermediate species that are not explained by the proposed mechanism. We must then modify the mechanism or discard it and propose a new one.

As an example, the reaction of nitrogen dioxide and carbon monoxide has been found to be second order with respect to NO_2 and zero order with respect to CO below 225°C.

$$NO_2(g) + CO(g) \longrightarrow NO(g) + CO_2(g) \quad \text{rate} = k[NO_2]^2$$

The balanced equation for the overall reaction shows the stoichiometry but *does not necessarily mean* that the reaction simply occurs by one molecule of NO_2 colliding with one molecule of CO. If the reaction really took place in that one step, then the rate would be first order in NO_2 and first order in CO, or rate = $k[NO_2][CO]$. The fact that the experimentally determined orders do not match the coefficients in the overall balanced equation tells us that *the reaction does not take place in one step*.

The following proposed two-step mechanism is consistent with the observed rate-law expression.

(1) $NO_2 + NO_2 \longrightarrow N_2O_4$ (slow)

(2) $N_2O_4 + CO \longrightarrow NO + CO_2 + NO_2$ (fast)

$NO_2 + CO \longrightarrow NO + CO_2$ (overall)

The rate-determining step of this mechanism involves a *bimolecular* collision between two NO_2 molecules. This is consistent with the rate expression involving $[NO_2]^2$. Because the CO is involved only after the slow step has occurred, the reaction rate would not depend on [CO] (i.e., the reaction would be zero order in CO) if this were the actual mechanism.

S TOP & THINK
A fundamental reaction step can often look similar to an overall balanced reaction equation, so they are not easy to distinguish. In general, you will be told if a given reaction is fundamental or elementary in nature. If you are not told that a reaction is a fundamental step, you need to be given additional experimental information in order to determine the kinetic orders.

S TOP & THINK
The *rate* of a reaction involves only the steps up to and including the rate-determining step. The *overall stoichiometry* includes all steps in a reaction.

▶ For clarity, reaction intermediates are shown on an orange background.

▶ Some reaction intermediates are so unstable that it is very difficult to prove experimentally that they exist.

In this proposed mechanism, N_2O_4 is formed in one step and is completely consumed in a later step. Such a species is called a **reaction intermediate**.

In other studies of this reaction, however, nitrogen trioxide, NO_3, has been detected as a transient (short-lived) intermediate. The mechanism now thought to be correct is

$$(1) \quad NO_2 + NO_2 \longrightarrow NO_3 + NO \qquad \text{(slow)}$$
$$(2) \quad NO_3 + CO \longrightarrow NO_2 + CO_2 \qquad \text{(fast)}$$
$$\overline{\quad NO_2 + CO \longrightarrow NO + CO_2 \qquad \text{(overall)}}$$

In this proposed mechanism, two molecules of NO_2 collide to produce one molecule each of NO_3 and NO. The reaction intermediate NO_3 then collides with one molecule of CO and reacts very rapidly to produce one molecule each of NO_2 and CO_2. Even though two NO_2 molecules are consumed in the first step, one is produced in the second step. The net result is that only one NO_2 molecule is consumed in the overall reaction.

▶ Nonproductive side reactions can produce observable intermediates (such as NO_3) that may not be part of the overall reaction to make products. This is one of the factors that can make kinetics very complicated.

Each of these proposed mechanisms meets both criteria for a plausible mechanism: (1) The steps add to give the equation for the overall reaction, and (2) the mechanism is consistent with the experimentally determined rate-law expression (in that two NO_2 molecules and no CO molecules are reactants in the slowest reaction step). The NO_3 that has been detected is evidence in favor of the second mechanism, but this does not unequivocally prove that mechanism; it may be possible to think of other mechanisms that would involve NO_3 as an intermediate and would also be consistent with the observed rate law.

You should be able to distinguish among various species that can appear in a reaction mechanism. So far, we have seen three kinds of species:

1. *Reactant*: More is consumed than is formed.
2. *Product*: More is formed than is consumed.
3. *Reaction intermediate*: Formed in earlier steps, then consumed in an equal amount in later steps.

The gas-phase reaction of NO and Br_2 is known to be second order in NO and first order in Br_2.

$$2NO(g) + Br_2(g) \longrightarrow 2NOBr(g) \quad \text{rate} = k[NO]^2[Br_2]$$

▶ Think how unlikely it is for three moving billiard balls to collide *simultaneously*.

A one-step collision involving two NO molecules and one Br_2 molecule would be consistent with the experimentally determined rate-law expression. However, the likelihood of all three molecules colliding simultaneously is far less than the likelihood of two colliding. *Routes involving only bimolecular collisions or unimolecular decompositions are thought to be far more favorable in reaction mechanisms.* The mechanism is believed to be

▶ Any fast step that precedes a slow step reaches equilibrium.

$$(1) \quad NO + Br_2 \rightleftharpoons NOBr_2 \qquad \text{(fast, equilibrium)}$$
$$(2) \quad NOBr_2 + NO \longrightarrow 2NOBr \qquad \text{(slow)}$$
$$\overline{\quad 2NO + Br \longrightarrow 2NOBr \qquad \text{(overall)}}$$

The first step involves the collision of one NO molecule (reactant) and one Br_2 molecule (reactant) to produce the intermediate species $NOBr_2$. The $NOBr_2$ can react rapidly, however, to re-form NO and Br_2. We say that this is an *equilibrium step*. Eventually another NO molecule (reactant) can collide with a short-lived $NOBr_2$ molecule and react to produce two NOBr molecules (product).

To analyze the rate law that would be consistent with this proposed mechanism, we again start with the slow (rate-determining) Step 2. Denoting the rate constant for this step as k_2, we could express the rate of this step as

$$\text{rate} = k_2[NOBr_2][NO]$$

However, $NOBr_2$ is a reaction intermediate, so its concentration at the beginning of the second step may not be easy to measure directly. Because $NOBr_2$ is formed in a fast equilibrium step, we can relate its concentration to the concentrations of the original

STOP & THINK
The rate-law expression of Step 2 (the rate-determining step) determines the rate law for the overall reaction. The overall rate law must not include the concentrations of any intermediate species formed in elementary reaction steps.

reactants. When a reaction or reaction step is at *equilibrium*, its forward (f) and reverse (r) rates are equal.

$$\text{rate}_{1f} = \text{rate}_{1r}$$

Because this is an elementary step, we can write the rate expression for both directions from the equation for the elementary step

▶ In a mechanism, each individual reaction step is a fundamental step.

$$k_{1f}[NO][Br_2] = k_{1r}[NOBr_2]$$

and then rearrange for $[NOBr_2]$.

$$[NOBr_2] = \frac{k_{1f}}{k_{1r}}[NO][Br_2]$$

When we substitute the right side of this equation for $[NOBr_2]$ in the rate expression for the rate-determining step, rate $= k_2[NOBr_2][NO]$, we arrive at the experimentally determined rate-law expression.

$$\text{rate} = k_2\left(\frac{k_{1f}}{k_{1r}}[NO][Br_2]\right)[NO] \quad \text{or} \quad \text{rate} = k[NO]^2[Br_2]$$

▶ The product and quotient of constants k_2, k_{1f}, and k_{1r} is another constant, k.

Similar interpretations apply to most other overall third- or higher-order reactions, as well as many lower-order reactions. When several steps are about equally slow, however, the analysis of experimental data is more complex. Fractional or negative reaction orders can result from complex multistep mechanisms.

One of the earliest kinetic studies involved the gas-phase reaction of hydrogen and iodine to form hydrogen iodide. The reaction was found to be first order in both hydrogen and iodine.

$$H_2(g) + I_2(g) \longrightarrow 2HI(g) \quad \text{rate} = k[H_2][I_2]$$

The mechanism that was accepted for many years involved the collision of single molecules of H_2 and I_2 in a simple one-step reaction. Current evidence indicates a more complex process, however. Most kineticists now accept the following mechanism.

(1) $I_2 \rightleftharpoons 2I$ (fast, equilibrium)
(2) $I + H_2 \rightleftharpoons H_2I$ (fast, equilibrium)
(3) $H_2I + I \longrightarrow 2HI$ (slow)

$H_2 + I_2 \longrightarrow 2HI$ (overall)

▶ Apply the algebraic approach described earlier to show that this mechanism is consistent with the observed rate-law expression.

In this case neither original reactant appears in the rate-determining step, but both appear in the rate-law expression. Each step is a fundamental reaction in itself. Transition state theory tells us that each step has its own activation energy. Because Step 3 is the slowest, we know that its activation energy is the highest, as shown in Figure 16-12.

In summary,

the experimentally determined reaction orders of reactants indicate the number of molecules of those reactants involved in (1) the slow step only, if it occurs first, or (2) the slow step *and* any fast equilibrium steps preceding the slow step.

16-8 Temperature: The Arrhenius Equation

The average kinetic energy of a collection of molecules is proportional to the absolute temperature. At a particular temperature, T_1, a definite fraction of the reactant molecules have sufficient kinetic energy, KE $> E_a$, to react to form product molecules on collision. At a higher temperature, T_2, a greater fraction of the molecules possess the necessary activation energy, and the reaction proceeds at a faster rate. This is depicted in Figure 16-13a.

Figure 16-12 A graphical representation of the relative energies of activation for a postulated mechanism for the gas-phase reaction

$$H_2 + I_2 \longrightarrow 2HI$$

Reaction intermediates are shown on an orange background.

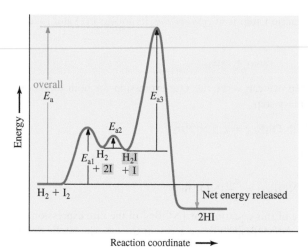

From experimental observations, Svante Arrhenius developed the mathematical relationship among activation energy, absolute temperature, and the specific rate constant of a reaction, k, at that temperature. The relationship, called the **Arrhenius equation**, is

* ▶ Recall that $e = 2.718$ is the base of *natural* logarithms (ln).

$$k = Ae^{-E_a/RT}$$

or, in logarithmic form,

$$\ln k = \ln A - \frac{E_a}{RT}$$

In this expression, A is a constant having the same units as the rate constant. It is equal to the fraction of collisions with the proper orientations when all reactant concentrations are one molar. R is the universal gas constant, expressed with the same energy units in its numerator as are used for E_a. For instance, when E_a is known in J/mol, the value $R = 8.314$ J/mol · K is appropriate. Here the unit "mol" is interpreted as "mole of reaction," as described in Chapter 15. One important point is the following: The greater the value of E_a, the smaller the value of k and the slower the reaction rate (other factors being equal). This is because fewer collisions take place with sufficient energy to get over a high-energy barrier (see Figure 16-13b).

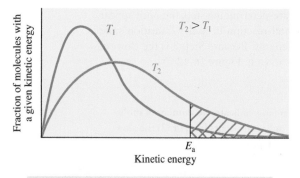

Ⓐ The effect of temperature on the number of molecules that have kinetic energies greater than E_a, the activation energy. At T_2, a higher fraction of molecules possess at least E_a, the activation energy. The area between the distribution curve and the horizontal axis is proportional to the total number of molecules present. The total area is the same at T_1 and T_2. The shaded areas represent the number of particles that exceed the energy of activation, E_a.

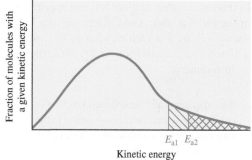

Ⓑ Consider two hypothetical reactions 1 and 2, where the activation energy of reaction 1 is less than that of reaction 2—that is, $E_{a1} < E_{a2}$. At any given temperature, a larger fraction of the molecules have energies that exceed E_{a1} than that exceed E_{a2}, so reaction 1 would have a higher specific rate constant, k, than reaction 2 at the same reactant concentrations.

Figure 16-13 Maxwell-Boltzmann distributions, shown here plotted in terms of kinetic energy. Such plots were introduced in Figure 12-9, where they were presented in terms of molecular speed.

Antimony powder reacts with bromine more rapidly at 75°C (*left*) than at 25°C (*right*).

The Arrhenius equation predicts that increasing T results in a faster reaction for the same E_a and concentrations.

| If T increases | $\Rightarrow$ | E_a/RT decreases | $\Rightarrow$ | $-E_a/RT$ increases | $\Rightarrow$ | $e^{-E_a/RT}$ increases | $\Rightarrow$ | k increases | $\Rightarrow$ | **Reaction speeds up** |

Let's look at how the rate constant varies with temperature for a given single reaction. Assume that the activation energy and the factor A do not vary with temperature. We can write the Arrhenius equation for two different temperatures. Then we subtract one equation from the other and rearrange the result to obtain

$$\ln \frac{k_2}{k_1} = \frac{E_a}{R}\left(\frac{1}{T_1} - \frac{1}{T_2}\right)$$

The advantage of setting up the equation this way is that the usually unknown A value is cancelled out, making it easier to solve for one of the other values.

Let's substitute some typical values into this equation. The activation energy for many reactions that occur near room temperature is about 50 kJ/mol. For such a reaction, a temperature increase from 300 K to 310 K would result in

$$\ln \frac{k_2}{k_1} = \frac{50,000\,\text{J/mol}}{(8.314\,\text{J/mol}\cdot\text{K})}\left(\frac{1}{300\,\text{K}} - \frac{1}{310\,\text{K}}\right) = 0.647$$

$$\frac{k_2}{k_1} = 1.91 \approx 2$$

Chemists sometimes use the rule of thumb that near room temperature the rate of a reaction approximately doubles with a 10°C rise in temperature. Such a "rule" must be used with care, however, because it obviously depends on the activation energy.

EXAMPLE 16-11 Arrhenius Equation

The specific rate constant, k, for the following first-order reaction is 9.16×10^{-3} s^{-1} at 0.0°C. The activation energy of this reaction is 88.0 kJ/mol. Determine the value of k at 2.0°C.

$$N_2O_5 \longrightarrow NO_2 + NO_3$$

Plan

First we tabulate the values, remembering to convert temperature to the Kelvin scale.

$E_a = 88,000\,\text{J/mol}$ $\qquad$ $R = 8.314\,\text{J/mol}\cdot\text{K}$

$k_1 = 9.16 \times 10^{-3}$ s^{-1} $\quad$ at $\quad$ $T_1 = 0.0°C + 273 = 273\,\text{K}$

$k_2 = \underline{\,?\,}$ $\qquad\qquad$ at $\quad$ $T_2 = 2.0°C + 273 = 275\,\text{K}$

A fever is your body's natural way of speeding up the immune system (and the rest of your metabolism) to more effectively fight an infection. Part of the old saying "feed a fever" makes a lot of sense from a kinetics viewpoint. Why—in terms of what you have learned in this chapter?

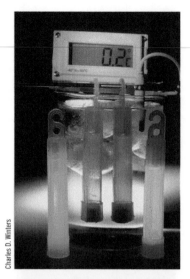

The reaction rate of chemiluminescent light sticks is strongly affected by temperature. In the top photo, the light sticks immersed in cold water (0.2°C) barely glow relative to those sitting outside the beaker at room temperature (22°C). In the bottom photo, the light sticks in the beaker are at 59°C and glow very brightly compared to the room temperature sticks. The light sticks at 59°C will "burn out," that is, the reactants that produce the light will be consumed much sooner due to the faster reaction at the higher temperature.

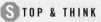 **STOP & THINK**
Remember to convert the answer from J/mol to kJ/mol because E_a values are usually listed in units of kJ/mol.

We use these values in the "two-temperature" form of the Arrhenius equation.

Solution

$$\ln \frac{k_2}{k_1} = \frac{E_a}{R} \left(\frac{1}{T_1} - \frac{1}{T_2} \right)$$

$$\ln \left(\frac{k}{9.16 \times 10^{-3} \, s^{-1}} \right) = \frac{88,000 \, \text{J/mol}}{8.314 \dfrac{\text{J}}{\text{mol} \cdot \text{K}}} \left(\frac{1}{273 \text{ K}} - \frac{1}{275 \text{ K}} \right) = 0.282$$

We can take inverse (natural) logarithms of both sides.

$$e^{\left[\ln\left(\frac{k_2}{9.16 \times 10^{-3} \, s^{-1}} \right) \right]} = e^{0.282}$$

$$\frac{k_2}{9.16 \times 10^{-3} \, s^{-1}} = 1.32$$

$$k_2 = 1.32(9.16 \times 10^{-3} \, s^{-1}) = 1.21 \times 10^{-2} \, s^{-1}$$

We see that a very small temperature difference, only 2°C, causes an increase in the rate constant (and hence in the reaction rate for the same concentrations) of about 32%. Such sensitivity of rate to temperature change makes the control and measurement of temperature extremely important in chemical reactions.

You should now work Exercise 55.

EXAMPLE 16-12   Activation Energy

The gas-phase decomposition of ethyl iodide to give ethylene and hydrogen iodide is a first-order reaction.

$$C_2H_5I \longrightarrow C_2H_4 + HI$$

At 600. K, the value of k is $1.60 \times 10^{-5} \, s^{-1}$. When the temperature is raised to 700. K, the value of k increases to $6.36 \times 10^{-3} \, s^{-1}$. What is the activation energy for this reaction?

Plan

We know k at two different temperatures. We solve the two-temperature form of the Arrhenius equation for E_a and evaluate.

Solution

$$k_1 = 1.60 \times 10^{-5} \, s^{-1} \text{ at } T_1 = 600. \text{ K} \qquad k_2 = 6.36 \times 10^{-3} \, s^{-1} \text{ at } T_2 = 700. \text{ K}$$

$$R = 8.314 \, \text{J/mol} \cdot K \qquad E_a = \underline{?}$$

We arrange the Arrhenius equation for E_a.

$$\ln \frac{k_2}{k_1} = \frac{E_a}{R} \left(\frac{1}{T_1} - \frac{1}{T_2} \right) \quad \text{so} \quad E_a = \frac{R \ln \dfrac{k_2}{k_1}}{\left(\dfrac{1}{T_1} - \dfrac{1}{T_2} \right)}$$

Substituting,

$$E_a = \frac{\left(8.314 \dfrac{\text{J}}{\text{mol} \cdot \text{K}} \right) \ln \left(\dfrac{6.36 \times 10^{-3} \, s^{-1}}{1.60 \times 10^{-5} \, s^{-1}} \right)}{\left(\dfrac{1}{600. \text{ K}} - \dfrac{1}{700. \text{ K}} \right)} = \frac{\left(8.314 \dfrac{\text{J}}{\text{mol} \cdot \text{K}} \right)(5.98)}{2.38 \times 10^{-4} \, \text{K}^{-1}} = \frac{2.09 \times 10^5 \, \text{J/mol}}{\text{or} \quad 209 \, \text{kJ/mol}}$$

You should now work Exercise 56.

The determination of E_a in the manner illustrated in Example 16-12 may be subject to considerable error because it depends on the measurement of k at only two temperatures. Any error in either of these k values would greatly affect the resulting value of E_a. A more reliable method that uses many measured values for the same reaction is based on a graphical approach. Let us rearrange the single-temperature logarithmic form of the Arrhenius equation and compare it with the equation for a straight line.

$$\underbrace{\ln k}_{y} = \underbrace{-\left(\frac{E_a}{R}\right)}_{m} \underbrace{\left(\frac{1}{T}\right)}_{x} + \underbrace{\ln A}_{b}$$

The value of the collision frequency factor, A, is very nearly constant over moderate temperature changes. Thus, $\ln A$ can be interpreted as the constant term in the equation (the intercept). The slope of the straight line obtained by plotting $\ln k$ versus $1/T$ equals $-E_a/R$. This allows us to determine the value of the activation energy from the slope (Figure 16-14). Exercises 59 and 60 use this method.

▶ Compare this approach to that described in the earlier Enrichment section for determining k.

16-9 Catalysts

Catalysts are substances that can be added to reaction mixtures to increase the rate of reaction. They provide alternative pathways that increase reaction rates by lowering activation energies.

The activation energy is lowered in all catalyzed reactions, as depicted in Figures 16-15 and 16-16. A catalyst does become involved in the reaction, and although it may be transformed via reaction steps during the catalysis, the starting catalyst is regenerated as the reaction

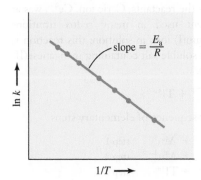

Figure 16-14 A graphical method for determining activation energy, E_a. At each of several different temperatures, the rate constant, k, is determined by methods such as those in Sections 16-3 and 16-4. A plot of $\ln k$ versus $1/T$ gives a straight line with negative slope. The slope of this straight line is $-E_a/R$. Use of this graphical method is often desirable because it partially compensates for experimental errors in individual k and T values.

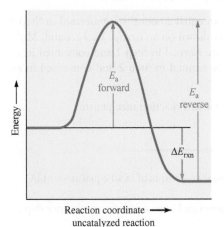

Reaction coordinate ⟶
uncatalyzed reaction

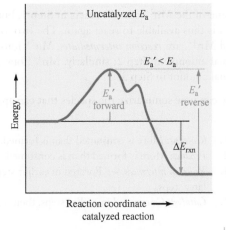

Reaction coordinate ⟶
catalyzed reaction

Figure 16-15 Potential energy diagrams showing the effect of a catalyst. The catalyst provides a different mechanism, corresponding to a lower-energy pathway, for the formation of the products. A catalyzed reaction typically occurs in several steps, each with its own barrier, but the overall energy barrier for the net reaction, E_a', is lower than that for the uncatalyzed reaction, E_a. The value of ΔE_{rxn} depends only on the states of the reactants and products (thermodynamics), so it is the same for either path.

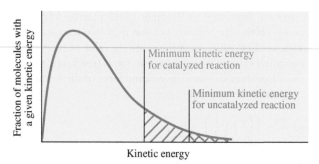

Figure 16-16 When a catalyst is present, the energy barrier is lowered. Thus, more molecules possess the minimum kinetic energy necessary for reaction. This is analogous to allowing more students to pass a course by lowering the requirements.

proceeds. The catalyst can then interact with more reactants to convert them into products. An effective catalyst can perform millions of such cycles. Because it is continually regenerated, a catalyst does not appear either as a reactant or as a product in the balanced equation for the reaction.

For constant T and the same concentrations,

Catalyst added $\Rightarrow$ E_a decreases $\Rightarrow$ E_a/RT decreases $\Rightarrow$ $-E_a/RT$ increases $\Rightarrow$ $e^{-E_a/RT}$ increases $\Rightarrow$ k increases $\Rightarrow$ **Reaction speeds up**

We can describe two categories of catalysts: (1) homogeneous catalysts and (2) heterogeneous catalysts.

Homogeneous Catalysis

A **homogeneous catalyst** exists in the same phase as the reactants. Ceric ion, Ce^{4+}, was at one time an important laboratory oxidizing agent used in many redox titrations (see Section 11.6). For example, Ce^{4+} oxidizes thallium(I) ions in solution; this reaction is catalyzed by the addition of a very small amount of a soluble salt containing manganese(II) ions, Mn^{2+}. The Mn^{2+} acts as a homogeneous catalyst.

$$2Ce^{4+} + Tl^+ \xrightarrow{Mn^{2+}} 2Ce^{3+} + Tl^{3+}$$

This reaction is thought to proceed by the following sequence of elementary steps.

$$
\begin{array}{llll}
Ce^{4+} & + Mn^{2+} & \longrightarrow Ce^{3+} & + Mn^{3+} & \text{step 1} \\
Ce^{4+} & + Mn^{3+} & \longrightarrow Ce^{3+} & + Mn^{4+} & \text{step 2} \\
Mn^{4+} & + Tl^+ & \longrightarrow Mn^{2+} & + Tl^{3+} & \text{step 3} \\
\hline
2Ce^{4+} & + Tl^+ & \longrightarrow 2Ce^{3+} & + Tl^{3+} & \text{overall}
\end{array}
$$

Some of the Mn^{2+} catalyst reacts in Step 1, but an equal amount is regenerated in Step 3 and is thus available to react again. The two ions shown on an orange background, Mn^{3+} and Mn^{4+}, are *reaction intermediates*. Mn^{3+} ions are formed in Step 1 and consumed in an equal amount in Step 2; similarly, Mn^{4+} ions are formed in Step 2 and consumed in an equal amount in Step 3.

We can now summarize the species that can appear in a reaction mechanism.

1. *Reactant:* More is consumed than is formed.
2. *Product:* More is formed than is consumed.
3. *Reaction intermediate:* Formed in earlier steps, then consumed in an equal amount in later steps.
4. *Catalyst:* Consumed in earlier steps, then regenerated in an equal amount in later steps.

Strong acids function as homogeneous catalysts in the acid-catalyzed hydrolysis of esters (a class of organic compounds—Section 23-14). Using ethyl acetate (a component of nail polish removers) as an example of an ester, we can write the overall reaction as follows.

▶ "Hydrolysis" means reaction with water.

$$CH_3\overset{\displaystyle O}{\overset{\|}{C}}-OCH_2CH_3(aq) + H_2O \xrightarrow{H^+} CH_3\overset{\displaystyle O}{\overset{\|}{C}}-OH(aq) + CH_3CH_2OH(aq)$$

ethyl acetate   acetic acid   ethanol

This is a thermodynamically favored reaction, but because of its high energy of activation, it occurs only very, very slowly when no catalyst is present. In the presence of strong acids, however, the reaction occurs more rapidly. In this acid-catalyzed hydrolysis, different intermediates with lower activation energies are formed. The sequence of steps in the *proposed* mechanism is shown at the bottom of this page.

We see that H^+ is a reactant in Step 1, but it is completely regenerated in Step 5. H^+ is therefore a catalyst. The species shown on orange backgrounds in Steps 1 through 4 are *reaction intermediates*. Ethyl acetate and water are the reactants, and acetic acid and ethanol are the products of the overall catalyzed reaction.

▶ All intermediates in this sequence of elementary steps are charged species, but this is not always the case.

Heterogeneous Catalysis

A **heterogeneous catalyst** is present in a different phase from the reactants. Such catalysts are usually solids, and they lower activation energies by providing surfaces on which reactions can occur. The first step in the catalytic process is usually *adsorption*, in which one or more of the reactants become attached to the solid surface. Some reactant molecules

▶ Groups of atoms that are involved in the change in each step are shown in blue. The catalyst, H^+, is shown in red. The intermediates are shown on an orange background.

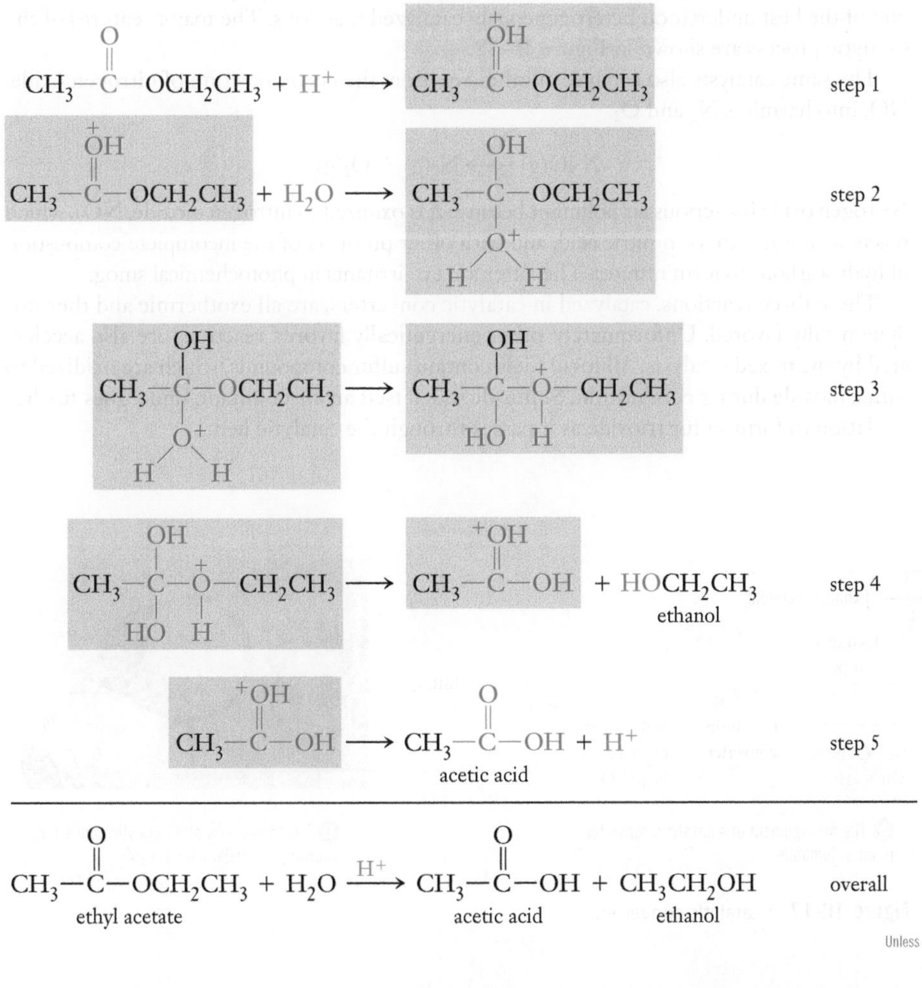

step 1

step 2

step 3

step 4
ethanol

step 5
acetic acid

$$CH_3\overset{\displaystyle O}{\overset{\|}{C}}-OCH_2CH_3 + H_2O \xrightarrow{H^+} CH_3\overset{\displaystyle O}{\overset{\|}{C}}-OH + CH_3CH_2OH$$

ethyl acetate   acetic acid   ethanol   overall

The petroleum industry uses numerous heterogeneous catalysts. Many of them contain highly colored compounds of transition metal ions. Several are shown here.

► Approximately 98% of all catalysts used in industry are heterogeneous. Therefore, when industrial chemists talk about a "catalyst," they usually mean a heterogeneous catalyst.

► At the high temperatures of the combustion of any fuel in air, nitrogen and oxygen combine to form nitrogen oxide.

► See the Chemistry in Use essay "Nitrogen Oxides and Photochemical Smog" in Chapter 28.

► Maintaining the continued efficiency of all three reactions in a "three-way" catalytic converter is a delicate matter. It requires control of such factors as the O_2 supply pressure and the order in which the reactants reach the catalyst. Modern automobile engines use microcomputer chips, based on an O_2 sensor in the exhaust stream, to control air valves.

may be held in particular orientations, or some bonds may be weakened; in other molecules, some bonds may be broken to form atoms or smaller molecular fragments. This causes *activation* of the reactants. As a result, *reaction* occurs more readily than would otherwise be possible. In a final step, *desorption*, the product molecules leave the surface, freeing reaction sites to be used again. Most heterogeneous catalysts are more effective as small particles, because they have relatively large surface areas.

Transition metals and their compounds function as effective catalysts in many homogeneous and heterogeneous reactions. Vacant *d* orbitals in many transition metal ions can accept electrons from reactants to form intermediates. These subsequently decompose to form products. Three transition metals, Pt, Pd, and Ni, are often used as finely divided solids to provide surfaces on which heterogeneous reactions can occur.

The catalytic converters (Figure 16-17) built into automobile exhaust systems contain two types of heterogeneous catalysts, powdered noble metals and powdered transition metal oxides. They catalyze the oxidation of unburned hydrocarbon fuel (reaction 1) and of partial combustion products such as carbon monoxide (reaction 2, shown in Figure 16-18).

1. $2C_8H_{18}(g) + 25O_2(g) \xrightarrow[\text{NiO}]{\text{Pt}} 16CO_2(g) + 18H_2O(g)$

      ↑

     isooctane (a component of gasoline)

2. $2CO(g) + O_2(g) \xrightarrow[\text{NiO}]{\text{Pt}} 2CO_2(g)$

It is desirable to carry out these reactions in automobile exhaust systems. Carbon monoxide is very poisonous. The latter reaction is so slow that a mixture of CO and O_2 gas at the exhaust temperature would remain unreacted for thousands of years in the absence of a catalyst! Yet the addition of only a small amount of a solid, finely divided transition metal catalyst promotes the production of up to a mole of CO_2 per minute. Because this reaction is a very simple but important one, it has been studied extensively by surface chemists. It is one of the best understood heterogeneously catalyzed reactions. The major features of the catalytic process are shown in Figure 16-18.

The same catalysts also catalyze another reaction, the decomposition of nitrogen oxide, NO, into harmless N_2 and O_2.

$$2NO(g) \xrightarrow[\text{NiO}]{\text{Pt}} N_2(g) + O_2(g)$$

Nitrogen oxide is a serious air pollutant because it is oxidized to nitrogen dioxide, NO_2, which reacts with water to form nitric acid, and with other products of the incomplete combustion of hydrocarbons to form nitrates. The latter are eye irritants in photochemical smog.

These three reactions, catalyzed in catalytic converters, are all exothermic and thermodynamically favored. Unfortunately, other energetically favored reactions are also accelerated by the mixed catalysts. All fossil fuels contain sulfur compounds, which are oxidized to sulfur dioxide during combustion. Sulfur dioxide, itself an air pollutant, undergoes further oxidation to form sulfur trioxide as it passes through the catalytic bed.

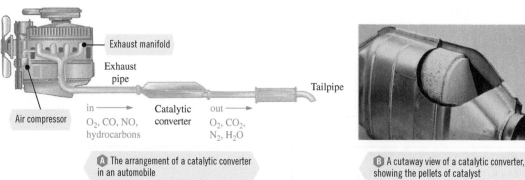

Exhaust manifold

Exhaust pipe

Tailpipe

Air compressor

in ⟶   Catalytic converter   out ⟶

O_2, CO, NO, hydrocarbons O_2, CO_2, N_2, H_2O

Ⓐ The arrangement of a catalytic converter in an automobile

Ⓑ A cutaway view of a catalytic converter, showing the pellets of catalyst

Figure 16-17 A catalytic converter.

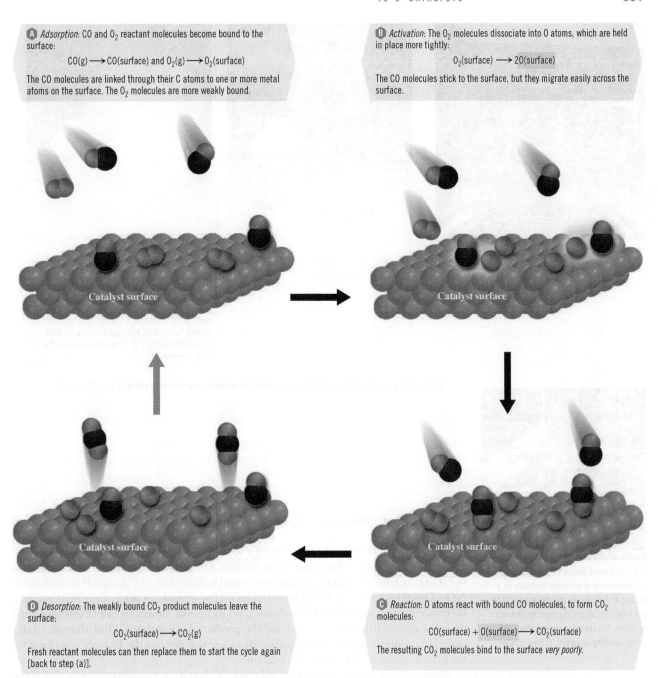

A *Adsorption*: CO and O_2 reactant molecules become bound to the surface:

$$CO(g) \longrightarrow CO(surface) \text{ and } O_2(g) \longrightarrow O_2(surface)$$

The CO molecules are linked through their C atoms to one or more metal atoms on the surface. The O_2 molecules are more weakly bound.

Catalyst surface

B *Activation*: The O_2 molecules dissociate into O atoms, which are held in place more tightly:

$$O_2(surface) \longrightarrow 2O(surface)$$

The CO molecules stick to the surface, but they migrate easily across the surface.

Catalyst surface

D *Desorption*: The weakly bound CO_2 product molecules leave the surface:

$$CO_2(surface) \longrightarrow CO_2(g)$$

Fresh reactant molecules can then replace them to start the cycle again [back to step (a)].

Catalyst surface

C *Reaction*: O atoms react with bound CO molecules, to form CO_2 molecules:

$$CO(surface) + O(surface) \longrightarrow CO_2(surface)$$

The resulting CO_2 molecules bind to the surface *very poorly*.

Catalyst surface

Figure 16-18 A schematic representation of the catalysis of the reaction

$$2CO(g) + O_2(g) \longrightarrow 2CO_2(g)$$

on a metallic surface.

$$2SO_2(g) + O_2(g) \xrightarrow[\text{NiO}]{\text{Pt}} 2SO_3(g)$$

Sulfur trioxide is probably a worse pollutant than sulfur dioxide, because SO_3 is the acid anhydride of strong, corrosive sulfuric acid. Sulfur trioxide reacts quickly with water vapor in the air, as well as in auto exhausts, to form sulfuric acid droplets. This problem must be overcome if the current type of catalytic converter is to see continued use. These same catalysts also suffer from the problem of being "poisoned"—that is, made inactive—by lead. Leaded fuels contain tetraethyl lead, $Pb(C_2H_5)_4$, and tetramethyl lead, $Pb(CH_3)_4$. Such fuels

A This reaction takes place more rapidly if the solution is heated.

B A very small amount of a transition metal oxide is added.

C This oxide catalyzes the decomposition reaction. The catalyzed reaction is rapid, so the exothermic reaction quickly heats the solution to the boiling point of water, forming steam. The temperature increase further accelerates the decomposition. We should never use a syringe with a metal tip to withdraw a sample from a 30% hydrogen peroxide solution, because the metal can dangerously catalyze the rapid decomposition.

A 30% hydrogen peroxide, H_2O_2, solution at room temperature decomposes very slowly to form O_2 and H_2O.

The bombardier beetle uses a catalyzed decomposition of hydrogen peroxide as a means of defense. An enzyme produced by the beetle catalyzes the rapid exothermic reaction. The resulting steam, along with other irritating chemical, is then ejected.

are not suitable for automobiles equipped with catalytic converters and are banned by U.S. law.

Reactions that occur in the presence of a solid catalyst, as on a metal surface (heterogeneous catalysis), often follow zero-order kinetics. For instance, the rate of decomposition of $NO_2(g)$ at high pressures on a platinum metal surface does not change if we add more NO_2. This is because only the NO_2 molecules on the surface can react. If the metal surface is completely covered with NO_2 molecules, no additional molecules can be *adsorbed* until the ones already there have reacted and the products have *desorbed*. Thus, the rate of the reaction is controlled only by the availability of reaction sites on the Pt surface, and not by the total number of NO_2 molecules available.

Some other important reactions that are catalyzed by transition metals and their oxides follow.

1. The Haber process for the production of ammonia (see Section 17-7).

$$N_2 + 3H_2 \xrightarrow[\text{high T, P}]{\text{Fe, Fe oxides}} 2NH_3$$

2. The contact process for the production of sulfur trioxide in the manufacture of sulfuric acid (see Section 28-11).

$$2SO_2 + O_2 \xrightarrow[400°C]{V_2O_5} 2SO_3$$

3. The chlorination of benzene (see Section 23-16).

$$C_6H_6 + Cl_2 \xrightarrow{FeBr_3} C_6H_5Cl + HCl$$
$$\text{benzene} \qquad \text{chlorobenzene}$$

4. The hydrogenation of unsaturated hydrocarbons (see Section 23-17).

$$RCH{=}CH_2 + H_2 \xrightarrow{\text{Pt}} RCH_2CH_3 \qquad R = \text{organic groups}$$

Enzymes as Biological Catalysts

Enzymes are proteins that act as catalysts for specific biochemical reactions in living systems. The reactants in enzyme-catalyzed reactions are called **substrates**. Thousands of vital processes in our bodies are catalyzed by many different enzymes. For instance, the enzyme carbonic anhydrase catalyzes the combination of CO_2 and water (the substrates), facilitating most of the transport of carbon dioxide in the blood. This combination reaction, ordinarily slow, proceeds rapidly in the presence of carbonic anhydrase; a single molecule of this enzyme can promote the conversion of more than 1 million molecules of carbon dioxide each second. Each enzyme is extremely specific, catalyzing only a few closely related reactions—or, in many cases, only one particular reaction—for only certain substrates. Modern theories of enzyme action attribute this to the requirement of very specific matching of shapes (molecular geometries) and polarities for a particular substrate to bind to a particular enzyme (Figure 16-19).

Enzyme-catalyzed reactions are important examples of zero-order reactions; that is, the rate of such a reaction is independent of the concentration of the substrate (provided some substrate is present).

$$\text{rate} = k$$

Each active site on an enzyme molecule can bind to only one substrate molecule at a time (or one pair, if the reaction links two reactant molecules), no matter how many other substrate molecules are available in the vicinity.

Ammonia is a very important industrial chemical that is used as a fertilizer and in the manufacture of many other chemicals. The reaction of nitrogen with hydrogen is a thermodynamically spontaneous reaction (product-favored), but without a catalyst it is very slow, even at high temperatures. The Haber process for its preparation involves the use of iron as a catalyst at 450°C to 500°C and high pressures.

$$N_2(g) + 3H_2(g) \xrightarrow{\text{Fe}} 2NH_3(g) \qquad \Delta G^0 = -194.7 \text{ kJ/mol (at 500°C)}$$

Even so, iron is not a very effective catalyst even under these extreme conditions.

When heated, a sugar cube (sucrose, melting point 185°C) melts but does not burn. A sugar cube rubbed in cigarette ash burns before it melts. The cigarette ash contains trace amounts of metal compounds that catalyze the combustion of sugar.

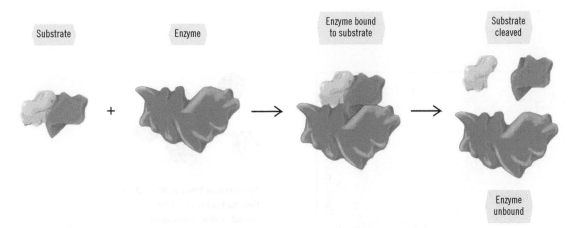

| Substrate | Enzyme | Enzyme bound to substrate | Substrate cleaved |

Enzyme unbound

Figure 16-19 A schematic representation of a simplified mechanism (lock-and-key) for enzyme reaction. The substrates (reactants) fit the active sites of the enzyme molecule much as keys fit locks. When the reaction is complete, the products do not fit the active sites as well as the reactants did. They separate from the enzyme, leaving it free to catalyze the reaction of additional reactant molecules. The enzyme is not permanently changed by the process. The illustration here is for a process in which a complex reactant molecule is split to form two simpler product molecules. The formation of simple sugars from complex carbohydrates is a similar reaction. Some enzymes catalyze the combination of simple molecules to form more complex ones.

CHEMISTRY IN USE

Ozone

Ozone, O_3, is such a powerful oxidizing agent that in significant concentrations it degrades many plastics, metals, and rubber, as well as both plant and animal tissues. We therefore try to minimize exposure to ozone in our immediate environment. In the upper atmosphere, however, ozone plays a very important role in the absorption of harmful radiation from the sun. Maintaining appropriate concentrations of ozone—minimizing its production where ozone is harmful and preventing its destruction where ozone is helpful—is an important challenge in environmental chemistry.

Ozone is formed in the upper atmosphere as some O_2 molecules absorb high-energy UV radiation from the sun and dissociate into oxygen atoms; these then combine with other O_2 molecules to form ozone.

$$O_2(g) + \text{UV radiation} \longrightarrow 2\ \boxed{O(g)} \quad \text{(step 1—occurs once)}$$
$$\underline{O_2(g) + \boxed{O(g)} \longrightarrow O_3(g) \quad \text{(step 2—occurs twice)}}$$
$$3O_2(g) + \text{UV radiation} \longrightarrow 2O_3(g) \quad \text{(net reaction)}$$

Although it also decomposes in the upper atmosphere, the ozone supply is continuously replenished by this process. Its concentration in the stratosphere ($\approx 7-31$ miles above the earth's surface) is about 10 ppm (parts per million), whereas it is only about 0.04 ppm near the earth's surface.

The high-altitude ozone layer is responsible for absorbing much of the dangerous ultraviolet light from the sun in the $20-30$ Å wavelength range.

$$O_3(g) + \text{UV radiation} \longrightarrow O_2(g) + \boxed{O(g)}$$
$$\text{(step 1—occurs once)}$$
$$\underline{O_2(g) + \boxed{O(g)} \longrightarrow O_3(g) \quad \text{(step 2—occurs once)}}$$
$$\text{No net reaction}$$

We see that each time this sequence takes place, it absorbs one photon of ultraviolet light; however, the process regenerates as much ozone as it uses up. Each stratospheric ozone molecule can thus absorb a significant amount of ultraviolet light. If this high-energy radiation reached the surface of the earth in higher intensity, it would be very harmful to plants and animals (including humans). It has been estimated that the incidence of skin cancer would increase by 2% for every 1% decrease in the concentration of ozone in the stratosphere.

Chlorofluorocarbons (CFCs) are chemically inert, nonflammable, nontoxic compounds that are superb solvents and have been used in many industrial processes; they are excellent coolants for air conditioners and refrigerators. Two CFCs that have been widely used are Freon-11 and Freon-12 (Freon is a DuPont trade name).

Cl
|
Cl—C—F
|
Cl
Freon-11 is CCl_3F

Cl
|
Cl—C—F
|
F
Freon-12 is CCl_2F_2

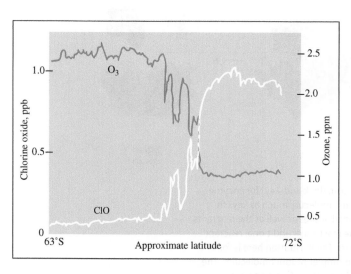

The compound known as HCFC-134, a fluorocarbon currently used in home and automobile air conditioners.

A plot that shows the decrease in [O_3] as [ClO] increases over Antarctica.

The CFCs are so unreactive that they do not readily decompose, that is, break down into simpler compounds, when they are released into the atmosphere. Over time the CFCs are carried into the stratosphere by air currents, where they are exposed to large amounts of ultraviolet radiation.

In 1974, Mario Molina and Sherwood Rowland of the University of California–Irvine demonstrated in their laboratory that when CFCs are exposed to high-energy ultraviolet radiation they break down to form chlorine *radicals*.

$$
\underset{\underset{\text{Cl}}{|}}{\overset{\overset{\text{F}}{|}}{\text{F}-\text{C}-\text{Cl}}} \xrightarrow[\text{radiation}]{\text{UV}} \underset{\underset{\text{Cl}}{|}}{\overset{\overset{\text{F}}{|}}{\text{F}-\text{C}\cdot}} + :\overset{..}{\underset{..}{\text{Cl}}}\cdot \quad \text{(a chlorine radical)}
$$

Molina and Rowland predicted that these very reactive radicals could cause problems by catalyzing the destruction of ozone in the stratosphere.

Each spring since 1979, researchers have observed a thinning of the ozone layer over Antarctica. Each spring (autumn in the Northern Hemisphere) beginning in 1983, satellite images have shown a "hole" in the ozone layer over the South Pole. During August and September 1987, a NASA research team flew a plane equipped with sophisticated analytical instruments into the ozone hole 25 times. Their measurements demonstrated that as the concentration of the chlorine oxide radicals, Cl — O, increased, the concentration of ozone decreased.

By September 1992, this *ozone hole* was nearly three times the area of the United States. In December 1994, three years of data from NASA's Upper Atmosphere Research Satellite (UARS) provided conclusive evidence that CFCs are primarily responsible for this destruction of the ozone layer. Considerable thinning of the ozone layer in the Northern Hemisphere has also been observed.

The following is a simplified representation of the *chain reaction* that is now believed to account for most of the ozone destruction in the stratosphere.

$$
:\overset{..}{\underset{..}{\text{Cl}}}\cdot + \text{O}_3 \longrightarrow \text{Cl}-\overset{..}{\underset{..}{\text{O}}}\cdot + \text{O}_2 \quad \text{(step 1)}
$$

$$
\text{Cl}-\overset{..}{\underset{..}{\text{O}}}\cdot + \text{O} \longrightarrow :\overset{..}{\underset{..}{\text{Cl}}}\cdot + \text{O}_2 \quad \text{(step 2)}
$$

$$
\overline{\text{O}_3(g) + \text{O} \longrightarrow 2\text{O}_2(g)} \quad \text{(net reaction)}
$$

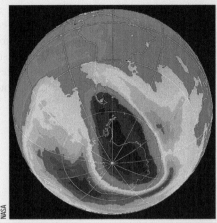

A computer-generated image of part of the Southern Hemisphere on October 17, 1994, reveals the ozone "hole" (*black and purple areas*) over Antarctica and the tip of South America. Relatively low ozone levels (*blue and green areas*) extend into much of South America as well as Central America. Normal ozone levels are shown in yellow, orange, and red. The ozone hole is not stationary but moves about as a result of air currents.

A sufficient supply of oxygen atoms, O, is available in the upper atmosphere for the second step to occur. The net reaction results in the destruction of a molecule of ozone. The chlorine radical that initiates the first step of this reaction sequence is regenerated in the second step, however, and so a single chlorine radical can act as a catalyst to destroy many thousands of O₃ molecules. Other well-known reactions also destroy ozone in the stratosphere, but the evidence shows conclusively that CFCs are the principal culprits.

Since January 1978, the use of CFCs in aerosol cans in the United States has been banned; increasingly strict laws prohibit the release into the atmosphere of CFCs from sources such as automobile air conditioners and discarded refrigerators. The Montreal Protocol, signed by 24 countries in 1989, called for reductions in production and use of many CFCs. International agreements have since called for a complete ban on CFC production. Efforts to develop suitable replacement substances and controls for existing CFCs continue. The good news is that scientists expect the ozone hole to decrease and possibly disappear during the 21st century *if* current international treaties remain in effect and *if* they are implemented throughout the world. These are two very large *ifs*.

Up to date information on stratospheric ozone can be found at the NASA website, ozonewatch.gsfc.nasa.gov/ and at other websites.

▶ The process is called nitrogen fixation. The ammonia can be used in the synthesis of many nitrogen-containing biological compounds such as proteins and nucleic acids.

▶ Transition metal ions are present in the active sites of some enzymes.

In contrast, the reaction between N_2 and H_2 to form NH_3 is catalyzed at room temperature and atmospheric pressure by a class of enzymes, called nitrogenases, that are present in some bacteria. Legumes are plants that support these bacteria; they are able to obtain nitrogen as N_2 from the atmosphere and convert it to ammonia.

In comparison with manufactured catalysts, most enzymes are tremendously efficient under very mild conditions. If chemists and biochemists could develop catalysts with a small fraction of the efficiency of enzymes, such catalysts could be a great boon to the world's health and economy. One of the most active areas of current chemical research involves attempts to discover or synthesize catalysts that can mimic the efficiency of naturally occurring enzymes such as nitrogenases. Such a development would be important in industry. Due to the high temperatures and pressures needed, an estimated 1% of the world's energy supply is used each year to run the Haber process. Developing better catalysts could decrease the cost of food grown with the aid of ammonia-based fertilizers. Ultimately this would help greatly to feed the world's growing population.

KEY TERMS

Activation energy   The kinetic energy that reactant molecules must have to allow them to reach the transition state so that a reaction can occur; also known as the *activation barrier*.

Arrhenius equation   An equation that relates the specific rate constant to activation energy and temperature.

Catalyst   A substance that increases the rate at which a reaction occurs. It is regenerated during the course of the reaction and can then interact with more reactants to convert them into products.

Chemical kinetics   The study of rates and mechanisms of chemical reactions and of the factors on which they depend.

Collision theory   A theory of reaction rates that states that effective collisions between reactant molecules must take place for reaction to occur.

Effective collision   A collision between molecules that results in reaction; one in which molecules collide with proper orientations and with sufficient energy to react.

Elementary step   An individual step in the mechanism by which a reaction occurs. For each elementary step, the reaction orders *do* match the reactant coefficients in that step.

Enzyme   A protein that acts as a catalyst in a biological system.

Fundamental step   See *Elementary step*.

Half-life of a reactant ($t_{1/2}$)   The time required for half of that reactant to be converted into product(s).

Heterogeneous catalyst   A catalyst that exists in a different phase (solid, liquid, or gas) from the reactants; the vast majority of heterogeneous catalysts are solids.

Homogeneous catalyst   A catalyst that exists in the same phase (liquid or gas) as the reactants.

Integrated rate equation   An equation that relates the concentration of a reactant remaining to the time elapsed; has different mathematical forms for different orders of reaction.

Method of initial rates   A method of determining the rate-law expression by carrying out a reaction with different initial concentrations and analyzing the resulting changes in initial rates.

Order of a reactant   The power to which the reactant's concentration is raised in the rate-law expression.

Order of a reaction   The sum of the powers to which all concentrations are raised in the rate-law expression; also called the overall order of a reaction.

Rate constant   An experimentally determined proportionality constant that is different for different reactions and that, for a given reaction, changes only with temperature or the presence of a catalyst; k in the rate-law expression, rate = $k[A]^x[B]^y$; also known as the *specific rate constant*.

Rate-determining step   The slowest elementary step in a reaction mechanism; the step that limits the overall rate of reaction.

Rate-law expression   An equation that relates the rate of a reaction to the concentrations of the reactants and the specific rate constant; rate = $k[A]^x[B]^y$. The exponents of reactant concentrations *do not necessarily* match the coefficients in the overall balanced chemical equation. The rate-law expression must be determined from experimental data; also known as the *rate law*.

Rate of reaction   The change in concentration of a reactant or product per unit time.

Reaction coordinate   The progress along the potential energy pathway from reactants to products.

Reaction intermediate   A species that is produced and then entirely consumed during a multistep reaction; usually short-lived.

Reaction mechanism   The sequence of fundamental steps by which reactants are converted into products.

Substrate   A reactant in an enzyme-catalyzed reaction.

Thermodynamically favorable (spontaneous) reaction   A reaction that occurs with a net release of free energy, G; a reaction for which ΔG is negative (see Section 15-16).

Transition state   A relatively high-energy state in which bonds in reactant molecules are partially broken and new ones are partially formed.

Transition state theory   A theory of reaction rates that states that reactants pass through high-energy transition states before forming products.

EXERCISES

. . . .

🐾 **Molecular Reasoning** exercises

▲ **More Challenging** exercises

General Concepts

1. Briefly summarize the effects of each of the four factors that affect rates of reactions.
2. 🐾 Describe the basic features of collision theory and transition state theory.
3. What is a rate-law expression? Describe how it is determined for a particular reaction.
4. Distinguish between reactions that are thermodynamically favorable and reactions that are kinetically favorable. What can be said about the relationship between the two?
5. 🐾 What is meant by the mechanism of a reaction? How does the mechanism relate to the order of the reaction?
6. What, if anything, can be said about the relationship between the coefficients of the balanced *overall* equation for a reaction and the powers to which concentrations are raised in the rate-law expression? To what are these powers related?
7. Express the rate of the reaction
$2N_2H_4(\ell) + N_2O_4(\ell) \longrightarrow 3N_2(g) + 4H_2O(g)$ in terms of: (a) $\Delta[N_2O_4]$; (b) $\Delta[N_2]$.
8. Express the rate of reaction in terms of the rate of change of each reactant and each product in the following equations.
   (a) $3ClO^-(aq) \longrightarrow ClO_3^-(aq) + 2Cl^-(aq)$
   (b) $2SO_2(g) + O_2(g) \longrightarrow 2SO_3(g)$
   (c) $C_2H_4(g) + Br_2(g) \longrightarrow C_2H_4Br_2(g)$
   (d) $(C_2H_5)_2(NH)_2 + I_2 \longrightarrow (C_2H_5)_2N_2 + 2HI$
   (liquid phase)
9. At a given time, N_2 is reacting with H_2 at a rate of $0.30\ M/min$ to produce NH_3. At that same time, what is the rate at which the other reactant is changing and the rate at which the product is changing?
$$N_2 + 3H_2 \longrightarrow 2NH_3$$
10. The following equation shows the production of NO and H_2O by oxidation of ammonia. At a given time, NH_3 is reacting at a rate of $1.20\ M/min$. At that same time, what is the rate at which the other reactant is changing and the rate at which each product is changing?
$$4NH_3 + 5O_2 \longrightarrow 4NO + 6H_2O$$
11. Why do large crystals of sugar burn more slowly than finely ground sugar?
12. Some fireworks are bright because of the burning of magnesium. Speculate on how fireworks might be constructed using magnesium. How might the sizes of the pieces of magnesium be important? What would

you expect to occur if the pieces that were used were too large? too small?

Charles Steele

Rate-Law Expression

13. A reaction has the experimental rate equation: Rate = $k[A]^2$. How will the rate change if the concentration of A is tripled? If the concentration of A is halved?
14. The rate expression for the following reaction at a certain temperature is rate = $k[NO]^2[O_2]$. Two experiments involving this reaction are carried out at the same temperature. In the second experiment the initial concentration of NO is halved, while the initial concentration of O_2 is doubled. The initial rate in the second experiment will be how many times that of the first?
$$2NO + O_2 \longrightarrow 2NO_2$$
15. The rate-law expression for the following reaction is found to be rate = $k[N_2O_5]$. What is the overall reaction order?
$$2N_2O_5(g) \longrightarrow 4NO_2(g) + O_2(g)$$
16. Use times expressed in seconds to give the units of the rate constant for reactions that are overall (a) first order; (b) second order; (c) third order; (d) of order $1\frac{1}{2}$.
17. Rate data were obtained at 25°C for the following reaction. What is the rate-law expression for this reaction?
$$A + 2B \longrightarrow C + 2D$$

Expt.	Initial [A] (mol/L)	Initial [B] (mol/L)	Initial Rate of Formation of C ($M \cdot min^{-1}$)
1	0.10	0.10	3.0×10^{-4}
2	0.30	0.30	9.0×10^{-4}
3	0.10	0.30	3.0×10^{-4}
4	0.20	0.40	6.0×10^{-4}

. . . .

18. Rate data were obtained for the following reaction at 25°C. What is the rate-law expression for the reaction?

$$2A + B + 2C \longrightarrow D + 2E$$

Expt.	Initial [A] (M)	Initial [B] (M)	Initial [C] (M)	Initial Rate of Formation of D (M · min⁻¹)
1	0.10	0.10	0.20	5.0×10^{-4}
2	0.20	0.30	0.20	1.5×10^{-3}
3	0.30	0.10	0.20	5.0×10^{-4}
4	0.40	0.30	0.60	4.5×10^{-3}

19. The reaction of $CO(g) + NO_2(g)$ is second-order in NO_2 and zeroth-order in CO at temperatures less than 500 K. (a) Write the rate law for the reaction. (b) How will the reaction rate change if the NO_2 concentration is halved? (c) How will the reaction rate change if the concentration of CO is doubled?

20. Listed below are the equations for three reactions and the rate-law expressions for them.

Reaction	Rate Law
$2NO(g) + 2H_2(g) \longrightarrow$ $N_2(g) + 2H_2O(g)$	rate $= k[NO]^2 [H_2]$
$2ICl(g) + H_2(g) \longrightarrow$ $2HCl(g) + I_2(g)$	rate $= k[ICl] [H_2]$
$H_2(g) + Br_2(g) \longrightarrow 2HBr(g)$	rate $= k[H_2] [Br_2]$

For which of these reactions will doubling the concentration of H_2 double the reaction rate? For which of these reactions will doubling the concentration of H_2 quadruple the reaction rate? For which of these reactions will doubling the concentration of H_2 have no effect on the rate of the reaction?

21. For a reaction involving the decomposition of Y, the following data are obtained:

Rate (mol/L · min)	0.228	0.245	0.202	0.158
[Y]	0.200	0.170	0.140	0.110

(a) Determine the order of the reaction.
(b) Write the rate expression for the decomposition of Y.
(c) Calculate k for the above experiment.

22. Rate data were collected for the following reaction at a particular temperature.

$$2ClO_2(aq) + 2OH^-(aq) \longrightarrow ClO_3^-(aq) + ClO_2^-(aq) + H_2O(\ell)$$

Expt.	Initial [ClO₂] (mol/L)	Initial [OH⁻] (mol/L)	Initial Rate of Formation of C (M s⁻¹)
1	0.012	0.012	2.07×10^{-4}
2	0.012	0.024	4.14×10^{-4}
3	0.024	0.012	8.28×10^{-4}
4	0.024	0.024	1.66×10^{-3}

(a) What is the rate-law expression for this reaction?
(b) Describe the order of the reaction with respect to each reactant and to the overall order. (c) What is the value, with units, for the specific rate constant?

23. The reaction

$$(C_2H_5)_2(NH)_2 + I_2 \longrightarrow (C_2H_5)_2N_2 + 2HI$$

gives the following initial rates.

Expt.	[(C₂H₅)₂(NH)₂]₀ (mol/L)	[I₂]₀ (mol/L)	Initial Rate of Formation of (C₂H₅)₂N₂
1	0.015	0.015	$3.15 \ M \cdot s^{-1}$
2	0.015	0.045	$9.45 \ M \cdot s^{-1}$
3	0.030	0.045	$18.9 \ M \cdot s^{-1}$

(a) Write the rate-law expression. (b) What is the value, with units, for the specific rate constant?

24. ▲ Given these data for the reaction $A + B \longrightarrow C$, write the rate-law expression.

Expt.	Initial [A] (M)	Initial [B] (M)	Initial Rate of Formation of C (M·s⁻¹)
1	0.10	0.20	5.0×10^{-6}
2	0.10	0.30	7.5×10^{-6}
3	0.20	0.40	4.0×10^{-5}

25. ▲ (a) Given these data for the reaction $A + B \longrightarrow C$, write the rate-law expression. (b) What is the value, with units, for the specific rate constant?

Expt.	Initial [A] (M)	Initial [B] (M)	Initial Rate of Formation of C (M·s⁻¹)
1	0.25	0.15	8.0×10^{-5}
2	0.25	0.30	3.2×10^{-4}
3	0.50	0.60	5.12×10^{-3}

26. ▲ (a) Given these data for the reaction $A + B \longrightarrow C$, write the rate-law expression. (b) What is the value, with units, for the specific rate constant?

Expt.	Initial [A] (M)	Initial [B] (M)	Initial Rate of Formation of C (M/s)
1	0.10	0.10	2.0×10^{-4}
2	0.10	0.20	8.0×10^{-4}
3	0.20	0.40	2.56×10^{-2}

🔴 **Molecular Reasoning** exercises   ▲ **More Challenging** exercises   Blue-Numbered exercises solved in Student Solutions Manual

27. ▲ Consider a chemical reaction between compounds A and B that is first order in A and first order in B. From the information shown here, fill in the blanks.

Expt.	Rate ($M \cdot s^{-1}$)	[A]	[B]
1	0.24	0.20 M	0.050 M
2	0.20	___ M	0.030 M
3	0.80	0.40 M	___ M

28. ▲ Consider a chemical reaction of compounds A and B that was found to be first order in A and second order in B. From the following information, fill in the blanks.

Expt.	Rate ($M \cdot s^{-1}$)	[A]	[B]
1	0.150	1.00 M	0.200 M
2	_____	2.00 M	0.200 M
3	_____	2.00 M	0.400 M

29. ▲ The rate of decomposition of NO_2 by the following reaction at a particular temperature is 5.4×10^{-5} mol NO_2/L·s when [NO_2] = 0.0110 mol/L.

$$2NO_2(g) \longrightarrow 2NO(g) + O_2(g)$$

(a) Assume that the rate law is rate = k[NO_2]. What rate of disappearance of NO_2 would be predicted when [NO_2] = 0.00550 mol/L? (b) Now assume that the rate law is rate = k[NO_2]^2. What rate of disappearance of NO_2 would be predicted when [NO_2] = 0.00550 mol/L? (c) The rate when [NO_2] = 0.00550 mol/L is observed to be 1.4×10^{-5} mol NO_2/L·s. Which rate law is correct? (d) Calculate the rate constant. (*Reminder*: Express the rate of reaction in terms of rate of disappearance of NO_2.)

Integrated Rate Equations and Half-Life

30. What is meant by the half-life of a reactant?

31. The rate law for the reaction of sucrose in water

$$C_{12}H_{22}O_{11} + H_2O \longrightarrow 2C_6H_{12}O_6$$

is rate = k[$C_{12}H_{22}O_{11}$]. After 2.57 hours at 25°C, 6.00 g/L of $C_{12}H_{22}O_{11}$ has decreased to 5.40 g/L. Evaluate k for this reaction at 25°C.

Sucrose (table sugar)

32. The rate constant for the decomposition of nitrogen dioxide

$$2NO_2 \longrightarrow 2NO + O_2$$

with a laser beam is 1.70 $M^{-1} \cdot min^{-1}$. Find the time, in seconds, needed to decrease 2.00 mol/L of NO_2 to 1.25 mol/L.

33. The second-order rate constant for the following gas-phase reaction is 0.0442 $M^{-1} \cdot s^{-1}$. We start with 0.135 mol C_2F_4 in a 2.00-liter container, with no C_4F_8 initially present.

$$2C_2F_4 \longrightarrow C_4F_8$$

(a) What will be the concentration of C_2F_4 after 1.00 hour? (b) What will be the concentration of C_4F_8 after 1.00 hour? (c) What is the half-life of the reaction for the initial C_2F_4 concentration given in part (a)? (d) How long will it take for half of the C_2F_4 that remains after 1.00 hour to disappear?

34. The decomposition reaction of carbon disulfide, CS_2, to carbon monosulfide, CS, and sulfur is first order with $k = 2.8 \times 10^{-7}$ s^{-1} at 1000°C.

$$CS_2 \longrightarrow CS + S$$

(a) What is the half-life of this reaction at 1000°C? (b) How many days would pass before a 2.00-gram sample of CS_2 had decomposed to the extent that 0.75 gram of CS_2 remained? (c) Refer to part (b). How many grams of CS would be present after this length of time? (d) How much of a 2.00-gram sample of CS_2 would remain after 45.0 days?

35. ▲ The first-order rate constant for the conversion of cyclobutane to ethylene at 1000.°C is 87 s^{-1}. (a) What is the half-life of this reaction at 1000.°C? (b) If you started with 4.00 g of cyclobutane, how long would it take to consume 2.50 g of it? (*Hint*: Write the ratio of concentrations, [A]$_0$/[A], in terms of mass, molecular weight, and volume.) (c) How much of an initial 1.00-g sample of cyclobutane would remain after 1.00 s?

$$H-\underset{\underset{H}{|}}{\overset{\overset{H}{|}}{C}}-\underset{\underset{H}{|}}{\overset{\overset{H}{|}}{C}}-H \longrightarrow 2 \; \underset{H}{\overset{H}{\diagdown}}C=C\underset{H}{\overset{H}{\diagup}}$$

cyclobutane　　　　ethylene

36. ▲ For the reaction

$$2NO_2 \longrightarrow 2NO + O_2$$

the rate equation is

rate = 1.4×10^{-10} $M^{-1} \cdot s^{-1}$ [NO_2]^2 at 25°C

(a) If 3.00 mol of NO_2 is initially present in a sealed 2.00-L vessel at 25°C, what is the half-life of the reaction? (b) Refer to part (a). What concentration and how many grams of NO_2 remain after 115 years? (c) Refer to part (b). What concentration of NO would have been produced during the same period of time?

t (seconds)	[HI] (mmol/L)
0.	5.46
250.	4.10
500.	2.73
750.	1.37

37. The first-order rate constant for the radioactive decay of radium-223 is 0.0606 day^{-1}. What is the half-life of radium-223?

38. Cyclopropane rearranges to form propene in a reaction that follows first-order kinetics. At 800. K, the specific rate constant for this reaction is 2.74×10^{-3} s^{-1}. Suppose we start with a cyclopropane concentration of 0.290 M. How long will it take for 99.0% of the cyclopropane to disappear according to this reaction?

cyclopropane → propene

39. The rate constant for the first-order reaction

$$N_2O_5 \longrightarrow 2NO_2 + \frac{1}{2} O_2$$

is 1.20×10^{-2} s^{-1} at 45°C, and the initial concentration of N_2O_5 is 0.01500 M. (a) How long will it take for the concentration to decrease to 0.00100 M? (b) How much longer will it take for a further decrease to 0.000900 M?

40. The familiar rule of thumb that the rate of a reaction roughly doubles for every 10°C rise in temperature (around room temperature) has been shown to be consistent with the Arrhenius equation, using a typical activation energy of 50 kJ/mol. How much would the rate of such a reaction change if the temperature of the reaction were changed from 90°C to 100°C or from 0°C to 10°C?

41. The thermal decomposition of ammonia at high temperatures was studied in the presence of inert gases. Data at 2000. K are given for a single experiment.

$$NH_3 \longrightarrow NH_2 + H$$

t (hours)	[NH$_3$] (mol/L)
0	8.000×10^{-7}
25	6.75×10^{-7}
50	5.84×10^{-7}
75	5.15×10^{-7}

Plot the appropriate concentration expressions against time to find the order of the reaction. Find the rate constant of the reaction from the slope of the line. Use the given data and the appropriate integrated rate equation to check your answer.

42. The following data were obtained from a study of the decomposition of a sample of HI on the surface of a gold wire. (a) Plot the data to find the order of the reaction, the rate constant, and the rate equation. (b) Calculate the HI concentration in mmol/L at 600. s.

43. The decomposition of SO_2Cl_2 in the gas phase,

$$SO_2Cl_2 \longrightarrow SO_2 + Cl_2$$

can be studied by measuring the concentration of Cl$_2$ as the reaction proceeds. We begin with [SO$_2$Cl$_2$]$_0$ = 0.250 M. Holding the temperature constant at 320.°C, we monitor the Cl$_2$ concentration, with the following results.

t (hours)	[Cl$_2$] (mol/L)
0.00	0.000
2.00	0.037
4.00	0.068
6.00	0.095
8.00	0.117
10.00	0.137
12.00	0.153
14.00	0.168
16.00	0.180
18.00	0.190
20.00	0.199

(a) Plot [Cl$_2$] versus t. (b) Plot [SO$_2$Cl$_2$] versus t. (c) Determine the rate law for this reaction. (d) What is the value, with units, for the specific rate constant at 320.°C? (e) How long would it take for 95% of the original SO$_2$Cl$_2$ to react?

44. ▲ At some temperature, the rate constant for the decomposition of HI on a gold surface is 0.080 $M \cdot$ s^{-1}.

$$2HI(g) \longrightarrow H_2(g) + I_2(g)$$

(a) What is the order of the reaction? (b) How long will it take for the concentration of HI to drop from 1.50 M to 0.15 M?

45. At body temperature, the first-order rate constant is 1.87×10^{-3} min^{-1} for the elimination of the cancer chemotherapy agent cisplatin. What would be the concentration of cisplatin in a patient 36 hours after the concentration had reached 4.75×10^{-3} M?

46. The first-order rate constant for the decomposition of an insecticide in moist soil is 3.0×10^{-3} d^{-1}. How long will it take for 75% of the insecticide to decompose?

Activation Energy, Temperature, and Catalysts

47. The following rearrangement reaction is first order:

$$CH_3NC \longrightarrow CH_3CN$$

In a table of kinetics data, we find the following values listed for this reaction: $A = 3.98 \times 10^{13}$ s^{-1}, $E_a = 160.$ kJ/mol. (a) Calculate the value of the specific rate constant at room temperature, 25°C. (b) Calculate the value of the specific rate constant at 115°C.

48. The following gas-phase decomposition reaction is first order:

$$C_2H_5Cl \longrightarrow C_2H_4 + HCl$$

In a table of kinetics data, we find the following values listed for this reaction: $A = 1.58 \times 10^{13}$ s^{-1}, $E_a = 237$ kJ/mol. (a) Calculate the value of the specific rate constant at room temperature, 25°C. (b) Calculate the value of the specific rate constant at 275°C.

49. Draw typical reaction energy diagrams for one-step reactions that release energy and that absorb energy. Distinguish between the net energy change, ΔE, and the activation energy for each kind of reaction. Indicate potential energies of products and reactants for both kinds of reactions.

50. Use graphs to illustrate how the presence of a catalyst can affect the rate of a reaction.

51. How do homogeneous catalysts and heterogeneous catalysts differ?

52. ♣ (a) Why should one expect an increase in temperature to increase the initial rate of reaction? (b) Why should one expect a reaction in the gaseous state to be faster than the same reaction in the solid state?

53. Assume that the activation energy for a certain reaction is 173 kJ/mol and the reaction is started with equal initial concentrations of reactants. How many times faster will the reaction occur at 40°C than at 10°C?

54. What is the activation energy for a reaction if its rate constant is found to triple when the temperature is raised from 600. K to 610. K?

55. For a gas-phase reaction, $E_a = 103$ kJ/mol, and the rate constant is 0.0850 min^{-1} at 273 K. Find the rate constant at 323 K.

56. The rate constant of a reaction is tripled when the temperature is increased from 298 K to 308 K. Find E_a.

57. The rate constant for the decomposition of N_2O

$$2N_2O(g) \longrightarrow 2N_2(g) + O_2(g)$$

is 2.6×10^{-11} s^{-1} at 300.°C and 2.1×10^{-10} s^{-1} at 330.°C. Calculate the activation energy for this reaction. Prepare a reaction coordinate diagram like Figure 16-10 using −164.1 kJ/mol as the ΔE_{rxn}.

58. For a particular reaction, $\Delta E = 51.51$ kJ/mol, $k = 8.0 \times 10^{-7}$ s^{-1} at 0.0°C, and $k = 8.9 \times 10^{-4}$ s^{-1} at 50.0°C. Prepare a reaction coordinate diagram like Figure 16-10 for this reaction.

59. ▲ You are given the rate constant as a function of temperature for the exchange reaction

$$Mn(CO)_5(CH_3CN)^+ + NC_5H_5 \rightarrow$$
$$Mn(CO)_5(NC_5H_5)^+ + CH_3CN$$

T (K)	k (min^{-1})
298	0.0409
305	0.0818
312	0.157

(a) Calculate E_a from a plot of log k versus $1/T$. (b) Use the graph to predict the value of k at 308 K. (c) What is the numerical value of the collision frequency factor, A, in the Arrhenius equation?

60. ▲ The rearrangement of cyclopropane to propene described in Exercise 38 has been studied at various temperatures. The following values for the specific rate constant have been determined experimentally.

T (K)	k (s^{-1})
600.	3.30×10^{-9}
650.	2.19×10^{-7}
700.	7.96×10^{-6}
750.	1.80×10^{-4}
800.	2.74×10^{-3}
850.	3.04×10^{-2}
900.	2.58×10^{-1}

(a) From the appropriate plot of these data, determine the value of the activation energy for this reaction. (b) Use the graph to estimate the value of k at 500 K. (c) Use the graph to estimate the temperature at which the value of k would be equal to 5.00×10^{-5} s^{-1}.

61. Biological reactions nearly always occur in the presence of enzymes as catalysts. The enzyme catalase, which acts on peroxides, reduces the E_a for the reaction from 72 kJ/mol (uncatalyzed) to 28 kJ/mol (catalyzed). By what factor does the reaction rate increase at normal body temperature, 37.0°C, for the same reactant (peroxide) concentration? Assume that the collision factor, A, remains constant.

62. ▲ The enzyme carbonic anhydrase catalyzes the hydration of carbon dioxide.

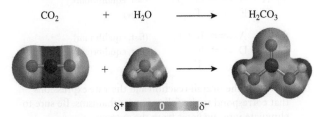

$$CO_2 \quad + \quad H_2O \quad \longrightarrow \quad H_2CO_3$$

This reaction is involved in the transfer of CO_2 from tissues to the lungs via the bloodstream. One enzyme molecule hydrates 10^6 molecules of CO_2 per second. How many grams of CO_2 are hydrated in one minute in 1 L by 1.0×10^{-6} M enzyme?

63. The following gas-phase reaction follows first-order kinetics.

$$ClO_2F \longrightarrow ClOF + O$$

The activation energy of this reaction is 186 kJ/mol. The value of k at 322°C is 6.76×10^{-4} s^{-1}. (a) What would be the value of k for this reaction at 25°C? (b) At what temperature would this reaction have a k value of 3.00×10^{-2} s^{-1}?

64. The following gas-phase reaction is first order.

$$N_2O_5 \longrightarrow NO_2 + NO_3$$

The activation energy of this reaction is 88 kJ/mol. The value of k at 0°C is 9.16×10^{-3} s^{-1}. (a) What would be the value of k for this reaction at room temperature, 25°C? (b) At what temperature would this reaction have a k value of 3.00×10^{-2} s^{-1}?

Reaction Mechanisms

65. Define reaction mechanism. Why do we believe that only bimolecular collisions and unimolecular decompositions are important in most reaction mechanisms?

66. For the reaction

$$CH_3-\underset{\underset{CH_3}{|}}{\overset{\overset{CH_3}{|}}{C}}-Br + OH^- \longrightarrow CH_3-\underset{\underset{CH_3}{|}}{\overset{\overset{CH_3}{|}}{C}}-OH + Br^-$$

the rate law is: Rate = $k[(CH_3)_3CBr]$.
Identify each mechanism that is compatible with the rate law.

(a) $(CH_3)_3CBr \longrightarrow (CH_3)_3C^+ + Br^-$ slow
      $(CH_3)_3C^+ + OH^- \longrightarrow (CH_3)_3COH$ fast
(b) $(CH_3)_3CBr + OH^- \longrightarrow (CH_3)_3COH + Br^-$
(c) $(CH_3)_3CBr + OH^- \longrightarrow (CH_3)_2(CH_2)CBr^- + H_2O$ slow
      $(CH_3)_2(CH_2)CBr^- \longrightarrow (CH_3)_2(CH_2)C + Br^-$ fast
      $(CH_3)_2(CH_2)C + H_2O \longrightarrow (CH_3)_3COH$ fast

67. Write the overall reaction and the rate expressions that correspond to the following reaction mechanisms. Be sure to eliminate intermediates from the answers.

(a) $A + B \rightleftharpoons C + D$ (fast, equilibrium)
    $C + E \longrightarrow F$ (slow)

(b) $A \rightleftharpoons B + C$ (fast, equilibrium)
    $C + D \rightleftharpoons E$ (fast, equilibrium)
       $E \longrightarrow F$ (slow)

68. Write the overall reaction and the rate expressions that correspond to the following mechanisms. Be sure to eliminate intermediates from the answers.

(a) $2A + B \rightleftharpoons D$ (fast, equilibrium)
    $D + B \longrightarrow E + F$ (slow)
       $F \longrightarrow G$ (fast)

(b) $A + B \rightleftharpoons C$ (fast, equilibrium)
    $C + D \rightleftharpoons F$ (fast, equilibrium)
       $F \longrightarrow G$ (slow)

69. The ozone, O_3, of the stratosphere can be decomposed by reaction with nitrogen oxide (commonly called nitric oxide), NO, from high-flying jet aircraft.

$$O_3(g) + NO(g) \longrightarrow NO_2(g) + O_2(g)$$

The rate expression is rate = $k[O_3][NO]$. Which of the following mechanisms are consistent with the observed rate expression?

(a) $NO + O_3 \longrightarrow NO_3 + O$ (slow)
    $NO_3 + O \longrightarrow NO_2 + O_2$ (fast)

    $O_3 + NO \longrightarrow NO_2 + O_2$ (overall)

(b) $NO + O_3 \longrightarrow NO_2 + O_2$ (slow)
                           (one step)

(c) $O_3 \longrightarrow O_2 + O$ (slow)
    $O + NO \longrightarrow NO_2$ (fast)

    $O_3 + NO \longrightarrow NO_2 + O_2$ (overall)

(d) $NO \longrightarrow N + O$ (slow)
    $O + O_3 \longrightarrow 2O_2$ (fast)
    $O_2 + N \longrightarrow NO_2$ (fast)

    $O_3 + NO \longrightarrow NO_2 + O_2$ (overall)

(e) $NO \rightleftharpoons N + O$ (fast, equilibrium)
    $O + O_3 \longrightarrow 2O_2$ (slow)
    $O_2 + N \longrightarrow NO_2$ (fast)

    $O_3 + NO \longrightarrow NO_2 + O_2$ (overall)

NO₂ sample

70. A proposed mechanism for the decomposition of ozone, $2O_3 \longrightarrow 3O_2$, is

$$O_3 \rightleftharpoons O_2 + O \quad \text{(fast, equilibrium)}$$
$$O + O_3 \longrightarrow 2O_2 \quad \text{(slow)}$$

Derive the rate equation for the net reaction.

71. A mechanism for the gas-phase reaction

$$H_2 + I_2 \longrightarrow 2HI$$

Charles D. Winters

🐾 **Molecular Reasoning** exercises   ▲ **More Challenging** exercises   Blue-Numbered exercises solved in Student Solutions Manual

Unless otherwise noted, all content on this page is © Cengage Learning.

was discussed in the chapter. (a) Show that this mechanism predicts the correct rate law, rate = $k[H_2][I_2]$.

I_2	$\rightleftharpoons$ 2I	(fast, equilibrium)
$I + H_2$	$\rightleftharpoons$ H_2I	(fast, equilibrium)
$H_2I + I$	$\longrightarrow$ 2HI	(slow)

(b) Identify any reaction intermediates in this proposed mechanism.

72. ▲ The combination of Cl atoms is catalyzed by $N_2(g)$. The following mechanism is suggested.

$N_2 + Cl$	$\rightleftharpoons$ N_2Cl	(fast, equilibrium)
$N_2Cl + Cl$	$\longrightarrow$ $Cl_2 + N_2$	(slow)

(a) Identify any reaction intermediates in this proposed mechanism. (b) Is this mechanism consistent with the experimental rate law, rate = $k[N_2][Cl]^2$?

73. The reaction between NO and Br_2 was discussed in Section 16-7. The following mechanism has also been proposed.

2NO	$\rightleftharpoons$ N_2O_2	(fast, equilibrium)
$N_2O_2 + Br_2$	$\longrightarrow$ 2NOBr	(slow)

Is this mechanism consistent with the observation that the reaction is second order in NO and first order in Br_2?

74. ▲ The following mechanism for the reaction between H_2 and CO to form formaldehyde, H_2CO, has been proposed.

H_2	$\rightleftharpoons$ 2H	(fast, equilibrium)
$H + CO$	$\longrightarrow$ HCO	(slow)
$H + HCO$	$\longrightarrow$ H_2CO	(fast)

(a) Write the balanced equation for the overall reaction. (b) The observed rate dependence is found to be one-half order in H_2 and first order in CO. Is this proposed reaction mechanism consistent with the observed rate dependence?

75. The reaction between nitrogen dioxide and ozone,

$$2NO_2 + O_3 \longrightarrow N_2O_5 + O_2$$

has been studied at 231 K. The experimental rate equation is rate = $k[NO_2][O_3]$. (a) What is the order of the reaction? (b) Is either of the following proposed mechanisms consistent with the given kinetic data? Show how you arrived at your answer.

(a)
$NO_2 + NO_2$	$\rightleftharpoons$ N_2O_4	(fast, equilibrium)
$N_2O_4 + O_3$	$\longrightarrow$ $N_2O_5 + O_2$	(slow)

(b)
$NO_2 + O_3$	$\longrightarrow$ $NO_3 + O_2$	(slow)
$NO_3 + NO_2$	$\longrightarrow$ N_2O_5	(fast)

Mixed Exercises

76. (a) What is the transition state in a reaction mechanism? (b) Are the energy of activation and the transition state related concepts? Explain. (c) How does the activation energy affect the rate of reaction?

77. Refer to the reaction and data in Exercise 63. Assume that we begin with 2.80 mol of ClO_2F in a 3.00-L container.

(a) How many moles of ClO_2F would remain after 2.00 minutes at 25°C? (b) How much time would be required for 99.0% of the ClO_2F to decompose at 25°C?

78. Refer to the reaction and data in Exercise 64. Assume that we begin with 2.80 mol of N_2O_5 in a 3.00-L container. (a) How many moles of N_2O_5 would remain after 2.00 minutes at 25°C? (b) How much time would be required for 99.0% of the N_2O_5 to decompose at 25°C?

79. The decomposition of gaseous dimethyl ether

$$CH_3OCH_3 \longrightarrow CH_4 + CO + H_2$$

follows first-order kinetics. Its half-life is 25.0 min at 500.°C. (a) Starting with 12.00 g of dimethyl ether at 500.°C, how many grams would remain after 150. minutes? (b) In part (a), how many grams would remain after 180. minutes? (c) In part (b), what fraction remains, and what fraction reacts? (d) Calculate the time, in minutes, required to decrease 24.0 mg of dimethyl ether to 2.40 mg.

80. The rate of the hemoglobin (Hb)–carbon monoxide reaction,

$$4Hb + 3CO \longrightarrow Hb_4(CO)_3$$

has been studied at 20°C. Concentrations are expressed in micromoles per liter (μmol/L).

Concentration (μmol/L)		Rate of Disappearance of Hb (μmol $\cdot$ L^{-1} $\cdot$ s^{-1})
[Hb]	[CO]	
3.36	1.00	0.941
6.72	1.00	1.88
6.72	3.00	5.64

(a) Write the rate equation for the reaction. (b) Calculate the rate constant for the reaction. (c) Calculate the rate at the instant when [Hb] = 1.50 and [CO] = 0.600 μmol/L.

81. How does an enzyme change the speed with which a reaction reaches equilibrium? Can an enzyme change the final equilibrium concentrations? Explain.

Conceptual Exercises

82. Some reactions occur faster than others due to differences in the shapes of the reactants. Use the collision theory to explain these observations.

83. How is it possible for two reactant molecules to collide with the correct orientation and still not react?

84. Write the net ionic equation for the following reaction. Construct a potential energy diagram, like that in Figure 16-10, for this reaction.

$$HCl(aq) + NaOH(aq) \longrightarrow NaCl(aq) + H_2O(\ell)$$

85. Starting with only two molecules of each reactant in a reaction that is first order in each reactant, show how the collision theory predicts that the rate of reaction will double if the amount of either reactant is doubled.

🔧 **Molecular Reasoning** exercises   ▲ **More Challenging** exercises   Blue-Numbered exercises solved in Student Solutions Manual

Unless otherwise noted, all content on this page is © Cengage Learning.

86. A sentence in an introductory chemistry textbook reads, "Dioxygen reacts with itself to form trioxygen, ozone, according to the following equation, $3O_2 \longrightarrow 2O_3$." As a student of chemistry, what would you write to criticize this sentence?

87. A stream of gaseous H_2 is directed onto finely divided platinum powder in the open air. The metal immediately glows white-hot and continues to do so as long as the stream continues. Explain.

88. Is the activation energy of a reaction expected to be higher or lower when the same reactants are in the gaseous state rather than the liquid or solid state? Explain.

89. Construct a diagram like that in Figure 16-10a. (a) Write a generic equation that would have such a potential energy diagram. (b) Is the reaction exothermic or endothermic? (c) Label the energy of activation. It is equal to the difference between what two points on your drawing?

90. Construct a diagram like that in Figure 16-10b. (a) Write a generic equation that would have such a potential energy diagram. (b) Is the reaction exothermic or endothermic? (c) Label the energy of activation. It is equal to the difference between what two points on your drawing?

91. In 1946 W. M. Grant and V. E. Kinsey reported on the rate of hydrolysis of di(2-chloroethyl)sulfide [$(ClCH_2CH_2)_2S$, commonly known as mustard gas] in aqueous solution:

$$(ClCH_2CH_2)_2S + H_2O \longrightarrow ClCH_2CH_2SCH_2CH_2OH + H^+ + Cl^-$$

What are some of the quantities that could be measured to observe the rate of this reaction?

92. When vinegar (5% acetic acid) is added to solid baking soda (sodium hydrogen carbonate), they react to produce sodium acetate, water, and carbon dioxide gas:

$$CH_3COOH(aq) + NaHCO_3(s) \longrightarrow NaCH_3COO(aq) + H_2O(\ell) + CO_2(g)$$

List four ways that the rate of this reaction can be increased.

93. According to the collision theory, the rate of a reaction can be expressed as the product of three factors:

$$\text{rate} = \text{collision frequency} \times \text{energy factor} \times \text{probability factor}$$

rate	number of effective collisions per unit volume per unit time
collision frequency	total number of collisions per unit volume per unit time
energy factor	fraction of the collisions with sufficient energy
probability factor	fraction of collisions with proper orientation

Which of these three factors are most affected by each of the following changes in the reaction conditions or the reacting substances? Increasing the temperature; switching to a similar substance that is a larger, heavier molecule; cutting large solid pieces into much smaller pieces; adding a catalyst; diluting a solution; increasing the pressure of a gas.

94. Listed below are the equations and the rate law expressions for the gas-phase decompositions of three compounds: N_2O_5, NO_2, and NH_3.

Reaction	Rate Law
$2N_2O_5(g) \longrightarrow 4NO_2(g) + O_2(g)$	rate = $k[N_2O_5]$
$2NO_2(g) \longrightarrow 2NO(g) + O_2(g)$	rate = $k[NO_2]^2$
$2NH_3(g) \xrightarrow{\text{Pt}} N_2(g) + 3H_2(g)$	rate = $k[NH_3]^2$

For each compound, start with an initial concentration of 0.80 M at 8:00 a.m. and a first half-life of 21 minutes. At what times will the concentration of each starting compound be 0.40 M? 0.20 M? 0.10 M?

Building Your Knowledge

95. The following explanation of the operation of a pressure cooker appears in a cookbook: "Boiling water in the presence of air can never produce a temperature higher than 212°F, no matter how high the heat source. But in a pressure cooker, the air is withdrawn first, so the boiling water can be maintained at higher temperatures." Support or criticize this explanation.

96. ▲ A cookbook gives the following general guideline for use of a pressure cooker. "For steaming vegetables, cooking time at a gauge pressure of 15 pounds per square inch (psi) is $\frac{1}{3}$ that at atmospheric pressure." Remember that gauge pressure is measured relative to the external atmospheric pressure, which is 15 psi at sea level. From this information, estimate the activation energy for the process of steaming vegetables. (*Hint:* Clausius and Clapeyron may be able to help you.)

97. For most reactions that involve an enzyme, the rate of product formation versus reactant concentration increases as reactant concentration increases until a maximum value is obtained, after which further increases do not yield increased rates. Using a description like that in Figure 16-19, describe how the reaction may be first order with respect to substrate but the amount of enzyme can also be a determining factor.

98. Using the mechanism and relative energy values shown in Figure 16-12, prepare Lewis formulas that illustrate the species that are likely to be present at each of the peaks and troughs in the graphical representation given in Figure 16-12. (*Hint:* You may need to label some bonds as being weaker, stretched, in the process of being formed, and so on.)

99. The activation energy for the reaction

$$2HI(g) \longrightarrow H_2(g) + I_2(g)$$

is 179 kJ/mol. Based upon Figure 16-12, construct a diagram for this multistep reaction. (*Hint:* Calculate ΔH^0 from values in Appendix K. How does ΔH^0 compare with ΔE for this reaction?) Show energy values if known.

🔷 **Molecular Reasoning** exercises   ▲ **More Challenging** exercises   Blue-Numbered exercises solved in Student Solutions Manual

100. ▲ The activation energy for the reaction between O_3 and NO is 9.6 kJ/mol.

$$O_3(g) + NO(g) \longrightarrow NO_2(g) + O_2(g)$$

(a) Use the thermodynamic quantities in Appendix K to calculate ΔH^0 for this reaction. (b) Prepare an activation energy plot similar to that in Figure 16-10 for this reaction. (*Hint*: How does ΔH^0 compare with ΔE for this reaction?)

Beyond the Textbook

NOTE: *Whenever the answer to an exercise depends on information obtained from a source other than this textbook, the source must be included as an essential part of the answer.*

101. Go to **en.wikipedia.org/wiki/Arrhenius_equation** or another website and locate information on the Arrhenius equation. For a particular reaction, how could one determine the value of *A* in the Arrhenius equation for that reaction?

101. Go to **www.chemguide.co.uk/physical/basicrates/ arrhenius.html** or another website and locate information on the Arrhenius equation. Describe a reaction found at this website that follows a first-order mechanism.

103. Use the *Handbook of Chemistry* and Physics or a website and locate information on kinetics, conversion factors, or rate constants. Which of the units for a second-order reaction found in this handbook or on another website is/are equivalent to the units used in this textbook?

104. Between 1957 and 1990, according to NASA, the average amount of ozone in the stratosphere dropped from about 320 to about 130 Dobson Units (a measure of thickness). In a very blunt calculation, the annual percentage drop over this period can be calculated as the quantity [(320 −130) / 320] divided by [1990 − 1957]. The 2007 CDC statistics show that 58, 094 Americans were diagnosed with cancerous skin melanomas that year. If the annual percent drop in ozone concentration calculated above held steady, how many additional new cases of skin cancer in the U.S. would have resulted for 2008 as a result of the drop?

105. Before the late 1920s (when Freon was invented), household refrigerators commonly ran on ammonia (NH_3) or sulfur dioxide (SO_2) gases. If you aren't already familiar with them, do a quick search on the safety hazards of these compounds. Would you rather have one of these old appliances in your kitchen, or a CFC-based fridge? Does your choice introduce conflict between what's best for the immediate environment and what's best for the global environment?

106. Another fluorinated hydrocarbon sold by DuPont is a common product you probably have in your home. What is it? (*Hint*: You won't stick to it.)

🞄 **Molecular Reasoning** exercises ▲ **More Challenging** exercises Blue-Numbered exercises solved in Student Solutions Manual

Unless otherwise noted, all content on this page is © Cengage Learning.

Chemical Equilibrium

17

Chemical equilibrium plays an important role in the operation of chemical plants where manufacturers want to optimize the amount of product. This can be done by adjusting reaction conditions such as temperature, pressure, and concentrations of reactants used, all key factors that affect chemical equilibrium.

littlewormy/Shutterstock.com

OBJECTIVES

After you have studied this chapter, you should be able to

▶ Explain the basic ideas of chemical equilibrium

▶ Explain what an equilibrium constant is and what it tells us

▶ Explain what a reaction quotient is and what it tells us

▶ Use equilibrium constants to describe systems at equilibrium

▶ Recognize the factors that affect equilibria and predict the resulting effects

▶ Use the equilibrium constant expressed in terms of partial pressures (K_P) and relate K_P to K_c

▶ Describe heterogeneous equilibria and write their equilibrium constants

▶ Use the relationships between thermodynamics and equilibrium

▶ Estimate equilibrium constants at different temperatures

17-1 Basic Concepts

Chemical reactions that can occur in either direction are called **reversible reactions**. Most reversible reactions do not go to completion. That is, even when reactants are mixed in stoichiometric quantities, they are not completely converted to products.

Reversible reactions can be represented in general terms as follows, where the capital letters represent formulas and the lowercase letters represent the stoichiometric coefficients in the balanced equation.

$$a\text{A} + b\text{B} \rightleftharpoons c\text{C} + d\text{D}$$

The double arrow ($\rightleftharpoons$) indicates that the reaction is reversible—that is, both the forward and reverse reactions occur simultaneously. In discussions of chemical equilibrium, the substances that appear on the left side of the balanced chemical equation are called the "reactants," and those on the right side are called the "products." In fact, the reaction can proceed in *either direction*. When A and B react to form C and D at the *same rate* at which C and D react to form A and B, the system is at *equilibrium*.

> **Chemical equilibrium** exists when two opposing reactions occur simultaneously at the same rate.

▶ The dynamic nature of chemical equilibrium can be proved experimentally by inserting radioactive atoms into a small percentage of reactant or product molecules. Even when the initial mixture is at equilibrium, radioactive atoms eventually appear both in reactant and in product molecules.

Chemical equilibria are **dynamic equilibria**; that is, individual molecules are continually reacting, even though the overall composition of the reaction mixture does not change. In a system at equilibrium, the equilibrium is said to lie toward the right if more C and D are present than A and B (product-favored), and to lie toward the left if more A and B are present (reactant-favored).

Consider a case in which the coefficients in the equation for a reaction are all 1. When substances A and B react, the rate of the forward reaction decreases as time passes because the concentrations of A and B decrease.

$$\text{A} + \text{B} \longrightarrow \text{C} + \text{D} \qquad (1)$$

As the concentrations of C and D build up, they start to re-form A and B.

$$\text{C} + \text{D} \longrightarrow \text{A} + \text{B} \qquad (2)$$

As more C and D molecules are formed, more can react, and so the rate of reaction between C and D increases with time. Eventually, the two reactions occur at the same rate, and the system is at equilibrium (Figure 17-1).

$$\text{A} + \text{B} \rightleftharpoons \text{C} + \text{D}$$

If a reaction begins with only C and D present, the rate of reaction (2) decreases with time, and the rate of reaction (1) increases with time until the two rates are equal.

STOP & THINK
In Chapter 7 the single line with an arrowhead on each end ↔ was introduced as a way of indicating resonance. We do not use it to indicate that a reaction is reversible. The double arrow $\rightleftharpoons$ is used to indicate a reversible reaction or an equilibrium.

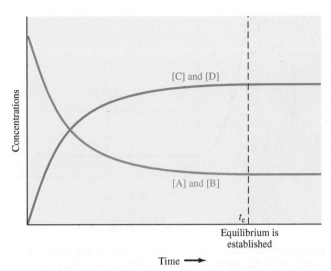

Figure 17-1 Variation in the concentrations of species present in the A + B $\rightleftharpoons$ C + D system as equilibrium is approached, beginning with equal concentrations of A and B only. For this reaction, production of products is favored. As we will see, this corresponds to a value of the equilibrium constant greater than 1. Brackets, [ ], represent the concentration, in moles per liter, of the species enclosed within them. The time axis may be in any convenient units—seconds, minutes, hours, and so on.

▶ Whether or not a reaction has reached equilibrium, chemists often use a single reaction arrow as a shortcut:

$$A + B \rightarrow C + D$$

The single reaction arrow does not necessarily mean that the reaction is not at equilibrium or that it goes completely to products. In this text, we are careful to show the equilibrium arrows $\rightleftharpoons$ when we wish to emphasize a reaction at equilibrium, as throughout Chapters 17 through 20.

The SO$_2$–O$_2$–SO$_3$ System

Consider the reversible reaction of sulfur dioxide with oxygen to form sulfur trioxide at 1500 K.

$$2SO_2(g) + O_2(g) \rightleftharpoons 2SO_3(g)$$

Suppose 0.400 mole of SO$_2$ and 0.200 mole of O$_2$ are injected into a closed 1.00-liter container. When equilibrium is established (at time t_e, Figure 17-2a), we find that 0.056 mole of SO$_3$ has formed and that 0.344 mole of SO$_2$ and 0.172 mole of O$_2$ remain unreacted. The amount of product at equilibrium does not increase further if the reaction is given additional time. These changes are summarized in the following reaction summary, using molarity units rather than moles. (They are numerically identical here because the volume of the reaction vessel is 1.00 liter.) The changes due to reaction are represented by the *changes* in concentrations.

▶ The numbers in this discussion come from experimental measurements.

▶ A setup such as this is called a *reaction summary*; we have used this description in Chapter 11 (Examples 11-1 and 11-2) and in Chapter 14 (Example 14-12). The ratio in the "change due to rxn" line is always determined by the coefficients in the balanced equation.

	$2SO_2(g)$	+	$O_2(g)$	$\rightleftharpoons$	$2SO_3(g)$
initial conc'n	0.400 M		0.200 M		0
change due to rxn	$-0.056\ M$		$-0.028\ M$		$+0.056\ M$
equilibrium conc'n	0.344 M		0.172 M		0.056 M

In a similar experiment, 0.500 mole of SO$_3$ is introduced alone into a closed 1.00-liter container. This time the reaction proceeds from *right to left* as the equation is written. The *changes in concentration* are in the same 2:1:2 ratio as in the previous case, as required by the coefficients of the balanced equation. The time required to reach equilibrium may be longer or shorter. When equilibrium is established (at time t_e, Figure 17-2b), 0.076 mole of SO$_3$, 0.212 mole of O$_2$, and 0.424 mole of SO$_2$ are present. These equilibrium amounts differ from those in the previous case, but they are related in an important way, as we will see in the next section.

	$2SO_2(g)$	+	$O_2(g)$	$\rightleftharpoons$	$2SO_3(g)$
initial conc'n	0		0		0.500 M
change due to rxn	$+0.424\ M$		$+0.212\ M$		$-0.424\ M$
equilibrium conc'n	0.424 M		0.212 M		0.076 M

The results of these experiments are summarized in the following table and in Figure 17-2.

	Initial Concentrations			Equilibrium Concentrations		
	[SO$_2$]	[O$_2$]	[SO$_3$]	[SO$_2$]	[O$_2$]	[SO$_3$]
Experiment 1	0.400 M	0.200 M	0 M	0.344 M	0.172 M	0.056 M
Experiment 2	0 M	0 M	0.500 M	0.424 M	0.212 M	0.076 M

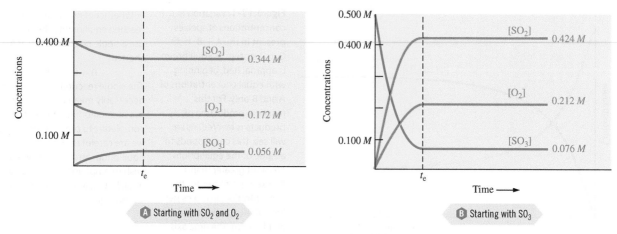

Figure 17-2 Establishment of equilibrium in the $2SO_2 + O_2 \rightleftharpoons 2SO_3$ system. (a) Beginning with stoichiometric amounts of SO_2 and O_2 and no SO_3. (b) Beginning with only SO_3 and no SO_2 or O_2. Greater changes in concentrations occur to establish equilibrium when starting with SO_3 than when starting with SO_2 and O_2. The equilibrium favors SO_2 and O_2.

17-2 The Equilibrium Constant

Suppose a reversible reaction occurs by a single elementary reaction step:

$$2A + B \rightleftharpoons A_2B$$

The rate of the forward reaction is $rate_f = k_f[A]^2[B]$; the rate of the reverse reaction is $rate_r = k_r[A_2B]$. In these expressions, k_f and k_r are the *specific rate constants* of the forward and reverse reactions, respectively. By definition, the two rates are equal *at equilibrium* ($rate_f = rate_r$). So we write

$$k_f[A]^2[B] = k_r[A_2B] \qquad \text{(at equilibrium)}$$

Dividing both sides of this equation by k_r and by $[A]^2[B]$ gives

$$\frac{k_f}{k_r} = \frac{[A_2B]}{[A]^2[B]}$$

> **S** **TOP & THINK**
> Remember that for *elementary* (or fundamental) reactions the coefficients do correspond to the orders of the reactant and product concentrations.

At any specific temperature, both k_f and k_r are constants, so k_f/k_r is also a constant.

This ratio is given a special name and symbol, the *equilibrium constant*, K_c or simply K.

> ▶ The subscript c refers to concentrations. The brackets, [ ], in this expression indicate *equilibrium concentrations* in moles per liter.

$$K_c = \frac{[A_2B]}{[A]^2[B]} \qquad \text{(at equilibrium)}$$

We can show that even if the overall reaction occurs by a multistep mechanism, the equilibrium constant is the product and ratio of the rate constants for each step of the mechanism. Regardless of the mechanism by which any reaction occurs, the concentrations of reaction intermediates cancel out and the equilibrium constant expression for the reaction has the same form. For a reaction in general terms, the equilibrium constant can always be written as follows:

$$\text{For } \underbrace{a\text{A} + b\text{B}}_{\text{reactants}} \rightleftharpoons \overbrace{c\text{C} + d\text{D}}^{\text{products}}, \qquad K_c = \frac{\overbrace{[\text{C}]_{eq}^{\,c}\,[\text{D}]_{eq}^{\,d}}^{\substack{\text{product} \\ \text{concentrations}}}}{\underbrace{[\text{A}]_{eq}^{\,a}\,[\text{B}]_{eq}^{\,b}}_{\substack{\text{reactant} \\ \text{concentrations}}}}$$

The **equilibrium constant, K_c,** is defined as the product of the *equilibrium concentrations* (in moles per liter) of the products, each raised to the power that corresponds to its coefficient in the balanced equation, divided by the product of the *equilibrium concentrations* of reactants, each raised to the power that corresponds to its coefficient in the balanced equation.

▶ Moles per liter is equal to molarity (*M*) or molar concentration.

Numerical values for K_c can be determined from experiments or from thermodynamic data (see Section 17-12). Some equilibrium constant expressions and their numerical values at 25°C are

$$N_2(g) + O_2(g) \rightleftharpoons 2NO(g) \qquad K_c = \frac{[NO]^2}{[N_2][O_2]} \qquad = 4.5 \times 10^{-31}$$

$$CH_4(g) + Cl_2(g) \rightleftharpoons CH_3Cl(g) + HCl(g) \qquad K_c = \frac{[CH_3Cl][HCl]}{[CH_4][Cl_2]} = 1.2 \times 10^{18}$$

$$N_2(g) + 3H_2(g) \rightleftharpoons 2NH_3(g) \qquad K_c = \frac{[NH_3]^2}{[N_2][H_2]^3} \qquad = 3.6 \times 10^8$$

▶ You must remember that the concentrations in the equilibrium constant expression are those at equilibrium.

Remember that calculations with K_c values always involve *equilibrium values* of concentrations.

The thermodynamic definition of the equilibrium constant involves activities rather than concentrations. The **activity** of a component of an ideal mixture is the ratio of its concentration or partial pressure to a standard concentration (1 *M* for solutions) or a standard pressure (1 atm for gases). For now, we can consider the activity of each species to be a dimensionless quantity whose numerical value can be determined as follows.

1. For any pure liquid or pure solid, the activity is taken as 1 (by definition).
2. For components of ideal solutions, the activity of each component is taken to be the ratio of its molar concentration to a standard concentration of 1 *M*, so the units cancel.
3. For gases in an ideal mixture, the activity of each component is taken to be the ratio of its partial pressure to a standard pressure of 1 atm, so again the units cancel.

▶ You can think of the activity as the molar concentration of a solution component or the partial pressure of a gas with the units omitted.

Because of the use of activities, *the equilibrium constant has no units; the values we put into K_c are numerically equal to molar concentrations but are dimensionless*, that is, they have no units. Therefore, calculations involving equilibrium are frequently carried out without units; we will follow that practice in this text.

The magnitude of K_c is a measure of the extent to which reaction occurs. For any balanced chemical equation, the value of K_c

1. is always the same at a given temperature.
2. changes if the temperature changes.
3. does not depend on the initial concentrations.

STOP & THINK
These are three very important ideas.

A value of K_c *much* greater than 1 indicates that the "numerator concentrations" (products) would be much greater than the "denominator concentrations" (reactants); this means that at equilibrium most of the reactants would be converted into products. For example, the reaction $CH_4(g) + Cl_2(g) \rightleftharpoons CH_3Cl(g) + HCl(g)$ shown earlier goes nearly to completion; earlier in this chapter and in Chapter 15, we called such a reaction spontaneous or "product-favored." On the other hand, if K_c is quite small, equilibrium is established when most of the reactants remain unreacted and only small amounts of products are formed. The reaction $N_2(g) + O_2(g) \rightleftharpoons 2NO(g)$ shown earlier reaches equilibrium with only a tiny amount of NO present; earlier in this chapter and in Chapter 15, we called such a reaction nonspontaneous or "reactant-favored."

For a given chemical reaction at a specific temperature, the product of the concentrations of the products formed by the reaction, each raised to the appropriate power, divided

▶ In the K_c expression, the power for each reactant and product is equal to its coefficient in the balanced chemical equation. Remember, however, that pure solids and liquids have activities = 1 and do not appear in the equilibrium expression.

by the product of the concentrations of the reactants, each raised to the appropriate power, always has the same value, that is, K_c. This does *not* mean that the individual equilibrium concentrations for a given reaction are always the same, but it does mean that this particular numerical combination of their values (K_c) is constant.

Consider again the SO_2—O_2—SO_3 equilibrium described in Section 17-1. We can use the *equilibrium* concentrations from either experiment to calculate the value of the equilibrium constant for this reaction at 1500 K.

From experiment 1: $2SO_2(g)$ + $O_2(g)$ $\rightleftharpoons$ $2SO_3(g)$
equilibrium conc'n   0.344 M 0.172 M 0.056 M

Substituting the numerical values (without units) into the equilibrium expression gives the value of the equilibrium constant.

$$K_c = \frac{[SO_3]^2}{[SO_2]^2[O_2]} = \frac{(0.056)^2}{(0.344)^2(0.172)} = 0.15$$

Alternatively,   From experiment 2: $2SO_2(g)$ + $O_2(g)$ $\rightleftharpoons$ $2SO_3(g)$
equilibrium conc'n   0.424 M 0.212 M 0.076 M

$$K_c = \frac{(0.076)^2}{(0.424)^2(0.212)} = 0.15$$

No matter what combinations of reactant and product concentrations we start with, the resulting *equilibrium* concentrations at 1500 K for this reversible reaction would always give the same value of K_c, 0.15.

For the reversible reaction written *as it is* with SO_2 and O_2 shown as reactants and SO_3 shown as the product, K_c is 0.15 at 1500 K.

EXAMPLE 17-1 Calculation of K_c

Some nitrogen and hydrogen are placed in an empty 5.00-liter container at 500°C. When equilibrium is established, 3.01 mol of N_2, 2.10 mol of H_2, and 0.565 mol of NH_3 are present. Evaluate K_c for the following reaction at 500°C.

$$N_2(g) + 3H_2(g) \rightleftharpoons 2NH_3(g)$$

Plan

We find the *equilibrium concentrations* by dividing the number of moles of each reactant and product by the volume, 5.00 liters. Then we substitute these equilibrium concentrations into the equilibrium constant expression.

Solution

The equilibrium concentrations are

$$[N_2] = 3.01 \text{ mol}/5.00 \text{ L} = 0.602 \ M$$
$$[H_2] = 2.10 \text{ mol}/5.00 \text{ L} = 0.420 \ M$$
$$[NH_3] = 0.565 \text{ mol}/5.00 \text{ L} = 0.113 \ M$$

We substitute these numerical values into the expression for K_c.

▶ Remember that the concentrations in K_c calculations are equilibrium values of molar concentration. But, since we should be using activities, which have no units, we do not carry concentration units along in the calculation.

$$K_c = \frac{[NH_3]^2}{[N_2][H_2]^3} = \frac{(0.113)^2}{(0.602)(0.420)^3} = 0.286$$

Thus, for the reaction of N_2 and H_2 to form NH_3 at 500°C, we can write

$$K_c = \frac{[NH_3]^2}{[N_2][H_2]^3} = \boxed{0.286}$$

The value of K_c from Example 17-1 is much different from the value given earlier for the same reaction at 25°C. For this reaction, products are favored at the lower temperature ($K_c = 3.6 \times 10^8$ at 25°C), whereas reactants are favored at the higher temperature ($K_c = 0.286$ at 500°C). You may recall from Chapter 16 that the rate constant, k, for any reaction increases with increasing temperature. But the value of K_c for different reactions responds differently to temperature changes; for some reactions, K_c increases with increasing temperature, whereas for others K_c decreases with increasing temperature, as we will see in Sections 17-6 and 17-13.

EXAMPLE 17-2 Calculation of K_c

We put 10.0 moles of N_2O into a 2.00-L container at some temperature, where it decomposes according to

$$2N_2O(g) \rightleftharpoons 2N_2(g) + O_2(g)$$

At equilibrium, 2.20 moles of N_2O remain. Calculate the value of K_c for the reaction.

Plan

We express all concentrations in moles per liter. The mole ratio from the balanced chemical equation allows us to find the *changes* in concentrations of the other substances in the reaction. We use the reaction summary to find the equilibrium concentrations to use in the K_c expression.

Solution

At equilibrium, 2.20 mol N_2O remain, so

$$\underline{?}\text{ mol } N_2O \text{ reacting} = 10.00 \text{ mol } N_2O \text{ initial} - 2.20 \text{ mol } N_2O \text{ remaining}$$
$$= 7.80 \text{ mol } N_2O \text{ reacting}$$

The initial $[N_2O] = (10.00 \text{ mol})/(2.00 \text{ L}) = 5.00\ M$; the concentration of N_2O that reacts is $(7.80 \text{ mol})/(2.00 \text{ L}) = 3.90\ M$. From the balanced chemical equation, each 2 mol N_2O that react produces 2 mol N_2 and 1 mol O_2, or a reaction ratio of

$$1 \text{ mol } N_2O \text{ reacting} : 1 \text{ mol } N_2 \text{ formed} : \tfrac{1}{2} \text{ mol } O_2 \text{ formed}$$

We can now write the reaction summary.

	$2N_2O(g) \rightleftharpoons$	$2N_2(g)$ +	$O_2(g)$
initial	5.00 M	0	0
change due to rxn	$-3.90\ M$	$+3.90\ M$	$+ \tfrac{1}{2}(3.90)$ $= 1.95\ M$
equilibrium	1.10 M	3.90 M	1.95 M

We put these equilibrium concentrations into the equilibrium constant expression and evaluate K_c.

$$K_c = \frac{[N_2]^2[O_2]}{[N_2O]^2} = \frac{(3.90)^2(1.95)}{(1.10)^2} = \boxed{24.5}$$

You should now work Exercises 24 and 28.

17-3 Variation of K_c with the Form of the Balanced Equation

The numerical value of K_c depends on how we write the balanced equation for the reaction. We wrote the equation for the reaction of SO_2 and O_2 to produce SO_3, and its equilibrium constant expression as

$$2SO_2(g) + O_2(g) \rightleftharpoons 2SO_3(g) \qquad \text{and} \qquad K_c = \frac{[SO_3]^2}{[SO_2]^2[O_2]} = 0.15$$

Suppose we write the equation for the same reaction in reverse. The equation and its equilibrium constant, written this way, are

$$2SO_3(g) \rightleftharpoons 2SO_2(g) + O_2(g) \quad \text{and} \quad K_c' = \frac{[SO_2]^2[O_2]}{[SO_3]^2} = \frac{1}{K_c} = \frac{1}{0.15} = 6.7$$

We see that K_c', the equilibrium constant for the equation written in reverse, is the *reciprocal* of K_c, the equilibrium constant for the original equation.

If the reaction is divided by a factor of 2, the new equilibrium expression is then written as:

▶ We see that K_c'' is the square root of K_c. $K_c^{1/2}$ means the square root of K_c.

$$SO_2(g) + \tfrac{1}{2}O_2(g) \rightleftharpoons SO_3(g) \quad \text{and} \quad K_c'' = \frac{[SO_3]}{[SO_2][O_2]^{1/2}} = K_c^{1/2} = 0.39$$

If an equation for a reaction is multiplied by any factor, n, then the original value of K_c is raised to the nth power. Thus, we must always know the balanced chemical equation that corresponds to the value of K_c for a chemical reaction.

EXAMPLE 17-3 Variation of the Form of K_c

You are given the following chemical equation and its equilibrium constant at a given temperature.

$$2HBr(g) + Cl_2(g) \rightleftharpoons 2HCl(g) + Br_2(g) \qquad K_c = 4.0 \times 10^4$$

Write the expression for the equilibrium constant and calculate its numerical value for each of the following at the same temperature.

(a) $4HBr(g) + 2Cl_2(g) \rightleftharpoons 4HCl(g) + 2Br_2(g)$

(b) $HBr(g) + \tfrac{1}{2}Cl_2(g) \rightleftharpoons HCl(g) + \tfrac{1}{2}Br_2(g)$

Plan

We recall the definition of the equilibrium constant. For the original equation,

▶ A coefficient of $\tfrac{1}{2}$ refers to $\tfrac{1}{2}$ of a mole, *not* $\tfrac{1}{2}$ of a molecule.

$$K_c = \frac{[HCl]^2[Br_2]}{[HBr]^2[Cl_2]} = 4.0 \times 10^4$$

Solution

(a) The original equation has been multiplied by 2, so K_c must be squared.

$$K_c' = \frac{[HCl]^4[Br_2]^2}{[HBr]^4[Cl_2]^2} \qquad K_c' = (K_c)^2 = (4.0 \times 10^4)^2 = 1.6 \times 10^9$$

(b) The original equation has been multiplied by $\tfrac{1}{2}$ (divided by 2), so K_c must be raised to the $\tfrac{1}{2}$ power. The value of K_c'' is the square root of the original K_c value.

$$K_c'' = \frac{[HCl][Br_2]^{1/2}}{[HBr][Cl_2]^{1/2}} = \sqrt{K_c} = \sqrt{4.0 \times 10^4} = 2.0 \times 10^2$$

You should now work Exercise 30.

17-4 The Reaction Quotient

▶ The reaction quotient is sometimes called the **mass action expression**.

The **reaction quotient, Q,** for the general reaction is given as follows.

$$\text{For } aA + bB \rightleftharpoons cC + dD, \qquad Q = \frac{[C]^c[D]^d}{[A]^a[B]^b} \quad \overset{\longleftarrow}{\underset{\longleftarrow}{}} \begin{array}{l}\text{not necessarily}\\\text{equilibrium}\\\text{concentrations}\end{array}$$

The reaction quotient has the same *form* as the equilibrium constant, but it involves specific values that are not *necessarily* equilibrium concentrations. If they *are* equilibrium concentrations, then $Q = K_c$. The concept of the reaction quotient is very useful. We can compare the magnitude of Q for any specified mixture of reactants and products with that of K_c for a reaction under given conditions to decide whether the forward or the reverse reaction must occur to a greater extent to establish equilibrium.

We can think of the reaction quotient, Q, as a measure of the progress of the reaction. When the mixture contains *only* reactants, the concentrations in the numerator are zero, so $Q = 0$. As the reaction proceeds to the right, the product concentrations (numerator) increase and the reactant concentrations (denominator) decrease, so Q would increase to an infinitely large value when all reactants have been consumed and only products remain. The value of K_c is a particular value of Q that represents equilibrium mixtures for the reaction.

If for any mixture $Q < K_c$, the forward reaction would occur to a greater extent than the reverse reaction for equilibrium to be established. This is because when $Q < K_c$, the numerator of Q is too small and the denominator is too large. To increase the numerator and to reduce the denominator, A and B must react to produce C and D. Conversely, if $Q > K_c$, the reverse reaction would occur to a greater extent than the forward reaction for equilibrium to be reached. When the value of Q reaches the value of K_c, the system is at equilibrium, so no further *net* reaction occurs.

▶ When the forward reaction occurs to a greater extent than the reverse reaction, we say that a *net* forward reaction has occurred.

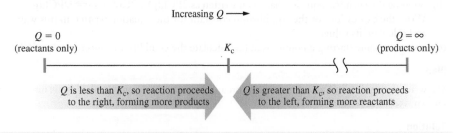

STOP & THINK
Once the balanced chemical equation is written, the substances on the left are called reactants and the ones on the right are products, no matter which way the reaction proceeds.

$Q < K_c$ Forward reaction predominates until equilibrium is established.
$Q = K_c$ System is at equilibrium.
$Q > K_c$ Reverse reaction predominates until equilibrium is established.

In Example 17-4 we calculate the value of Q and compare it with the *known* value of K_c to predict the direction of the reaction that leads to equilibrium.

EXAMPLE 17-4 The Reaction Quotient

At a very high temperature, $K_c = 65.0$ for the following reaction.

$$2HI(g) \rightleftharpoons H_2(g) + I_2(g)$$

The following concentrations were detected in a mixture. Is the system at equilibrium? If not, in which direction must the reaction proceed for equilibrium to be established?

$$[HI] = 0.500\ M, \qquad [H_2] = 2.80\ M, \qquad \text{and} \qquad [I_2] = 3.40\ M$$

▶ These concentrations could be present if we started with a mixture of HI, H_2, and I_2.

Plan

We substitute these concentrations into the expression for the reaction quotient to calculate Q. Then we compare Q with K_c to see whether the system is at equilibrium.

Solution

$$Q = \frac{[H_2][I_2]}{[HI]^2} = \frac{(2.80)(3.40)}{(0.500)^2} = 38.1$$

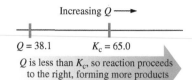

Increasing Q ⟶

$Q = 38.1$ $K_c = 65.0$

Q is less than K_c, so reaction proceeds to the right, forming more products

But $K_c = 65.0$, so $Q < K_c$. The system is *not* at equilibrium. For equilibrium to be established, the value of Q must *increase* until it equals K_c. This can occur only if the numerator *increases* and the denominator *decreases*. Thus, the forward (left-to-right) reaction must occur to a greater extent than the reverse reaction; that is, some HI must react to form more H_2 and I_2 to reach equilibrium.

You should now work Exercises 38, 40, and 42.

17-5 Uses of the Equilibrium Constant, K_c

▶ The equilibrium constant is a "constant" only if the temperature does not change.

We have seen (Section 17-2) how to calculate the value of K_c from one set of equilibrium concentrations. Once that value has been obtained, the process can be turned around to calculate equilibrium *concentrations* from the equilibrium *constant*.

EXAMPLE 17-5 Finding Equilibrium Concentrations

(a) The equation for the following reaction and the value of K_c at a given temperature are given. An equilibrium mixture in a 1.00-liter container contains 0.25 mol of PCl_5 and 0.16 mol of PCl_3. What equilibrium concentration of Cl_2 must be present?

$$PCl_3(g) + Cl_2(g) \rightleftharpoons PCl_5(g) \qquad K_c = 1.9$$

(b) Suppose the chemical equation had been written as $2PCl_3(g) + 2Cl_2(g) \rightleftharpoons 2PCl_5(g)$. Write the expression for the equilibrium constant for the equation written in this way, and calculate its value.

(c) Using the same starting amounts as in (a), calculate the equilibrium concentration of Cl_2.

Plan

We write the equilibrium constant expression and its value. Only one term, $[Cl_2]$, is unknown. We solve for it.

Solution

(a) Because the volume of the container is 1.00 liter, the molar concentration (mol/L) of each substance is numerically equal to the number of moles. The equilibrium constant expression and its numeric value are

$$K_c = \frac{[PCl_5]}{[PCl_3][Cl_2]} = 1.9$$

$$[Cl_2] = \frac{[PCl_5]}{K_c[PCl_3]} = \frac{(0.25)}{(1.9)(0.16)} = 0.82\ M$$

🅢 **TOP & THINK**

When the value of K_c and the equilibrium values of all but one concentration are known, the unknown concentration can be found.

(b) We saw in Section 17-3 that multiplying the chemical equation by 2 results in the value of K_c being squared. Thus for the equation $2PCl_3 + 2Cl_2 \rightleftharpoons 2PCl_5$, we have:

$$K_c' = \frac{[PCl_5]^2}{[PCl_3]^2[Cl_2]^2} = 1.9^2 = 3.6$$

(c) We insert the known equilibrium concentration values from (a) into the expression for K_c' and solve for the equilibrium concentration of $[Cl_2]$.

$$K_c' = \frac{[PCl_5]^2}{[PCl_3]^2[Cl_2]^2} = 3.6$$

$$[Cl_2]^2 = \frac{[PCl_5]^2}{K_c'[Cl_2]^2} = \frac{(0.25)^2}{(3.61)(0.16)^2} = 0.68$$

$$[Cl_2] = 0.82\ M$$

This is the same concentration that we calculated in (a). Correct equilibrium concentrations can be calculated from any representation of the balanced chemical equation, so long as we treat the equilibrium constant expression and its value in a consistent manner, as shown in Section 17-3.

Often we know the starting concentrations and want to know how much of each reactant and each product would be present at equilibrium. The next two examples illustrate this important kind of calculation.

EXAMPLE 17-6   Finding Equilibrium Concentrations

For the following reaction, the equilibrium constant is 49.0 at a certain temperature. If 0.400 mol each of A and B are placed in a 2.00-liter container at that temperature, what concentrations of all species are present at equilibrium?

$$A + B \rightleftharpoons C + D$$

▶ We have represented chemical formulas by single letters to simplify the notation in these calculations.

Plan

First we find the initial concentrations. Then we write the reaction summary and represent the equilibrium concentrations algebraically. Finally we substitute the algebraic representations of equilibrium concentrations into the K_c expression and find the equilibrium concentrations.

Solution

The initial concentrations are

$$[A] = \frac{0.400 \text{ mol}}{2.00 \text{ L}} = 0.200\ M \qquad [C] = 0\ M$$

$$[B] = \frac{0.400 \text{ mol}}{2.00 \text{ L}} = 0.200\ M \qquad [D] = 0\ M$$

We know that the reaction can only proceed to the right because only "reactants" are present. The reaction summary includes the values, or symbols for the values, of (1) initial concentrations, (2) changes in concentrations, and (3) concentrations at equilibrium.

Let x = moles per liter of A that react; then x = moles per liter of B that react and x = moles per liter of C and of D that are formed.

▶ The coefficients in the equation are all 1's, so the reaction ratio must be 1:1:1:1.

	A	+	B	$\rightleftharpoons$	C	+	D
initial	0.200 M		0.200 M		0 M		0 M
change due to rxn	$-x$ M		$-x$ M		$+x$ M		$+x$ M
at equilibrium	$(0.200 - x)$ M		$(0.200 - x)$ M		x M		x M

Now K_c is known, but concentrations are not. But the equilibrium concentrations have all been expressed in terms of the single variable x. We substitute the equilibrium concentrations (*not* the initial ones) into the K_c expression and solve for x.

$$K_c = \frac{[C][D]}{[A][B]} = 49.0$$

$$\frac{(x)(x)}{(0.200 - x)(0.200 - x)} = \frac{x^2}{(0.200 - x)^2} = 49.0$$

This quadratic equation has a perfect square on both sides. We solve it by taking the square roots of both sides of the equation and then rearranging for x.

$$\frac{x}{0.200 - x} = 7.00$$

$$x = 1.40 - 7.00x \qquad 8.00x = 1.40 \qquad x = \frac{1.40}{8.00} = 0.175$$

Now we know the value of x, so the equilibrium concentrations are

$$[A] = (0.200 - x)\ M = 0.025\ M; \qquad [C] = x\ M = 0.175\ M;$$

$$[B] = (0.200 - x)\ M = 0.025\ M; \qquad [D] = x\ M = 0.175\ M$$

S **TOP & THINK**
We see that the equilibrium concentrations of products are much greater than those of reactants because K_c is much greater than 1.

To check our answers, we use the equilibrium concentrations to calculate Q and verify that its value is equal to K_c.

$$Q = \frac{[C][D]}{[A][B]} = \frac{(0.175)(0.175)}{(0.025)(0.025)} = 49 \quad \text{Recall that } K_c = 49.0$$

The ideas developed in Example 17-6 may also be applied to cases in which the reactants are mixed in nonstoichiometric amounts. This is shown in Example 17-7.

EXAMPLE 17-7 Finding Equilibrium Concentrations

Consider the same system as in Example 17-6 at the same temperature. If 0.600 mol of A and 0.200 mol of B are mixed in a 2.00-liter container and allowed to reach equilibrium, what are the equilibrium concentrations of all species?

Plan

We proceed as we did in Example 17-6. The only difference is that now we have *nonstoichiometric* starting amounts of reactants.

Solution

As in Example 17-6, we let x = mol/L of A that react; then x = mol/L of B that react, and x = mol/L of C and D formed.

	A	+	B	$\rightleftharpoons$	C	+	D
initial	0.300 M		0.100 M		0 M		0 M
change due to rxn	$-x\ M$		$-x\ M$		$+x\ M$		$+x\ M$
equilibrium	$(0.300 - x)\ M$		$(0.100 - x)\ M$		$x\ M$		$x\ M$

The initial concentrations are governed by the amounts of reactants mixed together. But *changes in concentrations* due to reaction must still occur in the 1:1:1:1 ratio required by the coefficients in the balanced equation.

▶ The denominator of this equation is *not* a perfect square, so a different algebraic approach is required.

$$K_c = \frac{[C][D]}{[A][B]} = 49.0 \quad \text{so} \quad \frac{(x)(x)}{(0.300 - x)(0.100 - x)} = 49.0$$

We can arrange this quadratic equation into the standard form, $ax^2 + bx + c = 0$.

$$\frac{x^2}{0.0300 - 0.400x + x^2} = 49.0$$

$$x^2 = 1.47 - 19.6x + 49.0x^2$$

$$48.0x^2 - 19.6x + 1.47 = 0$$

Quadratic equations can be solved by use of the quadratic formula.

$$x = \frac{-b \pm \sqrt{b^2 - 4ac}}{2a}$$

In this case a = 48.0, b = −19.6, and c = 1.47. Substituting these values gives

$$x = \frac{-(-19.6) \pm \sqrt{-19.6)^2 - 4(48.0)(1.47)}}{2(48.0)}$$

$$= \frac{19.6 \pm 10.1}{96.0} = 0.309 \quad \text{or} \quad 0.099$$

Evaluating the quadratic formula always yields two roots. One root (the answer) has physical meaning. The other root, though mathematically correct, is extraneous; that is, it has no physical meaning. The value of x is defined as the number of moles of A per liter that react and the number of moles of B per liter that react. No more B can be consumed than was initially present (0.100 M), so $x = 0.309$ is the extraneous root. Thus, $x = 0.099$ is the root that has physical meaning. The equilibrium concentrations are

$$[A] = (0.300 - x)\,M = 0.201\,M; \qquad [B] = (0.100 - x)\,M = 0.001\,M;$$

$$[C] = [D] = x\,M = 0.099\,M$$

You should now work Exercises 34, 44, and 46.

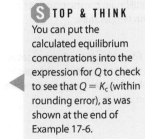

S TOP & THINK

You can put the calculated equilibrium concentrations into the expression for Q to check to see that $Q = K_c$ (within rounding error), as was shown at the end of Example 17-6.

Problem-Solving Tip   Solving Quadratic Equations— Which Root Shall We Use?

Quadratic equations can be rearranged into standard form.

$$ax^2 + bx + c = 0$$

All can be solved by the quadratic formula, which is

$$x = \frac{-b \pm \sqrt{b^2 - 4ac}}{2a} \qquad \text{(Appendix A)}$$

This formula gives *two* roots, both of which are *mathematically* correct. A foolproof way to determine which root of the equation has physical meaning is to substitute the value of the variable into the expressions for the equilibrium concentrations. For the extraneous root, one or more of these substitutions will lead to a negative concentration, which is physically impossible (there cannot be *less than none* of a substance present!). The correct root will give all positive concentrations. In Example 17-7, substitution of the extraneous root $x = 0.309$ would give $[A] = (0.300 - 0.309)\,M = -0.009\,M$ and $[B] = (0.100 - 0.309)\,M = -0.209\,M$. Either of these concentration values would be impossible, so we would know that 0.309 is an extraneous root. You should apply this test to subsequent calculations that involve solving a quadratic equation.

The following table summarizes Examples 17-6 and 17-7.

	Initial Concentrations (M)				Equilibrium Concentrations (M)			
	[A]	**[B]**	**[C]**	**[D]**	**[A]**	**[B]**	**[C]**	**[D]**
Example 17-6	0.200	0.200	0	0	0.025	0.025	0.175	0.175
Example 17-7	0.300	0.100	0	0	0.201	0.001	0.099	0.099

The data from the table can be substituted into the reaction quotient expression, Q, as a check (see margin). Even though the reaction is initiated by different relative amounts of reactants in the two cases, the ratios of equilibrium concentrations of products to reactants (each raised to the first power) agree within roundoff error.

▶ Check Example 17-6:

$$Q = \frac{[C][D]}{[A][B]} = \frac{(0.175)(0.175)}{(0.025)(0.025)}$$

$$Q = 49 = K_c$$

Check Example 17-7:

$$Q = \frac{(0.099)(0.099)}{(0.201)(0.001)}$$

$$Q = 49 = K_c$$

17-6 Disturbing a System at Equilibrium: Predictions

Once a reacting system has reached equilibrium, it remains at equilibrium until it is disturbed by some change of conditions. The guiding principle is known as **LeChatelier's Principle** (see Section 13-6).

▶ *LeChatelier* is pronounced "le-SHOT-lee-ay."

▶ Remember that the *value* of an equilibrium constant changes only with temperature.

▶ For reactions involving gases at constant temperature, changes in volume cause changes in pressure, and vice versa.

If a system at equilibrium is disturbed by changing its conditions (applying a stress), the system shifts in the direction that reduces the stress. Given sufficient time, a new state of equilibrium is established.

The reaction quotient, Q, helps us predict the direction of this shift. Three types of changes can disturb the equilibrium of a reaction.

1. Changes in concentration
2. Changes in pressure or volume (for reactions that involve gases)
3. Changes in temperature

We now study the effects of these types of stresses from a qualitative, or descriptive, point of view. In Section 17-8 we expand our discussion with quantitative examples.

Changes in Concentration

Consider the following system *starting at equilibrium*.

$$A + B \rightleftharpoons C + D \qquad K_c = \frac{[C][D]}{[A][B]}$$

When more of any reactant or product is *added* to the system, the value of Q changes so that it no longer matches K_c and the reaction is no longer at equilibrium. The stress due to the added substance is relieved by shifting the equilibrium in the direction that consumes some of the added substance, moving the value of Q back toward K_c. Let us compare the mass action expressions for Q and K_c. If more A or B is added, then $Q < K_c$, and the forward reaction occurs more rapidly and to a greater extent than the reverse reaction until equilibrium is reestablished. If more C or D is added, $Q > K_c$, and the reverse reaction occurs more rapidly and to a greater extent until equilibrium is reestablished.

Adding or removing reactants or products changes the value of Q. It *does not change* the value of K_c. Equilibrium is then reestablished by the reaction proceeding so as to change the value of Q back to the unaltered K_c value.

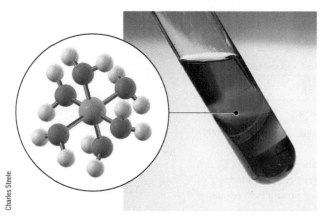 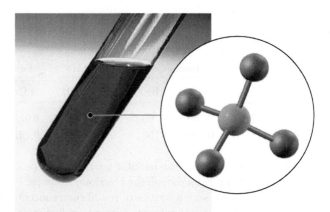

Charles Steele

Effects of changes in concentration on the equilibrium

$$[Co(OH_2)_6]^{2+} + 4Cl^- \rightleftharpoons [CoCl_4]^{2-} + 6H_2O$$

A solution of $CoCl_2 \cdot 6H_2O$ in isopropyl alcohol and water is purple (not shown) due to the mixture of $[Co(OH_2)_6]^{2+}$ (*pink*) and $[CoCl_4]^{2-}$ (*blue*). When we add concentrated HCl, the excess Cl^- shifts the reaction to the right (*blue, right photo*). Adding $AgNO_3(aq)$ removes some Cl^- by precipitation of $AgCl(s)$ and favors the reaction to the left (produces more $[Co(OH_2)_6]^{2+}$); the resulting solution is pink (*left photo*). Each model shows the structure of the cobalt complex species present in higher concentration; other ions and solvent molecules are not shown.

We can also view LeChatelier's Principle in the kinetic terms we used to introduce equilibrium. The rate of the forward reaction is proportional to the reactant concentrations raised to some powers:

$$\text{rate}_f = k_f[A]^x[B]^y$$

When we add more A to an equilibrium mixture, this rate increases so that it no longer matches the rate of the reverse reaction. As the reaction proceeds to the right, consuming some A and B and forming more C and D, the forward rate diminishes and the reverse rate increases until they are again equal. At that point, a new equilibrium condition has been reached, with more C and D than were present in the *original* equilibrium mixture. Not all of the added A has been consumed when the new equilibrium is reached, however.

If a reactant or product is *removed* from a system at equilibrium, the reaction that produces *that* substance occurs to a greater extent than its reverse. If some C or D is removed, then $Q < K$, and the forward reaction is favored until equilibrium is reestablished. If some A or B is removed, the reverse reaction is favored.

▶ The terminology used here is not as precise as we might like, but it is widely used. When we say that the equilibrium is "shifted to the left," we mean that the reaction to the left occurs to a greater extent than the reaction to the right.

Stress	Q	Direction of Shift of $A + B \rightleftharpoons C + D$
Increase concentration of A or B	$Q < K$	$\longrightarrow$ right
Increase concentration of C or D	$Q > K$	left $\longleftarrow$
Decrease concentration of A or B	$Q > K$	left $\longleftarrow$
Decrease concentration of C or D	$Q < K$	$\longrightarrow$ right

STOP & THINK
This tabulation summarizes a lot of useful information. Study it carefully.

When a new equilibrium condition is established, (1) the rates of the forward and reverse reactions are again equal, and (2) K_c is again satisfied by the new concentrations of reactants and products.

Practical applications of changes of this type are of great economic importance. Removing a product of a reversible reaction forces the reaction to produce more product than could be obtained if the reaction were simply allowed to reach equilibrium.

EXAMPLE 17-8 Disturbing a System at Equilibrium: Predictions

Given the following reaction at equilibrium in a closed container at 500°C, predict the effect of each of the following changes on the amount of NH_3 present at equilibrium: (a) forcing more H_2 into the system; (b) removing some NH_3 from the system.

$$N_2(g) + 3H_2(g) \rightleftharpoons 2NH_3(g)$$

Plan

We apply LeChatelier's Principle to each part of the question. We can reach the same conclusions by comparing Q with K, even though we don't have numerical values.

Solution

(a) LeChatelier's Principle: Adding a substance favors the reaction that uses up that substance (forward in this case).

More NH_3 is formed.

Q vs. K_c: At the start, the reaction is at equilibrium, so $Q = K_c$. Then we add more H_2, which *increases* $[H_2]$. Because $[H_2]$ is in the denominator of

$$Q = \frac{[NH_3]^2}{[N_2][H_2]^3}$$

Increasing $Q \longrightarrow$

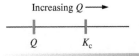

(a) Q is less than K_c, so reaction proceeds to the right, forming more products

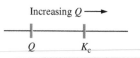

(b) Q is less than K_c, so reaction proceeds to the right, forming more products

this *decreases* the value of Q, making $Q < K_c$. To reestablish equilibrium, the reaction proceeds to the right.

$$\text{More NH}_3 \text{ is formed.}$$

(b) LeChatelier's Principle: Removing a substance favors the reaction that produces that substance (forward in this case).

$$\text{More NH}_3 \text{ is formed.}$$

Q vs. K_c: At the start, the reaction is at equilibrium, so $Q = K_c$. Then we remove some NH_3, which *decreases* $[NH_3]$. Because $[NH_3]$ is in the numerator of Q, this *decreases* the value of Q, making $Q < K_c$. To reestablish equilibrium, the reaction proceeds to the right.

$$\text{More NH}_3 \text{ is formed.}$$

At the new equilibrium, the concentration of any reactant or product that was added will still be greater than it was in the original equilibrium mixture. The concentration of any reactant or product that was removed will be less in the new equilibrium mixture because the reaction will have replaced only part of the removed material.

Another example of the shifting of equilibrium occurs in nature during the formation of stalactites and stalagmites in caves. Water containing carbon dioxide slowly passes through layers of limestone, $CaCO_3$. The limestone dissolves according to the following reaction:

$$CaCO_3(s) + CO_2(aq) + H_2O(\ell) \longrightarrow Ca^{2+}(aq) + 2HCO_3^-(aq)$$

Equilibrium is established when the dissolved limestone reaches the limit of its solubility:

$$Ca^{2+}(aq) + 2HCO_3^-(aq) \rightleftharpoons CaCO_3(s) + CO_2(aq) + H_2O(\ell)$$

When the solution reaches a cave, carbon dioxide slowly escapes from the solution, and the subsequent shift in the equilibrium causes $CaCO_3$ to be precipitated. Deposits of $CaCO_3$ produce the attractive cave formations.

Changes in Volume and Pressure

Changes in pressure have little effect on the concentrations of solids or liquids because they are only slightly compressible. Changes in pressure do cause significant changes in concentrations of gases, however. Such changes therefore affect the value of Q for reactions in which the number of moles of gaseous reactants differs from the number of moles of gaseous products. For an ideal gas,

$$PV = nRT \quad \text{or} \quad P = \left(\frac{n}{V}\right)(RT)$$

Stalactites and stalagmites are formed by the precipitation of calcium carbonate, $CaCO_3$, when gaseous CO_2 escapes from groundwater solutions.

The term (n/V) represents concentration, that is, mol/L. At constant temperature, n, R, and T are constants. Thus, if the volume occupied by a gas decreases, its partial pressure increases and its concentration (n/V) increases. If the volume of a gas increases, both its partial pressure and its concentration decrease.

Consider the following gaseous system at equilibrium.

$$A(g) \rightleftharpoons 2D(g) \quad K_c = \frac{[D]^2}{[A]}$$

At constant temperature, a decrease in volume (increase in pressure) increases the concentrations of both A and D. In the expression for Q, the concentration of D is squared and the concentration of A is raised to the first power. As a result, the numerator of Q increases more than the denominator as pressure increases. Thus, $Q > K_c$, and this equilibrium shifts to the left. Conversely, an increase in volume (decrease in pressure) shifts this reaction to the right until equilibrium is reestablished, because $Q < K_c$. We can summarize the effect of pressure (volume) changes on *this* gas-phase system at equilibrium.

Stress	Q^*	Direction of Shift of $A(g) \rightleftharpoons 2D(g)$
Volume decrease, pressure increase	$Q > K_c$	Toward smaller number of moles of gas (left for *this* reaction)
Volume increase, pressure decrease	$Q < K_c$	Toward larger number of moles of gas (right for *this* reaction)

*In Q for *this* reaction, there are more moles of gaseous product than gaseous reactant.

In general, for reactions that involve gaseous reactants or products, LeChatelier's Principle allows us to predict the following results.

1. If there is *no change in the total number of moles of gases in the balanced chemical equation*, a volume (pressure) change does not affect the position of equilibrium; Q is unchanged by the volume (or pressure) change.

2. If a *balanced chemical equation involves a change in the total number of moles of gases*, changing the volume (or pressure) of a reaction mixture changes the value of Q; it *does not change* the value of K_c. For such a reaction:

   (a) A decrease in volume (increase in pressure) shifts a reaction in the direction that produces the *smaller total number of moles of gas*, until Q again equals K_c.

   (b) An increase in volume (decrease in pressure) shifts a reaction in the direction that produces the *larger total number of moles of gas*, until Q again equals K_c.

The foregoing argument applies only when pressure changes are due to volume changes. It *does not apply* if the total pressure of a gaseous system is raised by merely pumping in an inert gas, for example, He. If the gas that is introduced is not involved in the reaction, the *partial pressure* of each reacting gas remains constant, so the system remains at equilibrium.

S **TOP & THINK**
Study this tabulation carefully. How would these conclusions change for a reaction in which there are more moles of gaseous reactants than moles of gaseous products?

▶ One practical application of these ideas is illustrated in the next section by the Haber process.

S **TOP & THINK**
Many students do not realize that as a different gas is pumped into a container, the partial pressures of any gases already present do not change.

EXAMPLE 17-9 Disturbing a System at Equilibrium: Predictions

(a) Given the following reaction at equilibrium in a closed container at 500°C, predict the effect of increasing the pressure by decreasing the volume.

$$N_2(g) + 3H_2(g) \rightleftharpoons 2NH_3(g)$$

(b) Make the same prediction for the following reaction at high temperature.

$$H_2(g) + I_2(g) \rightleftharpoons 2HI(g)$$

Plan

We apply LeChatelier's Principle to predict the effect on each reaction.

Solution

(a) Increasing the pressure favors the reaction that produces the smaller number of moles of gas (forward in this case).

More NH_3 is formed.

(b) This reaction involves the same number of moles of gas on both sides, so a pressure (volume) change does not disturb the equilibrium. There is no effect on the position of equilibrium.

Changes in Temperature

Consider the following exothermic reaction at equilibrium:

$$A + B \rightleftharpoons C + D + heat \qquad \text{(exothermic, gives off heat, } \Delta H \text{ is negative)}$$

Heat is produced by the forward (exothermic) reaction. Suppose we increase the temperature at constant pressure by adding heat to the system. This favors the reaction to the left, removing some of the extra heat. Lowering the temperature favors the reaction to the right as the system replaces some of the heat that was removed.

By contrast, for an endothermic reaction at equilibrium,

$$W + X + heat \rightleftharpoons Y + Z \qquad \text{(endothermic, absorbs heat, } \Delta H \text{ is positive)}$$

an increase in temperature at constant pressure favors the reaction to the right. A decrease in temperature favors the reaction to the left.

The *value of any equilibrium constant changes* as the temperature changes, as we will see in Section 17-13. Changing the temperature of a reaction at equilibrium thus causes Q to differ from K_c, but now this is because K_c has changed. The reaction then proceeds in the direction that moves Q toward the new value of K_c.

> The K_c values of exothermic reactions decrease with increasing T.
> The K_c values of endothermic reactions increase with increasing T.
> No other stresses affect the value of K_c.

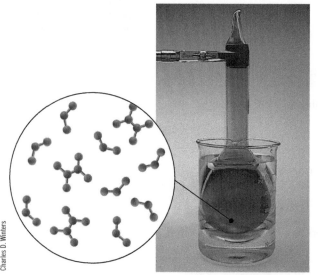
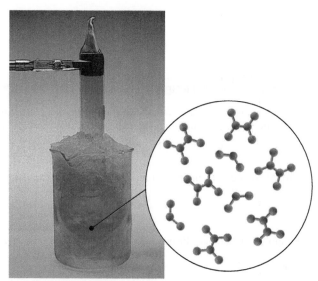

Charles D. Winters

The gas-phase equilibrium for the *exothermic* reaction

$$2NO_2(g) \rightleftharpoons N_2O_4(g) \qquad \Delta H^0_{rxn} = -57.2 \text{ kJ/mol rxn}$$

The two flasks contain the same *total mass* of gas. NO_2 is brown, whereas N_2O_4 is colorless. The higher temperature (50°C) of the flask on the left favors the reverse reaction; this mixture is more highly colored because it contains more NO_2. The flask on the right, at the temperature of ice water, contains less brown NO_2 gas. The lower temperature favors the formation of the N_2O_4. No matter how far the equilibrium shifts, both reactants and products will always be present.

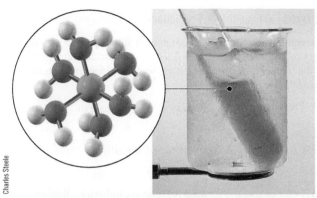
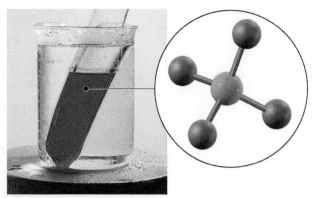

Charles Steele

Effect of temperature on the equilibrium of the *endothermic* **reaction:** $[Co(OH_2)_6]^{2+} + 4Cl^- + heat \rightleftharpoons [CoCl_4]^{2-} + 6H_2O$
We begin with a purple equilibrium mixture of the pink and blue complexes at room temperature (not shown). In hot water the forward reaction (endothermic) is favored and K_c is higher, so the solution is blue (*right photo*). At 0°C, the reverse reaction (exothermic) is favored and K_c is lower, so the solution is pink (*left photo*). Each model shows the structure of the cobalt complex species present in highest concentration; other ions and solvent molecules are not shown.

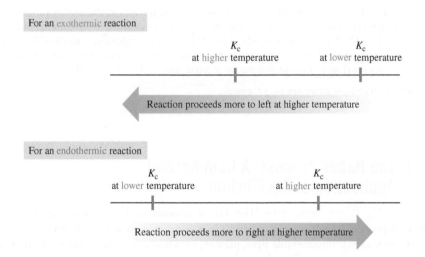

For an exothermic reaction

K_c at higher temperature K_c at lower temperature

Reaction proceeds more to left at higher temperature

For an endothermic reaction

K_c at lower temperature K_c at higher temperature

Reaction proceeds more to right at higher temperature

Addition of a Catalyst

Adding a catalyst to a system changes the rate of the reaction (see Section 16-9), but this *does not* shift the equilibrium in favor of either products or reactants. Because a catalyst affects the activation energy of *both* forward and reverse reactions equally, it changes both rate constants by the same factor, so their ratio, K_c, does not change.

> Adding a catalyst to a reaction at equilibrium has no effect on the amounts of reactants or products; it changes neither Q nor K_c.

The same equilibrium mixture is achieved with or without the catalyst, but the equilibrium is established more quickly in the presence of a catalyst.

Not all reactions attain equilibrium; they may occur too slowly, or else products or reactants may be continually added or removed. Such is the case with most reactions in biological systems. On the other hand, some reactions, such as typical acid–base neutralizations, achieve equilibrium very rapidly.

STOP & THINK
Theoretically, in closed systems all chemical reactions are reversible and can be characterized with equilibrium constants. You may wonder if the equilibrium reactions described are somehow different from those used in previous chapters where only a single reaction arrow was used. Many of those reactions have large K values and proceed to make mainly products. Equilibrium concentrations (and double reaction arrows) are most meaningful for reactions that have K values that are neither extremely large nor extremely small and have observable quantities of both reactants and products at equilibrium.

▶ Can you use the Arrhenius equation (Section 16-8) to show that lowering the activation energy barrier increases forward and reverse rates by the same factor?

EXAMPLE 17-10 Disturbing a System at Equilibrium: Predictions

Given the following reaction at equilibrium in a closed container at 500°C, predict the effect of each of the following changes on the amount of NH_3 present at equilibrium: (a) raising the temperature; (b) lowering the temperature; (c) introducing some platinum catalyst. (d) How would the value of K_c change in each case?

$$N_2(g) + 3H_2(g) \rightleftharpoons 2NH_3(g) \qquad \Delta H^0 = -92 \text{ kJ/mol rxn}$$

Plan

We apply LeChatelier's Principle to each part of the question individually.

Solution

(a) The negative value for ΔH tells us that the forward reaction is exothermic. Raising the temperature favors the endothermic reaction (reverse in this case).

Some NH_3 is used up.

(b) Lowering the temperature favors the exothermic reaction (forward in this case).

More NH_3 is formed.

(c) A catalyst does not favor either reaction.

It would have no effect on the amount of NH_3.

(d) In (a), there is less product (NH_3) present at equilibrium at the higher temperature, so the value of K_c is smaller at the higher temperature. In (b), there is more product (NH_3) present at equilibrium at the lower temperature, so the value of K_c is greater at the lower temperature. In (c), the change (introduction of a catalyst) causes no change in equilibrium concentrations, so K_c does not change.

You should now work Exercises 54, 57, and 58.

▶ As Germany prepared for World War I, Britain controlled the seas and thus the access to the natural nitrates in India and Chile that were needed to prepare explosives. Fritz Haber (1868–1934) developed the process to provide a cheaper and more reliable source of explosives. Currently the process is used mainly to produce fertilizers. Large amounts of ammonia are converted into ammonium sulfate or ammonium nitrate for use as fertilizers. Approximately 100 pounds of NH_3 is used for this purpose per person in the United States each year.

17-7 The Haber Process: A Commercial Application of Equilibrium

Nitrogen, N_2, is very unreactive. The Haber process is the economically important industrial process by which atmospheric N_2 is converted to ammonia, NH_3, a soluble, reactive compound. Innumerable dyes, plastics, explosives, fertilizers, and synthetic fibers are made from ammonia. The Haber process is an excellent practical application of LeChatelier Principle. It provides insight into kinetic and thermodynamic factors that influence reaction rates and the positions of equilibria. In this process the reaction between N_2 and H_2 to produce NH_3 is never allowed to reach equilibrium, but moves toward it.

$$N_2(g) + 3H_2(g) \rightleftharpoons 2NH_3(g) \qquad \Delta H^0 = -92 \text{ kJ/mol}$$

$$K_c = \frac{[NH_3]^2}{[N_2][H_2]^3} = 3.6 \times 10^8 \qquad \text{(at 25°C)}$$

The process is diagrammed in Figure 17-3. The reaction is carried out at about 450°C under pressures ranging from 200 to 1000 atmospheres. Hydrogen is obtained from coal gas or petroleum refining and nitrogen is obtained from liquefied air.

The value of K_c is 3.6×10^8 at 25°C. This very large value of K_c indicates that *if equilibrium were reached* at 25°C, virtually all of the N_2 and H_2 (mixed in a 1:3 mole ratio) would be converted into NH_3. At 25°C, the reaction occurs so slowly, however, that no measurable amount of NH_3 is produced within a reasonable time. Thus, the large equilibrium constant (a thermodynamic factor) indicates that the reaction proceeds toward the right almost completely. It tells us *nothing*, however, about how fast the reaction occurs (a kinetic factor).

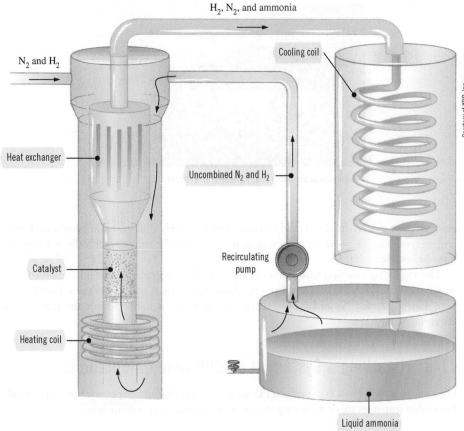

A nighttime photo of a large plant for the commercial production of ammonia, NH_3. Such an installation can produce up to 7000 metric tons of ammonia per day. There are nearly 100 such plants in the world.

Figure 17-3 A simplified representation of the Haber process for synthesizing ammonia.

There are four moles of gas on the left side of the equation and only two moles of gas on the right, so increasing the pressure favors the production of NH_3. The Haber process is therefore carried out at very high pressures, as high as the equipment will safely stand [see Example 17-9(a)].

The reaction is exothermic (ΔH^0_{rxn} is negative), so decreasing the temperature favors the *formation* of NH_3 [see Example 17-10(b)]. But the rates of both forward and reverse reactions would decrease greatly as the temperature decreases.

The addition of a catalyst of finely divided iron and small amounts of selected oxides speeds up both the forward and reverse reactions. This allows NH_3 to be produced not only more rapidly but at a lower temperature, which increases the overall production of NH_3 and extends the life of the equipment.

Table 17-1 shows the effects of changes in temperature and pressure on the equilibrium yield of NH_3, starting with 1:3 mole ratios of N_2:H_2. K_c changes by more than ten orders of magnitude, from 3.6×10^8 at 25°C to only 1.4×10^{-2} at 758°C. This tells us that the reaction proceeds *very far to the left* at high temperatures. Casual examination of the data might suggest that the reaction should be run at lower temperatures so that a higher

▶ In practice, the mixed reactants are compressed by special pumps and injected into the heated reaction vessel.

▶ Ten orders of magnitude is 10^{10}, that is,

$$1 \times 10^{10} = 10{,}000{,}000{,}000$$

Table 17-1 Effect of *T* and *P* on Yield of Ammonia

°C	K_c	Mole % NH_3 in Equilibrium Mixture		
		10 atm	100 atm	1000 atm
209	650	51	82	98
467	0.5	4	25	80
758	0.014	0.5	5	13

A major use of ammonia is as agricultural fertilizer.

percentage of the N_2 and H_2 would be converted into NH_3. The reaction occurs so slowly, however, even in the presence of a catalyst, that it cannot be run economically at temperatures below about 450°C.

The emerging reaction mixture is cooled and NH_3 (bp = −33.43°C) is removed as a liquid. This prevents the reaction from reaching equilibrium and favors the forward reaction. The unreacted N_2 and H_2 are recycled. Excess N_2 is used to favor the reaction to the right.

17-8 Disturbing a System at Equilibrium: Calculations

We can use equilibrium constants to calculate new equilibrium concentrations that result from adding species to, or removing species from, a system at equilibrium.

EXAMPLE 17-11 Disturbing a System at Equilibrium: Calculations

Some hydrogen and iodine are mixed at 229°C in a 1.00-liter container. When equilibrium is established, the following concentrations are present: [HI] = 0.490 M, [H_2] = 0.080 M, and [I_2] = 0.060 M. If an additional 0.300 mol of HI is then added, what concentrations will be present when the new equilibrium is established?

$$H_2(g) + I_2(g) \rightleftharpoons 2HI(g)$$

Plan

We use the original equilibrium concentrations to calculate the value of K_c. Then we determine the new concentrations after some HI has been added and calculate Q. Comparing the value of Q to K_c tells us which reaction is favored. Then we can represent the new equilibrium concentrations. We substitute these representations into the K_c expression and solve for the new equilibrium concentrations.

Solution

Calculate the value of K_c from the first set of equilibrium concentrations.

$$K_c = \frac{[HI]^2}{[H_2][I_2]} = \frac{(0.490)^2}{(0.080)(0.060)} = 50.$$

When we add 0.300 mol of HI to the 1.00-liter container, the [HI] instantaneously increases by 0.300 M.

	$H_2(g)$ +	$I_2(g)$ $\rightleftharpoons$	$2HI(g)$
original equilibrium	0.080 M	0.060 M	0.490 M
mol/L added	0 M	0 M	+0.300 M
new *initial* conc'n	0.080 M	0.060 M	0.790 M

Substitution of these *new initial* concentrations into the reaction quotient gives

$$Q = \frac{[HI]^2}{[H_2][I_2]} = \frac{(0.790)^2}{(0.080)(0.060)} = 130$$

Because $Q > K_c$, the reaction proceeds to the left to establish a new equilibrium. The new equilibrium concentrations can be determined as follows. Let x = mol/L of H_2 formed; so x = mol/L of I_2 formed, and $2x$ = mol/L of HI consumed.

	$H_2(g)$	+	$I_2(g)$	$\rightleftharpoons$	$2HI(g)$
new initial conc'n	0.080 M		0.060 M		0.790 M
change due to rxn	$+x\ M$		$+x\ M$		$-2x\ M$
new equilibrium	$(0.080 + x)\ M$		$(0.060 + x)\ M$		$(0.790 - 2x)\ M$

Substitution of these values into K_c allows us to evaluate x.

$$K_c = 50. = \frac{(0.790 - 2x)^2}{(0.080 + x)(0.060 + x)} = \frac{0.624 - 3.16x + 4x^2}{0.0048 + 0.14x + x^2}$$

$$0.24 + 7.0x + 50x^2 = 0.624 - 3.16x + 4x^2$$

$$46x^2 + 10.2x - 0.38 = 0$$

Solution by the quadratic formula gives x = 0.032 and -0.25.

Clearly, $x = -0.25$ is the extraneous root, because x cannot be less than zero in this case. This reaction does not consume a negative quantity of HI, because the reaction is proceeding toward the left. Thus, $x = 0.032$ is the root with physical meaning, so the new equilibrium concentrations are

$$[H_2] = (0.080 + x)\ M = (0.080 + 0.032)\ M = 0.112\ M$$

$$[I_2] = (0.060 + x)\ M = (0.060 + 0.032)\ M = 0.092\ M$$

$$[HI] = (0.790 - 2x)\ M = (0.790 - 0.064)\ M = 0.726\ M$$

In summary,

Original Equilibrium	Stress Applied	New Equilibrium
$[H_2]$ = 0.080 M		$[H_2]$ = 0.112 M
$[I_2]$ = 0.060 M	Add 0.300 M HI	$[I_2]$ = 0.092 M
$[HI]$ = 0.490 M		$[HI]$ = 0.726 M

You should now work Exercise 64.

▶ It is obvious that adding some HI favors the reaction to the left. If more than one substance is added to the reaction mixture, it might not be obvious which reaction will be favored. Calculating Q always lets us make the decision.

 STOP & THINK
Equal amounts of H_2 and I_2 must be formed by the *new progress* of the reaction.

▶ To "consume a negative quantity of HI" would be to form HI. The value $x = -0.25$ would lead to $[H_2] = (0.080 + x)M = (0.080 - 0.25)M = -0.17\ M$. A negative concentration is impossible, so $x = -0.25$ is the extraneous root.

STOP & THINK
We see that some of the additional HI is consumed, but not all of it. More HI remains after the new equilibrium is established than was present before the stress was imposed. The new equilibrium $[H_2]$ and $[I_2]$ are substantially greater than the original equilibrium concentrations, however.

We can also use the equilibrium constant to calculate new equilibrium concentrations that result from decreasing the volume (increasing the pressure) of a gaseous system that was initially at equilibrium.

EXAMPLE 17-12　Disturbing a System at Equilibrium: Calculations

At 22°C the equilibrium constant, K_c, for the following reaction is 4.66×10^{-3}. (a) If 0.800 mol of N_2O_4 were injected into a closed 1.00-liter container at 22°C, how many moles of each gas would be present at equilibrium? (b) If the volume were then halved (to 0.500 L) at constant temperature, how many moles of each gas would be present after the new equilibrium has been established?

$$N_2O_4(g) \rightleftharpoons 2NO_2(g) \qquad K_c = \frac{[NO_2]^2}{[N_2O_4]} = 4.66 \times 10^{-3}$$

Plan

(a) We are given the value of K_c and the initial concentration of N_2O_4. We write the reaction summary, which gives the representation of the equilibrium concentrations. Then we substitute these into the K_c expression and solve for the new equilibrium concentrations. (b) We obtain the *new initial* concentrations by adjusting the equilibrium concentrations from Part (a) for the volume change. Then we solve for the *new* equilibrium concentrations as we did in Part (a).

Solution

(a) Let x = mol/L of N_2O_4 consumed and $2x$ = mol/L of NO_2 formed.

	$N_2O_4(g)$	$\rightleftharpoons$	$2NO_2(g)$
initial	0.800 M		0 M
change due to rxn	$-x$ M		$+2x$ M
equilibrium	$(0.800 - x)$ M		$2x$ M

$$K_c = \frac{[NO_2]^2}{[N_2O_4]} = 4.66 \times 10^{-3} = \frac{(2x)^2}{0.800 - x} = \frac{4x^2}{0.800 - x}$$

$$3.73 \times 10^{-3} - 4.66 \times 10^{-3}x = 4x^2$$

$$4x^2 + 4.66 \times 10^{-3}x - 3.73 \times 10^{-3} = 0$$

▶ The value of x is the number of moles per liter of N_2O_4 that react. So x must be positive and cannot be greater than 0.800 M.

$$0 < x < 0.800 \ M$$

Solving by the quadratic formula gives $x = 3.00 \times 10^{-2}$ and $x = -3.11 \times 10^{-2}$. We use $x = 3.00 \times 10^{-2}$.

The original equilibrium concentrations are

$$[NO_2] = 2x \ M = 6.00 \times 10^{-2} \ M$$

$$[N_2O_4] = (0.800 - x) \ M = (0.800 - 3.00 \times 10^{-2}) \ M = 0.770 \ M$$

$$\underline{?} \text{ mol } NO_2 = 1.00 \text{ L} \times \frac{6.00 \times 10^{-2} \text{ mol } NO_2}{L} = 6.00 \times 10^{-2} \text{ mol } NO_2$$

$$\underline{?} \text{ mol } N_2O_4 = 1.00 \text{ L} \times \frac{0.770 \text{ mol } N_2O_4}{L} = 0.770 \text{ mol } N_2O_4$$

▶ LeChatelier's Principle tells us that a decrease in volume (increase in pressure) favors the production of N_2O_4. As before, we can confirm this by comparing Q with K_c.

$$Q = \frac{(0.120)^2}{1.54} = 9.35 \times 10^{-3}$$

$$Q > K_c$$

$$\therefore \longleftarrow \text{shift left}$$

(b) When the volume of the reaction vessel is halved, the concentrations are doubled, so the *new initial* concentrations of N_2O_4 and NO_2 are $2(0.770 \ M) = 1.54 \ M$ and $2(6.00 \times 10^{-2} \ M) = 0.120 \ M$, respectively.

	$N_2O_4(g)$	$\rightleftharpoons$	$2NO_2(g)$
new initial	1.54 M		0.120 M
change due to rxn	$+x$ M		$-2x$ M
new equilibrium	$(1.54 + x)$ M		$(0.120 - 2x)$ M

$$K_c = \frac{[NO_2]^2}{[N_2O_4]} = 4.66 \times 10^{-3} = \frac{(0.120 - 2x)^2}{1.54 + x}$$

Rearranging into the standard form of a quadratic equation gives

$$x^2 - 0.121x + 1.81 \times 10^{-3} = 0$$

▶ The root $x = 0.104$ would give a *negative* concentration for NO_2, which is impossible.

Solving as before gives $x = 0.104$ and $x = 0.017$.

The maximum value of x is 0.060 M, because $2x$ may not exceed the concentration of NO_2 that was present after the volume was halved. Thus, $x = 0.017 \ M$ is the root with physical significance. The new equilibrium concentrations in the 0.500-liter container are

$$[NO_2] = (0.120 - 2x) \ M = (0.120 - 0.034) \ M = 0.086 \ M$$

$$[N_2O_4] = (1.54 + x) \ M = (1.54 + 0.017) \ M = 1.56 \ M$$

$$? \text{ mol NO}_2 = 0.500 \text{ L} \times \frac{0.086 \text{ mol NO}_2}{\text{L}} = 0.043 \text{ mol NO}_2$$

$$? \text{ mol N}_2\text{O}_4 = 0.500 \text{ L} \times \frac{1.56 \text{ mol N}_2\text{O}_4}{\text{L}} = 0.780 \text{ mol N}_2\text{O}_4$$

In summary,

First Equilibrium	Stress	New Equilibrium
0.770 mol of N_2O_4	Decrease volume from	0.780 mol of N_2O_4
0.0600 mol of NO_2	1.00 L to 0.500 L	0.043 mol of NO_2

You should now work Exercise 66.

> **STOP & THINK**
> The concentrations [N_2O_4] and [NO_2] *both increase* because of the large decrease in volume. However, the *number of moles* of N_2O_4 increases, while the *number of moles* of NO_2 decreases. We predict this from LeChatelier's Principle.

Problem-Solving Tip There Are Several Ways to Solve Equilibrium Problems

When a stress is applied to a system originally at equilibrium, it is no longer at equilibrium. As we did in Example 17-12(b), we can apply the stress to the *old equilibrium values* and then treat these as the new "initial values." Alternatively, we could adjust the *original concentration values* to reflect the stress, and then treat these as the new "initial values." We could consider [N_2O_4]$_{initial}$ in Example 17-12(b) as the *original* 0.800 mol of N_2O_4 from part (a) in the *new* volume, 0.500 L. That is, [N_2O_4]$_{initial}$ = 0.800 mol/0.500 L = 1.60 M, with no NO_2 having yet been formed. [NO_2]$_{initial}$ = 0 M. From that starting point, the reaction would proceed to the *right*.

17-9 Partial Pressures and the Equilibrium Constant

It is often more convenient to measure pressures rather than concentrations of gases. Solving the ideal gas equation, $PV = nRT$, for pressure gives

$$P = \frac{n}{V}(RT) \quad \text{or} \quad P = M(RT)$$

> $\left(\dfrac{n}{V}\right)$ is $\left(\dfrac{\text{no. mol}}{\text{L}}\right)$

The pressure of a gas is directly proportional to its concentration (n/V). For reactions in which all substances that appear in the equilibrium constant expression are gases, we sometimes prefer to express the equilibrium constant in terms of partial pressures *in atmospheres* (K_P) rather than in terms of concentrations (K_c).

In general for a reaction involving gases,

$$a\text{A(g)} + b\text{B(g)} \rightleftharpoons c\text{C(g)} + d\text{D(g)} \qquad K_P = \frac{(P_C)^c(P_D)^d}{(P_A)^a(P_B)^b}$$

> ► K_P has no units for the same reasons that K_c has no units.

For instance, for the following reversible reaction,

$$\text{N}_2\text{(g)} + 3\text{H}_2\text{(g)} \rightleftharpoons 2\text{NH}_3\text{(g)} \qquad K_P = \frac{(P_{\text{NH}_3})^2}{(P_{\text{N}_2})(P_{\text{H}_2})^3}$$

EXAMPLE 17-13 Calculation of K_P

In an equilibrium mixture at 500°C, we find $P_{NH_3} = 0.147$ atm, $P_{N_2} = 6.00$ atm, and $P_{H_2} = 3.70$ atm. Evaluate K_P at 500°C for the following reaction.

$$N_2(g) + 3H_2(g) \rightleftharpoons 2NH_3(g)$$

Plan

We are given equilibrium partial pressures of all reactants and products. So we write the expression for K_P and substitute partial pressures in atmospheres into it.

Solution

$$K_P = \frac{(P_{NH_3})^2}{(P_{N_2})(P_{H_2})^3} = \frac{(0.147)^2}{(6.00)(3.70)^3} = 7.11 \times 10^{-5}$$

You should now work Exercises 70 and 73.

S TOP & THINK
One error that students sometimes make when solving K_P problems is to express pressures in torr. Remember that these pressures must be expressed in atmospheres.

17-10 Relationship Between K_P and K_c

If the ideal gas equation is rearranged, the molar concentration of a gas is

$$\left(\frac{n}{V}\right) = \frac{P}{RT} \quad \text{or} \quad M = \frac{P}{RT}$$

Substituting P/RT for n/V in the K_c expression for the N_2–H_2–NH_3 equilibrium gives the relationship between K_c and K_P for *this* reaction.

$$K_c = \frac{[NH_3]^2}{[N_2][H_2]^3} = \frac{\left(\frac{P_{NH_3}}{RT}\right)^2}{\left(\frac{P_{N_2}}{RT}\right)\left(\frac{P_{H_2}}{RT}\right)^3} = \frac{(P_{NH_3})^2}{(P_{N_2})(P_{H_2})^3} \times \frac{\left(\frac{1}{RT}\right)^2}{\left(\frac{1}{RT}\right)^4}$$

$$= K_P(RT)^2 \quad \text{or} \quad K_P = K_c(RT)^{-2}$$

In general the relationship between K_c and K_P is

$$K_P = K_c(RT)^{\Delta n} \quad \text{or} \quad K_c = K_P(RT)^{-\Delta n} \quad \Delta n = (n_{gas\ prod}) - (n_{gas\ react})$$

S TOP & THINK
Δn refers to the numbers of moles of gaseous substances in the balanced equation, *not* in the reaction vessel.

For reactions in which equal numbers of moles of gases appear on both sides of the equation, $\Delta n = 0$ and $K_P = K_c$.

In Example 17-1, we saw that for the ammonia reaction at 500°C (or 773 K), $K_c = 0.286$. We can describe this equilibrium in terms of partial pressures using K_P.

$$NH_3(g) + 3H_2(g) \rightleftharpoons 2NH_3(g) \quad \Delta n = 2 - 4 = -2$$
$$K_P = K_c(RT)^{\Delta n} = (0.286)[(0.0821)(773)]^{-2} = 7.10 \times 10^{-5}$$

This is essentially the same value we obtained in Example 17-13.

Problem-Solving Tip Be Careful About the Value of R

To decide which value of R to use when you convert between K_c and K_P, you can reason as follows. K_c involves molar concentrations, for which the units are mol/L; K_P involves pressures expressed in atm. Thus the appropriate value of R to use for these conversions must include these units. We use $0.08206 \frac{L \cdot atm}{mol \cdot K}$, rounded to the number of places appropriate to the problem.

For *gas-phase reactions*, we can calculate the amounts of substances present at equilibrium using either K_P or K_c. The results are the same by either method (when they are both expressed in the same terms). To illustrate, let's solve the following problem by both methods.

EXAMPLE 17-14   Calculations with K_c and K_P

We place 10.0 grams of $SbCl_5$ in a 5.00-liter container at 448°C and allow the reaction to attain equilibrium. How many grams of $SbCl_5$ are present at equilibrium? Solve this problem (a) using K_c and molar concentrations, and (b) using K_P and partial pressures.

$$SbCl_5(g) \rightleftharpoons SbCl_3(g) + Cl_2(g) \qquad K_c = 2.51 \times 10^{-2} \quad \text{and} \quad K_P = 1.48$$

(a) Plan (using K_c)

We calculate the initial concentration of $SbCl_5$, write the reaction summary, and represent the equilibrium concentrations; then we substitute into the K_c expression to obtain the equilibrium concentrations.

(a) Solution (using K_c)

Because we are given K_c, we use concentrations. The initial concentration of $SbCl_5$ is

$$[SbCl_5] = \frac{10.0 \text{ g } SbCl_5}{5.00 \text{ L}} \times \frac{1 \text{ mol}}{299 \text{ g}} = 0.00669 \; M \; SbCl_5$$

Let x = mol/L of $SbCl_5$ that react. In terms of molar concentrations, the reaction summary is

	$SbCl_5$	$\rightleftharpoons$	$SbCl_3$	+	Cl_2
initial	0.00669 M		0		0
change due to rxn	$-x \, M$		$+x \, M$		$+x \, M$
equilibrium	$(0.00669 - x) \, M$		$x \, M$		$x \, M$

$$K_c = \frac{[SbCl_3][Cl_2]}{[SbCl_5]}$$

$$2.51 \times 10^{-2} = \frac{(x)(x)}{0.00669 - x}$$

$$x^2 = 1.68 \times 10^{-4} - 2.51 \times 10^{-2}x$$

$$x^2 + 2.51 \times 10^{-2}x - 1.68 \times 10^{-4} = 0$$

Solving by the quadratic formula gives

$$x = 5.49 \times 10^{-3} \quad \text{and} \quad -3.06 \times 10^{-2} \text{ (extraneous root)}$$

$$[SbCl_5] = (0.00669 - x) \, M = (0.00669 - 0.00549) \, M = 1.20 \times 10^{-3} \, M$$

$$\underline{?} \text{ g } SbCl_5 = 5.00 \text{ L} \times \frac{1.20 \times 10^{-3} \text{ mol}}{L} \times \frac{299 \text{ g}}{\text{mol}} = 1.79 \text{ g } SbCl_5$$

Let us now solve the same problem *using K_P and partial pressures*.

(b) Plan (using K_P)

Calculate the initial partial pressure of $SbCl_5$, and write the reaction summary. Substitution of the representations of the equilibrium partial pressures into K_P gives their values.

(b) Solution (using K_P)

We calculate the initial *pressure* of $SbCl_5$ in atmospheres, using $PV = nRT$.

$$P_{SbCl_5} = \frac{nRT}{V} = \frac{\left[(10.0 \text{ g})\left(\frac{1 \text{ mol}}{299 \text{ g}}\right)\right]\left(0.0821 \frac{L \cdot atm}{mol \cdot K}\right)(721 \text{ K})}{5.00 \text{ L}} = 0.396 \text{ atm}$$

▶ The *partial pressure* of each substance is proportional to the *number of moles* of that substance.

Clearly, $P_{SbCl_3} = 0$ and $P_{Cl_2} = 0$ because only $SbCl_5$ is present initially. We write the reaction summary in terms of partial pressures in atmospheres, because K_P refers to pressures in atmospheres.

Let y = decrease in pressure (atm) of $SbCl_5$ due to reaction. In terms of partial pressures, the reaction summary is

	$SbCl_5$	$\rightleftharpoons$	$SbCl_3$ +	Cl_2
initial	0.396 atm		0	0
change due to rxn	$-y$ atm		$+y$ atm	$+y$ atm
equilibrium	$(0.396 - y)$ atm		y atm	y atm

$$K_P = \frac{(P_{SbCl_3})(P_{Cl_2})}{P_{SbCl_5}} = 1.48 = \frac{(y)(y)}{0.396 - y}$$

$$0.586 - 1.48y = y^2 \quad y^2 + 1.48y - 0.586 = 0$$

▶ If we did not know the value of K_P, we could calculate it from the known value of K_c, using the relationship $K_P = K_c(RT)^{\Delta n}$.

Solving by the quadratic formula gives

$$y = 0.325 \text{ and } -1.80 \text{ (extraneous root)}$$

$$P_{SbCl_5} = (0.396 - y) = (0.396 - 0.325) = 0.071 \text{ atm}$$

We use the ideal gas law, $PV = nRT$, to calculate the number of moles of $SbCl_5$.

$$n = \frac{PV}{RT} = \frac{(0.071 \text{ atm})(5.00 \text{ L})}{\left(0.0821 \dfrac{\text{L} \cdot \text{atm}}{\text{mol} \cdot \text{K}}\right)(721 \text{ K})} = 0.0060 \text{ mol } SbCl_5$$

▶ We see that within roundoff range the same result is obtained by both methods.

$$\underline{?} \text{ g } SbCl_5 = 0.0060 \text{ mol} \times \frac{299 \text{ g}}{\text{mol}} = \boxed{1.8 \text{ g } SbCl_5}$$

You should now work Exercises 75 and 78.

Charles Steele

▶ The decomposition reaction

$$2HgO(s) \rightleftharpoons 2Hg(\ell) + O_2(g)$$

The reaction is not at equilibrium here, because O_2 gas has been allowed to escape.

17-11 Heterogeneous Equilibria

Thus far, we have considered only equilibria involving species in a single phase, that is, **homogeneous equilibria.** **Heterogeneous equilibria** involve species in more than one phase. Consider the following reversible reaction at 25°C.

$$2HgO(s) \rightleftharpoons 2Hg(\ell) + O_2(g)$$

A solid, a liquid, and a gas are all present when equilibrium is established for this system. Neither solids nor liquids are significantly affected by changes in pressure. The fundamental definition of the equilibrium constant in thermodynamics is in terms of the activities of the substances involved.

For any pure solid or pure liquid, the activity is taken as 1 (Section 17-2), so terms for pure liquids and pure solids *do not* appear in the K expressions for heterogeneous equilibria.

Thus, for the reaction

$$2HgO(s) \rightleftharpoons 2Hg(\ell) + O_2(g) \qquad K_c = [O_2] \qquad \text{and} \qquad K_P = P_{O_2}$$

These equilibrium constant expressions indicate that equilibrium exists at a given temperature for *one and only one* concentration and one partial pressure of oxygen in contact with liquid mercury and solid mercury(II) oxide.

EXAMPLE 17-15 K_c and K_P for Heterogeneous Equilibrium

Write both K_c and K_P for the following reversible reactions.

(a) $2ZnS(s) + 3O_2(g) \rightleftharpoons 2ZnO(s) + 2SO_2(g)$

(b) $2NH_3(g) + H_2SO_4(\ell) \rightleftharpoons (NH_4)_2SO_4(s)$

(c) $S(s) + H_2SO_3(aq) \rightleftharpoons H_2S_2O_3(aq)$

Plan

We apply the definitions of K_c and K_P to each reaction.

Solution

(a) $K_c = \dfrac{[SO_2]^2}{[O_2]^3}$ $\qquad$ $K_p = \dfrac{(P_{SO_2})^2}{(P_{O_2})^3}$

(b) $K_c = \dfrac{1}{[NH_3]^2} = [NH_3]^{-2}$ $\qquad$ $K_P = \dfrac{1}{(P_{NH_3})^2} = (P_{NH_3})^{-2}$

(c) $K_c = \dfrac{[H_2S_2O_3]}{[H_2SO_3]}$ $\qquad$ K_P undefined; no gases involved

> **S TOP & THINK**
>
> Remember that concentrations or pressures for pure liquids and pure solids do not appear in the equilibrium constant expressions.

EXAMPLE 17-16 Heterogeneous Equilibria

The value of K_P is 27 for the thermal decomposition of potassium chlorate at a given high temperature. What is the partial pressure of oxygen in a closed container in which the following system is at equilibrium at the given temperature? (This can be a dangerous reaction.)

$$2KClO_3(s) \overset{heat}{\rightleftharpoons} 2KCl(s) + 3O_2(g)$$

Plan

Because two solids, $KClO_3$ and KCl, and only one gas, O_2, are involved, we see that K_P involves only the partial pressure of O_2, that is, $K_P = (P_{O_2})^3$.

Solution

We are given

$$K_P = (P_{O_2})^3 = 27$$

Let x atm $= P_{O_2}$ at equilibrium. Then we have

$$(P_{O_2})^3 = 27 = x^3 \qquad x = 3.0 \text{ atm}$$

This tells us that the partial pressure of oxygen at equilibrium is 3.0 atm.

You should now work Exercises 80 and 91.

17-12 Relationship Between ΔG_{rxn}^0 and the Equilibrium Constant

Let's consider in thermodynamic terms what may happen when two substances are mixed together at constant temperature and pressure. First, as a result of mixing, there is usually an increase in entropy due to the increase in disorder. If the two substances can react with each other, the chemical reaction begins, heat is released or absorbed, and the concentrations of the substances in the mixture change. An additional change in entropy, which depends on changes in the nature of the reactants and products, also begins to occur. The evolution or absorption of heat energy, the changes in entropy, and the changes in concentrations all continue until equilibrium is established. Equilibrium may be reached with large amounts of products formed, with virtually all of the reactants remaining unchanged, or at *any* intermediate combination of concentrations.

▶ Thermodynamic standard states are (1) pure solids or pure liquids at 1 atm, (2) solutions of one-molar concentrations, and (3) gases at partial pressures of 1 atm.

The standard free energy change for a reaction is ΔG^0_{rxn}. This is the free energy change that would accompany *complete* conversion of all reactants initially present at standard conditions to *all* products at standard conditions (see Section 15-16). The free energy change for any other concentrations or pressures is ΔG_{rxn} (no superscript zero). The two quantities are related by the equation

$$\Delta G_{rxn} = \Delta G^0_{rxn} + RT \ln Q$$

R is the universal gas constant, T is the absolute temperature, and Q is the reaction quotient. When a system is *at equilibrium*, $\Delta G_{rxn} = 0$ (see Section 15-16) and $Q = K$. Recall that the reaction quotient may represent nonequilibrium concentrations (or partial pressures) of products and reactants. As reaction occurs, the free energy of the mixture and the concentrations change until at equilibrium $\Delta G_{rxn} = 0$, and the concentrations of reactants and products satisfy the equilibrium constant. At that point, Q becomes equal to K (see Section 17-4). Then

$$0 = \Delta G^0_{rxn} + RT \ln K \quad \text{(at equilibrium)}$$

Rearranging gives

STOP & THINK

The energy units of R must match those of ΔG^0. We usually use

$R = 8.314 \text{ J/mol} \cdot \text{K}$

$$\Delta G^0_{rxn} = -RT \ln K$$

This equation shows the relationship between the standard free energy change and the *thermodynamic equilibrium constant*.

For the following generalized reaction, the **thermodynamic equilibrium constant** is defined in terms of the activities of the species involved.

$$a\text{A} + b\text{B} \rightleftharpoons c\text{C} + d\text{D} \quad K = \frac{(a_\text{C})^c (a_\text{D})^d}{(a_\text{A})^a (a_\text{B})^b}$$

where a_A is the activity of substance A, and so on (see Section 17-2). The mass action expression to which it is related involves concentration terms for species in solution and partial pressures for gases.

▶ We will encounter equilibrium calculations for solution and gaseous species in electrochemistry (see Chapter 21).

When the relationship $\Delta G^0_{rxn} = -RT \ln K$ is used with

1. all gaseous reactants and products, K represents K_P;
2. all solution reactants and products, K represents K_c;
3. a mixture of solution and gaseous reactants, K represents the thermodynamic equilibrium constant, and we do not make the distinction between K_P and K_c.

Figure 17-4 displays the relationships between free energy and equilibrium. The *left* end of each curve represents the total free energy of the reactants and the *right* end of each curve represents the total free energy of the products at standard state conditions. The difference between them is ΔG^0_{rxn}; like K, ΔG^0_{rxn} depends only on temperature and is a constant for any given reaction.

From the preceding equation that relates ΔG^0_{rxn} and K, we see that when ΔG^0_{rxn} is negative, $\ln K$ *must* be positive, and K is greater than 1. This tells us that products are favored over reactants at equilibrium. This case is illustrated in Figure 17-4a. When ΔG^0_{rxn} is positive, $\ln K$ *must* be negative, and K is less than 1. This tells us that reactants are favored over products at equilibrium (Figure 17-4b). In the *rare* case of a chemical reaction for which $\Delta G^0_{rxn} = 0$, then $K = 1$ and the numerator and the denominator must be equal in the equilibrium constant expression (i.e., $[\text{C}]^c[\text{D}]^d \cdots = [\text{A}]^a[\text{B}]^b \ldots$). These relationships are summarized as follows.

ΔG^0_{rxn}	K	**Product Formation**
$\Delta G^0_{rxn} < 0$	$K > 1$	Products favored over reactants at equilibrium
$\Delta G^0_{rxn} = 0$	$K = 1$	At equilibrium when $[\text{C}]^c[\text{D}]^d \ldots = [\text{A}]^a[\text{B}]^b \ldots$ (very rare)
$\Delta G^0_{rxn} > 0$	$K < 1$	Reactants favored over products at equilibrium

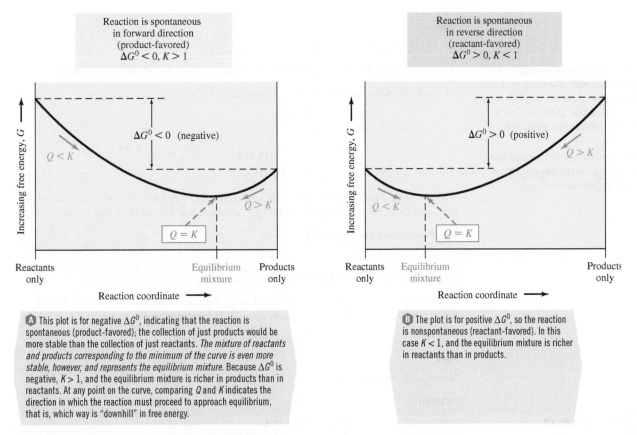

Figure 17-4 Variation in total free energy for a reversible reaction carried out at constant T. The *standard free energy change*, ΔG^0, represents the free energy change for the *complete* conversion of reactants into products (both at standard state concentrations: 1 M or 1 atm).

The direction of approach to equilibrium and the actual free energy change (ΔG_{rxn}) are *not* constants. They vary with the conditions and the initial concentrations. If the initial concentrations correspond to $Q < K$, equilibrium is approached from left to right on the curves in Figure 17-4, and the forward reaction predominates. If $Q > K$, equilibrium is approached from right to left, and the reverse reaction predominates.

The magnitude of ΔG^0_{rxn} indicates the *extent* to which a chemical reaction occurs under standard state conditions, that is, how far the reaction goes toward the formation of products before equilibrium is reached. The more negative the ΔG^0_{rxn} value, the larger is the value of K and the more favorable is the formation of products. We think of some reactions as going "to completion." These generally have *very negative* ΔG^0_{rxn} values. The more positive the ΔG^0_{rxn} value, the smaller is the value of K and the less favorable is the formation of products.

EXAMPLE 17-17 *K* Versus ΔG^0_{rxn}

Use the data in Appendix K to calculate K_P for the following reaction at 25°C.

$$2N_2O(g) \rightleftharpoons 2N_2(g) + O_2(g)$$

Plan

The temperature is 25°C, so we evaluate ΔG^0_{rxn} for the reaction from ΔG^0_f values in Appendix K. The reaction involves only gases, so K is K_P. This means that $\Delta G^0_{rxn} = -RT \ln K_P$. We solve for K_P.

S TOP & THINK

Remember (Section 15-16) that if the temperature is *not* 25°C, you must calculate ΔG^0 for the reaction from the equation

$$\Delta G^0 = \Delta H^0 - T\Delta S^0$$

using tabulated values of ΔH^0_f and S^0 as given in Appendix K. Review Examples 15-19 and 15-20.

Solution

$$\Delta G^0_{rxn} = [2\Delta G^0_{fN_2(g)} + \Delta G^0_{fO_2(g)}] - [2\Delta G^0_{fN_2O(g)}]$$

$$= [2(0) + 0] - [2(104.2)] = -208.4 \text{ kJ/mol, or } -2.084 \times 10^5 \text{ J/mol}$$

This is a gas-phase reaction, so ΔG^0_{rxn} is related to K_P by

$$\Delta G^0_{rxn} = -RT \ln K_P$$

$$\ln K_P = \frac{\Delta G^0_{rxn}}{-RT} = \frac{-2.084 \times 10^5 \text{ J/mol}}{-(8.314 \text{ J/mol·K})(298K)} = 84.1$$

$$K_P = e^{84.1} = 3.3 \times 10^{36}$$

The very large value of K_P tells us that the equilibrium lies *very* far to the right. This reaction is so slow at 25°C, however, that very little N_2O decomposes to N_2 and O_2 at that temperature.

You should now work Exercise 84.

▶ Units cancel when we express ΔG^0 in joules per mole. We interpret this as meaning "per mole of reaction"—that is, for the number of moles of each substance shown in the balanced equation.

▶ On some calculators, we evaluate e^x as follows: Enter the value of x, then press INV followed by ln x .

EXAMPLE 17-18 *K* Versus ΔG^0_{rxn}

In Examples 15-19 and 15-20 we evaluated ΔG^0_{rxn} for the following reaction at 25°C and found it to be $+173.1$ kJ/mol. Calculate K_P at 25°C for this reaction.

$$N_2(g) + O_2(g) \rightleftharpoons 2NO(g)$$

Plan

In this example we use $\Delta G^0_{rxn} = -RT \ln K_P$.

Solution

$$\Delta G^0_{rxn} = -RT \ln K_P$$

$$\ln K_P = \frac{\Delta G^0_{rxn}}{-RT} = \frac{1.731 \times 10^5 \text{ J/mol}}{-(8.314 \text{ J/mol·K})(298 \text{ K})}$$

$$= -69.9$$

$$K_P = e^{-69.9} = 4.4 \times 10^{-31}$$

This very small number indicates that at equilibrium almost no N_2 and O_2 are converted to NO at 25°C. For all practical purposes, the reaction does not occur at 25°C.

You should now work Exercise 86.

S TOP & THINK

The energy units in ΔG, often given in kJ, must be converted to J to match those of R.

A very important application of the relationships in this section is the use of measured *K* values to calculate ΔG^0_{rxn}.

EXAMPLE 17-19 *K* Versus ΔG^0_{rxn}

The equilibrium constant, K_P, for the following reaction is 5.04×10^{17} at 25°C. Calculate ΔG^0_{298} for the hydrogenation of ethylene to form ethane.

$$C_2H_4(g) + H_2(g) \rightleftharpoons C_2H_6(g)$$

Plan

We use the relationship between ΔG^0 and K_P.

Solution

$$\Delta G^0_{298} = -RT \ln K_P$$

$$= -(8.314 \text{ J/mol·K})(298 \text{ K}) \ln(5.04 \times 10^{17})$$

$$= -1.01 \times 10^5 \text{ J/mol}$$

$$= -101 \text{ kJ/mol}$$

You should now work Exercise 88(a).

17-13 Evaluation of Equilibrium Constants at Different Temperatures

Chemists have determined equilibrium constants for thousands of reactions. It would be an impossibly huge task to catalog such constants for each reaction at every temperature of interest. Fortunately, there is no need to do this. If we determine the equilibrium constant, K_{T_1}, for a reaction at one temperature, T_1, and also its ΔH^0, we can then estimate the equilibrium constant at a second temperature, T_2, using the **van't Hoff equation**.

$$\ln\left(\frac{K_{T_2}}{K_{T_1}}\right) = \frac{\Delta H^0}{R}\left(\frac{1}{T_1} - \frac{1}{T_2}\right)$$

Thus, if we know ΔH^0 for a reaction and K at a given temperature (say 298 K), we can use the van't Hoff equation to calculate the value of K at any other temperature.

▶ The van't Hoff equation assumes that ΔH^0 does not change with temperature. This is not strictly true, but the variation is generally quite small, so this equation gives good estimates of K_T.

▶ Compare the form of the van't Hoff equation to the Arrhenius equation (see Section 16-8) and to the Clausius–Clapeyron equation (see Section 13-9).

EXAMPLE 17-20 Evaluation of K_P at Different Temperatures

We found in Example 17-18 that $K_P = 4.4 \times 10^{-31}$ at 25°C (298 K) for the following reaction. $\Delta H^0 = 180.5$ kJ/mol for this reaction. Evaluate K_P at 2400. K.

$$N_2(g) + O_2(g) \rightleftharpoons 2NO(g)$$

Plan

We are given K_P at one temperature, 25°C, and the value of ΔH^0. We are given the second temperature, 2400. K. These data allow us to evaluate the right side of the van't Hoff equation, which gives us $\ln(K_{T_2}/K_{T_1})$. Because we know K_{T_1}, we can find the value for K_{T_2}.

Solution

Let $T_1 = 298$ K and $T_2 = 2400.$ K. Then

$$\ln\left(\frac{K_{T_2}}{K_{T_1}}\right) = \frac{\Delta H^0}{R}\left(\frac{1}{T_1} - \frac{1}{T_2}\right)$$

Let us first evaluate the right side of the equation.

$$\ln\left(\frac{K_{T_2}}{K_{T_1}}\right) = \frac{1.805 \times 10^5 \text{ J/mol}}{8.314 \text{ J/mol} \cdot \text{K}}\left(\frac{1}{298 \text{ K}} - \frac{1}{2400. \text{ K}}\right) = 63.8$$

Now take the inverse logarithm of both sides.

$$\frac{K_{T_2}}{K_{T_1}} = e^{63.8} = 5.1 \times 10^{27}$$

Solving for K_{T_2} and substituting the known value of K_{T_1}, we obtain

$$K_{T_2} = (5.1 \times 10^{27})(K_{T_1}) = (5.1 \times 10^{27})(4.4 \times 10^{-31}) = \boxed{2.2 \times 10^{-3}} \text{ at 2400. K}$$

You should now work Exercise 88(b–f).

▶ 2400 K is a typical temperature inside the combustion chambers of automobile engines. Large quantities of N_2 and O_2 are present during gasoline combustion because the gasoline is mixed with air.

S TOP & THINK
The positive value of ΔH^0 tells us that the reaction is endothermic. Thus, it is reasonable that the equilibrium constant is greater (2.2×10^{-3} compared to 4.4×10^{-31}) at the higher temperature of 2400. K.

⚙ **Problem-Solving Tip** Use the Correct K

The K values that appear in the van't Hoff equation represent the *thermodynamic equilibrium constant* (see Section 17-12). For a gas-phase reaction (such as that in Example 17-20), K represents K_P; if the value of K_c were given, we would have to convert it to K_P (see Section 17-10) before using the van't Hoff equation.

▶ The K_P value could be converted to K_c using the relationship $K_c = K_P(RT)^{-\Delta n}$ (see Section 17-10).

In Example 17-20 we see that K_{T_2} (K_P at 2400. K) is quite small, which tells us that the equilibrium favors N_2 and O_2 rather than NO. Nevertheless, K_{T_2} is very much larger than K_{T_1}, which is 4.4×10^{-31}. At 2400. K, significantly more NO is present at equilibrium relative to N_2 and O_2 than at 298 K. So automobiles emit small amounts of NO into the atmosphere that are sufficient to cause severe air pollution problems. Catalytic converters (see Section 16–9) are designed to catalyze the breakdown of NO into N_2 and O_2.

$$2NO(g) \rightleftharpoons N_2(g) + O_2(g)$$

This reaction is spontaneous. Catalysts do not shift the position of equilibrium. They favor neither consumption nor production of NO. They merely allow the system to reach equilibrium more rapidly. The time factor is very important because the NO stays in the automobile exhaust system for only a very short time.

KEY TERMS

Activity (of a component of an ideal mixture) A dimensionless quantity whose magnitude is equal to molar concentration in an ideal solution, equal to partial pressure (in atmospheres) in an ideal gas mixture, and defined as 1 for pure solids or liquids.

Chemical equilibrium A state of dynamic balance in which the rates of forward and reverse reactions are equal; there is no net change in concentrations of reactants or products while a system is at equilibrium.

Dynamic equilibrium An equilibrium in which processes occur continuously, with no *net* change.

Equilibrium constant (K) A dimensionless quantity that indicates the extent to which a reversible reaction occurs. K varies with temperature.

Heterogeneous equilibria Equilibria involving species in more than one phase.

Homogeneous equilibria Equilibria involving only species in a single phase, that is, all gases, all liquids, or all solids.

K_c Equilibrium constant with amounts of reactants and products expressed as molar concentrations.

K_P Equilibrium constant with amounts of reactants and products expressed as partial pressures.

LeChatelier's Principle If a stress (change of conditions) is applied to a system at equilibrium, the system shifts in the direction that reduces the stress, to move toward a new state of equilibrium.

Mass action expression For a reversible reaction,

$$aA + bB \rightleftharpoons cC + dD$$

the product of the molar concentrations of the products (species on the right), each raised to the power that corresponds to its coefficient in the balanced chemical equation, divided by the product of the concentrations of the reactants (species on the left), each raised to the power that corresponds to its coefficient in the balanced chemical equation. At equilibrium the mass action expression is equal to K; at other conditions, it is Q.

$$\frac{[C]^c[D]^d}{[A]^a[B]^b} = Q \text{ or, at equilibrium, } K_c$$

Reaction quotient (Q) The mass action expression under any set of conditions (not necessarily equilibrium); its magnitude relative to K determines the direction in which reaction must occur to establish equilibrium.

Reversible reactions Reactions that do not go to completion and occur in both the forward and reverse directions.

Thermodynamic equilibrium constant The equilibrium constant defined in terms of activities of the products and reactants.

van't Hoff equation The relationship between ΔH^0 for a reaction and its equilibrium constants at two different temperatures.

EXERCISES

🔵 **Molecular Reasoning** exercises

▲ **More Challenging** exercises

Blue-Numbered exercises are solved in the Student Solutions Manual

---

Basic Concepts

1. 🔵 Define and illustrate the following terms: (a) reversible reaction; (b) static equilibrium; (c) equilibrium constant.

2. Equilibrium constants do not have units. Explain.

3. 🔵 Distinguish between the terms *static equilibrium* and *dynamic equilibrium*. Which kind does a chemical equilibrium represent?

4. 🔵 (a) Describe three examples of static equilibrium. (b) Describe three examples of dynamic equilibrium (besides chemical equilibrium).

🔵 **Molecular Reasoning** exercises　　▲ **More Challenging** exercises　　Blue-Numbered exercises solved in Student Solutions Manual

5. 🐾 Explain the significance of (a) a very large value of K, (b) a very small value of K, and (c) a value of K of about 1.

6. 🐾 What can be said about the magnitude of the equilibrium constant in a reaction whose equilibrium lies far to the right? to the left?

7. 🐾 What is the relationship between equilibrium and the rates of opposing processes?

8. 🐾 What does the value of an equilibrium constant tell us about the time required for the reaction to reach equilibrium?

9. When giving the value of an equilibrium constant, it is necessary also to write the balanced chemical equation. Why? Give examples to illustrate your explanation.

10. (a) How is the equilibrium constant related to the forward and reverse rate constants? (b) Can the rate expressions for forward and reverse reactions be written from the balanced chemical equation? Explain. (c) Can the equilibrium constant expression be written from the balanced chemical equation? Explain.

11. (a) Sketch a set of curves similar to those in Figure 17-2 for concentration changes over time for a reaction

$$3A(g) + B(g) \rightleftharpoons C(g) + 2D(g)$$

assuming that K is much greater than 1. In each case, assume that A and B start at the same concentration and that no C or D are present. (b) Repeat part (a) for the case that K is much less than 1.

12. (a) Sketch a set of curves similar to those in Figure 17-2 for concentration changes over time for a reaction

$$2A(g) + B(g) \rightleftharpoons 2C(g)$$

assuming that K is much greater than 1. In each case, assume that A and B start at the same concentration and that no C is present. (b) Repeat part (a) for the case that K is much less than 1.

13. At some temperature, the reaction

$$N_2(g) + 3H_2(g) \rightleftharpoons 2NH_3(g)$$

has an equilibrium constant K_c numerically equal to 1. State whether each of the following is true or false, and explain why. (a) An equilibrium mixture must have the H_2 concentration three times that of N_2 and the NH_3 concentration twice that of H_2. (b) An equilibrium mixture must have the H_2 concentration three times that of N_2. (c) A mixture in which the H_2 concentration is three times that of N_2 *and* the NH_3 concentration is twice that of N_2 could be an equilibrium mixture. (d) A mixture in which the concentration of each reactant and each product is 1 M is an equilibrium mixture. (e) Any mixture in which the concentrations of all reactants and products are equal is an equilibrium mixture. (f) An equilibrium mixture must have equal concentrations of all reactants and products.

14. Why do we omit concentrations of pure solids and pure liquids from equilibrium constant expressions?

15. Consider the following compounds, in the states indicated, as possible reactants or products in a chemical reaction. Which of these compounds would be omitted from the equilibrium constant expression? Explain. $H_2O(s)$, $H_2O(\ell)$, $H_2O(g)$, $HCl(g)$, $HCl(aq)$, $NaHCO_3(s)$, $CH_3OH(\ell)$, $CH_3OH(aq)$, $Cl_2(aq)$, $N_2(g)$, $NH_3(\ell)$, $CO(g)$, and $Fe_2O_3(s)$.

16. Consider the following compounds, in the states indicated, as possible reactants or products in a chemical reaction. Which of these compounds would be omitted from the equilibrium constant expression? Explain. $CaCO_3(s)$, $H_2SO_4(\ell)$, $NaOH(s)$, $NaOH(aq)$, $O_2(g)$, $CH_3COOH(\ell)$, $CH_3COOH(aq)$, $HI(g)$, $I_2(s)$, $C(graphite)$, and $SO_3(g)$.

Equilibrium Constant Expression and Value of *K*

17. Write the expression for K_c for each of the following reactions:
(a) $CO(g) + H_2O(g) \rightleftharpoons CO_2(g) + H_2(g)$
(b) $SrCO_3(s) \rightleftharpoons SrO(s) + CO_2(g)$
(c) $2CHCl_3(g) + 3H_2(g) \rightleftharpoons 2CH_4(g) + 3Cl_2(g)$
(d) $H_2(g) + I_2(g) \rightleftharpoons 2HI(g)$
(e) $2NOCl(g) \rightleftharpoons 2NO(g) + Cl_2(g)$

18. Write the expression for K_c for each of the following reactions:
(a) $2H_2O(g) + 2SO_2(g) \rightleftharpoons 2H_2S(g) + 3O_2(g)$
(b) $4NH_3(g) + 5O_2(g) \rightleftharpoons 4NO(g) + 6H_2O(g)$
(c) $PCl_3(g) + Cl_2(g) \rightleftharpoons PCl_5(g)$
(d) $NaF(s) + H_2SO_4(\ell) \rightleftharpoons NaHSO_4(s) + HF(g)$
(e) $2SO_3(g) \rightleftharpoons 2SO_2(g) + O_2(g)$

19. Write the expression for K_c for each of the following reactions:
(a) $2CO(g) + O_2(g) \rightleftharpoons 2CO_2(g)$
(b) $2NO_2(g) \rightleftharpoons 2NO(g) + O_2(g)$
(c) $2HBr(g) \rightleftharpoons H_2(g) + Br_2(\ell)$
(d) $P_4(g) + 3O_2(g) \rightleftharpoons P_4O_6(s)$
(e) $N_2(g) + O_2(g) \rightleftharpoons 2NO(g)$

20. Write the expression for K_c for each of the following reactions:
(a) $2H_2O_2(g) \rightleftharpoons 2H_2O(g) + O_2(g)$
(b) $2ZnS(s) + 3O_2(g) \rightleftharpoons 2ZnO(s) + 2SO_2(g)$
(c) $NH_3(g) + HCl(g) \rightleftharpoons NH_4Cl(s)$
(d) $N_2O_4(g) \rightleftharpoons 2NO_2(g)$
(e) $2Cl_2(g) + 2H_2O(g) \rightleftharpoons 4HCl(g) + O_2(g)$

21. Write the expression for K_c for each of the following reactions:
(a) $TlCl_3(s) \rightleftharpoons TlCl(s) + Cl_2(g)$
(b) $CuCl_4^{2-}(aq) \rightleftharpoons Cu^{2+}(aq) + 4Cl^-(aq)$
(c) $2O_3(g) \rightleftharpoons 3O_2(g)$
(d) $4H_3O^+(aq) + 2Cl^-(aq) + MnO_2(s) \rightleftharpoons$
$Mn^{2+}(aq) + 6H_2O(\ell) + Cl_2(aq)$

22. On the basis of the equilibrium constant values, choose the reactions in which the *products* are favored.
(a) $NH_3(aq) + H_2O(\ell) \rightleftharpoons NH_4^+(aq) + OH^-(aq)$
$K = 1.8 \times 10^{-5}$
(b) $Au^+(aq) + 2CN^-(aq) \rightleftharpoons [Au(CN)_2]^-(aq)$
$K = 2 \times 10^{38}$
(c) $PbC_2O_4(s) \rightleftharpoons Pb^{2+}(aq) + C_2O_4^{2-}(aq)$ $K = 4.8$
(d) $HS^-(aq) + H^+(aq) \rightleftharpoons H_2S(aq)$ $K = 1.0 \times 10^7$

23. On the basis of the equilibrium constant values, choose the reactions in which the *reactants* are favored.
(a) $H_2O(\ell) \rightleftharpoons H^+(aq) + OH^-(aq)$ $K = 1.0 \times 10^{-14}$
(b) $[AlF_6]^{3-}(aq) \rightleftharpoons Al^{3+}(aq) + 6F^-(aq)$ $K = 2 \times 10^{-24}$
(c) $Ca_3(PO_4)_2(s) \rightleftharpoons 3Ca^{2+}(aq) + 2PO_4^{3-}(aq)$
$K = 1 \times 10^{-25}$
(d) $2Fe^{3+}(aq) + 3S^{2-}(aq) \rightleftharpoons Fe_2S_3(s)$ $K = 1 \times 10^{88}$

🐾 **Molecular Reasoning** exercises ▲ **More Challenging** exercises Blue-Numbered exercises solved in Student Solutions Manual

Calculation of K

24. The reaction between nitrogen and oxygen to form $NO(g)$ is represented by the chemical equation

$$N_2(g) + O_2(g) \rightleftharpoons 2NO(g)$$

Equilibrium concentrations of the gases at 1500 K are 6.4×10^{-3} mol/L for N_2, 1.7×10^{-3} mol/L for O_2, and 1.1×10^{-5} mol/L for NO. Calculate the value of K_c at 1500 K from these data.

25. At elevated temperatures, BrF_5 establishes the following equilibrium.

$$2BrF_5(g) \rightleftharpoons Br_2(g) + 5F_2(g)$$

The equilibrium concentrations of the gases at 1500 K are 0.0064 mol/L for BrF_5, 0.0018 mol/L for Br_2, and 0.0090 mol/L for F_2. Calculate the value of K_c.

26. At some temperature the reaction

$$PCl_3(g) + Cl_2(g) \rightleftharpoons PCl_5(g)$$

is at equilibrium when the concentrations of PCl_3, Cl_2, and PCl_5 are 10.0, 9.0, and 12.0 mol/L, respectively. Calculate the value of K_c for this reaction at that temperature.

27. For the reaction

$$CO(g) + H_2O(g) \rightleftharpoons CO_2(g) + H_2(g)$$

the value of the equilibrium constant, K_c, is 1.845 at a given temperature. We place 0.500 mole CO and 0.500 mole H_2O in a 1.00-L container at this temperature and allow the reaction to reach equilibrium. What will be the equilibrium concentrations of all substances present?

28. Given: $A(g) + B(g) \rightleftharpoons C(g) + 2D(g)$
One mole of A and one mole of B are placed in a 0.400-liter container. After equilibrium has been established, 0.20 mole of C is present in the container. Calculate the equilibrium constant, K_c, for the reaction.

29. The reaction

$$2NO_2(g) \rightleftharpoons N_2O_4(g)$$

has an equilibrium constant, K_c, of 170 at 25°C. If 2.0×10^{-3} mol of NO_2 is present in a 10.-L flask along with 1.5×10^{-3} mol of N_2O_4, is the system at equilibrium? If it is not at equilibrium, does the concentration of NO_2 increase or decrease as the system proceeds to equilibrium?

30. For the following equation, $K_c = 7.9 \times 10^{11}$ at 500. K.

$$H_2(g) + Br_2(g) \rightleftharpoons 2HBr(g)$$

(a) $\frac{1}{2}H_2(g) + \frac{1}{2}Br_2(g) \rightleftharpoons HBr(g)$ $K_c = ?$
(b) $2HBr(g) \rightleftharpoons H_2(g) + Br_2(g)$ $K_c = ?$
(c) $4HBr(g) \rightleftharpoons 2H_2(g) + 2Br_2(g)$ $K_c = ?$

31. Nitrosyl chloride, NOCl, decomposes to NO and Cl_2 at high temperatures.

$$2NOCl(g) \rightleftharpoons 2NO(g) + Cl_2(g)$$

Suppose you place 2.00 mol NOCl in a 1.0-L flask and raise the temperature to 462°C. When equilibrium has been established, 0.66 mol NO is present. Calculate the equilibrium constant, K_c, for the decomposition reaction from these data.

32. An equilibrium mixture contains 3.00 mol CO, 2.00 mol Cl_2, and 9.00 mol $COCl_2$ in a 50.-L reaction flask at 800. K. Calculate the value of the equilibrium constant, K_c, for the reaction $CO(g) + Cl_2(g) \rightleftharpoons COCl_2(g)$ at this temperature.

33. NO and O_2 are mixed in a container of fixed volume kept at 1000 K. Their initial concentrations are 0.0200 mol/L and 0.0300 mol/L, respectively. When the reaction

$$2NO(g) + O_2(g) \rightleftharpoons 2NO_2(g)$$

has come to equilibrium, the concentration of NO_2 is 2.2×10^{-3} mol/L. Calculate (a) the concentration of NO at equilibrium, (b) the concentration of O_2 at equilibrium, and (c) the equilibrium constant, K_c, for the reaction.

34. Antimony pentachloride decomposes in a gas-phase reaction at high temperatures.

$$SbCl_5(g) \rightleftharpoons SbCl_3(g) + Cl_2(g)$$

(a) At some temperature, an equilibrium mixture in a 5.00-L container is found to contain 6.91 g of $SbCl_5$, 16.45 g of $SbCl_3$, and 5.11 g of Cl_2. Evaluate K_c. (b) If 25.0 grams of $SbCl_5$ is placed in the 5.00-L container and allowed to establish equilibrium at the temperature in part (a), what will be the equilibrium concentrations of all species?

35. At standard temperature and pressure, the reaction indicated by the following equation has an equilibrium constant, K_c, equal to 0.021.

$$2HI(g) \rightleftharpoons H_2(g) + I_2(g)$$

Calculate the equilibrium constant, K_c, for the reverse equation.

36. The following reaction has an equilibrium constant, K_c, equal to 1538 at 1800.°C.

$$2NO(g) + O_2(g) \rightleftharpoons 2NO_2(g)$$

Calculate the equilibrium constant, K_c, for the reverse equation.

The Reaction Quotient, Q

37. Define the reaction quotient, Q. Distinguish between Q and K.
38. Why is it useful to compare Q with K? What is the situation when (a) $Q = K$? (b) $Q < K$? (c) $Q > K$?
39. ⚛ How does the form of the reaction quotient compare with that of the equilibrium constant? What is the difference between these two expressions?
40. ⚛ If the reaction quotient is larger than the equilibrium constant, what will happen to the reaction? What will happen if $Q < K$?
41. $K_c = 19.9$ for the reaction

$$Cl_2(g) + F_2(g) \rightleftharpoons 2ClF(g)$$

What will happen in a reaction mixture originally containing $[Cl_2] = 0.5$ mol/L, $[F_2] = 0.2$ mol/L, and $[ClF] = 7.3$ mol/L?

42. The concentration equilibrium constant for the gas-phase reaction

$$H_2CO \rightleftharpoons H_2 + CO$$

⚛ **Molecular Reasoning** exercises   ▲ **More Challenging** exercises   Blue-Numbered exercises solved in Student Solutions Manual

Unless otherwise noted, all content on this page is © Cengage Learning.

has the numerical value 0.50 at a given temperature. A mixture of H_2CO, H_2, and CO is introduced into a flask at this temperature. After a short time, analysis of a small sample of the reaction mixture shows the concentrations to be $[H_2CO] = 0.50$ M, $[H_2] = 0.80$ M, and $[CO] = 0.25$ M. Classify each of the following statements about this reaction mixture as true or false.

(a) The reaction mixture is at equilibrium.

(b) The reaction mixture is not at equilibrium, but no further reaction will occur.

(c) The reaction mixture is not at equilibrium, but will move toward equilibrium by using up more H_2CO.

(d) The forward rate of this reaction is the same as the reverse rate.

43. The value of K_c at 25°C for

$$C(graphite) + CO_2(g) \rightleftharpoons 2CO(g)$$

is 3.7×10^{-23}. Describe what will happen if 3.5 mol of CO and 3.5 mol of CO_2 are mixed in a 1.5-L graphite container with a suitable catalyst to make the reaction "go" at this temperature.

Uses of the Equilibrium Constant, K_c

44. For the reaction described by the equation

$$N_2(g) + C_2H_2(g) \rightleftharpoons 2HCN(g)$$

$K_c = 2.3 \times 10^{-4}$ at 300.°C. What is the equilibrium concentration of hydrogen cyanide if the initial concentrations of N_2 and acetylene (C_2H_2) were 3.5 mol/L and 2.5 mol/L, respectively?

45. The equilibrium constant, K_c, for the reaction

$$Br_2(g) + F_2(g) \rightleftharpoons 2BrF(g)$$

is 55.3. What are the equilibrium concentrations of all these gases if the initial concentrations of bromine and fluorine were both 0.240 mol/L?

46. $K_c = 96.2$ at 400. K for the reaction

$$PCl_3(g) + Cl_2(g) \rightleftharpoons PCl_5(g)$$

What is the concentration of Cl_2 at equilibrium if the initial concentrations were 0.24 mol/L for PCl_3 and 5.5 mol/L for Cl_2?

47. $K_c = 5.85 \times 10^{-3}$ at 25°C for the reaction

$$N_2O_4(g) \rightleftharpoons 2NO_2(g)$$

Twenty-two (22.0) grams of N_2O_4 is confined in a 5.00-L flask at 25°C. Calculate (a) the number of moles of NO_2 present at equilibrium, and (b) the percentage of the original N_2O_4 that is dissociated.

48. The reaction of iron and water vapor results in an equilibrium

$$3Fe(s) + 4H_2O(g) \rightleftharpoons Fe_3O_4(s) + 4H_2(g)$$

and an equilibrium constant, K_c, of 4.6 at 850.°C. What is the concentration of hydrogen present at equilibrium if the reaction is initiated with 24 g of H_2O and excess Fe in a 10.0-liter container?

49. The reaction of iron and water vapor results in an equilibrium

$$3Fe(s) + 4H_2O(g) \rightleftharpoons Fe_3O_4(s) + 4H_2(g)$$

and an equilibrium constant, K_c, of 4.6 at 850.°C. What is the concentration of water present at equilibrium if the reaction is initiated with 7.5 g of H_2 and excess iron oxide, Fe_3O_4, in a 15.0-liter container?

50. Carbon dioxide reacts with hot carbon in the form of graphite. The equilibrium constant, K_c, for the reaction is 10.0 at 850.°C.

$$CO_2(g) + C(graphite) \rightleftharpoons 2CO(g)$$

If 24.5 g of carbon monoxide is placed in a 2.50-L graphite container and heated to 850.°C, what is the mass of carbon dioxide at equilibrium?

51. Carbon dioxide reacts with hot carbon in the form of graphite. The equilibrium constant, K_c, for the reaction is 10.0 at 850.°C.

$$CO_2(g) + C(graphite) \rightleftharpoons 2CO(g)$$

If 25.0 g of carbon dioxide and 55.0 g of graphite are placed in a 2.50-L reaction vessel and heated to 850.°C, what is the mass of carbon monoxide at equilibrium?

52. A 75.7-gram sample of HI was placed in a 1.50-L reaction vessel and allowed to come to equilibrium as illustrated in the following equation

$$2HI(g) \rightleftharpoons H_2(g) + I_2(g)$$

The equilibrium constant, K_c, is 0.830. Calculate the concentration of each species present at equilibrium.

Factors That Influence Equilibrium

53. State LeChatelier's Principle. Which factors have an effect on a system at equilibrium? How does the presence of a catalyst affect a system at chemical equilibrium? Explain your answer.

54. What will be the effect of increasing the total pressure on the equilibrium conditions for (a) a chemical equation that has more moles of gaseous products than gaseous reactants, (b) a chemical equation that has more moles of gaseous reactants than gaseous products, (c) a chemical equation that has the same number of moles of gaseous reactants and gaseous products, and (d) a chemical equation in which all reactants and products are pure solids, pure liquids, or in an aqueous solution?

55. Suppose the following exothermic reaction is allowed to reach equilibrium.

$$A(g) + 3B(g) \rightleftharpoons 2C(g) + 3D(g)$$

Then we make each of the following changes and allow the reaction to reestablish equilibrium. Tell whether the amount of B present at the new equilibrium will be (i) greater than, (ii) less than, or (iii) the same as the amount of B before the change was imposed.

(a) The temperature is decreased while the volume is kept constant;
(b) more A is added;
(c) more C is added;
(d) a small amount of D is removed;
(e) the pressure is increased by decreasing the volume.

56. ☁ Suppose the following exothermic reaction is allowed to reach equilibrium.

$$A(g) + 3B(g) \rightleftharpoons 2C(g) + 3D(g)$$

Then we make each of the following changes and allow the reaction to reestablish equilibrium. For each change, tell whether the value of the equilibrium constant will be (i) greater than, (ii) less than, or (iii) the same as before the change was imposed.
(a) The temperature is increased while the volume is kept constant;
(b) more A is added;
(c) more C is added;
(d) a small amount of D is removed;
(e) the pressure is decreased by increasing the volume.

57. ☁ What would be the effect on an equilibrium mixture of Br_2, F_2, and BrF_5 if the total pressure of the system were decreased?

$$2BrF_5(g) \rightleftharpoons Br_2(g) + 5F_2(g)$$

58. ☁ What would be the effect on an equilibrium mixture of carbon, oxygen, and carbon monoxide if the total pressure of the system were decreased?

$$2C(s) + O_2(g) \rightleftharpoons 2CO(g)$$

59. ☁ A weather indicator can be made with a hydrate of cobalt(II) chloride, which changes color as a result of the following reaction.

$$\underset{\text{pink}}{[Co(OH_2)_6]Cl_2(s)} \rightleftharpoons \underset{\text{blue}}{[Co(OH_2)_4]Cl_2(s)} + 2H_2O(g)$$

Does a pink color indicate "moist" or "dry" air? Explain.

Cobalt(II) chloride hexahydrate

60. ☁ Predict whether the equilibrium for the photosynthesis reaction described by the equation

$$6CO_2(g) + 6H_2O(\ell) \rightleftharpoons C_6H_{12}O_6(s) + 6O_2(g)$$
$$\Delta H^0 = 2801.69 \text{ kJ/mol}$$

would (i) shift to the right, (ii) shift to the left, or (iii) remain unchanged if (a) $[CO_2]$ were decreased; (b) P_{O_2} were increased; (c) one half of the $C_6H_{12}O_6$ were removed; (d) the total pressure were decreased; (e) the temperature were increased; (f) a catalyst were added.

61. ☁ What would be the effect of increasing the temperature on each of the following systems at equilibrium?
(a) $H_2(g) + I_2(g) \rightleftharpoons 2HI(g)$ $\Delta H^0 = -9.45$ kJ/mol
(b) $PCl_5(g) \rightleftharpoons PCl_3(g) + Cl_2(g)$ $\Delta H^0 = 92.5$ kJ/mol
(c) $2SO_2(g) + O_2(g) \rightleftharpoons 2SO_3(g)$ $\Delta H^0 = -198$ kJ/mol
(d) $2NOCl(g) \rightleftharpoons 2NO(g) + Cl_2(g)$ $\Delta H^0 = 75$ kJ/mol
(e) $C(s) + H_2O(g) \rightleftharpoons CO(g) + H_2(g)$ $\Delta H^0 = 131$ kJ/mol

62. What would be the effect of increasing the pressure by decreasing the volume on each of the following systems at equilibrium?
(a) $2CO(g) + O_2(g) \rightleftharpoons 2CO_2(g)$
(b) $2NO(g) \rightleftharpoons N_2(g) + O_2(g)$
(c) $N_2O_4(g) \rightleftharpoons 2NO_2(g)$
(d) $Ni(s) + 4CO(g) \rightleftharpoons Ni(CO)_4(g)$
(e) $N_2(g) + 3H_2(g) \rightleftharpoons 2NH_3(g)$

63. Consider the system

$$4NH_3(g) + 3O_2(g) \rightleftharpoons 2N_2(g) + 6H_2O(\ell)$$
$$\Delta H = -1530.4 \text{ kJ}$$

How will the amount of ammonia at equilibrium be affected by
(a) removing $O_2(g)$?
(b) adding $N_2(g)$?
(c) adding water?
(d) expanding the container at constant pressure?
(e) increasing the temperature?

64. Given: $A(g) + B(g) \rightleftharpoons C(g) + D(g)$
(a) At equilibrium a 1.00-liter container was found to contain 1.60 moles of C, 1.60 moles of D, 0.40 mole of A, and 0.40 mole of B. Calculate the equilibrium constant for this reaction.
(b) If 0.20 mole of B and 0.20 mole of C are added to this system, what will the new *equilibrium* concentration of A be?

65. Given: $A(g) + B(g) \rightleftharpoons C(g) + D(g)$
When one mole of A and one mole of B are mixed and allowed to reach equilibrium at room temperature, the mixture is found to contain $\frac{2}{3}$ mole of C.
(a) Calculate the equilibrium constant.
(b) If two moles of A were mixed with two moles of B and allowed to reach equilibrium, how many moles of C would be present at equilibrium?

66. Given: $A(g) \rightleftharpoons B(g) + C(g)$
(a) When the system is at equilibrium at 200.°C, the concentrations are found to be: $[A] = 0.30 \, M$, $[B] = [C] = 0.25 \, M$. Calculate K_c.
(b) If the volume of the container in which the system is at equilibrium is suddenly doubled at 200.°C, what will be the new equilibrium concentrations?
(c) Refer back to part (a). If the volume of the container is suddenly halved at 200.°C, what will be the new equilibrium concentrations?

67. ▲ The equilibrium constant, K_c, for the dissociation of phosphorus pentachloride is 9.3×10^{-2} at 252°C. How many moles and grams of PCl_5 must be added to a 2.0-liter flask to obtain a Cl_2 concentration of 0.17 M?

$$PCl_5(g) \rightleftharpoons PCl_3(g) + Cl_2(g)$$

☁ **Molecular Reasoning** exercises   ▲ **More Challenging** exercises   Blue-Numbered exercises solved in Student Solutions Manual

68. At 25°C, K_c is 5.84×10^{-3} for the dissociation of dinitrogen tetroxide to nitrogen dioxide.

$$N_2O_4(g) \rightleftharpoons 2NO_2(g)$$

(a) Calculate the equilibrium concentrations of both gases when 4.00 grams of N_2O_4 is placed in a 2.00-liter flask at 25°C.
(b) What will be the new equilibrium concentrations if the volume of the system is suddenly increased to 3.00 liters at 25°C?
(c) What will be the new equilibrium concentrations if the volume is decreased to 1.00 liter at 25°C?

K in Terms of Partial Pressures

69. Write the K_P expression for each reaction in Exercise 17.
70. Under what conditions would K_c and K_P for a reaction be numerically equal? Are K_c and K_P numerically equal for any of the reactions in Exercises 17 and 18? Which ones?
71. 0.0100 mol of NH_4Cl and 0.0100 mol of NH_3 are placed in a closed 2.00-L container and heated to 603 K. At this temperature, all the NH_4Cl vaporizes. When the reaction

$$NH_4Cl(g) \rightleftharpoons NH_3(g) + HCl(g)$$

has come to equilibrium, 5.8×10^{-3} mol of HCl is present. Calculate (a) K_c and (b) K_P for this reaction at 603 K.
72. CO_2 is passed over graphite at 500. K. The emerging gas stream contains 4.0×10^{-3} mol percent CO. The total pressure is 1.00 atm. Assume that equilibrium is attained. Find K_P for the reaction

$$C(graphite) + CO_2(g) \rightleftharpoons 2CO(g)$$

73. At 425°C, the equilibrium partial pressures of H_2, I_2, and HI are 0.06443 atm, 0.06540 atm, and 0.4821 atm, respectively. Calculate K_P for the following reaction at this temperature.

$$2HI(g) \rightleftharpoons H_2(g) + I_2(g)$$

74. The equilibrium constant, K_P, for the reaction indicated by the following equation is 0.715 at 47°C.

$$N_2O_4(g) \rightleftharpoons 2NO_2(g)$$

Calculate the partial pressures of N_2O_4 and NO_2 in an experiment in which 3.3 moles of N_2O_4 is placed in a 5.0-L flask and allowed to establish equilibrium at 47°C.
75. The equilibrium constant, K_P, is 1.92 at 252°C for the decomposition reaction of phosphorus pentachloride indicated in the following equation.

$$PCl_5(g) \rightleftharpoons PCl_3(g) + Cl_2(g)$$

Calculate the partial pressures of all species present after 6.0 moles of PCl_5 is placed in an evacuated 4.0-liter container and equilibrium is reached at 252°C.
76. The following equilibrium partial pressures were measured at 750.°C: $P_{H_2} = 0.387$ atm, $P_{CO_2} = 0.152$ atm, $P_{CO} = 0.180$ atm, and $P_{H_2O} = 0.252$ atm. What is the value of the equilibrium constant, K_P, for the reaction?

$$H_2 + CO_2 \rightleftharpoons CO + H_2O$$

77. The dissociation of calcium carbonate has an equilibrium constant of $K_P = 1.16$ at 800°C.

$$CaCO_3(s) \rightleftharpoons CaO(s) + CO_2(g)$$

(a) What is K_c for the reaction?
(b) If you place 22.5 g of $CaCO_3$ in a 9.56-L container at 800°C, what is the pressure of CO_2 in the container?
(c) What percentage of the original 22.5-g sample of $CaCO_3$ remains undecomposed at equilibrium?
78. For the reaction

$$Br_2(g) \rightleftharpoons 2Br(g)$$

$K_P = 2550.$ at 4000 K. What is the value of K_c?
79. A stream of gas containing H_2 at an initial partial pressure of 0.200 atm is passed through a tube in which CuO is kept at 500. K. The reaction

$$CuO(s) + H_2(g) \rightleftharpoons Cu(s) + H_2O(g)$$

comes to equilibrium. For this reaction, $K_P = 1.6 \times 10^9$. What is the partial pressure of H_2 in the gas leaving the tube? Assume that the total pressure of the stream is unchanged.

Relationships Among *K*, Δ*G⁰*, Δ*H⁰*, and *T*

80. In the distant future, when hydrogen may be cheaper than coal, steel mills may make iron by the reaction

$$Fe_2O_3(s) + 3H_2(g) \rightleftharpoons 2Fe(s) + 3H_2O(g)$$

For this reaction, $\Delta H^0 = 96$ kJ/mol and $K_c = 8.11$ at 1000. K. (a) What percentage of the H_2 remains unreacted after the reaction has come to equilibrium at 1000. K? (b) Is this percentage greater or less if the temperature is decreased to below 1000. K?
81. What kind of equilibrium constant can be calculated from a ΔG^0 value for a reaction involving only gases?
82. What must be true of the value of ΔG^0 for a reaction if (a) $K \gg 1$; (b) $K = 1$; (c) $K \ll 1$?
83. A mixture of 3.00 mol of Cl_2 and 3.00 mol of CO is enclosed in a 5.00-L flask at 600.°C. At equilibrium, 3.3% of the Cl_2 has been consumed.

$$CO(g) + Cl_2(g) \rightleftharpoons COCl_2(g)$$

(a) Calculate K_c for the reaction at 600.°C. (b) Calculate ΔG^0 for the reaction at this temperature.
84. (a) Use the tabulated thermodynamic values of ΔH_f^0 and S^0 to calculate the value of K_P at 25°C for the gas-phase reaction

$$CO + H_2O \rightleftharpoons CO_2 + H_2$$

(b) Calculate the value of K_P for this reaction at 200.°C, by the same method as in part (a). (c) Repeat the calculation of part (a) using tabulated values of ΔG_f^0.

🔷 **Molecular Reasoning** exercises   ▲ **More Challenging** exercises   Blue-Numbered exercises solved in Student Solutions Manual

85. The equilibrium constant K_c for the reaction

$$H_2(g) + Br_2(g) \rightleftharpoons 2HBr(g)$$

is 1.6×10^5 at 1297 K and 3.5×10^4 at 1495 K. (a) Is ΔH^0 for this reaction positive or negative? (b) Find K_c for the reaction

$$\tfrac{1}{2}H_2(g) + \tfrac{1}{2}Br_2(g) \rightleftharpoons HBr(g)$$

at 1297 K. (c) Pure HBr is placed in a container of constant volume and heated to 1297 K. What percentage of the HBr is decomposed to H_2 and Br_2 at equilibrium?

86. The air pollutant sulfur dioxide can be partially removed from stack gases in industrial processes and converted to sulfur trioxide, the acid anhydride of commercially important sulfuric acid. Write the equation for the reaction, using the smallest whole-number coefficients. Calculate the value of the equilibrium constant for this reaction at 25°C from values of ΔG_f^0 in Appendix K.

87. The value of ΔH^0 for the reaction in Exercise 86 is -197.6 kJ/mol. (a) Predict qualitatively (i.e., without calculation) whether the value of K_P for this reaction at 500.°C would be greater than, the same as, or less than the value at room temperature (25°C). (b) Now calculate the value of K_P at 500.°C.

88. ▲ The following is an example of an alkylation reaction that is important in the production of isooctane (2,2,4-trimethylpentane) from two components of crude oil: isobutane and isobutene. Isooctane is an antiknock additive for gasoline. The thermodynamic equilibrium constant, K, for this reaction at 25°C is 4.3×10^6, and ΔH^0 is -78.58 kJ/mol.
(a) Calculate ΔG^0 at 25°C.
(b) Calculate K at 800.°C.
(c) Calculate ΔG^0 at 800.°C.
(d) How does the spontaneity of the forward reaction at 800.°C compare with that at 25°C?
(e) Why do you think the reaction mixture is heated in the industrial preparation of isooctane?
(f) What is the purpose of the catalyst? Does it affect the forward reaction more than the reverse reaction?

$$CH_3-\underset{\underset{CH_3}{|}}{\overset{\overset{CH_3}{|}}{C}}-H \ + \ CH_3-\overset{\overset{CH_3}{|}}{C}=CH_2 \ \underset{\text{catalyst}}{\overset{\text{heat}}{\longrightarrow}}$$

isobutane   **isobutene**

$$CH_3-\underset{\underset{CH_3}{|}}{\overset{\overset{CH_3}{|}}{C}}-CH_2-\underset{\underset{H}{|}}{\overset{\overset{CH_3}{|}}{C}}-CH_3$$

isooctane

89. ✿ Oxygen gas dissolves in water to a small extent according to the equilibrium:

$$O_2(g) \rightleftharpoons O_2(aq)$$

Fish extract the dissolved O_2. Without oxygen in the water, the fish will die. The heating of water in lakes and reservoirs has led to massive fish kills caused by suffocation. Without doing actual calculations, determine from these facts the sign of ΔH for the equilibrium as written above.

90. ▲ In the gas phase, acetic acid forms an equilibrium between the monomer and the dimer:

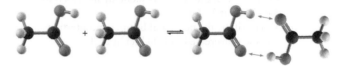

The equilibrium constant for the equation as written is 3.2×10^4 at 25°C. Is the monomer or dimer favored at equilibrium?

91. Describe, in general terms, each of the following reactions and also characterize each as product- or reactant-favored.
(a) $CO(g) + \tfrac{1}{2}O_2(g) \rightleftharpoons CO_2(g)$ $K_P = 1.2 \times 10^{45}$
(b) $H_2O(g) \rightleftharpoons H_2(g) + \tfrac{1}{2}O_2(g)$ $K_P = 9.1 \times 10^{-41}$
(c) $O_2(g) \rightleftharpoons 2O(g)$ $K_P = 1.2 \times 10^{-10}$

Mixed Exercises

92. At 700.°C, K_P is 1.50 for the reaction

$$C(s) + CO_2(g) \rightleftharpoons 2CO(g)$$

Suppose the total gas pressure at equilibrium is 1.00 atm. What are the partial pressures of CO and CO_2?

93. At $-10.$°C, the solid compound $Cl_2(H_2O)_8$ is in equilibrium with gaseous chlorine, water vapor, and ice. The partial pressures of the two gases in equilibrium with a mixture of $Cl_2(H_2O)_8$ and ice are 0.20 atm for Cl_2 and 0.00262 atm for water vapor. Find the equilibrium constant K_P for each of these reactions.
(a) $Cl_2(H_2O)_8(s) \rightleftharpoons Cl_2(g) + 8H_2O(g)$
(b) $Cl_2(H_2O)_8(s) \rightleftharpoons Cl_2(g) + 8H_2O(s)$
Why are your two answers so different?

94. A flask contains $NH_4Cl(s)$ in equilibrium with its decomposition products

$$NH_4Cl(s) \rightleftharpoons NH_3(g) + HCl(g)$$

For this reaction, $\Delta H = 176$ kJ/mol. How is the mass of NH_3 in the flask affected by each of the following disturbances? (a) The temperature is decreased. (b) NH_3 is added. (c) HCl is added. (d) NH_4Cl is added, with no appreciable change in the gas volume. (e) A large amount of NH_4Cl is added, decreasing the volume available to the gases.

95. The equilibrium constant for the reaction

$$H_2(g) + Br_2(\ell) \rightleftharpoons 2HBr(g)$$

is $K_P = 4.5 \times 10^{18}$ at 25°C. The vapor pressure of liquid Br_2 at this temperature is 0.28 atm.
(a) Find K_P at 25°C for the reaction

$$H_2(g) + Br_2(g) \rightleftharpoons 2HBr(g)$$

✿ **Molecular Reasoning** exercises   ▲ **More Challenging** exercises   Blue-Numbered exercises solved in Student Solutions Manual

Unless otherwise noted, all content on this page is © Cengage Learning.

(b) How will the equilibrium in part (a) be shifted by an increase in the volume of the container if (1) liquid Br_2 is absent, (2) liquid Br_2 is present? Explain why the effect is different in these two cases.

96. Given that K_P is 4.6×10^{-14} at 25°C for the reaction

$$2Cl_2(g) + 2H_2O(g) \rightleftharpoons 4HCl(g) + O_2(g)$$
$$\Delta H^0 = +115 \text{ kJ/mol}$$

calculate K_P and K_c for the reaction at 400.°C and at 800.°C.

97. $K_c = 19.9$ for the reaction

$$Cl_2(g) + F_2(g) \rightleftharpoons 2ClF(g)$$

What will happen in a reaction mixture originally containing $[Cl_2] = 0.200$ mol/L, $[F_2] = 0.300$ mol/L, and $[ClF] = 0.950$ mol/L?

98. A mixture of CO, H_2, CH_4, and H_2O is kept at 1133 K until the reaction

$$CO(g) + 3H_2(g) \rightleftharpoons CH_4(g) + H_2O(g)$$

has come to equilibrium. The volume of the container is 0.100 L. The equilibrium mixture contains 1.21×10^{-4} mol CO, 2.47×10^{-4} mol H_2, 1.21×10^{-4} mol CH_4, and 5.63×10^{-8} mol H_2O. Calculate K_P for this reaction at 1133 K.

99. What would the pressure of hydrogen be at equilibrium when $P_{WCl_6} = 0.012$ atm and $P_{HCl} = 0.10$ atm? $K_P = 1.37 \times 10^{21}$ at 900. K.

$$WCl_6(g) + 3H_2(g) \rightleftharpoons W(s) + 6HCl(g)$$

Conceptual Exercises

100. When the beam of the triple beam balance stops swinging, has the balance reached a dynamic or a static equilibrium? Explain.

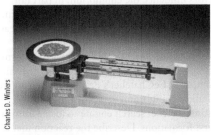

Triple beam balance

101. 🐾 The term "equilibrium" brings to mind the word "equal." What is the relationship between the two terms?

102. The masses of participants in a chemical equilibrium are not the same on both sides of the reaction. Does the equilibrium concept violate the Law of Conservation of Matter? Explain.

103. 🐾 A sample of benzoic acid, a solid carbon-containing acid, is in equilibrium with an aqueous solution of benzoic acid. A tiny quantity of D_2O, water containing the isotope ^2H, deuterium, is added to the solution. The solution is allowed to stand at constant temperature for several hours, after which some of the solid benzoic acid is removed and analyzed. The benzoic acid is found to contain a tiny quantity of deuterium, D, and the formula of the deuterium-containing molecules is C_6H_5COOD. Explain how this can happen. (*Hint*: Look at the ECP plot.)

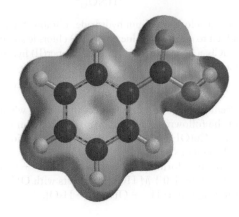

104. 🐾 Imagine yourself the size of atoms and molecules inside a beaker containing the following equilibrium mixture with a K greater than 1.

$$[Co(OH_2)_6]^{2+}(aq) + 4Cl^-(aq) \rightleftharpoons$$
$$\text{pink}$$
$$CoCl_4^{2-}(aq) + 6H_2O(\ell)$$
$$\text{blue}$$

Write a brief description of what you observe around you before and after additional water is added to the mixture.

105. Dinitrogen tetroxide dissociates into nitrogen dioxide:

$$N_2O_4(g) \rightleftharpoons 2NO_2(g)$$

At 110°C the concentrations of N_2O_4 and NO_2 are approximately equal. If 3.00 mol of dinitrogen tetroxide are placed in a sealed container at 110°C and allowed to come to equilibrium, how much NO_2 will be present at equilibrium?

106. In the process called esterification, an alcohol (like ethanol) reacts with a carboxylic acid (like propanoic acid), producing an ester and water:

$$C_2H_5OH(\ell) + C_3H_7COOH(\ell) \rightleftharpoons$$
$$C_3H_7COOC_2H_5(\ell) + H_2O(\ell)$$
ethanol   propanoic acid   ethyl propanoate   water

Two (2.00) moles each of ethanol and propanoic acid are combined and the system is allowed to come to equilibrium. At equilibrium, are the amounts of ethanol, propanoic acid, ethyl propanoate, and water 1.00 mol each? Would *any* ethanol be left? If the system takes 45 minutes to come to equilibrium, how long would it take for *all* of the ethanol and propanoic acid to be converted to ethyl propanoate and water?

🐾 **Molecular Reasoning** exercises ▲ **More Challenging** exercises Blue-Numbered exercises solved in Student Solutions Manual

Unless otherwise noted, all content on this page is © Cengage Learning.

107. Oh, no! Where's that scrap of paper I wrote the reaction for today's experiment on? What's this? (*unfolding*) Here's where I scribbled the equilibrium constant expression. Great . . . what use is that? Can you figure out what the equation is from this expression?

$$K_c = \frac{[Cu(NO_3)_2]^3[NO]^2}{[HNO_3]^8}$$

108. When colorless sodium hydroxide solution, NaOH, is added to clear, light blue copper(II) chloride solution, $CuCl_2$, a light blue cloud of solid copper(II) hydroxide, $Cu(OH)_2$, begins to precipitate:

$$Cu^{2+}(aq) + 2OH^-(aq) \rightleftharpoons Cu(OH)_2(s)$$

Predict what you would expect to observe when each of the following is added to this equilibrium system: more $Cu(OH)_2$ solution, more NaOH solution, 1 M aqueous ammonia, NH_3 solution [NH_3 reacts with Cu^{2+}, forming a deep blue ion: $Cu^{2+} + NH_3 \longrightarrow Cu(NH_3)_4^{2+}$], 0.1 M HCl [H^+ reacts with OH^-, forming water: $H^+ + OH^- \longrightarrow H_2O$].

Building Your Knowledge

109.  Hemoglobin, Hb, has four Fe atoms per molecule that, on the average, pick up roughly three molecules of O_2.

$$Hb(aq) + 3O_2(g) \rightleftharpoons Hb(O_2)_3(aq)$$

Discuss mountain (high altitude) or space sickness in terms of this equilibrium.

Astronauts incur the risk of high altitude sickness.

110. The van't Hoff equation can be written as

$$\ln\left(\frac{K_{T_2}}{K_{T_1}}\right) = \frac{\Delta H^0}{R}\left(\frac{1}{T_1} - \frac{1}{T_2}\right)$$

If $T_2 > T_1$ (the temperature of the reaction increases), how will $\frac{1}{T_2}$ compare with $\frac{1}{T_1}$, and what will be the sign of $\frac{1}{T_1} - \frac{1}{T_2}$? If ΔH^0 is negative (the reaction is exothermic) and knowing the value of R, what will be the sign of $\frac{\Delta H^0}{R}\left(\frac{1}{T_1} - \frac{1}{T_2}\right)$? Knowing that and the property $\ln\frac{x}{y} = \ln x - \ln y$, compare $\ln K_{T_2}$ with $\ln K_{T_1}$. How does K_{T_2} compare with K_{T_1}? Is that result consistent with the predictions of LeChatelier's Principle?

111. At 25°C, 550.0 g of deuterium oxide, D_2O (20.0 g/mol; density 1.10 g/mL), and 498.5 g of H_2O (18.0 g/mol; density 0.997 g/mL) are mixed. The volumes are additive. 47.0% of the H_2O reacts to form HDO. Calculate K_c at 25°C for the reaction

$$H_2O + D_2O \rightleftharpoons 2HDO$$

112. At its normal boiling point of 100.°C, the heat of vaporization of water is 40.66 kJ/mol. What is the equilibrium vapor pressure of water at 75°C?

113. Use the data in the preceding exercise to calculate the temperature at which the vapor pressure of water is 1.20 atm.

Beyond the Textbook

NOTE: *Whenever the answer to an exercise depends on information obtained from a source other than this textbook, the source must be included as an essential part of the answer.*

114. Use an internet search engine to locate information on a mission to Mars and equilibrium. (a) What chemical equilibrium is associated with the Mars mission? (b) What factors influence the equilibrium in part (a)?

115. Use an internet search engine to locate information on Max Bodenstein. What equilibrium constant or reaction did he study? What year did he publish his results?

116. Use an internet search engine to locate information on LeChatelier. State his principle in simple terms. Give an example of LeChatelier's Principle being applied to a non-chemical equilibrium.

117. Use an internet search engine to locate information on Karst formations. Describe the reaction that creates a Karst formation. Does that reaction ever establish an equilibrium?

Ionic Equilibria I: Acids and Bases

18

Ascorbic acid (*top*) and citric acid (*bottom*) are two of the weak acids present in oranges and other citrus fruits.

Charles D. Winters

OBJECTIVES

After you have studied this chapter, you should be able to

▶ Identify strong electrolytes and calculate concentrations of their ions

▶ Understand the autoionization of water

▶ Understand the pH and pOH scales

▶ Use ionization constants for weak monoprotic acids and bases

▶ Discuss the concepts of solvolysis and hydrolysis

▶ Describe how polyprotic acids ionize in steps and how to calculate concentrations of all species in solutions of polyprotic acids

▶ Apply acid–base equilibrium concepts to salts of strong bases and strong acids

▶ Apply acid–base equilibrium concepts to salts of strong bases and weak acids

▶ Apply acid–base equilibrium concepts to salts of weak bases and strong acids

▶ Apply acid–base equilibrium concepts to salts of weak bases and weak acids

▶ Apply acid–base equilibrium concepts to salts of small, highly charged cations

Aqueous solutions are very important. Nearly three-fourths of the earth's surface is covered with water. Enormous numbers of chemical reactions occur in the oceans and smaller bodies of water. Plant and animal fluids are mostly water. Life processes (chemical reactions) of all plants and animals occur in aqueous solutions or in contact with water. Before we were born, all of us developed in sacs filled with aqueous solutions, which protected and helped nurture us until we had developed to the point that we could live in the atmosphere.

18-1 A Review of Strong Electrolytes

In previous discussions we have seen that water-soluble compounds may be classified as either electrolytes or nonelectrolytes. **Electrolytes** are compounds that ionize (or dissociate into their constituent ions) to produce aqueous solutions that conduct an electric current. **Nonelectrolytes** exist as molecules in aqueous solution, and such solutions do not conduct an electric current.

Strong electrolytes are ionized or dissociated completely, or very nearly completely, in dilute aqueous solutions. Strong electrolytes include strong acids, strong bases, and most soluble salts. You should review the discussions of these substances in Sections 6-1 and 10-8. Some strong acids and strong bases are listed again in Table 18-1. See Section 6-1, Part 5, for the solubility guidelines for ionic compounds.

Concentrations of ions in aqueous solutions of strong electrolytes can be calculated directly from the molarity of the strong electrolyte, as the following example illustrates.

Table 18-1 Some Strong Acids and Strong Bases

Strong Acids	
HCl	HNO_3
HBr	$HClO_4$
HI	$(HClO_3)^*$
	H_2SO_4
Strong Bases	
LiOH	
NaOH	
KOH	$Ca(OH)_2$
RbOH	$Sr(OH)_2$
CsOH	$Ba(OH)_2$

*Chloric acid, $HClO_3$, is sometimes not included in a list of strong acids since it is not commonly encountered. However, its anion is much more common.

EXAMPLE 18-1 Calculation of Concentrations of Ions

Calculate the molar concentrations of Ba^{2+} and OH^- ions in 0.030 M barium hydroxide.

Plan

Write the equation for the dissociation of $Ba(OH)_2$, and construct the reaction summary. $Ba(OH)_2$ is a strong base that is completely dissociated.

Solution

From the equation for the dissociation of barium hydroxide, we see that *one* mole of $Ba(OH)_2$ produces *one* mole of Ba^{2+} ions and *two* moles of OH^- ions.

	(strong base)	$Ba(OH)_2(s)$	$\longrightarrow$	$Ba^{2+}(aq)$	+	$2OH^-(aq)$
initial		$0.030\,M$				
change due to rxn		$-0.030\,M$		$+0.030\,M$		$+2(0.030)\,M$
final		$0\,M$		$0.030\,M$		$0.060\,M$

$$[Ba^{2+}] = 0.030\,M \quad \text{and} \quad [OH^-] = 0.060\,M$$

You should now work Exercises 4 and 6.

▶ Recall that we use a single arrow ($\rightarrow$) to indicate that a reaction goes to completion, or nearly to completion, in the indicated direction.

18-2 The Autoionization of Water

Careful experiments on its electrical conductivity have shown that pure water ionizes to a very slight extent.

$$H_2O(\ell) + H_2O(\ell) \rightleftharpoons H_3O^+(aq) + OH^-(aq)$$

Because the H_2O is pure, its activity is 1, so we do not include its concentration in the equilibrium constant expression. This equilibrium constant is known as the **ion product for water** and is usually represented as K_w.

$$K_w = [H_3O^+][OH^-]$$

The formation of each H_3O^+ ion by the ionization of water is always accompanied by the formation of an OH^- ion. Thus, in *pure* water the concentration of H_3O^+ is *always* equal to the concentration of OH^-. Careful measurements show that, in pure water at 25°C,

$$[H_3O^+] = [OH^-] = 1.0 \times 10^{-7}\ \text{mol/L}$$

Substituting these concentrations into the K_w expression gives

$$K_w = [H_3O^+][OH^-] = (1.0 \times 10^{-7})(1.0 \times 10^{-7})$$
$$= 1.0 \times 10^{-14} \quad \text{(at 25°C)}$$

▶ Recall that we use a double arrow ($\rightleftharpoons$) to indicate that the reaction can go in either direction to reach equilibrium.

Although the expression $K_w = [H_3O^+][OH^-] = 1.0 \times 10^{-14}$ was obtained for pure water, *it is also valid for dilute aqueous solutions at 25°C*. This is one of the most useful relationships chemists have discovered. It gives a simple relationship between H_3O^+ and OH^- concentrations in *all* dilute aqueous solutions.

The *value* of K_w is different at different temperatures (Table 18-2), but the *relationship* $K_w = [H_3O^+][OH^-]$ is still valid.

In this text, we shall assume a temperature of 25°C for all calculations involving aqueous solutions unless we specify another temperature.

▶ Solutions in which the concentration of solute is less than about 1 mol/L are usually called dilute solutions.

EXAMPLE 18-2 Calculation of Ion Concentrations

Calculate the concentrations of H_3O^+ and OH^- ions in a 0.050 M HNO_3 solution.

Plan

Write the equation for the ionization of HNO_3, a strong acid, and construct the reaction summary, which gives the concentrations of H_3O^+ (and NO_3^-) ions directly. Then use the relationship $K_w = [H_3O^+][OH^-] = 1.0 \times 10^{-14}$ to find the concentration of OH^- ions.

Table 18-2 K_w at Some Temperatures

Temperature (°C)	K_w
0	1.1×10^{-15}
10	2.9×10^{-15}
25	1.0×10^{-14}
37*	2.4×10^{-14}
45	4.0×10^{-14}
60	9.6×10^{-14}

*Normal human body temperature.

▶ In the remainder of this textbook we will be working nearly exclusively with aqueous solutions, so we will often omit the "(aq)" and "(ℓ)."

Solution

The reaction summary for the ionization of HNO_3, a strong acid, is

(strong acid)	HNO_3	+	H_2O	$\longrightarrow$	H_3O^+	+	NO_3^-
initial	0.050 M				≈ 0 M		0 M
change due to rxn	−0.050 M				+0.050 M		+0.050 M
at equil	0 M				0.050 M		0.050 M

$$[H_3O^+] = [NO_3^-] = 0.050\ M$$

The $[OH^-]$ is determined from the equation for the autoionization of water and its K_w. The initial value for $[H_3O^+]$ is 0.050 M from the dissociation of the nitric acid. We assume that the initial concentration of OH^- from water is 1×10^{-7} M, which is negligible and is approximated as zero.

	$2H_2O$	$\rightleftharpoons$	H_3O^+	+	OH^-
initial			0.050 M		≈ 0 M
change due to rxn	$-2x$ M		$+x$ M		$+x$ M
at equil			$(0.050 + x)$ M		x M

$$K_w = [H_3O^+][OH^-]$$
$$1.0 \times 10^{-14} = (0.050 + x)(x)$$

Because the product $(0.050 + x)(x)$ is a very small number, we know that x must be very small. Thus, it will not matter whether we add x to 0.050; we can assume that $(0.050 + x) \approx 0.050$. This simplifies the math considerably and we can substitute this approximation into the equation and solve.

$$1.0 \times 10^{-14} = (0.050)(x) \quad \text{or} \quad x = \frac{1.0 \times 10^{-14}}{0.050} = 2.0 \times 10^{-13} M = [OH^-]$$

We see that the assumption that x is much smaller than 0.050 was a good one.

You should now work Exercise 14.

In solving Example 18-2 we assumed that *all* of the H_3O^+ (0.050 M) came from the ionization of HNO_3 and neglected the H_3O^+ formed by the ionization of H_2O. The ionization of H_2O produces only 2.0×10^{-13} M H_3O^+ and 2.0×10^{-13} M OH^- *in this solution*. Adding 2.0×10^{-13} M H_3O^+ from the ionization of water to the 0.050 M H_3O^+ from the nitric acid gives a total H_3O^+ concentration of 0.0500000000002 M. The difference is negligible relative to 0.050 M H_3O^+ from the nitric acid. Thus, we were justified in assuming that the $[H_3O^+]$ is derived solely from the strong acid. An easier way to carry out the calculation to find the $[OH^-]$ concentration is to write directly

$$K_w = [H_3O^+][OH^-] = 1.0 \times 10^{-14} \quad \text{or} \quad [OH^-] = \frac{1.0 \times 10^{-14}}{[H_3O^+]}$$

Then we substitute to obtain

$$[OH^-] = \frac{1.0 \times 10^{-14}}{0.050} = 2.0 \times 10^{-13}\ M$$

From now on, we shall use this more direct approach for such calculations.

When nitric acid is added to water, large numbers of H_3O^+ ions are produced. The large increase in $[H_3O^+]$ shifts the water ionization equilibrium far to the left (LeChatelier's Principle), and the $[OH^-]$ decreases.

$$H_2O(\ell) + H_2O(\ell) \rightleftharpoons H_3O^+(aq) + OH^-(aq)$$

In acidic solutions the H_3O^+ concentration is always greater than the OH^- concentration. We should not conclude that acidic solutions contain no OH^- ions. Rather, the $[OH^-]$ is always less than 1.0×10^{-7} M in such solutions. The reverse is true for basic solutions, in which the $[OH^-]$ is always greater than 1.0×10^{-7} M. By definition, "neutral" aqueous solutions at 25°C are solutions in which $[H_3O^+] = [OH^-] = 1.0 \times 10^{-7}$ M.

Solution	General Condition	At 25°C (M)	
acidic	$[H_3O^+] > [OH^-]$	$[H_3O^+] > 1.0 \times 10^{-7}$	$[OH^-] < 1.0 \times 10^{-7}$
neutral	$[H_3O^+] = [OH^-]$	$[H_3O^+] = 1.0 \times 10^{-7}$	$[OH^-] = 1.0 \times 10^{-7}$
basic	$[H_3O^+] < [OH^-]$	$[H_3O^+] < 1.0 \times 10^{-7}$	$[OH^-] > 1.0 \times 10^{-7}$

18-3 The pH and pOH Scales

The pH and pOH scales provide a convenient way to express the acidity and basicity of dilute aqueous solutions. The **pH** and **pOH** of a solution are defined as

$$pH = -\log [H_3O^+] \qquad or \qquad [H_3O^+] = 10^{-pH}$$
$$pOH = -\log [OH^-] \qquad or \qquad [OH^-] = 10^{-pOH}$$

Note that we write pH rather than pH_3O. At the time the pH concept was developed, H_3O^+ was represented as H^+. Various "p" terms are used in chemistry and other sciences. In general a lowercase "**p**" before a symbol means "negative logarithm of the symbol." Thus, pH is the negative logarithm of the H_3O^+ concentration, **pOH** is the negative logarithm of the OH^- concentration, and **pK** refers to the negative logarithm of an equilibrium constant. It is convenient to describe the autoionization of water in terms of **pK_w**.

▶ When dealing with pH, we always use the base-10 (common) logarithm, *not* the base-*e* (natural) logarithm. This is because pH is *defined* using base-10 logarithms.

$$pK_w = -\log K_w \qquad (= 14.0 \text{ at } 25°C)$$

EXAMPLE 18-3 Calculation of pH

Calculate the pH of a solution in which the H_3O^+ concentration is 0.050 mol/L.

Plan

We are given the value for $[H_3O^+]$; so we take the negative logarithm of this value.

Solution

$$[H_3O^+] = 0.050 \ M = 5.0 \times 10^{-2} \ M$$
$$pH = -\log [H_3O^+] = -\log [5.0 \times 10^{-2}] = 1.30$$

This answer contains only *two* significant figures. The "1" in 1.30 is *not* a significant figure; it comes from the power of ten.

You should now work Exercise 22.

STOP & THINK
In a pH value, only the digits after the decimal point are significant figures.

EXAMPLE 18-4 Calculation of H₃O⁺ Concentration from pH

The pH of a solution is 3.301. What is the concentration of H_3O^+ in this solution?

Plan

By definition, $pH = -\log [H_3O^+]$. We are given the pH, so we solve for $[H_3O^+]$.

Solution

From the definition of pH, we write

$$-\log [H_3O^+] = 3.301$$

▶ On most calculators the antilog function is indicated by a 10^x key (where $x = -3.301$ for this problem).

Multiplying through by -1 gives

$$\log [H_3O^+] = -3.301$$

Taking the inverse logarithm (antilog) of both sides of the equation gives

$$[H_3O^+] = 10^{-3.301} \quad \text{so} \quad [H_3O^+] = 5.00 \times 10^{-4} \, M$$

You should now work Exercise 24.

A convenient relationship between pH and pOH in *all dilute solutions at 25°C* can be easily derived. We start with the K_w expression.

$$[H_3O^+][OH^-] = 1.0 \times 10^{-14}$$

Taking the logarithm of both sides of this equation gives

$$\log [H_3O^+] + \log [OH^-] = \log (1.0 \times 10^{-14})$$

Multiplying both sides of this equation by -1 gives

$$(-\log [H_3O^+]) + (-\log [OH^-]) = -\log (1.0 \times 10^{-14})$$

or

▶ At any temperature, pH + pOH = pK_w.

$$pH + pOH = 14.00$$

We can now relate $[H_3O^+]$ and $[OH^-]$, as well as pH and pOH.

$$[H_3O^+][OH^-] = 1.0 \times 10^{-14} \quad \text{and} \quad pH + pOH = 14.00 \quad \text{(at 25°C)}$$

🛑 **STOP & THINK**

These are very important relationships. Remember them!

From this relationship, we see that pH and pOH can *both* be positive only if *both* are less than 14. If either pH or pOH is greater than 14, the other is obviously negative.

Please study carefully the following summary. It will be helpful.

Solution	General Condition	At 25°C
acidic	$[H_3O^+] > [OH^-]$	$[H_3O^+] > 1.0 \times 10^{-7} M > [OH^-]$
	pH < pOH	pH < 7.00 < pOH
neutral	$[H_3O^+] = [OH^-]$	$[H_3O^+] = 1.0 \times 10^{-7} M = [OH^-]$
	pH = pOH	pH = 7.00 = pOH
basic	$[H_3O^+] < [OH^-]$	$[H_3O^+] < 1.0 \times 10^{-7} M < [OH^-]$
	pH > pOH	pH > 7.00 > pOH

To develop familiarity with the pH and pOH scales, consider a series of solutions in which $[H_3O^+]$ varies from $10 \, M$ to $1.0 \times 10^{-15} \, M$. Obviously, $[OH^-]$ will vary from $1.0 \times 10^{-15} \, M$ to $10 \, M$ in these solutions. Table 18-3 summarizes these scales.

EXAMPLE 18-5   pH and pOH of an Acidic Solution

Calculate $[H_3O^+]$, pH, $[OH^-]$, and pOH for a 0.015 M HNO$_3$ solution.

Plan

We write the equation for the ionization of the strong acid, HNO$_3$, which gives us $[H_3O^+]$. Then we calculate pH. We use the relationships $[H_3O^+][OH^-] = 1.0 \times 10^{-14}$ and pH + pOH = 14.00 to find pOH and $[OH^-]$.

Solution

$$HNO_3 + H_2O \longrightarrow H_3O^+ + NO_3^-$$

Because nitric acid is a strong acid (it ionizes completely), we know that

$$[H_3O^+] = 0.015 \ M$$

$$pH = -\log [H_3O^+] = -\log (0.015) = -(-1.82) = 1.82$$

We also know that pH + pOH = 14.00. Therefore,

$$pOH = 14.00 - pH = 14.00 - 1.82 = 12.18$$

Because $[H_3O^+][OH^-] = 1.0 \times 10^{-14}$, $[OH^-]$ is easily calculated.

$$[OH^-] = \frac{1.0 \times 10^{-14}}{[H_3O^+]} = \frac{1.0 \times 10^{-14}}{0.015} = 6.7 \times 10^{-13} \ M$$

You should now work Exercise 25.

▶ All ions are hydrated in aqueous solution. We often omit the designations (ℓ), (g), (s), (aq), and so on.

S TOP & THINK

It is easy to confuse pH and pOH because pH values are far more commonly used. Here pOH = 12.18, which might initially lead someone thinking only about pH's to conclude that the solution is basic. But as one can see from the very small $[OH^-]$ value and pH of 1.82, the solution is definitely acidic.

pH Range for a Few Common Substances

Substance	pH Range
Gastric contents (human)	1.6–3.0
Soft drinks	2.0–4.0
Lemons	2.2–2.4
Vinegar	2.4–3.4
Tomatoes	4.0–4.4
Beer	4.0–5.0
Urine (human)	4.8–8.4
Milk (cow's)	6.3–6.6
Saliva (human)	6.5–7.5
Blood plasma (human)	7.3–7.5
Egg white	7.6–8.0
Milk of magnesia	10.5
Household ammonia	11–12

More acidic

More basic

Table 18-3 Relationships Among $[H_3O^+]$, pH, pOH, and $[OH^-]$

$[H_3O^+]$	pH			pOH	$[OH^-]$	
10^{-15}	15			−1	10^1	Increasing basicity
10^{-14}	14			0	1	
10^{-13}	13			1	10^{-1}	
10^{-12}	12			2	10^{-2}	
10^{-11}	11			3	10^{-3}	
10^{-10}	10			4	10^{-4}	
10^{-9}	9			5	10^{-5}	
10^{-8}	8			6	10^{-6}	
10^{-7}	7	H_3O^+ concentration	OH^- concentration	7	10^{-7}	Neutral
10^{-6}	6			8	10^{-8}	
10^{-5}	5			9	10^{-9}	
10^{-4}	4			10	10^{-10}	Increasing acidity
10^{-3}	3			11	10^{-11}	
10^{-2}	2			12	10^{-12}	
10^{-1}	1			13	10^{-13}	
1	0			14	10^{-14}	
10^1	−1			15	10^{-15}	

S TOP & THINK

Note that the pH can be 0 or a small negative number for concentrated strong acid solutions. A pH = −1.0 represents $[H^+]$ = 10.0 M, which is certainly possible for a concentrated strong acid. But pH = −2.0 indicates a solution with $[H^+]$ = 100.0 M, which is not possible.

EXAMPLE 18-6 Calculations Involving pH and pOH

Calculate $[H_3O^+]$, pH, $[OH^-]$, and pOH for a 0.015 M Ca(OH)$_2$ solution.

Plan

We write the equation for the ionization of the strong base Ca(OH)$_2$, which gives us $[OH^-]$. Then we calculate pOH. We use the relationships $[H_3O^+][OH^-] = 1.0 \times 10^{-14}$ and pH + pOH = 14.00 to find pH and $[H_3O^+]$.

Solution

$$Ca(OH)_2 \xrightarrow{H_2O} Ca^{2+} + 2OH^-$$

Because calcium hydroxide is a strong base (it dissociates completely), we know that

$$[OH^-] = 2 \times 0.015\ M = \boxed{0.030\ M}$$

$$pOH = -\log [OH^-] = -\log (0.030) = -(-1.52) = \boxed{1.52}$$

We also know that pH + pOH = 14.00. Therefore,

$$pH = 14.00 - pOH = 14.00 - 1.52 = \boxed{12.48}$$

Because $[H_3O^+][OH^-] = 1.0 \times 10^{-14}$, $[H_3O^+]$ is easily calculated.

$$[H_3O^+] = \frac{1.0 \times 10^{-14}}{[OH^-]} = \frac{1.0 \times 10^{-14}}{0.030} = \boxed{3.3 \times 10^{-13}\ M}$$

You should now work Exercises 26 and 37.

▶ A pH meter is much more accurate than an indicator for obtaining pH values.

The pH of a solution can be determined using a pH meter (Figure 18-1) or by the indicator method. Acid–base **indicators** are intensely colored complex organic compounds that are different colors in solutions of different pH (see Section 19-4). Many are weak acids or weak bases that are useful over rather narrow ranges of pH values. *Universal indicators* are mixtures of several indicators; they show several color changes over a wide range of pH values.

In the indicator method, we prepare a series of solutions of known pH (standard solutions). We add a universal indicator to each; solutions with different pH values turn different colors (Figure 18-2). We then add the same universal indicator to the unknown solution and compare its color to those of the standard solutions. Solutions with the same pH have the same color.

Universal indicator papers can also be used to determine pH. A drop of solution is placed on a piece of paper or a piece of the paper is dipped into a solution. The color of the paper is then compared with a color chart on the container to establish the pH of the solution.

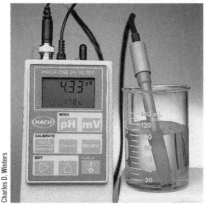

Charles D. Winters

Figure 18-1 A pH meter gives the pH of a solution directly. When the electrode is dipped into a solution, the meter displays the pH. The pH meter is based on the glass electrode. This sensing device generates a voltage that is proportional to the pH of the solution in which the electrode is placed. The instrument has an electrical circuit to amplify the voltage from the electrode and a meter that relates the voltage to the pH of the solution. Before being used, a pH meter should be calibrated with a series of solutions of known pH.

A Solutions containing a universal indicator. A universal indicator shows a wide range of colors as pH varies. The pH values are given by the black numbers. These solutions range from quite acidic (*upper left*) to quite basic (*lower right*).

Charles D. Winters

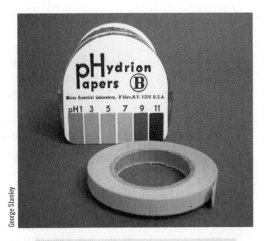

B Universal pH paper has been soaked in a universal indicator, allowing a wide range of pH's to be measured easily by comparing the test strip to the color scale on the container.

George Stanley

Figure 18-2

18-4 Ionization Constants for Weak Monoprotic Acids and Bases

We have discussed strong acids and strong bases. There are relatively few of these. Weak acids are much more numerous than strong acids. For this reason you were asked to learn the list of common strong acids (see Table 18-1). You may assume that other acids you encounter in this text will be weak acids. Table 18-4 contains names, formulas, ionization constants, and pK_a values for a few common weak acids; Appendix F contains a longer list of K_a values. *Weak* acids ionize only slightly in dilute aqueous solution. Our classification of acids as strong or weak is based on the *extent to which they ionize in dilute aqueous solution*.

Several weak acids are familiar to us. Vinegar is a 5% solution of acetic acid, CH_3COOH. Carbonated beverages are saturated solutions of carbon dioxide in water, which produces carbonic acid.

$$CO_2 + H_2O \rightleftharpoons H_2CO_3$$

Citrus fruits contain citric acid, $C_3H_5O(COOH)_3$. Some ointments and powders used for medicinal purposes contain boric acid, H_3BO_3. These everyday uses of weak acids suggest that there is a significant difference between strong and weak acids. The difference is that *strong acids ionize completely in dilute aqueous solution, whereas weak acids ionize only slightly*.

Let us consider the reaction that occurs when a weak acid, such as acetic acid, is dissolved in water. The equation for the ionization of acetic acid is

$$CH_3COOH(aq) + H_2O(\ell) \rightleftharpoons H_3O^+(aq) + CH_3COO^-(aq)$$

The equilibrium constant for this reaction could be represented as

$$K_c = \frac{[H_3O^+][CH_3COO^-]}{[CH_3COOH][H_2O]}$$

We should recall that the thermodynamic definition of K is in terms of activities. In dilute solutions, the activity of the (nearly) pure H_2O is essentially 1. The activity of each dissolved species is numerically equal to its molar concentration. Thus the **ionization constant** of a weak acid, K_a, does not include a term for the concentration of water.

The pH of some common substances is shown by a universal indicator. Refer to Figure 18-2 to interpret the indicator colors.

Charles Steele

▶ How can you tell that carbonated beverages are *saturated* CO_2 solutions?

▶ A **monoprotic acid** has only one ionizable H.

$$pK_a = -\log K_a$$

Table 18-4 Ionization Constants and pK_a Values for Some Weak Monoprotic Acids

Acid	Ionization Reaction	K_a at 25°C	pK_a
hydrofluoric acid	$HF + H_2O \rightleftharpoons H_3O^+ + F^-$	7.2×10^{-4}	3.14
nitrous acid	$HNO_2 + H_2O \rightleftharpoons H_3O^+ + NO_2^-$	4.5×10^{-4}	3.35
acetic acid	$CH_3COOH + H_2O \rightleftharpoons H_3O^+ + CH_3COO^-$	1.8×10^{-5}	4.74
hypochlorous acid	$HOCl + H_2O \rightleftharpoons H_3O^+ + OCl^-$	3.5×10^{-8}	7.45
hydrocyanic acid	$HCN + H_2O \rightleftharpoons H_3O^+ + CN^-$	4.0×10^{-10}	9.40

We often use HA as a general representation for a monoprotic acid and A^- for its conjugate base. The equation for the ionization of a weak acid can be written as

$$HA \rightleftharpoons H^+ + A^-$$

For example, for acetic acid, we can write either

$$K_a = \frac{[H^+][A^-]}{[HA]}$$

or

$$K_a = \frac{[H_3O^+][CH_3COO^-]}{[CH_3COOH]} = 1.8 \times 10^{-5}$$

▶ The value of K_a can be found in Table 18-4.

This expression tells us that in dilute aqueous solutions of acetic acid, the concentration of H_3O^+ multiplied by the concentration of CH_3COO^- and then divided by the concentration of *nonionized* acetic acid is equal to 1.8×10^{-5}.

Ionization constants for weak acids (and bases) must be calculated from *experimentally determined data*. Measurements of pH, conductivity, or depression of freezing point provide data from which these constants can be calculated.

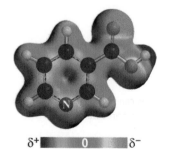

δ^+ ▬▬▬ 0 ▬▬▬ δ^-

▶ The structure of nicotinic acid is

Nicotinic acid, also called niacin, is a necessary vitamin in our diets. It is not physiologically related to nicotine.

EXAMPLE 18-7 Calculation of K_a and pK_a from Equilibrium Concentrations

Nicotinic acid is a weak monoprotic organic acid that we can represent as HA.

$$HA + H_2O \rightleftharpoons H_3O^+ + A^-$$

A dilute solution of nicotinic acid was found to contain the following concentrations at equilibrium at 25°C. What are the K_a and pK_a values? $[HA] = 0.049\ M$; $[H_3O^+] = [A^-] = 8.4 \times 10^{-4}\ M$.

Plan

We are given *equilibrium* concentrations, so we substitute these into the expression for K_a.

Solution

$$HA + H_2O \rightleftharpoons H_3O^+ + A^- \qquad K_a = \frac{[H_3O^+][A^-]}{[HA]}$$

$$K_a = \frac{(8.4 \times 10^{-4})(8.4 \times 10^{-4})}{0.049} = 1.4 \times 10^{-5}$$

The equilibrium constant expression is

$$K_a = \frac{[H_3O^+][A^-]}{[HA]} = 1.4 \times 10^{-5} \qquad pK_a = -\log(1.4 \times 10^{-5}) = 4.85$$

You should now work Exercise 30.

EXAMPLE 18-8 Calculation of K_a from Percent Ionization

In 0.0100 M solution, acetic acid is 4.2% ionized. Calculate its ionization constant.

Plan

We write the equation for the ionization of acetic acid and its equilibrium constant expression. Next we use the percent ionization to complete the reaction summary and then substitute into the K_a expression.

Solution

The equations for the ionization of CH_3COOH and its ionization constant are

$$CH_3COOH + H_2O \rightleftharpoons H_3O^+ + CH_3COO^- \quad \text{and} \quad K_a = \frac{[H_3O^+][CH_3COO^-]}{[CH_3COOH]}$$

Because 4.2% of the CH_3COOH ionizes,

$$M_{CH_3COOH} \text{ that ionizes} = 0.042 \times 0.0100 \ M = 4.2 \times 10^{-4} \ M$$

Each mole of CH_3COOH that ionizes forms one mole of H_3O^+ and one mole of CH_3COO^-. We represent this in the reaction summary.

	$CH_3COOH + H_2O \rightleftharpoons$	H_3O^+	+	CH_3COO^-
initial	0.0100 M	$\approx 0 \ M$		0 M
change	$-4.2 \times 10^{-4} \ M$	$+4.2 \times 10^{-4} \ M$		$+4.2 \times 10^{-4} \ M$
at equil	$9.58 \times 10^{-3} \ M$	$4.2 \times 10^{-4} \ M$		$4.2 \times 10^{-4} \ M$

Substitution of these values into the K_a expression gives the value for K_a.

$$K_a = \frac{[H_3O^+][CH_3COO^-]}{[CH_3COOH]} = \frac{(4.2 \times 10^{-4})(4.2 \times 10^{-4})}{9.58 \times 10^{-3}} = 1.8 \times 10^{-5}$$

You should now work Exercise 38.

S TOP & THINK

The acetic acid concentration at equilibrium is represented algebraically by (0.0100 $M - x$). We did not ignore the x value in this case because it was within two orders of magnitude of the number from which it was being subtracted. $0.0100 - 0.00042 = 0.00958$ or $9.58 \times 10^{-3} \ M$

EXAMPLE 18-9 Calculation of K_a from pH

The pH of a 0.115 M solution of chloroacetic acid, $ClCH_2COOH$, is measured to be 1.92. Calculate K_a for this weak monoprotic acid.

Plan

For simplicity, we represent $ClCH_2COOH$ as HA. We write the ionization equation and the expression for K_a. Next we calculate $[H_3O^+]$ for the given pH and complete the reaction summary. Finally, we substitute into the K_a expression.

Solution

The ionization of this weak monoprotic acid and its ionization constant expression may be represented as

$$HA + H_2O \rightleftharpoons H_3O^+ + A^- \quad \text{and} \quad K_a = \frac{[H_3O^+][A^-]}{[HA]}$$

We calculate $[H_3O^+]$ from the definition of pH.

$$pH = -\log[H_3O^+]$$
$$[H_3O^+] = 10^{-pH} = 10^{-1.92} = 0.012 \ M$$

We use the usual reaction summary as follows. At this point, we know the *original* [HA] and the *equilibrium* $[H_3O^+]$. From this information, we fill out the "change" line and then deduce the other equilibrium values.

	HA $+ H_2O \rightleftharpoons$	H_3O^+	+	A^-
initial	0.115 M	$\approx 0 \ M$		0 M
change due to rxn	$-0.012 \ M$	$+0.012 \ M$		$+0.012 \ M$
at equil	0.103 M	0.012 M		0.012 M

Some common household weak acids. A strip of paper impregnated with a universal indicator is convenient for estimating the pH of a solution.

Charles Steele

Ⓢ TOP & THINK
Note that because you were given the H_3O^+ concentration in the form of a pH, there are no x values to solve for, just the K_a value. This is, therefore, a very simple problem so long as you convert the pH to the H_3O^+ concentration and are careful to set up the balanced equation with the initial change due to rxn, and at equil values correctly.

Now that all concentrations are known, K_a can be calculated.

$$K_a = \frac{[H_3O^+][A^-]}{[HA]} = \frac{(0.012)(0.012)}{0.103} = 1.4 \times 10^{-3}$$

You should now work Exercise 40.

Problem-Solving Tip   Filling in Reaction Summaries

In Examples 18-8 and 18-9, the value of an equilibrium concentration was used to determine the change in concentration. You should become proficient at using a variety of data to determine values that are related via a chemical equation. Let's review what we did in Example 18-9. Only the equilibrium expression, initial concentrations, and the equilibrium concentration of H_3O^+ were known when we started the reaction summary. The following steps show how we filled in the remaining values in the order indicated by the numbered red arrows.

1. $[H_3O^+]_{equil} = 0.012\ M$, so we record this value
2. $[H_3O^+]_{initial} \approx 0$, so change in $[H_3O^+]$ due to rxn must be $+0.012\ M$
3. Formation of $0.012\ M\ H_3O^+$ consumes $0.012\ M\ HA$, so the change in $[HA] = -0.012\ M$
4. $[HA]_{equil} = [HA]_{orig} + [HA]_{chg} = 0.115\ M + (-0.012\ M) = 0.103\ M$
5. Formation of $0.012\ M\ H_3O^+$ also gives $0.012\ M\ A^-$
6. $[A^-]_{equil} = [A^-]_{orig} + [A^-]_{chg} = 0\ M + 0.012\ M = 0.012\ M$

At equilibrium, $[H_3O^+] = 0.012\ M$ so ①

	HA	$+ H_2O \rightleftharpoons$	H_3O^+	$+$	A^-
initial	0.115 M		0 M		0 M
change due to rxn	−0.012 M ③		+0.012 M ⑤		0.012 M
at equil	0.103 M ④		② 0.012 M		⑥ 0.012 M

Ionization constants are equilibrium constants for ionization reactions, so their values indicate the extents to which weak electrolytes ionize. At the same concentrations, acids with larger ionization constants ionize to greater extents (and are stronger acids) than acids with smaller ionization constants. From Table 18-4, we see that the order of decreasing acid strength for these five weak acids is

(strongest acid) $HF > HNO_2 > CH_3COOH > HOCl > HCN$ (weakest acid)

▶ Recall that in Brønsted–Lowry terminology, an acid forms its conjugate base by losing a H^+.

Conversely, in Brønsted–Lowry terminology (see Section 10-4), the order of increasing base strength of the anions of these acids (their conjugate bases) is

(weakest base) $F^- < NO_2^- < CH_3COO^- < OCl^- < CN^-$ (strongest base)

If we know the value of the ionization constant for a weak acid, we can calculate the concentrations of the species present in solutions of known initial concentrations.

EXAMPLE 18-10   Calculation of Concentrations from K_a

▶ We can write the formula for hypochlorous acid as HOCl rather than HClO to emphasize that its structure is H — O — Cl.

(a) Calculate the concentrations of the various species in $0.10\ M$ hypochlorous acid, HOCl. For HOCl, $K_a = 3.5 \times 10^{-8}$. (b) What is the pH of this solution?

Plan

We write the equation for the ionization of the weak acid and its K_a expression. Then we represent the *equilibrium* concentrations algebraically and substitute into the K_a expression. Then we solve for the concentrations and the pH.

Solution

(a) The equation for the ionization of HOCl and its K_a expression are

$$HOCl + H_2O \rightleftharpoons H_3O^+ + OCl^- \quad \text{and} \quad K_a = \frac{[H_3O^+][OCl^-]}{[HOCl]} = 3.5 \times 10^{-8}$$

We would like to know the concentrations of H_3O^+, OCl^-, and nonionized HOCl in solution. An algebraic representation of concentrations is required because there is no other obvious way to obtain the concentrations.

Let $x = $ mol/L of HOCl that ionizes. Then write the "change" line and complete the reaction summary.

	HOCl	+	H_2O	$\rightleftharpoons$	H_3O^+	+	OCl^-
initial	0.10 M				≈ 0 M		0 M
change due to rxn	$-x$ M				$+x$ M		$+x$ M
at equil	$(0.10 - x)$ M				x M		x M

Substituting these algebraic representations into the K_a expression gives

$$K_a = \frac{[H_3O^+][OCl^-]}{[HOCl]} = \frac{(x)(x)}{(0.10 - x)} = 3.5 \times 10^{-8}$$

This is a quadratic equation, but it is not necessary to solve it by the quadratic formula. The small value of the equilibrium constant, K_a, tells us that not very much of the original acid ionizes. Thus we can assume that $x \ll 0.10$. If x is small enough compared with 0.10, it will not matter (much) whether we subtract it, and we can assume that $(0.10 - x)$ is very nearly equal to 0.10. The equation then becomes

$$\frac{x^2}{0.10} \approx 3.5 \times 10^{-8} \quad x^2 \approx 3.5 \times 10^{-9} \quad \text{so} \quad x \approx 5.9 \times 10^{-5}$$

In our algebraic representation we let

$$[H_3O^+] = x\,M = 5.9 \times 10^{-5}\,M \; ; \; [OCl^-] = x\,M = 5.9 \times 10^{-5}\,M$$

$$[HOCl] = (0.10 - x)\,M = (0.10 - 0.000059)\,M = 0.10\,M$$

$$[OH^-] = \frac{K_w}{[H_3O^+]} = \frac{1.0 \times 10^{-14}}{5.9 \times 10^{-5}} = 1.7 \times 10^{-10}\,M$$

(b) pH $= -\log(5.9 \times 10^{-5}) = 4.23$

You should now work Exercise 42.

> **S**TOP & THINK
> We neglect the 1.0×10^{-7} mol/L of H_3O^+ produced by the ionization of *pure* water. Recall (Section 18-2) that the addition of an acid to water suppresses the ionization of H_2O, so $[H_3O^+]$ from H_2O is even less than $1.0 \times 10^{-7}\,M$.

> **S**TOP & THINK
> Because $x \approx 5.9 \times 10^{-5}$, it is more than two orders of magnitude smaller than 0.10. Thus our approximation of neglecting x in the expression $0.10 - x$ to simplify the algebra was a good one.

Problem-Solving Tip Simplifying Quadratic Equations

We often encounter quadratic or higher-order equations in equilibrium calculations. With modern programmable calculators, solving such problems by iterative methods is often feasible. But frequently a problem can be made much simpler by using some mathematical common sense.

When the linear variable (x) in a *quadratic* equation is added to or subtracted from a much larger number, it can often be disregarded if it is sufficiently small. A reasonable rule of thumb for determining whether the variable can be disregarded in equilibrium calculations is this: If the exponent of 10 in the K value is -3 (e.g., 10^{-3}) or less (-4, -5, -6, etc.), then the variable may be small enough to disregard when it is added to or subtracted from a number two orders of magnitude greater (e.g., 0.1 if $K_a = 0.001$ or 1×10^{-3}). Solve the problem neglecting x; then compare the value of x with the number it would have been added to (or subtracted from). If x is more than 5% of that number, the assumption was *not* justified, and you should solve the equation using the quadratic formula.

The pH of a Diet Coke is measured with a modern pH meter. Many carbonated soft drinks are acidic due to the dissolved CO_2 and phosphoric acid (both weak acids).

Let's examine the assumption as it applies to Example 18-10. Our quadratic equation is

$$\frac{(x)(x)}{(0.10 - x)} = 3.5 \times 10^{-8}$$

Because 3.5×10^{-8} is a very small K_a value, we know that the acid ionizes only slightly. Thus, x must be very small compared with 0.10, so we can write $(0.10 - x) \approx 0.10$.

The equation then becomes $\frac{x^2}{0.10} \approx 3.5 \times 10^{-8}$. To solve this, we rearrange and take the square roots of both sides. To check, we see that the result, $x \approx 5.9 \times 10^{-5}$, is only 0.059% of 0.10. This error is much less than 5%, so our assumption is justified. You may also wish to use the quadratic formula to verify that the answer obtained this way is correct to within roundoff error.

The preceding argument is purely algebraic. We could use our chemical intuition to reach the same conclusion. A small K_a value (10^{-3} or less) tells us that the extent of ionization is very small; therefore, nearly all of the weak acid exists as nonionized molecules. The amount that ionizes is insignificant compared with the concentration of nonionized weak acid.

From our calculations in Example 18-10, we can draw some conclusions. In a solution containing *only a weak monoprotic acid*, the concentration of H_3O^+ is equal to the concentration of the anion of the acid. The concentration of the dilute nonionized acid is approximately equal to the molarity of the solution.

EXAMPLE 18-11 Percent Ionization

Calculate the percent ionization of a 0.10 M solution of acetic acid.

Plan

Write the ionization equation and the expression for K_a. Follow the procedure used in Example 18-10 to find the concentration of acid that ionized. Then, substitute the concentration of acid that ionized into the expression for percent ionization. Percentage is defined as (part/whole) $\times$ 100%, so the percent ionization is

$$\% \text{ ionization} = \frac{[CH_3COOH]_{\text{ionized}}}{[CH_3COOH]_{\text{initial}}} \times 100\%$$

Solution

The equations for the ionization of CH_3COOH and its K_a are

$$CH_3COOH + H_2O \rightleftharpoons H_3O^+ + CH_3COO^- \qquad K_a = \frac{[H_3O^+][A^-]}{[HA]} = 1.8 \times 10^{-5}$$

We proceed as we did in Example 18-10. Let $x = [CH_3COOH]_{\text{ionized}}$.

	$CH_3COOH + H_2O \rightleftharpoons$	H_3O^+	$+ CH_3COO^-$
initial	0.10 M	$\approx 0 M$	0 M
change due to rxn	$-x M$	$+x M$	$+x M$
at equil	$(0.10 - x) M$	$x M$	$x M$

▶ We could write the initial $[H_3O^+]$ as $1.0 \times 10^{-7} M$. In very *dilute* solutions of weak acids, we might have to take this into account. In this acid solution, $(1.0 \times 10^{-7} + x) \approx x$.

Substituting into the ionization constant expression gives

$$K_a = \frac{[H_3O^+][A^-]}{[HA]} = \frac{(x)(x)}{(0.10 - x)} = 1.8 \times 10^{-5}$$

If we make the simplifying assumption that $(0.10 - x) \approx 0.10$, we have

$$\frac{x^2}{0.10} = 1.8 \times 10^{-5} \qquad x^2 = 1.8 \times 10^{-6} \qquad x = 1.3 \times 10^{-3}$$

This gives $[CH_3COOH]_{ionized} = x = 1.3 \times 10^{-3} M$. Now we can calculate the percent ionization for 0.10 M CH_3COOH solution.

$$\% \text{ ionization} = \frac{[CH_3COOH]_{ionized}}{[CH_3COOH]_{initial}} \times 100\% = \frac{1.3 \times 10^{-3}M}{0.10 M} \times 100\% = 1.3\%$$

Our assumption that $(0.10 - x) \approx 0.10$ is reasonable because $(0.10 - x) = (0.10 - 0.0013)$. Thus, x is approximately two orders of magnitude smaller than the 0.10 value it is being subtracted from (or only about 1% different than 0.10). When K_a for a weak acid is significantly greater than 10^{-3}, however, this assumption would introduce considerable error because the x value would be large enough to cause a significant change when subtracted from the initial concentration.

▶ Note that we did not need to solve explicitly for the equilibrium concentrations $[H_3O^+]$ and $[CH_3COO^-]$ to answer the question. From the setup, we see that these are both $1.3 \times 10^{-3} M$.

You should now work Exercise 48.

In dilute solutions, acetic acid exists primarily as nonionized molecules, as do all weak acids; there are relatively few hydronium and acetate ions. In 0.10 M solution, CH_3COOH is 1.3% ionized; for each 1000 molecules of CH_3COOH originally placed in the solution, there are 13 H_3O^+ ions, 13 CH_3COO^- ions, and 987 nonionized CH_3COOH molecules. For weaker acids of the same concentration, the number of molecules of nonionized acid would be even larger.

By now we should have gained some "feel" for the strength of an acid by looking at its K_a value. Consider 0.10 M solutions of HCl (a strong acid), CH_3COOH (Example 18-11), and HOCl (Example 18-10). If we calculate the percent ionization for 0.10 M HOCl (as we did for 0.10 M CH_3COOH in Example 18-11), we find that it is 0.059% ionized. In 0.10 M solution, HCl is very nearly completely ionized (100%). The data in Table 18-5 show that the $[H_3O^+]$ in 0.10 M HCl is approximately 77 times greater than that in 0.10 M CH_3COOH and approximately 1700 times greater than that in 0.10 M HOCl.

Many scientists prefer to use pK_a values rather than K_a values for weak acids. Recall that in general, "p" terms refer to negative logarithms. The pK_a value for a weak acid is just the negative logarithm of its K_a value.

$$pK_a = -\log K_a$$

EXAMPLE 18-12 pK_a Values

The K_a values for acetic acid and hydrofluoric acid are 1.8×10^{-5} and 7.2×10^{-4}, respectively. What are their pK_a values?

Plan

pK_a is defined as the negative logarithm of K_a, so we take the negative logarithm of each K_a.

Solution

For CH_3COOH,

$$pK_a = -\log K_a = -\log (1.8 \times 10^{-5}) = -(-4.74) = 4.74$$

For HF,

$$pK_a = -\log K_a = -\log (7.2 \times 10^{-4}) = -(-3.14) = 3.14$$

You should now work Exercise 50.

An inert solid has been suspended in the liquid to improve the quality of this photograph of a pH meter.

Table 18-5 Comparison of Extents of Ionization of Some Acids

Acid Solution	Ionization Constant	pK_a	$[H_3O^+]$	pH	Percent Ionization
0.10 M HCl	very large	—	0.10 M	1.00	≈ 100
0.10 M CH_3COOH	1.8×10^{-5}	4.74	0.0013 M	2.89	1.3
0.10 M HOCl	3.5×10^{-8}	7.46	0.000059 M	4.23	0.059

From Example 18-12, we see that the stronger acid (HF in this case) has the larger K_a value and the smaller pK_a value. Conversely, the weaker acid (CH_3COOH in this case) has the smaller K_a value and the larger pK_a value.

▶ A similar statement is true for weak bases; that is, a stronger base has the greater K_b value and the smaller pK_b value.

> The weaker an acid, the smaller its value of K_a, and the larger its value of pK_a.
>
> Similarly, the weaker a base, the smaller its value of K_b and the larger its value of pK_b.

Strong acids have large K_a values and *negative* pK_a values. HCl, for example, has a $K_a \approx 1 \times 10^8$ and a $pK_a \approx -8$.

EXAMPLE 18-13 Acid Strengths and K_a Values

Given the following list of weak acids and their K_a values, arrange the acids in order of (a) increasing acid strength and (b) increasing pK_a values.

Acid	K_a
HOCl	3.5×10^{-8}
HCN	4.0×10^{-10}
HNO$_2$	4.5×10^{-4}

Plan

(a) We see that HNO_2 is the strongest acid in this group because it has the largest K_a value. HCN is the weakest because it has the smallest K_a value. (b) We do not need to calculate pK_a values to answer the question. We recall that the weakest acid has the largest pK_a value and the strongest acid has the smallest pK_a value, so the order of increasing pK_a values is just the reverse of the order in Part (a).

Solution

Weakest to strongest acids:

Acid	K_a	pK_a
HCN	4.0×10^{-10}	9.40
HOCl	3.5×10^{-8}	7.46
HNO2	4.5×10^{-4}	3.35

Increasing acidity

(a) Increasing acid strength: $HCN < HOCl < HNO_2$
(b) Increasing pK_a values: $HNO_2 < HOCl < HCN$

You should now work Exercise 109.

▶ Do not confuse the strength of an acid with its reactivity. Acid strength refers to the extent of ionization of the acid and not to the reactions that it undergoes.

You may know that hydrofluoric acid dissolves glass. But HF is *not* a strong acid. The reaction of glass with hydrofluoric acid occurs because silicates react with HF to produce silicon tetrafluoride, SiF_4, or $[SiF_6]^{2-}$ in aqueous solution. This reaction tells us nothing about the acid strength of hydrofluoric acid, just that it has unusual reactivity towards SiO_2 (glass).

In Chapter 10 we studied the strengths of binary acids as a function of the position of the anion on the periodic table. We also studied the strengths of ternary acids containing the same central element as a function of the oxidation state of the central element. Those were empirical trends. We can now use K_a and pK_a values to make much more precise conclusions about compounds' acidity or basicity. The following example illustrates a correlation between K_a values and properties.

EXAMPLE 18-14 Comparison of the Acid Strength of Classes of Compounds

The structures of phenol, the smallest member of the group of compounds called phenols (C_6H_5OH), ethanol (CH_3CH_2OH), and water (HOH) are shown here.

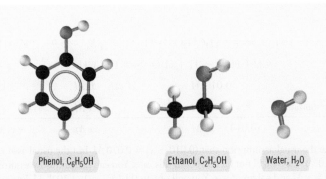

Phenol, C_6H_5OH Ethanol, C_2H_5OH Water, H_2O

► The circle in the phenol ring indicates delocalized π- bonding.

We see that each contains an $-O-H$ group, which can undergo an acid reaction. These three reactions and their pK_a values are

Water: $H-O-H + H_2O \rightleftharpoons H-O^- + H_3O^+$ $pK_w = 14.0$

Ethanol: $CH_3CH_2-O-H + H_2O \rightleftharpoons CH_3CH_2-O^- + H_3O^+$ $pK_a = 15.9$

Phenol: $C_6H_5-O-H + H_2O \rightleftharpoons C_6H_5-O^- + H_3O^+$ $pK_a = 10.0$

(a) Rank the three compounds in order of decreasing acidity. (b) Calculate the pH of 0.010 *M* solutions of ethanol and phenol and compare them to the acidity of pure water.

Plan

(a) Knowing the acidity, pK_a, of each compound, we can rank the compounds according to acidity. (b) From the equations for the ionization of each compound acting as an acid, we can write each K_a expression and substitute into it. Then we solve the resulting expression for the concentration of hydronium ion and determine the pH.

Solution

(a) Phenol has the lowest value for pK_a, so it is the strongest acid. Water is the next strongest acid of these three compounds. Ethanol has the largest pK_a, and, therefore, is the weakest acid.

► The lower the pK_a is, the larger the K_a and the stronger the compound when acting as an acid.

(b) *For water*:

$$H-O-H + H_2O \rightleftharpoons H-O^- + H_3O^+ \quad pK_a \text{ (or } pK_w) = 14.0$$

Therefore, $K_a = 1.0 \times 10^{-14}$

$$H-O-H + H_2O \rightleftharpoons H-O^- \quad + \quad H_3O^+$$

Due to autoionization $1.0 \times 10^{-7}\,M$ $1.0 \times 10^{-7}\,M$

$$[H_3O^+] = 1 \times 10^{-7} \quad pH = 7.00$$

For phenol:

	C_6H_5-O-H	$+ H_2O \rightleftharpoons$	$C_6H_5-O^-$	$+ H_3O^+$
initial	$0.010\,M$		$0\,M$	$1.0 \times 10^{-7}\,M$
change due to ionization	$-x\,M$		$+x\,M$	$+x\,M$
at equil	$(0.010 - x)\,M$		$x\,M$	$x + (1.0 \times 10^{-7})\,M$

$$pK_a = 10.0; \text{ therefore, } K_a = 1.0 \times 10^{-10}$$

Substituting into the ionization constant expression gives

$$1.0 \times 10^{-10} = \frac{x(x + (1.0 \times 10^{-7}))}{0.010 - x}$$

If we make the usual assumptions that $(0.010 - x) \approx 0.010\,M$ and that $x + (1.0 \times 10^{-7}) \approx x\,M$, we have

$$1.0 \times 10^{-10} = \frac{x^2}{0.010} \qquad x^2 = 1.0 \times 10^{-12} \qquad x = 1.0 \times 10^{-6}\,M$$

This gives $[H_3O^+] = 1.0 \times 10^{-6}$ $pH = 6.00$

[Solving by use of the quadratic formula yields $[H_3O^+] = 1.1 \times 10^{-6}\,M$ or $pH = 5.96$]

 TOP & THINK
It is not always *x* that is small enough to neglect! In the numerator, the number 1.0×10^{-7} is negligible compared to the variable *x*.

For ethanol:

$$CH_3CH_2O-H + H_2O \rightleftharpoons CH_3CH_2O^- + H_3O^+ \qquad pK_a = 15.9 \qquad K_a = 1.26 \times 10^{-16}$$

	$CH_3CH_2O-H + H_2O \rightleftharpoons$	$CH_3CH_2O^- +$	H_3O^+
initial	$0.010\ M$	$0\ M$	$1.0 \times 10^{-7}\ M$
change due to ionization	$-x\ M$	$+x\ M$	$+x\ M$
at equil	$(0.010 - x)\ M$	$x\ M$	$x + (1.0 \times 10^{-7})\ M$

We can make the usual assumption that $(0.010 - x) \simeq 0.010\ M$, but the usual assumption that $x + (1.0 \times 10^{-7}) \simeq x\ M$ will not be correct in this case. You can prove this by assuming that $x + (1.0 \times 10^{-7}) \simeq x\ M$ and solving for x. You will obtain $[H_3O^+] = 1 \times 10^{-9}\ M$. For very weak and very dilute acids (or bases), the quantity of hydrogen ion (or hydroxide ion) produced will be small compared to the quantity produced by the solvent, water. For very weak acids, the $[H_3O^+]$ at equilibrium can be considered to be that formed by pure water, $1.0 \times 10^{-7}\ M$, with a negligible amount added by the ionization of the ethanol. The acceptable assumption would be that

$$x + (1.0 \times 10^{-7}) \simeq 1.0 \times 10^{-7}\ M$$

$$K_a = 1.26 \times 10^{-16} = \frac{x\,(1.0 \times 10^{-7})}{0.010} \qquad x = 1.26 \times 10^{-11}$$

Substituting into the concentration values, we obtain

$$[H_3O^+] = (1.26 \times 10^{-11}) + (1.0 \times 10^{-7}) = 1.0 \times 10^{-7} \qquad pH = 7.00$$

We see that phenol, which is a stronger acid than water, results in an aqueous solution that has a pH less than 7 (an acidic solution). However, ethanol, which is a weaker acid than water, results in an aqueous solution that has the same acidity as pure water (pH = 7.0).

▶ The acidity of a molecule depends not just on the polarity of the O—H bond as shown by the greater $\delta-$ and $\delta+$ charge differential (red-blue color intensity) in the electrostatic charge potential plots, but also on the stabilization of the conjugate base (anion) after dissociation of the H^+. Phenol can stabilize its anion more effectively than ethanol due to the delocalized π-bonding in the C_6 ring system.

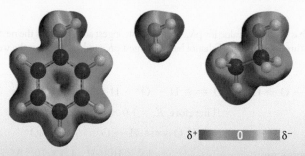

$\delta+$ ▬▬▬ 0 ▬▬▬ $\delta-$

You should now work Exercises 32 and 33.

Thus far we have focused our attention on acids. Very few common weak bases are soluble in water. Aqueous ammonia is the most frequently encountered example. From our earlier discussion of bonding in covalent compounds (see Section 8-8), we recall that there is one lone pair of electrons on the nitrogen atom in NH_3. When ammonia dissolves in water, it accepts H^+ from a water molecule in a reversible reaction (see Section 10-4). We say that NH_3 ionizes slightly when it undergoes this reaction. Aqueous solutions of NH_3 are basic because OH^- ions are produced.

$$:NH_3 + H_2O \rightleftharpoons NH_4^+ + OH^-$$

Amines are derivatives of NH_3 in which one or more H atoms have been replaced by organic groups, as the following structures indicate.

Ammonia, NH_3

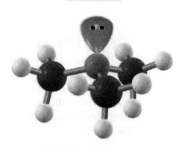

Trimethylamine, $(CH_3)_3N$

H—N̈—H	H₃C—N̈—H	H₃C—N̈—H	H₃C—N̈—CH₃
H	H	CH₃	CH₃
ammonia	methylamine	dimethylamine	trimethylamine
NH_3	CH_3NH_2	$(CH_3)_2NH$	$(CH_3)_3N$

18-4 IONIZATION CONSTANTS FOR WEAK MONOPROTIC ACIDS AND BASES

The nitrogen atom becomes increasingly basic with the addition of hydrocarbon groups such as CH_3 (methyl). These are considered electron-donating and increase the $\delta-$ charge on the nitrogen and the ability of the lone pair of electrons to remove H^+ from water.

Thousands of amines are known, and many are very important in biochemical processes. Low-molecular-weight amines are soluble weak bases. The ionization of trimethylamine, for example, forms trimethylammonium ions and OH^- ions.

The odor that we associate with fish is due to the presence of amines. This is one reason why lemon is often added to seafood. The citric acid (a weak acid) neutralizes the odor of the amines.

$$CH_3-\overset{..}{\underset{\underset{\displaystyle CH_3}{|}}{N}}-CH_3 + \textcircled{H}-OH \rightleftharpoons \left[CH_3-\overset{\overset{\displaystyle H}{|}}{\underset{\underset{\displaystyle CH_3}{|}}{N}}-CH_3\right]^+ + OH^-$$

trimethylamine
$(CH_3)_3N$
$\qquad\qquad$
trimethylammonium ion
$(CH_3)_3NH^+$

Now let us consider the behavior of ammonia in aqueous solutions. The reaction of ammonia with water and its ionization constant expression are

$$NH_3 + H_2O \rightleftharpoons NH_4^+ + OH^-$$

and

$$K_b = \frac{[NH_4^+][OH^-]}{[NH_3]} = 1.8 \times 10^{-5}$$

The fact that K_b for aqueous NH_3 has the same value as K_a for CH_3COOH is just a coincidence. It does tell us that in aqueous solutions of the same concentration, CH_3COOH and NH_3 are ionized to the same extent. Table 18-6 lists K_b and pK_b values for a few common weak bases. Appendix G includes a longer list of K_b values.

We use K_b and pK_b values for weak bases in the same way we used K_a and pK_a values for weak acids.

▶ The subscript "b" in K_b and pK_b indicates that the substance ionizes as a base. We do not include $[H_2O]$ in the K_b expression for the same reasons described for K_a.

$$pK_b = -\log K_b$$

Table 18-6 Ionization Constants and pK_b Values for Some Weak Bases

Base	Ionization Reaction	K_b at 25°C	pK_b
ammonia	$NH_3 + H_2O \rightleftharpoons NH_4^+ + OH^-$	1.8×10^{-5}	4.74
methylamine	$(CH_3)NH_2 + H_2O \rightleftharpoons (CH_3)NH_3^+ + OH^-$	5.0×10^{-4}	3.30
dimethylamine	$(CH_3)_2NH + H_2O \rightleftharpoons (CH_3)_2NH_2^+ + OH^-$	7.4×10^{-4}	3.13
trimethylamine	$(CH_3)_3N + H_2O \rightleftharpoons (CH_3)_3NH^+ + OH^-$	7.4×10^{-5}	4.13
pyridine	$C_5H_5N + H_2O \rightleftharpoons C_5H_5NH^+ + OH^-$	1.5×10^{-9}	8.82
aniline	$C_6H_5NH_2 + H_2O \rightleftharpoons C_6H_5NH_3^+ + OH^-$	4.2×10^{-10}	9.38

aniline

pyridine

EXAMPLE 18-15 pH of a Weak Base Solution

Calculate the $[OH^-]$, pH, and percent ionization for a 0.20 M aqueous NH_3 solution.

Plan

Write the equation for the ionization of aqueous NH_3 and represent the equilibrium concentrations algebraically. Then substitute into the K_b expression and solve for $[OH^-]$ and $[NH_3]_{ionized}$. Knowing $[OH^-]$, we calculate pOH, and from that we can find pH.

Solution

The equation for the ionization of aqueous ammonia and the algebraic representations of equilibrium concentrations follow. Let $x = [NH_3]_{ionized}$.

	NH_3	$+ H_2O \rightleftharpoons$	NH_4^+	$+ OH^-$
initial	0.20 M		0 M	$\approx 0\ M$
change due to rxn	$-x\ M$		$+x\ M$	$+x\ M$
at equil	$(0.20 - x)\ M$		$x\ M$	$x\ M$

▶ Since K_a and K_b values are usually known to only two significant figures, many answers in the following calculations will be limited to two significant figures. pH = 11.28, for example, has only two significant figures because the 11 relates to the exponent of ten.

▶ The value of x is only about 1% of the original concentration, so the assumption is justified.

Substitution into the ionization constant expression gives

$$K_b = \frac{[NH_4^+][OH^-]}{[NH_3]} = 1.8 \times 10^{-5} = \frac{(x)(x)}{(0.20 - x)}$$

Again, we can simplify this equation. The small value of K_b tells us that the base is only slightly ionized, so we can assume that $x \ll 0.20$, or $(0.20 - x) \approx 0.20$, and we have

$$\frac{x^2}{0.20} = 1.8 \times 10^{-5} \quad x^2 = 3.6 \times 10^{-6} \quad x = 1.9 \times 10^{-3}\ M$$

Then $[OH^-] = x = 1.9 \times 10^{-3}\ M$, pOH = 2.72, and pH = 14 - pOH = 11.28. $[NH_3]_{ionized} = x$, so the percent ionization may be calculated.

$$\% \text{ ionization} = \frac{[NH_3]_{ionized}}{[NH_3]_{initial}} \times 100\% = \frac{1.9 \times 10^{-3}}{0.20} \times 100\% = 0.95\% \text{ ionized}$$

You should now work Exercises 54 and 108.

Measurement of the pH of a solution of household ammonia.

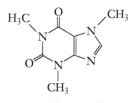

Caffeine is an example of a weak base.

EXAMPLE 18-16 Household Ammonia

The pH of a household ammonia solution is 11.50. What is its molarity? Assume that ammonia is the only base (or acid) present.

Plan

We are given the pH of an aqueous NH_3 solution. Use pH + pOH = 14.00 to find pOH, which we can convert to $[OH^-]$. Then complete the ionization reaction summary, and substitute the representations of equilibrium concentrations into the K_b expression.

Solution

At equilibrium pH = 11.50; we know that pOH = 2.50, so $[OH^-] = 10^{-2.50} = 3.2 \times 10^{-3}\ M$. This $[OH^-]$ results from the reaction, so we can write the change line. Then, letting x represent the *initial* concentration of NH_3, we can complete the reaction summary.

At equilibrium $[OH^-] = 3.2 \times 10^{-3}\ M$, so

	NH_3	$+ H_2O \rightleftharpoons$	NH_4^+	$+$	OH^-
Initial	$x\ M$		$0\ M$		$\approx 0\ M$
change	$-3.2 \times 10^{-3}\ M$		$+3.2 \times 10^{-3}\ M$		$+3.2 \times 10^{-3}\ M$
at equil	$(x - 3.2 \times 10^{-3})\ M$		$3.2 \times 10^{-3}\ M$		$3.2 \times 10^{-3}\ M$

Substituting these values into the K_b expression for aqueous NH_3 gives

$$K_b = \frac{[NH_4^+][OH^-]}{[NH_3]} = \frac{(3.2 \times 10^{-3})(3.2 \times 10^{-3})}{(x - 3.2 \times 10^{-3})} = 1.8 \times 10^{-5}$$

This suggests that $(x - 3.2 \times 10^{-3}) \approx x$. So we can approximate.

$$\frac{(3.2 \times 10^{-3})(3.2 \times 10^{-3})}{x} = 1.8 \times 10^{-5} \quad \text{and} \quad x = 0.57\ M\ NH_3$$

The solution is $0.57\ M\ NH_3$. Our assumption that $(x - 3.2 \times 10^{-3}) \approx x$ is justified.

You should now work Exercises 32, 54, and 56.

Problem-Solving Tip It Is Not Always *x* That Can Be Neglected

Students sometimes wonder about the approximation in Example 18-16, thinking that only *x* can be neglected. We can consider neglecting one term in an expression only when the expression involves *addition* or *subtraction*. The judgment we must make is whether *either* of the terms is sufficiently smaller than the other that ignoring it would not significantly affect the result. In Example 18-16, *x* represents the *initial* concentration of NH_3; 3.2×10^{-3} represents the concentration that ionizes, which cannot be greater than the original *x*. We know that NH_3 is a *weak* base ($K_b = 1.8 \times 10^{-5}$), so only a small amount of the original ionizes. We can safely assume that $3.2 \times 10^{-3} \ll x$, so $(x - 3.2 \times 10^{-3}) \approx x$, the approximation that we used in solving the example.

> **S TOP & THINK**
> We can *never* neglect *any* term in multiplication or division!

18-5 Polyprotic Acids

Thus far we have considered only *monoprotic* weak acids. Acids that can furnish *two* or more hydronium ions per molecule are called **polyprotic acids**. The ionizations of polyprotic acids occur stepwise, that is, one proton at a time. An ionization constant expression can be written for each step, as the following example illustrates. Consider phosphoric acid as a typical polyprotic acid. It contains three acidic hydrogen atoms and ionizes in three steps.

$$H_3PO_4 + H_2O \rightleftharpoons H_3O^+ + H_2PO_4^- \qquad K_{a1} = \frac{[H_3O^+][H_2PO_4^-]}{[H_3PO_4]} = 7.5 \times 10^{-3}$$

$$H_2PO_4^- + H_2O \rightleftharpoons H_3O^+ + HPO_4^{2-} \qquad K_{a2} = \frac{[H_3O^+][HPO_4^{2-}]}{[H_2PO_4^-]} = 6.2 \times 10^{-8}$$

$$HPO_4^{2-} + H_2O \rightleftharpoons H_3O^+ + PO_4^{3-} \qquad K_{a3} = \frac{[H_3O^+][PO_4^{3-}]}{[HPO_4^{2-}]} = 3.6 \times 10^{-13}$$

> ▶ Each K_a expression includes $[H_3O^+]$, so each K_a expression must be satisfied by the total concentration of H_3O^+ in the solution.

We see that K_{a1} is much greater than K_{a2}, and that K_{a2} is much greater than K_{a3}. This is generally true for polyprotic *inorganic* acids (Appendix F). Successive ionization constants often decrease by a factor of approximately 10^4 to 10^6, although some differences are outside this range. Large decreases in the values of successive ionization constants mean that each step in the ionization of a polyprotic acid occurs to a much lesser extent than the previous step. Thus, the $[H_3O^+]$ produced in the first step is very large compared with the $[H_3O^+]$ produced in the second and third steps. As we shall see, except in extremely dilute solutions of H_3PO_4, the concentration of H_3O^+ may be assumed to be that furnished by the first step in the ionization alone.

> ▶ Citrus fruits contain citric acid and ascorbic acid (vitamin C). Citric acid is an example of an organic triprotic weak acid. The structure of citric acid is:

$$\begin{array}{c} COOH \\ | \\ H_2C \\ | \\ HO-C-COOH \\ | \\ H_2C \\ | \\ COOH \end{array}$$

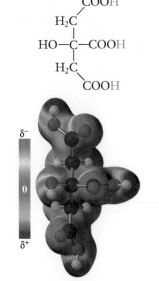

EXAMPLE 18-17 Solution of a Weak Polyprotic Acid

Calculate the concentrations of all species present in 0.10 *M* H_3PO_4.

Plan

Because H_3PO_4 contains three acidic hydrogens per formula unit, we show its ionization in three steps. For each step, write the appropriate ionization equation, with its K_a expression and value. Then represent the equilibrium concentrations from the *first* ionization step, and substitute into the K_{a1} expression. Repeat the procedure for the second and third steps *in order*.

Solution

First we calculate the concentrations of all species that are formed in the first ionization step. Let x = mol/L of H_3PO_4 that ionize; then $x = [H_3O^+]_{1st} = [H_2PO_4^-]$.

$$H_3PO_4 + H_2O \rightleftharpoons H_3O^+ + H_2PO_4^-$$
$$(0.10 - x)M \qquad x\,M \qquad x\,M$$

Substitution into the expression for K_{a1} gives

$$K_{a1} = \frac{[H_3O^+]_{1st}[H_2PO_4^-]}{[H_3PO_4]} = 7.5 \times 10^{-3} = \frac{(x)(x)}{(0.10 - x)}$$

> $x = -3.1 \times 10^{-2}$ is the extraneous root of the quadratic formula.

This equation must be solved by the quadratic formula because K_{a1} is too large to neglect x relative to 0.10 M. Solving gives the *positive* root $x = 2.4 \times 10^{-2}$. Thus, from the first step in the ionization of H_3PO_4,

$$x\, M = [H_3O^+]_{1st} = [H_2PO_4^-] = 2.4 \times 10^{-2}\, M$$

$$(0.10 - x)\, M = [H_3PO_4] = 7.6 \times 10^{-2}\, M$$

For the second step, we use the $[H_3O^+]$ and $[H_2PO_4^-]$ from the first step. Let y = mol/L of $H_2PO_4^{2-}$ that ionize; then $y = [H_3O^+]_{2nd} = [HPO_4^{2-}]$.

$$\begin{array}{cccccc}
H_2PO_4^- & + H_2O & \rightleftharpoons & H_3O^+ & + HPO_4^{2-} \\
(2.4 \times 10^{-2} - y)\, M & & & (2.4 \times 10^{-2} + y)\, M & y\, M
\end{array}$$

$$\underset{\text{from 1st step}}{\nearrow} \quad \underset{\text{from 2nd step}}{\nwarrow}$$

Substitution into the expression for K_{a2} gives

$$K_{a2} = \frac{[H_3O^+]_{1st\ \&\ 2nd}[HPO_4^{2-}]}{[H_2PO_4^-]} = 6.2 \times 10^{-8} = \frac{(2.4 \times 10^{-2} + y)(y)}{(2.4 \times 10^{-2} - y)}$$

Examination of this equation suggests that $y \ll 2.4 \times 10^{-2}$, so

$$\frac{(2.4 \times 10^{-2})(y)}{2.4 \times 10^{-2}} = 6.2 \times 10^{-8} \qquad y = 6.2 \times 10^{-8}\, M = [HPO_4^{2-}] = [H_3O^+]_{2nd}$$

We see that $[HPO_4^{2-}] = K_{a2}$ and $[H_3O^+]_{2nd} \ll [H_3O^+]_{1st}$. In general, in solutions of reasonable concentration of weak polyprotic acids for which $K_{a1} \gg K_{a2}$ and that contain no other electrolytes, *the concentration of the anion produced in the second ionization step is always equal to K_{a2}.*

For the third step, we use $[H_3O^+]$ from the *first* step and $[HPO_4^{2-}]$ from the *second* step. Let z = mol/L of HPO_4^{2-} that ionize; then $z = [H_3O^+]_{3rd} = [PO_4^{3-}]$.

$$\begin{array}{cccccc}
HPO_4^{2-} & + H_2O & \rightleftharpoons & H_3O^+ & + PO_4^{3-} \\
(6.2 \times 10^{-8} - z)\, M & & & (2.4 \times 10^{-2} + z)\, M & z\, M
\end{array}$$

$$\underset{\text{from 2nd step}}{\nearrow} \underset{\text{from 3rd step}}{\nwarrow} \qquad \underset{\text{from 1st step}}{\nearrow} \underset{\text{from 3rd step}}{\nwarrow}$$

$$K_{a3} = \frac{[H_3O^+][PO_4^{3-}]}{[HPO_4^{2-}]} = 3.6 \times 10^{-13} = \frac{(2.4 \times 10^{-2} + z)(z)}{(6.2 \times 10^{-8} - z)}$$

We make the usual simplifying assumption, and find that

$$z\, M = 9.3 \times 10^{-19}\, M = [PO_4^{3-}] = [H_3O^+]_{3rd}$$

> $[H_3O^+]_{2nd} = y = 6.2 \times 10^{-8}$ was disregarded in the second step and is also disregarded here.

The $[H_3O^+]$ found in the three steps can be summarized:

$$
\begin{array}{llll}
[H_3O^+]_{1st} & = 2.4 \times 10^{-2}\, M & = 0.024\, M \\
[H_3O^+]_{2nd} & = 6.2 \times 10^{-8}\, M & = 0.000000062\, M \\
[H_3O^+]_{3rd} & = 9.3 \times 10^{-19}\, M & = 0.00000000000000000093\, M \\
\hline
[H_3O^+]_{total} & = 2.4 \times 10^{-2}\, M & = 0.024\, M
\end{array}
$$

We see that the H_3O^+ furnished by the second and third steps of ionization is negligible compared with that from the first step.

You should now work Exercise 58.

We have calculated the concentrations of the species formed by the ionization of 0.10 M H_3PO_4. These concentrations are compared in Table 18-7. The concentration of $[OH^-]$ in 0.10 M H_3PO_4 is included. It was calculated from the known $[H_3O^+]$ using the ion product for water, $[H_3O^+][OH^-] = 1.0 \times 10^{-14}$.

Table 18-7 Concentrations of the Species in 0.100 M H_3PO_4 (Example 18-17)

Species	Concentration (mol/L)
H_3PO_4	$7.6 \times 10^{-2} = 0.076$
H_3O^+	$2.4 \times 10^{-2} = 0.024$
$H_2PO_4^-$	$2.4 \times 10^{-2} = 0.024$
HPO_4^{2-}	$6.2 \times 10^{-8} = 0.000000062$
OH^-	$4.2 \times 10^{-13} = 0.00000000000042$
PO_4^{3-}	$9.3 \times 10^{-19} = 0.00000000000000000093$

Nonionized H_3PO_4 is present in greater concentration than any other species in 0.10 M H_3PO_4 solution. The only other species present in significant concentrations are H_3O^+ and $H_2PO_4^-$. Similar statements can be made for other weak polyprotic acids for which the last K is very small.

Phosphoric acid is a typical *weak* polyprotic acid. Let us now describe solutions of sulfuric acid, a *very strong* polyprotic acid.

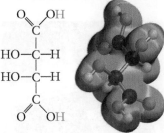

Tartaric acid is an example of a weak polyprotic acid. It contains two acidic protons. Tartaric acid and its salts are found in grapes.

EXAMPLE 18-18 Solution of a Strong Polyprotic Acid

Calculate concentrations of all species present in 0.10 M H_2SO_4. $K_{a2} = 1.2 \times 10^{-2}$.

Plan

Because the first ionization step of H_2SO_4 is complete, we read the concentrations for the first step from the balanced equation. The second ionization step is not complete, so we write the ionization equation, the K_{a2} expression, and the algebraic representations of equilibrium concentrations. Then we substitute into K_{a2} for H_2SO_4.

► The pH of solutions of most polyprotic acids is governed by the first ionization step.

Solution

As we pointed out, the first ionization step of H_2SO_4 is complete.

$$H_2SO_4 + H_2O \xrightarrow{100\%} H_3O^+ + HSO_4^-$$
$$0.10\ M \quad\quad \Rightarrow\ \ 0.10\ M \quad 0.10\ M$$

The second ionization step is not complete, however.

$$HSO_4^- + H_2O \rightleftharpoons H_3O^+ + SO_4^{2-} \quad \text{and} \quad K_{a2} = \frac{[H_3O^+][SO_4^{2-}]}{[HSO_4^-]} = 1.2 \times 10^{-2}$$

Let $x = [HSO_4^-]$ that ionizes. $[H_3O^+]$ is the sum of the concentrations produced in the first and second steps. So we represent the equilibrium concentrations as

$$HSO_4^- \quad + H_2O \rightleftharpoons H_3O^+ \quad + SO_4^{2-}$$
$$(0.10 - x)\ M \quad\quad\quad (0.10 + x)\ M \quad x\ M$$
$$\quad\quad\quad\quad\quad\quad\quad \uparrow \quad\quad\quad \uparrow$$
$$\quad\quad\quad\quad\quad\text{from 1st step} \quad \text{from 2nd step}$$

Substitution into the ionization constant expression for K_{a2} gives

$$K_{a2} = \frac{[H_3O^+][SO_4^{2-}]}{[HSO_4^-]} = 1.2 \times 10^{-2} = \frac{(0.10 + x)(x)}{0.10 - x}$$

A pH of 0.10 M H_3PO_4

B pH of 0.10 M H_2SO_4

pH comparison for equimolar solutions of two polyprotic acids.

Clearly, x cannot be disregarded because K_{a2} is too large. This equation must be solved by the quadratic formula, which gives $x = 0.010$ and $x = -0.12$ (extraneous). So $[H_3O^+]_{2nd} = [SO_4^{2-}] = 0.010\ M$. The concentrations of species in 0.10 M H_2SO_4 are

$$[H_2SO_4] \approx 0\ M \quad ; \quad [HSO_4^-] = (0.10 - x)\ M = 0.09\ M \quad ; \quad [SO_4^{2-}] = 0.010\ M$$
$$[H_3O^+] = (0.10 + x)\ M = 0.11\ M$$
$$[OH^-] = \frac{K_w}{[H_3O^+]} = \frac{1.0 \times 10^{-14}}{0.11} = 9.1 \times 10^{-14}\ M$$

In 0.10 M H_2SO_4 solution, the extent of the second ionization step is 10%.

You should now work Exercise 60.

In Table 18-8 we compare 0.10 M solutions of these two polyprotic acids. Their acidities are very different.

Table 18-8 Comparison of 0.10 M Solutions of Two Polyprotic Acids (Examples 18-17, 18-18)

	0.10 M H_3PO_4	0.10 M H_2SO_4
K_{a1}	7.5×10^{-3}	very large
K_{a2}	6.2×10^{-8}	1.2×10^{-2}
K_{a3}	3.6×10^{-13}	—
$[H_3O^+]$	$2.4 \times 10^{-2}\ M$	$0.11\ M$
pH	1.62	0.96
[nonionized molecules]	$7.6 \times 10^{-2}\ M$	$\approx 0\ M$

18-6 Solvolysis

Solvolysis is the reaction of a substance with the solvent in which it is dissolved. The solvolysis reactions that we will consider in this chapter occur in aqueous solutions, so they are called *hydrolysis* reactions. **Hydrolysis** is the reaction of a substance with water. Some hydrolysis reactions involve reaction with H_3O^+ or OH^- ions. One common kind of hydrolysis involves reaction of the anion of a *weak acid* (the conjugate base) with water to form *nonionized acid* molecules and OH^- ions. This upsets the H_3O^+/OH^- balance in water and produces basic solutions. This reaction is usually represented as

$$\underset{\substack{\text{anion of} \\ \text{weak acid}}}{A^-} + H_2O \rightleftharpoons \underset{\text{weak acid}}{HA} + OH^- \text{ (excess } OH^-\text{, so solution is basic)}$$

Recall that in

basic solutions	$[H_3O^+] < [OH^-]$ or $[OH^-] > 1.0 \times 10^{-7}\ M$
neutral solutions	$[H_3O^+] = [OH^-] = 1.0 \times 10^{-7}\ M$
acidic solutions	$[H_3O^+] > [OH^-]$ or $[H_3O^+] > 1.0 \times 10^{-7}\ M$

In Brønsted-Lowry terminology, anions of strong acids are extremely weak bases, whereas anions of weak acids are stronger bases (see Section 10-4). To refresh your memory, consider the following examples.

Nitric acid, a common strong acid, is essentially completely ionized in dilute aqueous solution. *Dilute* aqueous solutions of HNO_3 contain equal concentrations of H_3O^+ and NO_3^- ions. In dilute aqueous solution, nitrate ions show almost no tendency to react with H_3O^+ ions to form nonionized HNO_3; thus, NO_3^- is a very weak base.

$$HNO_3 + H_2O \xrightarrow{100\%} H_3O^+ + NO_3^-$$

► Examples of conjugate acid–base pairs.

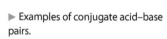

Acid		Conjugate Base
strong (HCl)	$\longrightarrow$	very weak (Cl^-)
weak (HCN)	$\longrightarrow$	stronger, but still weak (CN^-)

Base		Conjugate Acid
strong (OH^-)	$\longrightarrow$	very weak (H_2O)
weak (NH_3)	$\longrightarrow$	stronger, but still weak (NH_4^+)

Beckman Instruments

On the other hand, acetic acid (a weak acid) is only slightly ionized in dilute aqueous solution. Acetate ions have a strong tendency to react with H_3O^+ to form CH_3COOH molecules.

$$CH_3COOH + H_2O \rightleftharpoons H_3O^+ + CH_3COO^-$$

Hence, the CH_3COO^- ion is a stronger base than the NO_3^- ion, but it is still weak.

In dilute solutions, strong acids and strong bases are completely ionized or dissociated. In the following sections, we consider dilute aqueous solutions of salts. Based on our classification of acids and bases, we can identify four different kinds of salts.

1. Salts of strong bases and strong acids
2. Salts of strong bases and weak acids
3. Salts of weak bases and strong acids
4. Salts of weak bases and weak acids

18-7 Salts of Strong Bases and Strong Acids

We could also describe these as salts that contain the cation of a strong base and the anion of a strong acid. Salts derived from strong bases and strong acids give *neutral* solutions because neither the cation nor the anion reacts appreciably with H_2O. Consider an aqueous solution of NaCl, which is the salt of the strong base NaOH and the strong acid HCl. Sodium chloride is ionic in the solid state and dissociates into hydrated ions ($Na^+(aq)$ and $Cl^-(aq)$) in H_2O. H_2O ionizes slightly to produce equal concentrations of H_3O^+ and OH^- ions.

$$NaCl\ (solid) \xrightarrow[100\%]{H_2O} Na^+ + Cl^-$$

$$H_2O + H_2O \rightleftharpoons OH^- + H_3O^+$$

The salt of a strong acid and a strong base is often referred to as a *neutral salt.*

We see that aqueous solutions of NaCl contain four ions, Na^+, Cl^-, H_3O^+ and OH^-. The cation of the salt, Na^+, is such a weak acid that it does not react appreciably with water. The anion of the salt, Cl^-, is such a weak base that it does not react appreciably with water. Solutions of salts of strong bases and strong acids are therefore *neutral* because neither ion of such a salt reacts to upset the H_3O^+/OH^- balance in water.

18-8 Salts of Strong Bases and Weak Acids

When salts derived from strong bases and weak acids are dissolved in water, the resulting solutions are always basic. This is because anions of weak acids react with water to form hydroxide ions. Consider a solution of sodium acetate, $NaCH_3COO$, which is the salt of the strong base NaOH and the weak acid CH_3COOH (acetic acid). It is soluble and dissociates completely in water.

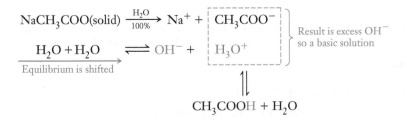

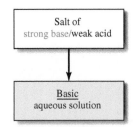

Though the salt of a strong base and a weak acid is not strictly included in the definition of a basic salt (see Section 10-9), many chemists refer to such a salt as a *weakly basic salt.*

Acetate ion is the conjugate base of a *weak* acid, CH_3COOH. Thus, it reacts to some extent with H_3O^+ to form CH_3COOH. As H_3O^+ is removed from the solution, causing more H_2O to ionize, an excess of OH^- builds up. So the solution becomes basic. The net

▶ This is like the reaction of a molecular weak base

$$NH_3 + H_2O \rightleftharpoons NH_4^+ + OH^-$$

(excess OH^- is produced; the solution becomes basic)

result of the preceding equations can be written as a single equation. This equation describes the *hydrolysis of acetate ions*.

$$CH_3COO^- + H_2O \rightleftharpoons CH_3COOH + OH^-$$

The equilibrium constant for this reaction is called a (base) *hydrolysis constant*, or K_b for CH_3COO^-.

$$K_b = \frac{[CH_3COOH][OH^-]}{[CH_3COO^-]} \qquad (K_b \text{ for } CH_3COO^-)$$

We can evaluate this equilibrium constant from other known expressions. Multiply the preceding expression by $[H_3O^+]/[H_3O^+]$ to give

$$K_b = \frac{[CH_3COOH][OH^-]}{[CH_3COO^-]} \times \frac{[H_3O^+]}{[H_3O^+]} = \frac{[CH_3COOH]}{[H_3O^+][CH_3COO^-]} \times \frac{[H_3O^+][OH^-]}{1}$$

We see that

$$K_b = \frac{1}{K_{a(CH_3COOH)}} \times \frac{K_w}{1} = \frac{K_w}{K_{a(CH_3COOH)}} = \frac{1.0 \times 10^{-14}}{1.8 \times 10^{-5}}$$

which gives

$$K_b = \frac{[CH_3COOH][OH^-]}{[CH_3COO^-]} = 5.6 \times 10^{-10}$$

We have calculated K_b, the hydrolysis constant for the acetate ion, CH_3COO^-, using the K_a value for the acid from which the anion was formed.

We can do the same kind of calculations for the anion of any weak monoprotic acid and find that $K_b = K_w/K_a$, where K_a refers to the ionization constant for the weak monoprotic acid from which the anion is derived.

This equation can be rearranged to

$$K_w = K_a K_b \qquad \text{(valid for } \textit{any conjugate acid–base pair} \text{ in aqueous solution)}$$

If either K_a or K_b is known, the other can be calculated.

STOP & THINK

Just as we can write a K_b expression to describe the extent of ionization of NH_3, we can do the same for the acetate ion. The reaction of NH_3 with H_2O is called an *ionization* because a neutral base reacts with water to produce the NH_4^+ and OH^- ions. It is also a *hydrolysis* reaction.

▶ Base hydrolysis constants, K_b values, for anions of weak acids can be determined experimentally. The values obtained from experiments agree with the calculated values. Please note that this K_b refers to a reaction in which the anion of a weak acid acts as a base.

▶ For conjugate acid–base pairs:

Small K_a　　　　　Large K_b
Large pK_a　　　　Small pK_b

Stronger conjugate acid　　Stronger conjugate base

Large K_a　　　　Small K_b
Small pK_a　　　　Large pK_b

EXAMPLE 18-19 K_b for the Anion of a Weak Acid

(a) Write the equation for the reaction of the base CN^- with water. (b) The value of the ionization constant for hydrocyanic acid, HCN, is 4.0×10^{-10}. What is the value of K_b for the cyanide ion, CN^-?

Plan

(a) The base CN^- accepts H^+ from H_2O to form the weak acid HCN and OH^- ions.
(b) We know that $K_a K_b = K_w$. So we solve for K_b and substitute into the equation.

Solution

(a) $CN^- + H_2O \rightleftharpoons HCN + OH^-$

(b) We are given $K_a = 4.0 \times 10^{-10}$ for HCN, and we know that $K_w = 1.0 \times 10^{-14}$.

$$K_{b(CN^-)} = \frac{K_w}{K_a} = \frac{1.0 \times 10^{-14}}{4.0 \times 10^{-10}} = 2.5 \times 10^{-5}$$

You should now work Exercises 74 and 76.

EXAMPLE 18-20 Calculations Based on Hydrolysis

Calculate $[OH^-]$, pH, and the percent hydrolysis for 0.10 M solutions of (a) sodium acetate, $NaCH_3COO$, and (b) sodium cyanide, $NaCN$.

Plan

We recall that all sodium salts are soluble and that all soluble ionic salts are completely dissociated in H_2O. We recognize that both $NaCH_3COO$ and $NaCN$ are salts of strong bases (which provide the cation) and weak acids (which provide the anion). The anions in such salts hydrolyze to give basic solutions. In the preceding text we determined that K_b for $CH_3COO^- = 5.6 \times 10^{-10}$, and in Example 18-19 we determined that K_b for $CN^- = 2.5 \times 10^{-5}$. As we have done before, we first write the appropriate chemical equation and equilibrium constant expression. Then we complete the reaction summary, substitute the algebraic representations of equilibrium concentrations into the equilibrium constant expression, and solve for the desired concentration(s).

> **S TOP & THINK**
> If we did not know either K_b value, we could use $K_{a(CH_3COOH)}$ to find $K_{b(CH_3COO^-)}$, or $K_{a(HCN)}$ to find $K_{b(CN^-)}$.

Solution

(a) The overall equation for the reaction of CH_3COO^- with H_2O and its equilibrium constant expression are

$$CH_3COO^- + H_2O \rightleftharpoons CH_3COOH + OH^-$$

$$K_b = \frac{[CH_3COOH][OH^-]}{[CH_3COO^-]} = 5.6 \times 10^{-10}$$

Let x = mol/L of CH_3COO^- that hydrolyzes. Then $x = [CH_3COOH] = [OH^-]$.

	CH_3COO^-	$+$	H_2O	$\rightleftharpoons$	CH_3COOH	$+$	OH^-
initial	0.10 M				0 M		$\approx 0\ M$
change due to rxn	$-x\ M$				$+x\ M$		$+x\ M$
at equil	$(0.10 - x)\ M$				$x\ M$		$x\ M$

Because the value of K_b (5.6×10^{-10}) is quite small, we know that the reaction does not go very far. We can assume $x \ll 0.10$, so $(0.10 - x) \approx 0.10$; this lets us simplify the equation to

$$5.6 \times 10^{-10} = \frac{(x)(x)}{0.10} \quad \text{so} \quad x = 7.5 \times 10^{-6}$$

$$x = 7.5 \times 10^{-6}\ M = [OH^-]; \text{ so pOH} = 5.12 \quad \text{and} \quad \boxed{pH = 8.88}$$

> ▶ pH = 14 − pOH

The 0.10 M $NaCH_3COO$ solution is distinctly basic.

$$\% \text{ hydrolysis} = \frac{[CH_3COO^-]_{hydrolyzed}}{[CH_3COO^-]_{initial}} \times 100\% = \frac{7.5 \times 10^{-6}\ M}{0.10\ M} \times 100\%$$

$$= 0.0075\% \text{ hydrolysis}$$

(b) Perform the same kind of calculation for 0.10 M $NaCN$. Let y = mol/L of CN^- that hydrolyzes. Then $y = [HCN] = [OH^-]$.

	CN^-	$+$	H_2O	$\rightleftharpoons$	HCN	$+$	OH^-
initial	0.10 M				0 M		$\approx 0\ M$
change due to rxn	$-y\ M$				$+y\ M$		$+y\ M$
at equil	$(0.10 - y)\ M$				$y\ M$		$y\ M$

$$K_b = \frac{[HCN][OH^-]}{[CN^-]} = 2.5 \times 10^{-5}$$

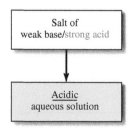

A The pH of 0.10 M NaCH$_3$COO is 8.88.

B The pH of 0.10 M NaCN is 11.20.

An inert solid has been suspended in the liquids to improve the quality of these photographs.

Substituting into this expression and realizing that $(0.10 - y) \approx 0.10$ gives

$$\frac{(y)(y)}{(0.10)} = 2.5 \times 10^{-5} \quad \text{so} \quad y = 1.6 \times 10^{-3} \, M$$

$$y = [OH^-] = 1.6 \times 10^{-3} M \quad pOH = 2.80 \quad \text{and} \quad pH = 11.20$$

The 0.10 M NaCN solution is even more basic (about 210 times) than the 0.10 M NaCH$_3$COO solution in Part (a).

$$\% \text{ hydrolysis} = \frac{[CN^-]_{\text{hydrolyzed}}}{[CN^-]_{\text{initial}}} \times 100\% = \frac{1.6 \times 10^{-3} \, M}{0.10 \, M} \times 100\%$$

$$= 1.6\% \text{ hydrolysis}$$

You should now work Exercises 79 and 80.

The 0.10 M solution of NaCN is considerably more basic than the 0.10 M solution of NaCH$_3$COO because CN$^-$ is a much stronger base than CH$_3$COO$^-$. This is expected because HCN is a much weaker acid than CH$_3$COOH; the K_b for CN$^-$ is much larger than the K_b for CH$_3$COO$^-$.

The percent hydrolysis for 0.10 M CN$^-$ (1.6%) is about 210 times greater than the percent hydrolysis for 0.10 M CH$_3$COO$^-$ (0.0075%). In Table 18-9 we compare 0.10 M solutions of CH$_3$COO$^-$, CN$^-$, and NH$_3$ (the familiar molecular weak base). We see that CH$_3$COO$^-$ is a much weaker base than NH$_3$, whereas CN$^-$ is a slightly stronger base than NH$_3$.

Table 18-9 Data for 0.10 M Solutions of NaCH$_3$COO, NaCN, and NH$_3$

	0.10 M NaCH$_3$COO	0.10 M NaCN	0.10 M aq NH$_3$
K_a for parent acid	1.8×10^{-5}	4.0×10^{-10}	
K_b for anion	5.6×10^{-10}	2.5×10^{-5}	K_b for NH$_3$ = 1.8×10^{-5}
$[OH^-]$	$7.5 \times 10^{-6} \, M$	$1.6 \times 10^{-3} \, M$	$1.3 \times 10^{-3} \, M$
% hydrolysis	0.0075%	1.6%	1.3% ionized
pH	8.88	11.20	11.11

18-9 Salts of Weak Bases and Strong Acids

The second common kind of hydrolysis reaction involves the reaction of the cation of a weak base with water to form nonionized molecules of the weak base and H$_3$O$^+$ ions. This upsets the H$_3$O$^+$/OH$^-$ balance in water, giving an excess of H$_3$O$^+$ and making such solutions *acidic*. Consider a solution of ammonium chloride, NH$_4$Cl, the salt of aqueous NH$_3$ and HCl.

▶ Ammonium chloride is an ionic salt that is soluble in water.

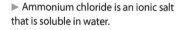

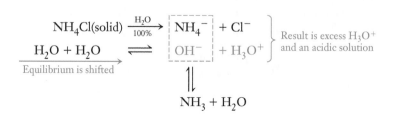

Ammonium ions from NH_4Cl react to some extent with OH^- to form nonionized NH_3 and H_2O molecules. This reaction removes OH^- from the system, so it causes more H_2O to ionize to produce an excess of H_3O^+ and an acidic solution. The net result of the preceding equations can be written as a single equation with its equilibrium constant expression.

$$NH_4^+ + H_2O \rightleftharpoons NH_3 + H_3O^+ \qquad K_a = \frac{[NH_3][H_3O^+]}{[NH_4^+]}$$

▶ Many chemists refer to the salt produced by the reaction of a weak base and a strong acid as a *weakly acidic salt*.

The expression $K_w = K_a K_b$ is valid for *any* conjugate acid-base pair in aqueous solution. We use it for the NH_4^+/NH_3 pair.

$$K_{a(NH_4^+)} = \frac{K_w}{K_{b(NH_3)}} = \frac{1.0 \times 10^{-14}}{1.8 \times 10^{-5}} = 5.6 \times 10^{-10} = \frac{[NH_3][H_3O^+]}{[NH_4^+]}$$

▶ Similar equations can be written for cations derived from other weak bases such as $CH_3NH_3^+$ and $(CH_3)_2NH_2^+$.

The fact that K_a for the ammonium ion, NH_4^+, is the same as K_b for the acetate ion should not be surprising. Recall that the ionization constants for CH_3COOH and aqueous NH_3 are equal (by coincidence). Thus, we expect CH_3COO^- to hydrolyze to the same extent as NH_4^+ does.

▶ If you wish to derive $K_a = K_w/K_b$ for this case, multiply the K_a expression by $[OH^-]/[OH^-]$ and simplify. This is similar to the derivation that was shown in detail for K_b in Section 18-8.

EXAMPLE 18-21 pH of a Soluble Salt of a Strong Acid and a Weak Base

Calculate the pH of a 0.20 *M* solution of ammonium nitrate, NH_4NO_3.

Plan

We recognize that NH_4NO_3 is the salt of a weak base, NH_3, and a strong acid, HNO_3, and that the cations of such salts hydrolyze to give acidic solutions. From our earlier calculations, we know that K_a for $NH_4^+ = 5.6 \times 10^{-10}$. We proceed as we did in Example 18-20.

▶ Ammonium nitrate is widely used as a fertilizer because of its high nitrogen content. It contributes to soil acidity.

Solution

The cation of the weak base reacts with H_2O. Let x = mol/L of NH_4^+ that hydrolyzes. Then $x = [NH_3] = [H_3O^+]$.

	NH_4^+	$+ H_2O \rightleftharpoons$	NH_3	$+ H_3O^+$
initial	0.20 *M*		0 *M*	≈0 *M*
change due to rxn	$-x$ *M*		$+x$ *M*	$+x$ *M*
at equil	$(0.20 - x)$ *M*		x *M*	x *M*

Substituting into the K_a expression gives

$$K_a = \frac{[NH_3][H_3O^+]}{[NH_4^+]} = \frac{(x)(x)}{0.20-x} = 5.6 \times 10^{-10}$$

Beckman Instruments

The pH of 0.20 *M* NH_4NO_3 solution is 4.96.

Making the usual simplifying assumption gives $x = 1.1 \times 10^{-5}$ *M* $= [H_3O^+]$ and pH = 4.96. The 0.20 *M* NH_4NO_3 solution is distinctly acidic.

You should now work Exercise 86.

18-10 Salts of Weak Bases and Weak Acids

Salts of weak bases and weak acids are the fourth class of salts. Most are soluble. Salts of weak bases and weak acids contain cations that would give acidic solutions and also contain anions that would give basic solutions. Will solutions of such salts be neutral, basic, or acidic? They may be any one of the three depending on the relative strengths of the weak molecular acid and weak molecular base from which each salt is derived. Thus, salts of this class may be divided into three types that depend on the relative strengths of their parent weak bases and weak acids.

CHEMISTRY IN USE

Taming Dangerous Acids with Harmless Salts

From yellowing paper in old books and newsprint to heartburn and environmental spills, many Americans encounter the effects of unwanted or excessive amounts of acid. Neutralizing unwanted acids with hydroxide bases might appear to be a good way to combat these acids. But even more effective chemicals exist that can neutralize acids without the risks posed by hydroxide bases. These acid-neutralizing chemicals are salts of weak acids and strong bases. Such salts can neutralize acids because hydrolysis makes their aqueous solutions basic. More importantly, there is a significant advantage in using relatively harmless salts such as sodium hydrogen carbonate (baking soda) rather than stronger bases such as sodium hydroxide (lye). For example, if we used too much sodium hydroxide to neutralize sulfuric acid spilled from a car battery, any excess lye left behind would pose an environmental and human health threat about equal to that of the spilled sulfuric acid. (Lye is the major ingredient in such commercial products as oven cleaners and Drāno.) We would not be concerned, however, if a little baking soda were left on the ground after the sulfuric acid from the car battery had been neutralized.

The same principle applies to acid indigestion. Rather than swallow lye (ugh!) or some other strong base to neutralize excess stomach acid, most people take antacids. Antacids typically contain salts such as calcium carbonate, sodium hydrogen carbonate (sodium bicarbonate), and magnesium carbonate, all of which are salts of weak acids. These salts hydrolyze to form hydroxide ions that reduce the degree of acidity in the stomach. Physicians also prescribe these and similar salts to treat peptic ulcers. The repeated use of antacids should always be under the supervision of a physician.

Salts of weak acids and strong bases can be used effectively against a major acid spill in much the same way they are used against sulfuric acid from a car battery or excess stomach acid. In a recent major acid spill, a tank car filled with nitric acid was punctured by the coupling of another rail car, spilling 22,000 gallons of concentrated nitric acid onto the ground. Several thousand residents living near the spill were evacuated. There were no fatalities or serious injuries, and there was no major environmental damage; resident firefighters neutralized the concentrated nitric acid by using airport snow blowers to spread *relatively* harmless sodium carbonate (washing soda) over the contaminated area.

Salts of weak acids and strong bases are also being used to combat the destructive aging process of paper. Think how serious this problem is for the Library of Congress, which loses 70,000 books each year to the acid-promoted decomposition of aging paper. Many of the twenty million books in the Library of Congress have a life expectancy of only 25 to 40 years. Paper ages because of the hydrolysis of aluminum sulfate. Aluminum sulfate has been used in the paper manufacturing process since the 1850s because it is an inexpensive sizing compound (it keeps ink from spreading out on paper). Aluminum sulfate is the salt of an insoluble weak base and a strong acid; it hydrolyzes in the water in paper (typically 4−7% H_2O) to give an acidic environment. The acid slowly eats away at cellulose fibers, which causes the paper to turn yellow and eventually disintegrate. To combat this aging, the Library of Congress individually treats its collections with solutions of salts of weak acids and strong bases at great cost. Meanwhile, the paper industry is fighting this aging process by increasing its output of alkaline paper. Some alkaline paper contains calcium carbonate, the same salt found in several brands of antacids. Calcium carbonate increases the pH of paper to between 7.5 and 8.5. Special manufacturing techniques produce calcium carbonate that is very fine and that has uniform particle size. Alkaline papers are expected to last about 300 years, in contrast to the average 25- to 40-year life expectancy of standard acidic paper.

Salts that hydrolyze to produce basic solutions can settle upset stomachs, prevent yellowing pages, and neutralize major and minor acid spills. A knowledge of hydrolysis is very useful and has many applications.

RONALD DELORENZO

Science Museum Library

ELEMENTS

⊙	Hydrogen	1	⊕ Strontian	46
◐	Azote	5	✱ Barytes	68
●	Carbon	54	I Iron	50
○	Oxygen	7	Z Zinc	56
⊘	Phosphorus	9	C Copper	56
⊕	Sulphur	13	L Lead	90
⊗	Magnesia	20	S Silver	190
⊖	Lime	24	G Gold	190
◫	Soda	28	P Platina	190
⦀	Potash	42	✷ Mercury	167

Salts of Weak Bases and Weak Acids for Which Parent K_b = Parent K_a

The most common salt of this type is ammonium acetate, NH_4CH_3COO, the salt of aqueous NH_3 and CH_3COOH. The ionization constants for both aqueous NH_3 and CH_3COOH are 1.8×10^{-5}. We know that ammonium ions react with water to produce H_3O^+.

$$NH_4^+ + H_2O \rightleftharpoons NH_3 + H_3O^+ \quad K_a = \frac{[NH_3][H_3O^+]}{[NH_4^+]} = 5.6 \times 10^{-10}$$

We also recall that acetate ions react with water to produce OH^-.

$$CH_3COO^- + H_2O \rightleftharpoons CH_3COOH + OH^-$$

$$K_b = \frac{[CH_3COOH][OH^-]}{[CH_3COO^-]} = 5.6 \times 10^{-10}$$

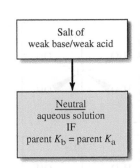

Because these K values are equal, the NH_4^+ produces just as many H_3O^+ ions as the CH_3COO^- produces OH^- ions. Thus, we predict that ammonium acetate solutions are neutral, and they are. There are very few salts, however, that have cations and anions with equal K values.

Salts of Weak Bases and Weak Acids for Which Parent K_b > Parent K_a

Salts of weak bases and weak acids for which K_b is greater than K_a are always basic because the anion of the weaker acid hydrolyzes to a greater extent than the cation of the stronger base.

Consider NH_4CN, ammonium cyanide. K_a for HCN (4.0×10^{-10}) is much smaller than K_b for NH_3 (1.8×10^{-5}), so K_b for CN^- (2.5×10^{-5}) is much larger than K_a for NH_4^+ (5.6×10^{-10}). This tells us that the CN^- ions hydrolyze to a much greater extent than do NH_4^+ ions, and so ammonium cyanide solutions are distinctly basic. Stated differently, CN^- is much stronger as a base than NH_4^+ is as an acid.

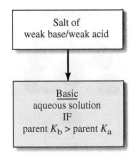

$$\left.\begin{array}{l} NH_4^+ + H_2O \rightleftharpoons NH_3 + H_3O^+ \\ CN^- + H_2O \rightleftharpoons HCN + OH^- \end{array}\right\} \rightarrow 2H_2O$$

The second reaction occurs to a greater extent; ∴ the solution is basic.

Salts of Weak Bases and Weak Acids for Which Parent K_b < Parent K_a

Salts of weak bases and weak acids for which K_b is less than K_a are acidic because the cation of the weaker base hydrolyzes to a greater extent than the anion of the stronger acid. Consider ammonium fluoride, NH_4F, the salt of aqueous ammonia and hydrofluoric acid.

K_b for aqueous NH_3 is 1.8×10^{-5} and K_a for HF is 7.2×10^{-4}. So the K_a value for NH_4^+ (5.6×10^{-10}) is slightly larger than the K_b value for F^- (1.4×10^{-11}). This tells us that NH_4^+ ions hydrolyze to a slightly greater extent than F^- ions. In other words, NH_4^+ is slightly stronger as an acid than F^- is as a base. Ammonium fluoride solutions are slightly acidic.

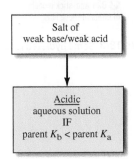

$$\left.\begin{array}{l} NH_4^+ + H_2O \rightleftharpoons NH_3 + H_3O^+ \\ F^- + H_2O \rightleftharpoons HF + OH^- \end{array}\right\} \rightarrow 2H_2O$$

The first reaction occurs to a greater extent; ∴ the solution is acidic.

EXAMPLE 18-22 Predicting Which Salts Are Acidic and Which Are Basic

Classify each of the following soil additives as an acidic salt, a neutral salt, or a basic salt: (a) $(NH_4)_2SO_4$; (b) NH_4NO_3; (c) Na_2CO_3.

Plan

We recognize these salts as being the products of acid/base reactions. We can categorize each parent acid and base as being stronger or weaker. We then predict the properties of each salt based upon the strength of its parent acid and parent base.

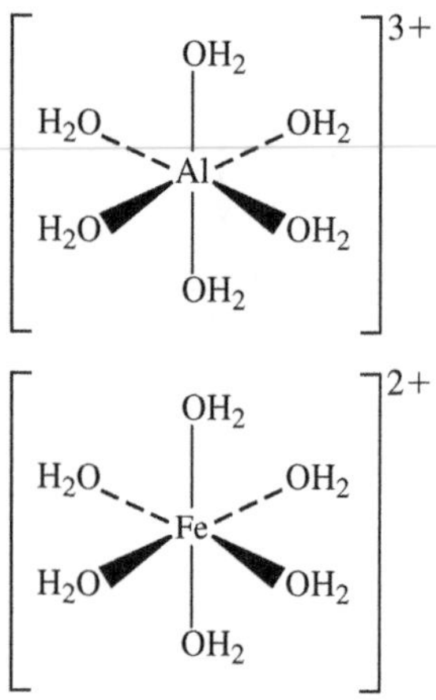

A Lewis structures of hydrated aluminum ions, $[Al(OH_2)_6]^{3+}$, and hydrated iron(II) ions, $[Fe(OH_2)_6]^{2+}$

B Ball-and-stick model of either of these ions

Figure 18-3 Some hydrated ions.

Solution

(a) $(NH_4)_2SO_4$ is the salt formed by the reaction of NH_3 (a weak base) and H_2SO_4 (a strong acid). Because H_2SO_4 is stronger as an acid than NH_3 is as a base, $(NH_4)_2SO_4$ is an acidic salt.

(b) NH_4NO_3 is the salt formed by the reaction of NH_3 (a weak base) and HNO_3 (a strong acid). Because HNO_3 is stronger as an acid than NH_3 is as a base, NH_4NO_3 is an acidic salt.

(c) Na_2CO_3 is the salt formed by the reaction of NaOH (a strong base) and H_2CO_3 (a weak acid). Because NaOH is stronger as a base then H_2CO_3 is as an acid, Na_2CO_3 is a basic salt.

You should now work Exercise 102.

18-11 Salts That Contain Small, Highly Charged Cations

Solutions of certain common salts of strong acids are acidic. For this reason, many homeowners apply iron(II) sulfate, $FeSO_4 \cdot 7H_2O$, or aluminum sulfate, $Al_2(SO_4)_3 \cdot 18H_2O$, to the soil around "acid-loving" plants such as azaleas, camelias, and hollies. You have probably tasted the sour, "acid" taste of alum, $KAl(SO_4)_2 \cdot 12H_2O$, a substance that is frequently added to pickles.

Each of these salts contains a small, highly charged cation and the anion of a strong acid. Solutions of such salts are acidic because these cations hydrolyze to produce excess hydronium ions. Consider aluminum chloride, $AlCl_3$, as a typical example. When solid anhydrous $AlCl_3$ is added to water, the water becomes very warm as the Al^{3+} ions become hydrated in solution. In many cases, the interaction between positively charged ions and the negative ends of polar water molecules is so strong that salts crystallized from aqueous solution contain definite numbers of water molecules. Salts containing Al^{3+}, Fe^{2+}, Fe^{3+}, and Cr^{3+} ions usually crystallize from aqueous solutions with six water molecules bonded (coordinated) to each metal ion. These salts contain the hydrated cations $[Al(OH_2)_6]^{3+}$, $[Fe(OH_2)_6]^{2+}$, $[Fe(OH_2)_6]^{3+}$, and $[Cr(OH_2)_6]^{3+}$, respectively, in the solid state. Such species also exist in aqueous solutions. Each of these species is octahedral, meaning that the metal ion (M^{n+}) is located at the center of a regular octahedron, and the O atoms in six H_2O molecules are located at the corners (Figure 18-3). In the metal–oxygen bonds of the hydrated cation, electron density is decreased around the O end of each H_2O molecule by the positively charged metal ion. This weakens the H—O bonds in coordinated H_2O molecules relative to the H—O bonds in noncoordinated H_2O molecules. Consequently, the coordinated H_2O molecules can donate H^+ to solvent H_2O molecules to form H_3O^+ ions. This produces acidic solutions (Figure 18-4).

The equation for the hydrolysis of hydrated Al^{3+} is written as follows.

$$[Al(OH_2)_6]^{3+} + H_2O \rightleftharpoons [Al(OH)(OH_2)_5]^{2+} + H_3O^+$$

$$K_a = \frac{[[Al(OH)(OH_2)_5]^{2+}][H_3O^+]}{[[Al(OH_2)_6]^{3+}]} = 1.2 \times 10^{-5}$$

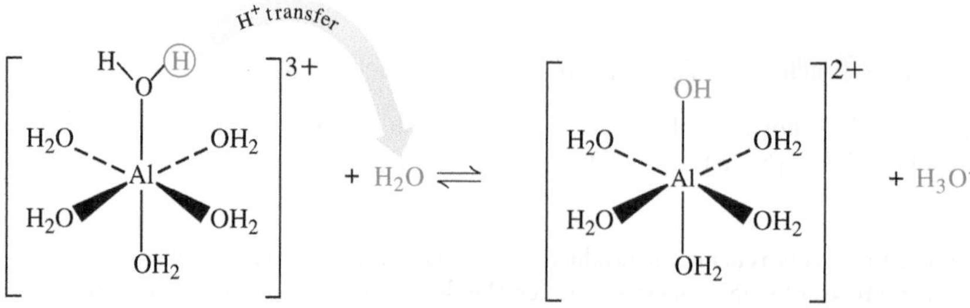

Figure 18-4 Hydrolysis of hydrated aluminum ions produces H_3O^+ by the transfer of a proton from a coordinated H_2O molecule to a noncoordinated one.

Removing one H^+ converts a coordinated water molecule to a coordinated hydroxide ion and decreases the positive charge on the hydrated species.

Hydrolysis of hydrated small, highly charged cations may occur beyond the first step. In many cases these reactions are quite complex. They may involve two or more cations reacting with each other to form large polymeric species. For most common hydrated cations, consideration of the first hydrolysis constant is adequate for our calculations.

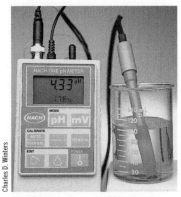

The blue color and the acidity of a 0.10 M $CuSO_4$ solution are both due to the hydrated Cu^{2+} ion.

EXAMPLE 18-23 Percent Hydrolysis

Calculate the pH and percent hydrolysis in 0.10 M $AlCl_3$ solution. $K_a = 1.2 \times 10^{-5}$ for $[Al(OH_2)_6]^{3+}$ (often abbreviated Al^{3+}).

Plan

We recognize that $AlCl_3$ produces a hydrated, small, highly charged cation that hydrolyzes to give an acidic solution. We represent the equilibrium concentrations and proceed as we did in earlier examples.

Solution

The equation for the reaction and its hydrolysis constant can be represented as

$$[Al(OH_2)_6]^{3+} + H_2O \rightleftharpoons [Al(OH)(OH_2)_5]^{2+} + H_3O^+$$

$$K_a = \frac{[[Al(OH)(OH_2)_5]^{2+}][H_3O^+]}{[[Al(OH_2)_6]^{3+}]} = 1.2 \times 10^{-5}$$

Let x = mol/L of $[Al(OH_2)_6]^{3+}$ that hydrolyzes. Then $x = [Al(OH)(OH_2)_5]^{2+} = [H_3O^+]$.

	$[Al(OH_2)_6]^{3+}$	+ $H_2O \rightleftharpoons$	$[Al(OH)(OH_2)_5]^{2+}$	+ H_3O^+
initial	0.10 M			
change due to rxn	$-x\,M$		$+x\,M$	$+x\,M$
at equil	$(0.10 - x)\,M$		$x\,M$	$x\,M$

$$\frac{(x)(x)}{(0.10 - x)} = 1.2 \times 10^{-5} \text{ so } x = 1.1 \times 10^{-3}$$

$[H_3O^+] = 1.1 \times 10^{-3}\ M$, pH = 2.96; the solution is quite acidic.

▶ Recall that x represents the concentration of Al^{3+} that hydrolyzes.

$$\% \text{ hydrolysis} = \frac{[Al^{3+}]_{hydrolyzed}}{[Al^{3+}]_{initial}} \times 100\% = \frac{1.1 \times 10^{-3}\ M}{0.10\ M} \times 100\% = 1.1\% \text{ hydrolyzed}$$

As a reference point, CH_3COOH is 1.3% ionized in 0.10 M solution (Example 18-11). In 0.10 M solution, $AlCl_3$ is 1.1% hydrolyzed. The acidities of the two solutions are very similar.

▶ The pH of 0.10 M $AlCl_3$ is 2.96. The pH of 0.10 M CH_3COOH is 2.89.

You should now work Exercise 92.

Smaller, more highly charged cations are stronger acids than larger, less highly charged cations (Table 18-10). This is because the smaller, more highly charged cations interact with coordinated water molecules more strongly.

For isoelectronic cations in the same period in the periodic table, the smaller, more highly charged cation is the stronger acid. (Compare K_a values for hydrated Li^+ and Be^{2+} and for hydrated Na^+, Mg^{2+}, and Al^{3+}.) For cations with the same charge from the same group in the periodic table, the smaller cation hydrolyzes to a greater extent. (Compare K_a values for hydrated Be^{2+} and Mg^{2+}.) If we compare cations of the same element in different oxidation states, the smaller, more highly charged cation is the stronger acid. (Compare K_a values for hydrated Fe^{2+} and Fe^{3+} and for hydrated Co^{2+} and Co^{3+}.)

Table 18-10 Ionic Radii and Hydrolysis Constants for Some Cations

Cation	Ionic Radius (Å)	Hydrated Cation	K_a
Li^+	0.90	$[Li(OH_2)_4]^+$	1×10^{-14}
Be^{2+}	0.59	$[Be(OH_2)_4]^{2+}$	1.0×10^{-5}
Na^+	1.16	$[Na(OH_2)_6]^+$	10^{-14}
Mg^{2+}	0.85	$[Mg(OH_2)_6]^{2+}$	3.0×10^{-12}
Al^{3+}	0.68	$[Al(OH_2)_6]^{3+}$	1.2×10^{-5}
Fe^{2+}	0.76	$[Fe(OH_2)_6]^{2+}$	3.0×10^{-10}
Fe^{3+}	0.64	$[Fe(OH_2)_6]^{3+}$	4.0×10^{-3}
Co^{2+}	0.74	$[Co(OH_2)_6]^{2+}$	5.0×10^{-10}
Co^{3+}	0.63	$[Co(OH_2)_6]^{3+}$	1.7×10^{-2}
Cu^{2+}	0.96	$[Cu(OH_2)_6]^{2+}$	1.0×10^{-8}
Zn^{2+}	0.74	$[Zn(OH_2)_6]^{2+}$	2.5×10^{-10}
Hg^{2+}	1.10	$[Hg(OH_2)_6]^{2+}$	8.3×10^{-7}
Bi^{3+}	0.74	$[Bi(OH_2)_6]^{3+}$	1.0×10^{-2}

Charles Steele

Pepto-Bismol contains $BiO(HOC_4H_6COO)$, bismuth subsalicylate, a *hydrolyzed* bismuth salt. Such salts "coat" polar surfaces such as glass and the lining of the stomach.

KEY TERMS

Amines   Derivatives of ammonia in which one or more hydrogen atoms has been replaced by organic groups.

Electrolytes   Compounds that ionize (or dissociate into their constituent ions) when dissolved in water to produce aqueous solutions that conduct an electric current.

Hydrolysis   The reaction of a substance with water.

Hydrolysis constant   An equilibrium constant for a hydrolysis reaction.

Indicator   An organic compound that exhibits different colors in solutions of different acidities.

Ion product for water   An equilibrium constant for the ionization of water,

$$K_w = [H_3O^+][OH^-] = 1.0 \times 10^{-14} \text{ at } 25°C$$

Ionization constant   An equilibrium constant for the ionization of a weak electrolyte.

Monoprotic acid   An acid that can form only one hydronium ion per molecule; may be strong or weak.

Nonelectrolytes   Compounds that do not ionize (or dissociate into their constituent ions) when dissolved in water to

produce aqueous solutions that will not conduct an electric current.

pH   The negative logarithm of the concentration (mol/L) of the H_3O^+ (or H^+) ion; the commonly used scale ranges from 0 to 14.

pK_a   The negative logarithm of K_a, the ionization constant for a weak acid.

pK_b   The negative logarithm of K_b, the ionization constant for a weak base.

pK_w   The negative logarithm of the ion product for water.

pOH   The negative logarithm of the concentration (mol/L) of the OH^- ion; the commonly used scale ranges from 14 to 0.

Polyprotic acid   An acid that can form two or more hydronium ions per molecule.

Solvolysis   The reaction of a substance with the solvent in which it is dissolved.

Strong electrolytes   Compounds that ionize (or dissociate into their constituent ions) completely, or nearly completely, when dissolved in water to produce aqueous solutions that conduct an electric current.

EXERCISES

🔵 **Molecular Reasoning** exercises

▲ **More Challenging** exercises

Blue-Numbered exercises are solved in the Student Solutions Manual

Review of Strong Electrolytes

1. List names and formulas for (a) seven strong acids; (b) six weak bases; (c) the common strong bases; (d) ten soluble ionic salts.

2. 🔵 (a) How are a strong acid and a weak acid similar? How are they different? (b) How are a strong base and a weak base similar? How are they different?

3. Which of the following are strong electrolytes: a Group 1A hydroxide, a Group 3A hydroxide, $Cu(OH)_2$, $Be(OH)_2$, H_3AsO_3, HI, HF, H_2SO_3, H_3PO_4, salts of a Group 1A metal?

4. Calculate the molarity of each of the following solutions. (a) 25.65 g of NaCl in 250. mL of solution; (b) 75.5 g of H_2SO_4 in 1.00 L of solution; (c) 0.126 g of phenol, C_6H_5OH, in 1.00 L of solution.

🔵 **Molecular Reasoning** exercises   ▲ **More Challenging** exercises   Blue-Numbered exercises solved in Student Solutions Manual

Unless otherwise noted, all content on this page is © Cengage Learning.

5. Square brackets, [ ], are often used in mathematical statements in chemistry. What is the meaning associated with square brackets?

6.  Calculate the concentrations of the constituent ions in solutions of the following compounds in the indicated concentrations: (a) 0.45 M HBr; (b) 0.045 M KOH; (c) 0.0112 M $CaCl_2$.

7. Calculate the concentrations of the constituent ions in solutions of the following compounds in the indicated concentrations: (a) 0.0105 M $Sr(OH)_2$; (b) 0.0105 M $HClO_3$; (c) 0.0105 M K_2SO_4.

8. Calculate the concentrations of the constituent ions in the following solutions: (a) 1.25 g of KOH in 1.50 L of solution; (b) 0.2505 g of $Ba(OH)_2$ in 250. mL of solution; (c) 1.26 g of $Ca(NO_3)_2$ in 100. mL of solution.

9. Calculate the concentrations of the constituent ions in the following solutions: (a) 1.57 g of $Al_2(SO_4)_3$ in 425 mL of solution; (b) 45.8 g of $CaCl_2 \cdot 6H_2O$ in 5.00 L of solution; (c) 18.4 g of HBr in 675 mL of solution.

The Autoionization of Water

10.  (a) Write a chemical equation showing the ionization of water. (b) Write the equilibrium constant expression for this equation. (c) What is the special symbol used for this equilibrium constant? (d) What is the relationship between [H$^+$] and [OH$^-$] in pure water? (e) How can this relationship be used to define the terms "acidic" and "basic"?

11.  Use K_w to explain the relationship between the hydronium ion concentration and the hydroxide ion concentration in aqueous solutions.

12. (a) Why is the concentration of OH$^-$ produced by the ionization of water neglected in calculating the concentration of OH$^-$ in a 0.060 M solution of NaOH? (b) Demonstrate that the concentration of OH$^-$ from H_2O can be neglected in the 0.060 M solution of NaOH.

13. Calculate the concentrations of OH$^-$ in the solutions described in Exercises 6(a), 7(b), and 9(c), and compare them with the OH$^-$ concentration in pure water.

14. Calculate the concentrations of H_3O^+ in the solutions described in Exercises 6(b), 7(a), and 8(b), and compare them with the H_3O^+ concentration in pure water.

15. Calculate [OH$^-$] that is in equilibrium with (a) [H_3O^+] = 6.3×10^{-4} mol/L (b) [H_3O^+] = 7.8×10^{-9} mol/L.

16. Calculate the concentration of water in moles/liter. Using the ratio of the concentrations of hydronium ions in pure water (1×10^{-7} M) to water itself, calculate the number of hydronium ions for every 10 billion water molecules.

The pH and pOH Scales

17. Write mathematical definitions for pH and pOH. What is the relationship between pH and pOH? How can pH be used to define the terms "acidic" and "basic"?

18. Find the pH and pOH of solutions with the following [H$^+$]. Classify each as acidic or basic. (a) 1.0 M; (b) 1.7×10^{-4} M; (c) 6.8×10^{-8} M; (d) 9.3×10^{-11} M.

19. (a) A sample of milk is found to have a pH of 6.50. What are the concentrations of H_3O^+ and OH$^-$ ions in this sample? (b) A sample of buttermilk has a pH of 4.50. Which sample is more acidic? (c) By what factor? (d) Activation of baking soda (as in the baking of biscuits) depends on an acidic environment. Which would be more effective in this process—milk or buttermilk? Why?

20. The normal pH of human blood ranges from 7.35 to 7.45. Calculate the concentrations of H_3O^+ and OH$^-$ ions in human blood that has a pH of 7.45.

21. Calculate the pH of (a) a 1.5×10^{-4} M solution of $HClO_4$, a strong acid, at 25°C and (b) 1.5×10^{-8} M solution of HCl at 25°C.

22. Calculate the pH of the following solutions: (a) 2.0×10^{-1} M HCl; (b) 0.050 M HNO_3; (c) 0.65 g·L^{-1} $HClO_4$; (d) 9.8×10^{-4} M NaOH.

23. Calculate the [H_3O^+], [OH$^-$], pH, and pOH in a 0.0548 M HCl solution.

24. A solution of HNO_3 has a pH of 3.52. What is the molarity of the solution?

25. Complete the following table. Is there an obvious relationship between pH and pOH? What is it?

Solution	[H₃O⁺]	[OH⁻]	pH	pOH
0.25 M HI	——	——	——	——
0.067 M RbOH	——	——	——	——
0.020 M $Ba(OH)_2$	——	——	——	——
0.00030 M $HClO_4$	——	——	——	——

🐾 **Molecular Reasoning** exercises   ▲ **More Challenging** exercises   Blue-Numbered exercises solved in Student Solutions Manual

Unless otherwise noted, all content on this page is © Cengage Learning.

26. Calculate the following values for each solution.

Solution	$[H_3O^+]$	$[OH^-]$	pH	pOH
(a) 0.085 M NaOH	___	___	___	___
(b) 0.075 M HCl	___	___	___	___
(c) 0.075 M Ca(OH)$_2$	___	___	___	___

27. 🔷 What is the relationship between base strength and the value of K_b? What is the relationship between acid strength and the value of pK_a?

28. 🔷 Predict which acid of each pair is the stronger acid. Briefly explain how you arrived at your answer. (a) H_3PO_4 or H_3AsO_4; (b) H_3AsO_3 or H_3AsO_4. (*Hint*: Review Chapter 10 and Appendix F.)

29. 🔷 Predict which acid of each pair is the stronger acid. Briefly explain how you arrived at your answer. (a) H_2O or H_2S; (b) HOBr or HOCl. (*Hint*: Review Chapter 10 and Appendix F.)

30. Write a chemical equation that represents the ionization of a weak acid, HA. Write the equilibrium constant expression for this reaction. What is the special symbol used for this equilibrium constant?

31. What is the relationship between the strength of an acid and the numerical value of K_a? What is the relationship between the acid strength and the value of pK_b?

32. Because K_b is larger for triethylamine

$$(C_2H_5)_3N(aq) + H_2O(\ell) \rightleftharpoons (C_2H_5)_3NH^+ + OH^-$$
$$K_b = 5.2 \times 10^{-4}$$

than for trimethylamine

$$(CH_3)_3N(aq) + H_2O(\ell) \rightleftharpoons (CH_3)_3NH^+ + OH^-$$
$$K_b = 7.4 \times 10^{-5}$$

an aqueous solution of triethylamine should have a larger concentration of OH^- ion than an aqueous solution of trimethylamine of the same concentration. Confirm this statement by calculating the $[OH^-]$ for 0.018 M solutions of both weak bases.

33. ▲ The equilibrium constant of the following reaction is 1.35×10^{-15}.

$$2D_2O \rightleftharpoons D_3O^+ + OD^-$$

D is deuterium, ^{2}H. Calculate the pD of pure deuterium oxide (heavy water). What is the relationship between $[D_3O^+]$ and $[OD^-]$ in pure D_2O? Is pure D_2O acidic, basic, or neutral?

34. (a) What is the pH of pure water at body temperature, 37°C? Refer to Table 18-2. (b) Is this acidic, basic, or neutral? Why?

35. Fill in the blanks in this table for the given solutions.

Sol'n	Temp. (°C)	Concentration (mol/L)		pH
		$[H_3O^+]$	$[OH^-]$	
(a)	25	1.0×10^{-4}	___	___
(b)	0	___	___	2.75
(c)	60	___	___	7.00
(d)	25	___	4.5×10^{-8}	___

36. Write a chemical equation that represents the equilibrium between water and a weak base, B. Write the equilibrium constant expression for this reaction. What is the special symbol used for this equilibrium constant?

37. Complete the following table by appropriate calculations.

$[H_3O^+]$	pH	$[OH^-]$	pOH
(a) ___	4.84	___	___
(b) ___	10.61	___	___
(c) ___	___	___	2.90
(d) ___	___	___	9.47

38. A 0.0750 M solution of a monoprotic acid is known to be 1.07% ionized. What is the pH of the solution? Calculate the value of K_a for this acid.

39. A 0.075 M aqueous solution of a weak, monoprotic acid is 0.85% ionized. Calculate the value of the ionization constant, K_a, for this acid.

40. A 0.10 M solution of chloroacetic acid, $ClCH_2CO_2H$, has a pH of 1.95. Calculate K_a for the acid.

41. The pH of a 0.35 M solution of uric acid is 2.17. What is the value of K_a for uric acid, a monoprotic acid?

42. Calculate the concentrations of all the species present in a 0.52 M benzoic acid solution. (See Appendix F.)

43. Find the concentrations of the various species present in a 0.45 M solution of HOBr. What is the pH of the solution? (See Appendix F.)

44. Hydrofluoric acid can be used to etch glass. Calculate the pH of a 0.38 M HF solution.

45. To what volume should 1.00 x 10^2 mL of any weak acid, HA, with a concentration of 0.20 M be diluted to double the percentage ionization?

46. Calculate the pH and pOH of a household ammonia solution that contains 2.05 moles of NH_3 per liter of solution.

47. Calculate the percent ionization in a 0.56 M NH_3 solution.

48. What is the percent ionization in a 0.0751 M solution of formic acid, HCOOH?

49. What is the percent ionization in (a) a 0.150 M CH_3COOH solution, (b) a 0.0150 M CH_3COOH solution?

50. The K_a values for two weak acids are 7.2×10^{-5} and 4.2×10^{-10}, respectively. What are their pK_a values?

51. What is the concentration of OI^- in equilibrium with $[H_3O^+] = 0.045$ mol/L and $[HOI] = 0.527$ mol/L?

52. Pyridine is 0.053% ionized in 0.00500 M solution. What is the pK_b of this monobasic compound?

53. A 0.068 M solution of benzamide has a pOH of 2.91. What is the value of pK_b for this monobasic compound?

54. In a 0.0100 M aqueous solution of methylamine, CH_3NH_2, the equilibrium concentrations of the species are $[CH_3NH_2] = 0.0080$ mol/L and $[CH_3NH_3^+] = [OH^-] = 2.0 \times 10^{-3}$ mol/L. Calculate K_b for this weak base.

$$CH_3NH_2(aq) + H_2O(\ell) \rightleftharpoons CH_3NH_3^+ + OH^-$$

55. What is the concentration of NH_3 in equilibrium with $[NH_4^+] = 0.010$ mol/L and $[OH^-] = 1.2 \times 10^{-5}$ mol/L?

56. Calculate $[OH^-]$, percent ionization, and pH for (a) 0.25 M aqueous ammonia, (b) 0.25 M methylamine solution.

🔷 **Molecular Reasoning** exercises ▲ **More Challenging** exercises Blue-Numbered exercises solved in Student Solutions Manual

Unless otherwise noted, all content on this page is © Cengage Learning.

57. Calculate [H_3O^+], [OH^-], pH, pOH, and percent ionization for 0.20 M aqueous ammonia.

Polyprotic Acids

58. 🞛 Calculate the concentrations of the various species in a 0.100 M H_3AsO_4 solution. Compare the concentrations with those of the analogous species in 0.100 M H_3PO_4 solution. (See Example 18-17 and Table 18-7.)

59. 🞛 Citric acid, the acid in lemons and other citrus fruits, has the structure

$$\begin{array}{c} CH_2COOH \\ | \\ HO{-}C{-}COOH \\ | \\ CH_2COOH \end{array}$$

which we may abbreviate as $C_3H_5O(COOH)_3$ or H_3A. It is a triprotic acid. Write the chemical equations for the three stages in the ionization of citric acid with the appropriate K_a expressions.

Charles D. Winters

60. Calculate the concentrations of H_3O^+, OH^-, $HSeO_4^-$, and SeO_4^{2-} in 0.15 M H_2SeO_4, selenic acid, solution.

61. Some kidney stones are crystalline deposits of calcium oxalate, a salt of oxalic acid, $(COOH)_2$. Calculate the concentrations of H_3O^+, OH^-, $COOCOOH^-$, and $(COO^-)_2$ in 0.12 M $(COOH)_2$. Compare the concentrations with those obtained in Exercise 60. How can you explain the difference between the concentrations of $HSeO_4^-$ and $COOCOOH^-$? between SeO_4^{2-} and $(COO^-)_2$?

62. Rust stains can be removed from painted surfaces with a solution of oxalic acid, $(COOH)_2$. Calculate the pH of a 0.045 M oxalic acid solution.

63. Calculate the pH and pOH of a carbonated soft drink that is 0.0035 M carbonic acid solution. Assume that there are no other acidic or basic components.

Charles D. Winters

Hydrolysis

64. 🞛 Define and illustrate the following terms clearly and concisely: (a) solvolysis; (b) hydrolysis.

65. Some anions, when dissolved, undergo no significant reaction with water molecules. What is the relative base strength of such an anion compared with water? What effect will dissolution of such anions have on the pH of the solution?

66. Some cations in aqueous solution undergo no significant reactions with water molecules. What is the relative acid strength of such a cation compared with water? What effect will dissolution of such cations have on the pH of the solution?

67. How can salts be classified conveniently into four classes? For each class, write the name and formula of a salt that fits into that category. Use examples other than those used in illustrations in this chapter.

Salts of Strong Bases and Strong Acids

68. 🞛 ▲ What determines whether the aqueous solution of a salt is acidic, basic, or neutral?

69. 🞛 Why do salts of strong bases and strong acids give neutral aqueous solutions? Use KNO_3 to illustrate. Write names and formulas for three other salts of strong bases and strong acids.

70. Which of the following are salts of a strong base and a strong acid? (a) Na_3PO_4; (b) K_2CO_3; (c) LiF; (d) $BaSO_4$; (e) $NaClO_3$.

71. Which of the following are salts of a strong base and a strong acid? (a) $Ba_3(PO_4)_2$; (b) $LiNO_3$; (c) NaI; (d) $CaCO_3$; (e) $KClO_4$.

Salts of Strong Bases and Weak Acids

72. Why do salts of strong bases and weak acids give basic aqueous solutions? Use sodium hypochlorite, $NaOCl$, to illustrate. (Clorox and some other commercial "chlorine bleaches" are 6% $NaOCl$.)

istockphoto.com/fjmdesign
<http://istockphoto.com/fjmdesign>

73. Some anions react with water to upset the H_3O^+/OH^- balance. What is the relative base strength of such an anion compared with water? What effect will dissolution of such anions have on the pH of the solution?

74. Calculate the equilibrium constant for the reaction of azide ions, N_3^-, with water.

🞛 **Molecular Reasoning** exercises   ▲ **More Challenging** exercises   Blue-Numbered exercises solved in Student Solutions Manual

Unless otherwise noted, all content on this page is © Cengage Learning.

75. Write names and formulas for three salts of strong bases and weak acids other than those that appear in Section 18-8.

76. Calculate hydrolysis constants for the following anions of weak acids: (a) NO_2^-; (b) OBr^-; (c) $HCOO^-$. What is the relationship between K_a, the ionization constant for a weak acid, and K_b, the hydrolysis constant for the anion of the weak acid? (See Appendix F.)

77. Calculate the equilibrium constant for the reaction of hypoiodite ions (OI^-) with water.

78. Calculate the pH of 1.5 M solutions of the following salts: (a) $NaCH_3COO$; (b) $KOBr$; (c) $LiCN$.

79. Calculate the pH of 0.75 M solutions of the following salts: (a) $NaNO_2$; (b) $NaOCl$; (c) $NaHCOO$.

80. (a) What is the pH of a 0.18 M solution of KOI? (b) What is the pH of a 0.18 M solution of KF?

Salts of Weak Bases and Strong Acids

81. 🦠 Why do salts of weak bases and strong acids give acidic aqueous solutions? Illustrate with NH_4NO_3, a common fertilizer.

82. Write names and formulas for four salts of weak bases and strong acids.

83. Use values found in Table 18-6 and in Appendix G to calculate hydrolysis constants for the following cations of weak bases: (a) NH_4^+; (b) $CH_3NH_3^+$, methylammonium ion; (c) $C_6H_5NH_3^+$, anilinium ion.

84. Use the values found in Table 18-6 and Appendix G to calculate hydrolysis constants for the following cations of weak bases: (a) $(CH_3)_2NH_2^+$, dimethylammonium ion; (b) $C_5H_5NH^+$, pyridinium ion; (c) $(CH_3)_3NH^+$, trimethylammonium ion.

85. Make a general statement relating parent base strength and extent of hydrolysis of the cations of Exercise 83 by using hydrolysis constants calculated in that exercise?

86. Calculate the pH of 0.26 M solutions of (a) NH_4NO_3, (b) $CH_3NH_3NO_3$, (c) $C_6H_5NH_3NO_3$.

Salts of Weak Bases and Weak Acids

87. 🦠 Why are some aqueous solutions of salts of weak acids and weak bases neutral, whereas others are acidic and still others are basic?

88. 🦠 In each of the following question parts write the names and formulas for three salts of a weak acid and a weak base that give (a) neutral, (b) acidic, and (c) basic aqueous solutions.

89. If both the cation and anion of a salt react with water when dissolved, what determines whether the solution will be acidic, basic, or neutral? Classify aqueous solutions of the following salts as acidic, basic, or neutral: (a) $NH_4F(aq)$; (b) $CH_3NH_3OI(aq)$.

Salts That Contain Small, Highly Charged Cations

90. Choose the hydrated cations that react with water to give acidic solutions. (a) $[Be(H_2O)_4]^{2+}$; (b) $[Al(H_2O)_6]^{3+}$; (c) $[Fe(H_2O)_6]^{3+}$; (d) $[Cu(H_2O)_6]^{2+}$. Write chemical equations for the reactions.

91. Why do some salts that contain cations related to insoluble bases (metal hydroxides) and anions related to strong acids give acidic aqueous solutions? Use $Fe(NO_3)_3$ to illustrate.

92. Calculate pH and percent hydrolysis for the following (see Table 18-10): (a) 0.15 M $Al(NO_3)_3$, aluminum nitrate; (b) 0.075 M $Co(ClO_4)_2$, cobalt(II) perchlorate; (c) 0.15 M $MgCl_2$, magnesium chloride.

93. ▲ Given pH values for solutions of the following concentrations, calculate hydrolysis constants for the hydrated cations: (a) 0.00050 M $CeCl_3$, cerium(III) chloride, pH = 5.99; (b) 0.10 M $Cu(NO_3)_2$, copper(II) nitrate, pH = 4.50; (c) 0.10 M $Sc(ClO_4)_3$, scandium perchlorate, pH = 3.44.

Mixed Exercises

94. A weak acid, HA, has a pK_a = 5.35. What is the concentration of the anion, A^-, in a 0.100 M solution?

95. Calculate the pH of the following solutions. (a) 0.0070 M $Ca(OH)_2$; (b) 0.25 M chloroacetic acid, $ClCH_2COOH$, $K_a = 1.4 \times 10^{-3}$; (c) 0.055 M pyridine, C_5H_5N.

96. 🦠 Classify aqueous solutions of the following salts as acidic, basic, or essentially neutral. Justify your choice. (a) $(NH_4)HSO_4$; (b) $(NH_4)_2SO_4$; (c) $LiCl$; (d) $LiBrO$; (e) $AlCl_3$.

97. Repeat Exercise 96 for (a) Na_2SO_4, (b) NH_4Cl, (c) KCl, (d) NH_4CN. (See Appendix F.)

98. 🦠 In aqueous solution some cations react with water to upset the H_3O^+/OH^- balance. What is the relative acid strength of such a cation compared with water? What effect will dissolution of these cations have on the pH of the solution?

99. 🦠 Some plants require acidic soils for healthy growth. Which of the following could be added to the soil around such plants to increase the acidity of the soil? Write equations to justify your answers. (a) $FeSO_4$; (b) Na_2SO_4; (c) $Al_2(SO_4)_3$; (d) $Fe_2(SO_4)_3$; (e) $BaSO_4$. Arrange the salts that give acidic solutions in order of increasing acidity. (Assume equal molarities of the salt solutions.)

100. 🦠 Some of the following salts are used in detergents and other cleaning materials because they produce basic solutions. Which of the following could *not* be used for this purpose? Write equations to justify your answers. (a) Na_2CO_3; (b) Na_2SO_4; (c) $(NH_4)_2SO_4$; (d) Na_3PO_4.

101. Calculate the pH of each of the following solutions: (a) 0.038 g of barium hydroxide in 450. mL of solution, (b) 0.050 g of hydrogen iodide in 750. mL of solution, (c) 0.00075 g of HCl in 1.00 L of solution.

102. For each of the following pairs, tell which solution would have the lower pH. Tell how you arrive at each answer. (a) 0.015 M NH_4Br, ammonium bromide, and 0.015 M NH_4NO_3, ammonium nitrate; (b) 0.015 M ammonium perchlorate, NH_4ClO_4, and 0.010 M ammonium fluoride,

---

🦠 **Molecular Reasoning** exercises   ▲ **More Challenging** exercises   Blue-Numbered exercises solved in Student Solutions Manual

Unless otherwise noted, all content on this page is © Cengage Learning.

NH$_4$F; (c) 0.010 M NH$_4$Cl and 0.050 M NH$_4$Cl. (*Hint*: Think before you calculate.)

103. 🔊 Arrange the following 0.1 M aqueous solutions in order of decreasing pH (highest to lowest).

Ba(NO$_3$)$_2$, HNO$_3$, NH$_4$NO$_3$, Al(NO$_3$)$_3$, NaOH

104. 🔊 Predict which base of each pair is the stronger base. Briefly explain how you arrived at your answer: (a) PH$_3$ or NH$_3$; (b) Br$^-$ or F$^-$; (c) ClO$_3^-$ or ClO$_2^-$; (d) HPO$_4^{2-}$ or PO$_4^{3-}$. (*Hint*: You may wish to review Chapter 10.)

Conceptual Exercises

105. How could we demonstrate that 0.012 M solutions of HCl and HNO$_3$ contain essentially no molecules of nonionized acid?

106. How could we demonstrate that 0.010 M solutions of HF and HNO$_2$ contain relatively few ions?

107. Carbonic acid, H$_2$CO$_3$, is diprotic and therefore has two ionization constants, $K_{a1} = 4.2 \times 10^{-7}$ and $K_{a2} = 4.8 \times 10^{-11}$. The pH of a carbonic acid solution can be calculated without using K_{a2}. Explain using a 0.100 M solution of carbonic acid.

108. Answer the following questions for 0.15 M solutions of the weak bases listed in Table 18-6. (a) In which solution is (i) the pH highest, (ii) the pH lowest, (iii) the pOH highest, (iv) the pOH lowest? (b) Which solution contains (i) the highest concentration of the cation of the weak base, (ii) the lowest concentration of the cation of the weak base?

109. Answer the following questions for 0.15 M solutions of the weak acids listed in Table 18-4. Which solution contains (a) the highest concentration of H$_3$O$^+$, (b) the highest concentration of OH$^-$, (c) the lowest concentration of H$_3$O$^+$, (d) the lowest concentration of OH$^-$, (e) the highest concentration of nonionized acid molecules, (f) the lowest concentration of nonionized acid molecules?

110. Write the balanced chemical equation for the ionization (or dissociation) that is believed to occur when each of the following compounds dissolves in water. Nitrous acid (HNO$_2$), barium hydroxide [Ba(OH)$_2$], hydrofluoric acid (HF), lithium hydroxide (LiOH), hydrocyanic acid (HCN), potassium hydroxide (KOH). Use a single arrow (⟶) to indicate a strong acid or strong base, one that ionizes completely (or to a very large extent), or a double arrow (⇌) to indicate a weak acid or weak base, one that ionizes partially (or to a small extent).

111. Listed below are five acids and their ionization constants. List their formulas in order of decreasing acid strength (i.e., strongest first).

benzoic acid (C$_6$H$_5$COOH), 6.3×10^{-5}
cyanic acid (HOCN), 3.5×10^{-4}
formic acid (HCOOH), 1.8×10^{-4}
hypobromous acid (HOBr), 2.5×10^{-9}
phenol (HC$_6$H$_5$O), 1.3×10^{-10}

112. "We have to get this problem of acid rain under control. We must do whatever it takes to get the pH down to zero!" Do you agree? Explain.

113. Of the following salts, which will produce an acidic solution? A neutral solution? A basic solution?

ammonium acetate (NH$_4$CH$_3$COO)
ammonium nitrate (NH$_4$NO$_3$)
ammonium sulfate [(NH$_4$)$_2$SO$_4$]
calcium sulfite (CaSO$_3$)
lithium hypochlorite (LiClO)
potassium chloride (KCl)

Building Your Knowledge

114. Ascorbic acid, C$_5$H$_7$O$_4$COOH, also known as vitamin C, is an essential vitamin for all mammals. Among mammals, only humans, monkeys, and guinea pigs cannot synthesize it in their bodies. K_a for ascorbic acid is 7.9×10^{-5}. Calculate [H$_3$O$^+$] and pH in a 0.110 M solution of ascorbic acid.

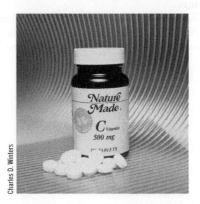

Charles D. Winters

115. Arrange the following common kitchen samples from most acidic to most basic.

carrot juice, pH 5.1   blackberry juice, pH 3.4
soap, pH 11.0   red wine, pH 3.7
egg white, pH 7.8   milk of magnesia, pH 10.5
sauerkraut, pH 3.5   lime juice, pH 2.0

116. ▲ The buildup of lactic acid in muscles causes pain during extreme physical exertion. The K_a for lactic acid, C$_2$H$_5$OCOOH, is 8.4×10^{-4}. Calculate the pH of a 0.110 M solution of lactic acid. Can you make a simplifying assumption in this case?

117. A 0.0100 molal solution of acetic acid freezes at $-0.01938°$C. Use this information to calculate the ionization constant for acetic acid. A 0.0100 molal solution is sufficiently dilute that it may be assumed to be 0.0100 molar without introducing a significant error.

118. The ion product for water, K_w, has the value 1.14×10^{-15} at 0.0 °C and 5.47×10^{-14} at 50. °C. Use the van't Hoff equation to estimate ΔH^0 for the ion product of water.

Beyond the Textbook

NOTE: *Whenever the answer to an exercise depends on information obtained from a source other than this textbook, the source must be included as an essential part of the answer.*

🔊 **Molecular Reasoning** exercises   ▲ **More Challenging** exercises   Blue-Numbered exercises solved in Student Solutions Manual

Unless otherwise noted, all content on this page is © Cengage Learning.

Use the *Handbook of Chemistry and Physics* or a suitable website to find answers to the following:

119. Find the approximate pH values of a number of foods, biological fluids, and so on. List the pH range of three foods or biological fluids found in this reference that are either not reported in the table in Section 18-3 of the textbook or have a different value reported in the table.

120. Locate the table of dissociation constants of organic acids in aqueous solutions and the table of dissociation constants of inorganic acids in aqueous solutions. In Section 18-5 of the textbook it is stated that for a polyprotic acid, "Successive ionization constants often decrease by a factor of approximately 10^4 to 10^6, although some differences are outside this range." Cite two inorganic examples that follow this rule of thumb. Cite one inorganic and two organic examples that are outside this range.

121. Nonmetal oxides are referred to as acid anhydrides. Use chemical equations to illustrate that via sequential reactions carbon dioxide added to water first hydrates and then dissociates to increase the acidity of the solution.

122. Use an internet search engine such as http://www.google.com to locate information about the medical condition achlorhydria. What acid and internal organ are involved? Does the normal presence of this acid explain one of the entries in the table "pH Range for a Few Common Substances" in Section 18-3?

123. Many medicines, vitamins, household cleaners, and lawn products contain acids and bases or their salts. Identify some of these products at your local supermarket. Are the acids and bases involved strong or weak? Some suggestions: Maalox®, Alka-Seltzer®.

124. Marble, a derivative of limestone (calcium carbonate), is often used for kitchen countertops. Marble countertops require extra care since they are easily etched by acidic substances such as lemon juice and vinegar. Write the unwanted chemical reaction between the marble and either vinegar or a generic acid.

⬤ **Molecular Reasoning** exercises   ▲ **More Challenging** exercises   Blue-Numbered exercises solved in Student Solutions Manual

Unless otherwise noted, all content on this page is © Cengage Learning.

Ionic Equilibria II: Buffers and Titration Curves

19

Anthocyanins are a general class of water-soluble pigments found in most plants. Every color but green has been recorded, with red, blue, and purple being the most common. Anthocyanins are natural indicators, as their colors vary with pH. Some plants can have different-colored flowers depending on the acidity of the soil they are grown in. Geraniums, for example, contain the anthocyanin pelargonin, which changes from orange-red (acidic) to blue (basic).

© Charles Volkland/AgeFotostock

OBJECTIVES

After you have studied this chapter, you should be able to

▶ Explain the common ion effect and give illustrations of its operation

▶ Recognize buffer solutions and describe their chemistry

▶ Describe how to prepare a buffer solution of a specified pH

▶ Carry out calculations related to buffer solutions and their action

▶ Explain what acid–base indicators are and how they function

▶ Describe what species are present at various stages of titration curves for (a) strong acids and strong bases, (b) weak acids and strong bases, and (c) weak acids and weak bases

▶ Carry out calculations based on titration curves for (a) strong acids and strong bases, and (b) weak acids and strong bases

In previous chapters, we calculated the acidity or basicity of separate aqueous solutions of strong acids, strong bases, weak acids, weak bases, and their salts. In this chapter, we will study (1) solutions that have both weak acids and weak bases present, (2) indicators, and (3) titration curves.

It can be a challenge to recognize the type of solution present. This can be even more difficult if the solution is formed by a partial or total neutralization reaction. To help you recognize the various solutions, a summary table (see Table 19-7) is included in Section 19-8. We encourage you to look at the table often as you progress through this chapter and as you review this and previous chapters.

19-1 The Common Ion Effect and Buffer Solutions

In laboratory reactions, in industrial processes, and in the bodies of plants and animals, it is often necessary to keep the pH nearly constant despite the addition of acids or bases. The oxygen-carrying capacity of the hemoglobin in your blood and the activity of the enzymes in your cells are very sensitive to the pH of your body fluids. A change in blood pH of 0.5 units (a change in $[H_3O^+]$ by a factor of about 3) can be fatal. Our bodies use a combination of compounds known as a *buffer system* to keep the pH within a narrow range.

▶ Buffer systems resist changes in pH.
▶ LeChatelier's Principle (Section 17-6) is applicable to equilibria in aqueous solution.

The operation of a buffer solution depends on the *common ion effect*, a special case of LeChatelier's Principle.

> When a solution of a weak electrolyte is altered by adding one of its ions from another source, the ionization of the weak electrolyte is suppressed. This behavior is termed the **common ion effect**.

Many types of solutions exhibit this behavior. Two of the most common kinds are

1. a solution of a weak acid plus a soluble ionic salt of the weak acid (e.g., CH_3COOH plus $NaCH_3COO$), and

2. a solution of a weak base plus a soluble ionic salt of the weak base (e.g., NH_3 plus NH_4Cl)

▶ You should review the discussion of conjugate acid–base pairs in Chapter 10 (Sections 10-4 and 10-5) and in Chapter 18 (Sections 18-6 to 18-10).

> A **buffer solution** contains a conjugate acid–base pair with both the acid and base in reasonable concentrations. The conjugate acid–base pair must be of weak to moderate strength. Strong acids and bases do not make buffer solutions. The acidic component reacts with added strong bases. The basic component reacts with added strong acids.

Weak Acids Plus Salts of Weak Acids

Consider a solution that contains acetic acid *and* sodium acetate, a soluble ionic salt of CH_3COOH. The $NaCH_3COO$ is completely dissociated into its constituent ions, but CH_3COOH is only slightly ionized.

$$NaCH_3COO \xrightarrow{H_2O} Na^+ + \boxed{CH_3COO^-} \quad \text{(to completion)}$$

$$CH_3COOH + H_2O \rightleftharpoons \boxed{H_3O^+ + CH_3COO^-} \quad \text{(reversible)}$$

Both CH_3COOH and $NaCH_3COO$ are sources of CH_3COO^- ions. The $NaCH_3COO$ is soluble, so it is completely dissociated and provides a high $[CH_3COO^-]$. This shifts the ionization equilibrium of CH_3COOH far to the left, as CH_3COO^- combines with H_3O^+ to form nonionized CH_3COOH and H_2O. The result is a drastic decrease in $[H_3O^+]$ in the solution (higher pH).

> Solutions that contain a weak acid plus a soluble salt of that weak acid are always less acidic than solutions that contain the same concentration of the weak acid alone.

In Example 18-11, we found that the H_3O^+ concentration in 0.10 *M* CH_3COOH is 1.3×10^{-3} mol/L (pH = 2.89). In Example 19-1, we will calculate the acidity of the same 0.10 *M* CH_3COOH solution after it is also made to be 0.20 *M* in the salt $NaCH_3COO$.

EXAMPLE 19-1 Weak Acid/Salt of Weak Acid Buffer Solution

Calculate the concentration of H_3O^+ and the pH of a buffer solution that is 0.10 *M* in CH_3COOH and 0.20 *M* in $NaCH_3COO$.

Plan

Write the appropriate equations for *both* $NaCH_3COO$ *and* CH_3COOH and the ionization constant expression for CH_3COOH. Then represent the *equilibrium* concentrations algebraically and substitute into the K_a expression.

Solution

The appropriate equations and ionization constant expression are

rxn 1 $\quad NaCH_3COO \longrightarrow Na^+ + CH_3COO^- \quad$ (to completion)

rxn 2 $\quad CH_3COOH + H_2O \rightleftharpoons H_3O^+ + CH_3COO^- \quad$ (reversible)

$$K_a = \frac{[H_3O^+][CH_3COO^-]}{[CH_3COOH]} = 1.8 \times 10^{-5}$$

This K_a expression is valid *for all solutions* that contain CH_3COOH. In a solution that contains both CH_3COOH and $NaCH_3COO$, the CH_3COO^- ions come from two sources. The ionization constant must be satisfied by the *total* CH_3COO^- concentration.

Because $NaCH_3COO$ is completely dissociated, the $[CH_3COO^-]$ *from* $NaCH_3COO$ will be 0.20 mol/L. Let $x = [CH_3COOH]$ that ionizes; then x is also equal to $[H_3O^+]$ and equal to $[CH_3COO^-]$ *from* CH_3COOH. The *total* concentration of CH_3COO^- is $(0.20 + x)$ *M*. The concentration of nonionized CH_3COOH is $(0.10 - x)$ *M*.

rxn 1 $\quad \begin{array}{ccc} NaCH_3COO & \longrightarrow & Na^+ + \boxed{CH_3COO^-} \\ 0.20\ M & & 0.20\ M \quad\ 0.20\ M \end{array}$

$\qquad\qquad\qquad\qquad\qquad\qquad\qquad\qquad \rightarrow$ Total $[CH_3COO^-]$

rxn 2 $\quad \begin{array}{ccc} CH_3COOH + H_2O & \rightleftharpoons & H_3O^+ + \boxed{CH_3COO^-} \\ (0.10-x)\ M & & x\ M \qquad x\ M \end{array}$ $\qquad = (0.20 + x)\ M$

S TOP & THINK

A solution of a *strong* acid and its salt (e.g., HCl/NaCl) has the acid nearly completely ionized, so the strong acid cannot be present in significant amounts. Therefore such a solution does not exhibit the common ion effect, nor can it act as a buffer solution. A similar reasoning applies to a solution of a *strong* base and its salt (e.g., NaOH/NaCl).

▶ This is an example of the common ion effect.

▶ We can represent these reactions with the abbreviated equations

$$NaA \longrightarrow Na^+ + A^-$$

$$HA \rightleftharpoons H^+ + A^-$$

Substitution into the ionization constant expression for CH_3COOH gives

$$K_a = \frac{[H_3O^+][CH_3COO^-]}{[CH_3COOH]} = \frac{(x)(0.20 + x)}{0.10 - x} = 1.8 \times 10^{-5}$$

The small value of K_a suggests that x is very small. This leads to two assumptions. It is reasonable to assume (1) that x (from the ionization of CH_3COOH) is small because CH_3COOH is a weak acid (rxn 2), and (2) that its ionization is further suppressed by the high concentration of CH_3COO^- formed by the soluble salt, $NaCH_3COO$ (rxn 1).

Assumption	Implication
$x \ll 0.20$, so $(0.20 + x) \approx 0.20$	Nearly all of the CH_3COO^- comes from $NaCH_3COO$ (rxn 1) and very little CH_3COO^- comes from ionization of CH_3COOH (rxn 2)
$x \ll 0.10$, so $(0.10 - x) \approx 0.10$	Very little of the CH_3COOH ionizes (rxn 2)

Introducing these assumptions gives

$$\frac{0.20\,x}{0.10} = 1.8 \times 10^{-5} \qquad \text{so} \qquad x = 9.0 \times 10^{-6}$$

$$x\,M = [H_3O^+] = 9.0 \times 10^{-6}\,M \qquad \text{so} \qquad pH = 5.05$$

This can be compared to $0.10\,M$ acetic acid, with $pH = 2.89$. The addition of the acetate anion, which is the conjugate base of acetic acid, raises the pH as one should qualitatively predict from acid–base reactions and the common ion effect (LeChatelier's Principle).

You should now work Exercise 10.

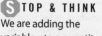

S TOP & THINK
We are adding the variable x to a quantity 0.20 that is more than two orders of magnitude greater than the K value, 1.8×10^{-5}, so it is safe to neglect x in the term $(0.20 + x)$. Similar reasoning lets us neglect x in the term $(0.10 - x)$. Review the Problem-Solving Tip "Simplifying Quadratic Equations" in Section 18-4.

▶ You can verify the validity of the assumptions by substituting the value $x = 9.0 \times 10^{-6}$ into the original K_a expression.

Let's calculate the percent ionization in the solution of Example 19-1.

$$\% \text{ ionization} = \frac{[CH_3COOH]_{\text{ionized}}}{[CH_3COOH]_{\text{initial}}} \times 100\%$$

$$= \frac{9.0 \times 10^{-6}\,M}{0.10\,M} \times 100\% = 0.0090\% \text{ ionized}$$

This compares with 1.3% ionization in $0.10\,M$ CH_3COOH (see Example 18-11). Table 19-1 compares these solutions. The third column shows that $[H_3O^+]$ is *140 times greater* in $0.10\,M$ CH_3COOH than in the solution to which 0.20 mol/L $NaCH_3COO$ has been added. This is due to the common ion effect.

The calculation of the pH of any solution containing significant amounts of both a weak acid and a salt of that weak acid may be carried out as we have done in Example 19-1.

James W. Morgenthaler

The two solutions of Table 19-1 in the presence of a universal indicator. The CH_3COOH solution is on the left.

Table 19-1 Comparison of $[H_3O^+]$ and pH in Acetic Acid and Sodium Acetate–Acetic Acid Solutions

Solution	% CH_3COOH Ionized	$[H_3O^+]$		pH
$0.10\,M$ CH_3COOH	1.3%	$1.3 \times 10^{-3}\,M$	2.89	
$0.10\,M$ CH_3COOH and $0.20\,M$ $NaCH_3COO$	0.0090%	$9.0 \times 10^{-6}\,M$	5.05	$\Delta pH = 2.16$

Alternatively, we may proceed as follows. We start by writing the equations for the ionization of the *weak monoprotic acid* and its K_a as we did previously.

$$HA + H_2O \rightleftharpoons H_3O^+ + A^- \quad \text{and} \quad \frac{[H_3O^+][A^-]}{[HA]} = K_a$$

▶ HA and A⁻ represent the weak acid and its conjugate base, respectively.

Solving this expression for $[H_3O^+]$ gives

$$[H_3O^+] = K_a \times \frac{[HA]}{[A^-]}$$

Consider a solution in which the weak acid and its anion (from an added salt) are both present at reasonable concentrations, such as greater than 0.050 M. Under these conditions, the concentration of the anion, $[A^-]$, in the solution can be assumed to be entirely due to the dissolved salt. With these restrictions, the preceding expression becomes

▶ These are the kinds of assumptions we made in Example 19-1.

$$[H_3O^+] = K_a \times \frac{[HA]}{[\text{conjugate base}]}$$

[HA] is the concentration of nonionized weak acid (in most cases, this is the total acid concentration), and [conjugate base] is the concentration of the anion from the dissolved salt.

Taking the logarithm of both sides of the preceding equation, we obtain

$$\log[H_3O^+] = \log K_a + \log\frac{[\text{acid}]}{[\text{conjugate base}]}$$

Multiplying by -1 gives

$$-\log[H_3O^+] = -\log K_a - \log\frac{[\text{acid}]}{[\text{conjugate base}]}$$

Recall that in Chapter 18 we defined $-\log[H_3O^+]$ (or $-\log[H^+]$) as pH and $-\log K_a$ as pK_a. The preceding equation can then be written as

▶ Remember that
$$-\log x = \log\frac{1}{x}$$

$$pH = pK_a + \log\frac{[\text{conjugate base (salt)}]}{[\text{acid}]} \qquad \text{(for acid/salt buffer)}$$

This equation is known as the **Henderson−Hasselbalch equation**. Workers in the biological sciences use it frequently. Because the amount of weak acid dissociated is very small, the values for [conjugate base] and [acid] are essentially their initial concentrations after mixing, ignoring the negligible reaction that they then undergo.

EXAMPLE 19-2 Weak Acid/Salt of Weak Acid Buffer Solution (via the Henderson−Hasselbalch Equation)

Use the Henderson−Hasselbalch equation to calculate the pH of the buffer solution in Example 19-1.

Plan

The Henderson−Hasselbalch equation is $pH = pK_a + \log\frac{[\text{conjugate base}]}{[\text{acid}]}$. The value for pK_a for acetic acid can be calculated from the value of K_a found in Example 19-1 and many other places. The solution in Example 19-1 is 0.10 M in CH_3COOH and 0.20 M in $NaCH_3COO$. The values used for [conjugate base] and [acid] are their initial concentrations after mixing but before reaction.

S **TOP & THINK**

For this example:
conjugate base = CH_3COO^-
　　　　　　　= $NaCH_3COO$
　　　　　　　= sodium acetate
　　　　　　　= salt of weak acid

Solution

The appropriate values needed for the Henderson-Hasselbalch equation are

$$pK_a = -\log K_a = -\log 1.8 \times 10^{-5} = 4.74$$

$$[\text{conjugate base}] = [CH_3COO^-] = [NaCH_3COO]_{\text{initial}} = 0.20\ M$$

$$[\text{acid}] = [CH_3COOH]_{\text{initial}} = 0.10\ M$$

$$pH = pK_a + \log\frac{[\text{conjugate base}]}{[\text{acid}]} = 4.74 + \log\frac{0.20}{0.10}$$

$$= 4.74 + \log 2.0 = 4.74 + 0.30 = 5.04$$

You should now work Exercise 11.

▶ The values $pK_a = 4.74$, log 2.0, and pH = 5.04 all have two significant figures.

Weak Bases Plus Salts of Weak Bases

Let us now consider the second common kind of buffer solution, containing a weak base and its salt. A solution that contains aqueous NH_3 and ammonium chloride, NH_4Cl, a soluble ionic salt of NH_3, is typical. The NH_4Cl is completely dissociated, but aqueous NH_3 is only slightly ionized.

rxn 1 $NH_4Cl \xrightarrow{H_2O} NH_4^+ + Cl^-$ (to completion)

rxn 2 $NH_3 + H_2O \rightleftharpoons NH_4^+ + OH^-$ (reversible)

Both NH_4Cl and aqueous NH_3 are sources of NH_4^+ ions. The soluble, completely dissociated NH_4Cl provides a high $[NH_4^+]$. This shifts the ionization equilibrium of aqueous NH_3 far to the left, as NH_4^+ ions combine with OH^- ions to form nonionized NH_3 and H_2O. The result is that $[OH^-]$ is decreased significantly.

Solutions that contain a weak base plus a soluble salt of that weak base are always less basic than solutions that contain the same concentration of the weak base alone.

EXAMPLE 19-3 Weak Base/Salt of Weak Base Buffer Solution

Calculate the concentration of OH^- and the pH of a solution that is 0.20 M in aqueous NH_3 and 0.10 M in NH_4Cl.

Plan

Write the appropriate equations for *both* NH_4Cl and NH_3 and the ionization constant expression for NH_3. Then represent the *equilibrium* concentrations algebraically and substitute into the K_b expression.

Solution

The appropriate equations and algebraic representations of concentrations are

rxn 1 $NH_4Cl \longrightarrow NH_4^+ + Cl^-$
         $0.10\ M$ $0.10\ M$ $0.10\ M$

rxn 2 $NH_3 + H_2O \rightleftharpoons NH_4^+ + OH^-$
         $(0.20 - x)\ M$ $x\ M$ $x\ M$ $\longrightarrow$ Total $[NH_4^+] = (0.10 + x)\ M$

Substitution into the K_b expression for aqueous NH_3 gives

$$K_b = \frac{[NH_4^+][OH^-]}{[NH_3]} = 1.8 \times 10^{-5} = \frac{(0.10 + x)(x)}{0.20 - x}$$

The small value of K_b suggests that x is very small. This leads to two assumptions. It is reasonable to assume (1) that x (from the ionization of NH_3) is small because NH_3 is a weak base (rxn 2), and (2) that its ionization is further suppressed by the high concentration of NH_4^+ formed by the soluble salt, NH_4Cl (rxn 1).

Assumption	Implication
$x \ll 0.10$, so $(0.10 + x) \approx 0.10$	Nearly all of the NH_4^+ comes from NH_4Cl (rxn 1), and very little NH_4^+ comes from ionization of NH_3 (rxn 2)
$x \ll 0.20$, so $(0.20 - x) \approx 0.20$	Very little of the NH_3 ionizes (rxn 2)

Introducing these assumptions gives

$$\frac{0.10x}{0.20} = 1.8 \times 10^{-5} M \qquad \text{so} \qquad x = 3.6 \times 10^{-5} M$$

$$x M = [OH^-] = 3.6 \times 10^{-5} M \qquad \text{so} \qquad pOH = 4.44 \qquad \text{and} \qquad pH = 14 - pOH = 9.56$$

You should now work Exercise 12.

In Example 18-15 we calculated $[OH^-]$ and pH for 0.20 M aqueous NH_3. Compare those results with the values obtained in Example 19-3 (Table 19-2). The concentration of OH^- is *53 times greater* in the solution containing only 0.20 M aqueous NH_3 than in the solution to which 0.10 mol/L NH_4Cl has been added. This is another demonstration of the common ion effect.

Table 19-2 Comparison of $[OH^-]$ and pH in Ammonia and Ammonium Chloride–Ammonia Solutions

Solution	% NH_3 Ionized	$[OH^-]$	pH	
0.20 M aq NH_3	0.95%	$1.9 \times 10^{-3} M$	11.28	$\Delta pH = -1.72$
0.20 M aq NH_3 and 0.10 M aq NH_4Cl	0.018%	$3.6 \times 10^{-5} M$	9.56	

We can derive a relationship for $[OH^-]$ in a solution containing a weak base, B, *plus* a salt that contains the cation, BH^+, of the weak base, just as we did for weak acids. In general terms the equation for the ionization of a monobasic weak base and its K_b expression are

$$B + H_2O \rightleftharpoons BH^+ + OH^- \quad \text{and} \quad \frac{[BH^+][OH^-]}{[B]} = K_b$$

Solving the K_b expression for $[OH^-]$ gives

$$[OH^-] = K_b \times \frac{[B]}{[BH^+]}$$

Taking the logarithm of both sides of the equation gives

$$\log[OH^-] = \log K_b + \log \frac{[B]}{[BH^+]}$$

STOP & THINK
The two-orders-of-magnitude guideline introduced in Chapter 18 suggests these assumptions. But we can *never* neglect the variable x when it stands alone as a multiplier! Adding or subtracting a very small number sometimes makes little difference in the result, but multiplying or dividing by a small number is always important!

▶ The percent ionization of NH_3 in this solution is
$$\frac{3.6 \times 10^{-5} M_{ionized}}{0.20 M_{original}} \times 100\% = 0.018\%$$

James W. Morgenthaler

The two solutions in Table 19-2 in the presence of a universal indicator. The NH_3 solution is on the left. Can you calculate the percentage of NH_3 that is ionized in these two solutions?

▶ B and BH^+ represent the weak base and its conjugate acid, respectively—for example, NH_3 and NH_4^+.

Multiplying by -1, substituting pK_b for $-\log K_b$ and rearranging gives another form of the *Henderson–Hasselbalch equation* for solutions containing a weak base plus a salt of the weak base.

$$pOH = pK_b + \log\frac{[BH^+]}{[B]} \quad \text{(for base/salt buffer)}$$

The Henderson–Hasselbalch equation is valid for solutions of weak bases plus salts of weak bases with univalent anions in reasonable concentrations. In general terms we can also write this equation as

$$pOH = pK_b + \log\frac{[\text{conjugate acid (salt)}]}{[\text{base}]} \quad \text{(for base/salt buffer)}$$

▶ You should solve Example 19-3 using this base/salt form of the Henderson–Hasselbalch equation.

19-2 Buffering Action

> A buffer solution is able to react with either H_3O^+ or OH^- ions, whichever is added, to keep the pH approximately the same.

Thus, a buffer solution resists changes in pH. When we add a modest amount of a strong base or a strong acid to a buffer solution, the pH changes very little. More concentrated buffer solutions can react with larger amounts of added acid or base before significant pH changes occur.

The two common kinds of buffer solutions are the ones we have just discussed—namely, solutions containing (1) a weak acid plus a soluble ionic salt of the weak acid and (2) a weak base plus a soluble ionic salt of the weak base.

Solutions of a Weak Acid and a Salt of the Weak Acid

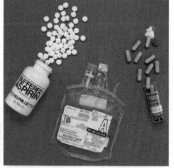

Maria G. Clarke

Three common examples of buffers. Many medications are buffered to minimize digestive upset. Most body fluids, including blood plasma, contain very efficient natural buffer systems. Buffer capsules are used in laboratories to prepare solutions of specified pH.

A solution containing acetic acid, CH_3COOH, and sodium acetate, $NaCH_3COO$, is an example of this kind of buffer solution. The acidic component is CH_3COOH. The basic component is $NaCH_3COO$ because the CH_3COO^- ion is the conjugate base of CH_3COOH. The operation of this buffer depends on the equilibrium

$$CH_3COOH + H_2O \rightleftharpoons H_3O^+ + CH_3COO^-$$
$$\quad \text{\textit{high} conc} \qquad\qquad\qquad\qquad \text{\textit{high} conc}$$
$$\quad \text{(from acid)} \qquad\qquad\qquad\qquad \text{(from salt)}$$

If we add a modest amount of a strong acid such as HCl to this solution, it produces H_3O^+. As a result of the added H_3O^+, the reaction shifts to the *left* as predicted by LeChatelier's Principle, to use up most of the added H_3O^+ and reestablish equilibrium. Because the $[CH_3COO^-]$ in the buffer solution is high, this can occur to a great extent. The net reaction is

$$H_3O^+ + CH_3COO^- \longrightarrow CH_3COOH + H_2O \quad (\approx 100\%)$$

or, written as a formula unit equation,

$$\underset{\text{added acid}}{HCl} + \underset{\text{base}}{NaCH_3COO} \longrightarrow \underset{\text{weaker acid}}{CH_3COOH} + \underset{\text{salt}}{NaCl} \quad (\approx 100\%)$$

▶ The net effect is to neutralize most of the H_3O^+ from HCl by forming nonionized CH_3COOH molecules. This slightly decreases the ratio $[CH_3COO^-]/[CH_3COOH]$, which governs the pH of the solution.

This reaction goes nearly to completion because CH_3COOH is a *weaker* acid than HCl; even when mixed from separate sources, its ions have a strong tendency to form nonionized CH_3COOH molecules rather than remain separate.

When a modest amount of a strong base, such as NaOH, is added to the CH_3COOH–$NaCH_3COO$ buffer solution, some of the OH^- can react directly with the acetic acid, CH_3COOH, which is present in a large amount. Additional OH^- can react with the relatively small amount of H_3O^+ present, greatly reducing $[H_3O^+]$. This disturbs the CH_3COOH ionization equilibrium.

$$CH_3COOH + H_2O \rightleftharpoons CH_3COO^- + H_3O^+$$

This equilibrium then shifts to the *right* to replace some of the depleted H_3O^+. Because the $[CH_3COOH]$ is high, this can occur to a great extent. The net result is the neutralization of most of the added OH^-.

$$\underset{\text{added base}}{OH^-} + \underset{\text{acid}}{CH_3COOH} \longrightarrow \underset{\text{weaker base}}{CH_3COO^-} + \underset{\text{water}}{H_2O} \quad (\approx 100\%)$$

or, written as a formula unit equation,

$$\underset{\text{added base}}{NaOH} + \underset{\text{acid}}{CH_3COOH} \longrightarrow \underset{\text{weaker base}}{NaCH_3COO} + \underset{\text{water}}{H_2O} \quad (\approx 100\%)$$

▶ The net effect is to neutralize most of the OH^- from NaOH. This *slightly* increases the ratio $[CH_3COO^-]$/$[CH_3COOH]$, which governs the pH of the solution.

EXAMPLE 19-4 Buffering Action

We add 0.010 mol of solid NaOH to 1.00 liter of a buffer solution that is 0.100 M in CH_3COOH and 0.100 M in $NaCH_3COO$. How much will $[H_3O^+]$ and pH change? Assume that there is no volume change due to the addition of solid NaOH.

Plan

Calculate $[H_3O^+]$ and pH for the original buffer solution. Then write the reaction summary that shows how much of the CH_3COOH is neutralized by NaOH (forming additional $NaCH_3COO$). Calculate $[H_3O^+]$ and pH for the resulting new buffer solution. Finally, calculate the change in pH.

Solution

For the 0.100 M CH_3COOH and 0.100 M $NaCH_3COO$ buffer solution, we can write

$$[H_3O^+] = K_a \times \frac{[\text{acid}]}{[\text{salt}]} = 1.8 \times 10^{-5} \times \frac{0.100}{0.100} = 1.8 \times 10^{-5}\ M;\ pH = 4.74$$

When solid NaOH is added, it reacts with CH_3COOH to form more $NaCH_3COO$.

	NaOH	+ CH₃COOH ⟶	NaCH₃COO + H₂O
start	0.010 mol	0.100 mol	0.100 mol
change due to rxn	−0.010 mol	−0.010 mol	+0.010 mol
after rxn	0 mol	0.090 mol	0.110 mol

The volume of the solution is still 1.00 liter, so we now have a buffer solution that is 0.090 M in CH_3COOH and 0.110 M in $NaCH_3COO$. In this solution,

$$[H_3O^+] = K_a \times \frac{[\text{acid}]}{[\text{salt}]} = 1.8 \times 10^{-5} \times \frac{0.090}{0.110} = 1.5 \times 10^{-5}\ M;\ pH = 4.82$$

The addition of 0.010 mol of solid NaOH to 1.00 liter of this buffer solution decreases $[H_3O^+]$ from $1.8 \times 10^{-5}\ M$ to $1.5 \times 10^{-5}\ M$ and increases pH from 4.74 to 4.82, a change of 0.08 pH unit, which is a very slight change. This is a change in $[OH^-]$ by a factor of only 1.2, from 5.5×10^{-10} to 6.6×10^{-10}, or a change of only 20%.

You should now work Exercise 24.

▶ We could have used the Henderson–Hasselbalch equation to solve for both pH values in this example.

$$pH = pK_a + \log \frac{[\text{conjugate base}]}{[\text{acid}]}$$

The $[H_3O^+]$ can then be obtained from the pH value.

▶ This is enough NaOH to neutralize 10% of the acid.

To emphasize the effectiveness of such a buffer solution, let's describe what happens when we add 0.010 mole of solid NaOH to one liter of 0.100 M CH_3COOH (pH = 2.89) to give a solution that is 0.090 M in CH_3COOH and 0.010 M in $NaCH_3COO$. The pH of the solution is 3.79, which is only 0.90 pH unit higher than that of the 0.100 M CH_3COOH solution.

By contrast, adding 0.010 mole of NaOH to enough pure H_2O to give one liter produces a 0.010 M solution of NaOH: $[OH^-] = 1.0 \times 10^{-2}\ M$ and pOH = 2.00. The pH of this solution is 12.00, an increase of 5.00 pH units above that of pure H_2O. This corresponds to decreasing $[H_3O^+]$ (and increasing $[OH^-]$) by a factor of 10^5, or 100,000.

▶ pH + pOH = 14

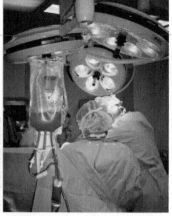

The pH of blood is controlled by naturally occurring buffers in the blood.

Table 19-3 Changes in pH Caused by Addition of Pure Base or Acid to One Liter of Solution

We Have 1.00 L of Original Solution	When We Add 0.010 mol NaOH(s)		When We Add 0.010 mol HCl(g)	
	pH *changes by*	$[H_3O^+]$ *decreases by a factor of*	pH *changes by*	$[H_3O^+]$ *increases by a factor of*
buffer solution (0.10 M NaCH$_3$COO and 0.10 M CH$_3$COOH)	+0.08 pH unit	1.2	−0.08 pH unit	1.2
0.10 M CH$_3$COOH	+0.90	7.9	−0.89	7.8
pure H$_2$O	+5.00	100,000	−5.00	100,000

In summary, 0.010 mole of NaOH

added to 1.00 L of the CH$_3$COOH/NaCH$_3$COO buffer, pH 4.74 $\longrightarrow$ 4.82
added to 1.00 L of 0.100 M CH$_3$COOH, pH 2.89 $\longrightarrow$ 3.78
added to 1.00 L of pure H$_2$O, pH 7.00 $\longrightarrow$ 12.00

In similar fashion we could calculate the effects of adding 0.010 mole of pure HCl(g) instead of pure NaOH to 1.00 liter of each of these three solutions. This would result in the following changes in pH:

added to 1.00 L of the CH$_3$COOH/NaCH$_3$COO buffer, pH 4.74 $\longrightarrow$ 4.66
added to 1.00 L of 0.100 M CH$_3$COOH, pH 2.89 $\longrightarrow$ 2.00
added to 1.00 L of pure H$_2$O, pH 7.00 $\longrightarrow$ 2.00

The results of adding NaOH or HCl to these solutions (Table 19-3) demonstrate the efficiency of the buffer solution. We recall that each change of 1 pH unit means that the $[H_3O^+]$ and $[OH^-]$ change *by a factor of 10*. In these terms, the effectiveness of the buffer solution in controlling pH is even more dramatic.

> **⑤ TOP & THINK**
> When a little strong acid is added to a solution of a weak acid, the $[H_3O]^+$ from the weak acid is negligible compared to the $[H_3O]^+$ from the strong acid.

Solutions of a Weak Base and a Salt of the Weak Base

An example of this type of buffer solution is one that contains the weak base ammonia, NH$_3$, and its soluble ionic salt ammonium chloride, NH$_4$Cl. The reactions responsible for the operation of this buffer are

$$NH_4Cl \xrightarrow{H_2O} NH_4^+ + Cl^- \quad \text{(to completion)}$$
$$NH_3 + H_2O \rightleftharpoons NH_4^+ + OH^- \quad \text{(reversible)}$$

high conc   *high* conc
(from base)   (from salt)

If a strong acid such as HCl is added to this buffer solution, the resulting H$_3$O$^+$ shifts the equilibrium reaction

$$2H_2O \rightleftharpoons H_3O^+ + OH^- \quad \text{(shifts *left*)}$$

strongly to the *left*. As a result of the diminished OH$^-$ concentration, the reaction

$$NH_3 + H_2O \rightleftharpoons NH_4^+ + OH^- \quad \text{(shifts *right*)}$$

shifts markedly to the *right*. Because the [NH$_3$] in the buffer solution is high, this can occur to a great extent. The net reaction is

$$H_3O^+ \quad + NH_3 \longrightarrow NH_4^+ + H_2O \quad (\approx 100\%)$$

added acid   base   weak acid   water

> ▶ The net effect is to neutralize most of the H$_3$O$^+$ from HCl. This slightly increases the ratio [NH$_4^+$]/[NH$_3$], which governs the pH of the solution.

When a strong base such as NaOH is added to the *original* buffer solution, it is neutralized by the more acidic component, NH_4^+ (the conjugate acid of ammonia).

$$NH_3 + H_2O \rightleftharpoons NH_4^+ + OH^- \qquad \text{(shifts \textit{left})}$$

Because the $[NH_4^+]$ is high, this can occur to a great extent. The result is the neutralization of OH^- by NH_4^+,

$$OH^- + NH_4^+ \longrightarrow NH_3 + H_2O \qquad (\approx 100\%)$$

or, as a formula unit equation,

$$\underset{\text{added base}}{NaOH} + \underset{\text{acid}}{NH_4Cl} \longrightarrow \underset{\text{weak base}}{NH_3} + \underset{\text{water}}{H_2O} + NaCl \qquad (\approx 100\%)$$

▶ The net effect is to neutralize most of the OH^- from NaOH. This slightly decreases the ratio $[NH_4^+]/[NH_3]$, which governs the pH of the solution.

Summary Changes in pH are minimized in buffer solutions because the basic component can react with added H_3O^+ ions (producing additional weak acid), while the acidic component can react with added OH^- ions (producing additional weak base).

19-3 Preparation of Buffer Solutions

Buffer solutions can be prepared by mixing other solutions. When solutions are mixed, the volume in which each solute is contained increases, so solute concentrations change. These changes in concentration must be considered. If the solutions are dilute, we may assume that their volumes are additive.

Buffer Preparation by Mixing of a Conjugate Acid–Base Pair

EXAMPLE 19-5 pH of a Buffer Solution

Calculate the concentration of H_3O^+ and the pH of a buffer solution prepared by mixing 200. mL of 0.10 M NaF and 100. mL of 0.050 M HF. $K_a = 7.2 \times 10^{-4}$ for HF.

Plan

Calculate the number of millimoles (or moles) of NaF and HF and then the molarity of each solute in the solution after mixing. Write the appropriate equations for both NaF and HF, represent the equilibrium concentrations algebraically, and substitute into the K_a expression for HF.

▶ Alternatively, we could substitute into the Henderson–Hasselbalch equation and solve for pH, and then convert to $[H_3O^+]$.

Solution

When two dilute solutions are mixed, we assume that their volumes are additive. The volume of the new solution will be 300. mL. Mixing a solution of a weak acid with a solution of its salt does not form any new species (though it does change their concentrations). So we have a straightforward buffer calculation. We calculate the number of millimoles (or moles) of each compound and the molarities in the new solution.

$$\left.\begin{array}{l} \underline{?}\ \text{mmol NaF} = 200.\ \text{mL} \times \dfrac{0.10\ \text{mmol NaF}}{\text{mL}} = 20.\ \text{mmol NaF} \\[2mm] \underline{?}\ \text{mmol HF} = 100.\ \text{mL} \times \dfrac{0.050\ \text{mmol HF}}{\text{mL}} = 5.0\ \text{mmol HF} \end{array}\right\} \text{in 300. mL}$$

The molarities of NaF and HF in the solution are

$$\frac{20.\ \text{mmol NaF}}{300.\ \text{mL}} = 0.067\ M\ \text{NaF} \qquad \text{and} \qquad \frac{5.0\ \text{mmol HF}}{300.\ \text{mL}} = 0.017\ M\ \text{HF}$$

 S TOP & THINK
It is very important to remember to use the total solution volume of 300 mL for the calculations in this example!

The appropriate equations and algebraic representations of concentrations are

$$
\begin{array}{ccccc}
\text{NaF} & \longrightarrow & \text{Na}^+ & + & \text{F}^- \\
0.067\,M & \Longrightarrow & 0.067\,M & & 0.067\,M \\
& & & & \hspace{1cm}\text{Total } [\text{F}^-] = (0.067 + x)\,M \\
\text{HF} & + \text{H}_2\text{O} \rightleftharpoons & \text{H}_3\text{O}^+ & + & \text{F}^- \\
(0.017 - x)\,M & & x\,M & & x\,M
\end{array}
$$

Substituting into the K_a expression for hydrofluoric acid gives

$$
K_a = \frac{[\text{H}_3\text{O}^+][\text{F}^-]}{[\text{HF}]} = \frac{(x)(0.067 + x)}{(0.017 - x)} = 7.2 \times 10^{-4}
$$

The two-orders-of-magnitude guideline suggests that x is negligible compared with 0.067 and 0.017 in this expression. (When in doubt, solve the equation using the simplifying assumption. Then decide whether the assumption was valid.) Assume that $(0.067 + x) \approx 0.067$ and $(0.017 - x) \approx 0.017$.

> ▶ Using the Henderson–Hasselbalch equation, we obtain:
>
> $$\text{pH} = -\log(7.2 \times 10^{-4}) + \log\frac{0.067}{0.017}$$
>
> $$\text{pH} = 3.14 + 0.596 = 3.74$$

$$
\frac{0.067x}{0.017} = 7.2 \times 10^{-4} \quad ; \quad x = 1.8 \times 10^{-4}\,M = [\text{H}_3\text{O}^+]
$$

$$
\text{pH} = 3.74 \text{ (Our assumption was valid)}
$$

You should now work Exercise 36.

We often need a buffer solution of a desired pH. One way to prepare such a solution is to add a salt of a weak base (or weak acid) to a solution of the weak base (or weak acid).

EXAMPLE 19-6 Buffer Preparation by Addition of a Salt

A solution is 0.10 M in aqueous NH_3. Calculate (a) the number of moles and (b) the number of grams of NH_4Cl that must be added to 500 mL of this solution to prepare a buffer solution with pH = 9.15. You may neglect the volume change due to addition of solid NH_4Cl.

Plan

Convert the given pH to the desired $[\text{OH}^-]$ by the usual procedure. Write the appropriate equations for the reactions of NH_4Cl and NH_3 and represent the equilibrium concentrations. Then, substitute into the K_b expression, and solve for the concentration of NH_4Cl required. This allows calculation of the amount of solid NH_4Cl to be added.

Solution

(a) Because the desired pH = 9.15, pOH = 14.00 − 9.15 = 4.85. So $[\text{OH}^-] = 10^{-\text{pOH}} = 10^{-4.85} = 1.4 \times 10^{-5}\,M\,\text{OH}^-$ desired. Let x mol/L be the necessary molarity of NH_4Cl. Because $[\text{OH}^-] = 1.4 \times 10^{-5}\,M$, this must be the $[\text{OH}^-]$ produced by ionization of NH_3. The equations and representations of equilibrium concentrations follow.

$$
\begin{array}{ccccc}
\text{NH}_4\text{Cl} & \xrightarrow{100\%} & \text{NH}_4^+ & + & \text{Cl}^- \\
x\,M & \Longrightarrow & x\,M & & x\,M \\
\text{NH}_3 & + \text{H}_2\text{O} \rightleftharpoons & \text{NH}_4^+ & + & \text{OH}^- \\
(0.10 - 1.4 \times 10^{-5})\,M & & 1.4 \times 10^{-5}\,M & & 1.4 \times 10^{-5}\,M
\end{array}
$$

$$
\text{Total } [\text{NH}_4^+] = (x + 1.4 \times 10^{-5})\,M
$$

Substitution into the K_b expression for aqueous ammonia gives

$$
K_b = \frac{[\text{NH}_4^+][\text{OH}^-]}{[\text{NH}_3]} = 1.8 \times 10^{-5} = \frac{(x + 1.4 \times 10^{-5})(1.4 \times 10^{-5})}{(0.10 - 1.4 \times 10^{-5})}
$$

NH$_4$Cl is 100% dissociated, so $x \gg 1.4 \times 10^{-5}$. Then $(x + 1.4 \times 10^{-5}) \approx x$.

$$\frac{(x)(1.4 \times 10^{-5})}{0.10} = 1.8 \times 10^{-5} \quad ; \quad x = 0.13 \; M = [\text{NH}_4^+] = M_{\text{NH}_4\text{Cl}}$$

Now we calculate the number of moles of NH$_4$Cl that must be added to prepare 500. mL (0.500 L) of buffer solution.

$$\underline{?} \; \text{mol NH}_4\text{Cl} = 0.500 \; \text{L} \times \frac{0.13 \; \text{mol NH}_4\text{Cl}}{\text{L}} = 0.065 \; \text{mol NH}_4\text{Cl}$$

(b) $\qquad \underline{?} \; \text{g NH}_4\text{Cl} = 0.065 \; \text{mol} \times \dfrac{53.5 \; \text{g NH}_4\text{Cl}}{\text{mol NH}_4\text{Cl}} = \boxed{3.5 \; \text{g NH}_4\text{Cl}}$

You should now work Exercise 39.

▶ Example 19-6 can be solved using the base/salt form of the Henderson–Hasselbalch equation as follows:

S TOP & THINK
Here x does *not* represent a *change* in concentration, but rather the initial concentration of NH$_4$Cl. We do *not* assume that $x \ll 1.4 \times 10^{-5}$, but rather the reverse.

$$\text{pOH} = \text{p}K_b + \log \frac{[\text{conjugate acid}]}{[\text{base}]}$$

$$4.85 = -\log (1.8 \times 10^{-5})$$
$$+ \log \frac{[\text{conjugate acid}]}{[0.10]}$$

$$4.85 = 4.74 + \log[\text{conjugate acid}] - (-1.00)$$

$$[\text{conjugate acid}] = 0.13 \; M = [\text{NH}_4^+]$$

A We add 3.5 grams of NH$_4$Cl to a 500-mL volumetric flask.

B We dissolve it in a little of the 0.10 M NH$_3$ solution.

C We then dilute to 500 mL with the 0.10 M NH$_3$ solution.

D When a universal indicator is added, the green color shows that the buffer solution has a lower pH than the original NH$_3$ solution (*blue*).

Charles Steele

Preparation of the buffer solution in Example 19-6.

We have used K_a or K_b expressions or the Henderson–Hasselbalch equation in its acid/salt or base/salt form to find the pH of buffered solutions. Either of these calculations involves a *ratio* of concentrations, for instance $\dfrac{[\text{conjugate base}]}{[\text{acid}]}$. Although the ratios are written in terms of molarities, it is not always necessary to use concentrations in the calculations. Both reagents are present in *a single* buffer solution, so the solution volume cancels from the molarity ratio. For example:

$$\frac{[\text{conjugate base}]}{[\text{acid}]} = \frac{\dfrac{\text{mol conjugate base}}{\text{L soln}}}{\dfrac{\text{mol acid}}{\text{L soln}}} = \frac{\text{mol conjugate base}}{\text{mol acid}}$$

Thus we see that a molarity ratio in the acid/salt Henderson–Hasselbalch equation can be treated as a mole (or millimole) ratio. A similar conclusion can be reached for the base/salt version of the Henderson–Hasselbalch equation or for the K_a or K_b expressions that were used in previous buffer calculations.

Thus the Henderson–Hasselbalch equation for buffers can be written in the following forms:

For an acid/salt buffer: $\text{pH} = \text{p}K_a + \dfrac{\text{mol conjugate base}}{\text{mol acid}} = \text{p}K_a + \dfrac{\text{mmol conjugate base}}{\text{mmol acid}}$

For a base/salt buffer: $\text{pOH} = \text{p}K_b + \dfrac{\text{mol conjugate acid}}{\text{mol base}} = \text{p}K_b + \dfrac{\text{mmol conjugate acid}}{\text{mmol base}}$

CHEMISTRY IN USE

Fun with Carbonates

Carbonates react with acids to produce carbon dioxide. This property of carbonates has been exploited in many ways, both serious and silly.

One of the giddiest applications of this behavior of carbonates is seen in Mad Dawg, a foaming bubble gum developed in the early 1990s. If you chew a piece of this gum, large quantities of foam are produced so that it is difficult to keep the colorful lather from oozing out of your mouth. The froth begins to form as your teeth mix saliva with the gum's ingredients (sodium hydrogen carbonate, citric acid, malic acid, food coloring, and flavoring).

How is this foam produced? When citric acid and malic acid dissolve in saliva, they produce hydrogen ions, which decompose the sodium hydrogen carbonate (baking soda) to produce carbon dioxide, a gas. These bubbles of carbon dioxide produce the foam. Large quantities of foam are produced because citric and malic acids taste sour, which stimulates salivation.

A common medical recipe for a similar combination of ingredients is found in Alka Seltzer tablets; these contain sodium hydrogen carbonate, citric acid, and aspirin. The acid and carbonate react in water to produce carbon dioxide, which gives the familiar fizz of Alka Seltzer.

Makeup artists add baking soda to cosmetics to produce monster-flesh makeup. When the hero throws acid (which is actually vinegar, a dilute solution of acetic acid) into the monster's face, the acetic acid reacts with sodium hydrogen carbonate to produce the disgustingly familiar scenes of "dissolving flesh" that we see in horror movies. The ability of baking soda to produce carbon dioxide delights children of all ages as it creates monsters in the movies.

Many early fire extinguishers utilized the reaction of sodium hydrogen carbonate with acids. A metal cylinder was filled with a solution of sodium hydrogen carbonate and water; a bottle filled with sulfuric acid was placed above the water layer. Invert-

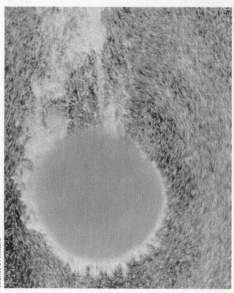

Charles D. Winters

Alka Seltzer™

ing the extinguisher activated it by causing the acid to spill into the carbonate solution. The pressure produced by gaseous carbon dioxide gas pushed the liquid contents out through a small hose.

Kitchen oven fires can usually be extinguished by throwing baking soda onto the flame. When heated, carbonates decompose to produce carbon dioxide, which smothers fires by depriving them of oxygen.

Chefs frequently use the heat-sensitive nature of carbonates to test the freshness of a box of baking soda. Pouring some boiling water over a little fresh baking soda results in active bubbling. Less active bubbling means the baking soda is unlikely to work well in a baking recipe.

RONALD DELORENZO

Buffer Preparation by Partial Neutralization

In practice, a common method of preparing a buffer solution is by *partial* neutralization of a weak acid solution by adding a strong base solution. For example,

$$HA + NaOH \longrightarrow NaA + H_2O \qquad (partial)$$

If an appreciable amount of the weak acid remains unneutralized, then this solution contains significant concentrations of a *weak acid* and *its conjugate base*, just as though we had added the salt from a separate solution; thus, it is a buffer solution. Example 19-7 illustrates the preparation of a buffer by this method.

EXAMPLE 19-7 Buffer Preparation by Partial Neutralization

Calculate the pH of a solution obtained by mixing 400. mL of a 0.200 M acetic acid solution and 100. mL of a 0.300 M sodium hydroxide solution.

Plan

Sodium hydroxide, NaOH, is a strong base, so it reacts with acetic acid, CH_3COOH, to form sodium acetate, $NaCH_3COO$. If an appreciable amount of excess acetic acid is still present after the sodium hydroxide has reacted, the excess acetic acid and the newly formed sodium acetate solution form a buffer solution.

▶ We are given the amounts of *both* reactants, so we recognize this as a *limiting reactant* problem. Review Section 3-3.

Solution

We first calculate how much of the weak acid has been neutralized. The numbers of millimoles of CH_3COOH and NaOH mixed are calculated as

$$\text{mmol } CH_3COOH = (0.200 \text{ mmol/mL}) \times 400. \text{ mL} = 80.0 \text{ mmol}$$

$$\text{mmol NaOH} = (0.300 \text{ mmol/mL}) \times 100. \text{ mL} = 30.0 \text{ mmol}$$

▶ Recall that M is equal to moles per liter or millimoles per milliliter.

Not enough NaOH is present to neutralize all of the CH_3COOH, so NaOH is the *limiting reactant*.

	NaOH	+ CH_3COOH	⟶	$NaCH_3COO$	+ H_2O
start	30.0 mmol	80.0 mmol		0	—
change	−30.0 mmol	−30.0 mmol		+30.0 mmol	—
after rxn	0.0 mmol	50.0 mmol		30.0 mmol	

Because $NaCH_3COO$ is a soluble salt, it provides 30.0 mmol CH_3COO^- to the solution. This solution contains a significant amount of CH_3COOH not yet neutralized *and* a significant amount of its conjugate base, CH_3COO^-. We recognize this as a buffer solution, so we can use the Henderson–Hasselbalch equation to find the pH.

$$pH = pK_a + \log\frac{\text{mmol conjugate base}}{\text{mmol acid}} = pK_a + \log\frac{\text{mmol } CH_3COO^-}{\text{mmol } CH_3COOH}$$

▶ We could also solve this problem using the K_a expression as we did in Example 19-5.

$$= 4.74 + \log\frac{30.0}{50.0} = 4.74 + \log(0.600) = 4.74 + (-0.222)$$

$$= 4.52$$

You should now work Exercises 54b to 54f.

More Complex Buffer Solutions

Living systems depend on buffers for pH control; most of these are complex buffers that contain a mixture of various acids and bases. Buffers can be made from weak acids/bases and the salts of *other* weak acids/bases. For example, we could make an acidic buffer from acetic acid and sodium bicarbonate ($NaHCO_3$, a salt of the weak acid carbonic acid). Or, we could make a basic buffer from NH_3 and $NH_2(CH_3)_2Br$ (the salt of the weak base dimethylamine, $HN(CH_3)_2$). But it is more difficult to calculate pH's or other concentration values from these mixed buffer systems. We have limited our discussion and detailed examples to buffers that contain a weak acid/base and the salt of that *same* acid/base.

19-4 Acid–Base Indicators

In Section 11-2, we described acid–base titrations and the use of indicators to tell us when to stop a titration. Detection of the end point in an acid–base titration is only one of the important uses of indicators.

An **indicator** is an organic dye; its color depends on the concentration of H_3O^+ ions, or the pH, in the solution. By the color an indicator displays, it "indicates" the acidity or basicity of a solution. Figure 19-1 displays solutions that contain three common indicators in solutions over the pH range 3 to 11. Carefully study Figure 19-1 and its legend.

▶ Phenolphthalein was the active component of the laxative Ex-Lax.

Bromthymol blue indicator is yellow in acidic solutions and blue in basic solutions.

The first indicators used were vegetable dyes. Litmus is a familiar example: red in acidic solution, blue in basic solution. Most of the indicators that we use in the laboratory today are synthetic compounds; that is, they have been made in laboratories by chemists. Phenolphthalein is the most common acid–base indicator. It is colorless in solutions of pH less than 8 ($[H_3O^+] > 10^{-8} M$) and turns bright pink as pH approaches 10.

Many acid–base indicators are weak organic acids. An organic indicator acid can be symbolized by HIn, where "In" represents various complex organic groups. Bromthymol blue is such an indicator. Its acid ionization constant is 7.9×10^{-8}. We can represent its ionization in dilute aqueous solution and its ionization constant expression as

$$HIn + H_2O \rightleftharpoons H_3O^+ + In^- \qquad K_a = \frac{[H_3O^+][In^-]}{[HIn]} = 7.9 \times 10^{-8}$$

color 1 ← for bromthymol blue → color 2
yellow   blue

HIn represents nonionized acid molecules, and In^- represents the anion (conjugate base) of HIn. The essential characteristic of an acid–base indicator is that HIn and In^- *must* have quite different colors. The relative amounts of the two species determine the color of the solution. Adding an acid favors the reaction to the left and gives more HIn molecules (color 1). Adding a base favors the reaction to the right and gives more In^- ions (color 2). The ionization constant expression can be rearranged.

$$\frac{[H_3O^+][In^-]}{[HIn]} = K_a \qquad so \qquad \frac{[In^-]}{[HIn]} = \frac{K_a}{[H_3O^+]}$$

This shows clearly how the $[In^-]/[HIn]$ ratio depends on $[H_3O^+]$ (or on pH) and the K_a value for the indicator. As a rule of thumb, when $[In^-]/[HIn] \geq 10$, color 2 (the color of In^-) is observed; conversely, when $[In^-]/[HIn] \leq \frac{1}{10}$, color 1 (the color of HIn) is observed.

Universal indicators are mixtures of several acid–base indicators that display a continuous range of colors over a wide range of pH values. Figure 18-2 shows concentrated solu-

Figure 19-1 Three common indicators in solutions that cover the pH range 3 to 11 (the black numbers).

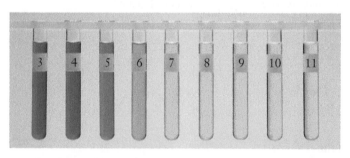

Ⓐ Methyl red is red at pH 4 and below; it is yellow at pH 7 and above. Between pH 4 and pH 7 it changes from red to red-orange, to orange, to yellow.

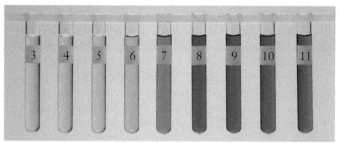

Ⓑ Bromthymol blue is yellow at pH 6 and below; it is blue at pH 8 and above. Between pH 6 and pH 8 it changes smoothly from yellow through green to blue.

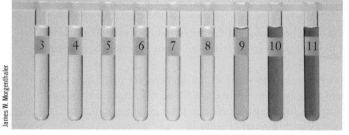

Ⓒ Phenolphthalein is colorless below pH 8 and bright pink above pH 10. It changes smoothly from colorless to bright pink in the pH range 8 to 10.

tions of a universal indicator in flat dishes so that the colors are very intense. The juice of red (purple) cabbage is a natural universal indicator (Figure 19-2).

One important use of universal indicators is in commercial indicator papers, which are small strips of paper impregnated with solutions of universal indicators. A strip of the paper is dipped into the solution of interest, and the color of the indicator on the paper indicates the pH of the solution. The photographs on page 717 illustrate the use of universal indicators to estimate pH. We shall describe the use of indicators in titrations more fully in Sections 19-5 and 19-6.

Anthocyanin pigments, which are found in many plants (see the geranium photo at the beginning of this chapter) are naturally occurring acid–base indicators. There are more than 400 organic molecules that belong to the anthocyanin family. Many red or purple foods, such as blueberries, raspberries, and grapes, contain anthocyanins. Some of these anthocyanins can also act as beneficial antioxidants in the diet. Antioxidants help to protect living cells against the damaging effects of free radicals and are thought to help inhibit the premalignant growth of some cancer cells.

Titration Curves

19-5 Strong Acid/Strong Base Titration Curves

A **titration curve** is a plot of pH versus the amount (usually volume) of acid or base added. It displays graphically the change in pH as acid or base is added to a solution and shows how pH changes during the course of the titration.

The **equivalence point** is the point at which chemically equivalent amounts of acid and base have reacted. The point at which the color of an indicator changes in a titration is known as the **end point**. It is determined by the K_a value for the indicator (see Section 19-4). Table 19-4 shows a few acid–base indicators and the pH ranges over which their colors change. Typically, color changes occur over a range of 1.5 to 2.0 pH units.

> Ideally, the end point and the equivalence point in a titration should coincide.

In practice, we try to select an indicator whose range of color change includes the equivalence point. We use the same procedures in both standardization and analysis to minimize any error arising from a difference between end point and equivalence point.

Review Section 11-2 before you consider the titration of 100.0 mL of a 0.100 M solution of HCl with a 0.100 M solution of NaOH. As we know, NaOH and HCl react in a 1:1 ratio. We calculate the pH of the solution at several stages as NaOH is added.

1. Before any NaOH is added to the 0.100 M HCl solution:

$$HCl + H_2O \xrightarrow{100\%} H_3O^+ + Cl^-$$

$$[H_3O^+] = 0.100\,M \quad \text{so} \quad pH = 1.00$$

The pH depends only on the initial concentration of strong acid.

2. After 20.0 mL of 0.100 M NaOH has been added:

	HCl	+	NaOH	⟶	NaCl	+ H₂O
start	10.0 mmol		2.0 mmol		0 mmol	
change	−2.0 mmol		−2.0 mmol		+2.0 mmol	
after rxn	8.0 mmol		0 mmol		2.0 mmol	

The concentration of unreacted HCl in the total volume of 120. mL is

$$M_{HCl} = \frac{8.00\ \text{mmol HCl}}{120.\ \text{mL}} = 0.067\ M\ \text{HCl}$$

$$[H_3O^+] = 6.7 \times 10^{-2}\,M \quad \text{so} \quad pH = 1.17$$

The pH depends on the amount of strong acid not yet neutralized.

3. After 50.0 mL of 0.100 M NaOH has been added (midpoint of the titration):

	HCl	+	NaOH	⟶	NaCl	+ H₂O
start	10.0 mmol		5.0 mmol		0 mmol	
change	−5.0 mmol		−5.0 mmol		+5.0 mmol	
after rxn	5.0 mmol		0 mmol		5.0 mmol	

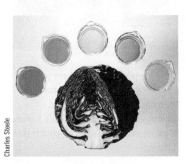

Figure 19-2 The juice of the red (purple) cabbage is a naturally occurring universal indicator. From left to right are solutions of pH 1, 4, 7, 10, and 13.

▶ Titrations are typically done with 50-mL or smaller burets. We have used 100. mL of solution in this discussion to simplify the arithmetic.

▶ If one prefers, these calculations can be done using moles and liters in the place of millimoles and milliliters.

Table 19-4 Range and Color Changes of Some Common Acid–Base Indicators

	pH Scale												
Indicators	**1**	**2**	**3**	**4**	**5**	**6**	**7**	**8**	**9**	**10**	**11**	**12**	**13**
methyl orange	red ⟵⟶			yellow									
methly red	red ⟵⟶				yellow								
bromthymol blue	yellow ⟵⟶					blue							
neutral red	red ⟵⟶					yellow							
phenolphthalein	colorless ⟵⟶						bright pink			colorless beyond 13.0			

Some household products. Each solution contains a few drops of a universal indicator. A color of yellow or red indicates a pH less than 7 (acidic). A green to purple color indicates a pH greater than 7 (basic).

▶ A strong base/strong acid salt does not undergo hydrolysis (Section 18–7).

$$M_{HCl} = \frac{5.00 \text{ mmol HCl}}{150. \text{ mL}} = 0.033 \, M \, HCl$$

$$[H_3O^+] = 3.3 \times 10^{-2} \, M \qquad \text{so} \qquad pH = 1.48$$

The pH depends on the amount of strong acid not yet neutralized.

4. After 100. mL of 0.100 M NaOH has been added:

	HCl	+	NaOH	⟶	NaCl	+ H₂O
start	10.0 mmol		10.0 mmol		0 mmol	
change	−10.0 mmol		−10.0 mmol		+10.0 mmol	
after rxn	0 mmol		0 mmol		10.0 mmol	

We have added enough NaOH to neutralize the HCl exactly, so *this is the equivalence point. A strong acid and a strong base react to give a neutral salt solution,* so $pH = 7.00$.

5. After 110.0 mL of 0.100 M NaOH has been added:

	HCl	+	NaOH	⟶	NaCl	+	H₂O
start	10.0 mmol		11.0 mmol		0 mmol		
change	−10.0 mmol		−10.0 mmol		+10.0 mmol		
after rxn	0 mmol		1.0 mmol		10.0 mmol		

The pH is determined by the excess NaOH.

$$M_{NaOH} = \frac{1.0 \text{ mmol NaOH}}{210. \text{ mL}} = 0.0048 \, M \, NaOH$$

$$[OH^-] = 4.8 \times 10^{-3} \, M \quad \text{so} \quad pOH = 2.32 \quad \text{and} \quad pH = 11.68$$

After the equivalence point, the pH depends only on the concentration of excess strong base.

Table 19-5 displays the data for the titration of 100.0 mL of 0.100 M HCl by 0.100 M NaOH solution. A few additional points have been included to show the shape of the curve better. These data are plotted in Figure 19-3a. This titration curve has a long "vertical section" over which the pH changes very rapidly with the addition of very small amounts of base. The pH changes very sharply from 3.60 (at 99.5 mL NaOH added) to 10.40 (at 100.5 mL of NaOH added) in the vicinity of the equivalence point (100.0 mL NaOH added). The midpoint of the vertical section (pH = 7.00) is the equivalence point.

For this kind of titration there are four distinct types of calculation, one for each of the four regions of the titration curves.

1. Before any strong base is added, the pH depends on the strong acid alone.

2. After some strong base has been added, but before the equivalence point, the remaining (excess) strong acid determines the pH.

3. At the equivalence point, the solution is neutral because it contains only a strong base/strong acid salt (no hydrolysis).

4. Beyond the equivalence point, excess strong base determines the pH.

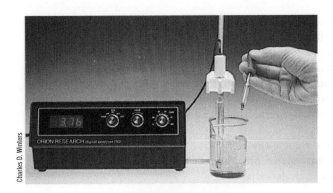

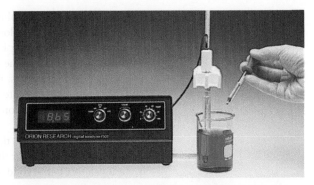

A Acidic solution just before the color change occurs; indicator is yellow.

B Solution after several *small drops* of NaOH are added. The solution is now basic and the indicator is blue. The equivalence point occurs at pH = 7, which is the middle of the bromthymol blue indicator color change (Table 19-4). There is a very large change in pH (3.76 to 8.65 for this solution) right around the equivalence point for strong acid/strong base titrations.

Titration of 0.100 *M* HCl with 0.100 *M* NaOH using bromthymol blue indicator.

Table 19-5 Titration Data for 100.0 mL of 0.100 *M* HCl Versus NaOH

mL of 0.100 *M* NaOH Added	mmol NaOH Added	mmol Excess Acid or Base	pH
0.0	0.00	10.0 H_3O^+	1.00
20.0	2.00	8.0	1.17
50.0	5.00	5.0	1.48
90.0	9.00	1.0	2.28
99.0	9.90	0.10	3.30
99.5	9.95	0.05	3.60
100.0	10.00	0.00 (eq. pt.)	7.00
100.5	10.05	0.05 OH^-	10.40
110.0	11.00	1.00	11.68
120.0	12.00	2.00	11.96

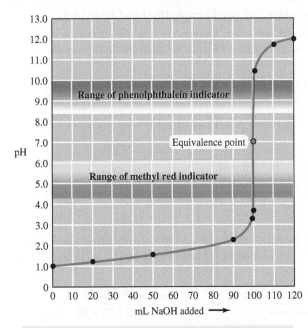

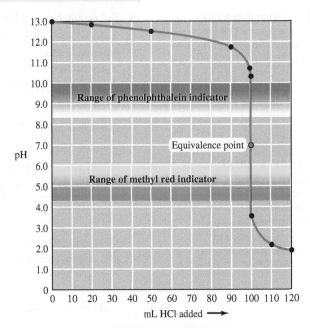

A The titration curve for 100. mL of 0.100 *M* HCl with 0.100 *M* NaOH. Note that the "vertical" section of the curve is quite long. The titration curves for other strong acids and bases are identical with this one *if* the same concentrations of acid and bases are used *and if* both are monoprotic.

B The titration curve for 100. mL of 0.100 *M* NaOH with 0.100 *M* HCl. This curve is similar to that in (a), but inverted.

Figure 19-3 Typical strong acid/strong base curves.

Ideally, the indicator color change should occur at pH = 7.00. For practical purposes, though, indicators with color changes in the pH range 4 to 10 can be used in the titration of strong acids and strong bases because the vertical portion of the titration curve is so long. Figure 19-3 shows the ranges of color changes for methyl red and phenolphthalein, two widely used indicators. Both fall within the vertical section of the NaOH/HCl titration curve, so either indicator would be suitable. When a strong acid is added to a solution of a strong base, the titration curve is inverted, but its essential characteristics are the same (see Figure 19-3b).

In Figure 19-3a, we see that the curve rises very slowly before the equivalence point. It then rises very rapidly near the equivalence point because there is no hydrolysis. The curve becomes almost flat beyond the equivalence point.

19-6 Weak Acid/Strong Base Titration Curves

When a weak acid is titrated with a strong base, the curve is quite different in two important ways. (i) Once the addition of strong base begins, the solution is buffered *before* the equivalence point. (ii) The solution is basic *at* the equivalence point because a salt of a weak acid and a strong base undergoes hydrolysis (see Section 18-8) to give a basic solution.

As before, we can separate the calculations for this kind of titration into four distinct types, corresponding to four regions of the titration curves.

1. Before any base is added, the pH depends on the weak acid alone.
2. After some base has been added, but before the equivalence point, a series of weak acid/salt buffer solutions determines the pH.
3. At the equivalence point, hydrolysis of the anion of the weak acid determines the pH.
4. Beyond the equivalence point, excess strong base determines the pH.

Consider the titration of 100.0 mL of 0.100 M CH$_3$COOH with 0.100 M NaOH solution. (The strong electrolyte is added to the weak electrolyte.)

▶ If it were not already known, we could calculate the pH of the weak acid solution in region 1 using the methods presented in Section 18-8.

1. Before any base is added, the pH is 2.89 (see Example 18-11 and Table 18-5).
2. As soon as some NaOH is added, but before the equivalence point, the solution is buffered because it contains both CH$_3$COOH and NaCH$_3$COO in significant concentrations.

$$\text{NaOH} + \text{CH}_3\text{COOH} \longrightarrow \text{NaCH}_3\text{COO} + \text{H}_2\text{O}$$
$$\text{lim amt} \qquad \text{excess}$$

For instance, after 20.0 mL of 0.100 M NaOH solution has been added, we have

	NaOH	+ CH$_3$COOH	⟶	NaCH$_3$COO + H$_2$O
start	2.00 mmol	10.00 mmol		0 mmol
change	−2.00 mmol	−2.00 mmol		+2.00 mmol
after rxn	0 mmol	8.00 mmol		2.00 mmol

STOP & THINK

The pH calculation in region 2 is that shown for partial neutralization in Example 19-7. Alternatively, we could use the K_a expression for these calculations as we did in Example 19-5.

We recognize that this is a buffer solution, so we can use the ratio (mmol conjugate base)/(mmol acid) in the Henderson–Hasselbalch equation as we did in Example 19-7.

$$\text{pH} = pK_a + \log\frac{\text{mmol conjugate base}}{\text{mmol acid}} = pK_a + \log\frac{\text{mmol CH}_3\text{COO}^-}{\text{mmol CH}_3\text{COOH}}$$
$$= 4.74 + \log\frac{2.00}{8.00} = 4.74 + \log(0.250) = 4.74 + (-0.602) = 4.14$$

After some NaOH has been added, the solution contains both NaCH$_3$COO and CH$_3$COOH, and so it is buffered until the equivalence point is reached. All points before the equivalence point are calculated in the same way.

3. At the equivalence point, the solution is 0.0500 M in NaCH$_3$COO.

	NaOH	+ CH$_3$COOH	$\longrightarrow$	NaCH$_3$COO	+ H$_2$O
start	10.0 mmol	10.0 mmol		0 mmol	
change	−10.0 mmol	−10.0 mmol		+10.0 mmol	
after rxn	0 mmol	0 mmol		10.0 mmol	

$$M_{\text{NaCH}_3\text{COO}} = \frac{10.0 \text{ mmol NaCH}_3\text{COO}}{200. \text{ mL}} = 0.0500 \ M \text{ NaCH}_3\text{COO}$$

The pH of a 0.0500 M solution of NaCH$_3$COO is 8.72 (Example 18-20 shows a similar calculation). The solution is distinctly basic at the equivalence point because of the hydrolysis of the acetate ion.

4. Beyond the equivalence point, the concentration of the excess NaOH determines the pH of the solution just as it did in the titration of a strong acid.

Figure 19-4 shows the titration curve for 100.0 mL of 0.100 M CH$_3$COOH titrated with a 0.100 M solution of NaOH, and Table 19-6 lists several points on the titration curve. This titration curve has a short vertical section (pH ≈ 7 to 10), so many indicators would not be suitable. Phenolphthalein is the indicator commonly used to titrate weak acids with strong bases (see Table 19-4).

The titration curves for weak bases and strong acids are similar to those for weak acids and strong bases except that they are inverted (recall that strong is added to weak). Figure 19-5 displays the titration curve for 100.0 mL of 0.100 M aqueous ammonia titrated with 0.100 M HCl solution.

▶ Just *before* the equivalence point, the solution contains relatively high concentrations of NaCH$_3$COO and relatively low concentrations of CH$_3$COOH. Just *after* the equivalence point, the solution contains relatively high concentrations of NaCH$_3$COO and relatively low concentrations of NaOH, both basic components. In both regions our calculations are only approximations. Exact calculations of pH in these regions are too complicated to discuss here.

Table 19-6 Titration Data for 100.0 mL of 0.100 M CH$_3$COOH with 0.100 M NaOH

mL 0.100 M NaOH Added		mmol Base Added	mmol Excess Acid or Base	pH
0.0 mL		0	10.0 CH$_3$COOH	2.89
20.0 mL		2.00	8.00	4.14
50.0 mL		5.00	5.00	4.74
75.0 mL	buffered	7.50	2.50	5.22
90.0 mL	region	9.00	1.00	5.70
95.0 mL		9.50	0.50	6.02
99.0 mL		9.90	0.10	6.74
100.0 mL		10.0	0 (equivalence point)	8.72
101.0 mL		10.1	0.10 OH$^-$	10.70
110.0 mL		11.0	1.0	11.68
120.0 mL		12.0	2.0	11.96

19-7 Weak Acid/Weak Base Titration Curves

In titration curves for weak acids and weak bases, pH changes near the equivalence point are too gradual for color indicators to be used. The solution is buffered both before and after the equivalence point. Figure 19-6 shows the titration curve for 100.0 mL of 0.100 M CH$_3$COOH solution titrated with 0.100 M aqueous NH$_3$. The pH at the equivalence point might not be 7.0, depending on which weak base/weak acid salt is formed. The calculation of values along a weak acid/weak base titration curve (other than the initial pH and the equivalence point) is beyond the scope of this text.

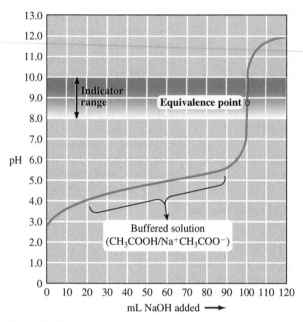

Figure 19-4 The titration curve for 100. mL of 0.100 M CH_3COOH with 0.100 M NaOH. The "vertical" section of this curve is much shorter than those in Figure 19-3 because the solution is buffered before the equivalence point. Methyl red indicator, which changes over the range 4.5 to 6.4, would obviously not be suitable for this titration.

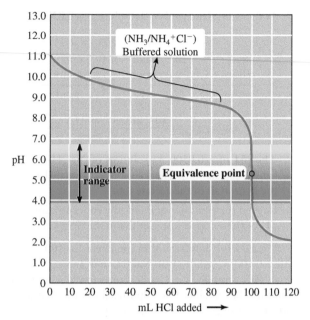

Figure 19-5 The titration curve for 100. mL of 0.100 M aqueous ammonia with 0.100 M HCl. The vertical section of the curve is relatively short because the solution is buffered before the equivalence point. The curve is very similar to that in Figure 19-4, but inverted. Phenolphthalein indicator, which changes over the range 8.0 to 10.0, should not be used for this titration.

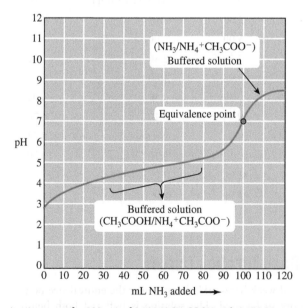

Figure 19-6 The titration curve for 100. mL of 0.100 M CH_3COOH with 0.100 M aqueous NH_3. At the equivalence point for *this* weak acid/weak base titration, the pH is 7.0, but this is not true for all such combinations (see Section 18-10). Because the solution is buffered before and after the equivalence point, the pH changes very gradually near the equivalence point. Color indicators cannot be used in such titrations. Instead, physical methods such as conductivity measurements or pH electrode-based instruments can be used to detect the end point.

Problem-Solving Tip Titration Curves

You can consider a titration curve in four parts.

1. **Initial solution** (before any titrant is added).

2. **Region before the equivalence point**. This may or may not be buffered. The solution is buffered in this region if the substance being titrated is a weak acid or weak base. A 1:1 buffer is created *halfway* to the equivalence point. At this halfway point of a weak acid/strong base titration, the pK_a of the weak acid is equal to the observed pH. At the half-way point of a weak base/strong acid titration, the pK_b of the weak base is equal to the observed pOH (or 14 - observed pH).

3. **Equivalence point**. Its location depends on the concentrations of the acid and the base solutions; its pH depends on the strengths of the acid and base.

4. **Region beyond the equivalence point**. This becomes nearly flat as more and more excess reactant is added. We often calculate only one or two points in this region.

Recognizing the four regions of a titration curve allows you to decide which kind of calculation is required.

19-8 Summary of Acid–Base Calculations

In this and the previous chapter, we have discussed several different types of acidic or basic solutions. Table 19-7 summarizes those many different types of solutions.

Table 19-7 A Review of the pH of Aqueous Solutions

Type of Aq. Solution	Example(s)	Resulting Chemistry or Type of Solution; Calculation	Section
Strong acid (represented as HX)	HNO_3 or HCl	*Complete* ionization; reaction goes to completion to form H_3O^+ $HX + H_2O \longrightarrow H_3O^+ + X^-$	18-1
Strong base [represented as MOH or $M(OH)_2$]	NaOH or $Ba(OH)_2$	*Complete* ionization; reaction goes to completion to form OH^- $MOH \longrightarrow M^+ + OH^-$ $M(OH)_2 \longrightarrow M^{2+} + 2OH^-$	18-1
Weak acid (represented as HA)	CH_3COOH or HCN	*Partial* ionization; $HA + H_2O \rightleftharpoons H_3O^+ + A^-$ Must solve equilibrium expression involving K_a for the acid, using *equilibrium* concentrations: $K_a = \dfrac{[H_3O^+][A^-]}{[HA]}$	18-4
Weak base (represented as B)	NH_3	*Partial* ionization; $B + H_2O \rightleftharpoons BH^+ + OH^-$ Must solve equilibrium expression involving K_b for the base, using *equilibrium* concentrations: $K_b = \dfrac{[BH^+][OH^-]}{[B]}$	18-4
Salt of *strong base* & *strong acid* (represented as MX)	$NaNO_3$ (salt of NaOH and HNO_3)	No hydrolysis—*neutral solution*	18-7
Salt of *strong base* & *weak acid* (represented as MA)	NaCN (salt of NaOH and HCN)	*Hydrolysis* of conjugate base (A^-) of *weak acid* $\Rightarrow$ *basic* solution $A^- + H_2O \rightleftharpoons HA + OH^-$ Solve weak base equilibrium for A^- using $K_{b\,for\,A^-} = \dfrac{K_w}{K_{a\,for\,HA}}$	18-8

Table 19-7 A Review of the pH of Aqueous Solutions—cont'd

Type of Aq. Solution	Example(s)	Resulting Chemistry or Type of Solution; Calculation	Section
Salt of *weak base* & *strong acid* (represented as BHX)	NH_4NO_3 (salt of NH_3 and HNO_3)	*Hydrolysis* of conjugate acid (BH^+) of *weak base* $\Rightarrow$ *acidic* solution $$BH^+ + H_2O \rightleftharpoons B + H_3O^+$$ Solve weak acid equilibrium for BH^+ using $K_{a\text{ for }BH^+} = \dfrac{K_w}{K_{b\text{ for }B}}$	18-9
Salt of *weak base* & *weak acid* (represented as BHA)	NH_4CN (salt of NH_3 and HCN)	*Hydrolysis* of conjugate base (A^-) of *weak acid* $\Rightarrow$ *basic* solution and *hydrolysis* of conjugate acid (BH^+) of *weak base* $\Rightarrow$ *acidic* solution. This solution can be *basic*, *neutral*, or *acidic*, depending on which hydrolysis occurs to a greater extent; use K_b and K_a values for the *hydrolysis reactions* to tell which ion will be the dominant factor. (We did not do calculations for this type of salt.)	18-10
Weak acid & *salt* of its conjugate base (represented as HA & MA)	$CH_3COOH + NaCH_3COO$ (or the product of the partial neutralization in a titration of a *weak acid* with a *strong base*)	Salt ionizes completely; $MA \longrightarrow M^+ + A^-$ Mixture of weak acid (HA) and its conjugate base (A^-) in significant concentrations gives a *buffer*. Use Henderson–Hasselbalch equation for acid/salt buffer (with ratio of initial concentrations, moles, or millimoles): $$pH = pK_a + \log\frac{[\text{conjugate base}]}{[\text{acid}]}$$	19-1, 19-2, 19-3
Weak base & *salt* of its conjugate acid (represented as B & BHX)	$NH_3 + NH_4Cl$ (or the product of the partial neutralization in a titration of a *weak base* with a *strong acid*)	Salt ionizes completely; $NH_4Cl \longrightarrow NH_4^+ + Cl^-$ Mixture of weak base (NH_3) and its conjugate acid (NH_4^+) in significant concentrations gives a *buffer*. Use Henderson–Hasselbalch equation for base/salt buffer (with ratio of initial concentrations, moles, or millimoles): $$pOH = pK_b + \log\frac{[\text{conjugate acid}]}{[\text{base}]}$$	19-1, 19-2, 19-3

KEY TERMS

Buffer solution A solution that resists changes in pH when strong acids or strong bases are added. A buffer solution contains an acid and its conjugate base, so it can react with added base or acid. Common buffer solutions contain either (1) a weak acid and a soluble ionic salt of the weak acid, *or* (2) a weak base and a soluble ionic salt of the weak base.

Common ion effect Suppression of ionization of a weak electrolyte by the presence in the same solution of a strong electrolyte containing one of the same ions as the weak electrolyte.

End point The point at which an indicator changes color and a titration should be stopped.

Equivalence point The point at which chemically equivalent amounts of reactants have reacted.

Henderson–Hasselbalch equation An equation that enables us to calculate the pH or pOH of a buffer solution directly.

For acid/salt buffer $pH = pK_a + \log\dfrac{[\text{conj. base}]}{[\text{acid}]}$

For base/salt buffer $pOH = pK_b + \log\dfrac{[\text{conj. acid}]}{[\text{base}]}$

Can also be written in terms of ratio of moles or millimoles.

Indicator (for acid–base reactions) An organic compound that exhibits different colors in solutions of different acidities; used to indicate the point at which reaction between an acid and a base is complete.

Titration A procedure in which one solution is added to another solution until the chemical reaction between the two solutes is complete; usually the concentration of one solution is known and that of the other is unknown.

Titration curve (for acid–base titration) A plot of pH versus volume of acid or base solution added.

EXERCISES

. . . .

🍄 **Molecular Reasoning** exercises

▲ **More Challenging** exercises

All exercises in this chapter assume a temperature of 25°C unless they specify otherwise. Values of K_a *and* K_b *can be found in Appendix F or will be specified in the exercise.*

Basic Ideas

1. (a) What is the relationship between pH, pOH, and pK_w? (b) What is the relationship between K_a and pH? (c) What is the relationship between K_a and pOH?

2. Define and give examples of (a) pK_a, (b) K_b, (c) pK_w, (d) pOH.

3. 🍄 Give the names and formulas of (a) five strong acids, (b) five weak acids, (c) four strong bases, (d) four weak bases.

4. 🍄 Write the balanced equation for an acid–base reaction that would produce each of the following salts; predict whether an aqueous solution of each salt is acidic, basic, or neutral. (a) $NaNO_3$; (b) Na_2S; (c) $Al_2(CO_3)_3$; (d) $Mg(CH_3COO)_2$; (e) $(NH_4)_2SO_4$.

5. 🍄 Write the balanced equation for an acid–base reaction that would produce each of the following salts; predict whether an aqueous solution of each salt is acidic, basic, or neutral. (a) CaF_2; (b) $ZnSO_4$; (c) $CaCl_2$; (d) K_3PO_4; (e) NH_4NO_3.

6. Under what circumstances can it be predicted that a neutral solution is produced by an acid–base reaction? (*Hint*: This question has more than one answer.)

The Common Ion Effect and Buffer Solutions

7. 🍄 Which of the following solutions are buffers? (Each solution was prepared by mixing and diluting appropriate quantities of the two solutes to yield the concentrations indicated.) Explain your decision for each solution. (a) 0.10 M HCN and 0.10 M NaCN; (b) 0.10 M NaCN and 0.10 M NaCl; (c) 0.10 M NH_3 and 0.10 M NH_4Br; (d) 0.10 M NaOH and 0.90 M KOH.

8. 🍄 Which of the following solutions are buffers? (Each solution was prepared by mixing and diluting appropriate quantities of the two solutes to yield the concentrations indicated.) Explain your decision for each solution. (a) 1.0 M CH_3CH_2COOH and 0.20 M $NaCH_3CH_2COO$; (b) 0.10 M NaOCl and 0.10 M HOCl; (c) 0.10 M NH_4Cl and 0.90 M NH_4Br; (d) 0.10 M NaCl and 0.20 M HF.

9. Which of these combinations would be the best to buffer the pH at approximately 9?
(a) $CH_3COOH/NaCH_3COO$
(b) HCl/NaCl
(c) NH_3/NH_4Cl

10. Calculate pH for each of the following buffer solutions. (a) 0.15 M HF and 0.20 M KF; (b) 0.040 M CH_3COOH and 0.025 M $Ba(CH_3COO)_2$.

11. The pK_a of HOCl is 7.45. Calculate the pH of a solution that is 0.0222 M HOCl and 0.0444 M NaOCl.

12. Calculate the concentration of OH^- and the pH for the following buffer solutions. (a) 0.25 M NH_3(aq) and 0.15 M NH_4NO_3; (b) 0.15 M NH_3(aq) and 0.20 M $(NH_4)_2SO_4$.

13. Calculate the concentration of OH^- and the pH for the following solutions. (a) 0.45 M NH_3(aq) and 0.35 M NH_4NO_3; (b) 0.10 M aniline, $C_6H_5NH_2$, and 0.25 M anilinium chloride, $C_6H_5NH_3Cl$.

14. ▲ Buffer solutions are especially important in our body fluids and metabolism. Write net ionic equations to illustrate the buffering action of (a) the $H_2CO_3/NaHCO_3$ buffer system in blood and (b) the NaH_2PO_4/Na_2HPO_4 buffer system inside cells.

15. Calculate the ratio of $[NH_3]/[NH_4^+]$ concentrations that gives (a) solutions of pH = 9.55 and (b) solutions of pH = 9.10.

16. 🍄 We prepare two solutions as follows. In solution A 0.50 mole of potassium acetate is added to 0.25 mole of acetic acid and diluted to a final volume of 1.00 liter. In solution B 0.25 mole of potassium acetate is added to 0.50 mole of acetic acid and diluted to 1.00 liter. (a) Which solution is expected to have the lower pH? (b) Explain how you can reach your conclusion without calculating the pH of each solution.

17. Compare the pH of a 0.30 M acetic acid solution to the pH of a solution composed of 0.30 M acetic acid to which 0.30 mole of sodium acetate per liter has been added.

18. (a) Calculate $[OH^-]$ and $[H_3O^+]$ in a 0.30 M aqueous ammonia solution. The K_b for aqueous ammonia is 1.8×10^{-5}. (b) Calculate $[OH^-]$ and $[H_3O^+]$ of a buffer solution that is 0.30 M in NH_3 and 0.40 M in NH_4Cl.

Buffering Action

19. 🍄 Briefly describe why the pH of a given buffer solution remains nearly constant when small amounts of acid or base are added. Over what pH range would a given buffer exhibit the best buffering action (nearly constant pH)?

20. Consider the ionization of formic acid, HCOOH.

$$HCOOH + H_2O \rightleftharpoons HCOO^- + H_3O^+$$

What effect does the addition of sodium formate, NaHCOO, have on the fraction of formic acid molecules that undergo ionization in aqueous solution?

. . . .

🍄 **Molecular Reasoning** exercises   ▲ **More Challenging** exercises   Blue-Numbered exercises solved in Student Solutions Manual

Unless otherwise noted, all content on this page is © Cengage Learning.

21. A solution is produced by dissolving 0.075 mole of formic acid and 0.065 mole of sodium formate in sufficient water to produce 1.00 L of solution. (a) Calculate the pH of this buffer solution. (b) Calculate the pH of the solution after an additional 0.020 mole of sodium formate is dissolved.

22. What is the pH of a solution that is 0.20 M in $HClO_4$ and 0.20 M $KClO_4$? Is this a buffer solution?

23. (a) Find the pH of a solution that is 0.50 M in formic acid and 0.30 M in sodium formate. (b) Find the pH after 0.050 mol HCl has been added to 1.00 liter of the solution.

24. One liter of 0.400 M NH_3 solution also contains 12.78 g of NH_4Cl. How much will the pH of this solution change if 0.142 mole of gaseous HCl is bubbled into it?

25. (a) Find the pH of a solution that is 0.11 M in nitrous acid and 0.15 M in sodium nitrite. (b) Find the pH after 0.14 mol of NaOH has been added to 1.00 liter of the solution.

26. (a) Find the pH of a solution that is 0.90 M in NH_3 and 0.80 M in NH_4Cl. (b) Find the pH of the solution after 0.10 mol of HCl has been added to 1.00 liter of the solution. (c) A solution was prepared by adding NaOH to pure water to give 1.00 liter of solution whose pH = 9.34. Find the pH of this solution after 0.10 mol of HCl has been added to it.

27. (a) Calculate the concentrations of CH_3COOH and CH_3COO^- in a solution in which their total concentration is 0.400 mol/L and the pH is 4.50. (b) If 0.100 mol of solid NaOH is added to 1.00 L of this solution, how much does the pH change?

28. ▲ A solution contains bromoacetic acid, $BrCH_2COOH$, and sodium bromoacetate, $NaBrCH_2COO$, with acid and salt concentrations that total 0.30 mol/L. If the pH is 3.50, what are the concentrations of the acid and the salt? $K_a = 2.0 \times 10^{-3}$ for $BrCH_2COOH$.

29. Calculate the concentration of propionate ion, $CH_3CH_2COO^-$, in equilibrium with 0.025 M CH_3CH_2COOH (propionic acid) and 0.015 M H_3O^+ from hydrochloric acid. $K_a = 1.3 \times 10^{-5}$ for CH_3CH_2COOH.

30. Calculate the concentration of $C_2H_5NH_3^+$ in equilibrium with 0.012 M $C_2H_5NH_2$ (ethylamine) and 0.0040 M OH^- ion from sodium hydroxide. $K_b = 4.7 \times 10^{-4}$ for ethylamine.

31. ▲ Buffer capacity is defined as the number of moles of a strong acid or strong base that are required to change the pH of one liter of the buffer solution by one unit. What is the buffer capacity of a solution that is 0.10 M in acetic acid and 0.10 M in sodium acetate?

Preparation of Buffer Solutions

32. A buffer solution of pH = 5.10 is to be prepared from propionic acid and sodium propionate. The concentration of sodium propionate must be 0.60 mol/L. What should be the concentration of the acid? $K_a = 1.3 \times 10^{-5}$ for CH_3CH_2COOH.

33. We need a buffer with pH = 9.25. It can be prepared from NH_3 and NH_4Cl. What must be the $[NH_4^+]/[NH_3]$ ratio?

34. What volumes of 0.125 M acetic acid and 0.065 M NaOH solutions must be mixed to prepare 1.00 L of a buffer solution of pH = 4.50 at 25°C?

35. One liter of a buffer solution is prepared by dissolving 0.115 mol of $NaNO_2$ and 0.070 mol of HCl in water. What is the pH of this solution? If the solution is diluted twofold with water, what is the pH?

36. One liter of a buffer solution is made by mixing 500. mL of 1.25 M acetic acid and 500. mL of 0.300 M calcium acetate. What is the concentration of each of the following in the buffer solution? (a) CH_3COOH; (b) Ca^{2+}; (c) CH_3COO^-; (d) H_3O^+; (e) What is the pH?

37. What must be the concentration of benzoate ion, $C_6H_5COO^-$, in a 0.0550 M benzoic acid, C_6H_5COOH, solution so that the pH is 5.10?

38. What must be the concentration of chloroacetic acid, $CH_2ClCOOH$, in a 0.015 M $NaCH_2ClCOO$ solution so that the pH is 2.75? $K_a = 1.4 \times 10^{-3}$ for $CH_2ClCOOH$.

39. What must be the concentration of NH_4^+ in a 0.075 M NH_3 solution so that the pH is 8.70?

Acid–Base Indicators

40. ✿ (a) What are acid–base indicators? (b) What are the essential characteristics of acid–base indicators? (c) What determines the color of an acid–base indicator in an aqueous solution?

41. K_a is 7.9×10^{-8} for bromthymol blue, an indicator that can be represented as HIn. HIn molecules are yellow, and In^- ions are blue. What color will bromthymol blue be in a solution in which (a) $[H_3O^+] = 1.0 \times 10^{-4}$ M, and (b) pH = 10.30?

42. ▲ The indicator metacresol purple changes from yellow to purple at pH 8.2. At this point it exists in equal concentrations as the conjugate acid and the conjugate base. What are K_a and pK_a for metacresol purple, a weak acid represented as HIn?

43. ✿ A series of acid–base indicators can be used to estimate the pH of an unknown solution. The solution was colorless with phenolphthalein, yellow with methyl orange, and yellow with methyl red. Use the values given in Table 19-4 to determine the possible range of pH values of the solution.

Charles D. Winters

44. ✿ A series of acid–base indicators can be used to estimate the pH of an unknown solution. The solution was colorless in phenolphthalein, blue in bromthymol blue, and yellow in methyl orange. Use the values given in Table 19-4 to determine the possible range of pH values of the solution.

✿ **Molecular Reasoning** exercises   ▲ **More Challenging** exercises   Blue-Numbered exercises solved in Student Solutions Manual

45.  Use Table 19-4 to choose one or more indicators that could be used to "signal" reaching a pH of (a) 3.5; (b) 7.0; (c) 10.3; (d) 8.0.

46. A solution of 0.020 M acetic acid is to be titrated with a 0.025 M NaOH solution. What is the pH at the equivalence point? Choose an appropriate indicator for the titration.

47. Demonstrate mathematically that neutral red is red in solutions of pH 3.00, whereas it is yellow in solutions of pH 10.00. HIn is red, and In⁻ is yellow. K_a is 2.0×10^{-7}.

Strong Acid/Strong Base Titration Curves

48. ▲ Make a rough sketch of the titration curve expected for the titration of a strong acid with a strong base. What determines the pH of the solution at the following points? (a) no base added; (b) half-equivalence point (the point at which one-half of the titrant volume required to reach the equivalence point has been added); (c) equivalence point; (d) excess base added. Compare your curve with Figure 19-3a.

For Exercises 49, 54, 65, and 66, calculate and tabulate [H₃O⁺], [OH⁻], pH, and pOH at the indicated points as we did in Table 19-5. In each case assume that pure acid (or base) is added to exactly 1 L of a 0.0100 molar solution of the indicated base (or acid). This simplifies the arithmetic because we may assume that the volume of each solution is constant throughout the titration. Plot each titration curve with pH on the vertical axis and moles of base (or acid) added on the horizontal axis.

49. Solid NaOH is added to 1.00 L of 0.0500 M HCl solution. Number of moles of NaOH added: (a) none; (b) 0.00500; (c) 0.01500; (d) 0.02500 (50% titrated); (e) 0.03500; (f) 0.04500; (g) 0.04750; (h) 0.0500 (100% titrated); (i) 0.0525; (j) 0.0600; (k) 0.0750 (50% excess NaOH). List the indicators in Table 19-4 that could be used in this titration.

50. A 25.00-mL sample of 0.1500 M HNO₃ is titrated with 0.100 M NaOH. Calculate the pH of the solution (a) before the addition of NaOH and after the addition of (b) 6.00 mL, (c) 15.00 mL, (d) 30.00 mL, (e) 37.44 mL, (f) 45.00 mL of NaOH solution.

51. A 36.30-mL sample of 0.245 M HNO₃ solution is titrated with 0.213 M KOH. Calculate the pH of the solution (a) before the addition of KOH and after the addition of (b) 6.15 mL, (c) 13.20 mL, (d) 26.95 mL, (e) 38.72 mL, (f) 42.68 mL of KOH solution.

52. A 22.0-mL sample of 0.145 M HCl solution is titrated with 0.106 M NaOH. Calculate the pH of the solution (a) before the addition of NaOH and after the addition of (b) 11.10 mL, (c) 24.0 mL, (d) 41.0 mL, (e) 54.4 mL, (f) 63.6 mL of NaOH solution.

Weak Acid/Strong Base Titration Curves

53.  Make a rough sketch of the titration curve expected for the titration of a weak monoprotic acid with a strong base. What determines the pH of the solution at the following points? (a) no base added; (b) half-equivalence point (the point at which one-half of the titrant volume required to reach the equivalence point has been added); (c) equivalence point; (d) excess base added. Compare your curve to Figure 19-4.

54. (See instructions preceding Exercise 49.) Solid NaOH is added to exactly 1 L of 0.0200 M CH₃COOH solution. Number of moles NaOH added: (a) none; (b) 0.00400; (c) 0.00800; (d) 0.01000 (50% titrated); (e) 0.01400; (f) 0.01800; (g) 0.01900; (h) 0.0200 (100% titrated); (i) 0.0210; (j) 0.0240; (k) 0.0300 (50% excess NaOH). List the indicators in Table 19-4 that could be used in this titration.

55. A 44.0-mL sample of 0.202 M CH₃COOH solution is titrated with 0.185 M NaOH. Calculate the pH of the solution (a) before the addition of any NaOH solution and after the addition of (b) 15.5 mL, (c) 20.0 mL, (d) 24.0 mL, (e) 27.2 mL, (f) 48.0 mL, (g) 50.2 mL of NaOH solution.

56. A 32.44-mL sample of 0.182 M CH₃COOH solution is titrated with 0.185 M NaOH. Calculate the pH of the solution (a) before the addition of any NaOH solution and after the addition of (b) 15.55 mL, (c) 20.0 mL, (d) 24.02 mL, (e) 27.2 mL, (f) 31.91 mL, (g) 33.12 mL of NaOH solution.

57. A solution contains an unknown weak monoprotic acid, HA. It takes 46.24 mL of NaOH solution to titrate 50.00 mL of the HA solution to the equivalence point. To another 50.00-mL sample of the same HA solution, 23.12 mL of the same NaOH solution is added. The pH of the resulting solution in the second experiment is 5.14. What are K_a and pK_a of HA?

58. ▲ Calculate the pH at the equivalence point of the titration of 100.0 mL of each of the following with 0.1500 M KOH: (a) 1.200 M acetic acid; (b) 0.1200 M acetic acid; (c) 0.06000 M acetic acid.

59. Describe, quantitatively, how the concentration of hydronium ion, [H₃O⁺], is calculated in the titration of a weak acid with a strong base: (a) before any base is added, (b) after some base has been added, but before the equivalence point, (c) at the equivalence point, and (d) beyond the equivalence point.

 **Molecular Reasoning** exercises   ▲ **More Challenging** exercises   Blue-Numbered exercises solved in Student Solutions Manual

Unless otherwise noted, all content on this page is © Cengage Learning.

Weak Base/Strong Acid Titration Curves

60. A 20.00-mL sample of 0.220 M triethylamine, $(CH_3CH_2)_3N$, is titrated with 0.544 M HCl.

$$K_b\ [(CH_3CH_2)_3N] = 5.2 \times 10^{-4}$$

(a) Write a balanced net ionic equation for the titration.
(b) How many mL of HCl are required to reach the equivalence point?
(c) Calculate $[(CH_3CH_2)_3N]$, $[(CH_3CH_2)_3NH^+]$, $[H^+]$, and $[Cl^-]$ at the equivalence point. (Assume that volumes are additive.)
(d) What is the pH at the equivalence point?

61. What is the pH of the solution that results from adding 25.0 mL of 0.12 M HCl to 25.0 mL of 0.43 M NH_3?

62. Suppose a small amount of a strong acid is added to an aniline buffer system $(C_6H_5NH_2/C_6H_5NH_3^+)$. What is the net ionic equation for the main reaction that occurs to largely neutralize the added acid?

63. If 1 M solutions of each of the following compounds were prepared, which one would have the highest pH?
(a) ammonium acetate (NH_4CH_3COO); (b) potassium chloride (KCl); (c) sodium cyanide (NaCN); (d) ammonium nitrate (NH_4NO_3); (e) hydrochloric acid (HCl).

Mixed Exercises

64. Give the name and formula of a compound that could be combined with each one of the following compounds to produce a buffer solution: HNO_2, NaCN, $(NH_4)_2SO_4$, NH_3, HCN, $(CH_3)_3N$.

65. ▲ (See instructions preceding Exercise 49.) Gaseous HCl is added to 1.00 L of 0.0100 M aqueous ammonia solution. Number of moles HCl added: (a) none; (b) 0.00100; (c) 0.00300; (d) 0.00500 (50% titrated); (e) 0.00700; (f) 0.00900; (g) 0.00950; (h) 0.0100 (100% titrated); (i) 0.0105; (j) 0.0120; (k) 0.0150 (50% excess HCl). List the indicators in Table 19-4, that could be used in this titration.

66. ▲ (See instructions preceding Exercise 49.) Gaseous NH_3 is added to 1.00 L of 0.0100 M HNO_3 solution. Number of moles NH_3 added: (a) none; (b) 0.00100; (c) 0.00400; (d) 0.00500 (50% titrated); (e) 0.00900; (f) 0.00950; (g) 0.0100 (100% titrated); (h) 0.0105; (i) 0.0130. What is the major difference between the titration curve for the reaction of HNO_3 and NH_3 and the other curves you have plotted? Consult Table 19-4. Can you suggest a satisfactory indicator for this titration?

67. Compare the pH of 0.54 M NaCl with the pH of 0.54 M NaOCl.

68. Compare the pH of 0.54 M NaCl with the pH of 0.54 M NH_4Cl.

Conceptual Exercises

69. Why do we need acid–base indicators that change at pH values other than 7?

70. The indicator indigo carmine changes from blue in acid to yellow in base. It is a weak acid with a K_a of 6×10^{-13}. At approximately what pH does indigo carmine change color?

71. In 100 mL of 0.100 M hypochlorous acid, HClO, the pH is 4.2, and the concentration of hypochlorite ion, ClO^-, is 6×10^{-5}. Describe what would happen if 100 mL of 0.100 M NaClO were added to it. (Include the effects, if any, on the pH and on the concentrations of HClO and ClO^-.)

72. The value of K_b for dimethylamine, $(CH_3)_2NH$, is 7.4×10^{-4}. Suppose dimethylammonium chloride, $(CH_3)_2NH_2Cl$, is dissolved in water. (a) Write the chemical equation for the dissociation of $(CH_3)_2NH_2Cl$ in water. (b) Write the chemical equation for the hydrolysis reaction which then occurs. (c) Describe the resulting solution as acidic, basic, or neutral.

73. When methylamine, CH_3NH_2, dissolves in water, it partially ionizes, producing the methylammonium ion, $CH_3NH_3^+$:

$$CH_3NH_2 + H_2O \rightleftharpoons CH_3NH_3^+ + OH^-$$

$$K_b = \frac{[CH_3NH_3^+][OH^-]}{[CH_3NH_2]}$$

What effect, if any, would dissolving methylammonium chloride $[CH_3NH_3Cl]$, a soluble salt, in this solution have on the equilibrium reaction?

74. Ammonium hypochlorite, NH_4ClO, is the salt of ammonia, NH_3, and hypochlorous acid, HClO. (a) What is the K_a for ammonium ion? (b) What is the K_b for hypochlorite ion? (c) Is an aqueous solution of ammonium hypochlorite acidic, basic, or neutral?

75. ☁ The pH of an equal molar acetic acid/sodium acetate buffer is 4.74. Draw a molecular representation of a small portion of this buffer solution. (You may omit the water molecules.) Draw another molecular representation of the solution after a very small amount of NaOH has been added.

76. ☁ Suppose you were asked on a laboratory test to outline a procedure to prepare a buffer solution of pH 8.0 using hydrocyanic acid, HCN. You realize that a pH of 8.0 is basic, and you find that the K_a of hydrocyanic acid is 4.0×10^{-10}. What is your response?

77. ☁ The odor of cooked fish is due to the presence of amines. This odor is lessened by adding lemon juice, which contains citric acid. Why does this work?

78. ☁ The *end point* of a titration is not the same as the *equivalence point* of a titration. Differentiate between these two concepts.

79. One function of our blood is to carry CO_2 from our tissues to our lungs. It is critical that the pH of our blood remains at 7.4 ± 0.05. Our blood is buffered to maintain that pH range, and the primary buffer system is composed of HCO_3^-/H_2CO_3. (a) Calculate the ratio of HCO_3^- to H_2CO_3 in blood if there were no other buffers present. (b) Determine whether in the presence of the HCO_3^-/H_2CO_3 system alone, our blood would be more effective as a buffer against the addition of acidic or basic solutions.

Building Your Knowledge

80. A 0.738-g sample of an unknown mixture composed only of NaCl and $KHSO_4$ required 36.8 mL of 0.115 M NaOH

☁ **Molecular Reasoning** exercises　　▲ **More Challenging** exercises　　Blue-Numbered exercises solved in Student Solutions Manual

Unless otherwise noted, all content on this page is © Cengage Learning.

for titration. What is the percent by mass of $KHSO_4$ in the mixture?

81. Acetylsalicylic acid, the active ingredient in aspirin, has a K_a value of 3.0×10^{-4}. We dissolve 0.0100 mole of acetylsalicylic acid in sufficient water to make 1.00 L of solution and then titrate it with 0.500 M NaOH solution. What is the pH at each of these points in the titration (a) before any of the NaOH solution is added, (b) at the equivalence point, (c) when a volume of NaOH solution has been added that is equal to half the amount required to reach the equivalence point?

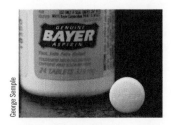

82. ▲ What is the pH of a solution that is a mixture of HOCl and HOI, each at 0.30 M concentration?

83. An unknown amount of water is mixed with 350. mL of a 6.0 M NaOH solution. A 75.0-mL sample of the resulting solution is titrated to neutrality with 52.5 mL of 6.00 M HCl. (a) Calculate the concentration of the diluted NaOH solution. (b) What was the concentration of the NaOH solution before it was diluted? (c) What volume of water was added? Assume that the volumes were additive.

84. A 3.5-L container of HCl had no concentration noted on its label. A 20.00-mL sample of this unknown HCl solution is titrated to a pH of 7.0 by 34.0 mL of 3.00 M NaOH solution. Determine the volume of this HCl solution required to prepare 1.5 L of 0.75 M HCl solution.

85. Calculate the pH at the equivalence point for the titration of a solution containing 150.0 mg of ethylamine, $C_2H_5NH_2$, with 0.1000 M HCl solution. The volume of the solution at the equivalence point is 250. mL. Select a suitable indicator. K_b for ethylamine appears in Exercise 30.

Beyond the Textbook

NOTE: *Whenever the answer to an exercise depends on information obtained from a source other than this textbook, the source must be included as an essential part of the answer.*

86. Search the web for buffered aspirin for pets. (a) For which animals did you find a buffered aspirin product being commercially marketed? (b) Which company(ies) markets a buffered roast-beef-flavored aspirin (VETRIN)? What is the dosage of aspirin in each VETRIN tablet? (c) What other dosages can you find?

87. Use a web search engine to find common misconceptions or mistakes involving buffers. What common errors do you find?

88. Use a web search engine to find lists of acid–base indicators. (a) Select an indicator that is not mentioned in this chapter and give its color and pH at each end of its transition range. (b) Determine what is meant by the statement "Vanilla extract has been described as an olfactory (acid–base) indicator."

89. 🜨 Locate a table of acid–base indicators on the web. What color will each solution be at the indicated pH when a small amount of the specified indicator is also present? (a) pH 3, congo red; (b) pH 5, congo red; (c) pH 6, methyl orange; (d) pH 9, thymolphthalein.

90. The flowers shown below are hydrangeas, which function as natural pH indicators. Use an internet search engine such as http://www.google.com to locate information about the relationship between hydrangea color and soil acidity. What metal (in ionic form) is responsible for the color difference? What chemicals are commonly used by gardeners to lower and raise soil pH?

Hydrangeas

🜨 **Molecular Reasoning** exercises   ▲ **More Challenging** exercises   Blue-Numbered exercises solved in Student Solutions Manual

Unless otherwise noted, all content on this page is © Cengage Learning.

Ionic Equilibria III: The Solubility Product Principle

20

The formation of caves and objects such as stalactites ("tight to ceiling") and stalagmites ("mites are on the ground") are based on acid–base and precipitation equilibria. CO_2 gas reacts with water to form carbonic acid (H_2CO_3), which is acidic enough to attack and dissolve limestone (calcium carbonate, $CaCO_3$), often the first stage of cave formation. Once a cave cavity is formed, the mineral solution dripping down from above can begin to precipitate the dissolved minerals (mainly calcium carbonate) to form the various features we see in caves.

Jeff Schultes/Shutterstock.com

OBJECTIVES

After you have studied this chapter, you should be able to

▶ Write solubility product constant expressions

▶ Explain how K_{sp} values are determined

▶ Use K_{sp} values in chemical calculations

▶ Recognize some common, slightly soluble compounds

▶ Describe fractional precipitation and how it can be used to separate ions

▶ Explain how simultaneous equilibria can be used to control solubility

▶ Describe some methods for dissolving precipitates

So far we have discussed mainly compounds that are quite soluble in water. Although most compounds dissolve in water to some extent, many are so slightly soluble that they are called "insoluble compounds." We shall now consider those that are only very slightly soluble. As a rule of thumb, compounds that dissolve in water to the extent of 0.020 mole/liter or more are classified as soluble. Refer to the solubility guidelines (see Table 6-4) as necessary.

Slightly soluble compounds are important in many natural phenomena. Our bones and teeth are mostly calcium phosphate, $Ca_3(PO_4)_2$, a slightly soluble compound. Many natural deposits of $Ca_3(PO_4)_2$ rock are mined and converted into agricultural fertilizer. Limestone caves have been formed by acidic water slowly dissolving away calcium carbonate, $CaCO_3$. Sinkholes are created when acidic water dissolves away most of the underlying $CaCO_3$. The remaining limestone can no longer support the weight above it, so it collapses, and a sinkhole is formed.

20-1 Solubility Product Constants

Charles D. Winters

The calcium carbonate, $CaCO_3$, in an eggshell has the same crystal structure as in the mineral calcite.

Suppose we add one gram of solid barium sulfate, $BaSO_4$, to 1.0 liter of water at 25°C and stir until the solution is *saturated*. Very little $BaSO_4$ dissolves; most of the $BaSO_4$ remains as undissolved solid. Careful measurements of conductivity show that one liter of a saturated solution of barium sulfate contains only 0.0025 gram of dissolved $BaSO_4$, no matter how much more solid $BaSO_4$ is added. The $BaSO_4$ that does dissolve is completely dissociated into its constituent ions.

$$BaSO_4(s) \xrightleftharpoons{H_2O} Ba^{2+}(aq) + SO_4^{2-}(aq)$$

(We usually omit H_2O over the arrows in such equations; the (aq) phase of the ions tells us that we are dealing with aqueous solution.)

In equilibria that involve slightly soluble compounds in water, the equilibrium constant is called the **solubility product constant, K_{sp}**. The activity of the pure solid $BaSO_4$ is one (see Section 17-11).

> The concentration of the solid is never included in the equilibrium constant expression.

For a saturated solution of $BaSO_4$ in contact with solid $BaSO_4$, we write

$$BaSO_4(s) \rightleftharpoons Ba^{2+}(aq) + SO_4^{2-}(aq) \qquad \text{and} \qquad K_{sp} = [Ba^{2+}][SO_4^{2-}]$$

The solubility product constant for $BaSO_4$ is the product of the concentrations of its constituent ions in a saturated solution.

> The *solubility product expression* for a compound is the product of the concentrations of its constituent ions, each raised to the power that corresponds to the number of ions in one formula unit of the compound. This product is constant at constant temperature for a saturated solution of the compound. This statement is the **solubility product principle**.

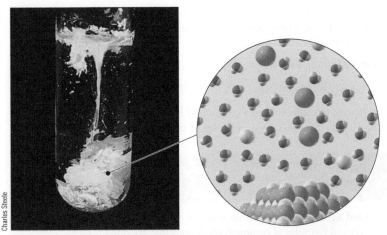

The dissolution process of ions going into solution. During precipitation, ions are producing more solid as the ions leave the solution. For a saturated solution, the dissolution and precipitation rates are equal. For the very slightly soluble salts discussed in this chapter, the ion concentrations are very low compared with the concentration in this illustration.

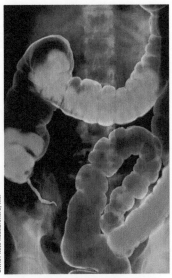

An X-ray photo of the gastrointestinal tract. Barium sulfate is the "insoluble" substance taken orally before stomach X-rays are made because the barium atoms absorb X-rays well. Even though barium ions are quite toxic, barium sulfate can still be taken orally without danger. The compound is so insoluble that it passes through the digestive system essentially unchanged.

The existence of a substance in the solid state can be indicated several ways. For example, $BaSO_4(s)$, $\overline{BaSO_4}$, and $BaSO_4\downarrow$ are sometimes used to represent solid $BaSO_4$. In this text we use the $\overline{(s)}$ notation for formulas of solid substances in equilibrium with their saturated aqueous solutions.

Consider dissolving slightly soluble calcium fluoride, CaF_2, in H_2O.

$$CaF_2(s) \rightleftharpoons Ca^{2+}(aq) + 2F^-(aq) \qquad K_{sp} = [Ca^{2+}][F^-]^2 = 3.9 \times 10^{-11}$$

We often shorten the term "solubility product constant" to "solubility product." Thus, the solubility products for barium sulfate, $BaSO_4$, and for calcium fluoride, CaF_2, are written as

$$K_{sp} = [Ba^{2+}][SO_4^{2-}] = 1.1 \times 10^{-10} \qquad K_{sp} = [Ca^{2+}][F^-]^2 = 3.9 \times 10^{-11}$$

The very small amount of solid zinc phosphate, $Zn_3(PO_4)_2$, that dissolves in water gives three zinc ions and two phosphate ions per formula unit.

$$Zn_3(PO_4)_2(s) \rightleftharpoons 3Zn^{2+}(aq) + 2PO_4^{3-}(aq) \qquad K_{sp} = [Zn^{2+}]^3[PO_4^{3-}]^2 = 9.1 \times 10^{-33}$$

Generally, we may represent the dissolution of a slightly soluble compound M_yX_z and its K_{sp} expression as

$$M_yX_z(s) \rightleftharpoons yM^{z+}(aq) + zX^{y-}(aq) \qquad \text{and} \qquad K_{sp} = [M^{z+}]^y[X^{y-}]^z$$

STOP & THINK
Remember to raise the concentration of each aqueous ion in the K_{sp} expression to the power of its coefficient in the balanced chemical equation.

Ⓐ Barium sulfate, $BaSO_4$, occurs in the mineral barite and has a relatively high density of 4.5 g/cm³.

Ⓑ Calcium fluoride, CaF_2, occurs in the mineral fluorite. The purple color is caused by defects in the crystal structure.

These two compounds are listed as "insoluble" in the solubility guidelines presented in Section 6-4. As we saw above, they are actually *very slightly soluble.*

In some cases, a compound contains more than two kinds of ions. Dissolution of the slightly soluble compound magnesium ammonium phosphate, $MgNH_4PO_4$, in water and its solubility product expression are represented as

$$MgNH_4PO_4(s) \rightleftharpoons Mg^{2+}(aq) + NH_4^+(aq) + PO_4^{3-}(aq)$$

$$K_{sp} = [Mg^{2+}][NH_4^+][PO_4^{3-}] = 2.5 \times 10^{-12}$$

Another way to express the solubility of a compound is **molar solubility**. This is defined as the number of moles that dissolve to give one liter of saturated solution.

Problem-Solving Tip You Must Recognize the Names and Formulas of Ions

To write the required equations for the dissolution and dissociation of salts, you must recognize the names and formulas of common, stable ions that are often present. Table 2-2 contains the names and formulas of several ions; since then you have learned others. Review these names and formulas carefully, and remember that formulas of ions must include their charges. Many of the ions you need to know are present in the salts listed in Appendix H.

20-2 Determination of Solubility Product Constants

▶ Unless otherwise indicated, solubility product constants and solubility data are given for 25°C (Appendix H).

If the solubility of a compound is known, the molar solubility and the value of its solubility product can be calculated.

EXAMPLE 20-1 Molar Solubility and Product Constants

One (1.0) liter of saturated barium sulfate solution contains 0.0025 gram of dissolved $BaSO_4$. (a) What is the molar solubility of $BaSO_4$? (b) Calculate the solubility product constant for $BaSO_4$.

Plan

▶ We frequently use statements such as "The solution contains 0.0025 gram of $BaSO_4$." What we mean is that 0.0025 gram of solid $BaSO_4$ *dissolves* to give a solution that contains equal concentrations of Ba^{2+} and SO_4^{2-} ions.

(a) We write the equation for the dissolution of $BaSO_4$ and the expression for its solubility product constant, K_{sp}. From the solubility of $BaSO_4$ in H_2O, we calculate its molar solubility and the concentrations of the ions. (b) This lets us calculate K_{sp}.

Solution

(a) In saturated solutions, solid and dissolved solute are in equilibrium. The equations for the dissolution of barium sulfate in water and its solubility product expression are

$$BaSO_4(s) \rightleftharpoons Ba^{2+}(aq) + SO_4^{2-}(aq) \qquad K_{sp} = [Ba^{2+}][SO_4^{2-}]$$

From the given solubility of $BaSO_4$ in H_2O, we can calculate its *molar solubility*.

▶ 1.1×10^{-5} mole of solid $BaSO_4$ dissolves to give a liter of saturated solution.

$$\frac{?\ \text{mol BaSO}_4}{L} = \frac{2.5 \times 10^{-3}\ \text{g BaSO}_4}{1.0\ L} \times \frac{1\ \text{mol BaSO}_4}{233\ \text{g BaSO}_4} = \frac{1.1 \times 10^{-5}\ \text{mol BaSO}_4/L}{\text{(dissolved)}}$$

(b) We know the molar solubility of $BaSO_4$. The dissolution equation shows that each formula unit of $BaSO_4$ that dissolves produces one Ba^{2+} ion and one SO_4^{2-} ion.

$$BaSO_4(s) \rightleftharpoons Ba^{2+}(aq) + SO_4^{2-}(aq)$$
$$1.1 \times 10^{-5}\ \text{mol/L} \Longrightarrow 1.1 \times 10^{-5}\ M \qquad 1.1 \times 10^{-5}\ M$$
$$\text{(dissolved)}$$

In a saturated solution $[Ba^{2+}] = [SO_4^{2-}] = 1.1 \times 10^{-5}\ M$. Substituting these values into the K_{sp} expression for $BaSO_4$ gives the calculated value of K_{sp}.

$$K_{sp} = [Ba^{2+}][SO_4^{2-}] = (1.1 \times 10^{-5})(1.1 \times 10^{-5}) = 1.2 \times 10^{-10}$$

▶ Calculated values may differ slightly from those found in tables due to data or rounding differences.

You should now work Exercise 6.

EXAMPLE 20-2 Molar Solubility and Solubility Product Constant

One (1.00) liter of a saturated solution of silver chromate at 25°C contains 0.0435 gram of dissolved Ag_2CrO_4. Calculate (a) its molar solubility and (b) its solubility product constant.

Plan

We proceed as in Example 20-1.

Solution

(a) The equation for the dissolution of silver chromate in water and its solubility product expression are

$$Ag_2CrO_4(s) \rightleftharpoons 2Ag^+(aq) + CrO_4^{2-}(aq) \quad \text{and} \quad K_{sp} = [Ag^+]^2[CrO_4^{2-}]$$

The molar solubility of silver chromate is calculated first.

$$\frac{?\ mol\ Ag_2CrO_4}{L} = \frac{0.0435\ g\ Ag_2CrO_4}{1.00\ L} \times \frac{1\ mol\ Ag_2CrO_4}{332\ g\ Ag_2CrO_4} = 1.31 \times 10^{-4}\ mol/L \atop \text{(dissolved)}$$

▶ 1.31×10^{-4} mole of solid Ag_2CrO_4 dissolves to give a liter of saturated solution.

(b) The equation for dissolution of Ag_2CrO_4 and its molar solubility give the concentrations of Ag^+ and CrO_4^{2-} ions in the saturated solution.

$$Ag_2CrO_4(s) \rightleftharpoons 2Ag^+(aq) + CrO_4^{2-}(aq)$$
$$1.31 \times 10^{-4}\ mol/L \Longrightarrow 2.62 \times 10^{-4}\ M \quad 1.31 \times 10^{-4}\ M$$
$$\text{(dissolved)}$$

Substitution into the K_{sp} expression for Ag_2CrO_4 gives the value of K_{sp}.

$$K_{sp} = [Ag^+]^2[CrO_4^{2-}] = (2.62 \times 10^{-4})^2(1.31 \times 10^{-4}) = 8.99 \times 10^{-12}$$

The calculated K_{sp} of Ag_2CrO_4 is 8.99×10^{-12}.

You should now work Exercise 8.

STOP & THINK

Remember to raise $[Ag^+]$ to the power of its coefficient in the balanced chemical equation.

The molar solubility and K_{sp} values for $BaSO_4$ and Ag_2CrO_4 are compared in Table 20-1. These data show that the molar solubility of Ag_2CrO_4 is greater than that of $BaSO_4$. The K_{sp} for Ag_2CrO_4, however, is less than the K_{sp} for $BaSO_4$ because the expression for Ag_2CrO_4 contains a *squared* term, $[Ag^+]^2$.

STOP & THINK

Values of solubility product constants are usually tabulated to only two significant figures. Note that although the Ag_2CrO_4 is more soluble (higher molar solubility), it has a smaller K_{sp} value relative to $BaSO_4$. One can only directly compare K_{sp} values for compounds with the same ion ratio.

Table 20-1 Comparison of Solubilities of $BaSO_4$ and Ag_2CrO_4

Compound	Molar Solubility	K_{sp}
$BaSO_4$	1.1×10^{-5} mol/L	$[Ba^{2+}][SO_4^{2-}] = 1.1 \times 10^{-10}$
Ag_2CrO_4	1.3×10^{-4} mol/L	$[Ag^+]^2[CrO_4^{2-}] = 9.0 \times 10^{-12}$

If we compare K_{sp} values for two 1:1 compounds, for example, AgCl and $BaSO_4$, the compound with the larger K_{sp} value has the higher molar solubility. The same is true for any two compounds that have the same ion ratio, for example, the 1:2 compounds CaF_2 and $Mg(OH)_2$ and the 2:1 compound Ag_2CO_3.

▶ $K_{sp(AgCl)} = 1.8 \times 10^{-10}$
$K_{sp(BaSO_4)} = 1.1 \times 10^{-10}$

The molar solubility of AgCl is only slightly greater than that of $BaSO_4$.

If two compounds have the *same ion ratio*, the one with the larger K_{sp} will have the greater *molar* solubility.

Appendix H lists some K_{sp} values. Refer to it as needed.

Problem-Solving Tip It's Important to Know the Difference Between Solubility, Molar Solubility, and Solubility Product Expression

Often you need to know which of the following is given or which is asked for. Be sure you have a clear understanding of their definitions.

▶ The *solubility* of a compound is the amount of the compound that dissolves in a specified volume of solution. Solubility is usually expressed as either grams per liter or grams per 100 mL.

▶ The *molar solubility* of a compound is the number of moles of the compound that dissolve to give one liter of saturated solution. It is expressed in moles per liter.

▶ The *solubility product expression* for a compound is the product of the molar concentrations of its constituent ions, each raised to the power that corresponds to the number of ions in one formula unit of the compound. Like all equilibrium constants, it has no dimensions.

20-3 Uses of Solubility Product Constants

When the solubility product for a compound is known, the solubility of the compound in H_2O can be calculated, as the following example illustrates.

EXAMPLE 20-3 Molar Solubilities from K_{sp} Values

▶ The values of K_{sp} are obtained from Appendix H.

Calculate the molar solubilities, concentrations of the constituent ions, and solubilities in grams per liter for (a) silver chloride, AgCl ($K_{sp} = 1.8 \times 10^{-10}$), and (b) zinc hydroxide, $Zn(OH)_2$ ($K_{sp} = 4.5 \times 10^{-17}$).

Plan

We are given the value for each solubility product constant. In each case we write the appropriate balanced equation, represent the equilibrium concentrations, and then substitute into the K_{sp} expression.

Solution

(a) The equation for the dissolution of silver chloride and its solubility product expression are

$$AgCl(s) \rightleftharpoons Ag^+(aq) + Cl^-(aq) \qquad K_{sp} = [Ag^+][Cl^-] = 1.8 \times 10^{-10}$$

Each formula unit of AgCl that dissolves produces one Ag^+ and one Cl^-. We let $x = $ mol/L of AgCl that dissolves, that is, the molar solubility.

$$AgCl(s) \rightleftharpoons Ag^+(aq) + Cl^-(aq)$$
$$x\,\text{mol/L} \Longrightarrow \quad x\,M \qquad x\,M$$

Substitution into the solubility product expression gives

$$K_{sp} = [Ag^+][Cl^-] = (x)(x) = 1.8 \times 10^{-10} \qquad x^2 = 1.8 \times 10^{-10} \qquad x = 1.3 \times 10^{-5}$$

$$x = \text{molar solubility of AgCl} = 1.3 \times 10^{-5}\,\text{mol/L}$$

One liter of saturated AgCl contains 1.3×10^{-5} mole of dissolved AgCl at 25°C. From the balanced equation, we know the concentrations of the constituent ions.

$$x = \text{molar solubility} = [Ag^+] = [Cl^-] = 1.3 \times 10^{-5}\,\text{mol/L} = 1.3 \times 10^{-5}\,M$$

Now we can calculate the mass of dissolved AgCl in one liter of saturated solution.

$$\frac{?\,\text{g AgCl}}{L} = \frac{1.3 \times 10^{-5}\,\text{mol AgCl}}{L} \times \frac{143\,\text{g AgCl}}{1\,\text{mol AgCl}} = 1.9 \times 10^{-3}\,\text{g AgCl/L}$$

A liter of saturated AgCl solution contains only 0.0019 g of dissolved AgCl.

(b) The equation for the dissolution of zinc hydroxide, $Zn(OH)_2$, in water and its solubility product expression are

$$Zn(OH)_2(s) \rightleftharpoons Zn^{2+}(aq) + 2OH^-(aq) \quad ; \quad K_{sp} = [Zn^{2+}][OH^-]^2 = 4.5 \times 10^{-17}$$

We let x = molar solubility, so $[Zn^{2+}] = x$ and $[OH^-] = 2x$, and we have

$$Zn(OH)_2(s) \rightleftharpoons Zn^{2+}(aq) + 2OH^-(aq)$$
$$x \text{ mol/L} \implies x\,M \qquad 2x\,M$$

Substitution into the solubility product expression gives

$$[Zn^{2+}][OH^-]^2 = (x)(2x)^2 = 4.5 \times 10^{-17}$$

$$4x^3 = 4.5 \times 10^{-17} \quad ; \quad x^3 = 11 \times 10^{-18} \quad ; \quad x = 2.2 \times 10^{-6}$$

$$x = \text{molar solubility of } Zn(OH)_2 = 2.2 \times 10^{-6} \text{ mol } Zn(OH)_2/L$$

$$[Zn^{2+}] = x = 2.2 \times 10^{-6}\,M \qquad \text{and} \qquad [OH^-] = 2x = 4.4 \times 10^{-6}\,M$$

We can now calculate the mass of dissolved $Zn(OH)_2$ in one liter of saturated solution.

$$\frac{?\text{ g } Zn(OH)_2}{L} = \frac{2.2 \times 10^{-6} \text{ mol } Zn(OH)_2}{L} \times \frac{99 \text{ g } Zn(OH)_2}{1 \text{ mol } Zn(OH)_2} = 2.2 \times 10^{-4} \text{ g } Zn(OH)_2/L$$

A liter of saturated $Zn(OH)_2$ solution contains only 0.00022 g of dissolved $Zn(OH)_2$.

You should now work Exercise 16.

 TOP & THINK
The $[OH^-]$ is twice the molar solubility of $Zn(OH)_2$ because each formula unit of $Zn(OH)_2$ produces two OH^- ions.

Problem-Solving Tip The Dissolution of a Slightly Soluble Base Is Not a K_b Problem

The K_{sp} expression describes the equilibrium between a slightly soluble compound and its ions; in Example 20-3(b) one of those ions is OH^-. A K_b expression describes the equilibrium between a *soluble* basic species, for example, the ammonia molecule or the acetate ion, and the products it forms in solution, including OH^-. Do you see why the dissolution of $Zn(OH)_2$ is not a K_b problem? But even though it is not formally a K_b problem, we can still determine the solution's pH (or pOH). We found that $[OH^-] = 4.4 \times 10^{-6}\,M$ in a *saturated* $Zn(OH)_2$ solution. From this we can calculate that pOH = 5.36 and pH = 8.64. A saturated $Zn(OH)_2$ solution is not very basic because $Zn(OH)_2$ is not very soluble in H_2O. The $[OH^-]$ is only about 44 times greater than it is in pure water.

The Common Ion Effect in Solubility Calculations

Recall from Section 19-1 that the common ion effect is a special case of LeChatelier's Principle. Thus, the common ion effect applies to solubility equilibria just as it does to other ionic equilibria. Silver acetate, $AgCH_3COO$, is a slightly soluble salt:

$$AgCH_3COO(s) \rightleftharpoons Ag^+(aq) + CH_3COO^-(aq)$$

If we add another salt that provides one of the product ions, say Ag^+ (from $AgNO_3$), this dissolution reaction is shifted to the *left* (LeChatelier's Principle). Example 20-4 illustrates this effect.

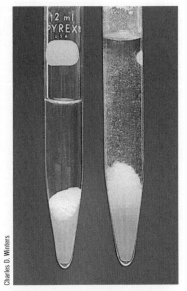

Charles D. Winters

The tube at the left contains a saturated solution of silver acetate, $AgCH_3COO$. When 1 M $AgNO_3$ is added to the tube, the equilibrium

$AgCH_3COO(s) \rightleftharpoons$

$Ag^+(aq) + CH_3COO^-(aq)$

shifts to the left and additional $AgCH_3COO$ precipitates. This demonstrates the common ion effect.

EXAMPLE 20-4 Molar Solubilities and the Common Ion Effect

For magnesium fluoride, MgF_2, $K_{sp} = 6.4 \times 10^{-9}$. (a) Calculate the molar solubility of magnesium fluoride in pure water. (b) Calculate the molar solubility of MgF_2 in 0.10 M sodium fluoride, NaF, solution. (c) Compare these molar solubilities.

Plan

For part (a), we write the appropriate chemical equations and solubility product expression, designate the equilibrium concentrations, and then substitute into the solubility product expression. For part (b), we recognize that NaF is a soluble ionic compound that is completely dissociated into its ions. MgF_2 is a slightly soluble compound. Both compounds produce F^- ions, so this is a common ion effect problem. We write the appropriate chemical equations and solubility product expression, represent the equilibrium concentrations, and substitute into the solubility product expression. For part (c), we compare the molar solubilities by calculating their ratio.

Solution

(a) We let x = molar solubility for MgF_2, a slightly soluble salt.

$$MgF_2(s) \rightleftharpoons Mg^{2+}(aq) + 2F^-(aq) \quad \text{(reversible)}$$
$$x \text{ mol/L} \Longrightarrow \quad x\, M \quad\quad 2x\, M$$

$$K_{sp} = [Mg^{2+}][F^-]^2 = 6.4 \times 10^{-9}$$

$$(x)(2x)^2 = 6.4 \times 10^{-9}$$

$$x = 1.2 \times 10^{-3}$$

molar solubility of MgF_2 in pure water = $1.2 \times 10^{-3}\, M$

(b) NaF is a soluble ionic salt, so $0.10\, M$ NaF contains $0.10\, M$ F^-:

$$NaF(s) \xrightarrow{H_2O} Na^+(aq) + F^-(aq) \quad \text{(complete)}$$
$$0.10\, M \Longrightarrow \quad 0.10\, M \quad\quad 0.10\, M$$

We let y = molar solubility for MgF_2, a slightly soluble salt.

$$MgF_2(s) \rightleftharpoons Mg^{2+}(aq) + 2F^-(aq) \quad \text{(reversible)}$$
$$y \text{ mol/L} \Longrightarrow \quad y\, M \quad\quad 2y\, M$$

The total $[F^-]$ is $0.10\, M$ from NaF *plus* $2y\, M$ from MgF_2, or $(0.10 + 2y)\, M$.

$$K_{sp} = [Mg^{2+}][F^-]^2 = 6.4 \times 10^{-9}$$

$$(y)(0.10 + 2y)^2 = 6.4 \times 10^{-9}$$

STOP & THINK

When the same ion is produced by a soluble and a slightly soluble salt, the amount from the slightly soluble salt can be neglected.

Very little MgF_2 dissolves, so y is small. This suggests that $2y \ll 0.10$, so $0.10 + 2y \approx 0.10$. Then

$$(y)(0.10)^2 = 6.4 \times 10^{-9} \quad \text{and} \quad y = 6.4 \times 10^{-7}$$

molar solubility of MgF_2 in $0.10\, M$ NaF = $6.4 \times 10^{-7}\, M$

(c) The ratio of molar solubility in water to molar solubility in $0.10\, M$ NaF solution is

$$\frac{\text{molar solubility (in } H_2O)}{\text{molar solubility (in NaF solution)}} = \frac{1.2 \times 10^{-3}\, M}{6.4 \times 10^{-7}\, M} = \frac{1900}{1}$$

The molar solubility of MgF_2 in $0.10\, M$ NaF ($6.4 \times 10^{-7}\, M$) is about 1900 times less than it is in pure water ($1.2 \times 10^{-3}\, M$).

You should now work Exercises 20 and 26.

ENRICHMENT

The Effects of Hydrolysis on Solubility

In Section 18-8 we discussed the hydrolysis of anions of weak acids. For example, we found that for CH_3COO^- and CN^- ions,

$$CH_3COO^- + H_2O \rightleftharpoons CH_3COOH + OH^-$$

$$K_b = \frac{[CH_3COOH][OH^-]}{[CH_3COO^-]} = 5.6 \times 10^{-10}$$

$$CN^- + H_2O \rightleftharpoons HCN + OH^- \qquad K_b = \frac{[HCN][OH^-]}{[CN^-]} = 2.5 \times 10^{-5}$$

We see that K_b for CN^-, the anion of a *very* weak acid, is much larger than K_b for CH_3COO^-, the anion of a much stronger acid. This tells us that in solutions of the same concentration, CN^- ions hydrolyze to a much greater extent than do CH_3COO^- ions. So we might expect that hydrolysis would have a much greater effect on the solubilities of cyanides such as AgCN than on the solubilities of acetates such as $AgCH_3COO$. It does.

Hydrolysis reduces the concentrations of anions of weak acids, such as F^-, CO_3^{2-}, CH_3COO^-, and CN^-, so its effect must be taken into account when we do very precise solubility calculations. Taking into account the effect of hydrolysis on solubilities of slightly soluble compounds is beyond the scope of this chapter, however.

The Reaction Quotient in Precipitation Reactions

Another application of the solubility product principle is the calculation of the maximum concentrations of ions that can coexist in solution. From these calculations, we can determine whether a **precipitate** will form in a given solution. The reaction quotient, Q (see Section 17-4), is useful in such decisions. We compare Q_{sp} with K_{sp}.

If $Q_{sp} < K_{sp}$	Forward process is favored
	No precipitation occurs; if solid is present, more solid can dissolve
If $Q_{sp} = K_{sp}$	Solution is *just* saturated
	Solid and solution are in equilibrium; neither forward nor reverse process is favored
If $Q_{sp} > K_{sp}$	Reverse process is favored
	Precipitation occurs to form more solid

> **STOP & THINK**
> Recall that the reaction quotient has the same algebraic form as the equilibrium constant, except that the concentrations in Q are not necessarily equilibrium concentrations. Remember also that solids do not appear in the expression for K_{sp} or for Q_{sp}.

EXAMPLE 20-5 Predicting Precipitate Formation

If 100. mL of 0.00075 M sodium sulfate, Na_2SO_4, is mixed with 50. mL of 0.015 M barium chloride, $BaCl_2$, will a precipitate form?

Plan

We are mixing solutions of two soluble ionic salts. First we find the amount of each solute at the instant of mixing. Next we find the molarity of each solute *at the instant of mixing*. Then we find the concentration of each ion in the *new* solution. Now we ask the question "*Could any combination of the ions in this solution form a slightly soluble compound?*" We consider the kinds of compounds mixed and determine whether a reaction could occur. Both Na_2SO_4 and $BaCl_2$ are soluble ionic salts. At the moment of mixing, the new solution contains a mixture of Na^+, SO_4^{2-}, Ba^{2+}, and Cl^- ions. We must consider the possibility of forming two new compounds, NaCl and $BaSO_4$.

i. Sodium chloride is a soluble ionic compound, so Na^+ and Cl^- do not combine to produce a precipitate in dilute aqueous solutions.

ii. $BaSO_4$, however, is only very slightly soluble, so the answer is "Yes, Ba^{2+} and SO_4^{2-} *might* form $BaSO_4$." We calculate Q_{sp} and compare it with K_{sp} to determine whether solid $BaSO_4$ is formed.

Solution

We find the *amount* of each solute at the instant of mixing.

$$\text{? mmol Na}_2\text{SO}_4 = 100.\text{ mL} \times \frac{0.00075 \text{ mmol Na}_2\text{SO}_4}{\text{mL}} = 0.075 \text{ mmol Na}_2\text{SO}_4$$

$$\text{? mmol BaCl}_2 = 50.\text{ mL} \times \frac{0.015 \text{ mmol BaCl}_2}{\text{mL}} = 0.75 \text{ mmol BaCl}_2$$

STOP & THINK
Recall that molarity can be expressed as mol/L or as mmol/mL. Because the amounts of dissolved ions in this problem are so small, it is convenient to express molarity as mmol/mL.

When *dilute* aqueous solutions are mixed, their volumes can be added to give the volume of the resulting solution.

$$\text{volume of mixed solution} = 100.\text{ mL} + 50.\text{ mL} = 150.\text{ mL}$$

Then we find the *molarity of each solute at the instant of mixing*.

$$M_{\text{Na}_2\text{SO}_4} = \frac{0.075 \text{ mmol Na}_2\text{SO}_4}{150.\text{ mL}} = 0.00050 \text{ } M \text{ Na}_2\text{SO}_4$$

$$M_{\text{BaCl}_2} = \frac{0.75 \text{ mmol BaCl}_2}{150.\text{ mL}} = 0.0050 \text{ } M \text{ BaCl}_2$$

Now we find the *concentration of each ion* in the new solution.

$$\text{Na}_2\text{SO}_4(s) \xrightarrow{100\%} 2\text{Na}^+(aq) + \text{SO}_4^{2-}(aq) \quad \text{(to completion)}$$
$$0.00050 \text{ } M \implies 0.0010 \text{ } M \quad 0.00050 \text{ } M$$

$$\text{BaCl}_2(s) \xrightarrow{100\%} \text{Ba}^{2+}(aq) + 2\text{Cl}^-(aq) \quad \text{(to completion)}$$
$$0.0050 \text{ } M \implies 0.0050 \text{ } M \quad 0.010 \text{ } M$$

Solid $BaSO_4$ will precipitate from the solution *if $Q_{sp} > K_{sp}$ for $BaSO_4$. K_{sp} for $BaSO_4$ is 1.1×10^{-10}. Substituting $[Ba^{2+}] = 0.0050 \text{ } M$ and $[SO_4^{2-}] = 0.00050 \text{ } M$ into the Q_{sp} expression for $BaSO_4$, we get

▶ Recall that Q has the same form as the equilibrium constant, in this case K_{sp}, but the concentrations are not necessarily equilibrium concentrations.

$$Q_{sp} = [Ba^{2+}][SO_4^{2-}] = (5.0 \times 10^{-3})(5.0 \times 10^{-4}) = 2.5 \times 10^{-6} \quad (Q_{sp} > K_{sp})$$

Because $Q_{sp} > K_{sp}$, solid $BaSO_4$ will precipitate until $[Ba^{2+}][SO_4^{2-}]$ just equals K_{sp} for $BaSO_4$.

You should now work Exercises 28 and 30.

Materials can also react in the solid state. When white solid potassium iodide, KI, and white solid lead(II) nitrate, $Pb(NO_3)_2$, are ground together, some yellow lead(II) iodide, PbI_2, forms.

Problem-Solving Tip Detection of Precipitates

The human eye is not a very sensitive detection device. As a rule of thumb, a precipitate can be seen with the naked eye if $Q_{sp} > K_{sp}$ by a factor of 1000. In Example 20-5,

Q_{sp} exceeds K_{sp} by a factor of $\dfrac{2.5 \times 10^{-6}}{1.1 \times 10^{-10}} = 2.3 \times 10^4 = 23,000$. We expect to be able

to see the $BaSO_4$ precipitate that is formed. Modern techniques enable us to detect smaller amounts of precipitates.

EXAMPLE 20-6 Initiation of Precipitation

What $[Ba^{2+}]$ is necessary to start the precipitation of $BaSO_4$ in a solution that is 0.0015 M in Na_2SO_4? Assume that the Ba^{2+} comes from addition of a solid soluble ionic compound such as $BaCl_2$. For $BaSO_4$, $K_{sp} = 1.1 \times 10^{-10}$.

Plan

These are the compounds in Example 20-5. We should recognize that Na_2SO_4 is a soluble ionic compound and that the molarity of SO_4^{2-} is equal to the molarity of the Na_2SO_4 solution. We are given K_{sp} for $BaSO_4$, so we solve for $[Ba^{2+}]$.

Solution

Because Na_2SO_4 is a soluble ionic compound, we know that $[SO_4^{2-}] = 0.0015\ M$. We can use K_{sp} for $BaSO_4$ to calculate the $[Ba^{2+}]$ required for Q_{sp} to just equal K_{sp}.

$$[Ba^{2+}][SO_4^{2-}] = 1.1 \times 10^{-10}$$

$$[Ba^{2+}] = \frac{1.1 \times 10^{-10}}{[SO_4^{2-}]} = \frac{1.1 \times 10^{-10}}{1.5 \times 10^{-3}} = 7.3 \times 10^{-8}\ M$$

Addition of enough $BaCl_2$ to give a barium ion concentration of $7.3 \times 10^{-8}\ M$ *just satisfies* K_{sp} for $BaSO_4$; that is, $Q_{sp} = K_{sp}$. Ever so slightly more $BaCl_2$ would be required for Q_{sp} to exceed K_{sp} and for precipitation of $BaSO_4$ to occur. Therefore

$$[Ba^{2+}] > 7.3 \times 10^{-8}\ M \qquad \text{(to initiate precipitation of BaSO}_4\text{)}$$

You should now work Exercise 29.

> **S TOP & THINK**
> For a 1:1 salt, the following qualitative check can be made. We can take the square root of the $BaSO_4$ K_{sp} value to get the solution concentrations of the Ba^{2+} and SO_4^{2-} ions for a saturated solution: $[Ba^{2+}] = [SO_4^{2-}] \approx 1 \times 10^{-5}\ M$. Because the $[SO_4^{2-}]$ from the Na_2SO_4 is just over two orders of magnitude larger than this ($1.5 \times 10^{-3}\ M$), the common ion concept should tell us that the $[Ba^{2+}]$ needs to be just over two orders of magnitude smaller (less than $1 \times 10^{-7}\ M$).

Often we wish to remove an ion from solution by forming an insoluble compound (as in water purification). We use K_{sp} values to calculate the concentrations of ions remaining in solution *after* precipitation has occurred.

EXAMPLE 20-7 Concentration of Common Ion

Suppose we wish to recover silver from an aqueous solution that contains a soluble silver compound such as $AgNO_3$ by precipitating insoluble silver chloride, AgCl. A soluble ionic compound such as NaCl can be used as a source of Cl^-. What is the minimum concentration of chloride ion needed to reduce the dissolved silver ion concentration to a maximum of $1.0 \times 10^{-9}\ M$? For AgCl, $K_{sp} = 1.8 \times 10^{-10}$.

Plan

We are given K_{sp} for AgCl and the required equilibrium $[Ag^+]$, so we solve for $[Cl^-]$.

Solution

The equations for the reaction of interest and the K_{sp} for AgCl are

$$AgCl(s) \rightleftharpoons Ag^+(aq) + Cl^-(aq) \qquad \text{and} \qquad [Ag^+][Cl^-] = 1.8 \times 10^{-10}$$

To determine the $[Cl^-]$ required to reduce the $[Ag^+]$ to $1.0 \times 10^{-9}\ M$, we solve the K_{sp} expression for $[Cl^-]$.

$$[Cl^-] = \frac{1.8 \times 10^{-10}}{[Ag^+]} = \frac{1.8 \times 10^{-10}}{1.0 \times 10^{-9}} = 0.18\ M\ Cl^-$$

To reduce the $[Ag^+]$ to $1.0 \times 10^{-9}\ M$ (0.00000011 g Ag^+/L), NaCl would be added until $[Cl^-] = 0.18\ M$ in the solution.

You should now work Exercise 32.

▶ The recovery of silver from the solutions used in developing and fixing black-and-white photographic film and prints presents just such a problem. Silver is an expensive metal, and the recovery is profitable. Moreover, if not recovered, the silver ions would constitute an undesirable pollutant in water supplies.

Silver chloride precipitates when chloride ions are added to a solution containing silver ions.

20-4 Fractional Precipitation; $Q_{sp} \geq K_{sp}$

▶ Cations such as Na^+ or K^+ are also present in this solution as spectator ions necessary to balance charge.

We sometimes wish to remove selected ions from solution while leaving others with similar properties in solution. This separation process is called **fractional precipitation**. Consider a solution that contains Cl^-, Br^-, and I^- ions. These halide ions are anions of elements in the same family in the periodic table. Although we expect them to have some similar properties, there are some differences in the solubility products for the silver halides:

Compound	Solubility Product
AgCl	1.8×10^{-10}
AgBr	3.3×10^{-13}
AgI	1.5×10^{-16}

S TOP & THINK
Because these compounds all have the same 1:1 ion ratio, we can directly compare the relative K_{sp} values and the solubilities of these compounds.

These K_{sp} values show that AgI is less soluble than AgBr and that AgBr is less soluble than AgCl. Silver fluoride is quite soluble in water. This is the expected trend based on the weaker interactions between water molecules and the larger halide anions that have their charge spread out over a larger surface area. This reduces the ability of the water molecules to solvate and pull the larger anions away from the solid.

EXAMPLE 20-8   Concentration Required to Initiate Precipitation

▶ NaCl, NaBr, NaI, AgNO₃, and NaNO₃ are soluble compounds that are completely dissociated in dilute aqueous solution. Thus, we do not use or need them in this problem.

Solid silver nitrate is slowly added to a solution that is 0.0010 M each in NaCl, NaBr, and NaI. Calculate the $[Ag^+]$ required to initiate the precipitation of each of the silver halides. For AgI, $K_{sp} = 1.5 \times 10^{-16}$; for AgBr, $K_{sp} = 3.3 \times 10^{-13}$; and for AgCl, $K_{sp} = 1.8 \times 10^{-10}$.

Plan

We are given a solution that contains equal concentrations of Cl^-, Br^-, and I^- ions, all of which form insoluble silver salts. Then we slowly add Ag^+ ions. We use each K_{sp} to determine the $[Ag^+]$ that must be exceeded to initiate precipitation of each salt as we did in Example 20-6.

▶ We ignore the extremely small change in volume caused by addition of solid AgNO₃.

Solution

We calculate the $[Ag^+]$ necessary to begin to precipitate each of the silver halides. The solubility product for AgI is

$$[Ag^+][I^-] = 1.5 \times 10^{-16}$$

$[I^-] = 1.0 \times 10^{-3}\,M$, so the $[Ag^+]$ that must be exceeded to start precipitation of AgI is

$$[Ag^+] = \frac{1.5 \times 10^{-16}}{[I^-]} = \frac{1.5 \times 10^{-16}}{1.0 \times 10^{-3}} = 1.5 \times 10^{-13}\,M$$

▶ We are not suggesting that a $1.5 \times 10^{-13}\,M$ solution of AgNO₃ be added. We are pointing out the fact that when sufficient AgNO₃ has been added to the solution to make $[Ag^+] > 1.5 \times 10^{-13}\,M$, AgI begins to precipitate.

Therefore, AgI will begin to precipitate when $[Ag^+] > 1.5 \times 10^{-13}\,M$.

Repeating this kind of calculation for silver bromide gives

$$[Ag^+][Br^-] = 3.3 \times 10^{-13}$$

$$[Ag^+] = \frac{3.3 \times 10^{-13}}{[Br^-]} = \frac{3.3 \times 10^{-13}}{1.0 \times 10^{-3}} = 3.3 \times 10^{-10}\,M$$

Thus, $[Ag^+] > 3.3 \times 10^{-10}\,M$ is needed to start precipitation of AgBr.

For the precipitation of silver chloride to begin,

$$[Ag^+][Cl^-] = 1.8 \times 10^{-10}$$

$$[Ag^+] = \frac{1.8 \times 10^{-10}}{[Cl^-]} = \frac{1.8 \times 10^{-10}}{1.0 \times 10^{-3}} = 1.8 \times 10^{-7}\,M$$

To precipitate AgCl, we must have $[Ag^+] > 1.8 \times 10^{-7}\,M$.

Problem-Solving Tip

Because of the simple 1:1 ion ratio, it is quite easy to estimate and check the answer without a calculator. The square root of the AgI K_{sp} value (simplified to 1×10^{-16}) gives $[Ag^+] = [I^-] \approx 1 \times 10^{-8}\ M$. The initial $[I^-]$ from the added NaI is $1 \times 10^{-3}\ M$, which is five orders of magnitude larger. The common ion concept and 1:1 ion ratio tell us that the $[Ag^+]$ needed to initiate precipitation should be five orders of magnitude smaller than $1 \times 10^{-8}\ M$, or about $1 \times 10^{-13}\ M$.

Similar estimations can be made for the AgBr and AgCl calculations to check their answers. The AgBr example is a little more complicated due to the odd power on the K_{sp} value (3.3×10^{-13}) that cannot simply be divided by two to estimate the square root of the power of 10. Changing this to 33×10^{-14} simplifies estimating the square root as approximately $6 \times 10^{-7} \approx [Ag^+] = [Br^-]$. The initial $[Br^-] = 1 \times 10^{-3}\ M$, which is a little more than three orders of magnitude larger than $6 \times 10^{-7}\ M$. Thus, the $[Ag^+]$ to initiate precipitation should be a little over three orders of magnitude smaller, or about $1 \times 10^{-10}\ M$.

This calculation tells us that when $AgNO_3$ is added slowly to a solution that is $0.0010\ M$ in each of NaI, NaBr, and NaCl, AgI precipitates first, AgBr precipitates second, and AgCl precipitates last.

We can also calculate the fraction of one salt that precipitates before another salt begins to form in a fractional precipitation experiment.

> As $AgNO_3$ is added to the solution containing Cl^-, Br^-, and I^- ions, some AgBr and AgCl may precipitate *locally*. As the solution is stirred, AgBr and AgCl redissolve as long as $[Ag^+]$ is not large enough to exceed their K_{sp} values in the *bulk* of the solution.

EXAMPLE 20-9 Fractional Precipitation

Refer to Example 20-8. (a) Calculate the percentage of I^- precipitated before AgBr precipitates. (b) Calculate the percentages of I^- and Br^- precipitated before Cl^- precipitates.

Plan

From Example 20-8 we know the $[Ag^+]$ that must be exceeded to initiate precipitation of each of three silver halides, AgI, AgBr, and AgCl. We use each of these values of $[Ag^+]$ with the appropriate K_{sp} expression, in turn, to find the concentration of each halide ion that remains in solution (unprecipitated). We express these halide ion concentrations as percent unprecipitated. Then we subtract each from exactly 100% to find the percentage of each halide that precipitates.

Solution

(a) In Example 20-8 we found that AgBr begins to precipitate when $[Ag^+] > 3.3 \times 10^{-10}\ M$. This value for $[Ag^+]$ can be substituted into the K_{sp} expression for AgI to determine $[I^-]$ remaining *unprecipitated* when AgBr begins to precipitate.

> In Example 20-7 we did a similar calculation, but we did not express the result in terms of the percentage of an ion precipitated.

$$[Ag^+][I^-] = 1.5 \times 10^{-16}$$

$$[I^-]_{unppt'd} = \frac{1.5 \times 10^{-16}}{[Ag^+]} = \frac{1.5 \times 10^{-16}}{3.3 \times 10^{-10}} = 4.5 \times 10^{-7}\ M$$

The percentage of I^- unprecipitated is

$$\%\ I^-{}_{unppt'd} = \frac{[I^-]_{unppt'd}}{[I^-]_{orig}} \times 100\% = \frac{4.5 \times 10^{-7}\ M}{1.0 \times 10^{-3}\ M} \times 100\%$$

$$= 0.045\%\ I^-\ \text{unprecipitated}$$

Therefore, 99.955% of the I^- precipitates *before* AgBr begins to precipitate.

(b) Similar calculations show that *just before* AgCl begins to precipitate, $[Ag^+] = 1.8 \times 10^{-7}\ M$, and the $[I^-]$ unprecipitated is calculated as in Part (a).

> We have subtracted 0.045% from *exactly* 100%. We have therefore *not* violated the rules for significant figures.

$$[Ag^+][I^-] = 1.5 \times 10^{-16}$$

$$[I^-]_{unppt'd} = \frac{1.5 \times 10^{-16}}{[Ag^+]} = \frac{1.5 \times 10^{-16}}{1.8 \times 10^{-7}} = 8.3 \times 10^{-10}\ M$$

Freshly precipitated AgCl is white (*left*), AgBr is very pale yellow (*center*), and AgI is yellow (*right*). Polarizabilities of these halide ions increase in the order $Cl^- < Br^- < I^-$. Colors of the silver halides become more intense in the same direction. Solubilities of the silver halides increase in the opposite direction.

The percentage of I^- unprecipitated just before AgCl precipitates is

$$\%I^-_{\text{unppt'd}} = \frac{[I^-]_{\text{unppt'd}}}{[I^-]_{\text{orig}}} \times 100\% = \frac{8.3 \times 10^{-10} M}{1.0 \times 10^{-3} M} \times 100\%$$

$$= 0.000083\% \ I^- \text{ unprecipitated}$$

Therefore, 99.999917% of the I^- precipitates before AgCl begins to precipitate.

A similar calculation for the amount of Br^- precipitated just before AgCl begins to precipitate gives

$$[Ag^+][Br^-] = 3.3 \times 10^{-13}$$

$$[Br^-]_{\text{unppt'd}} = \frac{3.3 \times 10^{-13}}{[Ag^+]} = \frac{3.3 \times 10^{-13}}{1.8 \times 10^{-7}} = 1.8 \times 10^{-6} M$$

$$\% \ Br^-_{\text{unppt'd}} = \frac{[Br^-]_{\text{unppt'd}}}{[Br^-]_{\text{orig}}} \times 100\% = \frac{1.8 \times 10^{-6} M}{1.0 \times 10^{-3} M} \times 100\%$$

$$= 0.18\% \ Br^- \text{ unprecipitated}$$

Thus, 99.82% of the Br^- precipitates before AgCl begins to precipitate.

You should now work Exercises 36 and 38.

We have described the precipitation reactions that occur when solid $AgNO_3$ is added slowly to a solution that is 0.0010 *M* in Cl^-, Br^-, and I^-. Silver iodide begins to precipitate first; 99.955% of the I^- precipitates before any solid AgBr is formed. Silver bromide begins to precipitate next; 99.82% of the Br^- and 99.999917% of the I^- precipitate before any solid AgCl forms. This shows that we can separate these ions very effectively by fractional precipitation.

20-5 Simultaneous Equilibria Involving Slightly Soluble Compounds

Many weak acids and bases react with many metal ions to form insoluble compounds. In such cases, we must take into account the weak acid or weak base equilibrium as well as the solubility equilibrium. The most common examples involve the reaction of metal ions with aqueous ammonia to form insoluble metal hydroxides.

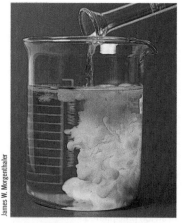

Pouring ammonium sulfide solution into a solution of cadmium nitrate gives a precipitate of cadmium sulfide.

$$(NH_4)_2S + Cd(NO_3)_2 \longrightarrow$$
$$CdS(s) + 2NH_4NO_3$$

Cadmium sulfide is used as a pigment in artists' oil-based paints.

EXAMPLE 20-10 Simultaneous Equilibria

We make a solution that is 0.10 *M* in magnesium nitrate, $Mg(NO_3)_2$, *and* 0.10 *M* in aqueous ammonia, a weak base. Will magnesium hydroxide, $Mg(OH)_2$, precipitate from this solution? K_{sp} for $Mg(OH)_2$ is 1.5×10^{-11}, and K_b for aqueous NH_3 is 1.8×10^{-5}.

Plan

We first write equations for the two *reversible* reactions and their equilibrium constant expressions. We note that $[OH^-]$ appears in both equilibrium constant expressions. From the statement of the problem, we know the concentration of Mg^{2+}. We use the K_b expression for aqueous NH_3 to find $[OH^-]$. Then we calculate Q_{sp} for $Mg(OH)_2$ and compare it with its K_{sp}.

Solution

Two equilibria and their equilibrium constant expressions must be considered.

$$Mg(OH)_2(s) \rightleftharpoons Mg^{2+}(aq) + 2OH^-(aq) \qquad K_{sp} = [Mg^{2+}][OH^-]^2 = 1.5 \times 10^{-11}$$

$$NH_3(aq) + H_2O(\ell) \rightleftharpoons NH_4^+(aq) + OH^-(aq) \qquad K_b = \frac{[NH_4^+][OH^-]}{[NH_3]} = 1.8 \times 10^{-5}$$

The $[OH^-]$ in 0.10 M aqueous NH_3 is calculated as in Example 18-14.

$$NH_3(aq) + H_2O(\ell) \rightleftharpoons NH_4^+(aq) + OH^-(aq)$$
$$(0.10 - x)\ M \qquad\qquad x\ M \qquad x\ M$$

$$\frac{[NH_4^+][OH^-]}{[NH_3]} = 1.8 \times 10^{-5} = \frac{(x)(x)}{(0.10 - x)} \quad ; \quad x = 1.3 \times 10^{-3}\ M = [OH^-]$$

Magnesium nitrate is a soluble ionic compound, so $[Mg^{2+}] = 0.10\ M$. Now that both $[Mg^{2+}]$ and $[OH^-]$ are known, we calculate Q_{sp} for $Mg(OH)_2$.

$$[Mg^{2+}][OH^-]^2 = Q_{sp}$$
$$(0.10)(1.3 \times 10^{-3})^2 = 1.7 \times 10^{-7} = Q_{sp}$$

$K_{sp} = 1.5 \times 10^{-11}$, so we see that $Q_{sp} > K_{sp}$.

Therefore, $Mg(OH)_2$ would precipitate, lowering $[Mg^{2+}]$ and $[OH^-]$ until $Q_{sp} = K_{sp}$.

You should now work Exercise 46.

S **TOP & THINK**

We have calculated the $[OH^-]$ produced by the ionization of 0.10 M aqueous NH_3. *This is the equilibrium concentration of OH^- in this solution. There is no reason to double this value when we put it into the K_{sp} expression!* But we do need to raise it to the coefficient power of 2 in the $Mg(OH)_2$ equilibrium expression.

Example 20-11 shows how we can calculate the concentration of weak base that is required to initiate precipitation of an insoluble metal hydroxide.

EXAMPLE 20-11   Simultaneous Equilibria

What concentration of aqueous ammonia is necessary to just start precipitation of $Mg(OH)_2$ from a 0.10 M solution of $Mg(NO_3)_2$? Refer to Example 20-10.

Plan

We have the same reactions and equilibrium constant expressions as in Example 20-10. We are given $[Mg^{2+}]$, so we use the K_{sp} expression of $Mg(OH)_2$ to calculate the $[OH^-]$ necessary to initiate precipitation. Then we find the molarity of aqueous NH_3 solution that would furnish the desired $[OH^-]$.

Solution

Two equilibria and their equilibrium constant expressions must be considered.

$$Mg(OH)_2(s) \rightleftharpoons Mg^{2+}(aq) + 2OH^-(aq) \qquad K_{sp} = 1.5 \times 10^{-11}$$
$$NH_3(aq) + H_2O(\ell) \rightleftharpoons NH_4^+(aq) + OH^-(aq) \qquad K_b = 1.8 \times 10^{-5}$$

We find the $[OH^-]$ necessary to initiate precipitation of $Mg(OH)_2$ when $[Mg^{2+}] = 0.10\ M$.

$$K_{sp} = [Mg^{2+}][OH^-]^2 = 1.5 \times 10^{-11}$$

$$[OH^-]^2 = \frac{1.5 \times 10^{-11}}{[Mg^{2+}]} = \frac{1.5 \times 10^{-11}}{0.10} = 1.5 \times 10^{-10} \qquad [OH^-] = 1.2 \times 10^{-5}\ M$$

Therefore, $[OH^-] > 1.2 \times 10^{-5}\ M$ to initiate precipitation of $Mg(OH)_2$.

Now we use the equilibrium of NH_3 as a weak base to find the $[NH_3]$ that will produce $1.2 \times 10^{-5}\ M\ OH^-$. Let x be the *original* $[NH_3]$.

$$NH_3(aq) + H_2O(\ell) \rightleftharpoons NH_4^+(aq) + OH^-(aq)$$
$$(x - 1.2 \times 10^{-5})\ M \qquad\qquad 1.2 \times 10^{-5}\ M \quad 1.2 \times 10^{-5}\ M$$

$$K_b = \frac{[NH_4^+][OH^-]}{[NH_3]} = 1.8 \times 10^{-5} = \frac{(1.2 \times 10^{-5})(1.2 \times 10^{-5})}{x - 1.2 \times 10^{-5}}$$

$$1.8 \times 10^{-5}\,x - 2.16 \times 10^{-10} = 1.44 \times 10^{-10}$$

$$1.8 \times 10^{-5}\,x = 3.6 \times 10^{-10} \qquad so \qquad x = 2.0 \times 10^{-5}\ M = [NH_3]_{orig}$$

S **TOP & THINK**

Remember that x represents the *original* concentration of the weak base NH_3. The small value of K_b tells us that 1.2×10^{-5} and x are of comparable magnitude, so neither can be disregarded in the term $(x - 1.2 \times 10^{-5})$.

The solution must be ever so slightly greater than $2.0 \times 10^{-5}\ M$ in NH_3 to initiate precipitation of $Mg(OH)_2$ in a 0.10 M solution of $Mg(NO_3)_2$.

A solution that contains a weak base can be buffered (by addition of a salt of the weak base) to decrease its basicity. Significant concentrations of some metal ions that form insoluble hydroxides can be kept in such solutions, as Example 20-12 illustrates.

EXAMPLE 20-12 Simultaneous Equilibria

What minimum number of moles of NH_4Cl must be added to 1.0 liter of solution that is 0.10 M in $Mg(NO_3)_2$ *and* 0.10 M in NH_3 to prevent precipitation of $Mg(OH)_2$?

Plan

These are the same compounds, in the same concentrations, that we used in Example 20-10. Because we know $[Mg^{2+}]$, we must find the maximum $[OH^-]$ that can exist in the solution *without exceeding* K_{sp} for $Mg(OH)_2$. Then we find the minimum concentration of NH_4Cl that is necessary to buffer the NH_3 solution to keep the $[OH^-]$ below this calculated value.

Solution

The buffering action of NH_4Cl in the presence of NH_3 decreases the concentration of OH^-. Again we have two equilibria.

> ▶ In Example 20-10, we found that $Mg(OH)_2$ will precipitate from a solution that is 0.10 M in $Mg(NO_3)_2$ and 0.10 M in NH_3.

$$Mg(OH)_2(s) \rightleftharpoons Mg^{2+}(aq) + 2OH^-(aq) \qquad K_{sp} = 1.5 \times 10^{-11}$$

$$NH_3(aq) + H_2O(\ell) \rightleftharpoons NH_4^+(aq) + OH^-(aq) \qquad K_b = 1.8 \times 10^{-5}$$

To find the *maximum $[OH^-]$ that can exist in solution without causing precipitation*, we substitute $[Mg^{2+}]$ into the K_{sp} for $Mg(OH)_2$.

$$[Mg^{2+}][OH^-]^2 = 1.5 \times 10^{-11}$$

> ▶ You may wish to refer to Example 19-6 to refresh your understanding of buffer solutions.

$$[OH^-]^2 = \frac{1.5 \times 10^{-11}}{[Mg^{2+}]} = \frac{1.5 \times 10^{-11}}{0.10} = 1.5 \times 10^{-10}$$

$$[OH^-] = 1.2 \times 10^{-5}\ M \qquad \text{(maximum } [OH^-] \text{ possible)}$$

To prevent precipitation of $Mg(OH)_2$ in *this* solution, $[OH^-]$ must be equal to or less than $1.2 \times 10^{-5}\ M$. We can use K_b for aqueous NH_3 to calculate the number of moles of NH_4Cl necessary to buffer 1.0 L of 0.10 M aqueous NH_3 so that $[OH^-] = 1.2 \times 10^{-5}\ M$. Let x = number of mol/L of NH_4Cl required.

$$
\begin{array}{ccc}
NH_4Cl(aq) \longrightarrow & NH_4^+(aq) & + Cl^-(aq) \quad \text{(to completion)} \\
x\,M \rightleftharpoons & x\,M & x\,M
\end{array}
$$

$$
\begin{array}{ccc}
NH_3(aq) + H_2O(\ell) \rightleftharpoons & NH_4^+(aq) & + OH^-(aq) \\
(0.10 - 1.2 \times 10^{-5})\,M & 1.2 \times 10^{-5}\,M & 1.2 \times 10^{-5}\,M
\end{array}
$$

We can assume that $(x + 1.2 \times 10^{-5}) \approx x$ and $(0.10 - 1.2 \times 10^{-5}) \approx 0.10$.

$$\frac{(x)(1.2 \times 10^{-5})}{0.10} = 1.8 \times 10^{-5}$$

> **S TOP & THINK**
> $x = 0.15\ M\ NH_4^+$ confirms that the approximation of $x + 1.2 \times 10^{-5} \approx x$ used above to simplify the algebra was a good one.

$$x = 0.15 \text{ mol of } NH_4^+ \text{ per liter of solution}$$

Addition of at least 0.15 mol of NH_4Cl to 1.0 L of 0.10 M aqueous NH_3 decreases $[OH^-]$ to $1.2 \times 10^{-5}\ M$ or lower. Then K_{sp} for $Mg(OH)_2$ is not exceeded in this solution, and so no precipitate would form.

You should now work Exercises 42 and 44.

Examples 20-10, 20-11, and 20-12 illustrate a very important point.

> *All relevant equilibria must be satisfied* when more than one equilibrium is required to describe a solution.

20-6 Dissolving Precipitates; $Q_{sp} < K_{sp}$

A precipitate dissolves when the concentrations of its ions are reduced so that K_{sp} is no longer exceeded, that is, when $Q_{sp} < K_{sp}$. The precipitate then dissolves until $Q_{sp} = K_{sp}$. Precipitates can be dissolved by several types of reactions. All involve removing ions from solution.

▶ Solubility product constants, like other equilibrium constants, are thermodynamic quantities. They tell us about the extent to which a given reaction can occur, but nothing about how fast that reaction occurs.

Converting an Ion to a Weak Electrolyte

Three specific illustrations follow.

1. *Converting OH^- to H_2O.* Insoluble $Al(OH)_3$ dissolves in acids. H^+ ions react with OH^- ions [from the saturated $Al(OH)_3$ solution] to form the weak electrolyte H_2O. This makes $Q_{sp} = [Al^{3+}][OH^-]^3 < K_{sp}$, so that the dissolution equilibrium shifts to the right and $Al(OH)_3$ dissolves.

$$Al(OH)_3(s) \rightleftharpoons Al^{3+}(aq) + 3OH^-(aq)$$
$$\underline{3H^+(aq) + 3OH^-(aq) \longrightarrow 3H_2O(\ell)}$$
overall rxn: $Al(OH)_3(s) + 3H^+(aq) \longrightarrow Al^{3+}(aq) + 3H_2O(\ell)$

2. *Converting NH_4^+ to NH_3.* Ammonium ions, from a salt such as NH_4Cl, dissolve insoluble $Mg(OH)_2$. The NH_4^+ ions combine with OH^- ions in the saturated $Mg(OH)_2$ solution. This forms the weak electrolytes NH_3 and H_2O. The result is that $Q_{sp} = [Mg^{2+}][OH^-]^2$, K_{sp}, and so the $Mg(OH)_2$ dissolves.

$$Mg(OH)_2(s) \rightleftharpoons Mg^{2+}(aq) + 2OH^-(aq)$$
$$\underline{2NH_4^+(aq) + 2OH^-(aq) \longrightarrow 2NH_3(aq) + 2H_2O(\ell)}$$
overall rxn: $Mg(OH)_2(s) + 2NH_4^+(aq) \longrightarrow Mg^{2+}(aq) + 2NH_3(aq) + 2H_2O(\ell)$

This process, dissolution of $Mg(OH)_2$ in an NH_4Cl solution, is the reverse of the reaction we considered in Example 20-10. There, $Mg(OH)_2$ precipitated from a solution of aqueous NH_3.

3. *Converting S^{2-} to H_2S.* Nonoxidizing acids dissolve most insoluble metal sulfides. For example, 6 M HCl dissolves MnS. The H^+ ions combine with S^{2-} ions to form H_2S, a gas that bubbles out of the solution. The result is that $Q_{sp} = [Mn^{2+}][S^{2-}] < K_{sp}$, and so the MnS dissolves.

$$MnS(s) \rightleftharpoons Mn^{2+}(aq) + S^{2-}(aq)$$
$$S^{2-}(aq) + H_2O(\ell) \longrightarrow HS^-(aq) + OH^-(aq)$$
$$\underline{2H^+(aq) + HS^-(aq) + OH^-(aq) \longrightarrow H_2S(g) + H_2O(\ell)}$$
overall rxn: $MnS(s) + 2H^+(aq) \longrightarrow Mn^{2+}(aq) + H_2S(g)$

Stalactite and stalagmite formations in limestone caves result from the dissolution and reprecipitation of calcium carbonate, $CaCO_3$.

 STOP & THINK
S^{2-}, like O^{2-}, does not exist in appreciable amounts in aqueous solutions. This is because it is basic enough to react with H_2O to produce HS^- and OH^-, as shown in the second reaction in part 3.

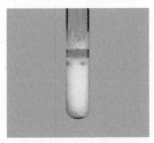

Manganese(II) sulfide, MnS, is salmon-colored. MnS dissolves in 6 M HCl. The resulting solution of $MnCl_2$ is pale pink.

Converting an Ion to Another Species by a Redox Reaction

Most insoluble metal sulfides dissolve in hot dilute HNO_3 because NO_3^- ions oxidize S^{2-} ions, or actually their hydrolysis product (HS^-), to elemental sulfur. This removes HS^- (and thus S^{2-}) ions from the solution and promotes the dissolving of more of the metal sulfide.

$$3HS^-(aq) + 2NO_3^-(aq) + 5H^+(aq) \longrightarrow 3S(s) + 2NO(g) + 4H_2O(\ell)$$

▶ It is sometimes convenient to ignore hydrolysis of S^{2-} ions in aqueous solutions. Leaving out the hydrolysis step may give the false impression that S^{2-} ions exist in solution. The overall net ionic equation for this redox reaction will be the same, however, with or without consideration of hydrolysis of the S^{2-} ions.

The elemental sulfur produced usually reacts with more HNO_3 to make H_2SO_4 via the following reaction.

$$S(s) + 2NO_3^-(aq) \longrightarrow 2NO(g) + SO_4^{2-}(aq)$$

Consider copper(II) sulfide, CuS, in equilibrium with its ions. This equilibrium lies far to the left. Removal of the S^{2-} or HS^- ions by oxidation to SO_4^{2-} favors the reaction to the right, and so CuS(s) dissolves in hot dilute HNO_3.

$$CuS(s) \rightleftharpoons Cu^{2+}(aq) + S^{2-}(aq)$$
$$S^{2-}(aq) + H_2O(\ell) \longrightarrow HS^-(aq) + OH^-(aq)$$
$$3HS^-(aq) + 3OH^-(aq) + 8H^+(aq) + 2NO_3^-(aq) \longrightarrow 3S(s) + 2NO(g) + 7H_2O(\ell)$$
$$S(s) + 2NO_3^-(aq) \longrightarrow 2NO(g) + SO_4^{2-}(aq)$$

Multiply the first, second, and fourth equations by 3, and then add and cancel like terms to give the net ionic equation for dissolving CuS(s) in hot dilute HNO_3.

$$3CuS(s) + 8NO_3^-(aq) + 8H^+(aq) \longrightarrow 3Cu^{2+}(aq) + 3SO_4^{2-}(aq) + 8NO(g) + 4H_2O(\ell)$$

Copper(II) sulfide, CuS, is black. As CuS dissolves in 6 M HNO_3, some NO is oxidized to brown NO_2 by O_2 in the air. The resulting solution of Cu^{2+} ions (SO_4^{2-} and NO_3^- counter anions) is blue.

Complex Ion Formation

The cations in many slightly soluble compounds can form *complex ions*. This often results in dissolution of the slightly soluble compound. Some metal ions share electron pairs donated by molecules and ions such as NH_3, CN^-, OH^-, F^-, Cl^-, Br^-, and I^- to form coordinate covalent bonds to metal ions. Coordinate covalent bonds are formed as these electron-donating groups (ligands) replace H_2O molecules from hydrated metal ions. The decrease in the concentration of the hydrated metal ion shifts the solubility equilibrium to the right.

▶ "Ligand" is the name given to an atom or a group of atoms bonded to the metal center in complex ions. Ligands are Lewis bases.

Many copper(II) compounds react with excess aqueous NH_3 to form the deep-blue complex ion $[Cu(NH_3)_4]^{2+}$.

$$Cu^{2+}(aq) + 4NH_3(aq) \rightleftharpoons [Cu(NH_3)_4]^{2+}(aq)$$

The dissociation of this complex ion is represented as

$$[Cu(NH_3)_4]^{2+}(aq) \rightleftharpoons Cu^{2+}(aq) + 4NH_3(aq)$$

The equilibrium constant for the dissociation of a complex ion is called its **dissociation constant, K_d**.

▶ As before, the outer brackets mean molar concentrations. The inner brackets are part of the formula of the complex ion.

$$K_d = \frac{[Cu^{2+}][NH_3]^4}{[[Cu(NH_3)_4]^{2+}]} = 8.5 \times 10^{-13}$$

Recall that $Cu^{2+}(aq)$ is really a hydrated ion, $[Cu(H_2O)_6]^{2+}$. The preceding reaction and its K_d expression are represented more accurately as

$$[Cu(NH_3)_4]^{2+} + 6H_2O \rightleftharpoons [Cu(H_2O)_6]^{2+} + 4NH_3$$

$$K_d = \frac{[[Cu(H_2O)_6]^{2+}][NH_3]^4}{[[Cu(NH_3)_4]^{2+}]} = 8.5 \times 10^{-13}$$

The more effectively a ligand competes with H_2O for a coordination site on the metal ions, the smaller K_d is. This tells us that in a comparison of complexes with the same number of ligands, the smaller the K_d value, the more stable the complex ion. Some complex ions and their dissociation constants, K_d, are listed in Appendix I.

Copper(II) hydroxide dissolves in an excess of aqueous NH_3 to form the deep-blue complex ion $[Cu(NH_3)_4]^{2+}$. This decreases the $[Cu^{2+}]$ so that $Q_{sp} = [Cu^{2+}][OH^-]^2 < K_{sp}$, and so the $Cu(OH)_2$ dissolves.

▶ For brevity, we often omit H_2O from formulas of hydrated ions. For example, we write $[Cu(H_2O)_6]^{2+}$ as Cu^{2+} or $Cu^{2+}(aq)$.

$$Cu(OH)_2(s) \rightleftharpoons Cu^{2+}(aq) \qquad + 2OH^-(aq)$$
$$Cu^{2+}(aq) + 4NH_3(aq) \rightleftharpoons [Cu(NH_3)_4]^{2+}(aq)$$

overall rxn: $\qquad \overline{Cu(OH)_2(s) + 4NH_3(aq) \rightleftharpoons [Cu(NH_3)_4]^{2+}(aq) \qquad + 2OH^-(aq)}$

Similarly, $Zn(OH)_2$ dissolves in excess NH_3 to form $[Zn(NH_3)_4]^{2+}$ ions.

$$Zn(OH)_2(s) + 4NH_3(aq) \rightleftharpoons [Zn(NH_3)_4]^{2+}(aq) + 2OH^-(aq)$$

Amphoteric hydroxides such as $Zn(OH)_2$ also dissolve in excess strong base by forming complex ions (Section 10-6).

$$Zn(OH)_2(s) + 2OH^-(aq) \rightleftharpoons [Zn(OH)_4]^{2-}(aq)$$

We see that we are able to shift equilibria (in this case, to dissolve $Zn(OH)_2$) by taking advantage of complex ion formation.

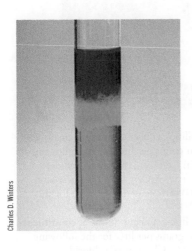

Charles D. Winters

Concentrated aqueous NH_3 was added *slowly* to a solution of copper(II) sulfate, $CuSO_4$. Unreacted blue copper(II) sulfate solution remains in the bottom part of the test tube. The light-blue precipitate in the middle is copper(II) hydroxide, $Cu(OH)_2$. The top layer contains deep-blue $[Cu(NH_3)_4]^{2+}$ ions that were formed as some $Cu(OH)_2$ dissolved in excess aqueous NH_3.

KEY TERMS

Common ion effect Suppression of ionization of a weak electrolyte by the presence in the same solution of a strong electrolyte containing one of the same ions as the weak electrolyte.

Complex ions Ions resulting from the formation of coordinate covalent bonds between simple cations and other ions or molecules (ligands).

Dissociation constant (K_d) The equilibrium constant that applies to the dissociation of a complex ion into a simple ion and coordinating species (ligands).

Fractional precipitation Removal of some ions from solution by precipitation while leaving other ions, with similar properties, in solution.

Molar solubility The number of moles of a solute that dissolve to produce a liter of saturated solution.

Precipitate A solid formed by mixing in solution the constituent ions of a slightly soluble compound.

Reaction quotient (in precipitation reactions) (Q_{sp}) The mass action expression that applies to the dissolution of a slightly soluble compound; this has the same algebraic form as that for K_{sp}, except that the concentrations are *not necessarily* equilibrium ones.

Solubility product constant (K_{sp}) The equilibrium constant that applies to the dissolution of a slightly soluble compound.

Solubility product principle The solubility product constant expression for a slightly soluble compound is the product of the concentrations of the constituent ions, each raised to the power that corresponds to the number of ions in one formula unit.

EXERCISES

 **Molecular Reasoning** exercises

▲ **More Challenging** exercises

Blue-Numbered exercises are solved in the Student Solutions Manual

Solubility Product

1. The solubility product constant values listed in Appendix H were determined at 25°C. How would those K_{sp} values change, if at all, with a change in temperature?
2. State the solubility product principle. What is its significance?
3. Why do we not include a term for the solid in a solubility product expression?
4. What do we mean when we refer to the molar solubility of a compound?
5. Write the solubility product expression for each of the following salts: (a) MgF_2; (b) $AlPO_4$; (c) $CuCO_3$; (d) Ag_3PO_4.
6. Write the solubility product expression for each of the following salts: (a) $Mn_3(AsO_4)_2$; (b) Hg_2I_2 [contains diatomic mercury(I) ions, Hg_2^{2+}]; (c) AuI_3; (d) $SrCO_3$.
7.  ▲ The K_{sp} value for $BaSO_4$ is calculated from the expression $K_{sp} = [Ba^{2+}][SO_4^{2-}]$, whereas the K_{sp} value for $Mg(OH)_2$ is calculated from the expression, $K_{sp} = [Mg^{2+}][OH^-]^2$. Explain why the hydroxide ion concentration is squared, but none of the other concentrations are squared.

Experimental Determination of K_{sp}

Values of K_{sp} calculated from the solubility data in these exercises may not agree exactly with the solubility products given in Appendix H because of rounding differences.

8. From the solubility data given for the following compounds, calculate their solubility product constants.
   (a) $SrCrO_4$, strontium chromate, 1.2 mg/mL
   (b) BiI_3, bismuth iodide, 7.7×10^{-3} g/L
   (c) $Fe(OH)_2$, iron(II) hydroxide, 1.1×10^{-3} g/L
   (d) SnI_2, tin(II) iodide, 10.9 g/L
9. From the solubility data given for the following compounds, calculate their solubility product constants.
   (a) CuBr, copper(I) bromide, 1.0×10^{-3} g/L
   (b) AgI, silver iodide, 2.8×10^{-8} g/10 mL
   (c) $Pb_3(PO_4)_2$, lead(II) phosphate, 6.2×10^{-7} g/L
   (d) Ag_2SO_4, silver sulfate, 5.0 mg/mL
10. Construct a table like Table 20-1 for the compounds listed in Exercise 8. Which compound has (a) the highest molar solubility, (b) the lowest molar solubility, (c) the largest K_{sp}, (d) the smallest K_{sp}?
11. Construct a table like Table 20-1 for the compounds listed in Exercise 9. Which compound has (a) the highest molar solubility, (b) the lowest molar solubility, (c) the largest K_{sp}, (d) the smallest K_{sp}?
12. A solution is produced by stirring 1.1000 gram of calcium fluoride in 1.00 liter of water at 25°C. Careful analysis shows that 0.0163 grams of calcium fluoride has dissolved. Calculate the K_{sp} for calcium fluoride based on these data.

13. ▲ Calculate the K_{sp} for zinc phosphate if 1.18×10^{-4} grams of zinc phosphate dissolved to make 2.5 liters of a saturated solution.

Uses of Solubility Product Constants

14. Seashells are calcium carbonate with traces of colored impurities. The solubility product for calcium carbonate is 4.8×10^{-9}. What is the solubility of calcium carbonate in grams per liter of solution?

Charles D. Winters

15. Calculate molar solubilities, concentrations of constituent ions, and solubilities in grams per liter for the following compounds at 25°C: (a) $Cd(CN)_2$, cadmium cyanide; (b) PbI_2, lead iodide; (c) $Sr_3(AsO_4)_2$, strontium arsenate; (d) Hg_2CO_3, mercury(I) carbonate [the formula for the mercury(I) ion is Hg_2^{2+}].
16. Calculate molar solubilities, concentrations of constituent ions, and solubilities in grams per liter for the following compounds at 25°C: (a) CuCl, copper(I) chloride; (b) $Ba_3(PO_4)_2$, barium phosphate; (c) PbF_2, lead(II) fluoride; (d) $Sr_3(PO_4)_2$, strontium phosphate.
17. What is the concentration of lead ions in one liter of saturated $PbCrO_4$ solution?
18. Barium sulfate is used to produce distinct X-rays of the gastrointestinal tract. What is the maximum mass of barium sulfate that can dissolve in 5.00 liters of water (a volume much greater than that of the average gastrointestinal tract)?
19. Calculate the molar solubility of CuBr in 0.010 M KBr solution.
20. Calculate the molar solubility of Ag_2SO_4 in 0.12 M K_2SO_4 solution.
21. Construct a table similar to Table 20-1 for the compounds listed in Exercise 15. Which compound has (a) the highest molar solubility; (b) the lowest molar solubility; (c) the

 **Molecular Reasoning** exercises   ▲ **More Challenging** exercises   Blue-Numbered exercises solved in Student Solutions Manual

Unless otherwise noted, all content on this page is © Cengage Learning.

highest solubility, expressed in grams per liter; (d) the lowest solubility, expressed in grams per liter?

22. Construct a table similar to Table 20-1 for the compounds listed in Exercise 16. Which compound has (a) the highest molar solubility; (b) the lowest molar solubility; (c) the highest solubility, expressed in grams per liter; (d) the lowest solubility, expressed in grams per liter?

23. Of the three compounds $CuCO_3$, $Ca(OH)_2$, and Ag_2CrO_4, which has (a) the highest molar solubility; (b) the lowest molar solubility; (c) the highest solubility, expressed in grams per liter; (d) the lowest solubility, expressed in grams per liter?

24. Of the three compounds, Ag_2CO_3, $AgCl$, and $Pb(OH)_2$, which has (a) the highest molar solubility; (b) the lowest molar solubility; (c) the highest solubility, expressed in grams per liter; (d) the lowest solubility, expressed in grams per liter?

25. What volume of water is required to dissolve 7.5 grams of copper(II) carbonate, $CuCO_3$?

26. Which has the greater molar solubility in 0.125 M K_2CrO_4 solution: $BaCrO_4$ or Ag_2CrO_4?

27. Will a precipitate form when 1.00 g of $AgNO_3$ is added to 50.0 mL of 0.050 M NaCl? If so, would you expect the precipitate to be visible?

28. Will a precipitate of $PbCl_2$ form when 5.0 g of solid $Pb(NO_3)_2$ is added to 1.00 L of 0.010 M NaCl? Assume that volume change is negligible.

29. Sodium bromide and lead nitrate are soluble in water. Will lead bromide precipitate when 1.03 g of NaBr and 0.332 g of $Pb(NO_3)_2$ are dissolved in sufficient water to make 1.00 L of solution?

30. Will a precipitate of $Cu(OH)_2$ form when 10.0 mL of 0.010 M NaOH is added to 1.00 L of 0.010 M $CuCl_2$?

31. A solution is 0.0100 M in Pb^{2+} ions. If 0.103 mol of solid Na_2SO_4 is added to 1.00 L of this solution (with negligible volume change), what percentage of the Pb^{2+} ions remain in solution?

32. A solution is 0.0100 M in Pb^{2+} ions. If 0.103 mol of solid NaI is added to 1.00 L of this solution (with negligible volume change), what percentage of the Pb^{2+} ions remain in solution?

33. ▲ A solution is 0.0100 M in $Ba(NO_3)_2$. If 0.103 mol of solid Na_3PO_4 is added to 1.00 L of this solution (with negligible volume change), what percentage of the Ba^{2+} ions remain in solution?

Fractional Precipitation

34. 🌺 What is fractional precipitation?

35. ▲ Solid Na_2SO_4 is added slowly to a solution that is 0.10 M in $Pb(NO_3)_2$ and 0.10 M in $Ba(NO_3)_2$. In what order will solid $PbSO_4$ and $BaSO_4$ form? Calculate the percentage of Ba^{2+} that precipitates just before $PbSO_4$ begins to precipitate.

36. ▲ To a solution that is 0.010 M in Cu^+, 0.010 M in Ag^+, and 0.010 M in Au^+, *solid* NaBr is added slowly. Assume that there is no volume change due to the addition of solid NaBr. (a) Which compound will begin to precipitate first? (b) Calculate $[Au^+]$ when AgBr just begins to precipitate.

What percentage of the Au^+ has precipitated at this point? (c) Calculate $[Au^+]$ and $[Ag^+]$ when CuBr just begins to precipitate.

37. ▲ A solution is 0.020 M in Pb^{2+} and 0.020 M in Ag^+. As Cl^- is introduced to the solution by the addition of solid NaCl, determine (a) which substance will precipitate first, AgCl or $PbCl_2$, and (b) the fraction of the metal ion in the first precipitate that remains in solution at the moment the precipitation of the second compound begins.

38. A solution is 0.050 M in K_2SO_4 and 0.050 M in K_2CrO_4. Solid $Pb(NO_3)_2$ is added slowly without changing the volume appreciably. (a) Which salt, $PbSO_4$ or $PbCrO_4$, will precipitate first? (b) What is $[Pb^{2+}]$ when the salt in part (a) begins to precipitate? (c) What is $[Pb^{2+}]$ when the other lead salt begins to precipitate? (d) What are $[SO_4^{2-}]$ and $[CrO_4^{2-}]$ when the lead salt in part (c) begins to precipitate?

39. Solid $Pb(NO_3)_2$ is added slowly to a solution that is 0.015 M each in NaOH, K_2CO_3, and Na_2SO_4. (a) In what order will solid $Pb(OH)_2$, $PbCO_3$, and $PbSO_4$ begin to precipitate? (b) Calculate the percentages of OH^- and CO_3^{2-} that have precipitated when $PbSO_4$ begins to precipitate.

40. Suppose you have three beakers that contain, respectively, 100 mL of each of the following solutions: (i) 0.0015 M KOH, (ii) 0.0015 M K_2CO_3, (iii) 0.0015 M KCN. (a) If solid zinc nitrate, $Zn(NO_3)_2$, were added slowly to each beaker, what concentration of Zn^{2+} would be required to initiate precipitation? (b) If solid zinc nitrate were added to each beaker until $[Zn^{2+}] = 0.0015$ M, what concentrations of OH^-, CO_3^{2-}, and CN^- would remain in solution, that is, unprecipitated? Neglect any volume change when solid is added.

41. ▲ Suppose you have three beakers that contain, respectively, 100 mL each of the following solutions: (i) 0.0015 M KOH, (ii) 0.0015 M K_2CO_3, (iii) 0.0015 M KI. (a) If solid lead nitrate, $Pb(NO_3)_2$, were added slowly to each beaker, what concentration of Pb^{2+} would be required to initiate precipitation? (b) If solid lead nitrate were added to each beaker until $[Pb^{2+}] = 0.0015$ M, what concentrations of OH^-, CO_3^{2-}, and I^- would remain in solution, that is, unprecipitated? Neglect any volume change when solid is added.

Simultaneous Equilibria

42. ▲ If a solution is made 0.080 M in $Mg(NO_3)_2$, 0.075 M in aqueous ammonia, and 3.5 M in NH_4NO_3, will $Mg(OH)_2$ precipitate? What is the pH of this solution?

43. ▲ If a solution is made 0.090 M in $Mg(NO_3)_2$, 0.090 M in aqueous ammonia, and 0.080 M in NH_4NO_3, will $Mg(OH)_2$ precipitate? What is the pH of this solution?

44. ▲ Calculate the solubility of CaF_2 in a solution that is buffered at $[H^+] = 0.0050$ M with $[HF] = 0.10$ M.

45. ▲ Calculate the solubility of AgCN in a solution that is buffered at $[H^+] = 0.000200$ M, with $[HCN] = 0.01$ M.

🌺 **Molecular Reasoning** exercises   ▲ **More Challenging** exercises   Blue-Numbered exercises solved in Student Solutions Manual

Unless otherwise noted, all content on this page is © Cengage Learning.

46. If a solution is 2.0×10^{-5} M in $Mn(NO_3)_2$ and 1.0×10^{-3} M in aqueous ammonia, will $Mn(OH)_2$ precipitate?

47. ▲ If a solution is 0.040 M in manganese(II) nitrate, $Mn(NO_3)_2$, and 0.080 M in aqueous ammonia, will manganese(II) hydroxide, $Mn(OH)_2$, precipitate?

48. Milk of magnesia is a suspension of the slightly soluble compound $Mg(OH)_2$ in water. (a) What is the molar solubility of $Mg(OH)_2$ in a 0.015 M NaOH solution? (b) What is the molar solubility of $Mg(OH)_2$ in a 0.015 M $MgCl_2$ solution?

Milk of magnesia

Charles D. Winters

49. How many moles of $Cr(OH)_3$ will dissolve in 555 mL of a solution with a pH of 5.00?

50. Determine whether a precipitate forms when a 0.00050 M solution of magnesium nitrate is brought to a pH of 8.70.

51. ▲ What concentration of NH_4NO_3 is necessary to prevent precipitation of $Mn(OH)_2$ in the solution of Exercise 47?

52. (a) What is the pH of a saturated solution of $Fe(OH)_2$? (b) What is the solubility in grams of $Fe(OH)_2$/100. mL of solution?

53. (a) What is the pH of a saturated solution of $Cu(OH)_2$? (b) What is the solubility in grams of $Cu(OH)_2$/100. mL of solution?

54. What is the maximum concentration of Zn^{2+} in a solution of pH 10.00?

55. A solution contains the metal ions listed below, all at the same concentration. The addition of thioacetamide, CH_3CSNH_2, produces sulfide ion, S^{2-}, causing the ions to precipitate as the sulfides. Each sulfide begins to precipitate at a different concentration of sulfide ion. As the concentration of sulfide ion increases (starting from 0), in what order will the metal sulfides precipitate? (Which one will start to precipitate first, second, third, and fourth?)
iron(II) ion, Fe^{2+} $K_{sp} = 4.9 \times 10^{-18}$ for FeS
lead(II) ion, Pb^{2+} $K_{sp} = 8.4 \times 10^{-28}$ for PbS
mercury(II) ion, Hg^{2+} $K_{sp} = 3.0 \times 10^{-53}$ for HgS
tin(II) ion, Sn^{2+} $K_{sp} = 1.0 \times 10^{-28}$ for SnS

56. If a solution is 4.3×10^{-4} M in magnesium nitrate, $Mg(NO_3)_2$, and 5.2×10^{-2} M in aqueous ammonia, NH_3, will $Mg(OH)_2$ precipitate? ($K_{sp} = 1.5 \times 10^{-11}$ for $Mg(OH)_2$, $K_b = 1.8 \times 10^{-5}$ for NH_3)

Dissolution of Precipitates and Complex Ion Formation

57. You will often work with salts of Fe^{3+}, Pb^{2+}, and Al^{3+} in the laboratory. (All are found in nature, and all are important economically.) If you have a solution containing these three ions, each at a concentration of 0.10 M, what is the order in which their hydroxides precipitate as aqueous NaOH is slowly added to the solution?

58. ☁ Explain, by writing appropriate equations, how the following insoluble compounds can be dissolved by the addition of a solution of nitric acid. (Carbonates dissolve in strong acids to form water and gaseous carbon dioxide.) What is the "driving force" for each reaction? (a) $Cu(OH)_2$; (b) $Sn(OH)_4$; (c) $ZnCO_3$; (d) $(PbOH)_2CO_3$.

59. ☁ Explain, by writing equations, how the following insoluble compounds can be dissolved by the addition of a solution of ammonium nitrate or ammonium chloride. (a) $Mg(OH)_2$; (b) $Mn(OH)_2$; (c) $Ni(OH)_2$.

60. ☁ The following insoluble sulfides can be dissolved in 3.0 M hydrochloric acid. Explain how this is possible, and write the appropriate equations. (a) MnS; (b) CuS.

61. ☁ The following sulfides are less soluble than those listed in Exercise 60 and can be dissolved in hot 6.0 M nitric acid, an oxidizing acid. Explain how, and write the appropriate balanced equations. (a) PbS; (b) CuS; (c) Bi_2S_3.

62. ☁ Why would MnS be expected to be more soluble in 0.10 M HCl solution than in water? Would the same be true for $Mn(NO_3)_2$?

63. ☁ ▲ For each pair, choose the salt that would be expected to be more soluble in acidic solution than in pure water, and justify your choice. (a) $Hg_2(CH_3COO)_2$ or Hg_2Br_2; (b) $Pb(OH)_2$ or PbI_2; (c) AgI or $AgNO_2$.

Mixed Exercises

64. We mix 25.0 mL of a 0.0030 M solution of $BaCl_2$ and 50.0 mL of a 0.050 M solution of NaF. (a) Find $[Ba^{2+}]$ and $[F^-]$ in the mixed solution at the instant of mixing (before any possible reaction occurs). (b) Would BaF_2 precipitate?

65. ▲ A concentrated, strong acid is added to a solid mixture of 0.025-mol samples of $Fe(OH)_2$ and $Cu(OH)_2$ placed in 1.0 L of water. At what values of pH will the dissolution of each hydroxide be complete? (Assume negligible volume change.)

66. A solution is 0.015 M in I^- ions and 0.015 M in Br^- ions. Ag^+ ions are introduced to the solution by the addition of solid $AgNO_3$. Determine (a) which compound will precipitate first, AgI or AgBr, and (b) the percentage of the halide ion in the first precipitate that is removed from solution before precipitation of the second compound begins.

67. Calculate the molar solubility of Ag_2SO_4 (a) in pure water, (b) in 0.010 M $AgNO_3$, and (c) in 0.010 M K_2SO_4.

68. Barium and all of its compounds are very toxic. For example, the estimated fatal dose of barium chloride ($BaCl_2$, solubility 375 g/L) is about 1 gram for a 70-kg human. Yet the barium contrast agent used in abdominal studies contains 450 grams of barium sulfate ($BaSO_4$, solubility 0.00246 g/L). Why would the use of barium

---

☁ **Molecular Reasoning** exercises   ▲ **More Challenging** exercises   Blue-Numbered exercises solved in Student Solutions Manual

carbonate ($BaCO_3$, solubility 0.02 g/L) as a contrast agent be tragic?

69. Solid sodium carbonate, Na_2CO_3, is added to a solution that is 0.0326 M in strontium nitrate, $Sr(NO_3)_2$, and 0.0270 M in barium nitrate, $Ba(NO_3)_2$. Calculate the percentage of strontium that has precipitated as strontium carbonate, $SrCO_3$, just before barium carbonate, $BaCO_3$, begins to precipitate. (K_{sp}: $SrCO_3 = 9.4 \times 10^{-10}$, $BaCO_3 = 8.1 \times 10^{-9}$)

Conceptual Exercises

70. In each of the following cases, decide whether a precipitate will form when mixing the indicated reagents, and write a balanced equation for the reaction.
    (a) $Na_2SO_4(aq) + Mg(NO_3)_2(aq)$
    (b) $K_3PO_4(aq) + FeCl_3(aq)$

71. ☁ (a) Are "insoluble" substances really insoluble? (b) What do we mean when we refer to insoluble substances?

72. ☁ Solubility product calculations are actually based on heterogeneous equilibria. Why are pure solids and liquids exempted from these calculations?

73. ☁ Draw a picture of a portion of a saturated silver chloride solution at the molecular level. Show a small amount of solid plus some dissociated ions. You need not show water or waters of hydration. Prepare a second drawing that includes the same volume of solution, but twice as much solid. Should your drawing include more, fewer, or the same number of silver ions?

74. ▲ The solubility product constants of silver chloride, $AgCl$, and silver chromate, Ag_2CrO_4, are 1.8×10^{-10} and 9.0×10^{-12}, respectively. Suppose that the chloride, $Cl^-(aq)$, and chromate, $CrO_4^{2-}(aq)$, ions are both present in the same solution at concentrations of 0.010 M each. A standard solution of silver ions, $Ag^+(aq)$, is dispensed slowly from a buret into this solution while it is stirred vigorously. Solid silver chloride is white, and silver chromate is red. What will be the concentration of $Cl^-(aq)$ ions in the mixture when the first tint of red color appears in the mixture?

75. Calculate the solubility of silver bromide, $AgBr$, in moles per liter, in pure water. Compare this value with the molar solubility of $AgBr$ in 225 mL of water to which 0.15 g of $NaBr$ has been added.

Building Your Knowledge

76. In the calculation of the molar solubility of a compound such as lead(II) chloride, $PbCl_2$, given its solubility product constant, K_{sp}, the concentration of the chloride ion, Cl^-, is exactly twice the concentration of the lead(II) ion, Pb^{2+}. In the solubility product constant expression, this "doubled" concentration is also squared. However, when lead(II) chloride is precipitated by the combining of solutions of lead(II) nitrate, $Pb(NO_3)_2$, and potassium chloride, KCl, the concentration of the chloride ion is not doubled before it is squared. Explain this difference.

77. The solubility product principle is generally valid only for saturated solutions in which the total concentration of ions of the slightly soluble compound is no more than about 0.01 M. Compounds with a K_{sp} less than 10^{-40} have extremely low solubility. Examine the table of solubility product constants in Appendix H. Which compound appears to be the most soluble? Calculate its molar solubility. Which compound appears to be the least soluble? Calculate its molar solubility.

78. A 1.00-L solution contains 0.010 M F^- and 0.010 M SO_4^{2-}. Solid barium nitrate is slowly added to the solution. (a) Calculate the $[Ba^{2+}]$ when $BaSO_4$ begins to precipitate. (b) Calculate the $[Ba^{2+}]$ when BaF_2 starts to precipitate. Assume no volume change occurs. K_{sp} values: $BaSO_4 = 1.1 \times 10^{-11}$, $BaF_2 = 1.7 \times 10^{-6}$.

79. How many moles of CO_3^{2-} must be added to 0.75 liter of a 0.10 M Sr^{2+} solution to produce a solution that is 1.0×10^{-6} M Sr^{2+}? How many moles of CO_3^{2-} are in the final solution, and how many moles of CO_3^{2-} are in the precipitate formed? Assume no volume change for the solution.

80. A fluoridated water supply contains 1 mg/L of F^-. What is the maximum amount of Ca^{2+}, expressed in grams per liter, that can exist in this water supply?

81. Many industrial operations require very large amounts of water as a coolant in heat exchange processes. Muddy or cloudy water is usually unsatisfactory because the dispersed solids may clog filters or deposit sediment in pipes and pumps. Murky water can be clarified on a large scale by adding agents to coagulate colloidal material, and then allowing the resulting solid to settle out in holding tanks or ponds before the clarified water is sent to plant intakes. Recent methods employ the addition of both calcium hydroxide and magnesium carbonate. If 56 g of $Ca(OH)_2$ and 45 g of $MgCO_3$ were added to 520. liters of water, would these compounds form a precipitate of calcium carbonate?

82. Magnesium carbonate is used in the manufacture of a high-density *magnesite* brick. This material is not well suited to general exterior use because the magnesium carbonate easily erodes. What percentage of 28 grams of surface-exposed $MgCO_3$ would be lost through the solvent action of 15 liters of water? Assume sufficient contact time for the water to become saturated with $MgCO_3$.

Beyond the Textbook

NOTE: *Whenever the answer to an exercise depends on information obtained from a source other than this textbook, the source must be included as an essential part of the answer.*

83. Search the web for an application of a relatively insoluble salt as a soil additive.

84. Use the *Handbook of Chemistry and Physics* or other suitable reference to find solubility data for inorganic compounds. What are some qualitative rules that relate to solubilities?

☁ **Molecular Reasoning** exercises   ▲ **More Challenging** exercises   Blue-Numbered exercises solved in Student Solutions Manual

Unless otherwise noted, all content on this page is © Cengage Learning.

85. Use the *Handbook of Chemistry and Physics* or other suitable reference to find solubility data for inorganic compounds. List five nickel salts that are listed as being insoluble in water but soluble in acid. Write the net ionic equation for the dissolution of each salt in a strong acid solution.

86. ▲ Suppose you had 1.00 L of the mixed halide solution described in Example 20-8. (a) Calculate the mass of solid $AgNO_3$ that you would need to add to the solution to accomplish the desired separations. (b) The minimum mass that can be weighed with a typical electronic balance is about 0.0001 g. Do you suppose that this would be a practical method to carry out this separation? (c) Can you suggest a more practical approach to this fractional precipitation? Show calculations to assess the practicality of your suggested approach.

87. The 23-foot Great Stalactite in Doolin Cave, County Clare, Ireland (below) has an estimated weight of 10 metric tons (1 metric ton = 1000 kg). Assume that this weight is entirely calcite, $CaCO_3$, which has a cold water

solubility of 0.0014 g/100 mL, and that the stalactite was formed by simple precipitation from a saturated solution. (a) How many liters of $CaCO_3$-saturated water deposited calcite from the cave ceiling to produce this massive formation? (b) From these data, what are the solubility product constant and the molar solubility for calcite?

88. Color change indicators have other uses than to monitor pH. Water-sensitive laboratory chemicals are often stored in the presence of the dessicant "Drierite" (white $CaSO_4(s)$, which attracts atmospheric water to form white $CaSO_4 \cdot 2H_2O$). "Indicating Drierite" also contains a small amount of colored $CoCl_2(s)$, which changes color upon hydration and signals that the storage atmosphere is no longer dry. Look up $CoCl_2$ and its hydrated salts $CoCl_2 \cdot 2H_2O(s)$ and $CoCl_2 \cdot 6H_2O(s)$ in the *CRC Handbook of Chemistry and Physics* or in various sources on the internet. (a) What colors are the "dry" (anhydrous) and "wet" forms of this salt? (b) Is the material in the photo below ready to use as a dessicant?

Courtesy of Doolin Cave

Doolin Cave

Courtesy of W. A. Hammond Drierite Co.

Commercially available indicating Drierite™

🐾 **Molecular Reasoning** exercises   ▲ **More Challenging** exercises   Blue-Numbered exercises solved in Student Solutions Manual

Unless otherwise noted, all content on this page is © Cengage Learning.

Electrochemistry

21

Nellis Air Force Base in Nevada has a giant solar panel power array that generates about 14 megawatts of power for the base. This provides about 25% of the electrical power needs for the base.

U.S. Air Force photo by Airman 1st Class Nadine Y. Barclay

OBJECTIVES

After you have studied this chapter, you should be able to

▶ Use the terminology of electrochemistry (terms such as "cell," "electrode," "cathode," "anode")

▶ Describe the differences between electrolytic cells and voltaic (galvanic) cells

▶ Recognize oxidation and reduction half-reactions, and know at which electrode each occurs for electrolytic cells and for voltaic cells

▶ Write half-reactions and overall cell reactions for electrolysis processes

▶ Use Faraday's Law of Electrolysis to calculate amounts of products formed, amounts of current passed, time elapsed, and oxidation state

▶ Describe the refining and plating of metals by electrolytic methods

▶ Describe the construction of simple voltaic cells from half-cells and a salt bridge, and understand the function of each component

▶ Write half-reactions and overall cell reactions for voltaic cells

▶ Write and interpret the shorthand notation for voltaic cells

▶ Compare various voltaic cells to determine the relative strengths of oxidizing and reducing agents

▶ Interpret standard reduction potentials

▶ Use standard reduction potentials, E^0, to calculate the potential of a standard voltaic cell, E^0_{cell}

▶ Use standard reduction potentials to identify the cathode and the anode in a standard cell

▶ Use standard reduction potentials to predict the spontaneity of a redox reaction

▶ Use standard reduction potentials to identify oxidizing and reducing agents in a cell or in a redox reaction

▶ Describe some corrosion processes and some methods for preventing corrosion

▶ Use the Nernst equation to relate electrode potentials and cell potentials to different concentrations and partial pressures

▶ Relate the standard cell potential (E^0_{cell}) to the standard Gibbs free energy change (ΔG^0) and the equilibrium constant (K_{eq})

▶ Distinguish between primary and secondary voltaic cells

▶ Describe the compositions and reactions of some useful primary and secondary cells (batteries)

▶ Describe the electrochemical processes involved in discharging and recharging a lead storage (automobile) battery

Electrochemistry deals with the chemical changes produced by electric current and with the production of electricity by chemical reactions. Many metals are purified or are plated onto jewelry by electrochemical methods. Digital watches, automobile starters, calculators, and pacemakers are just a few devices that depend on electrochemically produced power via batteries. Corrosion of metals is also an electrochemical process.

We learn much about chemical reactions from the study of electrochemistry. The amount of electrical energy consumed or produced can be measured quite accurately. All electrochemical reactions involve the transfer of electrons and are therefore *oxidation–reduction* reactions. The sites of oxidation and reduction are separated physically so that oxidation occurs at one location, and reduction occurs at the other. Electrochemical processes require some method of introducing a stream of electrons into a reacting chemical system and some means of withdrawing electrons. In most applications the reacting system is contained in a **cell**, and an electric current enters or exits by **electrodes**.

We classify electrochemical cells into two types.

1. **Electrolytic cells** are those in which electrical energy from an external source causes *nonspontaneous* chemical reactions to occur.

2. **Voltaic cells** are those in which *spontaneous* chemical reactions produce electricity and supply it to an external circuit.

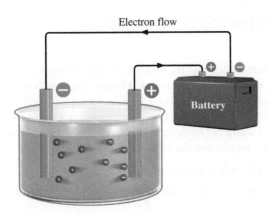

Electron flow

Battery

Figure 21-1 The motion of ions through a solution is an electric current. This accounts for ionic (electrolytic) conduction. Positively charged ions migrate toward the negative electrode, and negatively charged ions migrate toward the positive electrode. Here the rate of migration is greatly exaggerated for clarity. The ionic velocities are actually only slightly greater than random molecular speeds.

We will discuss several kinds of electrochemical cells. From experimental observations we deduce what is happening at each electrode and for the overall reaction chemistry. We then construct simplified diagrams of the cells.

21-1 Electrical Conduction

Electric current represents transfer of charge. Charge can be conducted through metals and through pure liquid electrolytes (that is, molten salts) or solutions containing electrolytes. The former type of conduction is called **metallic conduction**. It involves the flow of electrons with no similar movement of the atoms of the metal and no obvious changes in the metal (see Section 13-17). **Ionic**, or **electrolytic**, **conduction** is the conduction of electric current by the motion of ions through a solution or a pure liquid. Positively charged ions migrate toward the negative electrode, while negatively charged ions move toward the positive electrode. Both kinds of conduction, ionic and metallic, occur in electrochemical cells (Figure 21-1).

▶ A few organic salts are liquids at room temperature. These are called ionic liquids.

21-2 Electrodes

Electrodes are surfaces on which oxidation or reduction half-reactions occur. They may or may not participate in the reactions. Those that do not react are called **inert electrodes**. Regardless of the kind of cell, electrolytic or voltaic, the electrodes are identified as follows.

> The **cathode** is defined as the electrode at which *reduction* occurs as electrons are gained by some species. The **anode** is the electrode at which *oxidation* occurs as electrons are lost by some species.

Each of these can be either the positive or the negative electrode depending on the cell type.

Electrolytic Cells

In some electrochemical cells, *nonspontaneous* (reactant-favored) chemical reactions are forced to occur by the input of electrical energy. This process is called **electrolysis**. An electrolytic cell consists of a container for the reaction material with electrodes immersed in the reaction material and connected to a source of direct current. Inert electrodes are often used so that they do not become involved with the reaction.

S TOP & THINK
A simple way to remember the name of the electrode that is doing the reduction is to think about a "red cat"— that is, *reduction* always occurs at the *cathode*.

▶ *Lysis* means "splitting apart." In many electrolytic cells, compounds are split into their constituent elements, for example, the splitting of water, H_2O into H_2 and O_2 gases.

21-3 The Electrolysis of Molten Sodium Chloride (The Downs Cell)

▶ *Molten* NaCl, melting point = 801°C, is a clear, colorless liquid that looks like water.

Solid sodium chloride does not conduct electricity. Its ions vibrate about fixed positions, but they are not free to move throughout the crystal. Molten (melted) NaCl, however, is an excellent conductor because its ions are freely mobile. Consider a cell in which a source of direct current is connected by wires to two inert graphite electrodes (Figure 21-2a). They are immersed in a container of molten sodium chloride. When the current with a high enough voltage flows, we observe the following.

1. A pale green gas, which is chlorine, Cl_2, is liberated at one electrode.

▶ The Na metal remains liquid because its melting point is only 97.8°C. It floats because it is less dense than the molten NaCl.

2. Molten, silvery-white metallic sodium, Na, forms at the other electrode and floats on top of the molten sodium chloride.

From these observations we can deduce the processes of the cell. Chlorine must be produced by oxidation of Cl^- ions, and the electrode at which this happens must be the anode. Metallic sodium is produced by reduction of Na^+ ions at the cathode, where electrons are being forced into the cell.

▶ In this chapter, as in Chapters 6 and 11, we often use red type to emphasize reduction and blue type to emphasize oxidation.

$$2Cl^- \longrightarrow Cl_2(g) + 2e^- \qquad \text{(oxidation, anode half-reaction)}$$
$$\underline{2[Na^+ + e^- \longrightarrow Na(\ell)] \qquad \text{(reduction, cathode half-reaction)}}$$
$$\underbrace{2Na^+ + 2Cl^-}_{2NaCl(\ell)} \longrightarrow 2Na(\ell) + Cl_2(g) \quad \text{(overall cell reaction)}$$

The formation of metallic Na and gaseous Cl_2 from NaCl is *nonspontaneous* except at temperatures very much higher than 801°C. The direct current (dc) source must supply electrical energy to force this reaction to occur. Electrons are used in the cathode half-reaction (reduction) and produced in the anode half-reaction (oxidation). They therefore

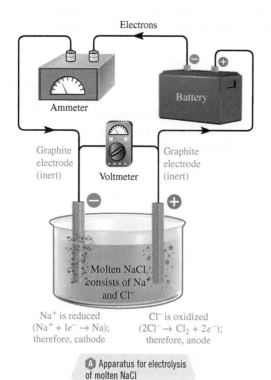

Na^+ is reduced
$(Na^+ + 1e^- \rightarrow Na)$;
therefore, cathode

Cl^- is oxidized
$(2Cl^- \rightarrow Cl_2 + 2e^-)$;
therefore, anode

Ⓐ Apparatus for electrolysis of molten NaCl

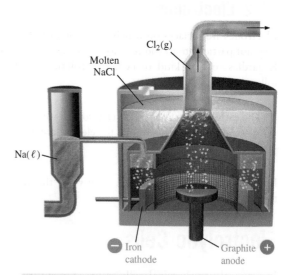

Ⓑ The Downs cell in which molten NaCl is commercially electrolyzed to Na(ℓ) and Cl_2(g). The liquid Na floats on the denser molten NaCl.

Figure 21-2 Electrolysis of molten sodium chloride, NaCl, to produce Na(ℓ) and Cl_2(g).

travel through the wire from *anode* to *cathode*. The dc source forces electrons to flow non-spontaneously from the positive electrode to the negative electrode. The anode is the positive electrode and the cathode the negative electrode *in all electrolytic cells*. Figure 21-2a is a simplified diagram of the cell.

Sodium and chlorine must not be allowed to come in contact with each other because they react spontaneously, rapidly, and explosively to form sodium chloride. Figure 21-2b shows the **Downs cell** that is used for the industrial electrolysis of sodium chloride. The Downs cell is expensive to run, mainly because of the cost of construction, the cost of the electricity, and the cost of heating the NaCl to melt it. Nevertheless, electrolysis of a molten sodium salt is the most practical means by which metallic Na can be obtained, owing to its extremely high reactivity. Once liberated by the electrolysis, the liquid Na metal is drained off, cooled, and cast into blocks. These must be stored in an inert environment (e.g., in mineral oil) to prevent reaction with atmospheric O_2 or water.

Electrolysis of NaCl in the Downs cell is the main commercial method of producing metallic sodium. The Cl_2 gas produced in the Downs cell is cooled, compressed, and marketed. This partially offsets the expense of producing metallic sodium, but most chlorine is produced by the cheaper electrolysis of aqueous NaCl.

21-4 The Electrolysis of Aqueous Sodium Chloride

Consider the electrolysis of a moderately concentrated solution of NaCl in water, using inert electrodes. The following experimental observations are made when a sufficiently high voltage is applied across the electrodes of a suitable cell.

1. H_2 gas is liberated at one electrode. The solution becomes basic in that vicinity.
2. Cl_2 gas is liberated at the other electrode.

Chloride ions are obviously being oxidized to Cl_2 in this cell, as they were in the electrolysis of molten NaCl. But Na^+ ions are not reduced to metallic Na. Instead, gaseous H_2 and aqueous OH^- ions are produced by reduction of H_2O molecules at the cathode. Water is more easily reduced than Na^+ ions. This is primarily because the reduction of Na^+ would produce the very active metal Na, whereas the reduction of H_2O produces the more stable products $H_2(g)$ and $OH^-(aq)$. The active metals Li, K, Ca, and Na (see Table 6-9) all react with water to produce $H_2(g)$ and the corresponding metal hydroxide [LiOH, KOH, $Ca(OH)_2$, and NaOH], so we do not expect these metals to be produced in aqueous solution. Later in this chapter (Section 21-14), we learn the quantitative basis for predicting which of several possible oxidations or reductions is favored. The half-reactions and overall cell reaction for this electrolysis are

$$2Cl^- \longrightarrow Cl_2 + 2e^- \qquad \text{(oxidation, anode)}$$
$$2H_2O + 2e^- \longrightarrow 2OH^- + H_2 \qquad \text{(reduction, cathode)}$$

$$\overline{2H_2O + 2Cl^- \longrightarrow 2OH^- + H_2 + Cl_2} \quad \begin{array}{l}\text{(overall cell reaction as} \\ \text{net ionic equation)}\end{array}$$

$$\underline{+ 2Na^+ \longrightarrow + 2Na^+} \qquad \text{(spectator ions)}$$

$$2H_2O + 2NaCl \longrightarrow 2NaOH + H_2 + Cl_2 \quad \begin{array}{l}\text{(overall cell reaction as} \\ \text{formula unit equation)}\end{array}$$

The cell is illustrated in Figure 21-3. As before, the battery forces the electrons to flow from the anode (+) through the wire to the cathode (−).

The overall cell reaction produces gaseous H_2 and Cl_2 and an aqueous solution of NaOH, called caustic soda. Solid NaOH is then obtained by evaporation of the solution. This is the most important commercial preparation of each of these substances. It is much less expensive than the electrolysis of molten NaCl because it is not necessary to heat the solution to the high temperatures needed to melt NaCl.

▶ The direction of *spontaneous* flow for negatively charged particles is from negative to positive.

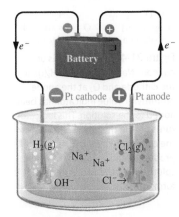

$2H_2O + 2e^- \rightarrow$ $\qquad$ $2Cl^- \rightarrow Cl_2(g) + 2e^-$
$H_2(g) + 2OH^-$ $\qquad\qquad$ Oxidation
Reduction

Figure 21-3 Electrolysis of aqueous NaCl solution. Although several reactions occur at both the anode and the cathode, the net result is the production of $H_2(g)$ and NaOH at the cathode and $Cl_2(g)$ at the anode. A few drops of phenolphthalein indicator were added to the solution. The solution turns pink at the cathode, where OH^- ions are formed and the solution becomes basic.

▶ We will omit the notation that indicates states of substances—(s), (ℓ), (g), and (aq)—except where states are not obvious. This is a common shortcut when writing equations.

▶ Not surprisingly, the fluctuations in commercial prices of these widely used industrial products—H_2, Cl_2, and NaOH—have often paralleled one another.

$$2(2H_2O + 2e^- \rightarrow$$
$$H_2(g) + 2OH^-)$$
Reduction

$$2H_2O \rightarrow$$
$$O_2(g) + 4H^+ + 4e^-$$
Oxidation

Figure 21-4 The electrolysis of aqueous Na₂SO₄ produces H₂(g) at the cathode and O₂ at the anode. Bromthymol blue indicator has been added to the solution. This indicator turns blue in the basic solution near the cathode (where OH⁻ is produced) and yellow in the acidic solution near the anode (where H⁺ is formed).

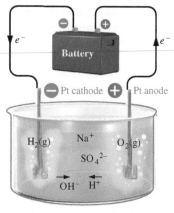

Michael Faraday (1791–1867) is considered the greatest experimental scientist of the 19th century. As a bookbinder's apprentice, he educated himself by extensive reading. Intrigued by his self-study of chemistry and by a lecture given by Sir Humphry Davy, the leading chemist of the day, Faraday applied for a position with Davy at the Royal Institution. He subsequently became director of that laboratory. His public lectures on science were very popular.

21-5 The Electrolysis of Aqueous Sodium Sulfate

In the electrolysis of aqueous sodium sulfate using inert electrodes, we observe the following.

1. Gaseous H_2 is produced at one electrode. The solution becomes basic around that electrode.

2. Gaseous O_2 is produced at the other electrode. The solution becomes acidic around that electrode.

As in the previous example, water is reduced in preference to Na^+ at the cathode. Observation 2 suggests that water is also preferentially oxidized relative to the sulfate ion, SO_4^{2-}, at the anode (Figure 21-4).

$$
\begin{array}{ll}
2(2H_2O + 2e^- \longrightarrow H_2 + 2OH^-) & \text{(reduction, cathode)} \\
2H_2O \longrightarrow O_2 + 4H^+ + 4e^- & \text{(oxidation, anode)} \\
\hline
6H_2O \longrightarrow 2H_2 + O_2 + \underbrace{4H^+ + 4OH^-}_{4H_2O} & \text{(overall cell reaction)} \\
2H_2O \longrightarrow 2H_2 + O_2 & \text{(net reaction)}
\end{array}
$$

The net result is the electrolysis of water. This occurs because H_2O is both more readily reduced than Na^+ and more readily oxidized than SO_4^{2-}. The ions of Na_2SO_4 conduct the current through the solution, but they take no part in the reaction.

21-6 Counting Electrons: Coulometry and Faraday's Law of Electrolysis

In 1832–1833, Michael Faraday's studies of electrolysis led to this conclusion.

> The amount of substance that undergoes oxidation or reduction at each electrode during electrolysis is directly proportional to the amount of electricity that passes through the cell.

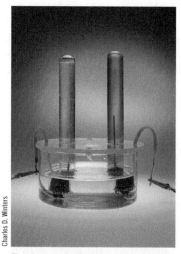

The electrolysis of the aqueous solution of KI, another Group 1A–Group 7A salt. At the cathode (*left*), water is reduced to H₂(g) and OH⁻ ions, turning the phenolphthalein indicator pink. The characteristic brownish color of aqueous I₂ appears at the anode (*right*).

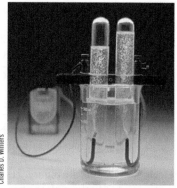

Electrolysis of water to produce O₂(g) and H₂(g). Na₂SO₄ was dissolved in the water to improve the conductivity. Can you identify which tube in this picture is collecting the H₂ gas and the name of the electrode producing it (anode or cathode)?

This is **Faraday's Law of Electrolysis**. A quantitative unit of electricity is now called the faraday.

> One **faraday** is the amount of electricity that corresponds to the gain or loss, and therefore the passage, of 6.022×10^{23} electrons, or *one mole* of electrons.

A smaller electrical unit commonly used in chemistry, physics, and electronics is the **coulomb (C)**. One coulomb is defined as the amount of charge that passes a given point when 1 ampere (A) of electric current flows for 1 second. One **ampere** of current equals 1 coulomb per second. One faraday is equal to 96,485 coulombs of charge.

▶ An ampere is usually called an "amp."

▶ For comparison, a 100-watt household incandescent light bulb uses a current of about 0.8 ampere.

$$1 \text{ ampere} = 1 \frac{\text{coulomb}}{\text{second}} \quad \text{or} \quad 1 \text{ A} = 1 \text{ C/s}$$

$$1 \text{ faraday} = 6.022 \times 10^{23} \, e^- = 96,485 \text{ C}$$

Table 21-1 shows the amounts of several elements produced during electrolysis by the passage of 1 faraday of electricity. The use of electrochemical cells to relate the amount of reactant or product to the amount of current passed is called **coulometry**.

Table 21-1 Amounts of Elements Produced at One Electrode in Electrolysis by 1 Faraday of Electricity

Half-Reaction	Number of e^- in Half-Reaction	Product (electrode)	Amount Produced
$Ag^+(aq) + e^- \longrightarrow Ag(s)$	1	Ag (cathode)	1 mol = 107.868 g
$2H^+(aq) + 2e^- \longrightarrow H_2(g)$	2	H_2 (cathode)	$\frac{1}{2}$ mol = 1.008 g
$Cu^{2+}(aq) + 2e^- \longrightarrow Cu(s)$	2	Cu (cathode)	$\frac{1}{2}$ mol = 31.773 g
$Au^{3+}(aq) + 3e^- \longrightarrow Au(s)$	3	Au (cathode)	$\frac{1}{3}$ mol = 65.656 g
$2Cl^-(aq) \longrightarrow Cl_2(g) + 2e^-$	2	Cl_2 (anode)	$\frac{1}{2}$ mol = 35.453 g = 11.2 L_{STP}
$2H_2O(\ell) \longrightarrow O_2(g) + 4H^+(aq) + 4e^-$	4	O_2 (anode)	$\frac{1}{4}$ mol = 8.000 g = 5.60 L_{STP}

EXAMPLE 21-1 Electrolysis

Calculate the mass of copper metal produced at the cathode during the passage of 2.50 amperes of current through a solution of copper(II) sulfate for 50.0 minutes.

Plan

The half-reaction that describes the reduction of copper(II) ions tells us the number of moles of electrons required to produce one mole of copper metal. Each mole of electrons corresponds to 1 faraday, or 9.65×10^4 coulombs, of charge. The product of current and time gives the number of coulombs.

$$\boxed{\begin{array}{c} \text{current} \\ \times \text{ time} \end{array}} \longrightarrow \boxed{\begin{array}{c} \text{no. of} \\ \text{coulombs} \end{array}} \longrightarrow \boxed{\begin{array}{c} \text{mol of } e^- \\ \text{passed} \end{array}} \longrightarrow \boxed{\begin{array}{c} \text{mass} \\ \text{of Cu} \end{array}}$$

▶ The amount of electricity in Examples 21-1 and 21-2 would be sufficient to light a 100-watt household incandescent light bulb for about 150 minutes, or 2.5 hours.

▶ When the number of significant figures in the calculation warrants, the value 96,485 coulombs is usually rounded to 96,500 coulombs (9.65×10^4 C).

Solution

The equation for the reduction of copper(II) ions to copper metal is

$$
\begin{array}{cccc}
Cu^{2+} & + & 2e^- & \longrightarrow & Cu & \quad \text{(reduction, cathode)} \\
1 \text{ mol} & & 2(6.02 \times 10^{23})e^- & & 1 \text{ mol} \\
63.5 \text{ g} & & 2(9.65 \times 10^4 \text{ C}) & & 63.5 \text{ g}
\end{array}
$$

CHEMISTRY IN USE

A Spectacular View of One Mole of Electrons

Early in our study of chemistry, we saw that atoms are made up of protons, neutrons, and electrons. We also discussed the incredibly large size of Avogadro's number, 6.022×10^{23}. Although individual atoms and molecules are invisible to the naked eye, one mole of atoms or molecules is easily detected. Because subatomic particles are even smaller than atoms and also invisible, you might never expect to see individual electrons. Let's consider the possibility, however, of seeing a faraday of charge. A faraday of charge contains Avogadro's number of electrons. Would this collection of 6.022×10^{23} electrons be visible? If so, what might it look like? It would look quite spectacular!

Throughout the 1980s, scientists carefully studied data collected during 5 million lightning flashes along the eastern United States. The data were collected by 36 instruments that were collectively known as the National Lightning Detection Network. The investigating scientists found that the electrical currents in lightning flashes over northern Florida measured about 45,000 amps, about double the 25,000-amp currents in lightning flashes over the New England states. This study showed that the amount of current flowing during lightning flashes was inversely proportional to the latitude (distance from the equator) of the storm.

One coulomb is the amount of charge that passes a point when a one-ampere current flows for one second (1 coulomb = 1 ampere·second). Thus, a current of 96,500 amps flowing for one second contains Avogadro's number of electrons, or one faraday of charge.

Garadd/Science Photo Library/Photo Researchers Inc

Measurements taken in northern Florida show that a typical two-second lightning strike over that section of the country would transfer approximately Avogadro's number of electrons between the clouds and the earth. So, for those living in northern Florida, a spectacular mental view of one mole of electrons can be obtained by visualizing a two-second lightning strike. Keep in mind that the average lightning strike lasts only a small fraction of a second, and that we can only have a mental view of a two-second lightning strike by extrapolation of what is seen in nature. Because New England lightning strikes produce only about half the current of lightning strikes over northern Florida, people in New England must try to imagine a four-second lightning strike.

RONALD DELORENZO

S TOP & THINK

Remember that 2.50 A = 2.50 C/s. Also remember to convert time into seconds when doing these calculations!

A family memento that has been electroplated with copper. To aid in electroplating onto nonconductors, such as shoes, the material is first soaked in a concentrated electrolyte solution to make it conductive.

We see that 63.5 grams of copper "plate out" for every 2 moles of electrons, or for every $2(9.65 \times 10^4$ coulombs$)$ of charge. We first calculate the number of coulombs passing through the cell.

$$\underline{?}\ C = 50.0\ \text{min} \times \frac{60\ s}{1\ \text{min}} \times \frac{2.50\ C}{s} = 7.50 \times 10^3\ C$$

We calculate the mass of copper produced by the passage of 7.50×10^3 coulombs.

$$\underline{?}\ g\ Cu = 7.50 \times 10^3\ C \times \frac{1\ mol\ e^-}{9.65 \times 10^4\ C} \times \frac{63.5\ g\ Cu}{2\ mol\ e^-} = 2.47\ g\ Cu \quad \text{(about the mass of a copper penny)}$$

Notice how little copper is deposited by this considerable current in 50 minutes.

You should now work Exercises 26 and 32.

Charles D. Winters

EXAMPLE 21-2 Electrolysis

What volume of oxygen gas (measured at STP) is produced by the oxidation of water at the anode in the electrolysis of copper(II) sulfate in Example 21-1?

Plan

We use the same approach as in Example 21-1. Here we relate the amount of charge passed to the number of moles, and hence the volume of O_2 gas produced at STP.

$$\boxed{\begin{array}{c}\text{current}\\ \times \text{ time}\end{array}} \longrightarrow \boxed{\begin{array}{c}\text{no. of}\\ \text{coulombs}\end{array}} \longrightarrow \boxed{\begin{array}{c}\text{mol of } e^-\\ \text{passed}\end{array}} \longrightarrow \boxed{\begin{array}{c}L_{STP}\\ \text{of } O_2\end{array}}$$

Solution

The equation for the oxidation of water and the equivalence between the number of coulombs and the volume of oxygen produced at STP are

$$2H_2O \longrightarrow O_2 + 4H^+ + 4e^- \qquad \text{(oxidation, anode)}$$

$$\begin{array}{ccc} & 1 \text{ mol} & 4(6.02 \times 10^{23})e^- \\ & 22.4 \text{ L}_{STP} & 4(9.65 \times 10^4 \text{ C}) \end{array}$$

The number of coulombs passing through the cell is 7.50×10^3 C. For every $4(9.65 \times 10^4$ coulombs) passing through the cell, one mole of O_2 (22.4 L at STP) is produced.

$$\underline{?}\ L_{STP}\ O_2 = 7.50 \times 10^3\ C \times \frac{1 \text{ mol } e^-}{9.65 \times 10^4\ C} \times \frac{1 \text{ mol } O_2}{4 \text{ mol } e^-} \times \frac{22.4\ L_{STP}\ O_2}{1 \text{ mol } O_2}$$

$$= 0.435\ L_{STP}\ O_2$$

You should now work Exercise 28.

Notice how little product is formed by what seems to be a lot of electricity. This suggests why electrolytic production of gases and metals is so costly.

21-7 Commercial Applications of Electrolytic Cells

Several elements are produced commercially by electrolysis. In Sections 21-3 to 21-5, we described some electrolytic cells that produce sodium (the Downs cell), chlorine, hydrogen, and oxygen. Electrolysis of molten compounds is also the common method of obtaining other Group 1A metals, 2A metals (except barium), and aluminum (see Section 26-3). Impure metals can also be refined electrolytically, as we will describe for copper in Section 26-8.

Metal-plated articles are common in our society. Jewelry and tableware are often plated with silver. Gold is plated onto jewelry and electrical contacts. Copper is plated onto many objects for decorative purposes (Figure 21-5). Automobiles formerly had steel bumpers plated with thin films of chromium. A chrome bumper required approximately 3 seconds of electroplating to produce a smooth, shiny surface only 0.0002 mm thick. When the metal atoms are deposited too rapidly, they are not able to form extended lattices. Rapid plating of metal results in rough, grainy, black surfaces. Slower plating produces smooth surfaces. "Tin cans" are steel cans plated electrolytically with tin; these are sometimes replaced by cans plated in a fraction of a second with an extremely thin chromium film.

Voltaic or Galvanic Cells

Voltaic, or **galvanic, cells** are electrochemical cells in which *spontaneous* (product-favored) oxidation–reduction reactions produce electrical energy. The two halves of the redox reaction are separated, requiring electron transfer to occur through an external circuit. In

Anodized aluminum is made by oxidizing a piece of aluminum in an electrolytic cell for a short period of time. The oxidation microscopically dissolves some of the Al surface and coats it with a protective and tough coating of aluminum oxide. This creates a matte finish that can be dyed different colors.

▶ These cells are named for Allesandro Volta (1745–1827) and Luigi Galvani (1737–1798), two Italian physicists of the 18th century.

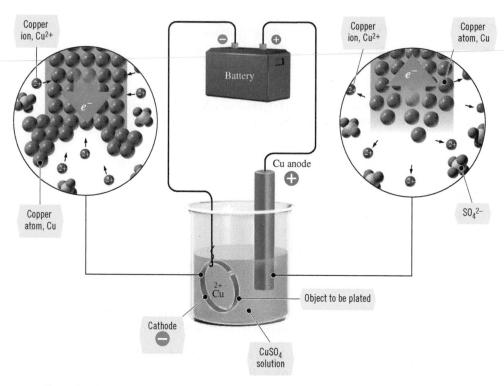

Figure 21-5 Electroplating with copper. The anode is pure copper, which dissolves during the electroplating process. This replenishes the Cu^{2+} ions that are removed from the solution as Cu plates out on the cathode.

this way, useful electrical energy is obtained. Everyone is familiar with some voltaic cells. The batteries commonly used in flashlights, portable radios, photographic equipment, and many toys and appliances are voltaic cells. Automobile batteries consist of voltaic cells connected in series so that their voltages add. We will first consider some simple laboratory cells used to measure the potential difference, or voltage, of a reaction under study. We will then look at some common voltaic cells.

21-8 The Construction of Simple Voltaic Cells

A **half-cell** contains the oxidized and reduced forms of an element, or other more complex species, in contact with one another. A common kind of half-cell consists of a piece of metal (the electrode) immersed in a solution of its ions. Consider two such half-cells in separate beakers (Figure 21-6). The electrodes are connected by a wire. A voltmeter can be inserted into the circuit to measure the potential difference between the two electrodes, or an ammeter can be inserted to measure the current flow. The electric current is the result of the spontaneous redox reaction that occurs. We measure the potential of the cell.

The circuit between the two solutions is completed by a **salt bridge**. This can be any medium through which ions can slowly pass. A salt bridge can be made by bending a piece of glass tubing into the shape of a "U," filling it with a hot saturated salt/5% agar solution, and allowing it to cool. The cooled mixture "sets" to the consistency of firm gelatin. As a result, the solution does not run out when the tube is inverted (see Figure 21-6), but the ions in the gel are still able to move. A salt bridge serves three functions.

1. It allows electrical contact between the two solutions.

2. It prevents mixing of the electrode solutions.

3. It maintains the electrical neutrality in each half-cell as ions flow into and out of the salt bridge.

▶ Neither a voltmeter nor an ammeter generates electrical energy. They are designed to measure electrical voltage or the amount of current flow.

▶ Agar is a porous, gelatinous material obtained from algae.

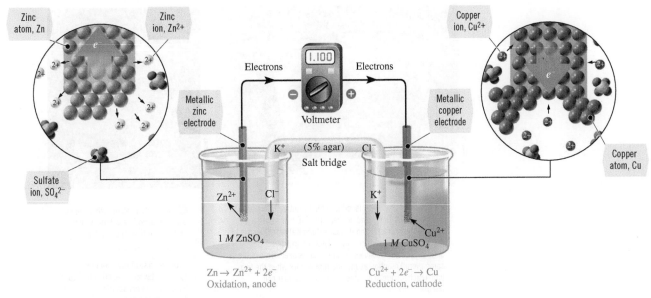

Figure 21-6 The zinc–copper voltaic cell utilizes the reaction

$$Zn(s) + Cu^{2+}(aq) \longrightarrow Zn^{2+}(aq) + Cu(s)$$

The standard potential of this cell is 1.100 volts.

> A cell in which all reactants and products are in their thermodynamic standard states (1 M for dissolved species and 1 atm partial pressure for gases) is called a **standard cell**.

▶ When we show a concentration in a *standard cell as* 1 *M*, it is assumed to be exactly 1 *M*.

21-9 The Zinc–Copper Cell

Consider a standard cell made up of two half-cells, one a strip of metallic Cu immersed in 1 M copper(II) sulfate solution and the other a strip of Zn immersed in 1 M zinc sulfate solution (see Figure 21-6). This cell is called the Daniell cell. The following experimental observations have been made about this cell.

1. The initial voltage is 1.100 volts.

2. The mass of the zinc electrode decreases. The concentration of Zn^{2+} *increases* in the solution around the zinc electrode as the cell operates.

3. The mass of the copper electrode increases. The concentration of Cu^{2+} *decreases* in the solution around this electrode as the cell operates.

The Zn electrode loses mass because some Zn metal is *oxidized* to Zn^{2+} ions, which go into solution. Thus the Zn electrode is the *anode*. At the *cathode*, Cu^{2+} ions are *reduced* to Cu metal. This plates out on the electrode, so its mass increases.

$$
\begin{array}{ll}
Zn \longrightarrow Zn^{2+} + 2e^- & \text{(oxidation, anode)} \\
\underline{Cu^{2+} + 2e^- \longrightarrow Cu} & \text{(reduction, cathode)} \\
Cu^{2+} + Zn \longrightarrow Cu + Zn^{2+} & \text{(overall cell reaction)}
\end{array}
$$

Electrons are released at the anode and consumed at the cathode. They therefore flow through the wire from anode to cathode, as in all electrochemical cells. In all *voltaic* cells, the electrons flow spontaneously from the negative electrode to the positive electrode. So, in contrast with electrolytic cells, the anode is negative and the cathode is positive. To maintain electroneutrality and complete the circuit, two Cl^- ions from the salt bridge migrate into the anode solution for every Zn^{2+} ion formed. Two K^+ ions migrate into the cathode solution to replace every Cu^{2+} ion reduced. Some Zn^{2+} ions from the anode vessel and some SO_4^{2-} ions from the cathode vessel also migrate into the salt bridge. Neither Cl^- nor K^+ ions are oxidized or reduced in preference to the zinc metal or Cu^{2+} ions.

▶ Reduction, however, still occurs at the cathode (recall the Red Cat) and oxidation at the anode.

▶ Compare the −/+, anode–cathode, and oxidation–reduction labels and the directions of electron flow in Figures 21-2a and 21-6.

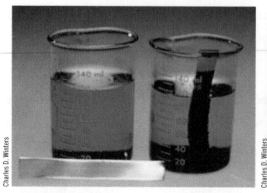

Charles D. Winters

Charles D. Winters

A A strip of zinc was placed in a blue solution of copper(II) sulfate, $CuSO_4$ (left-hand beaker). The Cu^{2+} in solution oxidizes the Zn metal and deposits Cu metal onto the strip of Zn, while the oxidized Zn^{2+} dissolved. The resulting zinc sulfate solution is colorless (right hand beaker). This is the same overall reaction as the one that occurs when the two half-reactions are separated in the zinc–copper cell (Figure 21-6).

B No reaction occurs when copper wire is placed in a colorless zinc sulfate solution. The reaction

$$Zn^{2+} + Cu(s) \longrightarrow Zn(s) + Cu^{2+}$$

is the *reverse* of the spontaneous (product-favored) reaction in Figure 21-6, so it is *nonspontaneous* (reactant-favored).

$Zn(s)/Cu^{2+}(aq)$ reaction.

As the redox reaction proceeds consuming the reactants, the cell voltage decreases. When the cell voltage reaches zero, the reaction has reached equilibrium and no further net reaction occurs. At this point, however, the Cu^{2+} ion concentration in the cathode cell is *not* zero. This description applies to any voltaic cell.

Voltaic cells can be represented as follows for the zinc–copper cell.

This is a shorthand notation for the cell reactions:

$$Zn \longrightarrow Zn^{2+} + 2e^-$$

and

$$Cu^{2+} + 2e^- \longrightarrow Cu$$

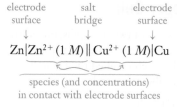

In this representation, a single line (|) represents an interface at which a potential develops, that is, an electrode. It is conventional to write the anode half-cell on the left in this notation.

The same reaction occurs when a piece of Zn is dropped into a solution of $CuSO_4$. The Zn dissolves and the blue color of Cu^{2+} ions disappears. Copper metal forms on the Zn and then settles to the bottom of the container. Because the oxidizing Cu^{2+} ions and reducing Zn metal are in direct contact with one another in this arrangement, the electron transfer is "short-circuited," so the spontaneous free energy available to perform work is wasted and simply heats the solution. Only by separating the two half-reactions into separate cells and connecting them with an external wire (circuit) and a salt bridge can one effectively use the electricity generated.

Problem-Solving Tip How to Tell the Anode from the Cathode

The correspondence between the names *anode* and *cathode* and the charge on the electrode is *different* for electrolytic cells than for voltaic (galvanic) cells. Students sometimes get confused by trying to remember which is which. Check the definitions of these two terms in Section 21-2. The surest way to name these electrodes is to determine what process takes place at each one. Remember the "red cat" mnemonic— *red*uction occurs at the *cat*hode, regardless of the type of electrochemical cell one is dealing with.

21-10 The Copper–Silver Cell

Now consider a similar standard voltaic cell consisting of a strip of Cu immersed in 1 M $CuSO_4$ solution and a strip of Ag immersed in 1 M $AgNO_3$ solution. A wire and a salt bridge complete the circuit. The following observations have been made.

1. The initial voltage of the cell is 0.462 volt.
2. The mass of the copper electrode decreases. The Cu^{2+} ion concentration *increases* in the solution around the copper electrode.
3. The mass of the silver electrode increases. The Ag^+ ion concentration *decreases* in the solution around the silver electrode.

In this cell the Cu electrode is the anode because Cu metal is oxidized to Cu^{2+} ions. The Ag electrode is the cathode because Ag^+ ions are reduced to metallic Ag (Figure 21-7).

$$\begin{array}{ll} Cu \longrightarrow Cu^{2+} + 2e^- & \text{(oxidation, anode)} \\ \underline{2(Ag^+ + e^- \longrightarrow Ag)} & \text{(reduction, cathode)} \\ Cu + 2Ag^+ \longrightarrow Cu^{2+} + 2Ag & \text{(overall cell reaction)} \end{array}$$

As before, ions from the salt bridge migrate to maintain electroneutrality. Some NO_3^- ions (from the cathode vessel) and some Cu^{2+} ions (from the anode vessel) also migrate into the salt bridge to keep the overall charge balanced. Two NO_3^- ions migrate for each Cu^{2+} ion.

Recall that in the zinc–copper cell the copper electrode was the *cathode*; now in the copper–silver cell the copper electrode has become the *anode*.

> Whether a particular electrode acts as an anode or a cathode in a voltaic cell depends on what the other electrode of the cell is.

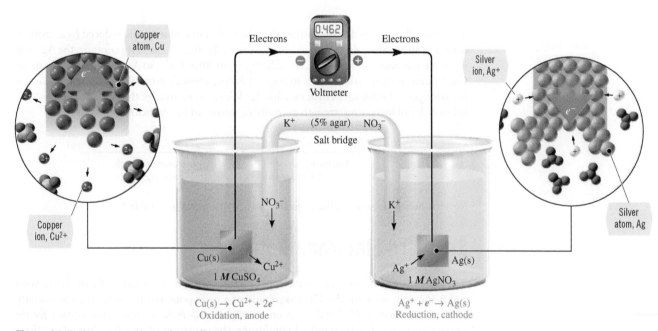

Figure 21-7 The copper–silver voltaic cell utilizes the reaction

$$Cu(s) + 2Ag^+(aq) \longrightarrow Cu^{2+}(aq) + 2Ag(s)$$

The standard potential of this cell is 0.462 volt. This standard cell can be represented as $Cu|Cu^{2+}$ (1 M)$||Ag^+$(1 M)$|Ag$.

▶ KCl is not used in this salt bridge because the Cl^- ions would react with Ag^+ ions to form insoluble AgCl(s).

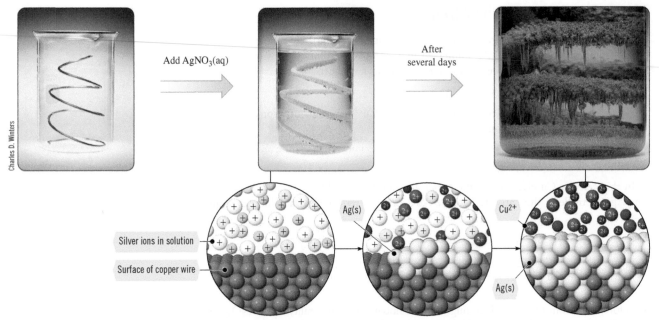

A spiral of copper wire is placed in a colorless solution of silver nitrate, $AgNO_3$. The silver is gradually displaced from solution and adheres to the wire. The resulting copper nitrate solution is blue.

$$2Ag^+ + Cu(s) \longrightarrow 2Ag(s) + Cu^{2+}$$

The same spontaneous (product-favored) reaction occurs when the two half-reactions are separated in the copper–silver cell (see Figure 21-7).

No reaction occurs when silver wire is placed in a blue copper sulfate solution. The reaction

$$Cu^{2+} + 2Ag(s) \longrightarrow Cu(s) + 2Ag^+$$

is the reverse of the spontaneous reaction shown here and in Figure 21-7; it has a *negative* E^0_{cell} and is *nonspontaneous* (reactant-favored).

The two cells we have described show that the Cu^{2+} ion is more easily reduced (is a stronger oxidizing agent) than Zn^{2+}, so Cu^{2+} oxidizes metallic zinc to Zn^{2+}. By contrast, the Ag^+ ion is more easily reduced (is a stronger oxidizing agent) than Cu^{2+}, so Ag^+ oxidizes Cu atoms to Cu^{2+}. Conversely, metallic Zn is a stronger reducing agent than metallic Cu, and metallic Cu is a stronger reducing agent than metallic Ag. We can now arrange the species we have studied in order of increasing strength as oxidizing agents and as reducing agents.

$$\underrightarrow{Zn^{2+} < Cu^{2+} < Ag^+}$$
Increasing strength
as oxidizing agents

$$\underrightarrow{Ag < Cu < Zn}$$
Increasing strength
as reducing agents

Results such as these form the basis for the activity series (see Table 6-9).

Standard Electrode Potentials

The potentials of the standard zinc–copper and copper–silver voltaic cells are 1.100 volts and 0.462 volt, respectively. The magnitude of a cell's potential measures the spontaneity of its redox reaction. *Higher (more positive) cell potentials indicate greater driving force for the reaction as written.* Under standard conditions, the oxidation of metallic Zn by Cu^{2+} ions has a greater tendency to go toward completion than does the oxidation of metallic Cu by Ag^+ ions. It is convenient to separate the total cell potential into the individual contributions of the two half-reactions. This lets us determine the relative tendencies of particular oxidation or reduction half-reactions to occur. Such information gives us a quantitative basis for specifying strengths of oxidizing and reducing agents. In the next several sections, we shall see how this is done for standard half-cells.

21-11 The Standard Hydrogen Electrode

Every oxidation must be accompanied by a reduction (i.e., the electrons must have somewhere to go). So it is impossible to determine experimentally the potential of any *single* electrode. We therefore establish an arbitrary standard. The conventional reference electrode is the **standard hydrogen electrode (SHE)**. This electrode contains a piece of metal electrolytically coated with a grainy black surface of inert platinum metal (finely divided metals such as Pt often appear black), immersed in a 1 M H^+ solution. Hydrogen, H_2, is bubbled at 1 atm pressure through a glass envelope over the platinized electrode (Figure 21-8).

By international convention, the standard hydrogen electrode is arbitrarily assigned a potential of *exactly* 0.0000 . . . volt.

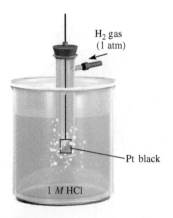

Figure 21-8 The standard hydrogen electrode (SHE). A molecular-level view of the operation of the SHE as a cathode is shown in Figure 21-9 and as an anode in Figure 21-10.

▶ The superscript in E^0 indicates thermodynamic standard-state conditions.

SHE Half-Reaction	E^0 (standard electrode potential)	
$H_2 \longrightarrow 2H^+ + 2e^-$	exactly 0.0000 . . . V	(SHE as anode)
$2H^+ + 2e^- \longrightarrow H_2$	exactly 0.0000 . . . V	(SHE as cathode)

As we shall see in Section 21-14, this arbitrary definition will provide a basis on which to describe the relative potential (and hence the relative spontaneity) of any half-cell process. We can construct a standard cell consisting of a standard hydrogen electrode and some other standard electrode (half-cell). Because the defined electrode potential of the SHE contributes exactly 0 volts to the sum, the voltage of the overall cell is then attributed entirely to the other half-cell.

21-12 The Zinc–SHE Cell

This cell consists of a SHE in one beaker and a strip of zinc immersed in 1 M zinc chloride solution in another beaker (Figure 21-9). A wire and a salt bridge complete the circuit. When the circuit is closed, the following observations can be made.

1. The initial potential of the cell is 0.763 volt.
2. As the cell operates, the mass of the zinc electrode decreases. The concentration of Zn^{2+} ions increases in the solution around the zinc electrode.
3. The H^+ concentration decreases in the SHE. Gaseous H_2 is produced.

We can conclude from these observations that the following half-reactions and cell reaction occur.

		E^0	
(oxidation, anode)	$Zn \longrightarrow Zn^{2+} + 2e^-$	0.763 V	
(reduction, cathode)	$2H^+ + 2e^- \longrightarrow H_2$	0.000 V	(by definition)
(cell reaction)	$Zn + 2H^+ \longrightarrow Zn^{2+} + H_2$ $E^0_{cell} = 0.763$ V		(measured)

The standard potential at the anode *plus* the standard potential at the cathode gives the standard cell potential. The potential of the SHE is 0.000 volts, and the standard cell potential is found to be 0.763 volt. So the standard potential of the zinc anode in this cell must be 0.763 volt. The $Zn|Zn^{2+}(1.0\ M)\|H^+(1.0\ M), H_2(1\ \text{atm})|Pt$ cell is depicted in Figure 21-9.

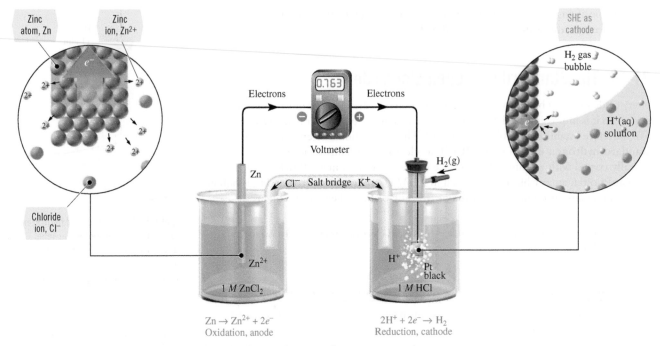

Figure 21-9 The Zn|Zn^{2+}(1 M)‖H$^+$(1 M); H$_2$(1 atm)|Pt cell, in which the following net reaction occurs.

$$Zn(s) + 2H^+(aq) \longrightarrow Zn^{2+}(aq) + H_2(g)$$

In this cell, the standard hydrogen electrode functions as the cathode.

Note that in *this* cell, H$^+$ is reduced to H$_2$, so the SHE is the *cathode*. At the other electrode, metallic zinc is oxidized to Zn^{2+}, so the zinc electrode is the *anode* in this cell.

The idea of "electron pressure" helps us to understand this process. The positive potential for the half-reaction

$$Zn \longrightarrow Zn^{2+} + 2e^- \qquad E^0_{oxidation} = +0.763 \text{ V}$$

says that this oxidation is *more favorable* than the corresponding oxidation of H$_2$,

$$H_2 \longrightarrow 2H^+ + 2e^- \qquad E^0_{oxidation} = 0.000 \text{ V}$$

▶ What we have informally called "electron pressure" is the tendency to undergo oxidation.

Before they are connected, each half-cell builds up a supply of electrons waiting to be released, thus generating an electron pressure. The process with the more *positive* E^0 value is favored, so we reason that the electron pressure generated at the Zn electrode is greater than that at the H$_2$ electrode. As a result, when the cell is connected, the electrons released by the oxidation of Zn flow through the wire *from the Zn electrode to the H$_2$ electrode*, where they are consumed by the reduction of H$^+$ ions. Oxidation occurs at the zinc electrode (anode), and reduction occurs at the hydrogen electrode (cathode).

21-13 The Copper–SHE Cell

Another voltaic cell consists of a SHE in one beaker and a strip of Cu metal immersed in 1 M copper(II) sulfate solution in another beaker. A wire and a salt bridge complete the circuit. For this cell, we observe the following (Figure 21-10).

1. The initial cell potential is 0.337 volt.

2. Gaseous hydrogen is consumed. The H$^+$ concentration *increases* in the solution of the SHE.

3. The mass of the copper electrode increases. The concentration of Cu^{2+} ions *decreases* in the solution around the copper electrode.

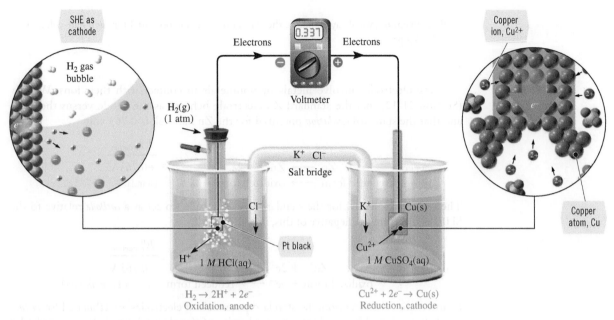

Figure 21–10 The standard copper–SHE cell,

$$Pt|H_2(1\ atm), H^+(1\ M)\|Cu^{2+}(1\ M)|Cu$$

In this cell, the standard hydrogen electrode functions as the anode. The net reaction is

$$H_2(g) + Cu^{2+}(aq) \longrightarrow 2H^+(aq) + Cu(s)$$

Thus, the following half-reactions and cell reaction occur.

		E^0	
(oxidation, anode)	$H_2 \longrightarrow 2H^+ + 2e^-$	0.000 V	(by definition)
(reduction, cathode)	$Cu^{2+} + 2e^- \longrightarrow Cu$	0.337 V	
(cell reaction)	$H_2 + Cu^{2+} \longrightarrow 2H^+ + Cu$ $\quad E^0_{cell} = 0.337\ V$		(measured)

The SHE functions as the *anode* in this cell, and Cu^{2+} ions oxidize H_2 to H^+ ions. The standard electrode potential of the copper half-cell is 0.337 volt as a *cathode* in the Cu–SHE cell.

> ▶ Recall that in the Zn–SHE cell the SHE was the *cathode*.

Again, we can think of $E^0_{oxidation}$ in the two half-cells as "electron pressures."

$$Cu \longrightarrow Cu^{2+} + 2e^- \qquad\qquad E^0_{oxidation} = -0.337\ V$$

$$H_2 \longrightarrow 2H^+ + 2e^- \qquad\qquad E^0_{oxidation} = 0.000\ V$$

> ▶ When we reverse the Cu half-reaction to write it as an oxidation, we must reverse its sign.

Now we see that the hydrogen electrode has the higher electron pressure (more positive, less negative E^0_{oxid}). When the cell is connected, electrons flow through the wire from the hydrogen electrode to the copper electrode. H_2 is oxidized to $2H^+$ (anode), and Cu^{2+} is reduced to Cu (cathode).

21-14 Standard Electrode Potentials

We can develop a series of representative electrode potentials by measuring the potentials of other standard electrodes versus the SHE in the way we have described for the standard Zn–SHE and standard Cu–SHE voltaic cells. The **standard electrode potential** for any half-cell is its potential with respect to the standard hydrogen electrode, measured at 25°C when the concentration of each ion in the solution is 1 M and the pressure of any gas involved is 1 atm.

> ▶ The activity series (see Table 6-9) is based on standard electrode potentials.

By convention, we always tabulate the standard electrode potential for each half-cell as a *reduction* process.

Many electrodes involve metals or nonmetals in contact with their ions. We saw (Section 21-12) that the standard Zn electrode behaves as the anode versus the SHE and that the standard *oxidation* potential for the Zn half-cell is 0.763 volt.

$$E^0_{\text{oxidation}}$$

(as anode) $Zn \longrightarrow Zn^{2+} + 2e^-$ +0.763 V

reduced form $\longrightarrow$ oxidized form + ne^- (standard *oxidation* potential)

The *reduction* potential for the standard zinc electrode (to act as a *cathode* relative to the SHE) is therefore the negative of this, or −0.763 volt.

$$E^0_{\text{reduction}}$$

(as cathode) $Zn^{2+} + 2e^- \longrightarrow Zn$ −0.763 V

oxidized form + $ne^- \longrightarrow$ reduced form   (standard *reduction* potential)

By international convention, the standard potentials of electrodes are tabulated for *reduction half-reactions*. These indicate the tendencies of the electrodes to behave as cathodes toward the SHE. Electrodes with positive E^0 values for reduction half-reactions act as cathodes versus the SHE. Those with negative E^0 values for reduction half-reactions act as anodes versus the SHE.

Electrodes with *Positive* $E^0_{\text{reduction}}$	Electrodes with *Negative* $E^0_{\text{reduction}}$
Reduction occurs *more readily* than the reduction of $2H^+$ to H_2.	Reduction is *more difficult* than the reduction of $2H^+$ to H_2.
Electrode acts as a *cathode* versus the SHE (which acts as the anode).	Electrode acts as an *anode* versus the SHE (which acts as the cathode).

The more positive the E^0 value for a half-reaction, the greater the tendency for the half-reaction to occur in the forward direction as written. Conversely, the more negative the E^0 value for a half-reaction, the greater the tendency for the half-reaction to occur in the reverse direction as written.

Table 21-2 lists standard reduction potentials for a few elements.

▶ The *oxidizing agent* is reduced.

1. The species on the *left* side are all cations of metals, hydrogen ions, or elemental nonmetals. These are all *oxidizing agents* (*oxidized forms* of the elements). Their strengths as oxidizing agents increase from top to bottom, that is, as the $E^0_{\text{reduction}}$ values become more positive. Fluorine, F_2, is the strongest oxidizing agent, and Li^+ is a very weak oxidizing agent.

▶ The *reducing agent* is oxidized.

2. The species on the *right* side are free metals, hydrogen, or anions of nonmetals. These are all *reducing agents* (*reduced forms* of the elements). Their strengths as reducing agents increase from bottom to top, that is, as the $E^0_{\text{reduction}}$ values become more negative. Metallic Li is a very strong reducing agent, and F^- is a very weak reducing agent.

The more positive the reduction potential, the stronger the species on the left is as an oxidizing agent and the weaker the species on the right is as a reducing agent.

Table 21-2 Standard Reduction Potentials in Aqueous Solution at 25°C

Element	Reduction Half-Reaction			Standard Reduction Potential E^0 (volts)	
Li	$Li^+ + e^-$	$\longrightarrow$	Li	−3.045	
K	$K^+ + e^-$	$\longrightarrow$	K	−2.925	
Ca	$Ca^{2+} + 2e^-$	$\longrightarrow$	Ca	−2.87	
Na	$Na^+ + e^-$	$\longrightarrow$	Na	−2.714	
Mg	$Mg^{2+} + 2e^-$	$\longrightarrow$	Mg	−2.37	
Al	$Al^{3+} + 3e^-$	$\longrightarrow$	Al	−1.66	
Zn	$Zn^{2+} + 2e^-$	$\longrightarrow$	Zn	−0.763	
Cr	$Cr^{3+} + 3e^-$	$\longrightarrow$	Cr	−0.74	
Fe	$Fe^{2+} + 2e^-$	$\longrightarrow$	Fe	−0.44	
Cd	$Cd^{2+} + 2e^-$	$\longrightarrow$	Cd	−0.403	
Ni	$Ni^{2+} + 2e^-$	$\longrightarrow$	Ni	−0.25	
Sn	$Sn^{2+} + 2e^-$	$\longrightarrow$	Sn	−0.14	
Pb	$Pb^2 + 2e^-$	$\longrightarrow$	Pb	−0.126	
H₂	$2H^+ + 2e^-$	$\longrightarrow$	H₂	0.000	(reference electrode)
Cu	$Cu^{2+} + 2e^-$	$\longrightarrow$	Cu	+0.337	
I₂	$I_2 + 2e^-$	$\longrightarrow$	$2I^-$	+0.535	
Hg	$Hg^{2+} + 2e^-$	$\longrightarrow$	Hg	+0.789	
Ag	$Ag^+ + e^-$	$\longrightarrow$	Ag	+0.799	
Br₂	$Br_2 + 2e^-$	$\longrightarrow$	$2Br^-$	+1.08	
Cl₂	$Cl_2 + 2e^-$	$\longrightarrow$	$2Cl^-$	+1.360	
Au	$Au^{3+} + 3e^-$	$\longrightarrow$	Au	+1.50	
F₂	$F_2 + 2e^-$	$\longrightarrow$	$2F^-$	+2.87	

(Left arrow label: Increasing strength as oxidizing agent; increasing ease of reduction. Right arrow label: Increasing strength as reducing agent; increasing ease of oxidation.)

21-15 Uses of Standard Electrode Potentials

The most important application of electrode potentials is the prediction of the spontaneity of redox reactions. Standard electrode potentials can be used to determine the spontaneity of redox reactions in general, whether or not the reactions can take place in electrochemical cells.

Suppose we ask the question: At standard conditions, will Cu^{2+} ions oxidize metallic Zn to Zn^{2+} ions, or will Zn^{2+} ions oxidize metallic copper to Cu^{2+}? One of the two possible reactions is spontaneous, and the reverse reaction is nonspontaneous. We must determine which one is spontaneous. We already know the answer to this question from experimental results (see Section 21-9), but let us demonstrate the procedure for predicting the spontaneous reaction from known values of standard electrode potentials.

1. Choose the appropriate half-reactions from a table of standard reduction potentials.

2. Write the equation for the half-reaction with the more positive (or less negative) E^0 value *for reduction*, along with its potential.

3. Write the equation for the other half-reaction *as an oxidation* and write its *oxidation potential*; to do this, reverse the tabulated reduction half-reaction and change the sign of E^0. (Reversing a half-reaction or a complete reaction also changes the sign of its potential.)

4. Balance the electron transfer. *We do not multiply the potentials by the numbers used to balance the electron transfer!* The reason is that each potential represents a *tendency* for a reaction process to occur relative to the SHE; this does not depend on *how many times* it occurs. An electric potential is an *intensive property*.

5. Add the reduction and oxidation half-reactions, and add the reduction and oxidation potentials. E^0_{cell} will be *positive* for the resulting overall cell reaction. This indicates that the reaction as written is *product-favored (spontaneous)*. A *negative* E^0_{cell} value would indicate that the reaction is *reactant-favored (nonspontaneous)*.

For the cell described here, the Cu^{2+}/Cu couple has the more positive reduction potential, so we keep it as the reduction half-reaction and reverse the other half-reaction. Following the steps outlined, we obtain the equation for the spontaneous (product-favored) reaction.

$$
\begin{array}{lll}
Cu^{2+} + 2e^- \longrightarrow Cu & +0.337\ V & \longleftarrow \text{reduction potential} \\
Zn \longrightarrow Zn^{2+} + 2e^- & +0.763\ V & \longleftarrow \text{oxidation potential} \\
\hline
Cu^{2+} + Zn \longrightarrow Cu + Zn^{2+} & E^0_{cell} = +1.100\ V &
\end{array}
$$

The positive E^0_{cell} value tells us that the forward reaction is spontaneous (product-favored) at standard conditions. So we conclude that copper(II) ions oxidize metallic zinc to Zn^{2+} ions as they are reduced to metallic copper. (Section 21-9 shows that the potential of the standard zinc-copper voltaic cell is 1.100 volts. This is the spontaneous reaction that occurs.)

The reverse reaction has a negative E^0 and is nonspontaneous.

nonspontaneous
reaction: $Cu + Zn^{2+} \longrightarrow Cu^{2+} + Zn$ $E^0_{cell} = -1.100$ volts

To make it occur, we would have to supply electrical energy with a potential difference greater than 1.100 volts. That is, this nonspontaneous reaction would have to be carried out in an *electrolytic cell*.

EXAMPLE 21-3   Predicting the Direction of Reactions

At standard conditions, will chromium(III) ions, Cr^{3+}, oxidize metallic copper to copper(II) ions, Cu^{2+}, or will Cu^{2+} oxidize metallic chromium to Cr^{3+} ions? Write the equation for the spontaneous (product-favored) reaction and calculate E^0_{cell} for this reaction.

Plan

We refer to the table of standard reduction potentials (see Appendix J) and choose the two appropriate half-reactions.

Solution

The copper half-reaction has the more positive reduction potential, so we write it first. Then we write the chromium half-reaction as an oxidation, balance the electron transfer, and add the two half-reactions and their potentials.

$$
\begin{array}{lll}
& & E^0 \\
\hline
3(Cu^{2+} + 2e^- \longrightarrow Cu) & \text{(reduction)} & +0.337\ V \\
2(Cr \longrightarrow Cr^{3+} + 3e^-) & \text{(oxidation)} & +0.74\ V \\
\hline
2Cr + 3Cu^{2+} \longrightarrow 2Cr^{3+} + 3Cu & E^0_{cell} = +1.08\ V &
\end{array}
$$

Because E^0_{cell} is positive, we know that the reaction is spontaneous (product-favored).

Cu^{2+} ions spontaneously oxidize metallic Cr to Cr^{3+} ions and are reduced to metallic Cu.

You should now work Exercise 58a.

S **TOP & THINK**

It is very important to change the sign of the E^0 potential whenever you reverse a half-reaction, as with the chromium equation shown here. But remember that we do *not* multiply the E^0 potentials by the factors used to balance the number of electrons when we add the half-reactions together to give the overall balanced equation.

Problem-Solving Tip The Sign of E^0 Indicates Spontaneity

For a reaction that is spontaneous at *standard conditions*, E^0_{cell} must be positive. A negative value of E^0_{cell} indicates that the reverse of the reaction written would be spontaneous at standard conditions.

21-16 Standard Electrode Potentials for Other Half-Reactions

In some half-cells, the oxidized and reduced species are both in solution as ions in contact with inert electrodes. For example, the standard iron(III) ion/iron(II) ion half-cell contains 1 M concentrations of the two ions. It involves the following half-reaction.

$$Fe^{3+} + e^- \longrightarrow Fe^{2+} \qquad E^0 = +0.771 \text{ V}$$

The standard dichromate ($Cr_2O_7^{2-}$) ion/chromium(III) ion half-cell consists of a 1 M concentration of each of the two ions in contact with an inert electrode. The balanced half-reaction in acidic solution (1.0 M H^+) is

$$Cr_2O_7^{2-} + 14H^+ + 6e^- \longrightarrow 2Cr^{3+} + 7H_2O \qquad E^0 = +1.33 \text{ V}$$

Standard reduction potentials for some other reactions are given in Table 21-3 and in Appendix J. These potentials can be used like those of Table 21-2.

▶ Platinum metal is often used as the inert electrode material. These two standard half-cells could be shown in shorthand notation as
Pt|Fe^{3+}(1 M), Fe^{2+}(1 M)   and
Pt|$Cr_2O_7^{2-}$(1 M), Cr^{3+}(1 M)

 TOP & THINK
If H^+ appears anywhere in the half-reaction, the solution is acidic. The presence of OH^- means that the solution is basic.

Table 21-3 Standard Reduction Potentials for Selected Half-Cells

Reduction Half-Reaction		Standard Reduction Potential E^0 (volts)
$Zn(OH)_4^{2-} + 2e^-$	$\longrightarrow Zn + 4OH^-$	−1.22
$Fe(OH)_2 + 2e^-$	$\longrightarrow Fe + 2OH^-$	−0.877
$2H_2O + 2e^-$	$\longrightarrow H_2 + 2OH^-$	−0.828
$PbSO_4 + 2e^-$	$\longrightarrow Pb + SO_4^{2-}$	−0.356
$NO_3^- + H_2O + 2e^-$	$\longrightarrow NO_2^- + 2OH^-$	+0.01
$Sn^{4+} + 2e^-$	$\longrightarrow Sn^{2+}$	+0.15
$AgCl + e^-$	$\longrightarrow Ag + Cl^-$	+0.222
$Hg_2Cl_2 + 2e^-$	$\longrightarrow 2Hg + 2Cl^-$	+0.27
$O_2 + 2H_2O + 4e^-$	$\longrightarrow 4OH^-$	+0.40
$NiO_2 + 2H_2O + 2e^-$	$\longrightarrow Ni(OH)_2 + 2OH^-$	+0.49
$H_3AsO_4 + 2H^+ + 2e^-$	$\longrightarrow H_3AsO_3 + H_2O$	+0.58
$Fe^{3+} + e^-$	$\longrightarrow Fe^{2+}$	+0.771
$ClO^- + H_2O + 2e^-$	$\longrightarrow Cl^- + 2OH^-$	+0.89
$NO_3^- + 4H^+ + 3e^-$	$\longrightarrow NO + 2H_2O$	+0.96
$O_2 + 4H^+ + 4e^-$	$\longrightarrow 2H_2O$	+1.229
$Cr_2O_7^{2-} + 14H^+ + 6e^-$	$\longrightarrow 2Cr^{3+} + 7H_2O$	+1.33
$Cl_2 + 2e^-$	$\longrightarrow 2Cl^-$	+1.360
$MnO_4^- + 8H^+ + 5e^-$	$\longrightarrow Mn^{2+} + 4H_2O$	+1.507
$PbO_2 + HSO_4^{2-} + 3H^+ + 2e^-$	$\longrightarrow PbSO_4 + 2H_2O$	+1.685

Increasing strength as oxidizing agent; increasing ease of reduction (left arrow, downward)

Increasing strength as reducing agent; increasing ease of oxidation (right arrow, upward)

EXAMPLE 21-4 Predicting the Direction of Reactions

In an acidic solution at standard conditions, will tin(IV) ions, Sn^{4+}, oxidize gaseous nitrogen oxide, NO, to nitrate ions, NO_3^-, or will NO_3^- oxidize Sn^{2+} to Sn^{4+} ions? Write the equation for the spontaneous (product-favored) reaction and calculate E^0_{cell} for this reaction.

Plan

We refer to the table of standard reduction potentials (Table 21-3) and choose the appropriate half-reactions.

TOP & THINK
Pay careful attention to the conditions (acidic or basic) and the half-cell reactions mentioned in the problem. For example, there also is a half-reaction in Table 21-3 with NO_3^- being reduced to NO_2^- in basic solution. This is not the correct half-reaction to use for this question.

Solution

The NO_3^-/NO reduction half-reaction has the more positive E^0 value, so we write it first and write the Sn^{4+}/Sn^{2+} half-reaction as an oxidation. We balance the electron transfer and add the two half-reactions to obtain the equation for the *spontaneous* reaction. Then we add the half-reaction potentials to obtain the overall cell potential.

$$
\begin{array}{ll}
 & E^0 \\
2(NO_3^- + 4H^+ + 3e^- \longrightarrow NO + 2H_2O) & \text{(reduction)} \quad +0.96 \text{ V} \\
3(Sn^{2+} \longrightarrow Sn^{4+} + 2e^-) & \text{(oxidation)} \quad -0.15 \text{ V} \\
\hline
2NO_3^- + 8H^+ + 3Sn^{2+} \longrightarrow 2NO + 4H_2O + 3Sn^{4+} & E^0_{cell} = +0.81 \text{ V}
\end{array}
$$

Because E^0_{cell} is positive for this reaction,

nitrate ions spontaneously oxidize tin(II) ions to tin(IV) ions and are reduced to nitrogen oxide in acidic solution.

You should now work Exercises 59 and 60.

Problem-Solving Tip Remember What We Mean by Standard Conditions

▶ Remember that temperature is not officially part of standard conditions. But most thermodynamic data are given at a temperature of 25°C.

When we say that a reaction takes place at *standard conditions*, we mean the following:

1. The temperature is the standard thermodynamic temperature, 25°C, unless stated otherwise.

2. All reactants and products are at *unit activity*. This means that

   a. Any solution species that takes part in the reaction is at a concentration of exactly 1 *M*;

   b. Any gas that takes part in the reaction is at a pressure of exactly 1 atm;

   c. Any other substance that takes part in the reaction is *pure*.

(When we say "takes part in the reaction," we mean either as a reactant or a product.) These are the same conditions that were described as *standard state conditions* for thermodynamic purposes (see Section 15-6). When one or more of these conditions is not satisfied, we must adjust our calculations for nonstandard conditions. We shall learn how to do this in Section 21-19.

Now that we know how to use standard reduction potentials, let's see how they explain the reaction that occurs in the electrolysis of aqueous NaCl. The first two electrolytic cells we considered involved *molten* NaCl and *aqueous* NaCl (see Sections 21-3 and 21-4). There was no doubt that in molten NaCl, metallic Na would be produced by reduction of Na^+, and gaseous Cl_2 would be produced by oxidation of Cl^-. But we found in aqueous NaCl that H_2O, rather than Na^+, was reduced. This is consistent with the less negative reduction potential of H_2O compared with Na^+.

$$
\begin{array}{ll}
 & E^0 \\
2H_2O + 2e^- \longrightarrow H_2 + 2OH^- & -0.828 \text{ V} \\
Na^+ + e^- \longrightarrow Na & -2.714 \text{ V}
\end{array}
$$

The more easily reduced species, H_2O, is reduced.

Electrode potentials measure only the relative *thermodynamic* likelihood for various half-reactions. In practice, kinetic factors can complicate matters. For instance, sometimes the electrode process is limited by the rate of diffusion of dissolved species to or from the electrode surface. At some cathodes, the rate of electron transfer from the electrode to a reactant is the rate-limiting step, and a higher voltage (called *overvoltage*) must be applied to accomplish the reduction. As a result of such factors, a half-reaction that is *thermodynamically* more favorable than some other process still might not occur at a significant rate. In the electrolysis of NaCl(aq), Cl^- is oxidized to Cl_2 gas (-1.360 V), instead of H_2O being oxidized to form O_2 gas (-1.229 V), because of the overvoltage of O_2 on Pt, the inert electrode.

21-17 Corrosion

Ordinary **corrosion** is the redox process by which metals are oxidized by oxygen, O_2, in the presence of moisture. There are other kinds, but this is the most common. The problem of corrosion and its prevention are of both theoretical and practical interest. Corrosion is responsible for the loss of billions of dollars annually in metal products. The mechanism of corrosion has been studied extensively. It is now known that the oxidation of metals occurs most readily at points of strain (where the metals are most "active"). Thus, a steel nail, which is mostly iron (see Section 26-7), first corrodes at the tip and head (Figure 21-11). A bent nail corrodes most readily at the bend.

A point of strain in a steel object acts as an anode where the iron is oxidized to iron(II) ions, and pits are formed (Figure 21-12).

$$Fe \longrightarrow Fe^{2+} + 2e^- \qquad \text{(oxidation, anode)}$$

The electrons produced then flow through the nail to areas exposed to O_2. These act as cathodes where oxygen is reduced to hydroxide ions, OH^-.

$$O_2 + 2H_2O + 4e^- \longrightarrow 4OH^- \qquad \text{(reduction, cathode)}$$

▶ Salt (NaCl) does not, in itself, do any active corrosion. It merely accelerates the corrosion of most metals, especially steel, by making the water layer more electrically conductive. This makes it easier for O_2 to oxidize the metal, via the electrically conducting saltwater, without having to diffuse through the water layer to come into physical contact with the metal surface.

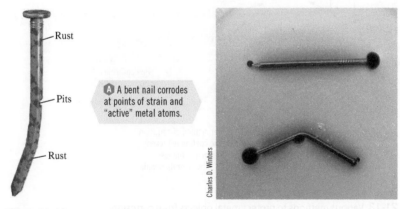

Ⓐ A bent nail corrodes at points of strain and "active" metal atoms.

Ⓑ Two nails were placed in an agar gel that contained phenolphthalein and potassium ferricyanide, $K_3[Fe(CN)_6]$. As the nails corrode, they produce Fe^{2+} ions at each end and at the bend (an oxidation, so these are anodic regions). Fe^{2+} ions react with $[Fe(CN)_6]^{3-}$ ions to form $Fe_3[Fe(CN)_6]_2$, an intensely blue-colored compound. The rest of each nail functions as the cathode, at which water is reduced to H_2 and OH^- ions. The OH^- ions turn phenolphthalein pink.

Charles D. Winters

Figure 21-11 Corrosion of steel (iron).

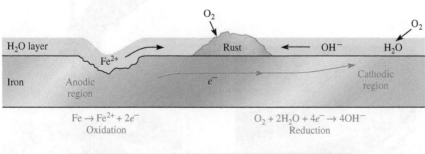

Overall process: $2Fe(s) + \frac{3}{2}O_2(aq) + xH_2O(\ell) \longrightarrow Fe_2O_3 \cdot xH_2O(s)$

Figure 21-12 The corrosion of iron. Pitting appears at the anodic region, where iron metal is initially oxidized to Fe^{2+} and then to Fe^{3+}. Rust appears at the cathodic region.

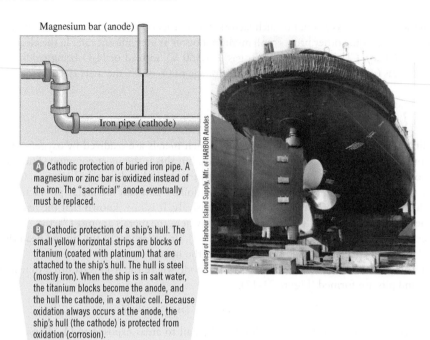

Magnesium bar (anode)

Iron pipe (cathode)

A Cathodic protection of buried iron pipe. A magnesium or zinc bar is oxidized instead of the iron. The "sacrificial" anode eventually must be replaced.

B Cathodic protection of a ship's hull. The small yellow horizontal strips are blocks of titanium (coated with platinum) that are attached to the ship's hull. The hull is steel (mostly iron). When the ship is in salt water, the titanium blocks become the anode, and the hull the cathode, in a voltaic cell. Because oxidation always occurs at the anode, the ship's hull (the cathode) is protected from oxidation (corrosion).

Courtesy of Harbour Island Supply, Mfr. of HARBOR Anodes

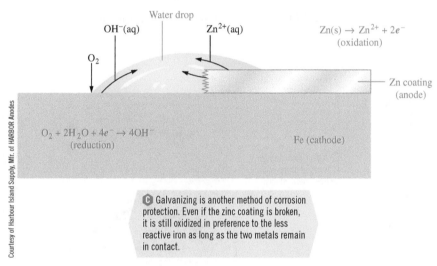

Water drop

$OH^-(aq)$ $Zn^{2+}(aq)$ $Zn(s) \rightarrow Zn^{2+} + 2e^-$
(oxidation)

O_2

Zn coating
(anode)

$O_2 + 2H_2O + 4e^- \rightarrow 4OH^-$
(reduction)

Fe (cathode)

Courtesy of Harbour Island Supply, Mfr. of HARBOR Anodes

C Galvanizing is another method of corrosion protection. Even if the zinc coating is broken, it is still oxidized in preference to the less reactive iron as long as the two metals remain in contact.

Figure 21-13 Various methods to protect metal objects from corrosion.

Charles D. Winters

Galvanized objects are made of steel that is coated with zinc to protect against corrosion.

Photodisc/Getty Images

Steel is plated with chromium for appearance as well as protection against corrosion.

Charles D. Winters

Corrosion is an undesirable electrochemical reaction with very serious economic consequences.

At the same time, the Fe^{2+} ions migrate through the moisture on the surface. The overall reaction is obtained by balancing the electron transfer and adding the two half-reactions.

$$2(Fe \longrightarrow Fe^{2+} + 2e^-) \quad \text{(oxidation, anode)}$$
$$O_2 + 2H_2O + 4e^- \longrightarrow 4OH^- \quad \text{(reduction, cathode)}$$
$$\overline{2Fe + O_2 + 2H_2O \longrightarrow 2Fe^{2+} + 4OH^-} \quad \text{(net reaction)}$$

The Fe^{2+} ions can migrate from the anode through the solution toward the cathode region, where they combine with OH^- ions to form iron(II) hydroxide. Iron is further oxidized by O_2 to the +3 oxidation state. The material we call rust is a complex hydrated form of iron(III) oxides and hydroxides with variable water composition; it can be represented as $Fe_2O_3 \cdot xH_2O$. The overall reaction for the rusting of iron is

$$2Fe(s) + \tfrac{3}{2}O_2(aq) + xH_2O(\ell) \longrightarrow Fe_2O_3 \cdot xH_2O(s)$$

▶ Note that water is needed for O_2 to oxidize iron. Dry O_2 will not, therefore, oxidize iron to form rust, nor does water without any O_2 have any ability to oxidize iron.

21-18 Corrosion Protection

There are several methods for protecting metals against corrosion. The most widely used are

1. Plating the metal with a thin layer of a less easily oxidized metal
2. Connecting the metal directly to a **sacrificial anode**, a piece of another metal that is more active and therefore is preferentially oxidized
3. Allowing a protective film, such as a metal oxide, to form naturally on the surface of the metal
4. Galvanizing, or coating steel with zinc, a more active metal
5. Applying a protective coating, such as paint

The thin layer of tin on tin-plated steel cans is less easily oxidized than iron, and it protects the steel underneath from corrosion. It is deposited either by dipping the can into molten tin or by electroplating. Copper is also less active than iron (see Table 21-2). It is sometimes deposited by electroplating to protect metals when food is not involved. Whenever the layer of tin or copper is breached, the iron beneath it corrodes even more rapidly than it would without the coating because of the adverse electrochemical cell that is set up.

Figure 21-13a shows an iron pipe connected to a strip of magnesium, a more active metal, to protect the iron from oxidation. The magnesium is preferentially oxidized. It is called a "sacrificial anode." Similar methods are used to protect bridges and the hulls of ships from corrosion. Other inexpensive active metals, such as zinc, are also used as sacrificial anodes. Galvanizing (coating the iron with zinc) combines these two approaches. Even if the zinc coating is broken so the iron is exposed, the iron is not oxidized as long as it is in contact with the more easily oxidized zinc (see Figure 21-13c).

Aluminum, a very active metal, reacts rapidly with O_2 from the air to form a surface layer of aluminum oxide, Al_2O_3, so thin that it is transparent. This very tough, hard substance is inert to oxygen, water, and most other corrosive agents in the environment. In this way, objects made of aluminum form their own protective layers and need not be treated further to inhibit corrosion. Iron oxide, unfortunately, does not form a protective coating on iron or steel. Instead it expands and flakes off, exposing fresh iron for further corrosion.

▶ Compare the potentials for the reduction half-reactions to see which metal is more easily oxidized. The more positive the reduction potential for a metal, the more stable the metal is as the free element and the harder it is to oxidize.

	$E^0_{reduction}$
$Mg^{2+} + 2e^- \longrightarrow Mg$	$-2.37\,V$
$Zn^{2+} + 2e^- \longrightarrow Zn$	$-0.763\,V$
$Fe^{2+} + 2e^- \longrightarrow Fe$	$-0.44\,V$
$Sn^{2+} + 2e^- \longrightarrow Sn$	$-0.14\,V$
$Cu^{2+} + 2e^- \longrightarrow Cu$	$+0.337\,V$

▶ Acid rain endangers structural aluminum by dissolving this Al_2O_3 coating.

Effect of Concentrations (or Partial Pressures) on Electrode Potentials

21-19 The Nernst Equation

Standard electrode potentials, designated E^0, refer to standard-state conditions. These standard-state conditions are one molar solutions for ions and one atmosphere pressure for gases with all solids and liquids in their standard states at 25°C. As any of the standard

S TOP & THINK
Remember that we refer to *thermodynamic* standard-state conditions and not standard temperature and pressure as in gas law calculations.

cells described earlier operates, and concentrations or pressures of reactants change, the observed cell voltage drops. Similarly, cells constructed with solution concentrations different from one molar, or gas pressures different from one atmosphere, cause the corresponding potentials to deviate from standard electrode potentials.

The **Nernst equation** is used to calculate electrode potentials and cell potentials for concentrations and partial pressures other than standard-state values.

▶ In this equation, the expression following the minus sign represents how much the *nonstandard* conditions cause the electrode potential to deviate from its standard value, E^0. The Nernst equation is normally presented in terms of base-10 logarithms, as we will do in this text.

$$E = E^0 - \frac{2.303 \, RT}{nF} \log Q$$

where

E = potential under the *nonstandard* conditions
E^0 = *standard* potential
R = gas constant, $8.314 \, \text{J/mol} \cdot \text{K}$
T = absolute temperature in K
n = number of moles of electrons transferred in the reaction or half-reaction
F = faraday, $96,485 \, \text{C/mol} \, e^- \times 1 \, \text{J/(V} \cdot \text{C)} = 96,485 \, \text{J/V} \cdot \text{mol} \, e^-$
Q = reaction quotient

The reaction quotient, Q, was introduced in Section 17-4. It involves a ratio of concentrations or pressures of products to those of reactants, each raised to the power indicated by the coefficient in the balanced equation. The Q expression used in the Nernst equation is the thermodynamic reaction quotient; it can include *both* concentrations and pressures. Substituting these values into the Nernst equation at 25°C gives

▶ At 25°C, the value of $\dfrac{2.303 \, RT}{F}$ is 0.0592; at any other temperature, this term must be recalculated. Can you show that this term has the units $\text{V} \cdot \text{mol}$?

$$E = E^0 - \frac{0.0592}{n} \log Q \qquad \text{(Note: in terms of base-10 log)}$$

In general, half-reactions for standard reduction potentials are written

$$x \, \text{Ox} + ne^- \longrightarrow y \, \text{Red}$$

"Ox" refers to the oxidized species and "Red" to the reduced species; x and y are their coefficients, respectively, in the balanced equation. The Nernst equation for any *cathode* half-cell (*reduction* half-reaction) is

$$E = E^0 - \frac{0.0592}{n} \log \frac{[\text{Red}]^y}{[\text{Ox}]^x} \qquad \text{(for reduction half-reaction)}$$

For the familiar half-reaction involving metallic zinc and zinc ions,

$$\text{Zn}^{2+} + 2e^- \longrightarrow \text{Zn} \qquad E^0 = -0.763 \, \text{V}$$

the corresponding Nernst equation is

▶ Metallic Zn is a pure solid, so its concentration does not appear in Q.

$$E = E^0 - \frac{0.0592}{2} \log \frac{1}{[\text{Zn}^{2+}]} \qquad \text{(for reduction)}$$

We substitute the E^0 value into the equation to obtain

$$E = -0.763 \, \text{V} - \frac{0.0592}{2} \log \frac{1}{[\text{Zn}^{2+}]}$$

EXAMPLE 21-5 The Nernst Equation for a Half-Cell Reaction

Calculate the potential, E, for the Fe^{3+}/Fe^{2+} electrode when the concentration of Fe^{2+} is exactly five times that of Fe^{3+}.

Plan

The Nernst equation lets us calculate potentials for concentrations other than one molar. The tabulation of standard reduction potentials gives us the value of E^0 for the reduction half-reaction. We use the balanced half-reaction and the given concentration ratio to calculate the value of Q. Then we substitute this into the Nernst equation, with n equal to the number of moles of electrons involved in the half-reaction.

Solution

The reduction half-reaction is

$$Fe^{3+} + e^- \longrightarrow Fe^{2+} \qquad E^0 = +0.771 \text{ V}$$

We are told that the concentration of Fe^{2+} is five times that of Fe^{3+}, or $[Fe^{2+}] = 5[Fe^{3+}]$. Calculating the value of Q,

$$Q = \frac{[\text{Red}]^y}{[\text{Ox}]^x} = \frac{[Fe^{2+}]}{[Fe^{3+}]} = \frac{5[Fe^{3+}]}{[Fe^{3+}]} = 5$$

The balanced half-reaction shows one mole of electrons, or $n = 1$. Putting values into the Nernst equation,

$$E = E^0 - \frac{0.0592}{n} \log Q = +0.771 - \frac{0.0592}{1} \log 5 = (+0.771 - 0.041) \text{ V}$$

$$= +0.730 \text{ V}$$

You should now work Exercise 82.

▶ The potential for this electrode is lower than E^0 for the standard Fe^{3+}/Fe^{2+} electrode, in which both concentrations are 1 *M*. This should make qualitative sense to you from having studied LeChatelier's Principle. The more products that are present relative to reactants, the less the driving force (lower potential) to make more products.

The Nernst equation can be applied to balanced equations for redox reactions. One approach is to correct the reduction potential for each half-reaction to take into account the nonstandard concentrations or pressures.

EXAMPLE 21-6 The Nernst Equation for an Overall Cell Reaction

A cell is constructed at 25°C as follows. One half-cell consists of the Fe^{3+}/Fe^{2+} couple in which $[Fe^{3+}] = 1.00 \text{ } M$ and $[Fe^{2+}] = 0.100 \text{ } M$; the other involves the MnO_4^-/Mn^{2+} couple in acidic solution in which $[MnO_4^-] = 1.00 \times 10^{-2} \text{ } M$, $[Mn^{2+}] = 1.00 \times 10^{-4} \text{ } M$, and $[H^+] = 1.00 \times 10^{-3} \text{ } M$. (a) Find the electrode potential for each half-cell with these concentrations, and (b) calculate the overall cell potential.

Plan

(a) We can apply the Nernst equation to find the reduction potential of each half-cell with the stated concentrations. (b) As in Section 21-15, we write the half-reaction with the more positive potential (*after* correction) along with its potential. We reverse the other half-reaction and change the sign of its E value. We balance the electron transfer and then add the half-reactions and their nonstandard, corrected potentials to find the overall cell potential.

Solution

(a) For the MnO_4^-/Mn^{2+} half-cell *as a reduction*,

$$MnO_4^- + 8H^+ + 5e^- \longrightarrow Mn^{2+} + H_2O \qquad E^0 = +1.507\ V$$

$$E = E^0 - \frac{0.0592}{n} \log \frac{[Mn^{2+}]}{[MnO_4^-][H^+]^8}$$

$$= +1.507\ V - \frac{0.0592}{5} \log \frac{1.00 \times 10^{-4}}{(1.00 \times 10^{-2})(1.00 \times 10^{-3})^8}$$

$$= +1.507\ V - \frac{0.0592}{5} \log (1.00 \times 10^{22}) = +1.507\ V - \frac{0.0592}{5} (22.0)$$

$$= +1.246\ V$$

(b) For the Fe^{3+}/Fe^{2+} half-cell *as a reduction*,

$$Fe^{3+} + e^- \longrightarrow Fe^{2+} \qquad E^0 = +0.771\ V$$

$$E = E^0 - \frac{0.0592}{n} \log \frac{[Fe^{2+}]}{[Fe^{3+}]} = +0.771\ V - \frac{0.0592}{1} \log \frac{0.100}{1.00}$$

$$= +0.771\ V - \frac{0.0592}{1} \log (0.100) = +0.771\ V - \frac{0.0592}{1} (-1.00)$$

$$= +0.830\ V$$

The corrected potential for the MnO_4^-/Mn^{2+} half-cell is greater than that for the Fe^{3+}/Fe^{2+} half-cell, so we reverse the latter, balance the electron transfer, and add.

	E (corrected)
$MnO_4^- + 8H^+ + 5e^- \longrightarrow Mn^{2+} + 4H_2O$	+1.246 V
$5(Fe^{2+} \longrightarrow Fe^{3+} + e^-)$	−0.830 V
$MnO_4^- + 8H^+ + 5Fe^{2+} \longrightarrow Mn^{2+}\ 4H_2O + 5Fe^{3+}$	$E_{cell} = 0.416\ V$

The cell in Example 21-6 can be represented in shorthand notation as

$$Pt\,|\,Fe^{2+}(0.100\ M),\ Fe^{3+}(1.00\ M)\,\|\,H^+(1.00 \times 10^{-3}\ M),\ MnO_4^-(1.00 \times 10^{-2}\ M),\ Mn^{2+}(1.00 \times 10^{-4}\ M)\,|\,Pt$$

The reaction as written is spontaneous (product-favored) under the stated conditions, with a potential of +0.416 volt *when the cell starts operation*. As the cell discharges and current flows, the product concentrations, $[Mn^{2+}]$ and $[Fe^{3+}]$, increase. At the same time, reactant concentrations, $[MnO_4^-]$, $[H^+]$, and $[Fe^{2+}]$, decrease. This increases the magnitude of $\log Q_{cell}$, so the correction factor becomes more negative. Thus, the overall E_{cell} *decreases* (the reaction becomes less favorable). Eventually the cell potential approaches zero (equilibrium), and the cell "runs down." The cell is completely run down ($E_{cell} = 0$) when the term $\frac{0.0592}{n} \log Q_{cell}$ is equal in magnitude to E_{cell}^0.

We can also find the cell potential for a nonstandard cell by first finding E_{cell}^0 for the overall standard cell reaction, and then using the Nernst equation to correct for nonstandard concentrations. The next example illustrates this approach.

EXAMPLE 21-7 The Nernst Equation for an Overall Cell Reaction

A cell is constructed at 25°C as follows. One half-cell consists of a chlorine/chloride, Cl_2/Cl^-, electrode with the partial pressure of $Cl_2 = 0.100$ atm and $[Cl^-] = 0.100\ M$. The other half-cell involves the MnO_4^-/Mn^{2+} couple in acidic solution with $[MnO_4^-] = 0.100\ M$, $[Mn^{2+}] = 0.100\ M$, and $[H^+] = 0.100\ M$. Apply the Nernst equation to the overall cell reaction to determine the cell potential for this cell.

Plan

First we determine the overall cell reaction and its *standard* cell potential, E^0_{cell}, as in Examples 21-3 and 21-4. Then we apply the Nernst equation to the overall cell.

Solution

The MnO_4^-/Mn^{2+} half-reaction has the more positive reduction potential, so we write it first. Then we write the Cl_2/Cl^- half-reaction as an oxidation, balance the electron transfer, and add the two half-reactions and their potentials to obtain the overall cell reaction and its E^0_{cell}.

$$ E^0$$

$$2(MnO_4^- + 8H^+ + 5e^- \longrightarrow Mn^{2+} + 4H_2O) \qquad +1.507\ V$$
$$5(2Cl^- \longrightarrow Cl_2 + 2e^-) \qquad -1.360\ V$$

$$\overline{2MnO_4^- + 16H^+ + 10Cl^- \longrightarrow 2Mn^{2+} + 8H_2O + 5Cl_2 \quad E^0_{cell} = +0.147\ V}$$

In the overall reaction, $n = 10$. We then apply the Nernst equation to this overall reaction by substituting appropriate concentration and partial pressure values. Because Cl_2 is a gaseous component, its term in the Nernst equation involves its partial pressure, P_{Cl_2}, in atm.

> ▶ Can you write the shorthand notation for this cell?

$$E_{cell} = E^0_{cell} - \frac{0.0592}{n} \log \frac{[Mn^{2+}]^2 (P_{Cl_2})^5}{[MnO_4^-]^2 [H^+]^{16} [Cl^-]^{10}}$$

$$= 0.147\ V - \frac{0.0592}{10} \log \frac{(0.100)^2 (0.100)^5}{(0.100)^2 (0.100)^{16} (0.100)^{10}}$$

$$= 0.147\ V - \frac{0.0592}{10} \log(1.00 \times 10^{21})$$

$$= 0.147\ V - \frac{0.0592}{10} (21.00) = \boxed{0.023\ V}$$

> ▶ When evaluating Q in the Nernst equation, remember that (a) molar concentrations are used for dissolved species, and (b) partial pressures of gases are expressed in atmospheres. Units are omitted.

You should now work Exercises 84 and 90.

The method illustrated in Example 21-7, applying the Nernst equation to the *overall* cell reaction, usually involves less calculation than correcting the separate half-reactions as in Example 21-6. We interpret our results as follows: The positive overall cell potentials in Examples 21-6 and 21-7 tell us that each of these cell reactions is spontaneous *in the direction written* for the concentrations given. If the resulting cell potential were negative, the *reverse* reaction would be favored at those concentrations. We could then reverse the equation for the overall cell reaction and change the sign of its potential to describe the spontaneous operation of the cell.

> ▶ Now solve Example 21-6 by applying the Nernst equation to the *overall* cell reaction to determine the cell potential.

Problem-Solving Tip *Be Careful of the Value of n*

How do you know what value of n to use? Remember that n must be the number of moles of electrons transferred in the *balanced* equation for the process to which you apply the Nernst equation.

1. For a *half-reaction*, n represents the number of moles of electrons in that half-reaction. In Example 21-6 we applied the Nernst equation to each half-reaction separately, so we used $n = 5$ for the half-reaction

$$MnO_4^- + 8H^+ + 5e^- \longrightarrow Mn^{2+} + H_2O$$

and we used $n = 1$ for the half-reaction

$$Fe^{3+} + e^- \longrightarrow Fe^{2+}$$

2. For an *overall reaction*, n represents the total number of moles of electrons transferred. In Example 21-7 we applied the Nernst equation to an *overall* reaction in which 10 moles of electrons was transferred from 10 moles of Cl^- to 2 moles of MnO_4^-, so we used the value $n = 10$.

21-20 Using Electrochemical Cells to Determine Concentrations

In Section 21-19, we used known concentrations to predict cell voltage. We can reverse this reasoning and apply the same ideas to *measure* the voltage of a cell and then use the Nernst equation to solve for an *unknown* concentration. The following example illustrates such an application.

EXAMPLE 21-8   The Nernst Equation

▶ This cell is similar to the zinc–hydrogen cell that we discussed in Section 21-12, except that the hydrogen concentration is not (necessarily) 1.00 *M*.

We construct an electrochemical cell at 25°C as follows. One half-cell is a standard Zn^{2+}/Zn cell, that is, a strip of zinc immersed in a 1.00 *M* Zn^{2+} solution; the other is a *nonstandard* hydrogen electrode in which a platinum electrode is immersed in a solution of *unknown* hydrogen ion concentration with gaseous hydrogen bubbling through it at a pressure of 1.000 atm. The observed cell voltage is 0.522 V. (a) Calculate the value of the reaction quotient Q. (b) Calculate $[H^+]$ in the second half-cell. (c) Determine the pH of the solution in the second half-cell.

Plan

We saw in Section 21-12 that the zinc–hydrogen cell operated with oxidation at the zinc electrode and reduction at the hydrogen electrode, with a *standard* cell potential of 0.763 V.

$$\text{overall:} \quad Zn + 2H^+ \longrightarrow Zn^{2+} + H_2 \qquad E^0_{cell} = 0.763 \text{ V}$$

(a) We rearrange the Nernst equation to solve for the reaction quotient, Q, from the measured cell voltage with $n = 2$. (b) We substitute concentrations and partial pressures in the expression for Q. Then we can solve for the only unknown, $[H^+]$. (c) The pH can be determined from the $[H^+]$ determined in Part (b).

Solution

(a)
$$E_{cell} = E^0_{cell} - \frac{0.0592}{n} \log Q$$

Substituting and solving for Q,

$$0.522 \text{ V} = 0.763 \text{ V} - \frac{0.0592}{2} \log Q$$

$$\frac{0.0592 \text{ V}}{2} \log Q = (0.763 - 0.522) \text{ V} = 0.241 \text{ V}$$

$$\log Q = \frac{(2)(0.241 \text{ V})}{0.0592 \text{ V}} = 8.14$$

$$Q = 10^{8.14} = 1.4 \times 10^8$$

(b) We write the expression for Q from the balanced overall equation, and solve for $[H^+]$.

$$Q = \frac{[Zn^{2+}]P_{H_2}}{[H^+]^2}$$

$$[H^+]^2 = \frac{[Zn^{2+}]P_{H_2}}{Q} = \frac{(1.00)(1.00)}{1.4 \times 10^8} = 7.1 \times 10^{-9}$$

$$[H^+] = 8.4 \times 10^{-5} \text{ } M$$

(c)
$$pH = -\log [H^+] = -\log (8.4 \times 10^{-5}) = 4.08$$

You should now work Exercises 85 and 86.

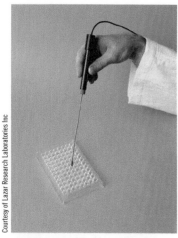

Microelectrodes have been developed to measure concentrations in very small volumes of solution.

A **pH meter** contains a voltaic cell, such as that in Example 21-8, that can be used to measure the [H⁺] concentration and thus the pH of an unknown solution. It is inconvenient to bubble hydrogen gas at a controlled pressure through a hydrogen electrode, so the routine use of this electrode in many environments is not practical. A typical commercial pH meter (Figure 21-14) incorporates a miniaturized pair of electrodes that are more portable and less fragile. One of these electrodes is a **glass electrode**, usually consisting of an AgCl-coated silver wire in contact with an HCl solution of known concentration (usually 1.00 M) in a thin-walled glass bulb (see Figure 21-14a). A **saturated calomel electrode** is often used as the second half-cell (reference electrode). This consists of a platinum wire in contact with a paste of liquid mercury and solid mercury(I) chloride, $Hg_2Cl_2(s)$, all immersed in a saturated solution of potassium chloride, KCl. When the glass electrode is placed in a solution, a potential is developed across the thin glass membrane; this potential depends on the [H⁺] concentration, and hence on the pH. The overall voltage of this cell thus measures the pH of the solution in contact with the glass electrode. Each change of one pH unit causes a voltage change of 0.0592 volts. The pH meter is designed to measure very small voltages, and its display is calibrated to give a direct readout of the pH of the solution. Meters of this general design have many routine but important applications in medicine, chemistry, biology, agriculture, environmental analysis, and numerous other areas. Glass electrodes can be made small enough to be implanted into blood vessels or even individual living cells.

Electrochemical procedures that use the principles illustrated here provide a convenient method for making many concentration measurements.

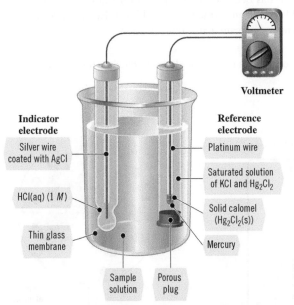

Ⓐ The glass electrode (*left*) is an Ag(s)/AgCl(s) half-cell immersed in a standard HCl solution that is enclosed by a thin glass membrane. This electrode develops a potential that is sensitive to the external pH relative to that in the internal HCl standard solution. The saturated calomel electrode is the reference electrode.

Ⓑ A portable pH meter that can be used in a wide variety of environments

Ⓒ The fragile tip of the glass electrode is usually surrounded by a protective open plastic cover.

Figure 21-14 The workings of a commercial pH meter.

ENRICHMENT

Concentration Cells

As we have seen, different concentrations of ions in a half-cell result in different half-cell potentials. We can use this idea to construct a **concentration cell**, in which both half-cells are composed of the same species but in different ion concentrations. Suppose we set up such a cell using the Cu^{2+}/Cu half-cell that we introduced in Section 21-9. We put copper electrodes into two aqueous solutions, one that is 0.10 M $CuSO_4$ and another that is 1.00 M $CuSO_4$. To complete the cell construction, we connect the two electrodes with a wire and join the two solutions with a salt bridge as usual (Figure 21-15). Now the relevant standard reduction half-reaction in either half-cell is

$$Cu^{2+} + 2e^- \longrightarrow Cu \qquad E^0 = +0.337 \text{ V}$$

Thus the Cu^{2+} ions in the more concentrated half-cell can be considered as the reactant, and those in the more dilute cell as the product.

$$Cu^{2+}(1.00\ M) \longrightarrow Cu^{2+}(0.10\ M)$$

The overall cell potential can be calculated by applying the Nernst equation to the overall cell reaction. We must first find E^0, the standard cell potential *at standard concentrations*; because the same electrode and the same type of ions are involved in both half-cells, this E^0_{cell} is always zero. Thus,

$$E_{cell} = E^0_{cell} - \frac{0.0592}{n} \log \frac{[\text{dilute solution}]}{[\text{concentrated solution}]}$$

$$= 0 - \frac{0.0592}{2} \log \frac{0.10}{1.00} = +0.030 \text{ V}$$

As the reaction proceeds, $[Cu^{2+}]$ decreases in the more concentrated half-cell and increases in the more dilute half-cell until the two concentrations are equal; at that point $E_{cell} = 0$, and equilibrium has been reached. This equilibrium $[Cu^{2+}]$ is the same concentration that would have been formed if we had just mixed the two solutions directly to obtain a solution of intermediate concentration.

In any concentration cell, the spontaneous reaction is always in the direction that tends to equalize the concentrations.

▶ The overall cell potential is positive; the reaction is spontaneous as written.

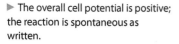

Figure 21-15 The concentration cell $Cu|Cu^{2+}(0.10\ M)\|Cu^{2+}(1.00\ M)|Cu$. The overall reaction lowers the $[Cu^{2+}]$ concentration in the more concentrated solution and increases it in the more dilute solution.

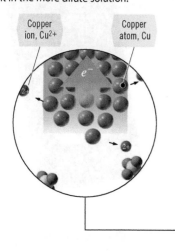

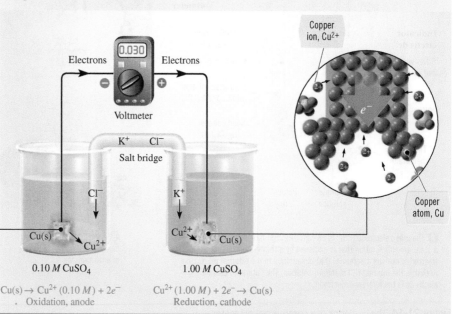

$$Cu(s) \rightarrow Cu^{2+}(0.10\ M) + 2e^-$$
Oxidation, anode

$$Cu^{2+}(1.00\ M) + 2e^- \rightarrow Cu(s)$$
Reduction, cathode

21-21 The Relationship of E^0_{cell} to ΔG^0 and K_{eq}

In Section 17-12 we studied the relationship between the standard Gibbs free energy change, ΔG^0, and the thermodynamic equilibrium constant, K_{eq}.

$$\Delta G^0 = -RT \ln K_{eq}$$

There is also a simple relationship between ΔG^0 and the standard cell potential, E^0_{cell}, for a redox reaction (reactants and products in standard states).

$$\Delta G^0 = -nFE^0_{cell}$$

ΔG^0 can be thought of as the *negative of the maximum electrical work* that can be obtained from a redox reaction. In this equation, n is the number of moles of electrons transferred in the overall process (mol e^-/mol rxn), and F is the faraday, 96,485 J/V $\cdot$ mol e^-.

Combining these relationships for ΔG^0 gives the relationship between E^0_{cell} values and equilibrium constants.

$$\underbrace{-nFE^0_{cell}}_{\Delta G^0} = \underbrace{-RT \ln K_{eq}}_{\Delta G^0}$$

After multiplying by -1, we can rearrange.

$$nFE^0_{cell} = RT \ln K_{eq} \quad \text{or} \quad E^0_{cell} = \frac{RT \ln K_{eq}}{nF} \quad \text{or} \quad \ln K_{eq} = \frac{nFE^0_{cell}}{RT}$$

If any one of the three quantities ΔG^0, K_{eq}, and E^0_{cell} is known, the other two can be calculated using these equations. It is usually much easier to determine K_{eq} for a redox reaction from electrochemical measurements than by measuring equilibrium concentrations directly, as described in Chapter 17. Keep in mind the following for all redox reactions *at standard conditions*.

Forward Reaction	ΔG^0	K_{eq}	E^0_{cell}	
spontaneous (product-favored)	$-$	>1	$+$	
at equilibrium	0	1	0	(all substances *at standard conditions*)
nonspontaneous (reactant-favored)	$+$	<1	$-$	

EXAMPLE 21-9   Calculation of ΔG^0 from Cell Potentials

Use tabulated standard electrode potentials to calculate the standard Gibbs free energy change, ΔG^0, in J/mol at 25°C for the following reaction.

$$3Sn^{4+} + 2Cr \longrightarrow 3Sn^{2+} + 2Cr^{3+}$$

Plan

We evaluate the standard cell potential as we have done before. Then we apply the relationship $\Delta G^0 = -nFE^0_{cell}$.

Solution

The standard reduction potential for the Sn^{4+}/Sn^{2+} couple is $+0.15$ volt; that for the Cr^{3+}/Cr couple is -0.74 volt. The equation for the reaction shows Cr being oxidized to Cr^{3+}, so the sign of the E^0 value for the Cr^{3+}/Cr couple is reversed. The overall reaction, the sum of the two half-reactions, has a cell potential equal to the sum of the two half-reaction potentials.

	E^0
$3(Sn^{4+} + 2\ e^- \longrightarrow Sn^{2+})$	$+0.15$ V
$2(Cr \longrightarrow Cr^{3+} + 3e^-)$	$-(-0.74$ V$)$
$3Sn^{4+} + 2Cr \longrightarrow 3Sn^{2+} + 2Cr^{3+}$	$E^0_{cell} = +0.89$ V

▶ Recall from Chapter 15 that ΔG^0 can be expressed in joules per *mole of reaction*. Here we ask for the number of joules of free energy change that corresponds to the reaction of 2 moles of chromium with 3 moles of tin(IV) to give 3 moles of tin(II) ions and 2 moles of chromium(III) ions.

STOP & THINK
The negative value of ΔG^0 tells us that the standard reaction is spontaneous (product-favored), consistent with the positive value of E^0_{cell}. This tells us nothing about the speed with which the reaction would occur.

The positive value of E^0_{cell} indicates that the forward reaction is spontaneous.

$$\Delta G^0 = -nFE^0_{cell} = -\left(\frac{6 \text{ mol } e^-}{\text{mol rxn}}\right)\left(\frac{9.65 \times 10^4 \text{ J}}{\text{V} \cdot \text{mol } e^-}\right)(+0.89 \text{ V})$$

$$= -5.2 \times 10^5 \text{ J/mol rxn} \quad \text{or} \quad -5.2 \times 10^2 \text{ kJ/mol rxn}$$

You should now work Exercise 101.

EXAMPLE 21-10   Calculation of K_{eq} from Cell Potentials

Use the standard cell potential to calculate the value of the equilibrium constant, K_{eq}, at 25°C for the following reaction.

$$2Cu + PtCl_6^{2-} \longrightarrow 2Cu^+ + PtCl_4^{2-} + 2Cl^-$$

Plan

We calculate E^0_{cell} for the reaction as written. Then we use it to calculate K_{eq}.

Solution

First we find the appropriate half-reactions. As the equation is written, Cu is oxidized to Cu^+, so we write the Cu^+/Cu couple as an oxidation and reverse the sign of its tabulated E^0 value. We balance the electron transfer and then add the half-reactions. The resulting E^0_{cell} value can be used to calculate the equilibrium constant, K_{eq}, for the reaction *as written*.

STOP & THINK
As the problem is stated, we must keep the equation as written. We must therefore accept either a positive or a negative value of E^0_{cell}. A negative value of E^0_{cell} would lead to $K_{eq} < 1$.

$$
\begin{array}{ll}
2(Cu \longrightarrow Cu^+ + e^-) & -(+0.521 \text{ V}) \\
PtCl_6^{2-} + 2e^- \longrightarrow PtCl_4^{2-} + 2Cl^- & +0.68 \text{ V} \\
\hline
2Cu + PtCl_6^{2-} \longrightarrow 2Cu^+ + PtCl_4^{2-} + 2Cl^- & E^0_{cell} = +0.16 \text{ V}
\end{array}
$$

Then we calculate K_{eq}.

$$\ln K_{eq} = \frac{nFE^0_{cell}}{RT} = \frac{(2)(9.65 \times 10^4 \text{ J/V} \cdot \text{mol})(+0.16 \text{ V})}{(8.314 \text{ J/mol} \cdot \text{K})(298 \text{ K})} = 12.5$$

$$K_{eq} = e^{12.5} = 2.7 \times 10^5$$

At equilibrium, $\quad K_{eq} = \dfrac{[Cu^+]^2[PtCl_4^{2-}][Cl^-]^2}{[PtCl_6^{2-}]} = 2.7 \times 10^5.$

The forward reaction is spontaneous, and the equilibrium lies far to the right.

You should now work Exercises 102 and 104.

Primary Voltaic Cells

▶ We use batteries as portable sources of electrical energy in many ways—in flashlights, cell phones, notebook computers, calculators, automobiles, etc. A **battery** is a voltaic cell (or a set of voltaic cells coupled together) that has been designed for practical use.

As any voltaic cell produces current (*discharges*), chemicals are consumed. **Primary voltaic cells** cannot be "recharged." Once the chemicals have been consumed, further chemical action is not possible. The electrolytes or electrodes (or both) cannot be regenerated by reversing the current flow through the cell using an external direct current source. The most familiar examples of primary voltaic cells are the ordinary "dry" cells that are used as energy sources in flashlights and other small appliances.

21-22 Dry Cells

The first dry cell was patented by Georges Leclanché (1839–1882) in 1866, and such cells are still in common use. The container of the **Leclanché cell**, made of zinc, also serves as one of the electrodes (Figure 21-16). The other electrode is a carbon rod in the center of the cell. The zinc container is lined with porous paper to separate it from the other materials of the cell. The rest of the cell is filled with a moist mixture (the cell is not really dry) of ammonium chloride (NH_4Cl), manganese(IV) oxide (MnO_2), zinc chloride ($ZnCl_2$), and a porous, inert filler. Dry cells are sealed to keep the moisture from evaporating. As the cell operates (the electrodes must be connected externally), the metallic Zn is oxidized to Zn^{2+}, and the liberated electrons flow along the container to the external circuit. Thus, the zinc electrode is the anode (negative electrode).

$$Zn \longrightarrow Zn^{2+} + 2e^- \quad \text{(oxidation, anode)}$$

The carbon rod is the cathode, at which ammonium ions are reduced.

$$2NH_4^+ + 2e^- \longrightarrow 2NH_3 + H_2 \quad \text{(reduction, cathode)}$$

Addition of the half-reactions gives the overall cell reaction

$$Zn + 2NH_4^+ \longrightarrow Zn^{2+} + 2NH_3 + H_2 \qquad E_{cell} = 1.6 \text{ V}$$

As H_2 is formed, it is oxidized by MnO_2 in the cell. This prevents collection of H_2 gas on the cathode, which would stop the reaction.

$$H_2 + 2MnO_2 \longrightarrow 2MnO(OH)$$

The ammonia produced at the cathode combines with zinc ions and forms a soluble compound containing the complex ions, $[Zn(NH_3)_4]^{2+}$.

$$Zn^{2+} + 4NH_3 \longrightarrow [Zn(NH_3)_4]^{2+}$$

This reaction reduces polarization due to the buildup of ammonia, and it keeps the concentration of Zn^{2+} from increasing substantially, which would decrease the cell potential by allowing it to diffuse away from the anode. Under heavy current conditions, NH_3 gas can build up in an ordinary dry cell slowing down the migration of the NH_4^+ ions to the cathode. This can cause a temporary reduction of the battery voltage and current, making

▶ The buildup of reaction products at an electrode is called **polarization** of the electrode.

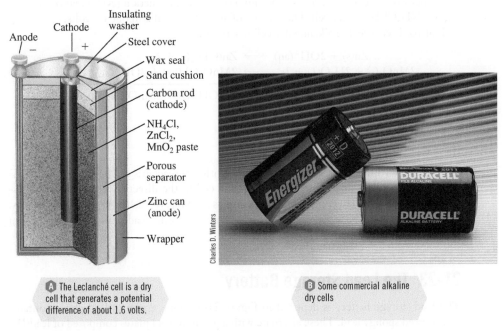

Anode — Cathode + Insulating washer
Steel cover
Wax seal
Sand cushion
Carbon rod (cathode)
NH_4Cl, $ZnCl_2$, MnO_2 paste
Porous separator
Zinc can (anode)
Wrapper

Charles D. Winters

A The Leclanché cell is a dry cell that generates a potential difference of about 1.6 volts.

B Some commercial alkaline dry cells

Figure 21-16 Dry cells are actually moist on the inside with a paste of the oxidizing agent.

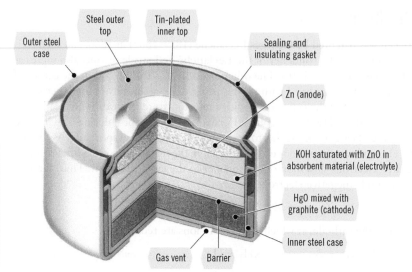

The mercury battery of the type frequently used in watches, calculators, and hearing aids is a primary cell. Although mercury in the water supply is known to cause health problems, no conclusive evidence has been found that the disposal of household batteries contributes to such problems. Nevertheless, manufacturers are working to decrease the amount of mercury in batteries. In recent years, the amount of mercury in alkaline batteries has decreased markedly; at the same time, the life of such batteries has increased dramatically.

it appear as if the battery is prematurely dying. On sitting, however, the NH_3 gas has time to react with the Zn^{2+} ions to form $[Zn(NH_3)_4]^{2+}$, which can diffuse away more readily from the cathode. Thus, a regular battery can often regain full power after sitting for a while after a period of heavy use. Once the reactant chemicals are sufficiently depleted, however, the battery is "dead."

Alkaline dry cells are similar to Leclanché dry cells except that (1) the electrolyte is basic (alkaline) because it contains KOH, and (2) the interior surface of the Zn container is rough; this gives a larger surface area. Alkaline cells have a longer shelf life than ordinary dry cells, and they stand up better under heavy use. Alkaline batteries give considerably better performance because they do not involve the gas production (NH_3) that occurs in a regular NH_4Cl-based dry cell. The voltage of an alkaline cell is about 1.5 volts.

During discharge, the alkaline dry cell reactions are

$$Zn(s) + 2OH^-(aq) \longrightarrow Zn(OH)_2(s) + 2e^- \qquad \text{(anode)}$$
$$2MnO_2(s) + 2H_2O(\ell) + 2e^- \longrightarrow 2MnO(OH)(s) + 2OH^-(aq) \qquad \text{(cathode)}$$

$$Zn(s) + 2MnO_2(s) + 2H_2O(\ell) \longrightarrow Zn(OH)_2(s) + 2MnO(OH)(s) \qquad \text{(overall)}$$

Secondary Voltaic Cells

In **secondary voltaic cells**, or *reversible cells*, the original reactants can be regenerated. This is done by passing a direct current through the cell in the direction opposite of the discharge current flow. This process is referred to as *charging*, or recharging, a cell or battery. The most common example of a secondary voltaic cell is the lead storage battery used in most automobiles.

21-23 The Lead Storage Battery

The lead storage battery is depicted in Figure 21-17. One group of lead plates contains compressed spongy lead. These alternate with a group of lead plates composed of lead(IV) oxide, PbO_2. The electrodes are immersed in a solution of about 40% sulfuric acid. When

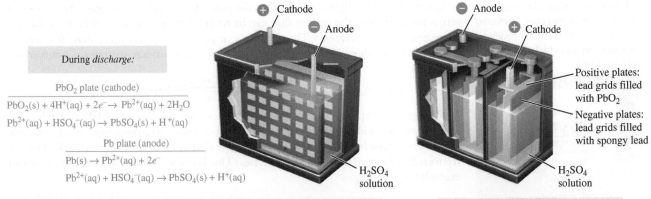

A A schematic representation of one cell of a lead storage battery. The reactions shown are those taking place during the *discharge* of the cell. Alternate lead grids are packed with spongy lead and lead(IV) oxide. The grids are immersed in a solution of sulfuric acid, which serves as the electrolyte. To provide a large reacting surface, each cell contains several connected grids, but for clarity only one of each is shown. Such a cell generates a voltage of about 2 volts.

B Six of these cells are connected together in series, although only three are shown, so that their voltages add to make a 12-volt battery. Although not shown, the cathodes are all connected in series, as are the anodes.

Figure 21-17 The lead storage battery, developed in 1859, is the oldest type of rechargeable battery. This is also called a lead-acid battery. Although not shown, the cathodes are all connected in series, as are the anodes.

the cell discharges, the spongy lead is oxidized to lead ions, and the lead plates accumulate a negative charge.

$$Pb \longrightarrow Pb^{2+} + 2e^- \quad \text{(oxidation)}$$

The lead ions then react with hydrogen sulfate ions from the sulfuric acid to form insoluble lead(II) sulfate. This coats the lead electrode.

$$Pb^{2+} + HSO_4^- \longrightarrow PbSO_4(s) + H^+ \quad \text{(precipitation)}$$

Thus, the net process at the anode *during discharge* is

$$Pb + HSO_4^- \longrightarrow PbSO_4(s) + H^+ + 2e^- \quad \text{(anode during discharge)}$$

The electrons travel through the external circuit and reenter the cell at the PbO_2 electrode, which is the cathode during discharge. Here, in the presence of hydrogen ions, the lead(IV) oxide is reduced to lead(II) ions, Pb^{2+}. These ions also react with HSO_4^- ions from the H_2SO_4 to form an insoluble $PbSO_4$ coating on the lead(IV) oxide electrode.

$$
\begin{array}{ll}
PbO_2 + 4H^+ + 2e^- \longrightarrow Pb^{2+} + 2H_2O & \text{(reduction)} \\
Pb^{2+} + HSO_4^- \longrightarrow PbSO_4(s) + H^+ & \text{(precipitation)} \\
\hline
PbO_2 + 3H^+ + HSO_4^- + 2e^- \longrightarrow PbSO_4(s) + 2H_2O & \text{(cathode during discharge)}
\end{array}
$$

The net cell reaction for discharge and its standard potential are obtained by adding the net anode and cathode half-reactions and their tabulated potentials. The tabulated E^0 value for the anode half-reaction is reversed in sign because it occurs as oxidation during discharge.

$$
\begin{array}{lcr}
 & & E^0 \\
Pb + HSO_4^- \longrightarrow PbSO_4(s) + H^+ + 2e^- & & -(-0.356\,\text{V}) \\
PbO_2 + 3H^+ + HSO_4^- + 2e^- \longrightarrow PbSO_4(s) \quad\quad + 2H_2O & & +1.685\,\text{V} \\
\hline
Pb + PbO_2 + \underbrace{2H^+ \;+ 2HSO_4^-}_{2H_2SO_4} \longrightarrow 2PbSO_4(s) \quad\quad + 2H_2O & & E^0_{cell} = +2.041\,\text{V}
\end{array}
$$

▶ The decrease in the concentration of sulfuric acid provides an easy method for measuring the degree of discharge because the density of the solution decreases accordingly. We simply measure the density of the solution with a hydrometer.

One cell creates a potential of about 2 volts. Automobile 12-volt batteries have six cells connected in series. The potential declines only slightly during use, because solid reactants are being consumed. As the cell is used, some H_2SO_4 is consumed, lowering its concentration.

▶ A *generator* supplies direct current (dc). An *alternator* supplies alternating current (ac), so an electronic rectifier is used to convert this to direct current for the battery.

Charles D. Winters

Rechargeable nicad batteries are used to operate many electrical devices.

▶ To see why a nicad battery produces a constant voltage, write the Nernst equation for its reaction. Look at Q.

▶ The efficiency of energy conversion of the fuel cell operation is 60–70% of the theoretical maximum (based on ΔG). This represents about twice the efficiency that can be realized from burning hydrogen in a heat engine coupled to a generator.

When a potential slightly greater than the potential the battery can generate is imposed across the electrodes, the current flow can be reversed. The battery can then be recharged by reversing all the reactions. The alternator or generator applies this potential when the engine is in operation. The reactions that occur in a lead storage battery are summarized as follows.

$$Pb + PbO_2 + 2H^+ + 2HSO_4^- \overset{\text{discharge}}{\underset{\text{charge}}{\rightleftharpoons}} 2PbSO_4(s) + 2H_2O$$

During many repeated charge–discharge cycles, some of the $PbSO_4$ falls to the bottom of the container and the H_2SO_4 concentration becomes correspondingly low. Eventually the battery cannot be recharged fully. It can be traded in for a new one, and the lead can be recovered and reused to make new batteries. This is one of the oldest and most successful examples of recycling.

21-24 The Nickel–Cadmium (Nicad) Cell

The nickel–cadmium (nicad) cell had widespread popularity from 1950 until recently because it can be recharged. It thus has a much longer useful life than ordinary (Leclanché) dry cells. Nicad batteries have been used in electronic toys, camcorders, and photographic equipment.

When the battery is delivering power the anode is cadmium, and the cathode is nickel(IV) oxide. The electrolytic solution is basic. The "discharge" reactions that occur in a nicad battery are

$$\begin{aligned} Cd(s) + 2OH^-(aq) &\longrightarrow Cd(OH)_2(s) + 2e^- &\text{(anode)} \\ NiO_2(s) + 2H_2O(\ell) + 2e^- &\longrightarrow Ni(OH)_2(s) + 2OH^-(aq) &\text{(cathode)} \\ \hline Cd(s) + NiO_2(s) + 2H_2O(\ell) &\longrightarrow Cd(OH)_2(s) + Ni(OH)_2(s) &\text{(overall)} \end{aligned}$$

The solid reaction product at each electrode adheres to the electrode surface. Hence, a nicad battery can be recharged by an external source of electricity; that is, the electrode reactions can be reversed. Because no gases are produced by the reactions in a nicad battery, the unit can be sealed. The voltage of a nicad cell is about 1.4 volts, slightly less than that of a Leclanché cell. The toxicity of cadmium and the limited number of recharges that a nicad battery can handle before deactivating are problems that newer battery technologies have addressed. Nickel-metal-hydride (NiMH) and lithium–ion batteries are two newer generation cells that are environmentally friendlier, have longer lifetimes, and higher power-to-weight ratios. The European Union is phasing out the use of nicad batteries, which has increased the use and lowered the price of NiMH and Li-ion batteries worldwide.

21-25 The Hydrogen–Oxygen Fuel Cell

Fuel cells are voltaic cells in which the reactants are continuously supplied to the cell and the products are continuously removed. The hydrogen–oxygen fuel cell (Figure 21-18) already has many applications. It is used in spacecraft to supplement the energy obtained from solar cells. Liquid H_2 is carried on board as a propellant. The boiled-off H_2 vapor that ordinarily would be lost is used in a fuel cell to generate electrical power.

Hydrogen (the fuel) is supplied to the anode compartment. Oxygen is fed into the cathode compartment. The diffusion rates of the gases into the cell are carefully regulated for maximum efficiency. Oxygen is reduced at the cathode, which consists of porous carbon impregnated with a finely divided Pt or Pd catalyst.

$$O_2 + 2H_2O + 4e^- \xrightarrow{\text{catalyst}} 4OH^- \quad \text{(cathode)}$$

The OH^- ions migrate through the electrolyte to the anode, an aqueous solution of a base. The anode is also porous carbon containing a small amount of catalyst (Pt, Ag, or CoO). Here H_2 is oxidized to H_2O.

$$H_2 + 2OH^- \longrightarrow 2H_2O + 2e^- \quad \text{(anode)}$$

Courtesy of Ford Motor Company

Nickel-metal-hydride battery technology is used in hybrid gas–electric vehicles such as the Ford Escape Hybrid. The NiMH batteries (330-volt) are used to power the electric drive motor in the transmission. The batteries are recharged when brakes are applied via regenerative braking.

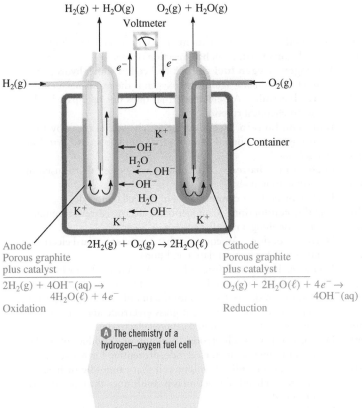

$H_2(g) + H_2O(g)$ $O_2(g) + H_2O(g)$

Voltmeter

$H_2(g) \rightarrow$ $\leftarrow O_2(g)$

e^- e^-

K^+

$\leftarrow OH^-$

H_2O

$\leftarrow OH^-$

$\leftarrow OH^-$

H_2O

$\leftarrow OH^-$

K^+ K^+

K^+

Container

$2H_2(g) + O_2(g) \rightarrow 2H_2O(\ell)$

Anode
Porous graphite
plus catalyst

$2H_2(g) + 4OH^-(aq) \rightarrow$
$4H_2O(\ell) + 4e^-$

Oxidation

Cathode
Porous graphite
plus catalyst

$O_2(g) + 2H_2O(\ell) + 4e^- \rightarrow$
$4OH^-(aq)$

Reduction

A The chemistry of a
hydrogen–oxygen fuel cell

B Space shuttle fuel cells are self-contained units, each measuring 14 × 15 × 45 in. and weighing 260 lb. Three of these are installed under the payload bay, just aft of the crew compartment, and are fueled by H_2 and O_2 from cryogenic tanks. Each fuel cell is capable of providing 12 kW continuously and up to 16 kW for short periods. A single unit can fully power the entire space shuttle in its day-to-day operation. The water produced is used for crew drinking and spacecraft cooling.

Figure 21-18 A hydrogen-oxygen fuel cell.

The net reaction is obtained from the two half-reactions.

$$
\begin{array}{ll}
O_2 + 2H_2O + 4e^- \longrightarrow 4OH^- & \text{(cathode)} \\
2(H_2 + 2OH^- \longrightarrow 2H_2O + 2e^-) & \text{(anode)} \\
\hline
2H_2 + O_2 \longrightarrow 2H_2O & \text{(net cell reaction)}
\end{array}
$$

The net reaction is the same as the burning of H_2 in O_2 to form H_2O, but combustion does not actually occur. Most of the chemical energy from the formation of H—O bonds is converted directly into electrical energy, rather than into heat energy as in combustion.

When the H_2/O_2 fuel cell is used aboard spacecraft, it is operated at a high enough temperature that the water evaporates at the same rate as it is produced. The vapor is then condensed to pure water.

Current research is aimed at modifying the design of fuel cells to lower their cost. Better catalysts would speed the reactions to allow more rapid generation of electricity and produce more power per unit volume. The H_2/O_2 cell is nonpolluting; the only substance released is H_2O. Catalysts have been developed that allow sunlight to decompose water into hydrogen and oxygen, which might be used to operate fuel cells, permitting the utilization of solar energy.

Gaseous H_2 is produced from H_2O at an illuminated photoelectrode. Light from the sun may soon be used to produce hydrogen, the ultimate clean-burning fuel.

Fuel cells have also been constructed using fuels other than hydrogen, such as methane or methanol. Biomedical researchers envision the possibility of using tiny fuel cells to operate pacemakers. The disadvantage of other power supplies for pacemakers, which are primary voltaic cells, is that their reactants are eventually consumed so that they require periodic surgical replacement. As long as the fuel and oxidizer are supplied, a fuel cell can—in theory, at least—operate forever. Eventually, tiny pacemaker fuel cells might be operated by the oxidation of blood sugar (the fuel) by the body's oxygen at a metal electrode implanted just below the skin.

KEY TERMS

Alkaline cell   A dry cell in which the electrolyte contains KOH.

Ampere   Unit of electric current; 1 ampere equals 1 coulomb per second.

Anode   The electrode at which oxidation occurs.

Battery   A voltaic cell (or a set of voltaic cells coupled together) that has been designed for practical use.

Cathode   The electrode at which reduction occurs.

Cathode protection   Protection of a metal against corrosion by making it a cathode (attaching it to a sacrificial anode of a more easily oxidized metal).

Cell potential   Potential difference, E_{cell}, between reduction and oxidation half-cells; may be at *nonstandard* conditions.

Concentration cell   A voltaic cell in which the two half-cells are composed of the same species but contain different ion concentrations.

Corrosion   Oxidation of metals in the presence of air and moisture.

Coulomb (C)   Unit of electric charge; the amount of charge that passes a given point when 1 ampere of electric current flows for 1 second.

Coulometry   The use of electrochemical cells to relate the amount of reactant or product to the amount of current passed through the cell.

Downs cell   An electrolytic cell for the commercial electrolysis of molten sodium chloride.

Dry cells   Ordinary batteries (voltaic cells) for flashlights, radios, and so on; many are Leclanché cells.

Electrochemistry   The study of the chemical changes produced by electric current and the production of electricity by chemical reactions.

Electrode potentials   Potentials, E, of half-reactions as reductions versus the standard hydrogen electrode.

Electrodes   Surfaces on which oxidation and reduction half-reactions occur in electrochemical cells.

Electrolysis   The process that occurs in electrolytic cells.

Electrolytic cell   An electrochemical cell in which electrical energy causes a nonspontaneous (reactant-favored) redox reaction to occur.

Electrolytic conduction   See *Ionic conduction*.

Electroplating   Plating a metal onto a (cathodic) surface by electrolysis.

Faraday   An amount of charge equal to 96,485 coulombs; corresponds to the charge associated with one mole of electrons, 6.022×10^{23} electrons.

Faraday's Law of Electrolysis   The amount of substance that undergoes oxidation or reduction at each electrode during electrolysis is directly proportional to the amount of electricity that passes through the cell.

Fuel cell   A voltaic cell in which the reactants (usually gases) are supplied continuously and products are removed continuously.

Galvanic cell   See *Voltaic cell*.

Glass electrode   An electrode consisting of an AgCl-coated silver wire in contact with an HCl solution of known standard concentration (usually $1.00\,M$) in a thin-walled glass bulb; when immersed in a solution, this electrode develops a potential that is sensitive to the relative $[H^+]$ concentrations (and hence to pH differences) of the internal standard solution and the outside solution.

Half-cell   The compartment in a voltaic cell in which the oxidation or reduction half-reaction occurs.

Hydrogen–oxygen fuel cell   A fuel cell in which hydrogen is the fuel (reducing agent) and oxygen is the oxidizing agent.

Inert electrode   An electrode that does not take part in the electrochemical reaction.

Ionic conduction   Conduction of electric current by ions through a pure liquid or a solution; also known as *Electrolytic conduction*.

Lead storage battery   A secondary voltaic cell that is used in most automobiles.

Leclanché cell   A common type of dry cell (battery).

Metallic conduction   Conduction of electric current through a metal or along a metallic surface.

Nernst equation   An equation that corrects standard electrode potentials for nonstandard conditions.

Nickel–cadmium cell (nicad battery)   A dry cell in which the anode is Cd, the cathode is NiO_2, and the electrolyte is basic.

pH meter   A device to measure the pH of a solution; typically consists of a pH-dependent glass electrode and a reference electrode (often a saturated calomel electrode).

Polarization of an electrode   Buildup of a product of oxidation or reduction at an electrode, preventing further reaction.

Primary voltaic cell   A voltaic cell that cannot be recharged; no further chemical reaction is possible once the reactants are consumed.

Sacrificial anode   A more active metal that is attached to a less active metal to protect the less active metal cathode against corrosion.

Salt bridge   A U-shaped tube containing an electrolyte that connects two half-cells of a voltaic cell.

Saturated calomel electrode   An electrode that consists of a platinum wire in contact with a paste of liquid mercury and solid mercury(I) chloride, $Hg_2Cl_2(s)$, all immersed in a saturated solution of potassium chloride, KCl; often used as the reference electrode in a pH meter.

Secondary voltaic cell   A voltaic cell that can be recharged; the original reactants can be regenerated by reversing the direction of current flow.

Standard cell   A cell in which all reactants and products are in their thermodynamic standard states ($1\,M$ for solution species and 1 atm partial pressure for gases).

Standard cell potential   The potential difference, E^0_{cell}, between standard reduction and oxidation half-cells.

Standard electrochemical conditions $1\,M$ concentration for solution species, 1 atm partial pressure for gases, and pure solids and liquids.

Standard electrode   A half-cell in which the oxidized and reduced forms of a species are present at unit activity: $1\,M$ solutions of dissolved species, 1 atm partial pressure of gases, and pure solids and liquids.

Standard electrode potential   By convention, the potential (E^0) of a half-reaction as a reduction relative to the standard hydrogen electrode, when all species are present at unit activity.

Standard hydrogen electrode (SHE)   An electrode consisting of a platinum electrode that is immersed in a $1\,M\,H^+$ solution and that has H_2 gas bubbled over it at 1 atm pressure; defined as the reference electrode, with a potential of *exactly* 0.0000 . . . volts.

Voltage Potential difference between two electrodes; a measure of the chemical potential for a redox reaction to occur.

Voltaic cell An electrochemical cell in which a spontaneous (product-favored) chemical reaction produces electricity; also known as a *galvanic cell*.

EXICISES

Molecular Reasoning exercises

▲ **More Challenging** exercises

Blue-Numbered exercises are solved in the Student Solutions Manual

Redox Review and General Concepts

1. (a) Define oxidation and reduction in terms of electron gain or loss. (b) What is the relationship between the numbers of electrons gained and lost in a redox reaction? (c) Do all electrochemical cells involve redox reactions?

2. Define and illustrate (a) oxidizing agent, and (b) reducing agent.

3. For each of the following unbalanced equations, (i) write the half-reactions for oxidation and for reduction, and (ii) balance the overall equation in acidic solution using the half-reaction method. (See Sections 11-4 and 11-5.)
   (a) $Hg^{2+} + Sn \longrightarrow Hg + Sn^{2+}$ (acidic solution)
   (b) $MnO_2 + Cl^- \longrightarrow Mn^{2+} + Cl_2$ (acidic solution)
   (c) $Sn^{2+} + O_2 \longrightarrow Sn^{4+} + H_2O$ (acidic solution)

4. For each of the following unbalanced equations, (i) write the half-reactions for oxidation and reduction, and (ii) balance the overall equation using the half-reaction method. (See Sections 11-4 and 11-5.)
   (a) $FeS + NO_3^- \longrightarrow$
   $NO + SO_4^{2-} + Fe^{2+}$ (acidic solution)
   (b) $Cr_2O_7^{2-} + Fe^{2+} \longrightarrow$
   $Cr^{3+} + Fe^{3+}$ (acidic solution)
   (c) $S^{2-} + Cl_2 + OH^- \longrightarrow$
   $SO_4^{2-} + Cl^- + H_2O$ (basic solution)

5. (a) Compare and contrast ionic conduction and metallic conduction. (b) What is an electrode? (c) What is an inert electrode?

6. Support or refute each of the following statements: (a) In any electrochemical cell, the positive electrode is the one toward which the electrons flow through the wire. (b) The cathode is the negative electrode in any electrochemical cell.

7. For each of the following unbalanced equations, (i) write the half-reactions for oxidation and reduction, (ii) identify the species that lose and the species that gain electrons, and (iii) write the balanced net ionic equation for the overall reaction. (See Sections 11-4 and 11-5.)
   (a) $Cr(s) + Au^{3+}(aq) \longrightarrow Cr^{2+}(aq) + Au(s)$
   (b) $NO_2^-(aq) + Cr_2O_7^{2-}(aq) \longrightarrow$
   $NO_3^-(aq) + Cr^{3+}(aq)$ (acidic solution)

(c) $N_2O_4(aq) + Br_2(aq) \longrightarrow$
   $NO_2^-(aq) + Br^-(aq)$ (basic solution)

8. Balance the following redox equations. All occur in basic solution. (See Sections 11-4 and 11-5.)
   (a) $Al(s) + OH^-(aq) \longrightarrow Al(OH)_4^-(aq) + H_2(g)$
   (b) $CrO_4^{2-}(aq) + SO_3^{2-}(aq) \longrightarrow Cr(OH)_3(s) + SO_4^{2-}(aq)$
   (c) $Zn(s) + Cu(OH)_2(s) \longrightarrow [Zn(OH)_4]^{2-}(aq) + Cu(s)$
   (d) $HS^-(aq) + ClO_3^-(aq) \longrightarrow S(s) + Cl^-(aq)$

Electrolytic Cells: General Concepts

9. (a) Solids such as potassium bromide, KBr, and sodium nitrate, $NaNO_3$, do not conduct electric current even though they are ionic. Why? Can these substances be electrolyzed as solids? (b) Support or refute the statement that the Gibbs free energy change, ΔG, is positive for any electrolysis reaction.

10. (a) Metallic magnesium cannot be obtained by electrolysis of aqueous magnesium chloride, $MgCl_2$. Why? (b) There are no sodium ions in the overall cell reaction for the electrolysis of aqueous sodium chloride. Why?

11. Consider the electrolysis of molten aluminum oxide, Al_2O_3, dissolved in cryolite, Na_3AlF_6, with inert electrodes. This is the Hall–Héroult process for commercial production of aluminum (Section 26-6). The following experimental observations can be made when current is supplied: (i) Silvery metallic aluminum is produced at one electrode. (ii) Oxygen, O_2, bubbles off at the other electrode. Diagram the cell, indicating the anode, the cathode, the positive and negative electrodes, the half-reaction occurring at each electrode, the overall cell reaction, and the direction of electron flow through the wire.

12. Do the same as in Exercise 11 for the electrolysis of molten calcium chloride with inert electrodes. The following observations are made when current is supplied: (i) Bubbles of pale green chlorine gas, Cl_2, are produced at one electrode. (ii) Silvery-white molten metallic calcium is produced at the other electrode.

13. Do the same as in Exercise 11 for the electrolysis of aqueous potassium sulfate, K_2SO_4. The following observations are made when current is supplied: (i) Bubbles of gaseous hydrogen are produced at one electrode, and the solution becomes more basic around that electrode. (ii) Bubbles of

Molecular Reasoning exercises ▲ **More Challenging** exercises **Blue-Numbered** exercises solved in Student Solutions Manual

Unless otherwise noted, all content on this page is © Cengage Learning.

gaseous oxygen are produced at the other electrode, and the solution becomes more acidic around that electrode.

14. In the electrolysis of aqueous sodium chloride, NaCl, shown below, what are the half-reactions that occur at each electrode?

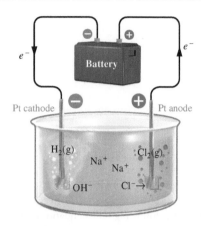

15. (a) Write the equation for the half-reaction when H_2O is reduced in an electrochemical cell. (b) Write the equation for the half-reaction when H_2O is oxidized in an electrochemical cell. (c) Write the equation for the electrolysis of water.

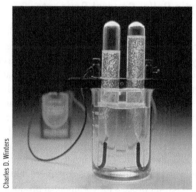

Electrolysis of water

16. Define (a) a coulomb, (b) an electric current, (c) an ampere, and (d) a faraday.
17. Calculate the number of electrons that have a total charge of 1 coulomb.
18. For each of the following cations, calculate (i) the number of faradays required to produce 1.00 mol of free metal and (ii) the number of coulombs required to produce 1.00 g of free metal. (a) Fe^{3+}; (b) Sn^{2+}; (c) Hg_2^{2+}
19. For each of the following cations, calculate (i) the number of faradays required to produce 1.00 mol of free metal, and (ii) the number of coulombs required to produce 1.00 g of free metal. (a) Fe^{2+}; (b) Au^{3+}; (c) K^+
20. In 400. min, 2.25 g of copper is obtained by electrolysis of a copper(I) acetate solution. (a) How many amperes is required for this experiment? (b) Using the same current and time, what mass of copper would be obtained from a copper(II) nitrate solution?

21. A mass of 1.20 g of silver is plated from a silver nitrate solution in 1.25 h. Calculate the (a) coulombs, (b) faradays, and (c) amperes necessary for this process.
22. Rhodium is an element that has the appearance of silver, but it does not tarnish like silver and, because it is very hard, does not become worn or scratched like silver. What mass of rhodium could be plated by electrolysis of a $Rh(NO_3)_3$ solution with a 0.755-A current for 15.0 min?
23. Hydrogen may be the fuel of the future for automobiles, according to some experts. Hydrogen can be isolated from water by electrolysis. Calculate the mass of hydrogen that is released when a 1.50-A current is passed through salt water for 12.75 h.
24. The mass of silver deposited on a spoon during electroplating was 0.976 mg. How much electric charge passed through the cell?
25. What mass of platinum could be plated onto a ring from the electrolysis of a platinum(II) salt with a 0.415-A current for 195 s?
26. What mass of silver could be plated onto a spoon from electrolysis of silver nitrate with a 2.78-A current for 45.0 min?
27. ☁ ▲ We pass enough current through a solution to plate out *one* mole of nickel metal from a solution of $NiSO_4$. In other electrolysis cells, this same current plates out *two* moles of silver from $AgNO_3$ solution but liberates only *one-half* mole of O_2 gas. Explain these observations.
28. ▲ A current is passed through 500. mL of a solution of CaI_2. The following electrode reactions occur:

anode: $2I^- \longrightarrow I_2 + 2e^-$
cathode: $2H_2O + 2e^- \longrightarrow H_2 + 2OH^-$

After some time, analysis of the solution shows that 41.5 mmol of I_2 has been formed. (a) How many faradays of charge have passed through the solution? (b) How many coulombs? (c) What volume of dry H_2 at STP has been formed? (d) What is the pH of the solution?
29. The cells in an automobile battery were charged at a steady current of 5.0 A for exactly 1.75 h. What masses of Pb and PbO_2 were formed in each cell? The overall reaction is

$$2PbSO_4(s) + 2H_2O(\ell) \longrightarrow Pb(s) + PbO_2(s) + 2H_2SO_4(aq)$$

30. ▲ The chemical equation for the electrolysis of a fairly concentrated brine solution is

$$2NaCl(aq) + 2H_2O(\ell) \longrightarrow Cl_2(g) + H_2(g) + 2NaOH(aq)$$

☁ **Molecular Reasoning** exercises   ▲ **More Challenging** exercises   Blue-Numbered exercises solved in Student Solutions Manual

What volume of gaseous chlorine would be generated at 752 torr and 15°C if the process were 85% efficient and if a current of 1.70 A flowed for 5.00 h?

31. An electrolytic cell contains 50.0 mL of a 0.165 M solution of $FeCl_3$. A current of 0.665 A is passed through the cell, causing deposition of Fe(s) at the cathode. What is the concentration of Fe^{3+}(aq) in the cell after this current has run for 20.0 min?

32. Suppose 250. mL of a 0.455 M solution of $CuCl_2$ is electrolyzed. How long will a current of 0.750 A have to run in order to reduce the concentration of Cu^{2+} to 0.167 M? What mass of Cu(s) will be deposited on the cathode during this time?

33. One method for recovering metals from their ores is by electrodeposition. For example, after passing a current of 18.0 amps through a molten tantalum salt for 38.0 min, workers at a metals processing plant isolated 15.4 g of pure tantalum. What was the oxidation state of tantalum in the molten material from which the metal was isolated?

34. ▲ Three electrolytic cells are connected in series; that is, the same current passes through all three, one after another. In the first cell, 1.20 g of Cd is oxidized to Cd^{2+}; in the second, Ag^+ is reduced to Ag; in the third, Fe^{2+} is oxidized to Fe^{3+}. (a) Find the number of faradays passed through the circuit. (b) What mass of Ag is deposited at the cathode in the second cell? (c) What mass of $Fe(NO_3)_3$ could be recovered from the solution in the third cell?

Voltaic Cells: General Concepts

35. What does voltage measure? How does it vary with time in a primary voltaic cell? Why?

36. ☁ (a) Why must the solutions in a voltaic cell be kept separate and not allowed to mix? (b) What are the functions of a salt bridge?

37. A voltaic cell containing a standard Fe^{3+}/Fe^{2+} electrode and a standard Ga^{3+}/Ga electrode is constructed, and the circuit is closed. Without consulting the table of standard reduction potentials, diagram and completely describe the cell from the following experimental observations. (i) The mass of the gallium electrode decreases, and the gallium ion concentration increases around that electrode. (ii) The ferrous ion, Fe^{2+}, concentration increases in the other electrode solution.

38. Repeat Exercise 37 for a voltaic cell that contains standard Co^{2+}/Co and Au^{3+}/Au electrodes. The observations are: (i) Metallic gold plates out on one electrode, and the gold ion concentration decreases around that electrode. (ii) The mass of the cobalt electrode decreases, and the cobalt(II) ion concentration increases around that electrode.

39. Write the shorthand notation for the cell described in Exercise 37.

40. Write the shorthand notation for the cell described in Exercise 38.

41. ☁ Appendix J lists selected reduction potentials in volts at 25°C. Why is it not necessary to list a mixture of reduction and oxidation potentials?

42. ▲ In Section 6-8 we learned how to predict from the activity series (Table 6-10) which metals replace which

others from aqueous solutions. From that table, we predict that nickel will displace silver. The equation for this process is

$$Ni(s) + 2Ag^+(aq) \longrightarrow Ni^{2+}(aq) + 2Ag(s)$$

Suppose we set up a voltaic cell based on this reaction. (a) What half-reaction would represent the reduction in this cell? (b) What half-reaction would represent the oxidation? (c) Which metal would be the anode? (d) Which metal would be the cathode? (e) Diagram this cell.

43. A cell is constructed by immersing a strip of iron in a 1.0 M $Fe(NO_3)_2$ solution and a strip of gold in a 1.0 M $AuNO_3$ solution. The circuit is completed by a wire and a salt bridge. As the cell operates, the strip of gold gains mass (only gold), and the concentration of gold ions in the solution around the gold strip decreases, while the strip of iron loses mass, and the concentration of iron ions increases in the solution around the iron strip. Write the equations for the half-reactions that occur at the cathode and at the anode.

44. ☁ When metallic copper is placed into aqueous silver nitrate, a spontaneous redox reaction occurs. No electricity is produced. Why?

45. Assume that a voltaic cell utilizes the redox reaction

$$2Al(s) + 3Ni^{2+}(aq) \longrightarrow$$
$$2Al^{3+}(aq) + 3Ni(s) \quad \text{(acidic solution)}$$

Potassium and nitrate ions may also be present. Draw this voltaic cell, and label the anode, cathode, electron flow, and ion flow.

46. ☁ Assume that a voltaic cell, proposed as a method for the purification of uranium, utilizes the redox reaction

$$3Mg(s) + 2U^{3+}(aq) \longrightarrow$$
$$3Mg^{2+}(aq) + 2U(s) \quad \text{(acidic solution)}$$

Potassium and nitrate ions may also be present. Draw this voltaic cell, and label the anode, cathode, electron flow, and ion flow.

Standard Cell Potentials

47. (a) What are standard electrochemical conditions? (b) Why are we permitted to assign arbitrarily an electrode potential of exactly 0 V to the standard hydrogen electrode?

48. What does the sign of the standard reduction potential of a half-reaction indicate? What does the magnitude indicate?

49. (a) What are standard reduction potentials? (b) What information is contained in tables of standard reduction potentials (Tables 21-2 and 21-3, Appendix J)? How is the information in such tables arranged?

50. Standard reduction potentials are 1.36 V for $Cl_2(g)/Cl^-$, 0.799 V for $Ag^+/Ag(s)$, 0.521 V for $Cu^+/Cu(s)$, 0.337 V for $Cu^{2+}/Cu(s)$, −0.44 V for $Fe^{2+}/Fe(s)$, −2.71 V for $Na^+/Na(s)$, and −2.925 V for $K^+/K(s)$. (a) Arrange the oxidizing agents in order of increasing strength. (b) Which of these oxidizing agents will oxidize Cu under standard-state conditions?

51. Standard reduction potentials are 1.455 V for the $PbO_2(s)/Pb(s)$ couple, 2.87 V for $F_2(g)/F^-$, 3.06 V for $F_2(g)/HF(aq)$, and 1.77 V for $H_2O_2(aq)/H_2O(\ell)$. Under standard-state conditions, (a) which is the strongest oxidizing agent, (b) which oxidizing agent(s) could oxidize lead to lead(IV) oxide, and (c) which oxidizing agent(s) could oxidize fluoride ion in an acidic solution?

52. Arrange the following less commonly encountered metals in an activity series from the most active to the least active: radium $[Ra^{2+}/Ra(s)$, $E^0 = -2.9\,V]$, rhodium $[Rh^{3+}/Rh(s)$, $E^0 = 0.80\,V]$, europium $[Eu^{2+}/Eu(s)$, $E^0 = -3.4\,V]$. How do these metals compare in reducing ability with the active metal lithium $[Li^+/Li(s)$, $E^0 = -3.0\,V]$, with hydrogen, and with platinum $[Pt^{2+}/Pt(s)$, $E^0 = 1.2\,V]$, which is a noble metal and one of the least active of the metals?

53. Using the list of standard electrode potentials, order the following by decreasing strength as an oxidizing agent (strongest first, weakest last): dichromate ion ($Cr_2O_7^{2-}$), hydrogen peroxide (H_2O_2), lead(IV) oxide (PbO_2), nitrate ion (NO_3^-), and permanganate ion (MnO_4^-).

54. Diagram the following cells. For each cell, write the balanced equation for the reaction that occurs spontaneously, and calculate the cell potential. Indicate the direction of electron flow, the anode, the cathode, and the polarity (+ or −) of each electrode. In each case, assume that the circuit is completed by a wire and a salt bridge. (a) A strip of magnesium is immersed in a solution that is 1.0 M in Mg^{2+}, and a strip of silver is immersed in a solution that is 1.0 M in Ag^+. (b) A strip of zinc is immersed in a solution that is 1.0 M in Zn^{2+}, and a strip of tin is immersed in a solution that is 1.0 M in Sn^{2+}.

55. Repeat Exercise 54 for the following cells. (a) A strip of chromium is immersed in a solution that is 1.0 M in Cr^{3+}, and a strip of gold is immersed in a solution that is 1.0 M in Au^{3+}. (b) A strip of aluminum is immersed in a solution that is 1.0 M in Al^{3+}, and a strip of lead is immersed in a solution that is 1.0 M in Pb^{2+}.

For Exercises 56, 57, 88, and 89, draw a cell similar to the one shown here with the anode on the left and the cathode on the right. Identify metal A, metal B, solution A, and solution B. Write the balanced equation for the reaction that occurs spontaneously. Calculate the cell potential and write the shorthand notation for the cell. Indicate the direction of electron flow, the anode, the cathode, and the polarity (+ or −) of each electrode.

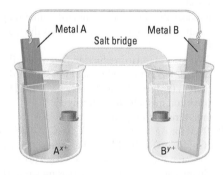

56. (See the directions above.) A standard magnesium and aluminum cell.

57. (See directions preceding Exercise 56.) A standard cadmium and zinc cell.

In answering Exercises 58–77, justify each answer by appropriate calculations. Assume that each reaction occurs at standard electrochemical conditions.

58. (a) Will Fe^{3+} oxidize Sn^{2+} to Sn^{4+} in acidic solution? (b) Will dichromate ions oxidize fluoride ions to free fluorine in acidic solution?

59. (See directions preceding Exercise 58.) (a) Will dichromate ions oxidize arsenous acid, H_3AsO_3, to arsenic acid, H_3AsO_4, in acid solution? (b) Will dichromate ions oxidize hydrogen peroxide, H_2O_2, to free oxygen, O_2, in acidic solution?

60. (See directions preceding Exercise 58.) (a) Will permanganate ions oxidize Cr^{3+} to $Cr_2O_7^{2-}$ in acidic solution? (b) Will sulfate ions oxidize arsenous acid, H_3AsO_3, to arsenic acid, H_3AsO_4, in acid solution?

61. (See directions preceding Exercise 58.) Calculate the standard cell potential, E^0_{cell}, for the cell described in Exercise 37.

62. (See directions preceding Exercise 58.) Calculate the standard cell potential, E^0_{cell}, for the cell described in Exercise 38.

63. (See directions preceding Exercise 58.) (a) Write the equation for the oxidation of $Zn(s)$ by $Br_2(\ell)$. (b) Calculate the potential of this reaction under standard-state conditions. (c) Is this a spontaneous reaction?

64. (See directions preceding Exercise 58.) For each of the following cells, (i) write the net reaction in the direction consistent with the way the cell is written, (ii) write the half-reactions for the anode and cathode processes, (iii) find the standard cell potential, E^0_{cell}, at 25°C, and (iv) state whether the standard cell reaction actually occurs as given or in the reverse direction.
(a) $Cr\,|\,Cr^{3+}\,\|\,Cu^{2+}\,|\,Cu$
(b) $Ag\,|\,Ag^+\,\|\,Cd^{2+}\,|\,Cd$

65. (See directions preceding Exercise 58.) Repeat Exercise 64 for the following cells:
(a) $Al\,|\,Al^{3+}\,\|\,Ce^{4+},\,Ce^{3+}\,|\,Pt$
(b) $Zn\,|\,Zn^{2+}\,\|\,Tl^+\,|\,Tl$

66. (See directions preceding Exercise 58.) Which of the following reactions are spontaneous in voltaic cells under standard conditions?
(a) $H_2(g) \longrightarrow H^+(aq) + H^-(aq)$
(b) $Zn(s) + 4CN^-(aq) + Ag_2CrO_4(s) \longrightarrow$
$\qquad\qquad Zn(CN)_4^{2-}(aq) + 2Ag(s) + CrO_4^{2-}(aq)$
(c) $MnO_2(s) + 4H^+(aq) + Sr(s) \longrightarrow$
$\qquad\qquad Mn^{2+}(aq) + 2H_2O(\ell) + Sr^{2+}(aq)$
(d) $Cl_2(g) + 2H_2O(\ell) + ZnS(s) \longrightarrow$
$\qquad\qquad 2HOCl(aq) + H_2S(aq) + Zn(s)$

67. (See directions preceding Exercise 58.) Consult a table of standard reduction potentials, and determine which of the following reactions are spontaneous under standard electrochemical conditions.
(a) $Mn(s) + 2H^+(aq) \longrightarrow H_2(g) + Mn^{2+}(aq)$
(b) $2Al^{3+}(aq) + 3H_2(g) \longrightarrow 2Al(s) + 6H^+(aq)$

Molecular Reasoning exercises   ▲ **More Challenging** exercises   Blue-Numbered exercises solved in Student Solutions Manual

Unless otherwise noted, all content on this page is © Cengage Learning.

(c) $2Cr(OH)_3(s) + 6F^-(aq) \longrightarrow$
$$2Cr(s) + 6OH^-(aq) + 3F_2(g)$$
(d) $Cl_2(g) + 2Br^-(aq) \longrightarrow Br_2(\ell) + 2Cl^-(aq)$

68. (See directions preceding Exercise 58.) Which of each pair is the stronger reducing agent? (a) Ag or H_2; (b) Sn or Pb; (c) Hg or Au; (d) Cl^- in acidic solution or Cl^- in basic solution; (e) HCl or H_2S; (f) Ag or Au

69. (See directions preceding Exercise 58.) Which of each pair is the stronger oxidizing agent? (a) Cu^+ or Ag^+; (b) Sn^{2+} or Sn^{4+}; (c) Fe^{2+} or Fe^{3+}; (d) I_2 or Br_2; (e) MnO_4^- in acidic solution or MnO_4^- in basic solution; (f) H^+ or Co^{2+}

70. ▲ (See directions preceding Exercise 58.) The element ytterbium forms both 2+ and 3+ cations in aqueous solution. $E^0 = -2.797$ V for $Yb^{2+}/Yb(s)$, and -2.267 V for $Yb^{3+}/Yb(s)$. What is the standard-state reduction potential for the Yb^{3+}/Yb^{2+} couple?

71. ▲ (See directions preceding Exercise 58.) The standard reduction potential for Cu^+ to $Cu(s)$ is 0.521 V, and for Cu^{2+} to $Cu(s)$ it is 0.337 V. Calculate the E^0 value for the Cu^{2+}/Cu^+ couple.

72. (See directions preceding Exercise 58.) Consider the suggestion for purifying uranium without an external energy source by setting up a voltaic cell with the reaction

$3Mg(s) + 2U^{3+}(aq) \longrightarrow$
$$3Mg^{2+}(aq) + 2U(s) \quad \text{(acidic solution)}$$

The standard reduction potentials are -1.798 for the uranium half-reaction and -2.37 for the magnesium half-reaction. (a) Will this setup work spontaneously? (b) Calculate the voltage produced by this cell as written.

73. (See directions preceding Exercise 58.) A reaction is proposed for a nickel–cadmium battery as $Ni(s) + Cd^{2+}(aq) \longrightarrow Ni^{2+}(aq) + Cd(s)$. (a) Is the reaction spontaneous as written? (b) Calculate the voltage produced by this voltaic cell as written.

74. (See directions preceding Exercise 58.) Selecting from half-reactions involving the following species, write the spontaneous reaction that will produce the voltaic cell with the highest voltage in an acidic solution: K^+, Ca^{2+}, Ni^{2+}, H_2O_2, and F_2.

75. (See directions preceding Exercise 58.) Propose the spontaneous reaction that will produce the voltaic cell with the highest voltage output by choosing from only reduction and oxidation potentials involving: Au^{3+} and Au^+, HgO and Hg, Ag_2O and Ag, S and S^{2-}, and SO_4^{2-} and SO_3^{2-}.

76. 🐾 (See directions preceding Exercise 58.) Tarnished silver is coated with a layer of $Ag_2S(s)$. The coating can be removed by boiling the silverware in an aluminum pan, with some baking soda or salt added to make the solution conductive. Explain this from the point of view of electrochemistry.

77. 🐾 ▲ (See directions preceding Exercise 58.) Describe the process of corrosion. How can corrosion of an easily oxidizable metal be prevented if the metal must be exposed to the weather?

Concentration Effects; Nernst Equation

For cell voltage calculations, assume that the temperature is 25°C unless stated otherwise.

78. ▲ Describe how the Nernst equation is important in electrochemistry. How would the Nernst equation be modified if we wished to use natural logarithms, ln? What is the value of the constant in the following equation at 25°C?

$$E = E^0 - \frac{\text{constant}}{n} \ln Q$$

79. Identify all of the terms in the Nernst equation. What part of the Nernst equation represents the correction factor for nonstandard electrochemical conditions?

80. By putting the appropriate values into the Nernst equation, show that it predicts that the voltage of a standard half-cell is equal to E^0. Use the Zn^{2+}/Zn reduction half-cell as an illustration.

81. Calculate the potential associated with the following half-reaction when the concentration of the cobalt(II) ion is 1.0×10^{-3} M.

$$Co(s) \longrightarrow Co^{2+} + 2e^-$$

82. Consider the voltaic cell
$2Ag^+(aq) + Cd(s) \longrightarrow 2Ag(s) + Cd^{2+}(aq)$
operating at 298 K.
(a) What is the E^0 for this cell?
(b) If $[Cd^{2+}] = 2.0$ M and $[Ag^+] = 0.25$ M, what is E_{cell}?
(c) If $E_{cell} = 1.25$ V and $[Cd^{2+}] = 0.100$ M, what is $[Ag^+]$?

83. The standard reduction potentials for the $H^+/H_2(g)$ and $O_2(g)$, $H^+/H_2O(\ell)$ couples are 0.0000 V and 1.229 V, respectively. (a) Write the half-reactions and the overall reaction, and calculate E^0 for the reaction

$$2H_2(g) + O_2(g) \longrightarrow 2H_2O(\ell)$$

(b) Calculate E for the cell when the pressure of H_2 is 2.85 atm and that of O_2 is 1.20 atm.

84. Consider the cell represented by the notation

$$Zn(s)\,|\,ZnCl_2(aq)\,\|\,Cl_2(g, 1\ atm);\ Cl^-(aq)\,|\,C$$

Calculate (a) E^0, and (b) E for the cell when the concentration of the $ZnCl_2$ is 0.35 mol/L and the Cl_2/Cl^- half-cell is at standard conditions.

85. What is the concentration of Ag^+ in a half-cell if the reduction potential of the Ag^+/Ag couple is observed to be 0.40 V?

86. What must be the pressure of fluorine gas in order to produce a half-cell reduction potential of the F_2/F^- couple of 2.70 V in a solution that contains 0.34 M F^-?

87. Calculate the cell potential of each of the following electrochemical cells at 25°C.
(a) $Sn(s)\,|\,Sn^{2+}(7.0 \times 10^{-3}\ M)\,\|\,Ag^+(0.110\ M)\,|\,Ag(s)$
(b) $Zn(s)\,|\,Zn^{2+}(0.500\ M)\,\|$
$$Fe^{3+}(7.2 \times 10^{-6}\ M),\ Fe^{2+}(0.20\ M)\,|\,Pt$$
(c) $Pt\,|\,H_2(1\ atm)\,|\,HCl(0.00880\ M)\,|\,Cl_2(1\ atm)\,|\,Pt$

🐾 **Molecular Reasoning** exercises   ▲ **More Challenging** exercises   Blue-Numbered exercises solved in Student Solutions Manual

Unless otherwise noted, all content on this page is © Cengage Learning.

88. (See the directions preceding Exercise 56.) Draw the cell represented by Exercise 87(a) and label its components.

89. (See the directions preceding Exercise 56.) Draw the cell represented by Exercise 87(c) and label its components.

90. Calculate the cell potential of each of the following electrochemical cells at 25°C.
    (a) $Pt\,|\,H_2(8.00\text{ atm}),\,H^+(1.00 \times 10^{-3}\ M)\,\|$
    $$Ag^+(0.00549\ M)\,|\,Ag(s)$$
    (b) $Pt\,|\,H_2(1.00\text{ atm}),\,H^+(pH = 5.97)\,\|$
    $$H^+(pH = 3.47),\,H_2(1.00\text{ atm})\,|\,Pt$$
    (c) $Pt\,|\,H_2(0.0361\text{ atm}),\,H^+(0.0175\ M)\,\|$
    $$H^+(0.0175\ M),\,H_2(5.98 \times 10^{-4}\text{ atm})\,|\,Pt$$

91. ▲ Find the potential of the cell in which identical iron electrodes are placed into solutions of $FeSO_4$ of concentration 1.5 mol/L and 0.15 mol/L.

92. What is the cell potential of a concentration cell that contains two hydrogen electrodes if the cathode contacts a solution with pH = 7.8 and the anode contacts a solution with $[H^+] = 0.05\ M$?

93. ▲ We construct a standard copper–cadmium cell, close the circuit, and allow the cell to operate. At some later time, the cell voltage reaches zero and the cell is "run down." Assume that the cell reaction is not limited by the mass of either electrode. (a) What will be the ratio of $[Cd^{2+}]$ to $[Cu^{2+}]$ when the cell is "run down"? (b) What will be their concentrations?

94. ▲ Repeat Exercise 93 for a standard zinc–nickel cell.

95. ▲ The cell potential for the cell $Zn(s) + 2H^+(\ ?\ M) \longrightarrow$ $Zn^{2+}(2.5\ M) + H_2(g)\ (4.5\text{ atm})$ is observed to be 0.445 V. What is the pH in the H^+/H_2 half-cell?

96. We wish to produce a 0.275-volt concentration cell using two hydrogen electrodes, both with hydrogen at a pressure of one atmosphere. One of the solutions has a pH of 1.5. Calculate the required pH of the other solution.

97. A concentration cell prepared using two hydrogen electrodes, both with the partial pressure of hydrogen being one atmosphere, produces 0.150 volts. The pH of one hydrogen electrode is 1.65; what is the pH of the other?

Relationships Among ΔG^0, E^0_{cell}, and K_{eq}

98. How are the signs and magnitudes of E^0_{cell}, ΔG^0, and K_{eq} related for a particular reaction? Why is the equilibrium constant K_{eq} related only to E^0_{cell} and not to E_{cell}?

99. In light of your answer to Exercise 98, how do you explain the fact that ΔG^0 for a redox reaction *does* depend on the number of electrons transferred, according to the equation $\Delta G^0 = -nFE^0_{cell}$?

100. Calculate E^0_{cell} from the tabulated standard reduction potentials for each of the following reactions in aqueous solution. Then calculate ΔG^0 and K_{eq} at 25°C from E^0_{cell}. Which reactions are spontaneous as written?
    (a) $MnO_4^- + 5Fe^{2+} \longrightarrow Mn^{2+} + 5Fe^{3+}$ (acidic solution)
    (b) $2Cu^+ \longrightarrow Cu^{2+} + Cu(s)$
    (c) $3Zn(s) + 2MnO_4^- + 4H_2O \longrightarrow$
    $$2MnO_2(s) + 3Zn(OH)_2(s) + 2OH^-$$

101. Calculate ΔG^0 (overall) and ΔG^0 per mole of metal for each of the following reactions from E^0 values: (a) Zinc dissolves in dilute hydrochloric acid to produce a solution that contains Zn^{2+}, and hydrogen gas is evolved.

(b) Chromium dissolves in dilute hydrochloric acid to produce a solution that contains Cr^{3+}, and hydrogen gas is evolved. (c) Silver dissolves in dilute nitric acid to form a solution that contains Ag^+, and NO is liberated as a gas. (d) Lead dissolves in dilute nitric acid to form a solution that contains Pb^{2+}, and NO is liberated as a gas.

102. For a certain cell, $\Delta G^0 = 25.0$ kJ. Calculate E^0 if n is (a) 1; (b) 2; (c) 4.
    Comment on the effect that the number of electrons exchanged has on the voltage of a cell.

103. Use tabulated reduction potentials to calculate the equilibrium constant for the reaction
    $$2I^- + Br_2(g) \rightleftharpoons I_2(s) + 2Br^-$$

104. Using the following half-reactions and E^0 data at 25°C:
    $$PbSO_4(s) + 2e^- \longrightarrow Pb(s) + SO_4^{2-} \qquad E^0 = -0.356\text{ V}$$
    $$PbI_2(s) + 2e^- \longrightarrow Pb(s) + 2I^- \qquad E^0 = -0.365\text{ V}$$

    calculate the equilibrium constant for the reaction
    $$PbSO_4(s) + 2I^- \rightleftharpoons PbI_2(s) + SO_4^{2-}$$

Practical Aspects of Electrochemistry

105. ☁ Distinguish among (a) primary voltaic cells, (b) secondary voltaic cells, and (c) fuel cells.

106. ☁ Sketch and describe the operation of (a) the Leclanché dry cell, (b) the lead storage battery, and (c) the hydrogen–oxygen fuel cell.

107. Why is the dry cell designed so that Zn and MnO_2 do not come into contact? What reaction might occur if they were in contact? How would this reaction affect the usefulness of the cell?

Charles D. Winters

108. ▲ People sometimes try to recharge dry cells, with limited success. (a) What reaction would you expect at the zinc electrode of a Leclanché cell in an attempt to recharge it? (b) What difficulties would arise from the attempt?

109. ☁ Briefly describe how a storage cell operates.

110. ☁ How does a fuel cell differ from a dry cell or a storage cell?

111. ☁ Does the physical size of a commercial cell govern the potential that it will deliver? What does the size affect?

112. Why is the electrode potential of the standard hydrogen electrode (SHE) 0.0000 volts?

113. What compound is produced by the chemical reaction that occurs in a hydrogen–oxygen fuel cell?

114. What commonly used metal forms a tough, transparent oxide surface layer that protects the metal from most

☁ **Molecular Reasoning** exercises   ▲ **More Challenging** exercises   Blue-Numbered exercises solved in Student Solutions Manual

Unless otherwise noted, all content on this page is © Cengage Learning.

corrosive agents in the environment and, thus, inhibits corrosion?

115. For a voltaic cell with the reaction

$$Pb(s) + Sn^{2+}(aq) \longrightarrow Pb^{2+}(aq) + Sn(s)$$

at what ratio of concentrations of lead and tin ions will $E_{cell} = 0$?

Mixed Exercises

116. Consider the electrochemical cell represented by $Mg(s)|$ $Mg^{2+} \| Fe^{3+} | Fe(s)$. (a) Write the half-reactions and the overall cell equation. (b) The standard reduction potential for $Fe^{3+}/Fe(s)$ is -0.036 V at 25°C. Determine the standard potential for the reaction. (c) Determine E for the cell when the concentration of Fe^{3+} is 10.0 mol/L and that of Mg^{2+} is 1.00×10^{-3} mol/L. (d) If a current of 150 mA is to be drawn from this cell for a period of 20.0 min, what is the minimum mass change of the magnesium electrode?

117. A sample of Al_2O_3 dissolved in a molten fluoride bath is electrolyzed using a current of 1.20 A. (a) What is the rate of production of Al in g/h? (b) The oxygen liberated at the positive carbon electrode reacts with the carbon to form CO_2. What mass of CO_2 is produced per hour?

118. ▲ The "life" of a certain voltaic cell is limited by the amount of Cu^{2+} in solution available to be reduced. If the cell contains 30.0 mL of 0.165 M $CuSO_4$, what is the maximum amount of electric charge this cell could generate?

119. A magnesium bar weighing 6.0 kg is attached to a buried iron pipe to protect the pipe from corrosion. An average current of 0.025 A flows between the bar and the pipe. (a) What reaction occurs at the surface of the bar? Of the pipe? In which direction do electrons flow? (b) How many years will be required for the Mg bar to be entirely consumed (1 year = 3.16×10^7 s)? (c) What reaction(s) will occur if the bar is not replaced after the time calculated in (b)?

120. (a) Calculate the ratio of ion concentrations of Mn^{2+} and Fe^{2+} necessary to produce a voltaic cell of 1.45 volts. The electrodes are solid manganese and iron. (b) Draw this voltaic cell. Indicate which electrode is the anode and which is the cathode, as well as the direction of electron flow.

121. (a) Calculate the ratio of ion concentrations of Mg^{2+} and Cu^{2+} necessary to produce a voltaic cell of 2.69 volts. The electrodes are solid magnesium and solid copper. (b) Draw this voltaic cell. Indicate which electrode is the anode and which is the cathode, as well as the direction of electron flow.

122. The production of uranium metal from purified uranium dioxide ore consists of the following steps:

$$UO_2(s) + 4HF(g) \longrightarrow UF_4(s) + 2H_2O(\ell)$$

$$UF_4(s) + 2Mg(s) \xrightarrow{\text{heat}} U(s) + 2MgF_2(s)$$

What is the oxidation number of U in (a) UO_2, (b) UF_4, and (c) U? Identify (d) the reducing agent, and (e) the substance reduced. (f) What current could the second reaction produce if 0.500 g of UF_4 reacted each minute? (g) What volume of HF(g) at 25°C and 10.0 atm would be

required to produce 0.500 g of U? (h) Would 0.500 g of Mg be enough to produce 0.500 g of U?

123. Which of each pair is the stronger oxidizing agent? (a) H^+ or Cl_2; (b) Zn^{2+} or Se in contact with acidic solution; (c) $Cr_2O_7^{2-}$ or Br_2 (acidic solution)

124. (a) Describe the process of electroplating. (b) Sketch and label an apparatus that a jeweler might use for electroplating silver onto jewelry. (c) A jeweler purchases highly purified silver to use as the anode in an electroplating operation. Is this a wise purchase? Why?

125. The same quantity of electric charge that deposited 0.612 g of silver was passed through a solution of a gold salt, and 0.373 g of gold was deposited. What is the oxidation state of gold in this salt?

Conceptual Exercises

126. In the lead storage battery, what substance is produced at both the anode and the cathode during discharge?

127. What metals found in commonly used voltaic cells have presented environmental concerns due to their toxicity?

128. Both I_2 and Br_2 are commercially prepared by oxidation of their halide binary salts from oceans or natural waters, and Cl_2 can be prepared by electrolyzing aqueous NaCl solutions. Why then is it not possible to prepare F_2 by electrolysis of an aqueous NaF solution?

129. Electroplating is performed by using a source of direct current. (a) Why can't an alternating current be used? (b) What would happen if an alternating current were used for electroplating?

130. Figure 21-5 is a schematic diagram of the electrolytic cell used to separate copper from the impurities, that is, zinc, iron, silver, gold, and platinum. Can this process be used to separate the metals left after the copper has been removed? Explain.

131. 🌦 A zinc–copper cell like that shown in Figure 21-6 is constructed, except that an inert platinum wire is used instead of the salt bridge. Will the cell still produce a potential?

Building Your Knowledge

132. An electrochemical cell was needed in which hydrogen and oxygen would react to form water. (a) Using the following standard reduction potentials for the couples given, determine which combination of half-reactions gives the maximum output potential:
$E^0 = -0.828$ V for $H_2O(\ell)/H_2(g)$, OH^-
$E^0 = 0.0000$ V for $H^+/H_2(g)$
$E^0 = 1.229$ V for $O_2(g)$, $H^+/H_2O(\ell)$
$E^0 = 0.401$ V for $O_2(g)$, $H_2O(\ell)/OH^-$
(b) Write the balanced equation for the overall reaction in (a).

133. (a) Given the following E^0 values at 25°C, calculate K_{sp} for cadmium sulfide, CdS.

$Cd^{2+}(aq) + 2e^- \longrightarrow Cd(s)$	$E^0 = -0.403$ V
$CdS(s) + 2e^- \longrightarrow Cd(s) + S^{2-}(aq)$	$E^0 = -1.21$ V

(b) Evaluate ΔG^0 at 25°C for the process

$$CdS(s) \rightleftharpoons Cd^{2+}(aq) + S^{2-}(aq)$$

🌦 **Molecular Reasoning** exercises   ▲ **More Challenging** exercises   Blue-Numbered exercises solved in Student Solutions Manual

Unless otherwise noted, all content on this page is © Cengage Learning.

134. Refer to tabulated reduction potentials. (a) Calculate K_{sp} for AgBr(s). (b) Calculate ΔG^0 for the reaction

$$AgBr(s) \rightleftharpoons Ag^+(aq) + Br^-(aq)$$

135. ▲ Under standard-state conditions, the following reaction is not spontaneous:

$$Br^- + 2MnO_4^- + H_2O(\ell) \longrightarrow$$
$$BrO_3^- + 2MnO_2(s) + 2OH^- \quad E^0 = -0.022\ V$$

The reaction conditions are adjusted so that $E = 0.120\ V$ by making $[Br^-] = [MnO_4^-] = 1.60\ mol/L$ and $[BrO_3^-] = 0.60\ mol/L$. (a) What is the concentration of hydroxide ions in this cell? (b) What is the pH of the solution in the cell?

136. Show by calculation that $E^0 = -1.662\ V$ for the reduction of Al^{3+} to Al(s), regardless of whether the equation for the reaction is written

(i) $\frac{1}{3}Al^{3+} + e^- \longrightarrow \frac{1}{3}Al(s) \quad \Delta G^0 = 160.4\ kJ/mol$
or
(ii) $Al^{3+} + 3e^- \longrightarrow Al(s) \quad \Delta G^0 = 481.2\ kJ/mol$

137. We wish to fill a balloon with H_2 at a pressure of 1.10 atm and a temperature of 25°C. The volume of the balloon when filled is 750 mL. How long must a current of 2.75 A be passed through the cell in order to produce this amount of H_2 by electrolysis of water?

138. In the electrolysis of water with copper electrodes (or with most other common metals), the amount of $O_2(g)$ produced is less than when one uses Pt electrodes. The amount of H_2 produced is the same for either electrode material. Explain.

Beyond the Textbook

NOTE: *Whenever the answer to an exercise depends on information obtained from a source other than this textbook, the source must be included as an essential part of the answer.*

139. Search the web to locate information on ion-selective electrodes. List five applications of ion-selective electrodes.

140. Use a reference book or search the web to locate a table of reduction potentials. List five half-reactions that are not in Appendix J and give the reduction potential of each.

141. Search the web to locate information on micro pH probes. The smallest probes that you found information about required an immersion depth of approximately how many millimeters?

142. Farrah Day's Beauty Salon advertises the removal of unwanted hair by "galvanic electrolysis." In this technique, a power supply is used to apply electrical current to the hair follicle using a tiny probe (electrode), generating tissue-destroying lye (NaOH) from body moisture and salts. (a) What's wrong with the name "galvanic electrolysis"? (b) Can you more accurately name the type of electrolytic cell involved? (c) What cell discussed in this chapter describes the electrochemical reactions taking place? Calculate the E^0_{cell} for the process.

🔹 **Molecular Reasoning** exercises   ▲ **More Challenging** exercises   Blue-Numbered exercises solved in Student Solutions Manual

Unless otherwise noted, all content on this page is © Cengage Learning.

Nuclear Chemistry

22

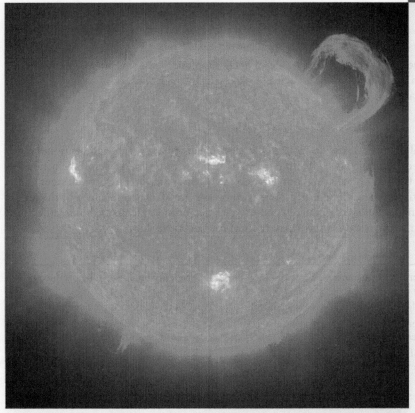

Our sun supplies energy to the earth from a distance of 93,000,000 miles. Like other stars, it is a giant nuclear fusion reactor. Much of its energy comes from the fusion of deuterium, ^2_1H, producing helium, ^4_2He.

NASA

OBJECTIVES

After you have studied this chapter, you should be able to

▶ Describe the makeup of the nucleus

▶ Describe the relationships between neutron–proton ratio and nuclear stability

▶ Tell what is meant by the band of stability

▶ Calculate mass deficiency and nuclear binding energy

▶ Describe the common types of radiation emitted when nuclei undergo radioactive decay

▶ Write and balance equations that describe nuclear reactions

▶ Predict the different kinds of nuclear reactions and describe how these reactions depend on positions of the nuclei relative to the band of stability

▶ Describe methods for detecting radiation

▶ Understand half-lives of radioactive elements

▶ Carry out calculations associated with radioactive decay

▶ Interpret decay series

▶ Tell about some uses of radionuclides, including the use of radioactive elements for dating objects

▶ Describe some nuclear reactions that are induced by bombardment of nuclei with particles

▶ Tell about nuclear fission and some of its applications, including nuclear reactors

▶ Tell about nuclear fusion and some prospects for and barriers to its use for the production of energy

Chemical properties are determined by electron distributions and are only indirectly influenced by atomic nuclei. Until now, we have discussed ordinary chemical reactions, so we have focused on electron configurations. **Nuclear reactions** involve changes in the composition of nuclei. These extraordinary processes are often accompanied by the release of tremendous amounts of energy and by transmutations of elements. Some differences between nuclear reactions and ordinary chemical reactions follow.

Nuclear Reaction	Chemical Reaction
1. Elements may be converted from one to another.	1. No new elements can be produced.
2. Particles within the nucleus are involved.	2. Only the electrons participate.
3. Tremendous amounts of energy are released or absorbed.	3. Relatively small amounts of energy are released or absorbed.
4. Rate of reaction is not influenced by external factors.	4. Rate of reaction depends on factors such as concentration, temperature, catalyst, and pressure.

Medieval alchemists spent lifetimes trying to convert other metals into gold without success. Years of failure and the acceptance of Dalton's atomic theory early in the 19th century convinced scientists that one element could not be converted into another. Then, in 1896, Henri Becquerel discovered "radioactive rays" (*natural* **radioactivity**) coming from a uranium compound. Ernest Rutherford's study of these rays showed that atoms of one element may indeed be converted into atoms of other elements by spontaneous nuclear disintegrations. Many years later, it was shown that nuclear reactions initiated by bombardment of nuclei with accelerated subatomic particles or other nuclei can also transform one element into another—accompanied by the release of radiation (*induced radioactivity*).

Becquerel's discovery led other researchers, including Marie and Pierre Curie, to discover and study new radioactive elements. Many radioactive isotopes, or **radioisotopes**, now have important medical, agricultural, and industrial uses.

Nuclear fission is the splitting of a heavy nucleus into lighter nuclei. **Nuclear fusion** is the combination of light nuclei to produce a heavier nucleus. These processes release huge amounts of energy that could satisfy a large portion of our future energy demands. Current research is aimed at surmounting the technological problems associated with safe and efficient use of nuclear fission reactors and with the development of controlled fusion reactors.

▶ In Chapter 4, we represented an atom of a particular isotope by its *nuclide symbol*. Radioisotopes are often called **radionuclides**.

Table 22-1 Fundamental Particles of Matter

Particle	Mass	Charge
Electron (e^-)	0.00054858 amu	1−
Proton (p or p^+)	1.0073 amu	1+
Neutron (n or n^0)	1.0087 amu	none

22-1 The Nucleus

In Chapter 4, we described the principal subatomic particles (Table 22-1). Recall that the neutrons and protons together constitute the nucleus, with the electrons occupying essentially empty space around the nucleus. The nucleus is only a minute fraction of the total volume of an atom; yet nearly all the mass of an atom resides in the nucleus. Thus, nuclei are extremely dense. It has been shown experimentally that nuclei of all elements have approximately the same density, 2.4×10^{14} g/cm³.

From an electrostatic point of view, it is amazing that positively charged protons can be packed so closely together. Yet many nuclei do not spontaneously decompose, so they must be stable. When Rutherford postulated the nuclear model of the atom in the early 20th century, scientists were puzzled by such a situation. Physicists have since detected many very short-lived subatomic particles (in addition to protons, neutrons, and electrons) as products of nuclear reactions. Well over 100 have been identified. A discussion of most of these particles is beyond the scope of a chemistry text. Furthermore, their functions are not entirely understood, but it is now thought that they help to overcome the proton–proton repulsions and to bind nuclear particles (**nucleons**) together. The attractive forces among nucleons appear to be important over only extremely small distances, about 10^{-13} cm.

22-2 Neutron–Proton Ratio and Nuclear Stability

The term "**nuclide**" is used to refer to different atomic forms of all elements. The term "isotope" applies only to different forms of the same element. Most naturally occurring nuclides have even numbers of protons and even numbers of neutrons; 157 nuclides fall into this category. Nuclides with odd numbers of both protons and neutrons are least common (there are only four), and those with odd–even and even–odd combinations are intermediate in abundance (Table 22-2). Furthermore, nuclides with certain "magic numbers" of protons and neutrons seem to be especially stable. Nuclides with a number of protons *or* a number of neutrons *or* a sum of the two equal to 2, 8, 20, 28, 50, 82, or 126 have unusual stability. Examples are $^4_2$He, $^{16}_8$O, $^{42}_{20}$Ca, $^{88}_{38}$Sr, and $^{208}_{82}$Pb. This suggests an energy level (shell) model for the nucleus analogous to the shell model of electron configurations.

Figure 22-1 is a plot of the number of neutrons (N) versus number of protons (Z) for the nuclides. For low atomic numbers, the most stable nuclides have equal numbers of protons and neutrons ($N = Z$). Above atomic number 20, the most stable nuclides have more neutrons than protons. Careful examination reveals an approximately stepwise shape to the plot due to the stability of nuclides with even numbers of nucleons.

Marie Sklodowska Curie (1867–1934) is the only person to have been honored with Nobel Prizes in both physics and chemistry. In 1903, Pierre (1859–1906) and Marie Curie and Henri Becquerel (1852–1908) shared the prize in physics for the discovery of natural radioactivity. Marie Curie also received the 1911 Nobel Prize in Chemistry for her discovery of radium and polonium and the compounds of radium. She named polonium for her native Poland. Marie's daughter, Irene Joliot-Curie (1897–1956), and Irene's husband, Frederick Joliot (1900–1958), received the 1935 Nobel Prize in Chemistry for the first synthesis of a new radioactive element.

▶ If enough nuclei could be gathered together to occupy one cubic centimeter, the total weight would be about 250 million tons!

▶ The nuclide symbol for an element (Section 4-7) is

$$^A_Z E$$

where E is the chemical symbol for the element, Z is its atomic number, and A is its mass number.

Table 22-2 Abundance of Naturally Occurring Nuclides

Number of protons	even	even	odd	odd
Number of neutrons	even	odd	even	odd
Number of such nuclides	157	52	50	4

Figure 22-1 A plot of the number of neutrons, N, versus the number of protons, Z, in nuclei. The stable nuclei (*green dots*) are located in an area known as the **band of stability**. All other nuclei in the white, pink, and blue regions are unstable and radioactive. No nuclei exist in the large gray shaded region. Most unstable, radioactive nuclei occur outside the band of stability. As atomic number increases, the N/Z ratio of the stable nuclei increases. Unstable nuclei above the band of stability are referred to as neutron-rich nuclei (*blue shading*); those below the band of stability are called neutron-poor nuclei (*pink shading*). Unstable (radioactive) nuclei decay by alpha emission, beta emission, positron emission, or electron capture. Lighter neutron-poor nuclei usually decay by positron emission or electron capture, either of which converts a proton into a neutron. Heavier neutron-poor nuclei usually decay by alpha emission, which decreases the neutron/proton ratio. Neutron-rich nuclei decay by beta emission, which transforms a neutron into a proton. Decay by alpha emission is by far the most predominant mode of decay for nuclei with atomic numbers beyond 83 (bismuth).

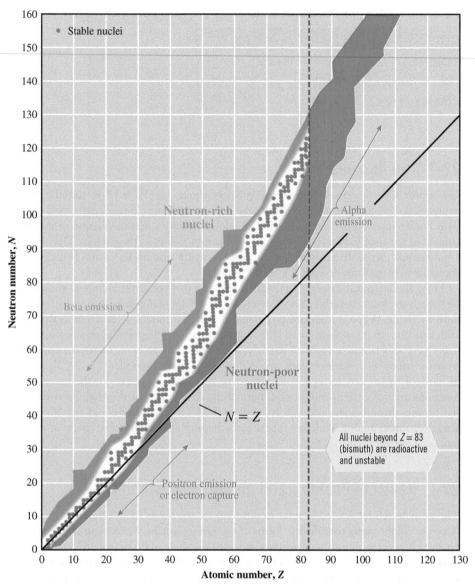

22-3 Nuclear Stability and Binding Energy

Experimentally, we observe that the masses of atoms other than $^1_1$H are always *less* than the sum of the masses of their constituent particles. We can now explain why this *mass deficiency* occurs. We also now know that the mass deficiency is in the nucleus of the atom and has nothing to do with the electrons; however, *because tables of masses of isotopes include the electrons, we shall also include them.*

The **mass deficiency**, Δm, for a nucleus is the difference between the sum of the masses of electrons, protons, and neutrons in the atom (the calculated mass) and the actual measured mass of the atom.

▶ Be sure you know how to find the numbers of protons, neutrons, and electrons in a specified atom. Review Section 4-7.

$$\Delta m = (\text{sum of masses of all } e^-, p^+, \text{ and } n^0) - (\text{actual mass of atom})$$

For most naturally occurring isotopes, the mass deficiency is only about 0.15% or less of the calculated mass of an atom.

EXAMPLE 22-1 Mass Deficiency

Calculate the mass deficiency for chlorine-35 atoms in amu/atom and in g/mol atoms. The actual mass of a chlorine-35 atom is 34.9689 amu.

Plan

We first find the numbers of protons, electrons, and neutrons in one atom. Then we determine the "calculated" mass as the sum of the masses of these particles. The mass deficiency is the actual mass subtracted from the calculated mass. This deficiency is commonly expressed either as mass per atom or as mass per mole of atoms.

Solution

Each atom of $^{35}_{17}Cl$ contains 17 protons, 17 electrons, and $(35 - 17) = 18$ neutrons. First we sum the masses of these particles.

protons:	17×1.0073 amu	$= 17.124$ amu	(masses from Table 22-1)
electrons:	17×0.00054858 amu $=$	0.0093 amu	
neutrons:	18×1.0087 amu	$= 18.157$ amu	

$$\text{sum} = 35.290 \text{ amu} \quad \longleftarrow \quad \text{calculated mass}$$

Then we subtract the actual mass from the "calculated" mass to obtain Δm.

$$\Delta m = 35.290 \text{ amu} - 34.9689 \text{ amu} = 0.321 \text{ amu} \qquad \text{mass deficiency (in one atom)}$$

We have calculated the mass deficiency in amu/atom. Recall (Section 4-9) that 1 gram is equal to 6.022×10^{23} amu. We can show that a number expressed in amu/atom is equal to the same number in g/mol of atoms:

$$\frac{? \text{ g}}{\text{mol}} = \frac{0.321 \text{ amu}}{\text{atom}} \times \frac{1 \text{ g}}{6.022 \times 10^{23} \text{ amu}} \times \frac{6.022 \times 10^{23} \text{ atoms}}{1 \text{ mol } ^{35}Cl \text{ atoms}}$$

$$= 0.321 \text{ g/mol of } ^{35}Cl \text{ atoms} \quad \longleftarrow \quad \text{(mass } \textit{deficiency} \text{ in a mole of Cl atoms)}$$

You should now work Exercises 14a and 16a, b.

What has happened to the mass represented by the mass deficiency? In 1905, Einstein set forth the Theory of Relativity, which stated that matter and energy are equivalent. An obvious corollary is that matter can be transformed into energy and energy into matter. The transformation of matter into energy occurs in the sun and other stars. It happened on earth when controlled nuclear fission was achieved in 1939 (Section 22-14). The reverse transformation, energy into matter, has not yet been accomplished on a large scale. Einstein's equation, which we encountered in Chapter 1, is $E = mc^2$. E represents the amount of energy released, m the mass of matter transformed into energy, and c the speed of light in a vacuum, 2.997925×10^8 m/s (usually rounded off to 3.00×10^8 m/s).

A mass deficiency represents the amount of matter that would be converted into energy and released if the nucleus were formed from initially separate protons and neutrons. This energy is the **nuclear binding energy (BE)**. It provides the powerful short-range force that holds the nuclear particles (protons and neutrons) together in a very small volume.

We can rewrite the Einstein relationship as

$$BE = (\Delta m)c^2$$

Specifically, if 1 mole of ^{35}Cl nuclei were to be formed from 17 moles of protons and 18 moles of neutrons, the resulting mole of nuclei would weigh 0.321 gram less than the original collection of protons and neutrons (Example 22-1).

Nuclear binding energies may be expressed in many different units, including kilojoules/mole of atoms, kilojoules/gram of atoms, and megaelectron volts/nucleon. Some useful equivalences are

▶ The abbreviation "M" stands for the prefix "mega," meaning 10^6 or "million" (Table 1-6).

$$1 \text{ megaelectron volt (MeV)} = 1 \text{ million electron volts} = 1.60 \times 10^{-13} \text{ J}$$
$$1 \text{ joule (J)} = 1 \text{ kg} \cdot \text{m}^2/\text{s}^2$$

Let's use the value of Δm for ^{35}Cl to calculate its nuclear binding energy.

EXAMPLE 22-2 Nuclear Binding Energy

Calculate the nuclear binding energy of ^{35}Cl in (a) kilojoules per mole of Cl atoms, (b) kilojoules per gram of Cl atoms, and (c) megaelectron volts per nucleon.

Plan

The mass deficiency that we calculated in Example 22-1 is related to the binding energy by the Einstein equation.

Solution

The mass deficiency is 0.321 g/mol = 3.21×10^{-4} kg/mol.

(a) $BE = (\Delta m)c^2 = \dfrac{3.21 \times 10^{-4} \text{ kg}}{\text{mol } ^{35}\text{Cl atoms}} \times (3.00 \times 10^8 \text{ m/s})^2 = 2.89 \times 10^{13} \dfrac{\text{kg} \cdot \text{m}^2/\text{s}^2}{\text{mol } ^{35}\text{Cl atoms}}$

$= 2.89 \times 10^{13} \text{ J/mol } ^{35}\text{Cl atoms} = \boxed{2.89 \times 10^{10} \text{ kJ/mol of } ^{35}\text{Cl atoms}}$

(b) From Example 22-1, the actual mass of ^{35}Cl is

$$\dfrac{34.9689 \text{ amu}}{^{35}\text{Cl atom}} \quad \text{or} \quad \dfrac{34.9689 \text{ g}}{\text{mol } ^{35}\text{Cl atoms}}$$

We use this mass to set up the needed conversion factor.

$$BE = \dfrac{2.89 \times 10^{10} \text{ kJ}}{\text{mol of } ^{35}\text{Cl atoms}} \times \dfrac{1 \text{ mol } ^{35}\text{Cl atoms}}{34.9689 \text{ g } ^{35}\text{Cl atoms}} = \boxed{8.26 \times 10^8 \text{ kJ/g } ^{35}\text{Cl atoms}}$$

(c) The number of nucleons in *one* atom of ^{35}Cl is 17 protons + 18 neutrons = 35 nucleons.

▶ The mass number, A, is equal to the number of nucleons in one atom (see Section 4-7).

$$BE = \dfrac{2.89 \times 10^{10} \text{ kJ}}{\text{mol of } ^{35}\text{Cl atoms}} \times \dfrac{1000 \text{ J}}{\text{kJ}} \times \dfrac{1 \text{ MeV}}{1.60 \times 10^{-13} \text{ J}} \times \dfrac{1 \text{ mol } ^{35}\text{Cl atoms}}{6.022 \times 10^{23} \, ^{35}\text{Cl atoms}} \times \dfrac{1 \, ^{35}\text{Cl atom}}{35 \text{ nucleons}}$$

$$= \boxed{8.57 \text{ MeV/nucleon}}$$

You should now work Exercises 14 and 16.

The nuclear binding energy of a mole of ^{35}Cl nuclei, 2.89×10^{13} J/mol, is an enormous amount of energy—enough to heat 6.9×10^7 kg ($\approx$ 76,000 tons) of water from 0°C to 100°C! Stated differently, this is also the amount of energy that would be required to separate 1 mole of ^{35}Cl nuclei into 17 moles of protons and 18 moles of neutrons. This has never been done.

▶ Some unstable radioactive nuclei do emit a single proton, a single neutron, or other subatomic particles as they decay in the direction of greater stability. None decomposes entirely into elementary particles.

Figure 22-2 is a plot of average binding energy per gram of nuclei versus mass number. It shows that nuclear binding energies (per gram) increase rapidly with increasing mass number, reach a maximum around mass number 50, and then decrease slowly. The nuclei with the highest binding energies (mass numbers 40 to 150) are the most stable. Large amounts of energy would be required to separate these nuclei into their component neu-

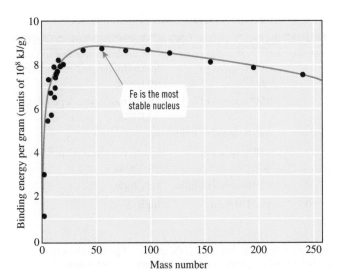

Figure 22-2 Plot of binding energy per gram versus mass number. Very light and very heavy nuclei are relatively unstable.

trons and protons. Even though these nuclei are the most stable ones, *all* nuclei are stable with respect to complete decomposition into protons and neutrons because all (except ^{1}H) nuclei have mass deficiencies. In other words, the energy equivalent of the loss of mass represents an associative force that is present in all nuclei except ^{1}H. It must be overcome to separate the nuclei completely into their subatomic particles.

22-4 Radioactive Decay

Nuclei whose neutron-to-proton ratios lie outside the stable region undergo spontaneous radioactive decay by emitting one or more particles or electromagnetic rays or both. The type of decay that occurs usually depends on whether the nucleus is above, below, or to the right of the band of stability (see Figure 22-1). Common types of **radiation** emitted in decay processes are summarized in Table 22-3.

The particles can be emitted at different kinetic energies. In addition, radioactive decay often leaves a nucleus in an excited (high-energy) state. Then the decay is followed by gamma ray emission.

$$(\text{excited nucleus}) \longrightarrow {}^A_Z E^* \longrightarrow {}^A_Z E + {}^0_0 \gamma$$

The energy of the **gamma ray** ($h\nu$) is equal to the energy difference between the ground and excited nuclear states. This is like the emission of lower-energy electromagnetic radiation that occurs as an atom in its excited electronic state returns to its ground state (see Section 4-13). Studies of gamma-ray energies strongly suggest that nuclear energy levels are quantized just as are electronic energy levels. This adds further support for a shell model for the nucleus.

The penetrating abilities of the particles and rays are proportional to their energies. Beta particles and positrons are about 100 times more penetrating than the heavier and slower-moving alpha particles. They can be stopped by a $\frac{1}{8}$-inch-thick (0.3 cm) aluminum plate. They can burn skin severely but cannot reach internal organs. Alpha particles have low penetrating ability and cannot damage or penetrate skin. They can damage sensitive internal tissue if inhaled, however. The high-energy gamma rays have great penetrating power and severely damage both skin and internal organs. They travel at the speed of light and can be stopped by thick layers of concrete or lead.

▶ Recall that the energy of electromagnetic radiation is $E = h\nu$, where h is Planck's constant and ν is the frequency.

Table 22-3 Common Types of Radioactive Emissions

Type and Symbol[a]	Identity	Mass (amu)	Charge	Velocity	Penetration
beta (β, β^-, $_{-1}^{0}\beta$, $_{-1}^{0}e$)	electron	0.00055	1−	≤90% speed of light	low to moderate, depending on energy
positron[b] ($_{+1}^{0}\beta$, $_{+1}^{0}e$)	positively charged electron	0.00055	1+	≤90% speed of light	low to moderate, depending on energy
alpha (α, $_{2}^{4}\alpha$, $_{2}^{4}$He)	helium nucleus	4.0026	2+	≤10% speed of light	low
proton ($_{1}^{1}p$, $_{1}^{1}$H)	proton, hydrogen nucleus	1.0073	1+	≤10% speed of light	low to moderate, depending on energy
neutron ($_{1}^{0}n$)	neutron	1.0087	0	≤10% speed of light	very high
gamma ($_{0}^{0}\gamma$) ray	high-energy electromagnetic radiation such as X-rays	0	0	speed of light	high

[a]The number at the upper left of the symbol is the number of nucleons, and the number at the lower left is the number of positive charges.
[b]On the average, a positron exists for about a nanosecond (1×10^{-9} second) before colliding with an electron and being converted into the corresponding amount of energy.

A technician cleans lead glass blocks that form part of the giant OPAL particle detector at CERN, the European center for particle physics near Geneva, Switzerland.

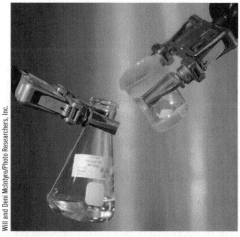

Robotics technology is used to manipulate highly radioactive samples safely.

22-5 Equations for Nuclear Reactions

In *chemical* reactions, atoms in molecules and ions are rearranged, but matter is neither created nor destroyed, and atoms are not changed into other atoms. In earlier chapters, we learned to write balanced chemical equations to represent chemical reactions. Such equations must show the same total number of atoms of each kind on both sides of the equation and the same total charge on both sides of the equation. In a nuclear reaction, a different kind of transformation occurs, one in which a proton can change into a neutron, a neutron can change into a proton, or particles can be captured by or ejected from the nucleus, but the total number of nucleons remains the same. This leads to two requirements for the equation for a nuclear reaction:

1. The sum of the mass numbers (the left superscript in the nuclide symbol) of the reactants must equal the sum of the mass numbers of the products; this maintains mass balance.

> 2. The sum of the atomic numbers (the left subscript in the nuclide symbol) of the reactants must equal the sum of the atomic numbers of the products; this maintains charge balance.

Because such equations are intended to describe only the changes in the nucleus, they do not ordinarily include ionic charges (which are due to changes in the arrangements of electrons). In the following sections, we will see many examples of such equations for nuclear reactions. For several of these, the check for mass balance and charge balance is illustrated in the margin.

22-6 Neutron-Rich Nuclei (Above the Band of Stability)

Nuclei in this region have too high a ratio of neutrons to protons. They undergo decays that *decrease* the ratio. The most common such decay is **beta emission**. A **beta particle** is an electron ejected *from the nucleus* when a neutron is converted into a proton.

$$_0^1n \longrightarrow {}_1^1p + {}_{-1}^0\beta$$

Thus, beta emission results in an increase of one in the number of protons (the atomic number) and a decrease of one in the number of neutrons, with no change in mass number. Examples of beta-particle emission are

$$_{88}^{228}\text{Ra} \longrightarrow {}_{89}^{228}\text{Ac} + {}_{-1}^0\beta \quad \text{and} \quad {}_6^{14}\text{C} \longrightarrow {}_7^{14}\text{N} + {}_{-1}^0\beta$$

The sum of the mass numbers on each side of the first equation is 228, and the sum of the atomic numbers on each side is 88. The corresponding sums for the second equation are 14 and 6, respectively.

▶ For beta emission by $_{88}^{228}\text{Ra}$:

mass balance (superscripts):
$$228 \longrightarrow 228 + 0$$
charge balance (subscripts):
$$88 \longrightarrow 89 + (-1) = 88$$

22-7 Neutron-Poor Nuclei (Below the Band of Stability)

Two types of decay for nuclei below the band of stability are **positron emission** or **electron capture** (*K* capture). Positron emission is most commonly encountered with artificially radioactive nuclei of the lighter elements. Electron capture occurs most often with heavier elements.

A **positron** has the mass of an electron, but a positive charge. Positrons are emitted when protons are converted to neutrons.

$$_1^1p \longrightarrow {}_0^1n + {}_{+1}^0\beta$$

Thus, positron emission results in a *decrease* by one in atomic number and an *increase* by one in the number of neutrons, with *no change* in mass number.

$$_{19}^{38}\text{K} \longrightarrow {}_{18}^{38}\text{Ar} + {}_{+1}^0\beta \quad \text{and} \quad {}_8^{15}\text{O} \longrightarrow {}_7^{15}\text{N} + {}_{+1}^0\beta$$

The same effect can be accomplished by electron capture (*K* capture), in which an electron from the *K* shell ($n = 1$) is captured by the nucleus. This is very different from an atom gaining an electron to form an ion.

$$_{47}^{106}\text{Ag} + {}_{-1}^0e \longrightarrow {}_{46}^{106}\text{Pd} \quad \text{and} \quad {}_{18}^{37}\text{Ar} + {}_{-1}^0e \longrightarrow {}_{17}^{37}\text{Cl}$$

Some nuclides, such as $_{11}^{22}\text{Na}$, undergo both electron capture and positron emission.

$$_{11}^{22}\text{Na} + {}_{-1}^0e \longrightarrow {}_{10}^{22}\text{Ne} \ (3\%) \quad \text{and} \quad {}_{11}^{22}\text{Na} \longrightarrow {}_{10}^{22}\text{Ne} + {}_{+1}^0\beta \ (97\%)$$

Some of the neutron-poor nuclei, especially the heavier ones, *increase* their neutron-to-proton ratios by undergoing **alpha emission**. **Alpha particles** are helium nuclei, $_2^4\text{He}$, consisting of two protons and two neutrons. Alpha particles carry a double positive charge,

▶ For positron emission by $_{19}^{38}\text{K}$:

mass: $38 \longrightarrow 38 + 0$
charge: $19 \longrightarrow 18 + 1 = 19$

▶ For electron capture by $_{18}^{37}\text{Ar}$:

mass: $37 + 0 \longrightarrow 37$
charge: $18 + (-1) \longrightarrow 17$

▶ For α-emission by $^{204}_{82}$Pb:

mass: $204 \longrightarrow 200 + 4 = 204$

charge: $82 \longrightarrow 80 + 2 = 82$

but charge is usually not shown in nuclear reactions. Alpha emission also results in an increase of the neutron-to-proton ratio. An example is the alpha emission of lead-204.

$$^{204}_{82}\text{Pb} \longrightarrow {}^{200}_{80}\text{Hg} + {}^{4}_{2}\alpha$$

22-8 Nuclei with Atomic Number Greater Than 83

▶ The only stable nuclide with atomic number 83 is $^{209}_{83}$Bi.

All nuclides with atomic number greater than 83 are beyond the band of stability and are radioactive. Many of these decay by alpha emission.

$$^{226}_{88}\text{Ra} \longrightarrow {}^{222}_{86}\text{Rn} + {}^{4}_{2}\alpha \quad \text{and} \quad {}^{210}_{84}\text{Po} \longrightarrow {}^{206}_{82}\text{Pb} + {}^{4}_{2}\alpha$$

The decay of radium-226 was originally reported in 1902 by Rutherford and Soddy. It was the first transmutation of an element ever observed. A few heavy nuclides also decay by beta emission, positron emission, and electron capture.

Some isotopes of uranium ($Z = 92$) and elements of higher atomic number, the **transuranium elements**, also decay by spontaneous nuclear fission. In this process, a heavy nuclide splits to form nuclides of intermediate mass and emits neutrons.

▶ For fission of $^{252}_{98}$Cf:

mass: $252 \longrightarrow$

$142 + 106 + 4(1) = 252$

charge: $98 \longrightarrow 56 + 42 + 4(0) = 98$

$$^{252}_{98}\text{Cf} \longrightarrow {}^{142}_{56}\text{Ba} + {}^{106}_{42}\text{Mo} + 4{}^{1}_{0}n$$

EXAMPLE 22-3 Equations for Nuclear Decay Reactions

Write the balanced equation for each of the following radioactive decay processes.

(a) ^{45}Ti decays by positron emission.

(b) ^{81}Kr decays by electron capture.

(c) ^{104}Ru decays by beta emission.

(d) ^{223}Ra decays by alpha emission.

Plan

For each process, write the nuclide symbols for the reactant species and for the particle that is captured (as a reactant) or that is emitted (as a product). Then use mass balance and charge balance to write the super- and subscripts for the product nuclide. Determine the atomic symbol for the product nuclide from its atomic number (the subscript).

Solution

Represent the unknown product nuclide as $^{y}_{x}\text{X}$.

(a) Ti (atomic number 22) emits a positron ($^{0}_{+1}\beta$ or $^{0}_{+1}e$)

$$^{45}_{22}\text{Ti} \longrightarrow {}^{y}_{x}\text{X} + {}^{0}_{+1}\beta$$

mass balance: $45 = y + 0; y = 45$
charge balance: $22 = x + 1; x = 21$ $\Big\} \Rightarrow {}^{45}_{21}\text{X}$

The element with atomic number 21 is scandium, Sc, so the complete equation is

$$^{45}_{22}\text{Ti} \longrightarrow {}^{45}_{21}\text{Sc} + {}^{0}_{+1}\beta$$

(b) Kr (atomic number 36) captures an electron ($^{0}_{-1}e$)

$$^{81}_{36}\text{Kr} + {}^{0}_{-1}e \longrightarrow {}^{y}_{x}\text{X}$$

mass balance: $81 + 0 = y; y = 81$
charge balance: $36 - 1 = x; x = 35$ $\Big\} \Rightarrow {}^{81}_{35}\text{X}$

The element with atomic number 35 is bromine, Br, so the complete equation is

$$^{81}_{36}\text{Kr} + {}^{0}_{-1}e \longrightarrow {}^{81}_{35}\text{Br}$$

(c) Ru (atomic number 44) emits a beta particle ($_{-1}^{0}\beta$ or $_{-1}^{0}e$)

$$^{104}_{44}\text{Ru} \longrightarrow {}^{y}_{x}\text{X} + {}^{0}_{-1}\beta$$

mass balance: $104 = y + 0; y = 104$ ⎫
charge balance: $44 = x - 1; x = 45$ ⎬ $\Rightarrow {}^{104}_{45}\text{X}$
⎭

The element with atomic number 45 is rhodium, Rh, so the complete equation is

$$^{104}_{44}\text{Ru} \longrightarrow {}^{104}_{45}\text{Rh} + {}^{0}_{-1}\beta$$

(d) Ra (atomic number 88) emits an alpha particle ($_{2}^{4}\alpha$)

$$^{223}_{88}\text{Ra} \longrightarrow {}^{y}_{x}\text{X} + {}^{4}_{2}\alpha$$

mass balance: $223 = y + 4; y = 219$ ⎫
charge balance: $88 = x + 2; x = 86$ ⎬ $\Rightarrow {}^{219}_{86}\text{X}$
⎭

The element with atomic number 86 is radon, Rn, so the complete equation is

$$^{223}_{88}\text{Ra} \longrightarrow {}^{219}_{86}\text{Rn} + {}^{4}_{2}\alpha$$

You should now work Exercises 34 and 35.

22-9 Detection of Radiation

Photographic Detection

Emanations from radioactive substances affect photographic plates just as ordinary visible light does. Becquerel's discovery of radioactivity resulted from the unexpected exposure of such a plate, wrapped in black paper, by a nearby enclosed sample of a uranium-containing compound, potassium uranyl sulfate. After a photographic plate has been developed and fixed, the intensity of the exposed area is related to the amount of radiation that has struck the plate. Quantitative detection of radiation by this method is difficult and tedious.

Detection by Fluorescence

Fluorescent substances can absorb high-energy radiation such as gamma rays and subsequently emit visible light. As the radiation is absorbed, the absorbing atoms jump to excited electronic states. The excited electrons return to their ground states through a series of transitions, some of which emit visible light. This method may be used for the quantitative detection of radiation, using an instrument called a **scintillation counter**.

Cloud Chambers

The original cloud chamber was devised by C. T. R. Wilson (1869–1959) in 1911. A chamber contains air saturated with vapor. Particles emitted from a radioactive substance ionize air molecules in the chamber. Cooling the chamber causes droplets of liquid to condense on these ions. The paths of the particles can be followed by observing the fog-like tracks produced. The tracks may be photographed and studied in detail. Figures 22-3 and 22-4 show a cloud chamber and a cloud chamber photograph, respectively.

Figure 22-3 A cloud chamber. The emitter is glued onto a pin stuck into a stopper that is mounted on the chamber wall. The chamber has some volatile liquid in the bottom and rests on dry ice. The cool air near the bottom becomes supersaturated with vapor. When an emission speeds through this vapor, ions are produced. These ions serve as sites about which the vapor condenses, forming tiny droplets, or fog.

Dry ice
−78°C

Science Source/Photo Researchers

Figure 22-4 Colored image obtained from a bubble chamber (a modern version of the cloud chamber) at the Stanford Linear Accelerator Center (SLAC), showing the collision between a proton and a high-energy photon (electromagnetic radiation). The collision point is marked by the arrow; the track of the photon is not visible because only charged particles leave tracks in a bubble chamber. Several particle paths are seen branching from the collision site heading toward the bottom of the image. The tight spirals elsewhere in the photo are caused by electrons whose trajectories are strongly curved by the applied magnetic field.

Gas Ionization Counters

▶ The Geiger counter can detect only β and γ radiation. The α-particles cannot penetrate the walls or window of the tube.

A common gas ionization counter is the **Geiger–Müller counter** (Figure 22-5). Radiation enters the tube through a thin window. Windows of different stopping powers can be used to admit only radiation of certain penetrating powers.

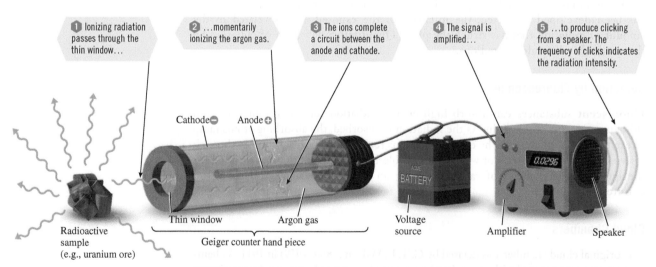

1 Ionizing radiation passes through the thin window...

2 ...momentarily ionizing the argon gas.

3 The ions complete a circuit between the anode and cathode.

4 The signal is amplified...

5 ...to produce clicking from a speaker. The frequency of clicks indicates the radiation intensity.

Cathode⊖   Anode⊕

0.0296

Thin window   Argon gas   Voltage source   Amplifier   Speaker

Radioactive sample (e.g., uranium ore)

Geiger counter hand piece

Figure 22-5 The principle of operation of a gas ionization counter. The center wire is positively charged, and the shell of the tube is negatively charged. When radiation enters through the window, it ionizes one or more gas atoms. The electrons are attracted to the central wire, and the positive ions are drawn to the shell. This constitutes a pulse of electric current, which is amplified and displayed on the meter or other readout.

22-10 Rates of Decay and Half-Life

Radionuclides have different stabilities and decay at different rates. Some decay nearly completely in a fraction of a second and others only after millions of years. The rates of all radioactive decays are independent of temperature and obey *first-order kinetics*. In Section 16-3, we saw that the rate of a first-order process is proportional to the concentration of only one substance. The rate law and the integrated rate equation for a first-order process (see Section 16-4) are

$$\text{rate of decay} = k[A] \quad \text{and} \quad \ln\left(\frac{A_0}{A}\right) = akt$$

Here A represents the amount of decaying radionuclide of interest remaining after some time t, and A_0 is the amount present at the beginning of the observation. The variable k is the rate constant, which is different for each radionuclide. Each atom decays independently of the others, so the stoichiometric coefficient a is *always* 1 for radioactive decay. We can therefore drop it from the calculations in this chapter and write the integrated rate equation as

▶ The reactant coefficient a is *always* 1 for radioactive decay.

$$\ln\left(\frac{A_0}{A}\right) = kt$$

Because A_0/A is a ratio, A_0 and A can represent either molar concentrations of a reactant or masses of a reactant. The rate of radioactive disintegrations follows first-order kinetics, so it is proportional to the amount of A present; we can write the integrated rate equation in terms of N, the number of disintegrations per unit time:

▶ The amounts A_0 and A are usually given as masses (grams, milligrams, micrograms, etc.), but can be expressed in any convenient units.

$$\ln\left(\frac{N_0}{N}\right) = kt$$

In nuclear chemistry, the decay rate is usually expressed in terms of the **half-life**, $t_{1/2}$, of the process. This is the amount of time required for half of the original sample to react. For a first-order process, $t_{1/2}$ is given by the equation

$$t_{1/2} = \frac{\ln 2}{k} = \frac{0.693}{k}$$

The isotope strontium-90 was introduced into the atmosphere by the atmospheric testing of nuclear weapons. Strontium is chemically similar to calcium, so strontium-90 now occurs with Ca in measurable quantities in milk, bones, and teeth as a result of its presence in food and water supplies. Strontium-90 is a radionuclide that undergoes beta emission with a half-life of 28.8 years. It may cause leukemia, bone cancer, and other related disorders. If we begin with a 16-μg sample of $^{90}_{38}$Sr, 8 μg will remain after one half-life of 28.8 years. After 57.6 years, 4 μg will remain; after 86.4 years, 2 μg; and so on (Figure 22-6).

Because all radioactive decay follows first-order kinetics, a similar plot for any radionuclide shows the same shape of *exponential decay curve*. About ten half-lives (288 years for $^{90}_{38}$Sr) must pass for any radionuclide to lose 99.9% of its radioactivity.

▶ In 1963, a treaty was signed by the United States, the Soviet Union, and the United Kingdom prohibiting the further testing of nuclear weapons in the atmosphere. Since then, strontium-90 has been disappearing from the air, water, and soil according to the curve in Figure 22-6. So the treaty has largely accomplished its aim up to the present.

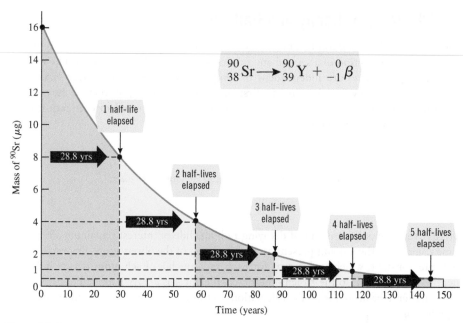

$$\ce{^{90}_{38}Sr -> ^{90}_{39}Y + ^{0}_{-1}\beta}$$

Figure 22-6 The decay of a 16-μg sample of $\ce{^{90}_{38}Sr}$, a radioactive nuclide with a half-life of 28.8 years.

▶ Gamma rays destroy both cancerous and normal cells, so the beams of gamma rays must be directed as nearly as possible at only cancerous tissue.

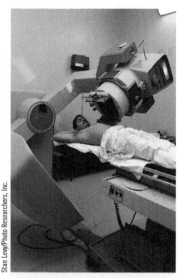

The γ-radiation from $\ce{^{60}Co}$ is used to treat cancers near the surface of the body.

EXAMPLE 22-4 Rate of Radioactive Decay

The "cobalt treatments" used in medicine to arrest certain types of cancer rely on the ability of gamma rays to destroy cancerous tissues. Cobalt-60 decays with the emission of beta particles and gamma rays, with a half-life of 5.27 years.

$$\ce{^{60}_{27}Co -> ^{60}_{28}Ni + ^{0}_{-1}\beta + ^{0}_{0}\gamma}$$

How much of a 3.42-μg sample of cobalt-60 remains after 30.0 years?

Plan

We determine the value of the specific rate constant, k, from the given half-life. This value is then used in the first-order integrated rate equation to calculate the amount of cobalt-60 remaining after the specified time.

Solution

We first determine the value of the specific rate constant.

$$t_{1/2} = \frac{0.693}{k} \quad \text{so} \quad k = \frac{0.693}{t_{1/2}} = \frac{0.693}{5.27 \text{ y}} = 0.131 \text{ y}^{-1}$$

This value can now be used to determine the ratio of A_0 to A after 30.0 years.

$$\ln\left(\frac{A_0}{A}\right) = kt = 0.131 \text{ y}^{-1}(30.0 \text{ y}) = 3.93$$

Taking the inverse ln of both sides, $\dfrac{A_0}{A} = 51$.

$$A_0 = 3.42 \ \mu\text{g, so}$$

$$A = \frac{A_0}{51} = \frac{3.42 \ \mu\text{g}}{51} = 0.067 \ \mu\text{g} \ \ce{^{60}_{27}Co} \quad \text{remains after 30.0 years.}$$

You should now work Exercise 54.

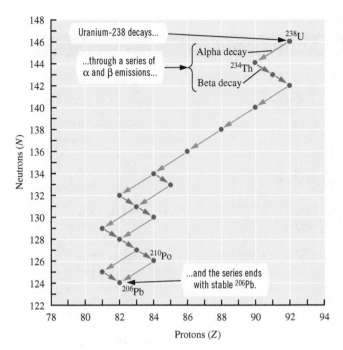

Figure 22-7 The ^{238}U decay series. Steps that occur by alpha emission are shown in red; beta emissions are shown in blue. See Table 22-4 for another representation of this and other decay series, including half-lives.

22-11 Decay Series

Many radionuclides cannot attain nuclear stability by only one nuclear reaction. Instead, they decay in a series of steps. For any particular decay step, the decaying nuclide is called the **parent** nuclide, and the product nuclide is the **daughter**.

Uranium-238 decays by alpha emission to thorium-234 in the first step of one series. Thorium-234 subsequently emits a beta particle to produce protactinium-234 in the second step. This series can be summarized as shown in Figure 22-7 and in Table 22-4a. The *net* reaction for the ^{238}U series is

$$^{238}_{92}\text{U} \longrightarrow\ ^{206}_{82}\text{Pb} + 8\ ^{4}_{2}\text{He} + 6\ ^{0}_{-1}\beta$$

"Branchings" are possible at various points in the chain. That is, two successive decays may be replaced by alternative decays, but they always result in the same final product. There are also decay series of varying lengths starting with some of the artificially produced radionuclides (see Section 22-13).

A few such series are known to occur in nature. Two begin with isotopes of uranium, ^{238}U and ^{235}U, and one begins with ^{232}Th. All three of these end with a stable isotope of lead (Z = 82). Table 22-4 outlines in detail the ^{238}U, ^{235}U, and ^{232}Th disintegration series, showing half-lives.

▶ Mass and charge balance for the decay series of $^{238}_{92}\text{U}$:

mass: 238 $\longrightarrow$
$$206 + 8(4) + 6(0) = 238$$
charge: 92 $\longrightarrow$
$$82 + 8(+2) + 6(-1) = 92$$

22-12 Uses of Radionuclides

Radionuclides have practical uses because they decay at known rates. Some applications make use of the radiation that is continuously emitted by radionuclides.

Radioactive Dating

The ages of articles of organic origin can be estimated by *radiocarbon dating*. The radioisotope carbon-14 is produced continuously in the upper atmosphere as nitrogen atoms capture cosmic-ray neutrons.

$$^{14}_{7}\text{N} + ^{1}_{0}n \longrightarrow\ ^{14}_{6}\text{C} + ^{1}_{1}\text{H}$$

Table 22-4 Emissions and Half-Lives of the Natural Radioactive Decay Series*

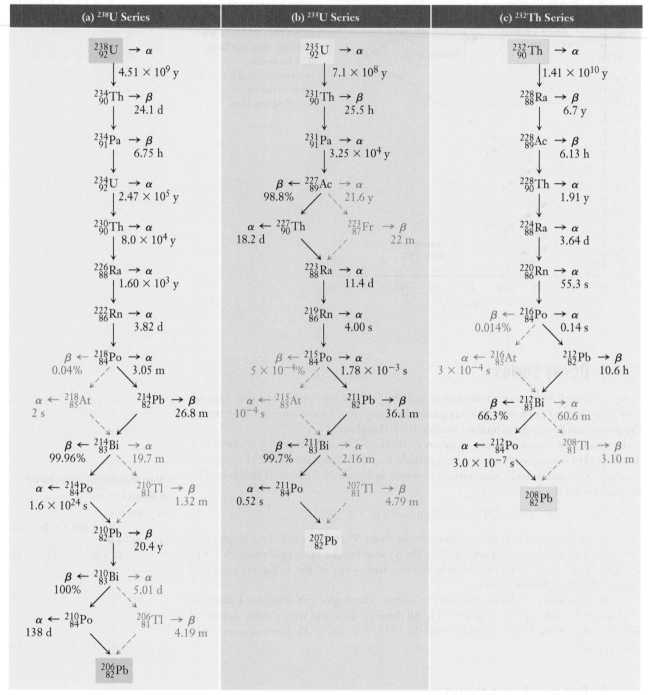

*Abbreviations are y, year; d, day; m, minute; and s, second. Less prevalent decay branches are shown in blue.

The carbon-14 atoms react with oxygen molecules to form $^{14}CO_2$. This process continually supplies the atmosphere with radioactive $^{14}CO_2$, which is removed from the atmosphere by photosynthesis. The intensity of cosmic rays is related to the sun's activity. As long as this remains constant, the amount of $^{14}CO_2$ in the atmosphere remains constant. $^{14}CO_2$ is incorporated into living organisms just as ordinary $^{12}CO_2$ is, so a certain fraction of all carbon atoms in living substances is carbon-14. This decays with a half-life of 5730 years.

$$^{14}_{6}C \longrightarrow \,^{14}_{7}N + \,^{0}_{-1}\beta$$

After death, the plant no longer carries out photosynthesis, so it no longer takes up $^{14}CO_2$. Other organisms that consume plants for food stop doing so at death. As the ^{14}C in dead tissue then decays, its β-emissions decrease with the passage of time. The activity per gram of carbon is a measure of the length of time elapsed since death. Comparison of ages of ancient trees calculated from ^{14}C activity with those determined by counting rings indicates that cosmic ray intensity has varied somewhat throughout history. The calculated ages can be corrected for these variations. The carbon-14 technique is useful only for dating objects less than 50,000 years old. Older objects have too little activity to be dated accurately.

The *potassium–argon* and *uranium–lead* methods are used for dating older objects. Potassium-40 decays to argon-40 with a half-life of 1.3 billion years.

$$^{40}_{19}K + ^{0}_{-1}e \longrightarrow ^{40}_{18}Ar$$

Because of its long half-life, potassium-40 can be used to date objects up to 1 billion years old by determination of the ratio of $^{40}_{19}K$ to $^{40}_{18}Ar$ in the sample. The uranium–lead method is based on the natural uranium-238 decay series, which ends with the production of stable lead-206. This method is used for dating uranium-containing minerals several billion years old because this series has an even longer half-life. All the ^{206}Pb in such minerals is assumed to have come from ^{238}U. Because of the very long half-life of $^{238}_{92}U$, 4.5 billion years, the amounts of intermediate nuclei can be neglected. A meteorite that was 4.6 billion years old fell in Mexico in 1969. Results of $^{238}U/^{206}Pb$ studies on such materials of extraterrestrial origin suggest that our solar system was formed several billion years ago.

In 1992, hikers in the Italian Alps found the remains of a man who had been frozen in a glacier for about 4000 years. This discovery was especially important because of the unusual preservation of tissues, garments, and personal belongings. Radiocarbon dating is used to estimate the ages of archaeological finds such as these.

EXAMPLE 22-5 Radiocarbon Dating

A piece of wood taken from a cave dwelling in New Mexico is found to have a carbon-14 activity (per gram of carbon) only 0.636 times that of wood cut today. Estimate the age of the wood. The half-life of carbon-14 is 5730 years.

Plan

As we did in Example 22-4, we determine the specific rate constant k from the known half-life. The time required to reach the present fraction of the original activity is then calculated from the first-order decay equation.

Solution

First we find the first-order specific rate constant for ^{14}C.

$$t_{1/2} = \frac{0.693}{k} \quad \text{or} \quad k = \frac{0.693}{t_{1/2}} = \frac{0.693}{5730 \text{ y}} = 1.21 \times 10^{-4} \text{ y}^{-1}$$

The present ^{14}C activity, N (disintegrations per unit time), is 0.636 times the original activity, N_0.

$$N = 0.636 \, N_0$$

We substitute into the first-order decay equation

$$\ln\left(\frac{N_0}{N}\right) = kt$$

$$\ln\left(\frac{N_0}{0.636 \, N_0}\right) = (1.21 \times 10^{-4} \text{ y}^{-1})t$$

We cancel N_0 and solve for t.

$$\ln\left(\frac{1}{0.636}\right) = (1.21 \times 10^{-4} \text{ y}^{-1})t$$

$$0.452 = (1.21 \times 10^{-4} \text{ y}^{-1})t \quad \text{or} \quad t = \boxed{3.74 \times 10^3 \text{ y (or 3740 y)}}$$

You should now work Exercises 58 and 62.

▶ Since more than 50% of the activity remains, it is reasonable that the item is less than 5730 years old.

EXAMPLE 22-6 Uranium–Lead Dating

A sample of uranium ore is found to contain 4.64 mg of ^{238}U and 1.22 mg of ^{206}Pb. Estimate the age of the ore. The half-life of ^{238}U is 4.51×10^9 years.

Plan

The original mass of ^{238}U is equal to the mass of ^{238}U remaining plus the mass of ^{238}U that decayed to produce the present mass of ^{206}Pb. We obtain the specific rate constant, k, from the known half-life. Then we use the ratio of original ^{238}U to remaining ^{238}U to calculate the time elapsed, with the aid of the first-order integrated rate equation.

Solution

First we calculate the amount of ^{238}U that must have decayed to produce 1.22 mg of ^{206}Pb, using the isotopic masses.

$$\underline{?} \text{ mg } ^{238}\text{U} = 1.22 \text{ mg } ^{206}\text{Pb} \times \frac{238 \text{ mg } ^{238}\text{U}}{206 \text{ mg } ^{206}\text{Pb}} = 1.41 \text{ mg } ^{238}\text{U}$$

Thus, the sample originally contained 4.64 mg + 1.41 mg = 6.05 mg of ^{238}U.
    We next evaluate the specific rate (disintegration) constant, k.

$$t_{1/2} = \frac{0.693}{k} \qquad \text{so} \qquad k = \frac{0.693}{t_{1/2}} = \frac{0.693}{4.51 \times 10^9 \text{ y}} = 1.54 \times 10^{-10} \text{ y}^{-1}$$

Now we calculate the age of the sample, t.

$$\ln\left(\frac{A_0}{A}\right) = kt$$

$$\ln\left(\frac{6.05 \text{ mg}}{4.64 \text{ mg}}\right) = (1.54 \times 10^{-10} \text{ y}^{-1})t$$

$$\ln 1.30 = (1.54 \times 10^{-10} \text{ y}^{-1})t$$

$$\frac{0.262}{1.54 \times 10^{-10} \text{ y}^{-1}} = t \qquad \text{or} \qquad t = 1.70 \times 10^9 \text{ years}$$

The ore is approximately 1.7 billion years old.

You should now work Exercise 76.

Medical Uses

The use of cobalt radiation treatments for cancerous tumors was described in Example 22-4. Several other nuclides are used in medicine as **radioactive tracers** (also called *radiopharmaceuticals*). Radioisotopes of an element have the same chemical properties as stable isotopes of the same element, so they can be used to "label" the presence of an element in compounds. A radiation detector can be used to follow the path of the element throughout the body. Modern computer-based techniques allow construction of an image of the area of the body where the radioisotope is concentrated. Salt solutions containing ^{24}Na can be injected into the bloodstream to follow the flow of blood and locate obstructions in the circulatory system. Thallium-201 tends to concentrate in healthy heart tissue, whereas technetium-99 concentrates in abnormal heart tissue. The two can be used together to survey damage from heart disease.

Iodine-123 concentrates in the thyroid gland, liver, and certain parts of the brain. This radioisotope is used to monitor goiter and other thyroid problems, as well as liver and brain tumors. One of the most useful radioisotopes in medical applications in recent years

▶ This isotope of iodine is also used in the treatment of thyroid cancer. Because of its preferential absorption in the thyroid gland, it delivers radiation where it is needed.

CHEMISTRY IN USE

Household Radon Exposure and Lung Cancer

Thanks to the Surgeon General's warnings on cigarette packages, most people are aware that smoking is the most common cause of lung cancer. But can you name the next most common? Few would guess that it's a noble gas, radon, and that this unhealthy element is almost certainly present in their home.

Radon has many unstable isotopes. The bad actor for humans is ^{222}Rn, a dense, odorless, colorless gas that is part of the ^{238}U decay chain (see Table 22-4). Radon is a major contributor to the background radiation dose experienced by the general population and the greatest natural source of radiation. (Other significant sources are cosmic radiation, particularly at high altitudes, and medical procedures.)

According to the World Health Organization, radon is responsible for between 3 and 14% of lung cancers. The Environmental Protection Agency (EPA) estimates that 21,000 cases of cancer in America each year are radon-related. The main damage to lung tissues is caused by the decay products ^{218}Po and ^{214}Po. When these species decay further, they emit alpha particles that harm DNA in the lung epithelium. Smokers are particularly at risk due to the elevated susceptibility of their lungs to cancer.

Radioactive decay is probably not high on your list of household worries, unless you happen to live next door to a missile silo or Three Mile Island. However, private homes are the locations of greatest radon health concern simply because people spend so much time in them. Uranium decay-chain elements occur naturally in geologic formations and thus ^{222}Rn is continually formed underground. It migrates upward through the pores and cracks of rocks and soils, and then into cracks, joints, and gaps in any buildings directly above. This migration is driven by the pressure differential between surface air (comparatively low) and the rocks and groundsoil (comparatively high). The radon is eventually concentrated in the building interiors and inhaled by the people living or working there.

The average indoor residential radon level in the United States is about 1.3 picocuries per liter (pCi/L), over three times the average outdoor level of ~0.4 pCi/L. While there is no safe level of radon, the "action level" for mitigation set by the EPA is 4 pCi/L. (One curie of radioactivity is equivalent to the quantity of a given nuclide that provides 3.700×10^{10} disintegrations per second). Radon levels are often evaluated at the time of a house purchase. However, testing can be carried out at any time by a professional or with a self-test kit.

Active soil depressurization (ASD) is the most common technique used to mitigate high radon levels in homes, and typically lowers concentrations by 50 to 99%. ASD relies on a simple, controlled alteration of the air-soil pressure differential. A suction point (open-ended pipe) is installed underneath the slab or floor of the house in a small excavated area where radon can collect. This pipe is connected by additional pipe to a continuously operating exhaust fan with a rooftop outlet. The vacuum effect of the fan pulls the radon-containing air from the excavated area into the pipe and ejects it above the roof, bypassing the interior of the house (and the lungs of its inhabitants).

LISA SAUNDERS BAUGH

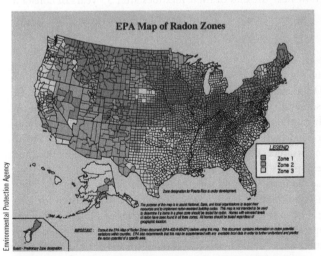

This EPA map shows predicted average indoor radon levels for all United States counties: Zone 1, >4 pCi/L; Zone 2, 2–4 pCi/L; Zone 3, <2 pCi/L. Homes in the lower zones are not necessarily safer; houses with elevated radon levels have been found in all three zones.

The white pipe in these photos is part of an Active Soil Depressurization system to mitigate radon in a New Jersey home. The U-tube manometer liquid level differential in the photo indicates that the exhaust pump is functioning properly.

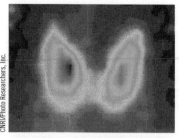

A scan of the radiation released by radioactive iodine concentrated in thyroid tissue gives an image of the thyroid gland.

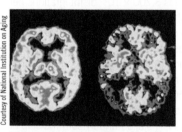

Positron emission tomography (PET) scans show differences in brain activity between a normal brain (*left*) and a brain affected by Alzheimer's disease (*right*). Glucose containing the radioactive isotope carbon-11 (^{11}C) was injected into the patients to obtain these images. When a carbon-11 isotope decays, it gives off a positron that immediately combines with an electron on another nearby atom. This matter–antimatter reaction converts these two particles into two high-energy gamma rays (180° apart) that are measured by a ring of detectors around the patient's head. Regions of the brain that are active consume more radioactive glucose and produce the brighter areas (red and yellow), while blue and black denote low or inactive portions.

▶ The procedure works because the female flies mate only once, so in an area highly populated with sterile males, the probability of a "productive" mating is very small.

is an isotope of technetium, an element that does not occur naturally on earth. This isotope, ^{99m}Tc, is produced by the decay of ^{99}Mo.

$$^{99}\text{Mo} \longrightarrow {}^{99m}\text{Tc} + {}_{-1}^{0}\beta$$

The "m" in the superscript ^{99m}Tc stands for "metastable." This isotope is formed at a high energy, and then slowly decays by emitting gamma radiation.

$$^{99m}\text{Tc} \longrightarrow {}^{99}\text{Tc} + \gamma$$

This decay has a half-life of 6.0 hours—long enough for the preparation of the radiopharmaceutical and subsequent imaging of the patient, but short enough to avoid a harmful radiation dose to the patient. Technetium is a transition metal located in the center of the periodic table. It can exist in several oxidation states from -1 to $+7$. These can undergo a diversity of reactions that allow ^{99m}Tc to be incorporated into a variety of chemical forms. One chemical compound of ^{99m}Tc can be used to image the bones, another to image the liver, another to image the brain, and so on.

Another form of imaging that uses positron emitters (see Section 22-6) is *positron emission tomography* (*PET*). Isotopes commonly used in this technique are short-lived positron emitters such as ^{11}C ($t_{1/2} = 20.4$ min), ^{13}N ($t_{1/2} = 9.98$ min), ^{15}O ($t_{1/2} = 2.05$ min), and ^{18}F ($t_{1/2} = 110$ min). The appropriate isotope is incorporated into a chemical that is normally taken up by the tissues that are being investigated, for instance carbon dioxide or glucose including ^{11}C or water including ^{15}O. This radioactive chemical can then be administered by inhalation or injection. The patient is then placed into a cylindrical gamma ray detector. When these radioisotopes decay, the emitted positron quickly encounters an electron and reacts in a matter–antimatter annihilation, to give off two gamma rays in opposite directions.

$$_{+1}^{0}\beta + {}_{-1}^{0}\beta \longrightarrow 2\gamma$$

The directions of emission of millions of such pairs of gamma rays detected over several minutes allow for a computer reconstruction of an image of the tissue containing the positron emitter.

The energy produced by the decay of plutonium-238 can be converted into electrical energy and used in batteries for special long-life applications. Some heart pacemakers in the mid to late 1970s used ^{238}Pu as an energy source, but they now use safer lithium battery technology. Voyager 1 and 2 space probes, launched in 1977, used batteries based on ^{238}PuO$_2$. These batteries are now near the end of their lifetimes after 35 years of constant operation.

Agricultural Uses

The pesticide DDT accumulates in fatty tissues, and is toxic to humans and animals repeatedly exposed to it. DDT persists in the environment for a long time. The DDT once used to control the screwworm fly was replaced by a radiological technique. Irradiating the male flies with gamma-rays alters their reproductive cells, sterilizing them. When great numbers of sterilized male flies are released in an infested area, they mate with females, that, of course, produce no offspring. This results in the reduction and eventual disappearance of the population.

Labeled fertilizers can also be used to study nutrient uptake by plants and to study the growth of crops. Gamma-irradiation of some foods allows them to be stored for longer periods without spoiling. For example, it retards the sprouting of potatoes and onions. In 1999, the FDA approved gamma-irradiation of red meat as a way to curb food-borne illnesses. In addition to reducing levels of *Listeria, Salmonella,* and other bacteria significantly, such irradiation is currently the only known way to completely eliminate the dangerous strain of *Escherichia coli* bacteria in red meat. Absorption of gamma-rays by matter produces no radioactive nuclides, so foods preserved in this way are *not* radioactive.

Moderate irradiation with gamma-rays from radioactive isotopes has kept the strawberries at the right fresh for 15 days, while those at the left are moldy. Such irradiation kills mold spores but does no damage to the food. The food does *not* become radioactive.

A weakly radioactive source such as americium is used in some smoke detectors. Radiation from the source ionizes the air to produce a weak current. Smoke particles interrupt the current flow by attracting the ions. This decrease in current triggers the alarm.

The Curiosity Mars rover is powered by a $^{238}PuO_2$ thermoelectric battery. The rover requires more power than solar cells could provide.

Industrial Uses

There are many applications of radiochemistry in industry and engineering. When great precision is required in the manufacture of strips or sheets of metal of definite thicknesses, the penetrating powers of various kinds of radioactive emissions are utilized. The thickness of the metal is correlated with the intensity of radiation passing through it. The flow of a liquid or gas through a pipeline can be monitored by injecting a sample containing a radioactive substance. Leaks in pipelines can also be detected in this way.

In addition to the ^{238}Pu-based heart pacemaker already mentioned, lightweight, portable power packs that use radioactive isotopes as fuel have been developed for other uses. Polonium-210, californium-242, and californium-244 have been used in such generators to power instruments for space vehicles and in polar regions. These generators can operate for years with only a small loss of power.

Research Applications

The pathways of chemical reactions can be investigated using radioactive tracers. When radioactive $^{35}S^{2-}$ ions are added to a saturated solution of cobalt sulfide in equilibrium with solid cobalt sulfide, the solid CoS becomes radioactive. This shows that sulfide ion exchange occurs between solid and solution in the solubility equilibrium.

$$CoS(s) \rightleftharpoons Co^{2+}(aq) + S^{2-}(aq) \qquad K_{sp} = 8.7 \times 10^{-23}$$

Photosynthesis is the process by which the carbon atoms in CO_2 are incorporated into glucose, $C_6H_{12}O_6$, in green plants.

$$6CO_2 + 6H_2O \xrightarrow[\text{chlorophyll}]{\text{sunlight}} C_6H_{12}O_6 + 6O_2$$

The process is far more complex than the net equation implies; it actually occurs in many steps and produces a number of intermediate products. By using labeled $^{14}CO_2$, which contains radioactive ^{14}C atoms, we can identify the intermediate molecules.

22-13 Artificial Transmutations of Elements

The first artificially induced nuclear reaction was carried out by Rutherford in 1915. He bombarded nitrogen-14 with alpha particles to produce an isotope of oxygen and a proton.

$$^{14}_{7}N + {}^{4}_{2}\alpha \longrightarrow {}^{1}_{1}H + {}^{17}_{8}O$$

Such reactions are often indicated in abbreviated form, with the bombarding particle and emitted subsidiary particles shown parenthetically between the parent and daughter nuclides.

$$^{14}_{7}\text{N} \, (^{4}_{2}\alpha, \, ^{1}_{1}p) \, ^{17}_{8}\text{O}$$

Several thousand different artificially induced reactions have been carried out with bombarding particles such as neutrons, protons, deuterons ($^{2}_{1}\text{H}$), alpha particles, and other small nuclei.

Bombardment with Positive Ions

A problem arises with the use of positively charged nuclei as projectiles. For a nuclear reaction to occur, the bombarding nuclei must actually collide with the target nuclei, which are also positively charged. Collisions cannot occur unless the projectiles have sufficient kinetic energy to overcome coulombic repulsion. The required kinetic energies increase with increasing atomic numbers of the target and of the bombarding particle.

▶ The first cyclotron was constructed by E. O. Lawrence (1901–1958) and M. S. Livingston at the University of California in 1930.

Particle accelerators called **cyclotrons** (atom smashers) and **linear accelerators** have overcome the problem of repulsion. A cyclotron (Figure 22-8) consists of two hollow, D-shaped electrodes called "dees." Both dees are in an evacuated enclosure between the poles of an electromagnet. The particles to be accelerated are introduced at the center in the gap between the dees. The dees are connected to a source of high-frequency alternating current that keeps them oppositely charged. The positively charged particles are attracted toward the negative dee. The magnetic field causes the path of the charged particles to curve 180° to return to the space between the dees. Then the charges are reversed on the dees, so the particles are repelled by the first dee (now positive) and attracted to the second. This repeated process is synchronized with the motion of the particles. They accelerate along a spiral path and eventually emerge through an exit hole oriented so that the beam hits the target atoms (Figure 22-9).

▶ The path of the particle is initially circular because of the interaction of the particle's charge with the electromagnet's field. As the particle gains energy, the radius of the path increases and the particle spirals outward.

▶ The first linear accelerator was built in 1928 by the German physicist Rolf Wideroe.

In a linear accelerator, the particles are accelerated through a series of tubes within an evacuated chamber (Figure 22-10). The odd-numbered tubes are at first negatively charged and the even ones positively charged. A positively charged particle is attracted toward the first tube. As it passes through that tube, the charges on the tubes are reversed so that the

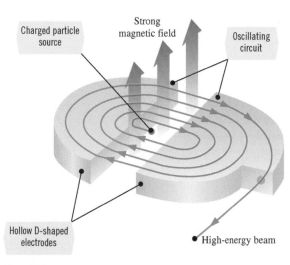

Charged particle source

Strong magnetic field

Oscillating circuit

Hollow D-shaped electrodes

High-energy beam

Figure 22-8 Schematic representation of a cyclotron.

Figure 22-9 The SLAC National Laboratory houses what used to be called the Stanford Linear Accelerator Center and is still commonly called that. The accelerator is 2 miles long and is the longest in the world.

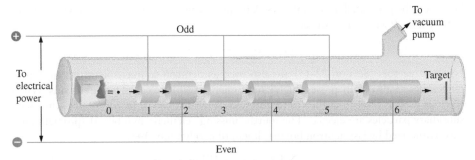

Figure 22-10 Diagram of an early type of linear accelerator. An alpha emitter is placed in the container at the left. Only those α-particles that happen to be emitted in line with the series of accelerating tubes can escape.

particle is repelled out of the first tube (now positive) and toward the second (negative) tube. Each time the particle nears the end of the second tube, the charges are again reversed. As this process is repeated, the particle is accelerated to very high velocities. The polarity is changed at constant frequency, so subsequent tubes are longer to accommodate the increased distance traveled by the accelerating particle per unit time. The bombardment target is located outside the last tube. If the initial polarities are reversed, negatively charged particles can also be accelerated. The longest linear accelerator, completed in 1966 at Stanford University, is about 2 miles long. It is capable of accelerating electrons to energies of nearly 20 GeV (see Figure 22-9).

▶ One gigaelectron volt (GeV) = 1×10^9 eV = 1.60×10^{-10} J. This is sometimes called 1 billion electron volts (BeV) in the United States.

Many nuclear reactions have been induced by such bombardment techniques. At the time of development of particle accelerators, there were a few gaps among the first 92 elements in the periodic table. Particle accelerators were used between 1937 and 1941 to synthesize three of the four "missing" elements: numbers 43 (technetium), 85 (astatine), and 87 (francium).

$$^{96}_{42}\text{Mo} + {}^2_1\text{H} \longrightarrow {}^{97}_{43}\text{Tc} + {}^1_0n$$

$$^{209}_{83}\text{Bi} + {}^4_2\alpha \longrightarrow {}^{210}_{85}\text{At} + 3\,{}^1_0n$$

$$^{230}_{90}\text{Th} + {}^1_1p \longrightarrow {}^{223}_{87}\text{Fr} + 2\,{}^4_2\alpha$$

▶ Mass and charge balance for alpha capture reaction
$$^{209}_{83}\text{Bi}\,(\alpha, 3n)\,{}^{210}_{85}\text{At}$$

mass: $209 + 4 \longrightarrow 210 + 3(1)$
charge: $83 + 2 \longrightarrow 85 + 3(0)$

Many hitherto unknown, unstable, artificial isotopes of known elements have also been synthesized so that their nuclear structures and behavior could be studied.

Neutron Bombardment

Neutrons bear no charge, so they are not repelled by nuclei as positively charged projectiles are. They do not need to be accelerated to produce bombardment reactions. Neutron

The CERN Large Hadron Collider, which spans the border between Switzerland and France, is more than 16 miles in circumference. The LHC recently provided evidence for the long-sought Higgs Boson subatomic particle.

beams can be generated in several ways. A frequently used method involves bombardment of beryllium-9 with alpha particles.

$$\ce{^{9}_{4}Be + ^{4}_{2}\alpha \longrightarrow ^{12}_{6}C + ^{1}_{0}n}$$

▶ Fast neutrons move so rapidly that they are likely to pass right through a target nucleus without reacting. Hence, the probability of a reaction is low, even though the neutrons may be very energetic.

Nuclear reactors (Section 22-15) are also used as neutron sources. Neutrons ejected in nuclear reactions usually possess high kinetic energies and are called **fast neutrons**. When they are used as projectiles they cause reactions, such as (n, p) or (n, α) reactions, in which subsidiary particles are ejected. The fourth "missing" element, number 61 (promethium), was synthesized by fast neutron bombardment of neodymium-142.

$$\ce{^{142}_{60}Nd + ^{1}_{0}n \longrightarrow ^{143}_{61}Pm + ^{0}_{-1}\beta}$$

Slow neutrons ("thermal" neutrons) are produced when fast neutrons collide with **moderators** such as hydrogen, deuterium, oxygen, or the carbon atoms in paraffin. These neutrons are more likely to be captured by target nuclei. Bombardments with slow neutrons can cause neutron-capture (n, γ) reactions.

$$\ce{^{200}_{80}Hg + ^{1}_{0}n \longrightarrow ^{201}_{80}Hg + ^{0}_{0}\gamma}$$

▶ Mass and charge balance for neutron capture reaction
$$\ce{^{6}_{3}Li}\ (n, \alpha)\ \ce{^{3}_{1}H}$$

mass: $6 + 1 \longrightarrow 3 + 4$
charge: $3 + 0 \longrightarrow 1 + 2$

Slow neutron bombardment also produces the ^{3}H isotope (tritium).

$$\ce{^{6}_{3}Li + ^{1}_{0}n \longrightarrow ^{3}_{1}H + ^{4}_{2}\alpha} \qquad (n, \alpha) \text{ reaction}$$

E. M. McMillan (1907–1991) discovered the first transuranium element, neptunium, in 1940 by bombarding uranium-238 with slow neutrons.

$$\ce{^{238}_{92}U + ^{1}_{0}n \longrightarrow ^{239}_{92}U + ^{0}_{0}\gamma}$$

$$\ce{^{239}_{92}U \longrightarrow ^{239}_{93}Np + ^{0}_{-1}\beta}$$

Several additional elements have been prepared by neutron bombardment or by bombardment of the nuclei so produced with positively charged particles. Some examples are

$$\left. \begin{array}{l} \ce{^{238}_{92}U + ^{1}_{0}n \longrightarrow ^{239}_{92}U + ^{0}_{0}\gamma} \\ \ce{^{239}_{92}U \longrightarrow ^{239}_{93}Np + ^{0}_{-1}\beta} \\ \ce{^{239}_{93}Np \longrightarrow ^{239}_{94}Pu + ^{0}_{-1}\beta} \end{array} \right\} \quad \text{plutonium}$$

$$\ce{^{239}_{94}Pu + ^{4}_{2}\alpha \longrightarrow ^{242}_{96}Cm + ^{1}_{0}n} \qquad \text{curium}$$

$$\ce{^{246}_{96}Cm + ^{12}_{6}C \longrightarrow ^{254}_{102}No + 4\,^{1}_{0}n} \qquad \text{nobelium}$$

$$\ce{^{252}_{98}Cf + ^{10}_{5}B \longrightarrow ^{257}_{103}Lr + 5\,^{1}_{0}n} \qquad \text{lawrencium}$$

22-14 Nuclear Fission

Isotopes of some elements with atomic numbers above 80 are capable of undergoing fission in which they split into nuclei of intermediate masses and emit one or more neutrons. Some fissions are spontaneous; others require that the activation energy be supplied by bombardment. A given nucleus can split in many different ways, liberating enormous amounts of energy. Some of the possible fissions that can result from bombardment of fissionable uranium-235 with fast neutrons follow. The uranium-236 is a short-lived intermediate.

$$\ce{^{235}_{92}U + ^{1}_{0}n \longrightarrow ^{236}_{92}U} \begin{cases} \ce{^{160}_{62}Sm + ^{72}_{30}Zn + 4\,^{1}_{0}n} + \text{energy} \\ \ce{^{146}_{57}La + ^{87}_{35}Br + 3\,^{1}_{0}n} + \text{energy} \\ \ce{^{141}_{56}Ba + ^{92}_{36}Kr + 3\,^{1}_{0}n} + \text{energy} \\ \ce{^{144}_{55}Cs + ^{90}_{37}Rb + 2\,^{1}_{0}n} + \text{energy} \\ \ce{^{144}_{54}Xe + ^{90}_{38}Sr + 2\,^{1}_{0}n} + \text{energy} \end{cases}$$

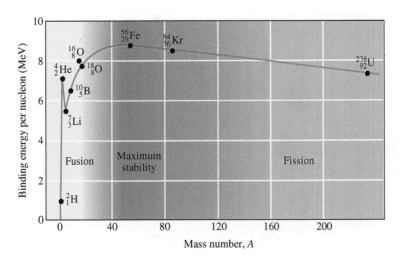

Figure 22-11 Variation in nuclear binding energy per nucleon with atomic mass. This plot shows the relative stability of the most stable isotopes of selected elements. The most stable nucleus is $^{56}_{26}$Fe, with a binding energy of 8.80 MeV per nucleon

$$1 \text{ MeV} = 1.60 \times 10^{-13} \text{ J}$$

Recall that the binding energy is the amount of energy that must be supplied to the nucleus to break it apart into subatomic particles. Figure 22-11 is a plot of binding energy per nucleon versus mass number. It shows that atoms of intermediate mass number have the highest binding energies per nucleon; therefore, they are the most stable. Thus, fission is an energetically favorable process for heavy atoms because atoms with intermediate masses and greater binding energies per nucleon are formed.

> The term *nucleon* refers to a nuclear particle, either a neutron or a proton.

Which isotopes of which elements undergo fission? Experiments with particle accelerators have shown that every element with an atomic number of 80 or more has one or more isotopes capable of undergoing fission, provided they are bombarded at the right energy. Nuclei with atomic numbers between 89 and 98 fission spontaneously with long half-lives of 10^4 to 10^{17} years. Nuclei with atomic numbers of 98 or more fission spontaneously with shorter half-lives of a few milliseconds to 60.5 days. One of the *natural* decay modes of the transuranium elements is via spontaneous fission. In fact, all known nuclides with *mass numbers* greater than 250 do this because they are too big to be stable. Most nuclides with mass numbers between 225 and 250 do not undergo fission spontaneously (except for a few with extremely long half-lives). They can be induced to undergo fission when bombarded with particles of relatively low kinetic energies. Particles that can supply the required activation energy include neutrons, protons, alpha particles, and fast electrons. For nuclei lighter than mass 225, the activation energy required to induce fission rises very rapidly.

> The word "fission" can be used either as a noun (meaning the process) or as a verb (meaning to undergo the process).

In Section 22-2, we discussed the stability of nuclei with even numbers of protons and even numbers of neutrons. We should not be surprised to learn that both ^{233}U and ^{235}U can be excited to fissionable states by slow neutrons much more easily than ^{238}U because they are less stable. It is so difficult to cause fission in ^{238}U that this isotope is said to be "nonfissionable."

Typically, two or three neutrons are produced per fission reaction. These neutrons can collide with other fissionable atoms to repeat the process. If sufficient fissionable material, the **critical mass**, is contained in a small enough volume, a sustained **chain reaction** can result. If too few fissionable atoms are present, most of the neutrons escape and no chain reaction occurs. Figure 22-12 depicts a fission chain reaction.

In an atomic bomb, two subcritical portions of fissionable material are brought together to form a critical mass. A nuclear fission explosion results. A tremendous amount of heat energy is released, as well as many radionuclides whose effects are devastating to life and the environment. The radioactive dust and debris are called *fallout*.

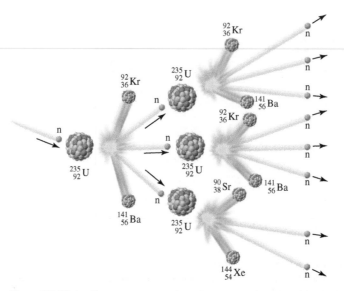

Figure 22-12 A self-propagating nuclear chain reaction. A stray neutron induces a single fission, liberating more neutrons. Each of them induces another fission, each of which is accompanied by release of two or three neutrons. The chain continues to branch in this way, very quickly resulting in an explosive rate of fission.

22-15 Nuclear Fission Reactors

In a nuclear fission reactor, the fission reaction is controlled by inserting materials to absorb some of the neutrons so that the mixture does not explode. The energy that is produced can be safely used as a heat source in a power plant.

Light Water Reactors

Most commercial nuclear power plants in the United States are "light water" reactors, moderated and cooled by ordinary water. Figure 22-13 is a schematic diagram of a light

Figure 22-13 A schematic diagram of a light water reactor plant. This design includes two closed loops of water. The water that carries heat from the reactor to the steam generator is in a closed loop and is not released to the environment.

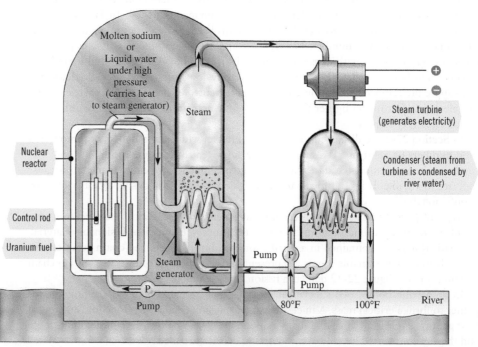

water reactor plant. The reactor core at the left replaces the furnace in which coal, oil, or natural gas is burned in a fossil fuel plant. Such a fission reactor consists of five main components: (1) fuel, (2) moderator, (3) control rods, (4) cooling system, and (5) shielding.

Fuel

Rods of U_3O_8 enriched in uranium-235 serve as the fuel. Unfortunately, uranium ores contain only about 0.7% $^{235}_{92}U$. Most of the rest is nonfissionable $^{238}_{92}U$. The enrichment is done in processing and reprocessing plants by separating gaseous $^{235}UF_6$ from $^{238}UF_6$, prepared from the ore. Separation by diffusion is based on the slower rates of diffusion of heavier gas molecules (see Section 12-14). Another separation procedure uses the ultracentrifuge.

A potentially more efficient method of enrichment would involve the use of sophisticated tunable lasers to ionize $^{235}_{92}U$ selectively and not $^{238}_{92}U$. The ionized $^{235}_{92}U$ could then be made to react with negative ions to form another compound, easily separated from the mixture. For this method to work, we must construct lasers capable of producing radiation monochromatic enough to excite one isotope and not the other—a difficult challenge.

Uranium is deposited on the negative electrode in the electrorefining phase of fuel reprocessing. The crystalline mass is about 97% LiCl and KCl. The remaining 3% uranium chloride is responsible for the amethyst color.

Moderator

The fast neutrons ejected during fission are too energetic to be absorbed efficiently by other nuclei. Thus, they must be slowed by collisions with atoms of comparable mass that do not absorb them, called *moderators*. The most commonly used moderator is ordinary water, although graphite is sometimes used. The most efficient moderator is helium, which slows neutrons but does not absorb them all. The next most efficient is **heavy water** (deuterium oxide, $^2_1H_2O$ or $^2_1D_2O$). This is so expensive that it has been used chiefly in research reactors. A Canadian-designed power reactor that uses heavy water is more neutron-efficient than light water reactors. This design is the basis of the many reactors in Canada.

Control Rods

Cadmium and boron are good neutron absorbers.

$$^{10}_5B + ^1_0n \longrightarrow ^7_3Li + ^4_2\alpha$$

The rate of a fission reaction is controlled by the use of movable control rods, usually made of cadmium or boron steel. They are automatically inserted in or removed from spaces between the fuel rods. The more neutrons absorbed by the control rods, the fewer fissions occur and the less heat is produced. Hence, the heat output is governed by the control system that operates the rods.

Cooling System

Two cooling systems are needed. First, the moderator itself serves as a coolant for the reactor. It transfers fission-generated heat to a steam generator. This converts water to steam. The steam then goes to turbines that drive generators to produce electricity. Another coolant (river water, seawater, or recirculated water) condenses the steam from the turbine, and the condensate is then recycled into the steam generator.

The danger of meltdown arises if a reactor is shut down quickly. The disintegration of radioactive fission products still goes on at a furious rate, fast enough to overheat the fuel elements and to melt them. So it is not enough to shut down the fission reaction. Efficient cooling must be continued until the short-lived isotopes are gone and the heat from their disintegration is dissipated. Only then can the circulation of cooling water be stopped.

The 1979 accident at Three Mile Island, near Harrisburg, Pennsylvania, was due to stopping the water pumps too soon *and* the inoperability of the emergency pumps. A combination of mechanical malfunctions, errors, and carelessness produced the overheating that damaged the fuel assembly. It did not and *could not explode*, although melting of the core material did occur. The 1986 accident at Chernobyl, in the USSR, involved a reactor

Nuclear powered submarines, like the USS Maryland shown here, typically use pressurized water reactors and highly enriched nuclear fuel to provide more power from a smaller reactor. One stealth issue with nuclear subs is the need to cool the reactor, which leaves a trail of warm water that rises to the surface.

▶ The neutrons are the worst problem of radiation. The human body contains a high percentage of H_2O, which absorbs neutrons very efficiently. The neutron bomb produces massive amounts of neutrons and so is effective against people, but it does not produce the long-lasting radiation or blast damage of the fission atomic bomb.

of a very different design and was far more serious. The effects of that disaster will continue for decades.

Shielding

It is essential that people and the surrounding countryside be adequately shielded from possible exposure to radioactive nuclides. The entire reactor is enclosed in a steel containment vessel. This is housed in a thick-walled concrete building. The operating personnel are further protected by a so-called biological shield, a thick layer of organic material made of compressed wood fibers. This absorbs the neutrons and beta and gamma rays that would otherwise be absorbed in the human body.

Breeder Reactors

The possibility of shortages in the known supply of $^{235}_{92}U$ has led to the development of **breeder reactors**, which can manufacture more fuel than they use. A breeder reactor is designed not only to generate electrical power but also to maximize neutron capture in the core by $^{238}_{92}U$. The fuel of a typical breeder reactor consists of the abundant but nonfissionable isotope $^{238}_{92}U$ mixed with $^{235}_{92}U$ or $^{239}_{94}Pu$, which produce neutrons when they undergo fission. Some of these neutrons are absorbed by $^{238}_{92}U$ to form $^{239}_{92}U$. This unstable uranium isotope soon leads, after two steps of beta emission, to $^{239}_{94}Pu$.

$$^{238}_{92}U + ^{1}_{0}n \longrightarrow ^{239}_{92}U \xrightarrow[t_{1/2} = 23.4 \text{ min}]{\beta \text{ decay}} ^{239}_{93}Np \xrightarrow[t_{1/2} = 2.35 \text{ days}]{\beta \text{ decay}} ^{239}_{94}Pu$$

This fissionable $^{239}_{94}Pu$ can then be used as fuel in a reactor.

For every $^{235}_{92}U$ or $^{239}_{94}Pu$ nucleus that undergoes fission, more than one neutron is captured by $^{238}_{92}U$ to produce $^{239}_{94}Pu$. Thus, the breeder reactor can produce more fissionable material than it consumes. After about 7 years, enough $^{239}_{94}Pu$ can be collected to fuel a new reactor *and* to refuel the original one.

Nuclear Power: Hazards and Benefits

Controlled fission reactions in nuclear reactors are of great use and have even greater potential. The fuel elements of a nuclear reactor have neither the composition nor the extremely compact arrangement of the critical mass of a bomb. Thus, no possibility of nuclear explosion exists. However, various dangers are associated with nuclear energy generation. The possibility of "meltdown" has been discussed with respect to cooling systems in light water reactors. Proper shielding precautions must be taken to ensure that the radionuclides produced are always contained within vessels from which neither they nor their radiations can escape. Long-lived radionuclides from spent fuel must be stored underground in heavy, shock-resistant containers until they have decayed to the point that they are no longer biologically harmful. As examples, strontium-90 ($t_{1/2}$ = 28 years) and plutonium-239 ($t_{1/2}$ = 24,000 years) must be stored for 280 years and 240,000 years, respectively, before they lose 99.9% of their activities. Critics of nuclear energy contend that the containers could corrode over such long periods, or burst as a result of earth tremors, and that transportation and reprocessing accidents could cause environmental contamination with radionuclides. They claim that river water used for cooling is returned to the rivers with too much heat (thermal pollution), thus disrupting marine life. (It should be noted, though, that fossil fuel electric power plants cause the same thermal pollution for the same amount of electricity generated.) The potential for theft also exists. Plutonium-239, a fissionable material, could be stolen from reprocessing plants and used to construct atomic weapons.

Proponents of the development of nuclear energy argue that the advantages far outweigh the risks. Nuclear energy plants do not pollute the air with oxides of sulfur, nitrogen, carbon, and particulate matter, as fossil fuel electric power plants do. The big advantage of nuclear fuels is the enormous amount of energy liberated per unit mass of fuel. At present, nuclear reactors provide about 20% of the electrical energy consumed in the United States.

U.S. Department of Energy/Photo Researchers, Inc.

Nuclear waste may take centuries to decompose, so we cannot afford to take risks in its disposal. Suggested approaches include casting it into ceramics, as shown here, to eliminate the possibility of the waste dissolving in groundwater. The encapsulated waste could then be deposited in underground salt domes. Located in geologically stable areas, such salt domes have held petroleum and compressed natural gas trapped for millions of years. The political problems of nuclear waste disposal are at least as challenging as the technological ones. Refer to the Chemistry in Use box entitled "Managing Nuclear Wastes" in this chapter.

In some parts of Europe, where natural resources of fossil fuels are scarcer, the utilization of nuclear energy is higher. For instance, in France and Belgium, more than 80% of electrical energy is produced from nuclear reactors. With rapidly declining fossil fuel reserves, it appears likely that nuclear energy and solar energy will become increasingly important. Intensifying public concerns about nuclear power, however, may mean that further growth in energy production using nuclear power in the United States must await technological developments to overcome the remaining hazards.

22-16 Nuclear Fusion

Fusion, the joining of light nuclei to form heavier nuclei, is favorable for the very light atoms. In both fission and fusion, the energy liberated is equivalent to the loss of mass that accompanies the reactions. Much greater amounts of energy per unit mass of *reacting atoms* are produced in fusion than in fission.

Spectroscopic evidence indicates that the sun is a tremendous fusion reactor consisting of 73% H, 26% He, and 1% other elements. Its major fusion reaction is thought to involve the combination of a deuteron, ^2_1H, and a triton, ^3_1H, at tremendously high temperatures to form a helium nucleus and a neutron with the release of a huge amount of energy.

$$^2_1\text{H} + {}^3_1\text{H} \longrightarrow {}^4_2\text{He} + {}^1_0n + 1.7 \times 10^{19}\,\text{kJ/mol rxn}$$

Thus, solar energy is actually a form of fusion energy. A fusion reaction releases a tremendous amount of energy. Compare the energy of the deuteron/triton fusion reaction with that released by the burning of the methane in natural gas (see Section 15-1).

$$CH_4(g) + 2O_2(g) \longrightarrow CO_2(g) + 2H_2O(\ell) + 890\,\text{kJ/mol rxn}$$

Fusion reactions are accompanied by even greater energy production per unit mass of reacting atoms than are fission reactions. They can be initiated only by extremely high temperatures, however. This is why fusion reactions are often referred to as **thermonuclear** reactions. The fusion of ^2_1H and ^3_1H occurs at the lowest temperature of any fusion reaction known, but even this is 40,000,000 K! Such temperatures exist in the sun and

The explosion of a thermonuclear (hydrogen) bomb releases tremendous amounts of energy. If we could learn how to control this process, we would have nearly limitless amounts of energy.

Nuclear fusion provides the energy of our sun and other stars. Development of controlled fusion as a practical source of energy requires methods to initiate and contain the fusion process. Here a very powerful laser beam has initiated a fusion reaction in a 1-mm target capsule that contained deuterium and tritium. In a 0.5-picosecond burst, 10^{13} neutrons were produced by the reaction $^2_1\text{H} + {}^3_1\text{H} \longrightarrow {}^4_2\text{He} + {}^1_0n$.

Managing Nuclear Wastes

Some may consider a career in managing nuclear waste as being just about the worst job anyone would ever want, but hundreds of technically trained people have spent years working to solve the problems associated with nuclear power. The major part of the continuing challenge is political. Nuclear power plants generate about 20% of the electricity in the United States. Most of the high-level nuclear waste (HLW) that is generated from nuclear power plants—in the form of spent nuclear fuel (SNF)—is generated where many people live, in the eastern half of the United States. The safest place for a repository is away from people, in a dry, remote location, probably in the western United States, where there are fewer people (and fewer votes!).

SNF constitutes about half of the HLW in the United States. The other half comes from the construction and existence of nuclear weapons. All HLW is a federal responsibility. About 90% of the radioactivity in nuclear waste is from HLW. The largest volume of nuclear waste is low-level waste (LLW), and that is mostly the responsibility of the state (or group of states) in which it is generated. LLW is rather awkwardly defined, being everything that is neither HLW nor defense waste and consists of wastes from hospitals; pharmaceutical labs; research labs; and the moon suits, tools, and the like from nuclear power plants. In the eastern United States, most of the LLW is in the form of the plastic beads that make up the ion-exchange resins used in nuclear power plants to clean various loops of water used in power production.

Plutonium wastes from the Los Alamos National Laboratory in northern New Mexico were trucked for the first time to the federal Waste Isolation Pilot Plant in Carlsbad in March 1999. The 600 pounds (270 kg) of waste consisted of plutonium-contaminated clothing and metal cans, packed in boxes and stainless steel containers. Most of the material was from the laboratory's manufacture of nuclear batteries used in NASA's deep space probes and will be buried in the depository carved out of ancient salt caverns about half a mile (0.8 km) below ground.

Most current attention is focused on SNF for two reasons. It is highly radioactive, and it can be seen as a "local" problem because it is made where electric customers live. Europe has reprocessing plants to recover the unused fissionable material for new fuel, but the United States disallowed the practice in the 1970s. This partially explains why spent fuel rods have been piling up at US nuclear plants.

Research focused on Yucca Mountain, Nevada, at the western edge of the National Test Site, for its suitability as a nuclear waste repository for SNF and some defense waste. In July 2002, after both houses of the US Congress voted to override a veto by the State of Nevada, President George W. Bush signed the bill making Yucca Mountain the central repository for the nation's nuclear waste. Many political leaders and residents of Nevada strongly opposed this plan, and they seriously question

that nuclear waste can be safely kept out of the human environment for 10,000 years, as is required under the federal Nuclear Waste Policy Act. Funding for the Yucca Mountain project was terminated in 2011 by Congress, but the controversy on where to store nuclear waste continues.

The numbers describing SNF are barely comprehensible to most people. The volume of all existing SNF could cover a large football stadium to a depth of 4 or 5 feet, but no sensible person would want to confine that much heat and radioactivity to one place. There are about 1000 isotopes of about 100 different elements in SNF, and most are radioactive. They decay into stable elements at different rates, giving off alpha, beta, and gamma emissions. It will take about 7000 years until the SNF will be only as radioactive as the rocks and minerals that make up our planet.

These fission products are housed in long titanium rods, each about the diameter of a pencil, that constitute the fuel assembly in a nuclear power plant. Workers wearing gloves can handle fuel assemblies before fissioning occurs. But after removal from a nuclear reactor, the fuel assembly is stored in a cooling pool of water beside the reactor for at least 10 years. On-site storage of the oldest fuel assemblies occurs in specially constructed concrete casks until the federal government takes ownership and finds a suitable place for it.

Other options that have been considered for HLW include outer space ejection and burial in deep ocean trenches. The consensus worldwide is that deep geological isolation is the best option.

For the student who likes a challenge, nuclear waste management is a good one. Here are a few key issues to study and discuss:

▶ *Transportation of the waste to its repository.* Should it be done by rail or by truck? Should there be public notification of the time of transport? Are there response measures in place in case of an accident?

▶ *The site's seismicity.* Will there be significant volcanic or seismic activity near the site in the next 10,000 years?

▶ *Hydrology.* Is there enough evidence to ensure that radionuclides will not seep into groundwater to any significant degree?

▶ *Weapons disarmament.* Should the plutonium from "disarmed" nuclear weapons eventually be turned into nuclear fuel or made useless immediately and buried with other HLW?

DONALD H. WILLIAMS
HOPE COLLEGE

other stars, but they are nearly impossible to achieve and contain on the earth. Thermonuclear bombs (called fusion bombs or hydrogen bombs) of incredible energy have been detonated in tests but, thankfully, never in war. In them, the necessary activation energy is supplied by the explosion of a fission bomb.

It is hoped that fusion reactions can be harnessed to generate energy for domestic power. Because of the tremendously high temperatures required, no currently known structural material can confine these reactions. At such high temperatures, all molecules dissociate and most atoms ionize, resulting in the formation of a new state of matter called a **plasma**. A very high temperature plasma is so hot that it melts and decomposes anything it touches, including structural components of a reactor. The technological innovation required to build a workable fusion reactor probably represents the greatest challenge ever faced by the scientific and engineering community.

▶ Plasmas have been called the fourth state of matter.

Recent attempts at the containment of lower-temperature plasmas by external magnetic fields have been successful, and they encourage our hopes. Fusion as a practical energy source, however, lies far in the future at best. The biggest advantages of its use would be that (1) the deuterium fuel can be found in virtually inexhaustible supply in the oceans; and (2) fusion reactions would produce only radionuclides of very short half-life, primarily tritium ($t_{1/2}$ = 12.3 years), so there would be no long-term waste disposal problem. If controlled fusion could be brought about, it could liberate us from dependence on uranium and fossil fuels.

U.S. Department of Energy/Photo Researchers, Inc.

The plasma in a fusion reactor must not touch the walls of its vacuum vessel, which would be vaporized. In the Tokamak fusion test reactor, the plasma is contained within a magnetic field shaped like a doughnut. The magnetic field is generated by D-shaped coils around the vacuum vessel.

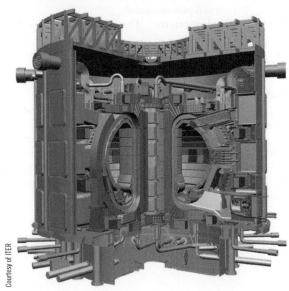

Courtesy of ITER

Proposed design for a new Tokamak reactor to be built in France. See www.iter.org/.

KEY TERMS

Alpha emission   Radioactive decay in which an alpha particle is given off from the nucleus.

Alpha particle (α)   A particle that consists of two protons and two neutrons; identical to a helium nucleus.

Artificial transmutation   An artificially induced nuclear reaction caused by bombardment of a nucleus with subatomic particles or small nuclei.

Band of stability   A band containing stable (nonradioactive) nuclides in a plot of number of neutrons versus number of protons (atomic number).

Beta emission   Radioactive decay in which an electron is ejected from the nucleus to convert a neutron into a proton.

Beta particle (β)   An electron emitted from the nucleus when a neutron decays to a proton and an electron.

Binding energy (nuclear binding energy)   The energy equivalent ($E = mc^2$) of the mass deficiency of an atom.

Breeder reactor   A fission reactor that produces more fissionable material than it consumes.

Chain reaction   A reaction that, once initiated, sustains itself and expands.

Cloud chamber   A device for observing the paths of speeding particles as vapor molecules condense on the ionized air molecules in their tracks.

Control rods   Rods of materials such as cadmium or boron steel that act as neutron absorbers (not merely moderators), used in nuclear reactors to control neutron fluxes and therefore rates of fission.

Critical mass   The minimum mass of a particular fissionable nuclide, in a given volume, that is required to sustain a nuclear chain reaction.

Cyclotron   A device for accelerating charged particles along a spiral path.

Daughter nuclide   A nuclide that is produced in a nuclear decay.

Electron capture   Absorption of an electron from the first energy level (*K shell*) by a proton as it is converted to a neutron; also known as *K capture; positron emission.*

Fast neutron   A neutron ejected at high kinetic energy in a nuclear reaction.

Fluorescence   Absorption of high-energy radiation by a substance and the subsequent emission of visible light.

Gamma ray (γ)   High-energy electromagnetic radiation.

Geiger-Müller counter   A type of gas ionization counter used to detect radiation.

Half-life of a radionuclide   The time required for half of a given sample to undergo radioactive decay.

Heavy water   Water containing deuterium, a heavy isotope of hydrogen, $_1^2\text{H}$.

Linear accelerator   A device used for accelerating charged particles along a straight-line path.

Mass deficiency (Δm)   The amount of matter that would be converted into energy if an atom were formed from constituent particles.

Moderator   A substance such as hydrogen, deuterium, oxygen, or paraffin capable of slowing fast neutrons upon collision.

Nuclear binding energy   The energy equivalent of the mass deficiency; energy released in the formation of an atom from subatomic particles. See *Binding energy.*

Nuclear fission   The process in which a heavy nucleus splits into nuclei of intermediate masses and one or more neutrons are emitted.

Nuclear fusion   The combination of light nuclei to produce a heavier nucleus.

Nuclear reaction   A reaction involving a change in the composition of a nucleus; it can evolve or absorb an extraordinarily large amount of energy.

Nuclear reactor   A system in which controlled nuclear fission reactions generate heat energy on a large scale. The heat energy is subsequently converted into electrical energy.

Nucleons   Particles comprising the nucleus; protons and neutrons.

Nuclides   Different atomic forms of all elements (in contrast to isotopes, which are different atomic forms of a single element).

Parent nuclide   A nuclide that undergoes nuclear decay.

Plasma   A physical state of matter that exists at extremely high temperatures, in which all molecules are dissociated and most atoms are ionized.

Positron   A nuclear particle with the mass of an electron but opposite charge.

Positron emission   See *Electron capture.*

Radiation   High-energy particles or rays emitted in nuclear decay processes.

Radioactive dating   A method of dating ancient objects by determining the ratio of amounts of a parent nuclide and one of its decay products present in an object and relating the ratio to the object's age via half-life calculations.

Radioactive tracer   A small amount of radioisotope that replaces a nonradioactive isotope of the element in a compound whose path (e.g., in the body) or whose decomposition products are to be monitored by detection of radioactivity; also known as a *radiopharmaceutical.*

Radioactivity   The spontaneous disintegration of atomic nuclei.

Radioisotope   A radioactive isotope of an element.

Radionuclide   A radioactive nuclide.

Scintillation counter   A device used for the quantitative detection of radiation.

Slow neutron   A fast neutron slowed by collision with a moderator.

Thermonuclear energy   Energy from nuclear fusion reactions.

Transuranium elements   The elements with atomic numbers greater than 92 (uranium); none occur naturally and all must be prepared by nuclear bombardment of other elements.

EXERCISES

- 🔵 **Molecular Reasoning** exercises
- ▲ **More Challenging** exercises

Nuclear Stability and Radioactivity

1. Define and compare nuclear fission and nuclear fusion. Briefly describe current uses of nuclear fission and fusion.
2. Differentiate between natural and induced radioactivity. Use the periodic table to identify the locations of those elements that are the result of induced radioactivity.
3. 🔵 How is a nuclear reaction different from a chemical reaction?
4. What is the equation that relates the equivalence of matter and energy? What does each term in this equation represent?
5. What is mass deficiency? What is binding energy? How are the two related?
6. What are nucleons? What is the relationship between the number of protons and the atomic number? What is the relationship among the number of protons, the number of neutrons, and the mass number?
7. Define the term "binding energy per nucleon." How can this quantity be used to compare the stabilities of nuclei?
8. Describe the general shape of the plot of binding energy per nucleon against mass number.
9. (a) Briefly describe a plot of the number of neutrons against the atomic number (for the stable nuclides). Interpret the observation that the plot shows a band with a somewhat step-like shape. (b) Describe what is meant by "magic numbers" of nucleons.
10. Potassium, $Z = 19$, has a series of naturally occurring isotopes: ^{39}K, ^{40}K, ^{41}K. Identify the isotope(s) of potassium that is/are most likely to be stable and which tend to decay.
11. Platinum, $Z = 78$, has a series of naturally occurring isotopes: ^{190}Pt, ^{192}Pt, ^{194}Pt, ^{195}Pt, ^{196}Pt, and ^{198}Pt. Identify the isotope(s) of platinum that is/are most likely to be stable and which tend to decay.
12. Indicate the type of emission and the decay product predicted for each unstable isotope listed in Exercise 10.
13. Indicate the type of emission and the decay product predicted for each unstable isotope listed in Exercise 11.
14. The actual mass of a ^{62}Ni atom is 61.9283 amu. (a) Calculate the mass deficiency in amu/atom and in g/mol for this isotope. (b) What is the nuclear binding energy in kJ/mol for this isotope?
15. The actual mass of a ^{108}Pd atom is 107.90389 amu. (a) Calculate the mass deficiency in amu/atom and in g/mol for this isotope. (b) What is the nuclear binding energy in kJ/mol for this isotope?

16. Calculate the following for ^{64}Zn (actual mass $= 63.9291$ amu): (a) mass deficiency in amu/atom; (b) mass deficiency in g/mol; (c) binding energy in J/atom; (d) binding energy in kJ/mol; (e) binding energy in MeV/nucleon.
17. Calculate the following for ^{49}Ti (actual mass $= 48.94787$ amu): (a) mass deficiency in amu/atom; (b) mass deficiency in g/mol; (c) binding energy in J/atom; (d) binding energy in kJ/mol; (e) binding energy in MeV/nucleon.
18. Calculate the nuclear binding energy in kJ/mol for each of the following: (a) $^{127}_{53}$I; (b) $^{81}_{35}$Br; (c) $^{35}_{17}$Cl. The atomic masses are 126.9044 amu, 80.9163 amu, and 34.96885 amu, respectively.
19. Repeat Exercise 18 for (a) $^{36}_{16}$S, (b) $^{39}_{19}$K, and (c) $^{24}_{12}$Mg. Their respective atomic masses are 35.96709 amu, 38.96371 amu, and 23.98504 amu. Which of these nuclides has the greatest binding energy per nucleon?
20. Compare the behaviors of α, β, and γ radiation (a) in an electrical field, (b) in a magnetic field, and (c) with respect to ability to penetrate various shielding materials, such as a piece of paper and concrete. What is the composition of each type of radiation?

Charles D. Winters

Geiger-Müller counter

21. 🔵 Why are α-particles that are absorbed internally by the body particularly dangerous?
22. Name some radionuclides that have medical uses and list the uses.
23. Describe how radionuclides can be used in (a) research, (b) agriculture, and (c) industry.
24. Name and describe four methods for detection of radiation.
25. Describe how (a) nuclear fission and (b) nuclear fusion generate more stable nuclei.
26. What evidence exists to support the theory that nucleons are arranged in "shells" or energy levels within the nucleus?

Nuclear Reactions

27. Consider a radioactive nuclide with a neutron/proton ratio that is larger than those for the stable isotopes of that element. What mode(s) of decay might be expected for this nuclide? Why?

28. Repeat Exercise 27 for a nuclide with a neutron/proton ratio that is smaller than those for the stable isotopes.

29. Calculate the neutron/proton ratio for each of the following radioactive nuclides, and predict how each of the nuclides might decay: (a) $^{13}_{5}B$ (stable mass numbers for B are 10 and 11); (b) $^{92}_{38}Sr$ (stable mass numbers for Sr are between 84 and 88); (c) $^{192}_{82}Pb$ (stable mass numbers for Pb are between 204 and 208).

30. What particle is emitted in the following nuclear reactions? Write an equation for each reaction.
    (a) Gold-198 decays to mercury-198.
    (b) Radon-222 decays to polonium-218.
    (c) Cesium-137 decays to barium-137.
    (d) Indium-110 decays to cadmium-110.

31. Write the symbols for the nuclide that is formed in each of the following radioactive decays (β refers to an e^-).
    (a) $^{125}Sn \xrightarrow{-\beta}$ (d) $^{147}Sm \xrightarrow{-\alpha}$
    (b) $^{13}C \xrightarrow{-n}$ (e) $^{184}Ir \xrightarrow{-p}$
    (c) $^{11}B \xrightarrow{-\gamma}$ (f) $^{40}_{19}K \xrightarrow{-\beta}$

32. Predict the kind of decays you would expect for the following radionuclides: (a) $^{60}_{27}Co$ (n/p ratio too high); (b) $^{20}_{11}Na$ (n/p ratio too low); (c) $^{222}_{86}Rn$; (d) $^{67}_{29}Cu$; (e) $^{238}_{92}U$; (f) $^{11}_{6}C$.

33. There are only four naturally occurring nuclides (out of 263) that have both an odd number of protons and an odd number of neutrons. They are also among the first seven elements and each of them has the same number of neutrons as protons. Give their nuclide symbols.

34. Fill in the missing symbols in the following nuclear bombardment reactions.
    (a) $^{96}_{42}Mo + ^{4}_{2}He \longrightarrow ^{100}_{43}Tc + \underline{?}$
    (b) $^{59}_{27}Co + ^{1}_{0}n \longrightarrow ^{56}_{25}Mn + \underline{?}$
    (c) $^{23}_{11}Na + \underline{?} \longrightarrow ^{23}_{12}Mg + ^{1}_{0}n$
    (d) $^{209}_{83}Bi + \underline{?} \longrightarrow ^{210}_{84}Po + ^{1}_{0}n$
    (e) $^{238}_{92}U + ^{16}_{8}O \longrightarrow \underline{?} + 5 ^{1}_{0}n$

35. Fill in the missing symbols in the following nuclear bombardment reactions.
    (a) $^{26}_{?}Mg + \underline{?} \longrightarrow ^{26}_{?}Al + ^{1}_{0}n$
    (b) $\underline{?} + ^{1}_{1}H \longrightarrow ^{29}_{14}Si + ^{0}_{0}\gamma$
    (c) $^{232}_{90}Th + \underline{?} \longrightarrow ^{240}_{96}Cm + 4 ^{1}_{0}n$
    (d) $^{40}_{18}Ar + \underline{?} \longrightarrow ^{43}_{?}K + ^{1}_{1}H$

36. Write the symbols for the daughter nuclides in the following nuclear bombardment reactions: (a) $^{60}_{28}Ni$ (n, p); (b) $^{98}_{42}Mo(^{1}_{0}n, \beta)$; (c) $^{35}_{17}Cl$ (p, α).

37. Write the symbols for the daughter nuclides in the following nuclear bombardment reactions: (a) $^{20}_{10}Ne$ (α, γ); (b) $^{15}_{7}N$ (p, α); (c) $^{10}_{5}B$ (n, α).

38. Write the nuclear equation for each of the following bombardment processes: (a) $^{14}_{7}N$ (α, p) $^{17}_{8}O$; (b) $^{106}_{46}Pd$ (n, p) $^{106}_{45}Rh$; (c) $^{23}_{11}Na$ (n, β^-) X. Identify X.

39. Repeat Exercise 38 for the following: (a) $^{113}_{48}Cd$ (n, γ) $^{114}_{48}Cd$; (b) $^{6}_{3}Li$ (n, α) $^{3}_{1}H$; (c) $^{2}_{1}H$ (γ, p) X. Identify X.

40. Fill in the mass number, atomic number, and symbol for the missing particle in each nuclear equation.
    (a) $^{242}_{94}Pu \longrightarrow ^{4}_{2}He + \underline{\quad}$
    (b) $\underline{\quad} \longrightarrow ^{32}_{16}S + ^{0}_{-1}e$
    (c) $^{252}_{98}Cf + \underline{\quad} \longrightarrow 3 ^{1}_{0}n + ^{259}_{103}Lr$
    (d) $^{55}_{26}Fe + \underline{\quad} \longrightarrow ^{55}_{25}Mn$
    (e) $^{15}_{8}O \longrightarrow \underline{\quad} + ^{0}_{+1}e$

41. Summarize the nuclear changes that occur for each of the following types of radioactive emissions by determining the algebraic expressions (using a or z) that would be placed in the boxes for the mass number and the atomic number that would complete the nuclear equation.
    beta: $^{a}_{z}X \to ^{\square}_{\square}Y + ^{0}_{-1}\beta$
    positron: $^{a}_{z}X \to ^{\square}_{\square}Y + ^{0}_{+1}\beta$
    alpha: $^{a}_{z}X \to ^{\square}_{\square}Y + ^{4}_{2}\alpha$
    proton: $^{a}_{z}X \to ^{\square}_{\square}Y + ^{1}_{1}\rho$
    neutron: $^{a}_{z}X \to ^{\square}_{\square}Y + ^{1}_{0}\eta$
    gamma: $^{a}_{z}X \to ^{\square}_{\square}Y + ^{0}_{0}\gamma$

42. Write the nuclear equations for the following processes: (a) $^{63}_{28}Ni$ undergoing β^- emission; (b) two deuterium atoms undergoing fusion to give $^{3}_{2}He$ and a neutron; (c) a nuclide being bombarded by a neutron to form $^{7}_{3}Li$ and an α-particle (identify the unknown nuclide); (d) $^{14}_{7}N$ being bombarded by a neutron to form three α-particles and an atom of tritium.

43. Write the nuclear equations for the following processes: (a) $^{220}_{86}Rn$ undergoing α decay; (b) $^{110}_{49}In$ undergoing positron emission; (c) $^{127}_{53}I$ being bombarded by a proton to form $^{121}_{54}Xe$ and seven neutrons; (d) tritium and deuterium undergoing fusion to form an α-particle and a neutron; (e) $^{95}_{42}Mo$ being bombarded by a proton to form $^{95}_{43}Tc$ and radiation (identify this radiation).

Radon test kit

Charles D. Winters

44. "Radioactinium" is produced in the actinium series from $^{235}_{92}U$ by the successive emission of an α-particle, a β^--particle, an α-particle, and a β^--particle. What are the symbol, atomic number, and mass number for "radioactinium"?

45. An alkaline earth element (Group 2A) is radioactive. It undergoes decay by emitting three α-particles in succession. In what periodic table group is the resulting element found?

46. A nuclide of element rutherfordium, $^{257}_{104}Rf$, is formed by the nuclear reaction of californium-249 and carbon-12, with the emission of four neutrons. This new nuclide rapidly decays by emitting an α-particle. Write the equation for each of these nuclear reactions.

47. Supply the missing information to complete each of the following equations.
   (a) $^{187}_{75}Re + \beta \longrightarrow \;?$
   (b) $^{243}_{95}Am + ^1_0n \longrightarrow \;^{244}_{?}Cm + \;? + \gamma$
   (c) $^{35}_{17}Cl + \;? \longrightarrow \;^{32}_{16}S + \;^4_2He$
   (d) $^{53}_{24}Cr + \;^4_2He \longrightarrow \;^1_0n + \;?$

48. Write balanced nuclear equations for:
   (a) the loss of an alpha particle by Th-230
   (b) the loss of a beta particle by lead-210
   (c) the fission of U-235 to give Ba-140, another nucleus, and an excess of two neutrons
   (d) the K-capture of Ar-37

49. Describe how (a) cyclotrons and (b) linear accelerators work.

Rates of Decay

50. What does the half-life of a radionuclide represent? How do we compare the relative stabilities of radionuclides in terms of half-lives?

51. Why must all radioactive decays be first order?

52. Describe the process by which steady-state (constant) ratios of carbon-14 to (nonradioactive) carbon-12 are attained in living plants and organisms. Describe the method of radiocarbon dating. What factors limit the use of this method?

53. If the half-life of a radioactive substance is 3.0 hours, how many hours would it take for 99.9% of a sample to decay?

54. The half-life of $^{11}_6C$ is 20.3 min. How long will it take for 92.5% of a sample to decay? How long will it take for 99.0% of the sample to decay?

55. The activity of a sample of tritium decreased by 5.5% over the period of a year. What is the half-life of ^3_1H?

56. A very unstable isotope of beryllium, ^8Be, undergoes emission with a half-life of 0.07×10^{-15} s (0.07 fs). How long does it take for 99.90% of a 2.0-μg sample of ^8Be to undergo decay?

57. The $^{14}_6C$ activity of an artifact from a burial site was 8.6/min · g C. The half-life of $^{14}_6C$ is 5730 years, and the current $^{14}_6C$ activity is 16.1/min · g C (that is, 16.1 disintegrations per minute per gram of carbon). How old is the artifact?

58. Gold-198 is used in the diagnosis of liver problems. The half-life of ^{198}Au is 2.69 days. If you begin with 2.8 μg of this gold isotope, what mass remains after 10.8 days?

59. Analysis of an ant found in a piece of amber provided 14.0 disintegrations of $^{14}_6C$/min · g C, whereas a living ant of the same species produces 16.0 disintegrations per minute per gram of carbon. Calculate the approximate age of the fossilized ant. The half-life of carbon-14 is 5730 years.

60. The practical limit for dating objects by carbon-14 is about 50,000 years because older objects have too little activity to be dated accurately. Given that the half-life of carbon-14 is 5730 years, approximately what fraction of the carbon-14 that was in an object 50,000 years ago is still carbon-14?

61. Strontium-90 is one of the harmful radionuclides that results from nuclear fission explosions. It decays by beta emission with a half-life of 28 years. How long would it take for 99.9% of a given sample released in an atmospheric test of an atomic bomb to disintegrate?

62. Carbon-14 decays by beta emission with a half-life of 5730 years. Assuming a particular object originally contained 8.50 μg of carbon-14 and now contains 0.80 μg of carbon-14, how old is the object?

Fission and Fusion

63. Briefly describe a nuclear fission process. What are the two most important fissionable materials?

64. What is a chain reaction? Why are nuclear fission processes considered chain reactions? What is the critical mass of a fissionable material?

65. Where have continuous nuclear fusion processes been observed? What is the main reaction that occurs in such sources?

66. The reaction that occurred in the first fusion bomb was ^7_3Li (p, α) X. (a) Write the complete equation for the process, and identify the product, X. (b) The atomic masses are 1.007825 amu for ^1_1H, 4.00260 amu for α, and 7.01600 amu for ^7_3Li. Find the energy for the reaction in kJ/mol.

67. Summarize how an atomic bomb works, including how the nuclear explosion is initiated.

68. Discuss the pros and cons of the use of nuclear energy instead of other, more conventional types of energy based on fossil fuels.

69. Describe and illustrate the essential features of a light water fission reactor.

70. How is fissionable uranium-235 separated from nonfissionable uranium-238?

71. Distinguish between moderators and control rods of nuclear reactors.

72. What are the major advantages and disadvantages of fusion as a potential energy source compared with fission? What is the major technological problem that must be solved to permit development of a fusion reactor?

73. ▲ Calculate the binding energy, in kJ/mol of nucleons, for the following isotopes: (a) $^{15}_8O$ with a mass of 15.00300 amu; (b) $^{16}_8O$ with a mass of 15.99491 amu; (c) $^{17}_8O$ with a mass of 16.99913 amu; (d) $^{18}_8O$ with a mass of 17.99915 amu; (e) $^{19}_8O$ with a mass of 19.0035 amu. Which of these would you expect to be most stable?

74. ▲ The first nuclear transformation (discovered by Rutherford) can be represented by the shorthand notation

🔺 **Molecular Reasoning** exercises   ▲ **More Challenging** exercises   Blue-Numbered exercises solved in Student Solutions Manual

Unless otherwise noted, all content on this page is © Cengage Learning.

$^{14}_{7}$N (α, p) $^{17}_{8}$O. (a) Write the corresponding nuclear equation for this process. The respective atomic masses are 14.00307 amu for $^{14}_{7}$N, 4.00260 amu for $^{4}_{2}$He, 1.007825 amu for $^{1}_{1}$H, and 16.99913 amu for $^{17}_{8}$O. (b) Calculate the energy change of this reaction in kJ/mol.

75. A proposed series of reactions (known as the carbon–nitrogen cycle) that could be important in the very hottest region of the interior of the sun is

$$^{12}C + {}^{1}H \longrightarrow A + \gamma$$
$$A \longrightarrow B + {}^{0}_{+1}e$$
$$B + {}^{1}H \longrightarrow C + \gamma$$
$$C + {}^{1}H \longrightarrow D + \gamma$$
$$D \longrightarrow E + {}^{0}_{+1}e$$
$$E + {}^{1}H \longrightarrow {}^{12}C + F$$

Identify the species labeled *A–F*.

76. ▲ The ultimate stable product of ^{238}U decay is ^{206}Pb. A certain sample of pitchblende ore was found to contain ^{238}U and ^{206}Pb in the ratio 67.8 atoms ^{238}U:32.2 atoms ^{206}Pb. Assuming that all the ^{206}Pb arose from uranium decay, and that no U or Pb has been lost by weathering, how old is the rock? The half-life of ^{238}U is 4.51×10^{9} years.

77. ▲ Calculate the energy released by the fission reaction

$$^{235}_{92}U \rightarrow {}^{90}_{38}Sr + {}^{144}_{58}Ce + {}^{1}_{0}n + 4{}^{0}_{-1}e$$

The actual masses of the species are, in amu, $^{235}_{92}U$, 235.0439; $^{90}_{38}Sr$, 89.8869; $^{144}_{58}Ce$, 143.8817; $^{1}_{0}n$, 1.0087; $^{0}_{-1}e$, 0.0005486.

78. ▲ Show by calculation which reaction produces the larger amount of energy per atomic mass unit of material reacting.

fission: $^{235}_{92}U + {}^{1}_{0}n \rightarrow {}^{94}_{40}Zr + {}^{140}_{58}Ce + 6{}^{0}_{-1}e + 2{}^{1}_{0}n$

fusion: $2{}^{2}_{1}H \rightarrow {}^{3}_{1}H + {}^{1}_{1}H$

The atomic masses are 235.0439 amu for $^{235}_{92}U$; 93.9061 amu for $^{94}_{40}Zr$; 139.9053 amu for $^{140}_{58}Ce$; 3.01605 amu for $^{3}_{1}H$; 1.007825 amu for $^{1}_{1}H$; 2.0140 amu for $^{2}_{1}H$; 1.0087 amu for $^{1}_{0}n$; 0.0005486 amu for $^{0}_{-1}e$.

79. ▲ Potassium-40 decays to ^{40}Ar with a half-life of 1.3×10^{9} years. What is the age of a lunar rock sample that contains these isotopes in the ratio 33.4 atoms ^{40}K: 66.6 atoms ^{40}Ar? Assume that no argon was originally in the sample and that none has been lost by weathering.

Conceptual Exercises

80. Both nuclear and conventional power plants produce wastes that can be detrimental to the environment. Discuss these wastes and the current problems created by them.

81. If the earth is 4.5×10^{9} years old and the amount of radioactivity in a sample becomes smaller with time, how is it possible for there to be any radioactive elements left on the earth that have half-lives of less than a few million years?

82. ● What common type of radioactive emission does not consist of matter?

83. ● How is the process of producing electricity from nuclear fission different from the production of electricity from burning coal?

Building Your Knowledge

84. Why is carbon-14 dating not accurate for estimating the age of material more than 50,000 years old?

85. Polonium-210 decays to Pb-206 by alpha emission. Its half-life is 138 days. What volume of helium at 25°C and 1.20 atm would be obtained from a 25.00-g sample of Po-210 left to decay for 75 hours?

Beyond the Textbook

NOTE: *Whenever the answer to an exercise depends on information obtained from a source other than this textbook, the source must be included as an essential part of the answer.*

86. Use an internet search engine (such as **http://www.google.com**) to locate information on uranium mining. What two countries produce nearly half of all uranium mined worldwide? What is the most common type of uranium mining? (*Hint:* Is it open pit, leaching, or something else?)

87. Use an internet search engine (such as **http://www.google.com**) to locate information on volatile uranium compounds. Isotopes of uranium are separated by taking advantage of the principle that gases of different masses diffuse at different rates. What is the volatile uranium compound used in the separation of its isotopes? What is "yellow cake"?

88. Open **www.woodrow.org/teachers/chemistry/institutes/1992** and under Nuclear Chemistry, click on Frederic Joliot-Curie, or locate a history of science site that describes the life of Frederic Joliot-Curie.

  (a) In addition to being married to each other, what was the relationship between Mme. Curie and Frederic Joliot-Curie?

  (b) What chemical evidence did Frederic Joliot-Curie provide to prove that transmutation had taken place?

89. Open **www.energy.gov** and enter "fusion" in the search window; go to the "fusion home page" or open a site associated with the U.S. Fusion Energy Sciences Program.

  (a) What is the mission of the U.S. Fusion Energy Sciences Program?

  (b) The U.S. Fusion Energy Sciences Program is a function of which federal department?

90. Use the *Handbook of Chemistry and Physics* to find the decay products and half-lives requested below.

  (a) Cm-244 decays in three steps, each emitting an alpha particle, to yield Th-232. Write the equation for each decay step.

  (b) What is the half-life of each product formed?

  (c) What is the half-life of Cm-244?

91. A full-page version of the Radon Zone Map is available at the EPA radon web site, http://www.epa.gov/radon/zonemap.html. What are the predicted average indoor radon levels for the counties where your college and home are located? If the numbers are low, does this mean you do not need to worry about radon exposure?

● **Molecular Reasoning** exercises   ▲ **More Challenging** exercises   Blue-Numbered exercises solved in Student Solutions Manual

Some Mathematical Operations

In chemistry we frequently use very large or very small numbers. Such numbers are conveniently expressed in *scientific*, or *exponential*, *notation*.

A-1 Scientific Notation

In scientific notation, a number is expressed as the *product of two numbers*. By convention, the first number, called the *digit term*, is between 1 and 10. The second number, called the *exponential term*, is an integer power of 10. Some examples follow.

$$10000 = 1 \times 10^4 \qquad 24327 = 2.4327 \times 10^4$$
$$1000 = 1 \times 10^3 \qquad 7958 = 7.958 \ \times 10^3$$
$$100 = 1 \times 10^2 \qquad 594 = 5.94 \ \times 10^2$$
$$10 = 1 \times 10^1 \qquad 98 = 9.8 \ \times 10^1$$
$$1 = 1 \times 10^0$$
$$1/10 = 0.1 = 1 \times 10^{-1} \qquad 0.32 = 3.2 \ \times 10^{-1}$$
$$1/100 = 0.01 = 1 \times 10^{-2} \qquad 0.067 = 6.7 \ \times 10^{-2}$$
$$1/1000 = 0.001 = 1 \times 10^{-3} \qquad 0.0049 = 4.9 \ \times 10^{-3}$$
$$1/10000 = 0.0001 = 1 \times 10^{-4} \qquad 0.00017 = 1.7 \ \times 10^{-4}$$

▶ Recall that, by definition, (any base)^0 = 1.

The exponent of 10 is the number of places the decimal point must be shifted to give the number in long form. A *positive exponent* indicates that the decimal point is *shifted right* that number of places. A *negative exponent* indicates that the decimal point is *shifted left*. When numbers are written in *standard scientific notation*, there is one nonzero digit to the left of the decimal point.

$$7.3 \times 10^3 = 73 \times 10^2 \qquad = 730 \times 10^1 \qquad = 7300$$
$$4.36 \times 10^{-2} = 0.436 \times 10^{-1} \ = 0.0436$$
$$0.00862 = 0.0862 \times 10^{-1} = 0.862 \times 10^{-2} = 8.62 \times 10^{-3}$$

In scientific notation the digit term indicates the number of significant figures in the number. The exponential term merely locates the decimal point and does not represent significant figures.

Problem-Solving Tip Know How to Use Your Calculator

Students sometimes make mistakes when they try to enter numbers into their calculators in scientific notation. Suppose you want to enter the number 4.36×10^{-2}. On most calculators, you would

1. Press 4.36

2. Press E, EE, or EXP, which stands for "times ten to the"

3. Press 2 (the magnitude of the exponent) and then ± or CHS (to change its sign)

▶ To be sure you know how to use your calculator, you should use it to check the answers to all calculations in this appendix.

A-1

The calculator display might show the value as $\boxed{4.36 \qquad -02}$ or as $\boxed{0.0436}$. Different calculators show different numbers of digits, which can sometimes be adjusted.

If you wished to enter -4.36×10^2, you would

1. Press 4.36, then press $\pm$ or CHS to change its sign,

2. Press E, EE, or EXP, and then press 2

The calculator would then show $\boxed{-4.36 \qquad 02}$ or $\boxed{-436.0}$.

Caution: Be sure you remember that the E, EE, or EXP button *includes* the "times 10" operation. An error that beginners often make is to enter "×10" explicitly when trying to enter a number in scientific notation. Suppose you mistakenly enter 3.7×10^2 as follows:

1. Enter 3.7

2. Press $\times$ and then enter 10

3. Press E, EE, or EXP and then enter 2

The calculator then shows the result as 3.7×10^3 or 3700—why? This sequence is processed by the calculator as follows: Step 1 enters the number 3.7; step 2 multiplies by 10, to give 37; step 3 multiplies this by 10^2, to give 37×10^2 or 3.7×10^3.

Other common errors include changing the sign of the exponent when the intent was to change the sign of the entire number (e.g., -3.48×10^4 entered as 3.48×10^{-4}).

When in doubt, carry out a trial calculation for which you already know the answer. For instance, multiply 300 by 2 by entering the first value as 3.00×10^2 and then multiplying by 2; you know the answer should be 600, and if you get any other answer, you know you have done something wrong. If you cannot find (or understand) the printed instructions for *your calculator*, your instructor or a classmate might be able to help.

Addition and Subtraction

In addition and subtraction all numbers are converted to the same power of 10, and then the digit terms are added or subtracted.

$$(4.21 \times 10^{-3}) + (1.4 \times 10^{-4}) = (4.21 \times 10^{-3}) + (0.14 \times 10^{-3}) = \boxed{4.35 \times 10^{-3}}$$

$$(8.97 \times 10^4) - (2.31 \times 10^3) = (8.97 \times 10^4) - (0.231 \times 10^4) = \boxed{8.74 \times 10^4}$$

Multiplication

The digit terms are multiplied in the usual way, the exponents are added algebraically, and the product is written with one nonzero digit to the left of the decimal.

▶ Two significant figures in answer.

$$(4.7 \times 10^7)(1.6 \times 10^2) = (4.7)(1.6) \times 10^{7+2} = 7.52 \times 10^9 = \boxed{7.5 \times 10^9}$$

▶ Two significant figures in answer.

$$(8.3 \times 10^4)(9.3 \times 10^{-9}) = (8.3)(9.3) \times 10^{4-9} = 77.19 \times 10^{-5} = \boxed{7.7 \times 10^{-4}}$$

Division

The digit term of the numerator is divided by the digit term of the denominator, the exponents are subtracted algebraically, and the quotient is written with one nonzero digit to the left of the decimal.

$$\frac{8.4 \times 10^7}{2.0 \times 10^3} = \frac{8.4}{2.0} \times 10^{7-3} = \boxed{4.2 \times 10^4}$$

▶ Three significant figures in answer.

$$\frac{3.81 \times 10^9}{8.412 \times 10^{-3}} = \frac{3.81}{8.412} \times 10^{[9-(-3)]} = 0.45292 \times 10^{12} = \boxed{4.53 \times 10^{11}}$$

Powers of Exponentials

The digit term is raised to the indicated power, and the exponent is multiplied by the number that indicates the power.

$$(1.2 \times 10^3)^2 = (1.2)^2 \times 10^{3 \times 2} = 1.44 \times 10^6 = \boxed{1.4 \times 10^6}$$

$$(3.0 \times 10^{-3})^4 = (3.0)^4 \times 10^{-3 \times 4} = 81 \times 10^{-12} = \boxed{8.1 \times 10^{-11}}$$

Electronic Calculators. *To square a number*: (1) enter the number and (2) touch the (x^2) button.

$$(7.3)^2 = 53.29 = \boxed{53} \qquad \text{(two sig. figs.)}$$

To raise a number y to power x: (1) enter the number; (2) touch the (y^x) button; (3) enter the power; and (4) touch the (=) button.

$$(7.3)^4 = 2839.8241 = \boxed{2.8 \times 10^3} \qquad \text{(two sig. figs.)}$$
$$(7.30 \times 10^2)^5 = 2.0730716 \times 10^{14} = \boxed{2.07 \times 10^{14}} \qquad \text{(three sig. figs.)}$$

Roots of Exponentials

If a calculator is not used, the exponent must be divisible by the desired root. The root of the digit term is extracted in the usual way, and the exponent is divided by the desired root.

$$\sqrt{2.5 \times 10^5} = \sqrt{2.5 \times 10^4} = \sqrt{25} \times \sqrt{10^4} = \boxed{5.0 \times 10^2}$$
$$\sqrt[3]{2.7 \times 10^{-8}} = \sqrt[3]{27 \times 10^{-9}} = \sqrt[3]{27} = \sqrt[3]{10^{-9}} = \boxed{3.0 \times 10^{-3}}$$

Electronic Calculators. *To extract the square root of a number*: (1) enter the number and (2) touch the $(\sqrt{x})$ button.

$$\sqrt{23} = 4.7958315 = \boxed{4.8} \qquad \text{(two sig. figs.)}$$

To extract some other root: (1) enter the number y; (2) touch the (INV) and then the (y^x) button; (3) enter the root to be extracted, x; and (4) touch the (=) button.

▶ These instructions are applicable to most calculators. If your calculator has other notation, consult the instructions for your calculator.

▶ On some calculator models, this function is performed by the $\sqrt[x]{y}$ button.

A-2 Logarithms

The logarithm of a number is the power to which a base must be raised to obtain the number. Two types of logarithms are frequently used in chemistry: (1) common logarithms (abbreviated log), whose base is 10, and (2) natural logarithms (abbreviated ln), whose base is $e = 2.71828$. The general properties of logarithms are the same no matter what base is used. Many equations in science were derived by the use of calculus, and these often involve natural (base e) logarithms. The relationship between $\log x$ and $\ln x$ is as follows.

$$\ln x = 2.303 \log x$$

▶ $\ln 10 = 2.303$

Finding Logarithms. The common logarithm of a number is the power to which 10 must be raised to obtain the number. The number 10 must be raised to the third power to equal 1000. Therefore, the logarithm of 1000 is 3, written $\log 1000 = 3$. Some examples follow.

Number	Exponential Expression	Logarithm
1000	10^3	3
100	10^2	2
10	10^1	1
1	10^0	0
1/10 = 0.1	10^{-1}	−1
1/100 = 0.01	10^{-2}	−2
1/1000 = 0.001	10^{-3}	−3

To obtain the logarithm of a number other than an integral power of 10, you must use either an electronic calculator or a logarithm table. On most calculators, you do this by (1) entering the number and then (2) pressing the (log) button.

$$\log 7.39 = 0.8686444 = \boxed{0.869}$$
$$\log 7.39 \times 10^3 = 3.8686 = \boxed{3.869}$$
$$\log 7.39 \times 10^{-3} = -2.1314 = \boxed{-2.131}$$

The number to the left of the decimal point in a logarithm is called the *characteristic*, and the number to the right of the decimal point is called the *mantissa*. The characteristic only locates the decimal point of the number, so it is usually not included when counting significant figures. The mantissa has as many significant figures as the number whose log was found.

To obtain the natural logarithm of a number on an electronic calculator, (1) enter the number and (2) press the (ln) or (ln x) button.

$$\ln 4.45 = 1.4929041 = \boxed{1.493}$$
$$\ln 1.27 \times 10^3 = 7.1468 \quad = \boxed{7.147}$$

► On some calculators, the inverse log is found as follows:
1. enter the value of the log
2. press the (2ndF) (second function) button
3. press (10^x)

On some calculators, the inverse natural logarithm is found as follows:
1. enter the value of the ln
2. press the (2ndF) (second function) button
3. press (e^x)

Finding Antilogarithms. Sometimes we know the logarithm of a number and must find the number. This is called finding the *antilogarithm* (or *inverse logarithm*). To do this on a calculator, (1) enter the value of the log; (2) press the (INV) button; and (3) press the (log) button.

$$\log x = 6.131; \qquad \text{so } x = \text{inverse log of } 6.131 = \boxed{1.352 \times 10^6}$$
$$\log x = -1.562; \qquad \text{so } x = \text{inverse log of } -1.562 = \boxed{2.74 \times 10^{-2}}$$

To find the inverse natural logarithm, (1) enter the value of the ln; (2) press the (INV) button; and (3) press the (ln) or (ln x) button.

$$\ln x = 3.552; \qquad \text{so } x = \text{inverse ln of } 3.552 = \boxed{3.49 \times 10^1}$$
$$\ln x = -1.248; \qquad \text{so } x = \text{inverse ln of } -1.248 = \boxed{2.87 \times 10^{-1}}$$

Calculations Involving Logarithms

Because logarithms are exponents, operations involving them follow the same rules as the use of exponents. The following relationships are useful.

$$\log xy = \log x + \log y \qquad \text{or} \qquad \ln xy = \ln x + \ln y$$
$$\log \frac{x}{y} = \log x - \log y \qquad \text{or} \qquad \ln \frac{x}{y} = \ln x - \ln y$$
$$\log x^y = y \log x \qquad \text{or} \qquad \ln x^y = y \ln x$$
$$\log \sqrt[y]{x} = \log x^{1/y} = \frac{1}{y} \log x \qquad \text{or} \qquad \ln \sqrt[y]{x} = \ln x^{1/y} = \frac{1}{y} \ln x$$

A-3 Quadratic Equations

Algebraic expressions of the form

$$ax^2 + bx + c = 0$$

are called **quadratic equations**. Each of the constant terms (a, b, and c) may be either positive or negative. All quadratic equations may be solved by the **quadratic formula**.

$$x = \frac{-b \pm \sqrt{b^2 - 4ac}}{2a}$$

To solve the quadratic equation $3x^2 - 4x - 8 = 0$, we use $a = 3$, $b = -4$, and $c = -8$. Substitution of these values into the quadratic formula gives

$$x = \frac{-(-4) \pm \sqrt{(-4)^2 - 4(3)(-8)}}{2(3)} = \frac{4 \pm \sqrt{16 + 96}}{6}$$
$$= \frac{4 \pm \sqrt{112}}{6} = \frac{4 \pm 10.6}{6}$$

The two roots of this quadratic equation are

$$x = \boxed{2.4} \qquad \text{and} \qquad x = \boxed{-1.1}$$

As you construct and solve quadratic equations based on the observed behavior of matter, you must decide which root has physical significance. Examination of the *equation that defines* x always gives clues about possible values for x. In this way you can tell which is extraneous (has no physical significance). Negative roots are often extraneous.

When you have solved a quadratic equation, you should always check the values you obtained by substituting them into the original equation. In the preceding example we obtained $x = 2.4$ and $x = -1.1$. Substitution of these values into the original quadratic equation, $3x^2 - 4x - 8 = 0$, shows that both roots are correct. Such substitutions often do not give a perfect check because some round-off error has been introduced.

A-4 Significant Figures

There are two kinds of numbers. *Numbers obtained by counting or from definitions are* **exact numbers**. They are known to be absolutely accurate. For example, the exact number of people in a closed room can be counted, and there is no doubt about the number of people. A dozen eggs is defined as exactly 12 eggs, no more, no fewer (Figure A-1).

Numbers obtained from measurements are not exact. Every measurement involves an estimate. For example, suppose you are asked to measure the length of this page to the nearest 0.1 mm. How do you do it? The smallest divisions (calibration lines) on a meter stick are 1 mm apart (see Figure 1-17). An attempt to measure to 0.1 mm requires estimation. If three different people measure the length of the page to 0.1 mm, will they get the same answer? Probably not. We deal with this problem by using significant figures.

Significant figures are digits believed to be correct by the person who makes a measurement. We assume that the person is competent to use the measuring device. Suppose one measures a distance with a meter stick and reports the distance as 343.5 mm. What does this number mean? In this person's judgment, the distance is greater than 343.4 mm but less than 343.6 mm, and the best estimate is 343.5 mm. The number 343.5 mm contains four significant figures. The last digit, 5, is a *best estimate* and is therefore doubtful, but it is considered to be a significant figure. In reporting numbers obtained from measurements, *we report one estimated digit, and no more.* Because the person making the measurement is not certain that the 5 is correct, it would be meaningless to report the distance as 343.53 mm.

To see more clearly the part significant figures play in reporting the results of measurements, consider Figure A-2a. Graduated cylinders are used to measure volumes of liquids when a high degree of accuracy is not necessary. The calibration lines on a 50-mL graduated cylinder represent 1-mL increments. Estimation of the volume of liquid in a 50-mL cylinder to within 0.2 mL ($\frac{1}{5}$ of one calibration increment) with reasonable certainty is possible. We might measure a volume of liquid in such a cylinder and report the volume as 38.6 mL, that is, to three significant figures.

Burets are used to measure volumes of liquids when higher accuracy is required. The calibration lines on a 50-mL buret represent 0.1-mL increments, allowing us to make estimates to within 0.02 mL ($\frac{1}{5}$ of one calibration increment) with reasonable certainty (Figure A-2b). Experienced individuals estimate volumes in 50-mL burets to 0.01 mL with considerable reproducibility. For example, using a 50-mL buret, we can measure out 38.56 mL (four significant figures) of liquid with reasonable accuracy.

Accuracy refers to how closely a measured value agrees with the correct value. **Precision** refers to how closely individual measurements agree with one another. Ideally, all measurements should be both accurate and precise. Measurements may be quite precise yet quite inaccurate because of some *systematic error*, which is an error repeated in each measurement. (A faulty balance, for example, might produce a systematic error.) Very accurate measurements are seldom imprecise.

Measurements are usually repeated to improve accuracy and precision. Average values obtained from several measurements are usually more reliable than individual measurements. Significant figures indicate how precisely measurements have been made (assuming the person who made the measurements was competent).

Some simple rules govern the use of significant figures.

1. Nonzero digits are always significant.

▶ An *exact* number may be thought of as containing an *infinite* number of significant figures.

▶ There is some uncertainty in all measurements.

▶ Significant figures indicate the *uncertainty* in measurements.

A A dozen eggs is exactly 12 eggs.

B A specific swarm of honeybees contains an *exact* number of live bees (though it would be difficult to count them, and any two *swarms* would be unlikely to contain the same exact number of bees).

Figure A-1

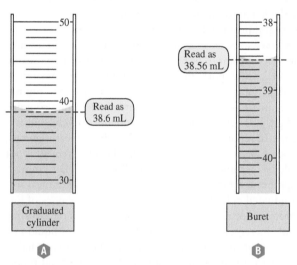

A Graduated cylinder

Read as 38.6 mL

B Buret

Read as 38.56 mL

Figure A-2 Measurement of the volume of water using two types of volumetric glassware. For consistency, we always read the bottom of the meniscus (the curved surface of the water). (a) A graduated cylinder is used to measure the amount of liquid *contained* in the glassware, so the scale increases from bottom to top. The level in a 50-mL graduated cylinder can usually be estimated to within 0.2 mL. The level here is 38.6 mL (three significant figures). (b) We use a buret to measure the amount of liquid *delivered* from the glassware, by taking the difference between an initial and a final volume reading. The level in a 50-mL buret can be read to within 0.02 mL. The level here is 38.56 mL (four significant figures).

For example, 38.56 mL has four significant figures; 288 g has three significant figures.

> **2.** Zeroes are sometimes significant, and sometimes they are not.
>   **a.** Zeroes at the *beginning* of a number (used just to position the decimal point) are never significant.

For example, 0.052 g has two significant figures; 0.00364 m has three significant figures. These could also be reported in scientific notation as 5.2×10^{-2} g and 3.64×10^{-3} m, respectively.

> **b.** Zeroes *between* nonzero digits are always significant.

For example, 2007 g has four significant figures; 6.08 km has three significant figures.

c. Zeroes at the *end* of a number that contains a decimal point are always significant.

For example, 38.0 cm has three significant figures; 440.0 m has four significant figures. These could also be reported as 3.80×10^1 cm and 4.400×10^2 m, respectively.

d. Zeroes at the *end* of a number that does not contain a decimal point may or may not be significant.

For example, does the quantity 24,300 km represent three, four, or five significant figures? Writing the number in this format gives insufficient information to answer the question. If both of the zeroes are used just to place the decimal point, the number should appear as 2.43×10^4 km (three significant figures). If only one of the zeroes is used to place the decimal point (i.e., the number was measured ±10), the number is 2.430 ± 10^4 km (four significant figures). If the number is actually known to be 24,300 ± 1, it should be written as 2.4300×10^4 km (five significant figures).

▶ When we wish to specify that all of the zeroes in such a number *are* significant, we may indicate this by placing a decimal point after the number. For instance, 130. grams represents a mass known to *three* significant figures, that is, 130 ± 1 gram.

3. Exact numbers can be considered as having an unlimited number of significant figures. This applies to defined quantities.

For example, in the equivalence 1 yard = 3 feet, the numbers 1 and 3 are exact, and we do not apply the rules of significant figures to them. The equivalence 1 inch = 2.54 centimeters is an exact one.

A calculated number can never be more precise than the numbers used to calculate it. The following rules show how to get the number of significant figures in a calculated number.

4. In addition and subtraction, the last digit retained in the sum or difference is determined by the position of the first doubtful digit.

Example A-1 Significant Figures (Addition and Subtraction)

(a) Add 37.24 mL and 10.3 mL. **(b)** Subtract 21.2342 g from 27.87 g.

Plan

We first check to see that the quantities to be added or subtracted are expressed in the same units. We carry out the addition or subtraction. Then we follow Rule 4 for significant figures to express the answer to the correct number of significant figures.

Solution

(a)
$$\begin{array}{r} 37.2\underline{4} \text{ mL} \\ +10.\underline{3} \text{ mL} \\ \hline 47.54 \text{ mL} \end{array}$$
is reported as 47.5 mL (calculator gives 47.54)

(b)
$$\begin{array}{r} 27.8\underline{7}\ \text{ g} \\ -21.2342 \text{ g} \\ \hline 6.6358 \text{ g} \end{array}$$
is reported as 6.64 g (calculator gives 6.6358)

🛑 **TOP & THINK**
The doubtful digits are underlined in Example A-1. In addition or subtraction, the last significant digit in the answer is determined by the position of the least doubtful digit. In (a), this is the first digit after the decimal point. In (b), this is the second digit after the decimal point.

5. In multiplication and division, an answer contains no more significant figures than the least number of significant figures used in the operation.

Example A-2 Significant Figures (Multiplication)

What is the area of a rectangle 1.23 cm wide and 12.34 cm long?

Plan

The area of a rectangle is its length times its width. We must first check to see that the width and length are expressed in the same units. (They are, but if they were not, we should first convert one to the units of the other.) Then we multiply the width by the length. We then follow Rule 5 for significant figures to find the correct number of significant figures. The units for the result are equal to the product of the units for the individual terms in the multiplication.

Solution

$$A = \ell \times w = (12.34 \text{ cm})(1.23 \text{ cm}) = 15.2 \text{ cm}^2$$

$$(\text{calculator result} = 15.1782)$$

Because three is the smallest number of significant figures used, the answer should contain only three significant figures. The number generated by an electronic calculator (15.1782) implies more accuracy than is justified; the result cannot be more accurate than the information that led to it. Calculators have no judgment, so you must exercise yours.

▶   12.34 cm
×   1.23 cm
          ‾‾‾‾‾‾‾
            3702
            2468
            1234
          ‾‾‾‾‾‾‾
          15.17   82 cm² = 15.2 cm²

▶ Rounding to an even number is intended to reduce the accumulation of errors in chains of calculations.

The step-by-step calculation in the margin demonstrates why the area is reported as 15.2 cm² rather than 15.1782 cm². The length, 12.34 cm, contains four significant figures, whereas the width, 1.23 cm, contains only three. If we underline each uncertain figure, as well as each figure obtained from an uncertain figure, the step-by-step multiplication gives the result reported in Example A-2. We see that there are only two certain figures (15) in the result. We report the first doubtful figure (.2), but no more. Division is just the reverse of multiplication, and the same rules apply.

In the three simple arithmetic operations we have performed, the number combination generated by an electronic calculator is not the "answer" in a single case! The correct result of each calculation, however, can be obtained by "rounding." The rules of significant figures tell us where to round.

In rounding off, certain conventions have been adopted. When the number to be dropped is less than 5, the preceding number is left unchanged (e.g., 7.34 rounds to 7.3). When it is more than 5, the preceding number is increased by 1 (e.g., 7.37 rounds to 7.4). When the number to be dropped is 5, the preceding number is set to the nearest even number (e.g., 7.45 rounds to 7.4, and 7.35 rounds to 7.4).

Problem-Solving Tip When Do We Round?

When a calculation involves several steps, we often show the answer to each step to the correct number of significant figures. *But we carry all digits in the calculator to the end of the calculation; then we round the final answer to the appropriate number of significant figures.* When carrying out such a calculation, it is safest to carry extra figures through all steps and then to round the final answer appropriately.

Electronic Configurations of the Atoms of the Elements

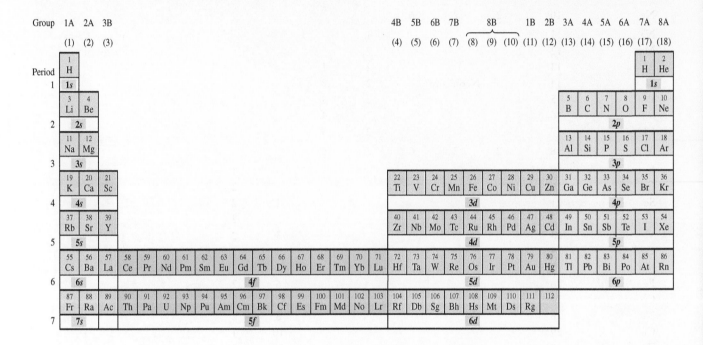

A periodic table colored to show the kinds of atomic orbitals (subshells) that are being filled in different parts of the periodic table. The atomic orbitals are given below the symbols of blocks of elements. The electronic structures of the A group elements are quite regular and can be predicted from their positions in the periodic table, but there are many exceptions in the *d* and *f* blocks. The populations of subshells are given in the table on pages A-6 and A-7.

Electron Configurations of the Atoms of the Elements

Element	Atomic Number	Populations of Subshells											
		1s	2s	2p	3s	3p	3d	4s	4p	4d	4f	5s
H	1	1										
He	2	2										
Li	3	2	1									
Be	4	2	2									
B	5	2	2	1								
C	6	2	2	2								
N	7	2	2	3								
O	8	2	2	4								
F	9	2	2	5								
Ne	10	2	2	6								
Na	11				1							
Mg	12				2							
Al	13				2	1						
Si	14	Neon core			2	2						
P	15				2	3						
S	16				2	4						
Cl	17				2	5						
Ar	18	2	2	6	2	6						
K	19							1				
Ca	20							2				
Sc	21						1	2				
Ti	22						2	2				
V	23						3	2				
Cr	24						5	1				
Mn	25						5	2				
Fe	26						6	2				
Co	27			Argon core			7	2				
Ni	28						8	2				
Cu	29						10	1				
Zn	30						10	2				
Ga	31						10	2	1			
Ge	32						10	2	2			
As	33						10	2	3			
Se	34						10	2	4			
Br	35						10	2	5			
Kr	36	2	2	6	2	6	10	2	6			
Rb	37											1
Sr	38											2
Y	39									1		2
Zr	40									2		2
Nb	41									4		1
Mo	42									5		1
Tc	43			Krypton core						5		2
Ru	44									7		1
Rh	45									8		1
Pd	46									10		
Ag	47									10		1
Cd	48									10		2

Electron Configurations of the Atoms of the Elements (continued)

Element	Atomic Number	Populations of Subshells									
		4d	4f	5s	5p	5d	5f	6s	6p	6d	7s
In	49	10		2	1						
Sn	50	10		2	2						
Sb	51	10		2	3						
Te	52	10		2	4						
I	53	10		2	5						
Xe	54	10		2	6						
Cs	55	10		2	6			1			
Ba	56	10		2	6			2			
La	57	10		2	6	1		2			
Ce	58	10	1	2	6	1		2			
Pr	59	10	3	2	6			2			
Nd	60	10	4	2	6			2			
Pm	61	10	5	2	6			2			
Sm	62	10	6	2	6			2			
Eu	63	10	7	2	6			2			
Gd	64	10	7	2	6	1		2			
Tb	65	10	9	2	6			2			
Dy	66	10	10	2	6			2			
Ho	67	10	11	2	6			2			
Er	68	10	12	2	6			2			
Tm	69	10	13	2	6			2			
Yb	70	10	14	2	6			2			
Lu	71	10	14	2	6	1		2			
Hf	72	10	14	2	6	2		2			
Ta	73	10	14	2	6	3		2			
W	74	10	14	2	6	4		2			
Re	75	10	14	2	6	5		2			
Os	76	10	14	2	6	6		2			
Ir	77	10	14	2	6	7		2			
Pt	78	10	14	2	6	9		1			
Au	79	10	14	2	6	10		1			
Hg	80	10	14	2	6	10		2			
Tl	81	10	14	2	6	10		2	1		
Pb	82	10	14	2	6	10		2	2		
Bi	83	10	14	2	6	10		2	3		
Po	84	10	14	2	6	10		2	4		
At	85	10	14	2	6	10		2	5		
Rn	86	10	14	2	6	10		2	6		
Fr	87	10	14	2	6	10		2	6		1
Ra	88	10	14	2	6	10		2	6		2
Ac	89	10	14	2	6	10		2	6	1	2
Th	90	10	14	2	6	10		2	6	2	2
Pa	91	10	14	2	6	10	2	2	6	1	2
U	92	10	14	2	6	10	3	2	6	1	2
Np	93	10	14	2	6	10	4	2	6	1	2
Pu	94	10	14	2	6	10	6	2	6		2
Am	95	10	14	2	6	10	7	2	6		2
Cm	96	10	14	2	6	10	7	2	6	1	2
Bk	97	10	14	2	6	10	9	2	6		2
Cf	98	10	14	2	6	10	10	2	6		2
Es	99	10	14	2	6	10	11	2	6		2
Fm	100	10	14	2	6	10	12	2	6		2
Md	101	10	14	2	6	10	13	2	6		2
No	102	10	14	2	6	10	14	2	6		2
Lr	103	10	14	2	6	10	14	2	6	1	2
Rf	104	10	14	2	6	10	14	2	6	2	2
Db	105	10	14	2	6	10	14	2	6	3	2
Sg	106	10	14	2	6	10	14	2	6	4	2
Bh	107	10	14	2	6	10	14	2	6	5	2
Hs	108	10	14	2	6	10	14	2	6	6	2
Mt	109	10	14	2	6	10	14	2	6	7	2
Ds	110	10	14	2	6	10	14	2	6	9	1
Rg	111	10	14	2	6	10	14	2	6	10	1

Krypton core

Common Units, Equivalences, and Conversion Factors

C-1 Fundamental Units of the SI System

The metric system was implemented by the French National Assembly in 1790 and has been modified many times. The International System of Units, or *le Système International* (SI), represents an extension of the metric system. It was adopted by the eleventh General Conference of Weights and Measures in 1960 and has also been modified since. It is constructed from seven base units, each of which represents a particular physical quantity (Table I).

The first five units listed in Table I are particularly useful in general chemistry. They are defined as follows.

▶ Alternatives for the standard mass have been proposed.

1. The *meter* is defined as the distance light travels in a vacuum in 1/299,792,458 second.

2. The *kilogram* represents the mass of a platinum–iridium block kept at the International Bureau of Weights and Measures at Sèvres, France.

3. The *second* was redefined in 1967 as the duration of 9,192,631,770 periods of a certain line in the microwave spectrum of cesium-133.

4. The *kelvin* is 1/273.16 of the temperature interval between absolute zero and the triple point of water.

5. The *mole* is the amount of substance that contains as many entities as there are atoms in exactly 0.012 kg of carbon-12 (12 g of ^{12}C atoms).

Table I SI Fundamental Units

Physical Quantity	Name of Unit	Symbol
length	meter	m
mass	kilogram	kg
time	second	s
temperature	kelvin	K
amount of substance	mole	mol
electric current	ampere	A
luminous intensity	candela	cd

Prefixes Used with Metric Units and SI Units
Decimal fractions and multiples of metric and SI units are designated by the prefixes listed in Table II. Those most commonly used in general chemistry are underlined.

Table II Traditional Metric and SI Prefixes

Factor	Prefix	Symbol	Factor	Prefix	Symbol
10^{12}	tera	T	10^{-1}	deci	d
10^9	giga	G	10^{-2}	centi	c
10^6	mega	M	10^{-3}	milli	m
10^3	kilo	k	10^{-6}	micro	μ
10^2	hecto	h	10^{-9}	nano	n
10^1	deka	da	10^{-12}	pico	p
			10^{-15}	femto	f
			10^{-18}	atto	a

C-2 Derived SI Units

In the International System of Units all physical quantities are represented by appropriate combinations of the base units listed in Table I. A list of the derived units frequently used in general chemistry is given in Table III.

Table III Derived SI Units

Physical Quantity	Name of Unit	Symbol	Definition
area	square meter	m^2	
volume	cubic meter	m^3	
density	kilogram per cubic meter	kg/m^3	
force	newton	N	$kg \cdot m/s^2$
pressure	pascal	Pa	N/m^2
energy	joule	J	$kg \cdot m^2/s^2$
electric charge	coulomb	C	$A \cdot s$
electric potential difference	volt	V	$J/(A \cdot s)$

Common Units of Mass and Weight

1 pound = 453.59 grams
1 pound = 453.59 grams = 0.45359 kilogram
1 kilogram = 1000 grams = 2.205 pounds
1 gram = 10 decigrams = 100 centigrams = 1000 milligrams
1 gram = 6.022×10^{23} atomic mass units
1 atomic mass unit = 1.6605×10^{-24} gram
1 short ton = 2000 pounds = 907.2 kilograms
1 long ton = 2240 pounds
1 metric tonne = 1000 kilograms = 2205 pounds

Common Units of Length

1 inch = 2.54 centimeters (exactly)
1 mile = 5280 feet = 1.609 kilometers
1 yard = 36 inches = 0.9144 meter
1 meter = 100 centimeters = 39.37 inches = 3.281 feet = 1.094 yards
1 kilometer = 1000 meters = 1094 yards = 0.6215 mile
1 Ångstrom = 1.0×10^{-8} centimeter = 0.10 nanometer = 1.0×10^{-10} meter = 3.937×10^{-9} inch

Common Units of Volume

1 quart = 0.9463 liter 1 liter = 1.056 quarts
1 liter = 1 cubic decimeter = 1000 cubic centimeters = 0.001 cubic meter
1 milliliter = 1 cubic centimeter = 0.001 liter = 1.056×10^{-3} quart
1 cubic foot = 28.316 liters = 29.902 quarts = 7.475 gallons

Common Units of Force* and Pressure

1 atmosphere = 760 millimeters of mercury = 1.01325×10^5 pascals = 1.01325 bar = 14.70 pounds per square inch
1 bar = 10^5 pascals = 0.98692 atm
1 torr = 1 millimeter of mercury
1 pascal = 1 kg/m · s² = 1 N/m²

*Force: 1 newton (N) = 1 kg · m/s², i.e., the force that, when applied for 1 second, gives a 1-kilogram mass a velocity of 1 meter per second.

Common Units of Energy

1 joule = 1×10^7 ergs
1 thermochemical calorie* = 4.184 joules = 4.184×10^7 ergs = 4.129×10^{-2} liter-atmospheres = 2.612×10^{19} electron volts
1 erg = 1×10^{-7} joule = 2.3901×10^{-8} calorie
1 electron volt = 1.6022×10^{-19} joule = 1.6022×10^{-12} erg = 96.487 kJ/mol†
1 liter-atmosphere = 24.217 calories = 101.325 joules = 1.01325×10^9 ergs
1 British thermal unit = 1055.06 joules = 1.05506×10^{10} ergs = 252.2 calories

*The amount of heat required to raise the temperature of one gram of water from 14.5°C to 15.5°C.

†Note that the other units are per particle and must be multiplied by 6.022×10^{23} to be strictly comparable.

Physical Constants

Quantity	Symbol	Traditional Units	SI Units
Acceleration of gravity	g	980.665 cm/s	9.80665 m/s
Atomic mass unit ($\frac{1}{12}$ the mass of ^{12}C atom)	amu or u	1.660539×10^{-24} g	1.660539×10^{-27} kg
Avogadro's number	N	$6.02214129 \times 10^{23}$ particles/mol	$6.02214129 \times 10^{23}$ particles/mol
Bohr radius	a_0	0.52918 Å 5.2918×10^{-9} cm	5.2918×10^{-11} m
Boltzmann constant	k	1.3807×10^{-16} erg/K	1.3807×10^{-23} J/K
Charge-to-mass ratio of electron	e/m	1.75882×10^8 coulomb/g	1.75882×10^{11} C/kg
Electronic charge	e	$1.60217656 \times 10^{-19}$ coulomb 4.8033×10^{-10} esu	$1.60217648 \times 10^{-19}$ C
Electron rest mass	m_e	9.1094×10^{-28} g 0.00054858 amu	9.1094×10^{-31} kg
Faraday constant	F	96,485 coulombs/eq 23.06 kcal/volt · eq	96,485 C/mol e^- 96,485 J/V · mol e^-
Gas constant	R	$0.08206 \dfrac{L \cdot atm}{mol \cdot K}$ $1.987 \dfrac{cal}{mol \cdot K}$	$8.3145 \dfrac{kPa \cdot dm^3}{mol \cdot K}$ 8.3145 J/mol · K
Molar volume (STP)	V_m	22.414 L/mol	22.414×10^{-3} m³/mol 22.414 dm³/mol
Neutron rest mass	m_n	1.67493×10^{-24} g 1.008665 amu	1.67493×10^{-27} kg
Planck constant	h	$6.62606957 \times 10^{-27}$ erg · s	$6.62606896 \times 10^{-34}$ J · s
Proton rest mass	m_p	1.6726×10^{-24} g 1.007277 amu	1.6726×10^{-27} kg
Rydberg constant	R_∞	3.2898×10^{15} cycles/s 2.17987×10^{-11} erg	1.09737×10^7 m⁻¹ 2.17987×10^{-18} J
Speed of light (in a vacuum)	c	$2.99792458 \times 10^{10}$ cm/s (186,281 miles/second)	2.99792458×10^8 m/s

$\pi = 3.1416$ $2.303\ R = 4.576$ cal/mol · K $= 19.15$ J/mol · K

$e = 2.71828$ $2.303\ RT$ (at 25°C) $= 1364$ cal/mol $= 5709$ J/mol

$\ln X = 2.303 \log X$

Some Physical Constants for a Few Common Substances

Specific Heats and Heat Capacities for Some Common Substances

Substance	Specific Heat (J/g · °C)	Molar Heat Capacity (J/mol · °C)
Al(s)	0.900	24.3
Ca(s)	0.653	26.2
Cu(s)	0.385	24.5
Fe(s)	0.444	24.8
Hg(ℓ)	0.138	27.7
H_2O(s), ice	2.09	37.7
H_2O(ℓ), water	4.18	75.3
H_2O(g), steam	2.03	36.4
C_6H_6(ℓ), benzene	1.74	136
C_6H_6(g), benzene	1.04	81.6
C_2H_5OH(ℓ), ethanol	2.46	113
C_2H_5OH(g), ethanol	0.954	420
$(C_2H_5)_2O$(ℓ), diethyl ether	3.74	172
$(C_2H_5)_2O$(g), diethyl ether	2.35	108

Heats of Transformation and Transformation Temperatures of Several Substances

Substance	mp (°C)	Heat of Fusion (J/g)	ΔH_{fus} (kJ/mol)	bp (°C)	Heat of Vaporization (J/g)	ΔH_{vap} (kJ/mol)
Al	658	395	10.6	2467	10520	284
Ca	851	233	9.33	1487	4030	162
Cu	1083	205	13.0	2595	4790	305
H_2O	0.0	334	6.02	100	2260	40.7
Fe	1530	267	14.9	2735	6340	354
Hg	−39	11	23.3	357	292	58.6
CH_4	−182	58.6	0.92	−164	—	—
C_2H_5OH	−117	109	5.02	78.3	855	39.3
C_6H_6	5.48	127	9.92	80.1	395	30.8
$(C_2H_5)_2O$	−116	97.9	7.66	35	351	26.0

Vapor Pressure of Water at Various Temperatures

Temperature (°C)	Vapor Pressure (torr)	Temperature (°C)	Vapor Pressure (torr)	Temperature (°C)	Vapor Pressure (torr)	Temperature (°C)	Vapor Pressure (torr)
−10	2.1	21	18.7	51	97.2	81	369.7
−9	2.3	22	19.8	52	102.1	82	384.9
−8	2.5	23	21.1	53	107.2	83	400.6
−7	2.7	24	22.4	54	112.5	84	416.8
−6	2.9	25	23.8	55	118.0	85	433.6
−5	3.2	26	25.2	56	123.8	86	450.9
−4	3.4	27	26.7	57	129.8	87	468.7
−3	3.7	28	28.3	58	136.1	88	487.1
−2	4.0	29	30.0	59	142.6	89	506.1
−1	4.3	30	31.8	60	149.4	90	525.8
0	4.6	31	33.7	61	156.4	91	546.1
1	4.9	32	35.7	62	163.8	92	567.0
2	5.3	33	37.7	63	171.4	93	588.6
3	5.7	34	39.9	64	179.3	94	610.9
4	6.1	35	42.2	65	187.5	95	633.9
5	6.5	36	44.6	66	196.1	96	657.6
6	7.0	37	47.1	67	205.0	97	682.1
7	7.5	38	49.7	68	214.2	98	707.3
8	8.0	39	52.4	69	223.7	99	733.2
9	8.6	40	55.3	70	233.7	100	760.0
10	9.2	41	58.3	71	243.9	101	787.6
11	9.8	42	61.5	72	254.6	102	815.9
12	10.5	43	64.8	73	265.7	103	845.1
13	11.2	44	68.3	74	277.2	104	875.1
14	12.0	45	71.9	75	289.1	105	906.1
15	12.8	46	75.7	76	301.4	106	937.9
16	13.6	47	79.6	77	314.1	107	970.6
17	14.5	48	83.7	78	327.3	108	1004.4
18	15.5	49	88.0	79	341.0	109	1038.9
19	16.5	50	92.5	80	355.1	110	1074.6
20	17.5						

Ionization Constants for Weak Acids at 25°C

Acid	Formula and Ionization Equation			K_a	pK_a
Acetic	CH_3COOH	$\rightleftharpoons$	$H^+ + CH_3COO^-$	1.8×10^{-5}	4.74
Arsenic	H_3AsO_4	$\rightleftharpoons$	$H^+ + H_2AsO_4^-$	$2.5 \times 10^{-4} = K_{a1}$	3.60
	$H_2AsO_4^-$	$\rightleftharpoons$	$H^+ + HAsO_4^{2-}$	$5.6 \times 10^{-8} = K_{a2}$	7.25
	$HAsO_4^{2-}$	$\rightleftharpoons$	$H^+ + AsO_4^{3-}$	$3.0 \times 10^{-13} = K_{a3}$	12.52
Arsenous	H_3AsO_3	$\rightleftharpoons$	$H^+ + H_2AsO_3^-$	$6.0 \times 10^{-10} = K_{a1}$	9.22
	$H_2AsO_3^-$	$\rightleftharpoons$	$H^+ + HAsO_3^{2-}$	$3.0 \times 10^{-14} = K_{a2}$	13.52
Benzoic	C_6H_5COOH	$\rightleftharpoons$	$H^+ + C_6H_5COO^-$	6.3×10^{-5}	4.20
Boric*	$B(OH)_3$	$\rightleftharpoons$	$H^+ + BO(OH)_2^-$	$7.3 \times 10^{-10} = K_{a1}$	9.14
	$BO(OH)_2^-$	$\rightleftharpoons$	$H^+ + BO_2(OH)^{2-}$	$1.8 \times 10^{-13} = K_{a2}$	12.74
	$BO_2(OH)^{2-}$	$\rightleftharpoons$	$H^+ + BO_3^{3-}$	$1.6 \times 10^{-14} = K_{a3}$	13.80
Carbonic	H_2CO_3	$\rightleftharpoons$	$H^+ + HCO_3^-$	$4.2 \times 10^{-7} = K_{a1}$	6.38
	HCO_3^-	$\rightleftharpoons$	$H^+ + CO_3^{2-}$	$4.8 \times 10^{-11} = K_{a2}$	10.32
Citric	$C_3H_5O(COOH)_3$	$\rightleftharpoons$	$H^+ + C_4H_5O_3(COOH)_2^-$	$7.4 \times 10^{-3} = K_{a1}$	2.13
	$C_4H_5O_3(COOH)_2^-$	$\rightleftharpoons$	$H^+ + C_5H_5O_5COOH^{2-}$	$1.7 \times 10^{-5} = K_{a2}$	4.77
	$C_5H_5O_5COOH^{2-}$	$\rightleftharpoons$	$H^+ + C_6H_5O_7^{3-}$	$7.4 \times 10^{-7} = K_{a3}$	6.13
Cyanic	$HOCN$	$\rightleftharpoons$	$H^+ + OCN^-$	3.5×10^{-4}	3.46
Formic	$HCOOH$	$\rightleftharpoons$	$H^+ + HCOO^-$	1.8×10^{-4}	3.74
Hydrazoic	HN_3	$\rightleftharpoons$	$H^+ + N_3^-$	1.9×10^{-5}	4.72
Hydrocyanic	HCN	$\rightleftharpoons$	$H^+ + CN^-$	4.0×10^{-10}	9.40
Hydrofluoric	HF	$\rightleftharpoons$	$H^+ + F^-$	7.2×10^{-4}	3.14
Hydrogen peroxide	H_2O_2	$\rightleftharpoons$	$H^+ + HO_2^-$	2.4×10^{-12}	11.62
Hydrosulfuric	H_2S	$\rightleftharpoons$	$H^+ + HS^2$	$1.0 \times 10^{-7} = K_{a1}$	7.00
	HS^-	$\rightleftharpoons$	$H^+ + S^{2-}$	$1.0 \times 10^{-19} = K_{a2}$	19.00
Hypobromous	$HOBr$	$\rightleftharpoons$	$H^+ + OBr^-$	2.5×10^{-9}	8.60

Acid	Formula and Ionization Equation			K_a	pK_a
Hypochlorous	HOCl	$\rightleftharpoons$	$H^+ + OCl^-$	3.5×10^{-8}	7.46
Hypoiodous	HOI	$\rightleftharpoons$	$H^+ + OI^-$	2.3×10^{-11}	10.64
Nitrous	HNO_2	$\rightleftharpoons$	$H^+ + NO_2^-$	4.5×10^{-4}	3.35
Oxalic	$(COOH)_2$	$\rightleftharpoons$	$H^+ + COOCOOH^-$	$5.9 \times 10^{-2} = K_{a1}$	1.23
	$COOCOOH^-$	$\rightleftharpoons$	$H^+ + (COO)_2{}^{2-}$	$6.4 \times 10^{-5} = K_{a2}$	4.19
Phenol	HC_6H_5O	$\rightleftharpoons$	$H^+ + C_6H_5O^-$	1.3×10^{-10}	9.89
Phosphoric	H_3PO_4	$\rightleftharpoons$	$H^+ + H_2PO_4^-$	$7.5 \times 10^{-3} = K_{a1}$	2.12
	$H_2PO_4^-$	$\rightleftharpoons$	$H^+ + HPO_4{}^{2-}$	$6.2 \times 10^{-8} = K_{a2}$	7.21
	$HPO_4{}^{2-}$	$\rightleftharpoons$	$H^+ + PO_4{}^{3-}$	$3.6 \times 10^{-13} = K_{a3}$	12.44
Phosphorous	H_3PO_3	$\rightleftharpoons$	$H^+ + H_2PO_3^-$	$1.6 \times 10^{-2} = K_{a1}$	1.80
	$H_2PO_3^-$	$\rightleftharpoons$	$H^+ + HPO_3{}^{2-}$	$7.0 \times 10^{-7} = K_{a2}$	6.15
Selenic	H_2SeO_4	$\rightleftharpoons$	$H^+ + HSeO_4^-$	Very large $= K_{a1}$	
	$HSeO_4^-$	$\rightleftharpoons$	$H^+ + SeO_4{}^{2-}$	$1.2 \times 10^{-2} = K_{a2}$	1.92
Selenous	H_2SeO_3	$\rightleftharpoons$	$H^+ + HSeO_3^-$	$2.7 \times 10^{-3} = K_{a1}$	2.57
	$HSeO_3^-$	$\rightleftharpoons$	$H^+ + SeO_3{}^{2-}$	$2.5 \times 10^{-7} = K_{a2}$	6.60
Sulfuric	H_2SO_4	$\rightleftharpoons$	$H^+ + HSO_4^-$	Very large $= K_{a1}$	
	HSO_4^-	$\rightleftharpoons$	$H^+ + SO_4{}^{2-}$	$1.2 \times 10^{-2} = K_{a2}$	1.92
Sulfurous	H_2SO_3	$\rightleftharpoons$	$H^+ + HSO_3^-$	$1.2 \times 10^{-2} = K_{a1}$	1.92
	HSO_3^-	$\rightleftharpoons$	$H^+ + SO_3{}^{2-}$	$6.2 \times 10^{-8} = K_{a2}$	7.21
Tellurous	H_2TeO_3	$\rightleftharpoons$	$H^+ + HTeO_3^-$	$2 \times 10^{-3} = K_{a1}$	2.70
	$HTeO_3^-$	$\rightleftharpoons$	$H^+ + TeO_3{}^{2-}$	$1 \times 10^{-8} = K_{a2}$	8.00

*Boric acid acts as a Lewis acid in aqueous solution.

Ionization Constants for Weak Bases at 25°C

Base	Formula and Ionization Equation				K_b	pK_b
Ammonia	NH_3	$+ H_2O \rightleftharpoons NH_4^+$		$+ OH^-$	1.8×10^{-5}	7.74
Aniline	$C_6H_5NH_2$	$+ H_2O \rightleftharpoons C_6H_5NH_3^+$		$+ OH^-$	4.2×10^{-10}	9.38
Dimethylamine	$(CH_3)_2NH$	$+ H_2O \rightleftharpoons (CH_3)_2NH_2^+$		$+ OH^-$	7.4×10^{-4}	3.13
Ethylenediamine	$(CH_2)_2(NH_2)_2$	$+ H_2O \rightleftharpoons (CH_2)_2(NH_2)_2H^+$		$+ OH^-$	$8.5 \times 10^{-5} = K_{b1}$	4.07
	$(CH_2)_2(NH_2)_2H^+$	$+ H_2O \rightleftharpoons (CH_2)_2(NH_2)_2H_2^{2+}$		$+ OH^-$	$2.7 \times 10^{-8} = K_{b2}$	7.57
Hydrazine	N_2H_4	$+ H_2O \rightleftharpoons N_2H_5^+$		$+ OH^-$	$8.5 \times 10^{-7} = K_{b1}$	6.07
	$N_2H_5^+$	$+ H_2O \rightleftharpoons N_2H_6^{2+}$		$+ OH^-$	$8.9 \times 10^{-16} = K_{b2}$	15.05
Hydroxylamine	NH_2OH	$+ H_2O \rightleftharpoons NH_3OH^+$		$+ OH^-$	6.6×10^{-9}	8.18
Methylamine	CH_3NH_2	$+ H_2O \rightleftharpoons CH_3NH_3^+$		$+ OH^-$	5.0×10^{-4}	3.30
Pyridine	C_5H_5N	$+ H_2O \rightleftharpoons C_5H_5NH^+$		$+ OH^-$	1.5×10^{-9}	8.82
Trimethylamine	$(CH_3)_3N$	$+ H_2O \rightleftharpoons (CH_3)_3NH^+$		$+ OH^-$	7.4×10^{-5}	4.13

Solubility Product Constants for Some Inorganic Compounds at 25°C

Substance	K_{sp}	Substance	K_{sp}
Aluminum compounds		*Chromium compounds*	
$AlAsO_4$	1.6×10^{-16}	$CrAsO_4$	7.8×10^{-21}
$Al(OH)_3$	1.9×10^{-33}	$Cr(OH)_3$	6.7×10^{-31}
$AlPO_4$	1.3×10^{-20}	$CrPO_4$	2.4×10^{-23}
Antimony compounds		*Cobalt compounds*	
Sb_2S_3	1.6×10^{-93}	$Co_3(AsO_4)_2$	7.6×10^{-29}
Barium compounds		$CoCO_3$	8.0×10^{-13}
$Ba_3(AsO_4)_2$	1.1×10^{-13}	$Co(OH)_2$	2.5×10^{-16}
$BaCO_3$	8.1×10^{-9}	$CoS (\alpha)$	5.9×10^{-21}
$BaC_2O_4 \cdot 2H_2O^*$	1.1×10^{-7}	$CoS (\beta)$	8.7×10^{-23}
$BaCrO_4$	2.0×10^{-10}	$Co(OH)_3$	4.0×10^{-45}
BaF_2	1.7×10^{-6}	Co_2S_3	2.6×10^{-124}
$Ba(OH)_2 \cdot 8H_2O^*$	5.0×10^{-3}	*Copper compounds*	
$Ba_3(PO_4)_2$	1.3×10^{-29}	$CuBr$	5.3×10^{-9}
$BaSeO_4$	2.8×10^{-11}	$CuCl$	1.9×10^{-7}
$BaSO_3$	8.0×10^{-7}	$CuCN$	3.2×10^{-20}
$BaSO_4$	1.1×10^{-10}	$Cu_2O (Cu^+ + OH^-)^\dagger$	1.0×10^{-14}
Bismuth compounds		CuI	5.1×10^{-12}
$BiOCl$	7.0×10^{-9}	Cu_2S	1.6×10^{-48}
$BiO(OH)$	1.0×10^{-12}	$CuSCN$	1.6×10^{-11}
$Bi(OH)_3$	3.2×10^{-40}	$Cu_3(AsO_4)_2$	7.6×10^{-36}
BiI_3	8.1×10^{-19}	$CuCO_3$	2.5×10^{-10}
$BiPO_4$	1.3×10^{-23}	$Cu_2[Fe(CN)_6]$	1.3×10^{-16}
Bi_2S_3	1.6×10^{-72}	$Cu(OH)_2$	1.6×10^{-19}
Cadmium compounds		CuS	8.7×10^{-36}
$Cd_3(AsO_4)_2$	2.2×10^{-32}	*Gold compounds*	
$CdCO_3$	2.5×10^{-14}	$AuBr$	5.0×10^{-17}
$Cd(CN)_2$	1.0×10^{-8}	$AuCl$	2.0×10^{-13}
$Cd_2[Fe(CN)_6]$	3.2×10^{-17}	AuI	1.6×10^{-23}
$Cd(OH)_2$	1.2×10^{-14}	$AuBr_3$	4.0×10^{-36}
CdS	3.6×10^{-29}	$AuCl_3$	3.2×10^{-25}
Calcium compounds		$Au(OH)_3$	1.0×10^{-53}
$Ca_3(AsO_4)_2$	6.8×10^{-19}	AuI_3	1.0×10^{-46}
$CaCO_3$	4.8×10^{-9}	*Iron compounds*	
$CaCrO_4$	7.1×10^{-4}	$FeCO_3$	3.5×10^{-11}
$CaC_2O_4 \cdot H_2O^*$	2.3×10^{-9}	$Fe(OH)_2$	7.9×10^{-15}
CaF_2	3.9×10^{-11}	FeS	4.9×10^{-18}
$Ca(OH)_2$	7.9×10^{-6}	$Fe_4[Fe(CN)_6]_3$	3.0×10^{-41}
$CaHPO_4$	2.7×10^{-7}	$Fe(OH)_3$	6.3×10^{-38}
$Ca(H_2PO_4)_2$	1.0×10^{-3}	Fe_2S_3	1.4×10^{-88}
$Ca_3(PO_4)_2$	1.0×10^{-25}	*Lead compounds*	
$CaSO_3 \cdot 2H_2O^*$	1.3×10^{-8}	$Pb_3(AsO_4)_2$	4.1×10^{-36}
$CaSO_4 \cdot 2H_2O^*$	2.4×10^{-5}	$PbBr_2$	6.3×10^{-6}

Substance	K_{sp}	Substance	K_{sp}
Lead compounds (cont.)		*Nickel compounds (cont.)*	
$PbCO_3$	1.5×10^{-13}	NiS (α)	3.0×10^{-21}
$PbCl_2$	1.7×10^{-5}	NiS (β)	1.0×10^{-26}
$PbCrO_4$	1.8×10^{-14}	NiS (γ)	2.0×10^{-28}
PbF_2	3.7×10^{-8}	*Silver compounds*	
$Pb(OH)_2$	2.8×10^{-16}	Ag_3AsO_4	1.1×10^{-20}
PbI_2	8.7×10^{-9}	AgBr	3.3×10^{-13}
$Pb_3(PO_4)_2$	3.0×10^{-44}	Ag_2CO_3	8.1×10^{-12}
$PbSeO_4$	1.5×10^{-7}	AgCl	1.8×10^{-10}
$PbSO_4$	1.8×10^{-8}	Ag_2CrO_4	9.0×10^{-12}
PbS	8.4×10^{-28}	AgCN	1.2×10^{-16}
Magnesium compounds		$Ag_4[Fe(CN)_6]$	1.6×10^{-41}
$Mg_3(AsO_4)_2$	2.1×10^{-20}	Ag_2O ($Ag^+ + OH^-$)[†]	2.0×10^{-8}
$MgCO_3 \cdot 3H_2O$*	4.0×10^{-5}	AgI	1.5×10^{-16}
MgC_2O_4	8.6×10^{-5}	Ag_3PO_4	1.3×10^{-20}
MgF_2	6.4×10^{-9}	Ag_2SO_3	1.5×10^{-14}
$Mg(OH)_2$	1.5×10^{-11}	Ag_2SO_4	1.7×10^{-5}
$MgNH_4PO_4$	2.5×10^{-12}	Ag_2S	1.0×10^{-49}
Manganese compounds		AgSCN	1.0×10^{-12}
$Mn_3(AsO_4)_2$	1.9×10^{-11}	*Strontium compounds*	
$MnCO_3$	1.8×10^{-11}	$Sr_3(AsO_4)_2$	1.3×10^{-18}
$Mn(OH)_2$	4.6×10^{-14}	$SrCO_3$	9.4×10^{-10}
MnS	5.1×10^{-15}	$SrC_2O_4 \cdot 2H_2O$*	5.6×10^{-8}
$Mn(OH)_3$	$\approx 1.0 \times 10^{-36}$	$SrCrO_4$	3.6×10^{-5}
Mercury compounds		$Sr(OH)_2 \cdot 8H_2O$*	3.2×10^{-4}
Hg_2Br_2	1.3×10^{-22}	$Sr_3(PO_4)_2$	1.0×10^{-31}
Hg_2CO_3	8.9×10^{-17}	$SrSO_3$	4.0×10^{-8}
Hg_2Cl_2	1.1×10^{-18}	$SrSO_4$	2.8×10^{-7}
Hg_2CrO_4	5.0×10^{-9}	*Tin compounds*	
Hg_2I_2	4.5×10^{-29}	$Sn(OH)_2$	2.0×10^{-26}
$Hg_2O \cdot H_2O$*		SnI_2	1.0×10^{-4}
($Hg_2^{2+} + 2OH^-$)[†]	1.6×10^{-23}	SnS	1.0×10^{-28}
Hg_2SO_4	6.8×10^{-7}	$Sn(OH)_4$	1.0×10^{-57}
Hg_2S	5.8×10^{-44}	SnS_2	1.0×10^{-70}
$Hg(CN)_2$	3.0×10^{-23}	*Zinc compounds*	
$Hg(OH)_2$	2.5×10^{-26}	$Zn_3(AsO_4)_2$	1.1×10^{-27}
HgI_2	4.0×10^{-29}	$ZnCO_3$	1.5×10^{-11}
HgS	3.0×10^{-53}	$Zn(CN)_2$	8.0×10^{-12}
Nickel compounds		$Zn_2[Fe(CN)_6]$	4.1×10^{-16}
$Ni_3(AsO_4)_2$	1.9×10^{-26}	$Zn(OH)_2$	4.5×10^{-17}
$NiCO_3$	6.6×10^{-9}	$Zn_3(PO_4)_2$	9.1×10^{-33}
$Ni(CN)_2$	3.0×10^{-23}	ZnS	1.1×10^{-21}
$Ni(OH)_2$	2.8×10^{-16}		

*$[H_2O]$ does not appear in equilibrium constants for equilibria in aqueous solution in general, so it does *not* appear in the K_{sp} expressions for hydrated solids.

[†]Very small amounts of oxides dissolve in water to give the ions indicated in parentheses. These solid hydroxides are unstable and decompose to oxides as rapidly as they are formed.

Dissociation Constants for Some Complex Ions

Dissociation Equilibrium		K_d
$[AgBr_2]^-$	$\rightleftharpoons$ Ag^+ + $2Br^-$	7.8×10^{-8}
$[AgCl_2]^-$	$\rightleftharpoons$ Ag^+ + $2Cl^-$	4.0×10^{-6}
$[Ag(CN)_2]^-$	$\rightleftharpoons$ Ag^+ + $2CN^-$	1.8×10^{-19}
$[Ag(S_2O_3)_2]^{3-}$	$\rightleftharpoons$ Ag^+ + $2S_2O_3^{2-}$	5.0×10^{-14}
$[Ag(NH_3)_2]^+$	$\rightleftharpoons$ Ag^+ + $2NH_3$	6.3×10^{-8}
$[Ag(en)]^+$	$\rightleftharpoons$ Ag^+ + en^*	1.0×10^{-5}
$[AlF_6]^{3-}$	$\rightleftharpoons$ Al^{3+} + $6F^-$	2.0×10^{-24}
$[Al(OH)_4]^-$	$\rightleftharpoons$ Al^{3+} + $4OH^-$	1.3×10^{-34}
$[Au(CN)_2]^-$	$\rightleftharpoons$ Au^+ + $2CN^-$	5.0×10^{-39}
$[Cd(CN)_4]^{2-}$	$\rightleftharpoons$ Cd^{2+} + $4CN^-$	7.8×10^{-18}
$[CdCl_4]^{2-}$	$\rightleftharpoons$ Cd^{2+} + $4Cl^-$	1.0×10^{-4}
$[Cd(NH_3)_4]^{2+}$	$\rightleftharpoons$ Cd^{2+} + $4NH_3$	1.0×10^{-7}
$[Co(NH_3)_6]^{2+}$	$\rightleftharpoons$ Co^{2+} + $6NH_3$	1.3×10^{-5}
$[Co(NH_3)_6]^{3+}$	$\rightleftharpoons$ Co^{3+} + $6NH_3$	2.2×10^{-34}
$[Co(en)_3]^{2+}$	$\rightleftharpoons$ Co^{2+} + $3en^*$	1.5×10^{-14}
$[Co(en)_3]^{3+}$	$\rightleftharpoons$ Co^{3+} + $3en^*$	2.0×10^{-49}
$[Cu(CN)_2]^-$	$\rightleftharpoons$ Cu^+ + $2CN^-$	1.0×10^{-16}
$[CuCl_2]^-$	$\rightleftharpoons$ Cu^+ + $2Cl^-$	1.0×10^{-5}
$[Cu(NH_3)_2]^+$	$\rightleftharpoons$ Cu^+ + $2NH_3$	1.4×10^{-11}
$[Cu(NH_3)_4]^{2+}$	$\rightleftharpoons$ Cu^{2+} + $4NH_3$	8.5×10^{-13}
$[Fe(CN)_6]^{4-}$	$\rightleftharpoons$ Fe^{2+} + $6CN^-$	1.3×10^{-37}
$[Fe(CN)_6]^{3-}$	$\rightleftharpoons$ Fe^{3+} + $6CN^-$	1.3×10^{-44}
$[HgCl_4]^{2-}$	$\rightleftharpoons$ Hg^{2+} + $4Cl^-$	8.3×10^{-16}
$[Ni(CN)_4]^{2-}$	$\rightleftharpoons$ Ni^{2+} + $4CN^-$	1.0×10^{-31}
$[Ni(NH_3)_6]^{2+}$	$\rightleftharpoons$ Ni^{2+} + $6NH_3$	1.8×10^{-9}
$[Zn(OH)_4]^{2-}$	$\rightleftharpoons$ Zn^{2+} + $4OH^-$	3.5×10^{-16}
$[Zn(NH_3)_4]^{2+}$	$\rightleftharpoons$ Zn^{2+} + $4NH_3$	3.4×10^{-10}

*The abbreviation "en" represents ethylenediamine, $H_2NCH_2CH_2NH_2$.

Standard Reduction Potentials in Aqueous Solution at 25°C

Acidic Solution	Standard Reduction Potential, E^0 (volts)
$Li^+(aq) + e^- \longrightarrow Li(s)$	-3.045
$K^+(aq) + e^- \longrightarrow K(s)$	-2.925
$Rb^+(aq) + e^- \longrightarrow Rb(s)$	-2.925
$Ba^{2+}(aq) + 2e^- \longrightarrow Ba(s)$	-2.90
$Sr^{2+}(aq) + 2e^- \longrightarrow Sr(s)$	-2.89
$Ca^{2+}(aq) + 2e^- \longrightarrow Ca(s)$	-2.87
$Na^+(aq) + e^- \longrightarrow Na(s)$	-2.714
$Mg^{2+}(aq) + 2e^- \longrightarrow Mg(s)$	-2.37
$H_2(g) + 2e^- \longrightarrow 2H^-(aq)$	-2.25
$Al^{3+}(aq) + 3e^- \longrightarrow Al(s)$	-1.66
$Zr^{4+}(aq) + 4e^- \longrightarrow Zr(s)$	-1.53
$ZnS(s) + 2e^- \longrightarrow Zn(s) + S^{2-}(aq)$	-1.44
$CdS(s) + 2e^- \longrightarrow Cd(s) + S^{2-}(aq)$	-1.21
$V^{2+}(aq) + 2e^- \longrightarrow V(s)$	-1.18
$Mn^{2+}(aq) + 2e^- \longrightarrow Mn(s)$	-1.18
$FeS(s) + 2e^- \longrightarrow Fe(s) + S^{2-}(aq)$	-1.01
$Cr^{2+}(aq) + 2e^- \longrightarrow Cr(s)$	-0.91
$Zn^{2+}(aq) + 2e^- \longrightarrow Zn(s)$	-0.763
$Cr^{3+}(aq) + 3e^- \longrightarrow Cr(s)$	-0.74
$HgS(s) + 2H^+(aq) + 2e^- \longrightarrow Hg(\ell) + H_2S(g)$	-0.72
$Ga^{3+}(aq) + 3e^- \longrightarrow Ga(s)$	-0.53
$2CO_2(g) + 2H^+(aq) + 2e^- \longrightarrow (COOH)_2(aq)$	-0.49
$Fe^{2+}(aq) + 2e^- \longrightarrow Fe(s)$	-0.44
$Cr^{3+}(aq) + e^- \longrightarrow Cr^{2+}(aq)$	-0.41
$Cd^{2+}(aq) + 2e^- \longrightarrow Cd(s)$	-0.403
$Se(s) + 2H^+(aq) + 2e^- \longrightarrow H_2Se(aq)$	-0.40
$PbSO_4(s) + 2e^- \longrightarrow Pb(s) + SO_4^{2-}(aq)$	-0.356
$Tl^+(aq) + e^- \longrightarrow Tl(s)$	-0.34
$Co^{2+}(aq) + 2e^- \longrightarrow Co(s)$	-0.28
$Ni^{2+}(aq) + 2e^- \longrightarrow Ni(s)$	-0.25
$[SnF_6]^{2-}(aq) + 4e^- \longrightarrow Sn(s) + 6F^-(aq)$	-0.25

Acidic Solution	Standard Reduction Potential, E^0 (volts)
$AgI(s) + e^- \longrightarrow Ag(s) + I^-(aq)$	−0.15
$Sn^{2+}(aq) + 2e^- \longrightarrow Sn(s)$	−0.14
$Pb^{2+}(aq) + 2e^- \longrightarrow Pb(s)$	−0.126
$N_2O(g) + 6H^+(aq) + H_2O + 4e^- \longrightarrow 2NH_3OH^+(aq)$	−0.05
$2H^+(aq) + 2e^- \longrightarrow H_2(g)$ (reference electrode)	0.000
$AgBr(s) + e^- \longrightarrow Ag(s) + Br^-(aq)$	0.10
$S(s) + 2H^+(aq) + 2e^- \longrightarrow H_2S(aq)$	0.14
$Sn^{4+}(aq) + 2e^- \longrightarrow Sn^{2+}(aq)$	0.15
$Cu^{2+}(aq) + e^- \longrightarrow Cu^+(aq)$	0.153
$SO_4^{2-}(aq) + 4H^+(aq) + 2e^- \longrightarrow H_2SO_3(aq) + H_2O$	0.17
$SO_4^{2-}(aq) + 4H^+(aq) + 2e^- \longrightarrow SO_2(g) + 2H_2O$	0.20
$AgCl(s) + e^- \longrightarrow Ag(s) + Cl^-(aq)$	0.222
$Hg_2Cl_2(s) + 2e^- \longrightarrow 2Hg(\ell) + 2Cl^-(aq)$	0.27
$Cu^{2+}(aq) + 2e^- \longrightarrow Cu(s)$	0.337
$[RhCl_6]^{3-}(aq) + 3e^- \longrightarrow Rh(s) + 6Cl^-(aq)$	0.44
$Cu^+(aq) + e^- \longrightarrow Cu(s)$	0.521
$TeO_2(s) + 4H^+(aq) + 4e^- \longrightarrow Te(s) + 2H_2O$	0.529
$I_2(s) + 2e^- \longrightarrow 2I^-(aq)$	0.535
$H_3AsO_4(aq) + 2H^+(aq) + 2e^- \longrightarrow H_3AsO_3(aq) + H_2O$	0.58
$[PtCl_6]^{2-}(aq) + 2e^- \longrightarrow [PtCl_4]^{2-}(aq) + 2Cl^-(aq)$	0.68
$O_2(g) + 2H^+(aq) + 2e^- \longrightarrow H_2O_2(aq)$	0.682
$[PtCl_4]^{2-}(aq) + 2e^- \longrightarrow Pt(s) + 4Cl^-(aq)$	0.73
$SbCl_6^-(aq) + 2e^- \longrightarrow SbCl_4^-(aq) + 2Cl^-(aq)$	0.75
$Fe^{3+}(aq) + e^- \longrightarrow Fe^{2+}(aq)$	0.771
$Hg_2^{2+}(aq) + 2e^- \longrightarrow 2Hg(\ell)$	0.789
$Ag^+(aq) + e^- \longrightarrow Ag(s)$	0.7994
$Hg^{2+}(aq) + 2e^- \longrightarrow Hg(\ell)$	0.855
$2Hg^{2+}(aq) + 2e^- \longrightarrow Hg_2^{2+}(aq)$	0.920
$NO_3^-(aq) + 3H^+(aq) + 2e^- \longrightarrow HNO_2(aq) + H_2O$	0.94
$NO_3^-(aq) + 4H^+(aq) + 3e^- \longrightarrow NO(g) + 2H_2O$	0.96
$Pd^{2+}(aq) + 2e^- \longrightarrow Pd(s)$	0.987
$AuCl_4^-(aq) + 3e^- \longrightarrow Au(s) + 4Cl^-(aq)$	1.00
$Br_2(\ell) + 2e^- \longrightarrow 2Br^-(aq)$	1.08
$ClO_4^-(aq) + 2H^+(aq) + 2e^- \longrightarrow ClO_3^-(aq) + H_2O$	1.19
$IO_3^-(aq) + 6H^+(aq) + 5e^- \longrightarrow \frac{1}{2}I_2(aq) + 3H_2O$	1.195
$Pt^{2+}(aq) + 2e^- \longrightarrow Pt(s)$	1.2
$O_2(g) + 4H^+(aq) + 4e^- \longrightarrow 2H_2O$	1.229
$MnO_2(s) + 4H^+(aq) + 2e^- \longrightarrow Mn^{2+}(aq) + 2H_2O$	1.23
$N_2H_5^+(aq) + 3H^+(aq) + 2e^- \longrightarrow 2NH_4^+(aq)$	1.24
$Cr_2O_7^{2-}(aq) + 14H^+(aq) + 6e^- \longrightarrow 2Cr^{3+}(aq) + 7H_2O$	1.33
$Cl_2(g) + 2e^- \longrightarrow 2Cl^-(aq)$	1.360
$BrO_3^-(aq) + 6H^+(aq) + 6e^- \longrightarrow Br^-(aq) + 3H_2O$	1.44
$ClO_3^-(aq) + 6H^+(aq) + 5e^- \longrightarrow \frac{1}{2}Cl_2(g) + 3H_2O$	1.47
$Au^{3+}(aq) + 3e^- \longrightarrow Au(s)$	1.50

Basic Solution	Standard Reduction Potential, E^0 (volts)
$MnO_4^-(aq) + 8H^+(aq) + 5e^- \longrightarrow Mn^{2+}(aq) + 4H_2O$	1.507
$NaBiO_3(s) + 6H^+(aq) + 2e^- \longrightarrow Bi^{3+}(aq) + Na^+(aq) + 3H_2O$	1.6
$Ce^{4+}(aq) + e^- \longrightarrow Ce^{3+}(aq)$	1.61
$2HOCl(aq) + 2H^+(aq) + 2e^- \longrightarrow Cl_2(g) + 2H_2O$	1.63
$Au^+(aq) + e^- \longrightarrow Au(s)$	1.68
$PbO_2(s) + SO_4^{2-}(aq) + 4H^+(aq) + 2e^- \longrightarrow PbSO_4(s) + 2H_2O$	1.685
$NiO_2(s) + 4H^+(aq) + 2e^- \longrightarrow Ni^{2+}(aq) + 2H_2O$	1.7
$H_2O_2(aq) + 2H^+(aq) + 2e^- \longrightarrow 2H_2O$	1.77
$Pb^{4+}(aq) + 2e^- \longrightarrow Pb^{2+}(aq)$	1.8
$Co^{3+}(aq) + e^- \longrightarrow Co^{2+}(aq)$	1.82
$F_2(g) + 2e^- \longrightarrow 2F^-(aq)$	2.87
$SiO_3^{2-}(aq) + 3H_2O + 4e^- \longrightarrow Si(s) + 6OH^-(aq)$	−1.70
$Cr(OH)_3(s) + 3e^- \longrightarrow Cr(s) + 3OH^-(aq)$	−1.30
$[Zn(CN)_4]^{2-}(aq) + 2e^- \longrightarrow Zn(s) + 4CN^-(aq)$	−1.26
$Zn(OH)_2(s) + 2e^- \longrightarrow Zn(s) + 2OH^-(aq)$	−1.245
$[Zn(OH)_4]^{2-}(aq) + 2e^- \longrightarrow Zn(s) + 4OH^-(aq)$	−1.22
$N_2(g) + 4H_2O + 4e^- \longrightarrow N_2H_4(aq) + 4OH^-(aq)$	−1.15
$SO_4^{2-}(aq) + H_2O + 2e^- \longrightarrow SO_3^{2-}(aq) + 2OH^-(aq)$	−0.93
$Fe(OH)_2(s) + 2e^- \longrightarrow Fe(s) + 2OH^-(aq)$	−0.877
$2NO_3^-(aq) + 2H_2O + 2e^- \longrightarrow N_2O_4(g) + 4OH^-(aq)$	−0.85
$2H_2O + 2e^- \longrightarrow H_2(g) + 2OH^-(aq)$	−0.828
$Fe(OH)_3(s) + e^- \longrightarrow Fe(OH)_2(s) + OH^-(aq)$	−0.56
$S(s) + 2e^- \longrightarrow S^{2-}(aq)$	−0.48
$Cu(OH)_2(s) + 2e^- \longrightarrow Cu(s) + 2OH^-(aq)$	−0.36
$CrO_4^{2-}(aq) + 4H_2O + 3e^- \longrightarrow Cr(OH)_3(s) + 5OH^-(aq)$	−0.12
$MnO_2(s) + 2H_2O + 2e^- \longrightarrow Mn(OH)_2(s) + 2OH^-(aq)$	−0.05
$NO_3^-(aq) + H_2O + 2e^- \longrightarrow NO_2^-(aq) + 2OH^-(aq)$	0.01
$O_2(g) + H_2O + 2e^- \longrightarrow OOH^-(aq) + OH^-(aq)$	0.076
$HgO(s) + H_2O + 2e^- \longrightarrow Hg(\ell) + 2OH^-(aq)$	0.0984
$[Co(NH_3)_6]^{3+}(aq) + e^- \longrightarrow [Co(NH_3)_6]^{2+}(aq)$	0.10
$N_2H_4(aq) + 2H_2O + 2e^- \longrightarrow 2NH_3(aq) + 2OH^-(aq)$	0.10
$2NO_2^-(aq) + 3H_2O + 4e^- \longrightarrow N_2O(g) + 6OH^-(aq)$	0.15
$Ag_2O(s) + H_2O + 2e^- \longrightarrow 2Ag(s) + 2OH^-(aq)$	0.34
$ClO_4^-(aq) + H_2O + 2e^- \longrightarrow ClO_3^-(aq) + 2OH^-(aq)$	0.36
$O_2(g) + 2H_2O + 4e^- \longrightarrow 4OH^-(aq)$	0.40
$Ag_2CrO_4(s) + 2e^- \longrightarrow 2Ag(s) + CrO_4^{2-}(aq)$	0.446
$NiO_2(s) + 2H_2O + 2e^- \longrightarrow Ni(OH)_2(s) + 2OH^-(aq)$	0.49
$MnO_4^-(aq) + e^- \longrightarrow MnO_4^{2-}(aq)$	0.564
$MnO_4^-(aq) + 2H_2O + 3e^- \longrightarrow MnO_2(s) + 4OH^-(aq)$	0.588
$ClO_3^-(aq) + 3H_2O + 6e^- \longrightarrow Cl^-(aq) + 6OH^-(aq)$	0.62
$2NH_2OH(aq) + 2e^- \longrightarrow N_2H_4(aq) + 2OH^-(aq)$	0.74
$OOH^-(aq) + H_2O + 2e^- \longrightarrow 3OH^-(aq)$	0.88
$ClO^-(aq) + H_2O + 2e^- \longrightarrow Cl^-(aq) + 2OH^-(aq)$	0.89

Selected Thermodynamic Values at 298.15 K

Species	ΔH_f^0 (kJ/mol)	S^0 (J/mol · K)	ΔG_f^0 (kJ/mol)
Aluminum			
Al(s)	0	28.3	0
AlCl$_3$(s)	−704.2	110.7	−628.9
Al$_2$O$_3$(s)	−1676	50.92	−1582
Barium			
BaCl$_2$(s)	−860.1	126	−810.9
BaSO$_4$(s)	−1465	132	−1353
Beryllium			
Be(s)	0	9.54	0
Be(OH)$_2$(s)	−907.1	—	—
Bromine			
Br(g)	111.8	174.9	82.4
Br$_2$(ℓ)	0	152.23	0
Br$_2$(g)	30.91	245.4	3.14
BrF$_3$(g)	−255.6	292.4	−229.5
HBr(g)	−36.4	198.59	−53.43
Calcium			
Ca(s)	0	41.6	0
Ca(g)	192.6	154.8	158.9
Ca^{2+}(g)	1920	—	—
CaC$_2$(s)	−62.8	70.3	−67.8
CaCO$_3$(s)	−1207	92.9	−1129
CaCl$_2$(s)	−795.0	114	−750.2
CaF$_2$(s)	−1215	68.87	−1162
CaH$_2$(s)	−189	42	−150
CaO(s)	−635.5	40	−604.2
CaS(s)	−482.4	56.5	−477.4
Ca(OH)$_2$(s)	−986.6	76.1	−896.8
Ca(OH)$_2$(aq)	−1002.8	76.15	−867.6
CaSO$_4$(s)	−1433	107	−1320
Carbon			
C(s, graphite)	0	5.740	0
C(s, diamond)	1.897	2.38	2.900
C(g)	716.7	158.0	671.3
CCl$_4$(ℓ)	−135.4	216.4	−65.27
CCl$_4$(g)	−103	309.7	−60.63
CHCl$_3$(ℓ)	−134.5	202	−73.72

Species	ΔH_f^0 (kJ/mol)	S^0 (J/mol · K)	ΔG_f^0 (kJ/mol)
CHCl$_3$(g)	−103.1	295.6	−70.37
CH$_4$(g)	−74.81	186.2	−50.75
C$_2$H$_2$(g)	226.7	200.8	209.2
C$_2$H$_4$(g)	52.26	219.5	68.12
C$_2$H$_6$(g)	−84.86	229.5	−32.9
C$_3$H$_8$(g)	−103.8	269.9	−23.49
C$_6$H$_6$(ℓ)	49.03	172.8	124.5
C$_8$H$_{18}$(ℓ)	−268.8	—	—
C$_2$H$_5$OH(ℓ)	−277.7	161	−174.9
C$_2$H$_5$OH(g)	−235.1	282.6	−168.6
CO(g)	−110.5	197.6	−137.2
CO$_2$(g)	−393.5	213.6	−394.4
CS$_2$(g)	117.4	237.7	67.15
COCl$_2$(g)	−223.0	289.2	−210.5
Cesium			
Cs$^+$(aq)	−248	133	−282.0
CsF(aq)	−568.6	123	−558.5
Chlorine			
Cl(g)	121.7	165.1	105.7
Cl$^-$(g)	−226	—	—
Cl$_2$(g)	0	223.0	0
HCl(g)	−92.31	186.8	−95.30
HCl(aq)	−167.4	55.10	−131.2
Chromium			
Cr(s)	0	23.8	0
(NH$_4$)$_2$Cr$_2$O$_7$(s)	−1807	—	—
Copper			
Cu(s)	0	33.15	0
CuO(s)	−157	42.63	−130
Fluorine			
F$^-$(g)	−322	—	—
F$^-$(aq)	−332.6	—	−278.8
F(g)	78.99	158.6	61.92
F$_2$(g)	0	202.7	0
HF(g)	−271	173.7	−273
HF(aq)	−320.8	—	−296.8

Species	ΔH_f^0 (kJ/mol)	S^0 (J/mol · K)	ΔG_f^0 (kJ/mol)	Species	ΔH_f^0 (kJ/mol)	S^0 (J/mol · K)	ΔG_f^0 (kJ/mol)
Hydrogen				$NH_4Cl(s)$	−314.4	94.6	−201.5
H(g)	218.0	114.6	203.3	$NH_4Cl(aq)$	−300.2	—	—
$H_2(g)$	0	130.6	0	$NH_4I(s)$	−201.4	117	−113
$H_2O(\ell)$	−285.8	69.91	−237.2	$NH_4NO_3(s)$	−365.6	151.1	−184.0
$H_2O(g)$	−241.8	188.7	−228.6	NO(g)	90.25	210.7	86.57
$H_2O_2(\ell)$	−187.8	109.6	−120.4	$NO_2(g)$	33.2	240.0	51.30
Iodine				$N_2O(g)$	82.05	219.7	104.2
I(g)	106.6	180.66	70.16	$N_2O_4(g)$	9.16	304.2	97.82
$I_2(s)$	0	116.1	0	$N_2O_5(g)$	11	356	115
$I_2(g)$	62.44	260.6	19.36	$N_2O_5(s)$	−43.1	178	114
ICl(g)	17.78	247.4	−5.52	NOCl(g)	52.59	264	66.36
HI(g)	26.5	206.5	1.72	$HNO_3(\ell)$	−174.1	155.6	−80.79
Iron				$HNO_3(g)$	−135.1	266.2	−74.77
Fe(s)	0	27.3	0	$HNO_3(aq)$	−206.6	146	−110.5
FeO(s)	−272	—	—	*Oxygen*			
Fe_2O_3(s, hematite)	−824.2	87.40	−742.2	O(g)	249.2	161.0	231.8
Fe_3O_4(s, magnetite)	−1118	146	−1015	$O_2(g)$	0	205.0	0
$FeS_2(s)$	−177.5	122.2	−166.7	$O_3(g)$	143	238.8	163
$Fe(CO)_5(\ell)$	−774.0	338	−705.4	$OF_2(g)$	23	246.6	41
$Fe(CO)_5(g)$	−733.8	445.2	−697.3	*Phosphorus*			
Lead				P(g)	314.6	163.1	278.3
Pb(s)	0	64.81	0	P_4(s, white)	0	177	0
$PbCl_2(s)$	−359.4	136	−314.1	P_4(s, red)	−73.6	91.2	−48.5
PbO(s, yellow)	−217.3	68.70	−187.9	$PCl_3(g)$	−306.4	311.7	−286.3
$Pb(OH)_2(s)$	−515.9	88	−420.9	$PCl_5(g)$	−398.9	353	−324.6
PbS(s)	−100.4	91.2	−98.7	$PH_3(g)$	5.4	210.1	13
Lithium				$P_4O_{10}(s)$	−2984	228.9	−2698
Li(s)	0	28.0	0	$H_3PO_4(s)$	−1281	110.5	−1119
LiOH(s)	−487.23	50	−443.9	*Potassium*			
LiOH(aq)	−508.4	4	−451.1	K(s)	0	63.6	0
Magnesium				KCl(s)	−436.5	82.6	−408.8
Mg(s)	0	32.5	0	$KClO_3(s)$	−391.2	143.1	−289.9
$MgCl_2(s)$	−641.8	89.5	−592.3	KI(s)	−327.9	106.4	−323.0
MgO(s)	−601.8	27	−569.6	KOH(s)	−424.7	78.91	−378.9
$Mg(OH)_2(s)$	−924.7	63.14	−833.7	KOH(aq)	−481.2	92.0	−439.6
MgS(s)	−347	—	—	*Rubidium*			
Mercury				Rb(s)	0	76.78	0
$Hg(\ell)$	0	76.02	0	RbOH(aq)	−481.16	110.75	−441.24
$HgCl_2(s)$	−224	146	−179	*Silicon*			
HgO(s, red)	−90.83	70.29	−58.56	Si(s)	0	18.8	0
HgS(s, red)	−58.2	82.4	−50.6	$SiBr_4(\ell)$	−457.3	277.8	−443.9
Nickel				SiC(s)	−65.3	16.6	−62.8
Ni(s)	0	30.1	0	$SiCl_4(g)$	−657.0	330.6	−617.0
$Ni(CO)_4(g)$	−602.9	410.4	−587.3	$SiH_4(g)$	34.3	204.5	56.9
NiO(s)	−244	38.6	−216	$SiF_4(g)$	−1615	282.4	−1573
Nitrogen				$SiI_4(g)$	−132	—	—
$N_2(g)$	0	191.5	0	$SiO_2(s)$	−910.9	41.84	−856.7
N(g)	472.704	153.19	455.579	$H_2SiO_3(s)$	−1189	134	−1092
$NH_3(g)$	−46.11	192.3	−16.5	$Na_2SiO_3(s)$	−1079	—	—
$N_2H_4(\ell)$	50.63	121.2	149.2	$H_2SiF_6(aq)$	−2331	—	—
$(NH_4)_3AsO_4(aq)$	−1268	—	—	*Silver*			
				Ag(s)	0	42.55	0

Species	ΔH_f^0 (kJ/mol)	S^0 (J/mol · K)	ΔG_f^0 (kJ/mol)	Species	ΔH_f^0 (kJ/mol)	S^0 (J/mol · K)	ΔG_f^0 (kJ/mol)
Sodium				$H_2SO_4(\ell)$	−814.0	156.9	−690.1
Na(s)	0	51.0	0	$H_2SO_4(aq)$	−907.5	17	−742.0
Na(g)	108.7	153.6	78.11	*Tin*			
$Na^+(g)$	601	—	—	Sn(s, white)	0	51.55	0
NaBr(s)	−359.9	—	—	Sn(s, grey)	−2.09	44.1	0.13
NaCl(s)	−411.0	72.38	−384	$SnCl_2(s)$	−350	—	—
NaCl(aq)	−407.1	115.5	−393.0	$SnCl_4(\ell)$	−511.3	258.6	−440.2
$Na_2CO_3(s)$	−1131	136	−1048	$SnCl_4(g)$	−471.5	366	−432.2
NaOH(s)	−426.7	—	—	$SnO_2(s)$	−580.7	52.3	−519.7
NaOH(aq)	−469.6	49.8	−419.2	*Titanium*			
Sulfur				$TiCl_4(\ell)$	−804.2	252.3	−737.2
S(s, rhombic)	0	31.8	0	$TiCl_4(g)$	−763.2	354.8	−726.8
S(g)	278.8	167.8	238.3	*Tungsten*			
$S_2Cl_2(g)$	−18	331	−31.8	W(s)	0	32.6	0
$SF_6(g)$	−1209	291.7	−1105	$WO_3(s)$	−842.9	75.90	−764.1
$H_2S(g)$	−20.6	205.7	−33.6	*Zinc*			
$SO_2(g)$	−296.8	248.1	−300.2	ZnO(s)	−348.3	43.64	−318.3
$SO_3(g)$	−395.6	256.6	−371.1	ZnS(s)	−205.6	57.7	−201.3
$SOCl_2(\ell)$	−206	—	—				
$SO_2Cl_2(\ell)$	−389	—	—				

Answers to Selected Even-Numbered Numerical Exercises

Chapter 1

30. **(a)** 423.<u>006</u> mL = 4.23006 × 10² mL (6 sig. fig.)
    (b) 0.001<u>073040</u> g = 1.073040 × 10⁻³ g (7 sig. fig.)
    (c) 1 <u>081.02</u> pounds = 1.08102 × 10³ pounds (6 sig. fig.)
32. **(a)** 50600 **(b)** 0.0004060 **(c)** 0.1610 **(d)** 0.000206 **(e)** 90000. **(f)** 0.0009000
34. 3.0 × 10⁴ cm³
36. **(a)** 0.4534 km **(b)** 3.63 × 10⁴ m **(c)** 4.87 × 10⁵ g **(d)** 1.32 × 10³ mL **(e)** 5.59 L **(f)** 6.251 × 10⁶ cm³
38. 82.42 cents/L
40. **(a)** 21.2 L **(b)** 2.11 pt **(c)** 0.4252 km/L
42. 57.3%
44. **(a)** 9.0 qt **(b)** 88.5 km/hr **(c)** 73 s
46. 1.65 g/cm³
48. **(a)** 42.2 cm³ **(b)** 3.48 cm **(c)** 1.37 in
50. 504 m
52. 3.2 × 10² g
54. **(a)** −9.4°C **(b)** 273.8 K **(c)** 130°F **(d)** 52.3°F
56. **(c)** 285.3°R
58. For Al, 660.4°C and 1221°F; for Ag, 961.9°C and 1763°F
60. 39.8°C, 312.1 K
62. 1440 J
64. **(a)** 9.12 × 10⁵ J **(b)** 11.8°C
66. 0.49 J/g°C
68. **(a)** 62.7 tons of ore **(b)** 69.3 kg of ore
70. 7.16 m
72. 110 mg drug (2 sig. fig.)
74. 500 mL (1 sig. fig.)
84. 4.79 × 10⁷ atoms

Chapter 2

24. 1.070
28. **(a)** 159.808 amu **(b)** 34.014 amu **(c)** 183.18 amu **(d)** 194.189 amu
30. **(a)** 34.08 amu **(b)** 137.3 amu **(c)** 52.46 amu **(d)** 127.91 amu
32. 364.7 g CCl₄, 0.3647 kg CCl₄
34. 1.82 × 10²⁵ H atoms
36. **(a)** 0.7340 mol NH₃
38. **(a)** 4.32 × 10²³ molecules CO₂ **(b)** 6.79 × 10²³ molecules N₂ **(c)** 1.53 × 10²³ molecules P₄ **(d)** 3.07 × 10²³ molecules P₂

44. 1.59 × 10⁻¹⁶ g CH₄
48. 78.25% Ag
50. **(a)** C₉H₉N **(b)** 131.2 g/mol
52. **(a)** C₁₃H₂₄N₄O₃S (FW = 316 g/mol) **(b)** same
54. C₈H₁₁O₃N
56. NaHCO₃
58. C₆H₁₄N₂O₂
60. C₂O₂
62. **(a)** 54.82% C, 5.624% H, 7.104% N, 32.45% O **(b)** 80.87% C, 11.70% H, 7.430% O **(c)** 63.15% C, 5.300% H, 31.55% O
64. Cu₃(CO₃)₂(OH)₂ is 55.31% Cu, Cu₂S is 79.84% Cu, CuFeS₂ is 34.63% Cu, CuS is 66.47% Cu, Cu₂O is 88.82% Cu, Cu₂CO₃(OH)₂ is 57.49% Cu
66. 0.1088 g C and 0.00647 g H giving 94.36% C and 5.61% H
68. 0.720 g CO₂
70. C₂H₆O
72. **(a)** 3.43 g O **(b)** 6.86 g O
74. **(a)** 9.02 g O **(b)** 13.5 g O
76. 498 g Hg
78. 209 g KMnO₄
80. 181 lb Cu₂S
82. **(a)** 352 g CuSO₄·H₂O **(b)** 296 g CuSO₄
84. 49.2 g Cr in ore, 75.7 g Cr recovered
86. **(a)** 73.4 lb MgCO₃ **(b)** 202 lb impurity **(c)** 21.2 lb Mg
88. **(a)** 63.92% **(b)** 47.6%
90. **(a)** 2.00 mol O₃ **(b)** 6.00 mol O **(c)** 96.0 g O₂ **(d)** 64.0 g O₂
92. C₁₄H₂₄O
96. **(a)** C₃H₅O₂ **(b)** C₆H₁₀O₄
102. ReO₂ (85.3358% Re⁺⁴), ReO₃ (79.5063% Re⁺⁶), Re₂O₃ (88.5833% Re⁺³), Re₂O₇ (76.8804% Re⁺⁷)
104. CH₃CH₂OH and CH₃OCH₃ will produce the most water
106. 261 g NaCl
108. ZnSO₄ is 35.95% more economical
110. **(a)** 844 g **(b)** 4.12 × 10³ g **(c)** 3.84 × 10³ g
112. C₃H₆O₂, C₃H₆O₂
114. 1230 mL ethanol
116. **(a)** 2.16 g/mL **(b)** 4.930 g/mL **(c)** 13.59 g/mL **(d)** 2.165 g/mL

Chapter 3

12. **(b)** 450. molecules H_2 **(c)** 300. molecules NH_3
14. **(b)** 5.2 mol HCl **(c)** 2.6 mol H_2O
16. 882 g $NaHCO_3$ (to 3 sig. fig.)
18. **(a)** 9.6 mol O_2 **(b)** 3.2 mol O_2 **(c)** 3.2 mol O_2
    (d) 3.2 mol O_2 **(e)** 13 mol O_2
20. **(a)** 8.00 mol O_2 **(b)** 6.40 mol NO **(c)** 9.60 mol H_2O
22. 178 g O_2
24. 87.22 g Fe_3O_4
26. **(b)** 55.80 g NaI
28. 79.9 g C_3H_8
30. **(c)** 211.59 g CO_2
32. 326.5 g superphosphate
34. **(a)** S_8 **(b)** 67.4 g S_2Cl_2 **(c)** 35.6 g Cl_2
36. 18.0 g $Ca_3(PO_4)_2$
38. **(a)** $AgNO_3$ **(b)** 14.9 g $BaCl_2$ **(c)** 52.7 g AgCl
40. 107 g PCl_5
42. **(a)** 2.27 g O_2 **(b)** 46.3%
44. 95.3%
46. 74.5%
48. **(a)** 349.7 g Fe **(b)** 47.61%
50. 82.6 g H_2TeO_3
52. 803.9 g $KClO_3$
54. **(b)** 2 mol N atoms **(c)** 106.4 g NH_3
56. 252 kg Zn
58. 149 g $(NH_4)_2SO_4$
60. 631 mL $(NH_4)_2SO_4$ soln
62. 0.866 M Na_3PO_4
64. 5.08×10^3 mL
66. **(a)** 20.0% $CaCl_2$ **(b)** 2.13 M $CaCl_2$
68. 0.00940 M $BaCl_2$ soln
70. 28.7 M HF soln
72. 0.250 L conc. HCl soln
76. 231 mL NaOH soln
78. 0.0408 L KOH soln
80. 8.41 mL HNO_3 soln
82. 0.00453 M $AlCl_3$ soln
84. 3.89% Fe_3O_4 in ore
86. 1.42 g Br_2
92. 3 mol Be, 0.966 g Be
94. 1.0% $SnCl_2$ < 1.0% $AlCl_3$ < 1.0% NaCl in molarity
96. 68.0 mL HNO_3
100. **(a)** 62.7% yield
104. **(a)** CH_3CH_2OH **(b)** 81.6% yield
106. 0.522 M NaCl
108. 623 g H_3PO_4
110. 0.0867 g AgCl
112. for Zn: \$65.4/mol H_2; for Al: \$36.0/mol H_2

Chapter 4

6. The charge on each droplet is a multiple of 1.22×10^{-19} coulombs
10. 2.3×10^{-14}
14. **(a)** mass % for e^- = 0.026265% **(b)** mass % for p = 48.228% **(c)** mass % for n = 51.745%
26. ^{69}Ga: 60.13%; ^{71}Ga: 39.87%

28. 87.61 amu
30. 55.85 amu
32. 69.17% ^{63}Cu
34. O, 15.999 amu; Cl, 35.453 amu
36. 73.3 amu
38. 52.0 amu
54. **(a)** 3.34×10^{14} s^{-1} **(b)** 6.79×10^{14} s^{-1}
    (c) 6.10×10^9 s^{-1} **(d)** 6.59×10^{18} s^{-1}
56. **(a)** 4.47×10^{14} s^{-1} **(b)** 2.96×10^{-19} J/photon **(c)** red
58. 5.85×10^{-19} J/photon, 352 kJ/mol
60. 2.5×10^{13} miles
64. 320 nm (violet/near ultraviolet)
68. 218 kJ/mol
70. 3.20×10^{15} s^{-1}
74. 2.53×10^{18} photons
76. **(a)** 1.59×10^{-14} m **(b)** 3.97×10^{-34} m
78. 1.88×10^3 m/s
122. **(a)** $n=3, \ell=1, m=-1,0,+1$ **(b)** $n=6, \ell=2, m=-2, -1,0,+1,+2$ **(c)** $n=3, \ell=1, m=-1,0,+1$ **(d)** $n=4, \ell=3, m=-3, -2, -1,0,+1,+2,+3$
132. 14.8 miles
136. 6 electrons
144. 3.76×10^{-36} kg
146. 216 kJ/mol
148. 3.35 m
150. **(a)** 8.43×10^{24} electrons **(b)** 1.00×10^{25} electrons

Chapter 5

46. **(a)** oxidation no. of P: in PCl_3, +3; in P_2O_5, +5; in P_4O_{10}, +5; in HPO_3, +5; in H_3PO_3, +3; in $POCl_3$, +5; in $H_4P_2O_7$, +5; in $Mg_3(PO_4)_2$, +5 **(b)** oxidation no. of Br: in Br^-, −1; in BrO^-, +1; in BrO_2^-, +3; in BrO_3^-, +5; in BrO_4^-, +7 **(c)** oxidation no. of Mn: in MnO, +2; in MnO_2, +4; in $Mn(OH)_2$, +2; in K_2MnO_4, +6; in $KMnO_4$, +7; in Mn_2O_7, +7 **(d)** oxidation no. of O: in OF_2, +2; in Na_2O, −2; in Na_2O_2, −1; in KO_2, −1/2
48. **(a)** oxidation no. of N: in N^{3-}, −3; in NO_2^-, +3; in NO_3^-, +5; in N_3^-, −1/3; in NH_4^+, −3 **(b)** oxidation no. of Cl: in Cl_2, 0; in HCl, −1; in HClO, +1; in $HClO_2$, +3; in $KClO_3$, +5; in Cl_2O_7, +7; in $Ca(ClO_4)_2$, +7, in PCl_5, −1
80. 122 pm
90. 1.05×10^{15} s^{-1}
92. 158 kJ

Chapter 6

128. **(a)** 0.294 mol **(b)** 0.353 mol, 0.0554 mol

Chapters 7 and 8

No numerical exercises in these chapters.

Chapter 9

20. **(a)** Ne_2^+, bo = 0.5 **(b)** Ne_2, bo = 0 **(c)** Ne_2^{2+}, bo = 1
22. **(a)** Li_2, bo = 1 **(b)** Li_2^+, bo = 0.5 **(c)** O_2^{2-}, bo = 1
24. **(a)** X_2, bo = 3 **(b)** X_2, bo = 0 **(c)** X_2^-, bo = 2.5
28. **(b)** N_2, bo = 3, N_2^-, bo = 2.5, N_2^+, bo = 2.5
30. NO^+, bo = 3

32. CN, bo = 2.5; CN$^+$, bo = 2.0; CN^{2+}, bo = 1.5; CN$^-$, bo = 3; CN^{2-}, bo = 2.5

46. NO, bo = 2.5; NO$^+$, bo = 3; NO$^-$, bo = 2. Bond lengths are inversely related to bond order.

Chapter 10

No numerical problems in this chapter.

Chapter 11

4. 0.115 M MgSO$_4$

6. 5.292 M H$_2$SO$_4$

8. 1.16 M NaCl

10. 5.49 M KI

12. 0.0375 M BaI$_2$

14. 1.35 M Na$_3$PO$_4$, 0.137 M NaOH

16. 0.857 M CH$_3$COOH

18. **(a)** 0.923 L NaOH soln, 0.222 L H$_3$PO$_4$ soln **(b)** 0.615 L NaOH soln, 0.222 L H$_3$PO$_4$ soln

24. 31.3 mL of CH$_3$COOH soln

26. 0.1969 M HNO$_3$

28. 0.07321 M NaOH

30. 18.0 mL HCl soln

34. 36.0% (COOH)$_2$·2H$_2$O

36. 0.150 g CaCO$_3$

38. 0.0310 M NaOH

40. 0.0539 M H$_3$PO$_4$, 32.5 mL H$_3$PO$_4$

42. 0.13 M H$_3$AsO$_4$, 14 mL H$_3$AsO$_4$

44. 0.233 M HCl

46. 32.7% Mg(OH)$_2$

52. oxidizing agents: **(a)** MnO$_4^-$ **(b)** Cr$_2$O$_7^{2-}$ **(c)** MnO$_4^-$ **(d)** Cr$_2$O$_7^{2-}$

62. 3.22 mL KMnO$_4$ soln

64. 4.26 mL KMnO$_4$ soln

66. **(a)** 0.09247 M I$_2$ **(b)** 0.3230 g As$_2$O$_3$

68. 3.5 mL NO$_3^-$ soln

70. 0.123 M KMnO$_4$

72. **(a)** 1.80 × 10^{-3} M MgNH$_4$PO$_4$ **(b)** 0.683 M NaCH$_3$COO **(c)** 2.60 × 10^{-4} M CaC$_2$O$_4$ **(d)** 0.0416 M (NH$_4$)$_2$SO$_4$

74. 7.60 mmol HCl, 12.5 mL HCl

76. 463 mL HCl soln

78. 17.2 mL H$_2$SO$_4$ soln

80. 30.94 mL NaOH soln

82. **(a)** 10.2 mL HI soln **(b)** 12.1 mL HI soln **(c)** 31.8 mL HI soln

88. 59.36% Fe

90. 69 g AgCl

92. 80. mmol HCl in stomach; 9.28 mmol HCl

Chapter 12

6. **(a)** 14.4 psi **(b)** 74.2 cm Hg **(c)** 29.2 in Hg **(d)** 98.9 kPa **(e)** 0.976 atm **(f)** 33.1 ft H$_2$O

10. 2.20 × 10^3 psi

14. 1.06 atm

16. **(a)** 18.2 atm **(b)** 0.274 L

18. 6.0 × 10^2 balloons

26. 0.81 L

28. **(a)** 1.441 L **(b)** increase by 41 cm

30. −78.5°C, 3.26 L; −195.8°C, 1.30 L; −268.9°C, 0.0713 L

34. 610. K or 337°C

40. 5.23 g/L

42. **(a)** #1 – Kr, #2 – Ne

46. 5.20 atm

48. **(a)** 1.75 × 10^8 L Cl$_2$, 6.18 × 10^6 ft^3 Cl$_2$ **(b)** 39.0 ft

50. 0.88 g He

52. 22.0 L/mol for SF$_6$, 21.7 L/mol for HF

54. 29.6 g/mol, 2% error

56. 30.0 g/mol, C$_2$H$_6$(g)

58. 46.7 amu/molecule

62. P_{total} = 92.8 atm, P_{CHCl_3} = 24.4 atm

64. X_{He} = 0.440, X_{Ar} = 0.299, X_{Xe} = 0.261

66. **(a)** 11.25 atm **(b)** 3.75 atm **(c)** 3.75 atm

68. 285 mL

70. **(a)** He, 4.00 atm; N$_2$, 1.00 atm **(b)** 5.00 atm **(c)** X_{He}, 0.800

72. 62.0 g NaN$_3$

74. 7.15 g SO$_2$

76. 4.45 g KClO$_3$

78. 4.67 L NH$_3$

80. 190. g KNO$_3$

82. 25.1% S by mass

84. **(a)** 2.83 L C$_8$H$_{18}$ **(b)** 44.8 min

90. **(b)** 1.17

98. **(a)** 0.821 atm **(b)** 0.805 atm **(c)** 18.1 atm, 13.6 atm

100. **(a)** 7.34 atm **(b)** 7.14 atm

102. liquid Fe: 8.12 cm^3/mol, solid Fe: 7.11 cm^3/mol

104. 0.13 g H$_2$O

106. 5000 L air (1 sig. fig.)

108. 121 g/mol

112. **(a)** 6130 g H$_2$O **(b)** 6150 mL H$_2$O

114. 4.90 atm, 4.78 atm, 2.4%

116. C$_2$N$_2$ (MW = 52.1 g/mol)

118. 38°F, 22°C, 40K

120. 1.106 g

128. 2.1 × 10^{15} atoms Xe

132. H$_2$ is the limiting reactant

134. 1.3 L O$_2$

136. 206 g MgSiO$_3$

138. **(a)** 2.11 × 10^{-6} M SO$_2$ **(b)** 4.74 × 10^{-5}

Chapter 13

36. bp = ~34°C.

40. **(a)** 43.4 kJ/mol

42. bp = ~ −1°C

44. 2.7 × 10^4 J

48. boiling Br$_2$ (9660 J)

50. 2.48 × 10^5 J

52. 52.6°C

54. 58.7°C

56. **(a)** 6.14 × 10^3 J **(b)** 19.2 g H$_2$O

58. for H$_2$O, 361.8 torr; for D$_2$O, 338.0 torr

60. **(c)** ΔH$_{vap}$ = 3.72 × 10^4 J/mol **(d)** 3.48 × 10^2 K or 75°C

62. 4.34 × 10^4 J or 43.4 kJ

64. 2.93 × 10^{-3} torr

88. **(a)** a/2 **(b)** 6 Cl^- ions **(c)** $a(2^{1/2})/2$ **(d)** 6 Na^+ ions

90. 1.00 g/cm^3

92. 22.4 g/cm^3

94. **(a)** 8 **(b)** tetrahedron with 4 nearest neighbors
    (c) $\sqrt{3}(a/4)$ **(d)** 1.545 Å **(e)** 3.515 g/cm^3

96. AW = $(1.37 \times 10^{-21}\ g)/(6.64 \times 10^{-24}\ mol)$ =
    206 g/mol, Pb (207.2 g/mol)

100. 1.542 Å

112. **(a)** true **(b)** false **(c)** false

114. CS_2 < acetone < ethanol

130. 63 g/mol, Cu

136. deviation at 0°C = 0.9%, deviation at 100°C = 0.8%

Chapter 14

18. X_{CH_4} (at 25°C) = 2.4×10^{-4}; X_{CH_4} (at 50°C) =
    1.7×10^{-4}; decreases

26. 0.886 m $C_{10}H_{16}O$; $X_{C_{10}H_{16}O}$ = 0.0393; 11.9%

28. 1.7 m C_6H_5COOH in C_2H_5OH

30. $X_{C_2H_5OH}$ = 0.322; X_{H_2O} = 0.678

32. 0.7767 M K_2SO_4, 0.8197 m K_2SO_4, 12.50% K_2SO_4,
    X_{H_2O} = 0.9854

34. **(a)** 1.48 m $C_6H_{12}O_6$

38. **(a)** 14.12 torr **(b)** 16.72 torr **(c)** 19.00 torr

40. $P_{acetone}$ = 172 torr, $P_{chloroform}$ = 148 torr

42. P_{total} = 320. torr, $X_{acetone}$ = 0.538, $X_{chloroform}$ = 0.462

44. **(a)** $P_{chloroform}$ = 65 torr **(b)** $P_{acetone}$ = 215 torr
    (c) P_{total} = 280 torr

46. 101.52°C

48. −5.52°C

50. **(a)** camphor **(b)** water **(c)** camphor **(d)** nitrobenzene

52. −18.6°C

54. 78.54°C

56. 1035°C

58. 170 g/mol

60. $C_2H_8O_2$ (64.08 g/mol)

62. **(a)** 70.% $C_{10}H_8$, 30.% $C_{14}H_{10}$ **(b)** 80.4°C

66. **(a)** 3 **(b)** 2 **(c)** 5 **(d)** 2

68. $CaCl_2$ < $KClO_3$ < CH_3COOH < CH_3OH (based on i)

70. 752 torr

72. 0.100 m Na_2SO_4 has lower fp

76. 100.247°C

78. 1%

80. i = 1.79, 79%

84. 0.571 atm

86. ΔT_f = 0.100°C, ΔT_b = 0.0276°C

88. 58.0 atm

90. **(a)** -1.02×10^{-4} °C **(b)** 1.35×10^{-3} atm or 1.02 torr
    (c) 1000% error **(d)** 10% error

98. bp = 100.44°C; π = 20.5 atm

100. **(a)** 60.5 g/mol **(b)** 117 g/mol

104. 37% lactose by mass

108. For water: 215 mL, 215 g, 11.9 mol; for ethanol:
     285 mL, 225 g, 4.88 mol; for glycerol: 630. g, 6.84 mol;
     X_{water} = 0.504, $X_{ethanol}$ = 0.207, $X_{glycerol}$ = 0.290

118. 1.0 m, 80% ionization; 2.0 m, 65% ionization; 4.0 m,
     17% ionization

126. 1.3 atm, 0.055 M

Chapter 15

16. **(a)** 2750 kJ **(b)** 58.1 g O_2

18. +98.9 kJ/mol rxn

30. −584 kJ/mol

32. −1015.4 kJ/mol rxn

34. +197.6 kJ/mol rxn

36. −289 kJ/mol rxn

38. **(a)** −124.0 kJ/mol rxn **(b)** −1656 kJ/mol rxn
    (c) +624.6 kJ/mol rxn

40. +46.36 kJ/g C_3H_8; +44.27 kJ/g C_8H_{18}

42. +333 kJ heat released

44. sucrose: −5430 kJ/mol, −15.9 kJ/g, −3.79 kcal/g;
    tristearin: −2900 kJ/mol, −3.3 kJ/g, −0.78 kcal/g

48. **(a)** −121 kJ/mol rxn **(b)** −103 kJ/mol rxn

50. −379 kJ/mol rxn

52. −93 kJ/mol HCl; −270 kJ/mol HF

54. 329 kJ/mol

56. 264 kJ/mol

58. 288 kJ/mol

60. 381 J/°C

62. 0.66 J/g·°C

64. **(a)** $+2.1 \times 10^3$ J **(b)** -1.0×10^2 kJ/mol rxn

66. −41.9 kJ/g $C_6H_6(l)$; −3270 kJ/mol $C_6H_6(l)$

68. −23.1 kJ; −460. kJ/mol rxn

70. -2.08×10^3 J/g $C_{10}H_{22}(l)$; −297 kJ/mol $C_{10}H_{22}(l)$

76. −336 J

78. **(a)** by surroundings **(b)** by system **(c)** by surroundings

80. **(a)** +983 J **(b)** 0 J

86. **(a)** ½ **(b)** ¼ **(c)** 1/1024

88. **(a)** 16 **(b)** 14 of 16 **(c)** 7/8 **(d)** 1/8

104. **(a)** −128.8 J/(mol rxn)·K **(b)** −182 J/(mol rxn)·K
     (c) +148.6 J/(mol rxn)·K

108. −432.8 J/(mol rxn)·K

110. −53.42 kJ/mol rxn

112. **(a)** ΔH^0 = −71.75 kJ/mol rxn, ΔS^0 = −268.0 J/
     (mol rxn)·K, ΔG^0 = +8.15 kJ/mol rxn
     (b) ΔH^0 = +14.7 kJ/mol rxn, ΔS^0 = +31.2 J/
     (mol rxn)·K, ΔG^0 = +5.4 kJ/mol rxn
     (c) ΔH^0 = −367.6 kJ/mol rxn, ΔS^0 = −11.62 J/
     (mol rxn)·K, ΔG^0 = −364.1 kJ/mol rxn

116. **(a)** ΔH^0 = −196.0 kJ/mol rxn, ΔG^0 = −233.6 kJ/
     mol rxn, ΔS^0 = +125.6 J/(mol rxn)·K

120. **(a)** spontaneous at all temperatures **(b)** spontaneous at
     T > 839.3 K **(c)** spontaneous at T > 980.7 K
     (d) spontaneous at T > 3000. K

122. **(a)** 97°C

126. ΔS^0 = −247 J/(mol rxn)·K, ΔH^0 = −88.2 kJ/mol rxn,
     product favored at T < 357 K

128. **(a)** −156 kJ/mol $C_6H_{12}(l)$ **(b)** −165 kJ/mol $C_6H_5OH(s)$

130. q = +10700 J; w = −788 J; ΔE = 9900 J

132. **(a)** 2.88 kJ/°C **(b)** 23.82°C

136. **(a)** sp. ht. = 0.13 J/g°C, tungsten, W (0.135 J/g°C)
     (b) yes, sp. ht = 0.26 J/g°C, molybdenum, Mo

142. 2.4 hr walking

144. for Pb: sp. ht = 0.13 J/g°C; molar heat capacity,
     27 J/mol°C

146. **(a)** 30.0 kJ/g **(b)** 7.17 kcal/g **(c)** 35.9 kcal

Chapter 16

10. O_2: 1.50 M/min, NO: 1.20 M/min, H_2O: 1.80 M/min
14. 1/2
18. Rate = $(2.5 \times 10^{-2}\ M^{-1}\cdot min^{-1})[B][C]$
22. Rate = $(1.2 \times 10^2\ M^{-2}\cdot s^{-1})[ClO_2]^2[OH^-]$
24. Rate = $(2.5 \times 10^{-3}\ M^{-2}\cdot s^{-1})[A]^2[B]$
26. Rate = $(20.\ M^{-4}\cdot s^{-1})[A]^3[B]^2$
28. 0.300 M/s, 1.20 M/s
32. 5.29 s
34. (a) 2.5×10^6 s (b) 41 days (c) 0.724 g Cs (d) 0.67 g
36. (a) 2.4×10^9 s or 76 yrs (b) 0.59 M NO_2, 54 g NO_2
    (c) 0.91 M NO reacted, 0.91 M NO produced
38. 1680 s or 28.0 min
40. k_2/k_1 = 1.6 for 90°C to 100°C, k_2/k_1 = 2.2 for 0°C to
    10°C
42. (a) zero order rxn, rate = k = 0.00273 $mM\cdot s^{-1}$,
    $[HI]$ = $[HI]_o - akt$ = 5.46 mM - (2)(0.00273 $mM\cdot s^{-1}$)t
    (b) 2.18 mmol/L
44. 8.4 s
46. 462 d
48. (a) 4.52×10^{-29} s^{-1} (b) 4.04×10^{-10} s^{-1}
54. 340 kJ/mol rxn
56. 84 kJ/mol rxn
58. 103 kJ/mol rxn
60. (a) 270. kJ/mol (b) 7.7×10^{-14} s^{-1} (c) 730. K
62. 2600 g CO_2/L·min
64. (a) 0.24 s^{-1} (b) 9°C
66. (a) yes (b) no (c) no
68. (a) Rate = $k[A]^2[B]^2$ (b) Rate = $k[A][B][D]$
70. Rate = $k[O_3]^2/[O_2]$
72. (a) N_2Cl (b) yes
74. (b) yes
78. (a) 8.7×10^{-13} mol N_2O_5 (b) 19 s
80. (a) Rate = $k[Hb][CO]$ (b) 0.280 L/μmol·s
    (c) 0.252 μmol/L·s
94. (1) 21 min, 42 min, 63 min (2) 21 min, 63 min, 150 min
    (3) same as (2)
96. 64 kJ/mol
100. −200. kJ/mol rxn

Chapter 17

24. K_c = 1.1×10^{-5}
26. K_c = 0.13
28. K_c = 0.12
30. (a) K_c' = 8.9×10^5 (b) K_c'' = 1.3×10^{-12}
    (c) K_c''' = 1.6×10^{-24}
32. K_c = 75
34. (a) K_c = 4.49×10^{-2} (b) $[SbCl_5]$ = 3.7×10^{-3} M,
    $[SbCl_3]$ = $[Cl_2]$ = 1.30×10^{-2} M
36. K_c' = 6.502×10^{-4}
42. (a) false (b) false (c) true (d) false
44. 0.044 M
46. 5.3 M
48. 0.079 M
50. 1 g (1 sig. fig.)
52. $[HI]$ = 0.140 M, $[H_2]$ = $[I_2]$ = 0.127 M
64. (a) K_c = 16 (b) 0.35 M

66. (a) K_c = 0.21 (b) $[A]$ = 0.12 M, $[B]$ = $[C]$ = 0.15 M
    (c) $[A]$ = 0.71 M, $[B]$ = $[C]$ = 0.39 M
68. (a) $[N_2O_4]$ = 0.0167 M, $[NO_2]$ = 9.90×10^{-3} M
    (b) $[N_2O_4]$ = 1.05×10^{-2} M, $[NO_2]$ = 7.82×10^{-3} M
    (c) $[N_2O_4]$ = 0.0360 M, $[NO_2]$ = 0.0145 M
72. K_p = 1.6×10^{-9}
74. $P_{N_2O_4}$ = 16 atm, P_{NO_2} = 3.3 atm
76. K_p = 0.771
78. K_c = 7.76
80. (a) 33.2%
84. (a) K_p = 1.1×10^5 (b) K_p = 2.3×10^2
    (c) K_p = 1.0×10^5
86. K_p = 7.0×10^{24}
88. (a) −37.9 kJ/mol rxn at 25°C (b) $K_{p\ 1073\ K}$ = 4.9×10^{-4}
    (c) ΔG° = +68.0 kJ/mol rxn at 800°C
92. P_{CO} = 0.685 atm, P_{CO_2} = 0.315 atm
96. At 400°C: K_p = 8.3×10^{-3}, K_c = 1.5×10^{-4}; at 800°C:
    K_p = 16, K_c = 0.18
98. K_p = 0.00432
112. P_{H_2O} = 0.39

Chapter 18

4. (a) 1.76 M NaCl (b) 0.770 M H_2SO_4 (c) 1.34×10^{-3}
   M C_6H_5OH
6. (a) $[H^+]$ = $[Br^-]$ = 0.45 M (b) $[K^+]$ = $[OH^-]$ = 0.045
   M (c) $[Ca^{2+}]$ = 0.0112 M, $[Cl^-]$ = 0.0224 M
8. (a) $[K^+]$ = $[OH^-]$ = 0.0149 M (b) $[Ba^{2+}]$ = 0.00585 M,
   $[OH^-]$ = 0.0117 M (c) $[Ca^{2+}]$ = 0.0768 M, $[NO_3^-]$ =
   0.154 M
14. 2.2×10^{-13} M; 4.76×10^{-13} M; 8.55×10^{-13} M;
    1.00×10^{-7} M
16. 55.6 M H_2O; 3.0×10^{-23} H_3O^+ ions
18. (a) 0.00, 14.00, acidic (b) 3.77, 10.23, acidic (c) 7.17,
    6.83, basic (d) 10.03, 3.97, basic
20. 3.5×10^{-8} M, 2.9×10^{-7} M at 25°C, 6.9×10^{-7} M at
    37°C
22. (a) 0.70 (b) 1.30 (c) 2.19 (d) 10.99
24. 3.0×10^{-4} M
26. (a) pH = 12.93 (b) pH = 1.12 (c) pH = 13.18
32. triethylamine: $[OH^-]$ = 2.8×10^{-3} M;
    trimethylamine: $[OH^-]$ = 1.1×10^{-3} M
34. (a) pH = 6.82 (b) neutral
38. pH = 3.096, K_a = 8.67×10^{-6}
40. K_a = 1.4×10^{-3}
42. $[C_6H_5COOH]$ = 0.51 M, $[H_3O^+]$ = $[C_6H_5COO^-]$ =
    5.7×10^{-3} M, $[OH^-]$ = 1.8×10^{-12} M
44. pH = 1.78
46. pH = 11.79
48. 4.9% ionized
50. For weak acid #1: pK_a = 4.14; for weak acid #2:
    pK_a = 9.38
52. pK_b = 8.82
54. K_b = 5.0×10^{-4}
56. (a) $[OH^-]$ = 2.1×10^{-3} M, 0.85% ionized,
    pH = 11.32 (b) $[OH^-]$ = 0.011 M, 4.4% ionized,
    pH = 12.04

58.

0.100 M H$_3$AsO$_4$ Solution		0.100 M H$_3$PO$_4$ Solution	
Species	Concentration (m)	Species	Concentration (m)
H$_3$AsO$_4$	0.095	H$_3$PO$_4$	0.076
H$_3$O$^+$	0.0050	H$_3$O$^+$	0.024
H$_2$AsO$_4^-$	0.0050	H$_2$PO$_4^-$	0.024
HAsO$_4^{2-}$	5.6×10^{-8}	HPO$_4^{2-}$	6.2×10^{-8}
OH$^-$	2.0×10^{-12}	OH$^-$	4.2×10^{-13}
AsO$_4^{3-}$	3.4×10^{-18}	PO$_4^{3-}$	9.3×10^{-19}

60. [H$_3$O$^+$] = 0.16 M, [OH$^-$] = 6.2×10^{-14} M, [HSeO$_4^-$] = 0.14 M, [SeO$_4^{2-}$] = 0.01 M

62. pH = 1.52

74. $K = 5.3 \times 10^{-10}$

76. **(a)** $K_b = 2.2 \times 10^{-11}$ **(b)** $K_b = 4.0 \times 10^{-6}$ **(c)** $K_b = 5.6 \times 10^{-11}$

78. **(a)** pH = 9.46 **(b)** pH = 11.39 **(c)** pH = 11.79

80. **(a)** pH = 11.94 **(b)** pH = 8.20

84. **(a)** $K_a = 1.4 \times 10^{-11}$ **(b)** $K_a = 6.7 \times 10^{-6}$ **(c)** $K_a = 1.4 \times 10^{-10}$

86. **(a)** pH = 4.92 **(b)** pH = 5.64 **(c)** pH = 2.60

92. **(a)** pH = 2.89, 0.87% hydrolysis **(b)** pH = 5.21, 8.2×10^{-3}% hydrolysis **(c)** pH = 6.17, 4.5×10^{-4}% hydrolysis

94. [A$^-$] = 6.7×10^{-4} M

114. [H$_3$O$^+$] = 2.9×10^{-3} M, pH = 2.54

116. pH = 2.04

118. +57 kJ/mol

Chapter 19

10. **(a)** pH = 3.27 **(b)** pH = 4.84

12. **(a)** [OH$^-$] = 3.0×10^{-5} M, pH = 9.48 **(b)** [OH$^-$] = 6.8×10^{-6} M, pH = 8.83

18. **(a)** [OH$^-$] = 2.3×10^{-3} M, [H$_3$O$^+$] = 4.3×10^{-12} M **(b)** [OH$^-$] = 1.4×10^{-5} M, [H$_3$O$^+$] = 7.4×10^{-10} M

22. pH = 0.70

24. pH decreases by 0.40 units (9.48 → 9.08)

26. **(a)** pH = 9.31 **(b)** pH = 9.20 **(c)** pH = 1.00

28. [BrCH$_2$COOH] = 4.1×10^{-2} M, [NaBrCH$_2$COO] = 0.26 M

30. [C$_2$H$_5$NH$_3^+$] = 0.0010 M

32. [CH$_3$CH$_2$COOH] = 0.36 M

34. 0.40 L NaOH, 0.60 L CH$_3$COOH

36. **(a)** 0.625 M CH$_3$COOH **(b)** 0.150 M Ca^{2+} **(c)** 0.300 M CH$_3$COO$^-$ **(d)** 3.8×10^{-5} M H$^+$ **(d)** pH = 4.42

38. 0.024 M ClCH$_2$COOH

42. $K_a = 6 \times 10^{-9}$, pK_a = 8.2

44. pH between 7.6 and 8

46. pH = 8.40, phenolphthalein

50. **(a)** pH = 0.8239 **(b)** pH = 0.9931 **(c)** pH = 1.250 **(d)** pH = 1.87 **(e)** pH = 4.0 **(f)** pH = 12.03

52. **(a)** pH = 0.839 **(b)** pH = 1.217 **(c)** pH = 1.85 **(d)** pH = 12.265 **(e)** pH = 12.529 **(f)** pH = 12.618

54. **(a)** pH = 3.22 **(b)** pH = 4.14 **(c)** pH = 4.57 **(d)** pH = 4.74 (halfway to the equivalence pt.) **(e)** pH = 5.11 **(f)** pH = 5.70 **(g)** pH = 6.02 **(h)** pH = 8.52 (at the equivalence pt.) **(i)** pH = 11.00 **(j)** pH = 11.60 **(k)** pH = 12.00

56. **(a)** pH = 2.74 **(b)** pH = 4.72 **(c)** pH = 4.97 **(d)** pH = 5.23 **(e)** pH = 5.51 **(f)** pH = 8.86 **(g)** pH = 11.54

58. **(a)** pH = 8.90 **(b)** pH = 8.79 **(c)** pH = 8.69

60. **(b)** 8.09 mL HCl **(c)** [(CH$_3$CH$_2$)$_3$N] = [H$_3$O$^+$] = 1.7×10^{-6} M, [(CH$_3$CH$_2$)$_3$NH$^+$] = [Cl$^-$] = 0.157 M

66. **(a)** pH = 2.00 **(b)** pH = 2.05 **(c)** pH = 2.22 **(d)** pH = 2.30 **(e)** pH = 3.00 **(f)** pH = 3.3 **(g)** pH = 5.62 **(h)** pH = 8.0 **(i)** pH = 8.73

68. pH (NaCl) = 7.00, pH (NH$_4$Cl) = 4.76

70. pH = 12.2

74. **(a)** K_a for NH$_4^+$ = 5.6×10^{-10} **(b)** K_b for ClO$^-$ = 2.9×10^{-7} **(c)** basic

80. 78.1% KHSO$_4$ by mass

82. pH = 4.00

84. 0.22 L HCl

Chapter 20

8. **(a)** $K_{sp} = 3.5 \times 10^{-5}$ **(b)** $K_{sp} = 7.7 \times 10^{-19}$ **(c)** $K_{sp} = 6.9 \times 10^{-15}$ **(d)** $K_{sp} = 1.01 \times 10^{-4}$

12. $K_{sp} = 3.65 \times 10^{-11}$

14. 6.9×10^{-3} g CaCO$_3$/L

16. **(a)** 4.4×10^{-4} mol CuCl/L; 4.4×10^{-4} M Cu$^+$; 4.4×10^{-4} M Cl$^-$; 0.043 g CuCl/L **(b)** 6.5×10^{-7} mol Ba$_3$(PO$_4$)$_2$/L; 2.0×10^{-6} M Ba^{2+}; 1.3×10^{-6} M PO$_4^{3-}$; 3.9×10^{-4} g Ba$_3$(PO$_4$)$_2$/L **(c)** 2.1×10^{-3} mol PbF$_2$/L; 2.1×10^{-3} M Pb^{2+}; 4.2×10^{-3} M F$^-$; 0.51 g PbF$_2$/L **(d)** 2.5×10^{-7} mol Sr$_3$(PO$_4$)$_2$/L; 7.4×10^{-7} M Sr^{2+}; 4.9×10^{-7} M PO$_4^{3-}$; 1.1×10^{-4} g Sr$_3$(PO$_4$)$_2$/L

18. 0.012 g BaSO$_4$

20. 6.0×10^{-3} mol Ag$_2$SO$_4$/L of 0.12 M K$_2$SO$_4$

28. precipitation does not occur

30. precipitation does occur

32. 0.013% Pb^{2+} in soln

36. **(a)** AuBr **(b)** 1.5×10^{-6} M Au$^+$; 99.985% Au$^+$ precipitated **(c)** 9.4×10^{-11} M Au$^+$; 6.2×10^{-7} M Ag$^+$

38. **(a)** PbCrO$_4$ **(b)** 3.6×10^{-13} M Pb^{2+} **(c)** 3.6×10^{-7} M Pb^{2+} **(d)** 0.050 M SO$_4^{2-}$, 5.0×10^{-8} M CrO$_4^{2-}$

40. **(a)** (i) 2.0×10^{-11} M Zn^{2+} (ii) 1.0×10^{-8} M Zn^{2+} (iii) 3.6×10^{-6} M Zn^{2+} **(b)** (i) 1.7×10^{-7} M OH$^-$ (ii) 1.0×10^{-8} M CO$_3^{2-}$ (iii) 7.3×10^{-5} M CN$^-$

42. pH = 7.59

44. 2.5×10^{-7} mol CaF$_2$/L

46. a precipitate will form but not be seen

48. **(a)** 6.7×10^{-8} mol Mg(OH)$_2$/L of 0.015 M NaOH **(b)** 1.6×10^{-5} mol Mg(OH)$_2$/L of 0.015 M MgCl$_2$

50. precipitation will not occur

52. **(a)** pH = 9.40 **(b)** 1.1×10^{-4} g Fe(OH)$_2$/100 mL

54. 4.5×10^{-9} M Zn^{2+}

56. a precipitate will form but not be seen

64. precipitation will not occur

66. 99.955% I$^-$ removed

74. 6.0×10^{-6} M Cl$^-$

78. **(a)** 1.1×10^{-9} M **(b)** 0.017 M

80. 0.6 g Ca^{2+}/L

82. 29% loss of MgCO$_3$

Chapter 21

18. (i) (a) 3 (b) 2 (c) 1
   (ii) (a) 5.18×10^3 C (b) 1.63×10^3 C (c) 481 C
20. (a) 0.142 A (b) 1.12 g Cu
22. 0.242 g Rh
24. 0.873 C
26. 8.39 g Ag
28. (a) 0.0830 faradays (b) 8.01×10^3 C (c) 0.930 L_{STP} H_2
   (d) pH = 13.220
30. 3.2 L Cl_2
32. 5.2 hr, 4.6 g Cu
34. (a) 0.0214 faradays (b) 2.31 g Ag (c) 5.18 g $Fe(NO_3)_3$
54. (a) +3.17 V (b) +0.62 V
56. 0.71 V
58. (a) yes (b) no
60. (a) yes (b) no
62. +1.78 V
64. (a) +1.08 V (b) −1.202 V
66. (a) no (b) yes (c) yes (d) no
68. (a) H_2 (b) Sn (c) Hg (d) Cl^- in base (e) H_2S (f) Ag
70. −1.207 V
72. (a) yes (b) +0.57 V
74. +5.80 V (K^+/K with F_2/F^-)
78. 0.0257
82. (a) 1.202 V (b) +1.16 V (c) 2.0 M
84. (a) +2.123 V (b) +2.143 V
86. $P_{F_2} = 2.2 \times 10^{-7}$ atm
88. +0.95 V
90. (a) +0.870 V (b) +0.148 V (c) +0.0527 V
92. +0.38 V
94. (a) $[Zn^{2+}]/[Ni^{2+}] = 2 \times 10^{17}$ (b) 1×10^{-17} M Ni^{2+};
   2.00 M Zn^{2+}
96. pH = 6.15
100. (a) $E^0_{cell} = +0.736$ V; $\Delta G^0 = -355$ kJ/mol rxn;
   $K = 1 \times 10^{62}$
   (b) $E^0_{cell} = +0.368$ V; $\Delta G^0 = -35.5$ kJ/mol rxn;
   $K = 1.6 \times 10^6$
   (c) $E^0_{cell} = +1.833$ V; $\Delta G^0 = -1061$ kJ/mol rxn;
   $K = 8.6 \times 10^{185}$
102. (a) −0.25 V (b) −0.13 V (c) −0.065 B
104. $K = 2$ (1 sig. fig.)
116. (b) +2.33 V (c) +2.44 V (d) 0.0227 g Mg
118. 955 C
120. $[Mn^{2+}]/[Fe^{2+}] = 10^{-24}$
122. (a) +4 (b) +4 (c) 0 (d) Mg(s) (e) $UF_4(s)$ (f) 10.2 A
   (g) 0.0206 L HF(g) (h) yes, 2.45 g U
134. (a) $K_{sp} = 10^{-12}$ (b) $\Delta G^o = +68$ kJ/mol rxn

Chapter 22

14. (a) 0.587 g/mol (b) 5.28×10^{10} kJ/mol of ^{62}Ni atoms
16. (a) 0.602 amu/atom (b) 0.602 g/mol (c) 9.00×10^{-11}
   J/atom (d) 5.42×10^{10} kJ/mol (e) 8.79 MeV/nucleon
18. (a) 1.04×10^{11} kJ/mol of ^{127}I atoms (b) 6.83×10^{10}
   kJ/mol of ^{81}Br atoms (c) 2.89×10^{10} kJ/mol of
   ^{35}Cl atoms
54. 67.5 min (90.0%), 88.0 min (95.0%)
56. 7×10^{-16} s
58. 0.17 μg
60. 0.00236

62. 1.98×10^4 yr
66. (b) $\Delta E = -1.68 \times 10^9$ kJ/mol rxn
74. (b) $\Delta E = +1.15 \times 10^8$ kJ/mol rxn
76. 2.52×10^9 yr
78. fission = -1.40×10^{-13} J/amu ^{235}U; fusion =
   -1.53×10^{-13} J/amu ^2H

Chapter 23

34. 2 isomers

Chapter 24

20. $[C_6H_5NH_2] = 0.12$ M; $[C_6H_5NH_3^+] = [OH^-] =$
   7.1×10^{-6} M; $[H_3O^+] = 1.4 \times 10^{-9}$ M
28. pH = 10.78
48. 24 isomers
52. 9; A-A, A-B, A-C, B-A, B-B, B-C, C-A, C-B, C-C
72. pH (sodium benzoate) = 8.64; more acidic

Chapter 25

4. (a) 4 (b) 4
6. (a) ox. no. +3 (b) +3 (c) +2 (d) +2
20. (a) ox. no. +2 (b) +3 (c) +1 (d) +2 (e) +2 (f) +3
32. (a) 2 isomers (b) 2 isomers
36. (a) 2 isomers (b) 2 isomers (c) 2 optical isomers
   (d) 3 isomers (e) 6 isomers
52. 8
58. $\Delta S^0_{rxn} = -153.5$ J/(mol rxn)·K
60. pH = 10.43
64. (a) 2.2×10^{-6} mol $Zn(OH)_2/L$ (b) 0.010 mol
   $Zn(OH)_2/L$ (c) $[Zn(OH)_4^{2-}] = 0.010$ M

Chapter 26

16. 39.997 g NaOH, 1.008 g H_2, 35.45 g Cl_2
40. 1.7×10^7 tons bauxite
46. (a) −15 kJ/mol rxn (b) +5 kJ/mol rxn (c) −25 J/
   (mol rxn)·K
50. 21.0 g Cu
52. bornite (63.33% Cu)
54. 104 tons SO_2
56. 33.4 tons C (coke)

Chapter 27

32. (a) −201.4 kJ/mol rxn (b) −138.9 kJ/mol rxn
   (c) −415.0 kJ/mol rxn
46. $Be(OH)_2$: $[OH^-] = 1.6 \times 10^{-7}$ M, pOH = 6.80,
   $Mg(OH)_2$: $[OH^-] = 3.1 \times 10^{-4}$ M, pOH = 3.51,
   $Ca(OH)_2$: $[OH^-] = 0.025$ M, pOH = 1.60, $Sr(OH)_2$:
   $[OH^-] = 0.086$ M, pOH = 1.06, $Ba(OH)_2$: $[OH^-] =$
   0.22 M; pOH = 0.67
48. 22.61% Cr
56. 384 g Co_3O_4
58. $\Delta H^0_{rxn} = -195.4$ kJ/mol Rb(s), $\Delta S^0_{rxn} = 29.4$ J/K per
   1 mol Rb(s), $\Delta G^0_{rxn} = -204.0$ kJ/mol Rb(s)
60. 0.200 mol CO_2, Q = Li in Li_2CO_3

Chapter 28

8. 2.19 g XeF_6
42. 2.04 tons H_2SO_4
44. $K_c = 2.7$
52. (a) 0 (b) +1 (c) +4 (d) +5 (e) +3
82. −140 kJ/mol rxn
86. 2.2×10^{25} g Si
88. 1.92 Å

Index of Equations

Bold entries in parentheses are chapter and section numbers, followed by the page number.

Arrhenius equation (single temperature)

$$k = Ae^{-E_a/RT}$$

(16-8), 644

Arrhenius equation (two temperatures)

$$\ln \frac{k_2}{k_1} = \frac{E_a}{R}\left(\frac{1}{T_1} - \frac{1}{T_2}\right)$$

(16-8), 645

Balmer–Rydberg equation

$$\frac{1}{\lambda} = R\left(\frac{1}{n_1^2} - \frac{1}{n_2^2}\right)$$

(4-13), 139

Bohr energy

$$E = -\frac{1}{n^2}\left(\frac{h^2}{8\pi^2 m a_0^2}\right) = -\frac{2.180 \times 10^{-18}\,\text{J}}{n^2}$$

(4-13), 142

Bohr radius

$$r = n^2 a_0 = n^2 \times 5.292 \times 10^{-11}\,\text{m}$$

(4-13), 142

Boiling point elevation

$$\Delta T_b = K_b m$$

(14-11), 524

Bond order

$$\text{bond order} = \frac{\left(\begin{array}{c}\text{number of bonding}\\\text{electrons}\end{array}\right) - \left(\begin{array}{c}\text{number of antibonding}\\\text{electrons}\end{array}\right)}{2}$$

(9-3), 335

Boyle's Law

$$PV = \text{constant} \qquad \text{(constant } n, T)$$ **(12-4)**, 406
$$P_1 V_1 = P_2 V_2 \qquad \text{(constant } n, T)$$ **(12-4)**, 408

Bragg equation (x-ray diffraction)

$$n\lambda = 2d\sin\theta$$

(13-14), 476

Buffers

$$\text{pH} = \text{p}K_a + \log\frac{[\text{conjugate base}]}{[\text{acid}]}$$ (for acid/salt buffer)

(19-1), 753

and

$$\text{pOH} = \text{p}K_b + \log\frac{[\text{conjugate acid}]}{[\text{base}]}$$ (for base/salt buffer)

(19-1), 756

Cell voltage, nonstandard conditions (Nernst equation)

$$E = E^0 - \frac{2.303\,RT}{nF}\log Q$$

(21-19), 828

or

$$E = E^0 - \frac{0.0592}{n}\log Q$$

Charles's Law

$$V = kT \qquad \text{(constant } n, P)$$ **(12-5)**, 410
$$\frac{V_1}{T_1} = \frac{V_2}{T_2} \qquad \text{(constant } n, P)$$ **(12-5)**, 411

Clausius–Clapeyron equation

$$\ln\left(\frac{P_2}{P_1}\right) = \frac{\Delta H_{\text{vap}}}{R}\left(\frac{1}{T_1} - \frac{1}{T_2}\right)$$

(13-9), 467

Combined Gas Law

$$\frac{P_1 V_1}{T_1} = \frac{P_2 V_2}{T_2} \qquad \text{(constant } n)$$

(12-7), 412

Concentrations:

Percent solute

$$\text{percent solute} = \frac{\text{mass of solute}}{\text{mass of solution}} \times 100\%$$ **(3-6)**, 98

Molarity, M

$$\text{molarity} = \frac{\text{number of moles of solute}}{\text{number of liters of solution}} \qquad \textbf{(3-6)}, 99$$

$$\text{molarity} = \frac{\text{number of millimoles of solute}}{\text{number of milliliters of solution}} \qquad \textbf{(11-1)}, 377$$

Molarity, dilution

$$V_1 M_1 = V_2 M_2 \qquad \text{(for dilution only)} \qquad \textbf{(3-7)}, 102$$

Molality, m

$$\text{molality} = \frac{\text{number of moles solute}}{\text{number of kilograms solvent}} \qquad \textbf{(14-8)}, 516$$

Mole fraction, X

$$X_A = \frac{\text{no. mol A}}{\text{total no. mol of all components}} \qquad \textbf{(12-11)}, 422$$

Coulomb's law (electrostatic attraction)

$$F \propto \frac{q^+ q^-}{d^2} \qquad \textbf{(5-3)}, 180$$
$$\textbf{(13-2)}, 453$$

Dalton's Law of partial pressures

$$P_{\text{total}} = P_A + P_B + P_C + \cdots \qquad \text{(constant } V, T)$$
$$\textbf{(12-11)}, 421$$

de Broglie equation

$$\lambda = \frac{h}{mv} \qquad \textbf{(4-14)}, 144$$

Density, D

$$\text{density} = \frac{\text{mass}}{\text{volume}} \quad \text{or} \quad D = \frac{m}{V} \qquad \textbf{(1-11)}, 26$$

Dilution

$$V_1 M_1 = V_2 M_2 \qquad \text{(for dilution only)} \qquad \textbf{(3-7)}, 102$$

Electrochemical potential vs. equilibrium constant

$$nFE^0_{\text{cell}} = RT \ln K \qquad \textbf{(21-21)}, 835$$

Electrochemical potential vs. standard free energy change

$$\Delta G^0 = -nFE^0_{\text{cell}} \qquad \textbf{(21-21)}, 835$$

Energy:

Matter−energy conversion
$$E = mc^2 \qquad \textbf{(1-1)}, 5$$

Photon energy
$$E = h\nu \quad \text{or} \quad E = \frac{hc}{\lambda} \qquad \textbf{(4-11)}, 137$$

Kinetic energy
$$E_{\text{kinetic}} = \tfrac{1}{2} mv^2 \qquad \textbf{(15-1)}, 553$$

Enthalpy change, ΔH

$$\Delta H = H_{\text{final}} - H_{\text{initial}} \qquad \text{or}$$
$$\Delta H = H_{\text{substances produced}} - H_{\text{substances consumed}} \qquad \textbf{(15-3)}, 556$$
$$\Delta H = \Delta E + P\,\Delta V \qquad \text{(constant } T \text{ and } P) \qquad \textbf{(15-11)}, 577$$
$$\Delta H = q_p \qquad \text{(constant } T \text{ and } P) \qquad \textbf{(15-11)}, 577$$

Enthalpy of reaction, ΔH^0_{rxn} (gas-phase reaction, estimation from bond energies)

$$\Delta H^0_{\text{rxn}} = \Sigma \text{B.E.}_{\text{reactants}} - \Sigma \text{B.E.}_{\text{products}} \qquad \textbf{(15-9)}, 569$$

Enthalpy of reaction, ΔH^0_{rxn} (Hess's Law), from enthalpies of formation

$$\Delta H^0_{\text{rxn}} = \Sigma n\, \Delta H^0_{\text{f products}} - \Sigma n\, \Delta H^0_{\text{f reactants}} \qquad \textbf{(15-8)}, 566$$

Enthalpy of reaction, ΔH^0_{rxn} (Hess's Law), from related reactions

$$\Delta H^0_{\text{rxn}} = \Delta H^0_a + \Delta H^0_b + \Delta H^0_c + \cdots \qquad \textbf{(15-8)}, 564$$

Enthalpy of solution, $\Delta H_{\text{solution}}$

$$\Delta H_{\text{solution}} = (\text{heat of solvation}) - (\text{crystal lattice energy}) \qquad \textbf{(14-2)}, 509$$

Entropy of reaction, ΔS^0_{rxn}

$$\Delta S^0_{\text{rxn}} = \Sigma n\, S^0_{\text{products}} - \Sigma n\, S^0_{\text{reactants}} \qquad \textbf{(15-14)}, 583$$

Equilibrium constant—Gibbs free energy relationship

$$\Delta G^0_{\text{rxn}} = -RT \ln K \qquad \textbf{(17-12)}, 696$$

Equilibrium constant in terms of concentrations, K_c

$$\textbf{(17-2)}, 670$$

$$K_c = \frac{\overbrace{[C]^c_{\text{eq}} [D]^d_{\text{eq}}}^{\text{product concentrations}}}{\underbrace{[A]^a_{\text{eq}} [B]^b_{\text{eq}}}_{\text{reactant concentrations}}}$$

Equilibrium constant in terms of pressures, K_P

$$K_P = \frac{(P_C)^c (P_D)^d}{(P_A)^a (P_B)^b} \qquad \textbf{(17-9)}, 691$$

Equilibrium constant, K_a, for weak acid HA

$$K_a = \frac{[H_3O^+][A^-]}{[HA]} \qquad \textbf{(18-4)}, 718$$

Equilibrium constant, K_b, for weak base B

$$K_b = \frac{[BH^+][OH^-]}{[B]} \qquad \textbf{(18-4)}, 727$$

Equilibrium constant, K_{sp}, for slightly soluble salt $M_y X_z$

For $M_y^{x+} X_z^{y-} (s) \rightleftharpoons yM^{x+}(aq) + zX^{y-}(aq)$,

$$K_{\text{sp}} = [M^{x+}]^y [X^{y-}]^z \qquad \textbf{(20-1)}, 781$$

Equilibrium constant, relationship between K_c and K_P

$$K_P = K_c(RT)^{\Delta n} \qquad \text{or} \qquad K_c = K_P(RT)^{-\Delta n}$$
$$\Delta n = (n_{\text{gas prod}}) - (n_{\text{gas react}}) \qquad \textbf{(17-10)}, 692$$

Equilibrium constant, thermodynamic

$$K = \frac{(a_C)^c(a_D)^d}{(a_A)^a(a_B)^b}$$ **(17-12)**, 696

Equilibrium constants for conjugate acid–base pair

$$K_a K_b = K_w$$ **(18-8)**, 734

First Law of Thermodynamics

$$\Delta E = q + w$$ **(15-10)**, 571

Formal charge (for Group A elements)

$$FC = (\text{group number}) - [(\text{number of bonds}) + (\text{number of unshared } e^-)]$$ **(7-7)**, 268

Formula weight (FW)

$$FW = \frac{\text{mass in grams}}{\text{no. of mol}}$$ **(2-6)**, 58

Free energy change, ΔG

$$\Delta G = \Delta H - T\,\Delta S \qquad (\text{constant } T \text{ and } P)$$ **(15-16)**, 592

Free energy–equilibrium constant relationship

$$\Delta G_{rxn}^0 = -RT \ln K$$ **(17-12)**, 696

Free energy of reaction, ΔG_{rxn}^0

$$\Delta G_{rxn}^0 = \Sigma\, n\, \Delta G_{f\,products}^0 - \Sigma\, n\, \Delta G_{f\,reactants}^0$$
(1 atm and 298 K *only*)   **(15-16)**, 592

Freezing point depression

$$\Delta T_f = K_f m$$ **(14-12)**, 525

Frequency vs. wavelength

$$\lambda \nu = c$$ **(4-11)**, 134

Half-life (first order, radioactive decay)

$$t_{1/2} = \frac{\ln 2}{k} = \frac{0.693}{k}$$ **(16-4)**, 627
    (22-10), 863

Henderson–Hasselbalch equations

$$pH = pK_a + \log \frac{[\text{conjugate base}]}{[\text{acid}]}$$ (for acid/salt buffer)
    (19-1), 753

and

$$pOH = pK_b + \log \frac{[\text{conjugate acid}]}{[\text{base}]}$$ (for base/salt buffer)
    (19-1), 756

Henry's Law (gas solubility)

$$C_{gas} = kP_{gas}$$ **(14-7)**, 515

Hess's Law, enthalpy of reaction, ΔH_{rxn}^0, from enthalpies of formation

$$\Delta H_{rxn}^0 = \Sigma\, n\, \Delta H_{f\,products}^0 - \Sigma\, n\, \Delta H_{f\,reactants}^0$$ **(15-8)**, 566

Hess's Law, enthalpy of reaction, ΔH_{rxn}^0, from enthalpies of related reactions

$$\Delta H_{rxn}^0 = \Delta H_a^0 + \Delta H_b^0 + \Delta H_c^0 + \cdots$$ **(15-8)**, 564

Ideal gas equation

$$PV = nRT$$ **(12-9)**, 415

Integrated rate equation, first order

$$\ln\left(\frac{[A]_0}{[A]}\right) = akt$$ **(16-4)**, 627

Integrated rate equation, second order

$$\frac{1}{[A]} - \frac{1}{[A]_0} = akt$$ **(16-4)**, 629

Integrated rate equation, zero order

$$[A] = [A]_0 - akt$$ **(16-4)**, 631

Internal energy

$$\Delta E = E_{final} - E_{initial} = E_{products} - E_{reactants} = q + w$$ **(15-10)**, 571
$$\Delta E = q_v$$ **(15-10)**, 573

Ion product for water, K_w

$$K_w = [H_3O^+][OH^-] = 1 \times 10^{-14} \qquad (\text{at } 25°C)$$ **(18-2)**, 711

K_a, ionization constant for weak acid HA

$$K_a = \frac{[H_3O^+][A^-]}{[HA]}$$ **(18-4)**, 718

K_b, ionization constant for weak base B

$$K_b = \frac{[BH^+][OH^-]}{[B]}$$ **(18-4)**, 727

K_{sp}, solubility product constant for slightly soluble salt $M_y X_z$

For $M_y^{x+} X_z^{y-} (s) \rightleftharpoons yM^{x+}(aq) + zX^{y-}(aq)$,
$$K_{sp} = [M^{x+}]^y[X^{y-}]^z$$ **(20-1)**, 781

K_w, ion product for water

$$K_w = [H_3O^+][OH^-] = 1 \times 10^{\times14} \qquad (\text{at } 25°C)$$ **(18-2)**, 711

Kinetic energy, average molecular, $\overline{KE}$

$$\overline{KE} \propto T$$ **(12-13)**, 429

Light, wavelength vs. frequency

$$\lambda \nu = c$$ **(4-11)**, 134

Mass number (A)

Mass number = number of protons + number of neutrons
= atomic number + neutron number **(4-7)**, 123

Molality, m

$$\text{molality} = \frac{\text{number of moles solute}}{\text{number of kilograms solvent}}$$ **(14-8)**, 516

Molarity, M

$$\text{molarity} = \frac{\text{number of moles of solute}}{\text{number of liters of solution}}$$ **(3-6)**, 99

$$\text{molarity} = \frac{\text{number of millimoles of solute}}{\text{number of milliliters of solution}}$$ **(11-1)**, 377

Mole fraction, X

$$X_A = \frac{\text{no. mol A}}{\text{total no. mol of all components}}$$ **(12-11)**, 422

Mole fraction, X, in gas mixture

$$X_A = \frac{P_A}{P_{total}}$$ **(12-11)**, 423

Moles, from mass

$$\text{mol} = \frac{\text{mass in grams}}{\text{formula weight}}$$ **(2-6)**, 56

Moles, solute in solution

moles solute = molarity × L solution **(3-6)**, 99
millimoles solute = molarity × mL solution **(11-1)**, 377

Nernst equation

$$E = E^0 - \frac{2.303\, RT}{nF} \log Q$$ **(21-19)**, 828

or

$$E = E^0 - \frac{0.0592}{n} \log Q$$

Nuclear binding energy

$BE = (\Delta m)c^2$ **(22-3)**, 855

Nuclear mass deficiency

Δm = (sum of masses of all e^-, p^+, and n^0)
− (actual mass of atom) **(22-3)**, 854

Osmotic pressure

$\pi = MRT$ **(14-15)**, 531

Percent solute

$$\text{percent solute} = \frac{\text{mass of solute}}{\text{mass of solution}} \times 100\%$$ **(3-6)**, 98

Percent yield

$$\text{percent yield} = \frac{\text{actual yield of product}}{\text{theoretical yield of product}} \times 100\%$$ **(3-4)**, 95

pH

pH = −log [H_3O^+]   or   [H_3O^+] = 10^{-pH} **(18-3)**, 713
pH + pOH = 14.00   (at 25°C) **(18-3)**, 714

pK_w

$pK_w = -\log K_w$ (= 14.0 at 25°C) **(18-3)**, 713

pOH

pOH = −log [OH^-]   or   [OH^-] = 10^{-pOH} **(18-3)**, 713

Quadratic formula

$$x = \frac{-b \pm \sqrt{b^2 - 4ac}}{2a}$$ **(17-5)**, 679

Radioactive decay

$$\ln\left(\frac{A_0}{A}\right) = kt \text{ or } \ln\left(\frac{N_0}{N}\right) = kt$$ **(22-10)**, 863

Raoult's Law

$P_{solvent} = X_{solvent} P^0_{solvent}$ **(14-9)**, 518

Raoult's Law, two volatile components

$P_{total} = X_A P^0_A + X_B P^0_B$ **(14-9)**, 519

Rate equation, integrated, first order

$$\ln\left(\frac{[A]_0}{[A]}\right) = akt$$ **(16-4)**, 627

Rate equation, integrated, second order

$$\frac{1}{[A]} - \frac{1}{[A]_0} = akt$$ **(16-4)**, 629

Rate equation, integrated, zero order

$[A] = [A]_0 - akt$ **(16-4)**, 631

Rate law expression

rate = $k[A]^x[B]^y \ldots$ **(16-3)**, 620

Rate of reaction

$$\text{rate of reaction} = -\frac{1}{a}\left(\frac{\Delta[A]}{\Delta t}\right) = -\frac{1}{b}\left(\frac{\Delta[B]}{\Delta t}\right)$$

$$= \frac{1}{c}\left(\frac{\Delta[C]}{\Delta t}\right) = \frac{1}{d}\left(\frac{\Delta[D]}{\Delta t}\right)$$ **(16-1)**, 615

Reaction quotient, Q

For $a\text{A} + b\text{B} \rightleftharpoons c\text{C} + d\text{D}$ **(17-4)**, 674

$$Q = \frac{[\text{C}]^c[\text{D}]^d}{[\text{A}]^a[\text{B}]^b} \longleftarrow \begin{array}{l} \text{not necessarily} \\ \text{equilibrium} \\ \text{concentrations} \end{array}$$

Second Law of Thermodynamics

$\Delta S_{\text{universe}} = \Delta S_{\text{system}} + \Delta S_{\text{surroundings}} > 0$ **(15-15)**, 590

Solubility product constant, K_{sp}, for slightly soluble salt $M_y X_z$

For $M_y^{x+} X_z^{y-}(\text{s}) \rightleftharpoons y M^{x+}(\text{aq}) + z X^{y-}(\text{aq})$,

$K_{\text{sp}} = [M^{x+}]^y [X^{y-}]^z$ **(20-1)**, 781

Specific gravity

$$\text{Sp. Gr} = \frac{D_{\text{substance}}}{D_{\text{water}}}$$ **(1-11)**, 28

Specific heat

$$\text{specific heat} = \frac{(\text{amount of heat in J})}{(\text{mass of substance in g})(\text{temperature change in °C})}$$
  (1-13), 32

Speed, average molecular, $\bar{u}$

$$\bar{u} \propto \sqrt{\frac{T}{\text{molecular weight}}}$$ **(12-13)**, 429

Speed, gas molecule (root mean square)

$$u_{\text{rms}} = \sqrt{\frac{3RT}{M}}$$ **(12-13)**, 433

Temperature conversions:

°C to K	$°C = K - 273.15°$	**(1-12)**, 30
K to °C	$K = °C + 273.15°$	**(1-12)**, 30
°C to °F	$°F = \left(x°C \times \dfrac{1.8°F}{1.0°C}\right) + 32°F$	**(1-12)**, 31
°F to °C	$°C = \dfrac{1.0°C}{1.8°F}(x°F - 32°F)$	**(1-12)**, 31

van der Waals equation

$$\left(P + \frac{n^2 a}{V^2}\right)(V - nb) = nRT$$ **(12-15)**, 437

van't Hoff equation

$$\ln\left(\frac{K_{T_2}}{K_{T_1}}\right) = \frac{\Delta H^0}{R}\left(\frac{1}{T_1} - \frac{1}{T_2}\right)$$ **(17-13)**, 699

van't Hoff factor, i

$$i = \frac{\Delta T_{\text{f(actual)}}}{\Delta T_{\text{f(if nonelectrolyte)}}} = \frac{K_f m_{\text{effective}}}{K_f m_{\text{stated}}} = \frac{m_{\text{effective}}}{m_{\text{stated}}}$$ **(14-14)**, 529

or more generally, as

$$i = \frac{\text{colligative property}_{\text{actual}}}{\text{colligative property}_{\text{if nonelectrolyte}}}$$

Vapor pressure lowering

$\Delta P_{\text{solvent}} = P^0_{\text{solvent}} - P_{\text{solvent}}$

$\Delta P_{\text{solvent}} = X_{\text{solute}} P^0_{\text{solvent}}$ **(14-9)**, 518

Wavelength vs. frequency

$\lambda \nu = c$ **(4-11)**, 134

Work

$w = fd$ **(15-10)**, 571

Work, due to change in number of moles of gas (e.g., phase change or chemical reaction)

$w = -(\Delta n)RT$ **(15-10)**, 574

Work, due to expansion

$w = -P \Delta V$ **(15-10)**, 573

Glossary/Index

Glossary terms, printed in **boldface**, are defined here as well as in the text (location indicated by boldface page numbers) and in Key Terms. Page numbers followed by i indicate illustrations or their captions; page numbers followed by t indicate tables.

formation of carboxylic acid derivatives, 968–969

oxidation of, 965–966

reactions, 961–962, 961–962t

Aldehyde A compound in which an alkyl or aryl group and a hydrogen atom are attached to a carbonyl group; general

formula is R—C—H; R may be H., R may be H, **918**–920, 919f, 919t, 921t, **943**

reactions, 963–965, 965–966

Aldose A monosaccharide that contains an aldehyde group, **971, 982**

Aliphatic hydrogens Hydrocarbons that do not contain aromatic rings, **890, 943**

Alkali metals Elements of Group 1A in the periodic table, except hydrogen, **132, 162, 240, 1053**

compounds and their uses, 1041–1042, 1041f

properties and occurrence, 1036–1037, 1037t

reactions of, 1037–1040, 1040t

reactions with halogens, 252–254

reactions with nonmetals, 254

Alkaline cell A dry cell in which the electrolyte contains KOH, **838, 842**

Alkaline earth metals Group 2A elements in the periodic table, **132, 162, 240, 1053**

compounds and their uses, 1043–1044, 1044f

properties and occurrence, 1042, 1042t

reactions of, 1042–1043, 1043f, 1043t

reactions with nonmetals, 255

Alkanes. *see* **Saturated hydrocarbons**

Alkenes Unsaturated hydrocarbons that contain a carbon-carbon double bond, 899–904, 899f, 901–902f, **943**

Alkylbenzene A compound containing an alkyl group bonded to a benzene ring, **907,** 909–910, 909f, **943,** 966–967

Alkyl group A group of atoms derived from an alkane by the removal of one hydrogen atom, **895, 943**

Alkyl hydrogen sulfates, 935

Alkynes Unsaturated hydrocarbons that contain a carbon-carbon triple bond, **905**–906, 905f, **943**

Allotropes Different structural forms of the same element, 44, 72, 487, **494**

Alloying Mixing of a metal with other substances (usually other metals) to modify its properties, **1030**

Alpha emission Radioactive decay in which an alpha particle is given off from the nucleus, 859–860, **882**

Alpha- (α-) particle A helium ion with a 2+ charge; an assembly of two protons and two neutrons, **162**

Alpha particle (α) A particle that consists of two protons and two neurons; identical to a helium nucleus, 857, 858t, **859**–860, **882**

Aluminum, 85–86, 134, 560–561, 1024–1026, 1025f, 1045–1046, 1045t, 1047

Aluminum chloride, 364

Aluminum chlorohydrate, 362f

Aluminum hydroxide, 354–355

Amide A compound containing the

O
‖
—C—N⟨ group, **929**–930, 929f, **943,** 969

Amines Derivatives of ammonia in which one or more hydrogen atoms has been replaced by organic groups, 726–727, **742, 921**–922, 922f, **943,** 1073, **1080**

reactions, 963

Amino group The —NR₂ group, where R can represent an H, alkyl, or aryl group, **943**

Ammine complexes Complex species that contain ammonia molecules bonded to metal ions, **993, 994t, 1010**

Ammonia, 192f, 390f, 792–793

ammine complexes, 993, 994t

buffer solutions, 755–756

Gay-Lussac's Law of Combining Volumes and, 428

neutralization with nitric acid, 361

pH of, 728

Ammonium chloride, 351f, 361f, 736, 754–755, 837

Ammonium ion, 300

Ammonium phosphate, 782

Ammonium sulfide, 207

Amorphous solid A noncrystalline solid with no well-defined, ordered structure, **475**–477, 478f, **494**

Ampere Unit of electric current; 1 ampere equals 1 coulomb per second, **842**

Amphiprotism The ability of a substance to exhibit amphoterism by accepting or donating protons, **354, 367**

Amphoteric oxide An oxide that shows some acidic and some basic properties, 195f, **201**

Amphoterism The ability of a substance to react with both acids and bases, **201, 354**–355, 355t, **367**

Amu. *see* **atomic mass unit (amu)**

Analytical chemistry, 3

Angstrom (Å) 10^{-10} meter, 10^{-1} nm, or 100 pm, 23, 177, 178f, **201**

Angular A term used to describe the molecular geometry of a molecule that has two atoms bonded to a central atom and one or more lone pairs on the central atom (AB_2U or AB_2U_2); also called V-shaped or bent, **322**

Angular momentum quantum number (λ) The quantum mechanical solution to a wave equation that designates the subshell, or set of orbitals (s, p, d, f), within a given main shell in which an electron resides, **148, 162**

Anhydrous Without water, 69–70, 72, 367

Anion A negatively charged ion; that is, an ion in which the atom or group of atoms has more electrons than protons, 48, 72, 251, **280**

Anode In a cathode-ray tube, the positive electrode; the electrode at which oxidation occurs, **162, 805, 842**

Anthocyanins, 749f, 765f

Antibonding electrons Electrons in antibonding orbitals, 335, 338, **343**

Antibonding molecular orbital A molecular orbital higher in energy than any of the atomic orbitals from which it is derived; when populated with electrons, lends instability to a molecule or ion. Denoted with an asterisk (*) superscript on its symbol, 331–334, **343**

Aqueous solution A solution in which the solvent is water, **97, 106, 208**–215. *see also* **Titration**

acid-base reactions in, 358–361

electrolytes and extent of ionization, 208–209, 209f

properties, 349

reactions in, 215–217, 215t

reversible reactions, 211–212, 213t

solubility guidelines for compounds in, 213–215

strong and weak acids, 209–211, 211f, 211t, 212f

strong bases, insoluble bases, and weak bases, 212, 213t

Aromatic hydrocarbons Benzene and similar condensed ring compounds; contain delocalized rings of electrons, **890, 943**

alkylbenzene, 907, 909–910, 909f

benzene, 906–907, 907f

Arrhenius, S., 349

Arrhenius equation An equation that related the specific rate constant to activation energy and temperature, 643–647, 644–646f, **656**

Arrhenius theory, 349–350

Artificial transmutation An artificially induced nuclear reaction caused by bombardment of a nucleus with subatomic particles or small nuclei, 871–874, 872–873f, **882**

Artists' pigments, 924

Aryl group The group of atoms remaining after a hydrogen atom is removed from an aromatic system, **943**

Ascorbic acid, 709f

Associated ions Short-lived species formed by the collision of dissolved ions of opposite charge, **528, 542**

Atmosphere (atm) A unit of pressure; the pressure that will support a column of mercury 760 mm high at 0°C; 760 torr., 404, 416, **438**

Atomic mass unit (amu) One-twelfth of the mass of an atom of the carbon-12 isotope; a unit used for stating atomic and formula weights, **51, 72, 162,** 420

Atomic number The integral number of protons in the nucleus; defines the identity of an element, 6, 35, 121–122, **132, 162, 240**

Atomic orbital The region or volume in space in which the probability of finding electrons is highest, **147, 148**–153, **162,** 297

energy level diagrams, 333–334, 334f

Atomic radius The radius of an atom, 177–179, 184f, **201**

Atomic spectra, 138–144

Atomic weight Weighted average of the masses of the constituent isotopes of an

13 Liquids and Solids

13-2. *Refer to Sections 13-2 and the Key Terms for Chapter 13.*

Hydrogen bonding is an especially strong dipole-dipole interaction between molecules in which one contains H in a highly polarized bond and the other contains a lone pair of electrons. The energy of a hydrogen bond is 4 to 5 times larger than a normal dipole-dipole interaction and roughly 10% of a covalent bond. It only occurs in systems where a hydrogen atom is directly bonded to a small, highly electronegative atom, such as N, O or F.

13-4. *Refer to Section 13-2 and Example 13-1.*

Permanent dipole-dipole forces can be found acting between the polar molecules of (c) NO and (d) SeF_4.

13-6. *Refer to Section 13-2, Example 13-1, and Exercise 13-4.*

Dispersion forces are the only important intermolecular forces of attraction operating between the nonpolar molecules of (a) molecular $AlBr_3$ and (b) PCl_5.

13-8. *Refer to Section 13-2, Example 13-1, and Solution to Exercise 13-12.*

The substances exhibiting strong hydrogen bonding in the liquid and solid states are:

(a) CH_3OH,

(d) $(CH_3)_2NH$ and

(e) CH_3NH_2.

13-10. *Refer to Section 13-2, Table 13-3 and Example 13-1.*

(a) The physical properties of ethyl alcohol (ethanol), $C_2H_6O \equiv CH_3CH_2OH$, are influenced mainly by hydrogen bonding since there is an H atom directly bonded to an O atom and lone pairs of electrons on the O atom, but are also affected by dispersion forces like any other molecule.

(b) Phosphine, PH_3, is a polar molecule. Refer to Exercise 13-14 for its structure. The intermolecular forces existing between the molecules are dispersion forces and dipole-dipole forces. One commonly says that dipole-dipole forces are more important. However, when one looks at Table 13-3, since PH_3 is larger than NH_3 and has no H-bonding, one can deduce that dispersion forces are more important than dipole-dipole interactions.

(c) Sulfur hexafluoride, SF_6, is a nonpolar molecule and therefore has only dispersion forces acting between its molecules.

13-12. *Refer to Section 13-2.*

Hydrogen bonding, usually occurring between molecules having an H atom directly bonded to a F, O, or N atom, is very strong compared with other dipole-dipole interactions. H bonding results from the attractions between the $\delta+$ atoms of one molecule (the H atoms) and the $\delta-$ atoms (usually F, O, and N atoms) of another molecule. The small sizes of the F, O, and N atoms, combined with their high electronegativities, concentrate the electrons in the molecule around the $\delta-$ atoms. This causes the H atom to behave somewhat like a bare proton. There is then a very strong attraction between the $\delta+$ H atom and a lone pair of electrons on a F, O, or N atom on another molecule.

13-14. *Refer to Section 13-2 and Example 13-5.*

The normal melting points and boiling points generally increase as the intermolecular forces between the molecules in the compounds increase.

silane, SiH_4

$$\begin{array}{c} H \\ H\!:\!\ddot{S}i\!:\!H \\ \ddot{H} \end{array}$$

SiH_4 is a nonpolar covalent molecule and has only dispersion forces acting between the molecules. PH_3 is a polar covalent molecule with dispersion forces and dipole-dipole interactions in operation. Since both molecules are about the same size as evidenced by having similar molecular weights, their dispersion forces are about the same. Therefore, the melting and boiling points of PH_3 ($-133°C$ and $-88°C$) are predicted to be higher than those of SiH_4 ($-185°C$ and $-112°C$) and they are.

phosphine, PH_3

$$\begin{array}{c} \ddot{} \\ H\!:\!\ddot{P}\!:\!H \\ \ddot{H} \end{array}$$

13-16. *Refer to Section 13-2, Exercise 13-2 Solution, and Example 13-1.*

(a) ammonia, NH_3

$$\begin{array}{c} H\text{-}\ddot{N}\text{-}H \\ | \\ H \end{array}$$

NH_3 has hydrogen bonds operating between its molecules, but not PH_3. NH_3 contains a hydrogen atom directly bonded to the small, highly electronegative atom, N.

phosphine, PH_3

$$\begin{array}{c} H\text{-}\ddot{P}\text{-}H \\ | \\ H \end{array}$$

(b) ethylene, C_2H_4

$$\begin{array}{c} H \qquad H \\ \diagdown C=C \diagup \\ H \diagup \qquad \diagdown H \end{array}$$

Hydrazine, N_2H_4, has hydrogen bonding, but not ethylene, C_2H_4, since N_2H_4 contains the small highly electronegative element, N, which is directly bonded to hydrogen atoms and has 2 lone pairs of electrons.

hydrazine, N_2H_4

$$\begin{array}{c} H\text{-}\ddot{N}\text{-}\ddot{N}\text{-}H \\ | \quad | \\ H \ H \end{array}$$

(c) hydrogen fluoride, HF

$$H\text{-}\ddot{F}\!:$$

Hydrogen fluoride, HF, has hydrogen bonding, but not hydrogen chloride, HCl, since in HF, the hydrogen atom is directly bonded to the small highly electronegative element, F.

hydrogen chloride, HCl

$$H\text{-}\ddot{C}l\!:$$

13-18. *Refer to Section 13-1 and Chapter 1.*

Copper metal: This solid at room temperature has a set volume and holds its shape. Its atoms are close together in an ordered crystalline structure and are vibrating in place.

Rubbing alcohol   This liquid at room temperature, which is either isopropyl alcohol (2-propanol, $CH_3CH(OH)CH_3$) or ethanol (CH_3CH_2OH), has a set volume, but takes on the shape of its container. Its molecules are close together but have no particular arrangement and are free to move past each other.

Nitrogen   This gas, N_2, at room temperature, has no set volume, taking on the volume of its container. Its molecules are relatively far apart and move independently of each other.

13-20. *Refer to Section 13-2 and Example 13-1.*

(a) NaF is composed of ions, while ClF, HF and F_2 are molecules.

(b) The ionic compound, NaF, is held together by ionic forces.
   ClF is a polar molecule and is held together by both dipole-dipole interactions and dispersion forces.
   HF is a very polar molecule and is held together by hydrogen bonding and dispersion forces.
   F_2 is a nonpolar molecule and only dispersion forces hold its molecules together.

(c) The relative strengths of the forces holding the particles together can be deduced from the relative boiling points. The higher the boiling point, the more energy that is required to separate the molecules or ions and the stronger the forces that must be holding the particles together. Therefore, in order of increasing force of attraction:

$$F_2 < ClF < HF \ll NaF$$

(d) Based on our answers to (b) and (c), we can say:

dispersion forces < dipole-dipole interactions < hydrogen bonding $\ll$ ion-ion interactions

Note: when molecules become very large, dispersion forces can become stronger than dipole-dipole interactions and even stronger than hydrogen bonding.

13-22. *Refer to Section 13-2 and Figure 13-5.*

(a) Ne, Ar, and Kr are nonpolar noble gases. Their boiling points are determined solely by dispersion forces, which in turn are dependent on an atom's size and polarizability. In order of increasing boiling point (i.e., increasing size),

$$Ne\,(-246°C) < Ar\,(-186°C) < Kr\,(-152°C)$$

(b) All three compounds, NH_3, H_2O and HF exhibit hydrogen bonding. Figure 13-5 gives the order of increasing boiling points as:

$$NH_3\,(-33°C) < HF\,(20°C) < H_2O\,(100°C)$$

Since the electronegativities of the elements follow the order, F > O > N, the charge separation for the bonds in these three molecules should follow the order, F-H > O-H > N-H. The same order also should be followed when *each* of these hydrogen atoms forms hydrogen bonding with a lone pair of electrons on a neighboring molecule.

The reason why H_2O has a higher boiling point than HF is because there is a larger number of H-bonds in the H_2O system. The limiting reactant for the formation of H-bonds in the NH_3 system is the number of lone pairs of electrons, while that in the HF system is the number of H atoms. In the H_2O system, all of the H atoms and lone pairs can participate in H-bond formation. In either the NH_3 or the HF system, 1 mol of the molecules could only give a maximum of 1 mole of H-bonds. But, in 1 mole of H_2O, a maximum of 2 moles of H-bonds could be obtained. This explains why the *total* H-bonding force is much higher in H_2O than in HF, as shown by the higher boiling point.

13-24. *Refer to Section 13-8.*

(a) The normal boiling point is the temperature at which the vapor pressure of a liquid is exactly equal to one atmosphere (760 torr) pressure.

(b) When the boiling point of a liquid is measured, the atmospheric pressure over the liquid must be specified since the boiling point of a liquid is the temperature at which its vapor pressure is exactly equal to the applied pressure. As the atmospheric pressure decreases, so does the boiling point of a liquid.

13-26. *Refer to Section 13-2.*

(a) CO_2 – dispersion forces only since the molecule is nonpolar
(b) NH_3 – hydrogen bonding and dispersion forces
(c) $CHCl_3$ – dipole-dipole interactions and dispersion forces
(d) CCl_4 – only dispersion forces since the molecule is nonpolar

13-28. *Refer to Section 13-4 and Figure 13-8.*

Surface tension is a measure of the inward intermolecular forces of attraction among liquid particles that must be overcome to expand the surface area. A measurable tension in the surface of the liquid is caused by an imbalance set up between the intermolecular forces operating at the surface and those operating within the liquid. As the temperature increases, the increased kinetic energy and greater movement of the particles in the liquid tend to counteract the intermolecular forces, resulting in the lowering of surface tension.

13-30. *Refer to Sections 13-3, 13-4, 13-6 and 13-7, and Table 13-6.*

There are similarities between the intermolecular attractions used to describe on a molecular level (1) viscosity, (2) surface tension, (3) the rate of evaporation and resulting vapor pressure of a liquid. For compounds in the liquid phase that have *strong* intermolecular forces of attraction operating between its molecules:

(1) the molecules cannot slide easily past each other, and the liquid has a *high* viscosity;

(2) the molecules at the surface of a liquid have an extra strong attraction for each other, and the liquid has a *high* surface tension, and

(3) the molecules are attracted strongly to each other in the liquid phase and do not evaporate as easily, effecting a *low* rate of evaporation and a *low* vapor pressure.

13-32. *Refer to Sections 13-2 and 13-7, and Table 13-3.*

The weaker the intermolecular forces between molecules in the liquid phase, the higher is the vapor pressure. Generally, hydrogen bonding and the cumulative dispersion forces in larger molecules are the most significant factors (see Table 13-3). In order of decreasing vapor pressure,

(a) $BiCl_3 > BiBr_3$ $BiCl_3$ is smaller than $BiBr_3$, has weaker dispersion forces and thus has a higher vapor pressure at the same temperature.

(b) $CO > CO_2$ CO is smaller, has weaker dispersion forces and a higher vapor pressure than CO_2. Note from Table 13-3 that even though CO is polar, its dipole moment is extremely low and its dipole-dipole interaction energy is nearly zero.

(c) $N_2 > NO$ Both are small molecules of comparable size, and therefore have comparable dispersion forces. NO however is polar and has permanent dipole-dipole interactions, whereas N_2 is nonpolar and has only dispersion forces. Therefore N_2 has the higher vapor pressure.

(d) $HCOOCH_3 > CH_3COOH$ Methyl formate ($HCOOCH_3$) and acetic acid (CH_3COOH) have the same molecular formula, $C_2H_4O_2$, and are about the same size, so both have similar dispersion forces. CH_3COOH can form hydrogen bonds, whereas $HCOOCH_3$ cannot. Therefore $HCOOCH_3$ has weaker intermolecular interactions and has a higher vapor pressure than CH_3COOH.

To confirm the above reasoning, use boiling points given below to indicate strengths of intermolecular forces. Remember, the lower the boiling point is, the weaker the intermolecular forces present in the liquid, and the higher the vapor pressure at a specific temperature.

Compound	B.P. (°C)	Compound	B.P. (°C)
$BiCl_3$	447	N_2	−195.8
$BiBr_3$	453	NO	−151.8
CO	−191.5	CH_3COOH	117.9
CO_2	−78.5 (sublimes)	$HCOOCH_3$	31.5

The order of increasing boiling points corresponds with the order of increasing temperatures when their vapor pressures are constant, e.g. 100 torr:

$$\text{butane} < \text{diethyl ether} < \text{1-butanol}$$

13-36. *Refer to Section 13-8.*

Vapor pressure curve for Cl_2O_7

The boiling point of Cl_2O_7 at 125 torr from the graph is about 34°C.

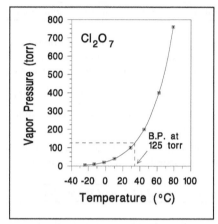

t (°C)	VP (torr)
−24	5
−13	10
−2	20
10	40
29	100
45	200
62	400
79	760

13-38. *Refer to Sections 13-2 and 12-11.*

The compounds that are expected to form hydrogen bonds in the liquid state are
  (c) HF,
  (d) CH_3CO_2H and
  (f) CH_3OH.

13-40. *Refer to Section 13-9.*

(a) at 37°C, the heat of vaporization of water = 2.41 kJ/g

at 37°C, ΔH°_{vap} (kJ/mol) $= 2.41\ \dfrac{kJ}{g} \times \dfrac{18.0\ g}{1\ mol} = $ **43.4 kJ/mol**

(b) The heat of vaporization at a certain temperature is the amount of heat required to change 1 gram of liquid to 1 gram of vapor at that temperature. The heat of vaporization is greater at 37°C than at 100°C because the average kinetic energy of the molecules is lower at the lower temperature. Therefore, more energy must be added per unit mass of the liquid to break the intermolecular forces between the molecules at the lower temperature.

13-42. *Refer to Section 13-8.*

Vapor pressure curve for $C_2H_4F_2$

The boiling point of $C_2H_4F_2$ at 200.
torr from the graph is about $-1°C$.

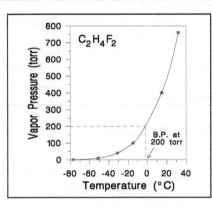

t	VP
(°C)	(torr)
−77.2	1
−51.2	10
−31.1	40
−15.0	100
14.8	400
31.7	760

13-44. *Refer to Sections 13-9 and 13-11, and the Key Terms for Chapter 13.*

This exercise involves 4 separate calculations:

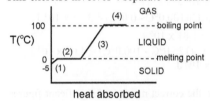

heat absorbed

$$
\begin{array}{ccccccccc}
& (1) & & (2) & & (3) & & (4) & \\
\text{ice} & \Rightarrow & \text{ice} & \Rightarrow & \text{water} & \Rightarrow & \text{water} & \Rightarrow & \text{steam} \\
-5.0°C & & 0.0°C & & 0.0°C & & 100.0°C & & 100.0°C
\end{array}
$$

$$? \text{ g H}_2\text{O} = 0.50 \text{ mol H}_2\text{O} \times \frac{18.0 \text{ g H}_2\text{O}}{1 \text{ mol H}_2\text{O}} = 9.0 \text{ g H}_2\text{O}$$

(1) heat required = mass **x** specific heat (s) **x** Δt = (9.0 g)(2.09 J/g·°C)(0.0°C - (−5.0°C)) = 94 J

(2) heat required = mass **x** heat of fusion = (9.0 g)(334 J/g) = 3.0 x 10^3 J

(3) heat required = mass **x** specific heat (ℓ) **x** Δt = (9.0 g)(4.184 J/g·°C)(100.0°C - 0.0°C) = 3.8 x 10^3 J

(4) heat required = mass **x** heat of vaporization = (9.0 g)(2260 J/g) = 2.0 x 10^4 J

Therefore, the total heat required = (1) + (2) + (3) + (4) = **2.7 x 10^4 J**

Note: Even though the individual calculations were each rounded to the correct number of significant figures, the final answer was obtained by rounding only at the very end.

13-46. *Refer to Section 13-6 and the Key Terms for Chapter 13.*

A dynamic equilibrium is a situation in which two (or more) opposing processes occur at the same rate so that no net change occurs. This is the kind of equilibrium that is established between two physical states of matter, e.g., between a liquid and its vapor, in which the rate of evaporation is equal to the rate of condensation in a closed container:

$$\text{liquid} \rightleftarrows \text{vapor.}$$

On a molecular level at equilibrium, molecules of liquid are escaping into the vapor and vapor molecules are condensing into the liquid. This is not a static situation. No net change occurs because the rates are the same. This can be demonstrated by observing a glass of water. In time, the water will evaporate and the water volume will decrease. By covering the glass with a glass or clear plastic plate, droplets of condensed water will be observed on the underside of the plate and the water volume will remain the same.

13-48. *Refer to Sections 13-9 and 13-11.*

Calculations of the heat required for each step:
(1) melting $Br_2(s)$ at m.p.: heat required = mass **x** heat of fusion = (50.0 g)(66.15 J/g) = 3310 J

(2) heating $Br_2(\ell)$ at m.p. $\rightarrow$ $Br_2(\ell)$ at b.p.:

heat required = mass x specific heat(ℓ) x Δt = $(50.0$ g$)(0.473$ J/g·°C$)(58.7°C$ - $-72°C)$ = 3090 J

(4) boiling $Br_2(\ell)$ at b.p.: heat required = mass x heat of vaporization = $(50.0$ g$)(193.21$ J/g$)$ = 9660 J

(5) heating $Br_2(g)$ at b.p. $\rightarrow$ $Br_2(g)$ at 100°C:

heat required = mass x specific heat(g) x Δt = $(50.0$ g$)(0.225$ J/g·°C$)(100.0°C$ - 58.7°C$)$ = 465 J

Therefore, the step needing the most energy is step 4, boiling the liquid bromine at its boiling point.

13-50. *Refer to Sections 13-9 and: 13-11.*

This exercise involves 5 separate calculations:

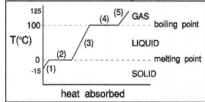

$$
\begin{array}{cccccccccc}
& (1) & & (2) & & (3) & & (4) & & (5) \\
\text{ice} & \Rightarrow & \text{ice} & \Rightarrow & \text{water} & \Rightarrow & \text{water} & \Rightarrow & \text{steam} & \Rightarrow & \text{steam} \\
-15.0°C & & 0.0°C & & 0.0°C & & 100.0°C & & 100.0°C & & 125.0°C
\end{array}
$$

(1) heat required = mass x specific heat (s) x Δt = $(80.0$ g$)(2.09$ J/g·°C$)(0.0°C$ - $(-15.0°C))$ = 111 J

(2) heat required = mass x heat of fusion = $(80.0$ g$)(334$ J/g$)$ = 2.67×10^4 J

(3) heat required = mass x specific heat (ℓ) x Δt = $(80.0$ g$)(4.184$ J/g·°C$)(100.0°C$ - 0.0°C$)$ = 3.35×10^4 J

(4) heat required = mass x heat of vaporization = $(80.0$ g$)(2260$ J/g$)$ = 1.81×10^5 J

(5) heat required = mass x specific heat (g) x Δt = $(80.0$ g$)(2.03$ J/g·°C$)(125.0°C$ - 100.0°C$)$ = 4.06×10^3 J

Therefore, the total heat required = (1) + (2) + (3) + (4) + (5) = **2.48×10^5 J**

Note: Even though the individual calculations were each rounded to the correct number of significant figures, the final answer was obtained by rounding only at the very end.

13-52. *Refer to Sections 1-14 and 13-9.*

Recall: When two substances are brought into contact with each other, the heat lost by one substance is equal in absolute value to the heat gained by the other.

$$|\text{heat gained by cold water}| = |\text{heat lost by hot water}|$$

$$|\text{mass x specific heat}(\ell) \text{ x } \Delta t|_{\text{cold}} = |\text{mass x specific heat}(\ell) \text{ x } \Delta t|_{\text{hot}}$$

$$|\text{mass x } \Delta t|_{\text{cold}} = |\text{mass x } \Delta t|_{\text{hot}}$$

$$(525 \text{ g})(t_{\text{final}} - 30.0°C) = (250. \text{ g})(100.°C - t_{\text{final}})$$

$$(525 \text{ x } t_{\text{final}}) - 15750 = 25000 - (250. \text{ x } t_{\text{final}})$$

$$775 \text{ x } t_{\text{final}} = 40750 \text{ (3 significant figures)}$$

$$t_{\text{final}} = \textbf{52.6°C}$$

13-54. *Refer to Section 13-9.*

Plan: (1) Determine if the final phase will be liquid or gas.

(2) Calculate the final temperature, t_{final} (°C).

(1) The amount of heat required to change the liquid water to steam at 100°C is

heat required $= |\text{mass x specific heat } (\ell) \text{ x } \Delta t|_{\text{water}} + |\text{mass x heat of vaporization}|$

$= (180. \text{ g})(4.184 \text{ J/g·°C})(100.°C - 0.°C) + (180. \text{ g})(2.26 \times 10^3 \text{ J/g})$

$= 4.82 \times 10^5$ J

The amount of heat released when the steam changes to liquid water at 100°C is

heat released = |mass $\times$ specific heat (g) $\times$ Δt|$_{\text{steam}}$ + |mass $\times$ heat of vaporization|
$\qquad$ = (18.0 g)(2.03 J/g·°C)(110.°C - 100.°C) + (18.0 g)(2.26 $\times 10^3$ J/g)
$\qquad$ = 4.10 $\times 10^4$ J

The heat required to convert water to steam is greater than the heat released when the steam condenses to water. Therefore, when the two systems are mixed, the liquid water will cause all of the steam to condense and the final temperature will be between 0°C and 100°C.

(2) $\qquad$ |amount of heat gained by water| = |amount of heat lost by steam|

$\qquad$ |mass $\times$ specific heat (ℓ) $\times$ Δt| = |mass $\times$ specific heat (g) $\times$ Δt| + |mass $\times$ heat of vaporization|

$\qquad\qquad\qquad\qquad$ + |mass $\times$ specific heat (ℓ) $\times$ Δt|$_{\text{water from steam}}$

(180. g)(4.184 J/g·°C)(t_{final} - 0.°C) = (18.0 g)(2.03 J/g·°C)(110.°C - 100.°C) + (18.0 g)(2.26 $\times 10^3$ J/g)
$\qquad\qquad\qquad\qquad$ + (18.0 g)(4.184 J/g·°C)(100.°C - t_{final})
(753 $\times$ t_{final}) - 0 = (370 J) + (4.07 $\times 10^4$ J) + (7.53 $\times 10^3$ J) - (75.3 $\times$ t_{final})
828 $\times$ t_{final} = 4.86 $\times 10^4$
t_{final} = **58.7°C**

13-56. *Refer to Sections 13-9 and 13-11.*

(a) This part of the exercise requires 2 separate calculations:

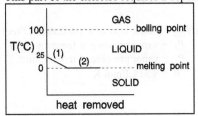

$\qquad\qquad\qquad\qquad$ (1) $\qquad$ (2)
$\qquad\qquad$ water $\Rightarrow$ water $\Rightarrow$ ice
$\qquad\qquad$ 25.0°C $\qquad$ 0.0°C $\qquad$ 0.0°C

(1) heat removed = mass $\times$ specific heat (ℓ) $\times$ Δt = (14.0 g)(4.184 J/g·°C)(25.0°C - 0.0°C) = 1.46 $\times 10^3$ J

(2) heat removed = mass $\times$ heat of fusion = (14.0 g)(334 J/g) = 4.68 $\times 10^3$ J
Therefore, the total heat removed = (1) + (2) = **6.14 $\times 10^3$ J**

(b) The mass of water at 100.0°C that could be cooled to 23.5°C by removing 6.14 $\times 10^3$ J of heat from the water sample is:

$\qquad$ heat removed from water in (a) = mass$_{\text{water}}$ $\times$ specific heat (ℓ) $\times$ Δt

$\qquad$ 6.14 $\times 10^3$ J = (mass$_{\text{water}}$)(4.184 J/g·°C)(100.0°C - 23.5°C)
$\qquad\qquad$ = (mass$_{\text{water}}$)(3.20 $\times 10^2$ J/g)
$\qquad$ mass$_{\text{water}}$ = **19.2 g**

13-58. *Refer to Section 13-9 and Example 13-4.*

We must use the Clausius-Clapeyron equation to find the vapor pressure at 80.00°C:

$$\ln\left(\frac{P_2}{P_1}\right) = \frac{\Delta H_{\text{vap}}}{R}\left(\frac{1}{T_1} - \frac{1}{T_2}\right)$$

$\qquad$ where $\qquad$ ΔH_{vap} = molar heat of vaporization (J/mol)
$\qquad\qquad\qquad$ P_1 = 760.00 torr
$\qquad\qquad\qquad$ P_2 = vapor pressure at 80.00°C
$\qquad\qquad\qquad$ R = 8.314 J/mol·K
$\qquad\qquad\qquad$ T_1 = normal boiling point of liquid (K)
$\qquad\qquad\qquad$ T_2 = 80.00°C + 273.15° = 353.15 K

(1) for water, H_2O: $\Delta H_{vap} = 40,656$ J/mol
$$T_1 = 100.00°C + 273.15° = 373.15 \text{ K}$$

Substituting,

$$\ln\left(\frac{P_2}{760.00 \text{ torr}}\right) = \frac{40656 \text{ J/mol}}{8.314 \text{ J/mol·K}}\left(\frac{1}{373.15 \text{ K}} - \frac{1}{353.15 \text{ K}}\right)$$
$$= (4890.)(0.0026799 - 0.0028317)$$
$$= -0.7424$$

$$\frac{P_2}{760.00 \text{ torr}} = 0.4760$$

$$P_2 = \textbf{361.8 torr}$$

(2) for heavy water, D_2O: $\Delta H_{vap} = 41,606$ J/mol
$$T_1 = 101.41°C + 273.15° = 374.56 \text{ K}$$

Substituting,

$$\ln\left(\frac{P_2}{760.00 \text{ torr}}\right) = \frac{41606 \text{ J/mol}}{8.314 \text{ J/mol·K}}\left(\frac{1}{374.56 \text{ K}} - \frac{1}{353.15 \text{ K}}\right)$$
$$= (5004)(0.0026698 - 0.0028317)$$
$$= -0.8101$$

$$\frac{P_2}{760.00 \text{ torr}} = 0.4448$$

$$P_2 = \textbf{338.0 torr}$$

13-60.   *Refer to Section 13-9.*

(a) The Clausius-Clapeyron equation is

$$\ln\left(\frac{P_2}{P_1}\right) = \frac{\Delta H_{vap}}{R}\left(\frac{1}{T_1} - \frac{1}{T_2}\right)$$

where ΔH_{vap} = molar heat of vaporization (J/mol)
   P_1 = vapor pressure at temperature, T_1 (K)
   P_2 = vapor pressure at temperature, T_2 (K)
   $R = 8.314$ J/mol·K

Expanding the equation we obtain:

$$\ln P_2 - \ln P_1 = \frac{\Delta H_{vap}}{R}\left(\frac{1}{T_1}\right) - \frac{\Delta H_{vap}}{R}\left(\frac{1}{T_2}\right)$$

$$\ln P_2 = \frac{-\Delta H_{vap}}{RT_2} + \left[\frac{\Delta H_{vap}}{R}\left(\frac{1}{T_1}\right) + \ln P_1\right]$$

If we let P_1 be a known vapor pressure of the substance at a particular temperature, T_1, and simplify by letting B stand for all the terms in the square brackets, we have

$$\ln P = \frac{-\Delta H_{vap}}{RT} + B$$

When ln P is plotted against $1/T$, we obtain a straight line with a slope of $-\Delta H_{vap}/R$ and y-intercept of B.

(b) Vapor pressure data for ethyl acetate, $CH_3COOC_2H_5$:

t (°C)	T (K)	1/T (K⁻¹)	P (torr)	ln P
-43.4	229.8	4.35×10^{-3}	1	0.0
-23.5	249.7	4.00×10^{-3}	5	1.6
-13.5	259.7	3.85×10^{-3}	10.	2.30
-3.0	270.2	3.70×10^{-3}	20.	3.00
9.1	282.3	3.54×10^{-3}	40.	3.69
16.6	289.8	3.45×10^{-3}	60.	4.09
27.0	300.2	3.33×10^{-3}	100.	4.61
42.0	315.2	3.17×10^{-3}	200.	5.30
59.3	332.5	3.01×10^{-3}	400.	5.99
?	?	?	760.	6.63

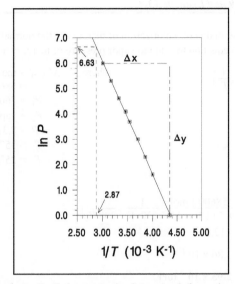

(c) The molar heat of vaporization of ethyl acetate is determined from the slope of the line. A linear regression fit to the data gives a slope of -4.471 with a coefficient of determination, $r^2 = 0.99958$, which indicates a very good fit. If you do not have access to a line-fitting program, the slope can be estimated by using two data points that are far apart as shown in the graph. This will work because the line fits the data well.

$$\text{slope} = \frac{-\Delta H_{vap}}{R} = \frac{\Delta y}{\Delta x} = \frac{\Delta \ln P}{\Delta \, 1/T} = \frac{(0.00 - 5.99)}{(4.35 \times 10^{-3} - 3.01 \times 10^{-3})} = -4.47 \times 10^3 \text{ K}$$

Therefore, $\Delta H_{vap} = -\text{slope} \times R = -(-4.47 \times 10^3 \text{ K})(8.314 \text{ J/mol·K}) = \textbf{3.72} \times \textbf{10}^4 \textbf{ J/mol}$

(d) The normal boiling point is the temperature at which the vapor pressure of ethyl acetate is 760 torr. From the graph, when $\ln P = \ln 760 = 6.63$,

$\qquad 1/T = 2.87 \times 10^{-3} \text{ K}^{-1}$ $\qquad\qquad$ and solving, $T = \textbf{3.48} \times \textbf{10}^2 \textbf{ K or 75°C}$

13-62. *Refer to Section 13-9 and Example 13-4.*

Use the Clausius-Clapeyron equation to solve for the molar heat of vaporization of isopropyl alcohol, ΔH_{vap}.

We know: $\qquad P_1 = 100.$ torr $\qquad\qquad T_1 = 39.5°C + 273.2° = 312.7$ K
$\qquad\qquad\qquad P_2 = 400.$ torr $\qquad\qquad T_2 = 67.8°C + 273.2° = 341.0$ K

$$\ln\!\left(\frac{P_2}{P_1}\right) = \frac{\Delta H_{vap}}{R}\!\left(\frac{1}{T_1} - \frac{1}{T_2}\right)$$

where $\quad \Delta H_{vap} = $ molar heat of vaporization (J/mol)
$\qquad\qquad\qquad P_1 = $ vapor pressure at temperature, T_1 (K)
$\qquad\qquad\qquad P_2 = $ vapor pressure at temperature, T_2 (K)
$\qquad\qquad\qquad R = 8.314$ J/mol·K

Plugging in, we have

$$\ln\!\left(\frac{400.\text{ torr}}{100.\text{ torr}}\right) = \frac{\Delta H_{vap}}{8.314 \text{ J/mol·K}}\left(\frac{1}{312.7 \text{ K}} - \frac{1}{341.0 \text{ K}}\right)$$

$$1.386 = \frac{\Delta H_{vap}}{8.314 \text{ J/mol·K}}\,(0.000265)$$

$$\Delta H_{vap} = \textbf{4.34} \times \textbf{10}^4 \textbf{ J/mol or 43.4 kJ/mol}$$

13-64. *Refer to Section 13-9 and Example 13-4.*

From Appendix E, the molar enthalpy of vaporization of mercury at the normal boiling point is 58.6 kJ/mol. Using the Clausius-Clapeyron equation to find the vapor pressure of mercury at 25°C, we have

$$\ln\left(\frac{P_2}{P_1}\right) = \frac{\Delta H_{vap}}{R}\left(\frac{1}{T_1} - \frac{1}{T_2}\right)$$

where
ΔH_{vap} = molar heat of vaporization (J/mol)
$\quad\quad\;\; = 58600$ J/mol
$P_1 = 760.$ torr
P_2 = vapor pressure at 25°C
$R = 8.314$ J/mol·K
$T_1 = 357°C + 273° = 630.$ K
$T_2 = 25°C + 273° = 298$ K

Substituting,

$$\ln\left(\frac{P_2}{760. \text{ torr}}\right) = \frac{58600 \text{ J/mol}}{8.314 \text{ J/mol·K}}\left(\frac{1}{630. \text{ K}} - \frac{1}{298 \text{ K}}\right)$$

$$= -12.5$$

$$\frac{P_2}{760. \text{ torr}} = 3.86 \times 10^{-6}$$

$$P_2 = \mathbf{2.93 \times 10^{-3}} \text{ torr}$$

13-66. *Refer to Section 13-13 and the Key Terms for Chapter 13.*

The critical point is the combination of critical temperature and critical pressure of a substance. The critical temperature is the temperature above which a gas cannot be liquefied, i.e., above this temperature, one cannot distinguish between a liquid and a gas. The critical pressure is the pressure required to liquefy a gas at its critical temperature. If the temperature is less than the critical temperature, the substance can be either a gas, liquid or solid, depending on the pressure.

13-68. *Refer to Section 13-13 and Figure 13-17b.*

(a) This point lies on the sublimation curve where the solid and gas phases are in equilibrium. So, both the solid and gaseous phases are present.

(b) This point is called the triple point where all three phases are in equilibrium with each other. So, the solid, liquid and gaseous phases are all present.

13-70. *Refer to Section 13-13 and Figure 13-17b.*

The melting point of carbon dioxide increases with increasing pressure, since the solid-liquid equilibrium line on its phase diagram slopes up and to the right. If the pressure on a sample of liquid carbon dioxide is increased at constant temperature, causing the molecules to get closer together, the liquid will solidify. This indicates that solid carbon dioxide has a higher density than the liquid phase. This is true for most substances. The notable exception is water.

13-72. *Refer to Section 13-13.*

(a) Phase diagram for butane:
    (not to scale)

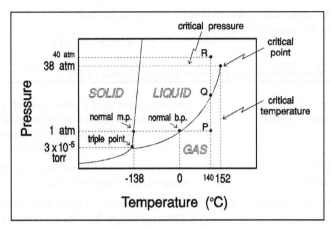

(b) As butane is compressed from 1 atm (point P) to 40 atm (point R) at 140°C, butane is converted from a gas to a liquid. Both phases are present simultaneously where the vapor pressure curve intersects the vertical isothermal line: $T = 140$°C indicated by point Q. At point Q, both phases are in equilibrium.

(c) Since 200°C is a temperature greater than the critical temperature of 152°C, there is no pressure at which two phases exist. At all pressures the liquid phase cannot be distinguished from the gas phase.

13-74. *Refer to Section 13-13 and the phase diagram for sulfur.*

(a) solid (monoclinic) (b) solid (rhombic) (c) solid (rhombic)
(d) solid (rhombic)   (e) vapor   (f) liquid

13-76. *Refer to Section 13-13.*

Ice, i.e., solid water, floats in liquid water because the solid state is less dense than the liquid state. However, like most other substances, solid mercury is more dense than liquid mercury and therefore, solid mercury sinks when placed in liquid mercury.

13-78. *Refer to Section 13-16.*

MoF_6 molecular solid   Pt   metallic solid
BN   covalent (network) solid   RbI ionic solid
Se_8 molecular solid

13-80. *Refer to Section 13-16.*

(a) SO_2F molecular solid   (d) Pb   metallic solid
(b) MgF_2 ionic solid   (e) PF_5 molecular solid
(c) W   metallic solid

13-82. *Refer to Sections 13-2 and 13-16.*

Melting points of ionic compounds increase with increasing ion-ion interactions which are functions of d, the distance between the ions, and q, the charge on the ions:

$$F \propto \frac{q^+q^-}{d^2}$$

Due to the differences in the number of charges on the cations (Na^+, Mg^{2+} and Al^{3+}), the following order of increasing melting points is predicted:

$$NaF \ < \ MgF_2 \ < \ AlF_3$$

13-84.   *Refer to Section 13-15, and Figures 13-25 and 13-29.*

Simple cubic lattice: The eight corners of a cubic unit cell are occupied by one kind of atom,
   (Example: CsCl)   while the center of the unit cell is occupied by another kind of atom. Note
                               that this is *not* the same as a body-centered cubic arrangement, because
                               the atom at the center of the cell is *not* the same as those at the corners.

Body-centered cubic lattice (bcc):   All eight corners *and* the point at the center of a cubic unit cell are
   (Example: Na)   occupied by a single kind of atom.

Face-centered cubic lattice (fcc):   The eight corners of a cubic unit cell as well as the central points on each
   (Example: Ni)   of the six faces of the cube are occupied by the same kind of atom.

CsCl   Na   Ni

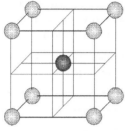 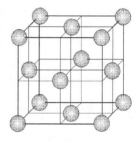

Unit cell 1: Cs^+ ◯ Cl^- ●   Na ◯   Ni ◯
Unit cell 2: Cs^+ ● Cl^- ◯

13-86.   *Refer to Section 13-15 and Figure 13-29.*

(1) For CsCl (simple cubic):   1 Cs^+ cation in center **x** 1 = 1 cation in the unit cell
                                 8 Cl^- anions in corners **x** 1/8 = 1 anion in unit cell

(2) For NaCl (face-centered cubic):
     (1 Na^+ cation in center **x** 1) + (12 Na^+ cations on edges **x** 1/4) = 4 cations in the unit cell
     (6 Cl^- anions on faces x 1/2) + (8 Cl^- anions on corners x 1/8) = 4 anions in the unit cell

(3) For ZnS (face-entered cubic):
     (4 Zn^{2+} cation in center **x** 1) = 4 cations in the unit cell
     (6 S^{2-} anions on faces x 1/2) + (8 S^{2-} anions on corners x 1/8) = 4 anions in the unit cell

---

13-88.   *Refer to Section 13-15, and Figures 13-28 and 13-29b.*

Consider the NaCl face-centered cubic structure with a unit
cell edge represented as *a*, shown here.

(a) The distance from Na^+ to its nearest neighbor is ***a*/2**.

(b) Each Na^+ ion has **6** equidistant nearest neighbors. They
     are **Cl^-** ions.

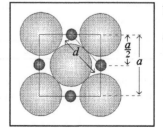

Na^+ ●

Cl^- ◯

(c) The distance, d, from Na^+ to nearest Na^+ is the length of a hypotenuse of an isosceles right triangle, with sides equal to $a/2$.

$$d^2 = (a/2)^2 + (a/2)^2$$

$$d = \sqrt{(a/2)^2 + (a/2)^2} = \sqrt{2(a/2)^2} = \sqrt{a^2/2} = \frac{a}{\sqrt{2}} \text{ or } \frac{a\sqrt{2}}{2}$$

(d) Each Cl^- ion has **6** equidistant nearest neighbors; they are $\mathbf{Na^+}$ ions.

13-90. *Refer to Section 13-16, Examples 13-9 and 13-10, and Exercise 13-84.*

Plan: (1) Calculate the volume of the unit cell, V.
(2) Calculate the mass, m, of Na atoms in the unit cell.
(3) Determine the density of Na:

$$\text{Density} = \frac{m_{\text{unit cell}}}{V_{\text{unit cell}}}$$

(1) let the length of the cube edge $= a = 4.24$ Å
$V_{\text{unit cell}} = a^3 = [4.24 \text{ Å} \times (1 \times 10^{-8} \text{ cm/Å})]^3 = 7.62 \times 10^{-23} \text{ cm}^3$

(2) A body-centered cubic unit cell contains 2 atoms in total:
in the corners: 8 Na atoms $\times$ 1/8 = 1 Na atom
in the center: 1 Na atom $\times$ 1   = 1 Na atom

$$? \ m_{\text{unit cell}} = 2 \text{ Na atoms} \times \frac{1 \text{ mol Na}}{6.02 \times 10^{23} \text{ atoms Na}} \times \frac{22.99 \text{ g Na}}{1 \text{ mol Na}} = 7.64 \times 10^{-23} \text{ g}$$

(3) Density of Na $= \dfrac{m}{V} = \dfrac{7.64 \times 10^{-23} \text{ g Na/unit cell}}{7.62 \times 10^{-23} \text{ cm}^3/\text{unit cell}} = \mathbf{1.00 \text{ g/cm}^3}$

The density of sodium as given by the *Handbook of Chemistry and Physics* is 0.97 g/cm³.

13-92. *Refer to Section 13-16, and Examples 13-9 and 13-10.*

Plan: (1) Calculate the mass of an Ir unit cell, m.
(2) Calculate the volume of an Ir unit cell, V.
(3) Calculate the density of Ir using

$$\text{Density} = \frac{\text{mass of unit cell } (m)}{\text{volume of unit cell } (V)}.$$

(1) $m_{\text{face-centered unit cell}}$ = mass of 4 Ir atoms

$$= 4 \text{ Ir atoms} \times \frac{1 \text{ mol Ir}}{6.022 \times 10^{23} \text{ Ir atoms}} \times \frac{192.2 \text{ g}}{1 \text{ mol Ir}}$$

$$= 1.277 \times 10^{-21} \text{ g}$$

(e) $V_{\text{cubic unit cell}}$ = (length of unit cell edge, a)³

For a face-centered cubic unit cell, the diagonal of a face is the hypotenuse of a right triangle and equals $4r$ where r is the radius of an Ir atom. From the Pythagorean Theorem,

$$(4r)^2 = a^2 + a^2$$
$$(4 \times 1.36 \text{ Å})^2 = 2a^2$$
$$a \text{ (Å)} = 3.85 \text{ Å}$$
$$a \text{ (cm)} = 3.85 \times 10^{-8} \text{ cm}$$

$$V_{\text{cubic unit cell}} = (a)^3 = (3.85 \times 10^{-8} \text{ cm})^3$$
$$= 5.71 \times 10^{-23} \text{ cm}^3$$

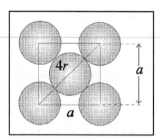

(3) Therefore,

$$\text{Density} = \frac{\text{mass of unit cell } (m)}{\text{volume of unit cell } (V)} = \frac{1.277 \times 10^{-21} \text{ g}}{5.71 \times 10^{-23} \text{ cm}^3} = \mathbf{22.4 \text{ g/cm}^3}$$

The Handbook of Chemistry and Physics (71st Ed.) gives the density of Ir as 22.421 g/cm^3.

13-94. *Refer to Section 13-16.*

(a) The cubic unit cell of diamond:

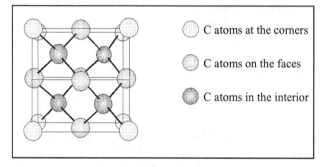

The unit cell contains 8 C atoms:

8 atoms at the corners **x** 1/8 = 1 atom
6 atoms on the faces **x** 1/2 = 3 atoms
4 atoms in the interior **x** 1 = 4 atoms

(b) Each C atom is at the center of a **tetrahedron** and hence has 4 nearest neighbors.

(c) Consider the lower right hand corner of diamond's unit cell:

Triangle BCD is in the plane of the base of the unit cell.
Line BC = line CD = a/4.
So, (line BD)^2 = (line BC)^2 + (line CD)^2

$$\text{line BD} = \sqrt{(a/4)^2 + (a/4)^2} = \frac{a}{2\sqrt{2}}$$

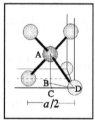

Triangle ABD is perpendicular to the base of the unit cell. Line AD is the distance from any carbon atom to its nearest neighbor.

$$\text{(line AD)}^2 = \text{(line AB)}^2 + \text{(line BD)}^2$$

$$\text{line AD} = \sqrt{\left(\frac{a}{4}\right)^2 + \left(\frac{a}{2\sqrt{2}}\right)^2} = \sqrt{3}\left(\frac{a}{4}\right)$$

(d) The unit cell edge is 3.567 Å. Therefore,

C-C bond length = line AD = $\sqrt{3}$(3.567 Å/4) = **1.545 Å**

(e) $V_{unit\ cell} = a^3 = [3.567\ \text{Å} \times (1 \times 10^{-8}\ \text{cm/Å})]^3 = 4.538 \times 10^{-23}\ \text{cm}^3$

$m_{unit\ cell} = 8\ \text{C atoms} \times \dfrac{1\ \text{mol}}{6.022 \times 10^{23}\ \text{atoms}} \times \dfrac{12.01\ \text{g}}{1\ \text{mol}} = 1.595 \times 10^{-22}\ \text{g}$

Density of $C_{diamond} = \dfrac{m}{V} = \dfrac{1.595 \times 10^{-22}\ \text{g}}{4.538 \times 10^{-23}\ \text{cm}^3} = \textbf{3.515 g/cm}^3$

The density of $C_{diamond}$ as given by the *Handbook of Chemistry and Physics* is 3.51 g/cm³.

13-96. *Refer to Section 13-16 and Example 13-11.*

Plan: (1) Calculate the volume of the unit cell, V.
   (2) Determine the mass of the unit cell, m, given the density.
   (3) Knowing the number of atoms in a face-centered cube, find the moles of atoms in the unit cell.
   (4) Determine the atomic weight of the unknown and its identity.

(1) let the length of the cube edge = a = 4.95 Å
   $V_{unit\ cell} = a^3 = [4.95\ \text{Å} \times (1 \times 10^{-8}\ \text{cm/Å})]^3 = 1.21 \times 10^{-22}\ \text{cm}^3$

(2) $m_{unit\ cell} = D \times V_{unit\ cell} = (11.35\ \text{g/cm}^3)(1.21 \times 10^{-22}\ \text{cm}^3) = 1.37 \times 10^{-21}\ \text{g}$ since $D = \dfrac{m}{V}$

(3) A face-centered cubic unit cell contains 4 atoms:
   in the corners   8 atoms × 1/8 = 1 atom
   on the faces   6 atoms × 1/2 = 3 atoms

   ? mol element in unit cell = 4 atoms × $\dfrac{1\ \text{mol atoms}}{6.02 \times 10^{23}\ \text{atoms}}$ = 6.64 × 10⁻²⁴ mol

(4) AW = $\dfrac{\text{g element}}{\text{mol element}} = \dfrac{m_{unit\ cell}}{mol_{unit\ cell}} = \dfrac{1.37 \times 10^{-21}\ \text{g}}{6.64 \times 10^{-24}\ \text{mol}} = \textbf{206 g/mol}$

   Therefore, the 4A element is **Pb (207.2 g/mol)**

13-98. *Refer to Section 13-14 and Figure 13-19.*

(a) Crystal diffraction studies use the x-ray region of the electromagnetic radiation.

(b) In the x-ray diffraction experiment, a monochromatic x-ray beam is attenuated by a system of slits and aimed at a crystal. The crystal is rotated to vary the angle of incidence. At certain angles which depend on the crystal's unit cell, x-rays are deflected and hit a photographic plate. After development the plate shows a set of symmetrically arranged spots. From the arrangement of the spots, the crystal structure can be determined.

(c) In order for diffraction to occur, the wavelength of the in-coming radiation must be about the same as the inter-nuclear separations in the crystal.

13-100. *Refer to Section 13-14.*

In x-ray diffraction, the Bragg equation is used:
   $n\lambda = 2d\sin\theta$ where n = 1 for the minimum diffraction angle
                λ = wavelength of Cu radiation
                θ = angle of incidence or diffraction angle, 19.98°
                d = spacing between parallel layers of Pt atoms, 2.256 Å

Solving for λ,

$$\lambda = \dfrac{2d\sin\theta}{n} = \dfrac{2 \times 2.256\ \text{Å} \times \sin 19.98°}{1} = \textbf{1.542 Å}$$

13-102. *Refer to Sections 13-16 and 13-17.*

When a metal is distorted (e.g., rolled into sheets or drawn into wire), new metallic bonds are formed and the environment around each atom is essentially unchanged. This can happen because the valence electrons of bonded metal atoms are only loosely associated with individual atoms, as though metal cations exist in a "cloud of electrons."

In ionic solids, the lattice arrangements of cations and anions are more rigid. When an ionic solid is distorted, it is possible for cation-cation and anion-anion alignments to occur. However, this will cause the solid to shatter due to electrostatic repulsions between ions of like charge.

13-104. *Refer to Section 13-17.*

According to band theory, the electrons in a substance must move within the conduction band in order to conduct electricity. The conduction band is a partially filled band or a band of vacant energy levels just higher in energy than a filled band. Metals conduct electricity because the highest energy electrons can easily move in a conduction band. For a metal, the electrical conductivity decreases as the temperature increases. The increase in temperature causes the metal ions to move more, which slows down the flow of electrons when an electric field is applied. A typical metalloid is a semiconductor. There is a small gap in energy between the highest-energy electrons and the conduction band. A semiconductor does not conduct electricity at low temperatures, but a small increase in temperature will excite some of the highest-energy electrons into the empty conduction band. So, electrical conductivity increases with increasing temperature for a metalloid.

13-106. *Refer to Sections 13-6 and 13-8.*

The first gas to evaporate would be the gas with the weakest intermolecular forces acting between the molecules.

Since a lower boiling point is an indicator of weaker intermolecular forces of attraction at work, N_2 (b.p. $-196°C$) would be the first gas to evaporate, then Ar (b.p. $-186°C$), and lastly, O_2 (b.p. $-183°C$).

13-108. *Refer to Section 13-3.*

Viscosity is the resistance to flow of a liquid. The stronger the intermolecular forces of attraction, the more viscous the liquid will be.

We expect the viscosity of ethylene glycol ($HOCH_2CH_2OH$) to be greater than that of ethanol (CH_3CH_2OH) because both molecules are about the same size, but ethylene glycol contains two sites where hydrogen bonding can occur whereas ethanol has only one.

13-110. *Refer to Sections 13-2 and 13-8.*

Consider the following three molecules with formula $C_2H_2Cl_2$:

(1) 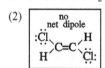 (2) 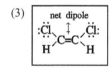 (3)

All of the compounds have dispersion forces of attraction operating between their molecules. Since the compounds are approximately the same size, their dispersion forces are about the same. Compounds (1) and (3) are polar and have permanent dipole-dipole interactions.

The compound with the lowest boiling point is the one with the weakest intermolecular forces of attraction. Compound (2) which is nonpolar and the same approximate size as the others should therefore have the lowest boiling point.

13-112. *Refer to Sections 13-7 and 13-8.*

(a) true

(b) false   The normal boiling point of a liquid is defined as the boiling point at 760 torr.

(c) false   As long as the temperature remains constant and some liquid remains, the vapor pressure of a liquid will remain the same. The escaping power of liquid molecules possessing sufficient kinetic energy to go into the gas phase stays the same regardless of the quantity of the liquid.

13-114. *Refer to Sections 13-2 and 13-9, and Exercise 13-113.*

The molar heat of vaporization, ΔH_{vap}, at the boiling point is a measure of the heat required to change 1 mole of a liquid into a gas. The stronger the intermolecular forces of attraction between the molecules, the greater will be the value of ΔH_{vap}. Since vapor pressure at a given temperature decreases with increasing intermolecular force, we expect the ΔH_{vap} values to be the reverse of the vapor pressure trend.

carbon disulfide < acetone < ethanol

This makes sense because carbon disulfide, CS_2, is a nonpolar molecule and has only dispersion forces at work holding the molecules together as a liquid. Acetone, CH_3COCH_3, is a larger polar molecule and has stronger dispersion forces (due to its larger size) as well as dipole-dipole forces present. Ethanol, CH_3CH_2OH, has dispersion forces and the very powerful hydrogen bonding present.

13-116. *Refer to Sections 13-13 and 13-14.*

(a) gas   (b) condense, liquid   (c) freeze, crystalline solid   (d) sublimation, gas

---

13-118. *Refer to Sections 13-2, 13-8, 13-10 and 13-12.*

(a) liquid (T = 0°C at STP)

(b) liquid; freezes (or solidifies or crystallizes); crystalline solid

(c) sublimation; gas; 24°C

13-120. *Refer to Section 13-8.*

When a liquid boils, the bubbles are filled with the liquid's vapor.

13-122. *Refer to Section 13-13.*

At room temperature, I would expect to find the sealed container with the iodine crystals filled with a lightly purplish gas (iodine vapor) and tiny crystals of iodine on the inside walls and top of the container where the iodine gas redeposited. The equilibrium involved here is

$$I_2(s) \rightleftarrows I_2(g)$$

In the sealed container of liquid water, I would expect to find droplets of water on the inside of the container where the water vapor condensed. The equilibrium involved in this case is:

$$H_2O(\ell) \rightleftarrows H_2O(g)$$

13-124. *Refer to Section 13-2 and Figure 13-3.*

The molecule, HBr, is polar since the electronegativities of H and Br are 2.1 and 2.8, respectively.

The arrows between the larger, more electronegative Br atom of one HBr molecule and the smaller, less electronegative H atom of an adjacent HBr molecule represent dipole-dipole attractions.

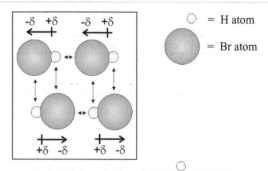

13-126. *Refer to Section 13-13.*

This question, "which freezes first, hot water or cold water," is a favorite of popular science magazines. This discussion is taken from the web site: http://math.ucr.edu/home/baez/physics/General/hot_water.html. Sir Francis Bacon, Descartes and even Aristotle are said to have remarked on it. There are five factors that can make the hot water freeze faster than expected.

(1) Evaporation: Evaporation occurs faster from warmer water. The mass of the hot water may decrease enough to make up for the greater temperature range it must cover to reach freezing. Also, there is a greater cooling effect from evaporation.

(2) Supercooling: The hot water sample may have a greater tendency to supercool, because it has less dissolved gas which can act as nucleation points for ice to form. Water that does not supercool may form a thin layer of ice at the surface which can insulate the rest of the water from the freezer and delay the freezing process.

(3) Convection: As water cools, convection currents within the sample occur as cooler, more dense water sinks, causing the water to circulate. However, water is most dense at 4°C, so at some point, colder liquid water will sink and a film of less dense ice will form at the top and insulate the sample from the cold. Some people speculate that the warmer sample will have more convection currents and will cool quicker.

(4) Dissolved Gases: Dissolved gases, such as oxygen and carbon dioxide, lower the freezing point of water. Cooler water has more dissolved gases and so the freezing point is slightly lower and the time that it would take to reach the slightly lower freezing point would be more.

(5) Conduction: When a hot water sample is put into a freezer, it has been observed that the sample container sometimes melts the ice-encrusted surface, which then allows for much better heat conduction than the frost on which the colder container rests. As a result, heat is drawn from the warmer container more rapidly.

13-128. *Refer to Sections 1-13, 13-9 and 13-11.*

The water droplets on the outside of a cold, unopened soda can came from condensed atmospheric water, just like dew forms on the morning grass.

13-130. *Refer to Sections 1-13, 13-9 and 13-11.*

Plan: (1) Find out the specific heat of the unknown metal in J/g·°C.
       (2) Using the Law of Dulong and Petit, find out the atomic weight of the metal and then identify it.

(1) |heat lost by the metal| = |heat gained by the water|

        |mass x specific heat x Δt|$_{metal}$ = |mass x specific heat(ℓ) x Δt|$_{water}$

    (100.2 g)(specific heat of metal)(99.9°C - 36.6°C) = (50.6 g)(4.184 J/g·°C)(36.6°C - 24.8°C)
            Solving, the specific heat of the metal = 0.394 J/g·°C

206

(2) $? AW = \dfrac{25 \text{ J/mol·°C}}{0.394 \text{ J/g·°C}} = 63$ g/mol

Therefore, the metal is likely to be **Cu**.

13-132. *Refer to Section 13-7.*

The vapor pressure of a liquid in equilibrium with its vapor cannot be treated like an ideal gas that obeys the gas laws; the equilibrium (liquid $\rightleftarrows$ vapor) controls the vapor pressure. As conditions are changed, the system adjusts itself until the system reaches equilibrium again; either the liquid which is present evaporates, or the vapor condenses.

In particular, if the temperature increases, more liquid evaporates becoming vapor to increase the vapor pressure. Mathematically, the saturated vapor pressure of a liquid increases exponentially instead of linearly with increasing temperature. Vapor pressure cannot be calculated from the ideal gas law: $P_1/T_1 = P_2/T_2$ since n, the number of moles of gas present, is not a constant but increases greatly with temperature.

13-134. *Refer to Section 13-2.*

(a) CH_3COOH

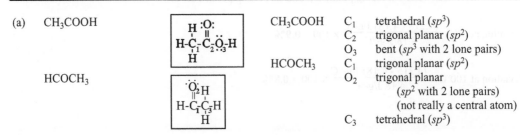

$HCOCH_3$

CH_3COOH	C_1	tetrahedral (sp^3)
	C_2	trigonal planar (sp^2)
	O_3	bent (sp^3 with 2 lone pairs)
$HCOCH_3$	C_1	trigonal planar (sp^2)
	O_2	trigonal planar
		(sp^2 with 2 lone pairs)
		(not really a central atom)
	C_3	tetrahedral (sp^3)

Both molecules in (a) are about the same size and are polar. Both molecules have dispersion and dipole-dipole intermolecular forces operating between their molecules. However, CH_3COOH has hydrogen bonds operating between its molecules, since CH_3COOH contains a hydrogen atom directly bonded to the small, highly electronegative atom, O; $HCOCH_3$ does not. Therefore, $HCOCH_3$ will have the lower boiling point.

(b) NHF_2

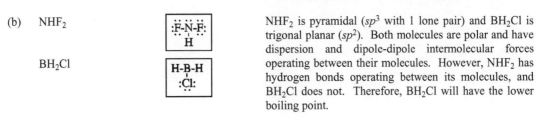

BH_2Cl

NHF_2 is pyramidal (sp^3 with 1 lone pair) and BH_2Cl is trigonal planar (sp^2). Both molecules are polar and have dispersion and dipole-dipole intermolecular forces operating between their molecules. However, NHF_2 has hydrogen bonds operating between its molecules, and BH_2Cl does not. Therefore, BH_2Cl will have the lower boiling point.

(c) CH$_3$CH$_2$OH

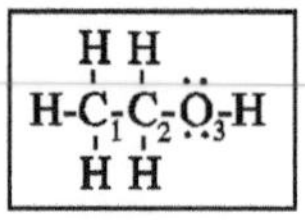

$$\begin{array}{|c|} \hline \text{H H} \\ \text{H-C}_1\text{-C}_2\text{-Ö}_3\text{H} \\ \text{H H} \\ \hline \end{array}$$

CH$_3$OCH$_3$

$$\begin{array}{|c|} \hline \text{H} \quad \text{H} \\ \text{H-C}_1\text{-Ö}_2\text{-C}_3\text{H} \\ \text{H} \quad \text{H} \\ \hline \end{array}$$

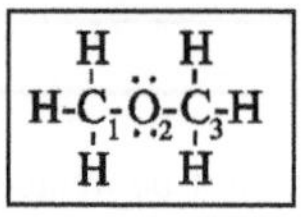

CH$_3$CH$_2$OH	C$_1$	tetrahedral (sp^3)
	C$_2$	tetrahedral (sp^3)
	O$_3$	bent (sp^3 with 2 lone pairs)
CH$_3$OCH$_3$	C$_1$	tetrahedral (sp^3)
	O$_2$	bent (sp^3 with 2 lone pairs)
	C$_3$	tetrahedral (sp^3)

Both molecules are about the same size and are polar. They have dispersion and dipole-dipole intermolecular forces operating between their molecules. However, CH$_3$CH$_2$OH has hydrogen bonds operating between its molecules, and CH$_3$OCH$_3$ does not. Therefore, CH$_3$OCH$_3$ will have the lower boiling point.

13-136. *Refer to Section 2-6.*

The specific heat of liquid water varies significantly with temperature. It is usually given as 4.18 J/g·°C but deviates the most near water's freezing point, 0°C and its boiling point, 100°C.

$$\% \text{ deviation at } 0°C = \frac{(4.218 - 4.18) \text{ J/g·°C}}{4.18 \text{ J/g·°C}} \times 100 = \textbf{0.9\%}$$

$$\% \text{ deviation at } 100°C = \frac{(4.215 - 4.18) \text{ J/g·°C}}{4.18 \text{ J/g·°C}} \times 100 = \textbf{0.8\%}$$

14 Solutions

14-2. *Refer to the Introduction to Chapter 14.*

	Type of Solution	Example	Solute	Solvent
(a)	solid dissolved in liquid	salt water	$NaCl(s)$	$H_2O(\ell)$
(b)	gas dissolved in gas	air (major components)	$O_2(g)$	$N_2(g)$
(c)	gas dissolved in liquid	$HCl(aq)$	$HCl(g)$	$H_2O(\ell)$
(d)	liquid dissolved in liquid	$CH_3COOH(aq)$	$CH_3COOH(\ell)$	$H_2O(\ell)$
(e)	solid dissolved in solid	brass	$Zn(s)$	$Cu(s)$

14-4. *Refer to Section 14-1 and Figure 14-1.*

Dissolution is favored when (a) solute-solute attractions and (b) solvent-solvent attractions are relatively small and (c) solvent-solute attractions are relatively large. Both processes (a) and (b) require energy, first to separate the solute particles from each other, then to separate the solvent molecules from each other. Process (c) releases energy as solute particles and solvent molecules interact. If the absolute value of heat absorbed in processes (a) and (b) is less than the absolute value of heat released in process (c), then the dissolving process is favored and releases heat.

14-6. *Refer to Section 14-1.*

There are two factors which control the spontaneity of a dissolution process: (1) the amount of heat absorbed or released and (2) the amount of increase in the disorder, or randomness of the system. All dissolution processes are accompanied by an increase in the disorder of both solvent and solute. Thus, their disorder factor is invariably favorable to solubility. Dissolution will always occur if the dissolution process is exothermic and the disorder term increases. Dissolution will occur when the dissolution process is endothermic if the disorder term is large enough to overcome the endothermicity, which opposes dissolution.

14-8. *Refer to Section 14-3.*

When two completely miscible, nonreactive liquids, A and B, are mixed in any proportions, the molecules of A and B will intermingle, and one phase only is always produced. When two completely immiscible liquids, C and D, are mixed in any proportions, two separate phases are always produced; one is pure C, the other is pure D.

However, when liquid E is slowly added to liquid F, with which it is only partially miscible, in the beginning only one phase is present ($V_E \ll V_F$). When the solubility of liquid E in liquid F is exceeded, two phases result. After a very large quantity of liquid E is added to liquid F ($V_E \gg V_F$), one phase again is present.

14-10. *Refer to Sections 14-2, 14-3 and 14-4.*

(a)	HCl in H_2O	high solubility since HCl will ionize in water, which is very polar, to form ions.
(b)	HF in H_2O	high solubility since both are polar covalent molecules which are capable of forming hydrogenbonds.
(c)	Al_2O_3 in H_2O	low solubility since both Al^{3+} and O^{2-} in the ionic solid have high charge-to-size ratios and therefore have larger lattice energies. The dissolution of Al_2O_3 is very endothermic and is not favored.

209

(d) SiO_2 in H_2O low solubility since SiO_2 is a covalent solid called quartz and H_2O is very polar.

(e) Na_2SO_4 in C_6H_{14} low solubility since Na_2SO_4 is an ionic solid and C_6H_{14} is a nonpolar solvent.

14-12.   *Refer to Sections 4-2, 14-2 and 14-3, and Exercise 14-10.*

(a) HCl in H_2O strong electrolyte

(b) HF in H_2O weak electrolyte

(c) Al_2O_3 in H_2O cannot be prepared in "reasonable" concentration

(d) SiO_2 in H_2O cannot be prepared in "reasonable" concentration

(e) Na_2SO_4 in C_6H_{14} cannot be prepared in "reasonable" concentration

---

14-14.   *Refer to Sections 14-1, 14-2 and 14-3.*

(a) The solubility of a solid in a liquid does not depend appreciably on pressure.

(b) The solubility of a liquid in a liquid also is essentially independent of pressure, because the interactions (solute:solute, solvent:solvent, and solute:solvent) are not affected by pressure.

14-16.   *Refer to Sections 14-2, 14-5 and 14-6.*

To determine if a sodium chloride solution is indeed saturated, one could add a known mass of solid sodium chloride, stir for an extended period of time, then quantitatively filter the remaining solid. If the weight of the dried NaCl is equal to what was originally added, the solution was saturated originally; if the weight is less, the solution was unsaturated. Another qualitative way would be to observe if a tiny added amount of NaCl would dissolve or not.

14-18.   *Refer to Section 14-7 and Exercise 14-17.*

Henry's Law states: $C_{gas} = X_{gas} = kP_{gas}$ where $C_{gas} = X_{gas}$ = mole fraction of gas at a certain temperature

k = Henry's Law constant

P_{gas} = partial pressure of gas above the solution

For CH_4 at 25°C:
$$X_{CH_4} = k\,P_{CH_4}$$
$$= (2.42 \times 10^{-5}\ atm^{-1})(10.\ atm)$$
$$X_{CH_4} = \mathbf{2.4 \times 10^{-4}}$$

For CH_4 at 50°C:
$$X_{CH_4} = k\,P_{CH_4}$$
$$= (1.73 \times 10^{-5}\ atm^{-1})(10.\ atm)$$
$$X_{CH_4} = \mathbf{1.7 \times 10^{-4}}$$

Therefore, the solubility of $CH_4(g)$ **decreases** with increasing temperature.

14-20.   *Refer to Section 14-2 and Table 14-1.*

The hydration energy of ions generally increases with increasing charge and decreasing size. The greater the hydration energy for an ion, the more strongly hydrated it is. In order of decreasing hydration energy:

(a) $Na^+ > Rb^+$ due to smaller size

(b) $Cl^- > Br^-$ due to smaller size

(c) $Fe^{3+} > Fe^{2+}$ due to higher charge and smaller size

(d) $Mg^{2+} > Na^+$ due to higher charge and smaller size

14-22. *Refer to Sections 14-1, 14-2 and 14-3.*

The dissolution of many ionic solids in liquids is endothermic (requires heat) due to the large solute-solute attractions that must be overcome relative to the solvent-solute attractions.
The mixing of two miscible liquids is exothermic (releases heat) since solute-solute attractions are less than the solvent-solute attractions.

14-24. *Refer to Section 3-6.*

If s = solubility of A in $\left(\dfrac{g\ A}{100.\ g\ H_2O}\right)$, then the maximum mass of A that will dissolve in 100. g of H_2O = s.
The solubility of A as a mass percent:

$$\text{solubility of A in}\left(\frac{g\ A}{100.\ g\ \text{solution}}\right) = \frac{g\ A}{g\ A + 100.\ g\ H_2O} \times 100\% = \frac{s}{s + 100} \times 100\%$$

14-26. *Refer to Section 14-8 and Example 14-1.*

(1) ? mol camphor, $C_{10}H_{16}O = \dfrac{45.0\ g}{152.2\ g/mol} = 0.296\ \text{mol}\ C_{10}H_{16}O$

? kg $C_2H_5OH = 425\ mL \times \dfrac{0.785\ g}{1\ mL} \times \dfrac{1\ kg}{1000\ g} = 0.334\ kg$

Therefore,

? m $C_{10}H_{16}O = \dfrac{\text{mol solute}}{\text{kg solvent}} = \dfrac{\text{mol}\ C_{10}H_{16}O}{\text{kg}\ C_2H_5OH} = \dfrac{0.296\ \text{mol}}{0.334\ kg} = \mathbf{0.886\ \textit{m}\ C_{10}H_{16}O}$

(2) ? mol $C_2H_5OH = 425\ mL \times \dfrac{0.785\ g}{1\ mL} \times \dfrac{1\ mol}{46.07\ g} = 7.24\ \text{mol}\ C_2H_5OH$

$X_{C_{10}H_{16}O} = \dfrac{\text{mol}\ C_{10}H_{16}O}{\text{mol}\ C_{10}H_{16}O + \text{mol}\ C_2H_5OH} = \dfrac{0.296\ \text{mol}}{0.296\ \text{mol} + 7.24\ \text{mol}} = \mathbf{0.0393}$ (no units)

(3) Recall that the mass of the solution equals the sum of the masses of the solute and the solvent.

mass % $C_{10}H_{16}O = \dfrac{g\ C_{10}H_{16}O}{g\ \text{soln}} \times 100\% = \dfrac{45.0\ g}{45.0\ g + 334\ g} \times 100\% = \mathbf{11.9\%\ C_{10}H_{16}O}$

14-28. *Refer to Section 14-8 and Example 14-1.*

We know: $m = \dfrac{\text{mol solute}}{\text{kg solvent}}$

? mol $C_6H_5COOH = \dfrac{56.5\ g\ C_6H_5COOH}{122\ g/mol} = 0.463\ \text{mol}\ C_6H_5COOH$

? kg $C_2H_5OH = 350\ mL\ C_2H_5OH \times \dfrac{0.789\ g\ C_2H_5OH}{1\ mL\ C_2H_5OH} \times \dfrac{1\ kg}{1000\ g} = 0.28\ kg\ C_2H_5OH$

Therefore, $m = \dfrac{0.463\ \text{mol}\ C_6H_5COOH}{0.28\ kg\ C_2H_5OH} = \mathbf{1.7\ \textit{m}\ C_6H_5COOH\ in\ C_2H_5OH}$

14-30. *Refer to Section 14-8 and Example 14-3.*

Plan: (1) Calculate the moles of each component.
(2) Calculate the mole fraction.

(1) $? \text{ mol C}_2\text{H}_5\text{OH} = \dfrac{55.0 \text{ g}}{46.1 \text{ g/mol}} = 1.19 \text{ mol C}_2\text{H}_5\text{OH}$

$? \text{ mol H}_2\text{O} = \dfrac{45.0 \text{ g}}{18.0 \text{ g/mol}} = 2.50 \text{ mol H}_2\text{O}$

(2) $X_{\text{C}_2\text{H}_5\text{OH}} = \dfrac{\text{mol C}_2\text{H}_5\text{OH}}{\text{mol C}_2\text{H}_5\text{OH} + \text{mol H}_2\text{O}} = \dfrac{1.19 \text{ mol}}{1.19 \text{ mol} + 2.50 \text{ mol}} = \textbf{0.322}$

$X_{\text{H}_2\text{O}} = \dfrac{\text{mol H}_2\text{O}}{\text{mol H}_2\text{O} + \text{mol C}_2\text{H}_5\text{OH}} = \dfrac{2.50 \text{ mol}}{2.50 \text{ mol} + 1.19 \text{ mol}} = \textbf{0.678}$

Alternative method: $X_{\text{H}_2\text{O}} = 1 - X_{\text{C}_2\text{H}_5\text{OH}} = 1.000 - 0.322 = \textbf{0.678}$

14-32. *Refer to Sections 3-6 and 14-8, and Examples 14-1 and 14-3.*

(1) $? \text{ mol K}_2\text{SO}_4 = \dfrac{12.50 \text{g}}{174.3 \text{ g/mol}} = 0.07172 \text{ mol K}_2\text{SO}_4$

$? \text{ mL soln} = \dfrac{\text{g soln}}{\text{density}} = \dfrac{100.00 \text{ g}}{1.083 \text{ g/mL}} = 92.34 \text{ mL}$

Therefore,

$M \text{ K}_2\text{SO}_4 = \dfrac{\text{mol K}_2\text{SO}_4}{\text{L soln}} = \dfrac{0.07172 \text{ mol}}{0.09234 \text{ L}} = \textbf{0.7767 } \boldsymbol{M} \textbf{ K}_2\textbf{SO}_4$

(2) $m \text{ K}_2\text{SO}_4 = \dfrac{\text{mol K}_2\text{SO}_4}{\text{kg H}_2\text{O}} = \dfrac{0.07172 \text{ mol}}{0.08750 \text{ kg}} = \textbf{0.8197 } \boldsymbol{m} \textbf{ K}_2\textbf{SO}_4$

since $\quad$ kg $\text{H}_2\text{O} =$ kg soln $-$ kg $\text{K}_2\text{SO}_4 = 0.10000$ kg $- 0.01250$ kg $= 0.08750$ kg

(3) mass % $\text{K}_2\text{SO}_4 = \dfrac{\text{g K}_2\text{SO}_4}{\text{g soln}} \times 100\% = \dfrac{12.50 \text{ g}}{100.00 \text{g}} \times 100\% = \textbf{12.50\% K}_2\textbf{SO}_4$

(4) $X_{\text{H}_2\text{O}} = \dfrac{\text{mol H}_2\text{O}}{\text{mol H}_2\text{O} + \text{mol K}_2\text{SO}_4} = \dfrac{\left(\dfrac{87.50 \text{ g}}{18.02 \text{ g/mol}}\right)}{\left(\dfrac{87.50 \text{ g}}{18.02 \text{ g/mol}} + \dfrac{12.50 \text{ g}}{174.3 \text{ g/mol}}\right)} = \dfrac{4.856 \text{ mol}}{4.856 \text{ mol} + 0.07172 \text{ mol}} = \textbf{0.9854}$

14-34. *Refer to Section 14-8, and Examples 14-1 and 14-2.*

(a) Plan: $\quad$ (1) Assuming 100 g of solution, calculate the moles of $\text{C}_6\text{H}_{12}\text{O}_6$. Note that the solution density is not required for the calculation.

$\qquad\qquad$ (2) Calculate the mass of water.

$\qquad\qquad$ (3) Determine the molality of $\text{C}_6\text{H}_{12}\text{O}_6$.

(1) Recall: % by mass $= \dfrac{\text{g solute}}{\text{g solution}} \times 100\%$, so solving for the mass of $\text{C}_6\text{H}_{12}\text{O}_6$ in 100 g soln, we have

$? \text{ g C}_6\text{H}_{12}\text{O}_6 \text{ in 100 g soln} = \dfrac{21.0\% \text{ C}_6\text{H}_{12}\text{O}_6 \times 100 \text{ g soln}}{100\%} = 21.0 \text{ g C}_6\text{H}_{12}\text{O}_6$

$? \text{ mol C}_6\text{H}_{12}\text{O}_6 = \dfrac{21.0 \text{ g}}{180.2 \text{ g/mol}} = 0.117 \text{ mol C}_6\text{H}_{12}\text{O}_6$

(2) $? \text{ g H}_2\text{O} = \text{g soln} - \text{g C}_6\text{H}_{12}\text{O}_6 = 100.0 \text{ g} - 21.0 \text{ g} = 79.0 \text{ g}$

(3) $m \text{ C}_6\text{H}_{12}\text{O}_6 = \dfrac{\text{mol C}_6\text{H}_{12}\text{O}_6}{\text{kg H}_2\text{O}} = \dfrac{0.117 \text{ mol}}{0.0790 \text{ kg}} = \textbf{1.48 } \boldsymbol{m} \textbf{ C}_6\textbf{H}_{12}\textbf{O}_6$

(b) Molality is independent of density and temperature since it is a measure of the number of moles of substance dissolved in 1 kilogram of solvent. Only concentration units that involve volume, e.g., molarity, are dependent on density and temperature. Therefore, the molality at a higher temperature would be the **same** as the molality at 20°C.

14-36. *Refer to Section 14-9.*

The vapor pressure of a liquid depends upon the ease with which the molecules are able to escape from the surface of the liquid. The vapor pressure of a liquid always decreases when nonvolatile solutes (ions or molecules) are dissolved in it, since after dissolution there are fewer solvent molecules at the surface to vaporize.

14-38. *Refer to Section 14-9 and Example 14-4.*

(a) From Raoult's Law,
$$\Delta P_{water} = X_{ethylene\ glycol}\, P^{\circ}_{water}$$

where ΔP_{water} = vapor pressure lowering of water, H_2O
P°_{water} = vapor pressure of pure water, 19.83 torr at 22°C
$X_{ethylene\ glycol}$ = mole fraction of ethylene glycol, $C_2H_6O_2$

$$\Delta P_{water} = 0.288 \times 19.83 = 5.71\ torr$$

$$P_{water} = P^{\circ}_{water} - \Delta P_{water} = 19.83\ torr - 5.71\ torr = \mathbf{14.12\ torr}$$

(b) Given: % ethylene glycol by mass = 39.0%
If we assume 100 g of solution, then the mass of ethylene glycol is 39.0 g and mass of water is 61.0 g.

$$X_{ethylene\ glycol} = \frac{mol\ C_2H_6O_2}{mol\ C_2H_6O_2 + mol\ H_2O} = \frac{\left(\dfrac{39.0\ g}{62.07\ g/mol}\right)}{\left(\dfrac{39.0\ g}{62.07\ g/mol} + \dfrac{61.0\ g}{18.02\ g/mol}\right)} = \frac{0.628\ mol}{0.628\ mol + 3.39\ mol} = 0.157$$

$$\Delta P_{water} = 0.157 \times 19.83 = 3.11\ torr$$
$$P_{water} = P^{\circ}_{water} - \Delta P_{water} = 19.83\ torr - 3.11\ torr = \mathbf{16.72\ torr}$$

(c) Given: 2.42 m ethylene glycol, $C_2H_6O_2$
If we assume 1.00 kg of water, then the solution contains 2.42 moles $C_2H_6O_2$

$$X_{ethylene\ glycol} = \frac{mol\ C_2H_6O_2}{mol\ C_2H_6O_2 + mol\ H_2O} = \frac{2.42\ mol}{2.42\ mol + \dfrac{1000.\ g}{18.02\ g/mol}} = \frac{2.42\ mol}{2.42\ mol + 55.5\ mol} = 0.0418$$

$$\Delta P_{water} = 0.0418 \times 19.83 = 0.829\ torr$$
$$P_{water} = P^{\circ}_{water} - \Delta P_{water} = 19.83\ torr - 0.829\ torr = \mathbf{19.00\ torr}$$

14-40. *Refer to Section 14-9, Example 14-5 and the figure accompanying Exercise 14-43.*

Plan: (1) Calculate the mole fraction of each component in the solution.
(2) By assuming the solution to be ideal, apply Raoult's Law to calculate the partial pressures of acetone and chloroform.

(1) Let X = mole fraction of the components
$$X_{acetone} = \frac{0.550\ mol\ acetone}{(0.550\ mol\ acetone) + (0.550\ mol\ chloroform)} = 0.500$$
$$X_{chloroform} = \frac{0.550\ mol\ chloroform}{(0.550\ mol\ acetone) + (0.550\ mol\ chloroform)} = 0.500\ (= 1 - X_{acetone})$$

(2) From Raoult's Law:

$P_{acetone} = X_{acetone} \, P^{\circ}_{acetone} = (0.500)(345 \text{ torr}) = \textbf{172 torr}$

$P_{chloroform} = X_{chloroform} \, P^{\circ}_{chloroform} = (0.500)(295 \text{ torr}) = \textbf{148 torr}$

Since the acetone-chloroform system is expected to show negative deviation from an ideal solution, $P_{acetone}$ should be less than 172 torr and $P_{chloroform}$ should be less than 148 torr.

14-42. *Refer to Sections 14-9 and 12-11, Examples 14-5 and 14-6, and Exercise 14-40 Solution.*

Plan: (1) Calculate the total pressure using Exercise 14-40 Solution: $P_{acetone} = 172$ torr, $P_{chloroform} = 148$ torr.

(2) Since the mole fraction of a component in a gaseous mixture equals the ratio of its partial pressure to the total pressure, the composition of the vapor above the solution can be calculated.

(1) $P_{total} = P_{acetone} + P_{chloroform} = 172 \text{ torr} + 148 \text{ torr} = \textbf{320. torr}$

(2) In the vapor, $\quad X_{acetone} = \dfrac{P_{acetone}}{P_{total}} = \dfrac{172 \text{ torr}}{320. \text{ torr}} = \textbf{0.538}$

$\quad\quad\quad\quad\quad\quad X_{chloroform} = \dfrac{P_{chloroform}}{P_{total}} = \dfrac{148 \text{ torr}}{320. \text{ torr}} = \textbf{0.462}$

14-44. *Refer to Section 14-9 and Exercise 14-43.*

Assuming real behavior for a chloroform/acetone solution in which the mole fraction of chloroform, $CHCl_3$, is 0.3 and using the dashed (curved) lines on the diagram in Exercise 14-43,

(a) $P_{chloroform} = \textbf{65 torr}$ (b) $P_{acetone} = \textbf{215 torr}$ (c) $P_{total} = \textbf{280 torr}$ ($P_{total} = P_{chloroform} + P_{acetone}$)

14-46. *Refer to Section 14-11, Example 14-7 and Table 14-2.*

Plan: (1) Find ΔT_b.

(2) Determine T_b for the ethylene glycol solution.

(1) From Table 14-2, K_b for $H_2O = 0.512 \text{ °C}/m$; B.P. = 100.00°C.

$\Delta T_b = K_b \, m = (0.512 \text{ °C}/m)(2.97 \; m) = 1.52°C$

(2) Boiling point of the ethylene glycol solution, $T_{b(soln)} = T_{b(solvent)} + \Delta T_b = 100.00°C + 1.52°C = \textbf{101.52°C}$

14-48. *Refer to Section 14-12, Examples 14-8 and 14-9, and Table 14-2.*

Plan: (1) Find ΔT_f.

(2) Determine T_f for the ethylene glycol solution.

(1) From Table 14-2, K_f for $H_2O = 1.86 \text{ °C}/m$; F.P. = 0.00°C.

$\Delta T_f = K_f \, m = (1.86 \text{ °C}/m)(2.97 \; m) = 5.52°C$

(2) Freezing point of the ethylene glycol solution, $T_{f(soln)} = T_{f(solvent)} - \Delta T_f = 0.00°C - 5.52°C = \textbf{–5.52°C}$

14-50. *Refer to Sections 14-11 and 14-12, and Table 14-2.*

Consider a 0.175 m solution of a nonvolatile nonelectrolyte in the solvents listed in Table 14-2.

(a) The greatest freezing point depression, ΔT_f, occurs in a **camphor** solution since $\Delta T_f = K_f \, m$ and this solvent has the largest value of K_f.

(b) The lowest freezing point, $T_{f(solution)}$, occurs in a solution of the non-electrolyte in **water**. Pure water freezes at 0°C, the lowest freezing point of all the listed solvents. The effect of freezing point depression on the freezing point of a 0.175 m solution is minor compared to the actual freezing points of the pure solvents.

(c) The greatest boiling point elevation, ΔT_b, occurs in a solution of **camphor** since $\Delta T_b = K_b m$ and this solvent has the largest value of K_b.

(d) The highest boiling point, $T_{b(solution)}$, for all the listed solvents occurs in a solution of **nitrobenzene** since it has the highest boiling point. The effect of boiling point elevation of a 0.175 m solution will be small and will not affect the rankings of solvents by boiling point.

14-52. *Refer to Section 14-12, Example 14-9 and Table 14-2.*

Plan: (1) Find ΔT_f.
(2) Determine T_f for the lemon juice.

(1) From Table 14-2, K_f for H_2O = 1.86 °C/m; F.P. = 0.00°C.

$\Delta T_f = K_f m = (1.86 \text{ °C}/m)(10.0 \ m) = 18.6°C$

(2) Freezing point of lemon juice, $T_{f(soln)} = T_{f(solvent)} - \Delta T_f = 0.00°C - 18.6°C = \mathbf{-18.6°C}$

14-54. *Refer to Section 14-11 and Example 14-7.*

Plan: (1) Calculate the molality of the nonelectrolyte solution.
(2) Find ΔT_b.
(3) Determine T_b for the nonelectrolyte solution.

(1) we know: molality, $m = \dfrac{\text{mol nonelectrolyte}}{\text{kg solvent}}$

? mol nonelectrolyte = 3.0 g $\times \dfrac{1 \text{ mol}}{137 \text{ g}} = 0.022$ mol

? kg solvent = 250.0 mL ethanol $\times \dfrac{0.789 \text{ g}}{1 \text{ mL}} \times \dfrac{1 \text{ kg}}{1000 \text{ g}} = 0.197$ kg ethanol

$m = \dfrac{0.022 \text{ mol nonelectrolyte}}{0.197 \text{ kg ethanol}} = 0.11 \ m$

(2) K_b for ethanol = 1.22 °C/m; B.P. = 78.41°C.

$\Delta T_b = K_b m = (1.22 \text{ °C}/m)(0.11 \ m) = 0.13°C$

(3) Boiling point of the ethyl alcohol solution, $T_{b(soln)} = T_{b(solvent)} + \Delta T_b = 78.41°C + 0.13°C = \mathbf{78.54°C}$

14-56. *Refer to Section 14-12 and Example 14-9.*

Plan: (1) Determine the molality of the solute, Zn, in the solid solution.
(2) Calculate the melting point (freezing point) of brass using $\Delta T_f = K_f m$.

(1) Assume 100. g of brass containing 12 g of Zn and 88 g of Cu.

$m \text{ Zn} = \dfrac{\text{mol solute}}{\text{kg solvent}} = \dfrac{\text{mol Zn}}{\text{kg Cu}} = \dfrac{12 \text{ g Zn}/65.4 \text{ g/mol}}{0.088 \text{ kg Cu}} = 2.1 \ m$

(2) $\Delta T_f = K_f m = 23 \text{ °C}/m \times 2.1 \ m = 48°C$

$T_{f(brass)} = T_{f(Cu)} - \Delta T_f = 1083°C - 48°C = \mathbf{1035°C}$

14-58. *Refer to Section 14-11 and Table 14-2.*

Plan: (1) Solve for the molality of the solution using $\Delta T_b = K_b m$
(2) Determine the moles of the solute
(3) Calculate the molecular weight.

From Table 14-2, for water: $\qquad T_f = 0°C \qquad K_f = 1.86 °C/m$

(1) $\quad m = \dfrac{\Delta T_b}{K_b} = \dfrac{0°C - -0.040°C}{1.86 °C/m} = 0.0215\ m\ C_{10}H_8$ (good to 2 significant figures)

(2) Recall: $m = \dfrac{\text{mol solute}}{\text{kg solvent}}$

So, ? mol solute = $(m)(\text{kg water}) = (0.0215\ m)(0.0500\ \text{kg}) = 0.00107$ g solute (good to 2 significant figures)

(3) $\quad MW = \dfrac{\text{g solute}}{\text{mol solute}} = \dfrac{0.180\ \text{g}}{0.00107} = \textbf{170 g/mol}$ (2 significant figures)

14-60. *Refer to Sections 14-13 and 2-9, Example 14-10, Exercise 2-52 Solution, and Appendix E.*

Plan: (1) Solve for the molality and the approximate formula weight of the nonelectrolyte, using $\Delta T_f = K_f\ m$.

(2) Determine the simplest (empirical) formula from the % composition data.

(3) Using the approximate formula weight of the simplest formula, determine the true molecular formula and the exact molecular weight.

From Appendix E, for benzene: $\quad T_f = 5.48°C \qquad K_f = 5.12 °C/m$

(1) $\quad m = \dfrac{\Delta T_f}{K_f} = \dfrac{0.53°C}{5.12 °C/m} = 0.10\ m$

(Note: use 0.1035 m to do the molecular weight calculation, remembering there are only 2 significant figures.)

Recall: $m = \dfrac{\text{mol solute}}{\text{kg solvent}} = \dfrac{\text{g solute/MW solute}}{\text{kg solvent}}$

Solving, ? MW solute $= \dfrac{\text{g solute}}{m \times \text{kg benzene}} = \dfrac{0.500\ \text{g}}{0.10\ m \times 0.0750\ \text{kg}} = 64$ g/mol (to 2 significant figures)

(2) Assume 100 g of sample containing 37.5 g C, 12.5 g H and 50.0 g O

$\quad$? mol C $= \dfrac{37.5\ \text{g}}{12.0\ \text{g/mol}} = 3.12$ mol $\qquad\qquad$ Ratio $= \dfrac{3.12}{3.12} = 1$

$\quad$? mol H $= \dfrac{12.5\ \text{g}}{1.008\ \text{g/mol}} = 12.4$ mol $\qquad\qquad$ Ratio $= \dfrac{12.4}{3.12} = 4$

$\quad$? mol O $= \dfrac{50.0\ \text{g}}{16.0\ \text{g/mol}} = 3.12$ mol $\qquad\qquad$ Ratio $= \dfrac{3.12}{3.12} = 1$

Therefore, the simplest formula is CH_4O (FW = 32.04 g/mol)

(3) $\quad n = \dfrac{\text{molecular weight}}{\text{simplest formula weight}} = \dfrac{64\ \text{g/mol}}{32\ \text{g/mol}} = 2$

The true molecular formula for the organic compound is $(CH_4O)_2 = \textbf{C}_2\textbf{H}_8\textbf{O}_2$ with a molecular weight (to 4 significant figures) of **64.08 g/mol**.

14-62. *Refer to Section 14-12 and Fundamental Algebra.*

(a) Plan: (1) Calculate the molality of the solution using $\Delta T_f = K_f\ m$

(2) Calculate the total number of moles of solutes.

(3) Calculate the masses of $C_{10}H_8$ and $C_{14}H_{10}$ and the % composition of each in the sample.

(1) From Appendix E, for benzene: $\qquad T_b = 80.1°C \qquad K_b = 2.53 °C/m$
$\qquad\qquad\qquad\qquad\qquad\qquad\qquad\qquad\ T_f = 5.48°C \qquad K_f = 5.12 °C/m$

$\quad m = \dfrac{\Delta T_f}{K_f} = \dfrac{5.48°C - 4.85°C}{5.12 °C/m} = 0.12\ m$ solutes

(2) Recall: $m = \dfrac{\text{mol solute}}{\text{kg solvent}}$

? mol solutes = 0.12 m × 0.360 kg = 0.043 mol solutes

(3) Let $\quad$ x = g $C_{10}H_8$ $\qquad$ and $\qquad$ $\dfrac{x}{128 \text{ g/mol}}$ = mol $C_{10}H_8$

$\qquad$ 6.00 g - x = g $C_{14}H_{10}$ $\qquad$ $\dfrac{6.00 \text{ g - x}}{178 \text{ g/mol}}$ = mol $C_{14}H_{10}$

Therefore,

the total moles of solute = mol $C_{10}H_8$ + mol $C_{14}H_{10}$ = $\dfrac{x}{128} + \dfrac{6.00 - x}{178}$

So, $\quad$ total moles = 0.043 mol = $\dfrac{x}{128} + \dfrac{6.00 - x}{178}$

When we multiply both sides of the equation by (128)(178), we obtain,

$$980 = (178x) + (768 - 128x)$$
$$50x = 212$$
$$x = 4.2 \text{ g } C_{10}H_8$$
$$6.00 - x = 1.8 \text{ g } C_{14}H_{10}$$

And, $\quad$ % $C_{10}H_8 = \dfrac{4.2 \text{ g } C_{10}H_8}{6.00 \text{ g sample}}$ × 100% = **70.%** $\qquad$ % $C_{14}H_{10} = \dfrac{1.8 \text{ g } C_{14}H_{10}}{6.00 \text{ g sample}}$ × 100% = **30.%**

(b) $\Delta T_b = K_b m_{total} = (2.53 \text{ °C}/m)(0.12 \text{ } m) = 0.30°C$

$T_{b(\text{solution})} = T_{b(\text{benzene})} + \Delta T_b = 80.1°C + 0.30°C = \textbf{80.4°C}$

14-64. *Refer to Section 14-14.*

The highest particle concentration is expected in 0.10 M $Al(NO_3)_3$. Theoretically, $Al(NO_3)_3$ dissociates completely into 4 ions whereas $Ca(NO_3)_2$ dissociates into 3 ions and $LiNO_3$ only dissociates into 2 ions, according to:

$$Al(NO_3)_3(aq) \rightarrow Al^{3+}(aq) + 3NO_3^-(aq)$$
$$Ca(NO_3)_2(aq) \rightarrow Ca^{2+}(aq) + 2NO_3^-(aq)$$
$$LiNO_3(aq) \rightarrow Li^+(aq) + NO_3^-(aq)$$

The solution with the greatest number of ions will conduct electricity the strongest. Therefore, **$Al(NO_3)_3$** is expected to be the best conductor of electricity among the three, even though there is some degree of association of ions.

14-66. *Refer to Section 14-14 and Table 14-3.*

The ideal value for the van't Hoff factor, i, for strong electrolytes at infinite dilution is the total number of ions present in a formula unit.

(a) Na_2SO_4 $\qquad$ i_{ideal} = **3**

(b) $NaOH$ $\qquad$ i_{ideal} = **2**

(c) $Al_2(SO_4)_3$ $\qquad$ i_{ideal} = **5**

(d) $CaSO_4$ $\qquad$ i_{ideal} = **2**

14-68. *Refer to Section 14-14 and Table 14-3.*

The value for the van't Hoff factor, i, for strong electrolytes in dilute solution approximates the total number of ions present in a formula unit. So, $i = 2$ for $KClO_3$ and $i = 3$ for $CaCl_2$. The value of i for the weak electrolyte such as CH_3COOH is between 1 and 2. The value of i for a nonelectrolyte such as CH_3OH is 1.

We know that $\Delta T_f = iK_f\, m$. For solutions of the same solvent at the same molality, the solute with the larger van't Hoff factor, i, has the lower freezing point. Therefore,

$$T_f \text{ for } CaCl_2 < T_f \text{ for } KClO_3 < T_f \text{ for } CH_3COOH < T_f \text{ for } CH_3OH$$

14-70. *Refer to Sections 14-9 and 14-14.*

Note: The van't Hoff factor, i, must be included in all calculations involving colligative properties, including vapor pressure lowering.

Plan: (1) Determine the total number of moles of ions in the solution.
  (2) Determine the mole fraction of water in the solution.
  (3) Calculate the vapor pressure above the solution.

(1) For NaCl, NaBr and NaI, the ideal van't Hoff factor, i_{ideal}, is 2.
  ? mol ions = (2 x mol NaCl) + (2 x mol NaBr) + (2 x mol NaI)

$$= \left(2 \times \frac{1.25 \text{ g NaCl}}{58.4 \text{ g/mol}}\right) + \left(2 \times \frac{1.25 \text{ g NaBr}}{103 \text{ g/mol}}\right) + \left(2 \times \frac{1.25 \text{ g NaI}}{150. \text{ g/mol}}\right)$$

$$= 0.0837 \text{ mol ions}$$

(2) $X_{water} = \dfrac{\text{mol } H_2O}{\text{mol ions + mol } H_2O} = \dfrac{(150. \text{ g}/18.0 \text{ g/mol})}{0.0837 \text{ mol} + (150. \text{ g}/18.0 \text{ g/mol})} = 0.990$

(3) At 100°C, the boiling point of water, the vapor pressure of pure water, $P^\circ_{water} = 760$ torr.
  $P_{water} = X_{water}\, P^\circ_{water} = 0.990 \times 760 \text{ torr} = \mathbf{752 \text{ torr}}$

14-72. *Refer to Section 14-14.*

We know that $\Delta T_f = iK_f\, m$. For solutions of the same solvent at the same molality, the solute with the larger van't Hoff factor, i, has the lower freezing point.

Since $Na_2SO_4(aq)$ will dissociate into 3 ions ($i_{ideal} = 3$) and $CaSO_4(aq)$ will dissociate into 2 ions ($i_{ideal} = 2$), **0.100 m Na$_2$SO$_4$** has the lower freezing point.

14-74. *Refer to Section 14-14.*

We know that $\Delta T_f = iK_f\, m$. For solutions of the same solvent at the same molality, the solute with the larger van't Hoff factor, i, has the higher boiling point. Since all are 1.1 m aqueous solutions:

(a) NaCl and LiCl solutions should boil at the same temperature, because both have $i_{ideal} = 2$.

(b) Na_2SO_4 ($i_{ideal} = 3$) solution boils at a higher temperature than that of NaCl ($i_{ideal} = 2$).

(c) Na_2SO_4 ($i_{ideal} = 3$) solution boils at a higher temperature than that of HCl ($i_{ideal} = 2$).

(d) HCl ($i_{ideal} = 2$) solution boils at a higher temperature than that of $C_6H_{12}O_6$ ($i_{ideal} = 1$, since it is a nonelectrolyte).

(e) $C_6H_{12}O_6$ ($i_{ideal} = 1$) and CH_3OH ($i_{ideal} = 1$) solutions should boil at the same temperature, since both are nonelectrolytes. However, CH_3OH is somewhat volatile, which would lower the boiling point.

(f) CH_3COOH solution ($1 < i < 2$ because acetic acid is a weak acid and only partially dissociates into its ions, H^+ and CH_3COO^-) boils at a higher temperature than that of CH_3OH ($i_{ideal} = 1$, since it is a nonelectrolyte). Note that both substances are somewhat volatile, which could affect their respective boiling points.

(g) KCl ($i_{ideal} = 2$) solution boils at a higher temperature than CH_3COOH solution ($1 < i < 2$, since acetic acid is a weak acid).

14-76. *Refer to Section 14-11, Example 14-7 and Table 14-2.*

Plan: (1) Calculate the molality of the urea solution.
(2) Find ΔT_b.
(3) Determine T_b for the aqueous solution.

(1) we know: molality, $m = \dfrac{\text{mol urea}}{\text{kg solvent}}$

? mol urea, $(NH_3)_2CO = 15.0 \text{ g} \times \dfrac{1 \text{ mol}}{62.08 \text{ g}} = 0.242$ mol

$m = \dfrac{0.242 \text{ mol } (NH_3)_2CO}{0.500 \text{ kg } H_2O} = 0.483 \ m$

(2) From Table 14-2, K_b for $H_2O = 0.512 \ °C/m$; B.P. = 100.000°C.
$\Delta T_b = iK_b m = (1)(0.512 \ °C/m)(0.483 \ m) = 0.247°C$

(3) Boiling point of urea solution, $T_{b(soln)} = T_{b(solvent)} + \Delta T_b = 100.000°C + 0.247°C = \textbf{100.247°C}$

14-78. *Refer to Section 14-14 and Example 14-12.*

Plan: (1) Calculate $m_{effective}$.
(2) Determine the % ionization of CH_3COOH.

(1) From Table 14-2, for water: $T_f = 0°C$, $K_f = 1.86 \ °C/m$
$m_{effective} = \dfrac{\Delta T_f}{K_f} = \dfrac{0.0000°C - (-0.188°C)}{1.86 \ °C/m} = 0.101 \ m$

(2) Let $\quad$ x = molality of CH_3COOH that ionizes
Then $\quad$ x = molality of H^+ and CH_3COO^- that formed

Consider:	CH_3COOH	$\rightleftarrows$	H^+	+	CH_3COO^-
Start	0.100 m		$\approx 0 \ m$		0 m
Change	- x m		+ x m		+ x m
Final	(0.100 - x) m		x m		x m

$M_{effective} = m_{CH_3COOH} + m_{H^+} + m_{CH_3COO^-}$
$0.101 = (0.100 - x) + x + x$
$0.101 = 0.100 + x$
$\quad\quad x = 0.001 \ m$

% ionization $= \dfrac{m_{ionized}}{m_{original}} \times 100\% = \dfrac{0.001 \ m}{0.100 \ m} \times 100\% = \textbf{1\%}$

14-80. *Refer to Section 14-14 and Example 14-12.*

Plan: (1) Find ΔT_f for the solution if CsCl had been a nonelectrolyte.
(2) Determine the van't Hoff factor, i.
(3) Calculate $m_{effective}$ for CsCl.
(4) Calculate the % dissociation.

(1) If CsCl were a nonelectrolyte
$\Delta T_f = K_f m = (1.86 \ °C/m)(0.121 \ m) = 0.225°C \quad\quad\quad\quad$ for water, $K_f = 1.86 \ °C/m$

(2) The van't Hoff factor, $i = \dfrac{\Delta T_{f(actual)}}{\Delta T_{f(nonelectrolyte)}} = \dfrac{0°C - (-0.403°C)}{0.225°C} = \textbf{1.79}$

(3) $m_{effective} = i \times m_{stated} = 1.79 \times 0.121 \ m = 0.217 \ m$

(4) Let   x = molality of CsCl that apparently dissociated
    Then   x = molality of Cs^+ and Cl^- that formed

Consider:	CsCl	$\rightleftharpoons$	Cs^+	+	Cl^-
Start	0.121 m		0 m		0 m
Change	- x m		+ x m		+ x m
Final	(0.121 - x) m		x m		x m

$m_{effective}$ = m_{CsCl} + m_{Cs^+} + m_{Cl^-}
   0.217 = (0.121 - x) + x + x
   0.217 = 0.121 + x
       x = 0.096 m

% ionization = $\dfrac{m_{dissociated}}{m_{original}}$ x 100% = $\dfrac{0.096\ m}{0.121\ m}$ x 100% = **79%**

14-82.   *Refer to Section 14-8.*

Assume 1 liter of solution.   ? g NaCl in 1 L soln = 1.00 x 10^{-4} mol NaCl x 58.4 g/mol = 5.84 x 10^{-3} g NaCl

The contribution of NaCl to the mass of 1 liter of water is nearly negligible. Since the density of water (1.00 g/mL) is essentially the same as the density of the solution:

$$M = \frac{1.00\ \text{x}\ 10^{-4}\ \text{mol NaCl}}{1\ \text{L soln}} \cong \frac{1.00\ \text{x}\ 10^{-4}\ \text{mol NaCl}}{1\ \text{kg}\ H_2O} = m$$

However, if acetonitrile (density at 20.0°C = 0.786 g/mL) were the solvent, 1 liter of solution which would be essentially pure solvent would have a mass of only 786 g. We see that molarity of NaCl in acetonitrile would *not* be equivalent to the molality, but would be less.

14-84.   *Refer to Section 14-15 and Example 14-13.*

osmotic pressure, $\pi = iMRT$ = (1)(0.0200 M)(0.0821 L·atm/mol·K)(75°C + 273°) = **0.571 atm**

14-86.   *Refer to Sections 14-11, 14-12 and 14-15, and Exercise 14-85.*

Plan: (1) Determine the molality of the aqueous solution.
         (2) Calculate ΔT_f and ΔT_b.

(1) Recall, osmotic pressure, $\pi = iMRT$

$$M = \frac{\pi}{iRT} = \frac{1.21\ \text{atm}}{(1)(0.0821\ \text{L·atm/mol·K})(273\ \text{K})} = 0.0540\ M$$

For dilute aqueous solution, $M \cong m$ (See Exercise 14-82 Solution).
Therefore, ? m = 0.0540 m

(2) From Table 14-2, for water: K_f = 1.86 °C/m K_b = 0.512 °C/m
$\Delta T_f = iK_f m$ = (1)(1.86 °C/m)(0.0540 m) = **0.100°C**
$\Delta T_b = iK_b m$ = (1)(0.512 °C/m)(0.0540 m) = **0.0276°C**

14-88.   *Refer to Section 14-15 and Table 14-3.*

From Table 14-3, the van't Hoff factor, i, for 1.00 m K_2CrO_4 is 1.95. Assume that i for 1.20 m is the same.
        $m_{effective}$ = i x m_{stated} = 1.95 x 1.20 m = 2.34 m

At this fairly concentrated concentration, we cannot say that $M \cong m$, so we must calculate the molarity using density of the solution given as 1.25 g/mL. Note: if a 1.20 m K_2CrO_4 soln contained 1000 g of water, there was also 1.20 mol (233 g) of K_2CrO_4 in the solution as well, hence the unit factor: 1000 g H_2O/1233 g soln.

$? \ M \ K_2CrO_4 \ soln = \dfrac{1.20 \ mol}{1000 \ g \ H_2O} \times \dfrac{1000 \ g \ H_2O}{1233 \ g \ soln} \times \dfrac{1.25 \ g \ soln}{1 \ mL \ soln} \times \dfrac{1000 \ mL \ soln}{1 \ L \ soln} = 1.22 \ M$ (use 1.2165 in next calc.)

osmotic pressure, $\pi = iMRT = (1.95)(1.22 \ M)(0.0821 \ L\cdot atm/mol\cdot K)(25°C + 273°) =$ **58.0 atm** (3 sig. fig.)

14-90. *Refer to Sections 14-12 and 14-15, Table 14-2, and Example 14-14.*

From Table 14-2, for water: $K_f = 1.86 \ °C/m$

(a) $m = \dfrac{mol \ substance}{kg \ solvent} = \dfrac{0.0110 \ g/2.00 \times 10^4 \ g/mol}{0.0100 \ kg} = 5.50 \times 10^{-5} \ m$

$\Delta T_f = iK_f m = (1)(1.86 \ °C/m)(5.50 \times 10^{-5} \ m) = 1.02 \times 10^{-4} \ °C$

$T_{f(soln)} = T_{f(water)} - \Delta T_f = 0°C - 1.02 \times 10^{-4} \ °C = \mathbf{-1.02 \times 10^{-4} \ °C}$

(b) For dilute solutions, $M \cong m$,
$\pi = MRT \cong mRT = (5.50 \times 10^{-5} \ m)(0.0821 \ L\cdot atm/mol\cdot K)(25°C + 273°) = \mathbf{1.35 \times 10^{-3} \ atm}$ or **1.02 torr**

(c) From the equation given in (a),
% error in $T_{f(soln)}$ = % error in ΔT_f = % error in MW

$? \ \% \ error \ in \ T_{f(soln)} = \dfrac{error \ in \ T_f}{T_f} \times 100\% = \dfrac{0.001°C}{1.02 \times 10^{-4} \ °C} \times 100\% = \mathbf{1000\%}$

Therefore, an error of only 0.001°C in the freezing point temperature corresponds to a 1000% error in the macromolecule's molecular weight.

(d) Since the osmotic pressure, $\pi = MRT = \dfrac{nRT}{V} = (mol/MW)\dfrac{RT}{V}$, % error in π = % error in MW

$? \ error \ in \ \pi = \dfrac{error \ in \ \pi}{\pi} \times 100\% = \dfrac{0.1 \ torr}{1.02 \ torr} \times 100\% = \mathbf{10\% \ error}$

Therefore, an error of 0.1 torr in osmotic pressure gives only a 10% error in molecular weight.

14-92. *Refer to Table 14-4 and the Key Terms for Chapter 14.*

(a) sol — a colloidal suspension of a solid dispersed in a liquid, e.g., detergents in water
(b) gel — a colloidal suspension of a solid dispersed in a liquid; a semirigid sol, e.g., jelly
(c) emulsion — a colloidal suspension of a liquid in a liquid, e.g., some cough medicines
(d) foam — a colloidal suspension of a gas in a liquid, e.g., bubbles in a bubble bath
(e) solid sol — a colloidal suspension of a solid in a solid, e.g., dirty ice
(f) solid emulsion — a colloidal suspension of a liquid dispersed in a solid, e.g., some kinds of sea ice containing pockets of brine
(g) solid foam — a colloidal suspension of a gas dispersed in a solid, e.g., marshmallows
(h) solid aerosol — a colloidal suspension of a solid in a gas, e.g., fine dust in air
(i) liquid aerosol — a colloidal suspension of a liquid in gas, e.g., insect spray

---

14-94. *Refer to Section 14-18 and the Key Terms for Chapter 14.*

Hydrophilic colloids are colloidal particles that attract water molecules, whereas hydrophobic colloids are colloidal particles that repel water molecules.

14-96. *Refer to Section 14-18.*

Soaps and detergents are both emulsifying agents. Solid soaps are usually sodium salts of long chain organic acids called fatty acids with the general formula, $R\text{-}COO^-Na^+$. On the other hand, synthetic detergents contain sulfonate, $-SO_3^-$, sulfate, $-SO_4^-$, or phosphate groups instead of carboxylate groups, $-COO^-$. "Hard" water contains Fe^{3+}, Ca^{2+} and/or Mg^{2+} ions, all of which displace Na^+ from soap molecules and give an undesirable precipitate coating. However, detergents do not form precipitates with the ions of "hard" water. A typical equation between soap and hard water that contains Ca^{2+} ions is shown below:

$$Ca^{2+}(aq) + 2RCOO^-Na^+(aq) \rightarrow (RCOO^-)_2Ca^{2+}(s) + 2Na^+(aq)$$

14-98. *Refer to Section 14-14 and Exercise 14-88 Solution.*

Plan: (1) Evaluate the van't Hoff factor, i, from the freezing point data.
(2) Calculate the boiling point of the solution.
(3) Calculate the osmotic pressure of the solution.

(1) ? m of $AlCl_3$ soln $= \dfrac{\text{mol } AlCl_3}{\text{kg solvent}} = \dfrac{1.56 \text{ g}/133.3 \text{ g/mol}}{0.0500 \text{ kg } H_2O} = 0.234 \; m \; AlCl_3$

we know: $\Delta T_f = iK_f m$, so

$i = \dfrac{\Delta T_f}{K_f m} = \dfrac{0°C \text{ - }(-1.61°C)}{(1.86°C/m)(0.234 \; m \; AlCl_3)} = 3.70$

(2) $\Delta T_b = iK_b m = (3.70)(0.512 \; °C/m)(0.234 \; m) = 0.443 \; °C$
$T_{b(soln)} = T_{b(solvent)} + \Delta T_b = 100.00°C + 0.443°C = \textbf{100.44°C}$ (to 2 decimal places)

(3) ? M $AlCl_3$ soln $= \dfrac{0.234 \text{ mol } AlCl_3}{1000 \text{ g } H_2O} \times \dfrac{50.0 \text{ g } H_2O}{51.56 \text{ g soln}} \times \dfrac{1.002 \text{ g soln}}{1 \text{ mL soln}} \times \dfrac{1000 \text{ mL soln}}{1 \text{ L soln}} = 0.227 \; M$

osmotic pressure, $\pi = iMRT = (3.70)(0.227 \; M)(0.0821 \; L\cdot atm/mol\cdot K)(25°C + 273°) = \textbf{20.5 atm}$

14-100. *Refer to Sections 14-12 and 14-13.*

From Appendix E, for water: $T_f = 0°C$ $K_f = 1.86 \; °C/m$
for benzene: $T_f = 5.48°C$ $K_f = 5.12 \; °C/m$

Plan: (1) Solve for the apparent molality of 1.00% CH_3COOH in water or benzene using $\Delta T_f = K_f m$ (assume $i = 1$).
(2) Calculate the apparent formula weight of CH_3COOH in each solvent.

(a) in aqueous solution:

(1) $m_{apparent} = \dfrac{\Delta T_f}{K_f} = \dfrac{0.310°C}{1.86 \; °C/m} = 0.167 \; m \; CH_3COOH \text{ in } H_2O$

(2) Recall: $m = \dfrac{\text{mol solute}}{\text{kg solvent}} = \dfrac{\text{g solute/MW}}{\text{kg solvent}}$

Assume 100.00 g of solution containing 1.00 g of CH_3COOH in 99.00 g of water.

? $FW_{apparent} = \dfrac{\text{g } CH_3COOH}{m_{apparent} \times \text{kg } H_2O} = \dfrac{1.00 \text{ g}}{0.167 \; m \times 0.09900 \text{ kg}} = \textbf{60.5 g/mol}$

The true molecular weight of acetic acid is 60.05 g/mol. Acetic acid is a weak acid and dissociates very slightly in water; the van't Hoff factor, i, is then only slightly larger than 1. The behavior of acetic acid approximates that of a nonelectrolyte in water.

(b) in benzene solution:

(1) $m_{apparent} = \dfrac{\Delta T_f}{K_f} = \dfrac{0.441°C}{5.12 \; °C/m} = 0.0861 \; m \; CH_3COOH \text{ in benzene}$

(2) Assume 100.00 g of solution containing 1.00 g of CH_3COOH in 99.00 g of benzene.

? $FW_{apparent} = \dfrac{\text{g } CH_3COOH}{m_{apparent} \times \text{kg benzene}} = \dfrac{1.00 \text{ g}}{0.0861 \; m \times 0.09900 \text{ kg}} = \textbf{117 g/mol}$

The apparent molecular weight, determined in benzene is almost twice as large as the true molecular weight. The reason is that in benzene, the acetic acid dimerizes due to hydrogen bonding. Most of the acetic acid exists as one particle called a dimer, consisting of 2 acetic acid molecules held together by hydrogen bonds.

14-102. *Refer to Section 14-18.*

(a) hydrophobic - hydrocarbons
(b) hydrophilic - starch
(c) The hydrophilic dispersion of starch in water is much easier to make and maintain. Using hydrogen bonding, the very polar water molecules surround and isolate each starch molecule from one another. The starch molecules cannot coalesce and therefore remain dispersed.

14-104. *Refer to Section 14-15.*

Plan: (1) Determine the total molarity of the drug-sugar mixture dissolved in water, using $\pi = MRT$ ($i = 1$).
(2) Determine the mass of lactose in the solution.
(3) Calculate the % lactose present.

(1) $M_{lactose} + M_{drug} = M_{total} = \dfrac{\pi}{RT} = \dfrac{(519/760)\ \text{atm}}{(0.0821\ \text{L·atm/mol·K})(25°C + 273)} = 0.0279\ M_{total}$

(2) Let   x = g lactose, $C_{12}H_{22}O_{11}$ (MW: 342 g/mol)
1 - x = g drug, $C_{21}H_{23}O_5N$ (MW: 369 g/mol)

$M_{total} = \dfrac{\text{mol lactose + mol drug}}{\text{L solution}}$

Substituting, $0.0279\ M = \left(\dfrac{\dfrac{x}{342\ \text{g/mol}} + \dfrac{1.00 - x}{369\ \text{g/mol}}}{0.100\ \text{L}} \right)$

$0.00279 = \dfrac{x}{342} + \dfrac{1.00 - x}{369}$

Multiplying both sides by the product, (342)(369), we obtain
$352 = 369x + 342(1.00 - x)$
$10. = 27x$
$x = 0.37\ \text{g lactose}$

(3) ? lactose by mass $= \dfrac{0.37\ \text{g lactose}}{1.00\ \text{g mixture}} \times 100\% = $ **37% lactose by mass**

14-106. *Refer to Sections 14-1 and 14-9.*

To have an ideal solution, the solvent-solvent, solute-solute and solute-solvent intermolecular forces should be as nearly identical as possible. Solution (c) consisting of $CH_4(\ell)$ dissolved in $CH_3CH_3(\ell)$ would be the most ideal.

14-108. *Refer to Sections 14-8.*

For water, H_2O: ? V = 200. mL + 300. mL x 0.05 = **215 mL**   (All calculations set to 3 sig. figs.)

? g = 215 mL x $\dfrac{0.998\ \text{g}}{1.00\ \text{mL}}$ = **215 g**   ? mol = 215 g x $\dfrac{1\ \text{mol}}{18.02\ \text{mol}}$ = **11.9 mol**

For ethanol, C_2H_5OH ? V = 300. mL x 0.95 = **285 mL**

? g = 285 mL x $\dfrac{0.789\ \text{g}}{1.00\ \text{mL}}$ = **225 g**   ? mol = 225 g x $\dfrac{1\ \text{mol}}{46.07\ \text{mol}}$ = **4.88 mol**

For glycerol, $C_3H_8O_3$? g = 500 mL x $\dfrac{1.26\ \text{g}}{1.00\ \text{mL}}$ = **630. g**   ? mol = 630. g x $\dfrac{1\ \text{mol}}{92.09\ \text{mol}}$ = **6.84 mol**

$X_{water} = \dfrac{\text{mol } H_2O}{\text{mol } H_2O + \text{mol } C_2H_5OH + \text{mol } C_3H_8O_3} = \dfrac{11.9\ \text{mol}}{11.9\ \text{mol} + 4.88\ \text{mol} + 6.84\ \text{mol}} = $ **0.504**

$$X_{\text{ethanol}} = \frac{\text{mol } C_2H_5OH}{\text{mol } H_2O + \text{mol } C_2H_5OH + \text{mol } C_3H_8O_3} = \frac{4.88 \text{ mol}}{11.9 \text{ mol} + 4.88 \text{ mol} + 6.84 \text{ mol}} = \mathbf{0.207}$$

$$X_{\text{glycerol}} = \frac{\text{mol } C_3H_8O_3}{\text{mol } H_2O + \text{mol } C_2H_5OH + \text{mol } C_3H_8O_3} = \frac{6.84 \text{ mol}}{11.9 \text{ mol} + 4.88 \text{ mol} + 6.84 \text{ mol}} = \mathbf{0.290}$$

Note: The sum of the mole fractions of all the solution components should add to 1.00. Rounding error is the reason why the mole fractions in this case add to 1.001.

I would use the moles and or the mole fraction to determine the identity of the solvent, since it is the value that is a direct measurement of the actual number of molecules present. In this case, water is definitely the solvent.

14-110. *Refer to Section 14-4.*

(1) $H_2S(g)$ dissolves appreciably in water because it is very polar and produces $H_2S(aq)$, hydrosulfuric acid, which ionizes as follows:

$$H_2S + H_2O \rightleftarrows HS^- + H_3O^+$$
$$HS^- + H_2O \rightleftarrows S^{2-} + H_3O^+$$

There are also lone pairs of electrons on S to which H_2O can hydrogen bond.

(2) $SO_2(g)$ dissolves appreciably in water because it reacts with water to produce sulfurous acid:

$$SO_2 + H_2O \rightarrow H_2SO_3$$

(3) $NH_3(g)$ dissolves appreciably in water because it hydrogen bonds with the water molecules.

14-112. *Refer to Chapter 14.*

When a solute is added to a solvent to create a solution, it

(a) **lowers** the vapor pressure,

(b) **raises** the boiling point,

(c) **lowers** the freezing point, and

(d) **raises** the osmotic pressure.

14-114. *Refer to Chapter 14.*

In chemistry, the word "dissolution" is used very specifically to describe the process by which a solute is dispersed by a solvent to form a solution. In popular or common usage, it is defined as the decomposition of a whole into fragments or parts.

When we speak of the dissolution of an estate, we are applying the common definition and we know that the estate will be broken up into parts. In chemistry, when we discuss the ease of solute dissolution, we are implying a full range of chemical and physical processes that take place when a solute dissolves.

14-116. *Refer to Section 14-14 and Table 14-3.*

The actual value for the van't Hoff factor can be larger than the ideal value if more particles are produced than expected, e.g., a molecular compound might be considered to have $i_{\text{ideal}} = 1$, but if it slightly ionizes, the i_{actual} would be larger than 1.

In another instance, suppose we thought we were making a solution of the molecule S_8 by weighing out a known mass. Sulfur also exists as S_2, S_4 and S_6. If these other allotropes were accidentally present in the solution, the observed i would be greater than expected, since there would be more particles present than expected.

A third case might involve dissolving an unstable compound in solution. If it decomposed into two or more soluble parts, the observed i would again be greater than expected.

14-118. *Refer to Section 14-8 and Example 14-12.*

Plan: (A) Calculate $m_{effective}$ at each concentration.
        (B) Determine the % ionization of the monoprotic acid.

(1) For 1.0 m acid:

(A) From Table 14-2, for water: $T_f = 0°C$, $K_f = 1.86 °C/m$

$$m_{effective} = \frac{\Delta T_f}{K_f} = \frac{0.0000°C - (-3.35°C)}{1.86 °C/m} = 1.80 \ m$$

(B) Let   x = molality of HA that ionizes
    Then   x = molality of H$^+$ and A$^-$ that formed

Consider: HA $\rightleftarrows$ H$^+$ +   A$^-$
Start   1.0 m ≈ 0 m 0 m
Change   - x m + x m + x m
Final   (1.0 - x) m x m x m

$$m_{effective} = m_{HA} + m_{H^+} + m_{A^-}$$
$$1.80 = (1.0 - x) + x + x$$
$$1.80 = 1.0 + x$$
$$x = 0.8 \ m$$

% ionization $= \dfrac{m_{ionized}}{m_{original}} \times 100\% = \dfrac{0.8 \ m}{1.0 \ m} \times 100\% = $ **80%**

(2) For 2.0 m acid:

(A) From Table 14-2, for water: $T_f = 0°C$, $K_f = 1.86 °C/m$

$$m_{effective} = \frac{\Delta T_f}{K_f} = \frac{0.0000°C - (-6.10°C)}{1.86 °C/m} = 3.28 \ m$$

(B) Let   x = molality of HA that ionizes
    Then   x = molality of H$^+$ and A$^-$ that formed

Consider: HA $\rightleftarrows$ H$^+$ +   A$^-$
Start   2.0 m ≈ 0 m 0 m
Change   - x m + x m + x m
Final   (2.0 - x) m x m x m

$$m_{effective} = m_{HA} + m_{H^+} + m_{A^-}$$
$$3.28 = (2.0 - x) + x + x$$
$$3.28 = 2.0 + x$$
$$x = 1.3 \ m$$

% ionization $= \dfrac{m_{ionized}}{m_{original}} \times 100\% = \dfrac{1.3 \ m}{2.0 \ m} \times 100\% = $ **65%**

(3) For 4.0 m acid:

(A) From Table 14-2, for water: $T_f = 0°C$, $K_f = 1.86 °C/m$

$$m_{effective} = \frac{\Delta T_f}{K_f} = \frac{0.0000°C - (-8.70°C)}{1.86 °C/m} = 4.68 \ m$$

(B) Let   x = molality of HA that ionizes
    Then   x = molality of H$^+$ and A$^-$ that formed

Consider: HA $\rightleftarrows$ H$^+$ +   A$^-$
Start   4.0 m ≈ 0 m 0 m
Change   - x m + x m + x m
Final   (4.0 - x) m x m x m

225

$$m_{\text{effective}} = m_{\text{HA}} + m_{\text{H}^+} + m_{\text{A}^-}$$
$$4.68 = (4.0 - x) + x + x$$
$$4.68 = 4.0 + x$$
$$x = 0.68 \ m$$

$$\% \text{ ionization} = \frac{m_{\text{ionized}}}{m_{\text{original}}} \times 100\% = \frac{0.68 \ m}{4.0 \ m} \times 100\% = \mathbf{17\%}$$

As you can see, the percent ionization decreases as the solution concentration increases.

Acid Concentration	Percent Ionization
1.0 m	80%
2.0 m	65%
4.0 m	17%

14-120. *Refer to Section 14-8.*

(a) The solution on the right is more concentrated. Concentration can be expressed as a ratio of the amount of solute to the amount of solution. The solution on the right has the same amount of solute as the solution on the left, but it is dissolved in less solution, so it is more concentrated.

(b) The 1.00 m solution is the one on the left, because it was prepared by dissolving 1.00 mol of K_2CrO_4 in 1.00 kg of water (1.00 liter = 1.00 kg since the density of water is 1.00 g/mL).

(c) The 1.00 M solution is the one on the right, because it was prepared by dissolving 1.00 mol of K_2CrO_4 to make 1.00 L of solution, as measured by the volumetric flask.

14-122. *Refer to Section 14-5.*

Rock candy consists of crystals of sugar on a stick or string. It could be made by first preparing a supersaturated solution of sugar.

(1) Heat some water to boiling in a glass or metal container with smooth sides and dissolve as much sugar as possible.

(2) Remove the hot saturated sugar solution carefully from the heat source and allow it to cool slowly. A supersaturated solution should result. A smooth container is critical so that the sugar doesn't start to crystallize before you're ready.

(3) Then put a stick or string with rough sides into the solution, perhaps with some sugar crystals, called seed crystals, imbedded in it. This will cause crystallization to begin and large sugar crystals should form on the stick or string in time.

14-124. *Refer to Sections 13-2 and 14-2.*

From the data given, DDT, a toxic pesticide now banned in the United States, is not very soluble in water, but is soluble in fat and fatty tissue. Therefore, it probably is not ionic, so therefore it does not have ion-ion interactions. It probably does not have hydrogen bonding in operation to any extent, because it is not very soluble in water, but is soluble in fat. So, most likely DDT has only dipole-dipole and dispersion forces in operation between its molecules.

In fact, DDT is a polar molecule and so there are dipole-dipole and dispersion forces present. Its official IUPAC name is 1,1,1-trichloro-2,2-bis(*p*-chlorophenyl)ethane and its structure is given here.

14-126. *Refer to Section 14-15 and Example 14-13.*

Plan: (1) Determine the osmotic pressure (in torr) required to force the sap to the top of the tree.
(2) Calculate the molarity of the sugar solution (sap).

(1) ? height of the tree (mm) = 40. ft $\times \dfrac{12\text{ in}}{1\text{ ft}} \times \dfrac{2.54\text{ cm}}{1\text{ in}} \times \dfrac{10\text{ mm}}{1\text{ cm}} = 1.2 \times 10^4$ mm

The pressure exerted by a column of sugar solution in the tree can be stated as 1.2×10^4 mm sugar solution. However, pressure is more often given in units of the height of an equivalent column of mercury in millimeters, called torr, which can be easily converted to atmospheres. The unit factor relating a column of mercury to a column of a liquid exerting the same pressure involves their densities. In this case since the density of the solution is only 1.10/13.6 that of mercury as determined from the relative densities, a 13.6 mm column of solution exerts the same pressure as a 1.10 mm column of mercury.

Thus,

? pressure (atm) = 1.2×10^4 mm sugar soln $\times \dfrac{1.10\text{ mm Hg}}{13.6\text{ mm sugar soln}} \times \dfrac{1\text{ torr}}{1\text{ mm Hg}} \times \dfrac{1\text{ atm}}{760\text{ torr}} = \mathbf{1.3\ atm}$

(2) Recall, osmotic pressure, $\pi = iMRT$

$$M = \frac{\pi}{iRT} = \frac{1.3\text{ atm}}{(1)(0.0821\text{ L·atm/mol·K})(273\text{ K} + 15^\circ\text{C})} = \mathbf{0.055\ }M$$

Note: If the numbers are not rounded off to 2 significant figures until the end, the answer is 0.054 M.

15 Chemical Thermodynamics

15-2. *Refer to the Key Terms for Chapters 1 and 15.*

(a) Heat is a form of energy that flows between two samples of matter due to their differences in temperature.
(b) Temperature is a measure of the intensity of heat, i.e., the hotness or coldness of an object.
(c) The system refers to the substances of interest in a process, i.e., it is the part of the universe that is under investigation.
(d) The surroundings refer to everything in the environment of the system of interest.
(e) The thermodynamic state of a system refers to a set of conditions that completely specifies all of the thermodynamic properties of the system.
(f) Work is the application of a force through a distance. For physical or chemical changes that occur at constant pressure, the work done on the system is $-P\Delta V$.

15-4. *Refer to Sections 1-1, 15-1 and 15-10, and Figure 15-1.*

(a) When heat is given off by a system, that system is labeled an exothermic process. Figure 15-1 describes such a process: the combustion of 1 mole of methane at 25°C:
$$CH_4(g) + 2O_2(g) \rightarrow CO_2(g) + 2H_2O(\ell) + 890 \text{ kJ}$$

(b) In this example, the system's volume at constant pressure decreases since the system initially has 3 moles of gas, but produces only 1 mole of gas, since volume is directly proportional to moles of gas at constant temperature and pressure. When gas is consumed in a process (e.g. in this process: 3 moles gaseous reactants → 1 mole gaseous product), the surroundings are doing work on the system.

Note: When we look at the volume of a system, we generally consider only the gases, since the volume of 1 mole of gas is so much greater than the volume of 1 mole of a liquid or solid, which are much more dense.

15-6. *Refer to Sections 1-1 and 15-1.*

An endothermic process absorbs heat energy from its surroundings; an exothermic process releases heat energy to its surroundings. If a reaction is endothermic in one direction, it is exothermic in the opposite direction. For example, the melting of 1 mole of ice water is an endothermic process requiring 6.02 kJ of heat:
$$H_2O(s) + 6.02 \text{ kJ} \rightarrow H_2O(\ell)$$
The reverse process, the freezing of 1 mole of liquid water, releasing 6.02 kJ of heat, is an exothermic process:
$$H_2O(\ell) \rightarrow H_2O(s) + 6.02 \text{ kJ}$$

15-8. *Refer to Section 15-1.*

According to the First Law of Thermodynamics, the total amount of energy in the universe is constant. When an incandescent light is turned on, electrical energy is converted mainly into light and heat energy. A small fraction of the energy is converted to chemical energy, which is why the filament eventually burns out.

15-10. *Refer to Sections 1-7 and 15-2.*

A state function is a variable that defines the state of a system; it is a function that is independent of the pathway by which a process occurs. Therefore, the change in a state function depends only on the initial and the final value, not on how that change occurred.

(a) Your bank balance is a state function, because it depends only on the difference between your deposits and withdrawals.

(b) The mass of a candy bar is a state function, since it is a constant wherever you are.

(c) However, your weight is not a state function. Weight depends on the gravitational attraction of your body to the center of the earth, which changes depending on where you are on the earth.

(d) The heat lost by perspiration during a climb up a mountain along a fixed path is not a state function, because it depends on the person - her/his size, build, metabolism and degree of fitness.

15-12.   *Refer to Sections 15-5, 15-6 and 15-17.*

(a) ΔH, the enthalpy change or heat of reaction, is the heat change of a reaction occurring at some constant pressure and temperature.

$\Delta H°$ is the standard enthalpy change of a reaction that occurs at 1 atm pressure. Unless otherwise stated, the reaction temperature is 25°C.

(b) As stated in (a), $\Delta H°_{rxn}$ is the standard enthalpy change of a reaction occurring at 1 atm pressure.

$\Delta H°_f$, the standard molar enthalpy of formation of a substance, is the enthalpy change for a reaction in which 1 mole of the substance in a specific state is formed from its elements in their standard states.

15-14.   *Refer to Sections 15-4 and 15-5.*

(i) Since the reaction is endothermic, (a) enthalpy increases, (b) $H_{product} > H_{reactant}$ and (c) ΔH is positive.

(ii) Since the reaction is exothermic, (a) enthalpy decreases, (b) $H_{reactant} > H_{product}$ and (c) ΔH is negative.

15-16.   *Refer to Section 15-5 and Example 15-5.*

Balanced equation: $CH_3OH(g) + \frac{3}{2}O_2(g) \rightarrow CO_2(g) + 2H_2O(\ell)$ $\qquad\qquad\qquad \Delta H$ $=$ -764 kJ/mol rxn

(a) Plan: heat evolved/mol rxn $\overset{(1)}{\Rightarrow}$ heat evolved/mol CH_3OH $\overset{(2)}{\Rightarrow}$ heat evolved/g CH_3OH $\overset{(3)}{\Rightarrow}$ heat evolved

? heat evolved (kJ) $= \dfrac{764\ kJ}{mol\ rxn} \times \dfrac{1\ mol\ rxn}{1\ mol\ CH_3OH} \times \dfrac{1\ mol\ CH_3OH}{32.0\ g\ CH_3OH} \times 115.0\ g\ CH_3OH =$ **2750 kJ evolved**

(b) Plan: heat evolved $\overset{(1)}{\Rightarrow}$ mol reaction $\overset{(2)}{\Rightarrow}$ mol O_2 $\overset{(3)}{\Rightarrow}$ g O_2

? g $O_2 = 925\ kJ \times \dfrac{1\ mol\ rxn}{764\ kJ} \times \dfrac{1.5\ mol\ O_2}{1\ mol\ rxn} \times \dfrac{32.0\ g\ O_2}{1\ mol\ O_2} =$ **58.1 g O_2**

15-18.   *Refer to Section 15-5 and Example 15-5.*

Balanced equation: $PbO(s) + C(s) \rightarrow Pb(s) + CO(g)$

Since the equation involves one mole of PbO, ΔH can be expressed in the units of kJ/mol PbO.

? heat supplied to the reaction $= \dfrac{5.95\ kJ}{13.43\ g\ PbO} \times \dfrac{223.2\ g\ PbO}{1\ mol\ PbO} = 98.9$ kJ/mol PbO

Therefore, since the heat is being added to the reaction, $\Delta H =$ **+98.9 kJ/mol rxn**

15-20.   *Refer to Section 15-3 and Appendix K.*

The standard molar enthalpy of formation, $\Delta H°_f$, is the amount of heat absorbed when 1 mole of the substance is produced from its elements in their standard states. At 25°C, $\Delta H°_f$ of liquid water is −285.8 kJ/mol and $\Delta H°_f$ of water vapor is −241.8 kJ/mol. This means that more heat is released when liquid water is formed from its elements, then when gaseous water is formed from its elements. So, the formation reaction of liquid water is more exothermic, which means that $H_2O(\ell)$ has a lower enthalpy than $H_2O(g)$. See the solution to Exercise 15-22.

15-22. *Refer to Sections 15-1 and 15-5.*

Consider the balanced reactions: (1) $CH_4(g) + 2O_2(g) \rightarrow CO_2(g) + 2H_2O(\ell)$ $\Delta H_1 = (-)$

(2) $CH_4(g) + 2O_2(g) \rightarrow CO_2(g) + 2H_2O(g)$ $\Delta H_2 = (-)$

The only difference between them is that Reaction (1) involves water in the liquid phase and Reaction (2) involves water as water vapor. Since more heat is released when $H_2O(g) \rightarrow H_2O(\ell)$, as shown in the adjacent diagram, Reaction (1) is more exothermic than Reaction (2).

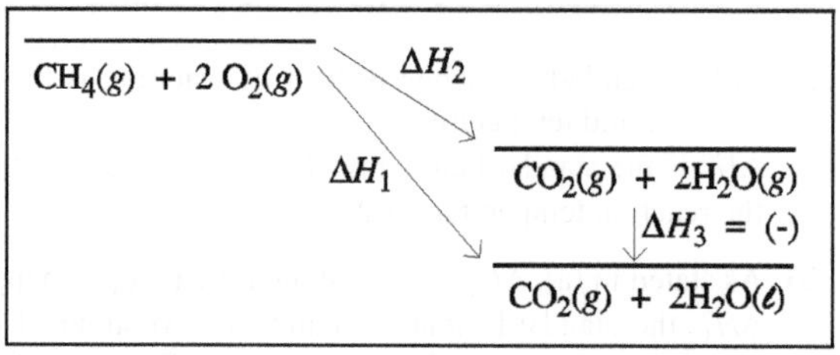

15-24. *Refer to Section 15-6 and the Key Terms for Chapter 15.*

The thermodynamic standard state of a substance is its most stable state under standard pressure (1 atm) and at some specific temperature (usually 25°C). "Thermodynamic" refers to the observation, measurement and prediction of energy changes that accompany physical changes or chemical reaction. "Standard" refers to the set conditions of 1 atm pressure and 25°C. The "state" of a substance is its phase: gas, liquid or solid. "Substance" is any kind of matter all specimens of which have the same chemical composition and physical properties.

15-26. *Refer to Section 15-7 and Appendix K.*

The standard molar enthalpy of formation, ΔH_f°, of elements in their standard states is zero. From the tabulated values of standard molar enthalpies in Appendix K, we can identify the standard states of elements.

(a) chlorine $Cl_2(g)$ (d) iodine $I_2(s)$

(b) chromium $Cr(s)$ (e) sulfur $S(s, \text{rhombic})$

(c) bromine $Br_2(\ell)$ (f) nitrogen $N_2(g)$

15-28. *Refer to Section 15-7, Example 15-6 and Appendix K.*

Hint: Use Appendix K to identify an element's standard state since its ΔH_f° value is equal to zero.

(a) $Ca(s) + O_2(g) + H_2(g) \rightarrow Ca(OH)_2(s)$ (e) $\frac{1}{4}P_4(s,\text{white}) + \frac{3}{2}H_2(g) \rightarrow PH_3(g)$

(b) $6C(s, \text{graphite}) + 3H_2(g) \rightarrow C_6H_6(\ell)$ (f) $3C(s,\text{graphite}) + 4H_2(g) \rightarrow C_3H_8(g)$

(c) $Na(s) + \frac{1}{2}H_2(g) + C(s,\text{graphite}) + \frac{3}{2}O_2(g) \rightarrow NaHCO_3(s)$ (g) $S(s,\text{rhombic}) \rightarrow S(g)$

(d) $Ca(s) + F_2(g) \rightarrow CaF_2(s)$ (h) $H_2(g) + \frac{1}{2}O_2(g) \rightarrow H_2O(\ell)$

15-30. *Refer to Section 15-5.*

The balanced equation for the standard molar enthalpy of formation of $Li_2O(s)$ is: $2Li(s) + \frac{1}{2}O_2(g) \rightarrow Li_2O(s)$

$$? \text{ kJ/mol } Li_2O(s) = \frac{146 \text{ kJ}}{3.47 \text{ g Li}} \times \frac{6.94 \text{ g Li}}{1 \text{ mol Li}} \times \frac{2 \text{ mol Li}}{1 \text{ mol Li}_2O} = 584 \text{ kJ/mol}$$

And so, $\Delta H^\circ_f \text{ }_{Li_2O(s)} = \textbf{−584 kJ/mol}$ since the reaction is exothermic.

15-32. *Refer to Section 15-8 and Examples 15-7 and 15-8.*

To obtain the desired equation,
(1) divide the first equation by 2 to give 2 moles of HCl on the reactant side,
(2) multiply the second equation by 2, giving 2 moles of HF on the product side. Then,
(3) reverse the third equation, so that H_2O, H_2 and $\frac{1}{2}O_2$ are eliminated when the modified equations are added together.

	ΔH°
$2HCl(g) + \frac{1}{2}O_2(g) \rightarrow H_2O(\ell) + Cl_2(g)$	−101.2 kJ/mol rxn
$H_2(g) + F_2(g) \rightarrow 2HF(\ell)$	−1200.0 kJ/mol rxn
$H_2O(\ell) \rightarrow H_2(g) + \frac{1}{2}O_2(g)$	+285.8 kJ/mol rxn
$2HCl(g) + F_2(g) \rightarrow 2HF(\ell) + Cl_2(g)$	**−1015.4 kJ/mol rxn**

15-34. *Refer to Section 15-8 and Examples 15-7 and 15-8.*

To obtain the desired equation,
(1) multiply the first equation by 2 to give 2 moles of SO_2 on the product side, then
(2) reverse the second equation and multiply by 2, giving 2 moles of SO_3 on the reactant side.

	ΔH°
$2S(s) + 2O_2(g) \rightarrow 2SO_2(g)$	−593.6 kJ/mol rxn
$2SO_3(g) \rightarrow 2S(s) + 3O_2(g)$	+791.2 kJ/mol rxn
$2SO_3(g) \rightarrow 2SO_2(g) + O_2(g)$	**+197.6 kJ/mol rxn**

15-36. *Refer to Section 15-8 and Examples 15-7 and 15-8.*

To obtain the desired hydrogenation equation,
(1) use the first equation as it is to give 2 moles of H_2 on the reactant side,
(2) use the second equation as it is to give 1 mole of C_3H_4 on the reactant side, then
(3) reverse the third equation to give 1 mole of C_3H_8 on the product side.

	ΔH°
$2H_2(g) + O_2(g) \rightarrow 2H_2O(\ell)$	−571.6 kJ/mol rxn
$C_3H_4(g) + 4O_2(g) \rightarrow 3\ CO_2(g) + 2H_2O(\ell)$	−1937 kJ/mol rxn
$3CO_2(g) + 4H_2O(\ell) \rightarrow C_3H_8(g) + 5O_2(g)$	+2220. kJ/mol rxn
$C_3H_4(g) + 2H_2(g) \rightarrow C_3H_8(g)$	**−289 kJ/mol rxn**

(a) Balanced equation: $NH_4NO_3(s) \rightarrow N_2O(g) + 2H_2O(\ell)$

$\Delta H^{\circ}_{rxn} = [\Delta H^{\circ}_{f\ N_2O(g)} + 2\Delta H^{\circ}_{f\ H_2O(\ell)}] - [\Delta H^{\circ}_{f\ NH_4NO_3(s)}]$

$\quad = [(1\ mol)(82.05\ kJ/mol) + (2\ mol)(-285.8\ kJ/mol)] - [(1\ mol)(-365.6\ kJ/mol)]$

$\quad = \textbf{-124.0 kJ/mol rxn}$

(b) Balanced equation: $2FeS_2(s) + \frac{11}{2}O_2(g) \rightarrow Fe_2O_3(s) + 4SO_2(g)$

$\Delta H^{\circ}_{rxn} = [\Delta H^{\circ}_{f\ Fe_2O_3(s)} + 4\Delta H^{\circ}_{f\ SO_2(g)}] - [2\Delta H^{\circ}_{f\ FeS_2(s)} + \frac{11}{2}\Delta H^{\circ}_{f\ O_2(g)}]$

$\quad = [(1\ mol)(-824.2\ kJ/mol) + (4\ mol)(-296.8\ kJ/mol)] - [(2\ mol)(-177.5\ kJ/mol) + (\frac{11}{2}mol)(0\ kJ/mol)]$

$\quad = \textbf{-1656 kJ/mol rxn}$

(c) Balanced equation: $SiO_2(s) + 3C(s,graphite) \rightarrow SiC(s) + 2CO(g)$

$\Delta H^{\circ}_{rxn} = [\Delta H^{\circ}_{f\ SiC(s)} + 2\Delta H^{\circ}_{f\ CO(g)}] - [\Delta H^{\circ}_{f\ SiO_2(s)} + 3\Delta H^{\circ}_{f\ C(s,graphite)}]$

$\quad = [(1\ mol)(-65.3\ kJ/mol) + (2\ mol)(-110.5\ kJ/mol)] - [(1\ mol)(-910.9\ kJ/mol) + (3\ mol)(0\ kJ/mol)]$

$\quad = \textbf{+624.6 kJ/mol rxn}$

(1) Balanced equation for combustion of propane: $C_3H_8(g) + 5O_2(g) \rightarrow 3CO_2(g) + 4H_2O(g)$

$\Delta H^{\circ}_{combustion} = [3\Delta H^{\circ}_{f\ CO_2(g)} + 4\Delta H^{\circ}_{f\ H_2O(g)}] - [\Delta H^{\circ}_{f\ C_3H_8(g)} + 5\Delta H^{\circ}_{f\ O_2(g)}]$

$\quad = [(3\ mol)(-393.5\ kJ/mol) + (4\ mol)(-241.8\ kJ/mol)]$

$\qquad\qquad - [(1\ mol)(-103.8\ kJ/mol) + (5\ mol)(0\ kJ/mol)]$

$\quad = -2043.9\ kJ/mol\ C_3H_8$

heat released (kJ/g) $= \dfrac{2043.9\ kJ}{1\ mol\ C_3H_8} \times \dfrac{1\ mol\ C_3H_8}{44.09\ g\ C_3H_8} = \textbf{46.36 kJ/g C}_\textbf{3}\textbf{H}_\textbf{8}$

(2) Balanced equation for combustion of octane: $C_8H_{18}(\ell) + \frac{25}{2}O_2(g) \rightarrow 8CO_2(g) + 9H_2O(g)$

$\Delta H^{\circ}_{combustion} = [8\Delta H^{\circ}_{f\ CO_2(g)} + 9\Delta H^{\circ}_{f\ H_2O(g)}] - [\Delta H^{\circ}_{f\ C_8H_{18}(\ell)} + \frac{25}{2}\Delta H^{\circ}_{f\ O_2(g)}]$

$\quad = [(8\ mol)(-393.5\ kJ/mol) + (9\ mol)(-241.8\ kJ/mol)]$

$\qquad\qquad - [(1\ mol)(-268.8\ kJ/mol) + (\frac{25}{2}mol)(0\ kJ/mol)]$

$\quad = -5055.4\ kJ/mol\ C_8H_{18}$

heat released (kJ/g) $= \dfrac{5055.4\ kJ}{1\ mol\ C_8H_{18}} \times \dfrac{1\ mol\ C_8H_{18}}{114.2\ g\ C_8H_{18}} = \textbf{44.27 kJ/g C}_\textbf{8}\textbf{H}_\textbf{18}$

Note: The sign convention for ΔH° tells the reader whether heat is being released or absorbed. However, when the question asks for "heat released" or "heat absorbed," the value of heat is a positive number. When the words "released" or "absorbed" are used, the sign convention is *not* used.

Balanced equation: $8Al(s) + 3Fe_3O_4(s) \rightarrow 4Al_2O_3(s) + 9Fe(s)$ $\qquad\qquad \Delta H^{\circ} = -3350.\ kJ/mol\ rxn$

Plan: (1) Determine the limiting reactant.

(2) Calculate the heat released based on the limiting reactant.

(1) $? \text{ mol Al} = \dfrac{27.6 \text{ g Al}}{27.0 \text{ g/mol}} = 1.02 \text{ mol Al}$ $\qquad ? \text{ mol Fe}_3\text{O}_4 = \dfrac{69.12 \text{ g Fe}_3\text{O}_4}{231.6 \text{ g/mol}} = 0.2984 \text{ mol Fe}_3\text{O}_4$

$\text{Required ratio} = \dfrac{8 \text{ mol Al}}{3 \text{ mol Fe}_3\text{O}_4} = 2.67$ $\qquad \text{Available ratio} = \dfrac{1.02 \text{ mol Al}}{0.2984 \text{ mol Fe}_3\text{O}_4} = 3.42$

Available ratio > Required ratio; Fe_3O_4 is the limiting reactant.

(2) $\Delta H° = 69.12 \text{ g Fe}_3\text{O}_4 \times \dfrac{1 \text{ mol Fe}_3\text{O}_4}{231.6 \text{ g Fe}_3\text{O}_4} \times \dfrac{-3350. \text{ kJ}}{3 \text{ mol Fe}_3\text{O}_4} = -333.3 \text{ kJ}$

Therefore, there are **+333 kJ** of heat released.

Note: The sign convention for $\Delta H°$ tells the reader whether heat is being released or absorbed. However, when the question asks for "heat released" or "heat absorbed," the value of heat is a positive number. When the words "released" or "absorbed" are used, the sign convention is *not* used.

15-44. *Refer to Section 15-9, Tables 15-2 and 15-3, and Examples 15-11 and 15-12.*

Balanced equations: oxidation of sucrose: $\qquad C_{12}H_{22}O_{11}(s) + 12O_2\,(g) \;\rightarrow\; 12CO_2(g) + 11H_2O(g)$

oxidation of tristearin: $\qquad C_{57}H_{110}O_6(s) + 163/2\,O_2\,(g) \;\rightarrow\; 57CO_2(g) + 55H_2O(g)$

Sucrose, $C_{12}H_{22}O_{11}$ contains 10 C-C bonds, 14 C-O bonds, 14 C-H bonds and 8 O-H bonds
Tristearin, $C_{57}H_{110}O_6$ contains 53 C-C bonds, 6 C-O bonds, 3 C=O bonds and 110 C-H bonds

Oxidation of 1 mol sucrose:

$\Delta H°_{rxn} = \Sigma\,\text{B.E.}_{reactants} - \Sigma\,\text{B.E.}_{products}$
$= [10\text{B.E.}_{C\text{-}C} + 14\text{B.E.}_{C\text{-}O} + 14\text{B.E.}_{C\text{-}H} + 8\text{B.E.}_{O\text{-}H} + 12\text{B.E.}_{O=O}] - [24\text{B.E.}_{C=O} + 22\text{B.E.}_{O\text{-}H}]$
$= [(10 \text{ mol})(346 \text{ kJ/mol}) + (14 \text{ mol})(358 \text{ kJ/mol}) + (14 \text{ mol})(413 \text{ kJ/mol}) + (8 \text{ mol})(463 \text{ kJ/mol})$
$\qquad + (12 \text{ mol})(498 \text{ kJ/mol})]$
$\qquad - [(24 \text{ mol})(799 \text{ kJ/mol})^* + (22 \text{ mol})(463 \text{ kJ/mol})]$
$= 23930 \text{ kJ} - 29360 \text{ kJ}$
$= \mathbf{-5430 \text{ kJ/mol sucrose}}$
$= \dfrac{-5430 \text{ kJ}}{1 \text{ mol sucrose}} \times \dfrac{1 \text{ mol}}{342.3 \text{ g}} = \mathbf{-15.9 \text{ kJ/g sucrose}}$
$= \dfrac{-5430 \text{ kJ}}{1 \text{ mol sucrose}} \times \dfrac{1 \text{ mol}}{342.3 \text{ g}} \times \dfrac{1 \text{ kcal}}{4.184 \text{ kJ}} = \mathbf{-3.79 \text{ kcal/g sucrose}}$

Oxidation of 1 mol tristearin:

$\Delta H°_{rxn} = \Sigma\,\text{B.E.}_{reactants} - \Sigma\,\text{B.E.}_{products}$
$= [53\text{B.E.}_{C\text{-}C} + 6\text{B.E.}_{C\text{-}O} + 110\text{B.E.}_{C\text{-}H} + 3\text{B.E.}_{C=O} + 163/2\text{B.E.}_{O=O}] - [114\text{B.E.}_{C=O} + 110\text{B.E.}_{O\text{-}H}]$
$= [(53 \text{ mol})(346 \text{ kJ/mol}) + (6 \text{ mol})(358 \text{ kJ/mol}) + (110 \text{ mol})(413 \text{ kJ/mol}) + (3 \text{ mol})(732 \text{ kJ/mol})$
$\qquad + (163/2 \text{ mol})(498 \text{ kJ/mol})]$
$\qquad - [(114 \text{ mol})(799 \text{ kJ/mol})^* + (110 \text{ mol})(463 \text{ kJ/mol})]$
$= 68100 \text{ kJ} - 71000 \text{ kJ}$
$= \mathbf{-2900 \text{ kJ/mol tristearin}}$
$= \dfrac{-2900 \text{ kJ}}{1 \text{ mol sucrose}} \times \dfrac{1 \text{ mol}}{891.5 \text{ g}} = \mathbf{-3.3 \text{ kJ/g tristearin}}$
$= \dfrac{-2900 \text{ kJ}}{1 \text{ mol sucrose}} \times \dfrac{1 \text{ mol}}{891.5 \text{ g}} \times \dfrac{1 \text{ kcal}}{4.184 \text{ kJ}} = \mathbf{-0.78 \text{ kcal/g tristearin}}$

Sucrose has the greater energy density, meaning that for 1 g of compound, more energy is released when sucrose is oxidized than when tristearin is oxidized.

* See extra information for the C=O bond in Table 15-3. C=O has different bond energies in CO_2 than in other compounds.

(a) For a reaction occurring in the gaseous phase, the net enthalpy change, ΔH°_{rxn}, equals the sum of the bond energies in the reactants minus the sum of the bond energies in the products:

$$\Delta H^\circ_{rxn} = \Sigma \text{ B.E.}_{reactants} - \Sigma \text{ B.E.}_{products}$$

If the products have higher bond energy and are therefore more stable than the reactants, the reaction is exothermic. If the opposite is true, the reaction is endothermic.

(b) Consider $O_2(g)$: $\Delta H^\circ_{f\ O_2(g)} = 0$ kJ/mol since the standard state of oxygen is $O_2(g)$

$$-\Sigma \text{ B.E.}_{O_2(g)} = -\text{B.E.}_{O=O} = -498 \text{ kJ/mol}$$

Therefore, one cannot say that $\Delta H^\circ_{f\ substance} = -\Sigma \text{ B.E.}_{substance}$. Bond energies are a measure of the energy involved in breaking of one mole of bonds in a gaseous substance to form gaseous atoms of the elements. The value of ΔH°_f is a measure of the energy involved in making one mole of the substance from its elements in their standard states. They differ in two major aspects: (1) In bond energy considerations, all the bonds are broken to give free atoms, while in ΔH°_f determinations, some bonds may still be maintained as diatomic or polyatomic free elements (e.g., $O_2(g)$ or $P_4(s)$). (2) The standard states of the elements are not necessarily the gaseous state. Moreover, the ΔH°_f equation is an exact calculation, but the bond energy equation is only an estimation of ΔH°_f because bond energies are average values from many different compounds.

(a) Balanced equation in terms of Lewis structures of the reactants and products:

$\Delta H^\circ_{rxn} = \Sigma \text{ B.E.}_{reactants} - \Sigma \text{ B.E.}_{products}$ in the gas phase
$= [\text{B.E.}_{C=C} + 4\text{B.E.}_{C-H} + \text{B.E.}_{Br-Br}] - [\text{B.E.}_{C-C} + 4\text{B.E.}_{C-H} + 2\text{B.E.}_{C-Br}]$
$= [(1 \text{ mol})(602 \text{ kJ/mol}) + (4 \text{ mol})(413 \text{ kJ/mol}) + (1 \text{ mol})(193 \text{ kJ/mol}]$
    $- [(1 \text{ mol})(346 \text{ kJ/mol}) + (4 \text{ mol})(413 \text{ kJ/mol}) + (2 \text{ mol})(285 \text{ kJ/mol})]$
$= \mathbf{-121 \text{ kJ/mol rxn}}$

(b) Balanced equation in terms of Lewis structures of the reactants and products:

$\Delta H^\circ_{rxn} = \Sigma \text{ B.E.}_{reactants} - \Sigma \text{ B.E.}_{products}$ in the gas phase
$= [2\text{B.E.}_{O-H} + \text{B.E.}_{O-O}] - [2\text{B.E.}_{O-H} + 1/2 \text{ B.E.}_{O=O}]$
$= [(2 \text{ mol})(463 \text{ kJ/mol}) + (1 \text{ mol})(146 \text{ kJ/mol})] - [(2 \text{ mol})(463 \text{ kJ/mol}) + (1/2 \text{ mol})(498 \text{ kJ/mol})]$
$= \mathbf{-103 \text{ kJ/mol rxn}}$

Balanced equation: $CCl_2F_2(g) + F_2(g) \rightarrow CF_4(g) + Cl_2(g)$

$\Delta H^\circ_{rxn} = \Sigma \text{ B.E.}_{reactants} - \Sigma \text{ B.E.}_{products}$ in the gas phase
$= [2\text{B.E.}_{C-Cl} + 2\text{B.E.}_{C-F} + \text{B.E.}_{F-F}] - [4\text{B.E.}_{C-F} + \text{B.E.}_{Cl-Cl}]$
$= [(2 \text{ mol})(339 \text{ kJ/mol}) + (2 \text{ mol})(485 \text{ kJ/mol}) + (1 \text{ mol})(155 \text{ kJ/mol})]$
            $- [(4 \text{ mol})(485 \text{ kJ/mol}) + (1 \text{ mol})(242 \text{ kJ/mol})]$
$= \mathbf{-379 \text{ kJ/mol rxn}}$

15-52. *Refer to Section 15-9, Table 15-2, Examples 15-11 and 15-12, and Appendix K.*

(1) Balanced equation for standard heat of formation of HCl: $\frac{1}{2}H_2(g) + \frac{1}{2}Cl_2(g) \rightarrow HCl(g)$

$\Delta H^{\circ}_{rxn} = \Sigma$ B.E.$_{reactants}$ - Σ B.E.$_{products}$ in the gas phase

       $= [\frac{1}{2}$ B.E.$_{H-H} + \frac{1}{2}$ B.E.$_{Cl-Cl}]$ - [B.E.$_{H-Cl}]$

       $= [(0.5 \text{ mol})(436 \text{ kJ/mol}) + (0.5 \text{ mol})(242 \text{ kJ/mol})]$ - $[(1 \text{ mol})(432 \text{ kJ/mol})]$

       $= \mathbf{-93 \text{ kJ/mol HCl}}$

For HCl(g), $\Delta H^{\circ}_f = -92.31$ kJ/mol

(2) Balanced equation for standard heat of formation of HF: $\frac{1}{2}H_2(g) + \frac{1}{2}F_2(g) \rightarrow HF(g)$

$\Delta H^{\circ}_{rxn} = \Sigma$ B.E.$_{reactants}$ - Σ B.E.$_{products}$ in the gas phase

       $= [\frac{1}{2}$ B.E.$_{H-H} + \frac{1}{2}$ B.E.$_{F-F}]$ - [B.E.$_{H-F}]$

       $= [(0.5 \text{ mol})(436 \text{ kJ/mol}) + (0.5 \text{ mol})(155 \text{ kJ/mol})]$ - $[(1 \text{ mol})(565 \text{ kJ/mol})]$

       $= \mathbf{-270 \text{ kJ/mol HF}}$

For HF(g), $\Delta H^{\circ}_f = -271$ kJ/mol

15-54. *Refer to Section 15-9 and Appendix K.*

The ΔH°_{rxn} of this reaction: $PCl_3(g) \rightarrow P(g) + 3Cl(g)$ is equal to 3 times the average P-Cl bond energy in $PCl_3(g)$ since this reaction involves the breaking of 3 P-Cl bonds.

$\Delta H^{\circ}_{rxn} = [\Delta H^{\circ}_{f\ P(g)} + 3\Delta H^{\circ}_{f\ Cl(g)}]$ - $[\Delta H^{\circ}_{f\ PCl_3(g)}]$

       $= [(1 \text{ mol})(314.6 \text{ kJ/mol}) + (3 \text{ mol})(121.7 \text{ kJ/mol})]$ - $[(1 \text{ mol})(-306.4 \text{ kJ/mol})]$

       $= 986$ kJ/mol rxn

Therefore, the average bond energy of an P-Cl bond in $PCl_3(g)$ is (986/3) kJ or **329 kJ**.

15-56. *Refer to Section 15-9, Appendix K, and Exercise 15-54.*

The ΔH°_{rxn} of the reaction: $PCl_5(g) \rightarrow P(g) + 5Cl(g)$
is equal to 5 times the average P-Cl bond energy in PCl_5, since this reaction involves the breaking of 5 P-Cl bonds.

$\Delta H^{\circ}_{rxn} = [\Delta H^{\circ}_{f\ P(g)} + 5\Delta H^{\circ}_{f\ Cl(g)}]$ - $[\Delta H^{\circ}_{f\ PCl_5(g)}]$

       $= [(1 \text{ mol})(314.6 \text{ kJ/mol}) + (5 \text{ mol})(121.7 \text{ kJ/mol})]$ - $[(1 \text{ mol})(-398.9 \text{ kJ/mol})]$

       $= 1322$ kJ/mol rxn

Therefore, the average bond energy of a P-Cl bond in $PCl_5(g)$ is (1322/5) kJ or **264 kJ**.
It takes less energy to break an average P-Cl bond in PCl_5 than one in PCl_3 because P is a relatively small atom and Cl is relatively large. It is more difficult to squeeze 5 atoms of Cl around a P than 3 atoms of Cl. Therefore, those 5 atoms of Cl in PCl_5 are not held as tightly and have weaker P-Cl bonds.

15-58. *Refer to Section 15-9, Table 15-2 and Example 15-12.*

$\Delta H^{\circ}_{rxn} = [5\text{B.E.}_{C-H} + \text{B.E.}_{C-C} + \text{B.E.}_{C-N} + 2\text{B.E.}_{N-H}]$ - $[4\text{B.E.}_{C-H} + \text{B.E.}_{C=C} + 3\text{B.E.}_{N-H}]$

Substituting,

     53.6 kJ $= [(5 \text{ mol})(413 \text{ kJ/mol}) + (1 \text{ mol})(346 \text{ kJ/mol}) + (1 \text{ mol})(\text{B.E.}_{C-N}) + (2 \text{ mol})(391 \text{ kJ/mol})]$

                 $- [(4 \text{ mol})(413 \text{ kJ/mol}) + (1 \text{ mol})(602 \text{ kJ/mol}) + (3 \text{ mol})(391 \text{ kJ/mol})]$

     53.6 kJ $= (1 \text{ mol})(\text{B.E.}_{C-N}) - 234$ kJ

     B.E.$_{C-N}$ = **288 kJ/mol**

Table 15-2 gives the bond energy for an average C-N bond as 305 kJ/mol.

235

15-60. *Refer to Sections 1-13 and 15-4, Example 15-1 and Exercise 1-60 Solution.*

Plan: (1) Determine the heat gained by the calorimeter.
(2) Find the heat capacity of the calorimeter (calorimeter constant).

(1) $|\text{heat lost}|_{\text{iron}} = |\text{heat gained}|_{\text{water}} + |\text{heat gained}|_{\text{calorimeter}}$

$|\text{specific heat} \times \text{mass} \times \Delta t|_{\text{iron}} = |\text{specific heat} \times \text{mass} \times \Delta t|_{\text{water}} + |\text{heat gained}|_{\text{calorimeter}}$

$(0.444 \text{ J/g}\cdot°\text{C})(93.3 \text{ g})(65.58°\text{C} - 19.68°\text{C}) = (4.184 \text{ J/g}\cdot°\text{C})(75.0 \text{ g})(19.68°\text{C} - 16.95°\text{C}) + |\text{heat gained}|_{\text{calorimeter}}$

$1.90 \times 10^3 \text{ J} = 8.57 \times 10^2 \text{ J} + |\text{heat gained}|_{\text{calorimeter}}$

Therefore, $|\text{heat gained}|_{\text{calorimeter}} = 1.90 \times 10^3 \text{ J} - 857 \text{ J} = 1.04 \times 10^3 \text{ J}$

(2) heat capacity of calorimeter (J/°C) $= \dfrac{|\text{heat gained}|_{\text{calorimeter}}}{\Delta T} = \dfrac{1.04 \times 10^3 \text{ J}}{19.68°\text{C} - 16.95°\text{C}} = \textbf{381 J/°C}$

15-62. *Refer to Sections 1-13 and 15-4, Example 15-1 and Exercise 1-60 Solution.*

$|\text{heat lost}|_{\text{metal}} = |\text{heat gained}|_{\text{water}} + |\text{heat gained}|_{\text{calorimeter}}$

$|\text{specific heat} \times \text{mass} \times \Delta t|_{\text{metal}} = |\text{specific heat} \times \text{mass} \times \Delta t|_{\text{water}} + |\text{calorimeter constant} \times \Delta t|_{\text{cal}}$

$(\text{Sp. Ht.})(36.5 \text{ g})(100.0°\text{C} - 32.5°\text{C}) = (4.184 \text{ J/g}\cdot°\text{C})(50.0 \text{ mL} \times 0.997 \text{ g/mL})(32.5°\text{C} - 25.0°\text{C})$
$+ (1.87 \text{ J/°C})(32.5°\text{C} - 25.0°\text{C})$

$(\text{Sp. Ht.})(2.46 \times 10^3 \text{ J}) = 1.6 \times 10^3 \text{ J} + 14 \text{ J}$

Specific heat of the metal $= \textbf{0.66 J/g}\cdot\textbf{°C}$

15-64. *Refer to Sections 15-4 and 15-5, Examples 15-2, 15-3 and 15-4, and Exercise 1-60 Solution.*

Balanced equation: $Pb(NO_3)_2(aq) + 2NaI(aq) \rightarrow PbI_2(s) + 2NaNO_3(aq)$

(a) $|\text{heat released}| = |\text{heat gained}|_{\text{soln}} + |\text{heat gained}|_{\text{calorimeter}}$

$= |\text{specific heat} \times \text{mass} \times \Delta t|_{\text{soln}} + |\text{heat capacity} \times \Delta t|_{\text{calorimeter}}$

$= (4.184 \text{ J/g}\cdot°\text{C})(200. \text{ g})(24.2°\text{C} - 22.6°\text{C})$
$+ (472 \text{ J/°C})(24.2°\text{C} - 22.6°\text{C})$

$= 1.3 \times 10^3 \text{ J} + 7.6 \times 10^2 \text{ J}$

$= \textbf{2.1} \times \textbf{10}^3 \textbf{ J}$

(b) This is a possible limiting reactant problem because amounts of both reactants are given. In this case, we are given stoichiometric amounts of both reactants.

$\text{mol Pb(NO}_3)_2 = \dfrac{6.62 \text{ g}}{331 \text{g/mol}} = 0.0200 \text{ mol}$

$\text{mol NaI} = \dfrac{6.00 \text{ g}}{149.9 \text{g/mol}} = 0.0400 \text{ mol}$

$\Delta H_{\text{rxn}} = \dfrac{-2.1 \times 10^3 \text{ J}}{0.0200 \text{ mol Pb(NO}_3)_2} \times \dfrac{1 \text{ mol Pb(NO}_3)_2}{1 \text{ mol rxn}} = \textbf{--1.0} \times \textbf{10}^5 \textbf{ J/mol rxn or --1.0} \times \textbf{10}^2 \textbf{ kJ/mol rxn}$

15-66. *Refer to Sections 15-4 and 15-10, and Examples 15-2 and 15-14.*

(a) $2C_6H_6(\ell) + 15 \, O_2(g) \rightarrow 12CO_2(g) + 6H_2O(\ell)$

(b) |heat released| = |heat gained|$_{water}$ + |heat gained|$_{calorimeter}$

= |specific heat x mass x Δt|$_{water}$ + |heat capacity x Δt|$_{calorimeter}$

= (4.184 J/g·°C)(945 g)(32.692°C - 23.640°C) + (891 J/°C)(32.692°C - 23.640°C)

= 3.58 x 10^4 J + 8.07 x 10^3 J

= 4.39 x 10^4 J or 43.9 kJ

Since heat is released in this reaction (the temperature of the water increased), ΔE is a negative quantity.

$$\Delta E = -\frac{43.9 \text{ kJ}}{1.048 \text{ g } C_6H_6(\ell)} = \textbf{–41.9 kJ/g } \textbf{C}_6\textbf{H}_6(\ell)$$

$$\Delta E = -\frac{43.9 \text{ kJ}}{1.048 \text{ g } C_6H_6(\ell)} \times \frac{78.11 \text{ g}}{1 \text{ mol}} = \textbf{–3270 kJ/mol } \textbf{C}_6\textbf{H}_6(\ell) \qquad \text{(to 3 significant figures)}$$

15-68.   *Refer to Sections 15-4 and 15-5, Examples 15-2, 15-3 and 15-4, and Exercise 1-60 Solution.*

Balanced equation: $Mg(s) + 2HCl(aq) \rightarrow MgCl_2(aq) + H_2(g)$

(1) |heat released| = |heat gained|$_{soln}$ + |heat gained|$_{calorimeter}$

= |specific heat x mass x Δt|$_{soln}$ + |heat capacity x Δt|$_{calorimeter}$

= (4.184 J/g·°C)[(100. mL x 1.10 g/mL) + 1.22 g](45.5°C - 23.0°C)

    + (562 J/°C)(45.5°C - 23.0°C)

= 1.05 x 10^4 J + 1.26 x 10^4 J

= **2.31 x 10^4 J**

Note: The mass of the solution equals the mass of the HCl solution plus the mass of the magnesium strip.

(2) This is a possible limiting reactant problem because amounts of both reactants are given. In this case, it is clear that Mg is the limiting reactant, since

$$\text{mol Mg} = \frac{1.22 \text{ g}}{24.3 \text{ g/mol}} = 0.0502 \text{ mol} \qquad\qquad \text{mol HCl} = 6.02 \text{ } M \times 0.100 \text{ L} = 0.612 \text{ mol}$$

$$\Delta H_{rxn} = \frac{-2.31 \times 10^4 \text{ J}}{0.0502 \text{ mol Mg}} \times \frac{1 \text{ mol Mg}}{1 \text{ mol rxn}} = \textbf{–4.60 x 10}^5 \textbf{ kJ or –460. kJ/mol rxn}$$

15-70.   *Refer to Sections 15-4 and 15-10, and Examples 15-2 and 15-14.*

Balanced equation: $2C_{10}H_{22}(\ell) + 31 \text{ } O_2(g) \rightarrow 20CO_2(g) + 22H_2O(\ell)$

|heat released| = |heat gained|$_{water}$ + |heat gained|$_{calorimeter}$

= |specific heat x mass x Δt|$_{water}$ + |heat capacity x Δt|$_{calorimeter}$

= (4.184 J/g·°C)(1250.0 g)(26.4°C - 24.6°C) + (2450 J/°C)(26.4°C - 24.6°C)

= 9400 J + 4400 J (each value has 2 significant figures)

= 13800 J (3 significant figures - see rules for adding numbers)

Since heat is released in this reaction (the temperature of the water increased), ΔE is a negative quantity.

$$\Delta E = -\frac{13800 \text{ J}}{6.620 \text{ g } C_{10}H_{22}(\ell)} = \textbf{–2.08 x 10}^3 \textbf{ J/g } \textbf{C}_{10}\textbf{H}_{22}(\ell)$$

$$\Delta E = -\frac{13800 \text{ J}}{6.620 \text{ g } C_{10}H_{22}(\ell)} \times \frac{142.3 \text{ g}}{1 \text{ mol}} \times \frac{1 \text{ kJ}}{1000 \text{ J}} = \textbf{–297 kJ/mol } \textbf{C}_{10}\textbf{H}_{22}(\ell)$$

15-72.   *Refer to Sections 15-10.*

(a) When heat is absorbed by a system or added to a system, q is "+." When heat is released or removed from a system, q is "–."

(b) When work is done on a system, w is "+." When work is done by a system, w is "–."

237

15-74. *Refer to Section 15-10 and Example 15-13.*

Balanced equation: $2NH_4NO_3(s) \rightarrow 2N_2(g) + 4H_2O(g) + O_2(g)$

(a) Work (w) is "–". The change in the moles of gas, Δn_{gas} ($= n_{gaseous\ products} - n_{gaseous\ reactants}$) is a positive value. The sign of the work term is opposite that of Δn_{gas} since $w = -P\Delta V = -\Delta n_{gas}RT$ at constant P and T, so work is "–".

(b) This reaction is responsible for many explosions, so intuitively we know that the system is doing work on the surroundings. The created gases of the system are expanding against the atmosphere and doing work on the surroundings.

15-76. *Refer to Section 15-10.*

For the system: $q = -175$ J, $w_{electrical} = +96$ J and $w_{PV} = -257$ J

$\Delta E = q + w_{total} = q + (w_{electrical} + w_{PV}) = -175\ J + [+96\ J + (-257\ J)] = \textbf{–336 J}$

15-78. *Refer to Section 15-10, Example 15-13 and Exercise 15-77.*

Plan: Evaluate $\Delta n_{gas} = n_{gaseous\ products} - n_{gaseous\ reactants}$. The sign of the work term is opposite that of Δn_{gas} since $w = -P\Delta V = -\Delta n_{gas}RT$ at constant P and T.

(a) $2SO_2(g) + O_2(g) \rightarrow 2SO_3(g)$

$\Delta n_{gas} = 2$ mol - 3 mol $= -1$ mol. Therefore, $w > 0$ and work is done by the surroundings on the system.

(b) $CaCO_3(s) \rightarrow CaO(s) + CO_2(g)$

$\Delta n_{gas} = 1$ mol - 0 mol $= +1$ mol. Therefore, work < 0, and work is done by the system on the surroundings.

(c) $CO_2(g) + H_2O(\ell) + CaCO_3(s) \rightarrow Ca^{2+}(aq) + 2HCO_3^-(aq)$

$\Delta n_{gas} = 0$ mol - 1 mol $= -1$ mol. Therefore, work > 0 and work is done on the system by the surroundings.

15-80. *Refer to Sections 15-10 and 15-11.*

(a) The balanced equation for the oxidation of 1 mole of HCl: $HCl(g) + 1/4\ O_2(g) \rightarrow 1/2\ Cl_2(g) + 1/2\ H_2O(g)$

$$work = -P\Delta V = -\Delta n_{gas}RT = -(n_{gaseous\ products} - n_{gaseous\ reactants})RT$$
$$= -(1\ mol - 5/4\ mol)(8.314\ J/mol \cdot K)(200°C + 273°)$$
$$= \textbf{+983 J}$$

Work is a positive number, therefore, work is done on the system by the surroundings. As the system "shrinks" from 5/4 mole of gas to 1 mole of gas, work is done on the system by the surroundings to decrease the volume. (Recall that $V \propto n$ at constant T and P.)

(b) The balanced reaction for the decomposition of 1 mole of NO: $NO(g) \rightarrow 1/2\ N_2(g) + 1/2\ O_2(g)$

$$work = -P\Delta V = -\Delta n_{gas}RT = -(1\ mol - 1\ mol)RT = \textbf{0 J}$$

There is no work done since the number of moles of gas, and hence the volume of the system, remains constant.

15-82. *Refer to the Introduction to Section 15-12, Sections 15-12 and 15-15.*

When fuel, e.g., gasoline, is burned, it first undergoes a physical change as it is converted from a liquid to the gaseous state. In the carburetor, the fuel is mixed with oxygen, and a spark ignites the mixture. The fuel then undergoes a chemical change as it reacts with oxygen gas to produce carbon dioxide and water. This reaction happens spontaneously. Let us consider gasoline as being primarily octane; the reaction in the engine is:

$$2C_8H_{18}(g) + 25\ O_2(g)\ \rightarrow\ 16CO_2(g) + 18H_2O(g) + heat$$

and it is exothermic, producing a great deal of heat. The Second Law of Thermodynamics states that in spontaneous changes, the universe tends toward a state of increasing entropy, $\Delta S_{universe} > 0$.

Does this make sense in this case? Absolutely. We are first going from a system containing 2 moles of liquid fuel to 2 moles of gaseous fuel - a big increase in entropy. Then before reaction we have 27 moles of gas, and after reaction we have a system containing 34 moles of gas. Entropy involves an increase in the relative positions of the molecules with respect to each other and the energies they can have. The entropy of this system has definitely increased after the combustion reaction has occurred.

15-84. *Refer to Section 15-14.*

The Third Law of Thermodynamics states that the entropy of a pure, perfect crystalline substance is zero at 0 K.

This means that all substances have some entropy (dispersal of energy and/or matter, i.e. disorder) except when the substance is a pure, perfect, motionless, vibrationless crystal at absolute zero Kelvin. This also implies that the entropy of a substance can be expressed on an absolute basis.

15-86. *Refer to Section 15-13.*

(a) The probability that a coin will come up heads in one flip is ½ = **0.5**.

(b) The probability that a coin comes up heads two times in a row is ½ x ½ = ¼ = **0.25**.

(c) The probability that the coin comes up heads 10 times in a row is $(½)^{10}$ = **1/1024 = 0.000977**.

15-88. *Refer to Section 15-13, Exercise 15-87 and Figure 15-13.*

Consider the following arrangements of molecules:

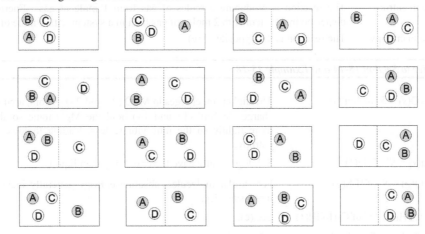

239

(a) A total of **16** different arrangements are possible.

(b) A mixture of unlike molecules in at least one of the flasks can be found in **14 out of 16** arrangements.

(c) The probability that at least one of the flasks contains a mixture of unlike molecules is 14/16 or **7/8**.

(d) The probability that the gases are not mixed is 2/16 or **1/8**.

15-90.   *Refer to Section 15-13.*

(a) heating glass to its softening temperature — entropy is increasing for the glass, so ΔS is positive

(b) sugar dissolving in coffee — entropy is increasing for both sugar and coffee, so ΔS is positive

(c) $CaCO_3$ is precipitating — entropy is decreasing for $CaCO_3$, so ΔS is negative

---

15-92.   *Refer to Section 15-14 and Table 15-4.*

There is an increase in entropy (dispersal of energy) in only the process (c) sublimation of dry ice, $CO_2(s) \rightarrow CO_2(g)$. In the other physical processes, the systems are becoming more ordered and the entropy is decreasing.

15-94.   *Refer to Section 15-14.*

When the volume occupied by one mole of Ar at 0°C is halved, there is a *decrease* in entropy (dispersal of energy), as signified by the negative sign of the entropy change, -5.76 J/(mol rxn)·K. In the smaller volume there are fewer energy levels available for the argon molecules to occupy and so, there is a decrease in entropy in the smaller volume.

15-96.   *Refer to Sections 15-13 and 15-14, and Table 15-4.*

(a) increase in entropy — When the NaCl dissolves, the ions disperse throughout the water. This allows the ions and the water molecules to transfer energy to each other. Dispersal of matter allows for more dispersal of energy.

(b) decrease in entropy — When some of the NaCl precipitates out as the saturated solution cools, there are fewer number of ways to distribute the same total energy.

(c) decrease in entropy — The solid phase is always more ordered than the liquid phase of a substance.

(d) increase in entropy — The gas phase is always more disordered than the liquid phase of a substance.

(e) increase in entropy — The reaction is producing 2 moles of gas from 1 mole of gas. Energy is more dispersed in a system with 2 moles of gas than in a system with 1 mole of gas.

(f) decrease in entropy — The reaction is the opposite of (e).

15-98.   *Refer to Section 15-14 and Example 15-18.*

(a) S° of MgO(s) < S° of NaF(s) — The higher ion charges in MgO (2+ and 2−) as compared to the ion charges in NaF (1+ and 1−) hold the MgO ionic solid together more tightly so the ions vibrate less, leading to lower absolute entropy.

(b) S° of Au(s) < S° of Hg(ℓ) — Solids generally have lower entropy than liquids.

(c) S° of H_2O(g) < S° of H_2S(g) — For similar molecules, absolute entropy generally increases with increasing size.

(d) S° of CH_3OH(ℓ) < S° of C_2H_5OH(ℓ) — See (c).

(e) S° of NaOH(s) < S° of NaOH(aq) — When substances are mixed, in this case, dissolved in water, the absolute entropy is always higher than either substance by itself.

15-100. *Refer to Section 15-14 and Example 15-17.*

Entropy increases and the change in entropy is positive when a reaction occurs
  (1) when there are more gaseous products than gaseous reactants ($\Delta n_{gas} > 0$) and
  (2) when there are more aqueous products than aqueous reactants if no gases are present.

(a) entropy change is negative   2 mol gaseous products $\rightarrow$ 1 mol gaseous product
(b) entropy change is negative   4 mol gaseous products $\rightarrow$ 2 mol gaseous product
(c) entropy change is positive   0 mol gaseous products $\rightarrow$ 1 mol gaseous product
(d) entropy change is negative   1/2 mol gaseous products $\rightarrow$ 0 mol gaseous product
(e) entropy change is negative   2 mol aqueous products $\rightarrow$ 0 mol gaseous product

15-102. *Refer to Sections 15-13 and 15-14.*

Consider the boiling of a pure liquid at constant pressure. (a) $\Delta S_{system} > 0$ (b) $\Delta H_{system} > 0$ (c) $\Delta T_{system} = 0$

15-104. *Refer to Section 15-13 and Example 15-15.*

(a) Balanced equation: $4HCl(g) + O_2(g) \rightarrow 2Cl_2(g) + 2H_2O(g)$

$$\Delta S^{\circ}_{rxn} = [2S^{\circ}_{Cl_2(g)} + 2S^{\circ}_{H_2O(g)}] - [4S^{\circ}_{HCl(g)} + S^{\circ}_{O_2(g)}]$$
$$= [(2 \text{ mol})(+223.0 \text{ J/mol·K}) + (2 \text{ mol})(+188.7 \text{ J/mol·K})]$$
$$- [(4 \text{ mol})(+186.8 \text{ J/mol·K}) + (1 \text{ mol})(+205.0 \text{ J/mol·K})]$$
$$= \mathbf{-128.8 \text{ J/(mol rxn)·K}}$$

The reaction is producing 4 moles of gas from 5 moles of gas. The energy and mass in the system is becoming less dispersed as the number of moles of gas decreases; entropy is decreasing and the change in entropy is expected to be negative.

(b) Balanced equation: $PCl_3(g) + Cl_2(g) \rightarrow PCl_5(g)$

$$\Delta S^{\circ}_{rxn} = [S^{\circ}_{PCl_5(g)}] - [S^{\circ}_{PCl_3(g)} + S^{\circ}_{Cl_2(g)}]$$
$$= [(1 \text{ mol})(+353 \text{ J/mol·K})] - [(1 \text{ mol})(+311.7 \text{ J/mol·K}) + (1 \text{ mol})(+223.0 \text{ J/mol·K})]$$
$$= \mathbf{-182 \text{ J/(mol rxn)·K}}$$

The reaction is producing 1 mole of gas from 2 moles of gas. For the same reasoning as shown in (a), the entropy is decreasing and the change in entropy is expected to be negative.

(c) Balanced equation: $2N_2O(g) \rightarrow 2N_2(g) + O_2(g)$

$$\Delta S^{\circ}_{rxn} = [2S^{\circ}_{N_2(g)} + S^{\circ}_{O_2(g)}] - [2S^{\circ}_{N_2O(g)}]$$
$$= [(2 \text{ mol})(+191.5 \text{ J/mol·K}) + (1 \text{ mol})(+205.0 \text{ J/mol·K})] - [(2 \text{ mol})(+219.7 \text{ J/mol·K})]$$
$$= \mathbf{+148.6 \text{ J/(mol rxn)·K}}$$

The reaction is producing 3 mole of gas from 2 moles of gas. The entropy is increasing and the change in entropy is expected to be positive.

15-106. *Refer to Sections 15-15 and 15-16, and Table 15-7.*

(a) always spontaneous: (iii) $\Delta H < 0, \Delta S > 0$
(b) always nonspontaneous: (ii) $\Delta H > 0, \Delta S < 0$
(c) spontaneous or nonspontaneous, depending on T and the magnitudes of ΔH and ΔS: (i)$\Delta H > 0, \Delta S > 0$
$$\text{(iv) } \Delta H < 0, \Delta S < 0$$

15-108. *Refer to Section 15-14, Example 15-16 and Appendix K.*

Balanced equation: $SiH_4(g) + 2O_2(g) \rightarrow SiO_2(s) + 2H_2O(\ell)$

$$\Delta S^{\circ}_{rxn} = [S^{\circ}_{SiO_2(s)} + 2S^{\circ}_{H_2O(\ell)}] - [S^{\circ}_{SiH_4(g)} + 2S^{\circ}_{O_2(g)}]$$

241

$$= [(1\ mol)(+41.84\ J/mol\cdot K) + (2\ mol)(+69.91\ J/mol\cdot K)] - [(1\ mol)(+204.5\ J/mol\cdot K)$$
$$+ (2\ mol)(+205.0\ J/mol\cdot K)]$$
$$= \textbf{-432.8 J/(mol rxn)·K}$$

15-110. *Refer to Sections 15-8 and 15-16.*

Since $\Delta G°$ is a state function like $\Delta H°$, we can use Hess's Law type of manipulations to determine the $\Delta G_f°$. The balanced equation representing the $\Delta G_f°$ of HBr(g) is: $\frac{1}{2}H_2(g) + \frac{1}{2}Br_2(\ell) \rightarrow HBr(g)$

	$\Delta G°$
$\frac{1}{2}Br_2(\ell) \rightarrow \frac{1}{2}Br_2(g)$	1.57 kJ
$H(g) + Br(g) \rightarrow HBr(g)$	−339.09 kJ
$\frac{1}{2}Br_2(g) \rightarrow Br(g)$	80.85 kJ
$\frac{1}{2}H_2(g) \rightarrow H(g)$	203.247 kJ
$\frac{1}{2}H_2(g) + \frac{1}{2}Br_2(\ell) \rightarrow HBr(g)$	**−53.42 kJ/mol rxn**

15-112. *Refer to Section 15-16, Example 15-20 and Appendix K.*

Plan: Calculate $\Delta H_{rxn}°$ and $\Delta S_{rxn}°$, then use the Gibbs free energy change equation , $\Delta G = \Delta H - T\Delta S$, to determine $\Delta G_{rxn}°$.

(a) Balanced equation: $3NO_2(g) + H_2O(\ell) \rightarrow 2HNO_3(\ell) + NO(g)$

$\Delta H_{rxn}° = [2\Delta H_{f\ HNO_3(\ell)}° + \Delta H_{f\ NO(g)}°] - [3\Delta H_{f\ NO_2(g)}° + \Delta H_{f\ H_2O(\ell)}°]$
$= [(2\ mol)(-174.1\ kJ/mol) + (1\ mol)(+90.25\ kJ/mol)]$
$\qquad - [(3\ mol)(+33.2\ kJ/mol) + (1\ mol)(-285.8\ kJ/mol)]$
$= \textbf{-71.75 kJ/mol rxn}$

$\Delta S_{rxn}° = [2S_{HNO_3(\ell)}° + S_{NO(g)}°] - [3S_{NO_2(g)}° + S_{H_2O(\ell)}°]$
$= [(2\ mol)(+155.6\ J/mol\cdot K) + (1\ mol)(+210.7\ J/mol\cdot K)]$
$\qquad - [(3\ mol)(+240.0\ J/mol\cdot K) + (1\ mol)(+69.91\ J/mol\cdot K)]$
$= \textbf{-268.0 J/(mol rxn)·K}$

$\Delta G_{rxn}° = \Delta H_{rxn}° - T\Delta S_{rxn}° = -71.75\ kJ - (298.15\ K)(-0.268\ kJ/K) = \textbf{+8.15 kJ/mol rxn}$

(b) Balanced equation: $SnO_2(s) + 2CO(g) \rightarrow 2CO_2(g) + Sn(s, white)$

$\Delta H_{rxn}° = [2\Delta H_{f\ CO_2(g)}° + \Delta H_{f\ Sn(s)}°] - [\Delta H_{f\ SnO_2(s)}° + 2\Delta H_{f\ CO(g)}°]$
$= [(2\ mol)(-393.5\ J/mol) + (1\ mol)(0\ kJ/mol)] - [(1\ mol)(-580.7\ kJ/mol) + (2\ mol)(-110.5\ kJ/mol)]$
$= \textbf{+14.7 kJ/mol rxn}$

$\Delta S_{rxn}° = [2S_{CO_2(g)}° + S_{Sn(s)}°] - [S_{SnO_2(s)}° + 2S_{CO(g)}°]$
$= [(2\ mol)(+213.6\ J/mol\cdot K) + (1\ mol)(+51.55\ J/mol\cdot K)]$
$\qquad - [(1\ mol)(+52.3\ J/mol\cdot K) + (2\ mol)(+197.6\ J/mol\cdot K)]$
$= \textbf{+31.2 J/(mol rxn)·K}$

$\Delta G_{rxn}° = \Delta H_{rxn}° - T\Delta S_{rxn}° = +14.7\ kJ - (298.15\ K)(+0.0312\ kJ/K) = \textbf{+5.4 kJ/mol rxn}$

(c) Balanced equation: $2Na(s) + 2H_2O(\ell) \rightarrow 2NaOH(aq) + H_2(g)$

$\Delta H_{rxn}° = [2\Delta H_{f\ NaOH(aq)}° + \Delta H_{f\ H_2(g)}°] - [2\Delta H_{f\ Na(s)}° + 2\Delta H_{f\ H_2O(\ell)}°]$
$= [(2\ mol)(-469.6\ kJ/mol) + (1\ mol)(0\ kJ/mol)] - [(2\ mol)(0\ kJ/mol) + (2\ mol)(-285.8\ kJ/mol)]$
$= \textbf{-367.6 kJ/mol rxn}$

242

$$\Delta S^{\circ}_{rxn} = [2S^{\circ}_{NaOH(aq)} + S^{\circ}_{H_2(g)}] - [2S^{\circ}_{Na(s)} + 2S^{\circ}_{H_2O(\ell)}]$$
$$= [(2 \text{ mol})(+49.8 \text{ J/mol·K}) + (1 \text{ mol})(+130.6 \text{ J/mol·K})]$$
$$- [(2 \text{ mol})(+51.0 \text{ J/mol·K}) + (2 \text{ mol})(+69.91 \text{ J/mol·K})]$$
$$= \mathbf{-11.62 \text{ J/(mol rxn)·K}}$$

$$\Delta G^{\circ}_{rxn} = \Delta H^{\circ}_{rxn} - T\Delta S^{\circ}_{rxn} = -367.6 \text{ kJ} - (298.15 \text{ K})(-0.01162 \text{ kJ/K}) = \mathbf{-364.1 \text{ kJ/mol rxn}}$$

15-114. *Refer to Sections 15-16 and 15-17.*

Recall: Gibbs free energy change equation: $\Delta G = \Delta H - T\Delta S$

(a) false   An exothermic reaction ($\Delta H < 0$) will be spontaneous ($\Delta G < 0$) only if either ΔS is positive, or, in the event ΔS is negative, the absolute value of $T\Delta S$ is smaller than that of ΔH.

(b) true   From the Gibbs free energy change equation; the $T\Delta S$ term has a negative sign in front.

(c) false   A reaction with $\Delta S_{sys} > 0$ will be spontaneous ($\Delta G < 0$) only if either ΔH is negative, or, in the event ΔH is positive, its absolute value is smaller than that of $T\Delta S$.

15-116. *Refer to Sections 15-16 and 15-17, and Appendix K.*

Balanced equation: $2H_2O_2(\ell) \rightarrow 2H_2O(\ell) + O_2(g)$

(a) $\Delta H^{\circ}_{rxn} = [2\Delta H^{\circ}_{f\,H_2O(\ell)} + \Delta H^{\circ}_{f\,O_2(g)}] - [2\Delta H^{\circ}_{f\,H_2O_2(\ell)}]$
$$= [(2 \text{ mol})(-285.8 \text{ kJ/mol}) + (1 \text{ mol})(0 \text{ kJ/mol})] - [(2 \text{ mol})(-187.8 \text{ kJ/mol})]$$
$$= \mathbf{-196.0 \text{ kJ/mol rxn}}$$

$\Delta G^{\circ}_{rxn} = [2\Delta G^{\circ}_{f\,H_2O(\ell)} + \Delta G^{\circ}_{f\,O_2(g)}] - [2\Delta G^{\circ}_{f\,H_2O_2(\ell)}]$
$$= [(2 \text{ mol})(-237.2 \text{ kJ/mol}) + (1 \text{ mol})(0 \text{ kJ/mol})] - [(2 \text{ mol})(-120.4 \text{ kJ/mol})]$$
$$= \mathbf{-233.6 \text{ kJ/mol rxn}}$$

$\Delta S^{\circ}_{rxn} = [2S^{\circ}_{H_2O(\ell)} + S^{\circ}_{O_2(g)}] - [2S^{\circ}_{H_2O_2(\ell)}]$
$$= [(2 \text{ mol})(+69.91 \text{ J/mol·K}) + (1 \text{ mol})(+205.0 \text{ J/mol·K})] - [(2 \text{ mol})(+109.6 \text{ J/mol·K})]$$
$$= \mathbf{+125.6 \text{ J/(mol rxn)·K}}$$

(b) Hydrogen peroxide, $H_2O_2(\ell)$, will be stable if $\Delta G^{\circ} > 0$ for the above balanced reaction at some temperature, i.e., if the above reaction is non-spontaneous. However, $\Delta H^{\circ}_{rxn} < 0$ and $\Delta S^{\circ}_{rxn} > 0$ for the decomposition of $H_2O_2(\ell)$ and the reaction is spontaneous ($\Delta G^{\circ} < 0$) for all temperatures. Hence, there is no temperature at which $H_2O_2(\ell)$ is stable at 1 atm.

15-118. *Refer to Section 15-17.*

Dissociation reactions, such as $HCl(g) \rightarrow H(g) + Cl(g)$, require energy to break bonds and therefore are endothermic with positive ΔH values. The ΔS values for such reactions are positive since 2 or more particles are being formed from 1 molecule, causing the system to become more energetically dispersed. Under the circumstances when ΔH and ΔS are both positive, the spontaneity of the reaction is favored at higher temperatures.

15-120. *Refer to Section 15-17, Examples 15-22 and 15-23, and Appendix K.*

Plan: Evaluate ΔH_{rxn} and ΔS_{rxn}. To assess the temperature range over which the reaction is spontaneous, use the signs of ΔH and ΔS and the Gibbs free energy change equation, $\Delta G = \Delta H - T\Delta S$. Assume that ΔH and ΔS are independent of temperature.

(a) Balanced equation: $CaCO_3(s) + H_2SO_4(\ell) \rightarrow CaSO_4(s) + H_2O(\ell) + CO_2(g)$

$\Delta H^\circ_{rxn} = [\Delta H^\circ_{f\,CaSO_4(s)} + \Delta H^\circ_{f\,H_2O(\ell)} + \Delta H^\circ_{f\,CO_2(g)}] - [\Delta H^\circ_{f\,CaCO_3(s)} + \Delta H^\circ_{f\,H_2SO_4(\ell)}]$

$\quad = [(1\ mol)(-1433\ kJ/mol) + (1\ mol)(-285.8\ kJ/mol) + (1\ mol)(-393.5\ kJ/mol)]$
$\qquad\qquad - [(1\ mol)(-1207\ kJ/mol) + (1\ mol)(-814.0\ kJ/mol)]$

$\quad = -91.3\ kJ/mol\ rxn$

$\Delta S^\circ_{rxn} = [S^\circ_{CaSO_4(s)} + S^\circ_{H_2O(\ell)} + S^\circ_{CO_2(g)}] - [S^\circ_{CaCO_3(s)} + S^\circ_{H_2SO_4(\ell)}]$

$\quad = [(1\ mol)(+107\ J/mol\cdot K) + (1\ mol)(+69.91\ J/mol\cdot K) + (1\ mol)(+213.6\ J/mol\cdot K)]$
$\qquad\qquad - [(1\ mol)(+92.9\ J/mol\cdot K) + (1\ mol)(+156.9\ J/mol\cdot K)]$

$\quad = +141\ J/(mol\ rxn)\cdot K$

Since ΔH is negative and ΔS is positive, the reaction is **spontaneous at all temperatures**.

(b) Balanced equation: $2HgO(s) \rightarrow 2Hg(\ell) + O_2(g)$

$\Delta H^\circ_{rxn} = [2\Delta H^\circ_{f\,Hg(\ell)} + \Delta H^\circ_{f\,O_2(g)}] - [2\Delta H^\circ_{f\,HgO(s)}]$

$\quad = [(2\ mol)(0\ kJ/mol) + (1\ mol)(0\ kJ/mol)] - [(2\ mol)(-90.83\ kJ/mol)]$

$\quad = +181.7\ kJ/mol\ rxn$

$\Delta S^\circ_{rxn} = [2S^\circ_{Hg(\ell)} + S^\circ_{O_2(g)}] - [2S^\circ_{HgO(s)}]$

$\quad = [(2\ mol)(+76.02\ J/mol\cdot K) + (1\ mol)(+205.0\ J/mol\cdot K)] - [(2\ mol)(+70.29\ J/mol\cdot K)]$

$\quad = +216.5\ J/(mol\ rxn)\cdot K$

At equilibrium, $\Delta G^\circ_{rxn} = 0 = \Delta H^\circ_{rxn} - T\Delta S^\circ_{rxn}$, and solving for T_{eq}

$T_{eq} = \dfrac{\Delta H_{rxn}}{\Delta S_{rxn}} = \dfrac{181.7\ kJ}{0.2165\ kJ/K} = 839.3\ K$

Since ΔH and ΔS are positive, the reaction is **spontaneous at $T > 839.3$ K**.

(c) Balanced equation: $CO_2(g) + C(s) \rightarrow 2CO(g)$

$\Delta H^\circ_{rxn} = [2\Delta H^\circ_{f\,CO(g)}] - [\Delta H^\circ_{f\,CO_2(g)} + \Delta H^\circ_{f\,C(s)}]$

$\quad = [(2\ mol)(-110.5\ kJ/mol)] - [(1\ mol)(-393.5\ kJ/mol) + (1\ mol)(0\ kJ/mol)]$

$\quad = +172.5\ kJ/mol\ rxn$

$\Delta S^\circ_{rxn} = [2S^\circ_{CO(g)}] - [S^\circ_{CO_2(g)} + S^\circ_{C(s)}]$

$\quad = [(2\ mol)(+197.6\ J/mol\cdot K)] - [(1\ mol)(+213.6\ J/mol\cdot K) + (1\ mol)(+5.740\ J/mol\cdot K)]$

$\quad = +175.9\ J/(mol\ rxn)\cdot K$

At equilibrium, $\Delta G^\circ_{rxn} = 0 = \Delta H^\circ_{rxn} - T\Delta S^\circ_{rxn}$, and solving for T_{eq}

$T_{eq} = \dfrac{\Delta H_{rxn}}{\Delta S_{rxn}} = \dfrac{172.5\ kJ}{0.1759\ kJ/K} = 980.7\ K$

Since ΔH and ΔS are positive, the reaction is **spontaneous at $T > 980.7$ K**.

(d) Balanced equation: $2Fe_2O_3(s) \rightarrow 4Fe(s) + 3O_2(g)$

$\Delta H^\circ_{rxn} = [4\Delta H^\circ_{f\,Fe(s)} + 3\Delta H^\circ_{f\,O_2(g)}] - [2\Delta H^\circ_{f\,Fe_2O_3(s)}]$

$\quad = [(4\ mol)(0\ kJ/mol) + (3\ mol)(0\ kJ/mol)] - [(2\ mol)(-824.2\ kJ/mol)]$

$\quad = +1648\ kJ/mol\ rxn$

$\Delta S^\circ_{rxn} = [4S^\circ_{Fe(s)} + 3S^\circ_{O_2(g)}] - [2S^\circ_{Fe_2O_3(s)}]$

$\quad = [(4\ mol)(+27.3\ J/mol\cdot K) + (3\ mol)(+205.0\ J/mol\cdot K)] - [(2\ mol)(+87.40\ J/mol\cdot K)]$

$\quad = +549.4\ J/(mol\ rxn)\cdot K$

$T_{eq} = \dfrac{\Delta H_{rxn}}{\Delta S_{rxn}} = \dfrac{1648\ kJ}{0.5494\ kJ/K} = 3000.\ K$

Since ΔH and ΔS are positive, the reaction is **spontaneous at $T > 3000.$ K**.

15-122. _Refer to Section 15-17, Appendix K and Example 15-21._

(a) The process is: $H_2O(\ell) \rightarrow H_2O(g)$

$\Delta H^{\circ}_{rxn} = \Delta H^{\circ}_{f\ H_2O(g)} - \Delta H^{\circ}_{f\ H_2O(\ell)} = (1\ mol)(-241.8\ kJ/mol) - (1\ mol)(-285.8\ kJ/mol) = +44.0\ kJ$

$\Delta S^{\circ}_{rxn} = S^{\circ}_{H_2O(g)} - S^{\circ}_{H_2O(\ell)} = (1\ mol)(+188.7\ J/mol{\cdot}K) - (1\ mol)(+69.91\ J/mol{\cdot}K) = +118.8\ J/K$

$T_{eq} = \dfrac{\Delta H_{rxn}}{\Delta S_{rxn}} = \dfrac{44.0\ kJ}{0.1188\ kJ/K} = 370\ K\ or\ \mathbf{97°C}$

(b) The known boiling point of water is, of course, 100°C. The discrepancy is because we assumed that the standard values of enthalpy of formation and entropy in Appendix K are independent of temperature. However, these tabulated values were determined at 25°C; we are using them to solve a problem at 100°C. Nevertheless, this assumption allows us to estimate the boiling point of water with reasonable accuracy.

15-124. _Refer to Sections 15-16 and 15-17, and Appendix K._

Balanced equation: $2NiO(s) \rightarrow 2Ni(s) + O_2(g)$

(1) The decomposition of $NiO(s)$ is product-favored (spontaneous) at 25°C if $\Delta G^{\circ}_{rxn} < 0$ at that temperature.
$\Delta G^{\circ}_{rxn} = -2\Delta H^{\circ}_{f\ NiO(s)} = -(2\ mol)(-216\ kJ/mol) = +432\ kJ$
Since $\Delta G^{\circ}_{rxn} > 0$, the reaction is reactant-favored, not product favored at 25°C.

(2) To determine how this reaction is affected by temperature, let's calculate ΔS°_{rxn} and ΔH°_{rxn}.
$\Delta H^{\circ}_{rxn} = -2\Delta H^{\circ}_{f\ NiO(s)} = -(2\ mol)(-244\ kJ/mol) = +488\ kJ/mol\ rxn$
$\Delta S^{\circ}_{rxn} = [2S^{\circ}_{Ni(s)} + S^{\circ}_{O_2(g)}] - [2S^{\circ}_{NiO(s)}]$
$\qquad = [(2\ mol)(+30.1\ J/mol{\cdot}K) + (1\ mol)(+205.0\ J/mol{\cdot}K)] - [(2\ mol)(+38.6\ J/mol{\cdot}K)]$
$\qquad = +188\ J/(mol\ rxn){\cdot}K$
$T_{eq} = \dfrac{\Delta H_{rxn}}{\Delta S_{rxn}} = \dfrac{+488\ kJ}{0.188\ kJ/K} = \mathbf{2.60 \times 10^3\ K\ or\ 2320°C}$

15-126. _Refer to Sections 15-16 and 15-17, and Appendix K._

Balanced equation: $C_2H_4(g) + H_2O(g) \rightarrow C_2H_5OH(\ell)$

ΔS°_{rxn} $= [S^{\circ}_{C_2H_5OH(\ell)}] - [S^{\circ}_{C_2H_4(g)} + S^{\circ}_{H_2O(g)}]$
$\qquad = [(1\ mol)(+161\ J/mol{\cdot}K)] - [(1\ mol)(+219.5\ J/mol{\cdot}K) + (1\ mol)(+188.7\ J/mol{\cdot}K)]$
$\qquad = \mathbf{-247\ J/(mol\ rxn){\cdot}K}$

Since $\Delta S^{\circ}_{rxn} < 0$, we know that the reaction is becoming more ordered, but we don't know if the reaction is spontaneous (product-favored) or not. We would also need to know ΔH°_{rxn}.

ΔH°_{rxn} $= [\Delta H^{\circ}_{f\ C_2H_5OH(\ell)}] - [\Delta H^{\circ}_{f\ C_2H_4(g)} + \Delta H^{\circ}_{f\ H_2O(g)}]$
$\qquad = [(1\ mol)(-277.7\ kJ/mol)] - [(1\ mol)(+52.26\ kJ/mol) + (1\ mol)(-241.8\ kJ/mol)]$
$\qquad = \mathbf{-88.2\ kJ/mol\ rxn}$

Since ΔH is negative and ΔS is negative, the reaction will be spontaneous at lower temperatures. Let's now find the temperature at which the reaction is at equilibrium:

ΔG°_{rxn} $= \Delta H^{\circ}_{rxn} - T\Delta S^{\circ}_{rxn} = 0$ at equilibrium
$T_{eq} = \dfrac{\Delta H_{rxn}}{\Delta S_{rxn}} = \dfrac{-88.2\ kJ}{-0.247\ kJ/K} = 357\ K,$

Therefore, the reaction is only product-favored at **temperatures below 357 K**.

245

Plan: Use Hess's Law and solve for ΔH_f° of the organic compound.

(a) Balanced equation: $C_6H_{12}(\ell) + 9O_2(g) \rightarrow 6CO_2(g) + 6H_2O(\ell)$

$\Delta H_{combustion}^\circ = [6\Delta H_f^\circ\, CO_2(g) + 6\Delta H_f^\circ\, H_2O(\ell)] - [\Delta H_f^\circ\, C_6H_{12}(\ell) + 9\Delta H_f^\circ\, O_2(g)]$

$-3920 \text{ kJ} = [(6 \text{ mol})(-393.5 \text{ kJ/mol}) + (6 \text{ mol})(-285.8 \text{ kJ/mol})]$
$\qquad\qquad\qquad\qquad - [(1 \text{ mol})\Delta H_f^\circ\, C_6H_{12}(\ell) + (9 \text{ mol})(0 \text{ kJ/mol})]$

$-3920 \text{ kJ} = -4075.8 \text{ kJ} - (1 \text{ mol})\Delta H_f^\circ\, C_6H_{12}(\ell)$

$\Delta H_f^\circ\, C_6H_{12}(\ell) = \mathbf{-156 \text{ kJ/mol } C_6H_{12}(\ell)}$

(b) Balanced equation: $C_6H_5OH(s) + 7O_2(g) \rightarrow 6CO_2(g) + 3H_2O(\ell)$

$\Delta H_{combustion}^\circ = [6\Delta H_f^\circ\, CO_2(g) + 3\Delta H_f^\circ\, H_2O(\ell)] - [\Delta H_f^\circ\, C_6H_5OH(s) + 7\Delta H_f^\circ\, O_2(g)]$

$-3053 \text{ kJ} = [(6 \text{ mol})(-393.5 \text{ kJ/mol}) + (3 \text{ mol})(-285.8 \text{ kJ/mol})]$
$\qquad\qquad\qquad\qquad - [(1 \text{ mol})\Delta H_f^\circ\, C_6H_5OH(s) + (7 \text{ mol})(0 \text{ kJ/mol})]$

$-3053 \text{ kJ} = -3218.4 \text{ kJ} - (1 \text{ mol})\Delta H_f^\circ\, C_6H_5OH(s)$

$\Delta H_f^\circ\, C_6H_5OH(s) = \mathbf{-165 \text{ kJ/mol } C_6H_5OH(s)}$

The vaporization process is: ethanol(ℓ) $\rightarrow$ ethanol(g)

$\Delta E = q + w$ where ΔE = change in internal energy
$\qquad\qquad\qquad\qquad\qquad\quad q$ = heat absorbed by the system
$\qquad\qquad\qquad\qquad\qquad\quad w$ = work done on the system

(1) The heat absorbed by the system, $q = \Delta H_{vap} \times \text{g ethanol} = +855 \text{ J/g} \times 12.5 \text{ g} = \mathbf{+10700 \text{ J}}$

(2) The work done on the system in going from a liquid to a gas,

$w = -P\Delta V = -P(V_{gas} - V_{liquid})$

where $V_{gas} = \dfrac{nRT}{P} = \dfrac{(12.5 \text{ g}/46.1 \text{ g/mol})(0.0821 \text{ L·atm/mol·K})(78.0°C + 273.15°)}{1.00 \text{ atm}} = 7.82 \text{ L}$

$V_{liquid} = 12.5 \text{ g ethanol} \times \dfrac{1.00 \text{ mL ethanol}}{0.789 \text{ g ethanol}} = 15.8 \text{ mL or } 0.0158 \text{ L}$

Therefore,

$w = -P\Delta V = -(1 \text{ atm})(7.82 \text{ L} - 0.02 \text{ L}) = -7.80 \text{ L·atm}$ (the negative value means the system is doing work)

To find a factor to convert L·atm to J, we can equate two values of the molar gas constant, R
$\qquad\qquad\qquad 0.0821 \text{ L·atm/mol·K} = 8.314 \text{ J/mol·K}$
$\qquad\qquad\qquad\qquad\qquad 1 \text{ L·atm} = 101 \text{ J}$

And so, $w = -7.80 \text{ L·atm} \times \dfrac{101 \text{ J}}{1 \text{ L·atm}} = \mathbf{-788 \text{ J}}$

(3) Finally, $\Delta E = q + w = 10700 \text{ J} + (-788 \text{ J}) = \mathbf{9900 \text{ J}}$

(a) heat gained by calorimeter $= 0.01520 \text{ g } C_{10}H_8 \times \dfrac{1 \text{ mol } C_{10}H_8}{128.16 \text{ g } C_{10}H_8} \times \dfrac{5156.8 \text{ kJ}}{1 \text{ mol } C_{10}H_8} = 0.6116 \text{ kJ}$

We know: $|\text{heat gained by calorimeter}| = |\text{heat capacity} \times \Delta t|$ where t is temperature in °C

Therefore, heat capacity $= \dfrac{|\text{heat gained by calorimeter}|}{|\Delta t|} = \dfrac{0.6116 \text{ kJ}}{0.212°C} = \mathbf{2.88 \text{ kJ/°C}}$

(b) |heat released in the reaction| = $0.1040 \text{ g } C_8H_{18} \times \dfrac{1 \text{ mol } C_8H_{18}}{114.22 \text{ g/mol}} \times \dfrac{5451.4 \text{ kJ}}{1 \text{ mol } C_8H_{18}} = 4.964 \text{ kJ}$

We also know:

|heat released in the reaction| = |heat gained by calorimeter|

Substituting, $4.964 \text{ kJ} = |\text{heat capacity} \times \Delta t|$

$= |2.88 \text{ kJ/°C} \times \Delta t|$

$\Delta t = 1.72°C$

Therefore, $t_{\text{final}} = t_{\text{initial}} + \Delta t = 22.102°C + 1.72°C = \mathbf{23.82°C}$.

15-134. *Refer to Sections 15-14, 15-15 and 15-16.*

When a rubber band is stretched: $\Delta H < 0$, since heat is released

$\Delta S < 0$, since the rubber band is becoming more ordered (more linear);

therefore, $\Delta G > 0$, since the process does not occur spontaneously

When the stretched rubber band is relaxed, the signs of the thermodynamic state functions change:

$\Delta H > 0$, since heat is absorbed (that's why your hand feels colder)

$\Delta S > 0$, since the rubber band is becoming more disordered; therefore

$\Delta G < 0$, since the process occurs spontaneously

The spontaneous process that occurs when the stretched rubber band is allowed to return to its original, random arrangement of polymer molecules, must be driven by the increase in the mass and energy dispersal of the system, since the reaction is endothermic ($\Delta H > 0$).

15-136. *Refer to Sections 1-13 and 15-4.*

(a) |heat lost|$_{\text{metal}}$ = |heat gained|$_{\text{water}}$

|specific heat $\times$ mass $\times \Delta t$|$_{\text{metal}}$ = |specific heat $\times$ mass $\times \Delta t$|$_{\text{water}}$

(specific heat of metal)(32.6 g)(99.83°C - 24.41°C) = (4.184 J/g·°C)(100.0 g)(24.41°C - 23.62°C)

(specific heat of metal)(2.46 $\times 10^3$) = 330

specific heat of metal = **0.13 J/g·°C**

Therefore, according to this calculation, the metal is **tungsten, W** (specific heat = 0.135 J/g·°C).

(b) |heat lost|$_{\text{metal}}$ = |heat gained|$_{\text{water}}$ + |heat gained|$_{\text{calorimeter}}$

|specific heat $\times$ mass $\times \Delta t$|$_{\text{metal}}$ = |specific heat $\times$ mass $\times \Delta t$|$_{\text{water}}$

+ |heat capacity $\times \Delta t$|$_{\text{calorimeter}}$

(specific heat of metal)(32.6 g)(99.83°C - 24.41°C) = (4.184 J/g·°C)(100.0 g)(24.41°C - 23.62°C)

+ (410 J/°C)(24.41°C - 23.62°C)

(specific heat of metal)(2.46 $\times 10^3$) = 330 + 320

specific heat of metal = 0.26 J/g·°C

Yes, the identification of the metal was different. When the heat capacity of the calorimeter is taken into account, the specific heat of the metal is 0.26 J/g·°C and the metal is identified as molybdenum, Mo (specific heat = 0.250 J/g·°C).

15-138. *Refer to Sections 15-13, 15-14. 15-16 and 15-17.*

(a) crystal growth from supersaturated solution:

$\Delta S < 0$ since the system is becoming more ordered; there are fewer number of ways to distribute the same total energy.

$\Delta G < 0$ since crystals spontaneously will form from a supersaturated solution

(b) sugar cube dissolving into hot tea

$\Delta S > 0$ since the system is becoming more disordered, i.e. the sugar molecules disperse throughout the tea., allowing the sugar molecules and the aqueous tea solution to transfer energy to each other. Dispersal of matter allows for more dispersal of energy.

$\Delta G < 0$ since the sugar cube easily and spontaneously dissolves into hot tea

(c) $H_2O(s) \rightarrow H_2O(\ell)$

$\Delta S > 0$ since the system is becoming more disordered; liquids always have higher entropy than solids

The sign of ΔG depends on the temperature. When $T > 0°C$, $\Delta G < 0$, since ice will spontaneously melt. When $T < 0°C$, $\Delta G > 0$, since liquid water will spontaneously freeze and when $T = 0°C$, $\Delta G = 0$, since that is the melting point of water and the reaction is at equilibrium.

15-140. *Refer to Section 15-10.*

Calculation: Activity Time Equivalent (min) $= \dfrac{\text{Food Fuel Value (kcal)}}{\text{Energy Output (kcal/min)}}$

Food	Fuel Value (kcal)	Activity Time Equivalent (min)				
		Sitting (1.7 kcal/min)	Walking (5.5 kcal/min)	Cycling (10 kcal/min)	Swimming (8.4 kcal/min)	Running (19 kcal/min)
Apple	100	59	18	10	12	5.3
Cola	105	62	19	11	13	5.5
Malted milk	500	290	91	50	60	26
Pasta	195	110	35	20	23	10
Hamburger	350	210	64	35	42	18
Steak	1000	590	180	100	120	53

15-142. *Refer to Section 15-3 and Fundamental Algebra.*

? kJ of energy found in 100. g protein $= 100.$ g protein $\times \dfrac{17 \text{ kJ}}{1 \text{ g protein}} = 1700$ kJ

? kJ of energy found in 100. g fat $= 100.$ g fat $\times \dfrac{39 \text{ kJ}}{1 \text{ g fat}} = 3900$ kJ

The difference in energy content is the amount of energy that must be burned up by walking instead of resting, so that the person doesn't gain weight:

    difference in energy content = 3900 - 1700 = 2200 kJ

? time required to walk instead of rest to burn off 2200 kJ $= \dfrac{\text{difference in energy content}}{\text{difference in utilization rate}}$

$$= \dfrac{(3900 - 1700) \text{ kJ}}{(1250 - 335) \text{ kJ/hr}}$$

$$= \dfrac{2200 \text{ kJ}}{915 \text{ kJ/hr}}$$

$$= \textbf{2.4 hr}$$

15-144. *Refer to Sections 1-13 and 15-4.*

$$|\text{heat lost}|_{\text{lead}} = |\text{heat gained}|_{\text{water}} + |\text{heat gained}|_{\text{calorimeter}}$$
$$|\text{specific heat} \times \text{mass} \times \Delta t|_{\text{lead}} = |\text{specific heat} \times \text{mass} \times \Delta t|_{\text{water}} + |\text{heat capacity} \times \Delta t|_{\text{calorimeter}}$$

(Sp. Ht. of Pb)(43.6 g)(100.0°C − 26.8°C) = (4.184 J/g·°C)(50.0 g)(26.8°C - 25.0°C)
$$+ (18.6 \text{ J/°C})(26.8°C - 25.0°C)$$

    (Specific heat of Pb)(3190) = 380 + 33

$$\text{Specific heat of Pb} = \textbf{0.13 J/g·°C}$$
$$\text{Molar heat capacity of Pb} = \text{0.13 J/g·°C} \times \text{207.2 g/mol} = \textbf{27 J/mol·°C}$$

15-146. *Refer to Section 15-4.*

(a) Heat gain by calorimeter $= (4572 \text{ J/°C})(27.93°C - 24.76°C) = 1.449 \times 10^4 \text{ J or } 14.49 \text{ kJ}$

Fuel value of butter $= \dfrac{14.49 \text{ kJ}}{0.483 \text{ g}} = \textbf{30.0 kJ/g}$

(b) Nutritional Calories/g butter $= \dfrac{30.0 \text{ kJ/g}}{4.184 \text{ kJ/kilocalorie}} = \textbf{7.17 kilocalorie/g}$

(c) Nutritional Calories/5.00 g pat of butter $= (7.17 \text{ kilocalorie/g}) \times 5.00 \text{ g} = \textbf{35.9 kilocalorie}$

16 Chemical Kinetics

16-2. *Refer to Sections 16-5 and 16-6.*

The collision theory of reaction rates states that molecules, atoms or ions must collide effectively in order to react. For an effective collision to occur, the reacting species must have (1) at least a minimum amount of energy in order to break old bonds and make new ones, and (2) the proper orientation toward each other.

Transition state theory complements collision theory. When particles collide with enough energy to react, called the activation energy, E_a, the reactants form a short-lived, high energy activated complex, or transition state, before forming the products. The transition state also could revert back to the reactants.

16-4. *Refer to the Introduction to Chapter 16.*

In Chapter 15, we learned that reactions which are thermodynamically favorable have negative ΔG values and occur spontaneously as written. However, thermodynamics cannot be used to determine the rate of a reaction. Kinetically favorable reactions must be thermodynamically favorable *and* have a low enough activation energy to occur at a reasonable rate at a certain temperature.

16-6. *Refer to Section 16-3.*

The coefficients of the balanced overall equation bear no necessary relationship to the exponents to which the concentrations are raised in the rate law expression. The exponents are determined experimentally and describe how the concentrations of each reactant affect the reaction rate. The exponents are related to the rate-determining (slow) step in a sequence of mainly unimolecular and bimolecular reactions called the mechanism of the reaction. It is the mechanism which lays out exactly the order in which bonds are broken and made as the reactants are transformed into the products of the reaction.

16-8. *Refer to Section 16-1 and Example 16-1.*

(a) $3ClO^-(aq) \rightarrow ClO_3^-(aq) + 2Cl^-(aq)$ rate of reaction $= -\dfrac{\Delta[ClO^-]}{3\Delta t} = \dfrac{\Delta[ClO^-]}{\Delta t} = \dfrac{\Delta[Cl^-]}{2\Delta t}$

(b) $2SO_2(g) + O_2(g) \rightarrow 2SO_3(g)$ rate of reaction $= -\dfrac{\Delta[SO_2]}{2\Delta t} = -\dfrac{\Delta[O_2]}{\Delta t} = \dfrac{\Delta[SO_3]}{2\Delta t}$

(c) $C_2H_4(g) + Br_2(g) \rightarrow C_2H_4Br_2(g)$ rate of reaction $= -\dfrac{\Delta[C_2H_4]}{\Delta t} = -\dfrac{\Delta[Br_2]}{\Delta t} = \dfrac{\Delta[C_2H_4Br_2]}{\Delta t}$

(d) $(C_2H_5)_2(NH)_2 + I_2 \rightarrow (C_2H_5)_2N_2 + 2HI$ rate of reaction $=$

$$-\dfrac{\Delta[(C_2H_5)_2(NH)_2]}{\Delta t} = -\dfrac{\Delta[I_2]}{\Delta t} = \dfrac{\Delta[(C_2H_5)_2N_2]}{\Delta t} = \dfrac{\Delta[HI]}{2\Delta t}$$

16-10. *Refer to Section 16-1 and Example 16-1.*

Balanced reaction: $4NH_3 + 5O_2 \rightarrow 4NO + 6H_2O$ rate of reaction $= -\dfrac{\Delta[NH_3]}{4\Delta t} = -\dfrac{\Delta[O_2]}{5\Delta t} = \dfrac{\Delta[NO]}{4\Delta t} = \dfrac{\Delta[H_2O]}{6\Delta t}$

Substituting,

$$\text{rate of reaction} = -\dfrac{\Delta[NH_3]}{4\Delta t} = \dfrac{1.20\ M\ NH_3}{4 \times 1\ \text{min}} = 0.300\ M/\text{min}$$

Therefore, rate of disappearance of $O_2 = -\dfrac{\Delta[O_2]}{\Delta t} = 5 \times$ rate of reaction $= 5 \times 0.300 \; M/min = \mathbf{1.50 \; M/min}$

rate of appearance of NO $= \dfrac{\Delta[NO]}{\Delta t} = 4 \times$ rate of reaction $= 4 \times 0.300 \; M/min = \mathbf{1.20 \; M/min}$

rate of appearance of $H_2O = \dfrac{\Delta[H_2O]}{\Delta t} = 6 \times$ rate of reaction $= 6 \times 0.300 \; M/min = \mathbf{1.80 \; M/min}$

16-12. *Refer to Section 16-2.*

Some fireworks are bright because of the burning of magnesium: $2Mg + O_2 \rightarrow 2MgO$.

This reaction gives off much energy as heat and light. The magnesium metal pieces could be interspersed in the body of the firework near the fuse. To get the best visual display, this oxidation reaction cannot go too fast or too slow. The speed of this reaction can be controlled by the size of magnesium pieces. If the pieces were large, the reaction would take place at a slower rate if it occurred at all due to the smaller surface area of magnesium exposed to the air. If the pieces were too small, the reaction would occur too quickly; there would be one large burst of light and the beauty of the sparks would be lost.

16-14. *Refer to Section 16-3 and Example 16-2.*

The simplest approach to this problem is to assume that the initial concentrations of NO and O_2 for the first experiment are each 1 M. Then for the second experiment, the initial concentration of NO is 1/2 M and that of O_2 is 2 M. Let us substitute these values into the rate-law expression, rate $= k[NO]^2[O_2]$

First experiment: rate $= k(1 \; M)^2(1 \; M) = k$
Second experiment: rate $= k(\frac{1}{2} \; M)^2(2 \; M) = \frac{1}{2} \, k$

Therefore, the rate of reaction in the second experiment would be **1/2** times that of the first experiment.

16-16. *Refer to Section 16-4.*

Plan: Use dimensional analysis and the rate-law expression to determine the units of k, the rate constant, in the following general equation: rate $(M/s) = k[A]^x$ where x = the overall order of the reaction
[A] = the reactant concentration (M)

	Overall Reaction Order	**Example**	**Units of k**
(a)	1	rate $= k[A]$	$(M/s)/M = s^{-1}$
(b)	2	rate $= k[A]^2$	$(M/s)/M^2 = M^{-1} \cdot s^{-1}$
(c)	3	rate $= k[A]^3$	$(M/s)/M^3 = M^{-2} \cdot s^{-1}$
(d)	1.5	rate $= k[A]^{1.5}$	$(M/s)/M^{1.5} = M^{-0.5} \cdot s^{-1}$

16-18. *Refer to Section 16-3 and Examples 16-3 and 16-4.*

The form of the rate-law expression: rate $= k[A]^x[B]^y[C]^z$

Step 1: rate dependence on [A]. Consider Experiments 1 and 3:

Method 1: By observation, [B] and [C] do not change; [A] increases by a factor of 3. However, the reaction rate does not change. Therefore, changing [A] does not affect reaction rate and the reaction is zero order with respect to A. In all subsequent determinations, the effect of A can be ignored.

Method 2: A mathematical solution is obtained by substituting the experimental values of Experiments 1 and 3 into rate-law expressions and dividing the latter by the former. Note: the calculations are easier when the experiment with the larger rate is in the numerator.

| Expt 3 | $\dfrac{5.0 \times 10^{-4}\ M/min}{5.0 \times 10^{-4}\ M/min}$ | $= \dfrac{k(0.30\ M)^x(0.10\ M)^y(0.20\ M)^z}{k(0.10\ M)^x(0.10\ M)^y(0.20\ M)^z}$ |
| Expt 1 | | |

$$1 = 3^x$$
$$x = 0$$

Step 2: rate dependence on [B]. Consider Experiments 1 and 2:

Method 1:   [B] changes by a factor of 3; [C] does not change; the reaction rate also changes by a factor of 3 ($= 1.5 \times 10^{-3}/5.0 \times 10^{-4}$). The reaction rate is directly proportional to [B] and y must be equal to 1. The reaction is first order with respect to B.

Method 2:

| Expt 2 | $\dfrac{1.5 \times 10^{-3}\ M/min}{5.0 \times 10^{-4}\ M/min}$ | $= \dfrac{k(0.20\ M)^0(0.30\ M)^y(0.20\ M)^z}{k(0.10\ M)^0(0.10\ M)^y(0.20\ M)^z}$ |
| Expt 1 | | |

$$3 = 3^y$$
$$y = 1$$

Step 3: rate dependence on [C]. Consider Experiments 2 and 4:

Method 1:   [B] does not change; [C] changes by a factor of 3; the reaction rate changes by a factor of 3 ($= 4.5 \times 10^{-3}/1.5 \times 10^{-3}$). The reaction rate is directly proportional to [C] and z must be equal to 1. The reaction is first order with respect to C.

Method 2:

| Expt 4 | $\dfrac{4.5 \times 10^{-3}\ M/min}{1.5 \times 10^{-3}\ M/min}$ | $= \dfrac{k(0.40\ M)^0(0.30\ M)^1(0.60\ M)^z}{k(0.20\ M)^0(0.30\ M)^1(0.20\ M)^z}$ |
| Expt 2 | | |

$$3 = 3^z$$
$$z = 1$$

The rate-law expression is: rate $= k[A]^0[B]^1[C]^1 = k[B][C]$. To calculate the value of k, substitute the values from any one of the experiments into the rate-law expression and solve for k. If we use the data from Experiment 1,

$$5.0 \times 10^{-4}\ M/min = k(0.10\ M)(0.20\ M)$$
$$k = 2.5 \times 10^{-2}\ M^{-1} \cdot min^{-1}$$

The rate-law expression is now: **rate $= (2.5 \times 10^{-2}\ M^{-1} \cdot min^{-1})[B][C]$**

16-20.   *Refer to Section 16-3 and Example 16-2.*

(1) To determine for which reaction will the rate double if $[H_2]$ is doubled, look for the reaction that is first order with respect to H_2. All three reactions are first order with respect to H_2.

(2) To determine for which reaction will the rate quadruple if $[H_2]$ is doubled, look for the reaction that is second order with respect to H_2. None of the reactions are second order with respect to H_2.

(3) To determine for which reaction will the rate is not affected if $[H_2]$ is doubled, look for the reaction that is zero order with respect to H_2. None of the reactions are zero order with respect to H_2.

16-22.   *Refer to Section 16-3, Examples 16-3 and 16-4, and Exercise 16-18 Solution.*

Balanced equation: $2ClO_2(aq) + 2OH^-(aq) \rightarrow ClO_3^-(aq) + ClO_2^-(aq) + H_2O(\ell)$

(a) The form of the rate-law expression: rate $= k[ClO_2]^x[OH^-]^y$

Step 1: rate dependence on $[ClO_2]$. Consider Experiments 1 and 3.

Method 1: By observation, $[OH^-]$ is constant and $[ClO_2]$ increases by a factor of 2 ($= 0.024/0.012$). The rate of reaction increases by a factor of 4 ($= 8.28 \times 10^{-4}/2.07 \times 10^{-4}$). The reaction rate increases as the square of $[ClO_2]$ and x equals 2. The reaction is second order with respect to ClO_2.

Method 2:

Expt 2	$\dfrac{8.28 \times 10^{-4}\ M/s}{2.07 \times 10^{-4}\ M/s}$	$= \dfrac{k(0.024\ M)^x(0.012\ M)^y}{k(0.012\ M)^x(0.012\ M)^y}$
Expt 1		

$$4 = 2^x$$
$$x = 2$$

Step 2: rate dependence on [OH$^-$]. Consider Experiments 1 and 2:

Method 1: By observation, [ClO$_2$] is constant and [OH$^-$] increases by a factor of 2 (= 0.024/0.012). The rate of reaction increases by a factor of 2 (= $4.14 \times 10^{-4}/2.07 \times 10^{-4}$). The reaction rate is directly proportional to [OH$^-$] and y equals 1. The reaction is first order with respect to OH$^-$.

Method 2:

Expt 2	$\dfrac{4.14 \times 10^{-4}\ M/s}{2.07 \times 10^{-4}\ M/s}$	$= \dfrac{k(0.012\ M)^1(0.024\ M)^y}{k(0.012\ M)^1(0.012\ M)^y}$
Expt 1		

$$2 = 2^y$$
$$y = 1$$

The rate-law expression is: rate $= k[ClO_2]^2[OH^-]^1 = k[ClO_2]^2[OH^-]$

(b) The reaction is second order with respect to ClO$_2$, first order with respect to OH$^-$ and third order overall.

(c) Using the data from Experiment 1 to calculate k, we have

$$2.07 \times 10^{-4}\ M/s = k(0.012\ M)^2(0.012\ M)$$
$$k = 1.2 \times 10^2\ M^{-2}{\cdot}s^{-1}$$

The rate-law expression is now: **rate $= (1.2 \times 10^2\ M^{-2}{\cdot}s^{-1})[ClO_2]^2[OH^-]$**

16-24   *Refer to Section 16-3, Examples 16-3 and 16-4, and Exercise 16-18 Solution.*

Balanced equation: A + B → C
The form of the rate-law expression: rate $= k[A]^x[B]^y$

Step 1: rate dependence on [B]. Consider Experiments 1 and 2:

Method 1: [A] does not change; [B] changes by a factor of 1.5 (= 0.30/0.20); reaction rate changes also by a factor of 1.5 (= $7.5 \times 10^{-6}/5.0 \times 10^{-6}$). The reaction rate is directly proportional to [B] and y equals 1. The reaction is first order with respect to B.

Method 2:

Expt 2	$\dfrac{7.5 \times 10^{-6}\ M/s}{5.0 \times 10^{-6}\ M/s}$	$= \dfrac{k(0.10\ M)^x(0.30\ M)^y}{k(0.10\ M)^x(0.20\ M)^y}$
Expt 1		

$$1.5 = 1.5^y$$
$$y = 1$$

Step 2: rate dependence on [A].
There is no pair of experiments in which [A] is changing and [B] is constant. Therefore, one may choose any 2 experiments in which [A] is varying and use Method 2. If we choose Exp. 1 and 3:

Expt 3	$\dfrac{4.0 \times 10^{-5}\ M/s}{5.0 \times 10^{-6}\ M/s}$	$= \dfrac{k(0.20\ M)^x(0.40\ M)^1}{k(0.10\ M)^x(0.20\ M)^1}$
Expt 1		

$$8 = (2)^x(2)^1$$
$$4 = 2^x$$
$$x = 2$$

Therefore, the reaction is second order with respect to A.

The rate-law expression is: rate $= k[A]^2[B]^1 = k[A]^2[B]$

Using the data from Experiment 1 to calculate k, we have

$$5.0 \times 10^{-6} \ M/s = k(0.10 \ M)^2(0.20 \ M)$$
$$k = 2.5 \times 10^{-3} \ M^{-2}\cdot s^{-1}$$

The rate-law expression is now: rate $= (2.5 \times 10^{-3} \ M^{-2}\cdot s^{-1})[A]^2[B]$

16-26   *Refer to Section 16-3, Examples 16-3 and 16-4, and Exercise 16-18 Solution.*

Balanced equation: $A + B \rightarrow C$

(a) The form of the rate-law expression: rate $= k[A]^x[B]^y$

   Step 1: rate dependence on [B]. Consider Experiments 1 and 2:

   Method 1: [A] does not change; [B] changes by a factor of 2 (= 0.20/0.10); reaction rate changes by a factor of 4 (= $8.0 \times 10^{-4}/2.0 \times 10^{-4}$). The reaction rate quadruples when the concentration doubles. Therefore, y equals 2. The reaction is second order with respect to B.

   Method 2:
   Expt 2 $\dfrac{8.0 \times 10^{-4} \ M/s}{2.0 \times 10^{-4} \ M/s} = \dfrac{k(0.10 \ M)^x(0.20 \ M)^y}{k(0.10 \ M)^x(0.10 \ M)^y}$
   Expt 1
   $$4.0 = 2.0^y$$
   $$y = 2$$

   Step 2: rate dependence on [A].
   There is no pair of experiments in which [A] is changing and [B] is constant. Therefore, one may choose any 2 experiments in which [A] is varying and use Method 2. If we choose Exp. 1 and 3:

   Expt 3 $\dfrac{2.56 \times 10^{-2} \ M/s}{2.0 \times 10^{-4} \ M/s} = \dfrac{k(0.20 \ M)^x(0.40 \ M)^2}{k(0.10 \ M)^x(0.10 \ M)^2}$
   Expt 1
   $$128 = (2)^x(4)^2$$
   $$8 = 2^x$$
   $$x = 3$$

   Therefore, the reaction is third order with respect to A.

   The rate-law expression is: rate $= k[A]^3[B]^2$

(b) Using the data from Experiment 1 to calculate k, we have

$$2.0 \times 10^{-4} \ M/s = k(0.10 \ M)^3(0.10 \ M)^2$$
$$k = 20. \ M^{-4}\cdot s^{-1}$$

   The rate-law expression is now: **rate $= (20. \ M^{-4}\cdot s^{-1})[A]^3[B]^2$**

16-28.   *Refer to Section 16-3.*

The rate-law expression: rate $= k[A][B]^2$

Plan: (1) Use the data for Experiment 1 and the rate-law expression to calculate the rate constant, k.
        (2) Substitute the given values into the complete rate-law expression to determine the reaction rate.

(1) Substituting, $0.150 \ M/s = k(1.00 \ M)(0.200 \ M)^2$
                            $k = 3.75 \ M^{-2}\cdot s^{-1}$

(2) Expt 2: rate $= (3.75 \ M^{-2}\cdot s^{-1})(2.00 \ M)(0.200 \ M)^2 = $ **0.300 M/s**
     Expt 3: rate $= (3.75 \ M^{-2}\cdot s^{-1})(2.00 \ M)(0.400 \ M)^2 = $ **1.20 M/s**

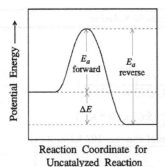

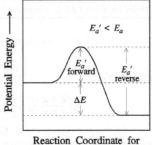

Reaction Coordinate for
Uncatalyzed Reaction

Reaction Coordinate for
Catalyzed Reaction

16-52. *Refer to Section 16-8 and Figure 16-13.*

(a) An increase in temperature does increase the initial rate of reaction. In order for a reaction to occur, the reactants must collide effectively. As the temperature increases, the velocity of the reactants increases and there are more collisions. Also, the reactant species must have a certain energy, called the activation energy, to produce products upon collision. When the temperature increases, more of the reactant species have that necessary amount of kinetic energy, so the reaction proceeds at a faster rate. Recall that the average kinetic energy of a container of molecules in the gas phase is directly proportional to the absolute temperature.

(b) A gas phase reaction is faster than the same reaction in the solid phase because the reacting species in the gaseous phase can move more quickly, have more collisions, resulting in a faster reaction rate.

16-54. *Refer to Section 16-8 and Example 16-12.*

The Arrhenius equation can be presented as:

$$\ln\left(\frac{k_2}{k_1}\right) = \frac{E_a}{R}\left(\frac{1}{T_1} - \frac{1}{T_2}\right)$$

where

k_2/k_1 = ratio of rate constants = 3.000
E_a = activation energy (J/mol)
R = 8.314 J/mol·K
T_1 = 600.0 K
T_2 = 610.0 K

Substituting,

$$\ln 3.000 = \frac{E_a}{8.314 \text{ J/mol·K}}\left(\frac{1}{600.0 \text{ K}} - \frac{1}{610.0 \text{ K}}\right)$$

$$1.0986 = \frac{E_a}{8.314 \text{ J/mol·K}}(2.7 \times 10^{-5} \text{ K}^{-1})$$

$$E_a = \mathbf{3.4 \times 10^5 \text{ J/mol rxn} \text{ or } 340 \text{ kJ/mol rxn}}$$

16-56. *Refer to Section 16-8 and Example 16-12.*

The Arrhenius equation can be presented as:

$$\ln\left(\frac{k_2}{k_1}\right) = \frac{E_a}{R}\left(\frac{1}{T_1} - \frac{1}{T_2}\right)$$

where

k_2/k_1 = ratio of rate constants = 3.000
E_a = activation energy (J/mol)
R = 8.314 J/mol·K
T_1 = 298 K
T_2 = 308 K

Substituting,

$$\ln 3.000 = \frac{E_a}{8.314 \text{ J/mol·K}}\left(\frac{1}{298 \text{ K}} - \frac{1}{308 \text{ K}}\right)$$

$$1.0986 = \frac{E_a}{8.314 \text{ J/mol·K}}(1.1 \times 10^{-4} \text{ K}^{-1})$$

$$E_a = \mathbf{8.4 \times 10^4 \text{ J/mol rxn} \text{ or } 84 \text{ kJ/mol rxn}}$$

For a particular reaction: $\Delta E^\circ = 51.51$ kJ/mol reaction

(1) From the Arrhenius equation:

$$\ln\!\left(\frac{k_2}{k_1}\right) = \frac{E_a}{8.314 \text{ J/mol·K}}\left(\frac{1}{273 \text{ K}} - \frac{1}{323 \text{ K}}\right)$$

$$\ln\!\left(\frac{8.9 \times 10^{-4} \text{ s}^{-1}}{8.0 \times 10^{-7} \text{ s}^{-1}}\right) = \frac{E_a}{8.314 \text{ J/mol·K}}\,(5.67 \times 10^{-4} \text{ K}^{-1})$$

$$E_a = \mathbf{1.03 \times 10^5 \text{ J/mol rxn}} \text{ or } \mathbf{103 \text{ kJ/mol rxn}}$$

(2) The reaction coordinate diagram for the reaction is:

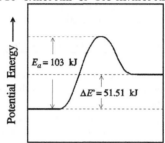

16-60. *Refer to Section 16-8.*

Plan: The Arrhenius equation can be rearranged: $\ln k = -\left(\dfrac{E_a}{R}\right)\!\left(\dfrac{1}{T}\right) + \ln A$.

Plot $\ln k$ against $1/T$.
The slope of the line $= -E_a/R$ and solve for E_a.

T (K)	$1/T$ (K^{-1})	k (s^{-1})	$\ln k$
600	1.67×10^{-3}	3.30×10^{-9}	-19.53
650	1.54×10^{-3}	2.19×10^{-7}	-15.33
700	1.43×10^{-3}	7.96×10^{-6}	-11.74
750	1.33×10^{-3}	1.80×10^{-4}	-8.623
800	1.25×10^{-3}	2.74×10^{-3}	-5.900
850	1.18×10^{-3}	3.04×10^{-2}	-3.493
900	1.11×10^{-3}	2.58×10^{-1}	-1.355

(a) By plotting the data as shown, we obtained
slope $= -E_a/R = -3.25 \times 10^4$ K
Therefore,
$E_a = -$ (slope) $\times R$
$= -(-3.25 \times 10^4$ K$) \times 8.314$ J/mol·K
$= \mathbf{2.70 \times 10^5}$ **J/mol or 270. kJ/mol**

(b) On the x axis, $1/T = 1/500$ K $= 0.00200$ K^{-1}
From the graph, we can estimate:
$y = \ln k = -30.2$
Therefore, $k = \mathbf{7.7 \times 10^{-14}}$ **s^{-1}** at 500 K

(c) On the y axis, $\ln k = \ln (5.00 \times 10^{-5}) = -9.9$
From the graph, we can estimate:
$x = 1/T = 0.00137$ K^{-1}
Therefore, $T = \mathbf{730}$ **K** when $k = 5.00 \times 10^{-5}$ s^{-1}

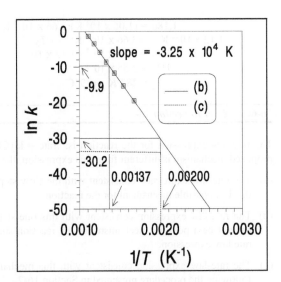

16-62.   *Refer to Sections 16-4 and 16-9.*

The hydration reaction of CO_2, $CO_2 + H_2O \rightarrow H_2CO_3$, is enzyme catalyzed. The rate of reaction does not depend on $[CO_2]$ or $[H_2O]$. This is deduced from the fact that it only takes 1 molecule of enzyme to react with 10^6 molecules of CO_2. Therefore, the reaction is zero-order with respect to CO_2.: rate $= k[CO_2]^0 = k$.

Plan: mol enzyme/L $\Rightarrow (1)$ molecules enzyme/L $\Rightarrow (2)$ hydration rate (molecules CO_2/L·s)
$\Rightarrow (3)$ hydration rate (molecules CO_2/L·min) $\Rightarrow (4)$ hydration rate (mol CO_2/L·min)
$\Rightarrow (5)$ hydration rate (g CO_2/L·min)

? g CO_2 hydrated/L·min $= \dfrac{1.0 \times 10^{-6} \text{ mol enzyme}}{1 \text{ liter}} \times \dfrac{6.02 \times 10^{23} \text{ molecules enzyme}}{1 \text{ mol enzyme}} \times \dfrac{10^6 \text{ molecules } CO_2}{1 \text{ molecule enzyme} \times 1\text{s}}$

$\times \dfrac{60 \text{ s}}{1 \text{ min}} \times \dfrac{1 \text{ mol } CO_2}{6.02 \times 10^{23} \text{ molecules } CO_2} \times \dfrac{44 \text{ g } CO_2}{1 \text{ mol } CO_2}$

$= \mathbf{2600}$ **g CO_2/L·min**

16-64.   *Refer to Section 16-8 and Example 16-11.*

(a) From the Arrhenius equation: $\ln\left(\dfrac{k_2}{k_1}\right) = \dfrac{E_a}{R}\left(\dfrac{1}{T_1} - \dfrac{1}{T_2}\right)$

Substituting,

$\ln \dfrac{k_2}{9.16 \times 10^{-3} \text{ s}^{-1}} = \dfrac{88 \times 10^3 \text{ J/mol}}{8.314 \text{ J/mol·K}}\left(\dfrac{1}{273 \text{ K}} - \dfrac{1}{298 \text{ K}}\right) = 3.3 \text{ (or 3.2526)}$

$\dfrac{k_2}{9.16 \times 10^{-3} \text{ s}^{-1}} = 26$

$k_2 = \mathbf{0.24}$ **s^{-1}**

(b) Substituting back into the Arrhenius equation and solving for T_2,

$\ln \dfrac{k_2}{k_1} = \dfrac{E_a}{R}\left(\dfrac{1}{T_1} - \dfrac{1}{T_2}\right)$

$\ln\left(\dfrac{3.00 \times 10^{-2} \text{ s}^{-1}}{9.16 \times 10^{-3} \text{ s}^{-1}}\right) = \dfrac{88 \times 10^3 \text{ J/mol}}{8.314 \text{ J/mol·K}}\left(\dfrac{1}{273 \text{ K}} - \dfrac{1}{T_2}\right)$

$$1.186 = (1.06 \times 10^4 \text{ K})(3.66 \times 10^{-3} \text{ K}^{-1} - 1/T_2)$$
$$1.12 \times 10^{-4} \text{ K}^{-1} = 3.66 \times 10^{-3} \text{ K}^{-1} - 1/T_2$$
$$1/T_2 = 3.66 \times 10^{-3} - 1.12 \times 10^{-4}$$
$$1/T_2 = 3.55 \times 10^{-3}$$
$$T_2 = \mathbf{282 \text{ K} \ \text{or} \ 9°C}$$

16-66. *Refer to Section 16-7.*

The rate-law expression for the reaction is: rate = k[$(CH_3)_3CBr$]. If the rate-law expression derived from a proposed mechanism is different from this expression, the mechanism cannot be the correct one.

(a) The rate-law expression consistent with the slow step of this mechanism is: rate = $k[(CH_3)_3CBr]$ and **yes**, this **is** a possible mechanism for the reaction.

(b) The rate-law expression consistent with this one-step mechanism is: rate = $k[(CH_3)_3CBr][OH^-]$ and this **cannot** be a possible mechanism for the reaction since it does not agree with the experimentally-derived rate law expression.

(c) The rate-law expression consistent with this mechanism is more complicated. It can be determined by following the procedure presented in Section 16-7:

From Step 2 (the slow step), rate = $k_2[(CH_3)_3(CH_2)CBr^-]$ where $(CH_3)_3(CH_2)CBr^-$ is an intermediate

Since $(CH_3)_3(CH_2)CBr^-$ is an intermediate, its concentration must be expressed in terms of the reactants, $(CH_3)_3CBr$ and OH^-.

For a fast, equilibrium step, we know:
$$\text{rate}_{1f} = \text{rate}_{1r}$$
From Step 1, $$k_{1f}[(CH_3)_3CBr][OH^-] = k_{1r}[(CH_3)_3(CH_2)CBr^-][H_2O]$$
$$[(CH_3)_3(CH_2)CBr^-] = \left(\frac{k_{1f}}{k_{1r}}\right)\frac{[(CH_3)_3CBr][OH^-]}{[H_2O]}$$

Substituting $$\text{rate} = k_2\left(\frac{k_{1f}}{k_{1r}}\right)\frac{[(CH_3)_3CBr][OH^-]}{[H_2O]}$$

And so, this **cannot** be a mechanism for this reaction, since it doesn't match the experimentally-derived rate law expression.

16-68. *Refer to Section 16-7.*

(a) Overall reaction: **2A + 2B → E + G**

Using the second step which is the slow, rate-determining step, we can write: rate = $k_2[D][B]$.

However, D is an intermediate; its concentration must be expressed in terms of the reactants, A and B. For a fast, equilibrium step, we know:
$$\text{rate}_{1f} = \text{rate}_{1r}$$
From Step 1, $$k_{1f}[A]^2[B] = k_{1r}[D]$$
$$[D] = \left(\frac{k_{1f}}{k_{1r}}\right)[A]^2[B]$$

Substituting into the rate law expression obtained in the slow, rate-determining step:
$$\text{rate} = k_2\left(\frac{k_{1f}}{k_{1r}}\right)[A]^2[B][B].$$
$$\mathbf{rate = k[A]^2[B]^2}$$

(b) **Overall reaction: A + B + D → G**   where C and F are intermediates

The rate-law expression consistent with this mechanism is more complicated. It can be determined by following the procedure presented in Section 16-7:

From Step 3 (the slow step), $\quad\quad\quad\quad$ rate $= k_3[F]$ $\quad\quad\quad$ where F is an intermediate

Since F is an intermediate, its concentration must ultimately be a function of the reactants, A, B and D.

For a fast, equilibrium step, we know:

Starting from the top in Step 1, $\quad\quad\quad$ rate$_{1f}$ = rate$_{1r}$

$$k_{1f}[A][B] = k_{1r}[C] \quad\quad\quad\quad \text{where C is an intermediate}$$

$$[C] = \left(\frac{k_{1f}}{k_{1r}}\right)[A][B]$$

From Step 2, $\quad\quad\quad\quad\quad\quad\quad k_{2f}[C][D] = k_{2r}[F]$

$$[F] = \left(\frac{k_{2f}}{k_{2r}}\right)\left(\frac{k_{1f}}{k_{1r}}\right)[A][B][D]$$

Substituting for [F] $\quad\quad\quad\quad$ rate $= k_3[F] = k_3\left(\frac{k_{2f}}{k_{2r}}\right)\left(\frac{k_{1f}}{k_{1r}}\right)[A][B][D]$

Therefore $\quad\quad\quad\quad\quad\quad\quad$ **rate $= k[A][B][D]$**

16-70. *Refer to Section 16-7.*

Overall reaction: $2O_3(g) \rightarrow 3O_2(g)$

The reaction mechanism: $\quad\quad\quad\quad O_3 \rightleftarrows O_2 + O \quad\quad\quad\quad$ (fast, equilibrium)

$\quad\quad\quad\quad\quad\quad\quad\quad\quad\quad\quad\quad\quad O + O_3 \rightarrow 2O_2 \quad\quad\quad\quad$ (slow)

From Step 2 (the slow step), $\quad\quad\quad$ rate $= k_2[O][O_3]$ $\quad\quad\quad$ where O is an intermediate

From the slow step, rate $= k_2[O][O_3]$. However, O is an intermediate and its concentration must be expressed in terms of the reactant, O_3.

For a fast, equilibrium step, we know:

$$\text{rate}_{1f} = \text{rate}_{1r}$$

$$k_{1f}[O_3] = k_{1r}[O_2][O]$$

$$[O] = \frac{k_{1f}}{k_{1r}}\frac{[O_3]}{[O_2]}$$

Substituting for [O] in the original rate equation, we have

$$\text{rate} = k_2[O][O_3] = k_2\left(\frac{k_{1f}}{k_{1r}}\frac{[O_3]}{[O_2]}\right)[O_3] \text{ or } \textbf{rate} = k\frac{[O_3]^2}{[O_2]}$$

16-72. *Refer to Section 16-7.*

The reaction mechanism: $\quad\quad\quad\quad N_2 + Cl \rightleftarrows N_2Cl \quad\quad\quad\quad$ (fast, equilibrium)

$\quad\quad\quad\quad\quad\quad\quad\quad\quad\quad\quad\quad N_2Cl + Cl \underset{k_2}{\rightarrow} Cl_2 + N_2 \quad\quad\quad$ (slow)

(a) The intermediate species is $\textbf{N}_2\textbf{Cl}$.

(b) From the slow step, rate $= k_2[N_2Cl][Cl]$. However, N_2Cl is an intermediate and its concentration must be expressed in terms of the reactants, N_2 and Cl. For a fast, equilibrium step, we know:

$$\text{rate}_{1f} = \text{rate}_{1r}$$

$$k_{1f}[N_2][Cl] = k_{1r}[N_2Cl]$$

$$[N_2Cl] = \frac{k_{1f}}{k_{1r}}[N_2][Cl]$$

Substituting for [N$_2$Cl] in the original rate equation, we have

$$\text{rate} = k_2[N_2Cl][Cl] = k_2\left(\frac{k_{1f}}{k_{1r}}[N_2][Cl]\right)[Cl] \text{ or } \text{rate} = k[N_2][Cl]^2$$

Yes, the mechanism is consistent with the experimental rate law, rate $= k[N_2][Cl]^2$.

The reaction mechanism:

$H_2 \rightleftarrows 2H$	(fast, equilibrium)
$H + CO \rightarrow HCO$	(slow)
$H + HCO \xrightarrow{k_2} H_2CO$	(fast)

(a) Balanced equation: $H_2 + CO \rightarrow H_2CO$

(b) From the slow step, rate = $k_2[H][CO]$. However, H is an intermediate and its concentration must be expressed in terms of the reactant, H_2. For a fast, equilibrium step, we know:

$$\text{rate}_{1f} = \text{rate}_{1r}$$
$$k_{1f}[H_2] = k_{1r}[H]^2$$
$$[H] = \left(\frac{k_{1f}}{k_{1r}}\right)^{1/2}[H_2]^{1/2}$$

Substituting for [H] in the original rate equation, we have

$$\text{rate} = k_2[H][CO] = k_2\left(\frac{k_{1f}}{k_{1r}}\right)^{1/2}[H_2]^{1/2}[CO] = k[H_2]^{1/2}[CO]$$

Yes, the mechanism is consistent with the observed rate dependence: rate = $k[H_2]^{1/2}[CO]$.

16-76. *Refer to Sections 16-6 and 16-9.*

(a) The transition state is a short-lived, high-energy intermediate state that the reactants must convert into before the products are formed.

(b) Yes, the activation energy and the transition state are related concepts. The activation energy, E_a, is the additional energy that must be absorbed by the reactants in their ground states to allow them to reach the transition state.

(c) The higher the activation energy, the more energy that is required by the reactants to form products and the slower is the overall reaction.

16-78. *Refer to Section 16-4 and Exercise 16-64 Solution.*

From Exercise 16-64, we know that the reaction, $N_2O_5 \rightarrow NO_2 + NO_3$, is first order with specific rate constant,
$k = 0.24 \ s^{-1}$ at 25°C.

(a) The integrated rate equation: $\ln\dfrac{A_0}{A} = akt$

where $A_0 = 2.80$ mol in 3.00 L (the volume is irrelevant in this question)
a = stoichiometric coefficient of N_2O_5

Substituting,
$$\ln\left(\frac{2.80 \text{ mol}}{? \text{ mol } N_2O_5}\right) = (0.24 \ s^{-1})(2.00 \text{ min} \times 60 \text{ s/min}) = 29 \ (28.8 \text{ to 3 sig. figs.})$$

$$\frac{2.80 \text{ mol}}{? \text{ mol } N_2O_5} = 3.2 \times 10^{12} \quad \text{(using 28.8 then rounding to 2 sig. figs.)}$$

$$? \text{ mol } N_2O_5 = \textbf{8.7} \times \textbf{10}^{-13} \textbf{ mol } \textbf{N}_2\textbf{O}_5 \textbf{ remaining after 2.00 min}$$

Note: Any small change in the value of k can greatly alter the answer.

(b) If 99.0% of N_2O_5 have decomposed, then 1.0% of N_2O_5 remains.

Substituting into the integrated rate equation, we have $\quad \ln\left(\dfrac{100.0\%}{1.0\%}\right) = (0.24 \ s^{-1})t \quad$ Solving, $t = \textbf{19 s}$

16-80. *Refer to Section 16-3.*

Balanced equation: $4Hb + 3CO \rightarrow Hb_4(CO)_3$ where Hb is hemoglobin

(a) The form of the rate-law expression: rate = $k[Hb]^x[CO]^y$

Step 1: rate dependence on [Hb]. Consider Experiments 1 and 2:

Method 1: By observation, [CO] does not change. [Hb] increases by a factor of 2 (= 6.72/3.36). The rate of disappearance of Hb also increases by a factor of 2 (= 1.88/0.941). The rate is directly proportional to [Hb] and x is equal to 1. The reaction is first order with respect to Hb.

Method 2:

Expt 2
Expt 1
$$\frac{1.88 \ \mu mol/L \cdot s}{0.941 \ \mu mol/L \cdot s} = \frac{k(6.72 \ \mu mol/L)^x(1.00 \ \mu mol/L)^y}{k(3.36 \ \mu mol/L)^x(1.00 \ \mu mol/L)^y}$$
$$2 = (2)^x$$
$$x = 1$$

Step 2: rate dependence on [CO]. Consider Experiments 2 and 3.

Method 1: By observation, [Hb] does not change. [CO] increases by a factor of 3 (= 3.00/1.00). The rate of disappearance of Hb also increases by a factor of 3 (= 5.64/1.88). The rate is directly proportional to [CO] and y is equal to 1. The reaction is first order with respect to CO.

Method 2:

Expt 3
Expt 2
$$\frac{5.64 \ \mu mol/L \cdot s}{1.88 \ \mu mol/L \cdot s} = \frac{k(6.72 \ \mu mol/L)^x(3.00 \ \mu mol/L)^y}{k(6.72 \ \mu mol/L)^x(1.00 \ \mu mol/L)^y}$$
$$3 = (3)^y$$
$$y = 1$$

The rate-law expression is: **rate** = $k[Hb]^1[CO]^1$ = **$k[Hb][CO]$**

(b) Substituting, the data from Experiment 1 into the rate-law expression to calculate k, we have,
$$0.941 \ \mu mol/L \cdot s = k(3.36 \ \mu mol/L)(1.00 \ \mu mol/L)$$
$$k = \mathbf{0.280 \ L/\mu mol \cdot s}$$

(c) Substituting into the complete rate-law expression, gives
$$\text{rate } = (0.280 \ L/\mu mol \cdot s)[Hb][CO]$$
$$= (0.280 \ L/\mu mol \cdot s)(1.50 \ \mu mol/L)(0.600 \ \mu mol/L)$$
$$= \mathbf{0.252 \ \mu mol/L \cdot s}$$

16-82. *Refer to Section 16-5.*

The shape of a reactant can affect the rate of a reaction. Collision theory tells us that in order for a reaction to occur there must be an effective collision. For a collision to be effective, the reacting species must
(1) possess at least a certain minimum energy, called the activation energy, and
(2) have the proper orientations toward each other at the time of collision. Because orientation is important, the shape of the molecule must also be important.

16-84. *Refer to Sections 4-9, 15-3, 15-11, and 16-6, Figure 16-10, and Appendix K.*

formula unit: $HCl(aq) + NaOH(aq) \rightarrow NaCl(aq) + H_2O(\ell)$

total ionic: $H^+(aq) + Cl^-(aq) + Na^+(aq) + OH^-(aq) \rightarrow Na^+(aq) + Cl^-(aq) + H_2O(\ell)$

net ionic: $H^+(aq) + OH^-(aq) \rightarrow H_2O(\ell)$

To find the ΔE of the reaction:

(1) $\Delta H^{\circ}_{rxn} = [\Delta H^{\circ}_{f\ NaCl(aq)} + \Delta H^{\circ}_{f\ H_2O(\ell)}] - [\Delta H^{\circ}_{f\ HCl(aq)} + \Delta H^{\circ}_{f\ NaOH(aq)}]$

$= [(1\ mol)(-407.1\ kJ/mol) + (1\ mol)(-285.8\ kJ/mol)]$

$- [(1\ mol)(-167.4\ kJ/mol) + (1\ mol)(-469.6\ kJ/mol)]$

$= -55.9\ kJ/mol\ rxn$

(2)   Because there are no gases involved in this reaction,

$\Delta E = \Delta H = -55.9\ kJ/mol\ rxn.$

This means that energy is released in this reaction. The potential energy diagram for the reaction is:

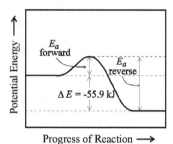

16-86.   *Refer to Section 16-7, Chapter 15 and Appendix K.*

Given the equation: $3O_2 \rightarrow 2O_3$, we definitely **cannot** say that O_2 reacts with itself to form O_3. This equation is the overall reaction, and does not give any information about the mechanism of the reaction, i.e. the order in which bonds are broken and formed to create the products from the reactant. It is only telling us that for every 3 moles of O_2 that react, 2 moles of O_3 are formed. There is no information in the equation as to how that happens. However, using thermodynamic tables and Hess's Law, we find that $\Delta H_{rxn} = +286\ kJ$ and $\Delta S_{rxn} = -137.4\ J/mol\ K$, so ΔG_{rxn} is positive at all temperatures and the reaction must be nonspontaneous at all temperatures.

16-88.   *Refer to Sections 13-9 and 16-8.*

The activation energy, E_a, is the additional energy that must be absorbed by the reactants in their ground states to allow them to reach the transition state. Consider the following three endothermic reactions; they are identical except that they occur in different phases:

$$A(g) + B(g) \rightarrow C(g) \qquad A(\ell) + B(\ell) \rightarrow C(\ell) \qquad A(s) + B(s) \rightarrow C(s)$$

We know that the potential energy of the reactants as gases is greater than that of the liquid reactants, and the potential energy of the solid reactants is the lowest. Let us assume that (1) the potential energy of the high-energy transition state for a reaction is independent of the phase of the reactants and products, and (2) the mechanism of the reaction doesn't change. Therefore, the activation energy of the gaseous reaction is expected to be less than the liquid-phase reaction, which is less than the solid-phase reaction.

So, we expect:

E_a for gas-phase rxn $< E_a$ for liquid-phase rxn $< E_a$ for solid-phase rxn

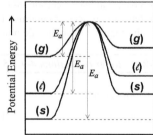

The potential energy diagram for the generic reaction shown in Figure 16-10b is shown at the right.

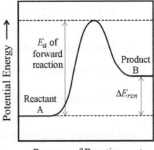

(a) The reaction can be expressed as: A → B.

(b) The reaction is endothermic because ΔE for the reaction is positive; the reaction absorbs energy as product B is formed.

(c) The activation energy of the forward reaction is shown as the difference between the potential energy of reactant, A, and the energy of the transition (excited) state.

Balanced reaction: $CH_3COOH(aq) + NaHCO_3(s) \rightarrow NaCH_3OO(aq) + H_2O(\ell) + CO_2(g)$

The reaction rate can be increased by (1) increasing the temperature, (2) grinding up the baking soda, thereby increasing the solid's surface area, (3) increasing the mixing rate when the acetic acid is added to the baking soda and (4) dissolving the baking soda in water before adding the vinegar. You might also try to find vinegar that has a concentration greater than 5% acetic acid.

(1) Balanced equation: $2N_2O_5(g) \rightarrow 4NO_2(g) + O_2(g)$

Since rate = $k[N_2O_5]$, we must use the first order integrated rate equation: : $\ln\left(\dfrac{A_0}{A_t}\right) = akt$ with $t_{1/2} = \dfrac{0.693}{ak}$

where a is the stoichiometric coefficient of N_2O_5

Solving for k in terms of half-life: $k = \dfrac{0.693}{a\,t_{1/2}} = \dfrac{0.693}{(2)(21\ \text{min})} = 0.0165\ \text{min}^{-1}$ (good to 2 significant figures)

(i) Substituting, $A_t = 0.40\ M$, $\ln\left(\dfrac{0.80\ M}{0.40\ M}\right) = (2)(0.0165\ \text{min}^{-1})t$, so $t =$ **21 min**

Alternatively, we have half of our sample remaining, so we must have waited one half-life, or 21 min.

(ii) Substituting, $A_t = 0.20\ M$, $\ln\left(\dfrac{0.80\ M}{0.20\ M}\right) = (2)(0.0165\ \text{min}^{-1})t$, so $t =$ **42 min**

Alternatively, we have 1/4 of our sample remaining, so we must have waited two half-lives, or 42 min.

(iii) Substituting, $A_t = 0.10\ M$, $\ln\left(\dfrac{0.80\ M}{0.10\ M}\right) = (2)(0.0165\ \text{min}^{-1})t$, so $t =$ **63 min**

Alternatively, we have 1/8 of our sample remaining, so we must have waited three half-lives, or 63 min.

(2) Balanced equation: $2NO_2(g) \rightarrow 2NO(g) + O_2(g)$

Since rate = $k[NO_2]^2$, we must use the second order integrated rate equation: $\dfrac{1}{[NO_2]} - \dfrac{1}{[NO_2]_0} = akt$

where a is the stoichiometric coefficient of NO_2

The relationship between $t_{1/2}$ and k for second order reactions is: $t_{1/2} = \dfrac{1}{ak[NO_2]_0}$,

so $k = \dfrac{1}{a\,t_{1/2}\,[NO_2]_0} = \dfrac{1}{(2)(21\ min)(0.80\ M)} = 0.0298\ M^{-1} min^{-1}$ (good to 2 significant figures)

(i) Substituting, $A_t = 0.40\ M$, $\dfrac{1}{0.40\ M} - \dfrac{1}{0.80\ M} = (2)(0.0298\ M^{-1} min^{-1})t$, so $t = $ **21 min**

Alternatively, we have half of our sample remaining, so we must have waited one half-life, or 21 min.

(ii) Substituting, $A_t = 0.20\ M$, $\dfrac{1}{0.20\ M} - \dfrac{1}{0.80\ M} = (2)(0.0298\ M^{-1} min^{-1})t$, so $t = $ **63 min**

(iii) Substituting, $A_t = 0.10\ M$, $\dfrac{1}{0.10\ M} - \dfrac{1}{0.80\ M} = (2)(0.0298\ M^{-1} min^{-1})t$, so $t = $ **150 min**

(3) Balanced equation: $2NH_3(g) \rightarrow N_2(g) + 3H_2(g)$

Since rate $= k[NH_3]^2$, the second order integrated rate equation: $\dfrac{1}{[NH_3]} - \dfrac{1}{[NH_3]_0} = akt$ and $t_{1/2} = \dfrac{1}{ak[NH_3]_0}$

where a is the stoichiometric coefficient of NO_2

so $k = \dfrac{1}{a\,t_{1/2}\,[NH_3]_0} = \dfrac{1}{(2)(21\ min)(0.80\ M)} = 0.0298\ M^{-1} min^{-1}$ (good to 2 significant figures)

(i) Substituting, $A_t = 0.40\ M$, $\dfrac{1}{0.40\ M} - \dfrac{1}{0.80\ M} = (2)(0.0298\ M^{-1} min^{-1})t$, so $t = $ **21 min**

Alternatively, we have half of our sample remaining, so we must have waited one half-life, or 21 min.

(ii) Substituting, $A_t = 0.20\ M$, $\dfrac{1}{0.20\ M} - \dfrac{1}{0.80\ M} = (2)(0.0298\ M^{-1} min^{-1})t$, so $t = $ **63 min**

(iii) Substituting, $A_t = 0.10\ M$, $\dfrac{1}{0.10\ M} - \dfrac{1}{0.80\ M} = (2)(0.0298\ M^{-1} min^{-1})t$, so $t = $ **150 min**

16-96. *Refer to Sections 13-9 and 16-8.*

Plan: (1) Use the Clausius-Clapeyron equation to calculate the steam temperature in the pressure cooker. Assume that ΔH_{vap} for H_2O is independent of temperature.

(2) Use the Arrhenius equation to calculate the activation energy for the process of steaming vegetables.

(1) From the Clausius-Clapeyron equation: $\ln\left(\dfrac{P_2}{P_1}\right) = \dfrac{\Delta H_{vap}}{R}\left(\dfrac{1}{T_1} - \dfrac{1}{T_2}\right)$

where
ΔH_{vap} = molar heat of vaporization for H_2O, 40.7 kJ/mol
P_1 = atmospheric pressure, 15 psi
P_2 = cooker pressure = P_1 + gauge pressure = (15 + 15) psi
T_1 = boiling point of water at 1 atm, 100.0°C
T_2 = steam temperature in the pressure cooker

Substituting, $\ln\left(\dfrac{30\ psi}{15\ psi}\right) = \dfrac{40.7 \times 10^3\ J/mol}{8.314\ J/mol \cdot K}\left(\dfrac{1}{373\ K} - \dfrac{1}{T_2}\right)$

$0.69 = (4.90 \times 10^3\ K)(2.68 \times 10^{-3}\ K^{-1} - 1/T_2)$

$1.4 \times 10^{-4}\ K^{-1} = (2.68 \times 10^{-3}\ K^{-1} - 1/T_2)$

$T_2 = 394\ K$

(2) From the Arrhenius equation: $\ln\left(\dfrac{k_2}{k_1}\right) = \dfrac{E_a}{R}\left(\dfrac{1}{T_1} - \dfrac{1}{T_2}\right)$

where
$T_1 = 373\ K$
$T_2 = 394\ K$
k_1 = rate constant for cooking vegetables at atmospheric pressure
k_2 = rate constant for cooking vegetables in the pressure cooker

$k_2/k_1 = 3$ since we assume the cooking process is 3x faster in the pressure cooker

Substituting, $\ln 3 = \dfrac{E_a}{8.314 \text{ J/mol·K}}\left(\dfrac{1}{373} - \dfrac{1}{394}\right)$

$E_a = \mathbf{6.4 \times 10^4}$ **J/mol or 64 kJ/mol**

16-98. *Refer to Section 16-7 and Figure 16-12.*

The overall reaction: $H_2(g) + I_2(g) \rightarrow 2HI(g)$ rate $= k[H_2][I_2]$

The reaction mechanism: (1) $I_2 \rightleftarrows 2I$ (fast, equilibrium)

(2) $I + H_2 \rightleftarrows H_2I$ (fast, equilibrium)

$\underline{(3) \quad H_2I + I \rightarrow 2HI \quad\quad\quad\text{(slow)}}$

$H_2 + I_2 \rightarrow 2HI$ (overall)

Step 1	**Reactant**	**Excited State (first peak)**	**Products (first trough)**
	$:\ddot{I}\!-\!\ddot{I}:$ (H-H present)	$:\ddot{I}\!-\!\ddot{I}:$ (H-H present)	2 $:\ddot{I}\cdot$ (H-H present)

Step 2	**Reactants**	**Excited State (second peak)**	**Product (second trough)**
	H-H and $:\ddot{I}\cdot$ ($:\ddot{I}\cdot$ present)	$\underset{\text{weaker bond}}{H\!-\!H\!-\!-\!:\ddot{I}\cdot}$ ($:\ddot{I}\cdot$ present)	H-H-$\ddot{I}:$ ($:\ddot{I}\cdot$ present) (9 e^- present) *

Step 3	**Reactants**	**Excited State (third peak)**	**Products (final)**
	$:\ddot{I}\cdot$ and H-H-$\ddot{I}:$ *	$:\ddot{I}\!-\!-\!-\!H\!-\!-\!-\!H\!-\!-\!-\!\ddot{I}:$	2 H-$\ddot{I}:$

* The extra weak H-H bond contains only one electron.

16-100. *Refer to Sections 15-3, 15-8 and 16-6, Appendix K and Figure 16-10.*

Balanced equation: $O_3(g) + NO(g) \rightarrow NO_2(g) + O_2(g)$ $E_a = 9.6$ kJ/mol reaction

(a) $\Delta H^\circ_{\text{rxn}} = [\Delta H^\circ_{f\,NO_2(g)} + \Delta H^\circ_{f\,O_2(g)}] - [\Delta H^\circ_{f\,O_3(g)} + \Delta H^\circ_{f\,NO(g)}]$

$= [(1 \text{ mol})(33.2 \text{ kJ/mol}) + (1 \text{ mol})(0 \text{ kJ/mol})] - [(1 \text{ mol})(143 \text{ kJ/mol}) + (1 \text{ mol})(90.25 \text{ kJ/mol})]$

$= \mathbf{-200.}$ **kJ/mol reaction**

(b)   For the reaction, $\Delta E^\circ = \Delta H^\circ$ since $\Delta n_{\text{gas}} = 0$.

Therefore, the activation energy plot for this reaction is:

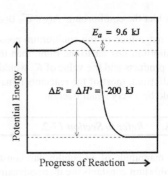

269

17 Chemical Equilibrium

17-2. *Refer to Section 17-2.*

Equilibrium constants do not have units because in the strict thermodynamic definition of the equilibrium constant, the *activity* of a component is used, not its concentration. The activity of a species in an ideal mixture is the ratio of its concentration or partial pressure to a standard concentration (1 *M*) or pressure (1 atm). Because activity is a ratio, it is unitless and the equilibrium constant involving activities is also unitless.

17-4. *Refer to Sections 17-1 and 17-2.*

Equilibrium can be defined as a state in which no observable changes occur as time goes by.

(a) Static equilibrium is an important concept in physics. A body is said to be "in static equilibrium" if a body at rest will stay at rest. (Note: this happens when its acceleration and angular acceleration are zero.) Examples of static equilibrium are (1) a ladder leaning against a wall, (2) a block of wood resting on a table and (3) a picture hanging on a wall.

(b) Dynamic equilibrium is a state in which no net change takes place because two opposing processes are occurring at the same time. For example (1) consider the movement of skiers on a busy day on the ski slopes when the number of skiers going up the hill on the chair lift equals the number of skiers skiing down the hill. If you took pictures during the day, the number of skiers at the top of the slope and the number at the bottom remain unchanged. (2) Another similar example would be children on a slide in the park. (3) A more scientific example is the evaporation of water in a closed container. Once the system is at equilibrium, the number of water molecules in the gaseous phase and the number of water molecules in the liquid phase remain unchanged. However, liquid molecules are constantly entering the gaseous phase while gaseous molecules are constantly condensing to form liquid molecules of water. We write this as:

$$H_2O(\ell) \rightleftarrows H_2O(g)$$

17-6. *Refer to Section 17-2.*

Consider the equilibrium: $A(g) \rightleftarrows B(g)$ $\qquad K_c = \dfrac{[B]}{[A]}$

The magnitude of the equilibrium constant, K_c, is a measure of the extent to which a reaction occurs. If the equilibrium lies far to the right, then this means that at equilibrium most of the reactants would be converted into products and the value of K_c would be much greater than one. If the equilibrium lies far to the left, then at equilibrium, most of the reactants remain unreacted and there are very little products formed. The value of K_c would be a very small fraction.

17-8. *Refer to Section 17-2.*

The magnitude of an equilibrium constant tells us *nothing* about how fast the system will reach equilibrium. Equilibrium constants are thermodynamic quantities, whereas the speed of a reaction is a kinetic quantity. The two are not related. Rather, an equilibrium constant is a measure of the extent to which a reaction occurs.

17-10. *Refer to Section 17-2.*

(a) The equilibrium constant is related to the specific rate constants of the forward and reverse reactions. Consider the following 1 step reaction: $A \rightleftarrows B$, where k_f is the specific rate constant of the forward reaction and k_r is the specific rate constant of the reverse reaction. At equilibrium, the rates of the forward reaction and the reverse reaction are equal.

$$\text{Rate}_f = \text{Rate}_r$$
$$k_f[A] = k_r[B]$$
$$K_c = \frac{k_f}{k_r} = \frac{[B]}{[A]}$$

K_c, the conventional equilibrium constant, is equal to the ratio of the forward rate constant divided by the reverse rate constant.

(b) The rate expressions for the forward and reverse reactions can NOT be written from the balanced chemical equation. You need the experimentally-derived rate law expression or the mechanism for the reaction. The only exception to this is if you know the reaction occurs in only 1 step - a 1-step mechanism, as in the example given in (a). In that case, the balanced chemical reaction is the rate-determining step (it's the only step) and you can determine the rate law expressions.

(c) However, the equilibrium constant expression (also called the mass action expression) can be written from the balanced chemical equation. Regardless of the mechanism by which a reaction occurs, the concentrations of reaction intermediates always cancel out and the equilibrium constant expression has the same form.

17-12. *Refer to Section 17-2 and Figure 17-2.*

(a) Consider the equilibrium: $2A(g) + B(g) \rightleftarrows 2C(g)$. Assuming that the concentrations of A and B are both 1 M initially, the *changes* that the concentrations undergo are related to the stoichiometric coefficients. When the reaction is at equilibrium at t_{eq}, the system has little A remaining. The change in [B] is 1/2 that of the [A], and the increase in the [C] is equal to twice the loss of [B]. In other words:

Let x = mol/L of B that react. Then 2x = mol/L of A that react, and 2x = mol/L of C that are produced.

	2A	+	B	$\rightleftarrows$	2C
initial	1.0 mol/L		1.0 mol/L		0 mol/L
change	- 2x mol/L		- x mol/L		+ 2x mol/L
at equilibrium	(1.0 - 2x) mol/L		(1.0 - x) mol/L		2x mol/L
at equilibrium (x = 0.45 M)	0.10 mol/L		0.55 mol/L		0.90 mol/L

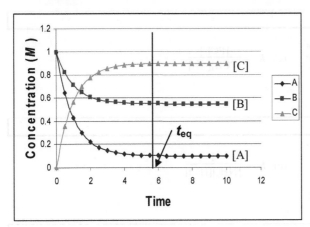

To see what is happens to [A], [B] and [C] when $K_c \gg 1$, let us assume that x = 0.45 M at equilibrium and work through this problem. From the equilibrium concentrations determined above, we can calculate K_c:

$$K_c = \frac{[C]^2}{[A]^2[B]} = \frac{(0.90)^2}{(0.10)^2(0.55)} = 150$$

Note the overall look of the graph. The concentrations of A and B decrease with time and the concentration of C increases with time as the system approaches equilibrium at time, t_{eq}. How much the reactants decrease and the products increase depends on the stoichiometric coefficients

(b) For the same equilibrium with $K_c \ll 1$, there will be more reactants and little products when the system reaches equilibrium.

	2A	+	B	$\rightleftarrows$	2C
initial	1.0 mol/L		1.0 mol/L		0 mol/L
change	- 2x mol/L		- x mol/L		+ 2x mol/L
at equilibrium	(1.0 - 2x) mol/L		(1.0 - x) mol/L		2x mol/L
at equilibrium (x = 0.05 M)	0.90 mol/L		0.95 mol/L		0.10 mol/L

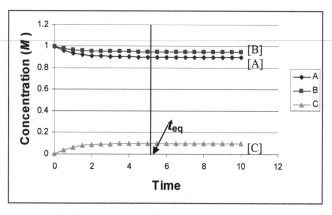

In this diagram, let us assume that x = 0.050 M at equilibrium. From the equilibrium concentrations, we can calculate K_c for this example.

$$K_c = \frac{[C]^2}{[A]^2[B]} = \frac{(0.10)^2}{(0.90)^2(0.95)} = 0.013$$

Note the overall look of the graph. The concentrations of the reactants, A and B, decrease slightly and the concentration of the product, C, increases slightly according to their stoichiometric coefficients as the system approaches equilibrium.

17-14. *Refer to Section 17-2 and Exercise 17-2 Solution.*

We omit concentrations of pure solids and pure liquids from equilibrium constant expressions because their activity is taken to be 1 and the thermodynamic equilibrium constant involves activities, rather than concentrations.

17-16. *Refer to Section 17-2 and Exercise 17-2 Solution.*

The concentrations of pure solids and pure liquids are omitted from the equilibrium constant expression because their activity is taken to be 1. Therefore, when writing equilibrium constant expressions, we would omit:

pure solids: $CaCO_3(s)$, $NaOH(s)$, $I_2(s)$ and $C(graphite)$
pure liquids: $H_2SO_4(\ell)$ and $CH_3COOH(\ell)$

17-18. *Refer to Section 17-2 and 17-11.*

(a) $K_c = \dfrac{[H_2S]^2[O_2]^3}{[H_2O]^2[SO_2]^2}$　(b) $K_c = \dfrac{[NO]^4[H_2O]^6}{[NH_3]^4[O_2]^5}$　(c) $K_c = \dfrac{[PCl_5]}{[PCl_3][Cl_2]}$

(d) $K_c = [HF]$　(e) $K_c = \dfrac{[SO_2]^2[O_2]}{[SO_3]^2}$

17-20. *Refer to Sections 17-2 and 17-11.*

(a) $K_c = \dfrac{[H_2O]^2[O_2]}{[H_2O_2]^2}$　(b) $K_c = \dfrac{[SO_2]^2}{[O_2]^3}$　(c) $K_c = \dfrac{1}{[NH_3][HCl]}$

(d) $K_c = \dfrac{[NO_2]^2}{[N_2O_4]}$　(e) $K_c = \dfrac{[HCl]^4[O_2]}{[Cl_2]^2[H_2O]^2}$

17-22. *Refer to Section 17-2.*

The products are favored in those reactions in which K > 1: (b), (c) and (d).

17-24. *Refer to Section 17-2 and Example 17-1.*

Balanced equation: $N_2(g) + O_2(g) \rightleftarrows 2NO(g)$　$K_c = \dfrac{[NO]^2}{[N_2][O_2]} = \dfrac{(1.1 \times 10^{-5})^2}{(6.4 \times 10^{-3})(1.7 \times 10^{-3})} = \mathbf{1.1 \times 10^{-5}}$

272

17-26. *Refer to Section 17-2 and Example 17-1.*

Balanced equation: $PCl_3(g) + Cl_2(g) \rightleftarrows PCl_5(g)$ $\quad K_c = \dfrac{[PCl_5]}{[PCl_3][Cl_2]} = \dfrac{(12)}{(10.)(9.0)} = \mathbf{0.13}$

17-28. *Refer to Section 17-2 and Example 17-2.*

Balanced equation: $A(g) + B(g) \rightleftarrows C(g) + 2D(g)$

Plan: (1) Determine the concentrations of the species of interest.
(2) Determine the concentrations of all species after equilibrium is reached.
(3) Calculate K_c.

(1) $[A]_{initial} = [B]_{initial} = \dfrac{1.00 \text{ mol}}{0.400 \text{ L}} = 2.50 \text{ } M$

$[C]_{equil} = \dfrac{0.20 \text{ mol}}{0.400 \text{ L}} = 0.50 \text{ } M$

(2)

	A	+	B	$\rightleftarrows$	C	+	2D
initial	2.50 M		2.50 M		0 M		0 M
change	- 0.50 M		- 0.50 M		+ 0.50 M		+ 1.00 M
at equilibrium	2.00 M		2.00 M		0.50 M		1.00 M

(3) $K_c = \dfrac{[C][D]^2}{[A][B]} = \dfrac{(0.50)(1.00)^2}{(2.00)(2.00)} = \mathbf{0.12}$

17-30. *Refer to Section 17-3 and Example 17-3.*

Balanced equation: $H_2(g) + Br_2(g) \rightleftarrows 2HBr(g)$ $\quad K_c = \dfrac{[HBr]^2}{[H_2][Br_2]} = 7.9 \times 10^{11}$

(a) $1/2 \text{ } H_2(g) + 1/2 \text{ } Br_2(g) \rightleftarrows HBr(g)$ $\quad K_c' = \dfrac{[HBr]}{[H_2]^{1/2}[Br_2]^{1/2}} = \sqrt{K_c} = \mathbf{8.9 \times 10^5}$

(b) $2HBr(g) \rightleftarrows H_2(g) + Br_2(g)$ $\quad K_c'' = \dfrac{[H_2][Br_2]}{[HBr]^2} = \dfrac{1}{K_c} = \mathbf{1.3 \times 10^{-12}}$

(c) $4HBr(g) \rightleftarrows 2H_2(g) + 2Br_2(g)$ $\quad K_c''' = \dfrac{[H_2]^2[Br_2]^2}{[HBr]^4} = \dfrac{1}{K_c^2} = \mathbf{1.6 \times 10^{-24}}$

17-32. *Refer to Sections 17-2, and Examples 17-1 and 17-2.*

Balanced equation: $CO(g) + Cl_2(g) \rightleftarrows COCl_2(g)$

At equilibrium,

$[CO] = \dfrac{3.00 \text{ mol}}{50. \text{ L}} = 0.060 \text{ } M$ $\quad [Cl_2] = \dfrac{2.00 \text{ mol}}{50. \text{ L}} = 0.040 \text{ } M$ $\quad [COCl_2] = \dfrac{9.00 \text{ mol}}{50. \text{ L}} = 0.18 \text{ } M$

$K_c = \dfrac{[COCl_2]}{[CO][Cl_2]} = \dfrac{(0.18)}{(0.060)(0.040)} = \mathbf{75}$

17-34. *Refer to Section 17-2 and Example 17-1.*

Balanced equation: $SbCl_5(g) \rightleftarrows SbCl_3(g) + Cl_2(g)$

(a) Plan: (1) Determine the equilibrium concentrations of $SbCl_5$, $SbCl_3$ and Cl_2 at some high temperature.
(2) Evaluate K_c at that temperature by substituting the equilibrium concentrations into the K_c expression.

(a) $[SbCl_5] = \dfrac{(6.91\ g/299\ g/mol)}{5.00\ L} = 4.62 \times 10^{-3}\ M$

$[SbCl_3] = \dfrac{(16.45\ g/228.11\ g/mol)}{5.00\ L} = 1.44 \times 10^{-2}\ M$

$[Cl_2] = \dfrac{(5.11\ g/70.9\ g/mol)}{5.00\ L} = 1.44 \times 10^{-2}\ M$

(2) $K_c = \dfrac{[SbCl_3][Cl_2]}{[SbCl_5]} = \dfrac{(1.44 \times 10^{-2})(1.44 \times 10^{-2})}{(4.62 \times 10^{-3})} = \mathbf{4.49 \times 10^{-2}}$

(b) Plan: (1) Determine the initial concentration of $SbCl_5$.

(2) Determine the equilibrium concentrations of $SbCl_5$, $SbCl_3$ and Cl_2 using the K_c expression.

(1) $[SbCl_5] = \dfrac{(25.0\ g/299\ g/mol)}{5.00\ L} = 0.0167\ M$

(2) Since $[Cl_2] = [SbCl_3] = 0$, the forward reaction will predominate. Some $SbCl_5$ will react and equal moles of $SbCl_3$ and Cl_2 will be produced.

Let x = moles per liter of $SbCl_5$ that react. Then
x = moles per liter of $SbCl_3$ produced = moles per liter of Cl_2 produced.

	$SbCl_5$	$\rightleftarrows$	$SbCl_3$	+	Cl_2
initial	0.0167 M		0 M		0 M
change	- x M		+ x M		+ x M
at equilibrium	(0.0167 - x) M		x M		x M

$K_c = \dfrac{[SbCl_3][Cl_2]}{[SbCl_5]} = \dfrac{(x)(x)}{(0.0167 - x)} = 4.49 \times 10^{-2}$

The quadratic equation: $x^2 + (4.49 \times 10^{-2})x - 7.50 \times 10^{-4} = 0$

Solving,

$x = \dfrac{-4.49 \times 10^{-2} \pm \sqrt{(4.49 \times 10^{-2})^2 - 4(1)(-7.51 \times 10^{-4})}}{2(1)} = \dfrac{-4.49 \times 10^{-2} \pm 7.08 \times 10^{-2}}{2}$

$= 1.30 \times 10^{-2}$ or -5.78×10^{-2} (discard)

Note: There are always two solutions when solving quadratic equations, but only one is meaningful in this type of chemical problem. The other solution, -5.79×10^{-2}, is discarded because a negative value for concentration has no physical meaning in this problem.

Therefore, at equilibrium: $[SbCl_5] = 0.0167 - x = \mathbf{3.7 \times 10^{-3}\ M}$

$[SbCl_3] = [Cl_2] = x = \mathbf{1.30 \times 10^{-2}\ M}$

17-36. *Refer to Section 17-3 and Example 17-3.*

Balanced equation: $2NO(g) + O_2(g) \rightleftarrows 2NO_2(g)$ $\qquad K_c = \dfrac{[NO_2]^2}{[NO]^2[O_2]} = 1538$

$\qquad\qquad\qquad 2NO_2(g) \rightleftarrows 2NO(g) + O_2(g)$ $\qquad K_c' = \dfrac{[NO]^2[O_2]}{[NO_2]^2} = \dfrac{1}{K_c} = \mathbf{6.502 \times 10^{-4}}$

17-38. *Refer to Section 17-4.*

Many systems are not at equilibrium. The mass action expression, also called the reaction quotient, Q, is a measure of how far a system is from equilibrium and in what direction the system must go to get to equilibrium. The reaction quotient has the same form as the equilibrium constant, K, but the concentration values put into Q are the actual values found in the system at that given moment.

(a) If $Q = K$, the system is at equilibrium.

(b) If $Q < K$, the system has greater concentrations of reactants than it would have if it were at equilibrium. The forward reaction will dominate until equilibrium is established.

(c) If $Q > K$, the system has greater concentrations of products than it would have if it were at equilibrium. The reverse reaction will dominate until equilibrium is reached.

17-40. *Refer to Section 17-4.*

If $Q > K$ for a reversible reaction, the reverse reaction occurs to a greater extent than the forward reaction until equilibrium is reached.

If $Q < K$, the forward reaction occurs to a greater extent than the reverse reaction until equilibrium is reached.

17-42. *Refer to Section 17-4 and Example 17-4.*

Balanced equation: $H_2CO \rightleftarrows H_2 + CO$　　　　　　　　$K_c = 0.50$

The reaction quotient, Q, for this reaction at the given moment is: $Q = \dfrac{[H_2][CO]}{[H_2CO]} = \dfrac{(0.80)(0.25)}{(0.50)} = 0.40$

Under the given conditions, $Q < K_c$.

(a) false   The reaction is not at equilibrium since $Q \neq K_c$.

(b) false   The reaction is not at equilibrium. However, the reaction will continue to proceed until equilibrium is reached.

(c) true   When $Q < K_c$, the system has more reactants and less products than it would have at equilibrium. Equilibrium will be reached by forming more H_2 and CO and using up more H_2CO.

(d) false   The forward rate of the reaction is more than that of the reverse reaction since $Q < K_c$. The forward and reverse rates are equal only at equilibrium.

17-44. *Refer to Section 17-5 and Example 17-7.*

Balanced equation: $N_2(g) + C_2H_2(g) \rightleftarrows 2HCN(g)$　　　　　　$K_c = 2.3 \times 10^{-4}$

Plan: Determine the equilibrium concentration of HCN using the K_c expression.

Let   x = moles per liter of N_2 that react. Then
　　 x = moles per liter of C_2H_2 that react, and 2x = moles per liter of HCN that are formed.

	N_2	+	C_2H_2	$\rightleftarrows$	2HCN
initial	3.5 *M*		2.5 *M*		0 *M*
change	- x *M*		- x *M*		+ 2x *M*
at equilibrium	(3.5 - x) *M*		(2.5 - x) *M*		2x *M*

$K_c = \dfrac{[HCN]^2}{[N_2][C_2H_2]} = \dfrac{(2x)^2}{(3.5 - x)(2.5 - x)} = 2.3 \times 10^{-4}$

Note: In the following calculations, the answers that are in parentheses contain too many significant figures, but are the ones used in the progressive calculation in order to minimize rounding errors.

Rearranging into a quadratic equation: $4.0x^2 + (1.38 \times 10^{-3})x - 2.01 \times 10^{-3} = 0$
Solving,

$x = \dfrac{-1.4 \times 10^{-3} \pm \sqrt{(1.38 \times 10^{-3})^2 - 4(4.0)(-2.01 \times 10^{-3})}}{(2)(4.0)} = \dfrac{-1.38 \times 10^{-3} \pm 0.179}{8.00} = 0.0222 \text{ or } -0.0225 \text{ (discard)}$

Note: There are always two solutions when solving quadratic equations, but only one is meaningful in this type of chemical problem. The other solution, −0.0225, is discarded because a negative value for concentration of HCN has no physical meaning in this problem.

Therefore, at equilibrium: [HCN] = 2x = **0.044 *M*** (two significant figures)

Balanced equation: $PCl_3(g) + Cl_2(g) \rightleftarrows PCl_5(g)$ $\hspace{2cm}$ $K_c = 96.2$

Plan: Determine the equilibrium concentration of Cl_2 using the equilibrium expression.

Let $\hspace{0.5cm}$ x = moles per liter of PCl_3 that react = moles per liter of Cl_2 that react. Then

$\hspace{1.3cm}$ x = moles per liter of PCl_5 that are formed.

	PCl_3	+	Cl_2	$\rightleftarrows$	PCl_5
initial	0.24 M		5.5 M		0 M
change	- x M		- x M		+ x M
at equilibrium	(0.24 - x) M		(5.5 - x) M		x M

$$K_c = \frac{[PCl_5]}{[PCl_3][Cl_2]} = \frac{(x)}{(0.24 - x)(5.5 - x)} = 96.2$$

Rearranging into a quadratic equation: $96.2x^2 - 553x + 127 = 0$

Solving, $x = \dfrac{553 \pm \sqrt{(553)^2 - 4(96.2)(127)}}{(2)(96.2)} = \dfrac{553 \pm 507}{192} = 0.24$ or 5.5 (discard)

Note: The solution, x = 5.5 is meaningless since it would result in a negative equilibrium concentration for PCl_3.

Therefore, at equilibrium: $[Cl_2] = 5.5 - x = $ **5.3 M**

Note: The equilibrium concentration of PCl_3 is not zero, but a small positive number. Solve for $[PCl_3]$ by plugging the calculated equilibrium concentrations of Cl_2 and PCl_5 into the equilibrium expression.

$$K_c = \frac{[PCl_5]}{[PCl_3][Cl_2]} = \frac{x}{[PCl_3](5.5 - x)} = \frac{0.24}{[PCl_3](5.3)} = 96.2 \hspace{1cm} \text{Solving, } [PCl_3] = 0.00047 \ M$$

Balanced equation: $3Fe(s) + 4H_2O(g) \rightleftarrows Fe_3O_4(s) + 4H_2(g)$ $\hspace{1cm}$ $K_c = 4.6$ at 850°C

Plan: $\hspace{0.5cm}$ (1) Determine the initial concentration of H_2O.

$\hspace{1.5cm}$ (2) Determine the equilibrium concentration of H_2 using the K_c expression.

$\hspace{0.8cm}$ Note: In the following calculations, the answers that are in parentheses contain too many significant figures, but are the ones used in the progressive calculation in order to minimize rounding errors.

(1) $[H_2O] = \dfrac{(24 \text{ g}/18.0 \text{ g/mol})}{10.0 \text{ L}} = 0.13 \ M$ $\hspace{0.3cm}$ (0.133 M - use this to minimize rounding errors, then round at the end)

(2) Since initially, $[H_2] = 0$, the forward reaction will predominate. Note that this is a heterogeneous equilibrium, so we can ignore the solids.

Let $\hspace{0.3cm}$ 4x = moles per liter of H_2O that react. Then, 4x = moles per liter of H_2 produced.

	$3Fe(s)$	+	$4H_2O(g)$	$\rightleftarrows$	$Fe_3O_4(s)$	+	$4H_2(g)$
initial			0.133 M				0 M
change			- 4x M				+ 4x M
at equilibrium			(0.133 - 4x) M				4x M

$$K_c = \frac{[H_2]^4}{[H_2O]^4} = \frac{(4x)^4}{(0.133 - 4x)^4} = 4.6$$

If we take the 4th root of both sides, we have: $\dfrac{4x}{0.133 - 4x} = \sqrt[4]{4.6} = 1.5 \ (1.465)$

276

Solving for x, we have: 4x = 0.195 - 5.86x (knowing our numbers are only good to 2 sig. figs.)

$$9.86x = 0.195$$
$$x = 0.020 \ (0.01978)$$

Therefore, at equilibrium: $[H_2O] = 0.133 - 4x = 0.054 \ M$ and $[H_2] = 4x = \mathbf{0.079 \ M}$

17-50.   *Refer to Section 17-5 and Example 17-7.*

Balanced equation: $CO_2(g) + C(graphite) \rightleftarrows 2CO(g)$ $K_c = 10.0$ at 850°C

Plan: (1) Determine the initial concentration of CO.
          (2) Determine the equilibrium concentration of CO_2 using the K_c expression.

(1) $[CO] = \dfrac{(24.5 \ g/28.0 \ g/mol)}{2.50 \ L} = 0.350 \ M$

(2) Since initially $[CO_2] = 0$, the reverse reaction will predominate. Note that this is a heterogeneous equilibrium, so we can ignore the solid graphite.

Let   2x = moles per liter of CO that react. Then, x = moles per liter of CO_2 produced.

	$CO_2(g)$	+	C(graphite)	$\rightleftarrows$	$2CO(g)$
initial	0 *M*				0.350 *M*
change	+ x *M*				- 2x *M*
at equilibrium	x *M*				(0.350 - 2x) *M*

$$K_c = \frac{[CO]^2}{[CO_2]} = \frac{(0.350 - 2x)^2}{(x)} = 10.0$$

Note: Even though the reverse reaction is favored, we still write the equilibrium expression for the reaction as originally written. This is a major point!

Rearranging into a quadratic equation: $4x^2 - 11.4x + 0.123 = 0$

Solving, $x = \dfrac{+11.4 \pm \sqrt{(11.4)^2 - 4(4)(0.123)}}{(2)(4)} = \dfrac{+11.4 \pm 11.3}{8.00} = 0.01$ or 2.8 (discard)

Note: There are always two solutions when solving quadratic equations, but only one is meaningful in this type of problem. The other solution, x = 2.8, is discarded because a negative value for the equilibrium concentration of CO would result and this has no physical meaning.

Therefore, at equilibrium: $[CO_2] = 0.01 \ M$

$$? \ g \ CO_2 = 2.50 \ L \times \frac{0.01 \ mol \ CO_2}{1 \ L} \times \frac{44 \ g \ CO_2}{1 \ mol \ CO_2} = \mathbf{1 \ g} \ (1 \ sig. \ fig.)$$

17-52.   *Refer to Section 17-5 and Example 17-6.*

Balanced equation: $2HI(g) \rightleftarrows H_2(g) + I_2(g)$ $K_c = 0.830$

Plan: (1) Determine the initial concentration of HI.
          (2) Determine the equilibrium concentrations of HI, H_2 and I_2 using the K_c expression.

(1) $[HI] = \dfrac{(75.7 \ g/127.9 \ g/mol)}{1.50 \ L} = 0.395 \ M \ (0.3946)$

(2) Since initially, $[H_2] = [I_2] = 0$, the forward reaction will predominate.

Let   2x = moles per liter of HI that react. Then, x = moles per liter of H_2 and I_2 produced.

	$2HI(g)$	$\rightleftarrows$	$H_2(g)$	+	$I_2(g)$
Initial	0.395 *M*		0 *M*		0 *M*
change	- 2x *M*		+ x *M*		+ x *M*
at equilibrium	(0.395 - 2x) *M*		x *M*		x *M*

$$K_c = \frac{[H_2][I_2]}{[HI]^2} = \frac{(x)^2}{(0.3946 - 2x)^2} = 0.830 \ \text{(We'll round at the end of the problem to 3 significant figures.)}$$

If we take the square root of both sides, we have: $\dfrac{x}{0.3946 - 2x} = \sqrt{0.830} = 0.9110$

Solving for x, we have: $x = 0.3595 - 1.822x$
$$2.822x = 0.3595$$
$$x = 0.1274$$
Therefore, at equilibrium: $[HI] = 0.3946 - 2x = \textbf{0.140 } \textbf{\textit{M}}$ and $[H_2] = [I_2] = x = \textbf{0.127 } \textbf{\textit{M}}$

17-54. *Refer to Section 17-6 and Example 17-9.*

When an equilibrium system involving gases is subjected to an increase in pressure resulting from a decrease in volume, the concentrations of the gases increase and there may or may not be a shift in the equilibrium. If there is the same number of moles of gas on each side of the equation, equilibrium is not affected. If the number of moles of gas on each side of the equation is different, the general rule is that such an increase in pressure shifts a system in the direction that produces the smaller number of moles of gas.

(a) shift to left   (b) shift to right   (c) equilibrium is unaffected   (d) equilibrium is unaffected

---

17-56. *Refer to Section 17-6, Examples 17-9 and 17-10, and Exercise 17-54 Solution.*

Balanced equation: $A(g) + 3B(g) \rightleftarrows 2C(g) + 3D(g) + heat$ (the reaction is exothermic)

(a) Whenever the temperature changes, K_c changes as well. In fact, the only variable that affects K_c is temperature. When the reaction is exothermic, adding heat causes (ii) K_c to decrease. (At that point, $Q_c > K_c$ and the reaction shifts to the left.)

(b) K_c (iii) stays the same when A is added since the temperature is constant. (The equilibrium will shift to the right.)

(c) K_c (iii) stays the same when more C is added since T is constant. (The equilibrium will shift to the left.)

(d) K_c (iii) stays the same when D is removed since T is constant. (The equilibrium will shift to the right.)

(e) K_c (iii) stays the same when the pressure is decreased by increasing the volume of the container since T is constant. (The equilibrium will shift to the right.)

17-58. *Refer to Section 17-6, Example 17-9 and Exercise 17-54 Solution.*

Balanced equation: $2C(s) + O_2(g) \rightleftarrows 2CO(g)$
If the total pressure were decreased, the equilibrium would **shift to the right** to create more gas molecules.

17-60. *Refer to Section 17-6 and Examples 17-9 and 17-10.*

Balanced equation: $6CO_2(g) + 6H_2O(\ell) \rightleftarrows C_6H_{12}O_6(s) + 6O_2(g)$ $\Delta H° = 2801.69$ kJ/mol rxn

(a) If $[CO_2]$ is decreased, (i) the equilibrium will shift to the left.

(b) If P_{O_2} is increased, (ii) the equilibrium will shift to the left.

(c) If one-half of $C_6H_{12}O_6(s)$ is removed, (iii) the equilibrium is unaffected since $C_6H_{12}O_6$ is a solid and does not appear in the equilibrium expression.

(d) If the total pressure is decreased, the equilibrium is unaffected since the total number of moles of gas is the same on both sides of the equation.

(e) If the temperature is increased, (i) the equilibrium will shift to the right since the forward reaction is endothermic ($\Delta H° > 0$) and the forward reaction will absorb more heat to minimize the effect of raising the temperature and adding heat.

(f) If a catalyst is added, (iii) the equilibrium is unaffected; equilibrium is simply reached at a faster rate.

When the pressure is increased by decreasing the volume, the equilibrium in question :

(a) shifts to right   (b) is not affected   (c) shifts to left

(d) shifts to right   (e) shifts to right

Balanced equation: $A(g) + B(g) \rightleftharpoons C(g) + D(g)$

(a) $K_c = \dfrac{[C][D]}{[A][B]} = \dfrac{(1.60 \text{ mol}/1.00 \text{ L})^2}{(0.40 \text{ mol}/1.00 \text{ L})^2} = \mathbf{16}$

(b) Since we are adding both reactant and product to the system, the value of Q_c must be evaluated to determine the direction of the reaction.

New $[B] = 0.40\ M + 0.20\ M = 0.60\ M$ $Q_c = \dfrac{[C][D]}{[A][B]} = \dfrac{(1.80\ M)(1.60\ M)}{(0.40\ M)(0.60\ M)} = 12$

New $[C] = 1.60\ M + 0.20\ M = 1.80\ M$

Since $Q_c < K_c$, the forward reaction proceeds.

Let   x = moles per liter of A or B that react *after* the addition of 0.20 moles per liter of A and C. Then
    x = moles per liter of C produced = moles per liter of D produced.

	A	+	B	$\rightleftharpoons$	C	+	D
initial	0.40 *M*		0.40 *M*		1.60 *M*		1.60 *M*
mol/L added	0 *M*		+ 0.20 *M*		+ 0.20 *M*		0 *M*
new system	0.40 *M*		0.60 *M*		1.80 *M*		1.60 *M*
change	- x *M*		- x *M*		+ x *M*		+ x *M*
at equil	(0.40 - x) *M*		(0.60 - x) *M*		(1.80 + x) *M*		(1.60 + x) *M*

$K_c = \dfrac{[C][D]}{[A][B]} = \dfrac{(1.80 + x)(1.60 + x)}{(0.40 - x)(0.60 - x)} = 16$ The quadratic equation: $15x^2 - 19.4x + 0.96 = 0$

Solving, $x = \dfrac{19.4 \pm \sqrt{(19.4)^2 - 4(15)(0.96)}}{2(15)} = \dfrac{19.4 \pm 17.9}{30} = 0.050$ or 1.24 (discard)

Therefore, the new equilibrium concentration of A is (0.40 - x) or **0.35 *M***

Balanced equation: $A(g) \rightleftharpoons B(g) + C(g)$

(a) $K_c = \dfrac{[B][C]}{[A]} = \dfrac{(0.25)^2}{0.30} = \mathbf{0.21}$

(b) If the volume is suddenly doubled, the initial concentrations will be halved and the system is no longer at equilibrium. We learned in Section 17-5 that the equilibrium will then shift to the side with the greater number of moles of gas, i.e., the right side.

Let   x = number of moles per liter of A that react *after* the volume is doubled.
    x = number of moles per liter of B produced = number of moles per liter of C produced.

	A	$\rightleftharpoons$	B	+	C
initial	0.30 *M*		0.25 *M*		0.25 *M*
new system	0.15 *M*		0.125* *M*		0.125* *M*
change	- x *M*		+ x *M*		+ x *M*
at equilibrium	(0.15 - x) *M*		(0.125* + x) *M*		(0.125* + x) *M*

*this number really should have 2 significant figures. The numbers will be rounded off at the end.

$$K_c = \frac{[B][C]}{[A]} = \frac{(0.125 + x)^2}{(0.15 - x)} = 0.21 \qquad \text{The quadratic equation: } x^2 + 0.46x - 0.016 = 0$$

Solving, $x = \dfrac{-0.46 \pm \sqrt{(0.46)^2 - 4(1)(-0.016)}}{2(1)} = \dfrac{-0.46 \pm 0.52}{2} = 0.03$ or -0.49 (discard)

Therefore, $\quad [A] = 0.15 - x = \mathbf{0.12 \; M}$
$\qquad\qquad\;\; [B] = [C] = 0.12 + 0.03 = \mathbf{0.15 \; M}$

(c) If the volume is suddenly halved, the initial equilibrium concentrations will be doubled and this system is no longer at equilibrium. The equilibrium will shift to the left, the side with the lesser number of moles of gas.

Let $\;\; x = $ number of moles per liter of B that react *after* the volume is halved, and
$\qquad\;\; x = $ number of moles per liter of C that react after the volume is halved. Then
$\qquad\;\; x = $ number of moles of A that are produced.

	A	$\rightleftharpoons$	B	+	C
initial	0.30 M		0.25 M		0.25 M
new system	0.60 M		0.50 M		0.50 M
change	+ x M		- x M		- x M
at equilibrium	(0.60 + x) M		(0.50 - x) M		(0.50 - x) M

$$K_c = \frac{[B][C]}{[A]} = \frac{(0.50 - x)^2}{(0.60 + x)} = 0.21 \qquad \text{The quadratic equation: } x^2 - 1.21x + 0.12 = 0$$

Solving, $x = \dfrac{1.21 \pm \sqrt{(-1.21)^2 - 4(1)(0.12)}}{2(1)} = \dfrac{1.21 \pm 0.99}{2} = 0.11$ or 1.1 (discard)

Therefore, $\quad [A] = 0.60 + x = \mathbf{0.71 \; M}$
$\qquad\qquad\;\; [B] = [C] = 0.50 - x = \mathbf{0.39 \; M}$

17-68. *Refer to Sections 17-6 and 17-8, Example 17-12, and Exercise 17-66 Solution.*

Balanced equation: $N_2O_4(g) \rightleftharpoons 2NO_2(g)$ $\qquad\qquad\qquad K_c = 5.84 \times 10^{-3}$

(a) $[N_2O_4]_{initial} = \dfrac{4.00 \text{ g}/92.0 \text{ g/mol}}{2.00 \text{ L}} = 0.0217 \; M$

Let $\;\; x = $ number of moles per liter of N_2O_4 that react. Then
$\qquad 2x = $ number of moles per liter of NO_2 that are produced.

	N_2O_4	$\rightleftharpoons$	$2NO_2$
initial	0.0217 M		0 M
change	- x M		+ 2x M
at equilibrium	(0.0217 - x) M		2x M

$$K_c = \frac{[NO_2]^2}{[N_2O_4]} = \frac{(2x)^2}{(0.0217 - x)} = 5.84 \times 10^{-3} \quad \text{The quadratic equation: } 4x^2 + (5.84 \times 10^{-3})x - 1.27 \times 10^{-4} = 0$$

Solving, $x = \dfrac{-5.84 \times 10^{-3} \pm \sqrt{(5.84 \times 10^{-3})^2 - 4(4)(-1.27 \times 10^{-4})}}{2(4)}$

$\qquad\qquad = \dfrac{-5.84 \times 10^{-3} \pm 4.55 \times 10^{-2}}{8} = 4.95 \times 10^{-3}$ or -6.41×10^{-3} (discard)

Therefore, at equilibrium $\qquad [N_2O_4] = 0.0217 - x = \mathbf{0.0167 \; M}$
$\qquad\qquad\qquad\qquad\qquad\;\; [NO_2] = 2x = \mathbf{9.90 \times 10^{-3} \; M}$

(b) When the volume is suddenly increased (2.00 L $\rightarrow$ 3.00 L), the concentrations of N_2O_4 and NO_2 are decreased by a factor of 2/3 and the equilibrium shifts to the right.

Let $\;\; x = $ number of moles per liter of N_2O_4 that react *after* the volume is increased. Then
$\qquad 2x = $ number of moles of NO_2 that are produced.

$[N_2O_4]_{new} = 1.67 \times 10^{-2} \ M \times 2/3 = 1.11 \times 10^{-2} \ M$
$[NO_2]_{new} = 9.90 \times 10^{-3} \ M \times 2/3 = 6.60 \times 10^{-3} \ M$

	N_2O_4	$\rightleftarrows$	$2 NO_2$
initial	$1.67 \times 10^{-2} \ M$		$9.90 \times 10^{-3} \ M$
new system	$1.11 \times 10^{-2} \ M$		$6.60 \times 10^{-3} \ M$
change	$- x \ M$		$+ 2x \ M$
at equilibrium	$(1.11 \times 10^{-2} - x) \ M$		$(6.60 \times 10^{-3} + 2x) \ M$

$K_c = \dfrac{[NO_2]^2}{[N_2O_4]} = \dfrac{(6.60 \times 10^{-3} + 2x)^2}{(1.11 \times 10^{-2} - x)} = 5.84 \times 10^{-3}$ The quadratic equation: $4x^2 + 0.0322x - 2.12 \times 10^{-5} = 0$

Solving, $x = \dfrac{-0.0322 \pm \sqrt{(0.0322)^2 - 4(4)(-2.12 \times 10^{-5})}}{2(4)}$

$= \dfrac{-0.0322 \pm 0.0371}{8} = 6.1 \times 10^{-4}$ or -8.66×10^{-3} (discard)

Therefore, at equilibrium $[N_2O_4] = 1.11 \times 10^{-2} - x = \mathbf{1.05 \times 10^{-2} \ M}$
$[NO_2] = 6.60 \times 10^{-3} + 2x = \mathbf{7.82 \times 10^{-3} \ M}$

(c) When the volume in (a) is suddenly halved (2.00 L → 1.00 L), the concentrations of N_2O_4 and NO_2 are doubled and the equilibrium shifts to the left side.

Let x = number of moles per liter of N_2O_4 that are produced *after* the volume is halved. Then
$2x$ = number of moles of NO_2 that are consumed.

	N_2O_4	$\rightleftarrows$	$2 NO_2$
initial	$1.67 \times 10^{-2} \ M$		$9.90 \times 10^{-3} \ M$
new system	$0.0334 \ M$		$0.0198 \ M$
change	$+ x \ M$		$- 2x \ M$
at equilibrium	$(0.0334 + x) \ M$		$(0.0198 - 2x) \ M$

$K_c = \dfrac{[NO_2]^2}{[N_2O_4]} = \dfrac{(0.0198 - 2x)^2}{(0.0198 + x)} = 5.84 \times 10^{-3}$ The quadratic equation: $4x^2 - 0.0850x + 1.97 \times 10^{-4} = 0$

Solving, $x = \dfrac{0.0850 \pm \sqrt{(-0.0850)^2 - 4(4)(1.97 \times 10^{-4})}}{2(4)}$

$= \dfrac{0.0850 \pm 0.0638}{8} = 2.65 \times 10^{-3}$ or 0.0186 (discard)

Therefore, at equilibrium $[N_2O_4] = 0.0334 + x = \mathbf{0.0360 \ M}$
$[NO_2] = 0.0198 - 2x = \mathbf{0.0145 \ M}$

17-70. *Refer to Sections 17-9 and 17-10.*

The values of K_p and K_c are numerically equal for reactions in which there are equal numbers of moles of gases on both sides of the equation, i.e., $\Delta n_{gas} = 0$. K_p and K_c are numerically equal for reactions 17(a), 17(c), 17(d) only.

17-72. *Refer to Section 17-9 and Example 17-13.*

Balanced equation: $C(graphite) + CO_2(g) \rightleftarrows 2CO(g)$

Plan: (1) Calculate the partial pressures of CO and CO_2.
    (2) Determine K_p.

(1) Since the CO_2 gas stream contains 4.0×10^{-3} mol percent CO, the mole fraction of CO is 4.0×10^{-5}.

P_{CO} = mole fraction CO $\times P_{total} = (4.0 \times 10^{-5})(1.00 \text{ atm}) = 4.0 \times 10^{-5}$ atm
P_{CO_2} = mole fraction $CO_2 \times P_{total} = 1.00$ (to 3 significant figures) $\times 1$ atm $= 1.00$ atm

(2) $K_p = \dfrac{(P_{CO})^2}{P_{CO_2}} = \dfrac{(4.0 \times 10^{-5} \text{ atm})^2}{1.00 \text{ atm}} = \mathbf{1.6 \times 10^{-9}}$

17-74. *Refer to Section 17-9 and Example 17-13.*

Balanced equation: $N_2O_4(g) \rightleftarrows 2NO_2(g)$

Plan: (1) Calculate the initial partial pressure of N_2O_4, $K_p = 0.715$ at $T = 47°C$ (320 K).
(2) Determine the partial pressures of N_2O_4 and NO_2 at equilibrium.

(1) $P_{N_2O_4} = M(RT) = (3.3 \text{ mol/5.0 L})(0.0821 \text{ L·atm/mol·K})(320 \text{ K}) = 17 \text{ atm}$ (17.3 atm - used in calculations)

(2) Let x = partial pressure of N_2O_4 that that reacted. Then
        2x = partial pressure of NO_2 at equilibrium.

	N_2O_4	$\rightleftarrows$	$2 NO_2$
initial	17.3 atm		0 atm
change	- x atm		+ 2x atm
at equilibrium	(17.3 - x) atm		2x atm

$K_p = \dfrac{[NO_2]^2}{[N_2O_4]} = \dfrac{(2x)^2}{(17.3 - x)} = 0.715$ The quadratic equation: $4x^2 + 0.715x - 12.4 = 0$

Solving, $x = \dfrac{-0.715 \pm \sqrt{(0.715)^2 - 4(4)(-12.4)}}{2(4)}$

$= \dfrac{-0.715 \pm 14.1}{8} = 1.67$ or -1.85 (discard)

Therefore, at equilibrium $P_{N_2O_4} = 17.3 - x = \mathbf{16 \text{ atm}}$ (answer limited to 2 sig. figs. by the data)
$P_{NO_2} = 2x = \mathbf{3.3 \text{ atm}}$

17-76. *Refer to Section 17-9 and Example 17-13.*

Balanced equation: $H_2(g) + CO_2(g) \rightleftarrows CO(g) + H_2O(g)$ $K_p = \dfrac{P_{CO} P_{H_2O}}{P_{H_2} P_{CO_2}} = \dfrac{(0.180 \text{ atm})(0.252 \text{ atm})}{(0.387 \text{ atm})(0.152 \text{ atm})} = \mathbf{0.771}$

17-78. *Refer to Section 17-10.*

Balanced equation: $Br_2(g) \rightleftarrows 2Br(g)$ $K_p = 2550.$ at 4000 K

$K_c = K_p(RT)^{-\Delta n} = 2550.(0.0821 \times 4000)^{-(2-1)} = \mathbf{7.76}$

17-80. *Refer to Sections 17-5 and 17-6.*

Balanced equation: $Fe_2O_3(s) + 3H_2(g) \rightleftarrows 2Fe(s) + 3H_2O(g)$ $K_c = 8.11$ at 1000 K $\Delta H = 96$ kJ/mol rxn

(a) Let   x = moles per liter of H_2 present initially and y = moles per liter of H_2 that react

	$Fe_2O_3(s)$	+	$3H_2$	$\rightleftarrows$	$2Fe(s)$	+	$3H_2O$
initial	-		x M		-		0 M
change			- y M				+ y M
at equilibrium			(x - y) M				y M

$K_c = \dfrac{[H_2O]^3}{[H_2]^3} = \dfrac{y^3}{(x-y)^3} = 8.11$ Taking the cube root of both sides: $\dfrac{y}{x-y} = 2.01$

$y = (x - y)2.01$
$y = 2.01x - 2.01y$
$3.01y = 2.01x$
$y = 0.668x$

Therefore, the percentage of H_2 that reacts = 66.8%

  the percentage of H_2 that remains unreacted = (100 - 66.8)% = **33.2%**

(b) At lower temperatures, the equilibrium will shift to the left since the reaction is endothermic ($\Delta H > 0$), and the percentage of H_2 that remains unreacted will be **greater**.

17-82. *Refer to Section 17-12.*

(a) If $K \gg 1$, the forward reaction is likely to be spontaneous, and ΔG°_{rxn} must be negative.

(b) If $K = 1$, then ΔG°_{rxn} is 0, and the reaction is at equilibrium when the concentrations of aqueous species are 1 M and the partial pressures of gaseous species are 1 atm, or when the numerator and denominator cancel.

(c) If $K \ll 1$, the forward reaction is likely to be nonspontaneous and ΔG°_{rxn} must be positive.

17-84. *Refer to Sections 17-12 and 15-15, Examples 17-17 and 17-18, and Appendix K.*

Balanced equation: $CO(g) + H_2O(g) \rightleftarrows CO_2(g) + H_2(g)$

(a) Plan: (1) Calculate ΔH°_{rxn} and ΔS°_{rxn} from data in Appendix K and substitute into the Gibbs-Helmholtz equation to determine ΔG°_{rxn}.

  (2) Calculate K_p, using $\Delta G^\circ = -RT \ln K_p$.

(1) $\Delta H^\circ_{rxn} = [\Delta H^\circ_{f\ CO_2(g)} + \Delta H^\circ_{f\ H_2(g)}] - [\Delta H^\circ_{f\ CO(g)} + \Delta H^\circ_{f\ H_2O(g)}]$

  $= [(1\ mol)(-393.5\ kJ/mol) + (1\ mol)(0\ kJ/mol)]$

    $- [(1\ mol)(-110.5\ kJ/mol) + (1\ mol)(-241.8\ kJ/mol)]$

  $= -41.2\ kJ/mol\ rxn$

$\Delta S^\circ_{rxn} = [S^\circ_{CO_2(g)} + S^\circ_{H_2(g)}] - [S^\circ_{CO(g)} + S^\circ_{H_2O(g)}]$

  $= [(1\ mol)(213.6\ J/mol\cdot K) + (1\ mol)(130.6\ J/mol\cdot K)]$

    $- [(1\ mol)(197.6\ J/mol\cdot K) + (1\ mol)(188.7\ J/mol\cdot K)]$

  $= -42.1\ J/K$

$\Delta G^\circ = \Delta H^\circ - T\Delta S^\circ = -41.2 \times 10^3\ J - (298\ K)(-42.1\ J/K) = -2.87 \times 10^4\ J/mol\ rxn$

(2) Substituting into $\Delta G^\circ = -RT \ln K_p$, we have

  $-2.87 \times 10^4\ J = -(8.314\ J/K)(298\ K)\ln K_p$

    $\ln K_p = 11.6$

    $K_p = \mathbf{1.1 \times 10^5}$

(b) If we assume that ΔH° and ΔS° for the reaction are independent of temperature, we can determine ΔG° at 200°C (473 K) by substituting into the Gibbs-Helmholtz equation.

  $\Delta G^\circ = \Delta H^\circ - T\Delta S^\circ = -41.2 \times 10^3\ J/mol - (473\ K)(-42.1\ J/mol\cdot K) = -2.13 \times 10^4\ J/mol\ rxn$

Then, $\Delta G^\circ = -RT \ln K_p$

  $-2.13 \times 10^4\ J = -(8.314\ J/K)(473\ K)\ln K_p$

    $\ln K_p = 5.42$

    $K_p = \mathbf{2.3 \times 10^2}$

(c) $\Delta G^\circ_{rxn} = [\Delta G^\circ_{f\ CO_2(g)} + \Delta G^\circ_{f\ H_2(g)}] - [\Delta G^\circ_{f\ CO(g)} + \Delta G^\circ_{f\ H_2O(g)}]$

  $= [(1\ mol)(-394.4\ kJ/mol) + (1\ mol)(0\ kJ/mol)]$

    $- [(1\ mol)(-137.2\ kJ/mol) + (1\ mol)(-228.6\ kJ/mol)]$

  $= \mathbf{-28.6\ kJ/mol\ rxn\ \ or\ \ -2.86 \times 10^4\ J/mol\ rxn}$

Substituting into $\Delta G^\circ = -RT \ln K_p$, we have

  $-2.86 \times 10^4\ J = -(8.314\ J/K)(298\ K)\ln K_p$

    $\ln K_p = 11.5$

    $K_p = \mathbf{1.0 \times 10^5}$

17-86. *Refer to Section 17-12, Examples 17-17 and 17-18, and Appendix K.*

Balanced equation: $2SO_2(g) + O_2(g) \rightleftarrows 2SO_3(g)$

$$\Delta G^\circ_{rxn} = [2\Delta G^\circ_f \,_{SO_3(g)}] - [2\Delta G^\circ_f \,_{SO_2(g)} + \Delta G^\circ_f \,_{O_2(g)}]$$
$$= [(2 \text{ mol})(-371.1 \text{ kJ/mol})] - [(2 \text{ mol})(-300.2 \text{ kJ/mol}) + (1 \text{ mol})(0 \text{ kJ/mol})]$$
$$= -141.8 \text{ kJ/mol rxn}$$

We know $\qquad\qquad \Delta G^\circ_{rxn} = -RT \ln K_p \qquad$ for a gas phase reaction

Substituting, $\qquad -141.8 \times 10^3 \text{ J} = -(8.314 \text{ J/K})(298.15 \text{ K})\ln K_p$

$$\ln K_p = 57.20$$
$$K_p = \mathbf{7.0 \times 10^{24}}$$

17-88. *Refer to Sections 17-12 and 17-13, and Examples 17-19 and 17-20.*

(a) $\quad \Delta G^\circ_{rxn} = -RT \ln K_p$
$$= -(8.314 \text{ J/K})(298.15 \text{ K})(\ln 4.3 \times 10^6)$$
$$= \mathbf{-3.79 \times 10^4 \text{ J/mol rxn} \ \ or \ \ -37.9 \text{ kJ/mol rxn at } 25°C}$$

(b) Using the van't Hoff equation,

$$\ln \frac{K_{p\,T_2}}{K_{p\,T_1}} = \frac{\Delta H^\circ}{R}\left(\frac{1}{T_1} - \frac{1}{T_2}\right)$$

$$\ln \frac{K_{p\,1073\,K}}{4.3 \times 10^6} = \frac{-78.58 \times 10^3 \text{ J}}{8.314 \text{ J/K}}\left(\frac{1}{298 \text{ K}} - \frac{1}{1073 \text{ K}}\right) = -22.9$$

$$\frac{K_{p\,1073\,K}}{4.3 \times 10^6} = e^{-22.9} = 1.13 \times 10^{-10}$$

$$K_{p\,1073\,K} = \mathbf{4.9 \times 10^{-4}}$$

(c) $\Delta G^\circ_{rxn} = -(8.314 \text{ J/K})(1073 \text{ K})(\ln 4.9 \times 10^{-4}) = \mathbf{+6.80 \times 10^4 \text{ J/mol rxn} \ \ or \ \ +68.0 \text{ kJ/mol rxn at } 800°C}$

(d) The forward reaction at 800°C is nonspontaneous (ΔG is positive) whereas at 25°C the forward reaction is spontaneous (ΔG is negative).

(e) The reaction mixture is heated to speed up the rate at which equilibrium is reached, not to shift the equilibrium toward more product. Heating actually decreases the amount of product present at equilibrium since the reaction is exothermic (ΔH is negative).

(f) The basic purpose of a catalyst is to speed up the reaction of interest. Recall from Chapter 16 that the presence of a catalyst increases the rates of both forward and reverse reactions to the same extent without affecting equilibrium. However, the catalyst increases the rate at which the reaction goes to equilibrium. Hopefully, the yield of desired product is adequate at equilibrium.

17-90. *Refer to Section 17-2.*

Balanced equation: $2CH_3COOH(g) \rightleftarrows (CH_3COOH)_2(g)$

Since the equilibrium constant at 25°C, 3.2×10^4, is much greater than 1, the **dimer** is favored at equilibrium.

17-92. *Refer to Section 17-9.*

Balanced equation: $C(s) + CO_2(g) \rightleftarrows 2CO(g)$

Given: $P_{CO} + P_{CO_2} = 1.00$ atm; $K_p = 1.50$

Let $\quad x$ = partial pressure of CO = P_{CO}. Then, $1.00 - x$ = partial pressure of $CO_2 = P_{CO_2}$

$$K_p = \frac{(P_{CO})^2}{P_{CO_2}} = \frac{x^2}{1.00 - x} = 1.50 \qquad\qquad \text{The quadratic equation: } x^2 + 1.50x - 1.50 = 0$$

Solving, $x = \dfrac{-1.50 \pm \sqrt{(1.50)^2 - 4(1)(-1.50)}}{2(1)} = \dfrac{-1.50 \pm 2.87}{2} = 0.685$ or -2.19 (discard)

Therefore, at equilibrium $\qquad P_{CO} = x = \textbf{0.685 atm}$

$\qquad\qquad\qquad\qquad\qquad\qquad P_{CO_2} = 1.00 - x = \textbf{0.315 atm}$

17-94. *Refer to Sections 17-6, 17-8 and 17-11, and Example 17-10.*

Balanced equation: $NH_4Cl(s) \rightleftarrows NH_3(g) + HCl(g)$ $\qquad\qquad \Delta H = 176$ kJ/mol rxn

(a) As the temperature decreases, the mass of NH_3 decreases because the reaction is endothermic ($\Delta H > O$).

(b) When more NH_3 is added, the total mass of NH_3 is initially the sum of the original equilibrium amount plus the additional NH_3. After equilibrium is reestablished by shifting to the left, the final mass of NH_3 is greater than the original equilibrium amount but less than the total mass present immediately after the additional NH_3 was added.

(c) When more HCl is added, the equilibrium shifts to the left and the mass of NH_3 decreases.

(d) When more solid NH_4Cl is added, the equilibrium is unaffected if the total gas volume is unchanged.

(e) When more solid NH_4Cl is added and the total gas volume decreases, the concentration of the gases will increase and the equilibrium will then shift to the left, thereby decreasing the concentrations of NH_3 and HCl back to their original values. However, although there is no change in the concentration of NH_3, the mass of NH_3 decreases due to the volume shrinkage.

17-96. *Refer to Sections 17-13 and 17-10, and Example 17-20.*

Balanced equation: $2Cl_2(g) + 2H_2O(g) \rightleftarrows 4HCl(g) + O_2(g)$ $\qquad \Delta H^\circ = +115$ kJ/mol rxn

$\qquad\qquad\qquad\qquad\qquad\qquad\qquad\qquad\qquad\qquad\qquad\qquad\qquad\qquad\qquad K_p = \textbf{4.6 x 10}^{-14}$ at 25°C

At 400.°C (673 K): Substituting into the van't Hoff equation, we have

$$\ln \frac{K_{p\,T_2}}{K_{p\,T_1}} = \frac{\Delta H^\circ}{R}\left(\frac{1}{T_1} - \frac{1}{T_2}\right)$$

$$\ln \frac{K_{p\,673\,K}}{4.6\text{ x }10^{-14}} = \frac{115\text{ x }10^3\text{ J}}{8.314\text{ J/K}}\left(\frac{1}{298\text{ K}} - \frac{1}{673\text{ K}}\right) = 25.9$$

$$\frac{K_{p\,673\,K}}{4.6\text{ x }10^{-14}} = e^{25.9} = 1.8\text{ x }10^{11}$$

$$K_{p\,673\,K} = \textbf{8.3 x 10}^{-3}$$

$$K_c = K_p(RT)^{-\Delta n} = (8.3\text{ x }10^{-3})(0.0821 \times 673)^{-(5-4)} = \textbf{1.5 x 10}^{-4}$$

At 800.°C (1073 K): Substituting into the van't Hoff equation, we have

$$\ln \frac{K_{p\,1073\,K}}{4.6\text{ x }10^{-14}} = \frac{115\text{ x }10^3\text{ J}}{8.314\text{ J/K}}\left(\frac{1}{298\text{ K}} - \frac{1}{1073\text{ K}}\right) = 33.5$$

$$K_{p\,1073\,K} = \textbf{16}$$

$$K_c = K_p(RT)^{-\Delta n} = (16)(0.0821 \times 1073)^{-(5-4)} = \textbf{0.18}$$

17-98. *Refer to Sections 17-10.*

Balanced equation: $CO(g) + 3H_2(g) \rightleftarrows CH_4(g) + H_2O(g)$

Plan: (1) Determine K_c at 1133 K.

$\qquad\quad$ (2) Calculate K_p.

(1) $K_c = \dfrac{[CH_4][H_2O]}{[CO][H_2]^3} = \dfrac{(1.21\text{ x }10^{-4}\text{ mol}/0.100\text{ L})(5.63\text{ x }10^{-8}\text{ mol}/0.100\text{ L})}{(1.21\text{ x }10^{-4}\text{ mol}/0.100\text{ L})(2.47\text{ x }10^{-4}\text{ mol}/0.100\text{ L})^3} = 37.4$

(2) $K_p = K_c(RT)^{\Delta n} = (37.4)(0.0821 \times 1133)^{(2-4)} = \textbf{0.00432}$

When the beam of a triple balance stops swinging, the balance has reached a static equilibrium. A body is said to be "in static equilibrium" if a body at rest will stay at rest. That statement is definitely true in this case.

17-102. *Refer to Section 17-1.*

The actual masses of reactant species do not equal the masses of product species at equilibrium.
For example, consider the equilibrium: $2A \rightleftarrows A_2$.

If K_c were very large, at equilibrium we would have essentially all A_2 and very little A. The Law of Conservation of Matter is not violated because wherever the system is, the total mass of the system remains constant: mass of A + mass of A_2 = constant.

Recall that for other reactions we have studied which we know obey the Law of Conservation of Matter, we assumed they went totally to completion. At the beginning of the reaction, there were only reactants and at the end, there were only products.

17-104. *Refer to Section 17-2 and Margin Figure in Section 17-6.*

Balanced equation: $[Co(OH_2)_6]^{2+}(aq) + 4Cl^-(aq) \rightleftarrows CoCl_4^{2-}(aq) + 6H_2O(\ell)$ $\qquad K_c = \dfrac{[CoCl_4^{2-}]}{[[Co(OH_2)_6]^{2+}][Cl^-]^4}$

Assuming $K>1$, initially in the beaker, I would "see" mostly the blue $CoCl_4^{2-}$ ions and fewer pink $[Co(OH_2)_6]^{2+}$ ions, along with colorless Cl^- ions. Overall, the solution would look blue.

When water is added, the reaction quotient, Q_c, would initially become greater than K_c since the concentrations of all the species would decrease. The system would then shift to the left toward the reactants to reach equilibrium, in much the same way as when working with gases and the volume of the container is increased, i.e. the concentrations of all species are decreased. In the gas case, the system shifts toward the side with the greater number of moles of gas. Here, the system shifts toward the side with the greater number of species. Water is not included as a relevant species since it does not appear in the equilibrium expression. The solution would become pinker, since the concentration of pink $[Co(OH_2)_6]^{2+}(aq)$ is increasing while the concentration of blue $CoCl_4^{2-}(aq)$ is decreasing.

17-106. *Refer to Sections 17-1 and 17-2.*

Balanced equation: $C_2H_5OH(\ell) + C_3H_7COOH(\ell) \rightleftarrows C_3H_7COOC_2H_5(\ell) + H_2O(\ell)$

Esterification is a homogeneous equilibrium reaction when all the species are liquids. The equilibrium constant value is commonly determined by following the concentration of acid with time by titration of solution aliquots with NaOH. As a homogeneous equilibrium, the concentrations of the liquids are not set to 1, but can be determined as M = moles/L of solution, assuming the liquids are ideal.

At equilibrium, it is very unlikely that the concentrations of each species will be 1.0 M, when starting with 2.0 M each of ethanol, $C_2H_5OH(\ell)$, and propanoic acid, $C_3H_7COOH(\ell)$. For this to happen, K_c would be equal to 1.

Yes, there would be some ethanol left, as there is always some amount of all species present at equilibrium.

If the system reached equilibrium after 45 minutes, then the forward reaction is occurring at the same rate as the reverse reaction. There will never be a time at which all the reactants will be converted to products.

Balanced equation: $Cu^{2+}(aq) + 2OH^-(aq) \rightleftarrows Cu(OH)_2(s)$ $\qquad K_c = \dfrac{1}{[Cu^{2+}][OH^-]^2}$

(1) more $Cu(OH)_2(s)$ is added: Nothing would happen.

(2) more NaOH solution is added: The system would shift to the right and more precipitate, $Cu(OH)_2(s)$, will form.

(3) $1\ M\ NH_3$ is added with $Cu^{2+}(aq) + 4NH_3(aq) \rightleftarrows [Cu(NH_3)_4]^{2+}(aq)$: What happens depends on the relative values of K_c for the formation of $Cu(OH)_2(s)$ and the K_c for the formation of $[Cu(NH_3)_4]^{2+}(aq)$. If one value of K_c is much larger than the other, then that equilibrium will prevail. Since in this problem, we don't know what the K_c values are, we can simply guess that the formation of the copper-ammonia complex would decrease the concentration of Cu^{2+} in solution and that some or all of the $Cu(OH)_2(s)$ might dissolve.

(4) addition of 0.10 M HCl: The acid would react with the free OH^- ions in solution, reducing the OH^- concentration. The system would shift to the left, forming more OH^- ions and some of the $Cu(OH)_2(s)$ would dissolve.

Consider the van't Hoff equation: $\ln\left(\dfrac{K_{T_2}}{K_{T_1}}\right) = \dfrac{\Delta H^\circ}{R}\left(\dfrac{1}{T_1} - \dfrac{1}{T_2}\right)$

If $T_2 > T_1$, then $1/T_2 < 1/T_1$ and $\left(\dfrac{1}{T_1} - \dfrac{1}{T_2}\right)$ will be a positive number.

Moreover, if the reaction is exothermic and $\Delta H^\circ < 0$, then $\dfrac{\Delta H^\circ}{R}\left(\dfrac{1}{T_1} - \dfrac{1}{T_2}\right)$ will be a negative value.

From mathematical rules, $\ln\left(\dfrac{K_{T_2}}{K_{T_1}}\right) = \ln K_{T_2} - \ln K_{T_1}$.

Therefore, $(\ln K_{T_2} - \ln K_{T_1}) < 0$, so $\ln K_{T_1} > \ln K_{T_2}$ and $K_{T_1} > K_{T_2}$

Does this agree with LeChatelier's Principle that for an exothermic reaction the system will shift to form more reactants at higher temperatures? Yes, it does. Recall, that that only temperature will affect the value of the equilibrium constant. More reactants are formed because the value of the equilibrium constant decreased when the temperature increased.

Plan: (1) Treat the vaporization process as a chemical reaction, then establish the relationship between K_{eq} and P_{H_2O}.
(2) Use the van't Hoff equation to evaluate P_{H_2O} at 75°C.

(1) For vaporization: $\qquad H_2O(\ell) \rightleftarrows H_2O(g) \qquad K_{eq} = P_{H_2O}$
At temperature $T_1 = 100°C = 373\ K$ $\qquad K_1 = P_{H_2O}$ at 373 K = 1 atm
At temperature $T_2 = 75°C = 348\ K$ $\qquad K_2 = P_{H_2O}$ at 348 K = unknown

287

(2) van't Hoff equation:

$$\ln\frac{K_2}{K_1} = \frac{\Delta H^{\circ}}{R}\left(\frac{1}{T_1} - \frac{1}{T_2}\right) \qquad \text{where } \Delta H^{\circ} = +40.66 \text{ kJ/mol}$$

$$\ln\frac{K_2}{1 \text{ atm}} = \frac{40660 \text{ J}}{8.314 \text{ J/K}}\left(\frac{1}{373 \text{ K}} - \frac{1}{348 \text{ K}}\right)$$

$$\ln\frac{K_2}{1 \text{ atm}} = -0.94$$

$$K_2 = P_{H_2O} \text{ at } 348 \text{ K} = \mathbf{0.39 \text{ atm}}$$

Note: In Appendix E, the vapor pressure of water, P_{H_2O}, at 348 K = 187.5 torr = 0.247 atm. The answers agree very well.

18 Ionic Equilibria I: Acids and Bases

(a) Strong and weak acids are electrolytes and ionize to some degree in water, producing H^+, according to Arrhenius and Brønsted-Lowry definitions. They have a sour taste, change the color of many indicators, react with metal oxides and metal hydroxides to form salts and water. They differ in the degree to which they ionize in aqueous solution. Strong acids ionize almost completely, whereas weak acids generally ionize less than 5%.

(b) Strong and weak bases are also electrolytes. Both produce OH^- ions. They generally have a bitter taste, have a slippery feeling, change the color of many indicators, and react with protic acids to form salts and sometimes water. Strong bases are soluble metal hydroxides (ionic compounds), like NaOH and dissociate completely into their ions. When strong bases react with protic acids, salt and water are produced. Two categories of weak bases are (1) the soluble organic base (covalent compounds), like NH_3 or CH_3NH_2, and (2) the conjugate base of a weak acid, such as CN^- or F^-. Both produce OH^- ions in solution by reacting with water in an equilibrium (see below). When weak organic bases react with protic acids, only salt is produced.

$$(1)\ \ NH_3(aq) + H_2O(\ell) \rightleftarrows NH_4^+(aq) + OH^-(aq) \qquad\qquad (2)\ \ CN^-(aq) + H_2O(\ell) \rightleftarrows HCN(aq) + OH^-(aq)$$

18-4.   *Refer to Sections 18-1 and 4-2, and Example 18-1.*

(a) $? \, M \, NaCl = \dfrac{\text{mol NaCl}}{\text{L soln}} = \dfrac{25.65 \text{ g}/58.44 \text{ g/mol}}{0.250 \text{ L}} = \textbf{1.76 } \boldsymbol{M} \textbf{ NaCl}$

(b) $? \, M \, H_2SO_4 = \dfrac{\text{mol } H_2SO_4}{\text{L soln}} = \dfrac{75.5 \text{ g}/98.1 \text{ g/mol}}{1.00 \text{ L}} = \textbf{0.770 } \boldsymbol{M} \textbf{ H}_2\textbf{SO}_4$

(c) $? \, M \, C_6H_5OH = \dfrac{\text{mol } C_6H_5OH}{\text{L soln}} = \dfrac{0.126 \text{ g}/94.1 \text{ g/mol}}{1.00 \text{ L}} = \textbf{1.34 x 10}^{-3} \, \boldsymbol{M} \textbf{ C}_6\textbf{H}_5\textbf{OH}$

18-6.   *Refer to Sections 18-1, 18-2 and 4-2, and Examples 18-1 and 18-2.*

These compounds are strong electrolytes.

(a) $0.45 \, M \, HBr(aq) \rightarrow 0.45 \, M \, H^+(aq) + 0.45 \, M \, Br^-(aq)$ $[H^+] = [Br^-] = \textbf{0.45 } \boldsymbol{M}$

(b) $0.045 \, M \, KOH(aq) \rightarrow 0.045 \, M \, K^+(aq) + 0.045 \, M \, OH^-(aq)$ $[K^+] = [OH^-] = \textbf{0.045 } \boldsymbol{M}$

(c) $0.0112 \, M \, CaCl_2(aq) \rightarrow 0.0112 \, Ca^{2+}(aq) + 0.0224 \, M \, Cl^-(aq)$ $[Ca^{2+}] = \textbf{0.0112 } \boldsymbol{M}; [Cl^-] = \textbf{0.0224 } \boldsymbol{M}$

18-8.   *Refer to Sections 18-1 and 18-2, and Example 18-2.*

These compounds are strong electrolytes.

(a) $? \, M \, KOH = \dfrac{(1.25 \text{ g}/56.1 \text{ g/mol})}{1.50 \text{ L}} = 0.0149 \, M$

	KOH	$\rightarrow$	K^+	$+$	OH^-
initial	0.0149 M		0 M		0 M
change	- 0.0149 M		+ 0.0149 M		+ 0.0149 M
final	0 M		0.0149 M		0.0149 M

Therefore, $[K^+] = \textbf{0.0149 } \boldsymbol{M}$ and $[OH^-] = \textbf{0.0149 } \boldsymbol{M}$

(b) $? M \ Ba(OH)_2 = \dfrac{(0.2505 \ g/171.3 \ g/mol)}{0.250 \ L} = 0.00585 \ M$ Therefore, $[Ba^{2+}] = \mathbf{0.00585 \ M}$; $[OH^-] = \mathbf{0.0117 \ M}$

(c) $? M \ Ca(NO_3)_2 = \dfrac{(1.26 \ g/164 \ g/mol)}{0.100 \ L} = 0.0768 \ M$ \qquad Therefore, $[Ca^{2+}] = \mathbf{0.0768 \ M}$; $[NO_3^-] = \mathbf{0.154 \ M}$

18-10. *Refer to Section 18-2.*

(a) $H_2O(\ell) + H_2O(\ell) \rightleftarrows H_3O^+(aq) + OH^-(aq)$

(b) $K_c = K_w = [H_3O^+][OH^-]$

(c) The equilibrium constant, K_w, is known as the ion product for water.

(d) In pure water, $[H_3O^+] = [OH^-]$. At 25°C, $[OH^-] = [H_3O^+] = 1.0 \times 10^{-7} \ M$.

(e) In acidic solutions, $[H_3O^+] > [OH^-]$. In basic solutions, $[OH^-] > [H_3O^+]$.

18-12. *Refer to Section 18-2.*

(a) A 0.060 M solution of NaOH does have 2 sources of OH^- ion: (1) OH^- from the complete dissociation of NaOH and (2) OH^- from the ionization of water. Since Source 1 dominates Source 2, the concentration of OH^- produced by the ionization of water is therefore neglected.

(b) From Source 1, $NaOH(aq) \rightarrow Na^+(aq) + OH^-(aq)$, $[OH^-] = 0.060 \ M$. To calculate $[OH^-]$ from Source 2, consider the ionization of water in a 0.060 M solution of NaOH.

Let x = moles per liter of H_3O^+ and OH^- produced by the ionization of water.

	$2H_2O(\ell)$	$\rightleftarrows$	$H_3O^+(aq)$	$+$	$OH^-(aq)$
initial			0 M		0.060 M
change			+ x M		+ x M
at equilibrium			x M		(0.060 + x) M

$K_w = [H_3O^+][OH^-] = x(0.060 + x) = 1.0 \times 10^{-14}$
We know that x $\ll$ 0.060. Let us make the approximation that $0.060 + x \approx 0.060$.
Then x x 0.060 = 1.0×10^{-14} and x = 1.7×10^{-13}. So, our approximation that x $\ll$ 0.060 is a good one.

Therefore, $[OH^-]$ from 0.060 M NaOH = 0.060 M
$[OH^-]$ from the ionization of water = $1.7 \times 10^{-13} \ M$, and can be neglected.

18-14. *Refer to Section 18-2, Example 18-3, and Exercises 18-6, 18-7 and 18-8 Solutions.*

In pure water at 25°C, $[H_3O^+] = 1.00 \times 10^{-7} \ M$.
To determine $[H_3O^+]$ in basic solution, use the K_w expression: $K_w = [H_3O^+][OH^-] = 1.00 \times 10^{-14}$.

Solution	$[OH^-]$	$[H_3O^+] = K_w/[OH^-]$
0.045 M KOH	0.045 M	$2.2 \times 10^{-13} \ M$
0.0105 M Sr(OH)$_2$	0.0210 M	$4.76 \times 10^{-13} \ M$
0.0585 M Ba(OH)$_2$	0.0117 M	$8.55 \times 10^{-13} \ M$
pure water	$1.00 \times 10^{-7} \ M$	$1.00 \times 10^{-7} \ M$

The $[H_3O^+]$ in basic solutions is much lower than that in pure water.

18-16. *Refer to Section 18-2 and Example 18-2.*

(1) $[H_2O] = \dfrac{mol \ H_2O}{L \ H_2O} = \dfrac{1000 \ g/18.0 \ g/mol}{1.00 \ L} = \mathbf{55.6 \ M}$ \qquad since the density of H_2O is 1.00 g/mL or 1000 g/L

(2) If we assume that we are working with 1.00 L of pure water which contains 55.6 moles of water and 1.0 x 10^{-7} mol of H_3O^+, then, we can calculate the number of H_3O^+ ions for every 1.0 x 10^{10} (10 billion) water molecules:

$$? \text{ H}_3\text{O}^+ \text{ ions} = (1.0 \times 10^{10} \text{ H}_2\text{O molecules}) \times \frac{1.0 \text{ mole H}_2\text{O molecules}}{6.02 \times 10^{23} \text{ H}_2\text{O molecules}} \times \frac{1.0 \times 10^{-7} \text{ mol H}_3\text{O}^+ \text{ ions}}{55.6 \text{ mol H}_2\text{O}}$$

$$= \mathbf{3.0 \times 10^{-23} \text{ H}_3\text{O}^+ \text{ ions}}$$

18-18. *Refer to Section 18-3 and Appendix A-2.*

When working with base 10 logarithms, the number of significant figures in the number (the antilogarithm) sets the number of significant digits in the mantissa (the decimal part of the logarithm). For example in (a), the number 1.0 has two significant figures, so there are 2 decimal places in the log of that number.

(a) $\text{pH} = -\log [\text{H}^+] = -\log (1.0) = \mathbf{0.00}$ $\text{pOH} = 14.00 - \text{pH} = \mathbf{14.00}$ **acidic**

(b) $\text{pH} = -\log (1.7 \times 10^{-4}) = \mathbf{3.77}$ $\text{pOH} = 14.00 - 3.77 = \mathbf{10.23}$ **acidic**

(c) $\text{pH} = -\log (6.8 \times 10^{-8}) = \mathbf{7.17}$ $\text{pOH} = 14.00 - 7.17 = \mathbf{6.83}$ **basic**

(d) $\text{pH} = -\log (9.3 \times 10^{-11}) = \mathbf{10.03}$ $\text{pOH} = 14.00 - 10.03 = \mathbf{3.97}$ **basic**

18-20. *Refer to Section 18-3, Examples 18-4 and 18-5, and Table 18-2.*

We know: $\text{pH} = -\log [\text{H}_3\text{O}^+] = 7.45$.

Therefore, $[\text{H}_3\text{O}^+] = \text{antilog} (-7.45) = \mathbf{3.5 \times 10^{-8} \textit{ M}}$

If we assume the blood is at 25°C, $[\text{OH}^-] = \dfrac{K_w}{[\text{H}_3\text{O}^+]} = \dfrac{1.00 \times 10^{-14}}{3.5 \times 10^{-8}} = \mathbf{2.9 \times 10^{-7} \textit{ M}}$

If we assume the blood is at 37°C (body temperature), $[\text{OH}^-] = \dfrac{K_w}{[\text{H}_3\text{O}^+]} = \dfrac{2.4 \times 10^{-14}}{3.5 \times 10^{-8}} = \mathbf{6.9 \times 10^{-7} \textit{ M}}$

18-22. *Refer to Sections 18-2 and 18-3, and Example 18-5.*

(a) $[\text{H}_3\text{O}^+] = [\text{HCl}] = 0.20 \textit{ M}$ since HCl is a strong acid $\text{pH} = -\log[\text{H}^+] = -\log (0.20) = \mathbf{0.70}$

(b) $[\text{H}_3\text{O}^+] = [\text{HNO}_3] = 0.050 \textit{ M}$ since HNO_3 is a strong acid $\text{pH} = -\log[\text{H}^+] = -\log (0.050) = \mathbf{1.30}$

(c) $[\text{H}_3\text{O}^+] = [\text{HClO}_4] = \dfrac{0.65 \text{ g HClO}_4}{1 \text{ L}} \times \dfrac{1 \text{ mol HClO}_4}{100.45 \text{ g HClO}_4} = 6.5 \times 10^{-3} \textit{ M}$ since HClO_4 is a strong acid

     $\text{pH} = -\log (6.5 \times 10^{-3} \textit{ M}) = \mathbf{2.19}$

(d) $[\text{OH}^-] = [\text{NaOH}] = 9.8 \times 10^{-4} \textit{ M}$ since NaOH is a strong soluble base

     $\text{pOH} = -\log[\text{OH}^-] = 3.01$; $\text{pH} = 14.00 - 3.01 = \mathbf{10.99}$

18-24. *Refer to Section 18-3 and Example 18-4.*

We know: $\text{pH} = -\log [\text{H}_3\text{O}^+] = 3.52$. Therefore, $[\text{H}_3\text{O}^+] = \text{antilog} (-3.52) = 3.0 \times 10^{-4} \textit{ M}$

Since HNO_3 is a strong acid, $[\text{HNO}_3] = [\text{H}_3\text{O}^+] = \mathbf{3.0 \times 10^{-4} \textit{ M}}$

18-26. *Refer to Section 18-3, and Examples 18-5 and 18-6.*

	Solution	$[\text{H}_3\text{O}^+]$	$[\text{OH}^-]$	pH	pOH
(a)	0.085 M NaOH	$1.2 \times 10^{-13} \textit{ M}$	0.085 M	12.93	1.07
(b)	0.075 M HCl	0.075 M	$1.3 \times 10^{-13} \textit{ M}$	1.12	12.88
(c)	0.075 M Ca(OH)$_2$	$6.7 \times 10^{-14} \textit{ M}$	0.15 M	13.18	0.82

18-28. *Refer to Section 10-7 and Appendix F.*

(a) H_3PO_4 is probably a slightly stronger acid than H_3AsO_4. P is probably slightly more electronegative than As, since electronegativity increases upward in a group, even though both are given an electronegativity of 2.1. The P atom would pull the electron density in the O-H bond away from the O-H bond toward itself a little more than the As atom would, making the O-H bond slightly weaker in H_3PO_4 and producing a slightly stronger acid. This deduction is verified by the K_a values in Appendix F: K_a for H_3PO_4 is 7.5 x 10^{-3} and K_a for H_3AsO_4 is 2.5 x 10^{-4}.

(b) H_3AsO_4 is a stronger acid than H_3AsO_3, because H_3AsO_4 has one more O atom. The extra oxygen atom with its high electronegativity, helps pull the electron density away from the H-O bond in H_3AsO_4, thereby weakening the H-O bond. The weaker the H-O bond, the easier it will break in solution and the stronger the acid will be. This deduction is verified by the data in Appendix F: K_a for H_3AsO_4 is 2.5 x 10^{-4} and K_a for H_3AsO_3 is 6.0 x 10^{-10}.

18-30. *Refer to Section 18-4.*

Balanced equation for the ionization of a weak acid: $HA(aq) + H_2O(\ell) \rightleftarrows H_3O^+(aq) + A^-(aq)$

In Chapter 17 (Section 17-11), we stated that for heterogeneous equilibria, terms for pure liquids and pure solids do not appear in K expressions because their activity values are essentially 1. In the discussion in Section 18-4, these terms are temporarily included. In dilute solutions, activities are equal to concentrations.

$$K = \frac{(H_3O^+)(A^-)}{(HA)(H_2O)} \quad \text{where (x) represents the activity of x} \qquad K_a = \frac{[H_3O^+][A^-]}{[HA](1)}$$

The symbol for the equilibrium constant is K_a. It is called the acid ionization constant.

18-32. *Refer to Section 18-4 and Examples 18-15 and 18-16.*

(1) For triethylamine, $(C_2H_5)_3N$

Balanced equation: $(C_2H_5)_3N + H_2O \rightleftarrows (C_2H_5)_3NH^+ + OH^-$ $\qquad K_b = 5.2 \times 10^{-4}$

Let $x = [(C_2H_5)_3N]$ that ionizes. Then $x = [(C_2H_5)_3NH^+]$ produced $= [OH^-]$ produced.

$K_b = \dfrac{[(C_2H_5)_3NH^+][OH^-]}{[(C_2H_5)_3N]} = 5.2 \times 10^{-4} = \dfrac{x^2}{0.018 - x} \approx \dfrac{x^2}{0.018}$ $\qquad$ Solving, $x = 3.1 \times 10^{-3}$

In this case, 3.1 x 10^{-3} is greater than 5% of 0.018 (= 9.0 x 10^{-4}). A simplifying assumption cannot be made and the original quadratic equation must be solved: $x^2 + (5.2 \times 10^{-4})x - 9.4 \times 10^{-6} = 0$

$$x = \frac{-(5.2 \times 10^{-4}) \pm \sqrt{(5.2 \times 10^{-4})^2 - 4(1)(-9.4 \times 10^{-6})}}{2(1)} = \frac{-5.2 \times 10^{-4} \pm 6.2 \times 10^{-3}}{2}$$

$$= 2.8 \times 10^{-3} \text{ or } -3.4 \times 10^{-3} \text{ (discard)}$$

Therefore, $[OH^-] = x =$ **2.8 x 10^{-3} M**

(2) For trimethylamine, $(CH_3)_3N$

Balanced equation: $(CH_3)_3N + H_2O \rightleftarrows (CH_3)_3NH^+ + OH^-$ $\qquad K_b = 7.4 \times 10^{-5}$

Let $x = [(CH_3)_3N]$ that ionizes. Then $x = [(CH_3)_3NH^+]$ produced $= [OH^-]$ produced.

$K_b = \dfrac{[(CH_3)_3NH^+][OH^-]}{[(CH_3)_3N]} = 7.4 \times 10^{-5} = \dfrac{x^2}{0.018 - x} \approx \dfrac{x^2}{0.018}$ $\qquad$ Solving, $x = 1.2 \times 10^{-3}$

In this case, 1.2 x 10^{-3} is greater than 5% of 0.018 (= 9.0 x 10^{-4}). A simplifying assumption cannot be made and the original quadratic equation must be solved: $x^2 + (7.4 \times 10^{-5})x - 1.3 \times 10^{-6} = 0$

$$x = \frac{-(7.4 \times 10^{-5}) \pm \sqrt{(7.4 \times 10^{-5})^2 - 4(1)(-1.3 \times 10^{-6})}}{2(1)}$$

$$= \frac{-7.4 \times 10^{-5} \pm 2.3 \times 10^{-3}}{2}$$

$$= 1.1 \times 10^{-3} \text{ or } -1.2 \times 10^{-3} \text{ (discard)}$$

Therefore, $[OH^-] = x =$ **1.1 x 10^{-3} M**

Yes, $[OH^-]$ in triethylamine(aq) is greater than $[OH^-]$ in trimethylamine(aq) for the same concentration.

Note: Even if the simplifying assumption may not always be applicable, it is a good idea to always use it in the beginning for several reasons. (1) It is quick and easy to use, and (2) it usually works. (3) Even when it does not yield the correct answer, it *usually* gives a reasonable estimate of the true value.

18-34. *Refer to Sections 18-2 and 18-3, and Table 18-2.*

(a) K_w at 37°C = 2.4 x 10^{-14} = $[H_3O^+][OH^-]$.   For pure water, $[H_3O^+] = [OH^-]$

$[H_3O^+] = \sqrt{K_w} = \sqrt{2.4 \times 10^{-14}} = 1.5 \times 10^{-7}$ at 37°C

pH = $-\log [H_3O^+]$ = $-\log (1.5 \times 10^{-7})$ = **6.82 at 37°C**

(b) Pure water at 37°C is **neutral** since $[H_3O^+] = [OH^-]$. The pH of a neutral aqueous solution is 7.0 only at 25°C.

18-36. *Refer to Section 18-4 and Exercise 18-30 Solution.*

Reaction of proton-accepting weak base, B, with water: $B(aq) + H_2O(\ell) \rightleftarrows BH^+(aq) + OH^-(aq)$

In Chapter 17 (Section 17-11), we stated that for heterogeneous equilibria, terms for pure liquids and pure solids do not appear in K expressions because their activity values are essentially 1. In the discussion in Section 18-4, these terms are temporarily included. In dilute solutions, activities are equal to concentrations.

$$K = \frac{(BH^+)(OH^-)}{(B)(H_2O)} \quad \text{where (x) represents the activity of x} \quad K_b = \frac{[BH^+][OH^-]}{[B](1)}$$

K_b, the ionization constant for bases, is used for calculations involving the equilibria of weak bases.

18-38. *Refer to Section 18-4 and Example 18-8.*

Balanced equation: $HX + H_2O \rightleftarrows H_3O^+ + X^-$

Since HX is 1.07% ionized, $[HX]_{reacted}$ = 0.0750 M x 0.0107 = 8.02 x 10^{-4} M

	HX	+	H_2O	$\rightleftarrows$	H_3O^+	+	X^-
initial	0.0750 M				$\approx 0\ M$		0 M
change	- 8.02 x 10^{-4} M				+ 8.02 x 10^{-4} M		+ 8.02 x 10^{-4} M
at equilibrium	0.0742 M				8.02 x 10^{-4} M		8.02 x 10^{-4} M

pH = $-\log (8.02 \times 10^{-4})$ = **3.096** $K_a = \dfrac{[H_3O^+][X^-]}{[HX]} = \dfrac{(8.02 \times 10^{-4})^2}{0.0742} = $ **8.67 x 10^{-6}**

18-40. *Refer to Section 18-4 and Example 18-9.*

Balanced equation: $ClCH_2COOH + H_2O \rightleftarrows H_3O^+ + ClCH_2COO^-$

Let   x = mol/L of $ClCH_2COOH$ that reacts. Then
       x = mol/L of H_3O^+ produced = mol/L of $ClCH_2COO^-$ produced.

Since pH = 1.95, $[H_3O^+]$ = x = antilog (-1.95) = 0.011 M

	$ClCH_2COOH$	+	H_2O	$\rightleftarrows$	H_3O^+	+	$ClCH_2COO^-$
initial	0.10 M				$\approx 0\ M$		0 M
change	- x M				+ x M		+ x M
at equilibrium	(0.10 - x) M				x M		x M

$K_a = \dfrac{[H_3O^+][ClCH_2COO^-]}{[ClCH_2COOH]} = \dfrac{x^2}{0.10 - x} = \dfrac{(0.011)^2}{0.10 - 0.011} = $ **1.4 x 10^{-3}**

18-42. *Refer to Section 18-4, Example 18-10 and Appendix F.*

Balanced equation: $C_6H_5COOH + H_2O \rightleftarrows H_3O^+ + C_6H_5COO^-$ $K_a = 6.3 \times 10^{-5}$

Let   x = mol/L of C_6H_5COOH that reacts. Then

     x = mol/L of H_3O^+ produced = mol/L of $C_6H_5COO^-$ produced.

	C_6H_5COOH	+	H_2O	$\rightleftarrows$	H_3O^+	+	$C_6H_5COO^-$
initial	0.52 M				≈ 0 M		0 M
change	- x M				+ x M		+ x M
at equilibrium	(0.52 - x) M				x M		x M

$$K_a = \frac{[H_3O^+][C_6H_5COO^-]}{[C_6H_5COOH]} = \frac{x^2}{0.52 - x} = 6.3 \times 10^{-5}$$

Assume that $0.52 - x \approx 0.52$. Then $x^2/0.52 = 6.3 \times 10^{-5}$ and $x = 5.7 \times 10^{-3}$. The simplifying assumption is justified since 5.7×10^{-3} is much less than 5% of 0.52 (= 0.026). However, in this example, it does make a slight difference in the concentration of C_6H_5COOH.

Therefore at equilibrium: $[C_6H_5COOH] = \mathbf{0.51}$ **M**

                     $[H_3O^+] = [C_6H_5COO^-] = x = \mathbf{5.7 \times 10^{-3}}$ **M**

                     $[OH^-] = K_w/[H_3O^+] = \mathbf{1.8 \times 10^{-12}}$ **M**

18-44. *Refer to Section 18-4, Example 18-10, and Appendix F.*

Balanced equation: $HF + H_2O \rightleftarrows H_3O^+ + F^-$ $K_a = 7.2 \times 10^{-4}$

Let   x = mol/L of HF that reacts. Then

     x = mol/L of H_3O^+ produced = mol/L of F^- produced.

	HF	+	H_2O	$\rightleftarrows$	H_3O^+	+	F^-
initial	0.38 M				≈ 0 M		0 M
change	- x M				+ x M		+ x M
at equilibrium	(0.38 - x) M				x M		x M

$$K_a = \frac{[H_3O^+][F^-]}{[HF]} = \frac{x^2}{0.38 - x} = 7.2 \times 10^{-4}$$

Assume that $0.38 - x \approx 0.38$. Then $x^2/0.38 = 7.2 \times 10^{-4}$ and $x = 0.017$. The simplifying assumption is justified since 0.017 is less than 5% of 0.38 (= 0.019).

Therefore at equilibrium: $[H_3O^+] = 0.017$ M   pH = **1.78**

18-46. *Refer to Section 18-4, Example 18-15, Table 18-6 and Appendix G.*

Balanced equation: $NH_3 + H_2O \rightleftarrows NH_4^+ + OH^-$ $K_b = 1.8 \times 10^{-5}$

Let $x = [NH_3]$ that ionizes. Then

     $x = [NH_4^+]$ produced = $[OH^-]$ produced.

	NH_3	+	H_2O	$\rightleftarrows$	NH_4^+	+	OH^-
initial	2.05 M				0 M		≈ 0 M
change	- x M				+ x M		+ x M
at equilibrium	(2.05 - x) M				x M		x M

$$K_b = \frac{[NH_4^+][OH^-]}{[NH_3]} = 1.8 \times 10^{-5} = \frac{x^2}{2.05 - x} \approx \frac{x^2}{2.05}$$

                                         Solving, $x = 6.1 \times 10^{-3}$ (6.07×10^{-3})

Since 6.1×10^{-3} is less than 5% of 2.05, the approximation is justified.

Therefore, $[OH^-] = 6.1 \times 10^{-3}$ M $pH = -\log \frac{K_w}{[OH^-]} = -\log\left(\frac{1.0 \times 10^{-14}}{6.1 \times 10^{-3}}\right) = \mathbf{11.79}$

         $pOH = -\log (6.07 \times 10^{-3}$ M$) = \mathbf{2.21}$ or   pH = 14.00 - pOH = 14.00 - 2.21 = **11.79**

Note: Keep all the significant figures and round at the end. Remember the number of decimal places in pH or pOH values are set by the number of significant figures in the [H$^+$] or [OH$^-$]; this is a result of working with logarithms.

18-48. *Refer to Section 18-4, Examples 18-10 and 18-11, and Appendix F.*

Balanced equation: $HCOOH + H_2O \rightleftarrows H_3O^+ + HCOO^-$ $\qquad$ $K_a = 1.8 \times 10^{-4}$

Let $x = [HCOOH]$ that ionizes. Then
$\qquad x = [H_3O^+]$ produced $= [HCOO^-]$ produced.

	HCOOH	+	H$_2$O	$\rightleftarrows$	H$_3$O$^+$	+	HCOO$^-$
initial	0.0751 *M*				≈ 0 *M*		0 *M*
change	- x *M*				+ x *M*		+ x *M*
at equilibrium	(0.0751 - x) *M*				x *M*		x *M*

$$K_a = \frac{[H_3O^+][HCOO^-]}{[HCOOH]} = \frac{x^2}{0.0751 - x} = 1.8 \times 10^{-4}$$

Assuming $0.0751 - x \approx 0.0751$, then $x^2/0.0751 = 1.8 \times 10^{-4}$ and $x = 3.7 \times 10^{-3}$. The simplifying assumption is justified since 3.7×10^{-3} is 4.9% of 0.0751.

$$\% \text{ ionization} = \frac{[HCOOH]_{ionized}}{[HCOOH]_{initial}} \times 100\% = \frac{3.7 \times 10^{-3} \ M}{0.0751 \ M} \times 100\% = \mathbf{4.9\%}$$

18-50. *Refer to Section 18-4 and Example 18-12.*

For Weak Acid #1 $\qquad pK_a = -\log K_a = -\log (7.2 \times 10^{-5}) = \mathbf{4.14}$
For Weak Acid #2 $\qquad pK_a = -\log K_a = -\log (4.2 \times 10^{-10}) = \mathbf{9.38}$

18-52. *Refer to Section 18-4.*

Balanced equation: $C_5H_5N(aq) + H_2O(\ell) \rightleftarrows C_5H_5NH^+(aq) + OH^-(aq)$

Since a 0.00500 *M* C_5H_5N solution is 0.053% ionized, $[C_5H_5N]_{reacted} = 0.00500 \ M \times 0.00053 = 2.7 \times 10^{-6} \ M$

	C$_5$H$_5$N	+	H$_2$O	$\rightleftarrows$	C$_5$H$_5$NH$^+$	+	OH$^-$
initial	0.00500 *M*				0 *M*		≈ 0 *M*
change	- 2.7 x 10^{-6} *M*				+ 2.7 x 10^{-6} *M*		+ 2.7 x 10^{-6} *M*
at equilibrium	0.00500 *M*				2.7 x 10^{-6} *M*		2.7 x 10^{-6} *M*

$$K_b = \frac{[C_5H_5NH^+][OH^-]}{[C_5H_5N]} = \frac{(2.7 \times 10^{-6})^2}{0.00500} = 1.5 \times 10^{-9}$$

$pK_b = -\log (1.5 \times 10^{-9}) = 8.82$ $\quad$ This agrees with the K_b value given in Appendix G and Table 18-6.

18-54. *Refer to Section 18-4.*

Balanced equation: $CH_3NH_2(aq) + H_2O(\ell) \rightleftarrows CH_3NH_3^+(aq) + OH^-(aq)$

$$K_b = \frac{[CH_3NH_3^+][OH^-]}{[CH_3NH_2]} = \frac{(2.0 \times 10^{-3})^2}{0.0080} = \mathbf{5.0 \times 10^{-4}}$$

This agrees with the K_b value given in Appendix G and Table 18-6.

18-56. *Refer to Section 18-4, Example 18-15, Table 18-6 and Appendix G.*

(a) Balanced equation: $NH_3 + H_2O \rightleftarrows NH_4^+ + OH^-$ $\qquad$ $K_b = 1.8 \times 10^{-5}$

Let $x = [NH_3]$ that ionizes. Then
$\qquad x = [NH_4^+]$ produced $= [OH^-]$ produced.

	NH_3	+	H_2O	$\rightleftarrows$	NH_4^+	+	OH^-
initial	0.25 M				0 M		$\approx 0\ M$
change	- x M				+ x M		+ x M
at equilibrium	(0.25 - x) M				x M		x M

$$K_b = \frac{[NH_4^+][OH^-]}{[NH_3]} = 1.8 \times 10^{-5} = \frac{x^2}{0.25 - x} \approx \frac{x^2}{0.25} \qquad \text{Solving, x} = 2.1 \times 10^{-3}$$

Since 2.1×10^{-3} is less than 5% of 0.25, the approximation is justified.

Therefore, $[OH^-] = \mathbf{2.1 \times 10^{-3}\ M}$

$$\% \text{ ionization} = \frac{[NH_3]_{\text{ionized}}}{[NH_3]_{\text{initial}}} \times 100\% = \frac{2.1 \times 10^{-3}\ M}{0.25\ M} \times 100\% = \mathbf{0.85\%}$$

$$pH = -\log \frac{K_w}{[OH^-]} = -\log\left(\frac{1.0 \times 10^{-14}}{2.1 \times 10^{-3}}\right) = \mathbf{11.32}$$

(b) Balanced equation: $CH_3NH_2 + H_2O \rightleftarrows CH_3NH_3^+ + OH^- \qquad K_b = 5.0 \times 10^{-4}$
Let   x = $[CH_3NH_2]$ that ionizes. Then
x = $[CH_3NH_3^+]$ produced = $[OH^-]$ produced.

	CH_3NH_2	+	H_2O	$\rightleftarrows$	$CH_3NH_3^+$	+	OH^-
initial	0.25 M				0 M		$\approx 0\ M$
change	- x M				+ x M		+ x M
at equilibrium	(0.25 - x) M				x M		x M

$$K_b = \frac{[CH_3NH_3^+][OH^-]}{[CH_3NH_2]} = 5.0 \times 10^{-4} = \frac{x^2}{0.25 - x} \approx \frac{x^2}{0.25} \qquad \text{Solving, x} = 0.011$$

Since 0.011 is less than 5% of 0.25, the approximation is justified.

Therefore, $[OH^-] = \mathbf{0.011\ M}$

$$\% \text{ ionization} = \frac{[CH_3NH_2]_{\text{ionized}}}{[CH_3NH_2]_{\text{initial}}} \times 100\% = \frac{0.011\ M}{0.25\ M} \times 100\% = \mathbf{4.4\%}$$

$$pH = -\log \frac{K_w}{[OH^-]} = -\log\left(\frac{1.0 \times 10^{-14}}{0.011}\right) = \mathbf{12.04}$$

18-58.   *Refer to Section 18-5, Example 18-17, Table 18-7 and Appendix F.*

Balanced equations: $H_3AsO_4 + H_2O \rightleftarrows H_3O^+ + H_2AsO_4^- \qquad K_{a1} = 2.5 \times 10^{-4}$
$\qquad\qquad\qquad\qquad\quad H_2AsO_4^- + H_2O \rightleftarrows H_3O^+ + HAsO_4^{2-} \qquad K_{a2} = 5.6 \times 10^{-8}$
$\qquad\qquad\qquad\qquad\quad HAsO_4^{2-} + H_2O \rightleftarrows H_3O^+ + AsO_4^{3-} \qquad K_{a3} = 3.0 \times 10^{-13}$

First Step:
Let x = $[H_3AsO_4]_{\text{ionized}}$.   Then $[H_3AsO_4]$ = (0.100 - x) M
$\qquad\qquad\qquad\qquad\qquad\qquad\qquad [H_3O^+] = [H_2AsO_4^-] = x\ M$

$$K_{a1} = \frac{[H_3O^+][H_2AsO_4^-]}{[H_3AsO_4]} = \frac{x^2}{(0.100 - x)} = 2.5 \times 10^{-4} \approx \frac{x^2}{0.100} \qquad \text{Solving, x} = 5.0 \times 10^{-3}$$

Since 5.0×10^{-3} is 5% of 0.100, the simplifying assumption is valid.

Therefore, $[H_3O^+] = [H_2AsO_4^-] = x = 5.0 \times 10^{-3}\ M$, $[H_3AsO_4] = 0.100 - x = 0.095\ M$

Second Step:
Let y = $[H_2AsO_4^-]_{\text{ionized}}$.   Then $[H_2AsO_4^-]$ = (5.0 $\times 10^{-3}$ - y) M
$\qquad\qquad\qquad\qquad\qquad\qquad\qquad [H_3O^+]$ = (5.0 $\times 10^{-3}$ + y) M
$\qquad\qquad\qquad\qquad\qquad\qquad\qquad [HAsO_4^{2-}]$ = y M

$$K_{a2} = \frac{[H_3O^+][HAsO_4^{2-}]}{[H_2AsO_4^-]} = \frac{(5.0 \times 10^{-3} + y)(y)}{(5.0 \times 10^{-3} - y)} = 5.6 \times 10^{-8} \approx \frac{(5.0 \times 10^{-3})y}{(5.0 \times 10^{-3})}$$
Solving, y = 5.6×10^{-8}

Therefore, because the simplifying assumptions are valid, $[H_3O^+] = [H_2AsO_4^-] = 5.0 \times 10^{-3}\ M$
$\qquad\qquad\qquad\qquad\qquad\qquad\qquad\qquad\qquad\qquad\qquad\qquad\qquad [HAsO_4^{2-}] = 5.6 \times 10^{-8}\ M$

Third Step:

Let $z = [HAsO_4^{2-}]_{ionized}$.   Then $[HAsO_4^{2-}] = (5.6 \times 10^{-8} - z)\ M$

$$[H_3O^+] = (5.0 \times 10^{-3} + z)\ M$$
$$[AsO_4^{3-}] = z\ M$$

$$K_{a3} = \frac{[H_3O^+][AsO_4^{3-}]}{[HAsO_4^-]} = \frac{(5.0 \times 10^{-3} + z)(z)}{(5.6 \times 10^{-8} - z)} = 3.0 \times 10^{-13} = \frac{(5.0 \times 10^{-3})z}{(5.6 \times 10^{-8})}$$ Solving, $z = 3.4 \times 10^{-18}$

Therefore, $[AsO_4^{3-}] = 3.4 \times 10^{-18}\ M$ because the simplifying assumptions are valid.

0.100 M H$_3$AsO$_4$ Solution		0.100 M H$_3$PO$_4$ Solution	
Species	Concentration (M)	Species	Concentration (M)
H_3AsO_4	0.095	H_3PO_4	0.076
H_3O^+	0.0050	H_3O^+	0.024
$H_2AsO_4^-$	0.0050	$H_2PO_4^-$	0.024
$HAsO_4^{2-}$	5.6×10^{-8}	HPO_4^{2-}	6.2×10^{-8}
OH^-	2.0×10^{-12}	OH^-	4.2×10^{-13}
AsO_4^{3-}	3.4×10^{-18}	PO_4^{3-}	9.3×10^{-19}

18-60. *Refer to Section 18-5, Example 18-18 and Appendix F.*

Balanced equations: $H_2SeO_4 + H_2O \rightleftarrows H_3O^+ + HSeO_4^-$ K_{a1} = very large

$HSeO_4^- + H_2O \rightleftarrows H_3O^+ + SeO_4^{2-}$ $K_{a2} = 1.2 \times 10^{-2}$

First Step:

Since K_{a1} is very large, H_2SeO_4 is a strong electrolyte and totally dissociates into H_3O^+ and $HSeO_4^-$.

Therefore, $[H_2SeO_4] \approx 0\ M$

$$[HSeO_4^-] = 0.15\ M$$
$$[H_3O^+] = 0.15\ M$$

Second Step:

Let $x = [HSeO_4^-]_{ionized}$.   Then $[HSeO_4^-] = (0.15 - x)\ M$

$$[H_3O^+] = (0.15 + x)\ M$$
$$[SeO_4^{2-}] = x\ M$$

$$K_{a2} = \frac{[H_3O^+][SeO_4^{2-}]}{[HSeO_4^-]} = \frac{(0.15 + x)(x)}{(0.15 - x)} = 1.2 \times 10^{-2} \approx \frac{(0.15)x}{(0.15)}$$ Solving, $x = 1.2 \times 10^{-2}$

In this case, 1.2×10^{-2} is greater than 5% of 0.15 (= 7.5×10^{-3}). A simplifying assumption cannot be made and the original quadratic equation must be solved: $x^2 + 0.16x - 1.8 \times 10^{-3} = 0$

$$x = \frac{-0.16 \pm \sqrt{(0.16)^2 - 4(1)(-1.8 \times 10^{-3})}}{2(1)} = \frac{-0.16 \pm 0.18}{2} = 0.01 \text{ or } -0.17 \text{ (discard)}$$

Therefore, $[H_3O^+] = 0.15 + x = \mathbf{0.16\ M}$

$$[OH^-] = K_w/[H_3O^+] = \mathbf{6.2 \times 10^{-14}\ M}$$
$$[HSeO_4^-] = 0.15 - x = \mathbf{0.14\ M}$$
$$[SeO_4^{2-}] = x = \mathbf{0.01\ M}$$

18-62. *Refer to Section 18-5, Example 18-17 and Appendix F.*

Balanced equations: $(COOH)_2 + H_2O \rightleftarrows H_3O^+ + COOCOOH^-$ $K_{a1} = 5.9 \times 10^{-2}$

$COOCOOH^- + H_2O \rightleftarrows H_3O^+ + (COO)_2^{2-}$ $K_{a2} = 6.4 \times 10^{-5}$

Let $x = [(COOH)_2]_{ionized}$.   Then $[(COOH)_2] = (0.045 - x)\ M$

$$[H_3O^+] = [COOCOOH^-] = x\ M$$

$$K_1 = \frac{[H_3O^+][COOCOOH^-]}{[(COOH)_2]} = \frac{x^2}{(0.045 - x)} = 5.9 \times 10^{-2} \approx \frac{x^2}{0.045}$$ Solving, $x = 0.052$

Since 0.052 is more than 100% of 0.045, the simplifying assumptions do not hold and we must solve the original quadratic equation: $x^2 + (5.9 \times 10^{-2})x - 2.7 \times 10^{-3} = 0$

$$x = \frac{-5.9 \times 10^{-2} \pm \sqrt{(5.9 \times 10^{-2})^2 - 4(1)(-2.7 \times 10^{-3})}}{2(1)} = \frac{-5.9 \times 10^{-2} \pm 0.12}{2} = 0.030 \text{ or } -0.090 \text{(discard)}$$

Therefore, $[(COOH)_2] = 0.045 - x = 0.045 - 0.030 = 0.015\ M$

$[H_3O^+] = [COOCOOH^-] = x = 0.030\ M$

$pH = -\log(0.030) = \textbf{1.52}$

Note: $[H_3O^+]$ from the second ionization is equal to the value of $K_{a2} = 6.4 \times 10^{-5}$. Therefore, the $[H_3O^+]$ furnished by the second ionization is negligible compared to that from the first ionization and was ignored.

18-64. *Refer to Section 18-6.*

(a) Solvolysis is the reaction of a substance with the solvent in which it is dissolved. Common solvents used include $H_2O(\ell)$, $NH_3(\ell)$, $H_2SO_4(\ell)$ and $CH_3COOH(\ell)$. There are many others. For example, glacial acetic acid, $CH_3COOH(\ell)$, is commonly used in non-aqueous titrations with weak acids:

$$C_6H_5NH_3^+ + CH_3COOH \rightleftarrows CH_3COOH_2^+ + C_6H_5NH_2$$
solute solvent aniline

(b) Hydrolysis is the reaction of a substance with the solvent, water, or its ions, OH^- and H_3O^+, e.g., the hydrolysis of the ion, CH_3COO^- to give a basic solution:

$$CH_3COO^- + H_2O \rightleftarrows CH_3COOH + OH^-$$
solute solvent

18-66. *Refer to Section 18-7.*

Some cations in aqueous solution, such as Na^+, K^+, and Ba^{2+}, undergo no real significant reaction with the water molecules in a solution. This is because their acid strengths are much less than the acid strength of water. In other words, these cations are such weak acids that they do not react with water. Because of this, when these cations are dissolved in water, the pH is unchanged.

18-68. *Refer to Sections 18-7, 18-8, 18-9 and 18-10.*

The pH of aqueous salt solutions depends on whether or not the ions produced by the dissociation of the salt will hydrolyze (react with water). If the cation of the salt hydrolyzes more than the anion, the solution is acidic. If the anion hydrolyzes more than the cation, the solution is basic. If the cation and the anion hydrolyze to the same extent or if neither hydrolyzes appreciably, the resulting solution is neutral.

18-70. *Refer to Section 18-7.*

Salts produced by a strong base reacting with a strong acid are

(d) $BaSO_4$ ($Ba(OH)_2$ with H_2SO_4) and

(e) $NaClO_3$ (NaOH with $HClO_3$).

Note: in dilute solution, H_2SO_4 is completely ionized and is considered to be a strong diprotic acid.

18-72. *Refer to Section 18-8.*

The solution of a salt derived from a strong base and weak acid is basic because the anion of a weak acid reacts with water (hydrolysis) to form hydroxide ions. Consider the soluble salt NaClO found in chlorine bleaches prepared by reacting NaOH, a strong base, and HClO, a weak acid. The salt dissociates completely in water and the conjugate base of the weak acid, ClO^-, hydrolyzes, producing OH^- ions.

$$NaClO(s) \xrightarrow[100\%]{H_2O} Na^+(aq) + ClO^-(aq)$$
$$ClO^-(aq) + H_2O(\ell) \rightleftarrows HClO(aq) + OH^-(aq)$$

18-74. *Refer to Section 18-8, Example 18-19 and Appendix F.*

Balanced equation: $N_3^-(aq) + H_2O(\ell) \rightleftarrows HN_3(aq) + OH^-(aq)$ (K_a for hydrazoic acid, $HN_3 = 1.9 \times 10^{-5}$)

$$K = K_b = \frac{[HN_3][OH^-]}{[N_3^-]} = \frac{K_w}{K_{a(HN_3)}} = \frac{1.0 \times 10^{-14}}{1.9 \times 10^{-5}} = \mathbf{5.3 \times 10^{-10}}$$

18-76. *Refer to Section 18-8, Example 18-19 and Appendix F.*

(a) for NO_2^-, $K_b = \dfrac{K_w}{K_{a(HNO_2)}} = \dfrac{1.0 \times 10^{-14}}{4.5 \times 10^{-4}} = \mathbf{2.2 \times 10^{-11}}$

(b) for BrO^-, $K_b = \dfrac{K_w}{K_{a(HBrO)}} = \dfrac{1.0 \times 10^{-14}}{2.5 \times 10^{-9}} = \mathbf{4.0 \times 10^{-6}}$

(c) for $HCOO^-$, $K_b = \dfrac{K_w}{K_{a(HCOOH)}} = \dfrac{1.0 \times 10^{-14}}{1.8 \times 10^{-4}} = \mathbf{5.6 \times 10^{-11}}$

The mathematical relationship between K_a, the ionization constant for a weak acid, and K_b, the base hydrolysis constant for the anion of the weak acid is $K_w = K_a \times K_b$.

The weaker the acid, the smaller is its K_a, the more its anion will hydrolyze, and the larger is K_b for the anion.

18-78. *Refer to Section 18-8, Example 18-20, and Appendix F.*

(a) Balanced equations: $NaCH_3COO \rightarrow Na^+ + CH_3COO^-$ (to completion)
                          $CH_3COO^- + H_2O \rightleftarrows CH_3COOH + OH^-$ (reversible)

Let $x = [CH_3COO^-]_{hydrolyzed}$ Then, $1.5 - x = [CH_3COO^-]$; $x = [CH_3COOH] = [OH^-]$

$K_b = \dfrac{K_w}{K_a} = \dfrac{1.0 \times 10^{-14}}{1.8 \times 10^{-5}} = 5.6 \times 10^{-10} = \dfrac{[CH_3COOH][OH^-]}{[CH_3COO^-]} = \dfrac{x^2}{1.5 - x} \approx \dfrac{x^2}{1.5}$ Solving, $x = 2.9 \times 10^{-5}$

$[OH^-] = 2.9 \times 10^{-5}$ M; pOH = 4.54; pH = 14 - 4.54 = **9.46**

(b) Balanced equations: $KOBr \rightarrow K^+ + OBr^-$ (to completion)
                          $OBr^- + H_2O \rightleftarrows HOBr + OH^-$ (reversible)

Let $x = [OBr^-]_{hydrolyzed}$ Then, $1.5 - x = [OBr^-]$; $x = [HOBr] = [OH^-]$

$K_b = \dfrac{K_w}{K_a} = \dfrac{1.0 \times 10^{-14}}{2.5 \times 10^{-9}} = 4.0 \times 10^{-6} = \dfrac{[HOBr][OH^-]}{[OBr^-]} = \dfrac{x^2}{1.5 - x} \approx \dfrac{x^2}{1.5}$ Solving, $x = 2.4 \times 10^{-3}$

$[OH^-] = 2.4 \times 10^{-3}$ M; pOH = 2.61; pH = 14 - 2.61 = **11.39**

(c) Balanced equations: $LiCN \rightarrow Li^+ + CN^-$ (to completion)
                          $CN^- + H_2O \rightleftarrows HCN + OH^-$ (reversible)

Let $x = [CN^-]_{hydrolyzed}$ Then, $1.5 - x = [CN^-]$; $x = [HCN] = [OH^-]$

$K_b = \dfrac{K_w}{K_a} = \dfrac{1.0 \times 10^{-14}}{4.0 \times 10^{-10}} = 2.5 \times 10^{-5} = \dfrac{[HCN][OH^-]}{[CN^-]} = \dfrac{x^2}{1.5 - x} \approx \dfrac{x^2}{1.5}$ Solving, $x = 6.1 \times 10^{-3}$

$[OH^-] = 6.1 \times 10^{-3}$ M; pOH = 2.21; pH = 14 - 2.21 = **11.79**

(a) Balanced equations: $KOI \rightarrow K^+ + OI^-$ (to completion)

$OI^- + H_2O \rightleftarrows HOI + OH^-$ (reversible)

Let x = $[OI^-]_{hydrolyzed}$ Then, 0.18 - x = $[OI^-]$; x = $[HOI]$ = $[OH^-]$

$K_b = \dfrac{K_w}{K_a} = \dfrac{1.0 \times 10^{-14}}{2.3 \times 10^{-11}} = 4.3 \times 10^{-4} = \dfrac{[HOI][OH^-]}{[OI^-]} = \dfrac{x^2}{0.18 - x} \approx \dfrac{x^2}{0.18}$ Solving, x = 8.8×10^{-3}

Since 8.8×10^{-3} is less than 5% of 0.18, the approximation is justified.

Therefore, $[OH^-] = 8.8 \times 10^{-3}$ *M*; pOH = 2.06; pH = 14 - 2.06 = **11.94**

(b) Balanced equations: $KF \rightarrow K^+ + F^-$ (to completion)

$F^- + H_2O \rightleftarrows HF + OH^-$ (reversible)

Let x = $[F^-]_{hydrolyzed}$ Then, 0.18 - x = $[F^-]$; x = $[HF]$ = $[OH^-]$

$K_b = \dfrac{K_w}{K_a} = \dfrac{1.0 \times 10^{-14}}{7.2 \times 10^{-4}} = 1.4 \times 10^{-11} = \dfrac{[HF][OH^-]}{[F^-]} = \dfrac{x^2}{0.18 - x} \approx \dfrac{x^2}{0.18}$ Solving, x = 1.6×10^{-6}

In this case, 1.6×10^{-6} is much less than 5% of 0.18 (= 9.0×10^{-3}). The simplifying assumption can be made.

Therefore, $[OH^-] = 1.6 \times 10^{-6}$ *M*; pOH = 5.80; pH = 14 - 5.80 = **8.20**

NH_4Cl	ammonium chloride	$CH_3NH_3NO_3$	methylammonium nitrate
C_5H_5NHBr	pyridinium bromide	$[(CH_3)_3NH]_2SO_4$	trimethylammonium sulfate

(a) for $(CH_3)_2NH_2^+$, $K_a = \dfrac{K_w}{K_{b((CH_3)_2NH)}} = \dfrac{1.0 \times 10^{-14}}{7.4 \times 10^{-4}} = \mathbf{1.4 \times 10^{-11}}$

(b) for $C_5H_5NH^+$, $K_a = \dfrac{K_w}{K_{b(C_5H_5N)}} = \dfrac{1.0 \times 10^{-14}}{1.5 \times 10^{-9}} = \mathbf{6.7 \times 10^{-6}}$

(c) for $(CH_3)_3NH^+$, $K_a = \dfrac{K_w}{K_{b((CH_3)_3N)}} = \dfrac{1.0 \times 10^{-14}}{7.4 \times 10^{-5}} = \mathbf{1.4 \times 10^{-10}}$

(a) Balanced equations: $NH_4NO_3 \rightarrow NH_4^+ + NO_3^-$ (to completion)

$NH_4^+ + H_2O \rightleftarrows NH_3 + H_3O^+$ (reversible)

Let x = $[NH_4^+]_{hydrolyzed}$ Then, 0.26 - x = $[NH_4^+]$; x = $[NH_3]$ = $[H_3O^+]$

$K_a = \dfrac{K_w}{K_{b(NH_3)}} = \dfrac{1.0 \times 10^{-14}}{1.8 \times 10^{-5}} = 5.6 \times 10^{-10} = \dfrac{[NH_3][H_3O^+]}{[NH_4^+]} = \dfrac{x^2}{0.26 - x} \approx \dfrac{x^2}{0.26}$

Solving, x = 1.2×10^{-5} Therefore, $[H_3O^+]$ = 1.2×10^{-5} *M*; pH = **4.92**

(b) Balanced equations: $CH_3NH_3NO_3 \rightarrow CH_3NH_3^+ + NO_3^-$ (to completion)

$CH_3NH_3^+ + H_2O \rightleftarrows CH_3NH_2 + H_3O^+$ (reversible)

Let x = $[CH_3NH_3^+]_{hydrolyzed}$ Then, 0.26 - x = $[CH_3NH_3^+]$; x = $[CH_3NH_2]$ = $[H_3O^+]$

$K_a = \dfrac{K_w}{K_{b(CH_3NH_2)}} = \dfrac{1.0 \times 10^{-14}}{5.0 \times 10^{-4}} = 2.0 \times 10^{-11} = \dfrac{[CH_3NH_2][H_3O^+]}{[CH_3NH_3^+]} = \dfrac{x^2}{0.26 - x} \approx \dfrac{x^2}{0.26}$

Solving, x = 2.3×10^{-6} Therefore, $[H_3O^+]$ = 2.3×10^{-6} *M*; pH = **5.64**

(c) Balanced equations: $C_6H_5NH_3NO_3 \rightarrow C_6H_5NH_3^+ + NO_3^-$ (to completion)

                                $C_6H_5NH_3^+ + H_2O \rightleftarrows C_6H_5NH_2 + H_3O^+$ (reversible)

Let $x = [C_6H_5NH_3^+]_{hydrolyzed}$ Then, $0.26 - x = [C_6H_5NH_3^+]$; $x = [C_6H_5NH_2] = [H_3O^+]$

$$K_a = \frac{K_w}{K_{b(C_6H_5NH_2)}} = \frac{1.0 \times 10^{-14}}{4.2 \times 10^{-10}} = 2.4 \times 10^{-5} = \frac{[C_6H_5NH_2][H_3O^+]}{[C_6H_5NH_3^+]} = \frac{x^2}{0.26 - x} \approx \frac{x^2}{0.26}$$

Solving, $x = 2.5 \times 10^{-3}$ Therefore, $[H_3O^+] = 2.5 \times 10^{-3}$ M; pH = **2.60**

18-88.   *Refer to Section 18-10 and Appendices F and G.*

(a) Salt of a weak acid and a weak base for which $K_a = K_b$ gives a neutral solution, e.g., ammonium acetate, NH_4CH_3COO. $K_{a(CH_3COOH)} = K_{b(NH_3)} = 1.8 \times 10^{-5}$.

(b) Salts of a weak acid and weak base for which $K_a > K_b$ gives an acidic solution, e.g., pyridinium fluoride, C_5H_5NHF. $K_{a(HF)} = 7.2 \times 10^{-4}$; $K_{b(C_5H_5N)} = 1.5 \times 10^{-9}$.

(c) A salts of a weak acid and weak base for which $K_b > K_a$ gives a basic solution, e.g., methylammonium cyanide, CH_3NH_3CN. $K_{a(HCN)} = 4.0 \times 10^{-10}$; $K_{b(CH_3NH_2)} = 5.0 \times 10^{-4}$.

18-90.   *Refer to Section 18-11 and Table 18-10.*

The cations that will reaction with water to form H^+ (or H_3O^+) ions are

(a) $[Be(OH_2)_4]^{2+} + H_2O \rightleftarrows [Be(OH)(OH_2)_3]^+ + H_3O^+$

(b) $[Al(OH_2)_6]^{3+} + H_2O \rightleftarrows [Al(OH)(OH_2)_5]^{2+} + H_3O^+$

(c) $[Fe(OH_2)_6]^{3+} + H_2O \rightleftarrows [Fe(OH)(OH_2)_5]^{2+} + H_3O^+$

(d) $[Cu(OH_2)_6]^{2+} + H_2O \rightleftarrows [Cu(OH)(OH_2)_5]^+ + H_3O^+$

18-92.   *Refer to Section 18-11, Example 18-23 and Table 18-10.*

(a) Balanced equation: $[Al(OH_2)_6]^{3+} + H_2O \rightleftarrows [Al(OH)(OH_2)_5]^{2+} + H_3O^+$ $K_a = 1.2 \times 10^{-5}$

Assume that the aluminum salt totally dissociated and all the Al^{3+} became $[Al(OH_2)_6]^{3+}$.

Let $x = [[Al(OH_2)_6]^{3+}]_{hydrolyzed}$ Then, $0.15 - x = [[Al(OH_2)_6]^{3+}]$; $x = [[Al(OH)(OH_2)_5]^{2+}] = [H_3O^+]$

$$K_a = \frac{[[Al(OH)(OH_2)_5]^{2+}][H_3O^+]}{[[Al(OH_2)_6]^{3+}]} = \frac{x^2}{0.15 - x} = 1.2 \times 10^{-5} \approx \frac{x^2}{0.15}$$ Solving, $x = 1.3 \times 10^{-3}$

Therefore, $[H_3O^+] = 1.3 \times 10^{-3}$ M; pH = **2.89**

% hydrolysis $= \dfrac{[Al(OH_2)_6]^{3+}_{hydrolyzed}}{[Al(OH_2)_6]^{3+}_{initial}} \times 100\% = \dfrac{1.3 \times 10^{-3}\ M}{0.15\ M} \times 100\% = $ **0.87%**

(b) Balanced equation: $[Co(OH_2)_6]^{2+} + H_2O \rightleftarrows [Co(OH)(OH_2)_5]^+ + H_3O^+$ $K_a = 5.0 \times 10^{-10}$

Assume that the cobalt(II) salt totally dissociated and all the Co^{2+} became $[Co(OH_2)_6]^{2+}$.

Let $x = [[Co(OH_2)_6]^{2+}]_{hydrolyzed}$ Then, $0.075 - x = [[Co(OH_2)_6]^{2+}]$; $x = [[Co(OH)(OH_2)_5]^+] = [H_3O^+]$

$$K_a = \frac{[[Co(OH)(OH_2)_5]^+][H_3O^+]}{[[Co(OH_2)_6]^{2+}]} = \frac{x^2}{0.075 - x} = 5.0 \times 10^{-10} \approx \frac{x^2}{0.075}$$ Solving, $x = 6.1 \times 10^{-6}$

Therefore, $[H_3O^+] = 6.1 \times 10^{-6}$ M; pH = **5.21**

% hydrolysis $= \dfrac{[[Co(OH_2)_6]^{2+}]_{hydrolyzed}}{[[Co(OH_2)_6]^{2+}]_{initial}} \times 100\% = \dfrac{6.1 \times 10^{-6}\ M}{0.075\ M} \times 100\% = $ **8.2 x 10⁻³ %**

(c) Balanced equation: $[Mg(OH_2)_6]^{2+} + H_2O \rightleftarrows [Mg(OH)(OH_2)_5]^+ + H_3O^+$ $K_a = 3.0 \times 10^{-12}$

Assume that the magnesium salt totally dissociated and all the Mg^{2+} became $[Mg(OH_2)_6]^{2+}$.

Let $x = [[Mg(OH_2)_6]^{2+}]_{hydrolyzed}$ Then, $0.15 - x = [[Mg(OH_2)_6]^{2+}]$; $x = [[Mg(OH)(OH_2)_5]^+] = [H_3O^+]$

$$K_a = \frac{[[Mg(OH)(OH_2)_5]^+][H_3O^+]}{[[Mg(OH_2)_6]^{2+}]} = \frac{x^2}{0.15 - x} = 3.0 \times 10^{-12} \approx \frac{x^2}{0.15}$$ Solving, $x = 6.7 \times 10^{-7}$

Therefore, $[H_3O^+] = 6.7 \times 10^{-7}$ M; pH = **6.17** (ignoring the H_3O^+ produced by the ionization of water)

$$\% \text{ hydrolysis} = \frac{[[Mg(OH_2)_6]^{2+}]_{hydrolyzed}}{[[Mg(OH_2)_6]^{2+}]_{initial}} \times 100\% = \frac{6.7 \times 10^{-7} \, M}{0.15 \, M} \times 100\% = \mathbf{4.5 \times 10^{-4} \, \%}$$

Note: To calculate the actual $[H_3O^+]$, let $x = [OH^-] = [H_3O^+]$ produced by the ionization of water.

Therefore $[H_3O^+]_{total}$ = $[H_3O^+]$ produced by hydrolysis + $[H_3O^+]$ produced by the ionization of water
$= 6.7 \times 10^{-7} + x$

We know that $K_w = 1.0 \times 10^{-14} = [H_3O^+][OH^-] = (6.7 \times 10^{-7} + x)(x) = (6.7 \times 10^{-7})x + x^2$
Solving the quadratic equation: $x^2 + (6.7 \times 10^{-7})x - 1.0 \times 10^{-14} = 0$, we have $x = 1.5 \times 10^{-8}$
Therefore, $[H_3O^+]_{total} = 6.7 \times 10^{-7} + x = 6.9 \times 10^{-7}$ M; pH = **6.16**

18-94.   Refer to Section 18-4 and Example 18-10.

Balanced equation: $HA + H_2O \rightleftarrows H_3O^+ + A^-$ $pK_a = 5.35$; $K_a = 4.5 \times 10^{-6}$

Let x = mol/L of HA that reacts. Then
x = mol/L of H_3O^+ produced = mol/L A^- produced.

	HA	+	H_2O	$\rightleftarrows$	H_3O^+	+	A^-
initial	0.100 M				≈ 0 M		0 M
change	- x M				+ x M		+ x M
at equilibrium	(0.100 - x) M				x M		x M

$$K_a = \frac{[H_3O^+][A^-]}{[HA]} = \frac{x^2}{0.100 - x} = 4.5 \times 10^{-6}$$

Assume that $0.100 - x \approx 0.100$. Then $x^2/0.100 = 4.5 \times 10^{-6}$ and $x = 6.7 \times 10^{-4}$. The simplifying assumption is justified since 6.7×10^{-4} is less than 5% of 0.100.

Therefore at equilibrium: $[A^-] = [H_3O^+] = \mathbf{6.7 \times 10^{-4} \, M}$

18-96.   Refer to Sections 18-7, 18-8, 18-9, and 18-11.

(a) $(NH_4)HSO_4$ acidic (salt of weak base and a strong acid)   (Section 18-9)
(b) $(NH_4)_2SO_4$ acidic (salt of a weak base and strong acid)   (Section 18-9)
(c) LiCl   neutral (salt of a strong base and strong acid)   (Section 18-7)
(d) LiBrO   basic (salt of a strong base and weak acid)   (Section 18-8)
(e) $AlCl_3$ acidic (salt of a small, highly charged cation)   (Section 18-11)

18-98.   Refer to Section 18-11.

If a cation reacts appreciably with water, its acid strength must be greater than that of water. The pH of the solution will be less than 7.

Balanced equations:

(a) $Na_2CO_3 \rightarrow 2Na^+ + CO_3^{2-}$

 $CO_3^{2-} + H_2O \rightleftarrows HCO_3^- + OH^- \qquad\qquad K_{b1} = 2.1 \times 10^{-4}$

 $HCO_3^- + H_2O \rightleftarrows H_2CO_3 + OH^- \qquad\qquad K_{b2} = 2.4 \times 10^{-8}$

(b) $Na_2SO_4 \rightarrow 2Na^+ + SO_4^{2-}$

 $SO_4^{2-} + H_2O \rightleftarrows HSO_4^- + OH^- \qquad\qquad K_{b1} = 8.3 \times 10^{-13}$

 $HSO_4^- + H_2O \rightleftarrows H_2SO_4 + OH^- \qquad\qquad K_{b2} = \text{very small}$

(c) $(NH_4)_2SO_4 \rightarrow 2NH_4^+ + SO_4^{2-}$

 $NH_4^+ + H_2O \rightleftarrows NH_3 + H_3O^+ \qquad\qquad K_a = 5.6 \times 10^{-10}$

 $SO_4^{2-} + H_2O \rightleftarrows HSO_4^- + OH^- \qquad\qquad K_{b1} = 8.3 \times 10^{-13}$

 $HSO_4^- + H_2O \rightleftarrows H_2SO_4 + OH^- \qquad\qquad K_{b2} = \text{very small}$

(d) $Na_3PO_4 \rightarrow 3Na^+ + PO_4^{3-}$

 $PO_4^{3-} + H_2O \rightleftarrows HPO_4^{2-} + OH^- \qquad\qquad K_{b1} = 2.8 \times 10^{-2}$

 $HPO_4^{2-} + H_2O \rightleftarrows H_2PO_4^- + OH^- \qquad\qquad K_{b2} = 1.6 \times 10^{-7}$

 $H_2PO_4^- + H_2O \rightleftarrows H_3PO_4 + OH^- \qquad\qquad K_{b3} = 1.3 \times 10^{-12}$

$(NH_4)_2SO_4$ definitely could not be used in cleaning materials since it produces an acidic solution, not a basic solution. Also, Na_2SO_4 cannot be used either since SO_4^{2-} is an extremely weak base (has a very small K_b).

18-102. *Refer to Sections 18-9 and 18-10.*

(a) NH_4Br and NH_4NO_3 are both salts derived from monoprotic strong acids and the weak base, NH_3. Since the concentration of NH_4^+ is the same in each solution, the pH values will be identical.

(b) NH_4ClO_4 is a salt derived from the monoprotic strong acid, $HClO_4$, and the weak base, NH_3. NH_4F is the salt derived from the monoprotic weak acid, HF, and the weak base, NH_3. The concentration of NH_4F (0.010 M) is less than the concentration of NH_4ClO_4 (0.015 M). After the salts dissociate, F^- is a weak base whereas ClO_4^- is too weak a base to react with water. Therefore, the solution of NH_4ClO_4 will have a lower pH (be more acidic) than the solution of NH_4F for two reasons: higher weak acid concentration (NH_4^+), and essentially no base present.

(c) The only difference between these two solutions of NH_4Cl is their concentration. Even though the 0.010 M solution of NH_4^+ will hydrolyze to a greater extent, the 0.050 M solution of NH_4Cl will be more acidic with a lower pH, because it has a higher concentration of the weak acid, NH_4^+.

18-104. *Refer to Sections 10-7, 18-4 and 18-9.*

(a) NH_3 is a stronger base than PH_3 because it can accept a proton more easily. Nitrogen is more electronegative than phosphorus and so the NH_4^+ ion is more stable than the PH_4^+ ion.

(b) F^- is a stronger base than Br^- since the conjugate acid of F^-, HF, is weaker than the conjugate acid of Br^-, HBr.

(c) ClO_2^- is a stronger base than ClO_3^- since the conjugate acid of ClO_2^-, $HClO_2$, is a weaker acid than the conjugate acid of ClO_3^-, $HClO_3$.

(d) PO_4^{3-} is a stronger base than HPO_4^{2-} since the conjugate acid of PO_4^{3-}, HPO_4^{2-}, is a weaker acid than the conjugate acid of HPO_4^{2-}, $H_2PO_4^-$.

Dilute aqueous solutions of weak acids, such as HF and HNO_2, contain relatively few ions because they are weak electrolytes, ionize only slightly into their ions and are therefore, poor conductors of electricity. This can be demonstrated using a conductivity apparatus as shown in Figure 6-1.

18-108. *Refer to Section 18-4, Example 18-14 and Table 18-6.*

(a) Recall that for a series of weak bases, as K_b increases, $[OH^-]$ increases, pOH decreases and pH increases.
   i. highest pH - dimethylamine   ii.   lowest pH - aniline

   iii. highest pOH - aniline   iv. lowest pOH - dimethylamine

(b) Consider the dissociation of a weak base: $B + H_2O \rightleftarrows BH^+ + OH^-$. As K_b increases, $[BH^+]$ increases.
   i. highest $[BH^+]$ - dimethylamine   ii. lowest $[BH^+]$ - aniline

18-110. *Refer to Sections 18-1 and 18-4.*

(1) $HNO_2 + H_2O \rightleftarrows H_3O^+ + NO_2^-$

(2) $Ba(OH)_2 \rightarrow Ba^{2+} + 2OH^-$

(3) $HF + H_2O \rightleftarrows H_3O^+ + F^-$

(4) $LiOH \rightarrow Li^+ + OH^-$

(5) $HCN + H_2O \rightleftarrows H_3O^+ + CN^-$

(6) $KOH \rightarrow K^+ + OH^-$

18-112. *Refer to Section 18-3.*

The quote, " We have to get this problem of acid rain under control. We must do whatever it takes to get the pH down to zero!" is scientifically in error. pH is defined as $-\log[H^+]$. When pH = 0, the $[H^+]$ is equal to 1.0 M, which is very acidic.

18-114. *Refer to Section 18-4 and Example 18-10.*

Balanced equation: $C_5H_7O_4COOH + H_2O \rightleftarrows H_3O^+ + C_5H_7O_4COO^-$ $K_a = 7.9 \times 10^{-5}$

Let x $= [C_5H_7O_4COOH]$ that ionizes. Then x $= [H_3O^+]$ produced $= [C_5H_7O_4COO^-]$ produced.

$$K_a = \frac{[H_3O^+][C_5H_7O_4COO^-]}{[C_5H_7O_4COOH]} = \frac{x^2}{0.110 - x} = 7.9 \times 10^{-5} \approx \frac{x^2}{0.110} \qquad \text{Solving, x} = 2.9 \times 10^{-3}$$

The simplifying assumption is justified since 2.9×10^{-3} is less than 5% of 0.110 ($= 5.5 \times 10^{-3}$).
Therefore, $[H_3O^+] = \mathbf{2.9 \times 10^{-3}}$ M; pH $= \mathbf{2.54}$

18-116. *Refer to Section 18-4.*

Balanced equation: $C_2H_5OCOOH + H_2O \rightleftarrows H_3O^+ + C_2H_5OCOO^-$ $K_a = 8.4 \times 10^{-4}$

Let x $= [C_2H_5OCOOH]$ that ionizes. Then x $= [H_3O^+]$ produced $= [C_2H_5OCOO^-]$ produced.

$$K_a = \frac{[H_3O^+][C_2H_5OCOO^-]}{[C_2H_5OCOOH]} = \frac{x^2}{0.110 - x} = 8.4 \times 10^{-4} \approx \frac{x^2}{0.110} \qquad \text{Solving, x} = 9.6 \times 10^{-3}$$

However, 9.6×10^{-3} is more than 5% of 0.110. A simplifying assumption *cannot* be made; we must solve the original quadratic equation: $x^2 + (8.4 \times 10^{-4})x - 9.2 \times 10^{-5} = 0$

$$x = \frac{-(8.4 \times 10^{-4}) \pm \sqrt{(8.4 \times 10^{-4})^2 - 4(1)(-9.2 \times 10^{-5})}}{2(1)} = \frac{-8.4 \times 10^{-4} \pm 1.9 \times 10^{-2}}{2}$$

$$= 9.1 \times 10^{-3} \text{ or } -9.9 \times 10^{-3} \text{ (discard)}$$

Therefore, $[H_3O^+] = 9.1 \times 10^{-3}$ M; pH = **2.04**

18-118. *Refer to Section 17-13.*

Using the van't Hoff equation, $\ln \dfrac{K_{w\ 323\ K}}{K_{w\ 273\ K}} = \dfrac{\Delta H^\circ}{R}\left(\dfrac{1}{T_1} - \dfrac{1}{T_2}\right)$ $T_1 = 0^\circ C$ or 273 K

$T_2 = 50^\circ C$ or 323 K

$$\ln \frac{5.47 \times 10^{-14}}{1.14 \times 10^{-15}} = \frac{\Delta H^\circ}{8.314\ J/K}\left(\frac{1}{273\ K} - \frac{1}{323\ K}\right)$$

$$3.87 = (\Delta H^\circ)(6.8 \times 10^{-5})$$

$$\Delta H^\circ = \textbf{+5.7} \times \textbf{10}^4 \textbf{ J/mol} \text{ or } \textbf{+57 kJ/mol}$$

Therefore, the reaction, $H_2O(\ell) + H_2O(\ell) \rightleftarrows H_3O^+(aq) + OH^-(aq)$, is endothermic with $\Delta H^\circ = $ **+57 kJ/mol**.

19 Ionic Equilibria II: Buffers and Titration Curves

19-2. *Refer to Sections 18-2, 18-3 and 18-4, and the Key Terms for Chapter 18.*

(a) pK_a is the negative logarithm of K_a, the ionization constant, for a weak acid.
Consider hydrofluoric acid, HF, reacting with water: $HF + H_2O \rightleftarrows H_3O^+ + F^-$

$$K_a = = \frac{[H_3O^+][F^-]}{[HF]} = 7.2 \times 10^{-4}, \text{ so } pK_a \text{ for HF is } -\log(7.2 \times 10^{-4}) = 3.14.$$

(b) K_b is the ionization constant for a weak base.
Consider ammonia, NH_3, reacting with water: $NH_3 + H_2O \rightleftarrows NH_4^+ + OH^-$

$$K_b = \frac{[NH_4^+][OH^-]}{[NH_3]} = 1.8 \times 10^{-5}$$

(c) pK_w is the negative logarithm of K_w, the ion product for water.
Water ionizes to a slight extent: $H_2O + H_2O \rightleftarrows H_3O^+ + OH^-$
$K_w = [H_3O^+][OH^-] = 1.0 \times 10^{-14}$ at 25°C, so pK_w at 25°C = 14.00.

(d) pOH is the negative logarithm of the molar concentration of the hydroxide ion in solution.
If $[OH^-] = 2.0 \times 10^{-3} M$, then pOH = $-\log(2.0 \times 10^{-3}) = 2.70$.

19-4. *Refer to Sections 18-7, 18-8 and 18-9.*

(a) $NaNO_3$ $HNO_3 + NaOH \rightarrow NaNO_3 + H_2O$ neutral (Section 18-7)

(b) Na_2S $H_2S + 2NaOH \rightarrow Na_2S + 2H_2O$ basic (Section 18-8)

(c) $Al_2(CO_3)_3$ $3H_2CO_3 + 2Al(OH)_3 \rightarrow Al_2(CO_3)_3 + 6H_2O$ basic (Section 18-8)

(d) $Mg(CH_3COO)_2$ $2CH_3COOH + Mg(OH)_2 \rightarrow Mg(CH_3COO)_2 + 2H_2O$ basic (Section 18-8)

(e) $(NH_4)_2SO_4$ $H_2SO_4 + 2NH_3 \rightarrow (NH_4)_2SO_4$ acidic (Section 18-9)

19-6. *Refer to Sections 18-7 and 18-10.*

A neutral salt solution is produced by an acid-base reaction in two cases:
(1) a strong acid reacting with a strong base, and
(2) a weak acid reacting with a weak base, where the K_a for the acid is the same value as the K_b for the base.

19-8. *Refer to Section 19-1.*

Buffer solutions are produced by mixing together solutions of a weak acid and its soluble, ionic salt or a weak base and its soluble, ionic salt in approximately the same concentrations. The concentration of one can be no more than ten times the concentration of the other.

(a) 1.0 M HCH_3CH_2COO and 0.20 M $NaCH_3CH_2COO$ is a **buffer** solution because it is composed of a weak acid (HCH_3CH_2COO) and its salt ($NaCH_3CH_2COO$) in appropriate concentrations.

(b) 0.10 M NaOCl and 0.10 M HOCl is a **buffer** solution because it is composed of a weak acid (HOCl) and its salt (NaOCl) in appropriate concentrations.

(c) 0.10 M NH_4Cl and 0.90 M NH_4Br is **not** a buffer solution because it is composed of two salts of the weak base, NH_3, with no NH_3 present.

(d) 0.10 M NaCl and 0.20 M HF is **not** a buffer solution because it is composed of a weak acid (HF) and a salt that does not contain F^- ion.

19-10. *Refer to Section 19-1, Examples 19-1 and 19-2, and Appendix F.*

(a) Balanced equations: $KF \rightarrow K^+ + F^-$ (to completion)

$HF + H_2O \rightleftarrows H_3O^+ + F^-$ (reversible) $\quad K_a = 7.2 \times 10^{-4}$

Since KF dissociates completely, $[F^-]$ from the salt = $[KF]_{initial}$ = 0.20 M

Let $\quad x = [HF]$ that ionizes. Then

$\quad x = [H_3O^+]$ produced from HF.

$\quad x = [F^-]$ produced from HF.

	HF	+	H₂O	⇄	H₃O⁺	+	F⁻
initial	0.15 M				≈ 0 M		0.20 M
change	- x M				+ x M		+ x M
at equilibrium	(0.15 - x) M				x M		(0.20 + x) M

$K_a = \dfrac{[H_3O^+][F^-]}{[HF]} = \dfrac{(x)(0.20+x)}{(0.15-x)} = 7.2 \times 10^{-4} \approx \dfrac{x(0.20)}{(0.15)} \qquad$ Solving, x = 5.4×10^{-4}

Therefore, $[H_3O^+] = 5.4 \times 10^{-4}$ M; pH = **3.27**

Alternatively, this problem can be solved using the Henderson-Hasselbalch equation:

$$pH = pK_a + \log \frac{[\text{conjugate base}]}{[\text{acid}]}$$

Substituting, pH = $3.14 + \log \dfrac{(0.20)}{(0.15)} = \mathbf{3.27}$

(b) Since $Ba(CH_3COO)_2$ dissociates totally, $[CH_3COO^-]$ from the salt = $2 \times [Ba(CH_3COO)_2]_{initial}$ = 0.050 M

Let $\quad x = [CH_3COOH]$ that ionizes. Then

$\quad x = [H_3O^+]$ produced from CH_3COOH

$\quad x = [CH_3COO^-]$ produced from CH_3COOH.

	CH₃COOH	+	H₂O	⇄	H₃O⁺	+	CH₃COO⁻
initial	0.040 M				≈ 0 M		0.050 M
change	- x M				+ x M		+ x M
at equilibrium	(0.040 - x) M				x M		(0.050 + x) M

$K_a = \dfrac{[H_3O^+][CH_3COO^-]}{[CH_3COOH]} = \dfrac{(x)(0.050+x)}{(0.040-x)} = 1.8 \times 10^{-5} \approx \dfrac{x(0.050)}{(0.040)} \qquad$ Solving, x = 1.4×10^{-5}

Therefore, $[H_3O^+] = 1.4 \times 10^{-5}$ M; pH = **4.84**

Alternatively, this problem can be solved using the Henderson-Hasselbalch equation:

$$pH = pK_a + \log \frac{[\text{conjugate base}]}{[\text{acid}]}$$

Substituting, pH = $4.74 + \log \dfrac{(0.050)}{(0.040)} = \mathbf{4.84}$

19-12. *Refer to Section 19-1, Example 19-3 and Appendix G.*

Balanced equations: $NH_4NO_3 \rightarrow NH_4^+ + NO_3^-$ (to completion)

$NH_3 + H_2O \rightleftarrows NH_4^+ + OH^-$ (reversible) $\quad K_b = 1.8 \times 10^{-5}$

(a) Since NH_4NO_3 is a soluble salt, $[NH_4^+]$ from the salt = $[NH_4NO_3]_{initial}$ = 0.15 M

Let $\quad x = [NH_3]$ that ionizes. Then $x = [NH_4^+]$ produced from NH_3 and $x = [OH^-]$ produced from NH_3.

	NH₃	+	H₂O	⇄	NH₄⁺	+	OH⁻
initial	0.25 M				0.15 M		≈ 0 M
change	- x M				+ x M		+ x M
at equilibrium	(0.25 - x) M				(0.15 + x) M		x M

$$K_b = \frac{[NH_4^+][OH^-]}{[NH_3]} = \frac{(0.15 + x)(x)}{(0.25 - x)} = 1.8 \times 10^{-5} \approx \frac{(0.15)(x)}{(0.25)} \qquad \text{Solving, } x = 3.0 \times 10^{-5}$$

Therefore, $[OH^-] = \mathbf{3.0 \times 10^{-5}}\ M$; pOH = 4.52; pH = **9.48**

Alternatively, using the Henderson-Hasselbalch equation,

$$pOH = pK_b + \log \frac{[\text{conjugate acid}]}{[\text{base}]} = -\log (1.8 \times 10^{-5}) + \log \frac{(0.25)}{(0.15)} = 4.57;\ pH = \mathbf{9.48}$$

(b) Since $(NH_4)_2SO_4$ is a soluble salt, $[NH_4^+]$ from the salt $= 2 \times [(NH_4)_2SO_4]_{\text{initial}} = 0.40\ M$

Using the Henderson-Hasselbalch equation,

$$pOH = pK_b + \log \frac{[\text{conjugate acid}]}{[\text{base}]} = -\log (1.8 \times 10^{-5}) + \log \frac{(0.40)}{(0.15)} = 5.17$$

$[OH^-] = \text{antilogarithm } (-5.17) = \mathbf{6.8 \times 10^{-6}}\ M \qquad\qquad pH = 14 - pOH = \mathbf{8.83}$

19-14. *Refer to Section 19-1.*

(a) Balanced equations: $\ NaHCO_3 \xrightarrow{H_2O} Na^+ + HCO_3^-$ (to completion)

                        $H_2CO_3 + H_2O \rightleftarrows H_3O^+ + HCO_3^-$ (reversible)

When a small amount of base is added to the buffer: $\ H_2CO_3 + OH^- \rightarrow HCO_3^- + H_2O$

When a small amount of acid is added to the buffer: $\ HCO_3^- + H_3O^+ \rightarrow H_2CO_3 + H_2O$

(b) Balanced equations: $\ NaH_2PO_2 \xrightarrow{H_2O} Na^+ + H_2PO_4^-$ (to completion)

                        $Na_2HPO_4 \xrightarrow{H_2O} 2Na^+ + HPO_4^{2-}$ (to completion)

                        $H_2PO_4^- + H_2O \rightleftarrows H_3O^+ + HPO_4^{2-}$ (reversible)

When a small amount of base is added to the buffer: $\ H_2PO_4^- + OH^- \rightarrow HPO_4^{2-} + H_2O$

When a small amount of acid is added to the buffer: $\ HPO_4^{2-} + H_3O^+ \rightarrow H_2PO_4^- + H_2O$

19-16. *Refer to Section 19-1.*

(a) Given: Solution A: 0.50 mol KCH_3COO + 0.25 mol CH_3COOH in 1.00 liter solution

            Solution B: 0.25 mol KCH_3COO + 0.50 mol CH_3COOH in 1.00 liter solution

Solution B is more acidic and will have the lower pH.

(b) Buffer solutions are solutions of conjugate acid-base pairs, as defined by Brønsted-Lowry theory. In this case, the acid is CH_3COOH and the conjugate base, CH_3COO^-, is provided by the salt. Fundamentally, the more acid the buffer solution has relative to the base, the more acidic the solution is overall. Solution A has half the amount of acid as conjugate base, whereas Solution B has twice the amount of acid as conjugate base. So, Solution B is more acidic and has the lower pH.

19-18. *Refer to Sections 18-4 and 19-1.*

(a) Balanced equation: $\ NH_3 + H_2O \rightleftarrows NH_4^+ + OH^-$ $K_b = 1.8 \times 10^{-5}$

Let $x = [NH_3]$ that ionizes. Then

      $x = [NH_4^+]$ produced $= [OH^-]$ produced.

	NH_3	$+$	H_2O	$\rightleftarrows$	NH_4^+	$+$	OH^-
initial	0.30 M				0 M		$\approx 0\ M$
change	$-\ x\ M$				$+\ x\ M$		$+\ x\ M$
at equilibrium	$(0.30 - x)\ M$				$x\ M$		$x\ M$

$$K_b = \frac{[NH_4^+][OH^-]}{[NH_3]} = 1.8 \times 10^{-5} = \frac{x^2}{0.30 - x} \approx \frac{x^2}{0.30} \qquad\qquad \text{Solving, } x = 2.3 \times 10^{-3}$$

Since 2.3×10^{-3} is less than 5% of 0.30, the approximation is justified.

Therefore, $[OH^-] = \mathbf{2.3 \times 10^{-3}}\ M$ $[H_3O^+] = \dfrac{K_w}{[OH^-]} = \left(\dfrac{1.0 \times 10^{-14}}{2.3 \times 10^{-3}}\right) = \mathbf{4.3 \times 10^{-12}}\ M$

(b) Balanced equations: $NH_4Cl \rightarrow NH_4^+ + Cl^-$ (to completion)

$NH_3 + H_2O \rightleftarrows NH_4^+ + OH^-$ (reversible) $K_b = 1.8 \times 10^{-5}$

Since NH_4Cl is a soluble salt, $[NH_4^+]$ from the salt $= [NH_4Cl]_{initial} = 0.40 \ M$

Let $x = [NH_3]$ that ionizes. Then $x = [NH_4^+]$ produced from NH_3 and $x = [OH^-]$ produced from NH_3.

	NH_3	+	H_2O	$\rightleftarrows$	NH_4^+	+	OH^-
initial	0.30 M				0.40 M		$\approx 0 \ M$
change	- x M				+ x M		+ x M
at equilibrium	(0.30 - x) M				(0.40 + x) M		x M

$K_b = \dfrac{[NH_4^+][OH^-]}{[NH_3]} = \dfrac{(0.40 + x)(x)}{(0.30 - x)} = 1.8 \times 10^{-5} \approx \dfrac{(0.40)(x)}{(0.30)}$ Solving, $x = 1.35 \times 10^{-5}$ (to 2 sig. fig.)

Therefore, $[OH^-] = \mathbf{1.4 \times 10^{-5} \ \textit{M}}$; $[H_3O^+] = \dfrac{K_w}{[OH^-]} = \left(\dfrac{1.0 \times 10^{-14}}{1.35 \times 10^{-5}}\right) = \mathbf{7.4 \times 10^{-10} \ \textit{M}}$

Alternatively, using the Henderson-Hasselbalch equation,

$pOH = pK_b + \log \dfrac{[\text{conjugate acid}]}{[\text{base}]} = -\log (1.8 \times 10^{-5}) + \log \dfrac{(0.40)}{(0.30)} = 4.87$; $[OH^-] = \mathbf{1.4 \times 10^{-5} \ \textit{M}}$

19-20.   *Refer to Section 19-2.*

Balanced equations: $NaHCOO \rightarrow Na^+ + HCOO^-$ (to completion)

$HCOOH + H_2O \rightleftarrows H_3O^+ + HCOO^-$ (reversible)

When the soluble salt sodium formate ($NaHCOO$) is added to a formic acid solution, the salt undergoes complete dissociation in water to produce the common ion, $HCOO^-$. The original equilibrium involving the weak acid shifts to the left. As a result, the fraction of $HCOOH$ molecules that undergo ionization in aqueous solution will be **less**.

19-22.   *Refer to Sections 18-3 and 19-1.*

Balanced equation: $HClO_4 + H_2O \rightarrow H_3O^+ + ClO_4^-$

$[H_3O^+] = [HClO_4] = 0.20 \ M$; $pH = -\log (0.15) = \mathbf{0.70}$

Perchloric acid, $HClO_4$, is a strong acid and dissociates completely into its ions, even in the presence of a supplier of common ion, $KClO_4$. A solution of 0.20 M $HClO_4$ and 0.20 M $KClO_4$ is *not* a buffer. There is no species present that could react with any added acid. A buffer must be a weak acid and its salt (its conjugate base) or a weak base and its salt (its conjugate acid).

19-24.   *Refer to Sections 19-1 and 19-2, Example 19-4 and Appendix G.*

From Section 19-1, we learned that if the concentrations of the weak acid or base and its salt are $\approx 0.05 \ M$ or greater, and the salt contains a univalent cation, then

for a weak acid buffer: $[H_3O^+] = \dfrac{[\text{acid}]}{[\text{salt}]} \times K_a = \dfrac{\text{mol acid}}{\text{mol salt}} \times K_a$

for a weak base buffer: $[OH^-] = \dfrac{[\text{base}]}{[\text{salt}]} \times K_b = \dfrac{\text{mol base}}{\text{mol salt}} \times K_b$

Original NH_3/NH_4^+ buffer: (K_b for $NH_3 = 1.8 \times 10^{-5}$)

$[NH_4Cl] = \dfrac{12.78 \ \text{g } NH_4Cl/L}{53.49 \ \text{g/mol}} = 0.2389 \ M$

$[OH^-] = \dfrac{[NH_3]}{[NH_4Cl]} \times K_b = \dfrac{0.400 \ M}{0.2389 \ M} \times (1.8 \times 10^{-5}) = 3.0 \times 10^{-5} \ M$;

$[H_3O^+] = \dfrac{K_w}{[OH^-]} = 3.3 \times 10^{-10} \ M$

$pH = 9.48$

New NH_3/NH_4^+ buffer: When 0.142 mol per liter of HCl is added to the original buffer presented in (a), it reacts with the base component of the buffer, NH_3, to form more of the acid component, NH_4^+ (the conjugate acid of NH_3). Since HCl is in the gaseous phase, there is no total volume change. A new buffer solution is created with a slightly more acidic pH. In this type of problem, always perform the acid-base limiting reactant problem first, then the equilibrium calculation.

	HCl	+	NH_3	$\rightarrow$	NH_4Cl
initial	0.142 mol		0.400 mol		0.239 mol
change	- 0.142 mol		- 0.142 mol		+ 0.142 mol
after reaction	0 mol		0.258 mol		0.381 mol

$$[OH^-] = \frac{mol\ NH_3}{mol\ NH_4Cl} \times K_b = \frac{0.258\ mol}{0.381\ mol} \times (1.8 \times 10^{-5}) = 1.2 \times 10^{-5}$$

$$[H_3O^+] = \frac{K_w}{[OH^-]} = 8.3 \times 10^{-10} \text{ and the new pH} = 9.08$$

The change in pH = final pH - initial pH = 9.08 - 9.48 = −0.40; **the pH decreases by 0.40 units**.

19-26. *Refer to Sections 19-1, 19-2 and 19-3, Example 19-4, Exercise 19-24 Solution, and Appendix G.*

(a) Original NH_3/NH_4^+ buffer: (K_b for $NH_3 = 1.8 \times 10^{-5}$)

$$[OH^-] = \frac{[NH_3]}{[NH_4Cl]} \times K_b = \frac{0.90\ M}{0.80\ M} \times (1.8 \times 10^{-5}) = 2.0 \times 10^{-5}\ M$$

$$[H_3O^+] = \frac{K_w}{[OH^-]} = 4.9 \times 10^{-10}\ M$$

pH = **9.31**

(b) New NH_3/NH_4^+ buffer: When 0.10 mol per liter of HCl is added to the original buffer presented in (a), it reacts with the base component of the buffer, NH_3, to form more of the acid component, NH_4^+ (the conjugate acid of NH_3). A new buffer solution is created with a slightly more acidic pH. In this type of problem, always perform the acid-base limiting reactant problem first, then the equilibrium calculation.

(1)

	HCl	+	NH_3	$\rightarrow$	NH_4Cl
initial	0.10 mol		0.90 mol		0.80 mol
change	- 0.10 mol		- 0.10 mol		+ 0.10 mol
after reaction	0 mol		0.80 mol		0.90 mol

(2) $$[OH^-] = \frac{mol\ NH_3}{mol\ NH_4Cl} \times K_b = \frac{0.80\ mol}{0.90\ mol} \times (1.8 \times 10^{-5}) = 1.6 \times 10^{-5}$$

$$[H_3O^+] = \frac{K_w}{[OH^-]} = 6.2 \times 10^{-10}$$

pH = **9.20**

(c) This is a simple strong acid/strong base neutralization problem.

Plan: (1) Find the concentration and the number of moles of NaOH from the pH of the solution.
(2) Perform the limiting reactant testing for the acid-base reaction.
(3) Determine the pH of the final solution.

(1) Since pH = 9.34, pOH = 14.00 - 9.34 = 4.66; $[OH^-] = 2.19 \times 10^{-5}\ M$
Therefore, 1.00 L would contain 2.19×10^{-5} mol NaOH

(2)

	NaOH	+	HCl	$\rightarrow$	NaCl
Initial	2.19×10^{-5} mol		0.10 mol		0 mol
Change	- 2.19×10^{-5} mol		- 2.19×10^{-5} mol		+ 2.19×10^{-5} mol
after reaction	0 mol		0.10 mol		2.19×10^{-5} mol

(3) The number of moles of HCl is essentially unaffected by the presence of 2.19×10^{-5} moles of NaOH. Therefore, $[H_3O^+] = [HCl] = 0.10$ mol/L; pH = **1.00**

Balanced equations: $NaBrCH_2COO \rightarrow Na^+ + BrCH_2COO^-$ (to completion)
$BrCH_2COOH + H_2O \rightleftarrows H_3O^+ + BrCH_2COO^-$ (reversible) $K_a = 2.0 \times 10^{-3}$

Since pH = 3.50, $[H_3O^+] = 3.2 \times 10^{-4} M$

Let $x = [BrCH_2COOH]$.   Then
$0.30 - x = [NaBrCH_2COO]$.

For a weak acid buffer: $[H_3O^+] = \dfrac{[BrCH_2COOH]}{[NaBrCH_2COO]} \times K_a$

$$3.2 \times 10^{-4} = \left(\frac{x}{0.30 - x}\right)(2.0 \times 10^{-3})$$

$$9.5 \times 10^{-5} - (3.2 \times 10^{-4})x = (2.0 \times 10^{-3})x$$

$$9.5 \times 10^{-5} = (2.3 \times 10^{-3})x$$

$$x = 4.1 \times 10^{-2}$$

Therefore, $[BrCH_2COOH] = x = \mathbf{4.1 \times 10^{-2}\ M}$
$[NaBrCH_2COO] = 0.30 - x = \mathbf{0.26\ M}$

Balanced equation: $C_2H_5NH_2 + H_2O \rightleftarrows C_2H_5NH_3^+ + OH^-$ $K_b = 4.7 \times 10^{-4}$

This is an example of the common ion effect; the common ion in this case is OH^-; $[OH^-]_{initial} = 0.0040\ M$

Let $x = [C_2H_5NH_2]$ that ionizes. Then
$x = [OH^-]$ produced from $C_2H_5NH_2$ and
$x = [C_2H_5NH_3^+]$ produced from $C_2H_5NH_2$.

	$C_2H_5NH_2$	+	H_2O	$\rightleftarrows$	$C_2H_5NH_3^+$	+	OH^-
initial	0.012 M				0 M		0.0040 M
change	- x M				+ x M		+ x M
at equilibrium	(0.012 - x) M				x M		(0.0040 + x) M

$K_b = \dfrac{[C_2H_5NH_3^+][OH^-]}{[C_2H_5NH_2]} = \dfrac{(x)(0.0040 + x)}{(0.012 - x)} = 4.7 \times 10^{-4} \approx \dfrac{x(0.0040)}{0.012}$ Solving, $x = 1.4 \times 10^{-3}$

However, x has the same order of magnitude as 0.0040, so the simplifying assumption does not hold. We must solve the original quadratic equation: $x^2 + 0.0045x - 5.6 \times 10^{-6} = 0$

$$x = \frac{-0.0045 \pm \sqrt{(0.0045)^2 - 4(1)(-5.6 \times 10^{-6})}}{2(1)} = \frac{-0.0045 \pm 0.0065}{2} = 0.0010 \text{ or } -0.0055 \text{ (discard)}$$

Therefore, $[C_2H_5NH_3^+] = x = \mathbf{0.0010\ M}$

Balanced equations: $NaCH_3CH_2COO + H_2O \rightarrow Na^+ + CH_3CH_2COO^-$ (to completion)
$CH_3CH_2COOH + H_2O \rightleftarrows H_3O^+ + CH_3CH_2COO^-$ (reversible) $K_a = 1.3 \times 10^{-5}$

In this buffer system: $[H_3O^+] = 7.9 \times 10^{-6}\ M$ since pH = 5.10
$[NaCH_3CH_2COO] = 0.60\ M$

$[acid] = [CH_3CH_2COOH] = \dfrac{[H_3O^+][salt]}{K_a} = \dfrac{(7.9 \times 10^{-6})(0.60)}{(1.3 \times 10^{-5})} = \mathbf{0.36\ M}$ since $K_a = \dfrac{[H_3O^+][salt]}{[acid]}$

Plan: (1) Perform the acid-base neutralization limiting reactant problem.
(2) Determine the volumes of acetic acid and sodium hydroxide that must be mixed without adding additional water by substituting into the modified K_a expression.

(1) Let V_A = volume (in liters) of 0.125 M CH$_3$COOH V_A x 0.125 M = initial moles of CH$_3$COOH
V_B = volume (in liters) of 0.065 M NaOH V_B x 0.065 M = initial moles of NaOH

In order to produce a buffer solution, NaOH must be consumed and is therefore the limiting reactant in the acid-base neutralization reaction.

	CH$_3$COOH	+	NaOH	$\rightarrow$	NaCH$_3$COO	+	H$_2$O
initial	0.125 M x V_A mol		0.065 M x V_B mol		0 mol		
change	- 0.065 M x V_B mol		- 0.065 M x V_B mol		+ 0.065 M x V_B mol		
after rxn	(0.125 $M \cdot V_A$ - 0.065 $M \cdot V_B$) mol		0 mol		0.065 M x V_B mol		

(2) For a weak acid buffer: pH = 4.50; [H$^+$] = 3.2 x 10^{-5}

$$[H^+] = K_a \times \frac{\text{mol CH}_3\text{COOH}}{\text{mol NaCH}_3\text{COO}}$$

Substituting,

$$3.2 \times 10^{-5} = (1.8 \times 10^{-5}) \times \frac{0.125\ M \cdot V_A - 0.065\ M \cdot V_B}{0.065\ M \times V_B}$$

$$1.8 = \frac{0.125\ V_A}{0.065\ V_B} - \frac{0.065\ V_B}{0.065\ V_B} = \frac{0.125\ V_A}{0.065\ V_B} - 1$$

$$2.8 = \frac{0.125\ V_A}{0.065\ V_B}$$

$$1.5 = \frac{V_A}{V_B}$$

$$V_A = 1.5\ V_B$$

Since $V_A + V_B$ = 1.00 L
Substituting, 1.5 $V_B + V_B$ = 1.00 L
2.5 V_B = 1.00 L
V_B = volume of NaOH = **0.40 L**
V_A = volume of CH$_3$COOH = 1.00 L - 0.40 L = **0.60 L**

19-36. *Refer to Section 19-3, Example 19-5 and Appendix F.*

Balanced equations: Ca(CH$_3$COO)$_2$ $\rightarrow$ Ca^{2+} + 2CH$_3$COO$^-$ (to completion)
CH$_3$COOH + H$_2$O $\rightleftarrows$ H$_3$O$^+$ + CH$_3$COO$^-$ (reversible) K_a = 1.8 x 10^{-5}

Recall: for dilution, $M_1V_1 = M_2V_2$. In this instance, the acetic acid solution and the calcium acetate solution are diluting each other.

$$[\text{CH}_3\text{COOH}]_{\text{initial}} = M_2 = \frac{M_1V_1}{V_2} = \frac{1.25\ M \times 500.\ \text{mL}}{500.\ \text{mL} + 500.\ \text{mL}} = 0.625\ M$$

$$[\text{Ca(CH}_3\text{COO)}_2]_{\text{initial}} = M_2 = \frac{M_1V_1}{V_2} = \frac{0.300\ M \times 500.\ \text{mL}}{500.\ \text{mL} + 500.\ \text{mL}} = 0.150\ M$$

Therefore, $[\text{CH}_3\text{COO}^-]_{\text{initial}}$ = 2 x $[\text{Ca(CH}_3\text{COO)}_2]_{\text{initial}}$ = 2 x 0.150 M = 0.300 M

Let x= [CH$_3$COOH] that ionizes. Then
x = [H$_3$O$^+$] produced from CH$_3$COOH = [CH$_3$COO$^-$] produced from CH$_3$COOH.

	CH$_3$COOH	+	H$_2$O	$\rightleftarrows$	H$_3$O$^+$	+	CH$_3$COO$^-$
initial	0.625 M				$\approx 0\ M$		0.300 M
change	- x M				+ x M		+ x M
at equilibrium	(0.625 - x) M				x M		(0.300 + x) M

$$K_a = \frac{[H_3O^+][CH_3COO^-]}{[CH_3COOH]} = \frac{(x)(0.300 + x)}{(0.625 - x)} = 1.8 \times 10^{-5} \approx \frac{x(0.300)}{(0.625)} \qquad \text{Solving, } x = 3.8 \times 10^{-5}$$

The simplifying assumption is justified and we have

(a) $[CH_3COOH] = \mathbf{0.625\ M}$

(b) $[Ca^{2+}] = [Ca(CH_3COO)_2]_{initial} = \mathbf{0.150\ M}$

(c) $[CH_3COO^-] = \mathbf{0.300\ M}$

(d) $[H^+] = \mathbf{3.8 \times 10^{-5}\ M}$

(e) $pH = -\log(3.8 \times 10^{-5}) = \mathbf{4.42}$

19-38.   *Refer to Section 19-3.*

Balanced equations: $NaClCH_2COO \rightarrow Na^+ + ClCH_2COO^-$ (to completion)

$ClCH_2COOH + H_2O \rightleftarrows H_3O^+ + ClCH_2COO^-$ (reversible) $K_a = 1.4 \times 10^{-3}$

$[ClCH_2COO^-]_{initial} = [NaClCH_2COO] = 0.015\ M$

Since pH = 2.75, $[H_3O^+] = 1.8 \times 10^{-3}\ M$

Therefore, $1.8 \times 10^{-3}\ M = [ClCH_2COOH]_{ionized} = [H_3O^+]_{produced\ from\ ClCH_2COOH}$
$= [ClCH_2COO^-]_{produced\ from\ ClCH_2COOH}$

Let $x = [ClCH_2COOH]_{initial}$

	$ClCH_2COOH$	+	H_2O	$\rightleftarrows$	H_3O^+	+	$ClCH_2COO^-$
initial	x M				$\approx 0\ M$		0.015 M
change	- 1.8 x 10⁻³ M				+ 1.8 x 10⁻³ M		+ 1.8 x 10⁻³ M
at equilibrium	(x - 1.8 x 10⁻³) M				1.8 x 10⁻³ M		0.017 M

$$K_a = \frac{[H_3O^+][ClCH_2COO^-]}{[ClCH_2COOH]} = \frac{(1.8 \times 10^{-3})(0.017)}{(x - 1.8 \times 10^{-3})} = 1.4 \times 10^{-3} \qquad \text{Solving, } x = 0.024$$

Therefore, $[ClCH_2COOH] = \mathbf{0.024\ M}$

19-40.   *Refer to Section 19-4.*

(a) Acid-base indicators are organic compounds which behave as weak acids or bases and exhibit different colors in solutions with different acidities.

(b) The essential characteristic of an acid-base indicator is that the conjugate acid-base pair must exhibit different colors. Consider the weak acid indicator, HIn. In solution, HIn dissociates slightly as follows:

$$HIn + H_2O \rightleftarrows H_3O^+ + In^-$$
$$\text{acid} \qquad\qquad\qquad \text{conjugate base}$$

HIn dominates in more acidic solutions with one characteristic color;   In⁻ dominates in more basic solutions with another color.

(c) The color of an acid-base indicator in an aqueous solution depends upon the ratio, $[In^-]/[HIn]$, which in turn depends upon $[H^+]$ and the K_a value of the indicator. A general rule of thumb: If $[In^-]/[HIn] < 0.1$, then the indicator will show its true acid color. If $[In^-]/[HIn] > 10$, then the indicator will show its true base color.

Balanced equation: $HIn + H_2O \rightleftarrows H_3O^+ + In^-$, where metacresol purple is represented by HIn
At pH 8.2, $[HIn] = [In^-]$; $[H_3O^+] = $ antilog $(-pH) = 6 \times 10^{-9}$ M

Substituting into the K_a expression, we have: $K_a = \dfrac{[H_3O^+][In^-]}{[HIn]} = \mathbf{6 \times 10^{-9}}$ Note: $[H^+] = K_a$ and pH =
$pK_a = \mathbf{8.2}$ pK_a at the endpoint, since $[In^-]$ and $[HIn]$ are equal and cancel out each other.

When a solution is colorless with phenolphthalein, pH < ~8. When a solution is blue in bromthymol blue, pH > ~7.6. When a solution is yellow in methyl orange, pH > ~4.4. Therefore, we know that this solution has a pH between ~7.6 and ~8.

Balanced equation: $CH_3COOH + NaOH \rightarrow NaCH_3COO + H_2O$

Method 1:   The resultant solution at the equivalence point of any acid-base reaction contains only salt and water.

To calculate the concentration of the salt solution: assume you have 1.00 L of 0.020 M acetic acid, so you have 1.00 L x 0.020 M = 0.020 mol CH_3COOH. Therefore, you will need to add 0.020 mol NaOH to make 0.020 mol of $NaCH_3COO$. The volume of NaOH that must be added to get to the equivalence point is

? L NaOH = 0.020 mol NaOH/0.025 M NaOH = 0.800 L   (rearrange M = mol/L to L = mol/M)

? total volume at the equivalence point = 1.00 L acetic acid + 0.80 L NaOH = 1.80 L soln

? $[NaCH_3COO]$ = 0.020 mol $NaCH_3COO$/1.80 L soln = 0.011 M $NaCH_3COO$

Method 2:   The pH is determined from the concentration of the salt. Even when the volume of solution is given, it is not necessary to use that information. We can calculate the concentration of a salt derived from a monoprotic acid and a base with one OH group by:

$$[salt] = \frac{M_A M_B}{M_A + M_B}$$ where M_A = molarity of the acid
M_B = molarity of the base

Derivation:
   Let V_A = volume (in L) of acid with molarity M_A V_B = volume (in L) of base with molarity M_B

   At the equivalence point, mol acid $(M_A V_A)$ = mol base $(M_B V_B)$ = mol salt produced

$$[salt] = \frac{\text{mol acid}}{\text{total volume}} = \frac{M_A V_A}{V_A + V_B} = \frac{M_A}{\left(1 + \frac{V_B}{V_A}\right)} = \frac{M_A}{\left(1 + \frac{M_A}{M_B}\right)} = \frac{M_A M_B}{M_A + M_B}$$ since $\frac{V_B}{V_A} = \frac{M_A}{M_B}$ in this case

Plan: (1) Calculate the concentration of $NaCH_3COO$.
        (2) Determine the pH.

(1) $[NaCH_3COO]$ = 0.011 M (see Method 1 above)

(2) The anion of the soluble salt hydrolyzes to form a basic solution: $CH_3COO^- + H_2O \rightleftarrows CH_3COOH + OH^-$

Let x = $[CH_3COO^-]_{hydrolyzed}$ Then, 0.011 - x = $[CH_3COO^-]$; x = $[CH_3COOH] = [OH^-]$

$K_b = \dfrac{K_w}{K_{a(CH_3COOH)}} = \dfrac{1.0 \times 10^{-14}}{1.8 \times 10^{-5}} = 5.6 \times 10^{-10} = \dfrac{[CH_3COOH][OH^-]}{[CH_3COO^-]} = \dfrac{x^2}{0.011 - x} \approx \dfrac{x^2}{0.011}$

Solving, x = 2.5 x 10^{-6}; $[OH^-]$ = 2.5 x 10^{-6} M; pOH = 5.60; pH = **8.40** with **phenolphthalein** as indicator.

Example of a titration of a strong acid with a strong base:

(a) When no base is added, the pH of the solution is determined by the initial strong acid concentration.

(b) At the point halfway to the equivalence point, only half of the base required to titrate all of the acid has been added. The strong base is the limiting reactant and the pH of the resulting solution is less than 7. It is calculated from the concentration of the remaining acid.

(c) At the equivalence point, only water and the salt of the strong acid and strong base are present. Since neither the cation nor the anion of the salt hydrolyzes appreciably, the pH is 7.

(d) Past the equivalence point, the strong acid is the limiting reactant and the pH of the solution is greater than 7. It is determined from the concentration of the excess strong base.

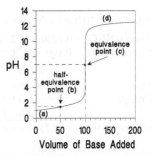

The graph compares well with Figure 19-3a.

Balanced equation: $HNO_3 + NaOH \rightarrow NaNO_3 + H_2O$
A 25.00 mL sample of 0.1500 M HNO_3 is titrated with 0.100 M NaOH.

(a) Initially: $[H^+] = [HNO_3] = 0.1500$ M; pH = **0.8239**

For the rest of the exercise, the plan is straightforward: for the neutralization reaction between HNO_3 and NaOH, perform the limiting reactant problem. The pH is determined from the concentration of excess HNO_3 or NaOH. Each calculation is totally independent of the other calculations.

(b) Addition of 6.00 mL of 0.100 M NaOH:

? mmol HNO_3 = 0.1500 M x 25.00 mL = 3.750 mmol HNO_3
? mmol NaOH = 0.100 M x 6.00 mL = 0.600 mmol NaOH

Before the equivalence point, NaOH is the limiting reactant. The pH is determined from the concentration of excess HNO_3 remaining. The salt produced, $NaNO_3$, is the salt of a strong acid and a strong base. It will not affect the pH of the solution.

	HNO_3	+	NaOH	$\rightarrow$	$NaNO_3$	+	H_2O
initial	3.750 mmol		0.600 mmol		0 mmol		
change	- 0.600 mmol		- 0.600 mmol		+ 0.600 mmol		
after reaction	3.150 mmol		0 mmol		0.600 mmol		

$$[H^+] = \frac{\text{mmol excess } HNO_3}{\text{total volume (mL)}} = \frac{3.150 \text{ mmol}}{(25.00 \text{ mL} + 6.00 \text{ mL})} = 0.1016 \ M; \ \text{pH} = \textbf{0.9931}$$

(c) Addition of 15.00 mL of 0.100 M NaOH:

? mmol HNO_3 = 0.1500 M x 25.00 mL = 3.750 mmol HNO_3
? mmol NaOH = 0.100 M x 15.00 mL = 1.50 mmol NaOH

	HNO_3	+	NaOH	$\rightarrow$	$NaNO_3$	+	H_2O
initial	3.750 mmol		1.50 mmol		0 mmol		
change	- 1.50 mmol		- 1.50 mmol		+ 1.50 mmol		
after reaction	2.25 mmol		0 mmol		1.50 mmol		

315

$$[\text{H}^+] = \frac{\text{mmol excess HNO}_3}{\text{total volume (mL)}} = \frac{2.25 \text{ mmol}}{(25.00 \text{ mL} + 15.00 \text{ mL})} = 0.0562 \ M; \ \text{pH} = \mathbf{1.250}$$

(d) Addition of 30.0 mL of 0.100 M NaOH:

? mmol HNO$_3$ = 0.1500 M x 25.00 mL = 3.750 mmol HNO$_3$
? mmol NaOH = 0.100 M x 30.00 mL = 3.00 mmol NaOH

	HNO$_3$	+	NaOH	→	NaNO$_3$	+	H$_2$O
Initial	3.750 mmol		3.00 mmol		0 mmol		
Change	- 3.00 mmol		- 3.00 mmol		+ 3.00 mmol		
after reaction	0.75 mmol		0 mmol		3.00 mmol		

$$[\text{H}^+] = \frac{\text{mmol excess HNO}_3}{\text{total volume (mL)}} = \frac{0.75 \text{ mmol}}{(25.00 \text{ mL} + 30.00 \text{ mL})} = 0.014 \ M; \ \text{pH} = \mathbf{1.87}$$

(e) Addition of 37.44 mL of 0.100 M NaOH:

? mmol HNO$_3$ = 0.1500 M x 25.00 mL = 3.750 mmol HNO$_3$
? mmol NaOH = 0.100 M x 37.44 mL = 3.74 mmol NaOH   (use 3.744 for next calculation)

	HNO$_3$	+	NaOH	→	NaNO$_3$	+	H$_2$O
Initial	3.750 mmol		3.74 mmol		0 mmol		
change	- 3.74 mmol		- 3.74 mmol		+ 3.74 mmol		
after reaction	0.01 mmol		0 mmol		3.74 mmol		

(use 0.006 for calculation)

$$[\text{H}^+] = \frac{\text{mmol excess HNO}_3}{\text{total volume (mL)}} = \frac{0.006 \text{ mmol}}{(25.00 \text{ mL} + 37.44 \text{ mL})} = 0.0001 \ M; \ \text{pH} = \mathbf{4.0}$$

(f) Addition of 45.00 mL of 0.100 M NaOH:

? mmol HNO$_3$ = 0.1500 M x 25.00 mL = 3.750 mmol HNO$_3$
? mmol NaOH = 0.100 M x 45.00 mL = 4.50 mmol NaOH

After the equivalence point, HNO$_3$ is the limiting reactant. The pH is determined from the concentration of excess NaOH.

	HNO$_3$	+	NaOH	→	NaNO$_3$	+	H$_2$O
initial	3.750 mmol		4.50 mmol		0 mmol		
change	- 3.750 mmol		- 3.750 mmol		+ 3.750 mmol		
after reaction	0 mmol		0.75 mmol		3.750 mmol		

$$[\text{OH}^-] = \frac{\text{mmol excess NaOH}}{\text{total volume (mL)}} = \frac{0.75 \text{ mmol}}{(25.00 \text{ mL} + 45.00 \text{ mL})} = 0.011 \ M; \ \text{pOH} = 1.97; \ \text{pH} = \mathbf{12.03}$$

19-52.   *Refer to Section 19-5 and Exercise 19-48 Solution.*

Balanced equation: HCl + NaOH $\rightarrow$ NaCl + H$_2$O
A 22.0 mL sample of 0.145 M HCl is titrated with 0.106 M NaOH.

(a) Initially: [H$^+$] = [HCl] = 0.145 M; pH = **0.839**

For the rest of the exercise, the plan is straightforward: for the neutralization reaction between HCl and NaOH, perform the limiting reactant problem. The pH is determined from the concentration of excess HCl or NaOH.

(b) Addition of 11.10 mL of 0.106 M NaOH:

? mmol HCl = 0.145 M x 22.0 mL = 3.19 mmol HCl
? mmol NaOH = 0.106 M x 11.10 mL = 1.18 mmol NaOH

Before the equivalence point, NaOH is the limiting reactant. The pH is determined from the concentration of excess HCl remaining. The salt produced, NaCl, is the salt of a strong acid and a strong base. It will not affect the pH of the solution.

	HCl	+	NaOH	→	NaCl	+	H₂O
initial	3.19 mmol		1.18 mmol		0 mmol		
change	- 1.18 mmol		- 1.18 mmol		+ 1.18 mmol		
after reaction	2.01 mmol		0 mmol		1.18 mmol		

$[H^+] = \dfrac{\text{mmol excess HCl}}{\text{total volume (mL)}} = \dfrac{2.01 \text{ mmol}}{(22.0 \text{ mL} + 11.10 \text{ mL})} = 0.0607 \ M; \ \text{pH} = \textbf{1.217}$

(c) Addition of 24.0 mL of 0.106 M NaOH:

? mmol HCl = 0.145 M x 22.0 mL = 3.19 mmol HCl
? mmol NaOH = 0.106 M x 24.0 mL = 2.54 mmol NaOH

	HCl	+	NaOH	→	NaCl	+	H₂O
initial	3.19 mmol		2.54 mmol		0 mmol		
change	- 2.54 mmol		- 2.54 mmol		+ 2.54 mmol		
after reaction	0.65 mmol		0 mmol		2.54 mmol		

$[H^+] = \dfrac{\text{mmol excess HCl}}{\text{total volume (mL)}} = \dfrac{0.65 \text{ mmol}}{(22.0 \text{ mL} + 24.0 \text{ mL})} = 0.014 \ M; \ \text{pH} = \textbf{1.85}$

(d) Addition of 41.0 mL of 0.106 M NaOH:

? mmol HCl = 0.145 M x 22.0 mL = 3.19 mmol HCl
? mmol NaOH = 0.106 M x 41.0 mL = 4.35 mmol NaOH

After the equivalence point, HCl is the limiting reactant. The pH is determined from the concentration of excess NaOH.

	HCl	+	NaOH	→	NaCl	+	H₂O
initial	3.19 mmol		4.35 mmol		0 mmol		
change	- 3.19 mmol		- 3.19 mmol		+ 3.19 mmol		
after reaction	0 mmol		1.16 mmol		3.19 mmol		

$[OH^-] = \dfrac{\text{mmol excess NaOH}}{\text{total volume (mL)}} = \dfrac{1.16 \text{ mmol}}{(22.0 \text{ mL} + 41.0 \text{ mL})} = 0.0184 \ M; \ \text{pOH} = 1.735; \ \text{pH} = \textbf{12.265}$

(e) Addition of 54.4 mL of 0.106 M NaOH:

? mmol HCl = 0.145 M x 22.0 mL = 3.19 mmol HCl
? mmol NaOH = 0.106 M x 54.4 mL = 5.77 mmol NaOH

	HCl	+	NaOH	→	NaCl	+	H₂O
initial	3.19 mmol		5.77 mmol		0 mmol		
change	- 3.19 mmol		- 3.19 mmol		+ 3.19 mmol		
after reaction	0 mmol		2.58 mmol		3.19 mmol		

$[OH^-] = \dfrac{\text{mmol excess NaOH}}{\text{total volume (mL)}} = \dfrac{2.58 \text{ mmol}}{(22.0 \text{ mL} + 54.4 \text{ mL})} = 0.0338 \ M; \ \text{pOH} = 1.471; \ \text{pH} = \textbf{12.529}$

(f) Addition of 63.6 mL of 0.106 M NaOH:

? mmol HCl = 0.145 M x 22.0 mL = 3.19 mmol HCl

? mmol NaOH = 0.106 M x 63.6 mL = 6.74 mmol NaOH

	HNO_3	+	NaOH	$\rightarrow$	$NaNO_3$	+	H_2O
initial	3.19 mmol		6.74 mmol		0 mmol		
change	- 3.19 mmol		- 3.19 mmol		+ 3.19 mmol		
after reaction	0 mmol		3.55 mmol		3.19 mmol		

$$[OH^-] = \frac{\text{mmol excess NaOH}}{\text{total volume (mL)}} = \frac{3.55 \text{ mmol}}{(22.0 \text{ mL} + 63.6 \text{ mL})} = 0.0415 \ M; \ pOH = 1.382; \ pH = \textbf{12.618}$$

19-54. *Refer to Section 19-6, Table 19-4 and Figure 19-4.*

The calculations for determining the pH at every point in the titration of 1 liter of 0.0200 M CH_3COOH with solid NaOH, assuming no volume change, can be divided into 4 types.

(1) Initially, the pH is determined by the concentration of the weak acid, CH_3COOH (K_a = 1.8 x 10^{-5}). Assuming the simplifying assumption works:

$$[H^+] = \sqrt{K_a[CH_3COOH]} = \sqrt{(1.8 \times 10^{-5})(0.0200)} = 6.0 \times 10^{-4} \ M; \ pH = 3.22$$

(2) Before the equivalence point, the pH is determined by the buffer solution consisting of the unreacted CH_3COOH and $NaCH_3COO$ produced by the reaction. Each calculation is a limiting reactant problem using the original concentration of CH_3COOH.

For example, at point (c) in the following table:

initial mmol CH_3COOH = 0.0200 M x 1000 mL = 20.0 mmol CH_3COOH

mmol of NaOH added = 0.00800 mol x 1000 mmol/mol = 8.00 mmol NaOH

	CH_3COOH	+	NaOH	$\rightarrow$	$NaCH_3COO$	+	H_2O
initial	20.0 mmol		8.00 mmol		0 mmol		
change	- 8.00 mmol		- 8.00 mmol		+ 8.00 mmol		
after reaction	12.0 mmol		0 mmol		8.00 mmol		

After the reaction, we have a 1 liter buffer solution consisting of 12.0 mmol CH_3COOH and 8.00 mmol $NaCH_3COO$.

$$[H_3O^+] = K_a \times \frac{\text{mmol } CH_3COOH}{\text{mmol } NaCH_3COO} = (1.8 \times 10^{-5}) \frac{12.0 \text{ mmol}}{8.00 \text{ mmol}} = 2.7 \times 10^{-5} \ M; \ pH = 4.57$$

Halfway to the equivalence point (i.e., when half of the required amount of base needed to reach the equivalence point is added), pH = pK_a. At point (d):

	CH_3COOH	+	NaOH	$\rightarrow$	$NaCH_3COO$	+	H_2O
initial	20.0 mmol		10.00 mmol		0 mmol		
change	- 10.00 mmol		- 10.00 mmol		+ 10.00 mmol		
after reaction	10.0 mmol		0 mmol		10.00 mmol		

$$[H_3O^+] = K_a \times \frac{\text{mmol } CH_3COOH}{\text{mmol } NaCH_3COO} = (1.8 \times 10^{-5}) \frac{10.0 \text{ mmol}}{10.00 \text{ mmol}} = 1.8 \times 10^{-5} \ M; \ pH = pK_a = 4.74$$

(3) At the equivalence point, there is no excess acid or base. The concentration of $NaCH_3COO$ determines the pH of the system (Refer to Exercise 19-46 Solution). At point (h):

	CH_3COOH	+	NaOH	$\rightarrow$	$NaCH_3COO$	+	H_2O
initial	20.0 mmol		20.0 mmol		0 mmol		
change	- 20.0 mmol		- 20.0 mmol		+ 20.0 mmol		
after reaction	0 mmol		0 mmol		20.0 mmol		

$$[NaCH_3COO] = \frac{0.0200\ mol}{1.00\ L} = 0.0200\ M$$

If aqueous NaOH had been added,
$[NaCH_3COO] = 0.0200\ mol/total\ volume\ (L)$

Then the anion of the salt hydrolyzes to produce a basic solution: $CH_3COO^- + H_2O \rightleftarrows CH_3COOH + OH^-$

Let x = $[CH_3COO^-]_{hydrolyzed}$ Then, 0.0200 - x = $[CH_3COO^-]$; x = $[CH_3COOH] = [OH^-]$

$$K_b = \frac{K_w}{K_{a(CH_3COOH)}} = 5.6 \times 10^{-10} = \frac{[CH_3COOH][OH^-]}{[CH_3COO^-]} = \frac{x^2}{0.0200 - x} \approx \frac{x^2}{0.0200}$$ Solving, x = 3.3×10^{-6}

$[OH^-]$ = x = $3.3 \times 10^{-6}\ M$; pOH = 5.48; pH = 8.52

(4) After the equivalence point, the pH is determined directly from the concentration of *excess* NaOH since CH_3COOH is now the limiting reactant. In the presence of the strong base, the effect of the weak base, CH_3COO^-, derived from the salt is negligible.

For example, at point (j):

	CH_3COOH	+	NaOH	$\rightarrow$	$NaCH_3COO$	+	H_2O
initial	20.0 mmol		24.0 mmol		0 mmol		
change	- 20.0 mmol		- 20.0 mmol		+ 20.0 mmol		
after reaction	0 mmol		4.0 mmol		20.0 mmol		

$$[OH^-] = [NaOH]_{excess} = \frac{0.0040\ mol}{1.00\ L} = 4.0 \times 10^{-3}\ M;\ pOH = 2.40;\ pH = 11.60$$

Note: If aqueous NaOH had been added, $[NaOH]_{excess} = 0.0040\ mol/total\ volume\ (L)$

Data Table:

	Mol NaOH Added	Type of Solution	$[H_3O^+]$ (M)	$[OH^-]$ (M)	pH	pOH
(a)	none	weak acid	6.0×10^{-4}	1.7×10^{-11}	3.22	10.78
(b)	0.00400	buffer	7.2×10^{-5}	1.4×10^{-10}	4.14	9.86
(c)	0.00800	buffer	2.7×10^{-5}	3.7×10^{-10}	4.57	9.43
(d)	0.01000	buffer	1.8×10^{-5}	5.6×10^{-10}	4.74 (= pK_a)	9.26
	(halfway to the equivalence point)					
(e)	0.01400	buffer	7.7×10^{-6}	1.3×10^{-9}	5.11	8.89
(f)	0.01800	buffer	2.0×10^{-6}	5.0×10^{-9}	5.70	8.30
(g)	0.01900	buffer	9.5×10^{-7}	1.1×10^{-8}	6.02	7.98
(h)	0.0200	salt	3.0×10^{-9}	3.3×10^{-6}	8.52	5.48
	(at the equivalence point)					
(i)	0.0210	strong base	1.0×10^{-11}	1.0×10^{-3}	11.00	3.0
(j)	0.0240	strong base	2.5×10^{-12}	4.0×10^{-3}	11.60	2.40
(k)	0.0300	strong base	1.0×10^{-12}	1.0×10^{-2}	12.00	2.00

Titration Curve: CH_3COOH vs. NaOH

An appropriate indicator would change color in the pH range, 7 - 10. From Table 19-4, **phenolphthalein** is the best indicator for this titration.

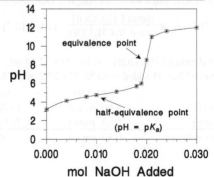

319

(a) Initially, the pH is determined by the concentration of the weak acid, CH_3COOH $(K_a = 1.8 \times 10^{-5})$. Assuming the simplifying assumption works:

$$[H^+] = \sqrt{K_a[CH_3COOH]} = \sqrt{(1.8 \times 10^{-5})(0.182)} = 1.8 \times 10^{-3} \, M; \quad pH = \mathbf{2.74}$$

Before the equivalence point, the pH is determined by the buffer solution consisting of the unreacted CH_3COOH and $NaCH_3COO$ produced by the reaction. Each calculation is a limiting reactant problem using the original concentration of CH_3COOH.

(b) initial mmol CH_3COOH = 0.182 *M* x 32.44 mL = 5.90 mmol CH_3COOH
mmol of NaOH added = 0.185 *M* x 15.55 mL = 2.88 mmol NaOH

	CH_3COOH	+	NaOH	$\rightarrow$	$NaCH_3COO$	+	H_2O
initial	5.90 mmol		2.88 mmol		0 mmol		
change	- 2.88 mmol		- 2.88 mmol		+ 2.88 mmol		
after reaction	3.02 mmol		0 mmol		2.88 mmol		

After the reaction, we have a total of 47.99 mL (= 32.44 + 15.55) buffer solution containing 3.02 mmol CH_3COOH and 2.88 mmol $NaCH_3COO$.

$$[H_3O^+] = K_a \times \frac{\text{mmol } CH_3COOH}{\text{mmol } NaCH_3COO} = (1.8 \times 10^{-5}) \frac{3.02 \text{ mmol}}{2.88 \text{ mmol}} = 1.9 \times 10^{-5} \, M; \quad pH = \mathbf{4.72}$$

(c) initial mmol CH_3COOH = 0.182 *M* x 32.44 mL = 5.90 mmol CH_3COOH
mmol of NaOH added = 0.185 *M* x 20.0 mL = 3.70 mmol NaOH

	CH_3COOH	+	NaOH	$\rightarrow$	$NaCH_3COO$	+	H_2O
initial	5.90 mmol		3.70 mmol		0 mmol		
change	- 3.70 mmol		- 3.70 mmol		+ 3.70 mmol		
after reaction	2.20 mmol		0 mmol		3.70 mmol		

After the reaction, we have a total of 52.4 mL (= 32.44 + 20.0) buffer solution containing 2.20 mmol CH_3COOH and 3.70 mmol $NaCH_3COO$.

$$[H_3O^+] = K_a \times \frac{\text{mmol } CH_3COOH}{\text{mmol } NaCH_3COO} = (1.8 \times 10^{-5}) \frac{2.20 \text{ mmol}}{3.70 \text{ mmol}} = 1.1 \times 10^{-5} \, M; \quad pH = \mathbf{4.97}$$

(d) initial mmol CH_3COOH = 0.182 *M* x 32.44 mL = 5.90 mmol CH_3COOH
mmol of NaOH added = 0.185 *M* x 24.02 mL = 4.44 mmol NaOH

	CH_3COOH	+	NaOH	$\rightarrow$	$NaCH_3COO$	+	H_2O
initial	5.90 mmol		4.44 mmol		0 mmol		
change	- 4.44 mmol		- 4.44 mmol		+ 4.44 mmol		
after reaction	1.46 mmol		0 mmol		4.44 mmol		

After the reaction, we have a total of 56.46 mL (= 32.44 + 24.02) buffer solution containing 1.46 mmol CH_3COOH and 4.44 mmol $NaCH_3COO$.

$$[H_3O^+] = K_a \times \frac{\text{mmol } CH_3COOH}{\text{mmol } NaCH_3COO} = (1.8 \times 10^{-5}) \frac{1.46 \text{ mmol}}{4.44 \text{ mmol}} = 5.9 \times 10^{-6} \, M; \quad pH = \mathbf{5.23}$$

(e) initial mmol CH_3COOH = 0.182 *M* x 32.44 mL = 5.90 mmol CH_3COOH
mmol of NaOH added = 0.185 *M* x 27.2 mL = 5.03 mmol NaOH

	CH_3COOH	+	NaOH	$\rightarrow$	$NaCH_3COO$	+	H_2O
initial	5.90 mmol		5.03 mmol		0 mmol		
change	- 5.03 mmol		- 5.03 mmol		+ 5.03 mmol		
after reaction	0.87 mmol		0 mmol		5.03 mmol		

After the reaction, we have a total of 59.6 mL (= 32.44 + 27.2) buffer solution containing 0.87 mmol CH_3COOH and 5.03 mmol $NaCH_3COO$.

$$[H_3O^+] = K_a \times \frac{\text{mmol } CH_3COOH}{\text{mmol } NaCH_3COO} = (1.8 \times 10^{-5}) \frac{0.87 \text{ mmol}}{5.03 \text{ mmol}} = 3.1 \times 10^{-6} \; M; \; pH = \textbf{5.51}$$

(f) initial mmol CH_3COOH = 0.182 M x 32.44 mL = 5.90 mmol CH_3COOH
mmol of NaOH added = 0.185 M x 31.91 mL = 5.90 mmol NaOH

	CH_3COOH	+	NaOH	→	$NaCH_3COO$	+	H_2O
initial	5.90 mmol		5.90 mmol		0 mmol		
change	- 5.90 mmol		- 5.90 mmol		+ 5.90 mmol		
after reaction	0 mmol		0 mmol		5.90 mmol		

After the reaction, we have a total of 64.35 mL (= 32.44 + 31.91) salt solution containing 0 mmol CH_3COOH and 5.90 mmol $NaCH_3COO$.

This is the equivalence point and there is no excess acid or base. The concentration of $NaCH_3COO$ determines the pH of the system (Refer to Exercise 19-46 Solution). At point (f):

$$[NaCH_3COO] = \frac{5.90 \text{ mmol}}{64.35 \text{ mL}} = 0.0917 \; M$$

Then the anion of the salt hydrolyzes to produce a basic solution: $CH_3COO^- + H_2O \rightleftarrows CH_3COOH + OH^-$

Let x = $[CH_3COO^-]_{\text{hydrolyzed}}$ Then, 0.0917 - x = $[CH_3COO^-]$; x = $[CH_3COOH]$ = $[OH^-]$

$$K_b = \frac{K_w}{K_{a(CH_3COOH)}} = 5.6 \times 10^{-10} = \frac{[CH_3COOH][OH^-]}{[CH_3COO^-]} = \frac{x^2}{0.0917 - x} \approx \frac{x^2}{0.0917} \qquad \text{Solving, x} = 7.2 \times 10^{-6}$$

$[OH^-]$ = x = $7.2 \times 10^{-6} \; M$; pOH = 5.14; pH = **8.86**

(g) initial mmol CH_3COOH = 0.182 M x 32.44 mL = 5.90 mmol CH_3COOH
mmol of NaOH added = 0.185 M x 33.12 mL = 6.13 mmol NaOH

	CH_3COOH	+	NaOH	→	$NaCH_3COO$	+	H_2O
initial	5.90 mmol		6.13 mmol		0 mmol		
change	- 5.90 mmol		- 5.90 mmol		+ 5.90 mmol		
after reaction	0 mmol		0.23 mmol		5.90 mmol		

After the equivalence point, the pH is determined directly from the concentration of *excess* NaOH since CH_3COOH is now the limiting reactant. In the presence of the strong base, the effect of the weak base, CH_3COO^-, derived from the salt is negligible.

After the reaction, we have a total of 65.56 mL (= 32.44 + 33.12) strong base solution containing 0.23 mmol NaOH and 5.90 mmol $NaCH_3COO$.

$$[OH^-] = [NaOH]_{\text{excess}} = \frac{0.23 \text{ mmol}}{65.56 \text{ mL}} = 3.5 \times 10^{-3} \; M; \; pOH = 2.46; \; pH = \textbf{11.54}$$

19-58. *Refer to Section 19-6, Figure 19-4, and Appendix F.*

Balanced equation: $CH_3COOH + KOH \rightarrow KCH_3COO + H_2O$

The resultant solution at the equivalence point of any acid-base reaction contains only salt and water. The pH is determined from the concentration of the salt.

(a) (1) initial mmol CH_3COOH = 1.200 M x 100.0 mL = 120.0 mmol CH_3COOH
So, at the equivalence point, mmol of KOH added = 120.0 mmol KOH, by definition

	CH_3COOH	+	KOH	→	KCH_3COO	+	H_2O
initial	120.0 mmol		120.0 mmol		0 mmol		
change	- 120.0 mmol		- 120.0 mmol		+ 120.0 mmol		
after reaction	0.0 mmol		0.0 mmol		120.0 mmol		

Volume of KOH needed to reach the equivalence point $= 120.0$ mmol $\times \dfrac{1 \text{ mL}}{0.1500 \text{ mmol}} = 800.0$ mL

Total volume at the equivalence point $= 100.0$ mL $+ 800.0$ mL $= 900.0$ mL

$[KCH_3COO] = \dfrac{100.0 \text{ mmol}}{900.0 \text{ mL}} = 0.1111 \; M$

(2) The anion of the soluble salt hydrolyzes to form a basic solution:

$$CH_3COO^- + H_2O \rightleftarrows CH_3COOH + OH^-$$

$$K_b = \dfrac{K_w}{K_{a(CH_3COOH)}} = \dfrac{1.0 \times 10^{-14}}{1.8 \times 10^{-5}} = 5.6 \times 10^{-10} = \dfrac{[CH_3OOH][OH^-]}{[CH_3COO^-]} = \dfrac{x^2}{0.1111 - x} \approx \dfrac{x^2}{0.1111}$$

Solving, $x = 7.9 \times 10^{-6}$; $[OH^-] = 7.9 \times 10^{-6} \; M$; pOH $= 5.10$; pH $= \mathbf{8.90}$

(b) (1) initial mmol $CH_3COOH = 0.1200 \; M \times 100.0$ mL $= 12.00$ mmol CH_3COOH

So, at the equivalence point, mmol of KOH added $= 12.00$ mmol KOH, by definition

	CH₃COOH	+	KOH	→	KCH₃COO	+	H₂O
initial	12.00 mmol		12.00 mmol		0 mmol		
change	- 12.00 mmol		- 12.00 mmol		+ 12.00 mmol		
after reaction	0.00 mmol		0.00 mmol		12.00 mmol		

Volume of KOH needed to reach the equivalence point $= 12.00$ mmol $\times \dfrac{1 \text{ mL}}{0.1500 \text{ mmol}} = 80.00$ mL

Total volume at the equivalence point $= 100.0$ mL $+ 80.00$ mL $= 180.0$ mL

$[KCH_3COO] = \dfrac{12.00 \text{ mmol}}{180.0 \text{ mL}} = 0.06667 \; M$

(2) $K_b = \dfrac{K_w}{K_{a(CH_3COOH)}} = \dfrac{1.0 \times 10^{-14}}{1.8 \times 10^{-5}} = 5.6 \times 10^{-10} = \dfrac{[CH_3OOH][OH^-]}{[CH_3COO^-]} = \dfrac{x^2}{0.06667 - x} \approx \dfrac{x^2}{0.06667}$

Solving, $x = 6.1 \times 10^{-6}$; $[OH^-] = 6.1 \times 10^{-6} \; M$; pOH $= 5.21$; pH $= \mathbf{8.79}$

(c) (1) initial mmol $CH_3COOH = 0.06000 \; M \times 100.0$ mL $= 6.000$ mmol CH_3COOH

So, at the equivalence point, mmol of KOH added $= 6.000$ mmol KOH, by definition

	CH₃COOH	+	KOH	→	KCH₃COO	+	H₂O
initial	6.000 mmol		6.000 mmol		0 mmol		
change	- 6.000 mmol		- 6.000 mmol		+ 6.000 mmol		
after reaction	0.000 mmol		0.000 mmol		6.000 mmol		

Volume of KOH needed to reach the equivalence point $= 6.000$ mmol $\times \dfrac{1 \text{ mL}}{0.1500 \text{ mmol}} = 40.00$ mL

Total volume at the equivalence point $= 100.0$ mL $+ 40.00$ mL $= 140.0$ mL

$[KCH_3COO] = \dfrac{6.000 \text{ mmol}}{140.0 \text{ mL}} = 0.04286 \; M$

(2) $K_b = \dfrac{[CH_3OOH][OH^-]}{[CH_3COO^-]} = 5.6 \times 10^{-10} = \dfrac{x^2}{0.04286 - x} \approx \dfrac{x^2}{0.04286}$

Solving, $x = 4.9 \times 10^{-6}$; $[OH^-] = 4.9 \times 10^{-6} \; M$; pOH $= 5.31$; pH $= \mathbf{8.69}$

19-60. *Refer to Section 19-2.*

(a) Balanced net ionic equation: $H^+(aq) + (CH_3CH_2)_3N(aq) \rightleftarrows (CH_3CH_2)_3NH^+(aq)$

(b) ? mL HCl $= 20.00$ mL $(CH_3CH_2)_3N \times \dfrac{0.220 \text{ mmol } (CH_3CH_2)_3N}{1 \text{ mL soln}} \times \dfrac{1 \text{ mmol HCl}}{1 \text{ mmoL } (CH_3CH_2)_3N}$

$\times \dfrac{1 \text{ mL HCl}}{0.544 \text{ mmol HCl}} = \mathbf{8.09 \text{ mL HCl}}$

(c) At the equivalence point, there is no excess acid or base, only salt and water. The pH of the salt solution, $(CH_3CH_2)_3NHCl$, is <7 since we have a solution of the salt of a weak base and a strong acid. The concentration of $(CH_3CH_2)_3NH^+$ determines the pH of the system.

? mmol HCl = mmol $(CH_3CH_2)_3N$ = 0.220 M $(CH_3CH_2)_3N$ x 20.00 mL = 4.40 mmol

	HCl	+	$(CH_3CH_2)_3N$	$\rightarrow$	$(CH_3CH_2)_3NH^+$
initial	4.40 mmol		4.40 mmol		0 mmol
change	- 4.40 mmol		- 4.40 mmol		+ 4.40 mmol
after reaction	0.00 mmol		0.00 mmol		4.40 mmol

$$[(CH_3CH_2)_3NH^+] = \frac{4.40 \text{ mmol}}{(20.00 \text{ mL} + 8.09 \text{ mL})} = 0.157 \ M$$

Then the cation of the salt hydrolyzes to produce an acidic solution:
$$(CH_3CH_2)_3NH^+ + H_2O \rightleftarrows (CH_3CH_2)_3N + H_3O^+$$

Let x = $[(CH_3CH_2)_3NH^+]_{hydrolyzed}$

Then, 0.157 - x = $[(CH_3CH_2)_3NH^+]$; x = $[(CH_3CH_2)_3N]$ = $[H_3O^+]$

$$K_a = \frac{K_w}{K_{b((CH_3CH_2)_3N)}} = \frac{1.0 \times 10^{-14}}{5.2 \times 10^{-4}} = 1.9 \times 10^{-11} = \frac{[(CH_3CH_2)_3N][H_3O^+]}{[(CH_3CH_2)_3NH^+]} = \frac{x^2}{0.157 - x} \approx \frac{x^2}{0.157}$$

Solving, x = 1.7×10^{-6}

$[(CH_3CH_2)_3N] = [H_3O^+]$ = x = $\mathbf{1.7 \times 10^{-6} \ M}$ $\qquad\qquad$ $[(CH_3CH_2)_3NH^+]$ = 0.157 - x = $\mathbf{0.157 \ M}$

$[Cl^-]$ = $\mathbf{0.157 \ M}$

(d) pH = **5.77**

19-62. *Refer to Section 19-2.*

Balanced equation: $H^+(aq) + C_6H_5NH_2(aq) \rightleftarrows C_6H_5NH_3^+(aq)$

The small amount of added acid will react with an equimolar amount of the base part of the buffer, $C_6H_5NH_2$, to form more of the conjugate acid part of the buffer.

19-64. *Refer to Section 19-1.*

Buffer solutions: HNO_2, nitrous acid, and **$NaNO_2$, sodium nitrite**

$\qquad\qquad\qquad$ NaCN, sodium cyanide, and **HCN, hydrocyanic acid**

$\qquad\qquad\qquad$ $(NH_4)_2SO_4$, ammonium sulfate, and **NH_3, ammonia**

$\qquad\qquad\qquad$ NH_3, ammonia, and **NH_4Cl, ammonium chloride**

$\qquad\qquad\qquad$ HCN, hydrocyanic acid, and **KCN, potassium cyanide**

$\qquad\qquad\qquad$ $(CH_3)_3N$, trimethylamine, and **$(CH_3)_3NHCl$, trimethylammonium chloride**

19-66. *Refer to Sections 19-6 and 19-8, Figure 19-5, Table 19-4, and Exercise 19-54 Solution.*

The calculations for determining the pH in the titration of 1.00 liter of 0.0100 M HNO_3 with gaseous NH_3 can be divided into 4 types:

(1) Initially, the pH is determined by the concentration of the strong acid, HNO_3.
$[H^+] = [HNO_3] = 0.0100 \ M$; pH = 2.00

(2) Before the equivalence point, the pH is essentially determined by the concentration of *excess* strong acid, HNO_3, since NH_3 is the limiting reagent. In the presence of the strong acid, the effect of the weak acid, NH_4^+, derived from the salt is negligible. For example, consider the limiting reactant problem for point (c) in the following table:

323

	HNO$_3$	+	NH$_3$	→	NH$_4$NO$_3$
initial	10.0 mmol		4.00 mmol		0 mmol
change	- 4.00 mmol		- 4.00 mmol		+ 4.00 mmol
after reaction	6.0 mmol		0 mmol		4.00 mmol

$[H^+] = [HNO_3]_{excess} = \dfrac{0.0060 \text{ mol}}{1.00 \text{ L}} = 6.0 \times 10^{-3} \ M; \ pH = 2.22$

(3) At the equivalence point, there is no excess acid or base. The pH of the salt solution, NH$_4$NO$_3$, is <7 since we have a solution of the salt of a weak base and a strong acid. The concentration of NH$_4$NO$_3$ determines the pH of the system. At point (g):

	HNO$_3$	+	NH$_3$	→	NH$_4$NO$_3$
initial	10.0 mmol		10.0 mmol		0 mmol
change	- 10.0 mmol		- 10.0 mmol		+ 10.0 mmol
after reaction	0 mmol		0 mmol		10.0 mmol

$[NH_4NO_3] = \dfrac{0.0100 \text{ mol}}{1.00 \text{ L}} = 0.0100 \ M$

Then the cation of the salt hydrolyzes to produce an acidic solution:
$$NH_4^+ + H_2O \rightleftharpoons NH_3 + H_3O^+$$

Let x = $[NH_4^+]_{hydrolyzed}$; then, 0.0100 - x = $[NH_4^+]$; x = $[NH_3]$ = $[H_3O^+]$

$K_a = \dfrac{K_w}{K_{b(NH_3)}} = 5.6 \times 10^{-10} = \dfrac{[NH_3][H_3O^+]}{[NH_4^+]} = \dfrac{x^2}{0.0100 - x} \approx \dfrac{x^2}{0.0100}$

Solving, x = 2.4×10^{-6}; $[H_3O^+]$ = x = $2.4 \times 10^{-6} \ M$; pH = 5.62

(4) After the equivalence point, the pH is determined by the buffer solution consisting of excess NH$_3$ and NH$_4^+$ produced by the neutralization reaction. For example, at point (i):

	HNO$_3$	+	NH$_3$	→	NH$_4$NO$_3$
initial	10.0 mmol		13.0 mmol		0 mmol
change	- 10.0 mmol		- 10.0 mmol		+ 10.0 mmol
after reaction	0 mmol		3.0 mmol		10.0 mmol

After the reaction, we have a 1 liter solution of NH$_3$ and the soluble salt, NH$_4$NO$_3$. OH$^-$ ion is produced by:

$$NH_3 + H_2O \rightleftharpoons NH_4^+ + OH^- \hspace{3cm} K_b = 1.8 \times 10^{-5}$$

$[OH^-] = K_b \times \dfrac{\text{mol NH}_3}{\text{mol NH}_4NO_3} = (1.8 \times 10^{-5}) \dfrac{3.0 \text{ mmol}}{10.0 \text{ mmol}} = 5.4 \times 10^{-6} \ M$

Therefore, pH = 8.73

Data Table:

	Mol NH$_3$ Added	Type of Solution	$[H_3O^+]$ (M)	$[OH^-]$ (M)	pH	pOH
(a)	none	strong acid	1.00×10^{-2}	1.00×10^{-12}	2.00	12.00
(b)	0.00100	strong acid	9.0×10^{-3}	1.1×10^{-12}	2.05	11.95
(c)	0.00400	strong acid	6.0×10^{-3}	1.7×10^{-12}	2.22	11.78
(d)	0.00500	strong acid	5.0×10^{-3}	2.0×10^{-12}	2.30	11.70
	(halfway to the equivalence point - has no real significance in this case)					
(e)	0.00900	strong acid	1.0×10^{-3}	1.0×10^{-11}	3.00	11.00
(f)	0.00950	strong acid	5×10^{-4}	2×10^{-11}	3.3	10.7
(g)	0.0100	salt	2.4×10^{-6}	4.2×10^{-9}	5.62	8.38
	(at the equivalence point)					
(h)	0.0105	buffer	1.1×10^{-8}	9×10^{-7}	8.0	6.0
(i)	0.0130	buffer	1.9×10^{-9}	5.4×10^{-6}	8.73	5.27

Titration Curve: HNO_3 vs. NH_3

The major difference between this titration curve of the HNO_3/NH_3 reaction and the other titrations is that the system is buffered after the equivalence point, not before the equivalence point. A satisfactory indicator for this titration is methyl red.

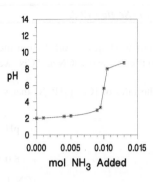

19-68. *Refer to Sections 18-7 and 18-9.*

(1) NaCl is the salt of a strong acid and a strong base, so its solution has a pH = **7.00**, no matter what its concentration is.

(2) NH_4Cl is the salt of a strong acid and a weak base, so its solution has a pH < 7.

Balanced equations: $NH_4Cl \rightarrow NH_4^+ + Cl^-$ (to completion)

$NH_4^+ + H_2O \rightleftarrows NH_3 + H_3O^+$ (reversible)

Let x = $[NH_4^+]_{hydrolyzed}$ Then, 0.54 - x = $[NH_4^+]$; x = $[NH_3]$ = $[H_3O^+]$

$K_a = \dfrac{K_w}{K_{b(NH_3)}} = \dfrac{1.0 \times 10^{-14}}{1.8 \times 10^{-5}} = 5.6 \times 10^{-10} = \dfrac{[NH_3][H_3O^+]}{[NH_4^+]} = \dfrac{x^2}{0.54 - x} \approx \dfrac{x^2}{0.54}$ Solving, x = 1.7×10^{-5}

Therefore, $[H_3O^+] = 1.7 \times 10^{-5}$ M; pH = **4.76**

19-70. *Refer to Section 19-4.*

For a weak acid indicator, indigo carmine, the acid form, HIn, is blue and the base form, In^-, is yellow.

$$HIn \rightleftarrows H^+ + In^- \qquad K_a = \dfrac{[H^+][In^-]}{[HIn]}$$

At the endpoint, when the color change occurs for the indicator, $[HIn] = [In^-]$. Then, $K_a = [H^+]$ and pK_a = pH. So, for the indicator, indigo carmine, pH at the endpoint = $pK_a = -\log(6 \times 10^{-13}) =$ **12.2.**

19-72. *Refer to Section 18-9.*

(a) dissociation: $(CH_3)_2NH_2Cl \rightarrow (CH_3)_2NH_2^+ + Cl^-$

(b) hydrolysis: $(CH_3)_2NH_2^+ + H_2O \rightleftarrows (CH_3)_2NH + H_3O^+$

(c) The resulting solution of the weak acid, dimethylammonium ion, will be **acidic.**

19-74. *Refer to Section 18-10.*

(a) for the ammonium ion, NH_4^+: $K_a = \dfrac{K_w}{K_{b(NH_3)}} = \dfrac{1.0 \times 10^{-14}}{1.8 \times 10^{-5}} = $ **5.6 x 10⁻¹⁰**

(b) for the hypochlorite ion, ClO^-: $K_b = \dfrac{K_w}{K_{a(HClO)}} = \dfrac{1.0 \times 10^{-14}}{3.5 \times 10^{-8}} = $ **2.9 x 10⁻⁷**

(c) A solution of NH_4ClO will be **basic** because the weak base, ClO^-, is stronger than the weak acid, NH_4^+, since K_b for ClO^- is larger than the K_a for NH_4^+.

Buffers made of weak acids work best at the pH where the pK_a = pH. At this pH, [weak acid] = [conjugate base]. So, a buffer made of HCN and CN$^-$ would work best at pH = -log (4.0×10^{-10}) = 9.40, not 8.0.

Solving for the ratio of [CN$^-$]/[HCN] that would give a pH of 8.0, we could use the Henderson-Hasselbalch equation:

$$pH = pK_a + \log \frac{[CN^-]}{[HCN]}$$

$$8.0 = 9.4 + \log \frac{[CN^-]}{[HCN]}$$

$$\log \frac{[CN^-]}{[HCN]} = -1.4$$

$$\frac{[CN^-]}{[HCN]} = 0.040$$

Since to be an effective buffer, their concentrations must be within 0.1 - 10 of each other, one really cannot make an effective buffer using HCN at pH = 8.0.

In a titration, the equivalence point is the point at which chemically equivalent amounts of reactants have reacted, whereas the end point is the point at which an indicator changes color and a titration should be stopped. So, a chemist needs to be careful when choosing an indicator in an acid-base reaction, to be certain that the pH at which the indicator changes color is close to the pH at the equivalence point of the titration.

Balanced equation: $NaOH + KHSO_4 \rightarrow NaKSO_4 + H_2O$

Plan: (1) Calculate the mass of $KHSO_4$ in the mixture
            (2) Determine the % by mass of $KHSO_4$

(1) ? g $KHSO_4$ = 0.0368 L NaOH x $\dfrac{0.115 \text{ mol NaOH}}{1 \text{ L NaOH}}$ x $\dfrac{1 \text{ mol } KHSO_4}{1 \text{ mol NaOH}}$ x $\dfrac{136.2 \text{ g } KHSO_4}{1 \text{ mol } KHSO_4}$ = 0.576 g $KHSO_4$

(2) ? % $KHSO_4$ by mass = $\dfrac{0.576 \text{ g } KHSO_4}{0.738 \text{ g mixture}}$ x 100 = **78.1%**

What we have is a solution of two weak acid, hypochlorous acid (HOCl) and hypoiodous acid (HOI):

$$HOCl + H_2O \rightleftarrows H_3O^+ + OCl^- \qquad K_a = 3.5 \times 10^{-8}$$
$$HOI + H_2O \rightleftarrows H_3O^+ + OI^- \qquad K_a = 2.3 \times 10^{-11}$$

Because the K_a for HOCl is more than 1000 times that of HOI, the pH in the solution is due only to the ionization of HOCl, following the same train of thought as for polyprotic acids.

$$K_a = \frac{[H_3O^+][OCl^-]}{[HOCl]} = \frac{x^2}{0.30 - x} = 3.5 \times 10^{-8}$$

Assume that 0.30 - x $\approx$ 0.30. Then $x^2/0.30 = 3.5 \times 10^{-8}$ and x = 1.0×10^{-4}.

The simplifying assumption is justified since 1.0×10^{-4} is less than 5% of 0.30 (= 0.015).
Therefore at equilibrium: [H_3O^+] = 1.0×10^{-4} M; pH = **4.00**

Balanced equation: $HCl + NaOH \rightarrow NaCl + H_2O$

Plan: (1) Determine the concentration of the acid solution.
           (2) Calculate the necessary volume by using dilution.

(1) At the equivalence point (pH = 7 at the equivalence point of a strong acid-strong base titration):

mmoles of acid = mmoles of base   (when the acid has 1 acidic H and the base has 1 basic OH)

$$M_{HCl} \times \text{mL of HCl} = M_{NaOH} \times \text{mL of NaOH}$$
$$M_{HCl} \times 20.00 \text{ mL} = 3.00 \ M \times 34.0 \text{ mL}$$
$$M_{HCl} = 5.10 \ M$$

(2) Solving for volume, using concepts for dilution:

$$M_1 V_1 = M_2 V_2$$
$$V_1 = \frac{M_2 V_2}{M_1}$$
$$= \frac{0.75 \ M \times 1.5 \text{ L}}{5.10 \ M}$$
$$= \textbf{0.22 L HCl}$$

Note: The fact that we had 3.5 L of the HCl solution did not enter into the calculations.

20 Ionic Equilibria III:
The Solubility Product Principle

20-2. *Refer to Section 20-1.*

The solubility product principle states that the solubility product expression for a slightly soluble compound is the product of the concentrations of its constituent ions, each raised to the power that corresponds to the number of ions in one formula unit of the compound. The quantity, K_{sp}, is constant at constant temperature for a saturated solution of the compound, when the system is at equilibrium. The significance of the solubility product is that it can be used to calculate the concentrations of the ions in solutions for such slightly soluble compounds.

20-4. *Refer to Section 20-1.*

The molar solubility of a compound is the number of moles of the compound that dissolve to produce one liter of saturated solution.

20-6. *Refer to Section 20-1.*

(a) $Mn_3(AsO_4)_2(s) \rightleftarrows 3Mn^{2+}(aq) + 2AsO_4^{3-}(aq)$ $\qquad$ $K_{sp} = [Mn^{2+}]^3[AsO_4^{3-}]^2$

(b) $Hg_2I_2(s) \rightleftarrows Hg_2^{2+}(aq) + 2I^-(aq)$ $\qquad$ $K_{sp} = [Hg_2^{2+}][I^-]^2$

(c) $AuI_3(s) \rightleftarrows Au^{3+}(aq) + 3I^-(aq)$ $\qquad$ $K_{sp} = [Au^{3+}][I^-]^3$

(d) $SrCO_3(s) \rightleftarrows Sr^{2+}(aq) + CO_3^{2-}(aq)$ $\qquad$ $K_{sp} = [Sr^{2+}][CO_3^{2-}]$

20-8. *Refer to Section 20-2, Examples 20-1 and 20-2, and Appendix H.*

Plan: (1) Calculate the molar solubility of the slightly soluble salt, which is the number of moles of the salt that will dissolve in 1 liter of solution.
(2) Determine the concentrations of the ions in solution.
(3) Substitute the ion concentrations into the K_{sp} expression to calculate K_{sp}.

(a) Balanced equation: $SrCrO_4(s) \rightleftarrows Sr^{2+}(aq) + CrO_4^{2-}(aq)$ $\qquad$ $K_{sp} = [Sr^{2+}][CrO_4^{2-}]$

(1) molar solubility (mol $SrCrO_4$/L) $= \dfrac{1.2 \text{ mg } SrCrO_4}{1 \text{ mL}} \times \dfrac{1000 \text{ mL}}{1 \text{ L}} \times \dfrac{1 \text{ g}}{1000 \text{ mg}} \times \dfrac{1 \text{ mol } SrCrO_4}{204 \text{ g } SrCrO_4}$

$\qquad\qquad = 5.9 \times 10^{-3}$ mol $SrCrO_4$/L (dissolved)

(2) $[Sr^{2+}] = [CrO_4^{2-}] = $ molar solubility $= 5.9 \times 10^{-3}$ M

(3) $K_{sp} = [Sr^{2+}][CrO_4^{2-}] = (5.9 \times 10^{-3})^2 = \mathbf{3.5 \times 10^{-5}}$ $\qquad$ (3.6×10^{-5} from Appendix H)

(b) Balanced equation: $BiI_3(s) \rightleftarrows Bi^{3+}(aq) + 3I^-(aq)$ $\qquad$ $K_{sp} = [Bi^{3+}][I^-]^3$

(1) molar solubility $= \dfrac{7.7 \times 10^{-3} \text{ g } BiI_3}{1 \text{ L}} \times \dfrac{1 \text{ mol } BiI_3}{590 \text{ g } BiI_3} = 1.3 \times 10^{-5}$ mol BiI_3/L (dissolved)

(2) $[Bi^{3+}] = $ molar solubility $= 1.3 \times 10^{-5}$ M
$[I^-] = 3 \times $ molar solubility $= 3.9 \times 10^{-5}$ M

(3) $K_{sp} = [Bi^{3+}][I^-]^3 = (1.3 \times 10^{-5})(3.9 \times 10^{-5})^3 = \mathbf{7.7 \times 10^{-19}}$ $\qquad$ (8.1×10^{-19} from Appendix H)

(c) Balanced equation: $Fe(OH)_2(s) \rightleftarrows Fe^{2+}(aq) + 2OH^-(aq)$ $\qquad$ $K_{sp} = [Fe^{2+}][OH^-]^2$

(1) molar solubility $= \dfrac{1.1 \times 10^{-3} \text{ g } Fe(OH)_2}{1 \text{ L}} \times \dfrac{1 \text{ mol } Fe(OH)_2}{89.9 \text{ g } Fe(OH)_2} = 1.2 \times 10^{-5}$ mol $Fe(OH)_2$/L (dissolved)

(2) $[Fe^{2+}]$ = molar solubility = 1.2×10^{-5} M
    $[OH^-]$ = 2 x molar solubility = 2.4×10^{-5} M
(3) $K_{sp} = [Fe^{2+}][OH^-]^2 = (1.2 \times 10^{-5})(2.4 \times 10^{-5})^2 = \textbf{6.9} \times \textbf{10}^{\textbf{-15}}$ (7.9×10^{-15} from Appendix H)

(d) Balanced equation: $SnI_2(s) \rightleftarrows Sn^{2+}(aq) + 2I^-(aq)$ $K_{sp} = [Sn^{2+}][I^-]^2$

(1) molar solubility = $\dfrac{10.9 \text{ g } SnI_2}{1 \text{ L}} \times \dfrac{1 \text{ mol } SnI_2}{372.5 \text{ g } SnI_2}$ = 2.93×10^{-2} mol SnI_2/L (dissolved)

(2) $[Sn^{2+}]$ = molar solubility = 2.93×10^{-2} M
    $[I^-]$ = 2 x molar solubility = 5.86×10^{-2} M

(3) $K_{sp} = [Sn^{2+}][I^-]^2 = (2.93 \times 10^{-2})(5.86 \times 10^{-2})^2 = \textbf{1.01} \times \textbf{10}^{\textbf{-4}}$ (1.0×10^{-4} from Appendix H)

20-10.   *Refer to Section 20-2, Table 20-1 and Exercise 20-8 Solution.*

Compound	Molar Solubility (M)	K_{sp} (calculated)
$SrCrO_4$	5.9×10^{-3}	$[Sr^{2+}][CrO_4^{2-}] = 3.5 \times 10^{-5}$
BiI_3	1.3×10^{-5}	$[Bi^{3+}][I^-]^3 = 7.7 \times 10^{-19}$
$Fe(OH)_2$	1.2×10^{-5}	$[Fe^{2+}][OH^-]^2 = 6.9 \times 10^{-15}$
SnI_2	2.93×10^{-2}	$[Sn^{2+}][I^-]^2 = 1.01 \times 10^{-4}$

(a) SnI_2 has the highest molar solubility.   (c) SnI_2 has the largest K_{sp}.
(b) $Fe(OH)_2$ has the lowest molar solubility.   (d) BiI_3 has the smallest K_{sp}.

20-12.   *Refer to Section 20-2, and Examples 20-1 and 20-2.*

Balanced equation: $CaF_2(s) \rightleftarrows Ca^{2+}(aq) + 2F^-(aq)$ $K_{sp} = [Ca^{2+}][F^-]^2$

Plan: (1) Determine the molar solubility of CaF_2 at 25°C.
         (2) Calculate K_{sp}.

(1) molar solubility = $\dfrac{0.0163 \text{ g } CaF_2}{1.00 \text{ L}} \times \dfrac{1 \text{ mol } CaF_2}{78.08 \text{ g } CaF_2}$ = 2.09×10^{-4} mol CaF_2/L (dissolved)

(2) $[Ca^{2+}]$ = molar solubility = 2.09×10^{-4} M; $[F^-]$ = 2 x molar solubility = 4.18×10^{-4} M
    $K_{sp} = [Ca^{2+}][F^-]^2 = (2.09 \times 10^{-4})(4.18 \times 10^{-4})^2 = \textbf{3.65} \times \textbf{10}^{\textbf{-11}}$ (3.9×10^{-11} from Appendix H)

20-14.   *Refer to Section 20-3, Example 20-3 and Appendix H.*

Balanced equation: $CaCO_3(s) \rightleftarrows Ca^{2+}(aq) + CO_3^{2-}(aq)$ $K_{sp} = 4.8 \times 10^{-9}$
Let x = molar solubility of $CaCO_3$. Then, $[Ca^{2+}] = [CO_3^{2-}] = x$
$K_{sp} = [Ca^{2+}][CO_3^{2-}] = x^2 = 4.8 \times 10^{-9}$ Solving, x = 6.9×10^{-5}
Therefore,   molar solubility = 6.9×10^{-5} mol $CaCO_3$/L (dissolved)

solubility (g/L) = $\dfrac{6.9 \times 10^{-5} \text{ mol } CaCO_3}{1 \text{ L}} \times \dfrac{100.09 \text{ g } CaCO_3}{1 \text{ mol } CaCO_3}$ = $\textbf{6.9} \times \textbf{10}^{\textbf{-3}}$ **g/L**

20-16.   *Refer to Section 20-3, Example 20-3 and Appendix H.*

(a) Balanced equation: $CuCl(s) \rightleftarrows Cu^+(aq) + I^-(aq)$ $K_{sp} = 1.9 \times 10^{-7}$
    Let x = molar solubility of CuCl. Then, $[Cu^+] = [Cl^-] = x$
    $K_{sp} = [Cu^+][Cl^-] = x^2 = 1.9 \times 10^{-7}$ Solving, x = 4.4×10^{-4}
    Therefore, molar solubility = **4.4 x 10⁻⁴ mol CuCl/L** (dissolved); $[Cu^+] = [Cl^-] = $ **4.4 x 10⁻⁴** M
    solubility (g/L) = $\dfrac{4.4 \times 10^{-4} \text{ mol CuCl}}{1 \text{ L}} \times \dfrac{99.00 \text{ g CuCl}}{1 \text{ mol CuCl}}$ = **0.043 g/L**

(b) Balanced equation: $Ba_3(PO_4)_2(s) \rightleftarrows 3Ba^{2+}(aq) + 2PO_4^{3-}(aq)$ $K_{sp} = 1.3 \times 10^{-29}$

Let x = molar solubility of $Ba_3(PO_4)_2$. Then, $[Ba^{2+}] = 3x$ and $[PO_4^{3-}] = 2x$

$K_{sp} = [Ba^{2+}]^3[PO_4^{3-}]^2 = (3x)^3(2x)^2 = 108x^5 = 1.3 \times 10^{-29}$ Solving, $x = 6.5 \times 10^{-7}$

Therefore,   molar solubility = **6.5×10^{-7} mol $Ba_3(PO_4)_2$/L** (dissolved)

$[Ba^{2+}] = 3x =$ **2.0×10^{-6} M**

$[PO_4^{3-}] = 2x =$ **1.3×10^{-6} M**

solubility (g/L) $= \dfrac{6.5 \times 10^{-7} \text{ mol } Ba_3(PO_4)_2}{1 \text{ L}} \times \dfrac{602 \text{ g } Ba_3(PO_4)_2}{1 \text{ mol } Ba_3(PO_4)_2} =$ **3.9×10^{-4} g/L**

(c) Balanced equation: $PbF_2(s) \rightleftarrows Pb^{2+}(aq) + 2F^-(aq)$ $K_{sp} = 3.7 \times 10^{-8}$

Let x = molar solubility of PbF_2. Then, $[Pb^{2+}] = x$ and $[F^-] = 2x$

$K_{sp} = [Pb^{2+}][F^-]^2 = (x)(2x)^2 = 4x^3 = 3.7 \times 10^{-8}$ Solving, $x = 2.1 \times 10^{-3}$

Therefore,   molar solubility = **2.1×10^{-3} mol PbF_2/L** (dissolved)

$[Pb^{2+}] = x =$ **2.1×10^{-3} M**

$[F^-] = 2x =$ **4.2×10^{-3} M**

solubility (g/L) $= \dfrac{2.1 \times 10^{-3} \text{ mol } PbF_2}{1 \text{ L}} \times \dfrac{245.2 \text{ g } PbF_2}{1 \text{ mol } PbF_2} =$ **0.51 g/L**

(d) Balanced equation: $Sr_3(PO_4)_2(s) \rightleftarrows 3Sr^{2+}(aq) + 2PO_4^{3-}(aq)$ $K_{sp} = 1.0 \times 10^{-31}$

Let x = molar solubility of $Sr_3(PO_4)_2$ Then, $[Sr^{2+}] = 3x$ and $[PO_4^{3-}] = 2x$

$K_{sp} = [Sr^{2+}]^3[PO_4^{3-}]^2 = (3x)^3(2x)^2 = 108x^5 = 1.0 \times 10^{-31}$ Solving, $x = 2.5 \times 10^{-7}$ (2.47×10^{-7} for calc.)

Therefore,   molar solubility = **2.5×10^{-7} mol $Sr_3(PO_4)_2$/L** (dissolved)

$[Sr^{2+}] = 3x =$ **7.4×10^{-7} M** (calculation done with 3 sig. fig. for x to minimize rounding

errors)

$[PO_4^{3-}] = 2x =$ **4.9×10^{-7} M**

solubility (g/L) $= \dfrac{2.5 \times 10^{-7} \text{ mol } Sr_3(PO_4)_2}{1 \text{ L}} \times \dfrac{452.8 \text{ g } Sr_3(PO_4)_2}{1 \text{ mol } Sr_3(PO_4)_2} =$ **1.1×10^{-4} g/L**

20-18.   *Refer to Section 20-3, Example 20-3 and Appendix H.*

Balanced equation: $BaSO_4(s) \rightleftarrows Ba^{2+}(aq) + SO_4^{2-}(aq)$ $K_{sp} = $ **1.1×10^{-10}**

Let x = molar solubility of $BaSO_4$. Then, $[Ba^{2+}] = [SO_4^{2-}] = x$

$K_{sp} = [Ba^{2+}][SO_4^{2-}] = x^2 = 1.1 \times 10^{-10}$ Solving, $x = 1.0 \times 10^{-5}$

Therefore,   molar solubility = 1.0×10^{-5} mol $BaSO_4$/L (dissolved)

solubility (g/L) $= \dfrac{1.0 \times 10^{-5} \text{ mol } BaSO_4}{1 \text{ L}} \times \dfrac{233.4 \text{ g } BaSO_4}{1 \text{ mol } BaSO_4} = 2.3 \times 10^{-3}$ g/L

mass of $BaSO_4$ that dissolves in 5.00 L = $(2.3 \times 10^{-3}$ g/L$)(5.00$ L$) =$ **0.012 g**

20-20.   *Refer to Section 20-3 and Example 20-4.*

This is a common ion effect problem similar to some buffer problems.

Balanced equations: $K_2SO_4(aq) \rightarrow 2K^+(aq) + SO_4^{2-}(aq)$ (to completion)

$Ag_2SO_4(s) \rightleftarrows 2Ag^+(aq) + SO_4^{2-}(aq)$ (reversible) $K_{sp} = 1.7 \times 10^{-5}$

Let   x = molar solubility of Ag_2SO_4 in 0.12 M K_2SO_4. Then,

$[Ag^+] = 2x$ (from Ag_2SO_4)

$[SO_4^{2-}] = x$ (from Ag_2SO_4) + 0.12 M (from K_2SO_4)

$K_{sp} = [Ag^+]^2[SO_4^{2-}] = (2x)^2(x + 0.12) = 1.7 \times 10^{-5} \approx (2x)^2(0.12)$ Solving, $x = 6.0 \times 10^{-3}$

Therefore, molar solubility = **6.0×10^{-3} mol Ag_2SO_4/L 0.12 M K_2SO_4 soln**

20-22. *Refer to Section 20-3, Table 20-1 and Exercise 20-16 Solution.*

Compound	K_{sp}	Molar Solubility (M)	Solubility (g/L)
CuCl	$[Cu^+][Cl^-] = 1.9 \times 10^{-7}$	4.4×10^{-4}	0.043
$Ba_3(PO_4)_2$	$[Ba^{2+}]^3[PO_4^{3-}]^2 = 1.3 \times 10^{-29}$	6.5×10^{-7}	3.9×10^{-4}
PbF_2	$[Pb^{2+}][F^-]^2 = 3.7 \times 10^{-8}$	2.1×10^{-3}	0.51
$Sr_3(PO_4)_2$	$[Sr^{2+}]^3[PO_4^{3-}]^2 = 1.0 \times 10^{-31}$	2.5×10^{-7}	1.1×10^{-4}

(a) **PbF_2** has the highest molar solubility.
(b) **$Sr_3(PO_4)_2$** has the lowest molar solubility.
(c) **PbF_2** has the highest solubility (g/L).
(d) **$Sr_3(PO_4)_2$** has the lowest solubility (g/L).

20-24. *Refer to Section 20-3, Example 20-3 and Appendix H.*

(1) Balanced equation: $Ag_2CO_3(s) \rightleftarrows 2Ag^+(aq) + CO_3^{2-}(aq)$ $\qquad$ $K_{sp} = 8.1 \times 10^{-12}$

Let x = molar solubility of Ag_2CO_3. Then, $[Ag^+] = 2x$ and $[CO_3^{2-}] = x$

$K_{sp} = [Ag^+]^2[CO_3^{2-}] = (2x)^2(x) = 4x^3 = 8.1 \times 10^{-12}$ $\qquad$ Solving, x = 1.3×10^{-4}

Therefore, molar solubility = 1.3×10^{-4} mol Ag_2CO_3/L (dissolved)
(Note: use 1.265×10^{-4} to minimize rounding errors in the following calculation)

$$\text{solubility (g/L)} = \frac{1.265 \times 10^{-4} \text{ mol } Ag_2CO_3}{1 \text{ L}} \times \frac{275.8 \text{ g } Ag_2CO_3}{1 \text{ mol } Ag_2CO_3} = 0.035 \text{ g/L}$$

(2) Balanced equation: $AgCl(s) \rightleftarrows Ag^+(aq) + Cl^-(aq)$ $\qquad$ $K_{sp} = 1.8 \times 10^{-10}$

Let x = molar solubility of AgCl. Then, $[Ag^+] = [Cl^-] = x$

$K_{sp} = [Ag^+][Cl^-] = x^2 = 1.8 \times 10^{-10}$ $\qquad$ Solving, x = 1.3×10^{-5}

Therefore, molar solubility = 1.3×10^{-5} mol AgCl/L (dissolved)

$$\text{solubility (g/L)} = \frac{1.3 \times 10^{-5} \text{ mol AgCl}}{1 \text{ L}} \times \frac{143.4 \text{ g AgCl}}{1 \text{ mol AgCl}} = 1.9 \times 10^{-3} \text{ g/L}$$

(3) Balanced equation: $Pb(OH)_2(s) \rightleftarrows Pb^{2+}(aq) + 2OH^-(aq)$ $\qquad$ $K_{sp} = 2.8 \times 10^{-16}$

Let x = molar solubility of $Pb(OH)_2$. Then, $[Pb^{2+}] = x$ and $[OH^-] = 2x$

$K_{sp} = [Pb^{2+}][OH^-]^2 = (x)(2x)^2 = 4x^3 = 2.8 \times 10^{-16}$ $\qquad$ Solving, x = 4.1×10^{-6}

Therefore, molar solubility = 4.1×10^{-6} mol $Pb(OH)_2$/L (dissolved)

$$\text{solubility (g/L)} = \frac{4.1 \times 10^{-6} \text{ mol } Pb(OH)_2}{1 \text{ L}} \times \frac{241.2 \text{ g } Pb(OH)_2}{1 \text{ mol } Pb(OH)_2} = 9.9 \times 10^{-4} \text{ g/L}$$

(a) **Ag_2CO_3** has the highest molar solubility.
(b) **$Pb(OH)_2$** has the lowest molar solubility.
(c) **Ag_2CO_3** has the highest solubility expressed in grams per liter.
(d) **$Pb(OH)_2$** has the lowest solubility expressed in grams per liter.

20-26. *Refer to Section 20-3, Example 20-4 and Appendix H.*

Plan: Determine separately the molar solubility of $BaCrO_4$ and Ag_2CrO_4 in 0.125 M K_2CrO_4. The salt with the higher molar solubility is more soluble in the K_2CrO_4 solution.

(1) Molar solubility of $BaCrO_4$ in 0.125 M K_2CrO_4:

Balanced equations: $K_2CrO_4(aq) \rightarrow 2K^+(aq) + CrO_4^{2-}(aq)$ $\qquad$ (to completion)
$\qquad\qquad\qquad\quad$ $BaCrO_4(s) \rightleftarrows Ba^{2+}(aq) + CrO_4^{2-}(aq)$ $\qquad$ (reversible) $\qquad$ $K_{sp} = 2.0 \times 10^{-10}$

Let   x = molar solubility of $BaCrO_4$ in 0.125 M K_2CrO_4. Then,
    $[Ba^{2+}]$ = x (from $BaCrO_4$)
    $[CrO_4^{2-}]$ = x (from $BaCrO_4$) + 0.125 M (from K_2CrO_4)

$K_{sp} = [Ba^{2+}][CrO_4^{2-}] = (x)(x + 0.125) = 2.0 \times 10^{-10} \approx (x)(0.125)$ Solving, x = 1.6×10^{-9}

Therefore, molar solubility = 1.6×10^{-9} mol $BaCrO_4$/L 0.125 M K_2CrO_4

(2) Molar solubility of Ag_2CrO_4 in 0.125 M K_2CrO_4:

Balanced equations: $K_2CrO_4(aq) \rightarrow 2K^+(aq) + CrO_4^{2-}(aq)$ (to completion)
                $Ag_2CrO_4(s) \rightleftarrows 2Ag^+(aq) + CrO_4^{2-}(aq)$ (reversible) $K_{sp} = 9.0 \times 10^{-12}$

Let   y = molar solubility of Ag_2CrO_4 in 0.125 M K_2CrO_4. Then,
    $[Ag^+]$ = 2y (from Ag_2CrO_4)
    $[CrO_4^{2-}]$ = y (from Ag_2CrO_4) + 0.125 M (from K_2CrO_4)

$K_{sp} = [Ag^+]^2[CrO_4^{2-}] = (2y)^2(y + 0.125) = 9.0 \times 10^{-12} \approx (2y)^2(0.125)$ Solving, y = 4.2×10^{-6}

Therefore, molar solubility = 4.2×10^{-6} mol Ag_2CrO_4/L 0.125 M K_2CrO_4

The molar solubility of **Ag_2CrO_4** in 0.125 M K_2CrO_4 is greater than that of $BaCrO_4$. We can therefore say that Ag_2CrO_4 is more soluble on a per mole basis. (Note: Ag_2CrO_4 is more soluble on a per gram basis also.)

20-28.   *Refer to Section 20-3, Example 20-5 and Appendix H.*

Balanced equations: $Pb(NO_3)_2(aq) + 2NaCl(aq) \rightarrow PbCl_2(s) + 2NaNO_3(aq)$ (Will precipitation occur?)
                $PbCl_2(s) \rightleftarrows Pb^{2+}(aq) + 2Cl^-(aq)$ (reversible) $K_{sp} = 1.7 \times 10^{-5}$

Plan: (1) Calculate the concentration of Pb^{2+} ions and Cl^- ions at the instant of mixing before combination occurs.
      (2) Determine the reaction quotient, Q_{sp}. If $Q_{sp} > K_{sp}$, then precipitation will occur.

(1) $[Pb^{2+}] = [Pb(NO_3)_2] = \dfrac{5.0 \text{ g } Pb(NO_3)_2/331 \text{ g/mol}}{1.00 \text{ L soln}} = 0.015 \ M \ Pb^{2+}$

   $[Cl^-] = [NaCl] = 0.010 \ M \ Cl^-$

(2) $Q_{sp} = [Pb^{2+}][Cl^-]^2 = (0.015)(0.010)^2 = 1.5 \times 10^{-6}$

   Since $Q_{sp} < K_{sp}$, precipitation will **not** occur.

20-30.   *Refer to Section 20-3, Example 20-5 and Appendix H.*

Balanced equations: $CuCl_2(aq) + 2NaOH(aq) \rightarrow Cu(OH)_2(s) + 2NaCl(aq)$ (Will precipitation occur?)
                $Cu(OH)_2(s) \rightleftarrows Cu^{2+}(aq) + 2OH^-(aq)$ (reversible) $K_{sp} = 1.6 \times 10^{-19}$

Plan: (1) Calculate the concentrations of Cu^{2+} ions and OH^- ions at the instant of mixing before reaction occurs using $M_1V_1 = M_2V_2$.
      (2) Determine the reaction quotient, Q_{sp}. If $Q_{sp} > K_{sp}$, a precipitate will form.

(1) $[Cu^{2+}] = [CuCl_2] = \dfrac{M_1V_1}{V_2} = \dfrac{0.010 \ M \times 1.00 \text{ L}}{(1.00 \text{ L} + 0.010 \text{ L})} = 9.9 \times 10^{-3} \ M$

   $[OH^-] = [NaOH] = \dfrac{M_1V_1}{V_2} = \dfrac{0.010 \ M \times 0.010 \text{ L}}{(1.00 \text{ L} + 0.010 \text{ L})} = 9.9 \times 10^{-5} \ M$

(2) $Q_{sp} = [Cu^{2+}][OH^-]^2 = (9.9 \times 10^{-3})(9.9 \times 10^{-5})^2 = 9.7 \times 10^{-11}$
   Since $Q_{sp} > K_{sp}$, a precipitate **will** form.

Balanced equations: $Pb^{2+}(aq) + 2NaI(aq) \rightarrow PbI_2(s) + 2Na^+(aq)$ (to completion)

$PbI_2(s) \rightleftarrows Pb^{2+}(aq) + 2I^-(aq)$ (reversible) $K_{sp} = 8.7 \times 10^{-9}$

Plan: (1) Do the limiting reactant problem to determine which reactant is in excess.
(2) If Pb^{2+} is the limiting reactant, calculate its concentration in the presence of the excess amount of NaI.
(3) Calculate %Pb^{2+} ions remaining in solution.

(1) ? mol Pb^{2+} = 0.0100 M x 1.00 L = 0.0100 mol Pb^{2+}
? mol NaI = 0.103 mol NaI

	$Pb^{2+}(aq)$	+	$2NaI(aq)$	$\rightarrow$	$PbI_2(s)$	+	$2Na^+(aq)$
initial	0.0100 mol		0.103 mol		0 mol		0 mol
change	- 0.0100 mol		- 0.020 mol		+ 0.0100 mol		+ 0.0200 mol
after reaction	0 mol		0.083 mol		0.0100 mol		0.0200 mol

Pb^{2+} is the limiting reactant; NaI is in excess.

(2) In the resulting solution, $[I^-]$ = $[NaI]$ = 0.083 mol/1.00 L = 0.083 M

Let x = molar solubility of PbI_2 in 0.083 M NaI. Then,
$[Pb^{2+}]$ = x (from PbI_2)
$[I^-]$ = 2x (from PbI_2) + 0.083 M (from NaI)

$K_{sp} = [Pb^{2+}][I^-]^2 = (x)(2x + 0.083)^2 = 8.7 \times 10^{-9} \approx (x)(0.083)^2$ Solving, x = 1.3 x 10^{-6}

Therefore, $[Pb^{2+}]$ = x = 1.3 x 10^{-6} M

(3) Therefore, % Pb^{2+} in solution = $\dfrac{1.3 \times 10^{-6} M}{0.0100 M}$ x 100% = **0.013%**

Fractional precipitation refers to a separation process whereby some ions are removed from solution by precipitation, leaving other ions with similar properties in solution.

Balanced equations: $Cu^+(aq) + NaBr(aq) \rightarrow CuBr(s) + Na^+(aq)$
$CuBr(s) \rightleftarrows Cu^+(aq) + Br^-(aq)$ (reversible) $K_{sp} = 5.3 \times 10^{-9}$

$Ag^+(aq) + NaBr(aq) \rightarrow AgBr(s) + Na^+(aq)$
$AgBr(s) \rightleftarrows Ag^+(aq) + Br^-(aq)$ (reversible) $K_{sp} = 3.3 \times 10^{-13}$

$Au^+(aq) + NaBr(aq) \rightarrow AuBr(s) + Na^+(aq)$
$AuBr(s) \rightleftarrows Au^+(aq) + Br^-(aq)$ (reversible) $K_{sp} = 5.0 \times 10^{-17}$

(a) Because the compounds to be precipitated are all of the same formular type, i.e., their cation to anion ratios are the same. The one with the smallest K_{sp} is the least soluble and will precipitate first. Hence, **AuBr** (K_{sp} = 5.0 x 10^{-17}) will precipitate first, then AgBr (K_{sp} = 3.3 x 10^{-13}), and finally CuBr (K_{sp} = 5.3 x 10^{-9}).

(b) AgBr will begin to precipitate when $Q_{sp(AgBr)} = K_{sp(AgBr)} = [Ag^+][Br^-]$. At this point,

$$[Br^-] = \frac{K_{sp(AgBr)}}{[Ag^+]} = \frac{3.3 \times 10^{-13}}{0.010} = 3.3 \times 10^{-11}\ M$$

At this concentration of $[Br^-]$, the $[Au^+]$ still in solution is governed by the K_{sp} expression for AuBr: $K_{sp(AuBr)} = [Au^+][Br^-]$.

$$[Au^+] = \frac{K_{sp(AuBr)}}{[Br^-]} = \frac{5.0 \times 10^{-17}}{3.3 \times 10^{-11}} = \mathbf{1.5 \times 10^{-6}\ M}$$

Therefore, the percentage of Au^+ that is still in solution when $[Br^-] = 3.3 \times 10^{-11}\ M$ is

$$\%\ Au^+\ \text{in solution} = \frac{[Au^+]}{[Au^+]_{initial}} \times 100\% = \frac{1.5 \times 10^{-6}\ M}{0.010\ M} \times 100\% = 0.015\%$$

$$\%\ Au^+\ \text{precipitated out} = 100.000\% - 0.015\% = \mathbf{99.985\%}$$

(c) CuBr will begin to precipitate when $Q_{sp(CuBr)} = K_{sp(CuBr)} = [Cu^+][Br^-]$

$$[Br^-] = \frac{K_{sp(CuBr)}}{[Cu^+]} = \frac{5.3 \times 10^{-9}}{0.010} = 5.3 \times 10^{-7}\ M$$

At this concentration of Br^-, $[Au^+]$ and $[Ag^+]$ in solution are governed by their K_{sp} expressions:

$$[Au^+] = \frac{K_{sp(AuBr)}}{[Br^-]} = \frac{5.0 \times 10^{-17}}{5.3 \times 10^{-7}} = \mathbf{9.4 \times 10^{-11}\ M}$$

$$[Ag^+] = \frac{K_{sp(AgBr)}}{[Br^-]} = \frac{3.3 \times 10^{-13}}{5.3 \times 10^{-7}} = \mathbf{6.2 \times 10^{-7}\ M}$$

20-38. *Refer to Section 20-4, Example 20-8 and 20-9, and Appendix H.*

Balanced equations: $K_2SO_4(aq) + Pb(NO_3)_2(aq) \rightarrow PbSO_4(s) + 2KNO_3(aq)$
$PbSO_4(s) \rightleftarrows Pb^{2+}(aq) + SO_4^{2-}(aq)$ $K_{sp} = 1.8 \times 10^{-8}$

$K_2CrO_4(aq) + Pb(NO_3)_2(aq) \rightarrow PbCrO_4(s) + 2KNO_3(aq)$
$PbCrO_4(s) \rightleftarrows Pb^{2+}(aq) + CrO_4^{2-}(aq)$ $K_{sp} = 1.8 \times 10^{-14}$

(a) Both $PbSO_4$ and $PbCrO_4$ are of the same molecular type, i.e., their cation to anion ratio is 1:1. **$PbCrO_4$** has the smaller K_{sp}, so it is less soluble and will precipitate first.

(b) $PbCrO_4$ will begin to precipitate when $Q_{sp(PbCrO_4)} = K_{sp(PbCrO_4)} = [Pb^{2+}][CrO_4^{2-}]$. At this point,

$$[Pb^{2+}] = \frac{K_{sp(PbCrO_4)}}{[CrO_4^{2-}]} = \frac{1.8 \times 10^{-14}}{0.050} = \mathbf{3.6 \times 10^{-13}\ M}$$

(c) $PbSO_4$ will begin to precipitate when $Q_{sp(PbSO_4)} = K_{sp(PbSO_4)} = [Pb^{2+}][SO_4^{2-}]$. At this point,

$$[Pb^{2+}] = \frac{K_{sp(PbSO_4)}}{[SO_4^{2-}]} = \frac{1.8 \times 10^{-8}}{0.050} = \mathbf{3.6 \times 10^{-7}\ M}$$

(d) When $PbSO_4$ begins to precipitate in (c), the concentration of SO_4^{2-} in solution is still the original concentration, **0.050 M**. The CrO_4^{2-} concentration can be calculated by substituting the concentration of Pb^{2+} obtained in (c) into the K_{sp} expression for $PbCrO_4$:

$$[CrO_4^{2-}] = \frac{K_{sp(PbCrO_4)}}{[Pb^{2+}]} = \frac{1.8 \times 10^{-14}}{3.6 \times 10^{-7}} = \mathbf{5.0 \times 10^{-8}\ M}$$

20-40. *Refer to Section 20-3, Example 20-6 and Appendix H.*

Balanced equations: (i) $2KOH(aq) + Zn(NO_3)_2(aq) \rightarrow Zn(OH)_2(s) + 2KNO_3(aq)$
$Zn(OH)_2(s) \rightleftarrows Zn^{2+}(aq) + 2OH^-(aq)$ $K_{sp} = 4.5 \times 10^{-17}$

(ii) $K_2CO_3(aq) + Zn(NO_3)_2(aq) \rightarrow ZnCO_3(s) + 2KNO_3(aq)$
$ZnCO_3(s) \rightleftarrows Zn^{2+}(aq) + CO_3^{2-}(aq)$ $K_{sp} = 1.5 \times 10^{-11}$

(iii) $2KCN(aq) + Zn(NO_3)_2(aq) \rightarrow Zn(CN)_2(s) + 2KNO_3(aq)$

$Zn(CN)_2(s) \rightleftarrows Zn^{2+}(aq) + 2CN^-(aq)$ $\hspace{2cm} K_{sp} = 8.0 \times 10^{-12}$

Plan: Substitute the value of K_{sp} and the given ion concentration into each K_{sp} equilibrium expression and solve for the other ion concentration. An ion concentration just greater than the calculated one will initiate precipitation.

(a) (i) $K_{sp} = [Zn^{2+}][OH^-]^2$
$4.5 \times 10^{-17} = [Zn^{2+}](0.0015)^2$
$[Zn^{2+}] = \mathbf{2.0 \times 10^{-11}\ \textit{M}}$

(ii) $K_{sp} = [Zn^{2+}][CO_3^{2-}]$
$1.5 \times 10^{-11} = [Zn^{2+}](0.0015)$
$[Zn^{2+}] = \mathbf{1.0 \times 10^{-8}\ \textit{M}}$

(iii) $K_{sp} = [Zn^{2+}][CN^-]^2$
$8.0 \times 10^{-12} = [Zn^{2+}](0.0015)^2$
$[Zn^{2+}] = \mathbf{3.6 \times 10^{-6}\ \textit{M}}$

(b) (i) $K_{sp} = [Zn^{2+}][OH^-]^2$
$4.5 \times 10^{-17} = (0.0015)[OH^-]^2$
$[OH^-] = \mathbf{1.7 \times 10^{-7}\ \textit{M}}$

(ii) $K_{sp} = [Zn^{2+}][CO_3^{2-}]$
$1.5 \times 10^{-11} = (0.0015)[CO_3^{2-}]$
$[CO_3^{2-}] = \mathbf{1.0 \times 10^{-8}\ \textit{M}}$

(iii) $K_{sp} = [Zn^{2+}][CN^-]^2$
$8.0 \times 10^{-12} = (0.0015)[CN^-]^2$
$[CN^-] = \mathbf{7.3 \times 10^{-5}\ \textit{M}}$

This problem assumes that the ions do not appreciably hydrolyze in water.

20-42. *Refer to Section 20-5, Examples 20-10 and 20-11, and Appendices G and H.*

Plan: Calculate the concentrations of Mg^{2+} and OH^- ions and determine Q_{sp}. If $Q_{sp} > K_{sp}$ for $Mg(OH)_2$, then precipitation will occur.

Two equilibrium must be considered: $\hspace{1cm} Mg(OH)_2(s) \rightleftarrows Mg^{2+}(aq) + 2OH^-(aq) \hspace{1cm} K_{sp} = 1.5 \times 10^{-11}$

$\hspace{5.5cm} NH_3(aq) + H_2O(\ell) \rightleftarrows NH_4^+(aq) + OH^-(aq) \hspace{0.5cm} K_b = 1.8 \times 10^{-5}$

We recognize that the given solution is a buffer. The NH_3/NH_4^+ equilibrium determines $[OH^-]$. Recall from Chapter 18:

$$[OH^-] = K_b \times \frac{[base]}{[salt]} = (1.8 \times 10^{-5}) \times \frac{0.075\ M}{3.5\ M} = 3.9 \times 10^{-7}\ M$$

Therefore, $\hspace{0.5cm} [OH^-] = 3.9 \times 10^{-7}\ M$

$\hspace{1.8cm} [Mg^{2+}] = [Mg(NO_3)_2] = 0.080\ M \hspace{1cm}$ since $Mg(NO_3)_2$ is a soluble salt

For $Mg(OH)_2$, $\hspace{0.5cm} Q_{sp} = [Mg^{2+}][OH^-]^2 = (0.080)(3.9 \times 10^{-7})^2 = 1.2 \times 10^{-14}$,

$\hspace{2.5cm} Q_{sp} < K_{sp}$, $Mg(OH)_2$ will **not** precipitate.

Since $[OH^-] = 3.9 \times 10^{-7}\ M$, pOH = 6.41; pH = **7.59**

20-44. *Refer to Section 20-5 and Appendices F and H.*

Balanced equations: $\hspace{0.3cm} CaF_2(s) \rightleftarrows Ca^{2+}(aq) + 2F^-(aq) \hspace{2.5cm} K_{sp} = 3.9 \times 10^{-11}$

$\hspace{2.6cm} HF(aq) + H_2O(\ell) \rightleftarrows H_3O^+(aq) + F^-(aq) \hspace{1.3cm} K_a = 7.2 \times 10^{-4}$

When equilibria are present in the same solution, all relevant equilibrium expressions must be satisfied:

$$K_{sp} = [Ca^{2+}][F^-]^2 \hspace{0.5cm} \text{and} \hspace{0.5cm} K_a = \frac{[H_3O^+][F^-]}{[HF]}$$

Here, the solid CaF_2 is allowed to dissolve in a solution where $[H_3O^+]$ is buffered at $0.00500\ M$ and $[HF] = 0.10\ M$, which is a source of common ion, F^-.

Let $\hspace{0.3cm} x$ = molar solubility of CaF_2. $\hspace{1cm}$ Then, $\hspace{1cm} x$ = mol/L of Ca^{2+} produced by dissolution of CaF_2

$\hspace{6.5cm} 2x$ = mol/L of F^- produced by dissolution of CaF_2

$\hspace{6.5cm} y$ = mol/L of F^- produced by dissociation of HF

Therefore, at equilibrium:

$CaF_2(s)$	$\rightleftarrows$	$Ca^{2+}(aq)$	+	$2F^-(aq)$
		$x\ M$		$(2x + y)\ M$

$HF(aq)$	+ $H_2O(\ell)$ $\rightleftarrows$	$H_3O^+(aq)$	+	$F^-(aq)$
$(0.10 - y)\ M$		$0.0050\ M$		$(2x + y)\ M$

Plan: (1) Use the K_a equilibrium expression to find the value of y in terms of x. Note: the H_3O^+ concentration is constant because it is buffered at 0.0050 M.

(2) Substitute for y in the K_{sp} expression to calculate x, the molar solubility.

(1) $K_a = \dfrac{[H_3O^+][F^-]}{[HF]} = \dfrac{(0.0050)(2x + y)}{(0.10 - y)} = 7.2 \times 10^{-4}$

Therefore,
$(0.0050)(2x + y) = (7.2 \times 10^{-4})(0.10 - y)$
$0.010x + 0.0050\,y = 7.2 \times 10^{-5} - 7.2 \times 10^{-4}\,y$
$0.010x + 0.0057y = 7.2 \times 10^{-5}$

$$y = \dfrac{7.2 \times 10^{-5} - 0.010x}{0.0057}$$

$$y = 0.0126 - 1.75x$$

(2) $K_{sp} = [Ca^{2+}][F^-]^2 = (x)(2x + y)^2 = 3.9 \times 10^{-11}$

Substituting for y,
$3.9 \times 10^{-11} = (x)(2x + (0.0126 - 1.75x))^2$
$= (x)(0.0126)^2$
$= x(1.59 \times 10^{-4})$
$x = 2.5 \times 10^{-7}$

(Note: The simplifying assumptions that 2x and 1.75x were small compared to 0.0126 were correct.)

Therefore, the molar solubility of CaF_2 in this system is **2.5 x 10^{-7} mol CaF$_2$/L**

The solubility of CaF_2 (g/L) $= \dfrac{2.5 \times 10^{-7} \text{ mol CaF}_2}{1 \text{ L soln}} \times \dfrac{78.08 \text{ g CaF}_2}{1 \text{ mol CaF}_2} =$ **2.0 x 10^{-5} g CaF$_2$/L**

20-46. *Refer to Section 20-5, Example 20-10, and Appendices G and H.*

Balanced equations: $Mn(OH)_2(s) \rightleftarrows Mn^{2+}(aq) + 2OH^-(aq)$ $\qquad K_{sp} = 4.6 \times 10^{-14}$
$NH_3(aq) + H_2O(\ell) \rightleftarrows NH_4^+(aq) + OH^-(aq)$ $\qquad K_b = 1.8 \times 10^{-5}$

Plan: Calculate the concentrations of Mn^{2+} and OH^- ions and determine Q_{sp} for $Mn(OH)_2$. If $Q_{sp} > K_{sp}$, then precipitation will occur.

$[Mn^{2+}] = [Mn(NO_3)_2] = 2.0 \times 10^{-5}\ M$
$[OH^-]$ is determined from the ionization of NH_3.

Let x = $[NH_3]$ that ionizes.
Then, $1.0 \times 10^{-3} - x = [NH_3]$; $\ x = [OH^-] = [NH_4^+]$

$K_b = \dfrac{[NH_3][OH^-]}{[NH_3]} = \dfrac{x^2}{1.0 \times 10^{-3} - x} = 1.8 \times 10^{-5} \approx \dfrac{x^2}{1.0 \times 10^{-3}}$ $\qquad$ Solving, $x = 1.3 \times 10^{-4}$

Since the value for x is greater than 5% of 1.0×10^{-3}, the simplifying assumption may not hold. When we solve the original quadratic equation, $x^2 + (1.8 \times 10^{-5})x - 1.8 \times 10^{-8} = 0$, $x = 1.3 \times 10^{-4}$. In this case, the simplifying assumption was adequate to 2 significant figures. Therefore, $[OH^-] = 1.3 \times 10^{-4}\ M$.

$Q_{sp} = [Mn^{2+}][OH^-]^2 = (2.0 \times 10^{-5})(1.3 \times 10^{-4})^2 = 3.4 \times 10^{-13}$

Thus, $Q_{sp} > K_{sp}$ by a factor of 7. Therefore **a precipitate will form** but will not be seen. To be seen, the Q_{sp} must be 1000 times larger than the K_{sp}.

20-48. *Refer to Section 20-3, Example 20-4 and Appendix H.*

(a) Balanced equations: $NaOH(aq) \rightarrow Na^+(aq) + OH^-(aq)$ $\qquad$ (to completion)
$Mg(OH)_2(s) \rightleftarrows Mg^{2+}(aq) + 2OH^-(aq)$ $\qquad$ (reversible) $\qquad K_{sp} = 1.5 \times 10^{-11}$

Let $\quad$ x = molar solubility of $Mg(OH)_2$ in 0.015 M NaOH. Then,
$[Mg^{2+}]$ = x (from $Mg(OH)_2$)
$[OH^-]$ = 2x (from $Mg(OH)_2$) + 0.015 M (from NaOH)

$K_{sp} = [Mg^{2+}][OH^-]^2 = (x)(2x + 0.015)^2 = 1.5 \times 10^{-11} \approx (x)(0.015)^2$ Solving, x = 6.7×10^{-8}

Therefore, molar solubility = **6.7×10^{-8} mol Mg(OH)$_2$/L 0.015 M NaOH soln**

(b) Balanced equations: $MgCl_2(aq) \rightarrow Mg^{2+}(aq) + 2Cl^-(aq)$ (to completion)
  $Mg(OH)_2(s) \rightleftarrows Mg^{2+}(aq) + 2OH^-(aq)$ (reversible) $K_{sp} = 1.5 \times 10^{-11}$

Let   x = molar solubility of Mg(OH)$_2$ in 0.015 M MgCl$_2$. Then,
  $[Mg^{2+}]$ = x (from Mg(OH)$_2$) + 0.015 M (from MgCl$_2$)
  $[OH^-]$ = 2x (from Mg(OH)$_2$)

$K_{sp} = [Mg^{2+}][OH^-]^2 = (x + 0.015)(2x)^2 = 1.5 \times 10^{-11} \approx (0.015)(2x)^2$ Solving, x = 1.6×10^{-5}

Therefore, molar solubility = **1.6×10^{-5} mol Mg(OH)$_2$/L 0.015 M MgCl$_2$ soln**

20-50.   *Refer to Section 20-3, Example 20-5 and Appendix H.*

Balanced equations: $Mg(NO_3)_2(aq) + 2OH^-(aq) \rightarrow Mg(OH)_2(s) + 2NO_3^-(aq)$ (Will precipitation occur?)
  $MgOH_2(s) \rightleftarrows Mg^{2+}(aq) + 2OH^-(aq)$ (reversible) $K_{sp} = 1.5 \times 10^{-11}$

Plan: (1) Calculate the concentration of Mg^{2+} ions and OH$^-$ ions at the instant of mixing before combination occurs.
  (2) Determine the reaction quotient, Q_{sp}. If $Q_{sp} > K_{sp}$, then precipitation will occur.

(1) $[Mg^{2+}]$ = $[Mg(NO_3)_2]$ = 0.00050 M
  Since pH = 8.70, pOH = 5.30, and $[OH^-]$ = 5.0×10^{-6} M

(2) $Q_{sp} = [Mg^{2+}][OH^-]^2 = (0.00050)(5.0 \times 10^{-6})^2 = 1.3 \times 10^{-14}$
  Since $Q_{sp} < K_{sp}$, precipitation will **not** occur.

20-52.   *Refer to Section 20-3, Example 20-3 and Appendix H.*

Balanced equation: $Fe(OH)_2(s) \rightleftarrows Fe^{2+}(aq) + 2OH^-(aq)$ $K_{sp} = 7.9 \times 10^{-15}$

(a) Plan: Calculate the molar solubility of Fe(OH)$_2$, then determine $[OH^-]$ and pH.

Let   x = molar solubility of Fe(OH)$_2$. Then
  x = moles/L of Fe^{2+}
  2x = moles/L of OH$^-$

$K_{sp} = [Fe^{2+}][OH^-]^2 = (x)(2x)^2 = 4x^3 = 7.9 \times 10^{-15}$ Solving, x = 1.3×10^{-5}

molar solubility = 1.3×10^{-5} mol Fe(OH)$_2$/L (dissolved)
$[OH^-]$ = 2x = 2.5×10^{-5} M; pOH = 4.60; pH = **9.40**

Note: Values for molar solubility and $[OH^-]$ were rounded to 2 significant figures after calculating them.

(b) solubility (g/100 mL) = $\dfrac{1.3 \times 10^{-5} \text{ mol Fe(OH)}_2}{1 \text{ L}} \times \dfrac{89.9 \text{ g}}{1 \text{ mol}} \times \dfrac{0.1 \text{ L}}{100 \text{ mL}} = $ **1.1×10^{-4} g/100 mL**

Note: Remember, do not divide by 100 mL; you want 100 mL to remain in the denominator.

20-54.   *Refer to Section 20-6 and Example 20-6.*

Balanced equation: $Zn(OH)_2(s) \rightleftarrows Zn^{2+}(aq) + 2OH^-(aq)$ $K_{sp} = 4.5 \times 10^{-17}$

Plan: (1) Calculate $[OH^-]$.
  (2) Substitute the value into the K_{sp} expression and solve for $[Zn^{2+}]$.

337

(1) Since pH = 10.00, pOH = 14.00 - 10.00 = 4.00 and [OH$^-$] = 1.0 x 10^{-4} M

(2) K_{sp} = [Zn^{2+}][OH$^-$]^2 [Zn^{2+}] = $\dfrac{[K_{sp}]}{[OH^-]^2}$ = $\dfrac{4.5 \times 10^{-17}}{(1.0 \times 10^{-4})^2}$ = **4.5 x 10^{-9} M**

20-56.   *Refer to Section 20-6 and Example 20-5.*

Balanced equations: Mg(OH)$_2$(s) $\rightleftarrows$ Mg^{2+}(aq) + 2OH$^-$(aq) K_{sp} = 1.5 x 10^{-11}

NH$_3$(aq) + H$_2$O(ℓ) $\rightleftarrows$ NH$_4$$^+$(aq) + OH$^-$(aq) K_b = 1.8 x 10^{-5}

Plan: Calculate the concentrations of Mg^{2+} and OH$^-$ ions and determine Q_{sp} for Mg(OH)$_2$. If Q_{sp} > K_{sp}, then precipitation will occur.

[Mg^{2+}] = [Mg(NO$_3$)$_2$] = 4.3 x 10^{-4} M
[OH$^-$] is determined from the ionization of NH$_3$.

Let x = [NH$_3$] that ionizes.
Then, 5.2 x 10^{-2} - x = [NH$_3$]; x = [OH$^-$] = [NH$_4$$^+$]

$K_b = \dfrac{[NH_3][OH^-]}{[NH_3]} = \dfrac{x^2}{5.2 \times 10^{-2} - x} = 1.8 \times 10^{-5} \approx \dfrac{x^2}{5.2 \times 10^{-2}}$ Solving, x = 9.7 x 10^{-4}

Since the value for x is less than 5% of 5.2 x 10^{-2}, the simplifying assumption does hold.
Therefore, [OH$^-$] = 9.7 x 10^{-4} M.

Q_{sp} = [Mg^{2+}][OH$^-$]^2 = (4.3 x 10^{-4})(9.7 x 10^{-4})^2 = 4.0 x 10^{-10}

Thus, Q_{sp} > K_{sp} by a factor of 27. Therefore **a precipitate will form** but will not be seen. To be seen, the Q_{sp} must be 1000 times larger than the K_{sp}.

20-58.   *Refer to Section 20-6.*

A slightly soluble compound will dissolve when the concentration of its ions in solution are reduced to such a level that Q_{sp} < K_{sp}. The following hydroxides and carbonates dissolve in strong acid, such as nitric acid.

(a)   Cu(OH)$_2$(s) $\rightleftarrows$ Cu^{2+}(aq) + 2OH$^-$(aq)
         $\dfrac{2H^+(aq) + 2OH^-(aq) \rightarrow 2H_2O(\ell)}{}$

         Cu(OH)$_2$(s) + 2H$^+$(aq) $\rightarrow$ Cu^{2+}(aq) + 2H$_2$O(ℓ)

The H$^+$ from the acid reacts with OH $^-$ lowering the concentration of OH$^-$ by forming H$_2$O, a weak electrolyte in an acid/base neutralization reaction. Whenever [OH$^-$] is low enough such that [Cu^{2+}][OH$^-$]^2 < K_{sp}, Cu(OH)$_2$(s) will dissolve.

(b)   Sn(OH)$_4$(s) $\rightleftarrows$ Sn^{4+}(aq) + 4OH$^-$(aq)
         $\dfrac{4H^+(aq) + 4OH^-(aq) \rightarrow 4H_2O(\ell)}{}$

         Sn(OH)$_4$(s) + 4H$^+$(aq) $\rightarrow$ Sn^{4+}(aq) + 4H$_2$O(ℓ)

The H$^+$ from the acid reacts with OH$^-$ and thus lowers the concentration of OH$^-$ in an acid/base neutralization reaction. Whenever [OH$^-$] is low enough such that [Sn^{4+}][OH$^-$]^4 < K_{sp}, Sn(OH)$_4$(s) will dissolve.

(c)   ZnCO$_3$(s) $\rightleftarrows$ Zn^{2+}(aq) + CO$_3$$^{2-}$(aq)
         $\dfrac{2H^+(aq) + CO_3^{2-}(aq) \rightarrow CO_2(g) + H_2O(\ell)}{}$

         ZnCO$_3$(s) + 2H$^+$(aq) $\rightarrow$ Zn^{2+}(aq) + CO$_2$(g) + H$_2$O(ℓ)

The H^+ from the acid removes CO_3^{2-} from solution in a reaction which forms $CO_2(g)$ and $H_2O(\ell)$. Whenever $[CO_3^{2-}]$ is low enough such that $[Zn^{2+}][CO_3^{2-}] < K_{sp}$, $ZnCO_3(s)$ will dissolve.

(d)
$$(PbOH)_2CO_3(s) \rightleftarrows 2Pb^{2+}(aq) + 2OH^-(aq) + CO_3^{2-}(aq)$$
$$\underline{4H^+(aq) + 2OH^-(aq) + CO_3^{2-}(aq) \rightarrow 3H_2O(\ell) + CO_2(g)}$$
$$(PbOH)_2CO_3(s) + 4H^+(aq) \rightarrow 2Pb^{2+}(aq) + 3H_2O(\ell) + CO_2(g)$$

The H^+ from the acid removes both OH^- and CO_3^{2-} ions from solution by forming $H_2O(\ell)$ and $CO_2(g)$. When $[OH^-]$ and $[CO_3^{2-}]$ are low enough such that $[Pb^{2+}]^2[OH^-]^2[CO_3^{2-}] < K_{sp}$, $(PbOH)_2CO_3(s)$ will dissolve.

20-60. *Refer to Section 20-6.*

Nonoxidizing acids dissolve some insoluble sulfides, including MnS and CuS. The H^+ ions react with S^{2-} ions to form gaseous H_2S, which bubbles out of the solutions. The Q_{sp} of the sulfide becomes less than the corresponding K_{sp} value and the metal sulfide dissolves.

(a)
$$MnS(s) \rightleftarrows Mn^{2+}(aq) + S^{2-}(aq)$$
$$\underline{2H^+(aq) + S^{2-}(aq) \rightarrow H_2S(g)}$$
$$MnS(s) + 2H^+(aq) \rightarrow Mn^{2+}(aq) + H_2S(g)$$

(b)
$$CuS(s) \rightleftarrows Cu^{2+}(aq) + S^{2-}(aq)$$
$$\underline{2H^+(aq) + S^{2-}(aq) \rightarrow H_2S(g)}$$
$$CuS(s) + 2H^+(aq) \rightarrow Cu^{2+}(aq) + H_2S(g)$$

20-62. *Refer to Sections 20-5 and 20-6, and Appendix F.*

Balanced equation: $MnS(s) \rightleftarrows Mn^{2+}(aq) + S^{2-}(aq)$

To make $MnS(s)$ more soluble, the above equilibrium must be shifted to the right. Applying LeChatelier's Principle, any process which will reduce either $[Mn^{2+}]$ or $[S^{2-}]$ will do this. In the presence of 0.10 M HCl (a strong acid), competing equilibria will lower $[S^{2-}]$ by producing the weak acids, HS^- and H_2S:

$$S^{2-} + H^+ \rightleftarrows HS^- \qquad K_{1'} = \frac{1}{K_{a2\,H_2S}} = \frac{1}{1.0 \times 10^{-19}} = 1.0 \times 10^{19}$$

$$HS^- + H^+ \rightleftarrows H_2S \qquad K_{2'} = \frac{1}{K_{a1\,H_2S}} = \frac{1}{1.0 \times 10^{-7}} = 1.0 \times 10^7$$

The equilibrium constants are very large, so the above equilibria are shifted far to the right, greatly reducing $[S^{2-}]$. In addition, $H_2S(g)$ will bubble out of solution when its solubility is exceeded, thus removing more S^{2-} from the system.

On the other hand, the soluble salt, $Mn(NO_3)_2$, would **not** become more soluble in the presence of 0.10 M HCl. The H^+ ions will not remove NO_3^- ions from solution, because HNO_3 is a strong acid and will ionize totally. The Cl^- ions do not remove Mn^{2+} from solution.

20-64. *Refer to Section 20-3, Example 20-5 and Appendix H.*

Balanced equations: $BaCl_2(aq) + 2NaF(aq) \rightarrow BaF_2(s) + 2NaCl(aq)$ (Will precipitation occur?)
$BaF_2(s) \rightleftarrows Ba^{2+}(aq) + 2F^-(aq)$ (reversible) $K_{sp} = 1.7 \times 10^{-6}$

Plan: (a) Calculate the concentration of Ba^{2+} ions and F^- ions at the instant of mixing before any reaction occurs, using $M_1V_1 = M_2V_2$.
(b) Determine the reaction quotient, Q_{sp}. If $Q_{sp} > K_{sp}$, then precipitation will occur.

(a) $[Ba^{2+}] = [BaCl_2] = \dfrac{M_1V_1}{V_2} = \dfrac{0.0030\ M \times 0.025\ L}{(0.025\ L + 0.050\ L)} = 1.0 \times 10^{-3}\ M$

$[F^-] = [NaF] = \dfrac{M_1V_1}{V_2} = \dfrac{0.050\ M \times 0.050\ L}{(0.025\ L + 0.050\ L)} = 3.3 \times 10^{-2}\ M$

(b) $Q_{sp} = [Ba^{2+}][F^-]^2 = (1.0 \times 10^{-3})(3.3 \times 10^{-2})^2 = 1.1 \times 10^{-6}$
Since $Q_{sp} < K_{sp}$, precipitation will **not** occur.

20-66. *Refer to Section 20-4, Examples 20-8 and 20-9, and Exercise 20-36 and 20-38 Solutions.*

Balanced equations: $AgBr(s) \rightleftarrows Ag^+(aq) + Br^-(aq)$ (reversible) $K_{sp} = 3.3 \times 10^{-13}$
$AgI(s) \rightleftarrows Ag^+(aq) + I^-(aq)$ (reversible) $K_{sp} = 1.5 \times 10^{-16}$

(a) Because the compounds to be precipitated are all of the same formula type, i.e., their cation to anion ratios are the same, the one with the smallest K_{sp} is the least soluble and will precipitate first. Hence, **AgI** ($K_{sp} = 1.5 \times 10^{-16}$) will precipitate first, then AgBr ($K_{sp} = 3.3 \times 10^{-13}$).

(b) AgBr will begin to precipitate when $Q_{sp(AgBr)} = K_{sp(AgBr)} = [Ag^+][Br^-]$. At this point,

$$[Ag^+] = \frac{K_{sp(AgBr)}}{[Br^-]} = \frac{3.3 \times 10^{-13}}{0.015} = 2.2 \times 10^{-11} \ M$$

At this concentration of $[Ag^+]$, the $[I^-]$ still in solution is governed by the K_{sp} expression for AgI: $K_{sp(AgI)} = [Ag^+][I^-]$.

$$[I^-] = \frac{K_{sp(AgI)}}{[Ag^+]} = \frac{1.5 \times 10^{-16}}{2.2 \times 10^{-11}} = 6.8 \times 10^{-6} \ M$$

Therefore, the percentage of I^- that is still in solution when AgBr begins to precipitate is

$$\% \ I^- \text{ in solution} = \frac{[I^-]}{[I^-]_{initial}} \times 100\% = \frac{6.8 \times 10^{-6} \ M}{0.015 \ M} \times 100\% = 0.045\%$$

and $\% \ I^-$ removed from solution $= 100.000\% - 0.045\% = \textbf{99.955\%}$

20-68. *Refer to general concepts in Chapter 20.*

It would be harmful to human health if a hospital worker mistakenly used 450 grams of $BaCO_3$ (solubility 0.02 g/L) rather than 450 grams of $BaSO_4$ (solubility 0.00246 g/L) as a contrast agent for X-raying the human intestinal tract. $BaCO_3$ is 10 times more soluble than $BaSO_4$ and would pose a considerable health risk, since Ba compounds are toxic. Also, $BaCO_3$ is even more soluble because of the effect of stomach acid.

20-70. *Refer to Section 20-3.*

(a) Balanced equation: $Na_2SO_4(aq) + Mg(NO_3)_2(aq) \rightarrow 2NaNO_3(aq) + MgSO_4(aq)$
Because both products are soluble compounds, there will not be a precipitate forming.

(b) Balanced equation: $K_3PO_4(aq) + FeCl_3(aq) \rightarrow 3KCl(aq) + FePO_4(s)$
This is a precipitation reaction, since one product, $FePO_4$, is not soluble.

20-72. *Refer to Sections 17-2, 17-11 and 20-1, and Appendix H.*

In the strict thermodynamic definition of the equilibrium constant, the activity of a component is used, not its concentration. The activity of a species in an ideal mixture is the ratio of its concentration or partial pressure to a standard concentration (1 M) or pressure (1 atm). The concentrations of pure solids and pure liquids are omitted from the equilibrium constant expression because their activity is taken to be 1.

20-74. *Refer to Section 20-4, and Examples 20-8 and 20-9.*

Balanced equations: $AgCl(s)$ (white) $\rightleftarrows Ag^+(aq) + Cl^-(aq)$ (reversible) $K_{sp} = 1.8 \times 10^{-10}$

$Ag_2CrO_4(s)$ (red) $\rightleftarrows 2Ag^+(aq) + CrO_4^{2-}(aq)$ (reversible) $K_{sp} = 9.0 \times 10^{-12}$

Because the compounds to be precipitated are not the same formular type, i.e., their cation to anion ratios are the not the same, you cannot tell just by looking at the values of K_{sp} which salt is more soluble..

The first tint of red color appears when Ag_2CrO_4 begins to precipitate: when $Q_{sp} = K_{sp} = [Ag^+]^2[CrO_4^{2-}]$.

At this point:
$$[Ag^+] = \sqrt{\frac{K_{sp(Ag_2CrO_4)}}{[CrO_4^{2-}]}} = \sqrt{\frac{9.0 \times 10^{-12}}{0.010}} = 3.0 \times 10^{-5} \, M$$

At this concentration of $[Ag^+]$, the $[Cl^-]$ still in solution is governed by the K_{sp} expression for AgCl:

$$K_{sp(AgCl)} = [Ag^+][Cl^-] \qquad [Cl^-] = \frac{K_{sp(AgCl)}}{[Ag^+]} = \frac{1.8 \times 10^{-10}}{3.0 \times 10^{-5}} = \mathbf{6.0 \times 10^{-6} \, M}$$

20-76. *Refer to Sections 20-2 and 20-3.*

Any solubility problem involving $PbCl_2$ involves the equilibrium:

$$PbCl_2(s) \rightleftarrows Pb^{2+}(aq) + 2Cl^-(aq)$$

and its corresponding equilibrium expression: $K_{sp} = [Pb^{2+}][Cl^-]^2$.

There are two main problem types when working with solubility that students confuse:

(1) a solid dissolving into water (or a solution containing a common ion) where one uses the molar solubility concept, and

(2) mixing two solutions to see if a precipitate occurs or not.

In problem type (1), there is a physical relationship between ions. For $PbCl_2$, the concentration of Cl^- ions in solution are twice the concentration of Pb^{2+} ions because they come from the same source, the solid $PbCl_2$. The molar solubility of $PbCl_2$, s, is equal to $[Pb^{2+}]$, so $[Cl^-] = 2s$, twice the molar solubility. So, $K_{sp} = [Pb^{2+}][Cl^-]^2 = (s)(2s)^2$. Some students get the mistaken impression that the concentration of Cl^- is doubled, but it is the molar solubility that must be doubled to equal the Cl^- concentration.

In problem type (2), there is no physical relationship between the ions that may or may not form a precipitate because they come from two different solutions. Here, one simply calculates the concentration of each ion after mixing, but just before precipitation begins. These actual concentrations are plugged into the Q_{sp} expression, which looks exactly like the K_{sp} expression. The Cl^- concentration isn't doubled, because it is what it is.

20-78. *Refer to Section 20-3.*

Balanced equations: $BaSO_4(s) \rightleftarrows Ba^{2+}(aq) + SO_4^{2-}(aq)$ $K_{sp} = 1.1 \times 10^{-11}$

$BaF_2(s) \rightleftarrows Ba^{2+}(aq) + 2F^-(aq)$ $K_{sp} = 1.7 \times 10^{-6}$

(a) $BaSO_4$ will begin to precipitate when $Q_{sp(BaSO_4)} = K_{sp(BaSO_4)} = [Ba^{2+}][SO_4^{2-}]$. At this point,

$$[Ba^{2+}] = \frac{K_{sp(BaSO_4)}}{[SO_4^{2-}]} = \frac{1.1 \times 10^{-11}}{0.010} = \mathbf{1.1 \times 10^{-9} \, M}$$

(b) BaF_2 will begin to precipitate when $Q_{sp(BaF_2)} = K_{sp(BaF_2)} = [Ba^{2+}][F^-]^2$. At this point,

$$[Ba^{2+}] = \frac{K_{sp(BaF_2)}}{[F^-]^2} = \frac{1.7 \times 10^{-6}}{(0.010)^2} = \mathbf{0.017 \, M}$$

20-80. *Refer to Section 20-3 and Appendix H.*

Balanced equation: $CaF_2(s) \rightleftarrows Ca^{2+}(aq) + 2F^-(aq)$ $\qquad\qquad K_{sp} = 3.9 \times 10^{-11}$

Plan: (1) Calculate $[F^-]$.
(2) Substitute the value into the K_{sp} expression and solve for $[Ca^{2+}]$.

(1) $[F^-] = \dfrac{1 \text{ mg F}^-}{1 \text{ L soln}} \times \dfrac{1 \text{ g F}^-}{1000 \text{ mg F}^-} \times \dfrac{1 \text{ mol F}^-}{19.0 \text{ g F}^-} = 5.3 \times 10^{-5} \ M$ (good to 1 significant figure; will round later)

(2) $K_{sp} = [Ca^{2+}][F^-]^2$

So, $[Ca^{2+}] = \dfrac{[K_{sp}]}{[F^-]^2} = \dfrac{3.9 \times 10^{-11}}{(5.3 \times 10^{-5})^2} = 1.4 \times 10^{-2} \ M$

? amount of Ca^{2+} (g/L) $= \dfrac{0.014 \text{ mol Ca}^{2+}}{1 \text{ L soln}} \times \dfrac{40.1 \text{ g Ca}^{2+}}{1 \text{ mol Ca}^{2+}} = \mathbf{0.6 \text{ g/L}}$ (1 significant figure)

20-82. *Refer to Section 20-3, Example 20-3 and Appendix H.*

Balanced equation: $MgCO_3(s) \rightleftarrows Mg^{2+}(aq) + CO_3^{2-}(aq)$ $\qquad\qquad K_{sp} = 4.0 \times 10^{-5}$

Plan: (1) Calculate the molar solubility of $MgCO_3$.
(2) Determine the mass of $MgCO_3$ that dissolves in 15 L of water to produce a saturated solution.
(3) Determine the percent loss of $MgCO_3$.

(1) Let x = molar solubility of $MgCO_3$. Then, $[Mg^{2+}] = [CO_3^{2-}] = x$

$K_{sp} = [Mg^{2+}][CO_3^{2-}] = x^2 = 4.0 \times 10^{-5}$

Solving, $x = 6.3 \times 10^{-3}$

Therefore, molar solubility = 6.3×10^{-3} mol $MgCO_3$/L (dissolved)

(2) ? g $MgCO_3$ dissolve in 15 L water $= \dfrac{6.3 \times 10^{-3} \text{ mol MgCO}_3}{1 \text{ L}} \times \dfrac{84.3 \text{ g MgCO}_3}{1 \text{ mol MgCO}_3} \times 15 \text{ L} = 8.0 \text{ g}$

(3) % loss of $MgCO_3 = \dfrac{MgCO_3 \text{ lost}}{\text{initial MgCO}_3} \times 100\% = \dfrac{8.0 \text{ g}}{28 \text{ g}} \times 100\% = \mathbf{29\%}$

21 Electrochemistry

21-2. *Refer to Section 4-7 and Example 4-5.*

In a redox reaction,

(a) oxidizing agents are the species that   (1) gain or appear to gain electrons,
   (2) are reduced, and
   (3) oxidize other substances.

(b) Reducing agents are the species that   (1) lose or appear to lose electrons,
   (2) are oxidized and
   (3) reduce other substances.

$$\overset{+3\ \ -2}{Fe_2O_3(s)} + \overset{+2\ -2}{3CO(g)} \rightarrow \overset{0}{2Fe(s)} + \overset{+4\ -2}{3CO_2(g)}$$

Consider the reaction:

Fe_2O_3 is the oxidizing agent because it contains Fe, which is being reduced from an oxidation state of +3 to 0. CO is the reducing agent because it contains C, which is being oxidized from an oxidation state of +2 to +4.

21-4. *Refer to Section 11-4.*

(a) oxidation: $3(FeS + 4H_2O \rightarrow SO_4^{2-} + Fe^{2+} + 8H^+ + 8e^-)$
    reduction: $8(3e^- + NO_3^- + 4H^+ \rightarrow NO + 2H_2O)$
    balanced equation: $3FeS + 8NO_3^- + 8H^+ \rightarrow 8NO + 3SO_4^{2-} + 3Fe^{2+} +$ $4H_2O$

(b) oxidation: $6(Fe^{2+} \rightarrow Fe^{3+} + e^-)$
    reduction: $6e^- + Cr_2O_7^{2-} + 14H^+ \rightarrow 2Cr^{3+} + 7H_2O$
    balanced equation: $6Fe^{2+} + Cr_2O_7^{2-} + 14H^+ \rightarrow 6Fe^{3+} + 2Cr^{3+} +$ $7H_2O$

(c) oxidation: $S^{2-} + 8OH^- \rightarrow SO_4^{2-} + 4H_2O + 8e^-$
    reduction: $4(2e^- + Cl_2 \rightarrow 2Cl^-)$
    balanced equation: $S^{2-} + 4Cl_2 + 8OH^- \rightarrow SO_4^{2-} + 8Cl^- +$ $4H_2O$

21-6. *Refer to Sections 21-2, 21-3 and 21-8.*

The cathode is defined as the electrode at which reduction occurs, i.e., where electrons are consumed, regardless of whether the electrochemical cell is an electrolytic or voltaic cell. In both electrolytic and voltaic cells, the electrons flow through the wire from the anode, where electrons are produced, to the cathode, where electrons are consumed. In an electrolytic cell, the dc source forces the electrons to travel nonspontaneously through the wire. Thus, the electrons flow from the positive electrode (the anode) to the negative electrode (the cathode). However, in a voltaic cell, the electrons flow spontaneously, *away* from the negative electrode (the anode) and toward the positive electrode (the cathode).

(a) The statement, "The positive electrode in any electrochemical cell is the one toward which the electrons flow through the wire," is **false**. It holds for voltaic cells, but not for electrolytic cells.

(b) The statement, "The cathode in any electrochemical cell is the negative electrode," is also **false**. It holds for any electrolytic cell, but not for a voltaic cell.

343

(a) oxidation: $2[4OH^-(aq) + Al(s) \rightarrow Al(OH)_4^-(aq) + 3e^-]$

reduction: $3[2e^- + 2H_2O(\ell) \rightarrow H_2(g) + 2OH^-(aq)]$

balanced equation: $2Al(s) + 2OH^-(aq) + 6H_2O(\ell) \rightarrow 2Al(OH)_4^-(aq) + 3H_2(g)$

(b) oxidation: $3[2OH^-(aq) + SO_3^{2-}(aq) \rightarrow SO_4^{2-}(aq) + H_2O(\ell) + 2e^-]$

reduction: $2[3e^- + CrO_4^{2-}(aq) + 4H_2O(\ell) \rightarrow Cr(OH)_3(s) + 5OH^-(aq)]$

balanced equation: $3SO_3^{2-}(aq) + 2CrO_4^{2-}(aq) + 5H_2O(\ell) \rightarrow 3SO_4^{2-}(aq) + 2Cr(OH)_3(s) + 4OH^-(aq)$

(c) oxidation: $4OH^-(aq) + Zn(s) \rightarrow [Zn(OH)_4]^{2-}(aq) + 2e^-$

reduction: $2e^- + Cu(OH)_2(s) \rightarrow Cu(s) + 2OH^-(aq)$

balanced equation: $Zn(s) + Cu(OH)_2(s) + 2OH^- \rightarrow [Zn(OH)_4]^{2-}(aq) + Cu(s)$

(d) oxidation: $3[OH^-(aq) + HS^-(aq) \rightarrow S(s) + H_2O(\ell) + 2e^-]$

reduction: $6e^- + ClO_3^-(aq) + 3H_2O(\ell) \rightarrow Cl^-(aq) + 6OH^-(aq)$

balanced equation: $3HS^-(aq) + ClO_3^-(aq) \rightarrow 3S(s) + Cl^-(aq) + 3OH^-(aq)$

(a) Magnesium metal is too reactive in water to be obtained by the electrolysis of $MgCl_2(aq)$. In other words, $H_2O(\ell)$ is more easily reduced to $OH^-(aq)$ and $H_2(g)$ than is $Mg^{2+}(aq)$ to $Mg(s)$. In electrochemical reactions, the species that is most easily reduced (or oxidized) will be reduced (or oxidized) first.

(b) Sodium ions do not appear in the overall cell reaction for the electrolysis of $NaCl(aq)$ because Na^+ ions are spectator ions and do not react. Since H_2O is more easily reduced than Na^+ ions, the reduction reaction involves H_2O:

reduction at cathode: $2e^- + 2H_2O(\ell) \rightarrow H_2(g) + 2OH^-(aq)$

oxidation at anode: $2Cl^-(aq) \rightarrow Cl_2(g) + 2e^-$

overall cell reaction: $2Cl^- + 2H_2O(\ell) \rightarrow H_2(g) + Cl_2(g) + 2OH^-(aq)$

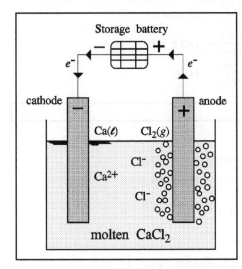

Electrolysis of molten calcium chloride:

oxidation:	$2Cl^-(molten) \rightarrow Cl_2(g) + 2e^-$
reduction:	$Ca^{2+}(molten) + 2e^- \rightarrow Ca(\ell)$
overall cell reaction:	$CaCl_2(\ell) \rightarrow Ca(\ell) + Cl_2(g)$

Electrolysis of aqueous sodium chloride:

reduction at cathode:	$2e^- + 2H_2O(\ell) \rightarrow H_2(g) + 2OH^-(aq)$
oxidation at anode:	$2Cl^-(aq) \rightarrow Cl_2(g) + 2e^-$
overall cell reaction:	$2Cl^- + 2H_2O(\ell) \rightarrow H_2(g) + Cl_2(g) + 2OH^-(aq)$

(a) A coulomb (C) is the amount of electrical charge that passes a given point when one ampere of current flows for one second.

(b) Electrical current is the motion of electrons or ions through a conducting medium.

(c) An ampere (A) is the practical unit of electrical current equal to the transfer of 1 coulomb per second. So, $1 A = 1 C/s$.

(d) A faraday of electricity corresponds to the charge on 6.022×10^{23} (1 mole) electrons, or 96,485 coulombs. It is the amount of electricity that reduces 1 equivalent weight of a substance at the cathode and oxidizes 1 equivalent weight of a substance at the anode.

(i) Recall that 1 faraday of electricity is equivalent to 1 mole of electrons passing through a system. Consider the general balanced half-reaction:

$$M^{n+} + ne^- \rightarrow M$$

The stoichiometry tells us that 1 mole of M requires n moles of electrons, hence n faradays of electricity.

Balanced Half-Reaction	**No. of Faradays/1 mol Free Metal**
(a) $Fe^{3+}(aq) + 3e^- \rightarrow Fe(s)$	3
(b) $Sn^{2+}(aq) + 2e^- \rightarrow Sn(s)$	2
(c) $Hg_2^{2+}(aq) + 2e^- \rightarrow 2Hg(\ell)$	1

(ii) The amount of charge required to deposit 1.00 g of each of the metals according to the reactions above:

(a) ? coulombs = 1.00 g Fe $\times \dfrac{1 \text{ mol Fe}}{55.85 \text{ g Fe}} \times \dfrac{3 \text{ mol } e^-}{1 \text{ mol Fe}} \times \dfrac{96500 \text{ C}}{1 \text{ mol } e^-} = \mathbf{5.18 \times 10^3 \text{ C}}$

(b) ? coulombs = 1.00 g Sn $\times \dfrac{1 \text{ mol Sn}}{118.7 \text{ g Sn}} \times \dfrac{2 \text{ mol } e^-}{1 \text{ mol Sn}} \times \dfrac{96500 \text{ C}}{1 \text{ mol } e^-} = \mathbf{1.63 \times 10^3 \text{ C}}$

(c) ? coulombs = 1.00 g Hg $\times \dfrac{1 \text{ mol Hg}}{200.6 \text{ g Hg}} \times \dfrac{2 \text{ mol } e^-}{2 \text{ mol Hg}} \times \dfrac{96500 \text{ C}}{1 \text{ mol } e^-} = \mathbf{481 \text{ C}}$

21-20. *Refer to Section 21-6 and Example 21-1.*

(a) Balanced half-reaction: $Cu^+ + e^- \rightarrow Cu$

Plan: g Cu $\Rightarrow$ mol Cu $\Rightarrow$ mol $e^- \Rightarrow$ coulombs $\Rightarrow$ amperes (= coulombs/sec)

? coulombs = 2.25 g Cu $\times \dfrac{1 \text{ mol Cu}}{63.55 \text{ g Cu}} \times \dfrac{1 \text{ mol } e^-}{1 \text{ mol Cu}} \times \dfrac{96500 \text{ C}}{1 \text{ mol } e^-} = 3420 \text{ coulombs}$

? amperes (coulombs/s) $= \dfrac{3420 \text{ C}}{400. \text{ min} \times (60 \text{ s/1 min})} = \mathbf{0.142 \text{ amperes}}$

(b) Balanced half-reaction: $Cu^{2+} + 2e^- \rightarrow Cu$

? g Cu = 400. min $\times \dfrac{60 \text{ s}}{1 \text{ min}} \times \dfrac{0.142 \text{ C}}{1 \text{ s}} \times \dfrac{1 \text{ mol } e^-}{96500 \text{ C}} \times \dfrac{1 \text{ mol Cu}}{2 \text{ mol } e^-} \times \dfrac{63.55 \text{ g Cu}}{1 \text{ mol Cu}} = \mathbf{1.12 \text{ g Cu}}$

(½ mass of Cu in (a))

21-22. *Refer to Section 21-6 and Example 21-1.*

Balanced half-reaction: $Rh^{3+} + 3e^- \rightarrow Rh$

? g Rh = 15.0 min $\times \dfrac{60 \text{ s}}{1 \text{ min}} \times \dfrac{0.755 \text{ C}}{1 \text{ s}} \times \dfrac{1 \text{ mol } e^-}{96500 \text{ C}} \times \dfrac{1 \text{ mol Rh}}{3 \text{ mol } e^-} \times \dfrac{102.9 \text{ g Rh}}{1 \text{ mol Rh}} = \mathbf{0.242 \text{ g Rh}}$

21-24. *Refer to Section 21-6.*

Balanced half-reaction: $Ag^+ + e^- \rightarrow Ag$

? coulombs = 0.976 mg Ag $\times \dfrac{1 \text{ g Ag}}{1000 \text{ mg Ag}} \times \dfrac{1 \text{ mol Ag}}{107.9 \text{ g Ag}} \times \dfrac{1 \text{ mol } e^-}{1 \text{ mol Ag}} \times \dfrac{96500 \text{ C}}{1 \text{ mol } e^-} = \mathbf{0.873 \text{ C}}$

21-26. *Refer to Section 21-6 and Example 21-1.*

Balanced half-reaction: $Ag^+ + e^- \rightarrow Ag$

? g Ag = 45.0 min $\times \dfrac{60 \text{ s}}{1 \text{ min}} \times \dfrac{2.78 \text{ C}}{1 \text{ s}} \times \dfrac{1 \text{ mol } e^-}{96500 \text{ C}} \times \dfrac{1 \text{ mol Ag}}{1 \text{ mol } e^-} \times \dfrac{107.9 \text{ g Ag}}{1 \text{ mol Ag}} = \mathbf{8.39 \text{ g Ag}}$

21-28. *Refer to Section 21-6 and Examples 21-1 and 21-2.*

Balanced half-reactions: anode $2I^- \rightarrow I_2 + 2e^-$
cathode $2H_2O + 2e^- \rightarrow H_2 + 2OH^-$

(a) The number of faradays passing through the cell is equivalent to the number of moles of electrons passing through the cell.

? faradays = 41.5×10^{-3} mol I_2 $\times \dfrac{2 \text{ mol } e^-}{1 \text{ mol } I_2} \times \dfrac{1 \text{ faraday}}{1 \text{ mol } e^-} = \mathbf{0.0830 \text{ faradays}}$

(b) ? coulombs = 0.0830 faradays $\times \dfrac{96500 \text{ C}}{1 \text{ faraday}} = \mathbf{8.01 \times 10^3 \text{ C}}$

(c) $? L_{STP}\ H_2 = 8.01 \times 10^3\ C \times \dfrac{1\ mol\ e^-}{96500\ C} \times \dfrac{1\ mol\ H_2}{2\ mol\ e^-} \times \dfrac{22.4\ L_{STP}\ H_2}{1\ mol\ H_2} = \mathbf{0.930\ L_{STP}\ H_2}$

Alternatively, $? L_{STP}\ H_2 = 0.0830\ faradays \times \dfrac{1\ mol\ H_2}{2\ faradays} \times \dfrac{22.4\ L_{STP}\ H_2}{1\ mol\ H_2} = \mathbf{0.930\ L_{STP}\ H_2}$

(d) Plan: (1) Determine the moles of OH^- formed.
(2) Calculate $[OH^-]$, pOH and pH.

(1) $? mol\ OH^- = 8.01 \times 10^3\ C \times \dfrac{1\ mol\ e^-}{96500\ C} \times \dfrac{2\ mol\ OH^-}{2\ mol\ e^-} = 0.0830\ mol\ OH^-$

(2) $[OH^-] = \dfrac{0.0830\ mol\ OH^-}{0.500\ L} = 0.166\ M\ OH^-$; pOH = 0.780; pH = **13.220**

21-30. *Refer to Section 21-6 and Example 21-2.*

Plan: (1) Determine the half-reaction involving Cl_2.
(2) Calculate the moles of Cl_2 produced at the experimental conditions at 83% efficiency.
(3) Calculate the volume of Cl_2 produced using the ideal gas law, $PV = nRT$.

(1) Balanced half-reaction: $2Cl^- \rightarrow Cl_2 + 2e^-$

(2) $? mol\ Cl_2 = 5.00\ hr \times \dfrac{3600\ s}{1\ hr} \times \dfrac{1.70\ C}{1\ s} \times \dfrac{1\ mol\ e^-}{96500\ C} \times \dfrac{1\ mol\ Cl_2}{2\ mol\ e^-} \times \dfrac{85}{100} = 0.13\ mol\ Cl_2$

(3) $V = \dfrac{nRT}{P} = \dfrac{(0.13\ mol)(0.0821\ L \cdot atm/mol \cdot K)(15°C + 273°)}{(752/760\ atm)} = \mathbf{3.2\ L\ Cl_2}$ (to 2 significant figures)

21-32. *Refer to Section 21-6.*

Balanced half-reaction: $Cu^{2+} + 2e^- \rightarrow Cu$

(1) Plan: $M,L\ CuCl_2\ soln \overset{(i)}{\Rightarrow} mol\ Cu^{2+}\ reacted \overset{(ii)}{\Rightarrow} mol\ e^-\ reacted \overset{(iii)}{\Rightarrow} time\ required$

(i) Original moles of $Cu^{2+} = (0.455\ M)(0.250\ L) = 0.114\ mol$
Final moles of $Cu^{2+} = (0.167\ M)(0.250\ L) = 4.18 \times 10^{-2}\ mol$
$? mol\ Cu^{2+}\ reacted = 0.114\ mol - 4.18 \times 10^{-2}\ mol = 0.072\ mol\ Cu^{2+}$

(ii) $? mol\ e^-\ reacted = 2 \times mol\ Cu^{2+}\ reacted = 2 \times 0.072\ mol = 0.14\ mol\ e^-$

(iii) $? time\ required\ (s) = 0.14\ mol\ e^- \times \dfrac{96500\ C}{1\ mol\ e^-} \times \dfrac{1\ amp\text{-}s}{1\ C} \times \dfrac{1}{0.750\ amp} = \mathbf{1.9 \times 10^4\ s \equiv 5.2\ hr}$

(2) $? mass\ of\ Cu = 0.072\ mol\ Cu \times \dfrac{63.55\ g\ Cu}{1\ mol\ Cu} = \mathbf{4.6\ g\ Cu}$

21-34. *Refer to Section 21-6 and Example 21-1.*

Balanced half-reactions: $Cd \rightarrow Cd^{2+} + 2e^-$ $Ag^+ + e^- \rightarrow Ag$ $Fe^{2+} \rightarrow Fe^{3+} + e^-$

(a) $? faradays = 1.20\ g\ Cd \times \dfrac{1\ mol\ Cd}{112.4\ g\ Cd} \times \dfrac{2\ mol\ e^-}{1\ mol\ Cd} \times \dfrac{1\ faraday}{1\ mol\ e^-} = \mathbf{0.0214\ faradays}$

(b) $? g\ Ag = 0.0214\ faradays \times \dfrac{1\ mol\ e^-}{1\ faraday} \times \dfrac{1\ mol\ Ag}{1\ mol\ e^-} \times \dfrac{107.9\ g\ Ag}{1\ mol\ Ag} = \mathbf{2.31\ g\ Ag}$

(c) $? g\ Fe(NO_3)_3 = 0.0214\ faraday \times \dfrac{1\ mol\ e^-}{1\ faraday} \times \dfrac{1\ mol\ Fe^{3+}}{1\ mol\ e^-} \times \dfrac{1\ mol\ Fe(NO_3)_3}{1\ mol\ Fe^{3+}} \times \dfrac{241.9\ g\ Fe(NO_3)_3}{1\ mol\ Fe(NO_3)_3}$
$= \mathbf{5.18\ g\ Fe(NO_3)_3}$

21-36. *Refer to the Introduction to Voltaic or Galvanic Cells, Section 21-8 and Figure 21-6.*

(a) In a voltaic cell, the solutions in the two half-cells must be kept separate in order to produce usable electrical energy since electricity is only produced when electron transfer is forced to occur through the external circuit. If the two half-cells were mixed, electron transfer would happen directly in the solution and could not be exploited to give electricity.

(b) A salt bridge in a voltaic or galvanic cell has three functions: it allows electrical contact between the two solutions; it prevents mixing of the electrode solutions; and it maintains electrical neutrality in each half-cell.

21-38. *Refer to Sections 21-8, 21-9 and 21-10, and Figure 21-6.*

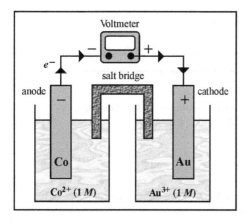

Voltaic cell:

oxidation at anode:	$3(Co \rightarrow Co^{2+} + 2e^-)$
reduction at cathode:	$2(Au^{3+} + 3e^- \rightarrow Au)$
overall cell reaction:	$3Co + 2Au^{3+} \rightarrow 3Co^{2+} + 2Au$

21-40. *Refer to Section 21-9 and Exercise 21-38.*

Shorthand notation: $Co|Co^{2+}(1\ M)||Au^{3+}(1\ M)/Au$

21-42. *Refer to Sections 21-8, 21-9 and 21-10.*

Balanced equation: $Ni(s) + 2Ag^+(aq) \rightarrow Ni^{2+}(aq) + 2Ag(s)$

(a) reduction half-reaction: $Ag^+(aq) + e^- \rightarrow Ag(s)$
(b) oxidation half-reaction: $Ni(s) \rightarrow Ni^{2+}(aq) + 2e^-$
(c) Ni is the anode.
(d) Ag is the cathode.
(e) Refer to cell diagram at right.

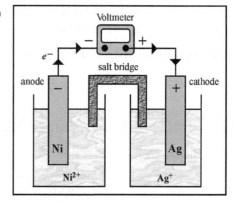

21-44. *Refer Exercise 21-36a Solution.*

No electricity is produced when $Cu(s)$ is placed into $AgNO_3(aq)$ even though a spontaneous redox reaction occurs:

$$Cu(s) + 2AgNO_3(aq) \rightarrow Cu(NO_3)_2(aq) + 2Ag(s)$$

The electron transfer at the solution-metal interface; it is not forced to occur through an external circuit where it would produce useful electrical energy.

21-46. *Refer to Sections 21-8, 21-9 and 21-10, and Figure 21-6.*

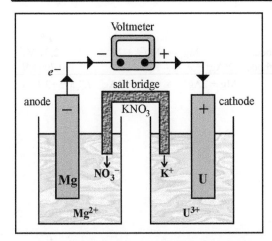

Voltaic cell:

oxidation at anode:	$3(Mg \rightarrow Mg^{2+} + 2e^-)$
reduction at cathode:	$2(U^{3+} + 3e^- \rightarrow U)$
overall cell reaction:	$3Mg + 2U^{3+} \rightarrow 3Mg^{2+} + 2U$

A note on ion flow: Because in the anodic half-cell, more positively-charged ions (Mg^{2+}) are being produced, more negatively-charged ions are required to keep the solution neutral. This is why the nitrate ions are flowing toward the anode. In the cathodic half-cell, positively-charged ions (U^{3+}) are being lost, so more positively-charged ions are needed to keep the solution neutral. This is why the potassium ions are flowing toward the cathode.

21-48. *Refer to Section 21-14 and the Introduction to Section 21-11.*

If the sign of the standard reduction potential, $E°$, of a half-reaction is positive, the half-reaction is the cathodic (reduction) reaction when connected to the standard hydrogen electrode (SHE). Half-reactions with more positive $E°$ values have greater tendencies to occur in the forward direction. Hence, the magnitude of a half-cell potential measures the spontaneity of the forward reaction.

If the $E°$ of a half-reaction is negative, the half-reaction is the anodic (oxidation) reaction when connected to the SHE. Half-reactions with more negative $E°$ values have greater tendencies to occur in the reverse direction.

21-50. *Refer to Section 21-14 and Table 21-2.*

(a) The substance that is the stronger oxidizing agent is the more easily reduced and has the more positive reduction potential. Therefore, in order of increasing strength,

K^+ (−2.925 V) < Na^+ (−2.71 V) < Fe^{2+} (−0.44 V) < Cu^{2+} (0.337 V) < Cu^+ (0.521 V) < Ag^+ (0.799 V) < Cl_2 (1.36 V)

(b) Under standard state conditions, both Cl_2 and Ag^+ can oxidize Cu, since their standard reduction potentials are more positive than those for Cu^+ and Cu^{2+}. Also, Cu^+ can oxidize Cu to Cu^{2+}.

21-52. *Refer to Section 21-15.*

The activity of a metal is based on how easily it oxidizes to positively-charged ions. Therefore, a more active metal loses electrons more readily, is more easily oxidized and is a better reducing agent. The strength of a reducing agent increases as its standard reduction potential becomes more negative.

most active   Eu (−3.4 V) > Ra (−2.9 V) > Rh (0.80 V)   least active

Only Eu is more active than Li (−3.0 V). Eu and Ra are more active than H_2 (0.00 V). All are more active than Pt (1.2 V).

(a) cell diagram:

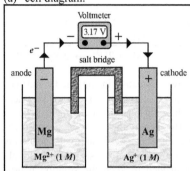

			$E°$
oxidation at anode:	Mg	$\rightarrow$ Mg^{2+} + 2e^-	+2.37 V
reduction at cathode:	2(Ag$^+$ + e^-	$\rightarrow$ Ag)	+0.7994 V

cell reaction: $\quad$ Mg + 2Ag$^+$ $\rightarrow$ Mg^{2+} + 2Ag $\quad$ $E°_{cell}$ = **+3.17 V**

(b) cell diagram:

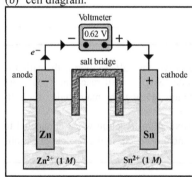

			$E°$
oxidation at anode:	Zn	$\rightarrow$ Zn^{2+} + 2e^-	+0.763 V
reduction at cathode:	Sn^{2+} + 2e^-	$\rightarrow$ Sn	−0.14 V

cell reaction: $\quad$ Zn + Sn^{2+} $\rightarrow$ Zn^{2+} + Sn $\quad$ $E°_{cell}$ = **+0.62 V**

For a standard magnesium and aluminum cell: Shorthand notation: Mg|Mg^{2+}(1 M)||Al^{3+}(1 M)|Al

(a) cell diagram:

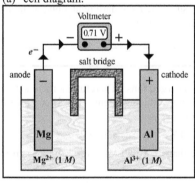

			$E°$
oxidation at anode:	3(Mg	$\rightarrow$ Mg^{2+} + 2e^-)	+2.37 V
reduction at cathode:	2(Al^{3+} + 3e^-	$\rightarrow$ Al)	+0.7994 V

cell reaction: $\quad$ 3Mg + 2Al^{3+} $\rightarrow$ 3Mg^{2+} + 2Al $\quad$ $E°_{cell}$ = **+0.71 V**

Plan: $\quad$ Calculate $E°_{cell}$ for each reaction as written. If $E°_{cell}$ is positive, the reaction is spontaneous and will go as written. If $E°_{cell}$ is negative, the reaction is nonspontaneous and will not go as written; the reverse reaction is spontaneous.

			$E°$
(a) reduction:		$2(Fe^{3+} + e^- \rightarrow Fe^{2+})$	+0.771 V
oxidation:		$Sn^{2+} \rightarrow Sn^{4+} + 2e^-$	−0.15 V
cell reaction:		$2Fe^{3+} + Sn^{2+} \rightarrow 2Fe^{2+} + Sn^{4+}$	$E°_{cell} = +0.62$ V

Yes, Fe^{3+} will oxidize Sn^{2+} to Sn^{4+} because the reaction is spontaneous ($E°_{cell} > 0$).

		$E°$
(b) reduction:	$Cr_2O_7^{2-} + 14H^+ + 6e^- \rightarrow 2Cr^{3+} + 7H_2O$	+1.33 V
oxidation:	$3(2F^- \rightarrow F_2 + 2e^-)$	−2.87 V
cell reaction:	$Cr_2O_7^{2-} + 14H^+ + 6F^- \rightarrow 2Cr^{3+} + 3F_2 + 7H_2O$	$E°_{cell} = -1.54$ V

No, $Cr_2O_7^{2-}$ ions cannot oxidize F^- ions to F_2 because the reaction is not spontaneous ($E°_{cell} < 0$).

21-60. *Refer to Sections 21-15 and 21-16, Examples 21-3 and 21-4, and Appendix J.*

Plan: Calculate $E°_{cell}$ for each reaction as written. If $E°_{cell}$ is positive, the reaction is spontaneous and will go as written. If $E°_{cell}$ is negative, the reaction is nonspontaneous and will not go as written; the reverse reaction is spontaneous.

		$E°$
(a) reduction:	$6(MnO_4^- + 8H^+ + 5e^- \rightarrow Mn^{2+} + 4H_2O)$	+1.51 V
oxidation:	$5(2Cr^{3+} + 7H_2O \rightarrow Cr_2O_7^{2-} + 14H^+ + 6e)$	−1.33 V
cell rxn:	$10\,Cr^{3+} + 11H_2O + 6MnO_4^- \rightarrow 5Cr_2O_7^{2-} + 22H^+ + 6Mn^{2+}$	$E°_{cell} = +0.18$ V

Yes, MnO_4^- ions can oxidize Cr^{3+} ions to $Cr_2O_7^{2-}$ ions since the reaction is spontaneous ($E°_{cell} > 0$).

		$E°$
(b) reduction (1):	$SO_4^{2-} + 4H^+ + 2e^- \rightarrow H_2SO_3 + H_2O$	+0.17 V
reduction (2):	$SO_4^{2-} + 4H^+ + 2e^- \rightarrow SO_2 + 2H_2O$	+0.20 V
oxidation:	$H_3AsO_3 + H_2O \rightarrow H_3AsO_4 + 2H^+ + 2e^-$	−0.58 V

No matter which sulfate reduction reaction is used, $E°_{cell}$ (= $E°_{cathode} + E°_{anode}$) < 0, where $E°_{cathode}$ is the standard reduction potential and $E°_{anode}$ is the standard oxidation potential. Therefore, **no**, SO_4^{2-} ions cannot oxidize H_3AsO_3 to H_3AsO_4.

21-62. *Refer to Section 21-15, Example 21-3 and Exercise 21-38 Solution.*

Refer to the cell diagram in Exercise 21-38 Solution.

		$E°$
reduction:	$2(Au^{3+} + 3e^- \rightarrow Au)$	+1.50 V
oxidation:	$3(Co \rightarrow Co^{2+} + 2e^-)$	+0.28 V
cell reaction:	$2Au^{3+} + 3Co \rightarrow 2Au + 3Co^{2+}$	$E°_{cell} = \textbf{+1.78 V}$

21-64. *Refer to Sections 21-9 and 21-15, and Appendix J.*

(a) Consider the voltaic cell: $Cr|Cr^{3+}||Cu^{2+}|Cu$

   (i) cell reaction: $2Cr + 3Cu^{2+} \rightarrow 2Cr^{3+} + 3Cu$

   (ii) oxidation half-reaction at anode: $Cr \rightarrow Cr^{3+} + 3e^-$ $E° = +0.74$ V

       reduction half-reaction at cathode: $Cu^{2+} + 2e^- \rightarrow Cu$ $E° = +0.337$ V

   (iii) $E°_{cell} = E°_{anode} + E°_{cathode} = +0.74$ V $+ (+0.337$ V$) = \textbf{+1.08 V}$

       Note: $E°_{cathode}$ is the standard reduction potential and $E°_{anode}$ is the standard oxidation potential.

   (iv) Yes, the standard reaction occurs as written since $E°_{cell} > 0$.

(b) Consider the voltaic cell: $Ag|Ag^+||Cd^{2+}|Cd$

    (i)   cell reaction: $2Ag + Cd^{2+} \rightarrow 2Ag^+ + Cd$

    (ii) oxidation half-reaction at anode: $Ag \rightarrow Ag^+ + e^-$ $E° = -0.7994$ V
           reduction half-reaction at cathode: $Cd^{2+} + 2e^- \rightarrow Cd$ $E° = -0.403$ V

    (iii) $E°_{cell} = E°_{cathode} + E°_{anode} = (-0.7994 \text{ V}) + (-0.403 \text{ V}) = \mathbf{-1.202}$ **V**
           Note: $E°_{cathode}$ is the standard reduction potential and $E°_{anode}$ is the standard oxidation potential.

    (iv) No, the standard reaction will not occur as written since $E°_{cell} < 0$; the reverse reaction will occur.

21-66.   *Refer to Sections 21-15 and 21-16, and Appendix J.*

Plan: Calculate $E°_{cell}$ for each reaction as written. If $E°_{cell}$ is positive, the reaction is spontaneous.

			E°
(a) reduction:	$H_2 + 2e^- \rightarrow 2H^-$		-2.25 V
oxidation:	$H_2 \rightarrow 2H^+ + 2e^-$		$+0.00$ V
cell reaction:	$2H_2 \rightarrow 2H^- + 2H^+$	$E°_{cell} =$	-2.25 V
or	$H_2 \rightarrow H^- + H^+$		

No, the reaction is non-spontaneous; $E°_{cell} < 0$.

			E°
(b) reduction:	$Ag_2CrO_4 + 2e^- \rightarrow 2Ag + CrO_4{}^{2-}$		$+0.446$ V
oxidation:	$Zn + 4CN^- \rightarrow Zn(CN)_4{}^{2-} + 2e^-$		$+1.26$ V
cell reaction:	$Ag_2CrO_4 + Zn + 4CN^- \rightarrow 2Ag + CrO_4{}^{2-} + Zn(CN)_4{}^{2-}$	$E°_{cell} =$	$+1.71$ V

Yes, the reaction is spontaneous; $E°_{cell} > 0$.

			E°
(c) reduction:	$MnO_2 + 4H^+ + 2e^- \rightarrow Mn^{2+} + 2H_2O$		$+1.23$ V
oxidation:	$Sr \rightarrow Sr^{2+} + 2e^-$		$+2.89$ V
cell reaction:	$MnO_2 + 4H^+ + Sr \rightarrow Mn^{2+} + 2H_2O + Sr^{2+}$	$E°_{cell} = +4.12$ V	

Yes, the reaction is spontaneous; $E°_{cell} > 0$.

			E°
(c) reduction:	$ZnS + 2e^- \rightarrow Zn + S^{2-}$		-1.44 V
oxidation:	$Cl_2 + 2H_2O \rightarrow 2HOCl + 2H^+ + 2e^-$		-1.63 V
cell reaction:	$ZnS + Cl_2 + 2H_2O \rightarrow Zn + 2HOCl + 2H^+ + S^{2-}$	$E°_{cell} = -3.07$ V	
or	$ZnS + Cl_2 + 2H_2O \rightarrow Zn + 2HOCl + H_2S$	since H_2S is a weak acid	

No, the reaction is non-spontaneous; $E°_{cell} < 0$.

21-68.   *Refer to Sections 21-15 and 21-16, and Appendix J.*

The substance that is the stronger reducing agent is the more easily oxidized. The reduced form of a species is a stronger reducing agent when the half-reaction has a more negative standard reduction potential. The stronger reducing agents are given below.

(a) H_2 (0.000 V) > Ag (0.7994 V)
(b) Sn (-0.14 V) > Pb (-0.126 V)
(c) Hg (0.855 V) > Au (1.68 V or 1.50 V)

(d) Cl^- in base (0.62 V or 0.89 V) > Cl^- in acid (1.36 V)
(e) H_2S (0.14 V) > HCl, i.e., Cl^- in acid (1.36 V) .
(f) Ag (0.7994 V) > Au (1.68 V or 1.50 V)

The half-reactions can be added together so that the desired half-reaction is obtained:

reduction half-reaction:	$Yb^{3+} + 3e^- \rightarrow Yb$	(1)
oxidation half-reaction:	$Yb \rightarrow Yb^{2+} + 2e^-$	(2)
net half-reaction:	$Yb^{3+} + e^- \rightarrow Yb^{2+}$	(3)

Since ΔG is a state function, we can write:

$$\Delta G^\circ_{rxn\,(3)} = \Delta G^\circ_{rxn\,(1)} + \Delta G^\circ_{rxn\,(2)}$$
$$\Delta G^\circ_{Yb^{3+}/Yb^{2+}} = \Delta G^\circ_{Yb^{3+}/Yb} + \Delta G^\circ_{Yb/Yb^{2+}}$$
$$-nFE^\circ_{Yb^{3+}/Yb^{2+}} = -nFE^\circ_{Yb^{3+}/Yb} + (-nFE^\circ_{Yb/Yb^{2+}})$$
$$-(1)E^\circ_{Yb^{3+}/Yb^{2+}} = -(3)(-2.267\ V) - (2)(2.797\ V)$$
$$E^\circ_{Yb^{3+}/Yb^{2+}} = \mathbf{-1.207\ V}$$

Plan: Calculate E°_{cell} for the reaction as written. If E°_{cell} is positive, the reaction is spontaneous.

		E°
reduction:	$2(U^{3+} + 3e^- \rightarrow U)$	$-1.798\ V$
oxidation:	$3(Mg \rightarrow Mg^{2+} + 2e^-)$	$+2.37\ V$
cell reaction:	$2U^{3+} + 3Mg \rightarrow 2U + 3Mg^{2+}$	$E^\circ_{cell} = +0.57\ V$

(a) **Yes**, the setup will work spontaneously because the reaction is spontaneous; $E^\circ_{cell} > 0$.

(b) $E^\circ_{cell} = \mathbf{+0.57\ V}$.

The standard reduction potentials in acidic solution for the species are listed below.

	E°
$K^+(aq) + e^- \rightarrow K(s)$	$-2.925\ V$
$Ca^{2+}(aq) + 2e^- \rightarrow Ca(s)$	$-2.87\ V$
$Ni^{2+}(aq) + 2e^- \rightarrow Ni(s)$	$-0.25\ V$
$O_2(g) + 2H^+(aq) + 2e^- \rightarrow H_2O_2(aq)$	$+0.682\ V$
$H_2O_2(aq) + 2H^+(aq) + 2e^- \rightarrow 2H_2O(\ell)$	$+1.77\ V$
$F_2(g) + 2e^- \rightarrow 2F^-(aq)$	$+2.87\ V$

The voltaic cell with the highest voltage will be the one connecting the K^+/K half-cell with the F_2/F^- half-cell:

		E°
reduction:	$F_2 + 2e^- \rightarrow 2F^-$	$+2.87\ V$
oxidation:	$2(K \rightarrow K^+ + e^-)$	$+2.925\ V$
cell reaction:	$F_2 + 2K \rightarrow 2F^- + 2K^+$	$E^\circ_{cell} = +5.80\ V$

The tarnish on silver, Ag_2S, can be removed by boiling the silverware in slightly salty water (to improve the water's conductivity) in an aluminum pan. The reaction is an oxidation-reduction reaction that occurs spontaneously, similar to the redox reaction occurring in a voltaic cell. The Ag in Ag_2S is reduced back to silver, while the Al in the pan is oxidized to Al^{3+}.

reduction reaction (at surface of the silverware): $3(Ag_2S(s) + 2e^- \rightarrow 2Ag(s) + S^{2-}(aq))$ 0.71 V

oxidation reaction (at surface of the aluminum pan): $2(Al(s) \rightarrow Al^{3+}(aq) + 3e^-)$ +1.66 V

overall reaction: $3Ag_2S(s) + 2Al \rightarrow 6Ag(s) + 3\,S^{2-}(aq) + 2Al^{3+}(aq)$ +0.96 V

The overall cell potential is +0.96 V, showing that the redox reaction is indeed spontaneous. The standard reduction potential for the half cell: $Ag_2S(s) + 2e^- \rightarrow 2Ag(s) + S^{2-}(aq)$ was obtained from the American Society for Metals (ASM) Handbook, available on the internet.

21-78.   *Refer to Section 21-19.*

The Nernst equation is used to calculate electrode potentials or cell potentials when the concentrations and partial pressures are other than standard state values. The Nernst equation using both base 10 and natural logarithms is given by:

$$E = E^\circ - \frac{2.303RT}{nF} \log Q$$

or

$$E = E^\circ - \frac{RT}{nF} \ln Q$$

where E = potential at nonstandard conditions (V)
E° = standard potential (V)
R = gas constant, 8.314 J/mol·K
T = absolute temperature (K); $T = °C + 273.15°$
F = Faraday's constant, 96485 J/V·mol e^-
n = number of moles of e^- transferred
Q = reaction quotient

Substituting at 25°C,   using base 10 logarithms: using natural logarithms:

$$E = E^\circ - \frac{(2.303)(8.314)(298.15)}{n(96485)} \log Q \qquad\qquad E = E^\circ - \frac{(8.314)(298.15)}{n(96485)} \ln Q$$

$$= E^\circ - \frac{0.0592}{n} \log Q \qquad\qquad\qquad = E^\circ - \frac{\mathbf{0.0257}}{n} \ln Q$$

21-80.   *Refer to Section 21-19 and Appendix J.*

Balanced reduction half-reaction: $Zn^{2+} + 2e^- \rightarrow Zn$ $E^\circ = -0.763$ V

For the standard half-cell, $[Zn^{2+}] = 1\ M$. Substituting these data into the Nernst equation, we have

$$E = E^\circ - \frac{0.0592}{n} \log \frac{1}{[Zn^{2+}]} = -0.763\ \text{V} - \frac{0.0592}{2} \log \frac{1}{1} = -0.763\ \text{V} \qquad \text{Note: } \log 1 = 0$$

Therefore, the Nernst equation predicts that the voltage of a standard half-cell equals E°.

21-82.   *Refer to Section 21-19, Example 21-7 and Appendix J.*

(a) E°

oxidation at anode: $Cd \rightarrow Cd^{2+} + 2e^-$ +0.403 V

reduction at cathode: $2(Ag^+ + e^- \rightarrow Ag)$ +0.7994 V

cell reaction: $Cd + 2Ag^+ \rightarrow Cd^{2+} + 2Ag$ $E^\circ_{cell} =$ **+1.202 V**

(b) $E_{cell} = E^\circ_{cell} - \dfrac{0.0592}{n} \log \dfrac{[Cd^{2+}]}{[Ag^+]^2} = +1.202\ \text{V} - \dfrac{0.0592}{2} \log \dfrac{(2.0)}{(0.25)^2} =$ **+1.16 V**

(c) $E_{cell} = E°_{cell} - \dfrac{0.0592}{n} \log \dfrac{[Cd^{2+}]}{[Ag^+]^2}$

$1.25 \text{ V} = +1.202 \text{ V} - \dfrac{0.0592}{2} \log \dfrac{(0.100)}{[Ag^+]^2}$

$0.05 \text{ V} = -\dfrac{0.0592}{2} \log \dfrac{(0.100)}{[Ag^+]^2}$

$-1.62 = \log \dfrac{(0.100)}{[Ag^+]^2} = \log(0.100) - 2\log[Ag^+]^2 = -1.00 - 2\log[Ag^+]$

$-0.62 = -2\log[Ag^+]$

$0.31 = \log[Ag^+]$

$[Ag^+] = \textbf{2.0 } \boldsymbol{M}$

21-84. *Refer to Section 21-19, Example 21-7 and Appendix J.*

		$\boldsymbol{E°}$
(a) oxidation half-reaction:	$Zn(s) \rightarrow Zn^{2+}(aq) + 2e^-$	+0.763 V
reduction half-reaction:	$Cl_2(g) + 2e^- \rightarrow 2Cl^-(aq)$	+1.360 V
cell reaction:	$Zn(s) + Cl_2(g) \rightarrow Zn^{2+}(aq) + 2Cl^-(aq)$	$E°_{cell} = $ **+2.123 V**

(b) $[Zn^{2+}] = [ZnCl_2] = 0.21 \ M;$ in the Cl_2/Cl^- halfcell, $[Cl^-] = 1.00 \ M;$ $P_{Cl_2} = 1.0$ atm

$E_{cell} = E°_{cell} - \dfrac{0.0592}{n} \log \dfrac{[Zn^{2+}][Cl^-]^2}{P_{Cl_2}} = +2.123 \text{ V} - \dfrac{0.0592}{2} \log \dfrac{(0.21)(1.00)^2}{(1.0)} = \textbf{+2.143 V}$

21-86. *Refer to Section 21-19, Example 21-7 and Appendix J.*

Balanced half-reaction: $F_2(g) + 2e^- \rightarrow 2F^-(aq)$ $E° = +2.87$ V

Applying the Nernst equation: $E = E° - \dfrac{0.0592}{n} \log \dfrac{[F^-]^2}{P_{F_2}}$

Substituting, $+2.70 \text{ V} = +2.87 \text{ V} - \dfrac{0.0592}{2} \log \dfrac{(0.34)^2}{P_{F_2}}$

$0.17 \text{ V} = \dfrac{0.0592}{2} \log \dfrac{0.12}{P_{F_2}}$

$5.7 = \log \dfrac{0.12}{P_{F_2}}$

Taking the antilogarithm of both sides, $5.5 \times 10^5 = \dfrac{0.12}{P_{F_2}}$

Therefore, $P_{F_2} = \textbf{2.2} \times \textbf{10}^{-7} \textbf{ atm}$

21-88. *Refer to Section 21-15 and Exercise 87a.*

For this non-standard cell: $\boldsymbol{E°}$

oxidation at anode:	$Sn \rightarrow Sn^{2+} + 2e^-$	+0.14 V
reduction at cathode:	$2(Ag^+ + e^- \rightarrow Ag)$	+0.7994 V
cell reaction:	$Sn + 2Ag^+ \rightarrow Sn^{2+} + 2Ag$ $E°_{cell} = $ +0.94 V	

Shorthand notation: $Sn|Sn^{2+}(7.0 \times 10^{-3} \ M)\|Ag^+(0.110 \ M)|Ag$

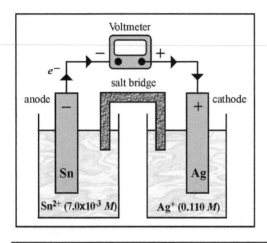

$$E_{cell} = E^{\circ}_{cell} - \frac{0.0592}{n} \log \frac{[Sn^{2+}]}{[Ag^+]^2}$$

$$= +0.94 \text{ V} - \frac{0.0592}{2} \log \frac{(7.0 \times 10^{-3})}{(0.110)^2}$$

$$= \mathbf{+0.95 \text{ V}}$$

21-90. *Refer to Section 21-19, Example 21-7 and Appendix J.*

0.95 V

Balanced reduction half-reactions:

(1)	$2H^+(aq) + 2e^- \rightarrow H_2(g)$	$E^{\circ} = 0.0000$ V
(2)	$Ag^+(aq) + e^- \rightarrow Ag(s)$	$E^{\circ} = 0.7994$ V

(a) Balanced equation: $H_2(g) + 2Ag^+(aq) \rightarrow 2H^+(aq) + 2Ag(s)$ $E^{\circ} = 0.7994$ V

$$E = E^{\circ} - \frac{0.0592}{n} \log \frac{[H^+]^2}{P_{H_2}[Ag^+]^2} = 0.7994 \text{ V} - \frac{0.0592}{2} \log \frac{(1.00 \times 10^{-3})^2}{(8.00)(5.49 \times 10^{-3})^2} = \mathbf{0.870 \text{ V}}$$

(b) Balanced equation: $H_2(1.00 \text{ atm}) + 2H^+(pH = 3.47) \rightarrow 2H^+(pH = 5.97) + H_2(1.00 \text{ atm})$

For pH = 5.97, $[H^+] = 1.07 \times 10^{-6}$ M
For pH = 3.47, $[H^+] = 3.39 \times 10^{-4}$ M

$$E = E^{\circ} - \frac{0.0592}{n} \log \frac{[H^+]^2 \, P_{H_2}}{P_{H_2}[H^+]^2} = 0.0000 \text{ V} - \frac{0.0592}{2} \log \frac{(1.07 \times 10^{-6})^2(1.00)}{(1.00)(3.39 \times 10^{-4})^2} = \mathbf{0.148 \text{ V}}$$

(c) Balanced equation: $H_2(0.0361 \text{ atm}) + 2H^+(0.0175 \text{ M}) \rightarrow 2H^+(0.0175 \text{ M}) + H_2(5.98 \times 10^{-4} \text{ atm})$

$$E = E^{\circ} - \frac{0.0592}{n} \log \frac{[H^+]^2 \, P_{H_2}}{P_{H_2}[H^+]^2} = 0.0000 \text{ V} - \frac{0.0592}{2} \log \frac{(0.0175)^2(5.98 \times 10^{-4})}{(0.0361)(0.0175)^2} = \mathbf{0.0527 \text{ V}}$$

21-92. *Refer to Section 21-20.*

Balanced half-reaction: $2H^+(aq) + 2e^- \rightarrow H_2(g)$ $E^{\circ} = 0.000$ V

In a concentration cell, we have 2 half cells containing the same ions and gases, only at different concentrations and/or partial pressures. Assume that the partial pressure of $H_2(g)$ in both cells is 1 atm. The spontaneous reaction occurring in the voltaic cell will proceed in the direction that will try to equalize the concentration of H^+ ion.

oxidation half-reaction:	$2H^+(0.05 \text{ M}) + 2e^- \rightarrow H_2(1 \text{ atm})$
reduction half-reaction:	$H_2(1 \text{ atm}) \rightarrow 2H^+(pH = 7.8) + 2e^-$
cell reaction:	$2H^+(0.05 \text{ M}) + H_2(1 \text{ atm}) \rightarrow 2H^+(pH = 7.8) + H_2(1 \text{ atm})$

For this concentration cell:

$$E_{cell} = E^\circ_{cell} - \frac{0.0592}{n} \log \frac{[H^+ (pH = 7.8)]^2 P_{H_2} (1 \text{ atm})}{([H^+] = 0.05 \ M)^2 P_{H_2} (1 \text{ atm})}$$

Substituting,

$$E_{cell} = 0 \text{ V} - \frac{0.0592}{2} \log \frac{(1.6 \times 10^{-8} \ M)^2 (1)}{(0.05)^2 (1)}$$

$$E_{cell} = 0 \text{ V} - \frac{0.0592}{2} \log (1.0 \times 10^{-13})$$

$$E_{cell} = \mathbf{+0.38 \text{ V}}$$

21-94. *Refer to Section 21-19, Exercise 21-93, Example 21-8 and Appendix J.*

(a) Plan: (1) Determine E°_{cell}.
　　　　　 (2) Use the Nernst equation to find the ratio of Zn^{2+} to Ni^{2+}.

			E°
(1) reduction half-reaction:	$Ni^{2+} + 2e^- \rightarrow Ni$		- 0.25 V
oxidation half-reaction:	$Zn \rightarrow Zn^{2+} + 2e^-$		+0.763 V
cell reaction:	$Ni^{2+} + Zn \rightarrow Ni + Zn^{2+}$	$E^\circ_{cell} =$	+0.513 V

(2) Using the Nernst equation:

$$E_{cell} = E^\circ_{cell} - \frac{0.0592}{n} \log \frac{[Zn^{2+}]}{[Ni^{2+}]}$$

Substituting,

$$0 = 0.513 \text{ V} - \frac{0.0592}{2} \log \frac{[Zn^{2+}]}{[Ni^{2+}]}$$

$$\log \frac{[Zn^{2+}]}{[Ni^{2+}]} = 17.33 \text{ or } 17.3 \text{ (3 significant figures)}$$

$$\frac{[Zn^{2+}]}{[Ni^{2+}]} = \mathbf{2 \times 10^{17}} \text{ (1 significant figures)}$$

(b) Since the cell starts at standard conditions, $[Ni^{2+}]_{initial} = [Zn^{2+}]_{initial} = 1.00 \ M$. Also, for every 1 mole of Zn^{2+} produced, there is 1 mole of Ni^{2+} lost. Therefore,

Let x = mol/L of Ni^{2+} that reacted. Then, x = mol/L of Zn^{2+} that were produced.

	Ni^{2+}	+	Zn	$\rightarrow$	Ni	+	Zn^{2+}
initial	1.00 *M*		-		-		1.00 *M*
change	- x *M*						+ x *M*
after reaction	(1.00 - x) *M*						(1.00 + x) *M*

Therefore, $[Zn^{2+}] + [Ni^{2+}] = (1.00 + x) \ M + (1.00 - x) \ M = 2.00 \ M$

We know from (a) that $[Zn^{2+}] = (2 \times 10^{17})[Ni^{2+}]$. Substituting for $[Zn^{2+}]$ and solving for $[Ni^{2+}]$,

$$2.00 \ M = (2 \times 10^{17})[Ni^{2+}] + [Ni^{2+}]$$
$$= [Ni^{2+}](2 \times 10^{17} + 1) \approx (2 \times 10^{17})[Ni^{2+}]$$
$$[Ni^{2+}] = \mathbf{1 \times 10^{-17} \ M}$$
$$[Zn^{2+}] = 2.00 \ M - [Ni^{2+}] = \mathbf{2.00 \ M} \text{ (to 3 significant figures)}$$

21-96. *Refer to Section 21-19.*

Balanced half-reaction: $2H^+(aq) + 2e^- \rightarrow H_2(g)$ 　　　　　 $E^\circ = 0.000 \text{ V}$

In a concentration cell, we have 2 half cells containing the same ions and gases, only at different concentrations and/or partial pressures. The way this question is worded, there may be 2 answers because the cell potential is a function of the square of the ratios of hydrogen ions in the two half cells.
Answer (1) - the cathodic half-cell is at pH = 1.5:

oxidation half-reaction:	$H_2\,(1\text{ atm}) \rightarrow 2H^+\,(\text{pH} = ?) + 2e^-$
reduction half-reaction:	$2H^+\,(\text{pH} = 1.5) + 2e^- \rightarrow H_2\,(1\text{ atm})$
cell reaction:	$2H^+\,(\text{pH} = 1.5) + H_2\,(1\text{ atm}) \rightarrow 2H^+\,(\text{pH} = ?) + H_2\,(1\text{ atm})$

For this concentration cell:

$$E_{\text{cell}} = E^\circ_{\text{cell}} - \frac{0.0592}{n} \log \frac{[H^+\,(?)]^2 P_{H_2}\,(1\text{ atm})}{[H^+\,(\text{pH} = 1.5)]^2 P_{H_2}\,(1\text{ atm})}$$

Substituting,

$$0.275\text{ V} = 0\text{ V} - \frac{0.0592}{2} \log \frac{[H^+]^2(1)}{(0.032)^2(1)}$$

$$\log \frac{[H^+]^2}{0.0010} = -9.29$$

Taking the antilogarithm,

$$\frac{[H^+]^2}{0.0010} = 5.1 \times 10^{-10}$$

Therefore,

$$[H^+] = 7.2 \times 10^{-7}\ M$$
$$\text{pH} = \mathbf{6.15}$$

Answer (2) - the anodic half-cell is at pH = 1.5:

oxidation half-reaction:	$H_2\,(1\text{ atm}) \rightarrow 2H^+\,(\text{pH} = 1.5) + 2e^-$
reduction half-reaction:	$2H^+\,(\text{pH} = ?) + 2e^- \rightarrow H_2\,(1\text{ atm})$
cell reaction:	$2H^+\,(\text{pH} = ?) + H_2\,(1\text{ atm}) \rightarrow 2H^+\,(\text{pH} = 1.5) + H_2\,(1\text{ atm})$

For this concentration cell:

$$E_{\text{cell}} = E^\circ_{\text{cell}} - \frac{0.0592}{n} \log \frac{[H^+\,(\text{pH} = 1.5)]^2 P_{H_2}\,(1\text{ atm})}{[H^+\,(\text{pH} = ?)]^2 P_{H_2}\,(1\text{ atm})}$$

Substituting,

$$0.275\text{ V} = 0\text{ V} - \frac{0.0592}{2} \log \frac{(0.032)^2(1)}{[H^+]^2(1)}$$

$$\log \frac{0.0010}{[H^+]^2} = -9.29$$

Taking the antilogarithm,

$$\frac{0.0010}{[H^+]^2} = 5.1 \times 10^{-10}$$

Therefore,

$$[H^+] = 1.4 \times 10^3\ M$$
$$\text{pH} = -3.14$$

As you can see, Answer (1) is the answer that makes sense. Answer (2) makes no sense, since it is impossible to have a solution with $[H^+] = 1.4 \times 10^3\ M$. So, the cathodic half-cell is at pH = 1.5 and the pH of the anodic half-cell must be **6.15**.

21-98. *Refer to Section 21-21.*

Because $\Delta G^\circ = -nFE^\circ_{\text{cell}}$ and $\Delta G^\circ = -RT \ln K$, the signs and magnitudes of E°_{cell}, ΔG° and K are related as shown in the following table for different types of reactions under standard state conditions.

Forward Reaction	E°_{cell}	ΔG°	K
spontaneous	+	−	>1
at equilibrium	0	0	1
non-spontaneous	−	+	<1

From the above equations, it is seen that the value of K is related to the value of ΔG° and E° of the cell, but not ΔG and E of the cell. E°, ΔG° and K are indicators of the thermodynamic tendency of an oxidation-reduction reaction to occur under standard conditions.

On the other hand, E and ΔG are related to the value of Q and are indicators of the spontaneity of a reaction under any given conditions. The reaction proceeds until $Q = K$ at which point $\Delta G = 0$ and $E_{\text{cell}} = 0$. Then:

$$\log K = \frac{nFE^\circ_{\text{cell}}}{2.303RT} \qquad \text{or} \qquad \ln K = \frac{nFE^\circ_{\text{cell}}}{RT}$$

			$E°$
(a)	reduction:	$MnO_4^- + 8H^+ + 5e^- \rightarrow Mn^{2+} + 4H_2O$	+1.507 V
	oxidation:	$5(Fe^{2+} \rightarrow Fe^{3+} + e^-)$	−0.771 V
	cell reaction:	$MnO_4^- + 8H^+ + 5Fe^{2+} \rightarrow Mn^{2+} + 4H_2O + 5Fe^{3+}$	$E°_{cell} =$ **+0.736 V**

The reaction is spontaneous as written under standard conditions since $E°_{cell} > 0$.

$\Delta G° = -nFE°_{cell} = -(5 \text{ mol } e^-)(96500 \text{ J/V·mol } e^-)(+0.736 \text{ V}) =$ **−3.55 x 10^5 J/mol rxn or −355 kJ/mol rxn**

At 25°C, $E°_{cell} = \dfrac{RT \ln K}{nF}$. So, $\ln K = \dfrac{nFE°_{cell}}{RT}$

Substituting, $\ln K = \dfrac{(5 \text{ mol})(9.65 \text{ x } 10^4 \text{ J/V·mol})(+0.736 \text{ V})}{(8.314 \text{ J/mol·K})(298 \text{ K})} = 143$ Solving, $K = e^{143}$ or **1 x 10^{62}**

			$E°$
(b)	reduction:	$Cu^+ + e^- \rightarrow Cu$	+0.521 V
	oxidation:	$Cu^+ \rightarrow Cu^{2+} + e^-$	−0.153 V
	cell reaction:	$2Cu^+ \rightarrow Cu + Cu^{2+}$	$E°_{cell} =$ **+0.368 V**

The reaction is spontaneous as written under standard conditions since $E°_{cell} > 0$.

$\Delta G° = -nFE°_{cell} = -(1 \text{ mol } e^-)(96500 \text{ J/V·mol } e^-)(0.368 \text{ V}) =$ **−3.55 x 10^4 J/mol rxn or −35.5 kJ/mol rxn**

at 25°C, $\ln K = \dfrac{nFE°_{cell}}{RT} = \dfrac{(1 \text{ mol})(9.65 \text{ x } 10^4 \text{ J/V·mol})(+0.368 \text{ V})}{(8.314 \text{ J/mol·K})(298 \text{ K})} = 14.3$

Solving, $K = e^{14.3}$ or **1.6 x 10^6**

			$E°$
(c)	reduction:	$2(MnO_4^- + 2H_2O + 3e^- \rightarrow MnO_2 + 4OH^-)$	+0.588 V
	oxidation:	$3(Zn + 2OH^- \rightarrow Zn(OH)_2 + 2e^-)$	+1.245 V
	cell reaction:	$2MnO_4^- + 3Zn + 4H_2O \rightarrow 2MnO_2 + 3Zn(OH)_2 + 2OH^-$	$E°_{cell} =$ **+1.833 V**

The reaction is spontaneous as written under standard conditions since $E°_{cell} > 0$.

$\Delta G° = -nFE°_{cell} = -(6 \text{ mol } e^-)(96485 \text{ J/V·mol } e^-)(1.833 \text{ V}) =$ **−1.061 x 10^6 J/mol rxn or −1061 kJ/mol rxn**

at 25°C, $\ln K = \dfrac{nFE°_{cell}}{RT} = \dfrac{(6 \text{ mol})(9.6485 \text{ x } 10^4 \text{ J/V·mol})(+1.833 \text{ V})}{(8.314 \text{ J/mol·K})(298.15 \text{ K})} = 428.1$

Solving, $K = e^{428.1}$ or **8.6 x 10^{185}**

Note (1): Since the $E°$ value has 4 significant figures, we must use values for the constants (F, R and T) that also have at least 4 significant figures.

Note (2): It is more difficult to find K in scientific notation because most calculators cannot handle numbers this big. So, use what you know about exponents to solve for K:

$e^{428.1} = e^{200.0} \text{ x } e^{228.1} = (7.2 \text{ x } 10^{86})(1.2 \text{ x } 10^{99}) = (7.2 \text{ x } 1.2)(10^{86} \text{ x } 10^{99}) = 8.6 \text{ x } 10^{185}$

21-102. *Refer to Section 21-21, Example 21-10, and Appendix J.*

We know that $\Delta G° = -nFE°_{cell}$, so $E°_{cell} = -\dfrac{\Delta G°}{nF}$

(a) $E°_{cell} = -\dfrac{\Delta G°}{nF} = -\dfrac{25000 \text{ J}}{(1 \text{ mol } e^-)(96485 \text{ J/V·mol } e^-)} = $ **−0.25 V**

(b) $E°_{cell} = -\dfrac{\Delta G°}{nF} = -\dfrac{25000 \text{ J}}{(2 \text{ mol } e^-)(96485 \text{ J/V·mol } e^-)} = $ **−0.13 V**

(c) $E°_{cell} = -\dfrac{\Delta G°}{nF} = -\dfrac{25000 \text{ J}}{(4 \text{ mol } e^-)(96485 \text{ J/V·mol } e^-)} = \mathbf{-0.065 \text{ V}}$

When $E°_{cell}$ is directly calculated from $\Delta G°$, the cell voltage is inversely proportional to the number of electrons passing through the cell.

21-104. *Refer to Section 21-21 and Example 21-10.*

		$E°$
reduction half-reaction:	$PbSO_4 + 2e^- \rightarrow Pb + SO_4{}^{2-}$	-0.356 V
oxidation half-reaction:	$Pb + 2I^- \rightarrow PbI_2 + 2e^-$	$+0.365$ V
cell reaction:	$PbSO_4 + 2I^- \rightarrow PbI_2 + SO_4{}^{2-}$	$E°_{cell} = +0.009$ V

$\ln K = \dfrac{nFE°_{cell}}{RT} = \dfrac{(2 \text{ mol})(96500 \text{ J/V·mol})(+0.009 \text{ V})}{(8.314 \text{ J/mol·K})(298 \text{ K})} = 0.7$ Solving, $K = \mathbf{2}$ (to 1 significant figure)

21-106. *Refer to Sections 21-22, 21-23 and 21-25.*

(a) The dry cell (Leclanchè cell) is shown in Figure 21-16. The container is made of zinc, which also acts as one of the electrodes. The other electrode is a carbon rod in the center of the cell. The cell is filled with a moist mixture of NH_4Cl, MnO_2, $ZnCl_2$ and a porous inert filler. The cell is separated from the zinc container by a porous paper. Dry cells are sealed to keep moisture from evaporating. As the cell operates, the Zn electrode is the anode and is oxidized to Zn^{2+} ions. The ammonium ion is reduced to give NH_3 and H_2 at the carbon cathode. The ammonia produced combines with Zn^{2+} ion and forms a soluble compound containing the complex ion, $Zn(NH_3)_4{}^{2+}$; H_2 is removed by being oxidized by MnO_2. This type of battery cannot be recharged.

(b) The lead storage battery is shown in Figure 21-17. It consists of a group of lead plates bearing compressed spongy lead alternating with a group of lead plates bearing lead(IV) oxide, PbO_2. The electrodes are immersed in a solution of about 40% sulfuric acid. When the cell discharges, the spongy lead is oxidized to give Pb^{2+} ions which then combine with sulfate ions to form insoluble $PbSO_4$, coating the anode. Electrons produced at the anode by oxidation of spongy lead travel through the external circuit to the cathode and reduce lead(IV) to lead(II) in the presence of H^+. The cathode also becomes coated with insoluble lead sulfate. The lead storage battery can be recharged by reversal of all reactions.

(c) The hydrogen-oxygen fuel cell is shown in Figure 21-18. Hydrogen (the fuel) is supplied to the anode compartment. Oxygen is fed into the cathode compartment. Oxygen is reduced at the cathode to OH^- ions. The OH^- ions migrate through the electrolyte, an aqueous solution of a base, to the anode, where H_2 is oxidized to H_2O. The net reaction of the cell is the same as the burning of hydrogen in oxygen to form water, but combustion does not occur. Rather, most of the chemical energy, produced from the destruction of H-H and O-O bonds and the formation of O-H bonds, is converted directly into electrical energy.

21-108. *Refer to Section 21-22, Figure 21-16, Exercise 21-106 Solution and Appendix J.*

(a) When attempting to recharge an Leclanchè cell (a dry cell), the electrodes are reversed; the zinc container which is the anode under normal operation becomes the cathode. The reaction expected is the reduction of Zn^{2+} to zinc metal:

$Zn^{2+} + 2e^- \rightarrow Zn$ $E° = -0.763$ V

(b) Recharging the battery means reversing the actual cell reaction to yield:

$H_2 + 2NH_3 + Zn^{2+} \rightarrow Zn + 2NH_4{}^+$ $E_{cell} = -1.6$ V

This is essentially an impossible task because each of the original products has been permanently removed from the system, especially the hydrogen gas.

(1) NH_3 and Zn^{2+} have reacted together to give a very stable zinc-ammonia complex:

$$Zn^{2+} + 4NH_3 \rightarrow Zn(NH_3)_4^{2+} \qquad K = \frac{1}{K_d} = 2.9 \times 10^9 \text{ (Appendix I)}$$

The zinc complex is more difficult than the free Zn^{2+} to reduce as deduced from the more negative standard reduction potential:

$$Zn(NH_3)_4^{2+} + 2e^- \rightarrow Zn + 4NH_3. \qquad E^\circ = -1.04 \text{ V}$$

(2) H_2 has reacted with MnO_2 to give the solid, $MnO(OH)$: $H_2 + 2MnO_2 \rightarrow 2MnO(OH)$.
Since there are essentially none of the original products, the recharging cannot occur.

21-110. *Refer to Sections 21-22, 21-23 and 21-25, and Exercise 21-106 Solution.*

A fuel cell is different from a dry cell or storage cell because:

(1) the reactant, the fuel (usually H_2) and oxygen are fed into the cell continuously and the products are constantly removed. Hence, the fuel cell creates electrical energy, but does not store it. It can operate indefinitely as long as fuel is available.
(2) The electrodes are made of an inert material such as platinum and do not react during the electrochemical process.
(3) Many fuel cells are non-polluting, e.g., the H_2/O_2 fuel cell whose only product is H_2O.

21-112. *Refer to Section 21-11.*

The half-cell reduction potential of the standard hydrogen electrode (SHE) was set arbitrarily to 0.000... V by international agreement. Since it is impossible to determine the potential of a single half-cell without comparing it to another, an arbitrary standard was established.

21-114. *Refer to Section 21-18.*

Aluminum is the metal that forms a resilient, transparent surface layer of its oxide, Al_2O_3. This layer protects the metal from most corrosive environmental agents.

21-116. *Refer to Sections 21-15, 21-19 and 21-6.*

Consider: $Mg(s)|Mg^{2+}(aq)||Fe^{3+}(aq)|Fe(s)$

(a) oxidation half reaction (at anode): $\qquad 3(Mg(s) \rightarrow Mg^{2+}(aq) + 2e^-) \qquad E^\circ = +2.37$ V
reduction half-reaction (at cathode): $\qquad 2(Fe^{3+}(aq) + 3e^- \rightarrow Fe(s)) \qquad E^\circ = -0.036$ V

overall cell reaction: $\qquad 3Mg(s) + 2Fe^{3+}(aq) \rightarrow 3Mg^{2+}(aq) + 2Fe(s)$

(b) $E^\circ_{cell} = E^\circ_{anode} + E^\circ_{cathode} = (+2.37 \text{ V}) + (-0.036 \text{ V}) = \textbf{+2.33 V}$
Note: $E^\circ_{cathode}$ is the standard reduction potential and E°_{anode} is the standard oxidation potential.

(c) $E = E^\circ - \dfrac{0.0592}{n} \log \dfrac{[Mg^{2+}]^3}{[Fe^{3+}]^2} = +2.33 \text{ V} - \dfrac{0.0592}{6} \log \dfrac{(1.00 \times 10^{-3})^3}{(10.0)^2} = +2.33 \text{ V} - (-0.109 \text{ V}) = \textbf{+2.44 V}$

(d) The minimum mass change of the magnesium electrode is the mass of Mg lost when 150 mA passes through the cell for 20.0 minutes.

$? \text{ g Mg} = 20.0 \text{ min} \times \dfrac{60 \text{ s}}{1 \text{ min}} \times \dfrac{0.150 \text{ C}}{1 \text{ s}} \times \dfrac{1 \text{ mol } e^-}{96500 \text{ C}} \times \dfrac{1 \text{ mol Mg}}{2 \text{ mol } e^-} \times \dfrac{24.30 \text{ g}}{1 \text{ mol Mg}} = \textbf{0.0227 g Mg}$

21-118. *Refer to Section 21-6.*

Balanced half-reaction: $Cu^{2+} + 2e^- \rightarrow Cu$

$? \text{ coulombs} = 0.0300 \text{ L soln} \times \dfrac{0.165 \text{ mol Cu}}{1 \text{ L soln}} \times \dfrac{2 \text{ mol } e^-}{1 \text{ mol Cu}} \times \dfrac{96500 \text{ C}}{1 \text{ mol } e^-} = \textbf{955 C}$

21-120. *Refer to Sections 21-19 and 21-20.*

		$E°$
oxidation half-reaction:	$Mn \rightarrow Mn^{2+} + 2e^-$	+1.18 V
reduction half-reaction:	$Fe^{2+} + 2e^- \rightarrow Fe$	−0.44 V
cell reaction:	$Mn + Fe^{2+} \rightarrow Mn^{2+} + Fe$	$E°_{cell} = +0.74 \text{ V}$

(a) Using the Nernst equation:

$$E_{cell} = E°_{cell} - \frac{0.0592}{n} \log \frac{[Mn^{2+}]}{[Fe^{2+}]}$$

Substituting,

$$1.45 \text{ V} = 0.74 \text{ V} - \frac{0.0592}{2} \log \frac{[Mn^{2+}]}{[Fe^{2+}]}$$

$$\log \frac{[Mn^{2+}]}{[Fe^{2+}]} = -24$$

$$\frac{[Mn^{2+}]}{[Fe^{2+}]} = \mathbf{10^{-24}}$$

(b) The anode is manganese and the cathode is iron. Since the electrons always flow from anode to cathode, they are flowing from the manganese anode to the iron cathode.

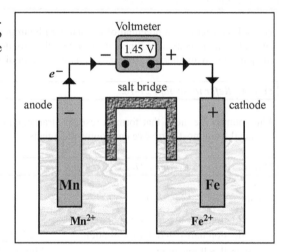

21-122. *Refer to Section 21-6.*

Balanced equations: $UO_2(s) + 4HF(g) \rightarrow UF_4(s) + 2H_2O(\ell)$

$\qquad\qquad\qquad\quad UF_4(s) + 2Mg(s) \rightarrow U(s) + 2MgF_2(s)$

(a) ox. no. U in $UO_2(s)$: +4

(b) ox. no. U in $UF_4(s)$: +4

(c) ox. no. U in $U(s)$: 0

(d) reducing agent: $Mg(s)$

(e) substance reduced (oxidizing agent): $UF_4(s)$

(f) U is being reduced from +4 oxidation number in $UF_4(s)$ to 0 in $U(s)$. Therefore, 4 moles of electrons are required to reduce 1 mole of $UF_4(s)$.

$$? \text{ coulombs/s} = \frac{0.500 \text{ g } UF_4}{1 \text{ min}} \times \frac{1 \text{ min}}{60 \text{ s}} \times \frac{1 \text{ mol } UF_4}{314 \text{ g } UF_4} \times \frac{4 \text{ mol } e^-}{1 \text{ mol } UF_4} \times \frac{96500 \text{ C}}{1 \text{ mol } e^-} = \mathbf{10.2 \text{ C/s} \text{ or } 10.2 \text{ A}}$$

(g) Plan: (1) Determine the number of moles of $HF(g)$.
    (2) Calculate the volume of $HF(g)$ using the ideal gas law, $PV = nRT$.

(1) $? \text{ mol HF} = 0.500 \text{ g U} \times \dfrac{1 \text{ mol U}}{238 \text{ g U}} \times \dfrac{1 \text{ mol } UF_4}{1 \text{ mol U}} \times \dfrac{4 \text{ mol HF}}{1 \text{ mol } UF_4} = 8.40 \times 10^{-3} \text{ mol HF}$

(2) $V = \dfrac{nRT}{P} = \dfrac{(8.40 \times 10^{-3} \text{ mol})(0.0821 \text{ L·atm/mol·K})(298 \text{ K})}{(10.0 \text{ atm})} = \mathbf{0.0206 \text{ L HF}(g)}$

(h) Plan: Determine the mass of U that can be prepared from 0.500 g Mg and compare.

$? \text{ g U} = 0.500 \text{ g Mg} \times \dfrac{1 \text{ mol Mg}}{24.30 \text{ g Mg}} \times \dfrac{1 \text{ mol U}}{2 \text{ mol Mg}} \times \dfrac{238.0 \text{ g U}}{1 \text{ mol U}} = 2.45 \text{ g U}$

Yes, 0.500 g Mg is more than enough to prepare 0.500 g U. In fact, 0.500 g Mg can ideally produce 2.45 g U.

21-124. *Refer to Section 21-7 and Figure 21-5.*

(a) Electroplating is a process that plates metal onto a cathodic surface by electrolysis.

(b) A simple silver electroplating apparatus for a jeweler consists of a dc generator (a battery) with the negative lead attached to the piece of jewelry (cathode) and the positive lead attached to a piece of silver metal (anode). The jewelry and the silver metal are both immersed in a beaker containing an aqueous solution of a silver salt such as $AgNO_3$. During electroplating, the Ag metal at the anode will be oxidized to Ag^+ ions, and the Ag^+ ions in solution will be reduced to Ag metal and plated onto the jewelry at the cathode.

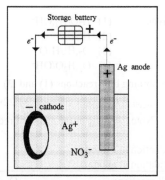

(c) Highly purified silver as the anode is not necessary in an electroplating operation. As the electrolytic cell operates, Ag and other metal impurities in a regular Ag anode oxidize to form metal cations in solution. However, only Ag^+ ions are reduced to Ag metal at the cathode because of its ease of reduction and higher concentration. This preference can be enhanced by setting the operating voltage just above the threshold required to electroplate silver. So, the extra cost of a highly purified silver anode is not necessary.

21-126. *Refer to Section 21-23.*

In the lead storage battery, insoluble lead sulfate, $PbSO_4(s)$, is produced at both the anode and cathode during cell discharge.

21-128. *Refer to Section 21-4, Table 21-2 and Appendix J.*

It is not possible to prepare F_2 by electrolysis of an aqueous NaF solution. In electrolysis, the most easily oxidized and reduced species are the ones involved. To prepare F_2, the oxidation of F^- would have to occur. However, water is more easily oxidized than is F^-, as seen by its position in the standard reduction potential chart (Appendix J and below). By inspection, H_2O is a stronger reducing agent than F^- because the reduction half-reaction has a less positive E°. So H_2O's oxidation is preferable to F^-'s oxidation. F_2 can be prepared from molten NaF, but not aqueous NaF.

$$O_2(g) + 4H^+(aq) + 4e^- \rightarrow 2H_2O(\ell) \qquad +1.229 \text{ V}$$
$$F_2(g) + 2e^- \rightarrow 2F^-(aq) \qquad +2.87 \text{ V}$$

363

In an electroplating process, once the concentration of the element of interest, e.g. Cu^{2+}, is sufficiently low, the impurity metal cations will start plating out of solution onto the object (cathode - source of electrons). The order of plating depends both on the reduction potentials of the metals and their concentrations. If we assume that the impurity concentrations are the same in the solution, then the most easily reduced species will be the first one to plate out. This is the metal with the most positive standard reduction potential. Of the metals given in the problem, the order in which the metals will plate out is:

(1) $Au^+ + e^- \rightarrow Au$ +1.68 V or
     $Au^{3+} + 3e^- \rightarrow Au$ +1.50 V

(2) $Pt^{2+} + 2e^- \rightarrow Pt$ +1.2 V

(3) $Ag^+ + e^- \rightarrow Ag$ +0.7994 V

(4) $Fe^{3+} + 3e^- \rightarrow Fe$ −0.036 V or
     $Fe^{2+} + 2e^- \rightarrow Fe$ −0.44 V

(5) $Zn^{2+} + 2e^- \rightarrow Zn$ −0.763 V

Yes, the electrolytic process can be used to individually separate the impurity metals if their reduction potentials are sufficiently different.

(a) Given:

			$E°$
(1) $H_2O/H_2,OH^-$	$2H_2O + 2e^- \rightarrow H_2 + 2OH^-$		−0.828 V
(2) H^+/H_2	$2H^+ + 2e^- \rightarrow H_2$		0.0000 V
(3) $O_2,H^+/H_2O$	$O_2 + 4H^+ + 4e^- \rightarrow 2H_2O$		+1.229 V
(4) $O_2,H_2O/OH^-$	$O_2 + 2H_2O + 4e^- \rightarrow 4OH^-$		+0.401 V

Combining half-reactions (1) and (3) would give the greatest voltage:

$$E°_{cell} = E°_{cathode} + E°_{anode} = +1.229 \text{ V} + 0.828 \text{ V} = +2.057 \text{ V}$$

Note: $E°_{cathode}$ is the standard reduction potential and $E°_{anode}$ is the standard oxidation potential.

(b) reduction at cathode: $O_2 + 4H^+ + 4e^- \rightarrow 2H_2O$

     oxidation at anode: $2(H_2 + 2OH^- \rightarrow 2H_2O + 2e^-)$

     cell reaction: $O_2 + 2H_2 + 4H^+ + 4OH^- \rightarrow 6H_2O$

     or $O_2 + 2H_2 + 4H_2O \rightarrow 6H_2O$

     or $O_2 + 2H_2 \rightarrow 2H_2O$

(a) Plan: The K_{sp} value for $AgBr(s)$ is the equilibrium constant for: $AgBr(s) \rightleftarrows Ag^+(aq) + Br^-(aq)$. It can be estimated from data in Appendix J. Choose the appropriate oxidation and reduction half-reactions that produce the above reaction and calculate $E°_{cell}$ and K_{sp} at 25°C.

		$E°$
reduction half-reaction:	$AgBr(s) + e^- \rightarrow Ag(s) + Br^-(aq)$	+0.10 V
oxidation half-reaction:	$Ag(s) \rightarrow Ag^+(aq) + e^-$	−0.7994 V
cell reaction:	$AgBr(s) \rightarrow Ag^+(aq) + Br^-(aq)$	$E°_{cell} = -0.70$ V

$$\ln K = \ln K_{sp} = \frac{nFE°_{cell}}{RT} = \frac{(1 \text{ mol})(96500 \text{ J/V·mol})(-0.70 \text{ V})}{(8.314 \text{ J/mol·K})(298 \text{ K})} = -27 \qquad \text{Solving, } K_{sp} = \mathbf{10^{-12}}$$

(From Appendix H, K_{sp} for AgBr = 3.3×10^{-13})

(b) $\Delta G° = -nFE°_{cell} = -(1 \text{ mol})(96500 \text{ 1/V·mol})(-0.70 \text{ V}) = \mathbf{+68,000 \text{ J/mol rxn} \text{ or } +68 \text{ kJ/mol rxn}}$

21-136. *Refer to Section 21-21.*

Balanced equations: (i) $\frac{1}{3}Al^{3+} + e^- \rightarrow \frac{1}{3}Al$ $\Delta G° = 160.4$ kJ/mol rxn

               (ii) $Al^{3+} + 3e^- \rightarrow Al$ $\Delta G° = 481.2$ kJ/mol rxn

(i) $E° = -\dfrac{\Delta G°}{nF} = -\dfrac{(+160400 \text{ J})}{(1 \text{ mol } e^-)(96485 \text{ J/V·mol } e^-)} = \mathbf{-1.662}$ **V**

(ii) $E° = -\dfrac{\Delta G°}{nF} = -\dfrac{(+481200 \text{ J})}{(3 \text{ mol } e^-)(96485 \text{ J/V·mol } e^-)} = \mathbf{-1.662}$ **V**

21-138. *Refer to Section 21-21.*

When water is electrolyzed with copper electrodes or using other common metals, the amount of $O_2(g)$ is less than when Pt electrodes are used, but the amount of $H_2(g)$ produced is independent of electrode material. Why does this happen? In electrolysis, the most easily oxidized species is oxidized and the most easily reduced species is reduced. If we compare Cu and H_2O by looking on the standard reduction potentials chart (data given below), we see that Cu is a stronger reducing agent than H_2O, because 0.337 V is less than 0.828 V. This means that Cu is more easily oxidized than water.

$$Cu^{2+}(aq) + 2e^- \rightarrow Cu(s) \qquad\qquad E° = 0.337 \text{ V}$$
$$O_2(g) + 4H^+(aq) + 4e^- \rightarrow 2H_2O(\ell) \qquad\qquad E° = 0.828 \text{ V}$$
$$Pt^{2+}(aq) + 2e^- \rightarrow Pt(s) \qquad\qquad E° = 1.23 \text{ V}$$

When Cu is used as an electrode, Cu will oxidize more easily than water and so, Cu^{2+} (from Cu) will be formed rather than O_2 (from H_2O), thereby lessening the amount of O_2 gas produced. You can also see that H_2O ($E° = 0.828$ V) is more easily oxidized than Pt ($E° = 1.23$ V), so a platinum electrode will not oxidize during the electrolysis of water as long as there is water present.

On the other hand, the only component present in the electrolysis cell that can be reduced is water, no matter what metal the electrode material is made from. So the formation of H_2 gas by the reduction of water is unaffected when platinum, Cu or most other common metals are used as electrodes.

22 Nuclear Chemistry

22-2. *Refer to the Introduction to Chapter 22, and Sections 22-1, 22-2 and 22-13.*

Natural radioactivity derives from spontaneous nuclear disintegrations. Induced radioactivity derives from the bombardment of nuclei with accelerated subatomic particles or other nuclei. Both cause atoms of one nuclide to be converted to another nuclide.

Using the elements mentioned in Section 22-13, induced radiation and the artificial transmutation of elements occur with both light elements, like the nonmetals ^{3}H, ^{12}C and ^{17}O as well has heavier elements, like ^{97}Tc, ^{112}Fr, ^{210}At and ^{239}U, which can be metals, metalloids or nonmetals. Transuranium elements, i.e. the elements with atomic numbers greater than 92 (uranium), must be prepared by nuclear bombardment of other elements.

22-4. *Refer to Sections 1-1 and 22-3.*

Einstein's equation relates matter and energy:

$$E = mc^2 \qquad \text{where } E = \text{amount of energy released}$$
$$m = \text{mass of matter transformed into energy}$$
$$c = \text{speed of light in a vacuum, } 3.00 \times 10^8 \text{ m/s}$$

If m is expressed in kg and c in m/s, the obtained E will be in units of J.

22-6. *Refer to Sections 22-1,5-5 and 5-7.*

Nucleons are the particles comprising the nucleus, i.e., it is a collective term for the protons and neutrons in a nucleus. The number of protons is the atomic number; the sum of the protons and the neutrons (the nucleons) is the mass number.

22-8. *Refer to Section 22-3 and Figure 22-11.*

The plot of binding energy per nucleon versus mass number for all the isotopes shows that binding energies/nucleon increase very rapidly with increasing mass number, reaching a maximum of 8.80 MeV per nucleon at mass number 56 for $^{56}_{26}$Fe, then decrease slowly.

22-10. *Refer to Sections 22-2 and 22-6.*

Potassium, with atomic number $Z = 19$, has three naturally occurring isotopes:

Isotope	Neutrons	Protons	Neutrons + Protons	n/p Ratio
^{39}K	20	19	39	1.05
^{40}K	21	19	40	1.11
^{41}K	22	19	41	1.16

The "magic numbers" which impart stability to a nucleus are 2, 8, 20, 28, 50, 82 or 122. The isotope, ^{39}K, has a magic number equal to its number of neutrons, so it is probably stable. The others have a larger neutron-to-proton ratio, making them neutron-rich nuclei, so ^{40}K and ^{41}K might be expected to decay by beta emission. In fact, both ^{39}K and ^{41}K are stable, and ^{40}K does decay by beta emission.

22-12. *Refer to Sections 22-5 and 22-6, and Exercise 22-10 Solution.*

According to predictions:

^{40}K   beta emission $^{40}_{19}$K $\rightarrow$ $^{40}_{20}$Ar + $^{0}_{-1}e$

^{41}K   beta emission $^{41}_{19}$K $\rightarrow$ $^{41}_{20}$Ar + $^{0}_{-1}e$ (However, ^{41}K is not unstable and does not decay.)

22-14. *Refer to Section 22-3, Table 22-1, Examples 22-1 and 22-2, and Appendix C.*

(a) One neutral atom of $^{62}_{28}$Ni contains 28 e^-, 28 p^+ and 34 n^0.

electrons: 28 x 0.00054858 amu   =   0.015 amu
protons: 28 x 1.0073 amu   = 28.204 amu
neutrons: 34 x 1.0087 amu   = 34.296 amu
                               sum = 62.515 amu

Δm = (sum of masses of e^-, p^+ and n^0) - (actual mass of a ^{62}Ni atom)
     = 62.515 amu - 61.9283 amu
     = 0.587 amu

Therefore, the mass deficiency for ^{62}Ni is **0.587 amu/atom or 0.587 g/mol**.

(b) Note: 1 joule = 1 kg x (1 m/s)^2

The nuclear binding energy, BE = $(\Delta m)c^2$
                         = $(0.587$ x 10^{-3} kg/mol$)(3.00$ x 10^8 m/s$)^2$
                         = 5.28 x 10^{13} kg·m^2/mol·s^2
                         = 5.28 x 10^{13} J/mol or **5.28 x 10^{10} kJ/mol of ^{62}Ni atoms**

22-16. *Refer to Section 22-3, Table 22-1, Examples 22-1 and 22-2, and Appendix C.*

(a) A neutral atom of $^{64}_{30}$Zn contains 30 e^-, 30 p^+ and 34 n^0.

electrons: 30 x 0.00054858 amu   =   0.016 amu
protons: 30 x 1.0073 amu   = 30.219 amu
neutrons: 34 x 1.0087 amu   = 34.296 amu
                               sum = 64.531 amu

Δm = (sum of masses of e^-, p^+ and n^0) - (actual mass of a ^{64}Zn atom)
     = 64.531 amu - 63.9291 amu
     = 0.602 amu

the mass deficiency, Δm, for ^{64}Zn is **0.602 amu/atom**

(b) This is equivalent to a mass deficiency of **0.602 g/mol**.

(c) Note: 1 J = 1 kg x (1 m/s)^2

The nuclear binding energy, $BE = (\Delta m)c^2$

$$= \left(\frac{0.602 \text{ g}}{1 \text{ mol}} \times \frac{1 \text{ kg}}{1000 \text{ g}} \times \frac{1 \text{ mol}}{6.02 \times 10^{23} \text{ atoms}}\right)(3.00 \times 10^8 \text{ m/s})^2$$

$$= 9.00 \times 10^{-11} \text{ kg·m}^2/\text{atom·s}^2$$

$$= \mathbf{9.00 \times 10^{-11} \text{ J/atom}}$$

(d) BE (kJ/mol)= $\dfrac{9.00 \times 10^{-11} \text{ J}}{1 \text{ atom}}$ x $\dfrac{6.02 \times 10^{23} \text{ atoms}}{1 \text{ mol}}$ x $\dfrac{1 \text{ kJ}}{1000 \text{ J}}$ = **5.42 x 10^{10} kJ/mol**

(e) Since there are 64 nucleons in a ^{64}Zn atom,

BE (MeV/nucleon) for ^{64}Zn= $\dfrac{9.00 \times 10^{-11} \text{ J}}{1 \text{ atom}}$ x $\dfrac{1 \text{ atom}}{64 \text{ nucleons}}$ x $\dfrac{1 \text{ MeV}}{1.60 \times 10^{-13} \text{ J}}$ = **8.79 MeV/nucleon**

(a) One neutral atom of $^{127}_{53}$I contains 53 e^-, 53 p^+ and 74 n^0.

electrons:	53 × 0.00054858 amu	=	0.029 amu
protons:	53 × 1.0073 amu	=	53.387 amu
neutrons:	74 × 1.0087 amu	=	74.644 amu
		sum =	128.060 amu

Δm = (sum of masses of e^-, p^+ and n^0) - (actual mass of a ^{127}I atom)
  = 128.060 amu - 126.9044 amu
  = 1.156 amu

Therefore, the mass deficiency for ^{127}I is 1.156 amu/atom or 1.156 g/mol.

Recall: 1 joule = 1 kg × (1 m/s)^2

The nuclear binding energy, $BE = (\Delta m)c^2$
    = (1.156 × 10^{-3} kg/mol)(3.00 × 10^8 m/s)^2
    = 1.04 × 10^{14} kg·m^2/mol·s^2
    = 1.04 × 10^{14} J/mol or **1.04 × 10^{11} kJ/mol of ^{127}I atoms**

(b) One neutral atom of $^{81}_{35}$Br contains 35 e^-, 35 p^+ and 46 n^0.

electrons:	35 × 0.00054858 amu	=	0.019 amu
protons:	35 × 1.0073 amu	=	35.256 amu
neutrons:	46 × 1.0087 amu	=	46.400 amu
		sum =	81.675 amu

Δm = 81.675 amu - 80.9163 amu = 0.759 amu/atom or 0.759 g/mol

$BE = (\Delta m)c^2$ = (0.759 × 10^{-3} kg/mol)(3.00 × 10^8 m/s)^2 = 6.83 × 10^{13} kg·m^2/mol·s^2
                = 6.83 × 10^{13} J/mol
                          or **6.83 × 10^{10} kJ/mol of ^{81}Br atoms**

(c) One neutral atom of $^{35}_{17}$Cl contains 17 e^-, 17 p^+ and 18 n^0.

electrons:	17 × 0.00054858 amu	=	0.0093 amu
protons:	17 × 1.0073 amu	=	17.124 amu
neutrons:	18 × 1.0087 amu	=	18.157 amu
		sum =	35.290 amu

Δm = 35.290 amu - 34.96885 amu = 0.321 amu/atom or 0.321 g/mol

$BE = (\Delta m)c^2$ = (0.321 × 10^{-3} kg/mol)(3.00 × 10^8 m/s)^2 = 2.89 × 10^{13} kg·m^2/mol·s^2
                = 2.89 × 10^{13} J/mol
                          or **2.89 × 10^{10} kJ/mol of ^{35}Cl atoms**

(a) In an electric field, an alpha (α) particle (a helium nucleus with a +2 charge) will be drawn toward the negative electrode, while a beta (β^-) particle (an electron with a -1 charge) will be drawn toward the positive electrode. Gamma (γ) radiation (very high energy electromagnetic radiation) will be unaffected by the electric field.

(b) The α particle and β^- particle will be drawn in opposite directions in a magnetic field and the γ radiation will be unaffected by a magnetic field.

(c) A piece of paper will reduce the α radiation significantly, but not the β^- or γ radiation; a thick concrete slab will prevent the α particles and β^- particles from passing, and most of the γ radiation.

There are many radionuclides that have medical uses, including:

(1) cobalt-60: This is used to arrest certain types of cancer. The technique uses the gamma rays produced in the decay of ^{60}Co to destroy cancerous tissue.

(2) plutonium-238: The energy produced in its decay is converted to electrical energy which powers heart pacemakers. Its relatively long half-life allows the device to be used for ten years before replacement.

Several radioisotopes are used as radioactive tracers (rediopharmaceuticals). These are injected into the body and allow physicians to study biological processes. These include

(3) sodium-24: This is used to follow the blood flow and locate stoppages in the circulatory system.

(4) thallium-201: This isotope helps locate healthy heart tissue.

(5) technetium-99: This metastable isotope has proven to be very useful and can be used to image the bones, liver, brain, and abnormal heart tissue.

(6) iodine-131: This concentrates in the thyroid gland, liver and certain parts of the brain. It is used to monitor goiter and other thyroid problems as well as liver and brain tumors.

Positron emission tomography (PET) is another form of imaging that uses positron emitters, such as ^{11}C, ^{13}N, ^{15}O and ^{18}F. These isotopes are incorporated into chemicals that are taken up by tissue. When the isotopes decay, the emitted positron reacts with a nearby electron, giving off 2 gamma rays, which are detected and an image of the tissue is created.

22-24.   *Refer to Section 22-9.*

(1) Photographic Detection: Radioactive substances affect photographic plates. Although the intensity of the affected spot is related to the amount of radiation, precise measurement by this method is tedious.

(2) Detection by Fluorescence: Fluorescent substances can absorb radiation and subsequently emit visible light. This is the basis for scintillation counting and can be used for quantitative detection.

(3) Cloud Chambers: A chamber containing air saturated with vapor is used. Radioactive particles ionize air molecules in the chamber. Cooling the chamber causes droplets of liquid to condense on these ions, giving observable fog-like tracks.

(4) Gas Ionization Counters: A common gas ionization counter is the Geiger-Müller counter where the electronic pulses derived from the ionization process are registered as counts. The instrument can be adjusted to detect only radiation with a desired penetrating power.

22-22.   *Refer to Section 22-2.*

Scientists have known that nuclides which have certain "magic numbers" of protons and neutrons are especially stable. Nuclides with a number of protons or a number of neutrons or a sum of the two equal to 2, 8, 20, 28, 50, 82 or 126 have unusual stability. Examples of this are $^{4}_{2}$He, $^{16}_{8}$O, $^{42}_{20}$Ca, $^{88}_{38}$Sr, and $^{208}_{82}$Pb. This suggests a shell (energy level) model for the nucleus similar to the shell model of electron configurations.

22-28.   *Refer to Section 22-7 and Figure 22-1.*

A nuclide with a neutron/proton ratio which is smaller than that for a stable isotope of the element can increase its ratio by undergoing:

(1) positron emission $^{1}_{1}p \rightarrow ^{1}_{0}n + ^{0}_{+1}\beta$

(2) electron capture (K capture) $^{1}_{1}p + ^{0}_{-1}e \rightarrow ^{1}_{0}n$

The net result of both processes is the loss of one proton and the gain of one neutron, thereby increasing the n/p ratio. Also, a heavier nuclide can undergo alpha emission to increase its n/p ratio.

22-30. *Refer to Section 22-5 and Table 22-3.*

(a) $^{198}_{79}\text{Au} \rightarrow {}^{198}_{80}\text{Hg} + \boxed{{}^{0}_{-1}\beta}$ beta particle

(c) $^{137}_{55}\text{Cs} \rightarrow {}^{137}_{56}\text{Ba} + \boxed{{}^{0}_{-1}\beta}$ beta particle

(b) $^{222}_{86}\text{Rn} \rightarrow {}^{218}_{84}\text{Po} + \boxed{{}^{4}_{2}\alpha}$ alpha particle

(d) $^{110}_{49}\text{In} \rightarrow {}^{110}_{48}\text{Cd} + \boxed{{}^{0}_{+1}\beta}$ positron

---

22-32. *Refer to Sections 22-6, 22-7 and 22-8, and Figure 22-1.*

(a) $^{60}_{27}\text{Co}$ (*n/p* ratio too high)   beta emission (neutron emission is less common)

(b) $^{20}_{11}\text{Na}$ (*n/p* ratio too low)   positron emission or electron capture (*K* capture)

(c) $^{222}_{86}\text{Rn}$ alpha emission

(d) $^{67}_{29}\text{Cu}$ (*n/p* ratio too high)   beta emission

(e) $^{238}_{92}\text{U}$ alpha emission

(f) $^{11}_{6}\text{C}$ (*n/p* ratio too low)   positron emission or electron capture (*K* capture)

22-34. *Refer to Sections 22-5, 22-6 and 22-13 and Table 22-3.*

In equations for nuclear reactions, the sums of the mass numbers and atomic numbers of the reactants must equal the sums for the products. Therefore,

(a) $^{96}_{42}\text{Mo} + {}^{4}_{2}\text{He} \rightarrow {}^{100}_{43}\text{Tc} + \boxed{{}^{0}_{+1}\beta}$ (b) $^{59}_{27}\text{Co} + {}^{1}_{0}n \rightarrow {}^{56}_{25}\text{Mn} + \boxed{{}^{4}_{2}\text{He}}$

(c) $^{23}_{11}\text{Na} + \boxed{{}^{1}_{1}\text{H}} \rightarrow {}^{23}_{12}\text{Mg} + {}^{1}_{0}n$ (d) $^{209}_{83}\text{Bi} + \boxed{{}^{2}_{1}\text{H}} \rightarrow {}^{210}_{84}\text{Po} + {}^{1}_{0}n$

(e) $^{238}_{92}\text{U} + {}^{16}_{8}\text{O} \rightarrow \boxed{{}^{249}_{100}\text{Fm}} + 5\,{}^{1}_{0}n$

22-36. *Refer to Sections 22-5 and 22-13.*

The equation for a nuclear reaction can be given in the following abbreviated form:
         parent nucleus (bombarding particle, emitted particle) daughter nucleus.

(a) $^{60}_{28}\text{Ni} + {}^{1}_{0}n \rightarrow \boxed{{}^{60}_{27}\text{Co}} + {}^{1}_{1}\text{H}$ (b) $^{98}_{42}\text{Mo} + {}^{1}_{0}n \rightarrow \boxed{{}^{99}_{43}\text{Tc}} + {}^{0}_{-1}\beta$

(c) $^{35}_{17}\text{Cl} + {}^{1}_{1}\text{H} \rightarrow \boxed{{}^{32}_{16}\text{S}} + {}^{4}_{2}\text{He}$

22-38. *Refer to Sections 22-5 and 22-13.*

The equation for a nuclear reaction can be given in the following abbreviated form:
         parent nucleus (bombarding particle, emitted particle) daughter nucleus.

(a) $^{14}_{7}\text{N} + {}^{4}_{2}\text{He} \rightarrow {}^{17}_{8}\text{O} + {}^{1}_{1}\text{H}$ (b) $^{106}_{46}\text{Pd} + {}^{1}_{0}n \rightarrow {}^{106}_{45}\text{Rh} + {}^{1}_{1}\text{H}$

(c) $^{23}_{11}\text{Na} + {}^{1}_{0}n \rightarrow \boxed{{}^{24}_{12}\text{Mg}} + {}^{0}_{-1}e$

22-40. *Refer to Sections 22-5, 22-6 and 22-13 and Table 22-3.*

In equations for nuclear reactions, the sums of the mass numbers and atomic numbers of the reactants must equal the sums for the products. Therefore,

(a) $^{242}_{94}Pu \rightarrow ^4_2He + \boxed{^{238}_{92}U}$

(b) $\boxed{^{32}_{15}P} \rightarrow ^{32}_{16}S + ^0_{-1}e$

(c) $^{252}_{98}Cf + 5\boxed{^2_1H} \rightarrow 3^1_0n + ^{259}_{103}Lr$

(d) $^{55}_{26}Fe + \boxed{^0_{-1}e} \rightarrow ^{55}_{25}Mn$

(e) $^{15}_8O \rightarrow \boxed{^{15}_7N} + ^0_{+1}e$

22-42. *Refer to Sections 22-5, 22-6 and 22-13.*

(a) $^{63}_{28}Ni \rightarrow ^{63}_{29}Cu + ^0_{-1}e$

(b) $2^2_1H \rightarrow ^3_2He + ^1_0n$

(c) $\boxed{^{10}_5B} + ^1_0n \rightarrow ^7_3Li + ^4_2He$

(d) $^{14}_7N + ^1_0n \rightarrow ^3_1H + 3^4_2He$

22-44. *Refer to Sections 22-5 and 22-8.*

Plan: Balance the nuclear reaction and identify the unknown "radioactinium."

$^{235}_{92}U \rightarrow \boxed{^{227}_{90}Th} + 2^4_2He + 2^0_{-1}\beta$

Therefore, "radioactinium" is the element thorium, **Th**, with atomic number **90** and mass number **227**.

22-46. *Refer to Sections 22-5, 22-8 and 22-13.*

Balanced nuclear reactions: (1) $^{249}_{98}Cf + ^{12}_6C \rightarrow ^{257}_{104}Rf + 4^1_0n$

(2) $^{257}_{104}Rf \rightarrow \boxed{^{253}_{102}No} + ^4_2He$

The element **nobelium**, No, is formed.

22-48. *Refer to Sections 22-5, 22-6 and 22-13.*

In equations for nuclear reactions, the sums of the mass numbers and atomic numbers of the reactants must equal the sums for the products. Therefore,

(a) $^{230}_{90}Th \rightarrow ^4_2He + ^{226}_{88}Ra$

(b) $^{210}_{82}Pb \rightarrow ^0_{-1}e + ^{210}_{83}Bi$

(c) $^{235}_{92}U \rightarrow ^{140}_{56}Ba + 2^1_0n + ^{93}_{36}Kr$

(d) $^{37}_{18}Ar + ^0_{-1}e \rightarrow ^{37}_{19}Cl$

22-50. *Refer to Section 22-10.*

The half-life of a radionuclide represents the amount of time required for half of the sample to decay. Relative stabilities of radionuclides are indicated by their half-life values. The shorter the half-life, the less stable is the radionuclide.

371

The radioisotope carbon-14 is produced continuously in the atmosphere as nitrogen atoms capture cosmic-ray neutrons:

$$^{14}_{7}N + ^{1}_{0}n \rightarrow ^{14}_{6}C + ^{1}_{1}H$$

The carbon-14 atoms react with O_2 to form $^{14}CO_2$. Like ordinary $^{12}CO_2$, it is removed from the atmosphere by living plants through the process of photosynthesis. As long as the cosmic-ray intensity remains constant, the amount of $^{14}CO_2$ and therefore its ratio to $^{12}CO_2$ in the atmosphere remains constant. Consequently, a certain fraction of carbon atoms in all living substances is carbon-14, a beta particle emitter with a half-life of 5730 years:

$$^{14}_{6}C \rightarrow ^{14}_{7}N + ^{0}_{-1}e$$

A steady state ratio of $^{14}C/^{12}C$ is maintained in living plants and organisms. After death the plant no longer carries out photosynthesis, so it no longer takes up $^{14}CO_2$. The radioactive emissions from the carbon-14 in dead tissue then decrease with the passage of time. The activity per gram of carbon in the sample in comparison with that in air gives a measure of the length of time elapsed since death. This is the basis of radiocarbon dating.

This technique is useful only when dating objects that are less than 50,000 years old (roughly 10 times the half-life of carbon-14). Older objects have too little activity to be accurately dated. This technique depends on cosmic-ray intensity being constant or at least predictable in order to keep the $^{14}C/^{12}C$ known throughout the time interval. Also, the sample must not be contaminated with organic matter having a different $^{14}C/^{12}C$ ratio.

22-54. *Refer to Section 22-10.*

For first order kinetics, $\ln\left(\dfrac{A_o}{A}\right) = kt$ and $t_{1/2} = \dfrac{0.693}{k}$ where A_o = initial amount of isotope
A = amount remaining after time, t
k = rate constant (units of time^{-1})
$t_{1/2}$ = half-life of isotope

Plan: (1) Calculate the rate constant, k, from the half-life of carbon-11.
(2) Assume the initial amount of carbon-11 is 100%. Calculate the time required for decay, t, using the first order rate equation.

(1) $k = \dfrac{0.693}{t_{1/2}} = \dfrac{0.693}{20.3 \text{ min}} = 0.0341 \text{ min}^{-1}$

(2) When 90.0% of the sample has decayed away, 10.0% of the sample remains.
Substituting into the first order rate equation,

$$\ln\left(\frac{100.0\%}{10.0\%}\right) = (0.0341 \text{ min}^{-1})t$$

$$2.303 = 0.0341t$$

$$t = \textbf{67.5 min}$$

When 95.0% of the sample has decayed, 5.0% of the sample remains. Substituting,

$$\ln\left(\frac{100.0\%}{5.0\%}\right) = (0.0341 \text{ min}^{-1})t$$

$$3.00 = 0.0341t$$

$$t = \textbf{88.0 min}$$

22-56. *Refer to Section 22-10 and Exercise 22-54 Solution.*

Balanced equation: $^{8}_{4}Be \rightarrow 2\,^{4}_{2}He$

Plan: (1) Determine the rate constant, k, from the half-life for ^{8}Be.
(2) Calculate the time required for 99.90% of ^{8}Be to decay.

(1) $k = \dfrac{0.693}{t_{1/2}} = \dfrac{0.693}{7 \times 10^{-17} \text{ s}} = 1 \times 10^{16} \text{ s}^{-1}$

(2) The first order rate equation: $\ln\left(\dfrac{A_o}{A}\right) = kt$

If 99.90% of ^{8}Be decayed away, then 0.10% remains. Note that the calculation does not depend on the initial amount of ^{8}Be. Substituting,

$$\ln\left(\dfrac{100.00\%}{0.10\%}\right) = (1 \times 10^{16}\ s^{-1})t$$

$$6.9 = (1 \times 10^{16}\ s^{-1})t$$

$$t = \mathbf{7 \times 10^{-16}\ s}\ \text{(to 1 significant figure)}$$

22-58. *Refer to Section 22-10.*

Plan: (1) Calculate the first order rate constant, k, from the half-life of gold-198.
        (2) Calculate the mass of gold-198 remaining.

(1) $k = \dfrac{0.693}{t_{1/2}} = \dfrac{0.693}{2.69\ d} = 0.258\ d^{-1}$

(2) The first order rate equation: $\ln\left(\dfrac{A_o}{A}\right) = kt$

Substituting,

$$\ln\left(\dfrac{2.8\ \mu g}{A_o}\right) = 0.258\ d^{-1} \times 10.8\ d$$

$$\ln\left(\dfrac{2.8\ \mu g}{A_o}\right) = 2.79$$

$$\left(\dfrac{2.8\ \mu g}{A_o}\right) = 16.2$$

$$A_o = \mathbf{0.17\ \mu g}$$

22-60. *Refer to Sections 22-10 and 22-12, and Example 22-5.*

Plan: (1) Calculate the first order rate constant, k, from the half-life of carbon-14.
        (2) Determine the fraction of C^{14} remaining, A/A_o, after 50,000 years.

(1) $k = \dfrac{0.693}{t_{1/2}} = \dfrac{0.693}{5730\ yr} = 1.21 \times 10^{-4}\ yr^{-1}$

(2) The first order rate equation: $\ln\left(\dfrac{A_o}{A}\right) = kt$

Substituting,

$$\ln\left(\dfrac{A_o}{A}\right) = 1.21 \times 10^{-4}\ yr^{-1} \times 50{,}000\ yr$$

$$\ln\left(\dfrac{A_o}{A}\right) = 6.05$$

$$\left(\dfrac{A_o}{A}\right) = 424$$

$$\left(\dfrac{A}{A_o}\right) = \mathbf{0.00236}$$

22-62. *Refer to Sections 22-10 and 22-12 and Example 22-5.*

Plan: (1) Calculate the first order rate constant, k, from the half-life of carbon-14.
        (2) Determine the age of the object.

(1) $k = \dfrac{0.693}{t_{1/2}} = \dfrac{0.693}{5730\ yr} = 1.21 \times 10^{-4}\ yr^{-1}$

(2) The first order rate equation: $\ln\left(\dfrac{A_o}{A}\right) = kt$

Substituting, $\ln\left(\dfrac{8.35\ \mu g}{0.76\ \mu g}\right) = (1.21 \times 10^{-4}\ \text{yr}^{-1})t$

$$2.40 = (1.21 \times 10^{-4})t$$

$$t = \mathbf{1.98 \times 10^4\ yr}$$

22-64. *Refer to Sections 22-14 and 22-15, and the Key Terms for Chapter 22.*

A chain reaction is a reaction that sustains itself once it has begun and may even expand. Normally, the limiting reactant is regenerated as a product to maintain the progress of the chain. Nuclear fission processes are considered chain reactions because the number of neutrons produced in the reaction equals or is greater than the number of neutrons absorbed by the fissioning nucleus.

For example:

$$\mathrm{^{235}_{92}U} + \mathrm{^{1}_{0}}n \rightarrow \left[\mathrm{^{236}_{92}U}\right] \rightarrow \mathrm{^{140}_{56}Ba} + \mathrm{^{93}_{36}Kr} + 3\ \mathrm{^{1}_{0}}n + \text{energy}$$

The critical mass of a fissionable material is the minimum mass of a particular fissionable nuclide in a set volume that is necessary to sustain a nuclear chain reaction.

22-66. *Refer to Sections 22-16 and 22-3.*

(a) $\mathrm{^{7}_{3}Li} + \mathrm{^{1}_{1}H} \rightarrow \boxed{\mathrm{^{4}_{2}He}} + \mathrm{^{4}_{2}He}$ where a proton is represented by $\mathrm{^{1}_{1}H}$

(b) Plan: (1) Determine the mass difference between the products and the reactants. The difference, Δm, is directly related to the energy involved in the reaction.
     (2) Calculate the amount of energy involved.

(1) Δm = mass of products − mass of reactants
     = (2 × mass of $\mathrm{^{4}_{2}He}$) − (mass of $\mathrm{^{7}_{3}Li}$ + mass of $\mathrm{^{1}_{1}H}$)
     = (2 × 4.00260 amu) − (7.01600 amu + 1.007825 amu)
     = −0.01862 amu or −0.01862 g/mol rxn

(2) ΔE = $(\Delta m)c^2$
     = $(-1.862 \times 10^{-5}\ \text{kg/mol rxn})(3.00 \times 10^8\ \text{m/s})^2$
     = $-1.68 \times 10^{12}\ \text{kg·(m/s)}^2/\text{mol rxn}$
     = $-1.68 \times 10^{12}\ \text{J/mol rxn}$ or $\mathbf{-1.68 \times 10^9\ kJ/mol\ rxn}$

Since $\Delta E < 0$, energy is being released in this fusion reaction, as expected.

22-68. *Refer to Section 22-15.*

The primary advantage of nuclear energy is that enormous amounts of energy are liberated per unit mass of fuel. Also, the air pollution (oxides of S, N, C and particulate matter) caused by fossil fuel electric power plants is not a problem with nuclear energy plants. In European countries, where fossil fuel reserves are scarce, most of the electricity is generated by nuclear power plants for these reasons.

There are, however, some disadvantages associated with nuclear power from controlled fission reactions. The radionuclides must be properly shielded to protect the workers and the environment from radiation and contamination. Spent fuel, containing long-lived radioisotopes, must be disposed of carefully using special containers placed underground in geologically inactive areas. This is because the radiation from the fuel is biologically dangerous and must be contained until the fuel has decayed to the point when it is no longer dangerous. The problem is that the time involved could be several hundred thousand years. If there is inadequate cooling in the reactor, there is the possibility of overheating the fuel and causing a "meltdown." This cooling water can cause biological damage to aquatic life if it is returned to the natural water system while it is still too warm. Finally, it is possible that Pu-239 could be stolen and used for bomb production.

In the future, when nuclear fusion power plants are in operation, most of these disadvantages will not be a concern. Fusion reactions produce only short-lived isotopes and so there would be no long-term storage problems. An added advantage is that there is a virtually inexhaustible supply of deuterium fuel in the world's oceans.

22-70. *Refer to Section 22-15.*

Uranium ores contain only about 0.7% ^{235}U which is fissionable. Most of the rest is nonfissionable ^{238}U. To enrich ^{235}U for use in nuclear power plants, the oxide is converted to UF_4 with HF and then oxidized to UF_6 by fluorine. The vapor of $^{235}UF_6$ and $^{238}UF_6$ is then subjected to repeated diffusion through porous barriers to concentrate $^{235}UF_6$ (Graham's Law). Gas centrifuges are now used for the concentration process which is also based upon the difference in masses of the two U isotopes.

22-72. *Refer to Section 22-16.*

The major advantages of fusion as a potential energy source are three-fold:
(1) Fusion reactions are accompanied by much greater energy production per unit mass of reacting atoms than fission reactions.
(2) The deuterium fuel for fusion reactions is present in a virtually inexhaustible supply in the world oceans.
(3) Fusion reactions produce only short-lived radionuclides; there would be no long-term waste-disposal problem.
The only disadvantage of fusion is that extremely high temperatures are required to initiate the fusion process. A structural material that can withstand the high temperatures (4×10^7 K or more) and contain the fusion reaction, does not as yet exist.

22-74. *Refer to Sections 22-12 and 22-3.*

(a) $^{14}_7N + ^4_2He \rightarrow ^{17}_8O + ^1_1H$

(b) Plan: (1) Determine the mass difference between the products and the reactants. The difference, Δm, is directly related to the energy involved in the reaction.
   (2) Calculate the amount of energy involved.

(1) Δm = mass of products - mass of reactants
   = (mass of $^{17}_8O$ + mass of ^1_1H) - (mass of $^{14}_7N$ + mass of ^4_2He)
   = (16.99913 amu + 1.007825 amu) - (14.00307 amu + 4.00260 amu)
   = 0.00128 amu or 0.00128 g/mol rxn

(2) $\Delta E = (\Delta m)c^2$
   = $(1.28 \times 10^{-6}$ kg/mol rxn$)(3.00 \times 10^8$ m/s$)^2$
   = 1.15×10^{11} kg·(m/s)^2/mol rxn
   = $+1.15 \times 10^{11}$ **J/mol rxn or +1.15 $\times 10^8$ kJ/mol rxn**

Since $\Delta E > 0$, energy is being absorbed in this reaction, as expected.

22-76. *Refer to Sections 22-10 and 22-12.*

Plan: (1) Calculate the first order rate constant, k, from the half-life of uranium-238.
   (2) Determine the age of the rock.
      Assume that all the ^{206}Pb came from ^{238}U.
      Because of the very long half-life of ^{238}U, 4.5 billion years,
         the amounts of intermediate nuclei can be neglected.

(1) $k = \dfrac{0.693}{t_{1/2}} = \dfrac{0.693}{4.51 \times 10^9 \text{ yr}} = 1.54 \times 10^{-10}$ yr^{-1}

(2) The first order rate equation: $\ln\left(\dfrac{N_o}{N}\right) = kt$

where N = number of ^{238}U atoms remaining
N_o = number of ^{238}U atoms originally present

Therefore, N_o = number of ^{238}U atoms remaining + number of ^{238}U atoms decayed
= number of ^{238}U atoms remaining + number of ^{206}Pb atoms produced

So, $\left(\dfrac{N_o}{N}\right) = \dfrac{67.8\ ^{238}U\ \text{atoms} + 32.2\ ^{206}Pb\ \text{atoms}}{67.8\ ^{238}U\ \text{atoms}} = \left(\dfrac{100.0}{67.8}\right)$

Substituting, $\ln\left(\dfrac{100.0}{67.8}\right) = (1.54 \times 10^{-10}\ \text{yr}^{-1})t$

$0.389 = (1.54 \times 10^{-10})t$

$t = \mathbf{2.52 \times 10^{9}\ yr}$

22-78.   *Refer to Sections 22-14 and 22-3.*

Balanced equations: fission $^{235}_{92}U + ^{1}_{0}n \rightarrow ^{94}_{40}Zr + ^{140}_{58}Ce + 6\ ^{0}_{-1}e + 2\ ^{1}_{0}n$
fusion $2\ ^{2}_{1}H \rightarrow ^{3}_{1}H + ^{1}_{1}H$

Plan: (1)   Determine the mass difference between the products and the reactants. The difference, Δm, is directly related to the energy involved in the reaction.
(2)   Calculate the amount of energy involved.

(1)   fission: Δm = mass of products - mass of reactants
= [mass of $^{94}_{40}Zr$ + mass of $^{140}_{58}Ce$ + (6 $\times$ mass of $^{0}_{-1}e$) + (2 $\times$ mass of $^{1}_{0}n$)]
- [mass of $^{235}_{92}U$ + mass of $^{1}_{0}n$]
= [93.9061 amu + 139.9053 amu + (6 $\times$ 0.000549 amu) + (2 $\times$ 1.0087 amu)]
- [235.0439 amu + 1.0087 amu)
= −0.2205 amu or −0.2205 g/mol rxn

fusion: Δm = [mass of $^{3}_{1}H$ + mass of $^{1}_{1}H$] - [2 $\times$ mass of $^{2}_{1}H$]
= [3.01605 amu + 1.007825 amu] - [2 $\times$ 2.0140 amu]
= −0.00412 amu or −0.00412 g/mol rxn

(2)   fission: $\Delta E = (\Delta m)c^2 = (-2.205 \times 10^{-4}\ \text{kg/mol rxn})(3.00 \times 10^8\ \text{m/s})^2 = -1.98 \times 10^{13}\ \text{kg·(m/s)}^2/\text{mol rxn}$
$= -1.98 \times 10^{13}\ \text{J/mol rxn}$

$\Delta E\ (\text{J/amu}\ ^{235}U) = \dfrac{-1.98 \times 10^{13}\ \text{J}}{1\ \text{mol rxn}} \times \dfrac{1\ \text{mol rxn}}{1\ \text{mol}\ ^{235}U} \times \dfrac{1\ \text{mol}\ ^{235}U}{6.02 \times 10^{23}\ \text{atoms}\ ^{235}U} \times \dfrac{1\ \text{atom}\ ^{235}U}{235.0439\ \text{amu}}$
$= -1.40 \times 10^{-13}\ \text{J/amu}\ ^{235}U$

fusion: $\Delta E = (\Delta m)c^2 = (-4.12 \times 10^{-6}\ \text{kg/mol rxn})(3.00 \times 10^8\ \text{m/s})^2 = -3.71 \times 10^{11}\ \text{kg·(m/s)}^2/\text{mol rxn}$
$= -3.71 \times 10^{11}\ \text{J/mol rxn}$

$\Delta E\ (\text{J/amu}\ ^{2}H) = \dfrac{-3.71 \times 10^{11}\ \text{J}}{1\ \text{mol rxn}} \times \dfrac{1\ \text{mol rxn}}{2\ \text{mol}\ ^{2}H} \times \dfrac{1\ \text{mol}\ ^{2}H}{6.02 \times 10^{23}\ \text{atoms}\ ^{2}H} \times \dfrac{1\ \text{atom}\ ^{2}H}{2.0140\ \text{amu}}$
$= -1.53 \times 10^{-13}\ \text{J/amu}\ ^{2}H$

Therefore, the above **fusion** process produces about 10% more energy per amu of material than fission. However, the above fusion reaction is not a typical one because it involves the production of two particles from two particles of similar size. In general, fusion processes produce much more energy than fission processes on a per unit mass basis.

22-80. *Refer to Section 22-15.*

Both nuclear and conventional power plants produce environmentally-sensitive waste. Both use cooling water which when put into streams and rivers while still at elevated temperatures can cause significant damage to the biota.

Conventional power plants can pollute the air with particulate matter and the oxides of sulfur, nitrogen, and carbon, causing acid rain and other problems. However, with proper scrubbing and filtering at the source, this pollution has been greatly reduced.

Nuclear power produces spent fuel that contains radionuclides that will emit radiation for hundreds and thousands of years. At present, they are being stored underground indefinitely in heavy, shock-proof containers. These containers could be stolen or may corrode with time, or leak as a result of earthquakes and tremors. Transportation and reprocessing accidents could cause environmental contamination. One solution is for the United States to go to breeder reactors, as has been done in other countries, to reduce the level and amount of radioactive waste.

22-82. *Refer to Section 22-4 and Table 22-3.*

The common type of radioactive emission that does not consist of matter is the gamma ray.

22-84. *Refer to Section 22-12.*

One of the limitations of radiocarbon dating artifacts is due to the half-life of the carbon-14, 5730 years. In radiochemistry, a good rule of thumb is the following: when an element decays for more than about 10 times its half-life, there is very little left to measure accurately. In the case of C-14, that time is 10 x 5730 yr or 57300 years.

What fraction of a C-14 sample remains after 10 half-lives? The answer is that only $(\frac{1}{2})^{10} = 0.00098$ or 0.098% of the C-14 in the original sample remains, and this is generally too little activity to measure accurately.

Location of Commonly Used Information

Element Colors for Models

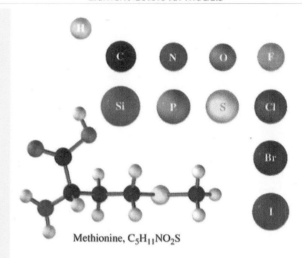

Methionine, $C_5H_{11}NO_2S$

Color Scale for Electrical Charge Potential (ECP) Surfaces

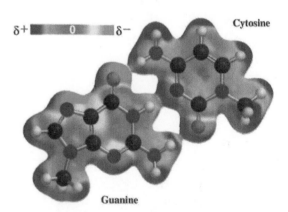

Hydrogen bonding between two DNA base pairs

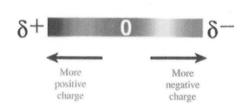